ANNUAL REVIEW OF BIOCHEMISTRY

ANNUAL REVIEW OF BIOCHEMISTRY

VOLUME 63, 1994

CHARLES C. RICHARDSON, *Editor*
Harvard Medical School

JOHN N. ABELSON, *Associate Editor*
California Institute of Technology

ALTON MEISTER, *Associate Editor*
Cornell University Medical College

CHRISTOPHER T. WALSH, *Associate Editor*
Harvard Medical School

ANNUAL REVIEWS INC. 4139 EL CAMINO WAY P.O. BOX 10139 PALO ALTO, CALIFORNIA 94303-0897

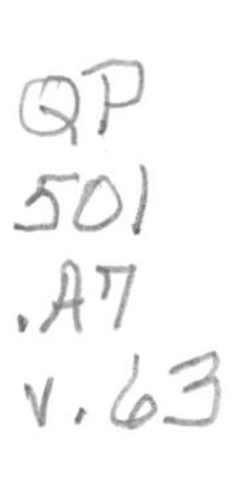

ANNUAL REVIEWS INC.
Palo Alto, California, USA

International Standard Serial Number: 0066–4154
International Standard Book Number: 0–8243–0863-8
Library of Congress Catalog Card Number: 32-25093

♾ The paper used in this publication meets the minimum requirements of American National Standard for Information Sciences—Permanence of Paper for Printed Library Materials, ANSI Z39.48-1984.

Typesetting by Kachina Typesetting Inc., Tempe, Arizona; John Olson, President; Jeannie Kaarle, Typesetting Coordinator; and by the Annual Reviews Inc. Editorial Staff

PRINTED AND BOUND IN THE UNITED STATES OF AMERICA

Annual Review of Biochemistry
Volume 63 (1994)

CONTENTS

(*Continued*)

SOME RELATED ARTICLES IN OTHER *ANNUAL REVIEWS*

From the *Annual Review of Biophysics and Biomolecular Structure*, Volume 23 (1994)

Membrane Proteins: From Sequence to Structure, G. von Heijne
Global Statistics of Protein Sequences: Implications for the Origin, Evolution, and Prediction of Structure, S. H. White
Conformational and Thermodynamic Properties of Supercoiled DNA, A. V. Vologodskii and N. R. Cozzarelli
The β-Ribbon DNA Recognition Motif, S. E. V. Phillips
G-Quartet Structures in Telomeric DNA, J. R. Williamson
DNA Branched Junctions, N. C. Seeman and N. R. Kallenbach
Protein Structure–Based Drug Design, P. J. Whittle and T. L. Blundell
Polypeptide Interactions with Molecular Chaperones and Their Relationships to In Vivo Protein Folding, S. J. Landry and L. M. Gierasch

From the *Annual Review of Cell Biology*, Volume 10 (1994)

Feedback Controls in Cell Cycle, T. A. Weinert
Protein Tyrosine Kinases, A. Hunter
Serine/Threonine Protein Phosphatases: New Avenues for Cell Regulation, S. Shenolikar
Structure of Actin-Binding Proteins: Insights about Function at Atomic Resolution, T. D. Pollard, S. Almo, S. Quirk, V. Vinson, and E. E. Lattman
Prenylation G Proteins, J. Glomset
Signal Sequence Recognition and Protein Targetting to the Endoplasmic Reticulum Membrane, P. Walter and A. E. Johnson
The Role of Nuclear Factor kappa B and Related Proteins in the Control of Transcription, U. Siebenlist, K. D. Brown, and G. Franzoso
Amyloids, D. Selkoe

From the *Annual Review of Genetics*, Volume 27 (1993)

DNA Repair Genes and Proteins of Saccharomyces cerevisiae, L. Prakash, S. Prakash, and P. Sung
The Function of Heat-Shock Proteins in Stress Tolerance: Degradation and Reactivation of Damaged Proteins, D. A. Parsell and S. Lindquist

From the *Annual Review of Immunology*, Volume 12 (1994)

Structure of Peptides Associated with Class I and Class II MHC Molecules, V. H. Engelhard
Assembly, Transport, and Function of MHC Class II Molecules, P. Cresswell
Making Antibodies by Phage Display Technology, G. Winter, A. D. Griffiths, R. E. Hawkins, and H. R. Hoogenboom

 (*Continued*)

From the *Annual Review of Medicine,* Volume 45 (1994)

Molecular Basis of Hereditary Disorders of Connective Tissue, D. J. Tilstra and P. H. Byers
Cellular and Molecular Biology of Alzheimer's Disease and Animal Models, D. L. Price and S. S. Sisodia

From the *Annual Review of Microbiology,* Volume 47 (1993)

Regulation of the Heat-Shock Response in Bacteria, T. Yura, H. Nagai, and H. Mori

From the *Annual Review of Neuroscience,* Volume 17 (1994)

Cloned Glutamate Receptors, M. Hollmann and S. Heinemann
Nitric Oxide and Synaptic Function, E. M. Schuman and D. V. Madison

From the *Annual Review of Nutrition,* Volume 14 (1994)

Inborn Errors of Fructose Metabolism, G. Van den Berghe
Nitric Oxide Synthases: Why So Complex?, B. S. Siler Masters

From the *Annual Review of Physiology,* Volume 56 (1994)

Modulation of Ion Channels by Protein Phosphorylation and Dephosphorylation, I. B. Levitan
Structure and Function of Cyclic Nucleotide-Dependent Protein Kinases, S. H. Francis and J. D. Corbin

From the *Annual Review of Plant Physiology and Plant Molecular Biology,* Volume 45 (1994)

The Role of Carbonic Anhydrase in Photosynthesis, M. R. Badger and G. D. Price
Molecular Biology of Carotenoid Biosynthesis in Plants, G. E. Bartley, P. A. Scolnik, and G. Giuliano
The Plant Mitochondrial Genome: Physical Structure, Information Content, RNA Editing, and Gene Migration to the Nucleus, W. Schuster and A. Brennicke

J. MURRAY LUCK (1899–1993)

We note with sadness the death of J. Murray Luck, professor emeritus of chemistry, Stanford University, on August 26, 1993. Dr. Luck was the founding Editor of the *Annual Review of Physiology* (1939) as well as the *Annual Review of Biochemistry* (1932).

In addition to his research activities as a member of the chemistry department at Stanford from 1926 to 1965, where he taught and inspired more than 2500 students, Dr. Luck created the nonprofit enterprise that now publishes the authoritative, highly cited Annual Reviews in 26 fields in the biological and biomedical, social and behavioral, and physical sciences. The continuing importance of Annual Reviews in the scientific research process stands as a living tribute to Murray Luck's unique combination of gifts as scholar, teacher, and entrepreneur, as well as to his unswerving commitment to the highest ideals of intellectual quality and service.

In his own inimitable style, Dr. Luck recounted the origins of the Annual Reviews in his Foreword to *The Excitement and Fascination of Science* (1965) and in "Confessions of a biochemist," his prefatory chapter in the *Annual Review of Biochemistry*, Volume 50 (1981), which we recommend to our readers. All of the Annual Reviews remain an enduring legacy to generations of researchers and students from this versatile, accomplished, unselfish, and very wise man. With great admiration and affection, the Editorial Committee dedicates this volume to his memory.

ANNUAL REVIEWS INC. is a nonprofit scientific publisher established to promote the advancement of the sciences. Beginning in 1932 with the *Annual Review of Biochemistry,* the Company has pursued as its principal function the publication of high quality, reasonably priced *Annual Review* volumes. The volumes are organized by Editors and Editorial Committees who invite qualified authors to contribute critical articles reviewing significant developments within each major discipline. The Editor-in-Chief invites those interested in serving as future Editorial Committee members to communicate directly with him. Annual Reviews Inc. is administered by a Board of Directors, whose members serve without compensation.

Osamu Hayaishi

Annu. Rev. Biochem. 1994. 63:1–24

TRYPTOPHAN, OXYGEN, AND SLEEP

Osamu Hayaishi
Osaka Bioscience Institute, 6-2-4 Furuedai, Suita, Osaka 565, Japan

KEY WORDS: amino acid, poly(ADP ribose), oxygenase, prostaglandin D_2, prostaglandin E_2

CONTENTS

PROLOGUE

My father, Jitsuzo Hayaishi, was born in 1882 in Miyazu, a small town about 50 miles north of Kyoto where he practiced medicine for some years. He must have been a very courageous and ambitious person to go to America and start his medical training all over again. He took a boat to California, crossed an unimaginably enormous expanse of land by train, and ended his journey in Baltimore, where he was admitted to the University of Maryland School of Medicine. After he received his M.D. and had passed the National Board Examinations, he returned to California to practice medicine. Soon he met and married my mother, Mitsu Uchida, who was visiting her brother in San Francisco, and settled in Stockton, where I was born in 1920. A

0066-4154/94/0701-0001$05.00

year later my father decided to go to Berlin to study immunology for two years at the Robert Koch Institute. We returned to Japan in 1923, settling in Osaka, where my father spent the rest of his life practicing medicine. A conscientious doctor, devoted to clinical practice, he was always respected and trusted by his patients. His only hobby was reading the new medical journals and books. Perhaps his sincere attitude toward medical science and the scholarly atmosphere at home were what led me to a career in medicine and basic research.

I graduated from the Osaka University School of Medicine in 1942 during World War II, and served as a medical officer in the Japanese Navy for three years. With the war's end in 1945, I returned to Osaka to find that the city had been almost totally demolished. The home of my parents and family had completely vanished. Food was scarce, and supplies of most commodities, gas, electricity, and even water, were extremely limited. I then thought of joining my parents in their hometown, Miyazu, and helping my father in his medical practice, as he was then already 64 years old. But my father had decided to return to Osaka to open a small hospital and to look after patients during this time of food shortages and very poor hygienic conditions. Inspired by his courage and foresight, I decided to remain in Osaka. Had I chosen the practice of medicine, as had many of my classmates, I would have been economically very comfortable even in those difficult times.

Instead, perhaps because I inherited my father's adventurous spirit, I was rash enough to undertake laboratory work and joined the Department of Bacteriology, located in the Institute of Microbial Diseases, at the Osaka University School of Medicine. I was very fortunate to be surrounded by many excellent and ambitious young colleagues, including Teishiro Seki, Masami Suda, and Tsunehisa Amano. Despite the disastrous events during and after World War II, everyone in the department was eager to read the new journals and catch up with the latest developments in the United States and Europe. Morale was high and people were enthusiastic. Reading *Bacterial Metabolism* by Marjory Stephenson and *Dynamic Aspects of Biochemistry* by Ernest Baldwin impressed and inspired me and aroused my interest in biochemistry.

TRYST WITH TRYPTOPHAN

Osaka University was a center of biochemical research in Japan, and Yashiro Kotake and his associates were well known for their research on tryptophan metabolism. Kotake was responsible for the isolation and characterization of kynurenine, the key intermediate of tryptophan metabolism in both mammals and microorganisms. He kindly provided us with several grams of tryptophan, kynurenine, and other related chemicals, and encouraged us to use these valuable compounds for our experiments. At that time, research funds were extremely limited, facilities were poor, chemicals were scarce, and there were no laboratory animals. To overcome some of these difficulties, I decided to use microorganisms. So I went to the backyard of the Institute, took some soil samples, mixed them with tryptophan and water in a test tube, and waited. It was a very simple experiment, but after a few days, I observed some cloudiness in the supernatant. Thus, by this enrichment culture technique, I was able to isolate several strains of soil bacteria that could grow on tryptophan as their sole source of carbon and nitrogen.

It was already known then, mainly through the efforts of Kotake and his associates, that in mammals tryptophan was degraded through kynurenine to kynurenic acid, anthranilic acid, and xanthurenic acid as the end metabolites. However, in my soil isolates, I was able to demonstrate in collaboration with Masami Suda, Yoshiharu Oda, and others that in microorganisms anthranilic acid was further metabolized to catechol, then to muconic acid, and ultimately oxidized to CO_2, ammonia, and H_2O (Figure 1). Encouraged by this unexpected discovery, I proceeded to isolate and purify the enzyme that catalyzed the oxidative cleavage of catechol to produce *cis,cis*-muconic acid as the reaction product (1). It appeared to me that this was the first clear-cut demonstration of the enzymatic cleavage of the benzene ring, namely, the enzymatic conversion of an aromatic into an aliphatic compound. This enzyme was a nonheme iron protein with a molecular weight of approximately 90,000. Because two atoms of oxygen were consumed, I took it for granted that oxygen atoms were incorporated into the substrate and named the enzyme pyrocatechase instead of catechol oxidase (Figure 2). I chose this name because several catechol oxidases had already been reported at that time, but these enzymes seemed to be grossly different from my pyrocatechase. They did not cleave the aromatic structure of catechol and they were copper proteins rather than nonheme iron proteins. The catalytic mechanism of pyrocatechase remained unclarified for some

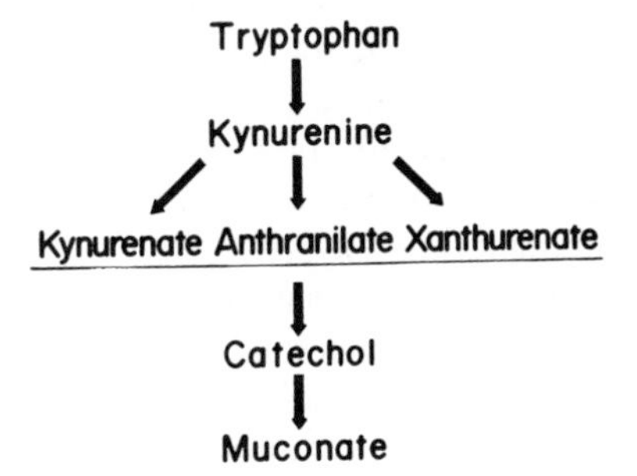

Figure 1 Major metabolic pathways of tryptophan (~1950)

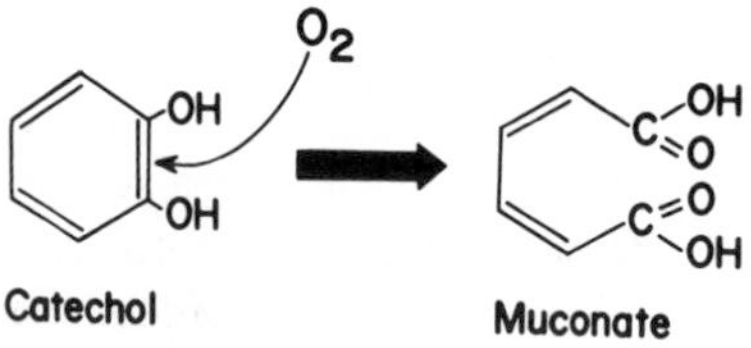

Catechol Muconate

Figure 2 Pyrocatechase-catalyzed reaction

years. In fact, in the *Enzymologie* textbook edited by Otto Hoffman-Ostenhof and published in 1954, pyrocatechase was described as "not yet classified" in the category of oxidoreductases (2).

Being a very naive and inexperienced biochemist, I assumed that the oxygen consumed by the pyrocatechase-catalyzed reaction must have been incorporated into catechol, as might be surmised by Lavoisier's principles of biological oxidation, formulated in the 18th century. He postulated the addition of oxygen atoms to the substrate in oxidations and the removal of oxygen from the oxide in reductions. I was startled to learn, however, that at the start of the present century, the enzymatic incorporation of molecular oxygen into substrates had been ruled out by Heinrich Wieland, the 1927 Nobel laureate in chemistry. In 1932, he wrote a book *On the Mechanism of Oxidation,* which was then considered the "bible" of biological oxidation (3).

In his famous dehydrogenation theory, he excluded direct addition of the oxygen molecule to the substrate in a biological oxidation system. According to this theory, biological oxidation processes entail the transfer of hydrogen atoms, or their equivalent, from the substrate molecule to an appropriate acceptor, such as a coenzyme or one of various dyes. Oxygen molecules may, in some instances, serve as immediate electron acceptors, and enzymes that catalyze such reactions are termed "oxidases." Thus, according to Wieland, molecular oxygen accepts hydrogen atoms and is reduced to water or hydrogen peroxide. However, it is never incorporated into the substrate, as had been proposed by Lavoisier and his contemporaries some 200 years earlier. According to Wieland, when the overall reaction appears to be the addition of oxygen, the oxygen atoms are always derived from a water molecule rather than molecular (or atmospheric) oxygen. For example, when aldehydes are converted to acids, they are hydrated first and then dehydrogenated. Whereas the sum appeared to be the addition of oxygen to the substrate aldehyde to form acid, this oxygen atom, it was postulated, is really derived from water rather than from molecular oxygen.

Wieland thought the direct addition of molecular oxygen to a substrate to be completely irrelevant to biological oxidation. Nevertheless, I considered the possibility that pyrocatechase may incorporate molecular oxygen into the substrate rather than dehydrogenate it, because oxygen could not be replaced by known coenzymes or electron acceptor dyes. Furthermore, orthobenzoquinone, the dehydrogenated product of catechol, did not serve as an intermediate in the overall reaction. Unfortunately, it was not possible to carry out experiments to prove my hypothesis, because the facilities were not available then in Japan. About this time, I married Takiko Satani in 1946, and we welcomed the arrival of our daughter Mariko in 1947.

PILGRIMAGE TO THE UNITED STATES

One day early in the summer of 1949, I received a letter from David E. Green inviting me to join his group as a postdoctoral fellow. I knew that Green had spent may years in Cambridge, England and had just returned to the United States to become Director of the Enzyme Institute at the University of Wisconsin in Madison. I was delighted, but hesitated to accept the offer, because Japan was still an occupied country and I feared that hostility against the Japanese might still exist among the American people. My starting salary as an instructor at Osaka University was 60 yen a month (approximately 17 US cents, not dollars!), whereas the William Waterman Fellowship in Enzyme Chemistry that Green had arranged for me was $250 a month, a 1500-fold difference! Furthermore, research facilities and working conditions were presumably much better in the United States. After pondering over Green's letter for some time and discussing this matter with my father and friends, I finally accepted the offer.

Shortly thereafter, I received an airplane ticket from Green and boarded a luxurious double-decker, four-engine Boeing B-377 from Tokyo airport. I felt as if I was going up to heaven from hell. It took almost 36 hours to fly from Tokyo to San Francisco, with three stops for refueling. I changed planes in San Francisco, then Chicago, and finally arrived at the Madison airport. The time was mid-November, 1949. With a heavy snowfall in progress, awaiting me were two young Americans, Bernard Katchman and Ephraim Kaplan. I felt extremely fortunate to have the opportunity to become good friends with Bernie and Eph, the kindest and most considerate human beings I had ever met. In spite of the language barrier and the ethnic and religious difference, they treated me like a brother, teaching me things in the lab and instructing me in the matters of daily life. Without their help, I would almost certainly have been lonesome, desperate, and homesick.

Soon I was able to adjust to and enjoy the American way of life in this beautiful university town and made many friends, among them Philip and Muriel Feigelson, Henry Lardy, Philip Cohen, and Takeru Higuchi. Yet I was uneasy because my experiments were contradicting Green's cyclophorase hypothesis. At the Federation meetings in April of 1950, I heard a talk by Arthur Kornberg and was greatly impressed by his presentation.

It was not a long lecture, only 10 minutes, but I felt as if I were hit by a bolt of lightening. I wanted very much to work under his guidance. Arthur kindly suggested that I apply for an National Institutes of Health (NIH) fellowship, but I could not wait for the final decision and left Madison in September to spend four months with Roger Stanier at the University of California at Berkeley. There again I was very fortunate to meet a number of outstanding scientists, including C. B. van Niel, H. A. Barker, Michael Doudoroff, and William Hassid, all helpful, kind, and welcoming. During my stay in Roger's laboratory, we both worked very hard and were able to publish six papers on tryptophan metabolism in bacteria, which appeared in the *Journal of Biological Chemistry, Journal of Bacteriology,* and *Science* (4–9). It was undoubtedly one of the most productive periods of my life.

In December 1950, my wife and daughter joined me and we moved from California to Bethesda, Maryland to join Arthur's group at the National Institute of Arthritis and Metabolic Diseases. Again I was very lucky, since this was the beginning of the so-called "Golden Age" at the NIH. In the Enzyme Section, which was headed by Arthur, a number of young active scientists were working very hard, including Bernard Horecker and Leon Heppel. Every day, we had a luncheon seminar, which was also attended by Herbert Tabor and Alan Mehler. Sometimes Earl and Thressa Stadtman, Chris Anfinsen, Alton Meister, Sidney Udenfriend, Bernhard Witkop, and other outstanding scientists joined us. It was a stimulating and friendly atmosphere, and because of my friendship with Arthur, I felt confident and comfortable.

For two years at the NIH, I worked with Arthur on bacterial degradation of uracil and characterized the intermediate products, malonyl CoA and acetyl CoA (10, 11). Feodor Lynen came to visit me at the NIH on several occasions, because of his interest in acetyl CoA. This was the beginning of a long association with Fitzy, not only scientifically rewarding, but also cemented by a most heart-warming friendship.

About two years after my arrival in Bethesda, Arthur decided to move to St. Louis to assume the chairmanship of the Department of Microbiology at the Washington University School of Medicine. He asked me to come with him as an assistant professor, his first academic appointment. Again I hesitated somewhat. As it was only my third year in the United States, it seemed an almost overwhelming task for me to teach medical students in English and at the same time help Arthur renovate the department and reorganize the teaching schedule. After a long talk with Arthur, I finally accepted the offer and moved to St. Louis in late 1952.

I spent the summer of 1954 as a visiting scientist at the NIH, collaborating with Herbert Tabor, who had isolated a metabolite of histamine from rat urine but was having some difficulty with its identification. From the elementary analysis, it appeared to be a conjugate of ribose and imidazoleacetic acid, yet it resisted acid and alkaline hydrolysis. From Herman Kalckar I obtained a partially purified riboside hydrolase of *Lactobacillus*. As predicted, the compound Herb had isolated from urine was unequivocally identified as the riboside of imidazoleacetic acid (12, 13). With an offer to be Chief of the Section on Toxicology in the Laboratory of Pharmacology and Toxicology at the National Institute of Arthritis and Metabolic Diseases, I decided to accept and return to Bethesda in late 1954.

DISCOVERY OF OXYGENASE

In reorganizing the research program of the Section on Toxicology, I immediately thought of pyrocatechase, my first love. I decided to explore the possibility that pyrocatechase could incorporate isotopically labeled molecular oxygen into catechol. When I spoke about this idea to several friends at the NIH, almost everyone was skeptical. After all, dehydrogenation was the dominant theory in the field of biological oxidation. The hottest topic in this field at that time was oxidative phosphorylation, which had been studied by many big names, including Severo Ochoa, Ephraim Racker, William Slater, Britton Chance, and Albert Lehninger.

Nevertheless, I was determined to carry out this crucial experiment. Again I was lucky, because I was able to collaborate with Simon Rothberg, a skillful mass spectroscopist at the NIH, and Masayuki Katagiri, an excellent enzymologist from Japan. I used the heavy isotope, ^{18}O, present either as molecular oxygen or as water. Since $H_2{}^{18}O$ was not commercially available in the United States at that time, I had to write to David Samuel at the Weizmann Institute in Israel, who kindly supplied me with a highly concentrated preparation of $H_2{}^{18}O$. The product, muconic acid, was isolated and analyzed for its ^{18}O content by mass spectrometry. Were pyrocatechase a dehydrogenase or an oxidase, the incorporated oxygen should, according to Wieland's theory, have been derived from water. Contrary to that theory and the generally held belief, our results clearly demonstrated that the incorporated oxygen was derived exclusively from molecular oxygen.

Despite our clear-cut results, our data were at first brushed off by our peers and greeted with skepticism. As I was completing a manuscript and about to mail it to the *Journal of Biological Chemistry,* Alan Mehler told me that Howard Mason and his associates in Oregon had just demonstrated that a mushroom phenolase complex could catalyze the incorporation of one atom of molecular oxygen into 3,4-dimethylphenol to form 4,5-dimethylcatechol. Mason's paper (14), along with my communication to the *Journal of the American Chemical Society* (15), appeared in 1955 as "Letters to the Editor." These findings, along with subsequent work by Konrad Bloch and others, established that "oxygen fixation" did indeed occur in biological systems, and the concept of "biological oxygenation" was introduced. In essence, we had revived the old concept of Lavoisier, and I proposed that we call this new type of oxidative enzyme an "oxygenase." In principle oxygenases can be classified into two categories: dioxygenases—which incorporate two atoms of molecular oxygen per substrate molecule, as exemplified by pyrocatechase—and monooxygenases. The latter incorporate one atom of oxygen per substrate molecule, the other atom being reduced to water at the expense of a reducing equivalent such as NADH or NADPH; tryptophan 5-monooxygenase, which produces 5-hydroxytryptophan from tryptophan, is a typical example.

In 1956, I was asked to organize the first symposium on oxygenases at the American Chemical Society meeting in Atlantic City, and in this capacity I invited Mason, Bloch, Mika Hayano, and others. By then, biological oxygenation was generally accepted, and the new field of oxygenases had emerged. In 1957, two major things happened to me. One was a visit from Hoffman-Ostenhof from Vienna, who was planning the International Congress of Biochemistry in that city in 1958. Recognizing the chemical and physiological significance of oxygenases, he asked me to organize and chair a colloquium at the Congress. The other remarkable event was a visit from Kikuo Ogiu, Chairman of the Department of Pharmacology of the Kyoto University Faculty of Medicine. He conveyed an invitation from Ko Hirasawa, Dean of the Faculty of Medicine, to come to Kyoto to be the Chairman and Professor of the Department of Medical Chemistry. This department of biochemistry, the second oldest in Japan, had been founded in 1899 by Torasaburo Araki, the last student of Felix Hoppe-Seyler. It was considered the most presitigious center for biochemistry in Japan. Such an appointment, to a 37-year-old graduate of Osaka University, was unprecedented in Japanese academic circles, and I felt very much honored and flattered. It was the most crucial decision I was called on to make in my scientific career, and I finally decided to return to Japan.

When I was about to leave Bethesda, I met Jesse Greenstein of the National Cancer Institute, one of the pioneers in the biochemistry of cancer. Although he knew me only casually, he urged me to apply for an NIH research grant and kindly volunteered to write a strong recommendation letter. Encouraged by his suggestion, I applied and was fortunate to receive substantial NIH support for a number of years after my return to Japan. This funding, together with other grants from the Jane Coffin Child's Memorial Fund, the Rockefeller Foundation, and various Japanese government agencies, especially the Ministry of Education, Science and Culture, enabled me to rebuild and refurnish the old department at Kyoto University and to continue my work with many young scientists.

KYOTO UNIVERSITY, MY SECOND ALMA MATER

When I moved to Kyoto in February 1958, the first graduate student who applied to join me was Yasutomi Nishizuka, in later years famous for his work on protein kinase C. Soon the Department of Medical Chemistry was deluged with applicants from all over Japan. These postdoctoral fellows and graduate students were all highly motivated, bright, and hard working. The most important thing I had to teach them was to be creative researchers rather than encyclopedic scholars. Although my salary as the Chairman and Professor of one of the most prestigious departments in the medical school in Japan was about one thirteenth (7.7%) of what I earned at the NIH, the work was rewarding and spirits were high.

In the summer of that year, I attended the Fourth International Congress of Biochemistry in Vienna and chaired the Colloquium on Oxygenases. After the Congress, I spent a month touring Europe with Masami Suda, visiting many biochemistry laboratories in England, Germany, France, Denmark, Sweden, Italy, and Switzerland, meeting many scientists whom I had known only from the literature and correspondence. When I visited the Karolinska Institute in Stockholm, Sune Bergström, who had recently returned from Lund to chair the Department of Biochemistry, greeted me warmly and showed me around the Institute. He told me about his pioneering work on prostaglandins, which revealed their versatile and fascinating biological functions, and suggested the involvement of oxygenase reactions in their biosynthesis. This was my first exposure to prostaglandins and the start of a long association with Bergström, Bengt Samuelsson, Sten Orrenius, Peter Reichard, and their colleagues at the Karolinska Institute, and Lars Ernster of Stockholm University. I was also greeted very warmly by Otto Warburg in Berlin, who was happy to hear that tryptophan pyrrolase was a hemoprotein dioxygenase and to discuss the nature of so-called "active oxygen."

Tryptophan pyrrolase cleaves the indole ring of tryptophan, and was shown in 1957 to incorporate two atoms of molecular oxygen into the tryptophan substrate to form formylkynurenine (Figure 3). To reflect this finding, it was renamed tryptophan 2,3-dioxygenase. This enzyme is a hemoprotein, and we later demonstrated that the enzyme-substrate-oxygen complex is the obligatory intermediate in the reaction. Various forms of chemically active oxygen have been reported, including singlet oxygen, superoxide, and hydroxyl radicals, but these do not seem to be involved as dissociable intermediates per se in most oxygenase-catalyzed reactions.

In 1964, I presented a plenary lecture at the Sixth International Congress of Biochemistry in New York and proposed that the enzymatically activated form of oxygen in the tryptophan 2,3-dioxygenase-catalyzed reaction resides

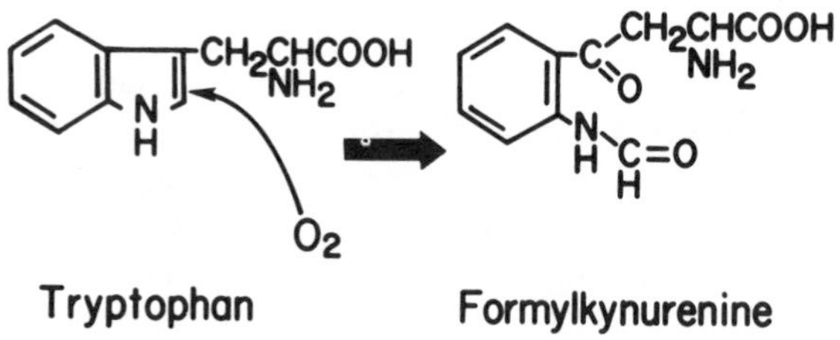

Tryptophan   Formylkynurenine

Figure 3 Tryptophan pyrrolase (2,3-dioxygenase)

in the ternary complex of enzyme heme:oxygen:substrate. The session was chaired by John T. Edsall, who, I recall, made very warm and encouraging remarks at the end of my presentation. This prediction was amply substantiated by the subsequent work of Yuzuru Ishimura and others in my laboratory with tryptophan dioxygenase and later by Fusao Hirata with indoleamine dioxygenase. Furthermore, such a ternary complex was later demonstrated to be an intermediate in the P450-catalyzed monooxygenase reaction studied by Ron Estabrook and associates and also by Irwin Gunsalus and coworkers.

At first, these oxygenase reactions were generally thought to be rather unusual and limited to primitive organisms, such as soil bacteria and mushrooms. It took several years before the universal existence of oxygenases was confirmed in many laboratories (16, 17). During the next decade, a number of oxygenases were isolated from animals, plants, and microorganisms (18), and some of them were crystallized by Mitsuhiro Nozaki, Hitoshi Fujisawa, Shozo Yamamoto, and others in my laboratory (19). These studies elucidated numerous metabolic pathways of physiological importance and demonstrated the ubiquitous presence of oxygenase reactions. For example, the two most important metabolic pathways for tryptophan in mammals, one leading to formation of pyridine nucleotide coenzymes and poly(ADP-ribose), and the other to formation of various indoleamines, are initiated by two well-known oxygenases, tryptophan 2,3-dioxygenase and tryptophan 5-monooxygenase, respectively.

In the 1940s, Roger Stanier isolated several strains of soil microorganisms that metabolized tryptophan via kynurenic acid. When I was in his laboratory in 1950, the pathway through anthranilate and catechol was referred to as the "aromatic pathway," while the pathway through kynurenic acid was referred to as the "quinolinic pathway." After returning to Kyoto in 1958, I carried out detailed studies on both pathways in collaboration with many researchers, including Shigeru Kuno, Yasutomi Nishizuka, Shosaku Numa, Arata Ichiyama, Hitoshi Fujisawa, Siroh Senoh, Yutaka Kojima, and others.

A large number of intermediates and enzymes responsible for the steps in these pathways were isolated and characterized.

We then proceeded to investigate the metabolic pathways of tryptophan in mammalian liver. Two pathways were delineated, one leading to the biosynthesis of NAD, and the other, the glutarate pathway, leading to the complete combustion of the carbon skeleton of tryptophan. Again, intermediates were isolated and characterized, and the enzymes responsible for each step were purified in collaboration with Nishizuka, Numa, Senoh, Ichiyama, Atsushi and Teruko Nakazawa, Hiroshi Okamoto, Takashi Murachi, and others.

DISCOVERY OF POLY- AND MONO(ADP-RIBOSYL)ATION

During the course of these studies, we were able, in collaboration with Nishizuka, Tasuku Honjo and Kunihiro Ueda, and subsequently with Hiroto Okayama, Yutaka Shizuta, and Ronald Reeder, to demonstrate that NAD, the well-known coenzyme of many dehydrogenases, can serve as an ADP-ribosyl donor in a unique type of covalent modification of proteins. In the presence of the appropriate enzyme, either a single ADP-ribosyl moiety can be covalently attached to an acceptor protein, or in some instances, this moiety can be polymerized to poly ADP-ribose. For example, in 1968, the ADP-ribosyl moiety of NAD was shown to be transferred to Elongation Factor 2. As a consequence, Factor 2 was inactivated and protein synthesis was inhibited. The reaction, catalyzed by diphtheria toxin, accounts for its toxicity. The first demonstration of this unique covalent modification of proteins had thus been made (20). Later studies in other laboratories have shown that other bacterial toxins, such as pertussis, botulinum, and cholera, catalyze a similar reaction. The biosynthesis and the structure of poly(ADP-ribose) were investigated independently in three laboratories: Paul Mandel's laboratory in Strassbourg, Takashi Sugimura's laboratory in Tokyo, and ours in Kyoto (21). It has been almost 30 years since the discovery of this unique and interesting polymer in the nucleus of mammalian cells, but its true physiological function remains unknown; several lines of circumstantial evidence indicate its involvement in DNA repair, in development, and in carcinogenesis (22).

TRYPTOPHAN REVISITED

With regard to other aspects of tryptophan metabolism, we discovered and studied two new enzymes during the period from 1967 to 1983: indoleamine 2,3-dioxygenase (IDO), and tryptophan side-chain α,β-oxidase (TSO). In

1937, Kotake and Noboru Ito had reported that rabbits fed D-tryptophan excreted D-kynurenine in their urine. In as much as liver tryptophan pyrrolase (TPO) is specific for the L-isomer, these authors postulated the existence of another enzyme for cleavage of the pyrrole moiety of D-tryprophan. In 1963, Kiyoshi Higuchi, Kuno, and I isolated and purified a new enzyme from rabbit intestine that catalyzed such a reaction. In contrast to TPO, which was found only in liver, this new type of TPO was found in many tissues other than the liver. The liver enzyme is specific for L-tryptophan, whereas the intestinal enzyme acts on both L- and D-tryptophan, as well as on L- and D-5-hydroxytryptophan and several indoleamines. The intestinal enzyme has been therefore referred to as indoleamine 2,3-dioxygenase.

Interestingly, IDO was shown to require and utilize the superoxide anion, O_2^-, a univalently reduced form of oxygen, in addition to molecular oxygen. The evidence for participation of the superoxide anion in the IDO-catalyzed reaction can be summarized as follows: First, O_2^- is required and utilized by the enzyme; the O_2^- can be provided as the K salt of O_2^- or by enzymatic generation of O_2^- by the xanthine oxidase system. Secondly, scavengers of O_2^-, such as superoxide dismutase or tyron, inhibit the reaction. Finally, the heavy oxygen isotope, ^{18}O, is incorporated into the product of the reaction when carried out in the presence of $K_2{}^{18}O$.

In 1979, in collaboration with Ryotaro Yoshida, Yoshihiro Urade, and coworkers, we demonstrated that IDO activity in mouse lung increased more than 100-fold over the basal level in response to infection with a virus such as influenza (23). When mice were exposed to the virus, the specific activity of IDO in mouse lung started to increase after a 5-day lag period and peaked within 10 days; the activity then gradually decreased and returned to normal levels in about 3 weeks. When this complex response was studied in greater detail by Yoshida and coworkers, we could show that IDO activity in human lung slices was induced by γ-interferon, while α- and β-interferons were hardly active. Further studies in my laboratory, together with those in the laboratories of Ryo Kido, Elmer Pfefferkorn, and others, revealed that when virus, bacteria, or parasites invade host tissues, interferons are generated and IDO activity is elevated. It is also well known that O_2^- is generated by granulocytes under these conditions. As a consequence, tryptophan, an essential amino acid, is degraded by IDO and O_2^-. and its depletion inhibits the growth of viruses, bacteria, parasites, and tumor cells.

Our preliminary studies with IDO, the participation of O_2^-, and the possible role of IDO in inflammatory processes were first announced at the First International Study Group for Tryptophan Research (ISTRY) meeting in Padova, Italy in 1974, organized by the late Luigi Musajo, Graziella Allegri, and their associates. Presumably, in recognition of this effort, I received in

1988 the degree of Honoris Causa from the University of Padova, the second oldest university in Europe.

The other new enzyme, tryptophan side-chain α,β-oxidase (TSO), isolated and purified from *Pseudomonas,* catalyzes a novel tryptophan degradation (24). In collaboration with Katsuji Takai, Shu Narumiya, Flora Zavala, and others, we showed that TSO catalyzes the oxidation of the side chain of tryptophan, forming one equivalent each of 3-indoleglycolaldehyde, ammonia, and CO_2. TSO shows a broad substrate specifity and oxidizes not only L- and D-tryptophan, but also a variety of indole deriatives, such as indole ethanol and methanol. Interestingly, a number of peptides containing tryptophan also serve as substrates for TSO. The product of enzymatic oxidation by TSO of a tryptophan residue was identified to be α,β-didehydrotryptophan. Because many peptides (e.g. somatostatin, ACTH, glucagon, LHRF) and proteins are substrates for such a modification, TSO has proved to be a useful method for modification of tryptophan residues in peptide chemistry.

ACTIVE RETIREMENT

In 1983, I retired from Kyoto University, as is mandatory at the age of 63. During my 25-year tenure, the second Department of Medical Chemistry was created in 1968, and Shosaku Numa was installed to head this department. He had been a protégé of Feodor Lynen, who came often to Japan and visited us in Kyoto. Being the son-in-law of Wieland, he was always appreciative of my discovery and subsequent work on oxygenases. In 1976, at the Tenth International Congress of Biochemistry held in Hamburg, I presided over the General Assembly of the International Union of Biochemistry (IUB), and recommended that Lynen become the President-Elect. Shockingly, he succumbed soon thereafter to an aortic aneurysm. A colorful, kind, and gregarious person, an exemplar of Bavarian "Gemütlichkeit," he earned the respect and affection of everyone he met. On the occasion of a visit to his laboratory in Münich, I had met Harland Wood, who was spending a sabbatical leave there, and started another long and wonderful friendship, which was tragically lost with Harland's death in 1992.

During my time at Kyoto University, I had been responsible for the guidance of more than 500 staff members, postdoctoral fellows, graduate students, and visiting scientists from all over Japan and numerous foreign countries. Furthermore, concurrent with my appointment at Kyoto University, I served as Professor at the Department of Biochemistry of Osaka University School of Medicine (1961–1963), and also as Professor at the Department of Physiological Chemistry and Nutrition, Faculty of Medicine of the University of Tokyo (1970–1974). Through these appointments, I had the pleasure of meeting many outstanding scientists and greatly enjoyed the association with a large number of able young students in both universities.

Upon the occasion of my retirement, David E. Green sent me this greeting.

> The fortunate ones are thrice blessed. The first blessing is to have a profession that can lead to organized discoveries. The second blessing is to share these discoveries with many colleagues. The third blessing is to make a discovery that improves the quality of life.
>
> Most scientists have the first blessing. A few come close to the second blessing. A rare few, who are among our greatest scientists, receive all three blessings. Osamu Hayaishi is among the handful in the world that have reached these three great honors. We all honor him for this triumph of triumphs.

I doubt that I deserved then or now such august compliments, but I do consider myself extremely fortunate to have lived and participated in a most exciting era of biochemistry and to have shared the excitement with so many friends and with young, able collaborators from all over the world.

In 1983, I was appointed President of the Osaka Medical College at Takatsuki City near Osaka. At about the same time, the Research Development Corporation of Japan (JRDC), a subsidiary of the Science and Technology Agency of the Japanese government, appointed me to be leader of the Hayaishi Bioinformation Transfer Project. This project, part of Exploratory Research for Advanced Technology (ERATO), was initiated by the Japanese government in 1981 to foster the creation of advanced technologies and promote future interdisciplinary scientific activities. With the generous funding from ERATO, I was able to organize a team of some 30 scientists and to launch a new project investigating the metabolism and functions of prostaglandins and related compounds in the central nervous system of mammals. With this I found an unexpected opportunity for new directions in my research career.

SABBATICAL LEAVES

Although most Japanese universities lack a sabbatical system, professors are usually allowed some leave provided that the teaching and administrative responsibilities are assumed by others in the department. During my tenure at Kyoto University, I spent six months in the Laboratory of Clinical Biochemistry at the National Heart Institute with Sidney Udenfriend in 1962. While in Udenfriend's laboratory, I was able to associate with many excellent scientists, including Julius Axelrod, Bernhard Witkop, and Herbert Weissbach. Herb and I discovered an ADP-stimulated threonine deaminase from *Clostridium tetanomorphum,* one of the first examples of a positive feedback regulation. On my way home to Japan, I visited Jacques Monod and Francois

Jacob at the Pasteur Institute in Paris. When they heard my seminar, they both exclaimed "allostery," the term that they had just coined for the negative feedback phenomenon.

In 1968, the late Sidney Colowick and Victor Najjar invited me to spend three months as a visiting professor in the microbiology department at Vanderbilt University School of Medicine in Nashville, supported by a special grant from the NIH. At the same time, similar appointments were held by Stanford Moore in the biochemistry department and by Paul Greengard in the pharmacology department. Greengard, Colowick, and I were able to show the reversibility of the adenylate cyclase reaction utilizing the crystalline enzyme preparation that Masaharu Hirata and I had purified from *Brevibacterium* in Kyoto. The results clearly indicated the "high-energy" bond nature of the cyclic phosphodiester linkage (25). During that three-month stay in the "southern capital," we made many new friends—Earl Sutherland, Stanley Cohen, Oscar Touster, Rollo and Jane Park, Leon Cunningham, Tadashi Inagami, and Michio Ui—and thoroughly enjoyed their hospitality. It was also a wonderful opportunity to learn about the southern states, which, according to Colowick, is the best way to become acquainted with and appreciate US history and tradition.

In 1973, when the Ninth International Congress of Biochemistry was held in Stockholm, I was nominated President of the IUB. Partly because I have always considered myself a bench biochemist and ignorant of international affairs and partly because of the language problem, I was hesitant at first to undertake the appointment. I was eventually persuaded by Harland Wood, who telephoned me several times at the NIH during my stays there as a Fogarty Scholar (1972–1974). It was during this period in Bethesda, where Takiko and I spent two to three months in each of three summers, that I would write or edit books, organize meetings, and do some collaborative work with scientists both inside and outside the NIH. This was a wonderful way to recharge batteries, learn new things, and make many new friends.

As the successor to Hugo Theorell to preside over the IUB, I worked hard to make the organization more prosperous, active, and meaningful to every biochemist in the world. I was very fortunate to have had many excellent helpers and advisors, above all, Harland Wood, William Whelan, the hard-working and excellent general secretary, and Kunio Yagi. I felt very much honored that the IUB decided to create an Osamu Hayaishi lectureship supported by an endowment from Keizo Saji, President of the Suntory Company, a long-time friend. The first lecture of this series was presented by Jeremy Knowles at the Twelfth International Congress of Biochemistry in Perth in 1982.

In 1987, the Osaka Bioscience Institute was created by the City of Osaka in commemoration of its centennial anniversary, and I was appointed as the director of this new institute, where I have been able to continue my work on prostaglandins in the central nervous system (CNS).

THE MYSTERY OF SLEEP

Prostaglandin (PG) was discovered in semen by Raphael Kurzrok and Charles Lieb (with the technical assistance of Sarah Ratner!) in 1930, but the unique structure of this family of compounds was not elucidated until the early 1960s by Bergström and coworkers. During the next 30 years, research in the prostaglandin field progressed exponentially. Various PGs were shown to be ubiquitously distributed throughout almost all types of cells in mammalian organs and tissues and to play vital roles in the regulation of a variety of physiological functions and pathological conditions, such as contraction and relaxation of smooth muscles, aggregation and disaggregation of blood platelets, inflammation, and pain. In 1982, the Nobel Prize in Medicine or Physiology was awarded to three pioneers in this area: Bergström, Samuelsson, and John Vane. However, the abundance, metabolism, and function of PGs in the CNS was poorly understood. To quote from the *Biochemical Basis of Neuropharmacology,* "The mammalian CNS contains predominantly the PGF series, with small amounts of a PGE in some cases. None of the available evidence indicates that prostaglandins act as typical CNS neurotransmitters" (26).

In the late 1970s, several groups, including my own, reported the presence of a relatively large amount of PGD_2 in the brains of rats and other mammals, including humans. Until that time, PGD_2 had long been considered a minor and biologically inactive prostanoid or perhaps even as a non-enzymatic decomposition product of PGH_2. However, these publications indicated that PGD_2 is unique among the PGs in being present in relatively high concentrations in the mammalian brain. Subsequently, Takao Shimizu, Kikuko Watanabe, Urade, and others found that PGD_2 is actively synthesized and metabolized by specific enzymes in both neurons and glial cells.

To determine the neural functions of PGD_2, Yasuyoshi and Yumiko Watanabe studied the intracerebral distribution of its binding protein, the putative receptor, by autoradiography with $[^3H]PGD_2$ combined with computer-assisted image processing and color coding. The binding protein for PGD_2 was located mainly in the gray matter, namely the neuron-rich areas. It was highly concentrated in specific regions, such as the olfactory bulb, cingulate cortex, occipital cortex, hippocampus, hypothalamus, and preoptic area, indicative of PGD_2 involvement in certain specific neural functions.

The preoptic area has long been proposed as a center of sleep and temperature regulation on the basis of neuroanatomical and electrophysiological experiments. However, the biochemical mechanisms involved in the induction of sleep had never been elucidated. While studying the effect of PGD_2 on brain temperature, Ryuji Ueno and I, in collaboration with the late Teruo Nakayama and Youzou Ishikawa of Osaka University, happened to observe that when saline was injected into the preoptic area of control rats under conditions of partial sleep deprivation, the rats remained awake most of the time. However, when a few nanomols of PGD_2 were micro-injected into the preoptic areas, the wakefulness period decreased almost 50% and the amount of sleep increased more than fivefold. The site of action was confined to the preeoptic area, whereas a PGD_2 injection into other areas (e.g. posterior hypothalamus, cerebral cortex, locus ceruleus, thalamus) had no significant effect. The response was dose-dependent and specific for PGD_2, indicating that PGD_2 may be a natural or physiological sleep-promoting substance in the rat brain (27).

Sleep is one of the most important and yet least understood biological phenomena of the brain. Although we spend almost one-third of our lifetime sleeping and repeat the sleep-wake cycle every day and night, the biochemical mechanism of sleep-wake regulation has thus far eluded us. During the past several decades, an intensive search for endogenous sleep-regulating substances has been carried out in laboratories throughout the world, but the results have not been convincing. Having decided to focus our research on the mystery of sleep, I consulted Yasuji Katsuki, who kindly introduced me to Shojiro Inoué of the Tokyo Medical and Dental University. With his help, we employed the more sophisticated continuous infusion sleep bioassay system, originally developed by Inoué and Kazuki Honda. Eventually we could show that PGD_2 and E_2 are probably two of the major endogenous sleep-regulating substances. PGD_2 induces sleep and PGE_2 promotes wakefulness in the rat, dog, and monkey, and likely in humans as well. The site of PGD_2 action appears to be localized to the sleep center in or near the preoptic area. PGE_2 action is on the wake center in or near the posterior hypothalamus. The REM sleep center in the brain stem also appears to be under the control of PGD_2 and E_2, although further experiments are required to pinpoint the exact sites of their actions (28, 29).

Recently, we discovered that inorganic quadrivalent selenium compounds are potent, specific, and noncompetitive inhibitors of brain prostaglandin D synthase, with a K_i value of about 10 μM. In contrast, hexavalent compounds and organic selenium compounds are not inhibitory at all. Other enzymes in the arachidonate cascade system are not inhibited. This inhibition can be reversed by sulfhydryl compounds, such as glutathione or dithiothreitol (DTT). When selenium chloride was slowly infused into the third ventricle of a rat, sleep was inhibited in a time- and dose-dependent manner. After about two hours, both slow-wave sleep and REM sleep were almost fully inhibited and the rat was completely and continuously awake (30). The effect was reversible: When the infusion was interrupted, sleep was restored. Furthermore, the effect was reversed by sulfhydryl compounds such as DTT and reduced glutathione, as in the case of in vitro experiments. These results clearly show that PGD_2 is involved in the regulation of physiological sleep and that PGD synthase is the key enzyme in control of natural sleep.

In 1990, when I presented a lecture on our work on sleep at the Royal Society in London, I met Victor Pentreath of the University of Salford, United Kingdom, who told me that the sleep seen in African sleeping sickness patients is probably caused by PGD_2. He and colleagues determined the amount of PGD_2 and PGE_2 and also $IL_1\alpha$, another putative sleep substance in the cerebrospinal fluid (CSF) in sleeping sickness patients. The amount of PGD_2 was specifically, significantly, and progressively elevated in the CSF, indicating that sleep in the terminal stage of these patients is caused mainly, if not exclusively, by PGD_2 (31). Roberts and coworkers reported that patients with systemic mastocytosis enter into deep sleep after the episodic production of large amounts of PGD_2 by their mast cells (32). All told, these results indicate that PGD_2 induces sleep in humans under certain pathological conditions.

As regards the waking effect of PGE_2, in order to find out if PGE_2 is indeed involved in the regulation of the sleep-wake cycle, Hitoshi Matsumura assessed the effect of AH 6809, a PGE_2 antagonist in the smooth muscle and pain systems. If PGE_2 induces wakefulness or inhibits sleep under physiological conditions, such a PGE_2 antagonist should counteract the effect of endogenous PGE_2; that is, it should increase the amount of sleep or decrease the time of wakefulness. When AH 6809 was infused slowly into the third ventricle of a rat during the night, the amount of slow-wave sleep was increased by 22% over the control, and REM sleep by 89%. These data clearly indicate that PGE_2 is involved in the maintenance of the waking state under physiological conditions (33).

In 1988, while visiting at Stanford University, I was introduced to William Dement, Director of the Sleep Disorder Center, who maintained a large colony of narcoleptic dogs. In collaboration with Dement and his colleagues, Seiji Nishino and I were able to demonstrate that the treatment of cataplexy in these narcoleptic dogs with PGE_2 or PGE_2 methyl ester induced a dose-dependent reduction of this manifestation of pathological REM sleep.

Narcolepsy is a genetic disorder of sleep, characterized by an uncontrollable desire for sleep. This sudden attack of sleep is usually triggered by emotions such as anger, joy, or even appetite. When such a dog sees his favorite food or female, he almost immediately undergoes paralysis and sleeps. To study this phenomenon more quantitatively, Dement and coworkers developed a unique biological assay system called the "food-elicited cataplexy" test. In this test, the elapsed time to finish eating 12 pieces of food and the number of attacks of cataplexy during this period are determined. When the PGE_2 methyl ester, which penetrates the blood-brain barrier much more easily than PGE_2, was injected intravenously, the number of attacks as well as the total elapsed time decreased significantly compared with the control dog, indicating that PGE_2 inhibits sleep or induces wakefulness in a narcoleptic dog. In principle, therefore, PGE_2 and its derivatives could be a potentially useful drug for prevention or treatment in narcoleptic patients (34).

During the past 10 years or so, I have again been fortunate to be associated with and helped by a large number of established leading scientists in the sleep research field, colleagues who have been extremely helpful and encouraging to me, an amateur sleep researcher. Among them I am especially grateful to Shojiro Inoué, Michel Jouvet, William Dement, Michael Chase, Victor Pentreath, and Alexander Borbély for their collaboration, hospitality, and kindnesses.

Recently, the molecular biological approach to the sleep problem has yielded several unexpected findings that may lead to an unusual new hypothesis about the mechanism of sleep regulation. PGD synthase catalyzes the isomerization of PGH_2 to PGD_2 (Figure 4). This enzyme is a monomeric protein with a molecular weight of approximately 26,000. In order to study the properties and the regulatory mechanisms of PGD synthase, and possibly to find a new specific inhibitor of this enzyme, Urade and I purified the enzyme from human and rat brains, isolated cDNAs encoding PGD synthase, and determined the nucleotide and amino-acid sequences of the rat brain enzyme and two isozymes of the human enzyme. A computer-assisted homology search in databases of protein primary structures showed PGD synthase to be a member of the lipocalin superfamily (35), small secretory proteins that bind and transport small lipophilic molecules widely distributed in the animal kingdom (e.g. mammals, birds, reptiles, insects, and crustacea). The only known exception is PGD synthase, which is an enzyme rather than a lipid transporter. The gene structure of the members of this family has also been delineated, and that of the rat enzyme was remarkably analogous to the gene structures of several other members of this family, in the number and sizes of exons and phase of splicing of introns. Based on these results, a phylogenetic tree of the lipocalin superfamily was constructed by Hiroyuki Toh of the Protein Engineering Research Institute, Osaka.

Because of the high evolutionary divergence of the lipocalin superfamily, homology of the amino-acid sequences is rather weak. Yet, the tertiary structure is well conserved, forming a remarkably similar β-barrel, as revealed by X-ray crystallographic studies on retinal-binding protein, β-lactoglobulin,

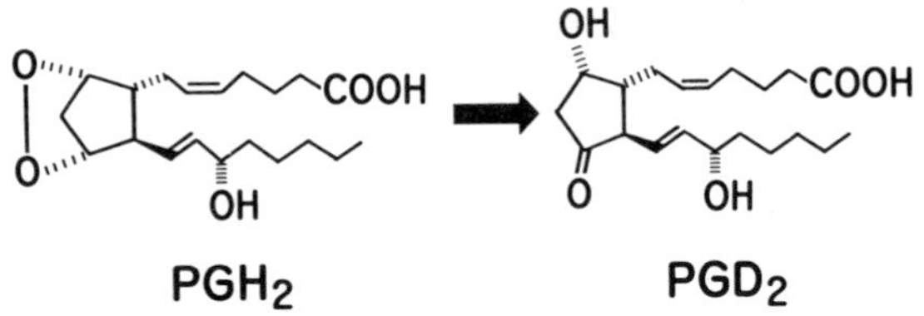

Figure 4 Enzymatic conversion of PGH_2 to PGD_2 by PGD synthase

and insecticyanin. The deduced tertiary structure of the brain PGD synthase indicates that the free SH group of cysteine at residue 65 is unique to the human and rat PGD synthases in the lipocalin family. The residue is located in the hydrophobic pocket and is thought to be in the active site of the enzyme. As mentioned earlier, inorganic quadrivalent selenium compounds are potent, specific, and noncompetitive inhibitors of brain PGD synthase, presumably interacting with this free sulfhydryl group.

The autoradiograms obtained after in situ hybridization with ^{35}S-labeled antisense RNA for rat brain PGD synthase showed that the mRNA for the enzyme was intensely expressed in the choroid plexus and the leptomeninges throughout the brain, and in the spinal cord of the adult rat. Positive signals indicating mRNA encoding the enzyme were also observed in oligodendrocytes, but hardly detected in neurons throughout the CNS (36).

Immunohistochemical studies with monoclonal and polyclonal antibodies against the enzyme revealed immunoreactivity in the leptomeninges and choroid plexus. The immunoreactivity for the enzyme was also found in oligodendrocytes, confirming our previously reported data, but these positive cells were diffusely and sporadically distributed in various parts of the brain. Immunoreactivity was occasionally observed in somata and dendrites of stellate neurons, apical dendrites of pyramidal neurons, and microglia-like cells in the superficial part of the cortex. However, no positive signals for the mRNA of the enzyme were detected in these cells.

The PGD synthase activity (nmol/min/mg of protein) was remarkably high in the isolated leptomeninges (14.2) and choroid plexus (7.0) of the adult rat brain, as compared with the activity in the whole brain (2.9).

During the past 10 years, we have shown that PGD_2 is the major endogenous sleep-regulating substance in the brains of rats, monkeys, and probably humans, and that its site of action is in or near the preoptic area. However, the major site of synthesis of PGD_2 has not yet been clearly demonstrated. In the study described above, PGD synthase was shown to be predominantly expressed in the leptomeninges, choroid plexus, and, to a lesser extent, in oligodendrocytes in the adult rat brain. PGD synthase-like immunoreactivity was also demonstrated mainly in these tissues and cells. Furthermore, the enzyme activity was shown to be localized in these tissues. The experimental evidence therefore strongly indicates that PGD synthase is present mainly, if not exclusively, in choroid plexus, leptomeninges, and oligodendrocytes, and that PGD_2 is produced mostly in these tissues and cells.

Earlier, we characterized PGD synthase as a membrane-associated enzyme, because the enzyme is N-glycosylated, has a putative signal sequence, and by immuno-electron microscopy is located on the rough-surfaced endoplasmic reticulum and outer nuclear membrane of oligodendrocytes. These characteristics are also shared by various secretory proteins. According to Kikuko Watanabe, rat CSF contains a significant PGD synthase activity (200–300 nmol/min/mg protein). Hoffmann et al (37) recently demonstrated that the protein termed β-trace, a major constituent of human CSF, shows a high degree of homology with the rat and human PGD synthases, indicating that β-trace is structurally identical to PGD synthase. Furthermore, Achen et al (38) recently demonstrated that the major protein secreted by amphibian choroid plexus has the highest homology (41% identity and 84% similarity) with the rat and human PGD synthase among lipocalins so far identified.

As mentioned earlier, immunoreactivity for PGD synthase is sometimes observed in certain neurons and microglia-like cells in the superficial part of the cortex, although the mRNA for the enzyme is not detected in these cells. This difference might be due to the greater ease of detection of immunoreactivity, despite the use of a highly sensitive method for in situ hybridization. Alternatively, these results may be interpreted to mean that the enzyme was originally produced in the cells of leptomeninges and choroid plexus, and possibly oligodendrocytes, and then secreted and transported through the CSF to the neurons.

Several lines of experimental evidence presented by us (36) and others pose intriguing and provocative hypotheses that need to be critically evaluated. Are the major sites of production of PGD_2, the leptomeninges, choroid plexus, and oligodendrocytes? If so, how is the process controlled? Alternatively, the PGD synthase may be secreted and transported to some other sites, possibly glial cells and neurons, where the supply of its substrate is presumably more abundant. Although the choroid plexus and leptomeninges have long been believed to play a role in the mechanical support and chemical homeostasis of the brain, a more active function as a neuroendocrine pathway for communication within the CNS has also been suggested recently. For example, the choroid plexus and walls of the ventricles contain serotonergic-binding proteins (39–41), and glucose utilization increases in the choroid plexus during slow-wave sleep (42). The presence of the sleep-promoting substance(s) in the CSF has been shown by a number of investigators (for a review, see 43), and suggests a humoral, rather than a neural, mechanism for sleep regulation. In their pioneering work on "sleep substances," Rene Legendre and Henri Pieron in 1907 deprived adult dogs of total sleep for 150 to 293 hours, took their CSF, blood, or serum, injected it into normal dogs, and showed that sleep-promoting substances accumulated in the CSF or serum under these conditions (44). Independently, Kuniomi Ishimori in Japan made similar observations with dogs in 1909 (45). Subsequently, several other authors also reported the presence of a sleep-promoting substance(s) in the CSF and urine, and some of them were chemically identified. However, their site of synthesis has not been clearly delineated. Our results (36) may provide some experimental basis for a rather provocative hypothesis that PGD_2 is not a typical and classical neurotransmitter but rather a "neurohormone" that circulates through the CSF in the ventricles, subarachnoidal space, and superior sagittal sinus. Alternatively, the PGD synthase itself may serve as a transporter or a carrier protein for PGD_2. Or else, PGD synthase may relocate into other parts of the brain tissue and produce PGD_2 locally.

EPILOGUE

Life is interesting because it is so unpredictable. During the past 50 years, my research interests have wandered about to several places, mostly guided by serendipity though sometimes by the sagacity of experience. Now I have come to my final destination, sleep. (I hope that readers of this essay have stayed awake to this point.) I am determined to continue my effort to challenge and solve this formidable problem.

I have always been fortunate to be associated with so many outstanding, kind, and helpful people, both scientists and nonscientists, whose names are too numerous to mention here, but to whom I express my deepest gratitude. Finally, I dedicate this article to Takiko, for her many years of faithful support and devotion and for providing a happy family life with our daughter, Mariko, Masashi Akizuki, her husband, and their sons, Masato and Shuji. I am deeply indebted to Dr. Arthur Kornberg for critical and thorough reading of this manuscript and numerous helpful comments. I also thank Drs. L. Frye, F. I. Tsuji, and S. Sri Kantha for help in the preparation of this manuscript, and Miss Junko Kawahara for secretarial assistance.

Literature Cited

1. Hayaishi O, Hashimoto Z. 1950. *J. Biochem.* 37:371–74
2. Hoffman-Ostenhof O. 1954. *Enzymologie*. Vienna: Springer-Verlag
3. Wieland H. 1932. *On the Mechanism of Oxidation*. New Haven, Conn: Yale Univ. Press
4. Stanier RY, Hayaishi O. 1951. *Science* 114:326–30
5. Stanier RY, Hayaishi O, Tsuchida M. 1951. *J. Bacteriol.* 62:355–66
6. Stanier RY, Hayaishi O. 1951. *J. Bacteriol.* 62:367–75
7. Hayaishi O, Stanier RY. 1951. *J. Bacteriol.* 62:691–701
8. Tsuchida M, Hayaishi O, Stanier RY. 1952. *J. Bacteriol.* 64:49–54
9. Hayaishi O, Stanier RY. 1952. *J. Biol. Chem.* 195:735–40
10. Hayaishi O, Kornberg A. 1952. *J. Biol. Chem.* 197:717–32
11. Hayaishi O, Kornberg A. 1954. *J. Biol. Chem.* 206:647–63
12. Hayaishi O, Tabor H, Hayaishi T. 1954. *J. Am. Chem. Soc.* 76:5570–71
13. Tabor H, Hayaishi O. 1955. *J. Am. Chem. Soc.* 77:505–6
14. Mason HS, Fowlks WL, Peterson E. 1955. *J. Am. Chem. Soc.* 77: 2914–15
15. Hayaishi O, Katagiri M, Rothberg S. 1955. *J. Am. Chem. Soc.* 77:5450–51
16. Hayaishi O. 1962. *Annu. Rev. Biochem.* 31:25–46
17. Hayaishi O. 1962. In *Oxygenases,* ed. O Hayaishi, p. 1. New York: Academic
18. Hayaishi O. 1969. *Annu. Rev. Biochem.* 38:21–44
19. Hayaishi O. 1966. *Bacteriol. Rev.* 30: 720–31
20. Honjo T, Nishizuka Y, Hayaishi O, Kato I. 1968. *J. Biol. Chem.* 243:3553–55
21. Hayaishi O, Ueda K. 1977. *Annu. Rev. Biochem.* 46:95–116
22. Hayaishi O, Ueda K. 1982. In *ADP-Ribosylation Reactions—Biology and Medicine*, ed. O Hayaishi, K Ueda, pp. 3–16. New York: Academic
23. Yoshida R, Urade Y, Tokuda M, Hayaishi O. 1979. *Proc. Natl. Acad. Sci. USA* 76:4084–86
24. Takai K, Ushiro H, Noda Y, Narumiya S, Tokuyama T, Hayaishi O. 1977. *J. Biol. Chem.* 252:2648–56
25. Hayaishi O, Greengard P, Colowick SP. 1971. *J. Biol. Chem.* 246:5840–43
26. Cooper JR, Bloom F, Roth RH. 1974. *Biochemical Basis of Neuropharmacology*, pp. 242–43. Oxford: Oxford Univ. Press
27. Ueno R, Ishikawa Y. Nakayama T, Hayaishi O. 1982. *Biochem. Biophys. Res. Commun.* 109:576–82
28. Hayaishi O. 1988. *J. Biol. Chem.* 263:14593–96
29. Hayaishi O. 1991. *FASEB J.* 5:2575–81
30. Matsumura H, Takahata R, Hayaishi O. 1991. *Proc. Natl. Acad. Sci. USA* 88:9046–50
31. Pentreath VW, Rees K, Owolabi OA, Philip KA, Doua R. 1990. *Trans. R. Soc. Trop. Med. Hyg.* 84:795–99
32. Roberts JL II, Sweetman BJ, Lewis RA, Austen KF, Oats JA. 1980. *N. Engl. J. Med.* 303:1400–4
33. Matsumura H, Honda K, Choi WS, Inoué S, Sakai T, Hayaishi O. 1989. *Proc. Natl. Acad. Sci. USA* 86:5666–69
34. Nishino S, Mignot E, Fruhstorfer B, Dement WC, Hayaishi O. 1989. *Proc. Natl. Acad. Sci. USA* 96:2483–87
35. Nagata A, Suzuki Y, Igarashi M, Eguchi N, Toh H, et al. 1991. *Proc. Natl. Acad. Sci. USA* 88:4020–24
36. Urade Y, Kitahama K, Ohishi H, Kaneko T, Mizuno N, Hayaishi O. 1993. *Proc. Natl. Acad. Sci. USA* 90:9070–74
37. Hoffmann A, Conradt HS, Gross G, Nimtz M, Lottspeich F, Wurster U. 1993. *J. Neurochem.* 61:451–56
38. Achen MG, Harms PJ, Thomas T, Richardson SJ, Wettenhall REH, Schreiber G. 1992. *J. Biol. Chem.* 267:23170–74
39. Moskowitz MA, Liebmann JE, Reinhard JF Jr, Schlosberg A. 1979. *Brain Res.* 169:590–94
40. Yagaloff KA, Hartig PR. 1985. *J. Neurosci.* 5:3178–83

41. Conn PJ, Sanders-Bush E, Hoffman BJ, Hartig PR. 1986. *Proc. Natl. Acad. Sci. USA* 83:4086–88
42. Bobillier P, Seguin S, Petitjean F, Buda C, Salvert D, et al. 1982. *Brain Res.* 240:359–63
43. Inoué S. 1989. *Biology of Sleep Substances*. Roca Raton, FL: CRC
44. Legendre R, Pieron H. 1907. *C. R. Soc. Biol.* 62:312–14
45. Ishimori K. 1909. *Tokyo Igakkai Zasshi* 23:429–57

Annu. Rev. Biochem. 1994. 63:25–61

STEROID 5α-REDUCTASE: TWO GENES/TWO ENZYMES

David W. Russell[1] *and Jean D. Wilson*

University of Texas Southwestern Medical Center, 5323 Harry Hines Boulevard, Dallas, Texas 75235

KEYWORDS: androgens, prostate, membrane proteins, NADPH-dependent reductases, inhibitors

---

CONTENTS

[1]Department of Molecular Genetics. To whom correspondence should be addressed

0066-4154/94/0701-0025$05.00

INTRODUCTION

Virilization in mammals is mediated by two steroid hormones, testosterone and dihydrotestosterone. Both hormones bind to a typical steroid hormone receptor, the androgen receptor, and activate genes containing androgen-responsive DNA sequences. Early studies implicated testosterone as the major androgenic hormone and postulated that dihydrotestosterone was an inactive metabolite of testosterone.

The notion that dihydrotestosterone is in fact a potent androgen with physiological roles distinct from those of testosterone came from two observations. First, androgen target tissues contained an enzyme activity (steroid 5α-reductase) capable of reducing testosterone to dihydrotestosterone (1), and second, the product of this enzyme accumulated in the nuclei of responsive cells, such as those of the rat ventral prostate (1, 2). The subsequent study of an inborn error of male phenotypic sexual differentiation, now termed steroid 5α-reductase 2 deficiency, provided formal genetic proof of the crucial role of dihydrotestosterone in androgen action (3, 4).

Males with this genetic disease have a biochemical defect in the synthesis of dihydrotestosterone in the embryo, which in turn leads to a developmental defect in the formation of the external genitalia and the prostate (3, 4). They exhibit a striking phenotype in which the internal genitalia (epididymis, seminal vesicles, vas deferens) are normal, but the external genitalia resemble those of the female. In addition, these subjects appear to have less baldness and acne. The facts that dihydrotestosterone mediates growth of the prostate and that individuals who lacked 5α-reductase failed to develop a prostate led to the development of therapeutic inhibitors of the enzyme. These drugs are used in the treatment of endocrine disorders whose underlying etiology requires dihydrotestosterone action.

Steroid 5α-reductase thus plays an important role in male developmental biology, physiology, and pharmacology. In the current review, we survey recent advances in the study of steroid 5α-reductase made possible by the isolation of cDNAs and genes that encode this enzyme. The genetics and clinical features of steroid 5α-reductase 2 deficiency are reviewed elsewhere (5).

HISTORICAL PERSPECTIVE

Background

The first androgenic hormone to be isolated and characterized was androsterone, a 5α-reduced 19-carbon steroid that was isolated by Butenant in 1931 from 25,000 liters of urine from adult men (reviewed in 6). This

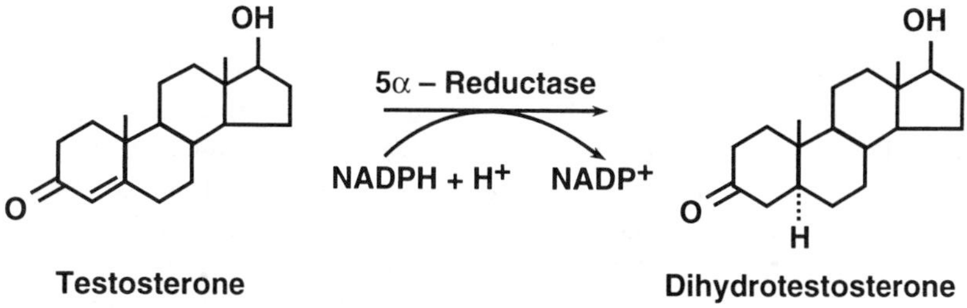

OH
5α – Reductase
NADPH + H+ NADP+
O
Testosterone
OH
O
H
Dihydrotestosterone

Other substrates: 20α-hydroxy-preg-4-en-3-one
17α-hydroxy-progesterone
Epitestosterone
Progesterone
Androstenedione

Figure 1 The enzymatic reaction catalyzed by steroid 5α-reductase.

steroid is a potent androgen in bioassay systems and was assumed to be the male hormone until Ernst Laquer and his colleagues demonstrated in 1935 that the androgen secreted by the testis is in fact testosterone, a 19-carbon steroid with a 4,5 double bond (Figure 1) (reviewed in 6). It was generally assumed thereafter that the metabolism of testosterone to androsterone served to inactivate and promote the excretion of the hormone. Indeed, a variety of substrates with a 3-oxo-$\Delta^{4,5}$ structure in the steroid A-ring are 5α-reduced by steroid 5α-reductase (Figure 1).

The 5α-reductase enzyme was initially characterized in the 1950s in rat liver slices based on its ability to convert deoxycorticosterone to 5α-reduced metabolites (7, 8). Subsequent work by Tomkins and others showed that the enzyme was present in the particulate fraction, utilized reduced pyridine nucleotide as a cofactor, and was capable of metabolizing a variety of steroid substrates (9–12), but it was uncertain in these early studies whether a single enzyme or multiple enzymes are responsible for the 5α-reduction of steroids.

The 5α-reduction of steroid substrates renders their 3-oxo groups more susceptible to reduction by 3α- and 3β-hydroxysteroid dehydrogenases and to sulfation and glucuronylation. The latter modifications reduce the affinity of the steroid for binding proteins, make it more hydrophilic, and facilitate its excretion (13). For this reason, 5α-reductase was thought to participate in the catabolism of steroid substrates. However, research in the 1960s showed that 5α-reductase would not catalyze the back reaction (dehydrogenation of reduced steroids) (14), implying that 5α-reduction might be a regulatory step. Furthermore, it was documented that dihydrotestosterone, the 5α-reduced precursor of androsterone formation, is a more potent androgen than testosterone in bioassays involving the prostate (15) and that the administration of radiolabeled testosterone to rats resulted in a time-dependent accumulation of dihydrotestosterone in the nuclei of ventral prostate cells (1, 2). Finally, dihydrotestosterone was shown to bind preferentially to specific nuclear (androgen) receptor proteins (16, 17). This body of data indicated that the 5α-reduction of testosterone is a crucial step in androgen action and focused attention on 5α-reductase.

Definitive evidence for the key role of 5α-reductase was subsequently obtained from two lines of investigation. First, developmental studies showed that the activity of 5α-reductase in mammalian embryos was highest in the primordia of the prostate and external genitalia prior to their virilization, but very low in wolffian duct structures (18–20), suggesting that the reaction is crucial for formation of the normal male phenotype during embryogenesis. Second, genetic studies showed that a rare disorder of male sexual differentiation, originally termed pseudovaginal perineoscrotal hypospadias (21), was caused by mutations in 5α-reductase (3, 4). The analysis of enzyme activity in skin slices (3), and of the urinary and serum steroids in these subjects (4), showed a generalized defect in the conversion of testosterone to dihydrotestosterone. This disease was subsequently referred to as steroid 5α-reductase deficiency.

The phenotype of 5α-reductase deficiency was faithfully exhibited by cells cultured from affected individuals (22). Thus, fibroblasts grown from genital skin possessed a 5α-reductase activity with an acidic pH optimum that was absent in the genetic disease (23). In certain patients, a biochemical defect (unstable enzyme, reduced affinity for testosterone or NADPH cofactor) was demonstrated in the acidic pH optimum enzyme (24). Additional studies revealed a second activity with an alkaline pH optimum in both genital and nongenital fibroblasts that was normal in the disease (25).

Expression Cloning of Steroid 5α-Reductase cDNAs

Further insight into 5α-reductase deficiency and the existence of multiple enzymes was hampered by the extreme insolubility of the protein. Between 1971 and 1991, many attempts were made to purify 5α-reductase from both rat and human sources (26–34). In these studies, several detergents were used successfully to solubilize the enzyme from cell membranes; however, enzyme activity was rapidly lost upon chromatography.

This purification impasse was broken in 1989, when the technique of expression cloning in *Xenopus laevis* oocytes (35) was used to isolate a cDNA from rat liver encoding a 5α-reductase isozyme (36). In these experiments, clutches of oocytes were injected with female rat liver mRNA and allowed to express the mRNA at 19°C for a period of two to three days. Thereafter, the incubation temperature was increased to 37°C (the optimum assay temperature for the rat liver 5α-reductase), and radioactive testosterone was added to the media. Subsequent thin-layer chromatography assays revealed that oocytes injected with liver mRNA had acquired the ability to synthesize dihydrotestosterone, whereas mock-injected eggs had not. These initial results indicated that the *Xenopus* expression cloning system could be used to assay for the presence of the liver 5α-reductase mRNA. In simultaneous studies, Farkash et al (37) also showed that 5α-reductase mRNA could be expressed in *Xenopus* oocytes.

The full-length cDNA encoding the rat liver 5α-reductase isolated with the *Xenopus* system (36) was subsequently used to isolate a human homolog of the rat cDNA by cross-hybridization screening of a prostate cDNA library (38). When compared to the major 5α-reductase activity present in human prostate, the cDNA and its encoded enzyme possessed several puzzling features (39). First, the cDNA-encoded enzyme was inhibited weakly by finasteride, a powerful 4-azasteroid inhibitor of 5α-reductase enzyme activity in the prostate. Second, the cDNA produced a 5α-reductase with an alkaline pH optimum, unlike the predominant acidic pH optimum enzyme activity of prostate. Third, when the gene that specified the cDNA was isolated and analyzed in subjects with 5α-reductase deficiency, no mutations were detected in the coding sequence. Finally, genetic methods were used to exclude the cloned locus as a candidate disease gene in seven unrelated families with 5α-reductase deficiency (39).

These findings strongly suggested the existence of a second human 5α-reductase gene. To isolate a cDNA encoding this putative second enzyme, an expression cloning strategy was devised in cultured human cells (40). Preliminary experiments revealed that expression vectors containing the first 5α-reductase cDNA could be diluted into a pool of 10,000 irrelevant cDNAs and still produce an enzyme activity in transfected 293 cells (a human embryonic kidney cell line) that was threefold over background. With this information, pools consisting of approximately 10,000 independent human prostate cDNAs cloned in an expression vector were transfected into 293 cells and assayed for 5α-reductase activity. Screening of 200 pools identified two candidate collections that expressed a low level of enzyme activity in multiple transfection experiments. When one such pool was subdivided into aliquots of roughly 1000 clones, it was shown in biochemical and pharmacological studies that the encoded enzyme activity was different from the first human 5α-reductase cDNA.

Subsequent division of the active cDNA pool into progressively smaller groups resulted in the isolation of a single 5α-reductase clone that was different from the first cDNA (40). The 5α-reductase enzyme activity encoded by the second cDNA had an acidic pH optimum and was sensitive to finasteride. In addition, the gene from which the cDNA was derived was mutated in subjects with 5α-reductase deficiency. In later studies, polymerase chain reaction cloning was used to isolate the rat homolog of the second human 5α-reductase cDNA (41). In both species, the second cDNA encoded a different protein that catalyzed the same biochemical reaction (5α-reduction of steroid substrates) as that encoded by the first cDNA. Enzymes that meet this general criterion are referred to as isozymes (42). In the case of 5α-reductase, we number the two isozymes based on the chronological order in which their cDNAs were isolated and refer to them as the type 1 and type 2 isozymes, cDNAs, or genes.

Several points are of general interest concerning expression cloning of the 5α-reductase cDNAs. First, the approaches used in these studies are readily adaptable to the isolation of other cDNAs encoding trace enzymes that are difficult to purify, but for which there is a biochemical assay. For example, a novel steroid 17β-hydroxysteroid dehydrogenase isozyme has been isolated from a prostate cDNA expression library by similar methods (43). Second, expression cloning allows the isolation of related gene products that would otherwise be difficult to isolate by cross-hybridization methods. Finally, the conclusions reached by Moore et al in the 1970s (23, 25) were correct regarding the existence of distinct human 5α-reductase enzymes with acidic and alkaline pH optima and the presence of mutations in the acidic pH optimum enzyme in 5α-reductase deficiency.

BIOCHEMICAL PROPERTIES OF STEROID 5α-REDUCTASE

Protein Structure

The primary structures of the rat and human 5α-reductase isozymes were determined from their respective cDNAs. As shown in Figure 2, they are hydrophobic proteins composed of 254–260 amino acids with predicted molecular weights of 28,000–29,000. There are no consensus sequences for N-linked glycosylation (Asn-X-Ser/Thr) or for O-linked glycosylation (Ser/Thr/Pro-rich regions). An average of 37% of the residues have side chains commonly found buried in the hydrophobic interior of globular proteins (Cys, Ile, Leu, Met, Phe, Val) (44). These hydrophobic amino acids are distributed throughout the enzyme, and do not give rise to clear-cut transmembrane regions in hydropathy plots. This structural feature suggests that the 5α-reductase isozymes are intrinsic membrane proteins deeply embedded in the lipid bilayer, and it explains the need for detergents to solubilize the enzyme in earlier purification attempts. The hydrophobic amino acid content may also underlie the aberrant electrophoretic mobilities in sodium dodecyl sulfate-containing polyacrylamide gels that have been reported for the 5α-reductase isozymes. In these gels, the isozymes migrate with molecular weights of 21,000–27,000 instead of the predicted 28,000–29,000 (36, 45, 46). In vitro translation studies indicate that the 5α-reductase isozymes do not have cleavable signal sequences (36).

Amino acid sequence identities between the 5α-reductase isozymes are indicated by capital letters in Figure 2. Sequence identity is 60% between type 1 isozymes and 77% between type 2 isozymes. When the two classes are compared, they average 47% identity, indicating that the rat and human type 1 isozymes are true homologs, as are the type 2 isozymes. Several sequences of amino acids are identical among all of the 5α-reductase isozymes (Figure 2), and oligonucleotide primers made to these sequences have proven useful in cDNA cloning of 5α-reductases using the polymerase chain reaction (40).

Database comparisons have so far revealed five proteins with extended sequence identities with the 5α-reductases (Figure 2). These include a protein of unknown function encoded by a partial cDNA isolated from the nematode *Caenorhabditis elegans* (47), the Epstein Barr virus terminal protein possibly involved in maintaining the virus's latent infection cycle (48), a rat protein of unknown function referred to as SC2 (49), a tobacco chloroplast NADH-ubiquinone oxidoreductase chain 5 homolog (50), and a portion of the reverse transcriptase (*pol*) gene of the Cas-Br-E murine leukemia virus (51). The significance of these homologies is unclear. In the cases of the *C. elegans,* Epstein Barr virus, and SC2 proteins, the identities may reflect a simple conservation of a hydrophobic transmembrane region. However, the fact that each of these proteins shares the same carboxyl-terminal end with the 5α-reductases (Figure 2) implies a functional conservation.

pH Optima

When examined in transfected cell lysates, the type 1 isozymes have broad pH optima that span the alkaline range (pH 6–8.5) (38, 41), while the type 2 isozymes have narrow acidic pH optima centered around 5.0 (40, 41). Representative pH curves for the two isozymes are illustrated in Figure 3*A*. The alkaline and acidic pH optima are thus diagnostic for the type 1 and type 2 isozymes, respectively. This feature can often be used to assign a 5α-reductase activity in a human or rat tissue to a particular isozyme. For example, cultured human genital skin fibroblasts express an acidic pH optimum enzyme (23), and would thus be predicted to express the type 2 isozyme. This prediction has been confirmed by RNA blotting using type 2 cDNA probes and by immunoblotting using selective antipeptide antiserum (46). Similarly, nongenital skin fibroblasts express an alkaline pH optimum 5α-reductase enzyme activity (25), and

```
h5R1(1-46)       MatatgvAEerLLaaLAYLqcAV--gcavfArnRqtnSvYGRH--aLPSh
r5R1(1-42)       M----eldELcLLdmLvYLeGfm--AfvsivglRsvgSPYGRysPqwP--
h5R2(1-43)       M-----QvqCQQSPVLAGSATLV--ALGALALyVaKPSgYGKHTESLKpA
r5R2(1-43)       M-----QiVChQvPVLAGSATLa--TmGTLiLcLgKPasYGKHTESVsSg
EBVTP(254-299)   vlvlivdAVLQLSP-LlGavTvVsmTLllLAfvLwlsSPgGlgT--L-gA
CasBrE(209-253)  tvwvrrhqtknLePrwkGpyT-V--lLtTpALkVdgiSawv-HaahVK-A
ndhF(213-262)    iynevdflfvtlcaVLlf-AGAV--AksAqfLhVwlPdamegpTPSLihA

h5R1(47-96)      rlRVPARAAWVVQELPSLALPLYqYAsEsApRLrsaPNciLLAMFLVHYg
r5R1(43-92)      giRVPARPAWFiQELPSmAwPLYeYiRpaAaRLgnLPNrVLLAMFLiHYv
h5R2(44-92)      ATRLPARAAWFLQELPSFAVPaGILARQPL-SLFGPPGTVLLGLFCVHYF
r5R2(44-92)      vpfLPARiAWFLQELPSFvVsVGmLAwQPr-SLFGPPGnVLLALFsaHYF
EBVTP(300-343)   A-lLtlaAA--LalLaSLiLgtlnLttmfL--LmlLwtlVvL-LiCsscs
SC2(137-139)                                                    HYi
CasBrE(254-278)  ATtsPARtAWkVQr--Sqn-PLkILsREPs*
ndhF(263-307)    ATmV-A-AgiFLavalpLfrvipyimyli---svigiiTVLLGatlalaq

h5R1(97-145)     HRcLIYPFLMR-GGKPMPLlactmAimFCTcNGYLQSRYLSHCAVYaDDW
r5R1(93-141)     qRTLVFPvLiR-GGKPtlLVtFvLAflFCTfNGYvQSRYLSqfAVYaEDW
h5R2(93-140)     HRTFVYSLLnR-G-RPYPAiLILRgTAFCTGNGvLQgYYLiYCAEYPDgW
r5R2(93-140)     HRTFIYSLLtR-G-RPfPAVLFLRATAFCiGNGlLQAYYLvYCAEYPEeW
EBVTP(344-393)   scpLskiLLaR-l-flYaLaLlLLAsALiaGgsiLQfksLSstefiPnlf
SC2(140-188)     kRlLetlFvhRfshgtMPLrnIfknctyywGfaawmAYYinH-plYtppT
```

Figure 2 Sequence identities between steroid 5α-reductase isozymes and other proteins. Identities are indicated by capital letters in the single-letter amino acid code. Termination codons are indicated by asterisks. Dots are placed every 10 residues above the first sequence. Amino acid numbers are indicated in the lefthand margin. Only identities between the 5α-reductase isozymes and other proteins are capitalized. Circled amino acids are mutated in subjects with 5α-reductase 2 deficiency. The boxed amino acid at position 89 in the human 5α-reductase type 2 isozyme is polymorphic with V (valine). Key: h5R1, human 5α-reductase type 1 isozyme (38);

```
h5R1(146-189)    VTDPRFLiGFgLWLtGMLINIHSDHILRNLRKPGdTGYKIPRGG------
r5R1(142-185)    VThPcFLtGFALWLvGMvINIHSDHILRNLRKPGETGYKIPRGG------
h5R2(141-184)    YTDiRFSLGVFLFILGMGINIHSDYILRQLRKPGEisYRIPqGG------
r5R2(141-184)    YTDvRFSfGVFLFILGMGINIHSDYtLRQLRKPGEviYRIPRGG------
Ce5α                                       sgnprtsyskslmeIhfls------
EBVTP(394-429)   cmlllivaG-iLFILaiLtewgS-----gnRtyG--pvfmclGG------
SC2(189-238)     YgvqqvkLalAiFgicqLgNfsihmaLRdLRpaGsktrKIPyptknpftw

h5R1(190-239)    LFEYVtAANYFGEImEWCGYALASWSVQGAAFAFFTFCFLsgRAKeHHeW
r5R1(186-235)    LFEYVSAANYFGElvEWCGFALASWSLQGvvFALFTLstLltRAKqHHqW
h5R2(185-234)    LFTYVSGANFLGEIIEWIGYALATWSLPAlAFAFFsLCFLGLRAFhHHRF
r5R2(185-234)    LFTYVSGANFLGEIIEWIGYALATWSVPAfAFAFFTLCFLGmqAFyHHRF
Ce5α             ssnYVScpNYtyEvasWIfFsimvqSLPAiiFttagFaqmaiwAqgkHRn
EBVTP(430-474)   LlTmVaGAvwL---tvmsntlLsaWiLtA-gFliF-LigfaLfgvirccr
SC2(239-288)     LFllVScpNYtyEvgsWIGFAimTqcVPvAlFsLvgFtqmtiwAKgkHRs

h5R1(240-259)    YLrKFEeYPKSRKILIPFVL*
r5R1(236-255)    YheKFEDYPKfRKIIIPFLF*
h5R2(235-254)    YLKMFEDYPKSRKALIPFIF*
r5R2(235-254)    YLKMFkDYPKSRKALIPFIF*
Ce5α             YLKeFpDYPKnRKAIvPFVL*
EBVTP(475-497)   YccyycltleSeerpptpyrntv*
SC2(289-308)     YLKeFrDYPplRmpIIPFLL*
```

r5R1, rat 5α-reductase type 1 isozyme (36); h5R2, human 5α-reductase type 2 isozyme (40); r5R2, rat 5α-reductase type 2 isozyme (41); Ce5α, *Caenorhabditis elegans* expressed sequence tag (wEST00337) (47); EBVTP, Epstein-Barr virus terminal protein (48); CasBrE, Cas-Br-E murine leukemia virus polymerase polyprotein (51); ndhF, tobacco chloroplast NADH-ubiquinone oxidoreductase chain 5 homolog (50); SC2, rat protein of unknown function (49).

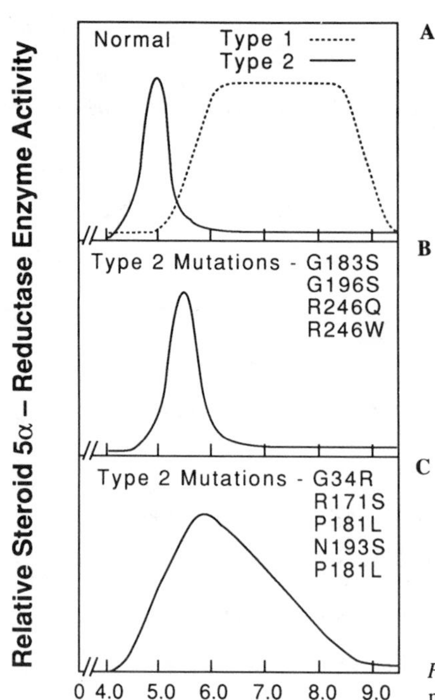

Figure 3 pH profiles of *A*. normal and *B*, *C*. mutant steroid 5α-reductase isozymes.

blotting studies have confirmed that these cells express the type 1 mRNA and isozyme (46). It is not known if the distinct pH optima will be a characteristic of the isozymes in all species.

The biochemical basis of the acidic pH optimum of the type 2 isozyme is puzzling. This feature is characteristic of the isozyme from rats, mice (unpublished observations), and humans, and is thus evolutionarily conserved. The pH optimum of an enzyme is usually the consequence of the ionic state of charged amino acids in the active site (52). This fact, together with the symmetry of the pH profile (Figure 3*A*), suggests the involvement of a small number of amino acids in determining the acidic pH optimum. However, nine different mutations that map throughout the protein alter the pH optimum (see below).

The acidic pH optimum of the type 2 isozyme could reflect a location of the protein within an acidic subcellular compartment such as the endosome or lysosome. However, the activity of the type 2 isozyme in transfected Chinese hamster ovary (CHO) cells is not affected by compounds (chloroquine, NH_4Cl, monensin) that neutralize the pH of acidic subcellular compartments (45). In addition, immunocytochemical studies show that both the type 1 and type 2 isozymes are present in the endoplasmic reticulum of CHO cells, a compartment with a neutral pH (see below).

Further studies with transfected CHO cells have suggested that the type 2 isozyme may have a neutral pH optimum in its native state. If these cells are treated with digitonin under conditions in which the plasma membrane is permeabilized but the endoplasmic reticulum membrane is left intact, the type 2 isozyme exhibits equivalent activity at both acidic and neutral pH (45). However, physical lysis methods (disruption with a polytron or perforating the plasma membrane via nitrocellulose filter overlay) result in a type 2 isozyme with activity only at pH 5.0. The isozyme would thus appear to function at a neutral pH within the cell and to shift to a pH 5.0 active form upon cell lysis. The shift in pH requirement may reflect a conformational change in the isozyme.

Kinetic Constants

Further evidence to support a neutral pH optimum for the type 2 isozyme in intact cells comes from kinetic studies with both transfected and non-transfected cells and cell lysates. Transfection of the type 1 isozyme from rats or humans produces an enzyme in cell lysates with micromolar affinity (apparent K_m = 1–5 μM) for substrates such as testosterone, androstenedione, and progesterone (38). A similar value is obtained for the human type 1 isozyme in intact CHO cells using testosterone as a substrate (45). The apparent testosterone K_m is in the submicromolar range (0.1–1.0 μM) at V_{max} (i.e. pH 5.0) for the human type 2 isozyme (40, 45). The apparent K_m for NADPH cofactor as determined at V_{max} is in the low micromolar range (3–10 μM) for both isozymes (45).

Surprisingly, the apparent K_m values at pH 7.0 for the type 2 isozyme in cell lysates, permeabilized cells, and intact cells are in the low nanomolar range (4–50 nM) (45, 53). The type 2 isozyme is most efficient between pH 6.0 and 7.0 on the basis of data from V_{max}/K_m versus pH plots (45, 53). The observations that the type 2 isozyme has a higher affinity for steroid substrates and is a more efficient enzyme at neutral pH in cell lysates again suggest that this isozyme acts at neutral pH within the cell.

Overexpression and Purification

Kinetic studies to date have been carried out with crude cellular extracts and are subject to nonspecific effects of endogenous inhibitors, substrate accessibility constraints, nonhomogeneous lipid environment perturbations, and numerous other problems. As noted above, several attempts have been made to purify one or more of the 5α-reductase isozymes from rat and human tissues. These attempts have not met with success due to the apparent lability of the enzyme. In addition, most tissues and cultured cell lines, especially those from the human, express low levels of the isozymes (specific activities < 0.1 nmol dihydrotestosterone/min/mg protein). A clear exception to this observation is female rat liver (33, 54), a tissue that expresses large amounts of the type 1 isozyme (specific activity = 4–20 nmol dihydrotestosterone/min/mg protein) (36, 41).

The cloning of cDNAs should allow high-level expression in heterologous cells and the subsequent purification of a 5α-reductase isozyme. To this end, the expression of one or more of the cDNAs has been reported in *Escherichia coli* (55), the yeast *Saccharomyces cerevisiae* (56), *Xenopus* oocytes (36, 37), Simian COS cells (38), human embryonic kidney cells (40), and CHO cells (45). Several groups have also succeeded in expressing the human cDNAs in the baculovirus system. In our hands, infected insect cells express a large amount of immunoreactive 5α-reductase isozyme (RC Wigley and DW Russell, unpublished observations) and may thus be appropriate hosts for subsequent purification attempts. Reaching this goal may be facilitated by the use of affinity chromatography on inhibitor columns (see below).

Intracellular Turnover and Subcellular Localization

In pulse-chase experiments with transfected CHO cell lines, both human 5α-reductase isozymes have long half-lives (20–30 hours), which are not altered by the presence of micromolar concentrations of two 4-azasteroid inhibitors. One of these 4-azasteroids (finasteride) is thought to bind irreversibly to the human 5α-reductases (see below). These results suggest that the enzyme is not regulated at the level of protein degradation, at least not in cultured hamster fibroblasts. In agreement with this hypothesis, no alterations in steady-state protein level or enzyme activity could be detected in synchronized CHO cells at various phases of the cell cycle (AE Thigpen, unpublished observations). Nor has evidence been found for posttranslational modification of the isozymes (e.g. phosphorylation, fatty acylation, isoprenylation).

Indirect fluorescent immunocytochemistry indicated that both human 5α-reductase isozymes reside in the endoplasmic reticulum (45), presumably embedded in the lipid bilayer of this organelle. The subcellular localization of 5α-reductase was previously shown to differ depending on the tissue source of the enzyme. Thus, enzyme activity sedimented with the nuclear fraction of human (57) and rat (26, 27) prostate cells, but with the endoplasmic reticulum fraction of liver cells (27). These results have been confirmed by immunohistochemical studies in both the human (type 2 isozyme) and rat prostate (type 1 isozyme), in which the subcellular distribution of the antigen is perinuclear, and in rat and human liver (type

1 and type 2 isozymes, respectively), in which a reticular distribution is detected (58; RI Silver, EL Wiley, AE Thigpen, JM Guileyardo, JD McConnell, DW Russell, manuscript in preparation). The different subcellular localization of 5α-reductase in liver and prostate may reflect a difference in the proliferation of the endoplasmic reticulum, since this organelle is continuous with and extends outwards from the nuclear membrane. Alternatively, the difference may have a regulatory underpinning, as the cells of the prostate are androgen dependent whereas those of the liver are not. A perinuclear localization of the enzyme might facilitate subsequent binding of product by nuclear androgen receptors.

GENETICS OF STEROID 5α-REDUCTASE

Gene Structure

Both human 5α-reductase isozyme genes contain five exons separated by four intervening sequences (59–61). The positions of the introns are essentially identical in the two genes (arrows, Figure 2), suggesting that the two isozymes arose by a primordial gene duplication. Preliminary characterization of the mouse and rat 5α-reductase genes indicates that this general structure is conserved in these species as well. Although the structure is shared, the two genes are located on separate chromosomes in both human and mouse (59, 60). The gene encoding the human type 1 isozyme (gene symbol *SRD5A1*) is located on the distal short arm of chromosome 5 (band p15), whereas the type 2 isozyme gene (symbol *SRD5A2*) is located in band p23 of chromosome 2 (60). A processed pseudogene (gene symbol *SRD5AP1*), apparently derived from an mRNA transcript of the type 1 gene, has been isolated and mapped to the long arm (bands q24-qter) of the X chromosome. The pseudogene contains a nonsense codon in place of that specifying amino acid 147 and is, therefore, not believed to encode a functional protein (59). It is not known if an mRNA is transcribed from this pseudogene.

Three biallelic DNA polymorphisms have been identified in the human type 1 gene, including a silent change in codon 30 of exon 1 (CGG versus CGC, both encoding arginine), which creates or destroys a *Hin*fI site, respectively. Alanine 116 is encoded by the sequence GCA or GCG, and this polymorphism affects an *Nsp*I site in exon 2. Threonine 160 is specified by either an ACG or an ACA codon in exon 3, and this alteration can be detected by single-stranded DNA conformation analysis. Although not definitively established, the frequency of the two alleles of each of the three polymorphisms appears to be roughly 0.5 (59).

Three polymorphisms have so far been found in the human 5α-reductase type 2 gene. A comparison of the two reported gene structures (60, 61) revealed two of these. A variable number of dinucleotide (AT) repeats are present in a segment of exon 5 specifying the 3′-untranslated region of the mRNA (62). The most frequent allele (present on 96% of chromosomes analyzed) possesses no AT repeats, while two variant alleles contain 9 and 18 copies of the dinucleotide and some adjacent sequence differences. The frequency of the variant alleles was very low (< 0.02) in a screen of 172 chromosomes. It is not known whether the presence or absence of the repeats in the 3′-untranslated region of the mRNA alters its stability or translation. A second polymorphism alters the protein coding region at codon 89, which can be either GTA (valine) or CTA (leucine) (Figure 2). The biochemical consequences of this alteration have not been determined. A third polymorphism, revealed in a genetic screen of the *SRD5A2* locus, involves a C/T variation within the first few nucleotides of intron 1 of the gene.

A major impetus for the characterization of these and other polymorphisms in the 5α-reductase genes is their potential usefulness in inheritance studies. Several human disorders with an apparent genetic component are postulated to involve dihydrotestosterone action, including male pattern baldness, hirsutism, prostate cancer, benign prostatic hyperplasia, and acne. An involvement of one or more of the 5α-reductase genes could be deduced by following the inheritance of *SRD5A1* and *SRD5A2* alleles in families in which the putative genetic trait is segregating. The biallelic nature and the low frequency of the polymorphisms described so far limits their genetic informativeness in this type of study, and additional polymorphic alleles will have to be isolated to overcome this drawback. Nevertheless, the polymorphisms have proven useful in excluding the *SRD5A1* gene in 5α-reductase deficiency and in studies of the inheritance of this disease.

Human Steroid 5α-Reductase 2 Deficiency

The power of 5α-reductase as a paradigm of developmental biology comes from the study of naturally occurring mutations in the gene that disrupt male sexual differentiation. The development of the male phenotype in mammals can be divided into three temporal stages, beginning with the establishment of chromosomal sex at the time of fertilization (reviewed in 5, 63, 64). Thereafter, gonadal sex is determined by the expression of a key regulatory gene on the Y chromosome (the testis-determining gene or *SRY*) (65). Expression of this gene transforms a bipotential gonad into a fetal testis capable of synthesizing testosterone and other hormones required for the third phase of sexual development, the establishment of male phenotypic sex. In this last stage, testosterone acts in concert with the product of the androgen receptor gene to initiate the formation of the internal male reproductive structures such as the epididymis, the seminal vesicle, and the vas deferens. In the embryonic urogenital tract, testosterone is converted into dihydrotestosterone, which in turn binds to the androgen receptor and drives the differentiation of the external genitalia (penis and scrotum) and the prostate gland from these anlages.

The outlines of this complex developmental process have been deduced from animal experimentation (66) and the study of naturally occurring mutations in key genes that alter the male phenotype. For example, in XY individuals, mutations that eliminate androgen receptor activity lead to subjects that have testes and synthesize testosterone and dihydrotestosterone but fail to develop male internal or external reproductive structures (67). In the case of 5α-reductase deficiency, the failure to synthesize dihydrotestosterone leads to alterations in the development of the external genitalia and prostate but does not affect other steps in the developmental pathway (5).

As described above, a series of studies suggested that mutations in the type 2 gene caused 5α-reductase deficiency (25, 39). This hypothesis was proven with the cloning of a human type 2 cDNA and a series of Southern blotting experiments in which a large deletion was shown to be present in the genes of a related group of subjects with 5α-reductase deficiency (40). Since this time, 29 different mutations in the type 2 gene have been identified and studied at the clinical, genetic, and biochemical levels. The clinical and genetic aspects of 5α-reductase deficiency are reviewed in a separate paper (5). Here, we focus on the biochemical insight into 5α-reductase provided by the mutations.

Mutations in the Steroid 5α-Reductase 2 Gene

Mutations in the 5α-reductase type 2 gene have been identified by standard methods of molecular genetic analysis (40, 60, 68; WC Wigley, JS Prihoda, I Mowszowicz, BB Mendonca, MI New, JD Wilson, DW Russell, manuscript submitted). As of September of 1993, these studies have revealed two deletion mutations, two nonsense mutations, one splicing defect, and 24 missense mutations (Figure 4). Mutations have been detected in 37 unrelated families, and in these approximately 65% of the affected individuals are true homozygotes and 35% are compound heterozygotes. The relative abundance of compound heterozygotes is unexpected given the rarity of the disease, and suggests that the carrier frequency may be relatively high in the general population. In several instances, the same mutation is found in individuals with a common ancestry, presumably reflecting a founder gene effect. For example, the mutation *Q126R* is found in subjects of presumed Portuguese ancestry from Brazil, Belgium, Louisiana, and New York. Similarly, the mutation 359, ΔTC is present in several families from Malta. There are two good examples of recurrent mutations in this collection,

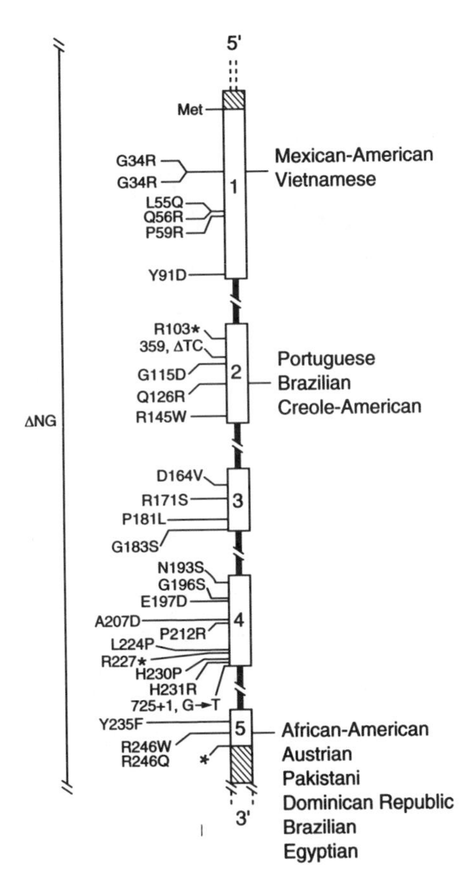

Figure 4 Mutations in steroid 5α-reductase type 2 gene found in subjects with 5α-reductase deficiency. Exons are indicated by large open boxes. Introns are indicated by connecting black lines and are not drawn to scale. Mutations are indicated (normal amino acid followed by position followed by mutant amino acid) in the single-letter amino acid code on the left of the gene schematic. The symbol Δ indicates a deletion. 725+1, G→T represents a splice junction mutation. Mutations representing a founder gene or potential recurrent mutations are indicated on the right of the schematic.

the *R246W* and *R246Q* mutations in exon 5. These two alleles arise from mutation at either position of a CG dinucleotide and are found in six different ethnic groups (Figure 4).

In most affected individuals, the disease is inherited in an autosomal recessive fashion (68). However, in four well-documented subjects, only a single mutation has been found (5). These findings raise the interesting possibility that certain mutations may be dominant, but an alternate explanation is that the second mutation lies in a region of the gene not yet screened. Despite extensive characterization of the individual mutations, we have not detected a correlation between the severity of the manifestation and the particular 5α-reductase mutation inherited by an affected individual.

Functional Domains Revealed by Mutation Analysis

The biochemical consequences of 22 of the missense mutations have been characterized by re-creating individual mutations in an expressible cDNA followed by transfection analysis in cultured mammalian cells. As indicated in Table 1, all of the mutations affect the V_{max} of the enzyme. Twelve of the missense mutations result in a protein with no detectable enzyme activity, whereas the remaining ten give rise to proteins with severely decreased, but measurable enzyme activity.

The last-mentioned group of mutations can be divided into two classes: those that affect the ability of the enzyme to bind testosterone substrate and

Table 1 Biochemical parameters of mutations in 5α-reductase type 2

Class	Mutation	V_{max} (nmol/min/mg)	Apparent testosterone K_m (μM)	Apparent NADPH K_m (μM)	pH Optimum	Protein half-life (h)
Normal	—	2–5	0.5–1.0	10–20	5.0	≥20
No enzyme activity						
	L55Q, P59R, Y91D, Q126R, E197D, A207D, P212R, L224P, H230P					≤6
	Q56R, G115D					15–20
	D164V					≥20
Testosterone-binding abnormality						
	G34R	0.5	10–12	8–15	5.2–6.0	≥20
	H231R	0.7	13–14	18–25	5.4–6.5	≥20
NADPH-binding abnormality						
	R145W	0.04	0.9–1.2	480–580	5.0	6–20
	R171S	1.2	1.2–2.3	320–340	5.6–6.0	6–20
	P181L	0.005	1.9–2.2	210–530	6.0–6.5	≤6
	G183S	0.4	0.8–1.8	144–316	5.3	6–20
	N193S	0.4	1.2–1.8	184–325	5.5–6.5	≤6
	G196S	0.06	0.5–1.0	150–180	5.2	≤6
	R246W	0.06	0.5–1.0	600–650	5.4	≤6
	R246Q	0.002	0.9–1.4	200–400	5.4	≤6

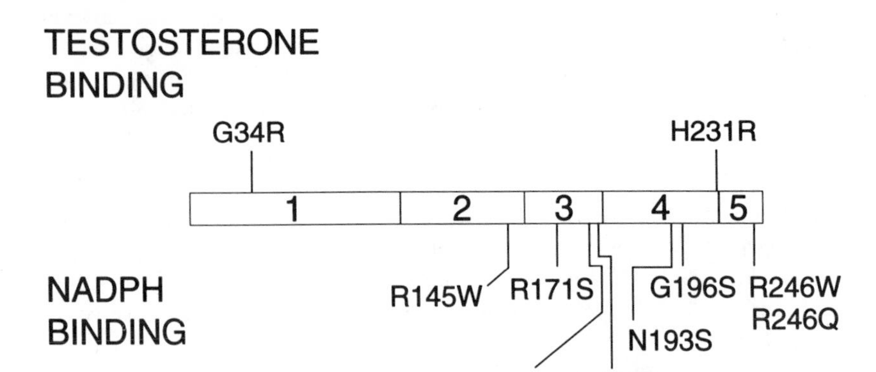

Figure 5 Biochemical effects of mutations in the steroid 5α-reductase type 2 gene found in subjects with 5α-reductase deficiency.

those that decrease the affinity for the NADPH cofactor (Table 1). Mutations that affect testosterone binding (by increasing the apparent K_m) map to the two ends of the type 2 isozyme, whereas those that affect NADPH binding map throughout the last half of the protein (Figure 5). At face value, the locations of the mutations suggest a nonlinear arrangement of the amino acids that make up the two functional domains of the enzyme. The fact that mutations alter either substrate or cofactor binding, but not both, further suggests that the amino acid determinants of these domains may be distinct.

No consensus steroid-binding domain can be deduced from the current mutational analysis, but a determinant of finasteride sensitivity has been mapped in the type 1 isozyme to amino acids 26–29, which are adjacent to the *G34R* mutation of the type 2 isozyme (69). The *G34R* mutation decreases the type 2 isozyme's affinity for substrate. Since finasteride can act as a competitor of substrate binding (see below), these findings suggest that an extended sequence of amino acids forms a substrate-binding domain at the amino terminus of the protein.

The 5α-reductase isozymes utilize only NADPH as a cofactor, preferentially inserting the 4*S*-hydrogen of the nicotinamide ring into the 5α configuration of the steroid substrate (14, 70). Despite this dependence on NADPH, the amino acid sequences of the rat and human isozymes do not contain consensus adenine dinucleotide-binding sequences or NADPH selectivity residues identified in other reductase enzymes (71). This absence suggests that the cofactor-binding domain of the 5α-reductases may represent a novel structure, a hypothesis in keeping with the diversity and location of mutations that affect NADPH binding in the type 2 isozyme (Figure 5).

This domain is presumably conserved between the 5α-reductase isozymes, as all but one (an arginine at position 145) of the mutated amino acids that affect cofactor binding are identical among the four sequenced isozymes (circled amino acids, Figure 2).

Kinetic studies of the rat 5α-reductase type 1 suggest an ordered bi-bi mechanism of catalysis in which NADPH is the first substrate to bind and $NADP^+$ is the last product to leave the enzyme (33). In agreement with these results, binding of radiolabeled 4-azasteroid inhibitors required the presence of NADPH (29), and numerous studies indicate that NADPH stabilizes enzyme activity in some tissue extracts. Similarly, mutant enzymes that have a lower affinity for NADPH have a shorter half-life in both fibroblasts (e.g. 24) and in transfected cells (Table 1). These results strengthen the conclusion that mutations affecting NADPH cofactor binding directly disrupt this domain of the enzyme.

A surprising finding from these studies is that all but one (*R145W*) of the mutations that diminish enzyme activity also change the pH optimum of the type 2 isozyme. The observed change can be subtle, as indicated by the roughly one-half unit increase in this parameter exhibited by the *G196S* mutation (Figure 3*B*). Other mutations, such as *R171S*, drastically alter the pH profile to a much broader and more alkaline curve (Figure 3*C*). Mutations that alter either substrate or cofactor affinity render the pH optimum more alkaline, suggesting that the effect is on conformation rather than on catalysis. It seems likely that a conformation-dependent residue (presumably with an acidic pK_a) or ion pair required for catalysis is being affected by even subtle substitutions in the type 2 isozyme. Further mutagenesis and structural studies are needed to uncover the biochemical basis of these observations.

REGULATION OF STEROID 5α-REDUCTASE

The regulation of 5α-reductase expression has been examined in several biological systems. The interpretation of data reported in early studies is complicated by the existence of two isozymes, as in most cases it was assumed that only a single 5α-reductase activity was being measured. Using the initial studies as guides and employing the same experimental systems, the regulation of 5α-reductase has been re-examined using gene-specific probes and isozyme-selective antibodies. Current studies have focused on establishing which tissues and cell types express a given 5α-reductase isozyme and the impact of hormonal manipulation on expression. Implicit in these studies is an attempt to determine why there are two isozymes and what unique and overlapping roles they play in androgen and steroid hormone metabolism.

Tissue Distribution of Steroid 5α-Reductase Isozymes

Tissue distribution surveys have been carried out for the rat (41), human (46), and mouse (M Mahendroo, DW Russell, unpublished observations). One or more of the 5α-reductase isozymes is detected in many tissues, and unique expression patterns exist in each of the species (Table 2). For example, the type 1 isozyme is present in the liver of the mouse and rat, whereas both isozymes are present in human liver. A unique sexual dimorphism in 5α-reductase expression also exists in the rat liver: Females express 10–20 times more type 1 isozyme activity and mRNA than do males (36, 37, 41, 54). This sexually dimorphic expression is thought to be due to differences in the pattern of growth hormone secretion between male and female rats (72; but see 73). In the ventral prostate of the rat, both isozymes are detected, whereas the type 2 isozyme is the predominant, if not exclusive, isozyme in the human and mouse prostate. These differences highlight the fact that findings in one species cannot always be broadly extrapolated.

Table 2 Tissue distribution of 5α-reductase isozymes

Species	Tissue	5α-Reductase isozyme[a]		Method of detection[b]
		Type 1	Type 2	
Rat	Ventral prostate	+	+	mRNA, protein
	Epididymis	+	+++	mRNA, protein
	Seminal vesicle, vas deferens, testes	+	+	mRNA
	Liver	+++	−	mRNA, protein
	Adrenal, brain,[c] colon, intestine, kidney	++	+	mRNA
	Heart	−	−	mRNA
	Lung	++	−	mRNA
	Muscle, spleen, stomach, ovary	+	−	mRNA
Human	Prostate, epididymis, seminal vesicle, genital skin	−	++	mRNA, protein
	Testis, ovary, adrenal, brain,[d] kidney	−	−	mRNA, protein
	Liver	+	+	mRNA, protein
	Nongenital skin	++	−	mRNA, protein

[a] A + indicates that the mRNA or protein was detected. The number of + signs is an approximate indication of the amount of 5α-reductase isozyme detected. A − sign indicates that no 5α-reductase isozyme was detected.

[b] Key: mRNA, isozyme mRNA detected by blot hybridization; protein, isozyme detected by immunoblotting. 5α-Reductase enzyme activity has been reported in all tissues.

[c] Whole brain.

[d] Cerebellum, hypothalamus, medulla oblongata, pituitary, pons.

They also make it difficult to assign distinct physiological roles to one or the other of the isozymes.

In the rat, the levels of mRNA for a given isozyme appear to correlate well with the levels of protein determined by immunoblotting and enzyme activity (41). In contrast, several tissues in the human express the type 1 isozyme mRNA but do not appear to have detectable protein (39, 46). Whether this finding is indicative of translational regulation or an unusual mRNA structure that precludes translation, or instability of the protein, is not known. The sequences of the 5′-untranslated regions of the 5α-reductase isozyme mRNAs are different, and these mRNAs may therefore be translated with different efficiencies. Along these lines, there are tissue-specific differences in the lengths of the rat type 1 mRNA (41), and multiple mRNAs are detected for all human and rat isozymes in blot hybridization experiments (41, 46). The molecular basis of or physiological reason for these differences has not yet been elucidated.

Cell-Type-Specific Expression of Steroid 5α-Reductase

As summarized in Table 3, in situ mRNA hybridization studies (58) and immunohistochemical analyses (RI Silver, EL Wiley, AE Thigpen, JM Guileyardo, JD McConnell, and DW Russell, manuscript submitted) have revealed cell-specific expression patterns and gradients of 5α-reductase isozyme expression within tissues. An informative example of cell specialization is found in the rat ventral prostate, in which basal epithelial cells express the type 1 isozyme and stromal cells express the type 2 isozyme (58). Basal epithelial cells are a less differentiated type of prostate cell that are thought to be precursors of the more abundant, terminally differentiated lumenal epithelial cells (74). Basal epithelial cells persist and retain a proliferative capacity in the castrate animal, whereas the lumenal epithelial cells die following castration. In the basal cells, the 5α-reductase type 1 isozyme may be responsible for synthesizing dihydrotestosterone that acts in an autocrine manner to stimulate their differentiation into lumenal epithelial cells. Alternatively, dihydrotestosterone may act in a paracrine fashion to stabilize or stimulate the division of the adjacent androgen-dependent lumenal epithelium (58).

Expression of the type 2 isozyme in the stroma of the prostate is consistent with the central role played by these cells in the development of the gland. Reconstitution studies have indicated that the stroma provides a signal required by the epithelium to differentiate into a functional prostate (75). This signal is derived in part from dihydrotestosterone, as individuals with 5α-reductase deficiency fail to develop a normal prostate (76). The available evidence suggests that the dihydrotestosterone signal involved in differentiation acts directly on the stroma rather than on the epithelium (58).

Table 3 Cell-type-specific expression of 5α-reductase isozymes

Species	Tissue	Cell type	Isozymes	Method of detection[a]	Comments
Rat	Ventral prostate	Basal epithelial cells	Type 1	mRNA, protein	
	Ventral prostate	Stromal cells	Type 2	mRNA	
	Epididymis	Principle (epithelial) cells	Type 1 and 2	mRNA, protein	Gradient: head >> tail
	Liver	Hepatocytes	Type 1	mRNA, protein	Gradient: portal triad >> central vein
Human	Prostate	Stromal cells	Type 2	protein	
	Epididymis	Principle (epithelial) cells	Type 2	protein	No apparent gradient
	Liver	Hepatocytes	Type 1 and 2	protein	No apparent gradient
	Seminal vesicle	Stromal cells	Type 2	protein	
	Genital skin	Fibroblasts	Type 2	protein	
	Nongenital skin	Sebaceous cells	Type 1	protein	

[a] 5α-Reductase RNA and protein were detected in tissue sections by in situ mRNA hybridization and immunohistochemical staining, respectively.

In contrast to the one isozyme–one cell type of expression pattern in the rat ventral prostate, the epithelial or principal cells of the rat epididymis express both isozymes (58). These cells also exhibit a gradient of isozyme expression in which the epithelial cells proximal to the testis express very large amounts of enzyme activity and mRNA, whereas those distal to the testis express lower levels. This gradient of expression was first reported by Viger & Robaire (77, 78) for the type 1 isozyme, and was subsequently demonstrated for the type 2 isozyme (41). At the cellular level, high-level expression begins in the epithelial cells at their junction with the efferent ducts of the testis (58). The biological role of the 5α-reductase gradient in the epididymis is not known. The gradient may simply reflect proximity to the androgen-secreting testis, or it may play a role in initiating maturation of sperm as they travel down the epididymal tubules.

A gradient of expression is also detected for the type 1 isozyme in liver hepatocytes (Table 3). Here, very high levels of type 1 isozyme and mRNA are present in hepatocytes surrounding the portal triad, and this expression decreases to almost undetectable levels in hepatocytes surrounding the central veins (58). This expression pattern mirrors the maturation of the hepatocytes and is similar to that reported for certain cytochrome P-450 enzymes and for albumin. Again, the physiological significance of this gradient may be important for steroid catabolism or hepatocyte maturation in the liver.

In the human, the cell-type-specific expression patterns are in general similar to those of the rat, but neither the epididymal nor liver gradients of 5α-reductase expression are present (RI Silver, EL Wiley, AE Thigpen, JM Guileyardo, JD McConnell, and DW Russell, manuscript submitted). In addition, expression of the type 1 isozyme has not been demonstrated in the human prostate (Table 3). This absence may reflect a sensitivity problem, as a low level of expression of the type 1 isozyme would be missed by available detection methods, or a functional difference, since the human gland is composed of a much larger proportion of stromal cells than is the rat ventral prostate. The resulting higher level of expression of the type 2 isozyme in the stroma of the human gland may provide all of the dihydrotestosterone required for prostate growth and maintenance.

Androgen Regulation

5α-reductase expression is regulated by androgen in a number of tissues and species. The best-studied tissue is the rat ventral prostate, an androgen-dependent tissue with remarkable experimental properties. Castration of male rats causes a marked regression in the size and weight of the ventral prostate (79). Regression is caused by apoptosis of the lumenal epithelial cells (80), and can be readily reversed or prevented by administration of testis extracts (79). Studies in the 1970s showed that castration and re-administration of testosterone led to a marked induction of 5α-reductase enzyme activity (81), suggesting that dihydrotestosterone was the active androgen in this process. It was subsequently shown that treatment of intact rats with inhibitors of 5α-reductase (see below) also led to prostate regression, and that this effect could be overcome by administration of dihydrotestosterone (82).

The induction of 5α-reductase enzyme activity in the ventral prostate of castrate rats given testosterone is accompanied by a large increase ($\geq$10-fold) in the mRNA for both the type 1 and type 2 isozymes (36, 41). It is not known whether this increase in steady-state mRNA levels is a consequence of mRNA stabilization or an increase in the transcription of the 5α-reductase genes. However, castration and re-administration of androgens do not change the cell-type-specific expression patterns of the two isozymes (58).

Surprisingly, dihydrotestosterone appears to be the active androgen that enhances 5α-reductase mRNA and enzyme activity in the ventral prostate (83). This conclusion was reached in experiments in which a 5α-reductase inhibitor was given to castrated animals together with testosterone. As expected, this regimen blocked growth of the gland, but also abolished the induction of 5α-reductase enzyme activity and severely attenuated induction of the mRNA (41, 83). The requirement for dihydrotestosterone results in an unusual situation in which the product of the enzyme is responsible for regulating the expression of the gene that encodes the enzyme. This so-called feed-forward regulation (41) is the exact opposite of the usual feedback regulation in which the product of an enzyme negatively regulates expression of the gene.

Feed-forward regulation or autoregulation has been described for several genes that play crucial roles in development. Examples include the *Drosophila fushi tarazu* (84), *engrailed* (85), and *Sex-lethal* genes (86), which are involved in pattern formation and sex determination in the embryo, the *cI* gene of bacteriophage lambda (87), which is involved in lysis versus lysogeny decisions, the *CLN1* and *CLN2* genes of yeast (88), which are involved in cell cycle control, and several transcription factors such as GATA-1 (89), which is involved in differentiation of certain hematopoietic cells. The common thread among these genes is that the cell requires a large amount of the gene product in a short period of time, either to flip a developmental switch, or as in the case of 5α-reductase, to synthesize a hormone that acts as a morphogen.

The observation that dihydrotestosterone is much more efficient at inducing 5α-reductase gene expression in the regenerating ventral prostate than is testosterone may have broader implications for androgen receptor action. Thus, it seems likely that there may be three classes of androgen-responsive genes: those that respond to the androgen receptor coupled with testosterone, those that require the androgen receptor coupled to dihydrotestosterone, and those that respond to either ligand-bound receptor. Whether these differences in receptor-ligand responses are a consequence of the DNA sequence of the androgen-responsive element or are due to the interaction of the receptor with different transcriptional adaptor proteins remains to be determined. Nevertheless, this hypothesis might explain the existence of two androgens with distinct physiological roles.

Ontogeny of Steroid 5α-Reductase Expression

The expression of the 5α-reductase during development has been most closely examined in humans. Early studies indicated that the expression of 5α-reductase in the urogenital sinus and urogenital tubercle of the embryonic urogenital tract preceded formation of the external genitalia and prostate (20), suggesting a cause and effect relationship for the involvement of dihydrotestosterone in phenotypic sexual differentiation. Immunoblotting experiments indicate that most of this enzyme activity in the early embryo can be attributed to the type 2 isozyme (46). From these findings and from the phenotype of the genetic disease, it would appear that 5α-reductase type 2 is responsible for embryonic virilization of the external genitalia and prostate in men. In contrast, both the type 1 and type 2 mRNAs are present beginning around day 13 of development of the rat embryo (DM Berman, DW Russell, unpublished observations), suggesting that in this species, both isozymes may play a role in male phenotypic sexual differentiation.

The type 1 isozyme is thought to be expressed at birth in both the liver and nongenital skin of humans (46). Expression in the liver persists throughout life, whereas expression in the skin continues through age 2 to 3, and then decreases to an undetectable level. During or shortly after puberty, the type 1 isozyme again is detectable in nongenital skin, including scalp, and remains present in these tissues throughout adult life. In adult scalp, most of the type 1 isozyme is found in the sebaceous gland (Table 3). The expression of the type 2 isozyme also increases in the liver and nongenital skin at approximately the time of birth, and like the type 1 isozyme, persists in the liver. Expression of the type 2 isozyme in nongenital skin continues through age 2 to 3, is extinguished thereafter, and does not appear to be re-expressed at puberty. Expression of the type 2 isozyme in male accessory reproductive tissues such as the prostate, epididymis, and seminal vesicles, and in the genital skin, is sustained throughout life (46).

The increase in the type 1 isozyme in the skin at puberty and its continuous presence in the liver may explain a curious clinical feature of subjects with 5α-reductase deficiency. A majority of affected individuals with this disease virilize to varying extents at puberty, and dihydrotestosterone is demonstrable, albeit in small amounts, in all affected individuals including those that do not synthesize a functional type 2 enzyme (5). Since the liver and skin constitute a large proportion of body mass (~25%), and since these two organs express a substantial amount of the type 1 isozyme (46), it seems reasonable to conclude that the observed virilization is due to the synthesis of dihydrotestosterone by the type 1 isozyme. If this hypothesis is true, then dihydrotestosterone can act as a true circulating hormone as well as an autocrine or paracrine mediator of androgen action. Other genetic and pharmacological reasons also support this notion (5).

The temporal patterns of expression of the 5α-reductase isozymes in humans raise several interesting issues. The mechanisms of the increase and decrease in isozyme levels are not known, nor is the role these phenomena play in several disorders that involve dihydrotestosterone, such as acne, hirsutism, and male pattern baldness. Men who lack the type 2 isozyme appear to have less temporal hair regression, an observation that suggests a role for this isozyme in hair loss. However, this hypothesis is not consistent with the finding that the type 1 isozyme (which is normal in 5α-reductase defiency) is the predominant 5α-reductase in the scalp of post-pubertal males (and females). It seems likely that the expression of one isozyme may influence the expression of the other in tissues that express both; however, such binary interactions have not yet been demonstrated. Finally, there must be hormonal control of isozyme expression to allow an integration of androgen synthesis with other endocrine systems. The elucidation of this control will be of great future interest.

5α-Reduction of Other Steroid Hormones

Virtually all steroid hormones with a $\Delta^{4,5}$,3-oxo structure, including glucocorticoids, progestogens, mineralocorticoids, and androgens, can be 5α-reduced by tissue slices or homogenates and are excreted, in part, as 5α-reduced metabolites. Work by Tomkins and others (10–12) suggested on indirect grounds that a separate enzyme might be responsible for the 5α-reduction of steroids of each of these classes, but subsequent genetic (4) and enzymological evidence (90) indicated that a single enzyme has the capacity to 5α-reduce many (or all) steroid hormones; indeed, the fact that excretion of all 5α-reduced steroids is impaired in individuals with 5α-reductase deficiency was clear evidence that the type 2 enzyme has broad enzymatic capacity, and the progestogens progesterone and 20α-hydroxy-preg-4-en-3-one appear to be the best endogenous substrates for the enzyme (90). This assumption was confirmed in studies with the expressed type 1 and type 2 isozymes, each of which has the capacity to 5α-reduce a variety of steroid hormones (38, 41, 45).

Whether 5α-reduction of steroid hormones other than androgens plays a physiological role in hormone action or is only a step in metabolic turnover is not entirely clear. 5α-Dihydroprogesterone is a major hormone in the circulation of both normal cycling women (91) and of pregnant women (92); a specific role for 5α-dihydroprogesterone has been suggested in the shell gland of the chicken (93) and in the feedback control of gonadotropin secretion in the rat (94). 5α-Dihydrocortisol is present in the aqueous humor of the eye, is synthesized in the lens of the eye, and may play a role in the regulation of aqueous humor formation (95). 5α-Dihydroaldosterone is a potent antinatriuretic agent with somewhat different physiologic effects than aldosterone itself (96, 97); its formation in the kidney (98) is enhanced by restriction of dietary sodium intake (99), in keeping with a role of the metabolite in the conservation of sodium. In summary, the 5α-reduction pathway may be of general significance in both the metabolism and action of steroid hormones of several classes.

PHARMACOLOGY OF STEROID 5α-REDUCTASE

With the recognition that 5α-reductase and dihydrotestosterone play important roles in androgen action came the idea that inhibitors of the enzyme might have therapeutic value (19). The clinical situations in which inhibitors would be useful became apparent with the demonstration of 5α-reductase activity in the skin and adult prostate and the characterization of 5α-reductase deficiency (3, 4, 100–103). Thus, the absence of a normal prostate in men with this disease and the demonstration that dihydrotestosterone is the active androgen in the human prostate at all ages suggested that 5α-reductase inhibitors would be useful in the treatment of benign prostatic hyperplasia. This syndrome is marked by urinary obstruction and occurs in a majority of men over the age of 55 (104). 5α-Reductase-deficient men appear to have less male pattern baldness and almost no acne, suggesting that inhibitors may also be useful in the treatment of these common afflictions. Finally, hirsutism (inappropriate growth of facial and body hair in women) is mediated by dihydrotestosterone, and the early stages of prostate cancer in some men are androgen dependent (105).

The 5α-reductase enzyme is not inhibited by product; thus the first inhibitors were steroids that mimicked testosterone substrate and in many cases were substrates themselves (i.e. not true inhibitors) (106, 107). The expanding possible therapeutic applications of 5α-reductase inhibitors attracted the interest of the major pharmaceutical research laboratories in the 1970s and led to the development of several potent inhibitors. Since the development of these inhibitors was reviewed by Metcalf et al in 1989 (108) has come the formal proof of two 5α-reductase isozymes in humans (38, 40), the observation that both 5α-reductase isozymes contribute to serum dihydrotestosterone levels (5, 46), the realization that many inhibitors exhibit isozyme selectivity (38–41, 109, 110), and the approval of the inhibitor finasteride for the treatment of benign prostatic hyperplasia (111).

Figure 6 illustrates several different classes of inhibitors and their isozyme selectivities for the human 5α-reductases. In addition to their therapeutic effects, these drugs serve as powerful tools for elucidating the biology of 5α-reductase and the role of dihydrotestosterone in androgen action.

Steroidal Inhibitors

The largest class of 5α-reductase inhibitors described to date are steroid derivatives. Of these, the 4-azasteroids such as 4-MA and finasteride (Figure 6) have been the most extensively studied. The compound 4-MA was initially reported by Brooks et al (82), who demonstrated that this drug inhibits 5α-reductase enzyme activity in rat prostate. 4-MA had a very low affinity for the androgen receptor and was thus not expected to produce undesirable anti-androgen effects such as impotence, impairment of muscle growth, or gynecomastia (112). However, the drug was subsequently shown to be an inhibitor of another steroid-metabolizing enzyme (3β-hydroxysteroid dehydrogenase) (113) and to cause hepatotoxicity.

Extensive structure-activity studies by Rasmusson et al led to the development of the 4-azasteroid finasteride (114, 115), which differs from 4-MA by the presence of a double bond in the A ring, the absence of a 4-methyl substitution, and the presence of a t-butyl amide group at the 17 position (Figure 6). Finasteride is a potent inhibitor of both the rat and human prostatic 5α-reductase enzyme activities (116) and has an interesting mechanism of inhibition (see below).

Frye et al (117) have synthesized an extensive series of 6-azasteroids, some of which are potent inhibitors of both human 5α-reductase isozymes (e.g. GI157669X, Figure 6). In addition to their heterocyclic B ring, these compounds are unique in having a $\Delta^{4,5}$ bond in the A ring, which may more faithfully mimic the natural steroid substrates of 5α-reductase.

Metcalf, Levy, Holt, and colleagues (118–121) have reported several additional classes of steroidal inhibitors, such as SKF 105657, with modifications at the 3-position of the A ring (Figure 6). These compounds lack a heterocyclic A ring but share substituents at the 17 position of the D ring with the 4-azasteroids. Many of them are postulated to mimic an enzyme-bound enolate and accordingly have different mechanisms of inhibition compared to the 4-azasteroids (108).

Nonsteroidal Inhibitors

Several nonsteroidal inhibitors of 5α-reductase have been described, including certain benzoylaminophenoxybutanoic acid derivatives (e.g. ONO-3805, Figure 6). The ONO-3805 compound was reported in 1988 to be a potent

INHIBITOR	K_i, HUMAN 5α-REDUCTASE 1	K_i, HUMAN 5α-REDUCTASE 2
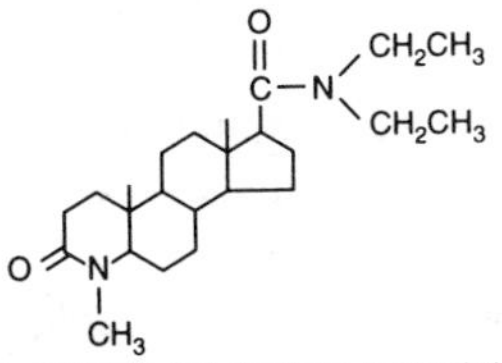 4-MA, 17β-(N,N,-diethyl)carbamoyl-4-methyl-4-aza-5α-androstan-3-one	8 nM	4 nM
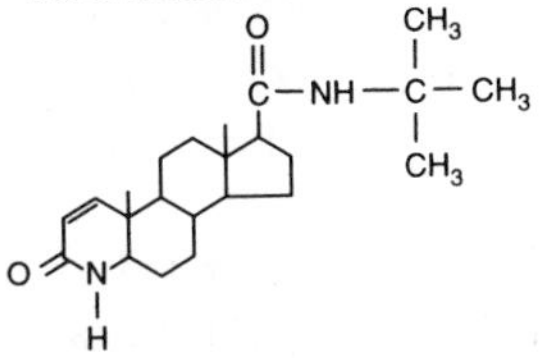 Finasteride, 17β-(N,t,-butyl)carbamoyl-4-aza-5α-androst-1-en-3-one	325 nM	12 nM
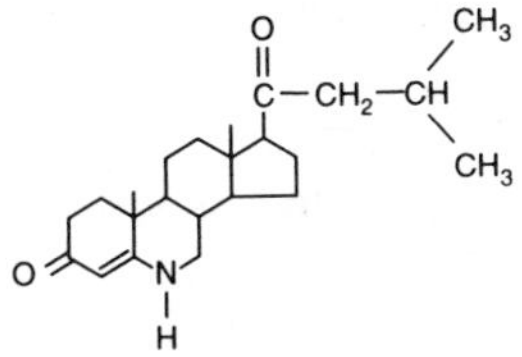 GI157669X, 17β-(i,-butyl)carbonyl-6-aza-androst-4-en-3-one	9 nM	0.7 nM
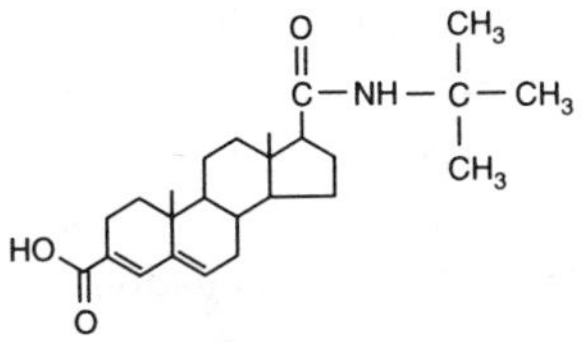 SKF 105657, 17β-(N,t,-butyl)carboxamide-androst-3,5-diene-3-carboxylic acid	425 nM	1.0 nM
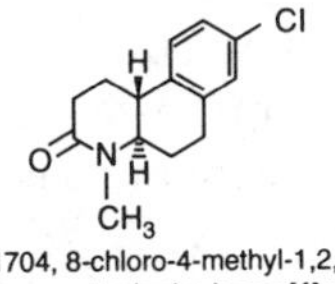 LY191704, 8-chloro-4-methyl-1,2,3,4,4a,5,6,10b-octaahydro-benzo[f]quinolin-3(2H)-one	3 nM	20 μM
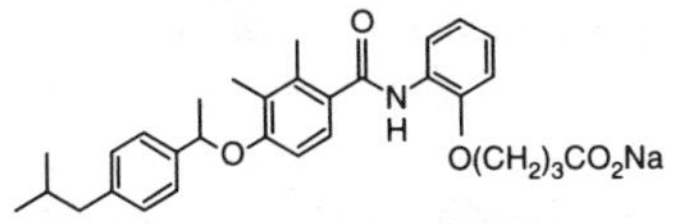 ONO-3805, 4-[2-[4-(4-isobutylbenzyloxy)-2,3-dimethylbenzoylamino]phenoxy] butanoic acid	84 nM	40 nM

Figure 6 Inhibitors of and inhibition constants for the human steroid 5α-reductase isozymes. K_i values for GI157669X were supplied by S Frye, and for SKF 105657 by M Levy. IC_{50} values for ONO-3805 were supplied by G Harris. See text for additional details.

inhibitor of rat prostatic 5α-reductase activity (122), and also inhibits the human 5α-reductases. Additional derivatives of this general class have been described by Blagg et al (123, 124).

A second nonsteroidal class of inhibitors are benzoquinolinones such as LY191704 (Figure 6). This series of tricyclic compounds was isolated as selective inhibitors of the human 5α-reductase type 1 isozyme (110, 125). They share clear structural features with the 4-azasteroids, including the presence of a heterocyclic ring containing nitrogen and a 3-oxo group, but lack the characteristic fourth ring of steroids. The fact that three classes of 5α-reductase inhibitors (the 4-azasteroids, 6-azasteroids, and benzoquinolinones) all contain heterocyclic nitrogen rings may explain why riboflavin is a weak inhibitor of 5α-reductase (126).

Polyunsaturated fatty acids, but not their mono-unsaturated or saturated counterparts, are also moderately potent inhibitors of 5α-reductase enzyme activity (127). For example, linoleic acid (9,12-$C_{18:2}$) inhibits the individual rat and human 5α-reductase isozymes at concentrations between 0.3 and 5 μM, whereas oleic acid (9-$C_{18:1}$) and stearic acid (C_{18}) inhibit the isozymes only at levels above their critical micelle concentration (K Normington, personal communication). As described above, 5α-reductase is a membrane-bound enzyme that appears to require the unique environment of the lipid bilayer for activity. Attempts to solubilize the protein with ionic and non-ionic detergents are frequently successful; however, subsequent chromatography results in rapid loss of enzyme activity. The addition of phospholipids to solubilized enzyme preparations frequently stimulates activity (31, 34, 128), again suggesting a role for lipids in modulating enzyme activity. The inhibition of 5α-reductase enzyme activity by polyunsaturated fatty acids may therefore reflect another form of this modulation.

Isozyme Selectivity of Inhibitors

In characterizing the above inhibitors, several interesting observations were made regarding the existence of possible isozymes of 5α-reductase. For example, a marked species specificity exists in the ability of some steroids to inhibit prostatic 5α-reductase enzyme activity (116, 121). In addition, within the ventral prostate of the rat, a single inhibitor showed a differential ability to inhibit 5α-reductase depending on the age of the animal (129). In retrospect, the interpretation of these and other studies was complicated by the fact that the prostates (or any other tissues or cell lines) of different species express alternate isozymes and by the different pH optima of the isozymes. Thus, if human prostate extracts were used in an assay carried out in an acidic pH buffer, then the type 2 isozyme was being scored. However, if rat prostatic or liver extracts were used at a more neutral pH, then the type 1 isozyme was being assayed.

With the cloning and expression of type 1 cDNAs (36, 38), it became possible to compare directly the 5α-reductase expressed by the cloned cDNA with those present in various tissue extracts. These comparisons revealed unambiguous evidence for pharmacological differences between the type 1 and type 2 isozymes (39) and were a major impetus to seek the second isozyme (see above). Isozyme selectivity is readily demonstrated using extracts derived from cells transfected with the individual cDNAs (Figure 6), and in certain instances inhibitor preference can be quite striking (e.g. for LY191704 and the human type 1 isozyme, Figure 6, and 21,21-pentamethylene-4-aza-5α-pregn-1-ene-3,20-dione for the rat type 2 isozyme, Ref. 41).

The structural basis for isozyme selectivity has been studied in some detail for the inhibitor finasteride. In short-term assays, this 4-azasteroid is 100-fold more potent against the rat type 1 isozyme versus the human type 1 isozyme (38). By constructing and analyzing chimeric 5α-reductase enzymes composed of different portions of the human and rat type 1 isozymes, this difference in sensitivity was traced to a four-amino-acid segment encoded within exon 1 (69). Since finasteride is a small molecule (M_r = 374) and inhibited the normal rat and human type 1 isozymes as well as the chimeric enzymes in a competitive fashion with substrate, it was postulated that these four amino acids also define a portion of the substrate-binding domain. Structure-activity studies comparing different inhibitors, isozymes, and 5α-reductases should provide additional insight into the functional domains of the enzyme as well as structure-activity relationships in the inhibitors.

Given the number of 5α-reductase inhibitors described (and undescribed), there almost certainly exist compounds that are selective for any isozyme from any species. In the future, the use of these isozyme-selective inhibitors in vivo will prove invaluable in assigning physiological roles to a given 5α-reductase. To date, studies have used inhibitors such as finasteride and other 4-azasteroids that block both 5α-reductase isozymes (83, 130–133) and thus cannot be interpreted at the isozyme level.

Mechanisms of Inhibition

As would be expected from the diversity of chemical structures, different 5α-reductase inhibitors manifest different mechanisms of inhibition. For the most part, the 4-azasteroids are competitive with substrate (116, 134), whereas some steroids and the nonsteroidal compounds exhibit noncompetitive or uncompetitive kinetics in the presence of substrate or cofactor. For example, the compound SKF 105657 is uncompetitive with substrate and is thought to bind an enzyme-$NADP^+$ intermediate (33). This mechanism of inhibition is common among steroids with negatively charged A-ring modifications (118–121).

The effect of kinetic mechanism on efficacy is well illustrated by finasteride. In human males, a single 0.5 mg dose of this drug can decrease serum dihydrotestosterone levels by 65% for a period of 5–7 days (135–137). This remarkable efficacy was initially attributed to a low nanomolar K_i for the type 2 isozyme, a competitive mechanism of inhibition with substrate, and the inhibitor's selectivity. Subsequent studies have shown that finasteride is in fact a time-dependent inactivator of 5α-reductase type 2 (and presumably also of type 1) (53). Finasteride has a slow K_{on} rate but an even slower dissociation rate ($T_{1/2} > 30$ days), operationally leading to irreversible inhibition (53). The time-dependent inhibition of 5α-reductase by finasteride means that IC_{50} or K_i values determined in short-term assays (e.g. those in Figure 6) are a gross underestimate of the true inhibition constants. Assays in which low concentrations of finasteride are used together with extended incubation times reveal an apparent K_i for the human type 1 isozyme of about 1 nM versus a low picomolar K_i for the type 2 isozyme.

Indirect evidence in support of irreversible finasteride binding also comes from immunoprecipitation experiments using conformation-dependent antibodies against the type 2 isozyme (45). Immunoprecipitation of this isozyme was blocked in the presence of non-ionic detergents and finasteride; however, this block could be reversed if the reaction was carried out in the presence of ionic detergents. The latter more harsh solubilization conditions presumably removed the bound finasteride and allowed recognition of the isozyme by the conformation-dependent antibodies, but the alternate explanation that the ionic detergent exposed an epitope but did not remove the drug could not be ruled out (45). The presence of 4-MA in the immunoprecipitation reactions did not interfere with antibody recognition, suggesting that this property was unique to finasteride. Time-dependent inhibition of the human type 2 isozyme is also observed with certain of the 6-azasteroids (S Frye, personal communication) and has been reported for at least two other steroids (108, 138).

Taken together, the results suggest that finasteride is a time-dependent inactivator of the type 2 isozyme and provide insight into why the drug is so efficacious in vivo. It is not yet clear if inactivation is the result of a covalent attachment of the drug or whether binding causes an irreversible conformational change in the enzyme. However, the presence of the $\Delta^{1,2}$ bond in the 4-azasteroids is crucial for time-dependent inactivation. This bond may be a target for Michael addition catalyzed by a nucleophilic group on the enzyme leading to formation of a covalent end-product. Future studies using radiolabeled finasteride should allow dissection of this interaction and determination of whether finasteride is itself a substrate for 5α-reductase. The unusual pH optimum of the type 2 isozyme may also be linked to the observed irreversible inhibition.

Physiological Consequences of Inhibition

Finasteride was approved in 1992 for the treatment of benign prostatic hyperplasia in the United States (111). Extensive studies in male humans and other species indicate that finasteride lowers the serum and intraprostatic concentrations of dihydrotestosterone, causing apoptosis of epithelial cells in the prostate and subsequent reduction in gland size (137). The drug has few if any untoward side effects, and is used as an alternative to surgical treatment of benign prostatic hyperplasia. With time, finasteride may also be prescribed to prevent the development of prostatic hyperplasia.

Finasteride is a selective but not specific inhibitor of the human type 2 isozyme of 5α-reductase (Figure 6), and would thus be expected to have a more profound effect on tissues such as the prostate that express this isozyme (Table 2). It is unclear how effectively the drug inhibits the type 1 isozyme in vivo. Most studies show a residual serum dihydrotestosterone level of about 40% after long-term finasteride therapy (139). Similarly, individuals with a known genetic defect in the type 2 gene can have serum dihydrotestosterone levels in the low normal range (5). The residual hormone in both cases may be synthesized by the type 1 isozyme that is relatively insensitive to finasteride inhibition and whose gene is normal in the genetic disease. Alternatively, finasteride may not have access to the type 1 isozyme in the skin, or the pharmacokinetics of finasteride coupled with a slow rate of time-dependent inactivation of the type 1 isozyme may decrease efficacy. The possibility of a third 5α-reductase isozyme that is resistant to finasteride must also be kept in mind. Such findings underscore the need to develop pan-specific inhibitors of the isozymes, and may explain why not all subjects with benign prostatic hyperplasia respond equally well to finasteride therapy (111, 139). Similarly, other disorders involving dihydrotestosterone (male pattern baldness, acne, hirsutism, prostate cancer) may require isozyme-specific inhibitors and the development of alternate routes of administration. Additional inhibitors may include antisense oligonucleotides, or even oligonucleotides themselves, as the human type 1 isozyme has been shown to bind to the DNA sequence 5′-CGCTTCCGGGA-3′ (140).

CONCLUSION

The cloning of cDNAs encoding distinct 5α-reductase isozymes has provided a number of tools for dissecting the central roles played by this enzyme and its products in androgen and steroid hormone action. These tools have so far been used to gain preliminary insight into the biochemistry, genetics, regulation, and pharmacology of 5α-reductase. In the future, they will be used to answer a host of interesting questions about the physiological roles of the isozymes. Included in these questions will be the tertiary structure of the protein and a further definition of functional domains in the molecule, insight into the regulatory proteins that control the temporal and cell-type-specific expression patterns of the 5α-reductase genes, the consequences of targeted mutation of the type 1 and/or type 2 genes in the mouse, a more complete definition of the physiological roles of the two isozymes, the role of 5α-reduction in the physiology of other steroid hormones, and the development and clinical use of new inhibitors of 5α-reductase.

ACKNOWLEDGMENTS

We are grateful to the members of our laboratories for their hard work and unflagging interest. We thank Steven Frye, Mark Levy, and Georgianna Harris for providing unpublished information. Research in our laboratories was supported by grants from the National Institutes of Health (GM43753 and DK03892), the Perot Family Foundation, and the Robert A. Welch Foundation (I-0971).

Literature Cited

1. Bruchovsky N, Wilson JD. 1968. *J. Biol. Chem.* 243:2012–21
2. Anderson KM, Liao S. 1968. *Nature* 219:277–79
3. Walsh PC, Madden JD, Harrod MJ, Goldstein JL, MacDonald PC, Wilson JD. 1974. *N. Engl. J. Med.* 291:944–49
4. Imperato-McGinley J, Guerrero L, Gautier T, Peterson RE. 1974. *Science* 186:1213–15
5. Wilson JD, Griffin JE, Russell DW. 1993. *Endocrine Rev.* 14:577–93
6. Tausk M. 1984. *Discov. Pharmacol.* 2:307–18
7. Schneider JJ, Horstmann PM. 1951. *J. Biol. Chem.* 191:327–38
8. Schneider JJ. 1952. *J. Biol. Chem.* 199:235–44
9. Forchielli E, Dorfman RI. 1952. *J. Biol. Chem.* 223:443–48
10. Tomkins GM. 1957. *J. Biol. Chem.* 225:13–24
11. McGuire JS, Tomkins GM. 1960. *J. Biol. Chem.* 235:1634–38
12. McGuire JS, Hollis VW, Tomkins GM. 1960. *J. Biol. Chem.* 235:3112–16
13. Bondy PK. 1981. In *Williams Textbook of Endocrinology,* ed. JD Wilson, DW Foster, pp. 816–90. Philadelphia: Saunders
14. Wilson JD. 1975. *Handb. Physiol.* 5:491–508
15. Saunders FJ. 1963. In *Biology of the Prostate and Related Tissue,* ed. EP Vollmer, pp. 139–59. Washington, DC: US Gov. Print. Off.
16. Mainwaring WIP. 1969. *J. Endocrinol.* 45:531–41
17. Fang S, Anderson KM, Liao S. 1969. *J. Biol. Chem.* 244:6584–95
18. Wilson JD, Lasnitzki I. 1971. *Endocrinology* 89:659–68
19. Wilson JD. 1972. *N. Engl. J. Med.* 287:1284–91
20. Siiteri PK, Wilson JD. 1974. *J. Clin. Endocrinol. Metab.* 38:113–25
21. Nowakowski H, Lenz W. 1961. *Rec. Prog. Hormone Res.* 17:53–95
22. Wilson JD. 1975. *J. Biol. Chem.* 250:3498–504
23. Moore RJ, Griffin JE, Wilson JD. 1975. *J. Biol. Chem.* 250:7168–72
24. Leshin M, Griffin JE, Wilson JD. 1972. *J. Clin. Invest.* 247:685–91
25. Moore RJ, Wilson JD. 1976. *J. Biol. Chem.* 251:5895–900

26. Frederiksen DW, Wilson JD. 1971. *J. Biol. Chem.* 246:2584–93
27. Moore RJ, Wilson JD. 1972. *J. Biol. Chem.* 247:958–67
28. Moore RJ, Wilson JD. 1974. *Biochemistry* 13:450–56
29. Liang T, Heiss CE, Ostrove S, Rasmusson GH, Cheung A. 1983. *Endocrinology* 112:1460–68
30. Houston B, Chisholm GD, Habib FK. 1985. *J. Steroid Biochem.* 22:461–67
31. Ichihara K, Tanaka C. 1987. *Biochem. Int.* 15:1005–11
32. Enderle-Schmitt U, Neuhaus C, Aumüller G. 1989. *Biochim. Biophy. Acta* 987:21–28
33. Levy MA, Brandt M, Greway AT. 1990. *Biochemistry* 29:2808–15
34. Sargent NSE, Habib FK. 1991. *J. Steroid Biochem. Mol. Biol.* 38:73–77
35. Noma Y, Sideras P, Naito T, Bergstedt-Lindquist S, Azuma C, et al. 1986. *Nature* 319:640–46
36. Andersson S, Bishop RW, Russell DW. 1989. *J. Biol. Chem.* 264:16249–55
37. Farkash Y, Soreq H, Orly J. 1988. *Proc. Natl. Acad. Sci. USA* 85:5824–28
38. Andersson S, Russell DW. 1990. *Proc. Natl. Acad. Sci. USA* 87:3640–44
39. Jenkins EP, Andersson S, Imperato-McGinley J, Wilson JD, Russell DW. 1992. *J. Clin. Invest.* 89:293–300
40. Andersson S, Berman DM, Jenkins EP, Russell DW. 1991. *Nature* 354: 159–61
41. Normington K, Russell DW. 1992. *J. Biol. Chem.* 267:19548–54
42. Markert CL, Moller F. 1959. *Proc. Natl. Acad. Sci. USA* 45:753–63
43. Wu L, Einstein M, Geissler WM, Chan HK, Elliston KO, Andersson S. 1993. *J. Biol. Chem.* 268:12964–69
44. Fletterick RJ, Schroer T, Matela RJ. 1985. *Protein Structure,* pp. 39–63. Palo Alto, CA: Blackwell
45. Thigpen AE, Cala KM, Russell DW. 1993. *J. Biol. Chem.* 268:17404–12
46. Thigpen AE, Silver RI, Guileyardo JM, Casey ML, McConnell JD, Russell DW. 1993. *J. Clin. Invest.* 92: 903–10
47. McCombie WR, Adams MD, Kelley JM, FitzGerald MG, Utterback TR, et al. 1992. *Nature Genet.* 1:124–31
48. Laux G, Perricaudet M, Farrell PJ. 1988. *EMBO J.* 7:769–74
49. Johnston IG, Rush SJ, Gurd JW, Brown IR. 1992. *J. Neurosci. Res.* 32:159–66
50. Shinozaki K, Ohme M, Tanaka M, Wakasugi T, Hayashida N, et al. 1986. *EMBO J.* 5:2043–49
51. Rassart E, Nelbach L, Jolicoeur P. 1986. *J. Virol.* 60:910–19
52. Fersht A. 1985. *Enzyme Structure and Mechanism,* pp. 155–75. New York: Freeman
53. Faller B, Farley D, Nick H. 1993. *Biochemistry* 32:5705–10
54. Yates FE, Herbst AL, Urquhart J. 1958. *Endocrinology* 63:887–902
55. Harris GS, Azzolina BA. 1990. *Fed. Am. Soc. Exp. Biol. J.* 4:2717 (Abstr.)
56. Ordman AB, Farley D, Meyhack B, Nick H. 1991. *J. Steroid Biochem. Mol. Biol.* 39:487–92
57. Hudson RW. 1981. *J. Steroid Biochem.* 14:579–84
58. Berman DM, Russell DW. 1993. *Proc. Natl. Acad. Sci. USA* 90:9359–63
59. Jenkins EP, Hsieh C-L, Milatovich A, Normington K, Berman DM, et al. 1991. *Genomics* 11:1102–12
60. Thigpen AE, Davis DL, Milatovich A, Mendonca BB, Imperato-McGinley J, et al. 1992. *J. Clin. Invest.* 90:799–809
61. Labrie F, Sugimoto Y, Luu-The V, Simard J, Lachance Y, et al. 1992. *Endocrinology* 131:1571–73
62. Davis DL, Russell DW. 1993. *Human Mol. Genet.* 2:820
63. Wilson JD. 1978. *Annu. Rev. Physiol.* 40:279–306
64. Wilson JD. 1984. *The Harvey Lect. Ser.* 79:145–70
65. Sinclair AH, Berta P, Palmer MS, Hawkins JR, Griffiths BL, et al. 1990. *Nature* 346:240–44
66. Jost A. 1970. *Philos. Trans. R. Soc. London* 259:119–30
67. Griffin JE, McPhaul MJ, Russell DW, Wilson JD. 1994. In *Metabolic Basis Inherited Disease,* ed. CR Scriver, AL Beaudet, WS Sly, D Valle. New York: McGraw-Hill. In press
68. Thigpen AE, Davis DL, Gautier T, Imperato-McGinley J, Russell DW. 1992. *N. Engl. J. Med.* 327:1216–19
69. Thigpen AE, Russell DW. 1992. *J. Biol. Chem.* 267:8577–83
70. Björkhem I, Buchman M, Byström S. 1992. *J. Biol. Chem.* 267:19872–75
71. Perham RN, Scrutton NS, Berry A. 1991. *BioEssays* 13:515–25
72. Gustafsson J-Å, Mode A, Norstedt G, Skett P. 1983. *Annu. Rev. Physiol.* 45:51–60
73. Bullock P, Gemzik B, Johnson D, Thomas P, Parkinson A. 1991. *Proc. Natl. Acad. Sci. USA* 88:5227–31
74. Verhagen APM, Aalders TW, Ramaekers FCS, Debruyne FMJ, Schalken JA. 1988. *Prostate* 13:25–38
75. Cunha GR, Donjacour AA, Cooke PS,

Mee S, Bigsby RM, et al. 1987. *Endocrine Rev.* 8:338–62
76. Imperato-McGinley J, Gautier T, Zirinsky K, Hom T, Palomo O, et al. 1992. *J. Clin. Endocrinol. Metab.* 75:1022–26
77. Viger RS, Robaire B. 1991. *Endocrinology* 128:2407–14
78. Viger RS, Robaire B. 1992. *Endocrinology* 131:1534–40
79. Moore CR, Price D, Gallagher TF. 1930. *Am. J. Anat.* 45:71–107
80. Raff MC. 1992. *Nature* 356:397–400
81. Moore RJ, Wilson JD. 1973. *Endocrinology* 93:581–92
82. Brooks JR, Baptista EM, Berman C, Ham EA, Hichens M, et al. 1981. *Endocrinology* 109:830–36
83. George FW, Russell DW, Wilson JD. 1991. *Proc. Natl. Acad. Sci. USA* 88:8044–47
84. Ish-Horowicz D, Pinchin SM, Ingham PW, Gyurkovics HG. 1989. *Cell* 57: 223–32
85. Heemskerk J, DiNardo S, Kostriken R, O'Farrell PH. 1991. *Nature* 352: 404–10
86. Bell LR, Horabin JI, Schedl P, Cline TW. 1991. *Cell* 65:229–39
87. Ptashne M. 1986. *A Genetic Switch: Gene Control and Phage λ*. Cambridge, MA: Blackwell. 128 pp.
88. Cross FR, Tinkelenberg AH. 1991. *Cell* 65:875–83
89. Tsai S-F, Strauss E, Orkin SH. 1991. *Genes Dev.* 5:919–31
90. Fisher LK, Kogut MD, Moore RJ, Goebelsmann U, Weitman JJ, et al. 1978. *J. Clin. Endocrinol. Metab.* 47:653–64
91. Milewich L, Gomez-Sanchez C, Crowley G, Porter JC, Madden JD, MacDonald PC. 1977. *J. Clin. Endocrinol. Metab.* 45:617–22
92. Milewich L, Gomez-Sanchez C, Madden JD, MacDonald PC. 1975. *Gynecol. Invest.* 6:291–306
93. Morgan MD, Wilson JD. 1970. *J. Biol. Chem.* 245:3781–89
94. Putnam-Roberts C, Brann DW, Mahesh VB. 1992. *J. Steroid Biochem.* 42:875–82
95. Weinstein BI, Kandalaft N, Ritch R, Gordon GG, Southren AL. 1991. *Invest. Ophthal. Visual Sci.* 32:2130–35
96. Morris DJ, Brem AS. 1987. *Am. Physiol. Soc.* 252:365–73
97. Morris DJ, Souness GW, Saccoccio NA, Harnik M. 1989. *Steroids* 53:21–36
98. McDermott M, Latif SA, Morris DJ. 1983. *J. Steroid Biochem.* 19:1205–11
99. Gorsline J, Latif SA, Morris DJ. 1988. *Am. J. Hypertension* 1:272–75
100. Gloyna RE, Wilson JD. 1969. *J. Clin. Endocrinol. Metab.* 29:970–77
101. Gloyna RE, Siiteri PK, Wilson JD. 1970. *J. Clin. Invest.* 246:1746–53
102. Siiteri P, Wilson JD. 1979. *J. Clin. Invest.* 49:1737–45
103. Voigt W, Fernandez EP, Hsia SL. 1970. *J. Biol. Chem.* 245:5594–99
104. Walsh PC. 1992. In *Campbell's Urology*, ed. PC Walsh, AB Retik, TA Stamey, ED Vaughan Jr, pp. 1007–27. Philadelphia: Saunders
105. Stamey TA, McNeal JE. 1992. See Ref. 104, pp. 1159–221
106. Voigt W, Hsia SL. 1973. *Endocrinology* 92:1216–22
107. Hsia SL, Voigt W. 1974. *J. Invest. Dermatol.* 62:224–27
108. Metcalf BW, Levy MA, Holt DA. 1989. *Trends Pharmacol. Sci.* 10:491–95
109. Harris G, Azzolina BA, Baginsky W, Cimis G, Rasmusson GH, et al. 1992. *Proc. Natl. Acad. Sci. USA* 89:10787–791
110. Hirsch KS, Jones CD, Audia JE, Andersson S, McQuaid LA, et al. 1993. *Proc. Natl. Acad. Sci. USA* 90:5277–81
111. Gormley GJ, Stoner E, Bruskewitz RC, Imperato-McGinley J, Walsh PC, et al. 1992. *N. Engl. J. Med.* 327: 1187–91
112. Liang T, Heiss CE, Cheung AH, Reynolds GF, Rasmusson GH. 1984. *J. Biol. Chem.* 259:734–39
113. Brandt M, Levy MA. 1989. *Biochemistry* 28:140–48
114. Rasmusson GH, Reynolds GF, Utne T, Jobson RB, Primka RL, et al. 1984. *J. Med. Chem.* 27:1690–701
115. Rasmusson GH, Reynolds GF, Steinberg NG, Walton E, Patel GF, et al. 1986. *J. Med. Chem.* 29:2298–15
116. Liang T, Cascieri MA, Cheung AH, Reynolds GF, Rasmusson GH. 1985. *Endocrinology* 117:571–79
117. Frye SV, Cribbs CM, Haffner CD, Maloney PR, Andrews RC. 1993. *World Patent No. WO93/13124*
118. Holt DA, Levy MA, Oh H-J, Erb JM, Heaslip JI, et al. 1990. *J. Med. Chem.* 33:943–50
119. Holt DA, Levy MA, Ladd DL, Oh H-J, Erb JM, et al. 1990. *J. Med. Chem.* 33:937–42
120. Holt DA, Oh H-J, Levy MA, Metcalf BW. 1991. *Steroids* 56:4–7
121. Levy MA, Metcalf BW, Brandt M, Erb JM, Oh H-J, et al. 1991. *Biorg. Chem.* 19:245–60

122. Nakai H, Terashima H, Arai Y. 1988. *Eur. Patent No. 0291245A2*
123. Blagg J, Cooper K, Spargo PL. 1993. *World Patent No. WO93/01050*
124. Blagg J, Cooper K, Spargo PL. 1993. *World Patent No. WO93/02051*
125. Jones CD, Audia JE, Lawhorn DE, McQuaid LA, Neubauer BL, et al. 1993. *J. Med. Chem.* 36:421–23
126. Nakayama O, Yagai M, Kiyoto S, Okuhara M, Kohsaka M. 1990. *J. Antibiotics* 43:1615–16
127. Liang T, Liao S. 1992. *Biochem. J.* 285:557–62
128. Cooke GM, Robaire B. 1985. *J. Biol. Chem.* 260:7489–95
129. Martini L, Zoppi S, Motta M. 1986. *J. Steroid Biochem.* 24:177–82
130. Imperato-McGinley J, Binienda Z, Arthur A, Mininberg DT, Vaughan ED Jr, Quimby FW. 1985. *Endocrinology* 16:807–12
131. Rittmaster RS, Magor KE, Manning AP, Norman RW, Lazier CB. 1991. *Mol. Endocrinol.* 5:1023–29
132. Lamb JC, English H, Levandoski PL, Rhodes GR, Johnson RK, Isaacs JT. 1992. *Endocrinology* 130:685–94
133. Imperato-McGinley J, Sanchez RS, Spencer JR, Yee B, Vaughan ED. 1992. *Endocrinology* 131:1149–56
134. Liang T, Heiss CE. 1981. *J. Biol. Chem.* 256:7998–8005
135. Vermeulen A, Giagulli VA, Deschepper P, Buntinx A, Stoner E. 1989. *Prostate* 14:45–53
136. Rittmaster RS, Stoner E, Thompson DL, Nance D, Lasseter KC. 1989. *J. Andrology* 10:259–62
137. McConnell JD, Wilson JD, George FW, Geller J, Pappas F, Stoner E. 1992. *J. Clin. Endocrinol. Metab.* 74:505–8
138. Blohm T, Metcalf BW, Sjoerdsma A, Schatzman GL. 1980. *Biochem. Biophys. Res. Commun.* 95:273–80
139. The Finasteride Study Group. 1993. *Prostate* 22:291–99
140. Gaston K, Fried M. 1992. *Nucleic Acids Res.* 20:6297–301

Annu. Rev. Biochem. 1994. 63:63–100

A MOLECULAR DESCRIPTION OF SYNAPTIC VESICLE MEMBRANE TRAFFICKING

Mark K. Bennett and Richard H. Scheller[1]

Department of Molecular and Cellular Physiology and Howard Hughes Medical Institute, Stanford University Medical Center, Stanford, California 94305

KEY WORDS: exocytosis, endocytosis, protein targeting, secretion, membrane fusion

---

CONTENTS

[1]Corresponding author. Address as above.

0066-4154/94/0701-0063$05.00

INTRODUCTION

Synaptic transmission, the process by which neurons communicate with their target cells, is essential for nearly all neuronal processes, from simple reflex pathways through the complex information processing carried out by higher brain regions. An understanding of the cellular and molecular mechanisms that underlie this process is key to a more complete understanding of neuronal function, both in normal and pathological conditions. Synaptic transmission is most commonly mediated by the release of chemical messengers (neurotransmitters) from the presynaptic neuron and the subsequent recognition of these neurotransmitters by specific receptors on the appropriate postsynaptic target cell. Fundamental insight into the presynaptic mechanisms involved in synaptic transmission at the frog neuromuscular junction was provided by classical electrophysiological experiments carried out by Katz and his colleagues (1–3). In these experiments, it was demonstrated that neurotransmitter is released from the presynaptic nerve terminal in discrete packets or "quanta" by a mechanism requiring the stimulation-dependent influx of calcium into the nerve terminal. The morphological correlates of the electrophysiologically defined quanta, 50 nm–diameter electron-translucent membranous vesicles abundant in the presynaptic nerve terminal, were termed synaptic vesicles (4). Although some nonvesicular release of neurotransmitter has been documented (5), it is widely accepted that synaptic vesicles are responsible for both the storage (by active uptake) and calcium-regulated release (by exocytotic fusion with the plasma membrane) of neurotransmitter at most synapses (for reviews see 6–9).

Given the central position occupied by synaptic vesicles in presynaptic physiology, they have not surprisingly been the topic of intensive investigation. Synaptic vesicles are abundant in nervous tissue and, owing to their unique physical properties, have been extensively purified and biochemically characterized (for recent reviews see 6, 10). A number of well-characterized proteins either integral to or peripherally associated with the synaptic vesicle membrane have been identified. A schematic diagram of the structures and topologies of the synaptic vesicle proteins that are discussed in the course of this review is presented in Figure 1*a*.

OVERVIEW OF THE SYNAPTIC VESICLE LIFECYCLE

The lifecycle of a synaptic vesicle consists of a series of membrane-trafficking events centered around the exocytotic release of neurotransmitter, as outlined in Figure 1*b*. The lifecycle begins with the biogenesis of the synaptic vesicle. This stage involves the synthesis of synaptic vesicle components in the cell body (*Ia*), nerve-terminal targeting mediated by axonal transport (*Ib*), and possibly vesicle formation in the nerve terminal by an endocytic process (*Ic*). Within the nerve terminal, the newly formed synaptic vesicle undergoes a maturation process that includes the active uptake of neurotransmitter from the cytoplasm (*IIa*) and reversible anchoring to cytoskeletal elements (*IIb*). The next stage of the lifecycle, exo/endocytic cycling, begins with the docking and calcium-regulated fusion of the synaptic vesicle with the presynaptic plasma membrane (*IIIa*), resulting in neurotransmitter release. This step is followed by the endocytic recovery (*IIIb*) and intraterminal recycling (*IIIc*) of synaptic vesicle components. After multiple rounds of exo/endocytic cycling, the components of the synaptic vesicle membrane are ultimately either turned over in the nerve terminal or targeted to the cell body for degradation (*IV*).

The synaptic vesicle lifecycle, and particularly the exo/endocytic cycling process, distinguishes the release of neurotransmitter from other regulated secretory pathways in several ways that are of critical importance for synaptic function. First, the fusion of synaptic vesicles with the presynaptic membrane in response to stimulation is extremely rapid, thereby allowing a high degree of temporal information to be transmitted across the synapse. Second, the endocytic recycling of synaptic vesicles is very efficient, enabling the presynaptic neuron to maintain the faithful release of neurotransmitter over a wide range of stimulation conditions. In contrast, dense-core granules containing neuropeptides and catecholamines in neural and endocrine cells release their contents more slowly and require recycling to the Golgi complex for vesicle regeneration. A third distinguishing feature of the synaptic vesicle lifecycle is the importance of its regulation. Since the mechanisms underlying the synaptic vesicle lifecycle are likely, at least in part, to determine the efficiency of neurotransmitter release, they represent strong candidates for presynaptic regulatory processes that might be involved in generating both the short- and long-term changes in synaptic efficacy thought to form the basis of learning and memory.

In this review, we describe the synaptic vesicle lifecycle with an emphasis on the molecular mechanisms that mediate each step. The protein constituents of the synaptic vesicle membrane are likely to be intimately associated with these molecular mechanisms. Additional soluble and plasma membrane–associated components are likely to be required at distinct stages of the synaptic vesicle lifecycle. In recent years, a great deal of progress has been made in defining the molecular mechanisms that underlie synaptic vesicle membrane trafficking. This progress has come not only from the biochemical characterization of the components of the synaptic vesicle membrane, but also from the biophysical characterization of neurotransmitter release and membrane fusion processes, as well as genetic and biochemical dissection of the constitutive secretory pathway in non-neuronal cells. These results

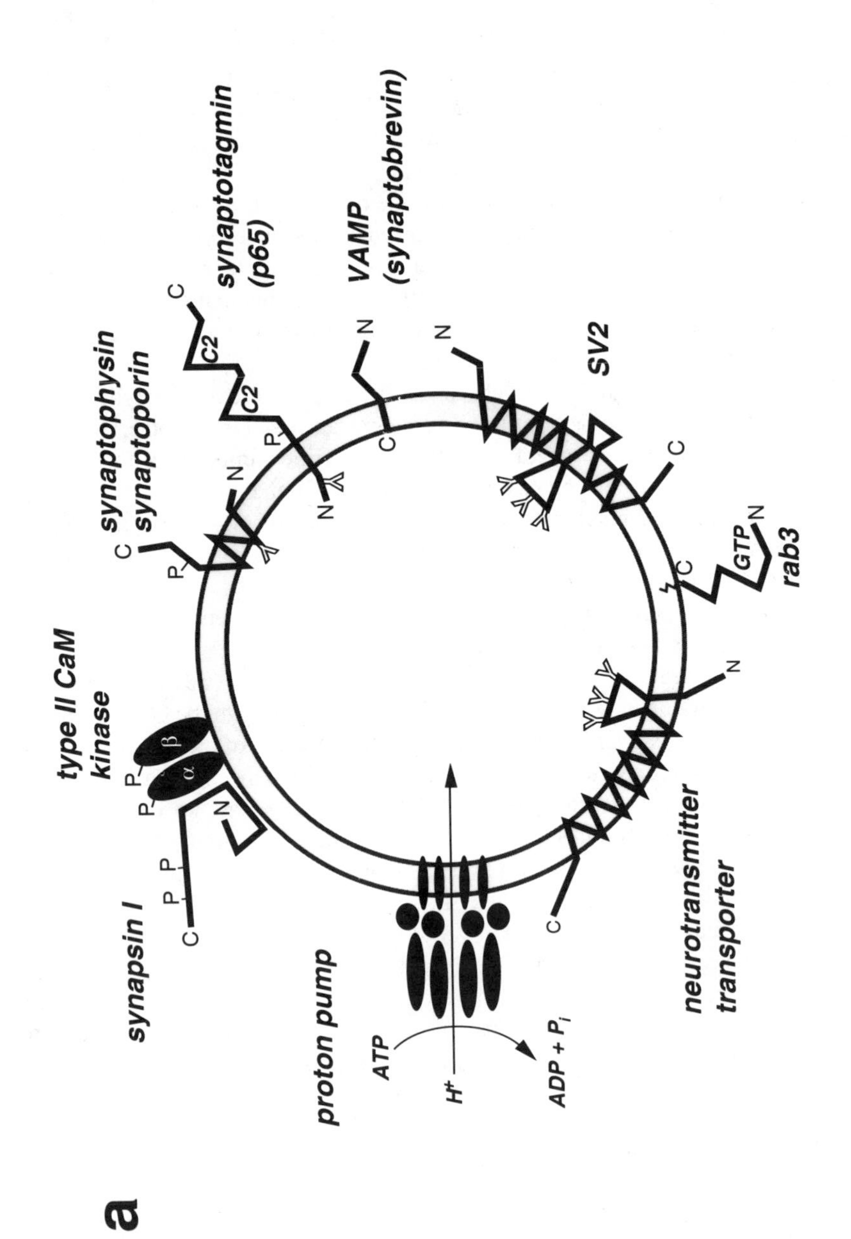
a
synapsin I
type II CaM kinase
α
β
synaptophysin
synaptoporin
synaptotagmin
(p65)
C2
C2
VAMP
(synaptobrevin)
SV2
GTP
rab3
neurotransmitter
transporter
proton pump
ATP
H+
ADP + P_i

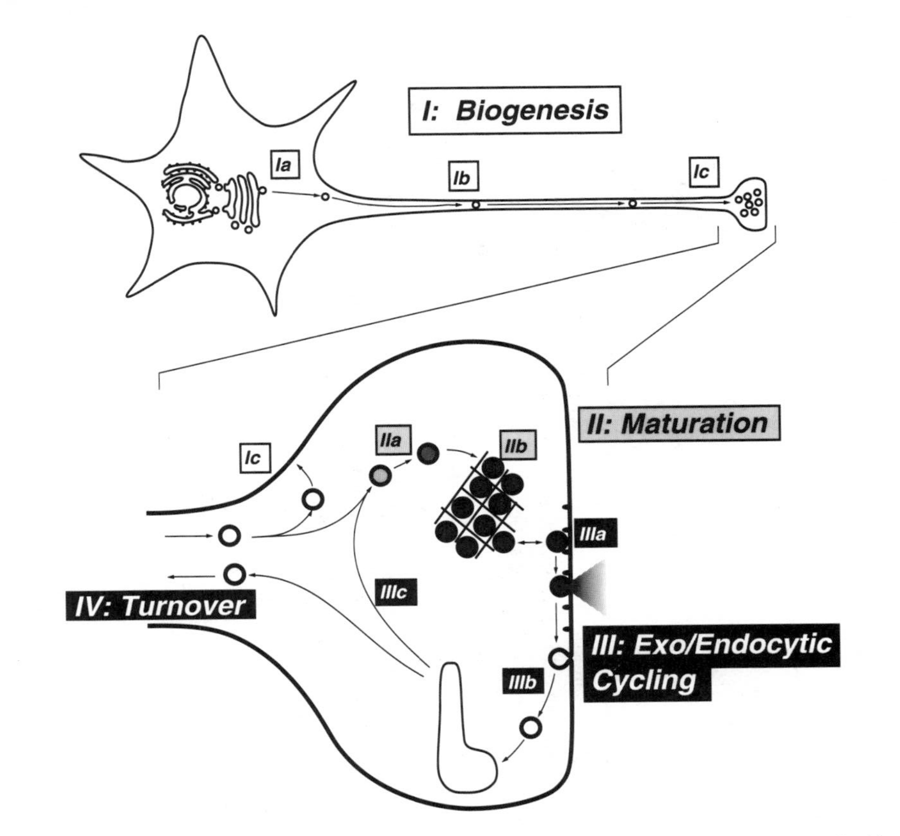

Figure 1 The synaptic vesicle: proteins and lifecycle. Panel *a*. Schematic representation of the synaptic vesicle proteins that are described in detail in the course of this review. Panel *b*. Overview of synaptic vesicle lifecycle consisting of stages of biogenesis (*I*), maturation (*II*), exo/endocytic cycling (*III*), and turnover (*IV*). See text for details.

suggest that synaptic vesicle targeting and membrane fusion involve a combination of general components that are also involved in constitutive membrane trafficking and unique components that confer specific regulatory properties to the process of neurotransmitter release.

SYNAPTIC VESICLE BIOGENESIS

Protein Synthesis

The integral membrane components of the synaptic vesicle membrane are synthesized in the cell body and initially follow a pathway that is common to all components of the secretory pathway, that is, insertion into the membrane of the endoplasmic reticulum (ER) followed by transport through the Golgi complex to the trans-Golgi network (TGN). The steady-state distribution of synaptic vesicle proteins includes proteins within the biosynthetic pathway, mature vesicles in the nerve terminals, and proteins that are targeted for degradation. While most vesicle protein is concentrated at synaptic sites, a significant amount of several vesicle proteins is detected in the region of the Golgi complex (11–14), reflecting the abundance of these proteins in transit through the biosynthetic and degradative pathways. Two synaptic vesicle proteins, the peripheral protein synapsin I and the lipid-anchored protein rab3A (Figure 1*a*), are not localized to the Golgi region (11, 15, 16), suggesting that their association with vesicle membranes occurs after exit from the TGN at an as-yet-undetermined location.

Axonal Transport

The majority of synaptic vesicle markers in mature neurons are localized to the nerve terminal. This steady-state localization likely results from selective targeting to, followed by selective retention within, the nerve terminal. The selective targeting of synaptic vesicle components to the nerve terminal is mediated by axonal transport, as evidenced by the accumulation of synaptic vesicle membrane markers on the proximal side of a block of axoplasmic transport (17). It is likely that this axonal transport is mediated by an anterograde microtubule motor protein of the kinesin family, since kinesin also accumulates on the proximal side of an axoplasmic block in association with membranes (18–20), some of which may be synaptic vesicles or their precursors (21, 22). The anterograde movement of fluorescently labeled synaptic vesicles following their introduction into squid axoplasm (23) and nerve terminals (24) is consistent with a role for kinesin in synaptic vesicle membrane transport. Further evidence for the role of a kinesin family member in synaptic vesicle localization is provided by mutations in the kinesin-related *unc-104* gene in *Caenorhabditis elegans*

(25). The *unc* mutants, initially identified because of their uncoordinated movements, are often associated with neuromuscular defects. While nerve terminals in the *unc-104* mutants displayed a reduced number of synaptic vesicles, clusters of vesicles morphologically similar to synaptic vesicles accumulated in the cell body. It remains to be determined whether the vesicles accumulating in the cell body of the *unc-104* mutants represent constitutive transport vesicles or synaptic vesicles, and whether they are derived directly from the TGN or are generated by endocytosis. Further analysis of these vesicles could provide important insight into the site and mechanism of initial synaptic vesicle formation. Interestingly, mutations in the kinesin heavy-chain gene in *Drosophila* (26) and *C. elegans* (27) exhibited defects in certain axonal properties, but synaptic vesicle targeting appeared normal. This may reflect some functional redundancy within the growing family of kinesin-related proteins (28). Upon arrival in the nerve terminal, selective retention of synaptic vesicles is likely to contribute to the maintenance of their nerve-terminal localization. One mechanism for this selective retention is the interaction between synaptic vesicles and the actin cytoskeleton to be described in a later section.

Synaptic Vesicle Formation

While the biosynthetic steps of protein synthesis and axonal transport are well established, the initial site of synaptic vesicle formation remains unclear. One possibility is that an immature synaptic vesicle forms directly from the TGN. Alternatively, synaptic vesicle formation could occur by endocytosis following constitutive delivery of synaptic vesicle components to the nerve terminal. In this case, initial vesicle formation might occur by a mechanism similar or identical to that responsible for the endocytosis and recycling of mature synaptic vesicles in the nerve terminal following regulated exocytosis. The site of vesicle formation has been addressed by both morphological and biochemical characterization of axonally transported organelles, as well as in neurons and neuroendocrine cells grown in culture.

Ultrastructural analysis of the membranes accumulating on the proximal side of a low-temperature block of membrane transport in the mouse saphenous nerve has revealed a predominance of tubulovesicular structures and axoplasmic reticulum, but very few synaptic vesicle profiles (29), suggesting that synaptic vesicles do not form directly form the TGN. Immunocytochemical localization of the synaptic vesicle protein SV2 in the electromotor axons of the electric fish *Torpedo marmorata* identified a variety of immunopositive membranes, including tubulovesicular and multivesicular bodies, electron-dense granules, and synaptic vesicle–like profiles (12). These results suggest that some synaptic vesicles may be generated directly from the TGN, although the anterograde axonal transport of the other SV2-positive organelles is also possible. Unfortunately, it is not possible by morphological analysis alone to determine if the synaptic vesicle–like profiles have the biochemical and physical properties characteristic of synaptic vesicles. A vesicle population (VP_0) containing biosynthetically labeled proteoglycans characteristic of synaptic vesicles has been isolated from electromotor axons and partially characterized (30). The VP_0 vesicles displayed physical and biochemical properties similar to those of mature synaptic vesicles, and the radiolabeled proteoglycan was sequentially chased into populations of reserve (VP_1) and actively recycling vesicles (VP_2). Although these results suggest a TGN origin for synaptic vesicles, it was not demonstrated that the VP_0 vesicles excluded nerve-terminal plasma membrane components, and the possibility that the VP_0 vesicles were generated during homogenization from the axonal tubulovesicular structures observed by electron microscopy was not addressed. In the rabbit optic nerve, pulse-chase experiments have been used to address whether the synaptic vesicle protein synaptophysin and the plasma membrane glucose transporter GLUT1 are localized to the same vesicles (21). While some overlap of the two markers was found on sedimentation gradients, immunoprecipitation of synaptophysin-positive membrane suggested that synaptophysin and GLUT1 were in different vesicles. Part of the difficulty in addressing the biosynthesis of synaptic vesicles in the nervous system is the inaccessibility of most preparations to experimental manipulation. This has prompted the use of neuroendocrine cell lines, in particular PC12 cells, and isolated neurons in culture to address the biogenesis of synaptic vesicles.

Several types of endocrine cells and endocrine-derived cell lines generate a population of small synaptic-like vesicles, sharing a number of molecular and physical properties with synaptic vesicles (11, 31, 32). These synaptic-like vesicles, which may serve a paracrine function (32), have provided an experimental model for analysis of the synaptic vesicle lifecycle (for review see 33, 34). The biogenesis of synaptic-like vesicles has been extensively studied in PC12 cells, a pheochromocytoma-derived cell line that displays a number of neural and endocrine properties, utilizing the synaptic vesicle protein synaptophysin as a marker. Newly synthesized synaptophysin is delivered to the cell surface in constitutive secretory vesicles, independent of the pathway of dense-core secretory granule biogenesis (35, 36). Pulse-labeled synaptophysin undergoes multiple rounds of endocytic and exocytic cycling, apparently passing through an early endosome compartment, before its eventual appearance, with 3 h of chase, in the synaptic-like vesicle fraction (36). The majority of the synaptophysin (90%) that accumulated in the synaptic-like vesicles had been present on the surface 15 min after the pulse labeling (as marked by cell-surface biotinylation), consistent with previous demonstration of an endocytic origin for these organelles (37, 38). At least a portion of the synaptic-like vesicles can form directly from an endosomal compartment (39). These results suggest that an early endosome compartment, rather than the TGN, plays an important role in the initial formation of synaptic-like vesicles, at least in PC12 cells. This raises the possibility that synaptic vesicle formation is similar to the formation of other early endosome–derived vesicles in non-neuronal cells, including transcytotic (40) and water channel–containing vesicles (41) in epithelial cells, and vesicles containing the insulin-regulated glucose transporter GLUT4 in adipocytes and muscle cells (42).

Several lines of evidence suggest that synaptic vesicle biogenesis in neurons may follow a pathway similar to that of synaptic-like vesicles in PC12 cells. First, as previously discussed, it has been difficult to demonstrate conclusively that synaptic vesicles bud directly from the TGN. Second, endocytic compartments are localized to nerve terminals (43–46), where synaptic vesicles are capable of recycling by endocytosis following stimulated exocytosis (discussed in a later section). Finally, studies on developing hippocampal neurons grown in culture have demonstrated that synaptic vesicle proteins are efficiently targeted to the axonal processes (even prior to development of synapses) where the vesicles undergo multiple rounds of exo/endocytic cycling (47). Although this exo/endocytic cycling is similar to that of newly synthesized synaptophysin in PC12 cells, the involvement of an intermediate early endosomal compartment has not been established. Interestingly, the exo/endocytic cycling in hippocampal cultures occurs in the absence of extracellular calcium (47), as does the apparently constitutive secretion of neurotransmitter from growing neuronal processes in culture (48, 49). The continuous exo/endocytic recycling of synaptic vesicles (and accompanying neurotransmitter release if they are filled) observed in immature neurons could be functionally important for several developmental processes, including axonal growth, target recognition, and synapse formation. The switch from constitutive release in developing axons to the regulated release that is characteristic of mature synapses might simply involve a modification of the calcium regulation of the exocytotic machinery induced by contact with the target cell (50, 51).

In addition to the axonal targeting of synaptic vesicle proteins in cultured hippocampal neurons, a portion of several of these proteins is also detected in somatodendritic regions with a distribution similar to that of the transferrin receptor, a marker for early endosomes (46, 52). The steady-state distribution of synaptophysin in PC12 cells also suggests significant overlap with early-endosomal markers (52–54). Treatment of hippocampal neuron cultures with the fungal metabolite brefeldin A (BFA) resulted in a change in the transferrin receptor distribution from a punctate to a tubular pattern (46), consistent with the disruption of early endosomes caused by BFA in non-neuronal cells (55). The somatodendritic synaptophysin precisely colocalized with transferrin receptor-positive tubules following BFA treatment, while the axonal staining pattern was unaffected. These results demonstrate that somatodendritic early endosomes are different from axonal early endosomes, perhaps reflecting the specialization of the latter for highly efficient recycling of synaptic vesicles. In contrast to the behavior of synaptophysin, the somatodendritic distribution of five other synaptic vesicle markers was unaffected by BFA treatment (46). This result suggests that synaptophysin may have a function in early endosomes as well as in synaptic vesicles. Alternatively, the synaptophysin that remains in a BFA-sensitive endosome may simply reflect inefficient targeting to the membranes containing the other synaptic vesicle markers. The slow time course of synaptophysin appearance in the synaptic-like vesicles of PC12 cells (36) is consistent with either of these possibilities. These observations raise concerns about the use of synaptophysin as a marker for synaptic vesicle biogenesis.

Synaptic Vesicle Protein Sorting

While the results to date favor the endocytic pathway for vesicle biogenesis, conclusive evidence in neurons is lacking. It is clear that synaptic vesicle membrane proteins must be sorted from other membrane proteins at one or more stages of the synaptic vesicle lifecycle. If vesicle biogenesis and recycling involve an endosomal intermediate, then two levels of sorting are required. In this case, initial sorting from resident plasma-membrane proteins during endocytosis would be followed by sorting from other proteins in the endocytic pathway during synaptic vesicle formation. How is this protein sorting accomplished? One possibility is that each synaptic vesicle protein has independent targeting information that directs it to the appropriate location. Analysis of multiple synaptic vesicle proteins has failed to reveal any primary sequence motif that could be considered a targeting sequence. Although it is possible that a common sorting signal is generated by secondary or tertiary folding (56, 57), it is also possible that synaptic vesicle protein sorting involves the formation of specific membrane-protein complexes mediated by protein-protein interactions in which only one component contains sorting information. Synaptic vesicle proteins can be recovered in multimeric complexes following detergent solubilization (58), consistent with the possibility that protein-protein interactions play a role in sorting.

One method that has been exploited to address the presence of targeting signals on synaptic vesicle proteins is expression in non-neuronal cells. When synaptophysin is expressed in non-neuronal cells, it is localized to an early endosome compartment (37, 38, 52, 53, 59, 60), consistent with the targeting of synaptophysin to early endosomes in PC12 cells and hippocampal neurons discussed previously. The endosomal targeting of synaptophysin requires its carboxyl-terminal cytoplasmic domain (61), which contains multiple tyrosine residues that could serve as signals for endocytosis (57). In contrast to synaptophysin, the localization of synaptotagmin to the plasma membrane and SV2 to an unidentified intracellular compartment following expression in CHO cells (60, 62) demonstrates that not all synaptic vesicle proteins are targeted to early endosomes. Although the steady-state levels of synaptotagmin in the plasma membrane of neurons are low (47), it has been detected by subcellular fractionation of adrenal chromaffin cells in a plasma membrane–enriched fraction (63). The fact that synaptotagmin expressed in fibroblasts is localized to the plasma membrane suggests either that synaptotagmin lacks a signal for endocytosis, that a neural-specific modification of synaptotagmin is required for its endocytosis, or that a neural-specific endocytic pathway, perhaps involving nervous system–specific clathrin light-chain or assembly proteins (64–66), is required for synaptotagmin endocytosis. If synaptotagmin truly lacks an endocytosis signal, then its exo/endocytic cycling behavior in neurons may be mediated by interactions with other (vesicle) proteins. Interestingly, coexpression of synaptotagmin, synaptophysin, and SV2 in CHO cells did not alter their individual distributions (60). This suggests that interactions between these proteins either do not occur under the conditions of coexpression or are not involved in protein targeting.

The mechanism that is responsible for synaptic vesicle protein sorting may be coupled to the mechanism that maintains the uniform size of synaptic vesicles. Since vesicle diameter is one factor that determines the total transmitter content, the regulation of vesicle size is likely to be tightly regulated. So far it has not been possible to generate synaptic vesicle–sized organelles by expressing synaptic vesicle proteins individually (37, 38, 53, 60) or in combinations (60) in several non-neuronal cells. This suggests that the generation of a synaptic vesicle–sized organelle requires the expression of all (or most) synaptic vesicle membrane proteins, and/or other nervous system–specific factors involved in vesicle budding (64–66). It is possible that synaptic vesicle formation involves strict control of membrane protein stoichiometry, perhaps through the assembly of a multimeric complex of vesicle proteins (58), and that this stoichiometry influences vesicle size. If this is the case, then one explanation for the apparent inefficient sorting of synaptophysin from an early endosomal compartment in PC12 cells and hippocampal neurons might be its presence in excess of that required for proper vesicle stoichiometry.

SYNAPTIC VESICLE MATURATION

Once the synaptic vesicle is generated, either biosynthetically or during endocytic recycling, it is subject to two processes prior to neurotransmitter release that we have grouped together as maturation. These two processes, neurotransmitter uptake and cytoskeletal anchoring, are not well defined in the context of the synaptic vesicle lifecycle, but are nonetheless important for an understanding of synaptic vesicle function.

Neurotransmitter Uptake

The mechanism used by synaptic vesicles to take up and store small neurotransmitter molecules has been studied pharamacologically and biochemically, both in isolated vesicles and by reconstitution in proteoliposomes (67). At least four distinct vesicular transport systems have been identified with these techniques: one each for biogenic amines, acetylcholine, and glutamate, and a fourth for both gamma-amino butyric acid (GABA) and glycine. Each of these transport systems is dependent on the electrochemical gradient generated by the vacuolar proton pump for its activity. The vacuolar proton pump is a multimeric protein complex containing at least nine subunits that is abundant in synaptic vesicles, clathrin-coated vesicles, chromaffin granules, and a variety of other intracellular organelles (for reviews see 68, 69). The proton pump consists of both peripheral cytoplasmically oriented subunits required for ATP binding and hydrolysis and integral membrane subunits responsible for proton translocation. The translocation of a proton into the lumen of the synaptic vesicle generates an electrochemical gradient that is used to drive the uptake of neurotransmitter. The two components of the electrochemical gradient, pH and potential, are differentially utilized by vesicular transport systems. While the glutamate transporter is preferentially driven by the potential gradient, the amine transporter is driven largely by the pH gradient, and the GABA/glycine transporter utilizes both (70). The stage in the synaptic vesicle lifecycle at which neurotransmitter uptake occurs has not been determined. It is likely that the coexistence of an active transporter and proton pump in the same membrane is sufficient for transmitter uptake. Indeed, transmitter uptake into an early endosomal fraction has been demonstrated (39).

The proteins responsible for the vesicular transport of neurotransmitters have begun to be identified through molecular biological techniques. A novel expression-cloning strategy, selection for resistance to the drug N-methyl-4-phenylpyridinium (MPP^+), resulted in the isolation of a cDNA clone for the chromaffin granule amine transporter from chromaffin cells (71, 72). A second, highly homologous amine transporter (62% amino acid identity) was subsequently isolated from brain by low-stringency screening with the chromaffin cell cDNA (71) and by a different expression-cloning strategy (73). When expressed in CHO or CV-1 cells, the two classes of amine transporter display the transport and pharmacological properties expected of the amine transporters from chromaffin granules or synaptic vesicles. Both classes of cDNA encode proteins with 12 predicted membrane-spanning segments, a common structural feature of transporters, but with no homology to plasma membrane neurotransmitter transporters. Limited sequence homology, particularly over the first six membrane-spanning regions, with bacterial drug-resistance transporters was observed. This is of interest since the bacterial transporters utilize the same energy source (proton gradient) and transport substances with the same topology (out of the cytoplasm) as the vesicular amine transporters. The existence of two vesicular amine transporters was not expected based on previous transport studies, but may reflect the different membrane-trafficking pathways followed by chromaffin granules and synaptic vesicles. Interestingly, in PC12 cells, both an amine transporter (presumably derived from the chromaffin granule) and an acetylcholine transporter are active in an early endosomal compartment (39). However, only the acetylcholine transporter is packaged into synaptic-like vesicles (39, 74), suggesting that differential sorting of the two transporters occurs in the endosomal compartment.

In contrast to the expression-cloning strategy used for the isolation of the amine transporter cDNA clones, a molecular candidate for the vesicular acetylcholine transporter has been identified through the characterization of the *C. elegans unc-17* mutant (75). The protein encoded by the *unc-17* gene is predicted to span the membrane 12 times, and displays 37–39% amino acid identity with the two vesicular amine transporters. Evidence that the product is the vesicular acetylcholine transporter includes its restricted expression in cholinergic neurons, colocalization with the synaptic vesicle protein synaptotagmin, and accumulation in cell soma in axonal transport–deficient *unc-104* mutants. However, it has not yet been directly demonstrated that the *unc-17* gene product is required for acetylcholine transport. The identification of related vesicular transporters for biogenic amines and acetylcholine makes it likely that the amino acid transporters will be part of the same family. The isolation of cDNA clones encoding two forms of the synaptic vesicle protein SV2 has revealed that it too is predicted to span the membrane 12 times (76–78), suggestive of a transporter function. Indeed, the first six transmembrane domains of SV2 display homology with bacterial nutrient and drug transporters, as well as mammalian plasma-membrane glucose transporters. However, no significant homology with other vesicular transporters characterized at a molecular level was identified. This observation, in combination with the widespread pattern of SV2 expression (77, 79), makes it unlikely that SV2 is a specific neurotransmitter transporter. It remains to be determined whether SV2 is involved in the transport of non-neurotransmitter substances found in vesicles such as ATP or chloride, or participates in some other aspect of synaptic vesicle function, such as membrane fusion.

A biochemical and pharmacological approach has been utilized to characterize the active transport of acetylcholine into cholinergic synaptic vesicles from the electric fish *Torpedo californica* (80). Through the use of the pharmacological agent vesamicol, a noncompetitive inhibitor of acetylcholine transport, a partial purification of the acetylcholine transporter/vesamicol receptor has been described (81). The vesamicol receptor behaves as a very heterogeneous complex that is associated with a synaptic vesicle proteoglycan and the epitopes recognized by monoclonal antibodies SV1 and SV2. It has recently been shown that SV2 is a keratan sulfate proteoglycan that in certain cell types, such as cholinergic neurons, also shares the SV1 epitope (82). The enrichment of SV2 during the purification of the vesamicol receptor raises the possibility that SV2 may be a subunit or regulator of vesicular transporters, independent of neurotransmitter type. The relationship between the vesamicol receptor and the *unc-17* gene product remains to be determined.

Cytoskeletal Anchoring

Within the presynaptic nerve terminal, synaptic vesicles are clustered near the active zone, a region of the presynaptic plasma membrane located immediately opposite the postsynaptic cell, where vesicle fusion and neurotransmitter release occurs (83). While some of the synaptic vesicles are found in close apposition to the presynaptic plasma membrane, the majority are localized a short distance away in the nerve-terminal cytoplasm. This observation, in combination with the extremely rapid kinetics of neurotransmitter release, has led to the proposal that at least two distinct pools of vesicles exist: a readily releasable pool adjacent to the plasma membrane and capable of rapid exocytotic fusion, and a reserve pool that can be recruited into the releasable pool in response to nerve stimulation. Recent studies of synaptic vesicle recycling have demonstrated that newly recycled vesicles are functionally equivalent to the pre-existing pool in the nerve terminal and that the entire pool of vesicles can be rapidly turned over with moderate stimulation (84, 85). These results suggest that exchange between the reserve and releasable pools is a very dynamic process. Quick-freeze deep-etch analysis of the presynaptic terminal has revealed that synaptic vesicles are connected with each other, with the actin cytoskeleton, and with the plasma membrane by a variety of filamentous elements (86–88), suggesting that the intermixing of vesicle pools requires restructuring of this network.

A likely candidate for mediating the cross-linking of synaptic vesicles with the actin cytoskeleton is the synapsin family of peripheral synaptic vesicle–associated proteins (89, 90). The best-characterized member of the synapsin family, both structurally and functionally, is synapsin I, a major physiological substrate for calmodulin-dependent protein kinase II (CaM kinase II) in the nerve terminal. Phosphorylation by CaM kinase II has been found to negatively regulate several of the activities of synapsin I, including the ability to bind to synaptic vesicles and to initiate and bundle actin filaments (91, 92–95). Although the relative physiological importance of each of these effects has not been established, the phosphorylation state of synapsin I has been found to influence neurotransmitter release, perhaps by regulating intraterminal vesicle dynamics. In several preparations (96–98), dephosphorylated synapsin I inhibits neurotransmitter release, phospho-synapsin I is without effect, and CaM kinase II potentiates release. Based on these observations, a model has been proposed in which synapsin I regulates the availability of synaptic vesicles for exocytosis (99). Prior to stimulation, synapsin I cross-links the reserve pool of synaptic vesicles both to each other and to the actin cytoskeleton, a proposal that is supported by morphological studies (87, 88). Upon nerve stimulation, an influx of calcium results in CaM kinase II–mediated phosphorylation of synapsin I, causing its dissociation from synaptic vesicles and actin filaments, thereby allowing the vesicle to join the readily releasable pool. Interestingly, the binding site for synapsin I on the synaptic vesicle is the α-subunit of CaM kinase II (100), perhaps allowing rapid and efficient regulation of synapsin I phosphorylation. In addition to a role in controlling the availability of synaptic vesicles for release, membrane-cytoskeletal interactions could contribute to the maintenance of synaptic vesicle localization within the nerve terminal. A role in the regulation of neurotransmitter release for synapsin II, which shares many structural features with synapsin I but is not phosphorylated by CaM kinase II, has not been demonstrated.

In adrenal chromaffin cells, another system that has been widely utilized to study regulated secretion, chromaffin granules have also been found to exist in readily releasable and reserve pools (101). Although much slower than the fusion of synaptic vesicles, chromaffin granule fusion is of sufficient speed to suggest that some granules must be in close apposition with the plasma membrane (102). As with synaptic vesicles, the availability of fusion-competent vesicles appears to be regulated by the actin cytoskeleton in a calcium-dependent manner (for review see 103), but by a synapsin I–independent mechanism.

BIOPHYSICAL CHARACTERIZATION OF SYNAPTIC VESICLE DOCKING AND FUSION

Biophysical analysis of regulated secretion (as well as other membrane fusion events) has provided important insight into the possible underlying molecular mechanisms. In this section we first discuss the kinetics and calcium sensitivity of secretion, which suggests that neurotransmitter release is likely to occur in close proximity to voltage-gated calcium channels in the presynaptic plasma membrane. We then focus on the biophysical characterization of an intermediate in the process of membrane fusion, the fusion pore, which may be relevant for the mechanism of neurotransmitter release.

Kinetics and Calcium Sensitivity of Release

The squid giant synapse has provided a good experimental model for the study of presynaptic release mechanisms, in part because of its accessibility to experimental manipulation during simultaneous pre- and postsynaptic physiological measurements. This system has provided important clues as to the location and sensitivity of the calcium sensor that triggers fusion. Analysis of the time course of release has demonstrated that the delay between the influx of calcium into the nerve terminal and the initiation of neurotransmitter release is ~200 μs (104). This speed suggests that much of the machinery required for vesicle fusion is preassembled or primed at release sites. Perturbation of the process of neurotransmitter release by presynaptic injection of different calcium chelators has demonstrated that the calcium-binding step occurs very rapidly, on the order of tens of microseconds, and that the amplitude of the calcium signal that triggers fusion is 100 μM or greater (105). These measurements suggest that the calcium sensors involved in regulating vesicle fusion will bind calcium rapidly and with low affinity, thereby producing a close temporal correlation between calcium channel activity and the regulation of the fusion machinery. Furthermore, the nonlinear dependence of secretion on calcium influx indicates that the binding of multiple calcium ions to one or more calcium receptors may be required to trigger release (106).

Modeling of the calcium dynamics in the vicinity of calcium channels suggests that the high concentrations required to trigger fusion could only occur in close proximity to a single calcium channel or clusters of channels. Experimental support for the existence of such microdomains of high calcium concentration has been described (107). The observation that increasing the number of activated calcium channels (by broadening the presynaptic action potential with a potassium channel blocker) causes a linear increase in the amount of transmitter released indicates that very little overlap occurs between the microdomains generated by separate calcium channels (108). If this is the case, then the calcium sensor for exocytosis is likely to be localized to within a few tens of nanometers of the presynaptic calcium channel, at least in squid. The suggestion that vesicle fusion occurs in close proximity to calcium channels is supported by freeze-fracture analysis of the presynaptic plasma membrane showing that vesicle fusion occurs near well-ordered arrays of intramembranous particles in the presynaptic plasma membrane (109). That these intramembranous particles may represent calcium channels (110) is supported by their apparent disruption by auto-antibodies against calcium channels in patients with Lambert-Eaton myasthenic syndrome (111). At least one class of presynaptic calcium channel that is involved in the regulation of neurotransmitter release, the ω-conotoxin-sensitive N-type calcium channel (112), has been localized to the active zone at the frog neuromuscular junction (113, 114), although not at an ultrastructural level. An active zone localization of other types of calcium channels involved in regulation of neurotransmitter secretion, such as the ω-agatoxin-sensitive P-type channel (115, 116), has yet to be demonstrated. It is possible that the mechanism that maintains the active zone localization and close proximity of synaptic vesicles and calcium channels involves a physical association between the synaptic vesicle docking or fusion machinery and the presynaptic calcium channels. Experimental support for this possibility is presented in the next section. The involvement of intracellular cytoskeletal elements and/or synaptic basal lamina components in this colocalization is also possible.

Although secretion from neuroendocrine and endocrine cells is at least tenfold slower than is synaptic vesicle fusion (102, 117), similar local high concentrations of calcium have been shown to be important (101, 118–121). This suggests that the two pathways may utilize similar docking and fusion mechanisms.

Fusion Pore

Capacitance measurements can be utilized to monitor the process of membrane fusion with very high temporal resolution (122). Studies of the fusion of large granules with the plasma membrane in mast cells (123) and intercellular fusion mediated by influenza hemagglutinin (124, 125) have demonstrated that the earliest detectable event in membrane fusion is the formation of an aqueous pore spanning the fusing bilayers. The pore is approximately the size of a large ion channel, and can, on occasion, be observed to repeatedly open and close, suggesting that pore opening is reversible. Some of the mast-cell granule contents can be secreted during the transient fusion events (126). In addition, net lipid flux through the fusion pore into the granule readily occurs during reversible opening, suggesting that at least a portion of the pore is lined with lipid (127). However, lipid flux during viral hemagglutinin-mediated fusion occurs only slowly, and the early stages of this pore have therefore been proposed to be entirely protein lined (128). Whether this represents a fundamental difference in the mechanism of fusion in the two systems remains to be determined. The molecular correlate of the fusion pore has yet to be identified, although several models have been proposed. One model is that a gap junction–like structure forms between the two membranes, which opens (and reversibly closes) and eventually dissociates during pore dilation and full bilayer fusion (129). It has been suggested that the kinetics of synaptic vesicle fusion may require such a preassembled pore complex. Alternatively, the pore may be wholly lipidic (130). In this case the role of proteins in the fusion process would be to bring the appropriate membrane into close apposition, overcoming hydration forces that are inhibitory to membrane fusion. A third possibility is that the fusion pore is lined with both protein and lipid, with proteins forming a hydrophobic bridge to allow lipid flow between the two bilayers (131). Models for the mechanism of fusion generated by the influenza hemagglutinin, which requires a pH-induced conformational change that exposes an amphipathic fusion peptide, are most like the third possibility (132). It is useful to keep these models in mind when considering the synaptic proteins whose functions may include fusion.

BIOCHEMICAL CHARACTERIZATION OF SYNAPTIC VESICLE DOCKING AND FUSION

Genetic and biochemical analyses of the constitutive secretory pathways in the yeast *Saccharomyces cerevisiae* and in animal cells have revealed a number of similarities with synaptic vesicle membrane trafficking (133, 134). This suggests that common mechanisms may underlie most membrane-trafficking events. The components of the yeast secretory pathway that have counterparts in the nerve terminal are illustrated schematically in Figure 2 (see also Table 1). Some of these proteins, based on studies in both yeast and animal cells, appear to be involved at multiple vesicular transport steps, suggesting that they are general components of the transport vesicle docking and fusion machinery. Other proteins are members of families with a distinct isoform involved at each stage of the secretory pathway. Such proteins may ensure that a particular transport vesicle recognizes and fuses with only the appropriate target membrane, a regulatory event that is common to all membrane-trafficking steps. An additional level of regulation involving calcium is required in the case of synaptic vesicle fusion, probably utilizing proteins that are not shared with constitutive trafficking steps. Because of the high level of conservation among different membrane-trafficking pathways, the regulatory role of calcium in synaptic vesicle fusion may be to relieve an inhibition or block of the constitutive fusion machinery (133, 135).

In this section we discuss the proteins that are likely to be involved in synaptic vesicle docking and fusion, with frequent reference to similarities

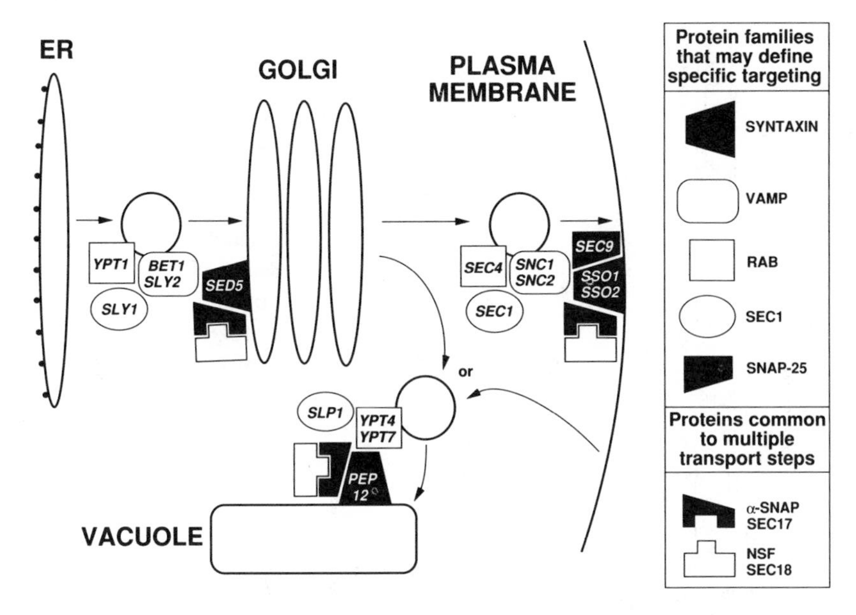

Figure 2 Similarities between yeast and nerve-terminal membrane-trafficking pathways. Schematic diagram illustrating protein families that may function in defining the specificity of vesicular trafficking events both in the nerve terminal and at various stages of the yeast secretory pathway. See text for details.

with the constitutive secretory pathway in yeast and animal cells (Figure 2 and Table 1). Since the molecular machinery involved in synaptic vesicle docking and fusion is likely to include protein components localized to both the vesicle and plasma membrane, as well as soluble factors, we organize our discussion around these three classes of proteins. In no case has it been possible to assign a function exclusively in docking, or exclusively in fusion to a particular protein. It may be that the two events are mechanistically coupled in a cascade of specific reactions, beginning with docking and culminating in membrane fusion, involving, at least in part, the same set of proteins throughout. Recent experiments that support this possibility, and a current working model, are discussed at the end of this section.

Synaptic Vesicle Proteins

VAMP VAMP (136, 137) or synaptobrevin (9, 138), is an 18-kDa protein anchored to the cytoplasmic surface of the synaptic vesicle by a carboxyl-terminal transmembrane domain (Figure 1*a*). Evidence that VAMP may play a role in synaptic vesicle docking or fusion is provided by the

Table 1 Similarity between yeast and nerve-terminal membrane-trafficking machinery

		Yeast membrane trafficking		Higher eukaryote membrane trafficking	
Family	Predicted localization	Gene	Transport step	Nerve terminal	Other homologs
Protein families that may contribute to docking and fusion specificity					
rab	transport vesicle	*SEC4*	Golgi to plasma membrane	rab3A	yes
		YPT1	ER to Golgi		
		YPT7	Endosome to vacuole		
VAMP	transport vesicle	*SNC1* & *SNC2*	Golgi to plasma membrane	VAMP	yes
		BET1/SLY12	ER to Golgi		
		SEC22/SLY2	ER to Golgi		
syntaxin	target membrane	*SSO1* & *SSO2*	Golgi to plasma membrane	syntaxin 1A and 1B	yes
		SED5	ER to Golgi		
		PEP12	Golgi to vacuole		
SEC1	soluble?	*SEC1*	Golgi to plasma membrane	*unc-18* & *rop1*	no
		SLY1	ER to Golgi		
		SLP1	Golgi to vacuole		
SNAP-25	target membrane	*SEC9*	Golgi to plasma membrane	SNAP-25	no
Proteins required at multiple transport steps					
—	soluble	*SEC18*	several	NSF	—
—	soluble	*SEC17*	several	α-SNAP	—

observation that VAMP is a substrate for the zinc-endoproteases associated with tetanus toxin and botulinum toxins type B and F, each of which is able to block neurotransmitter release (139–141). In addition, VAMP is recovered in a 20S particle along with syntaxin, SNAP-25, NSF, and α/γ-SNAP (134). Each of these proteins, as is discussed below, has been implicated in the regulation of membrane trafficking. Further support of a role for VAMP in membrane trafficking is provided by studies of the yeast secretory pathway. Several proteins have been identified in yeast that have domain structures similar to, and sequence homology with, VAMP (Figure 2 and Table 1). The yeast proteins most like VAMP are the products of the *SNC1* and *SNC2* genes (142, 143). The *SNC* gene products, which share 79% amino acid sequence identity with each other and 30–40% identity with mammalian VAMP proteins, are localized to a population of post-Golgi vesicles that accumulate in *sec6* mutants. Disruption of both *SNC* genes results in a block of invertase secretion and the accumulation of post-Golgi transport vesicles. These results indicate that Snc1p and Snc2p are required for the proper targeting or fusion of Golgi-derived transport vesicles with the plasma membrane. Similar conclusions have been drawn from studies of several other yeast proteins that share structural features and topology with VAMP, but limited sequence homology (Figure 2). Included in this group are the products of the genetically interacting *SEC22/SLY2, BET1/-SLY12,* and *BOS1* genes (144–146), each of which displays defects in ER-to-Golgi membrane trafficking. These mutants accumulate vesicles that may be involved in ER-to-Golgi transport (147, 148), and at least two of the products, Sec22p and Bos1p, are colocalized with ER-derived transport vesicles generated in an in vitro assay (145). Interestingly, each of these genes has been found to interact with the *YPT1* gene, the product of which is a low-molecular-weight GTP-binding protein of the rab family (see below), suggesting a possible interaction between the rab and VAMP families of proteins. The identification of multiple VAMP-like proteins in yeast suggests that these proteins may be involved in defining the specificity of vesicular transport. In support of this possibility is the identification of several VAMP-related proteins in higher eukaryotes that may be involved in vesicular trafficking (149–151). Interestingly, one of the most highly conserved regions of VAMP is predicted to form an amphipathic α-helix similar to that formed by the viral fusion peptide (J White, unpublished observation). The functional significance of this observation remains to be determined.

RAB3A The rab family of low-molecular-weight GTP-binding proteins is involved in the regulation of a wide range of membrane-trafficking pathways in yeast and higher eukaryotes (for review see 152; WE Balch, this volume). This growing protein family is now known to contain more than 20 members, many of which are localized to specific intracellular membrane compartments (153). One member of this family is rab3A (Figure 1*a*), which is attached to the synaptic vesicle membrane by a geranylgeranylated cysteine residue near its C terminus (154, 155). Biochemical and genetic studies suggest that the rab family may mediate vectorial membrane trafficking by a process that involves the cyclical binding and hydrolysis of GTP. For example, rab5, which is localized to the early endosome, regulates early endosome fusion in vitro (156) and in vivo (157). In yeast, three members of the rab family, Ypt1p (158), Sec4p (159), and Ypt7p (160), have been identified (Figure 2 and Table 1). The *YPT1* and *SEC4* gene products are required for the targeting of ER-derived transport vesicles to the Golgi, and Golgi-derived transport vesicles to the plasma membrane, respectively (145, 159, 161). As with the VAMP family, it is likely that the rab proteins are involved in regulating the specificity of vesicular targeting. However, the yeast proteins are not solely responsible for specificity, since chimeric Sec4p/Ypt1p molecules can be produced that function in both ER-to-Golgi and Golgi-to-plasma membrane transport (162, 163). The downstream effectors of the rab proteins have not been identified, although some candidates have been characterized. For example, *SEC4* interacts with two genes, *SEC8* and *SEC15*, the products of which form a 19.5S particle partially associated with the plasma membrane (164). To date, no similar complex has been identified in higher eukaryotes. Several proteins have been identified that regulate the binding and hydrolysis of GTP by rab3A, including a GTPase-activating protein (165), a guanine nucleotide-releasing factor (166), and a guanine nucleotide dissociation inhibitor (167). Each of these proteins could potentially influence the function of rab3A in the nerve terminal. Interestingly, rabphillin (168), a protein of unknown function that interacts with only the GTP-bound form of rab3A, displays homology to the synaptic vesicle protein synaptotagmin.

Several groups have addressed the function of rab proteins by examining the effects of a soluble peptide that corresponds to the effector domain of rab3A (169). The effects on membrane targeting and fusion have been either stimulatory (170–173) or inhibitory (169, 174), depending on the preparation or assay utilized. Further work will be needed to determine if these varied effects are actually mediated by the rab effector. Recent experiments suggest that the binding and hydrolysis of GTP by one form of rab3 is required for the proper targeting of secretory vesicle to, or selective retention at, release sites (175). Furthermore, the nonhydrolyzable GTP analog GTPγS has a wide variety of effects on intracellular membrane trafficking, including an inhibition of neurotransmitter release from the presynaptic nerve terminal at a stage after synaptic vesicle docking with the plasma membrane (176).

It is not currently known whether this effect is mediated by rab3A, or other GTP-binding proteins.

SYNAPTOTAGMIN Synaptotagmin (also known as p65; Figure 1*a*) is a synaptic vesicle protein that consists of an amino-terminal intravesicular domain, a single transmembrane domain, and a cytoplasmic domain composed of two repeats homologous to the C2 regulatory domain of protein kinase C (177; 178). Synaptotagmin forms a homo-oligomer that binds calcium in a complex with negatively charged phospholipids (179). This observation, along with the fact that no synaptotagmin homolog has yet been identified in the constitutive secretory pathway, suggests that synaptotagmin may function as a calcium sensor that triggers exocytosis. The apparent low affinity for calcium and the multiple binding sites for calcium displayed by synaptotagmin are consistent with the properties expected of the calcium trigger for vesicle fusion. Support for a role of synaptotagmin in synaptic vesicle docking or fusion includes direct interactions with syntaxin (58) and neurexin (180), and either direct or indirect interactions with N-type calcium channels (181, 182). The interactions with these three presynaptic plasma membrane proteins could potentially position synaptotagmin, and the vesicle, in an optimal position to respond to calcium. Functional data supportive of a role for synaptotagmin in regulated secretion include the inhibition of secretion generated by microinjection of either synaptotagmin peptides into the squid giant synapse (183), or soluble synaptotagmin fragments and anti-synaptotagmin antibodies into PC12 cells (184). However, genetic studies on PC12 cells, as well as in *Drosophila* and *C. elegans,* have demonstrated that synaptotagmin is not absolutely required for regulated neurotransmitter release (185–187). This may reflect the ability of another protein, such as rabphillin (168), to compensate partially for the loss of synaptotagmin. Alternatively, synaptotagmin may not be an essential component of the machinery required for vesicle docking and fusion, but rather may function as a calcium-sensitive negative regulator of the constitutive fusion machinery (133, 135). This possibility is supported by recent electrophysiological recordings from synaptotagmin-deficient *Drosophila,* which revealed an increase in spontaneous neurotransmitter release and a decrease in evoked release (A DiAntonio and T Schwarz, unpublished observations).

SYNAPTOPHYSIN Synaptophysin (188–190), and the related synaptoporin (191), are synaptic vesicle proteins that span the membrane four times (Figure 1*a*). Evidence that synaptophysin may function in the process of neurotransmitter release has been provided by experiments using *Xenopus* embryos and oocytes. Microinjection of an anti-synaptophysin antibody into a blastomere of an early embryo resulted in a reduced frequency of spontaneous transmitter release and a reduced amplitude of evoked transmitter release, as assayed in cultured neuromuscular junctions two days following injection (192). Synaptophysin is also required for the regulated secretion of glutamate from *Xenopus* oocytes injected with total cerebellar RNA (193). Purified synaptophysin behaves as a homo-oligomeric complex (194, 195), which upon reconstitution into artificial bilayers can form a transmembrane ion channel (196). Ion channels have been electrophysiologically identified in a number of synaptic vesicle and secretory granule preparations (197–199). One possible function for such channels is the formation of a fusion pore during exocytosis. Preliminary support for this possibility is provided by the observation that an antibody directed against synaptophysin is able to block both a calcium-dependent ion channel present in neurosecretory granules and secretion from a permeabilized nerve-terminal preparation (200). If synaptophysin functions as part of a gap junction–like fusion pore, then it is likely to do so in conjunction with a partner in the plasma membrane with similar properties. A plasma membrane protein that interacts with synaptophysin has been identified, but not molecularly characterized (201). Other roles for the putative synaptophysin channel are also possible. Although no synaptophysin-related protein has been identified in the yeast secretory pathway, a non-neuronal form of synaptophysin has been identified (202), raising the possibility that the function of synaptophysin in neurotransmitter secretion may be shared in other membrane-trafficking pathways. Of the synaptic vesicle proteins characterized to date, synaptophysin is the least conserved phylogenetically. In fact no synaptophysin homolog has yet been identified in *Drosophila* or *C. elegans,* suggesting that the function of synaptophysin is not under as much evolutionary pressure as are those of other synaptic vesicle proteins.

Plasma Membrane Proteins

SYNTAXINS Several components of the presynaptic plasma membrane that may constitute part of the synaptic vesicle docking and fusion machinery have been identified. Two such proteins are syntaxin 1A and 1B, a pair of homologous, cytoplasmically oriented 35-kDa proteins with carboxyl-terminal membrane anchors (182, 203). The syntaxins were initially identified by their ability to interact with the synaptic vesicle protein synaptotagmin. An interaction between syntaxins and ω-conotoxin-sensitive N-type calcium channels has also been detected (182, 204, 205). These observations suggest that syntaxins might be involved in the docking of synaptic vesicles near voltage-gated calcium channels in the presynaptic plasma membrane.

Whether syntaxin is associated with other types of calcium channels implicated in the regulation of neurotransmitter release, such as P-type channels, has not been determined. Microinjection experiments demonstrated that an antibody against syntaxin as well as soluble syntaxin fragments are able to block calcium-regulated secretion from PC12 cells (206), suggesting that syntaxin has an important role in vesicle docking or fusion. Furthermore, syntaxin has been identified as a substrate for the zinc-endoprotease associated with botulinum neurotoxin type C (C Montecucco, unpublished observation; R Jahn, unpublished observation), raising the possibility that the inhibition of neurotransmitter release caused by this toxin is related to syntaxin cleavage. Recent studies of the yeast secretory pathway have demonstrated that syntaxin is a member of a family of proteins that may be involved in defining the specificity of vesicular trafficking (Figure 2 and Table 1). Three genes have been identified in yeast that encode proteins with a carboxyl-terminal membrane anchor and display significant homology to syntaxin. The yeast proteins with the highest level of homology to syntaxin are the products of the *SSO1* and *SSO2* genes (S Keränen, personal communication), two genes isolated as multicopy suppressors of the late-acting secretory mutant *sec1*. Another syntaxin-related yeast protein, Pep12p, is required for proper targeting of hydrolases to the vacuole (K Becherer and E Jones, personal communication), and a third, Sed5p, is required at an early stage of the secretory pathway, between the ER and Golgi (207). The identification of several syntaxin-related proteins in yeast has now been extended with the identification of a syntaxin family in rat (206), raising the possibility that syntaxins function as target membrane-specific receptors for the docking and fusion of transport vesicles. This possibility is supported by the observation that syntaxin is recovered in a 20S particle with other proteins implicated in vesicular trafficking, including VAMP and the general factors NSF and α/γ-SNAP (134).

SNAP-25 SNAP-25 was initially identified as a brain-specific cDNA clone that encodes a 25-kDa protein localized to the plasma membrane in the nerve terminal (208). It has subsequently been demonstrated that SNAP-25 is an abundant axonally transported, [^{3}H]palmitate-labeled protein (209). Since SNAP-25 lacks hydrophobic sequences sufficient to span the membrane, palmitylation is likely to be responsible for its membrane association. SNAP-25 is the third membrane component, along with VAMP and syntaxin, that participates in the formation of a 20S particle with the soluble membrane-trafficking factors NSF and α/γ-SNAP (134). Further evidence for a role of SNAP-25 in membrane trafficking includes the cleavage of SNAP-25 by botulinum neurotoxin type A (C Montecucco, unpublished observation; R Jahn, unpublished observation) and the structural homology between

SNAP-25 and the late-acting yeast secretory mutant *sec9* (210). A role for SNAP-25 in constitutive membrane delivery in neurons is suggested by the inhibition of axonal growth by SNAP-25 antisense oligonucleotides (211). It remains to be determined whether other forms of SNAP-25 are expressed in yeast or higher eukaryotes, and to what extent SNAP-25 might be involved in defining the specificity of membrane trafficking.

NEUREXINS Another plasma-membrane protein that interacts with synaptotagmin is the α-latrotoxin receptor (212). α-latrotoxin, a component of black widow spider venom, binds to its plasma membrane–localized receptor and induces the efficient and calcium-independent fusion of synaptic vesicles with the presynaptic plasma membrane (213). The α-latrotoxin receptor is a member of a family of proteins known as the neurexins (214), each of which is composed of a large extracellular domain, a single transmembrane domain, and a short intracellular region that is responsible for the interaction with synaptotagmin (180). At least three neurexin genes undergo extensive alternative RNA splicing, resulting in a large number of isoforms. Based on this diversity, as well as homology of the neurexin extracellular domain to laminin A and agrin, it has been proposed that these molecules may serve as cell recognition signals during synaptogenesis. Although the functional significance of the interaction between neurexin and synaptotagmin is not clear, one possibility is that an α-latrotoxin-induced change in neurexin conformation is transduced to the synaptic vesicle fusion machinery through synaptotagmin.

GAP-43 GAP-43 (also known as B50, neuromodulin, and F1) is an abundant protein associated with neuronal growth cones for which a wide variety of functions have been proposed (215). Like SNAP-25, GAP-43 is attached to the plasma membrane by fatty acylation (216). The possibility that GAP-43 is involved in neurotransmitter release is suggested by the inhibition of secretion from permeabilized nerve terminals caused by antibodies directed against GAP-43 (217). In addition, the elimination of GAP-43 by expression of antisense constructs in PC12 cells causes an increase in basal secretion and decrease in evoked secretion (218). These results suggest that GAP-43, like synaptotagmin, may act as a negative regulator of secretion at resting calcium concentrations.

Soluble Proteins

SEC1 The product of the *SEC1* gene in yeast is one of several homologous proteins implicated in the regulation of different stages of the secretory pathway (Figure 2 and Table 1). Sec1p is a hydrophilic protein required for proper transport from the Golgi to the plasma membrane (219). The products of two other genes, *SLY1* (144) and *SLP1* (220), each related to Sec1p, are involved in transport from the ER to the Golgi and from the Golgi to the vacuole, respectively. It is of interest that two of these yeast genes, *SEC1* and *SLY1*, interact genetically with *SSO1* and *YPT1*, respectively, the products of which are proposed to play a role in defining transport specificity. Although no mammalian homologs in the *SEC1* family have been described, a *SEC1*-like gene, *rop1*, has been identified in *Drosophila* and is predominantly expressed in the nervous system (221). In addition, the *C. elegans unc-18* gene encodes a SEC1-like protein that is involved in cholinergic neurotransmission (222). Further characterization of the SEC1 family is required to determine if they are soluble or membrane-associated proteins, and what role they might play in the transport vesicle docking and fusion machinery.

NSF AND SNAPS NSF (NEM-sensitive factor) was initially identified and purified as a soluble factor required for reconstitution of intra-Golgi membrane transport in vitro (223). Biochemical and molecular characterization of NSF has demonstrated that it is a homo-tetrameric ATP-binding protein with a low level of ATPase activity (224, 225). NSF and its yeast homolog Sec18p are required for vesicular transport at many, but not all, stages of the secretory and endocytic pathways (224, 226–228). The accumulation of vesicles in an in vitro Golgi transport assay following NSF depletion (223) and in *sec18* mutant yeast (148, 161) suggests that NSF may be a general component of the transport-vesicle docking and fusion machinery. The attachment of NSF to membranes, a requirement for its activity, is dependent on other soluble factors known as SNAPs (soluble NSF attachment proteins), of which three classes (α, β, and γ) have been identified (229). α-SNAP is functionally and structurally related to the yeast *SEC17* gene product, while β-SNAP represents a brain-specific isoform of α-SNAP (230, 231). Addition of detergent-solubilized brain membranes to NSF and α/γ-SNAP results in their assembly into a particle that sediments at 20S, the stability of which is controlled by ATP hydrolysis (232). The components of the 20S particle that were derived from the brain membranes include the synaptic vesicle protein VAMP, and the synaptic plasma membrane proteins syntaxin and SNAP-25 (134). These proteins have been termed SNAREs, for SNAP receptors, although it remains to be determined which of the components interacts with α/γ-SNAP. These results indicate that a common machinery may underlie both constitutive and regulated vesicle docking and fusion reactions. As discussed below, further dissection of the 20S particle has provided important insight into the molecular events that precede membrane fusion and neurotransmitter release (233).

OTHER SOLUBLE PROTEINS A number of soluble proteins have been identified that may be involved in the regulated secretory pathway of PC12 cells and bovine chromaffin cells. For example, reconstitution of secretion from permeabilized PC12 cells has been utilized to purify two factors that act sequentially to promote calcium-dependent secretion (234). One of these proteins, involved in an ATP-dependent and calcium-independent "priming reaction," has been identified as a phosphatidylinositol transfer protein (235). A phospholipid transfer protein has also been found to be important for the export of proteins from the Golgi complex in yeast (236). These results emphasize the importance of lipids as potential regulators of membrane-trafficking pathways. A second factor, p145, is required for the reconstitution of a calcium-dependent step in secretion from permeabilized PC12 cells (237). Another set of soluble factors that may participate in regulated secretion, as well as constitutive membrane-trafficking events, are the annexins (for reviews, see 238, 239). Annexin II, as well as two other soluble factors (Exo1 and Exo2), has been found to stimulate calcium-dependent secretion from permeabilized adrenal chromaffin cells (240, 241). A synthetic peptide corresponding to a domain conserved between annexin II and Exo1 has been found to block the stimulation of exocytosis produced by these proteins (242). Interestingly, this same peptide has been found to inhibit the binding of both protein kinase C and synaptotagmin to a set of proteins that may act as receptors for the C2 domains shared by these proteins (243). The identification of multiple proteins involved in the regulation of secretion from chromaffin cells may reflect several calcium-dependent steps that precede exocytosis, as has been identified electrophysiologically (101). It remains to be determined whether these soluble proteins play a role in neurotransmitter release from the nerve terminal.

Model of Synaptic Vesicle Docking and Fusion

From the above discussion, it is clear that a large number of proteins are likely to be involved in regulating the process of synaptic vesicle docking and fusion. It is possible that the 20S particle, containing NSF, α/γ-SNAP, VAMP, syntaxin, and SNAP-25, represents an intermediate in the docking and fusion machinery that is common to both regulated and constitutive membrane-trafficking events. This possibility is supported by the fact that each of the constituents has a counterpart in the yeast secretory pathway (Figure 2 and Table 1). Recent dissection of the assembly and disassembly of the 20S particle has provided important insight into its role in vesicle docking and fusion (233). These experiments demonstrate that a stoichiometric complex of VAMP, syntaxin, and SNAP-25, as well as substoichiometric amounts of synaptotagmin, can be detected in the absence of NSF and α/γ-SNAP. Upon its addition, α-SNAP is able to displace syn-

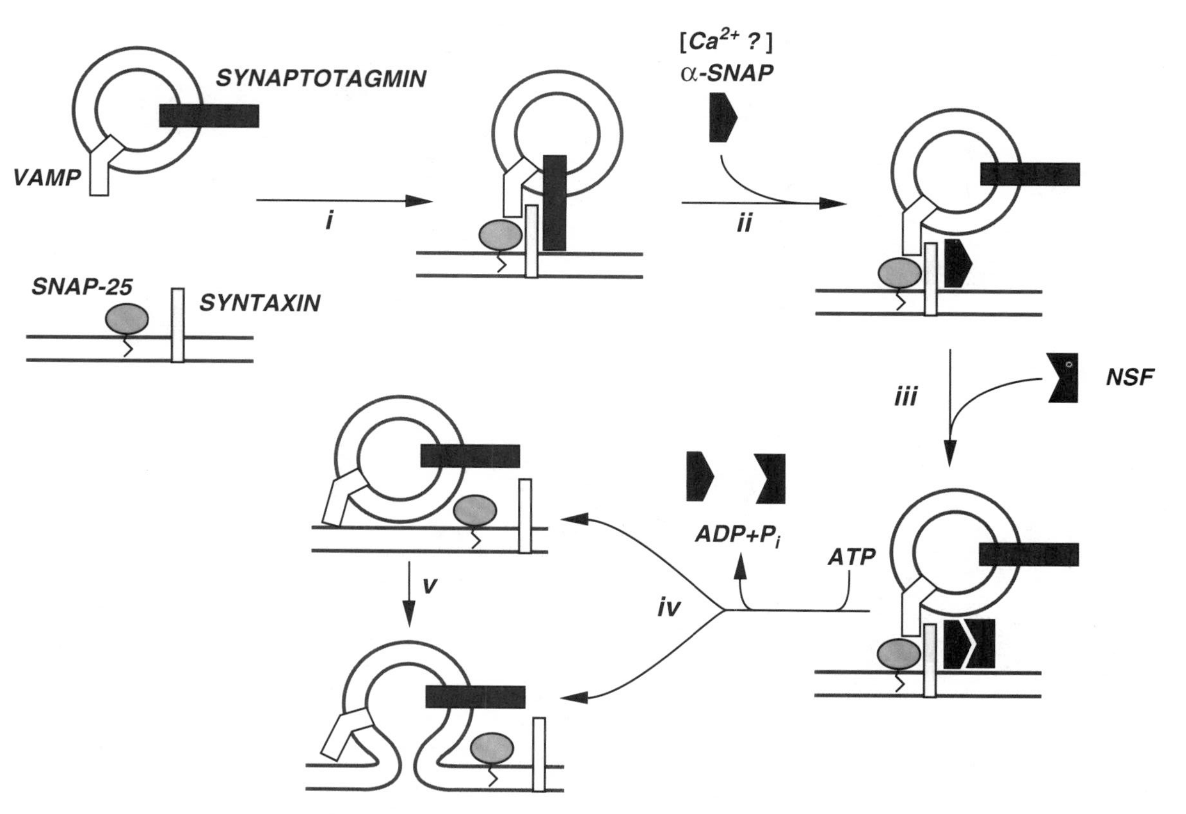

Figure 3 Model of synaptic vesicle docking and fusion. See text for discussion.

aptotagmin from the complex, thereby allowing the recruitment of NSF and the formation of the full 20S particle. The addition of ATP and its subsequent hydrolysis results in the near complete disassembly of the 20S particle. A model for synaptic vesicle docking and fusion that incorporates these findings, referred to as the SNARE hypothesis, is presented in Figure 3. The first stage is independent of NSF and α/γ-SNAP and involves the formation of a complex between VAMP and synaptotagmin from the vesicle membrane and syntaxin and SNAP-25 from the plasma membrane (step *i*). An attractive prediction of the SNARE hypothesis is that an interaction between VAMP and syntaxin isoforms provides specificity to the docking and fusion reactions at each stage of the secretory pathway. The role of synaptotagmin in the vesicle-docking complex may be to act as a negative regulator of subsequent steps that would, in its absence, proceed rapidly toward fusion (i.e. in a constitutive transport step). The potential competition between α-SNAP and synaptotagmin for a common binding site, perhaps on syntaxin (182, 244), is a candidate mechanism for this negative regulation. Once the synaptotagmin has been displaced by α-SNAP (step *ii*), NSF can bind (step *iii*) and hydrolyze ATP (step *iv*), resulting in the disruption of the 20S particle. The potential conformational changes generated by ATP hydrolysis may either lead directly (step *iv*) or indirectly (step *v*) to membrane fusion. Although many mechanistic questions remain unanswered, the SNARE hypothesis represents a useful starting point for further experimentation. It will be particularly important to better define the events upstream (step *i*) and downstream (steps *iv* and *v*) of the NSF/SNAP cycle, including the potential role of rab proteins in vesicle docking and fusion. Furthermore, the role of calcium in triggering vesicle fusion is not adequately accounted for by the available data or the current model. Synaptotagmin has been proposed to act as a calcium sensor, yet the association of synaptotagmin with the vesicle-docking complex is not affected by calcium (233). It is possible that calcium could have some effect on the complex under physiological conditions, since synaptotagmin binds calcium only in the presence of acidic phospholipids (179). If calcium does promote, through synaptotagmin, the assembly of the NSF/SNAP complex, it remains to be determined whether the subsequent ATP hydrolysis and complex disassembly could occur rapidly enough for a direct role in synaptic vesicle fusion.

SYNAPTIC VESICLE RECYCLING

Endocytosis

After exocytosis, the components of the synaptic vesicle membrane are efficiently recycled. It has been suggested that this recycling occurs by the reversal of the exocytic process (245, 246), perhaps by closure of the fusion pore before vesicle membrane is incorporated into the plasma membrane. However, several lines of evidence suggest that the recycling of synaptic vesicles, at least partially, proceeds by an endocytic mechanism similar to that previously described in the discussion of vesicle biogenesis. First, most morphological evidence is consistent with a mechanism of full fusion followed by selective recovery via clathrin-coated vesicles and an endosomal compartment (43, 247), at least following intense stimulation. The absence of morphologically distinct recycling structures during mild or moderate stimulation conditions may reflect the high level of efficiency of an endocytic pathway that is fully dedicated to synaptic vesicle recycling. Strong evidence in favor of an endocytic recycling pathway is provided by the *Drosophila* temperature-sensitive paralytic mutant *shibire* (248, 249). The *shibire* mutants display a temperature-sensitive defect in endocytosis that is evident in a variety of cell types. The morphological result of this defect, and the basis of the behavioral effect, is the complete depletion of synaptic vesicles from the nerve terminal. The *shibire* gene product is the *Drosophila* homolog of the mammalian protein dynamin (250, 251). It has recently been demonstrated that dynamin is involved in the formation of clathrin-coated vesicles in mammalian cells, both in vitro and in vivo (252, 253). Additional evidence in favor of the endocytic pathway for synaptic vesicle recycling is provided by the observation that 95% of nerve terminal–derived coated vesicles, and 80% of whole-brain coated vesicles, contain synaptic vesicle proteins (254). It remains to be determined whether brain-specific isoforms of clathrin light chain (64), β-adaptin (66), or the accessory protein AP3 (65) are preferentially involved in nerve-terminal membrane trafficking. While integral components of the vesicle membrane follow the endocytic pathway, rab3A is not found in nerve-terminal coated vesicles (254), and may recycle via a soluble intermediate (255). Experiments utilizing endocytic dye tracers have addressed the time course of synaptic vesicle recycling. It has been estimated that the endocytic stage of recycling occurs with a $t_{1/2}$ of 60 s, while an additional 30 s are required for the newly endocytosed material to reach a releasable vesicle pool (256, 257).

Protein Turnover

After an indeterminate number of rounds of exo/endocytic cycling, the components of the synaptic vesicle must be turned over. Very little is known about the half-life of most synaptic vesicle proteins. In hippocampal neuronal cultures, an antibody against the lumenal domain of synaptotagmin briefly added to the cells to monitor exo/endocytic cycling remains concentrated in terminals even after six days (47), suggesting that synaptotagmin is turned over slowly. Similarly, in PC12 cells, the half-life of synaptophysin has been reported to be greater than two days (37). When protein turnover does occur, two mechanisms are possible. Several of the integral components of the synaptic vesicle membrane are returned to the cell body by retrograde axonal transport prior to degradation, as indicated by their accumulation at the distal side of a block of axoplasmic transport (17). Other components of the synaptic vesicle, such as synapsin I, appear not be retrogradely transported, and are likely turned over in the nerve terminal.

CONCLUDING REMARKS

Activity-dependent changes in synaptic efficacy are believed to underlie the processes of learning and memory (258). The synaptic vesicle lifecycle provides many presynaptic targets for regulatory mechanisms that might result in changes in synaptic efficacy. Short-term regulation of the amount of transmitter stored in the vesicle, the availability of vesicles to release sites, or the calcium sensitivity of the fusion machinery could each dramatically influence the level of neurotransmitter release. Long-term or stable modulation of synaptic properties might involve the expression of different combinations of members of the small gene families to which each synaptic vesicle protein belongs (6, 10). It is possible that the different family members will have distinct functional properties. If this were the case, through a combinatorial mechanism, secretory machinery with quite different properties could be generated using a relatively small number of building blocks. In addition, most of the components of the synaptic vesicle membrane have been found to be phosphorylated, at least in vitro (90). As we learn more about the functions of synaptic proteins, an understanding of how those functions are regulated will follow. The regulation of the cytoskeletal anchoring of synaptic vesicles by synapsin I is a prime example.

The study of synaptic vesicle membrane trafficking has progressed rapidly in the past two years, in part because of its convergence with studies of the constitutive secretory pathways of non-neuronal cells. The continued combination of these approaches promises further advances of our understanding of the mechanism of neurotransmitter release in the near future.

ACKNOWLEDGMENT

We would like to thank the members of our lab for numerous discussions, and in particular Beverly Wendland, Nicole Calakos, and Jonathan Pevsner for helpful comments on the manuscript.

Literature Cited

1. Fatt P, Katz B. 1952. *J. Physiol.* 117:109–28
2. Del Castillo J, Katz B. 1954. *J. Physiol.* 124:553–59
3. Katz B, Miledi R. 1967. *J. Physiol.* 189:533–44
4. DeRobertis EDP, Bennett HS. 1955. *J. Biophys. Biochem. Cytol.* 1:47–58
5. Tauc L. 1982. *Physiol. Rev.* 62:857–93
6. Kelly RB. 1993. *Cell* 72:43–53
7. Rash JE, Walrond JP, Morita M. 1988. *J. Electron Microsc. Tech.* 10:153–85
8. Trimble WS, Linial M, Scheller RH. 1991. *Annu. Rev. Neurosci.* 14:93–122
9. Südhof TC, Baumert M, Perin MS, Jahn R. 1989. *Neuron* 2:1475–81
10. Südhof TC, Jahn R. 1991. *Neuron* 6:665–77
11. Navone F, Jahn R, Gioia GD, Stukenbrok H, Greengard P, De Camilli P. 1986. *J. Cell Biol.* 103:2511–27
12. Janetzko A, Zimmermann H, Volknandt W. 1989. *Neuroscience* 32:65–77
13. Baumert M, Takei K, Hartinger J, Burger PM, Fischer von Mollard G, et al. 1990. *J. Cell Biol.* 110:1285–94
14. Tixier-Vidal A, Faivre-Bauman A, Picart R, Wiedenmann B. 1988. *Neuroscience* 3:847–61
15. Fletcher TL, Cameron P, De Camilli P, Banker G. 1991. *J. Neurosci.* 11: 1617–26
16. Matteoli M, Takei K, Cameron R, Hurlbut P, Johnston PA, et al. 1991. *J. Cell Biol.* 115:625–33
17. Dahlström AB, Czernik AJ, Li J-Y. 1992. *Mol. Neurobiol.* 6:157–77
18. Hirokawa N, Sato-Yoshitake R, Kobayashi N, Pfister KK, Bloom GS, Brady ST. 1991. *J. Cell Biol.* 114:295–302
19. Morin PJ, Johnson RJ, Fine RE. 1993. *Biochim. Biophys. Acta* 1146:275–81
20. Dahlström AB, Pfister KK, Brady ST. 1991. *Acta Physiol. Scand.* 141:469–76
21. Morin PJ, Liu N, Johnson RJ, Leeman SE, Fine RE. 1991. *J. Neurochem.* 56:415–27
22. Leopold PL, McDowall AW, Pfister KK, Bloom GS, Brady ST. 1992. *Cell Motil. Cytoskel.* 23:19–33
23. Schroer TA, Brady ST, Kelly RB. 1985. *J. Cell Biol.* 101:568–72
24. Llinas R, Sugimori M, Lin J-W, Leopold PL, Brady ST. 1989. *Proc. Natl. Acad. Sci. USA* 86:5656–60
25. Hall DH, Hedgecock EM. 1991. *Cell* 65:837–47
26. Gho M, McDonald K, Ganetzky B, Saxton WM. 1992. *Science* 258:313–16
27. Hall DH, Plenefisch J, Hedgecock EM. 1991. *J. Cell Biol.* 115:389a (Abstr.)
28. Brown SS. 1993. *Curr. Opin. Cell Biol.* 5:129–34
29. Tsukita S, Ishikawa H. 1980. *J. Cell Biol.* 84:513–30
30. Kiene M-L, Stadler H. 1987. *EMBO J.* 6:2209–15
31. Navone F, Gioia GD, Jahn R, Browning M, Greengard P, De Camilli P. 1989. *J. Cell Biol.* 109:3425–33
32. Reetz A, Solimena M, Matteoli M, Folli F, Takei K, De Camilli P. 1991. *EMBO J.* 10:1275–84
33. De Camilli P, Jahn R. 1990. *Annu. Rev. Physiol.* 52:625–45
34. Régnier-Vigouroux A, Huttner WB. 1993. *Neurochem. Res.* 18:59–64
35. Cutler DF, Cramer LP. 1990. *J. Cell Biol.* 110:721–30
36. Régnier-Vigouroux A, Tooze SA, Huttner WB. 1991. *EMBO J.* 10:3589–601
37. Johnston PA, Cameron PL, Stukenbrok H, Jahn R, De Camilli P, Südhof TC. 1989. *EMBO J.* 8:2863–72
38. Clift-O'Grady L, Linstedt AD, Lowe AW, Grote E, Kelly RB. 1990. *J. Cell Biol.* 110:1693–703
39. Bauerfeind R, Régnier-Vigouroux A, Flatmark T, Huttner WB. 1993. *Neuron* 11:105–21
40. Bomsel M, Mostov K. 1991. *Curr. Opin. Cell Biol.* 3:647–53
41. van der Goot FG, Seigneur A, Guay-Woodford L, Zeidel ML. 1992. *J. Membr. Biol.* 128:133–39
42. Slot JW, Geuze HJ, Gigengack S, James DE, Lienhard GE. 1991. *Proc, Natl. Acad. Sci. USA* 88:7815–19
43. Heuser JE, Reese TS. 1973. *J. Cell Biol.* 57:315–44
44. Sulzer D, Holtzman E. 1989. *J. Neurocytol.* 18:529–40
45. Parton RG, Simons K, Dotti CG. 1992. *J. Cell Biol.* 119:123–37
46. Mundigl O, Matteoli M, Daniels L, Thomas-Reetz A, Metcalf A, et al. 1993. *J. Cell Biol.* 122:1207–21
47. Matteoli M, Takei K, Perin MS, Südhof TC, De Camilli P. 1992. *J. Cell Biol.* 117:849–61
48. Young SH, Poo M-m. 1983. *Nature* 305:634–37
49. Hume RI, Role LW, Fischbach GD. 1983. *Nature* 305:632–34
50. Xie Z-P, Poo M-m. 1986. *Proc. Natl. Acad. Sci. USA* 83:7069–73
51. Zoran MJ, Doyle RT, Haydon PG. 1991. *Neuron* 6:145–51
52. Cameron PL, Südhof TC, Jahn R, De

Camilli P. 1991. *J. Cell Biol.* 115:151–64
53. Linstedt AD, Kelly RB. 1991. *Neuron* 7:309–17
54. Lah JJ, Burry RW. 1993. *J. Neurocytol.* 22:92–101
55. Wood SA, Brown WJ. 1992. *J. Cell Biol.* 119:273–85
56. Baranski TJ, Faust PL, Kornfeld S. 1990. *Cell* 63:281–91
57. Trowbridge IS. 1991. *Curr. Opin. Cell Biol.* 3:634–41
58. Bennett MK, Calakos N, Kreiner T, Scheller RH. 1992. *J. Cell Biol.* 116:761–75
59. Leube RE, Wiedenmann B, Franke WW. 1989. *Cell* 59:433–46
60. Feany MB, Yee A, Delvy ML, Buckley KM. 1993. *J. Cell Biol.* 123:575–84
61. Linstedt AD, Kelly RB. 1991. *J. Physiol., Paris* 85:90–96
62. Feany MB, Buckley KM. 1993. *Nature* 364:537–40
63. Fournier S, Trifaró J-M. 1988. *J. Neurochem.* 51:1599–609
64. Brodsky FM, Hill BL, Acton SL, Nathke I, Wong DH, et al. 1991. *Trends Biochem. Sci.* 16:208–13
65. Zhou S, Tannery NH, Yang J, Puszkin S, Lafer EM. 1993. *J. Biol. Chem.* 268:12655–62
66. Ponnambalam S, Robinson MS, Jackson AP, Peiperl L, Parham P. 1990. *J. Biol. Chem.* 265:4814–20
67. Edwards RH. 1992. *Curr. Opin. Neurobiol.* 2:586–94
68. Nelson N. 1992. *J. Exp. Biol.* 172:19–27
69. Forgac M. 1992. *J. Exp. Biol.* 172:155–69
70. McMahon HT, Nicholls DG. 1991. *Biochim. Biophys. Acta* 1059:243–64
71. Liu Y, Peter D, Roghani A, Schuldiner S, Prive GG, et al. 1992. *Cell* 70:539–51
72. Liu Y, Roghani A, Edwards RH. 1992. *Proc. Natl. Acad. Sci. USA* 89:9074–78
73. Erickson JD, Eiden LE, Hoffman BJ. 1992. *Proc. Natl. Acad. Sci. USA* 89:10993–97
74. Blumberg D, Schweitzer ES. 1992. *J. Neurochem.* 58:801–10
75. Alfonso A, Grundahl K, Duerr JS, Han H-P, Rand JB. 1993. *Science* 261:617–19
76. Bajjalieh SM, Peterson K, Shinghal R, Scheller RH. 1992. *Science* 257:1271–73
77. Bajjalieh S, Peterson K, Linial M, Scheller RH. 1993. *Proc. Natl. Acad. Sci. USA* 90:2150–54
78. Feany MB, Lee S, Edwards RH, Buckley KM. 1992. *Cell* 70:861–67
79. Buckley K, Kelly RB. 1985. *J. Cell Biol.* 100:1284–94
80. Parsons SM, Prior C, Marshall IG. 1993. *Int. Rev. Neurobiol.* 35:279–89
81. Bahr BA, Parsons SM. 1992. *Biochemistry* 31:5763–69
82. Scranton TW, Iwata M, Carlson SS. 1993. *J. Neurochem.* 61:29–44
83. Heuser JE, Reese TS. 1977. In *Handbook of Physiology; The Nervous System,* ed. ER Kandel, 1(Part I):99–136. Bethesda, Md: Am. Physiol. Soc.
84. Searl T, Prior C, Marshall IG. 1991. *J. Physiol.* 444:99–116
85. Betz WJ, Bewick GS. 1991. *Science* 255:200–3
86. Gotow T, Miyaguchi K, Hashimoto PH. 1991. *Neuroscience* 40:587–98
87. Landis DMD, Hall AK, Weinstein LA, Reese TS. 1988. *Neuron* 1:201–9
88. Hirokawa N, Sobue K, Kanda K, Harada A, Yorifuji H. 1989. *J. Cell Biol.* 108:111–26
89. Valtorta F, Benfenati F, Greengard P. 1992. *J. Biol. Chem.* 267:7195–98
90. Greengard P, Valtorta F, Czernik AJ, Benfenati F. 1993. *Science* 259:780–85
91. Huttner WB, Schiebler W, Greengard P, De Camilli P. 1983. *J. Cell Biol.* 96:1374–88
92. Schiebler W, Jahn R, Doucet J-P, Rothlein J, Greengard P. 1986. *J. Biol. Chem.* 261:8383–90
93. Bähler M, Greengard P. 1987. *Nature* 326:704–7
94. Valtorta F, Greengard P, Fesce R, Chieregatti E, Benfenati F. 1992. *J. Biol. Chem.* 267:11281–88
95. Benfenati F, Valtorta F, Chieregatti E, Greengard P. 1992. *Neuron* 8:377–86
96. Llinás R, Gruner JA, Sugimori M, McGuinness TL, Greengard P. 1991. *J. Physiol.* 436:257–82
97. Hackett JT, Cochran SL, Greenfield LJ, Brosius DC, Ueda T. 1990. *J. Neurophysiol.* 63:701–6
98. Nichols RA, Chilcote TJ, Czernik AJ, Greengard P. 1992. *J. Neurochem.* 58:783–85
99. Benfenati F, Valtorta F, Greengard P. 1991. *Proc. Natl. Acad. Sci. USA* 88:575–79
100. Benfenati F, Valtorta F, Rubenstein JL, Gorelick FS, Greengard P, Czernik AJ. 1992. *Nature* 359:417–20
101. Neher E, Zucker RS. 1993. *Neuron* 10:21–30
102. Chow RH, von Rüden L, Neher E. 1992. *Nature* 356:60–63
103. Trifaro JM, Vitale ML, Rodriguez Del

Castillo A. 1992. *Eur. J. Pharmacol.* 225:83–104
104. Llinas R, Steinberg IZ, Walton K. 1981. *Biophys. J.* 33:323–51
105. Adler EM, Augustine GJ, Duffy SN, Charlton MP. 1991. *J. Neurosci.* 11: 1496–507
106. Augustine GJ, Charlton MP. 1986. *J. Physiol.* 381:619–40
107. Llinás R, Sugimori M, Silver RB. 1992. *Science* 256:677–79
108. Augustine GJ. 1990. *J. Physiol.* 431: 343–64
109. Heuser JE, Reese TS, Landis DMD. 1974. *J. Neurocytol.* 3:109–31
110. Pumplin DW, Reese TS, Llinás R. 1981. *Proc. Natl. Acad. Sci. USA* 78:7210–13
111. Vincent A, Lang B, Newsom-Davis J. 1989. *Trends Neurosci.* 12:496–502
112. Tsien RW, Ellinor PT, Horne WA. 1991. *Trends Pharmacol. Sci.* 12:349–54
113. Cohen MW, Jones OT, Angelides KJ. 1991. *J. Neurosci.* 11:1032–39
114. Torri Tarelli F, Passafaro M, Clementi F, Sher E. 1991. *Brain Res.* 547:331–34
115. Turner TJ, Adams ME, Dunlap K. 1992. *Science* 258:310–13
116. Uchitel OD, Protti DA, Sanchez V, Cherksey BD, Sugimori M, Llinás R. 1992. *Proc. Natl. Acad. Sci. USA* 89:3330–33
117. Thomas P, Wong JG, Almers W. 1993. *EMBO J.* 12:303–6
118. Verhage M, McMahon HT, Ghijsen WE, Boomsma F, Scholten G, et al. 1991. *Neuron* 6:517–24
119. Thomas P, Wong JG, Lee AK, Almers W. 1993. *Neuron* 11:93–104
120. Neher E, Augustine GJ. 1992. *J. Physiol.* 450:273–301
121. Augustine GJ, Neher E. 1992. *J. Physiol.* 450:247–71
122. Fernandez JM, Neher E, Gomperts BD. 1984. *Nature* 312:453–55
123. Breckenridge LJ, Almers W. 1987. *Nature* 328:814–17
124. Spruce AE, Iwata A, White JM, Almers W. 1989. *Nature* 342:555–58
125. Spruce AE, Iwata A, Almers W. 1991. *Proc. Natl. Acad. Sci. USA* 88:3623–27
126. Alvarez de Toledo G, Fernandez-Chacon R, Fernandez JM. 1993. *Nature* 363:554–58
127. Monck JR, Alvarez de Toledo G, Fernandez JM. 1990. *Proc. Natl. Acad. Sci. USA* 87:7804–8
128. Tse FW, Iwata A, Almers W. 1993. *J. Cell Biol.* 121:543–52
129. Almers W, Tse FW. 1990. *Neuron* 4:813–18
130. Monck JR, Fernandez JM. 1992. *J. Cell Biol.* 119:1395–404
131. Zimmerberg J, Curran M, Cohen FS. 1991. *Ann. NY Acad. Sci.* 635:307–17
132. White JM. 1992. *Science* 258:917–23
133. Bennett MK, Scheller RH. 1993. *Proc. Natl. Acad. Sci. USA* 90:2559–63
134. Söllner T, Whiteheart SW, Brunner M, Erdjument-Bromage H, Geromanos S, et al. 1993. *Nature* 362:318–24
135. Popov SV, Poo M-m. 1993. *Cell* 73:1247–49
136. Trimble WS, Cowan DM, Scheller RH. 1988. *Proc. Natl. Acad. Sci. USA* 85:4538–42
137. Elferink LA, Trimble WS, Scheller RH. 1989. *J. Biol. Chem.* 264:11061–64
138. Baumert M, Maycox PR, Navone F, De Camilli P, Jahn R. 1989. *EMBO J.* 8:379–84
139. Schiavo G, Benfenati F, Poulain B, Rossetto O, Polverino de Laureto P, et al. 1992. *Nature* 359:832–35
140. Schiavo G, Shone CC, Rossetto O, Alexander FCG, Montecucco C. 1993. *J. Biol. Chem.* 268:11516–19
141. Link E, Edelmann L, Chou JH, Binz T, Yamasaki S, et al. 1992. *Biochem. Biophys. Res. Commun.* 189:1017–23
142. Gerst JE, Rodgers L, Riggs M, Wigler M. 1992. *Proc. Natl. Acad. Sci. USA* 89:4338–42
143. Protopopov V, Govindan B, Novick P, Gerst JE. 1993. *Cell* 74:855–61
144. Dascher C, Ossig R, Gallwitz D, Schmitt HD. 1991. *Mol. Cell. Biol.* 11:872–85
145. Lian JP, Ferro-Novick S. 1993. *Cell* 73:735–45
146. Newman AP, Groesch ME, Ferro-Novick S. 1992. *EMBO J.* 11:3609–17
147. Shim J, Newman AP, Ferro-Novick S. 1991. *J. Cell Biol.* 113:55–64
148. Kaiser CA, Schekman R. 1990. *Cell* 61:723–33
149. Cain CC, Trimble WS, Lienhard GE. 1992. *J. Biol. Chem.* 267:11681–84
150. Chin A, Burgess RW, Wong BR, Schwarz TL, Scheller RH. 1993. *Gene* 131:175–81
151. McMahon HT, Ushkaryov YA, Edelmann L, Link E, Binz T, et al. 1993. *Nature* 364:346–49
152. Pfeffer SR. 1992. *Trends Cell Biol.* 2:41–46
153. Chavrier P, Parton RG, Hauri HP, Simons K, Zerial M. 1990. *Cell* 62: 317–29
154. Fischer von Mollard G, Mignery GA, Baumert M, Perin MS, Hanson TJ, et al. 1990. *Proc. Natl. Acad. Sci. USA* 87:1988–92

155. Farnsworth CC, Kawata M, Yoshida Y, Takai Y, Gelb MH, Glomset JA. 1991. *Proc. Natl. Acad. Sci. USA* 88:6196–200
156. Gorvel JP, Chavrier P, Zerial M, Gruenberg J. 1991. *Cell* 64:915–25
157. Bucci C, Parton RG, Mather IH, Stunnenberg H, Simons K, et al. 1992. *Cell* 70:715–28
158. Segev N, Mulholland J, Botstein D. 1988. *Cell* 52:915–24
159. Salminen A, Novick PJ. 1987. *Cell* 49:527–38
160. Wichmann H, Hengst L, Gallwitz D. 1992. *Cell* 71:1131–42
161. Rexach MF, Schekman RW. 1991. *J. Cell Biol.* 114:219–29
162. Dunn B, Stearns T, Botstein D. 1993. *Nature* 362:563–65
163. Brennwald P, Novick P. 1993. *Nature* 362:560–63
164. Bowser R, Müller H, Govindan B, Novick P. 1992. *J. Cell Biol.* 118: 1041–56
165. Burstein ES, Linko-Stentz K, Lu Z, Macara IG. 1991. *J. Biol. Chem.* 266:2689–92
166. Burstein ES, Macara IG. 1992. *Proc. Natl. Acad. Sci. USA* 89:1154–58
167. Sasaki T, Kikuchi A, Araki S, Hata Y, Isomura M, et al. 1990. *J. Biol. Chem.* 265:2333–37
168. Shirataki H, Kaibuchi K, Sakoda T, Kishida S, Yamaguchi T, et al. 1993. *Mol. Cell. Biol.* 13:2061–68
169. Plutner H, Schwaninger R, Pind S, Balch WE. 1990. *EMBO J.* 9:2375–83
170. Padfield PJ, Balch WE, Jamieson JD. 1992. *Proc. Natl. Acad. Sci. USA* 89:1656–60
171. Oberhauser AF, Monck JR, Balch WE, Fernandez JM. 1992. *Nature* 360:270–73
172. Edwardson JM, MacLean CM, Law GJ. 1993. *FEBS Lett.* 320:52–56
173. Richmond J, Haydon PG. 1993. *FEBS Lett.* 326:124–30
174. Davidson JS, Eales A, Roeske RW, Millar RP. 1993. *FEBS Lett.* 326:219–21
175. Ngsee JK, Fleming AM, Scheller RH. 1993. *Mol. Biol. Cell* 4:747–56
176. Hess SD, Doroshenko PA, Augustine GJ. 1993. *Science* 259:1169–72
177. Perin MS, Fried VA, Mignery GA, Jahn R, Südhof TC. 1990. *Nature* 345:260–63
178. Wendland B, Miller KG, Schilling J, Scheller RH. 1991. *Neuron* 6:993–1007
179. Brose N, Petrenko AG, Südhof TC, Jahn R. 1992. *Science* 256: 1021–25
180. Hata Y, Davletov B, Petrenko AG, Jahn R, Südhof TC. 1993. *Neuron* 10:307–15
181. Leveque C, Hoshino T, David P, Shoji-Kasai Y, Leys K, et al. 1992. *Proc. Natl. Acad. Sci. USA* 89:3625–29
182. Bennett MK, Calakos N, Scheller RH. 1992. *Science* 257:255–59
183. Bommert K, Charlton MP, DeBello WM, Chin GJ, Betz H, Augustine GJ. 1993. *Nature* 363:163–65
184. Elferink LA, Peterson MR, Scheller RH. 1993. *Cell* 72:153–59
185. Shoji-Kasai Y, Yoshida A, Sato K, Hoshino T, Ogura A, et al. 1992. *Science* 256:1820–23
186. DiAntonio A, Parfitt KD, Schwarz TL. 1993. *Cell* 73:1281–90
187. Nonet ML, Grundahl K, Meyer BJ, Rand JB. 1993. *Cell* 73:1291–305
188. Südhof TC, Lottspeich F, Greengard P, Mehl E, Jahn R. 1987. *Science* 238:1142–44
189. Buckley KM, Floor E, Kelly RB. 1987. *J. Cell Biol.* 105:2447–56
190. Leube RE, Kaiser P, Seiter A, Zimbelmann R, Franke WW, et al. 1987. *EMBO J.* 6:3261–68
191. Knaus P, Marqueze-Pouey B, Scherer H, Betz H. 1990. *Neuron* 5:453–62
192. Alder J, Xie Z, Valtorta F, Greengard P, Poo M-m. 1992. *Neuron* 9:759–68
193. Alder J, Lu B, Valtorta F, Greengard P, Poo M-m. 1992. *Science* 257:657–61
194. Rehm H, Wiedenmann B, Betz H. 1986. *EMBO J.* 5:535–41
195. Johnston PA, Südhof TC. 1990. *J. Biol. Chem.* 265:8869–73
196. Thomas L, Hartung K, Langosch D, Rehm H, Bamberg E, et al. 1988. *Science* 242:1050–53
197. Rahamimoff R, DeRiemer SA, Sakmann B, Stadler H, Yakir N. 1988. *Proc. Natl. Acad. Sci. USA* 85:5310–14
198. Sato M, Inoue K, Kasai M. 1992. *Biophys. J.* 63:1500–5
199. Lee CJ, Dayanithi G, Nordmann JJ, Lemos JR. 1992. *Neuron* 8:335–42
200. Lemos JR, Lee CJ, Dayanithi G, Nordmann JJ. 1992. *Soc. Neurosci. Abstr.* 18:576
201. Thomas L, Betz H. 1990. *J. Cell Biol.* 111:2041–52
202. Zhong C, Hayzer DJ, Runge MS. 1992. *Biochim. Biophys. Acta* 1129: 235–38
203. Inoue A, Obata K, Akagawa K. 1992. *J. Biol. Chem.* 267:10613–19
204. Morita T, Mori H, Sakimura K, Mishina M, Sekine Y, et al. 1992. *Biomed. Res.* 13:357–64

205. Yoshida A, Oho C, Omori A, Kuwahara R, Ito T, Takahashi M. 1992. *J. Biol. Chem.* 267:24925–28
206. Bennett MK, Garcia-Arraras JE, Elferink LA, Peterson K, Fleming AM, et al. 1993. *Cell* 74:863–73
207. Hardwick KG, Pelham HRB. 1992. *J. Cell Biol.* 119:513–21
208. Oyler GA, Higgins GA, Hart RA, Battenberg E, Billingsley M, et al. 1989. *J. Cell Biol.* 109:3039–52
209. Hess DT, Slater TM, Wilson MC, Skene JHP. 1992. *J. Neurosci.* 12:4634–41
210. De Camilli P. 1993. *Nature* 364:387–88
211. Osen-Sand A, Catsicas M, Staple J, Jones KA, Ayala G, et al. 1993. *Nature* 364:445–48
212. Petrenko AG, Perin MS, Davletov BA, Ushkaryov YA, Geppert M, Südhof TC. 1991. *Nature* 353:65–68
213. Petrenko AG. 1993. *FEBS Lett.* 325:81–85
214. Ushkaryov YA, Petrenko AG, Geppert M, Südhof TC. 1992. *Science* 257:50–56
215. Skene JHP. 1989. *Annu. Rev. Neurosci.* 12:127–56
216. Skene JH, Virag I. 1989. *J. Cell Biol.* 108:613–24
217. Dekker LV, DeGraan PNE, Pijnappel P, Oestreicher AB, Gispen WH. 1991. *J. Neurochem.* 56:1146–53
218. Ivins KJ, Neve KA, Feller DJ, Fidel SA, Neve RL. 1993. *J. Neurochem.* 60:626–33
219. Aalto MK, Ruohonen L, Hosono K, Keränen S. 1991. *Yeast* 7:643–50
220. Wada Y, Kitamoto K, Kanbe T, Tanaka K, Anraku Y. 1990. *Mol. Cell. Biol.* 10:2214–23
221. Salzberg A, Cohen N, Halachmi N, Kimchie Z, Lev Z. 1993. *Development* 117:1309–19
222. Hosono R, Hekimi S, Kamiya Y, Sassa T, Murakami S, et al. 1992. *J. Neurochem.* 58:1517–25
223. Malhotra V, Orci L, Glick BS, Block MR, Rothman JE. 1988. *Cell* 54:221–27
224. Rothman JE, Orci L. 1992. *Nature* 355:409–15
225. Tagaya M, Wilson DW, Brunner M, Arango N, Rothman JE. 1993. *J. Biol. Chem.* 268:2662–66
226. Wilson DW, Wilcox CA, Flynn GC, Chen E, Kuang WJ, et al. 1989. *Nature* 339:355–59
227. Graham TR, Emr SD. 1991. *J. Cell Biol.* 114:207–18
228. Riezman H. 1993. *Trends Cell Biol.* 3:273–77
229. Clary DO, Griff IC, Rothman JE. 1990. *Cell* 61:709–21
230. Griff IC, Schekman R, Rothman JE, Kaiser CA. 1992. *J. Biol. Chem.* 267:12106–15
231. Whiteheart SW, Griff IC, Brunner M, Clary DO, Mayer T, et al. 1993. *Nature* 362:353–55
232. Wilson DW, Whiteheart SW, Wiedmann M, Brunner M, Rothman JE. 1992. *J. Cell Biol.* 117:531–38
233. Söllner T, Bennett MK, Whiteheart SW, Scheller RH, Rothman JE. 1993. *Cell* 75:409–18
234. Hay JC, Martin TFJ. 1992. *J. Cell Biol.* 119:139–51
235. Hay JC, Martin TFJ. 1993. *Nature.* Submitted
236. Bankaitis VA, Aitken JR, Cleves AE, Dowhan W. 1990. *Nature* 347:561–62
237. Walent JH, Porter BW, Martin TFJ. 1992. *Cell* 70:765–75
238. Creutz CE. 1992. *Science* 258:924–30
239. Gruenberg J, Emans N. 1993. *Trends Cell Biol.* 3:224–27
240. Ali SM, Geisow MJ, Burgoyne RD. 1989. *Nature* 340:313–15
241. Morgan A, Burgoyne RD. 1992. *Nature* 355:833–36
242. Roth D, Morgan A, Burgoyne RD. 1993. *FEBS Lett.* 320:207–10
243. Mochly-Rosen D, Miller KG, Scheller RH, Khaner H, Lopez J, Smith BL. 1992. *Biochemistry* 31:8120–24
244. Whiteheart SW, Brunner M, Wilson DW, Wiedmann M, Rothman JE. 1992. *J. Biol. Chem.* 267:12239–43
245. Torri-Tarelli F, Haimann C, Ceccarelli B. 1987. *J. Neurocytol.* 16:205–14
246. Ceccarelli B, Hurlbut WP. 1980. *J. Cell Biol.* 87:297–303
247. Miller TM, Heuser JE. 1984. *J. Cell Biol.* 98:685–98
248. Kosaka T, Ikeda K. 1983. *J. Cell Biol.* 97:499–507
249. Koenig JH, Ikeda K. 1989. *J. Neurosci.* 9:3844–60
250. Chen MS, Obar RA, Schroeder CC, Austin TW, Poodry CA, et al. 1991. *Nature* 351:583–86
251. van der Bliek AM, Meyerowitz EM. 1991. *Nature* 351:411–14
252. van der Bliek AM, Redelmeier TE, Damke H, Tisdale EJ, Meyerowitz EM, Schmid SL. 1993. *J. Cell Biol.* 122:553–63
253. Herskovits JS, Burgess CC, Obar RA, Vallee RB. 1993. *J. Cell Biol.* 122:565–78
254. Maycox PR, Link E, Reetz A, Morris SA, Jahn R. 1992. *J. Cell Biol.* 118:1379–88

255. Fischer von Mollard G, Sudhof TC, Jahn R. 1991. *Nature* 349:79–81
256. Ryan TA, Reuter H, Wendland B, Schweitzer FE, Tsien RW, Smith SJ. 1993. *Neuron*. In press
257. Betz WJ, Bewick GS. 1993. *J. Physiol.* 460:287–309
258. Hawkins RD, Kandel ER, Siegelbaum SA. 1993. *Annu. Rev. Neurosci.* 16: 625–65

Annu. Rev. Biochem. 1994. 63:101–32

STRUCTURE AND FUNCTION OF G PROTEIN–COUPLED RECEPTORS

Catherine D. Strader, Tung Ming Fong*, Michael R. Tota*, and Dennis Underwood***

Department of Molecular Pharmacology and Biochemistry*, and Department of Molecular Systems**, Merck Research Laboratories, Rahway, New Jersey 07065

Richard A. F. Dixon

Texas Biotechnology Corporation, 7000 Fannin, Suite 1920, Houston, Texas 77030

KEY WORDS: adrenergic, neurokinin, signal transduction, mutagenesis, modeling

---

CONTENTS

INTRODUCTION

A variety of cell-surface receptors mediate their intracellular actions by a pathway that involves activation of one or more guanine nucleotide–binding regulatory proteins (G proteins). These receptors form a large and functionally diverse superfamily. Receptors that belong to this class respond to a variety of hormone and neurotransmitter agonists, ranging from small

0066-4154/94/0701-0101$05.00

biogenic amines, such as epinephrine and histamine, and retinals—to peptides, such as substance P and bradykinin—to large glycoprotein hormones, such as luteinizing hormone and parathyroid hormone. The signaling pathway common to this family of receptors is initiated by the binding of the agonist to its specific receptor on the surface of the cell. This interaction causes a conformational change in the receptor and allows it to interact with the heterotrimeric G protein in the cell membrane, forming a high-affinity agonist–receptor–G protein complex. The receptor–G protein interaction catalyzes guanine nucleotide exchange on the α subunit of the G protein, leading to the dissociation of the G protein complex into its α and $\beta\gamma$ subunits. The GTP-bound form of the G protein is the activated form, which dissociates from the receptor and activates an effector protein, modulating levels of intracellular second messengers and thereby transducing the signal. The dissociation of the G protein from the receptor also reduces the affinity of the receptor for the agonist. At least 15 different G protein α subunits, 5 β subunits, and 5 γ subunits have been identified to date (see Ref. 1 for review). These G proteins mediate the stimulation or inhibition of a diverse collection of effector enzymes and ion channels, including adenylyl cyclase, guanylyl cyclase, phospholipases C and A_2, and Ca^{2+} and K^+ channels. Thus, stimulation of different cell-surface G protein–coupled receptors with specific hormone agonists results in changes in levels of a wide variety of second-messenger molecules.

The molecular exploration of the superfamily of G protein–coupled receptors has revealed the existence of many previously unsuspected receptor subtypes, which bind related agonists with subtly different pharmacology that had not been distinguished by classical pharmacological approaches. For example, cross-hybridization and cloning based on sequence similarity has revealed the existence of 5 subtypes of the muscarinic receptor (2), 9 adrenergic receptors (3), and more than 14 subtypes of serotonin receptor (4); each of these receptor classes was previously thought to contain only 2–4 members from traditional pharmacological characterization of endogenous receptors in animal tissues. The discovery of this large variety of receptor subtypes, often with different tissue or cellular distributions, has opened up a rich array of potential new pharmacological interventions without the side effects inherent in less specific therapies.

RECEPTOR STRUCTURE

Despite the wide variety of agonists that stimulate the diverse second-messenger pathways activated by the family of G protein–coupled receptors, these receptors share considerable structural homology, reflecting their common mechanism of action. Cloning and sequence determination of more than 100 members of this large family of receptor proteins show that these receptors are characterized by seven hydrophobic stretches of 20–25 amino acids, predicted to form transmembrane α helices, connected by alternating extracellular and intracellular loops [shown schematically for the β-adrenergic receptor (βAR) in Figure 1]. Most of the primary sequence homology

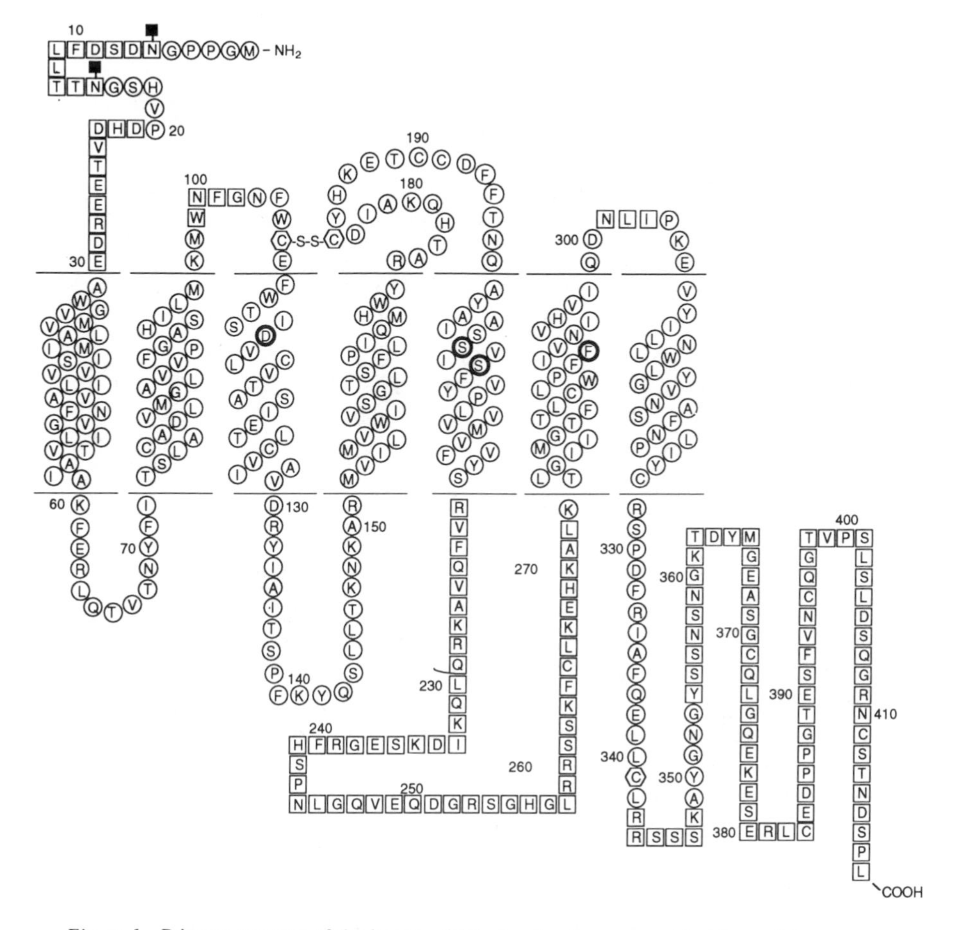

Figure 1 Primary structure of the hamster βAR showing the proposed topology of the seven transmembrane helices. The extracellular domain is at the top of the figure. The locations of Asp113, Ser204, Ser207, and Phe290 are indicated by bold circles. Glycosylation sites are indicated with solid boxes. The boundaries of the cell membrane are represented by dashed lines. Residues shown by deletion mutagenesis not to be required for ligand binding are shown in squares. Conserved Cys residues are shown in hexagons, and all other residues are shown in circles. Reprinted with permission from the *American Journal of Respiratory Cell and Molecular Biology*, 1989, Vol. 1, No. 2, p. 82.

among this family of receptors is contained within the hydrophobic transmembrane domain, with the hydrophilic loop regions being more divergent. In fact, the receptors may be subclassified by the degree of primary sequence homology within the transmembrane domains. The primary sequence identity in the transmembrane domain of these receptors ranges from 85–95% for species homologs of a given receptor, to 60–80% for related subtypes of the same receptor, to 35–45% for other members of the same family, down to 20–25% for unrelated G protein–coupled receptors. This classification approximates the relationships among the various classes of agonists that act at these receptors, suggesting structure-function relationships in this region of the receptor that can be further probed by genetic and pharmacological analysis.

G protein–coupled receptors are glycoproteins, with heterogeneous patterns of glycosylation that contribute to the anomalous migration of these proteins on polyacrylamide gels. All of the receptors cloned to date have at least one consensus sequence for N-linked glycosylation (Asn-X-Ser/Thr) in the extracellular domain. These putative glycosylation sites are usually located near the N terminus of the protein, although occasionally there are potential sites in the second extracellular loop. Site-directed mutagenesis of the β_2-adrenergic receptor, the NK1 neurokinin receptor, and the lutropin receptor, as well as biochemical analysis of rhodopsin, has shown that the potential glycosylation sites near the N termini are actually glycosylated in the native proteins (5–8). Prevention of receptor glycosylation, either by inhibition of glycosylation enzymes in the cell or by site-directed mutagenesis of the potential glycosylation sites on the receptor, results in a decrease in the level of expression of the β_2AR or rhodopsin on the cell surface, as well as changes in the appearance and migration of the receptor bands on polyacrylamine gels. However, the lack of glycosylation does not seem to have a dramatic effect on the ligand-binding or functional activity of those receptor molecules that are expressed in the plasma membrane.

G protein–coupled receptors contain a number of conserved cysteine residues, some of which appear to play a role in receptor structure. There are two highly conserved Cys residues in the second and third extracellular loops of the receptors (Cys106 and Cys184 in the β_2AR, Figure 1). Substitution of either of these Cys residues with Val (9, 10) by site-directed mutagenesis results in a destabilization of the tertiary structure of both rhodopsin and the βAR, accompanied by alterations in the binding characteristics of the receptors, suggesting a role for these residues in maintaining the active conformation of the receptor. Biochemical analysis of rhodopsin showed that these two Cys residues are involved in an intramolecular disulfide bridge that links the second and third intracellular loops of the receptors, thus constraining the conformation of the extracellular domain of the protein (11). The invariance of these Cys residues in most G protein–coupled receptors suggests that this intramolecular disulfide bridge is a structural feature common to the entire family of receptors.

An additional highly conserved Cys residue is found within the C-terminal tail of many G protein–coupled receptors (Cys341 in Figure 1). This residue has been shown to be palmitylated in the α- and β-adrenergic receptors (12, 13) and rhodopsin (14). The function of the palmityl substituent is not clear at present, but it has been speculated that the palmityl group may anchor a part of the cytoplasmic tail of the receptor to the plasma membrane, thus controlling the tertiary structure of this region of the receptor. Substitution of this Cys residue with Ser in rhodopsin (11), with Leu in the βAR (9), or with Ala in the α_{2a} receptor (13), removing the potential for palmitylation at this site, did not appear to affect the overall structure or the function of these receptors. However, substitution of this Cys with Gly in the βAR resulted in a loss of the ability of the receptor to activate G proteins (12), suggesting that the nature of the side chain (hydrophobic or palmitylated Cys) at this position is critical for receptor function.

MOLECULAR MODELING OF G PROTEIN–COUPLED RECEPTORS

A number of G protein–coupled receptor models have been published (15–31) since rhodopsin was first modeled in 1986 by R Feldmann and P Hargrave (personal communication) and by Findlay & Pappin (32). These models are all based on the high-resolution, electron cryo-microscopy structure of bacteriorhodopsin determined by Henderson and coworkers (33). Bacteriorhodopsin is a seven-transmembrane-domain protein from *Halobacterium halobium*. It is not a G protein–coupled receptor, but it is functionally related to rhodopsin and therefore proposed to be structurally homologous to the G protein–coupled receptors. However, comparison of bacteriorhodopsin with the recent structure of rhodopsin (33), which is coupled to the G protein transducin, indicates that there may be differences in the way that the helices are packed between the two transmembrane proteins. Baldwin has recently proposed a model of the G protein–coupled receptors that is based on the topological restraints of the loops, the proposed amphipathicity and residue variability of various helices, and the packing arrangement suggested by the low-resolution rhodopsin structure (34). Efforts have also been directed toward the development of amino acid similarity matrix methods, which characterize the salient features of residues in the proposed transmembrane, helical regions of the receptors (18, 34–36). These approaches provide a fingerprint of the helical regions of each G protein–coupled receptor and enable clustering of the known receptor sequences into families and subtypes. The importance of these methods is underscored by the apparent lack of a significant degree of sequence homology between bacteriorhodopsin and the eukaryotic G protein–coupled receptors. Interestingly, an alternate proposal of sequence alignment shows higher homology (up to 40% similarity) between bacteriorhodopsin and the G protein–coupled receptors, in which the sequential ordering of the helices is permitted to vary (26).

Donnelly's three-dimensional model of the human β_2AR (18) was used initially as the basis for construction of the G protein–coupled receptor models. It is clear from an examination of the periodicity in the position of conserved vs nonconserved residues in the putative transmembrane domains (36) that those residues that are conserved would be predicted to face along one edge of an α-helix. The classification of conserved and nonconserved residues is based on the observation from X-ray data of the substitution probability for each residue type in different environments: solvent accessible, protein buried, and lipid facing. Each position in an alignment across a set of G protein–coupled receptors shows a characteristic variability in residue type according to the type of environment that this position faces. The periodicity of this variability was shown by Donnelly and coworkers to be indicative of an α-helix. A recent review by Hibert and coworkers emphasizes the importance and the limitations of the approaches used in developing the G protein–coupled receptor models (37). Using this kind of information, with the high-resolution electron cryo-microscopy map of bacteriorhodopsin as a footprint (33), one can organize the membrane-spanning portion of the receptor by packing the helices such that the character of each face matches the appropriate environment (D Underwood, unpublished data). The side chains can then be oriented according to known preferences of amino acid side chains.

A similar three-dimensional model of the human neurokinin-1 receptor (NK1R) was also constructed (D Underwood, unpublished data), based on the sequence alignment between the β_2AR and the NK1R. The sequence alignments were performed by, in the main, pairwise hand alignment using the graphical tools available in the modeling program QUANTA (38). Portions of the alignment were produced using the Needleman & Wunch algorithm (39) but, in general, the automatic methods gave less than satisfactory results, due to the low primary sequence homology between the receptors and the variability in length and sequence of the loop regions. The structural model was created from the generated alignment using the coordinates of the β_2AR model. Each of the side-chain positions was placed according to the following considerations: All atoms of the side chain in common with atoms in the template (β_2AR) were placed accordingly, and those side-chain atoms that were not in common with the template were built according to known side-chain preferences. The loop regions were added to the helices by searching the library of protein structures generated from the Brookhaven Protein Data Bank (40) for protein fragments that most closely matched the geometric requirement of bridging from one helix to the next. The loop regions, which are by definition lacking in secondary structure, are well solvated and are the least well-defined region of the G protein–coupled receptor model. Both receptor models were refined using the CHARMm forcefield (41). It is difficult to define the end of the helices and the beginning of the more variable, hydrophilic loop regions to complete the structural model of the receptor. However, the presence of charged and polar residues appears to herald the transition from membrane-buried helices to hydrophilic loop structure. Clearly, in a dynamic sense, the end of a helix and the beginning of a loop need not be distinct boundaries, since helices will tend to ravel and unravel in response to local environmental changes. With the exception of the third intracellular loop, the hydrophilic loops of the receptors are relatively short (~10 amino acids), consistent with a seven-helix bundle forming the transmembrane core of the receptor. The addition of a disulfide bridge between the conserved Cys residues on the second and third extracellular domains further constrains the flexibility of the loops. These receptor models provide a framework from which to explore the functional domains of the receptors, as described below.

LOCALIZATION OF THE LIGAND-BINDING DOMAIN

The family of G protein–coupled receptors binds and is activated by a wide variety of hormone agonists, ranging in size from small molecules to large glycoproteins. Much of the current emphasis on receptor research is focused on determining the structures of the binding domains for these structurally divergent ligands, using genetic, biochemical, and biophysical approaches. The most widely studied of these receptor-binding domains is that of rhodopsin, which contains a retinal chromophore covalently attached via a Schiff base formed between the aldehyde group of retinal and the ε-amino group in the side chain of Lys296 in the seventh transmembrane helix of rhodopsin (42). Upon exposure to light, the covalently attached retinal undergoes a conversion from the 9-*cis* to the all-*trans* conformation, and the associated series of conformational changes in rhodopsin triggers the association with the G protein transducin. Thus, light-activated retinal serves as an agonist to initiate the rhodopsin-mediated signal-transduction cascade. Because of its spectroscopic properties, rhodopsin has been widely characterized biophysically and biochemically. The bound retinal has been determined to be buried approximately 22 Å into the membrane bilayer, with the long axis of retinal parallel to the plane of the membrane and the plane of the chromophore perpendicular to the plane of the membrane (43).

When the β-adrenergic receptor was first cloned and the structural similarity to rhodopsin noted, it was suggested that the conserved transmembrane domains of these receptors might serve as a common site of agonist binding, even though covalent attachment of the ligand is unique to rhodopsin (44). Deletion analysis of each of the extracellular and cytoplasmic loops of the βAR provided experimental evidence to implicate the transmembrane domain in ligand binding, since the hydrophilic extramembraneous segments of the receptor could be deleted without adversely affecting the binding of agonists or antagonists (45). However, the deletion approach for identifying regions of the receptor that could be removed without affecting the binding properties (and thus could be ruled out as contributing to the binding site) proved of limited usefulness in identifying binding domains within the transmembrane core of the receptor. Deletion of any of the hydrophobic transmembrane domains of the βAR, either singly or in pairs, resulted in a loss of receptor protein expressed at the surface of the cell, demonstrating the important structural roles of these regions. Thus, the individual residues in these regions that are involved in ligand binding ultimately had to be defined by single-amino-acid substitutions in the transmembrane core of the receptor.

Some clues as to regions of the receptors that are critical for binding have also arisen from a chimeric receptor approach to mutagenesis. In this approach, chimeras are formed between two related G protein–coupled receptors by splicing together complementary helical regions from the two receptors to form, for example, a chimeric receptor containing the N-terminal–to–helix 3 region from one receptor subtype and the helix 4–to–C-terminal region from another. By moving the "splice junction" to different positions in the receptor sequence, one can obtain information as to which helical domains are the most critical for the binding of specific ligands. This approach has been useful for locating general regions involved in the ligand-binding domains of various receptors, which can then be further characterized by specific point mutations in those regions. Thus, the helix 7 region of α- and β-adrenergic receptors was discovered to be critical for selective antagonist binding to these receptor subfamilies by this approach (46). Chimeric receptors were also useful in localizing the general regions of the β_1 and β_2 receptors involved in selective agonist binding (47, 48), in localizing the binding site for nonpeptide antagonists to the NK1 and NK2 neurokinin (49–52) and AT1 angiotensin receptors (T. Schwartz, personal communication), in localizing the regions of the NK1 receptor involved in determining species specificity (51, 52), and in localizing the G protein activation sites for the muscarinic and β-adrenergic receptors. However, structural effects can make these mutants difficult to interpret, since formation of a fully active chimeric receptor requires that all the interhelical contacts responsible for the tertiary structure of the receptor be maintained in the chimeric protein. Thus, chimeric receptors frequently display a reduced binding affinity or decreased level of expression that probably signifies a disruption of the tertiary structure of the receptor, and further information must be obtained from single-amino-acid substitutions, which may be less likely to cause an overall disruption of the structure of the protein.

THE LIGAND-BINDING DOMAIN OF BIOGENIC AMINE RECEPTORS

For the βAR, the design of single-amino-acid replacements with which to identify residues important for the binding of β-adrenergic ligands was aided by the long history of medicinal chemistry research that had gone into designing β-adrenergic agonists and antagonists (53). These studies had generated a consensus pharmacophore for the βAR, which could be superimposed on the new information from deletion mutagenesis that the ligand-binding site of this receptor resides within the transmembrane domain, resulting in the "pharmacophore map" of the binding site shown in Figure 2. β-adrenergic agonists and antagonists are biogenic amines, and the basic amine moiety is the source of much of the binding energy of these compounds. On this basis, it seemed likely that there is an acidic group in the binding pocket of the receptor that would provide a counterion for the protonated amine of the ligand. Such a salt bridge could provide up to 10

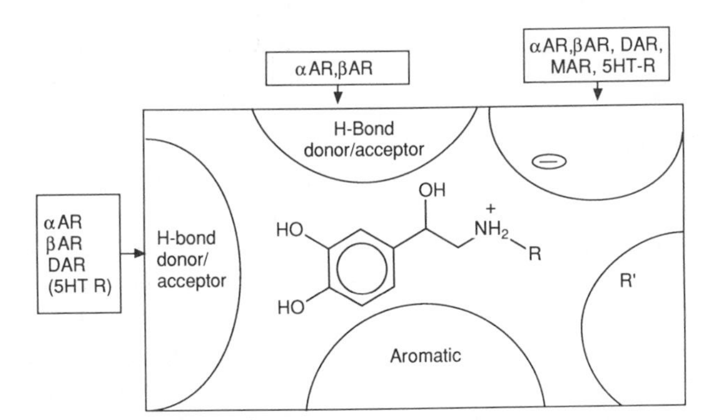

Figure 2 Pharmacophore map of the catecholamine-binding site of the βAR. The transmembrane helices are represented by circles; the nature of the interactions, as deduced from the structure-activity relationships of β-adrenergic agonists and antagonists, is indicated. Reprinted with permission from *Trends in Pharmacological Sciences,* 1989. December supplement, p. 28.

kcal/mol of binding energy within an essentially hydrophobic environment. Likewise, the β-hydroxyl group of the ligand is important for the stereoselective binding of both agonists and antagonists to the receptor, suggesting the existence of a hydrogen bond donor or acceptor in the binding site to interact with this moiety. The catechol ring of the agonist is critical for agonist binding and activation of the receptor, with both hydroxyl groups of the catechol being essential for full agonist responsiveness. Thus, one might expect to find a pair of hydrogen bond donors or acceptors in the receptor that can interact with the catechol hydroxyl groups, as well as some type of aromatic side chain to interact with the phenyl ring itself. The amine end of the β-adrenergic ligands has been less thoroughly mapped, but clearly plays a role in the subtype specificity of β-adrenergic receptors. The β_1 receptor subtype binds epinephrine (which has a methyl group at this position) and norepinephrine (in which R = H) with equal affinity, while the β_2 receptor subtype prefers epinephrine to norepinephrine and the β3 receptor binds norepinephrine with higher affinity than epinephrine.

The utility of pharmacophore maps such as that shown in Figure 2 is further enhanced by the homology in the transmembrane domains among G protein–coupled receptors. Assuming that these transmembrane domains evolved to bind the specific agonist hormones, then one would expect a certain pattern of conservation of key binding residues across the family of receptors. Thus, the putative acidic counterion for the amine group of the β-adrenergic agonists might be conserved among all biogenic amine G protein–coupled receptors, while the amino-acid side chains that interact with the aromatic catechol moiety should be conserved among catecholamine receptors, but not with other biogenic amine receptors of this class.

Single-amino-acid replacements of acidic amino acids in the transmembrane domain of the βAR with neutral amino-acid residues have implicated the side chain of Asp113 in the third transmembrane helix as the counterion for the amine groups of β-adrenergic agonists and antagonists (Figure 3*A*). Thus, substitution of this Asp with a Glu residue, moving the position of the carboxylate anion, resulted in a 100-fold decrease in ligand affinity (54, 55). Substitution of this residue with an Asn or Ser residue, removing the charge from the side chain, resulted in a 10,000-fold decrease in agonist and antagonist affinity (56). Interestingly, these mutant receptors could still be fully activated by epinephrine, indicating that Asp113 plays a role in ligand binding but not in receptor activation. In designing site-directed mutagenesis experiments, it is important to try to distinguish between structural effects of an amino acid substitution and direct effects of the substitution on receptor-ligand interactions. One way of making this distinction is to alter the ligand in parallel with mutating the receptor. If the effect of the mutation is to decrease binding affinity by specifically disrupting a critical interaction with the ligand, then it should be possible to form a new, specific interaction between the mutant receptor and an appropriately modified ligand. Thus, if Asp113 in the wild-type βAR forms an ion pair with the amine group of the ligand, then one would expect that the Ser113 mutant receptor should be activated by compounds that can form a hydrogen bond at this position, rather than an ion pair. This was in fact the case for this mutant receptor. The Ser113-substituted βAR was fully activated by catechol esters and ketones, carbonyl-containing hydrogen bond acceptors that did not activate the wild-type βAR at all (56).

Sequence alignment of the family of G protein–coupled receptors revealed that an aspartate residue was found at the position in helix 3 corresponding to that of Asp113 of the βAR in all G protein–coupled receptors that bind biogenic amines, but not in other G protein–coupled receptors, strongly suggesting that interacting with amines is the main function of this aspartate residue. Subsequent mutagenesis studies of other biogenic amine receptors have confirmed that this Asp residue also plays a critical role in ligand binding to the α_2 adrenergic (57), muscarinic (58), and histamine receptors (59). Interestingly, Lys181 at the analogous position of the ETB endothelin receptor is important for high-affinity binding of peptide agonists and antagonists to that receptor (60). Mutagenesis of rhodopsin, in which the retinal chromophore is anchored to the receptor by a covalent interaction with Lys296 in helix 7, showed that the counterion for this Schiff base is provided by the side chain of Glu113, which is located in helix 3 just one helical turn above the position of Asp113 in the βAR, again suggesting that this region of the receptor plays a key role in determining ligand specificity (42).

Another Asp residue, Asp79 in helix 2, is one of the most highly conserved residues throughout the family of G protein–coupled receptors. Substitution of this Asp with Ala in the β_2AR resulted in a decrease in agonist affinity and a reduction in agonist efficacy with no effect on antagonist interactions, consistent with a role for this residue in promoting receptor activation (54). A similar effect upon substitution of this Asp residue has been described for other G protein–coupled receptors, including the α_2 adrenergic (61), lutropin (62), AT_1 angiotensin (63), 5HT-2 serotonin (64), and NK1 neurokinin (R Huang, H Yu, C Strader, T Fong, manuscript in preparation). Thus, the conserved Asp in helix 2 appears to be involved in transduction of the agonist-binding signal to stimulate an interaction with the appropriate G protein.

A similar approach, combining receptor mutagenesis with structural modification of the ligand, has indicated that the aromatic catechol moiety of β-adrenergic agonists binds to a pocket involving residues in the fifth and sixth transmembrane helices of the βAR (Figure 3). Substitution of either

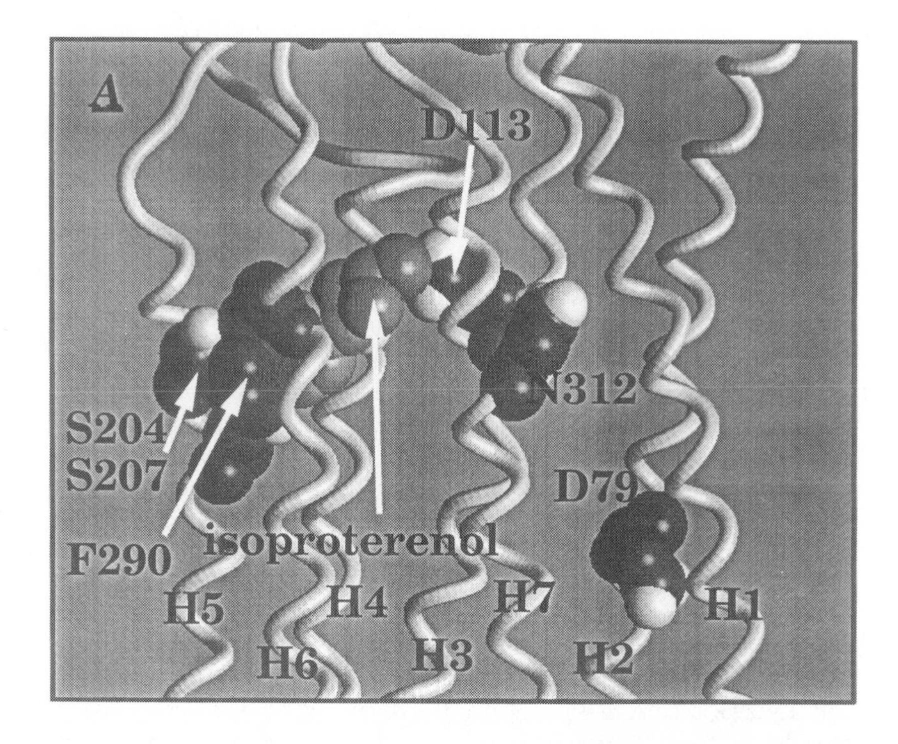

Ser204 or Ser207 in helix 5 of the receptor with an Ala residue, thereby removing the hydroxyl side chain from that position in the receptor, resulted in an approximately 10-fold decrease in the affinity of catecholamine agonists for the receptor, and a 50% reduction in agonist efficacy (65). A similar pattern of reduced affinity and efficacy was observed when either of the catechol hydroxyl groups was removed from the ligand by substitution with a proton. Examination of the additivity of the effects of substitution of the *meta* or *para* hydroxyl groups on the ligand with the effects of substitution of Ser204 or Ser207 in the receptor gave a pattern that would be consistent with the existence of two hydrogen bonds linking the catechol moiety of the ligand to helix 5 of the receptor: one specific interaction between the side chain of Ser204 and the *meta* hydroxyl group of the ligand, and a second hydrogen bond between the side chain of Ser207 and the *para* hydroxyl group of the ligand. Apparently the formation of both of these hydrogen bonds is essential for full agonist activation of the receptor, while formation of the "incorrect" hydrogen bond pair results in ligand binding but not in receptor activation. These Ser residues would be predicted to lie one helical turn apart on helix 5, so that simultaneous binding of the agonist to both Ser residues would severely constrain its conformation in the

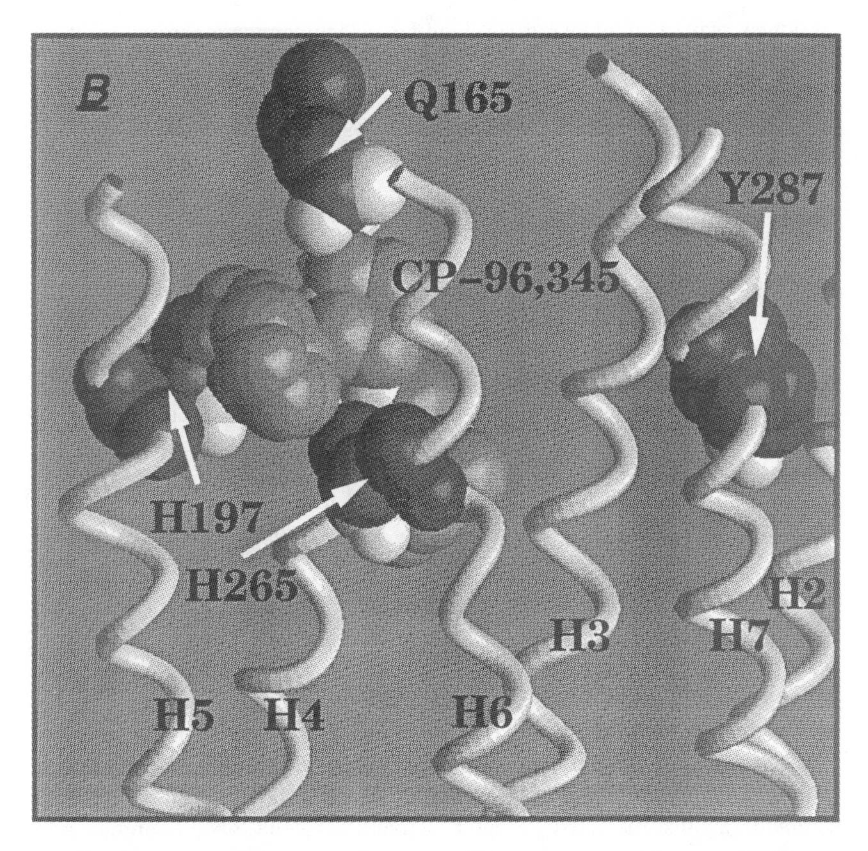

Figure 3 Molecular models of (*A*) β_2AR and (*B*) NK1R, showing a side view of the binding site. Isoproterenol is docked in the binding site of the βAR and CP-96,345 in the NK1R. The transmembrane helices are numbered (1–7) at the bottom of each panel, and the beginnings of the extracellular loops can be seen at the top. Residues determined by mutagenesis to be involved in ligand binding are shown in "CPK" format.

receptor. Subsequent mutagenesis of the βAR has suggested a role for Phe290 in the sixth transmembrane helix in stabilizing the interaction of the aromatic catechol-containing ring with the receptor (66).

Examination of the pattern of conservation of these residues across the family of G protein–coupled receptors suggests that they co-evolved with the catechol ring: Ser residues are located at positions analogous to those of Ser204 and Ser207, and a Phe residue at the position analogous to Phe290, in all G protein–coupled receptors that bind catecholamines, but are not present in other G protein–coupled receptors. Mutagenesis of the α_2 adrenergic and D2 dopamine receptors has confirmed that these Ser residues play a similar role in binding catecholamine agonists to those receptors, as well. In addition, when the histamine receptor was cloned and its primary sequence determined, it was noted that this receptor contains an Asp and a Thr residue at positions analogous to Ser204 and Ser207 of the βAR, respectively. Substitution of Asp186 and Thr190 in the H2 histamine receptor with Ala demonstrated that these two amino-acid side chains are critical for binding histamine to the receptor, and suggest a specific interaction with the imidazole ring of histamine (59). Interestingly, substitution of Lys199 in this region of the AT1 angiotensin receptor with Gln decreased the affinity of the receptor for peptide agonists and antagonists, indicating that this region may be important in the binding of peptides to G protein–coupled receptors, as well (67). Mutagenesis studies of the muscarinic receptor have implicated Thr231 and Thr234 in the analogous region of helix 5 in the binding and activation by acetylcholine and carbamylcholine, perhaps by an interaction with the ester moiety (68). Taken together, these results indicate that residues within this region of helices 5 and 6 make up a specific binding pocket for the gamma substituents of biogenic amines and are critical determinants of the agonist specificity of their receptors. In light of the identification of the third intracellular loop of G protein–coupled receptors, which connects helices 5 and 6, as a critical determinant of receptor activation (see below), it is interesting to speculate that the simultaneous binding of the ligand to residues in helices 5 and 6 of the receptor can affect the conformation of this region of the receptor in such a way as to promote the interaction of the third intracellular loop with the G protein. Thus, this region appears to be optimally located to promote agonist-mediated receptor activation. None of the amino acid substitutions in this region affected the binding of antagonists to the βAR, consistent with a role for this region in agonist activation, rather than as a general binding site for both agonists and antagonists.

Figure 3*A* shows a low-energy model of isoproterenol (T Halgren, personal communication) docked into the region between helices 3, 4, 5, and 6, with the side chains of the receptor adjusted so that the postulated interactions with Asp113, Ser204, Ser207, and Phe290 could occur. Examination of this model would suggest that the hydroxyl side chain of Ser165 in the fourth transmembrane helix is ideally located to interact with the β-hydroxyl group on the ethanolamine chain of epinephrine (Figure 3*A*). This β-hydroxyl group is critical for the binding of agonists and antagonists to the βAR, with the R isomer binding with 100-fold higher affinity than the S isomer. The postulated interaction between the side chain of Ser165 and the β-hydroxyl group of the ligand could not be tested directly by site-directed mutagenesis: Substitution of Ser165 with Ala resulted in a processing error in the βAR such that it was not expressed on the cell surface (65). Such processing defects probably indicate an effect of the amino acid substitution on the folding of the protein, suggesting that the residue in question is critical for helical packing in the transmembrane domain of the receptor, and preventing further probing of that position by mutagenesis (9). However, the apparent optimal positioning of Ser165 in the binding site of the receptor according to molecular modeling, in addition to the failure to identify any other potential hydrogen bond donor or acceptor to interact with this group despite a large mutagenesis effort (65, 66), has led us to postulate that the side chain of Ser165 forms a hydrogen bond with the β-hydroxyl group of adrenergic ligands in the binding site of the receptor.

As mentioned above, a chimeric receptor approach was useful in localizing the region of the β-adrenergic receptor responsible for the subtype specificity of endogenous agonists. Construction of a series of hybrid receptors in which increasingly long N-terminal stretches of the β_1 receptor were replaced by the analogous regions of the β_2 receptor showed that the relative potencies of epinephrine and norepinephrine "switched" from the β_1 phenotype to the β_2 as the helix 4 region was replaced (47). Subsequent construction of more specific hybrids confirmed that replacement of helices 4–5 of the β_2 receptor with those from the β_1 receptor was sufficient to cause an increased affinity for norepinephrine and a decreased affinity for epinephrine (48). However, no single-residue replacement in this region was able to account for the reversal of subtype specificity observed with the chimeric receptor. Thus, the binding pocket for the N-methyl substituent of epinephrine appears to involve either direct or conformational effects of multiple residues in the helix 4–5 region of the receptor.

Less is known about the residues involved in the antagonist-binding site of the βAR. Structure activity relationships derived during the development of β-blocking drugs have shown that antagonist activity is correlated with an increase in the hydrophobicity of the aromatic portion of the compound and an increased separation between the aromatic nucleus and the amine moiety, achieved by substitution of a phenoxypropanolamine for the phenethanolamine chain. The phenoxy oxygen of the antagonists appears to interact with the side chain of Asn312 in helix 7 of the βAR (Figure 3*A*). This Asn residue is conserved among biogenic amine receptors that bind phenoxypropanolamine antagonists with high affinity, including the β-adrenergic and 5HT-1a receptors, but is replaced by a Phe or Val residue in other serotonin and α-adrenergic receptors. Introduction of an Asn residue in place of the endogenous Phe residue at this position of the α_2 receptor increases the affinity of the receptor for phenoxypropanolamine antagonists such as cyanopindolol (69). Similarly, substitution of a Val residue for the analogous Asn of the 5HT-1a receptor decreases its affinity for this class of compounds (70).

As mentioned above, a large part of the binding energy of the antagonists arises from the formation of the ion pair between the amine moiety of the antagonist and the carboxylate side chain of Asp113 in the receptor. We observed that the replacement of Asp113 with a Glu residue, while decreasing the affinity of the receptor for adrenergic ligands, also resulted in the appearance of partial agonist activity for some β-adrenergic antagonists (71). This observation suggests that the ability of a compound to bind simultaneously to the carboxylate side chain at position 113 and to the catechol-binding nucleus in helices 5 and 6 might be a key determinant of agonist activity at the βAR. According to this hypothesis, the most potent antagonists would be those that are able to block the binding of the agonist to this "activation nucleus," while forming high-affinity binding interactions with other residues within the hydrophobic core of the receptor.

THE LIGAND-BINDING DOMAIN OF PEPTIDE RECEPTORS

Structure of the Peptide-Binding Site

As discussed in the previous section, small molecules such as biogenic amines bind to the transmembrane domain of their receptors. However, many of the G protein–coupled receptors are peptide receptors whose natural ligands range from tripeptides, such as thyrotropin-releasing hormone, to large 20–40-amino-acid peptides, such as glucagon. Because these peptide agonists are substantially larger than biogenic amines, it seems likely that the peptide-binding sites of G protein–coupled receptors might be composed of a larger surface area than that contained within the transmembrane domain. Molecular modeling of G protein–coupled receptors suggests that the extracellular domains may provide the additional points of contact with the peptides. As with the biogenic amine receptors, analysis of the structure-activity relationships determined for the ligands can provide clues with which to focus mutagenesis-based approaches to define the ligand-binding domains of these peptide receptors. For example, structural analysis of the neurokinin peptide family [including the undecapeptide substance P (SP), and the structurally related dodecapeptides neurokinin A and neurokinin B] has indicated that a core structure encompassing the conserved C terminus (FXGLM-NH_2) is required for agonist activity at all three NK receptor subtypes (72). The remaining N-terminal residues are divergent and define the subtype specificity of the peptides. It has therefore been hypothesized that the divergent N termini of the neurokinin peptides are responsible for the receptor subtype specificity, while the conserved C terminus mediates their common activation mechanism.

As a first step toward elucidating the role of the extracellular domains of the NK1R for SP binding, deletion and chimeric mutagenesis approaches were applied to the NK1R (73, 74). Substitution or deletion of the first, second, and third extracellular segments of the NK1R resulted in a substantial reduction in peptide-binding affinities. This observation is in contrast to the β_2AR, where deletion of most of the extracellular residues did not affect the binding of biogenic amines (45). Subsequent single-residue substitutions in the NK1R identified three residues (Asn23, Gln24, and Phe25) in the first extracellular segment and two residues (Asn96 and His108) in the second extracellular segment that are required for the high-affinity binding of neurokinin peptides (73). These residues are not involved in nonpeptide antagonist binding to the receptor. Only one of these five residues is conserved among the three neurokinin receptor subtypes. Because substitutions of divergent amino acids did not affect the relative affinities of the three neurokinin peptides to the receptor, a model in which the differential peptide-binding affinity of the NK1R originates from interactions between divergent residues of the neurokinin peptides and divergent residues of the receptor appears to be oversimplified.

The construction of chimeric NK1/NK3 receptors with multiresidue substitutions in the third extracellular segment of the NK1R substantially reduced the affinity of the NK1R for neurokinin peptides. However, individual substitution of each of the divergent residues in this region with Ala did not cause large changes in peptide affinity (R Huang, H Yu, C Strader, T Fong, in preparation). These data indicate that the first and second extracellular segments are critical for the binding of neurokinin peptides, while the third and fourth extracellular loops appear to play only a minor role in peptide binding. The involvement of the extracellular domain in peptide binding has also been demonstrated for other peptide or glycoprotein hormone receptors. For example, the N-terminal region of the thyrotropin receptor is required for the high-affinity binding of its agonists (74, 75). The lutropin receptor has a long N-terminal region (338 amino acids) which, when expressed in the absence of the transmembrane domain, exhibits high affinity and specificity for luteinizing hormone (76). In addition, substitution of the extracellular segments of the formyl peptide receptor has been found to affect the high-affinity binding of formyl peptides (77). The divergent residues in the N-terminal region of two interleukin-8 receptor subtypes have been identified as determinants of subtype specificity, but these residues do not affect the binding affinity of interleukin-8 itself significantly (78). Thus, in contrast to the biogenic amine receptors, the extracellular domains appear to play a role in the interactions of peptides with G protein–coupled receptors.

To determine the role of the transmembrane domain in peptide binding to the NK1R, site-directed mutagenesis has been used to probe the contribution of putative interior residues. These studies have demonstrated that three residues (Asn85, Asn89, and Tyr92) in the second transmembrane domain and Tyr287 in the seventh transmembrane domain of the NK1R are involved in the binding of all neurokinin peptides (R Huang, C Strader, and T Fong, manuscript in preparation). The residues in helix 2 are not involved in nonpeptide antagonist binding, while Tyr287 is involved in the binding of at least one nonpeptide antagonist, the perhydroisoindole RP67580. Based on the binding affinities of several SP analogs for the wild-type and the N85A mutant of the NK1R, Asn85 has been proposed to interact with the C-terminal amide of SP. The amino acid requirement at positions 85 and 89 is very stringent in that Asn, but not Ser or Gln, confers high affinity for peptides, suggesting the existence of hydrogen-bonding interactions between the bound peptides and these side chains in a rigidly constrained environment. On the other hand, substitution of Tyr92 or Tyr287 with other amino acids indicates that Tyr92 is involved in an aromatic interaction with a residue(s) on the neurokinin peptides, while Tyr287 appears to be involved in both hydrogen-bonding and aromatic interactions with the peptide agonists.

It is interesting that the four residues (Asn85, Asn89, Tyr92, and Tyr287) in the transmembrane domain of the NK1R that have been identified as contributing to peptide-binding affinity are conserved among the three neurokinin receptor subtypes. These residues appear to interact with all neurokinin peptides, suggesting that these amino acids are critical for high-affinity binding, rather than for determination of subtype selectivity. Taken together, the results from analysis of extracellular and transmembrane residues indicate that the residues important for peptide binding can be divided into two groups. One group, consisting of both conserved and divergent residues, appears to interact with all three neurokinins and is likely to be the major determinant of binding affinity. The other group consists of divergent residues that interact either directly or indirectly with a specific peptide, and each residue in this group appears to contribute a small amount of binding energy to the overall peptide affinity.

Chimeric receptor studies of the NK1R have demonstrated that the peptide selectivity can be altered in the chimeric receptors, suggesting the locations of domains that are critical for subtype specificity (50, 73, 79). However, no single-residue substitutions, either of divergent or conserved amino acids, have been found that can change the receptor subtype specificity for the neurokinin peptides (R Huang, H Yu, C Strader, and T Fong, manuscript in preparation), although small changes in affinity have been observed (39, 53). Two—mutually non-exclusive—explanations for the failure of single-residue substitutions to alter NKR subtype specificity may be considered.

First, peptide selectivity may be determined by direct interactions with a large number of divergent residues in the receptors. Second, peptide selectivity may be determined by the degree of conformational compatibility between each peptide and the binding pocket of the receptor. The second possibility is supported by the observation that the solution conformations of different neurokinin peptides are not identical (80, 81). Furthermore, analysis of conformationally constrained SP analogs suggests that different peptide conformations can be recognized by different neurokinin receptor subtypes (82). Biophysical analysis of receptor-peptide complexes will be required to address these possibilities directly.

Heterologous expression of the cloned NK1R demonstrates that its major second-messenger pathway is the phosphatidyl inositol (PI) hydrolysis pathway. Site-directed mutagenesis experiments have identified two residues in the transmembrane region of the NK1R that are critical for receptor activation. Substitution of Glu78 in the second transmembrane domain or Tyr205 in the fifth transmembrane domain with Ala reduces agonist-binding affinity without affecting antagonist affinity, and results in a complete loss of activity in stimulating PI hydrolysis; i.e. peptides can still bind to the E78A or Y205A mutant receptors but are unable to activate the receptor (R Huang, H Yu, C Strader, and T Fong, manuscript in preparation). Because of the close proximity of Glu78 to the residues in the second transmembrane domain (Asn85, Asn89, Tyr92) determined to be critical for peptide binding, this region of the NK1R may serve, at least in part, to trigger the agonist-induced conformational changes in the receptor responsible for receptor activation.

An Asp/Glu residue at the analogous position in the second transmembrane domain and a Tyr/Phe residue at the analogous position in the fifth transmembrane domain are conserved in many G protein–coupled receptors. As mentioned above, the acidic residue in the second transmembrane domain has been found to be involved in activation of other G protein–coupled receptors. The role of the Tyr/Phe residue in the fifth transmembrane domain in the activation of other G protein–coupled receptors has not been explored. However, given the highly conserved nature of an aromatic residue at this position, it seems likely that this region will prove to be important for activation of other receptors, as well. In this light, it is interesting that the region identified as being critical for catecholamine-mediated activation of the adrenergic receptors is in helices 5 and 6, while residues in helices 2 and 5 have been implicated in SP-mediated NK1R activation. Perhaps the helix 2 and helix 5 regions play a common role in mediating receptor activation throughout the family of G protein–coupled receptors, with the mechanisms of conformational triggering by the various peptide and non-peptide agonists varying for different members of this large receptor family.

The Binding Site for Nonpeptide Antagonists

The discovery of morphine as an agonist of the opiate receptor has long been considered as proof that it is possible to develop nonpeptide ligands for peptide receptors. Recent discoveries of nonpeptide antagonists for the CCK receptor, the angiotensin II receptor, and the NK1R further affirm that possibility (83–91). However, the development of nonpeptide antagonists for other G protein–coupled receptors remains a challenge. It is important to understand the molecular basis of nonpeptide antagonist-receptor interactions in order to develop new generations of antagonists for these receptors.

It has been well documented that biogenic amines bind to the transmembrane domain of their receptors. Since the nonpeptide SP antagonists are substantially smaller than SP and approach the size of biogenic amines, it also seemed likely that the nonpeptide antagonists bind to the transmembrane domain of the NK1R. Extensive mutagenesis studies on the extracellular and transmembrane domains have confirmed that the extracellular domains of the NK1R play a minor role in nonpeptide antagonist binding. Most of the recently developed nonpeptide NK1 antagonists show pronounced species specificity, with the quinuclidine CP-96,345 showing selectivity for the human receptor over the rat and the perhydroisoindole RP67580 showing selectivity for the rat receptor (85, 86). This degree of selectivity is remarkable, since the rat and human receptors are 95% identical, differing by only 22 amino acids, only eight of which lie within the transmembrane domain. Construction of chimeric human/rat NK1 receptors (51, 52), followed by single-residue substitutions (51), revealed that the exchange of two divergent residues in the transmembrane domain (Val116 for Ile in helix 3 and Ile290 for Ser in helix 7) is sufficient to reverse the species selectivity of CP-96,345 and RP67580. However, these residues do not appear to interact directly with the antagonists, because different combinations of homologous substitutions result in similar changes in antagonist-binding affinity (51, 52, 93). Furthermore, the magnitude of the changes in affinities observed upon substitution of these amino acids (10–100-fold) is too large to be accounted for by the direct van der Waals contacts that might be postulated to occur with either of these residues. Rather, these results suggest that these phylogenetically divergent residues affect the local conformation of the antagonist-binding pocket, which is composed of conserved residues in the transmembrane core (51). A similar conclusion may also be drawn from the observation that a single Val319-to-Leu substitution in the sixth transmembrane domain of the CCK_B receptor could account for the observed 20-fold difference in the binding affinity of benzodiazepine antagonists between the dog and human CCK_B receptors (94).

Four residues (Gln165, His197, His265, and Tyr287) have been identified as critical sites of contact for nonpeptide antagonists with the NK1R. Based on various amino-acid substitutions at position 197 and the analysis of a series of CP-96,345 analogs, His197 in the fifth transmembrane domain of the NK1R has been identified as interacting with the benzhydryl moiety of CP-96,345, possibly via an amino-aromatic interaction (95). His265 in the sixth transmembrane domain of the NK1R appears to lie in close proximity to the substituted benzyl moiety of bound quinuclidine antagonists (95a). Analysis of the chemical nature of the residues that can be substituted at position 265 suggests that His265 forms a hydrogen bond with RP67580. The interaction at this position is species specific: His265 of the human NK1R does not interact with CP-96,345, while His265 of the rat NK1R does. These data clearly demonstrate that the local conformation of the antagonist-binding pocket in the rat and human NK1Rs is slightly different, suggesting that the intermolecular distances between the antagonists and the residues in the binding site vary between the human and rat NK1Rs. In support of this hypothesis, modification of the substituted benzyl moiety of CP-96,345 results in new antagonists that are sensitive to the chemical nature of the side chain at position 265 in the human NK1R. SP binding is not affected by amino acid substitutions for either His197 or His265, indicating that these residues bind to small-molecule antagonists but do not interact with the peptide agonists. Gln165 in the fourth transmembrane domain of the NK1R appears to interact with the C3 heteroatom of quinuclidine antagonists such as CP-96,345 (Figure 3; T Fong, M Cascieri, H Yu, C Swain, and C Strader, manuscript in preparation). Gln165 also contributes slightly to the binding of perhydroisoindole antagonists such as RP67580, but the exact molecular interaction with that class of compounds has not yet been determined. Tyr287 in the seventh transmembrane domain has been shown to be required for peptide binding to the NK1R. This residue also appears to participate in an aromatic interaction with RP67580, although it does not interact with the quinuclidine class of antagonists.

Analysis of chimeric NK1/NK2 and NK1/NK3 receptors has also implicated the regions at the top of transmembrane domains 5 and 6 in the binding of nonpeptide antagonists to both the NK1 and NK2 receptors. Substitution of this region of the NK1R into the NK3R resulted in the creation of a binding site for CP-96,345 in the NK3R (50). This region of the receptor is near the locations of His197 and His265, which are both conserved among the three neurokinin receptor subtypes. Since CP-96,345 is more than 10,000-fold selective for the NK1 receptor subtype, these data suggest the additional involvement of nearby divergent residues in the high-affinity binding of CP-96,345 to the NK1R, either through direct interactions or through a conformational effect of the chimeric receptor. The analogous regions (helices 5–7) of the NK2R and the AT1 angiotensin receptor have also been implicated by chimeric receptor studies in the high-affinity binding of small-molecule antagonists (50; T Schwartz, personal communication).

Based on the available data on antagonist binding and the effects of amino-acid substitutions on binding affinity, a model of a complex between CP-96,345 and the NK1R has been developed. As shown in Figure 3*B*, CP-96,345 can be docked into the transmembrane region of the molecular model of the NK1R, where it interacts with Gln165 and His197, with His265 located nearby. Other antagonists (e.g. RP67580) utilize additional residues (such as Tyr287) for their binding. The general features of this model are applicable to the binding of other SP antagonists having a different core structure, and are consistent with chimeric receptor studies suggesting that NK1 and NK2 antagonists bind to similar regions of the NK1R and NK2R, respectively (50).

The data described above are consistent with the β_2AR studies indicating that the agonist-binding site and the antagonist-binding site are overlapping but not identical. Among all the key residues identified in the binding site of the NK1R, few appear to be directly involved in the binding of both the peptide agonists and the small-molecule antagonists. On the other hand, kinetic and thermodynamic analyses of ligand-binding interactions, as well as pharmacological analysis of functional efficacy, indicate that these compounds are competitive antagonists of SP. The simplest explanation is that pharmacologically defined competitive relationships arise from a mutual exclusion effect and do not require that the peptides and antagonists make the same intermolecular contacts with the receptor. It is also important to note that, although the set of residues found thus far to be critical for peptide binding (located in helix 2, helix 7, and extracellular segments) is different from the set of residues found to be critical for CP-96,345 binding (located in helices 4, 5, and 6), the proposed arrangement of the seven helices of G protein–coupled receptors is such that it appears unlikely that both a peptide and an antagonist could bind to the transmembrane domain simultaneously. The close proximity of a bound neurokinin peptide to the fifth and sixth transmembrane domains of the NK1R is also supported indirectly by the observation that His198 and His267 of the NK2R (equivalent to His197 and His265 in the NK1R) are required for the binding of neurokinin peptides to that receptor subtype (R Huang, H Yu, C Strader, and T Fong, manuscript in preparation).

BIOPHYSICAL ANALYSIS OF G PROTEIN–COUPLED RECEPTORS

Biophysical studies on G protein–coupled receptors have been limited to date, due to the low abundance of these receptors, the difficulty in solubilizing them from their native membranes in an active form, and difficulties in purification. Because of its relative ease of purification, and the exploitation of the endogenous chromophore-ligand retinal, rhodopsin has been the target of many physical studies. Biophysical studies on rhodopsin and bacteriorhodopsin have supplied most of the physical measurements that have been used to build the structural model of G protein–coupled receptors. The results from a variety of studies, including small-angle X-ray diffraction, neutron diffraction, circular dichroism, polarized infrared spectroscopy, and Raman spectroscopy, were used in building a seven-transmembrane, α-helical model of rhodopsin (19, 42, 92, 96). Perhaps the best "model" for rhodopsin and other G protein–coupled receptors is bacteriorhodopsin. Early electron-diffraction studies defined the orientation of the transmembrane regions of bacteriorhodopsin (96). Subsequent high-resolution electron cryo-microscopy has resulted in a structure with 3.5-Å resolution parallel to the membrane plane (33). These data have been used to refine the orientation of the helices and the position of the chromophore in the protein. Recently a 9-Å resolution electron diffraction structure was obtained for rhodopsin (97). However, there is not yet high-resolution X-ray diffraction data for rhodopsin or bacteriorhodopsin and there are certainly no physical data approaching this resolution for any of the other G protein–coupled receptors. Models for the structure of G protein–coupled receptors have arisen from a combination of structural data derived from bacteriorhodopsin and rhodopsin and mutagenesis data on the receptors themselves. It is important to try to collect physical data on additional G protein–coupled receptors.

We have utilized the fluorescent βAR antagonist carazolol to characterize the physical properties of the ligand-binding domain of the β_2AR. The fluorescence of receptor-bound carazolol was found to be unaltered when exposed to solvent-based collisional quenchers, indicating that the bound antagonist is not exposed to the solvent. Sodium nitrite, which would quench carazolol fluorescence through space (or protein) by energy transfer over a distance of up to 11 Å, also showed no effect on the fluorescence of bound carazolol, indicating that the ligand-binding site for the βAR is buried deep (at least 11 Å) beneath the surface of the receptor (98). Carazolol also displayed some modest environment sensitivity, and has indicated that the binding site is hydrophobic and constrained. Thus, the fluorescence experiments provide direct physical evidence that the binding domain of the βAR is located in the transmembrane core of the protein. These results are consistent with mutagenesis data that has demonstrated that residues predicted to be buried in the transmembrane domain (Asp113, Ser204, Ser207, and Phe 290) are critical for ligand binding to the βAR. Similar experiments on rhodopsin utilizing diffusion-enhanced fluorescence energy transfer have indicated that 11-*cis* retinal is buried 22 Å from the intradiscal surface of the membrane (43).

Thus, it appears that G protein–coupled receptors that bind biogenic amines and retinal are similar in that both proteins accommodate their activating ligands in a pocket located close to the center of the lipid bilayer, formed by the seven transmembrane helices. However, many G protein–coupled receptors, such as the NK1R, bind not small ligands but peptides or small proteins. This presents a conceptual challenge, as any model encompassing the binding of larger ligands must begin to depart from the framework supplied by rhodopsin and bacteriorhodopsin. Indeed, site-directed mutagenesis experiments have indicated that the peptide agonist-binding domain for the SP receptor is much more diffuse than that of biogenic amine receptors. Fluorescently labeled peptides may be useful in exploring subdomains of the ligand-binding sites of this class of receptors. Furthermore, it will be interesting to place fluorescent probes near proposed ligand-binding domains or in regions that are believed to be important for receptor–G protein coupling.

Recent work utilizing several fluorescent derivatives of the formyl peptide (fMLP) receptor has indicated that the receptor contains at least two microenvironments. A pentapeptide analog of fMLP, labeled at the C-terminal Lys with fluorescein, bound to the receptor with the fluorescein in a polar environment, while the labeled tetrapeptide analog placed the fluorescein in a hydrophobic or aromatic microenvironment (99). This type of approach, in combination with site-directed mutagenesis, will contribute to defining the expanded binding site that must be present to accommodate the peptide.

Future biophysical studies of the family of G protein–coupled receptors will not be limited to static determinations of structure, but will also probe the conformational changes associated with ligand binding and receptor activation. A knowledge of the basis of receptor action mandates an understanding of the conformational changes that allow the binding of a ligand to the extracellular and transmembrane domains of the receptor to affect protein-protein interactions on the intracellular surface. In addition to using high-resolution "static" methods, such as X-ray diffraction, it will be necessary to utilize techniques such as fluorescence and nuclear magnetic resonance (NMR) that can report more dynamic processes.

Fluorescein-labeled formyl peptides have also been used for kinetic analysis of ligand interactions (100). Rapid kinetic measurements were used to monitor receptor–G protein interactions, and a rate constant of ≥ 5 sec^{-1} was determined for the G protein–mediated conformational changes in the receptor induced after nucleotide binding. Since this conformational change is much faster than the rate of GDP release, these data would imply that the high-affinity state of the receptor must be coupled to a G protein containing an empty nucleotide-binding site (101). We have recently prepared Lys3-fluorescein-labeled SP and used this probe to characterize the kinetics of the interaction between SP and the NK1R. The addition of guanine nucleotide induced a rapid conversion of the receptor from the high- to the low-affinity state, consistent with an empty guanine nucleotide–binding site on the G protein (M Tota and C Strader, manuscript in preparation). By using fluorescently labeled agonists in combination with fluorescent guanine nucleotide analogs, it will be possible to examine the rapid conformational changes that occur during the agonist activation cycle.

The progress toward the application of physical techniques to the study of G protein–coupled receptors will be enhanced by the overexpression of these receptors using systems such as baculovirus-infected Sf9 insect cells. Molecular biology should help speed this progress, not only by overexpression of the proteins and by allowing the engineering of the receptors such that they can be more easily purified, but also by facilitating the placement of physical probes on the receptors. However, even with the availability of large amounts of pure protein, the characterization of G protein–coupled receptors may be slow, as there have been only a few examples of high-resolution structural data obtained from membrane-bound proteins. Therefore, genetic and biophysical analyses, together with molecular modeling, will continue to be important in understanding the mechanism of action of this class of receptors.

THE ACTIVATION OF G PROTEINS BY RECEPTORS

The seven-transmembrane-domain receptors all rely on the large family of G proteins to mediate the intracellular flow of information to the effector molecules. The original paradigm that a receptor activates a single effector pathway within a particular cell has given way to the current concept of great promiscuity among individual receptors, G proteins, and effectors. It is now clear that a single receptor can activate several different pathways in a given cell, although the predominant pathway may vary from one cell type to another. This observation is best exemplified by the α_2 receptor, which can couple to four different G proteins in the same cell (102–106). A second tier of heterogeneity is exemplified by the βAR, which stimulates two second-messenger pathways (adenylyl cyclase and Ca^{2+} channels), both through activation of $G_{s\alpha}$ (107). Much previous confusion concerning second-messenger pathway activation by a given receptor has been reconciled by these observations.

Much of the research into the interaction between receptors and G proteins has centered on localizing the sites of contact, the sequences required for activation, sequence elements controlling specificity, and elements involved in desensitization and uncoupling. Several reviews have covered this subject in detail for both the receptors (22, 108) and the G proteins (1), and only the highlights are presented here. As illustrated in Figure 1, most of the receptors in this class are buried in the membrane, exposing only a small surface to the cytoplasm. This surface is composed predominantly of the ends of the transmembrane helices and the connecting loops. It is now clear from several lines of evidence that much of this surface is involved in the interaction with the G proteins. In particular, mutational studies have implicated residues at the cytoplasmic ends of all helices and the portions of the loops closest to the membrane as critical for overall interaction and specificity determination.

Initial studies with the βAR demonstrated that removal of residues from either end of the third intracellular domain uncoupled the receptor from G_s (45). In addition, mutations within the second intracellular domain and at the proximal end of the C-terminal tail reduced the efficiency of coupling (45, 46, 109). Subsequently, point mutations and chimeric substitutions within these same regions continued to point to their involvement in direct interaction with the G protein (46, 110–116). So far, this theme has been demonstrated for the biogenic amine receptors (adrenergic, muscarinic, serotonergic), peptide receptors (angiotensin) (117), and rhodopsin (42).

Other lines of evidence point to the importance of the intracellular loops in activating the G proteins. Several investigators have used synthetic peptides that correspond to sequences within various receptors to block the interaction with the G protein or to stimulate directly different G proteins (118–120). One interesting group of peptides that has been studied extensively is the mastoparans. These peptides directly activate members of the G_i family (118). Biochemical and biophysical studies of these peptides have suggested that the amphiphilic nature of the peptides may be essential to their action (121). In addition, modeling studies on the βAR have suggested that the regions of the third cytoplasmic loop that are important for G protein activation are amphiphilic helices (66). Recent mutational studies on the βAR suggest that the hydrophobic residues in these regions, which may determine the secondary structure, contribute most significantly to the ability of the receptor to activate the G protein (122). Additional mutational studies on the muscarinic receptor suggest that the insertion and substitution of residues that would affect the amphipathic nature of the region also affect the ability of the receptor to couple with its G protein (123). These results indicate that while specific residues and sequences are important for mediating the interaction, the secondary structure of the region is also of great importance.

Although there has been much speculation as to how the receptor transmits its signal to the G protein, the exact mechanisms are still unknown (1). Recent data have suggested that the receptor exists in several dynamic states, as reflected by changes in agonist affinity (1, 101, 107, 124). When agonists bind, the receptor shifts to a state with higher affinity for agonists and induces a conformational change in the G protein, triggering its activation. Once this activation occurs, the G protein is released and the receptor reverts to a state with low affinity for agonists. Thus, not only does the receptor send information to the G protein, but the G protein feeds back information to the receptor, inducing a conformational switch. This feedback response, or "guanine nucleotide effect," is very sensitive and is often a better indicator of receptor–G protein coupling than is stimulation of the effector. The molecular basis for the high-affinity state has recently been explored by mutational studies of the adrenergic receptors. Several years ago, point mutations in the third intracellular loop of the βAR were reported that increased the affinity of agonists but not antagonists for the uncoupled state of the receptor (109). It was suggested at the time that this effect might be due to a relaxed ability of the mutant receptor to enter the activated state. More recently, two sets of β_2AR and α_2 receptor mutants having changes in the third intracellular loops have been identified that exhibit increased affinity toward agonists in the coupled and uncoupled states, but not toward antagonists (124, 125). In addition, these mutations demonstrate an increase in the basal (ligand-independent) activation of the G protein and a supersensitivity to agonists, suggesting that these mutant receptors are constitutively activated. In all of these cases, the data imply that the native receptor is conformationally constrained in the unliganded state and that agonists induce a change that results in a higher affinity for both the ligand and for the G protein. The nature of these conformational changes remains to be determined.

Taken together, these results support a model in which the intracellular ends of the transmembrane domains and the portions of the intracellular loops connecting them are held together in an inactive conformation in the resting state of the receptor. Upon binding agonist, a conformational change occurs in the receptor that "opens" the intracellular surface, allowing the G protein to interact with that portion of the receptor, most likely involving the amino-terminal end of the third intracellular loop, the second intracellular loop, and the C terminus. While the C-terminal portion of the third intracellular loop is important for G protein interaction, it may be more critical for maintaining the resting state of the receptor (as suggested from the data on constitutively activated receptors). The observation from molecular modeling that the α-helical regions would be predicted to continue from the transmembrane domains into the loops is consistent with such a mechanism. Examination of these models suggests that the binding of agonists may, in a well-defined and predictable way, affect the manner in which the various components of the receptor interact, thereby changing the way the receptor interacts with the G protein. These changes may compose the signal required to elicit the exchange of GDP for GTP, which ultimately puts the second-messenger system into action.

Other components of the system also influence the specificity of the receptor–G protein interaction. While it is clear that βγ subunits of the G protein are critical for regulating the interaction, virtually nothing is known about the sequence elements involved or the exact function of these proteins (1). Likewise, the roles of arrestin and the G protein–coupled receptor kinases in regulating the interaction, beyond a simple blocking function in homologous desensitization, have yet to be determined. The possible existence of a large complex consisting of the receptor, various G proteins, effectors, kinases, and arrestin is only now beginning to be realized. Future investigations will probe the roles of each of these components in receptor activation and regulation.

CONCLUSIONS

Cloning and primary sequence analysis have revealed that the family of G protein–coupled receptors shares a high degree of structural similarity that reflects the common mechanism of action of these pharmacologically very different receptors. The endogenous ligands for these receptors vary widely in structure, as do the effector systems activated by the various receptors. However, the modeling and mutagenesis data suggest that the ligand-binding and signal-transduction domains of G protein–coupled receptors are highly homologous, even for receptors whose ligands do not appear to be related. In this light, it is instructive to compare the ligand-binding domains of the βAR and NK1R, receptors whose endogenous ligands are completely unrelated and whose inherent structural similarity was not appreciated before the molecular biological characterization of their respective genes. The βAR and NK1R share only 17% primary sequence identity (31% within the transmembrane domain). However, an examination of the binding sites for small molecules within these receptors reveals striking similarities, as shown in Figures 3 *A* and *B*. His265, which is critical for anchoring the benzyl aromatic substituent of the quinuclidine antagonists to the NK1R, is at a location in helix 6 analogous to that of Phe290, which anchors the aromatic catechol ring of catecholamine agonists to the βAR. His197, which binds the benzhydryl substituent of the antagonists to the NK1 receptor, is located in helix 5 one helical turn above the position of Ser204 and Ser207, which bind the catechol hydroxyl groups of catecholamines to the βAR. These similarities suggest the existence of a general small-molecule-binding site in this region of G protein–coupled receptors, perhaps a residual biogenic amine or retinal site, which can potentially be exploited for the design of small-molecule antagonists for this class of receptors. Of interest along these lines is the recent discovery of small molecules, based on a derivatized glucose core, that bind to both peptide and nonpeptide G protein–coupled receptors (126).

The realization of the breadth of the superfamily of G protein–coupled receptors, especially the discovery of a multitude of "new" receptor subtypes that are homologous but distinct from those previously characterized pharmacologically, has opened possibilities for therapeutic specificity by permitting the targeting of specific receptor subtypes in tissues of interest. The integration of genetic analysis with molecular modeling allows the initial mapping of the ligand-binding sites of these receptors: biophysical analysis will ultimately provide higher-resolution structural data. The combination of these approaches with the synthesis and modeling of small molecules will ultimately lead to the design of more potent and specific medicines that act by stimulating or inhibiting members of the G protein–coupled receptor superfamily.

Literature Cited

1. Conklin BR, Bourne HR. 1993. *Cell* 73:631–41
2. Peralta EG, Ashkenazi A, Winslow JW, Smith DH, Ramachandran J, Capon DJ. 1987. *EMBO J.* 6:3923–29
3. Dohlman HG, Thorner J, Caron MG, Lefkowitz RJ. 1991. *Annu. Rev. Biochem.* 60:653–88
4. Humphrey PPA, Hartig P, Hoyer D. 1993. *Trends Pharmacol. Sci.* 14:233–36
5. Rands E, Candelore MR, Cheung AH, Hill WS, Strader CD, Dixon RAF. 1990. *J. Biol. Chem.* 265:10759–64
6. Kage R, Hershey AA, Krause JE, Boyd ND, Leeman SE. 1991. *Soc. Neurosci. Abstr.* 17:805
7. Fong TM, Yu H, Huang RRC, Strader CD. 1992. *Biochemistry* 31:1806–11
8. Liu XB, Davis D, Segaloff DL. 1993. *J. Biol. Chem.* 268:1513–16
9. Dixon RAF, Sigal IS, Candelore MR, Register RB, Scattergood W, et al. 1987. *EMBO J.* 11:3269–75
10. Dohlman HG, Caron MG, DeBlasi A, Frielle T, Lefkowitz RJ. 1990. *Biochemistry* 29:2335–42
11. Karnik SS, Sakmar TP, Chen HB, Khorana HG. 1988. *Proc. Natl. Acad. Sci. USA* 85:8459–63
12. O'Dowd BF, Hnatowich M, Caron MG, Lefkowitz RJ, Bouvier, M. 1989. *J. Biol. Chem.* 264:7564–69
13. Kennedy ME, Limbird LE. 1993. *J. Biol. Chem.* 268:8003–11
14. Papac DI, Thornburg KR, Bullesbach EE, Crouch RK, Knapp DR. 1992. *J. Biol. Chem.* 267:16889–94
15. Burbach JPH, Meijer OC. 1992. *Eur. J. Pharmacol.-Mol. Pharmacol.* 227:1–18
16. Chou K-C, Carlacci L, Maggiora GM, Parodi LA, Schulz MW. 1992. *Protein Sci.* 1:810–27
17. Cronet P, Sander C, Vriend G. 1993. *Protein Eng.* 6:59–64
18. Donnelly D, Johnson MS, Blundell TL, Saunders J. 1989. *FEBS Lett.* 251:109–16
19. Findlay J, Eliopoulos E. 1990. *Trends Pharmacol. Sci.* 11:492–99
20. Hibert MF, Trumpp-Kallmeyer S, Bruinvels A, Hoflack J. 1991. *Mol. Pharmacol.* 40:8–15
21. Ijzerman AP, Van Galen JM, Jacobson

KA. 1992. *Drug Design Discov.* 9:49–67
22. Kobilka B. 1992. *Annu. Rev. Neurosci.* 15:87–114
23. Lewell XQ. 1992. *Drug Design Discov.* 9:29–48
24. Maloney-Huss K, Lybrand TP. 1992. *J. Mol. Biol.* 225:859–71
25. Nodvall G, Hacksell U. 1993. *J. Med. Chem.* 36:967–76
26. Pardo L, Ballesteros JA, Osman R, Weinstein H. 1992. *Proc. Natl. Acad. Sci. USA* 89:4009–12
27. Sankararamakrishnan R, Vishveshwara S. 1992. *J. Biomol. Struct. Dyn.* 9: 1073–95
28. Sankararamakrishnan R, Vishveshwara S. 1993. *Proteins: Struct. Funct. Genet.* 15:26–41
29. Trumpp-Kallmeyer S, Hoflack J, Bruinvels A, Hibert M. 1992. *J. Med. Chem.* 35:3448–62
30. Yamamoto Y, Kamiya K, Terao S. 1993. *J. Med. Chem.* 36:820–25
31. Zhang D, Weinstein H. 1993. *J. Med. Chem.* 36:934–38
32. Findlay JBC, Pappin DJC. 1986. *Biochem. J.* 238:625–42
33. Henderson R, Baldwin JM, Ceska TA, Zemlin F, Beckmann E, Downing KH. 1990. *J. Mol. Biol.* 213:899–929
34. Baldwin JM. 1993. *EMBO J.* 12:1693–97
35. Attwood TK, Eliopoulos EE, Findlay JBC. 1991. *Gene* 98:153–59
36. Overington J, Donnelly D, Johnson MS, Sali A, Blundell TL. 1992. *Protein Sci.* 1:216–26
37. Hilbert MF, Trumpp-Kallmeyer S, Hoflack J, Bruinvels A. 1993. *Trends Pharmacol. Sci.* 14:7–12
38. IQUANTA 3.3. 1992. Molecular Simulations Inc., Waltham, MA
39. Needleman SB, Wunch CD. 1970. *J. Mol. Biol.* 48:443–53
40. Bernstein FC, Koetzle TF, Williams GJB, Meyer EF, Brice MD, et al. 1977. *J. Mol. Biol.* 112:535–42
41. Brooks BB, Bruccoleri RE, Olafson BD, States DJ, Swaminathan S, Karplus M. 1983. *J. Comput. Chem.* 4:187–217
42. Khorana HG. 1992. *J. Biol. Chem.* 267:1–4
43. Thomas DD, Stryer L. 1982. *Mol. Biol.* 154:145–57
44. Dixon RAF, Kobilka BK, Strader DJ, Benovic JL, Dohlman HG, et al. 1986. *Nature* 321:75–79
45. Dixon RAF, Sigal IS, Rands E, Register RB, Candelore MR, et al. 1987. *Nature* 326:73–77
46. Kobilka BK, Kobilka TS, Daniel K, Regan JW, Caron MG, Lefkowitz RJ. 1988. *Science* 240:1310–16
47. Frielle T, Daniel KW, Caron MG, Lefkowitz RJ. 1988. *Proc. Natl. Acad. Sci. USA* 85:9494–98
48. Dixon RAF, Hill WS, Candelore MR, Rands E, Diehl RE, et al. 1989. *Proteins* 6:267–74
49. Gether U, Johanson TE, Snider RM, Lowe JA, Nakanishi S, et al. 1993. *Nature* 362:345–48
50. Gether U, Yokota Y, Edmonds-Alt X, Breliere JC, Lowe JA, et al. 1993. *Proc. Natl. Acad. Sci. USA* 90:6194–98
51. Fong TM, Yu H, Strader CD. 1992. *J. Biol. Chem.* 267:25668–71
52. Sachais BS, Snider RM, Lowe JA III, Krause JE. 1993. *J. Biol. Chem.* 268: 2319–23
53. Main BG, Tucker H. 1985. *Prog. Med. Chem.* 22:122–64
54. Strader CD, Sigal IS, Register RB, Candelore MR, Rands E, Dixon RAF. 1987. *Proc. Natl. Acad. Sci. USA* 84:4384–88
55. Strader CD, Sigal IS, Candelore MR, Rands E, Hill WS, Dixon RAF. 1988. *J. Biol. Chem.* 263:10267–71
56. Strader CD, Gaffney T, Sugg EE, Candelore MR, Keys R, et al. 1991. *J. Biol. Chem.* 266:5–8
57. Wang CD, Buck MA, Fraser CM. 1991. *Mol. Pharmacol.* 40:168–79
58. Fraser CM, Wang CD, Robinson DA, Gocayne JD, Venter JC. 1989. *Mol. Pharmacol.* 36:840–47
59. Gantz I, Del Valle J, Wang L, Tashiro T, Munzert G, et al. 1992. *J. Biol. Chem.* 267:20840–43
60. Zhu G, Wu L-S, Mauzy C, Egloff AM, Mirzadegan T, Chung F-ZJ. 1992. *Cell. Biochem.* 50:159–64
61. Horstman DA, Brandon S, Wilson AL, Guyer CA, Cragoe EJ, Limbird LE. 1990. *J. Biol. Chem.* 265:21590–95
62. Ji I, Ji TH. 1991. *J. Biol. Chem.* 266:14953–57
63. Bihoreau C, Monnot C, Davies E, Teutsch B, Bernstein KE, et al. 1993. *Proc. Natl. Acad. Sci. USA* 90:5133–37
64. Wang C, Gallaher TK, Shih JC. 1993. *Mol. Pharmacol.* 43:931–40
65. Strader CD, Candelore MR, Hill WS, Sigal IS, Dixon RAF. 1989. *J. Biol. Chem.* 264:13572–78
66. Dixon RAF, Sigal IS, Strader CD. 1988. *Cold Spring Harbor Symp. Quant. Biol.* 53:487–97
67. Yamano Y, Ohyama K, Chaki S, Guo DF, Inagami T. 1992. *Biol. Biophys. Res. Commun.* 187:1426–31
68. Wess J, Gdula D, Brann MR. 1991. *EMBO J.* 10:3729–34

69. Suryanarayana S, Daunt DA, VonZastrow M, Kobilka BK. 1991. *J. Biol. Chem.* 266:15488–92
70. Guan XM, Peroutka SJ, Kobilka BK. 1992. *Mol. Pharmacol.* 41:695–99
71. Strader CD, Candelore MR, Hill WS, Dixon RAF, Sigal IS. 1989. *J. Biol. Chem.* 264:16470–77
72. Regoli D, Drapeau G, Dion S, D'Orleans-Juste P. 1989. *Pharmacology* 38:1–15
73. Fong TM, Huang RRC, Strader CD. 1992. *J. Biol. Chem.* 267:25664–67
74. Nagayama Y, Wadsworth HL, Chazenbalk GD, Russo D, Seto P, Rapport B. 1991. *Proc. Natl. Acad. Sci. USA* 88:902–5
75. Wadsworth HL, Chazenbalk GD, Nagayama Y, Russo D, Rapport B. 1990. *Science* 249:1423–25
76. Tsai-Morris CH, Buczko E, Wang W, Dufau ML. 1990. *J. Biol. Chem.* 265:19385–88
77. Perez HD, Holmes R, Vilander LR, Adams RR, Manzana W, et al. 1993. *J. Biol. Chem.* 268:2292–95
78. LaRosa GJ, Thomas KM, Kaufmann ME, Mark R, White M, et al. 1992. *J. Biol. Chem.* 267:25402–6
79. Yokota Y, Akazawa C, Ohkubo H, Nakanishi S. 1992. *EMBO J.* 11:3585–91
80. Levian-Teilelbaum D, Kolodny N, Chorev M, Selinger Z, Gilon C. 1989. *Biopolymers* 28:51–64
81. Savian G, Temussi PA, Motta A, Maggi CA, Rovero P. 1991. *Biochemistry* 30:10175–81
82. Cascieri MA, Chichi GG, Freidinger RM, Colton CD, Perlow DS, et al. 1986. *Mol. Pharmacol.* 29:34–38
83. Chang RSL, Lotti VJ, Monaghan RL, Birnbaum J, Stapley EO, et al. 1985. *Science* 230:177–79
84. Chiu AT, McCall DE, Price WA, Wong PC, Carini DJ, et al. 1990. *J. Pharmacol. Exp. Ther.* 252:711–18
85. Snider RM, Constantine JW, Lowe JA III, Longo KP, Lebel WS, et al. 1991. *Science* 251:435–37
86. Garret C, Carruette A, Fardin V, Moussaoui S, Peyronel J-F, et al. 1991. *Proc. Natl. Acad. Sci. USA* 88:10208–12
87. Lowe JA III, Drozda SE, Snider RM, Longo KP, Zorn SH, et al. 1992. *J. Med. Chem.* 35:2591–2600
88. Cascieri MA, Ber E, Fong TM, Sadowski S, Bansal A, et al. 1992. *Mol. Pharmacol.* 42:458–63
89. Appell KC, Fragale BJ, Loscig J, Singh S, Tomczuk BE. 1992. *Mol. Pharmacol.* 41:772–78
90. Seward EM, Swain CJ, Merchant K, Owen SN, Sabin V, et al. 1993. *Bioorg. Med. Chem. Lett.* 3:1361–66
91. MacLeod AM, Merchant KJ, Swain CJ, Baker R, Cascieri MA, Sadowski S. 1993. *J. Med. Chem.* In press
92. Pappin DJC, Eliopoulos E, Brett M, Findlay JBC. 1984. *Int. J. Macromol.* 6:73–76
93. Pradier L, Emile L, Habert E, Mercken L, Le Guern L, et al. 1992. *Soc. Neurosci. Abstr.* 18:400
94. Beinborn M, Lee YM, McBribe EW, Quinn S, Kopin AS. 1993. *Nature* 362:348–50
95. Fong TM, Cascieri MA, Yu H, Bansal A, Swain C, Strader CD. 1993. *Nature* 362:350–53
95a. Fong T, Yu H, Cascieri M, Underwood D, Swain C, Strader C. 1993. *J. Biol. Chem.* In press
96. Henderson R, Unwin PNT. 1975. *Nature* 257:28–32
97. Gebhard FX, Schertler CV, Henderson R. 1993. *Nature* 326:770–72
98. Tota M, Strader CD. 1990. *J. Biol. Chem.* 265:16891–97
99. Fay SP, Domalewski MD, Skar LA. 1993. *Biochemistry* 32:1627–31
100. Fay SP, Posner RG, Swann WN, Sklar LA. 1991. *Biochemistry* 30:5066–75
101. Neubig RR, Skar LA. 1993. *Mol. Pharmacol.* 43:734–40
102. Conklin BR, Farfel Z, Lustig KD, Julius D, Bourne HR. 1993. *Nature* 363:274–76
103. Eason MG, Kurose H, Holt BD, Raymond JR, Liggett SB. 1992. *J. Biol. Chem.* 267:15795–801
104. Federman AD, Conklin BR, Schrader KA, Reed RR, Bourne HR. 1992. *Nature* 356:159–61
105. Wong YH, Federman A, Pace AM, Zachary I, Evan T, et al. 1991. *Nature* 351:63–65
106. Wong YH, Conklin BR, Bourne HR. 1992. *Science* 255:339–42
107. Birnbaumer L, Abramowitz J, Brown AM. 1990. *Biochem. Biophys. Acta* 1031:163–224
108. Ostrowski J, Kjelsberg MA, Caron MG, Lefkowitz RJ. 1992. *Annu. Rev. Pharmacol. Toxicol.* 32:267–83
109. Strader CD, Dixon RAF, Cheung AH, Candelore MR, Blake AD, Sigal IS. 1987. *J. Biol. Chem.* 34:16439–43
110. O'Dowd BF, Hnatowich M, Regan JW, Leader WM, Caron MG, Lefkowitz RJ. 1988. *J. Biol. Chem.* 263: 15985–92
111. Wess J, Bonner TL, Dorje F, Brann MR. 1990. *Mol. Pharmacol.* 38:517–23

112. Wong SKF, Parker EM, Ross EM. 1990. *J. Biol. Chem.* 265:6219–24
113. Liggett SB, Caron MG, Lefkowitz RJ, Hnatowich M. 1991. *J. Biol. Chem.* 266:4816–21
114. Cotecchia S, Exum S, Caron MG, Lefkowitz RJ. 1990. *Proc. Natl. Acad. Sci. USA* 87:2896–900
115. Cotecchia S, Ostrowski J, Kjelsberg MA, Caron MG, Lefkowitz RJ. 1992. *J. Biol. Chem.* 267:1633–39
116. Lechleiter J, Hellmiss R, Duerson K, Ennulat D, David N, et al. 1990. *EMBO J.* 9:4381–90
117. Ohyaman K, Yamano Y, Chaki S, Kondo T, Inagami T. 1992. *Biochem. Biophys. Res. Commun.* 189:677–83
118. Higashijima T, Burnier J, Ross EM. 1990. *J. Biol. Chem.* 265:14176–86
119. Cheung AH, Huang RR, Graziano MP, Strader CD. 1991. *FEBS Lett.* 279:277–80
120. Konig B, Arendt A, McDowell JH, Kahlert M, Hargrave PA, Hoffman KP. 1989. *Proc. Natl. Acad. Sci. USA* 86:6878–82
121. Sukumar M, Higashijima T. 1992. *J. Biol. Chem.* 267:21421–24
122. Cheung AH, Huang RR, Strader CD. 1992. *Mol. Pharmacol.* 41:1061–65
123. Duerson K, Carroll R, Clapham D. 1993. *FEBS Lett.* 324:103–8
124. Saman P, Cotecchia S, Costa T, Lefkowitz RJ. 1993. *J. Biol. Chem.* 268:4625–36
125. Kjelsberg MA, Cotecchia S, Ostrowski J, Caron MG, Lefkowitz RJ. 1992. *J. Biol. Chem.* 267:1430–33
126. Hirschmann R, Nicolaou KC, Pietranico S, Salvino J, Leahy EM, et al. 1992. *J. Am. Chem. Soc.* 114:9217–18

Annu. Rev. Biochem. 1994. 63:133–73

THE RETROVIRAL ENZYMES

Richard A. Katz and Anna Marie Skalka
Institute for Cancer Research, Fox Chase Cancer Center, Philadelphia, Pennsylvania 19111

KEY WORDS: retroviral protease, reverse transcriptase, ribonuclease H, retroviral integrase, HIV-1 inhibitors

---

CONTENTS

INTRODUCTION

Retroviruses encode enzymes that are assembled into the viral particle and catalyze essential steps in the infectious cycle. All retroviruses encode at

0066-4154/94/0701-0133$05.00

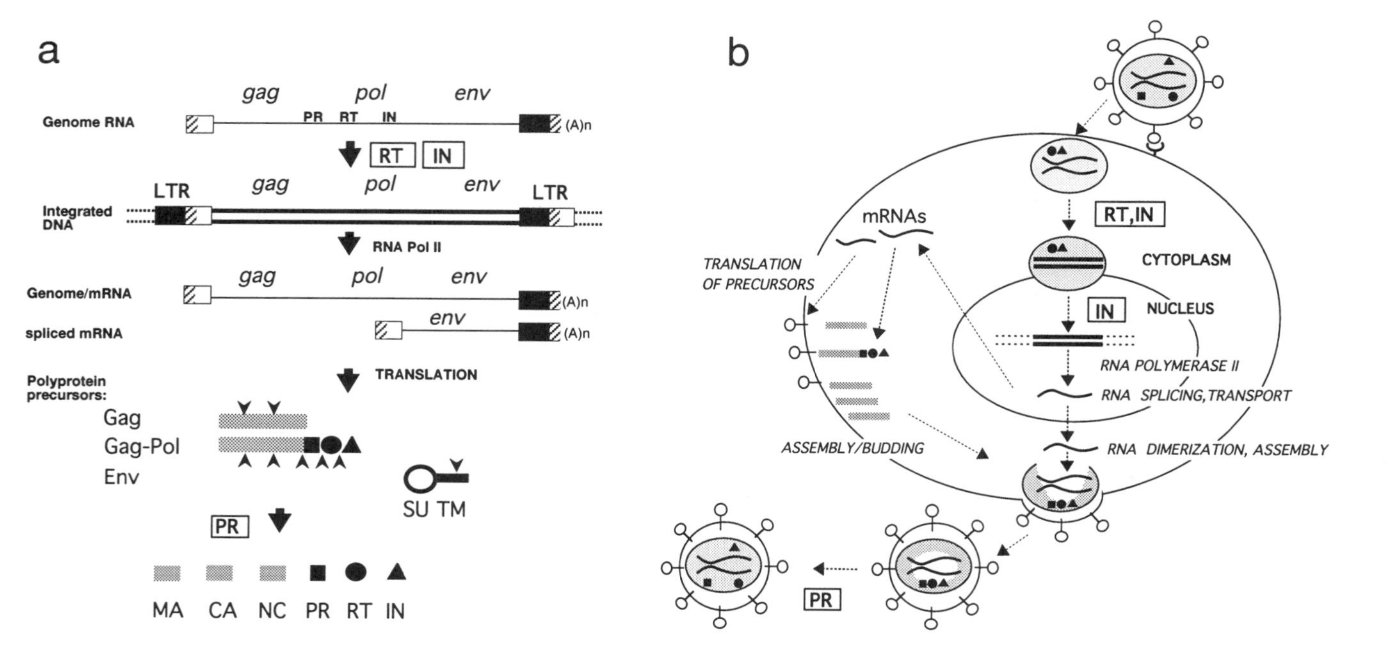

Figure 1 Common steps in the retroviral replication cycle. Depicted is a prototypic life cycle of a C-type retrovirus. Both simple (e.g. ASLV, MLV) and more complex (e.g. HIV) retroviruses encode the three replication genes *gag, pol,* and *env*. The complex retroviruses encode additional regulatory proteins not included here. *a*. Steps in replication. *Top:* Retroviral single-stranded RNA genome (thin line). Shown are the three genes *gag, pol,* and *env,* and the region encoding the three enzymes: PR, RT, and IN. Regions containing *cis*-acting sequences that function in replication are shown as rectangles: U3, black; U5, open; R, hatched. RT and IN copy the viral RNA into DNA (thick lines) and integrate it into the host-cell chromosome (dashed lines), respectively. *Middle:* Host RNA polymerase II transcribes the integrated viral DNA, and the viral mRNAs are translated into polyprotein precursors. *Bottom:* The mature viral structural proteins (MA, matrix; CA, capsid; NC, nucleocapsid; as well as PR, RT, and IN) are produced by PR-mediated proteolytic processing (arrows). The Env precursor, which gives rise to the surface (SU) and transmembrane (TM) proteins of the viral envelope, is also shown. In some retroviruses, PR cleaves near the C terminus of TM. *b*. The retroviral replication cycle. Sites of action of retroviral enzymes are indicated by boxed abbreviations. *Top:* Virion is shown containing the two copies of the viral RNA genome, mature PR, RT, and IN (symbols as shown in part *a*). After receptor binding and penetration, virion RT and IN produce the double-stranded DNA copy and integrate it into the host chromosome. The viral proteins are delivered to the membrane assembly site in the form of polyprotein precursors. *Bottom:* After assembly and budding, an immature, noninfectious virion is produced with a "hollow" morphology. Condensation of the core and infectivity are associated with PR-mediated cleavage of the precursor polypeptides.

least four activities: (*a*) protease (PR), (*b*) the polymerase and (*c*) ribonuclease H (RNase H) activities of reverse transcriptase (RT), and (*d*) integrase (IN). These activities are required for proteolytic maturation of the viral particle (PR), reverse transcription of the viral RNA into DNA (polymerase and RNase H), and integration of the viral DNA into the host-cell chromosome (IN). Although there has been a strong and productive academic interest in these enzymes over the past two decades (see Refs. 1–4 for background on retroviruses), the worldwide AIDS epidemic and the emergence of other human diseases with retroviral etiology have accelerated the pace of research in this area. Basic research with avian sarcoma-leukosis viruses (ASLVs) and murine leukemia viruses (MLVs) has provided a strong foothold; however much of our current knowledge comes from work on the enzymes of human immunodeficiency virus-1 (HIV-1).

This review summarizes recent insights into biochemical and structural aspects of the retroviral enzymes, with a focus on HIV. We have also provided background information, as well some historical perspective. There has been tremendous progress in some areas, especially protein structure. However, it is clear that a full understanding of the roles of these enzymes will ultimately require the reconstitution of higher-order systems, which more accurately reproduce the host-cell environment.

A summary of the retroviral replication cycle is shown in Figure 1. For details, we direct the reader to several excellent reviews (3–5). Other recent reviews have focused on aspects of RT (6, 7), PR (8–10), and IN (7, 11–13a). We also recommend a recent monograph that covers RT in depth (14).

PROTEASE

General Features and Roles

As is typical of many RNA viruses, retroviral translation products are long polyprotein precursors, which are specifically cleaved to release "mature" proteins by a protease that is included as a zymogen in the same precursor. Economical features of this scheme are that several viral proteins may be expressed from the same mRNA in stoichiometrically correct amounts, and that one signal on the precursor can direct several proteins to the assembly site (Figure 1). Since retroviruses rely on their own proteases for maturation, these enzymes are attractive targets for antivirals (see Ref. 15).

Sequences encoding the protease (PR) are always found in the same genetic location, i.e. between those of the core structural components (in the *gag* gene) and the enzymes RT and IN (in the *pol* gene) (Figure 2). Most are translated by frame-shifting, some by readthrough suppression

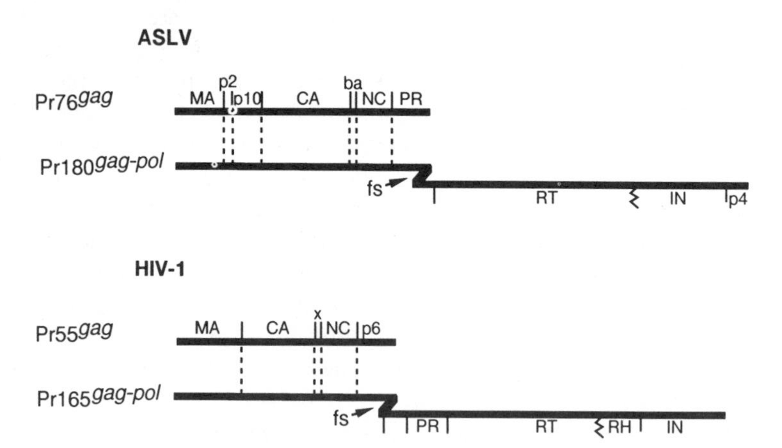

Figure 2 Linear representation of the polyprotein precursors of ASLV and HIV-1. Nomenclature is as in Figure 1; RH = ribonuclease H domain of RT present only in the larger subunit of HIV-1 RT. Proteins of unknown function are identified by their sizes (kDa). The arrow and "fs" show the location of the frameshift that allows synthesis of the Gag-Pol polyproteins. Vertical lines show locations of the main PR processing sites. The jagged vertical line shows the location of the partial processing site that produces the smaller subunits of RT.

mechanisms. Much of our current understanding of the structure and activity of retroviral PRs comes from detailed analysis of bacterially expressed and chemically synthesized enzymes of ASLVs (124 amino acids) and HIV-1 (99 amino acids).

The retroviral PRs belong to the aspartyl protease class, which includes well-characterized cell-derived enzymes such as pepsin, cathepsin D, chymosin, and renin. All contain the conserved active-site triplet, Asp-Thr/Ser-Gly. The cell-derived enzymes are approximately twice the size of the retroviral PRs and contain two interacting domains, each of which contributes one conserved triplet to form the catalytic site. Each retroviral PR monomer contains a single triplet, and the active form of this enzyme is a dimer. It is supposed that the ancestral form of the pepsins was also a dimer and that the single-chain, cellular enzymes arose as a consequence of gene duplication. A K_d of 440 ± 52 nM (at low ionic strength and in the absence of ligands) at 37°C was most recently reported for the HIV-1 PR dimer (16). Like the pepsins, the retroviral PRs are maximally active at acidic pH (4.5–6.5); however, they do show significant activity at a neutral pH of 7 to 7.5 (17), a property that may be explained, at least in part, by amino-acid differences around the active site (17, 18). Like the aspartyl proteases, PRs are also active in high ionic reaction conditions in vitro (e.g. 1–2 M NaCl; 19), which may also reflect the hydrophobic character of the substrate (see below). However, the retroviral PRs are not nearly as efficient as the cell-derived enzymes; k_{cat} values calculated for the best target sites under optimal conditions, ~0.4 s^{-1} for ASLV (20) and ~6.8 s^{-1} (21) for HIV-1 PRs, are very much lower than the highest value, ~400 s^{-1} reported for pepsin (22).

Natural Substrates and Specificity

The exact sites of PR cleavage have been identified for a number of viral polyproteins by comparing the amino-acid sequences of the precursors, deduced from the proviral DNA sequence, with the N- and C-terminal amino-acid sequences of mature viral proteins determined by direct protein-sequencing techniques. Comparison of these cleavage sites shows no unique primary sequence of amino acids within a substrate polyprotein or among polyproteins. However, several amino acids occur repeatedly, and the sequences are quite hydrophobic (see, for example, Figure 3*c*). An extensive cataloging of sequences from a number of retroviral polyproteins (23) has shown the strongest conservation for the amino acid at the N-terminal side of the scissile bond (P1), which is always hydrophobic and unbranched at the β carbon, and more moderate conservation at other positions (designated P4 to P1, and P1′ to P3′ for residues on the N- and C-terminal sides of the target site, respectively, cf Figure 3*c*). The existence of such shared general features (see Ref. 24) is consistent with the earlier observed partial cross-reactivities between PRs from different retroviruses using Gag polyprotein precursors as substrates.

Crystallographic analyses indicate that substrate target sequences must also be located in regions that are flexible or extended in order to be accessible to these enzymes (see below). This also limits target-site selection, and may help to explain an early observation that denatured nonviral proteins are more readily cleaved by PR than are their native forms. The requirement for accessibility may also provide a clue to the mechanism that controls the partial cleavages that occur during processing of RT in ASLV and HIV-1 polyprotein precursors (see section on RT).

Peptide Substrates

The discovery that small synthetic peptides could function as substrates and inhibitors paved the way for detailed study of PR specificity. For both ASLV and HIV-1 PRs, the minimal peptide length required for recognition and cleavage was shown to be seven amino acids (see Figure 3*c*). However, some peptides of six amino acids (P3 to P3′) or less, which are not cleaved, can act as competitive inhibitors (25–27). Although both enzymes cleave peptides representing individual target sequences in their natural substrates with different efficiencies, the HIV-1 PR is approximately an order of

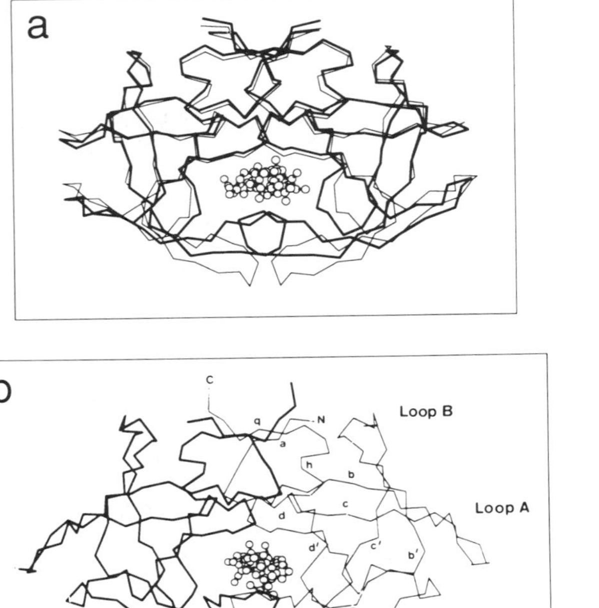

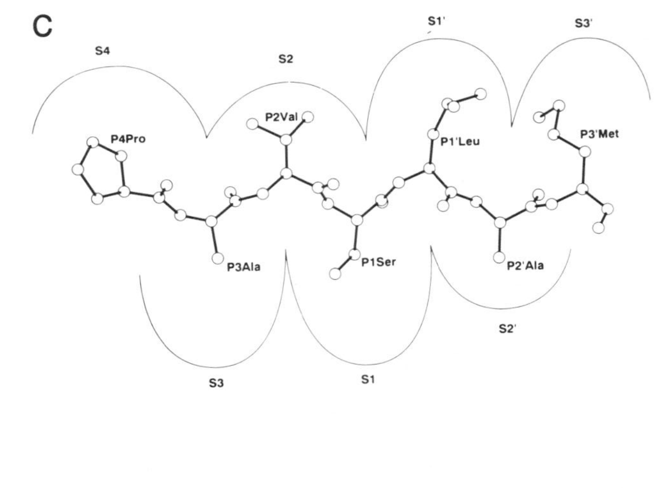

Figure 3 Structures of retroviral PRs complexed with inhibitors. *a*. Comparison of the Cα-backbones of empty (light lines) and inhibitor-complexed (heavy lines) HIV-1 PR. The inhibitor is shown in a ball-and-stick representation in the binding site. The extensive movement that must occur upon substrate binding, especially in the region of the flaps (at the bottom), is inferred by the difference in position of the Cα-backbones. *b*. Structure of the dimer of RSV PR with modeled flaps and substrate. In the RSV crystal structure, the "flaps" are not resolved and are presumed to be disordered. This model was derived by alignment with the HIV-1 structures (shown in part *a*), which allows similar configurations at the flap interfaces if "extra" residues are assigned to small loops in the turns at the opposite end of the flap structures (see Ref. 48). Subunits are distinguished by heavy and light lines; common structural elements are indicated by small letters on the "light" subunit (45a). *c*. Extended configuration of substrate (NCPR border of ASLV Gag) as it is presumed to lie within the active-site cleft with side chains interacting at distinct enzyme subsites (S4→S3'). This figure was kindly provided by I Weber.

magnitude more active than ASLV PR. The high concentration of the ASLV PR in virions presumably compensates for its lower catalytic activity. The best peptide target sites seem to correspond to those that are cleaved first in the normal polyprotein precursors. Thus, this susceptibility difference may reflect a requirement for stepwise processing.

Several groups have designed modifications of the original peptide assays, which used HPLC, TLC, and sometimes radioactivity to detect cleavage products. One of the first of these employed peptide substrates with N-terminal proline, and fluorescamine to detect the production of a new primary N-terminal amine during cleavage (28). Other, improved fluorogenic or chromogenic assays have greatly facilitated kinetic and mutagenesis studies of the PRs (29–36).

Inhibitors

The aspartyl protease inhibitor pepstatin A, although poorly reactive with the PRs, was used early on to demonstrate that the retroviral enzymes belonged to the pepsin family. The pepstatin-PR interaction has also been exploited for affinity purification of the HIV-1 and HIV-2 enzymes (37).

Many renin inhibitors, developed as antihypertensive drugs, mimic transition-state intermediates, and some bind tightly to the active site of HIV-1 PR. In addition, symmetrical peptidyl inhibitors containing two N termini and no C terminus were also shown to have activity against HIV-1 PR. Examples of the structures and range of activities of these peptidomimetic inhibitors have been provided in a recent review by Wlodawer & Erickson (10). Several derivatives of these compounds, which are active in the nM range, are potent inhibitors of HIV replication in tissue culture. A number of the peptidelike active-site inhibitors are in clinical trials for treatment of AIDS.

Computer-assisted drug design has been used recently by scientists at DuPont Merck to develop a novel nonpeptidyl inhibitor of HIV-1 PR, which is reported to be quite potent in tissue-culture assays (38). Crystallographic analyses show that this cyclic urea-containing compound binds to the active site of the enzyme and has the unique property of replacing a water molecule that is ordinarily bound between peptidyl inhibitors and the enzyme flaps (see below) (39).

The PR dimer is stabilized by a four-stranded β sheet interaction between the N and C termini of the two PR subunits, which accounts for >50% of their intermolecular contacts (40, 41). Compounds that interfere with this interaction have been tested as specific PR inhibitors. HIV-1 replication in MT4 tissue-culture cells can be inhibited by acylated octapeptides that represent the N and C termini of PR (42) and also, a tetrapeptide that mimics the C terminus can cause dissociative inhibition of HIV-1 PR in vitro (43). A 13-amino-acid peptide representing both termini of PR exhibits concentration-independent PR inhibition and prevents assembly of active PR dimers in re-folding experiments (44). These peptides are relatively weak inhibitors (K_is in the 50–200 mM range), but they could be valuable as lead compounds and as probes of enzyme folding and structure.

Three-Dimensional Structure

Crystal structures for virion-derived ASLV PR and bacterially expressed HIV-1 PR were reported, almost simultaneously, early in 1989 (44a, 44b). The general features of the two were quite similar, and both confirmed that these enzymes are members of the aspartyl protease family (see also Ref. 45). The recent review by Wlodawer & Erickson (10) provides a detailed description of the structural features of HIV PR as revealed by crystallographic analysis of PRs and PR-inhibitor complexes.

As with the pepsins, the active-site loops of the PR dimers are found at the base of a large cleft, which constitutes the substrate-binding pocket (Figure 3*a* and *b*). A water molecule, bound between the catalytic Asp residues, presumably takes part in the hydrolytic cleavage of the scissile bond (cf Figure 4). Projecting over the cleft are arms or "flaps," one from each monomer, as opposed to the single prominent flap found in the pepsinlike enzymes.

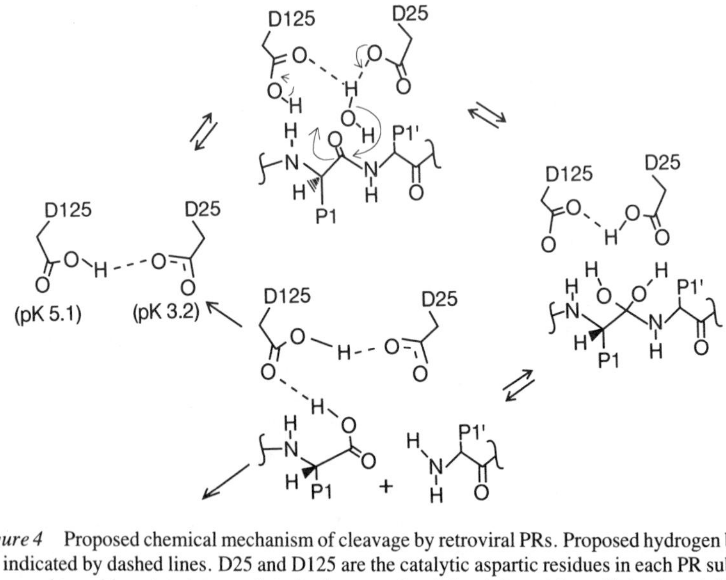

Figure 4 Proposed chemical mechanism of cleavage by retroviral PRs. Proposed hydrogen bonds are indicated by dashed lines. D25 and D125 are the catalytic aspartic residues in each PR subunit. Presumed transition-state intermediate is shown at the right. Adopted from Hyland et al (52).

Crystallographic analyses of HIV-1 PR complexed with inhibitors suggest that substrate peptides are embedded in the enzyme in an extended β-sheet. Each amino-acid side chain lies in a separate pocket, or subsite (S), formed by PR residues on alternating sides of the inhibitor (see Figure 3*c*). All of the NH_2 and CO groups of the inhibitors' backbone, from P3 to P3′, form hydrogen bonds with residues of the enzyme either at the base of the active-site clefts or with the flaps that have formed antiparallel β strands enclosing the inhibitors. Another water molecule participates in the interaction of inhibitor and flaps, and there appears to be some distortion between the P1 and P1′ residues of the bound inhibitor that may contribute to the hydrolysis reaction with cleavable substrates. The uncomplexed enzyme is a symmetric dimer; the complexed enzyme has asymmetry imposed, which can be fixed by comparison of the position of the flaps with respect to the direction (N→C-terminal) of inhibitor binding (46). Although somewhat unexpected, binding asymmetry has also been observed with a symmetrical inhibitor of HIV-1 PR, presumably because optimization of some interactions occurs at the expense of others (47).

One of the most striking features revealed by comparison of the structure of uncomplexed and inhibitor-complexed HIV-1 PR is the "hinge" movement by which the subunits appear to tighten down on the inhibitor in the active-site cleft. There is also a significant conformational change in the flaps, which occupy very different positions in the uncomplexed and complexed enzymes (Figure 3*a*). The substantial conformational restrictions imposed upon enzyme and substrate in the complex have implications for enzyme function. As noted above, the requirement for extended β conformation implies that to be susceptible, target sites must be located in flexible portions of 8–10 residues in substrate proteins. Furthermore, interactions between residues in alternating, but spatially adjacent, positions (i.e. P3 and P1, or P2 and P1′, Figure 3*c*) could affect the ability of potential targets to adopt optimal configurations (21). Finally, after cleavage, each half of a target site will only be bound to one subunit of the enzyme, whose flexibility may allow independent motion that facilitates release of the products.

Recently a model for ASLV PR complexed with substrate, based on the HIV-1 structures, has been produced, in which the limitations on acceptable amino acids at each position in the substrate are described (48; Figure 3*c*). Calculation of the probability of encountering a target heptamer in a random polypeptide that might be accommodated in this succession of subsites is once in approximately 1000 amino acids. Although future work may modify the descriptions of each subsite, this rough estimate is similar to that which may be calculated on the basis of the observed sequence complexities in the natural targets of ASLV, and reinforces the notion that the precise cleavage specificity of the PRs arises from the additivity of interactions at each subsite (49).

Catalytic Mechanism

The structural similarities between the pepsinlike enzymes and the retroviral PRs suggested that the viral and cellular enzymes utilized similar enzymatic mechanisms (50). Through isotope exchange, pH rate studies, and analysis of solvent kinetic isotope effects on HIV-1 PR, Hyland et al (51, 52) have obtained experimental support for the general acid–general base mechanism first proposed for the cell-derived aspartyl proteases (53). Additional details were also elaborated by these investigators, including identification of the probable rate-limiting step: collapse of an enzyme-bound amide hydrate intermediate (Figure 4). The refined chemical-reaction model derived from these analyses agrees well with that deduced from earlier crystallographic and modeling studies of Jaskólski et al (46). The model also helps to explain the relative effectiveness of different inhibitors, and may aid in the design of more potent derivatives in the future.

Mutational Analyses

Site-directed mutagenesis of molecular clones that produce viral enzymes is a powerful tool for investigating the roles of specific amino acids. Not surprisingly, a variety of replacements of the catalytic Asp residue inactivate the retroviral PRs. A mutant in which the catalytic Asp of ASLV PR was changed to Ser is inactive, but appears to be properly folded (28, 54). This mutant could be especially useful for substrate-binding studies. Using saturation mutagenesis, Loeb et al (55) identified three regions where consecutive residues were extremely sensitive to nonconservative substitution. These regions included 11 residues surrounding the catalytic Asp (a.a. 23–33), 6 amino acids in the region of the flap (a.a. 48–53), and 16 amino acids that are part of a hydrophobic domain that interacts with the active-site loop (a.a. 75–90). These studies also identified temperature-sensitive mutants that have been mapped to seven separate locations in the enzyme (56).

Kinetic analyses with an extensive series of peptide substrates have identified residues that might contribute to the unique specificities and the different efficiencies of ASLV and HIV-1 PRs (17, 28, 57). Studies with equine infectious anemia virus (EIAV) (58) have extended these comparisons. Using a similar strategy, Tözsér et al (21) were able to delineate some distinguishing features of substrate-enzyme interactions for HIV-1 and HIV-2 PRs. These authors suggest that the Pro at the P1′ position may facilitate a conformational change required in passage through the transition state.

A series of mutations have been made in residues predicted to form substrate–side-chain–binding pockets in attempts to alter ASLV specificity to resemble that of HIV-1 PR (48), and each has been tested using a set of systematically altered target peptides derived from the ASLV NC-PR cleavage site (Figure 3*c*; CE Cameron et al, personal communication). Following a similar line of reasoning, Konvalinka et al (59) changed five amino acids in ASLV PR (A100L, V104T, A105P, G106V, S107D) to resemble those of HIV-1 PR, and produced an enzyme with apparently higher catalytic activity, altered specificity, and higher pH stability, features resembling HIV-1 PR.

Because of the high mutation rate of retroviruses, mutants that are resistant to peptidelike PR inhibitors can be expected to arise with relatively high frequency, although a recent report suggests that their appearance might be delayed compared to RT inhibitor resistance (60). Recent analyses of resistant mutants produced by passage of HIV-1 in tissue culture show amino-acid changes (e.g. of R8, V31, and V82) (AH Kaplan and R Swanstrom, personal communication; 61, 62, 62a) consistent with side-chain interactions presumed to be important in the active site and substrate-binding subsites (57). The significance of other observed changes (M46, G48, L90, and L97) is not readily apparent and requires further analysis.

The close proximity of the C terminus of one PR subunit to the N terminus of the second (cf Figure 3) makes it possible to join the two by inserting DNA sequences encoding linkers of 0–4 amino acids between tandem copies of the PR open reading frame (63, 63a, 64). Such constructs have been used to introduce flap mutations in one subunit of ASLV PR and to make ASLV-HIV chimeras, both of which were inactive (17). Chimeras containing HIV-1 and HIV-2 subunits are, however, fully active (64a).

One of the first crystal structures for HIV-1 PR was obtained from a chemically synthesized product that contained L-α-amino-n-butyric acid, in place of the two Cys residues, to reduce synthetic difficulties and to ease handling (40). Chemical synthesis of HIV-1 PR entirely with D amino acids has produced a "mirror image" enzyme with chiral specificity for D peptide substrates (65). Schnölzer & Kent (66) have recently reported a method for linking synthetic peptides, each corresponding to approximately half of an HIV-1 monomer (residues 1–68 and 69–99), to reconstitute active enzyme.

Control of Activation

Processing of precursor polyproteins is ordinarily delayed until after virus release. One model for PR activation suggests that aggregation of the precursors at budding sites promotes PR dimerization, which in turn, activates PR. The observation that mutations that prevent aggregation of precursors also prevent processing (67–72) is consistent with this model. In fact, PR activation has recently been used in genetic experiments to screen for mutations that affect precursor polyprotein multimerization (73). However, in D-type retroviruses, processing is also delayed until after budding, despite the fact that core assembly is completed inside of the cell (70). Also, HIV-1 *gag* mutants that assemble intracellularly as well as at the plasma membrane show processing only in particles that are released into the medium (74). These results suggest a mechanism that delays dimerization or PR activation until nascent virion particles are separated from their host cell. Vogt et al (75) have shown that activation of ASLV PR does not depend on changes in Ca^{2+} ion concentration, but other possible mechanisms—such as pH changes, alterations in the concentration of other ions or molecules, or conformational rearrangements—remain to be explored. Genetic and biochemical studies with ASLV suggest that sequences at or near the P2-P10 border in the Gag precursor may participate in control of particle release and PR activation (CE Cameron, JW Wills, and J Leis, personal communication).

Normal control of PR activation appears to fail in situations where PR-containing precursors are highly expressed: under cell-free conditions (76), in heterologous cells or systems (77–80), or even during replication in natural host cells (81). Premature activation has also been observed with polyprotein precursors that contain PR sequences that have been engineered for higher than normal catalytic activity (AM Skalka and J Leis, unpublished observations). In such circumstances the polyproteins are processed intracellularly and, as a consequence, particle assembly is defective. Some of the ASLV- and HIV-1-linked dimers exhibited catalytic activities that were even higher than the unlinked wild-type dimers (64, 82). Polyprotein precursors containing such linked PR dimers are processed prematurely (67, 82). Thus, it seems unlikely that host cells contain inhibitors of PR, which might account for the normal delay in its activation. In addition to blocking normal viral assembly, intracellular activation of retroviral PR is also toxic to the host cell (67, 81–83). The exact mechanism of cytotoxicity is unknown, but many investigators have reported PR activity on cellular components (see Ref. 84 for a compilation).

Processing of precursors within budding virions appears to be initiated with cleavages that release the PR subunits. For ASLV, a mutation at the NC-PR cleavage site in the Gag precursor blocks processing of the polyprotein precursor, and such a precursor cannot efficiently cleave another Gag polyprotein precursor in *trans* (85). Fusion of HIV-1 PR with other viral or nonviral sequences can drastically reduce this enzyme's activity, and recent mutagenesis studies with HIV-1 Gag-Pol precursors suggest that the precursor form of the PR is inefficient (86). Furthermore, cleavages near the PR termini are known to be important for HIV-1 PR subunit release and optimal activity (87). Structural considerations suggest that rather than the initial activating cleavages occurring intramolecularly, precursor PR dimers may have very low activity that releases PR from other precursors in *trans*. A cascade should ensue, leading to release of more subunits and finally to complete precursor processing.

Other Functions?

Based on studies with EIAV, Roberts & Oroszlan (88) have suggested that late PR cleavage of NC proteins, which occurs after an infecting virion enters a new host cell, may be important for early steps in replication. Consistent with this proposal, a synthetic peptide inhibitor of HIV-1 PR (UK-88, 947) was reported to block synthesis of viral DNA if added to cell cultures one hour before infection (89). A similar test, of the effects of the inhibitor Ro31-8959 on early steps in HIV-1 replication, failed to support the EIAV results (61). However, recent, more extensive studies of these two compounds (S Oroszlan, personal communication) show dose-dependent inhibitions of HIV-1 replication in single cycles of infection. Thus, the significance of this early replication effect deserves further study.

Gag precursor cleavage by PR leads to morphological changes within the virion that are required for stable genome RNA dimer formation (89a). Two recent reports suggest that non-enzymatic features of PR proteins can affect both particle assembly (90) and the release and stability of virus particles (91). Such features remain to be explored in future studies, which may provide other clues to the mechanism of virus morphogenesis and maturation.

REVERSE TRANSCRIPTASE

The discovery of reverse transcriptase within the retroviral particle by Temin & Mitzutani and by Baltimore demonstrated that genetic information could flow from RNA to DNA. This seminal observation led to exploitation of this enzyme for biotechnology, and revealed a new pathway for cellular DNA transposition. In this section we focus on the elaborate process by which the multifunctional RT copies information from the retroviral single-stranded RNA genome to form double-stranded DNA during the retroviral replication cycle. This requires polymerization activities, as well as other highly specialized functions. It is not known to what extent, if any, subviral architecture or cellular proteins contribute to the reverse transcription process in vivo.

General Features and Roles

Retroviral RTs exhibit three enymatic activities: an RNA-dependent DNA polymerase (RDDP), Ribonuclease H (RNase H), and a DNA-dependent DNA polymerase (DDDP) (see Refs. 6 and 14 for reviews). These are employed in copying the plus-strand RNA genome to produce a minus

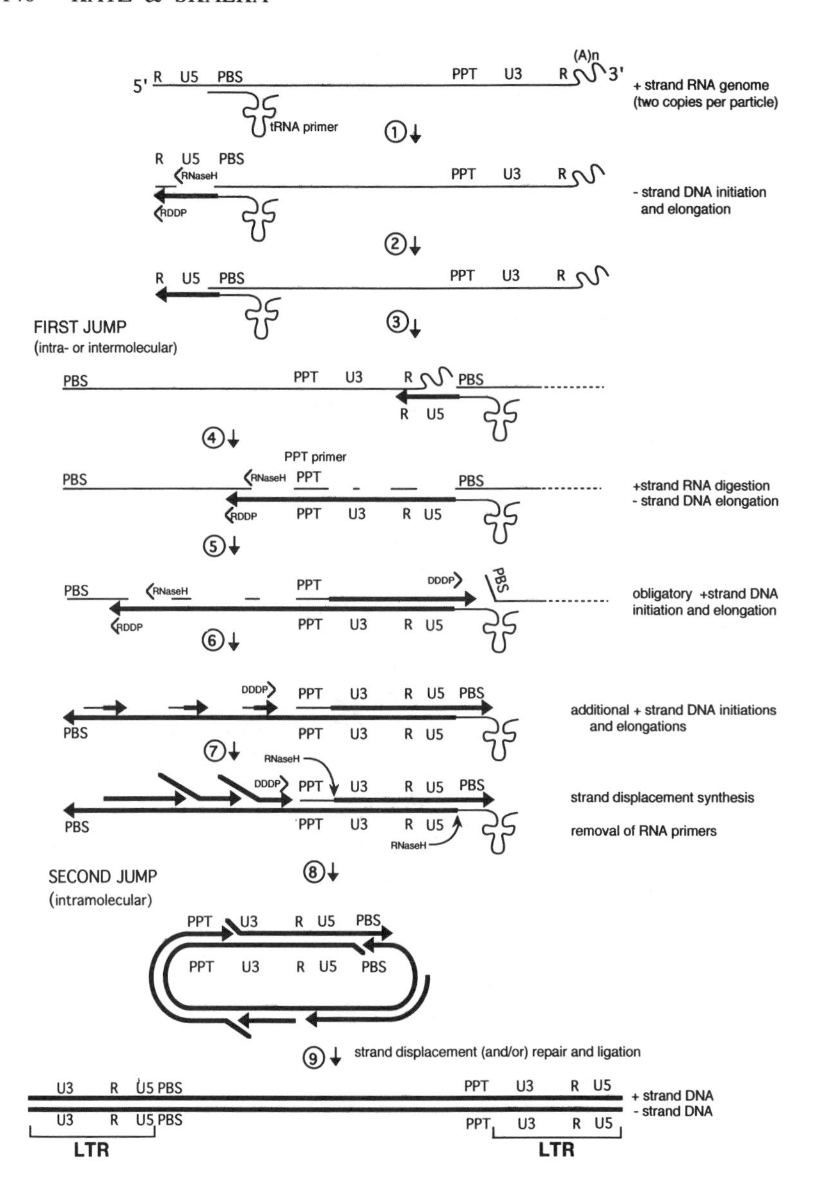

5' R U5 PBS PPT U3 R (A)n 3'
+ strand RNA genome (two copies per particle)
tRNA primer
1
- strand DNA initiation and elongation
RNaseH
RDDP
2
3
FIRST JUMP (intra- or intermolecular)
4
PPT primer
+strand RNA digestion
- strand DNA elongation
5
obligatory +strand DNA initiation and elongation
DDDP
6
additional + strand DNA initiations and elongations
7
strand displacement synthesis
removal of RNA primers
8
SECOND JUMP (intramolecular)
9
strand displacement (and/or) repair and ligation
+ strand DNA
- strand DNA
LTR
LTR

strand of DNA, removal of the RNA template, and synthesis of the plus strand of DNA using the minus-strand DNA as a template, respectively. It has been assumed that the same polymerase active site is used for both RNA- and DNA-dependent synthesis. Figure 5 shows a current model for the reverse transcription process.

The RTs of retroviruses from different vertebrate species display characteristic properties in vitro with respect to metal requirements, template preferences, processivity, and error rates. They also show considerable diversity with respect to subunit composition (Figure 6). The prototype ASLV RT is a heterodimer composed of a β chain of 95 kDa and an α chain of 63 kDa. These two chains have common N termini that contain both the polymerase and RNase H domains. The C-terminal region that is removed during processing to form the α chain corresponds to the IN domain, which likely functions as a separate peptide of 32 kDa. In contrast, the MLV RT is isolated as a monomer of 80 kDa, containing polymerase and RNase H domains, while the IN domain is released as a separate peptide of 46 kDa. However, recent evidence suggests that MLV RT may dimerize on the template RNA (92). The HIV-1 RT is a heterodimer consisting of large (p66) and small (p51) subunits with common N termini. Both subunits contain the polymerase domain; the C-terminal fragment that is removed from p66 by asymmetric processing corresponds to most of the RNase H domain. The IN domain is released from both subunits. Thus, although

←

Figure 5 The reverse transcription process. Two copies of the 7–10 kb RNA genome are present in each retroviral particle. Poly rA tail is shown [(A)n]. Minus-strand DNA synthesis starts near the 5′ end of the plus-strand RNA genome using a specific host tRNA as primer. The tRNA is annealed via at least 18 nt at its 3′ end to a site on the RNA genome denoted the primer-binding site (PBS). *Step 1:* Synthesis proceeds to the 5′ end of the RNA genome through the U5 region, ca. 100 nt, ending at the R region, which is terminally redundant in the RNA, forming the "minus-strand strong stop DNA." *Step 2:* The RNA portion of the RNA-DNA hybrid product is digested by the RNase H activity of RT, thus exposing the single-strand DNA product. *Step 3:* This exposure facilitates hybridization with the R region at the 3′ end of the same, or the second RNA genome, a reaction known as the "first jump." *Step 4:* When minus-strand elongation passes a specific region near the 3′ end of the RNA genome, characterized by a polypurine tract (PPT), a unique plus-strand RNA primer is formed by RT RNase H cleavage at its borders. Plus-strand synthesis then continues back to the U5 region using the minus-strand DNA as a template. *Step 5:* Meanwhile, minus-strand synthesis continues through the genome using the plus-strand RNA as a template, and removing the RNA template in its wake via RNase H activity. *Step 6:* The RNase digestion products formed are presumed to provide additional primers for plus-strand synthesis at a number of internal locations along the minus-strand DNA. *Step 7:* PPT-initiated plus-strand DNA synthesis stops after copying the annealed portion of the tRNA to generate the plus-strand DNA form of the PBS, forming the "plus-strand strong stop" product. The tRNA is then removed by RNase H activity of RT. *Step 8:* This may facilitate annealing to the PBS complement on the minus-strand DNA, providing the complementarity for the "second jump." DNA synthesis then continues. *Step 9:* Strand displacement synthesis by RT to the PBS and PPT ends, and/or repair and ligation of a circular intermediate, produces a linear duplex with long terminal repeats (LTRs).

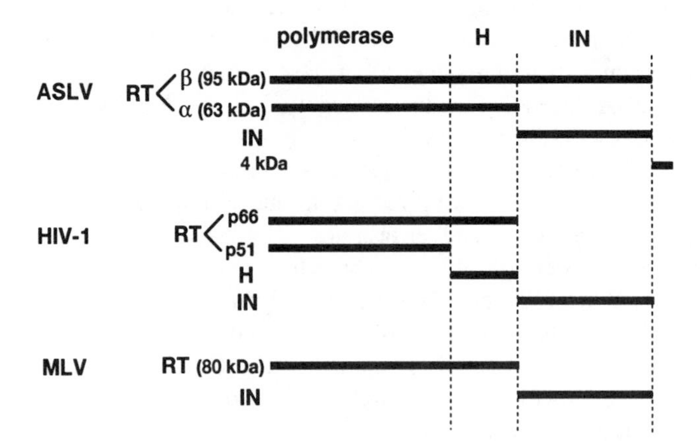

Figure 6 Domain and subunit relationships of reverse transcriptases from different retroviruses.

asymmetric processing is a frequent feature of RTs, the composition of the asymmetric units can differ. It remains to be seen whether there is a unifying structure-function relationship between domains or subunits.

Biochemical Properties of RT

It is estimated that there are 50–100 RT molecules per virion, and it is unclear whether one or more than one enzyme molecule contributes to forming a single DNA copy. The biochemical capabilities of RT have been studied in vitro using purified RT and model template-primers. Kinetic studies with HIV-1 RT reveal an ordered reaction pathway similar to that of other polymerases, with binding of the template-primer as a first step, followed by binding of the dXTP substrate (93). Processivity measurements on HIV-1 RT vary depending on the template; processivity on the best template, poly(rA), is greater than 300 nt, at an elongation rate of 10–15 nucleotides per second (94). This elongation rate is similar to that of other eukaryotic DNA polymerases. Processivity on natural templates and homopolymers other than poly(rA) is low as compared with other replicative polymerases (94). The probability of termination decreases considerably after addition of the first nucleotide, suggesting that initiation and processive synthesis are distinct steps (93). The RNase H domain may also contribute to the processivity of polymerase (92).

The current model for reverse transcription predicts two "jumps" between the ends of templates (see Ref. 95 for a review) (Figure 5). The first and second jumps are from RNA and DNA templates, respectively, and are thought to be essential for replication. The jumps from an RNA template may depend on RNase H activity (discussed below) to expose the end of the product strand DNA such that it can hybridize to the "acceptor" template (96, 97). In vitro studies using model substrates suggest that the jumps may not require dissociation of the RT and primer (94) and may be associated with high processivity (98). Since a prototype DNA polymerase, such as the Klenow fragment of *Escherichia coli* DNA polymerase I, is unable to carry out such jumps (98), some novel aspect of RT structure may be responsible.

Retroviruses, like other RNA viruses, mutate at a much higher rate than cellular genes (see Ref. 99 for a review). It is believed that a high error rate in nucleic acid replication results in a genetically diverse population of viruses, with individual members poised for expansion in response to selective pressures. This can result in the rapid appearance of drug-resistant mutants or antigenic variants that can escape immune surveillance, as in the case of HIV infections. RTs lack editing functions and have been shown to be error prone in vitro, and therefore are presumed to contribute to the high mutation rate in vivo (see Refs. 100 and 101 for reviews). However, retroviral replication is dependent on host RNA polymerase II, which also lacks editing functions.

In vitro, RT errors include misinsertions and rearrangements. Their frequency is dependent on the source of RT (e.g. HIV vs MLV), the template sequence (e.g. RNA vs DNA, homopolymers vs natural templates), local sequence context, and the composition of the mismatch (100, 101). For example, misincorporation by HIV-1 RT can be as high as 1/30 at some positions, while at other positions it can be as low as $1/10^6$. HIV-1 RT appears to be more error prone than MLV RT in vitro. Using MLV, a comparison between the in vivo and in vitro mutation rates on the same target sequence suggests that there may be in vivo cofactors that increase RT fidelity (102).

In vitro studies indicate that several novel properties of RTs may contribute to error rates. First, ASLV and HIV-1 RTs are proficient at extending certain mismatched terminal bases (103, 104). Efficient extension facilitates incorporation of the mismatched base into the DNA strand. In addition to a direct misincorporation mechanism (nontemplated addition), there is experimental support for "dislocation-mediated" substitution, where temporary basepair slipping within homopolymeric runs extrudes a base on the template strand. A "correct" base is then inserted, but mispairing occurs when the template and product strands realign. Single-nucleotide deletions may occur in a similar way. Bebenek et al (105) present evidence that these are major mechanisms with HIV-1 RT in vitro. Lastly, errors may be related to the loose association of RT with the template, a necessity for jumping (106, 107).

A fifth retroviral enzyme, dUTPase, is found in a subset of retroviruses and may function to reduce errors in retroviral DNA synthesis by minimizing dUTP incorporation (108). This enzyme is encoded between RT and IN or within *gag,* and is carried within the virus particle. With EIAV, the gene is dispensable for replication in dividing cells, but is required in nondividing macrophages, where endogenous dUTPase levels may be low (109).

RTs exhibit strand displacement activity during plus-strand synthesis, which may be important for production of a complete plus strand after multiple initiations have occurred; however, strand displacement synthesis is not unique to RTs (Figure 5, see Ref. 110 for a review). One model for retroviral recombination is based on this activity.

Activation of RT polymerase after budding (Figure 1) presumably protects the cell from deleterious effects of uncontrolled RT activity and also prevents the viral RNA from being reverse transcribed before assembly. There are differing reports as to whether RT is completely inactive when embedded in a larger protein, including the Gag-Pol precursor (summarized in Ref. 111). In the ASLV system, virions containing unprocessed Gag-Pol precursor show little RT activity (111). In contrast, some RT activity can be detected in association with the HIV-1 Gag-Pol precursor (112). These results may reflect real differences between the viruses.

Roles of HIV-1 RT Subunits

HIV-1 RT purified from virions is a heterodimer composed of a 66-kDa subunit (p66) and a 51-kDa subunit (p51) (113) (Figure 7*a*). Several in vitro studies have provided evidence that the enzyme functions as such (114, 115). The bacterially expressed p66 subunit alone is active, but studies with the p51-like chains have produced conflicting results (summarized in Ref. 116). Restle et al (117) showed that p66 and p51 subunits are active as homodimers, but are inactive as monomers; thus different homodimer content may account for variations in reported activities of p51 subunits.

The role of each subunit within the heterodimer has been investigated by reconstituting HIV RT in which one of the two subunits has been inactivated by mutation of a catalytic residue (see below) (118). Results showed that mutation of the p51 subunit has little effect, whereas mutation of the p66 subunit inactivates the heterodimer. Thus, the catalytic-site residues present on p51 do not seem to contribute directly to activity. Similar results have been reported by Hostomsky et al (119), who found that p51 alone (likely a homodimer form) has detectable activity, which is negated when p51 is present in a heterodimer with inactive p66. Their interpretation (which is now supported by structural data) was that the conformation of p51 is altered in the heterodimer. Experiments using HIV-1/HIV-2 mixed heterodimers

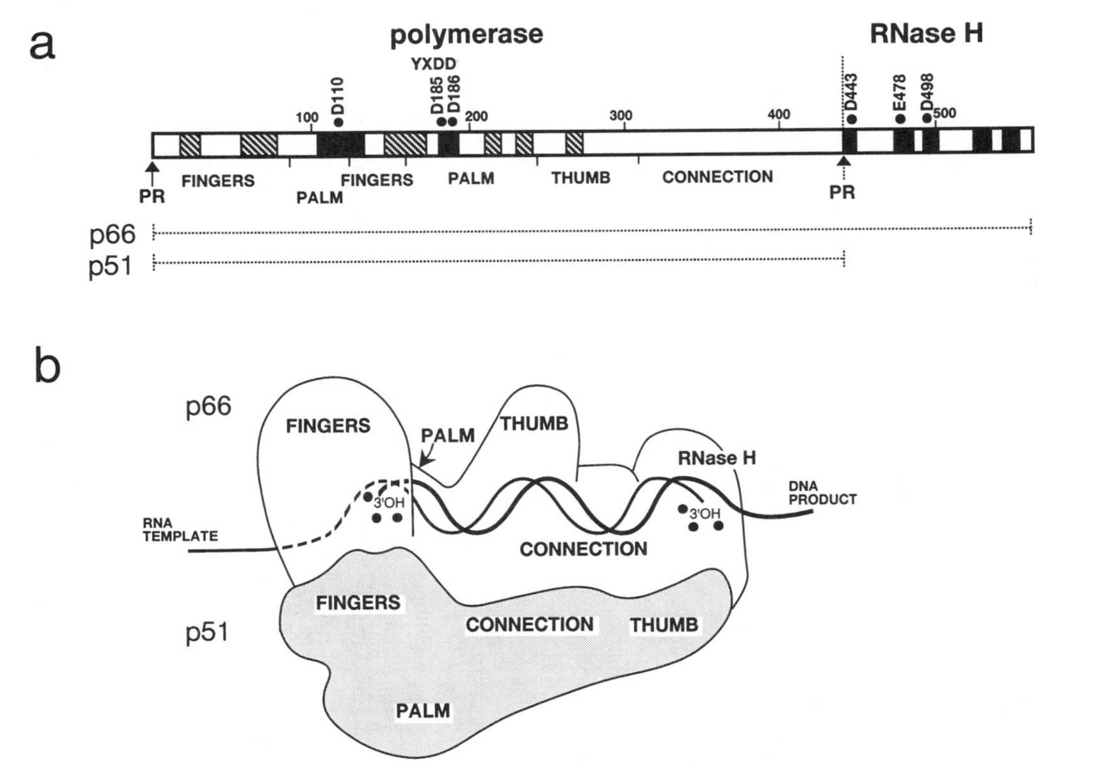

Figure 7 Conserved residues and structure of HIV-1 RT. The most highly conserved residues, which are thought to be components of the two active sites, are indicated by filled circles. *a*. Linear map of RT (adapted from Jacobo-Molina and Arnold, Ref. 129). The sequences included in the two subunits are shown by dashed lines below. Polymerase domain: Hatched areas indicate regions of homology within the RT family. Filled areas indicate regions of homology with all polymerases. RNase H domain: Filled areas indicate regions with homology to *E. coli* RNase H. Dashed arrow indicates partial processing by PR. *b*. Outline of the structure of HIV-1 RT (adapted from Kohlstaedt et al, Ref. 121). Filled circles correspond to active-site residues shown in part *a*. The RNA template–DNA product duplex is shown lying in a cleft. Dashed lines indicate duplex passing behind the fingers. The polymerase active site (left) shows the 3'OH of the primer surrounded by active-site residues (see Jacobo-Molina et al, Ref. 123). The putative RNase H active site (right) shows conserved residues surrounding the 3' end of template strand that is being degraded.

and a specific inhibitor of HIV-1 RT (a "TIBO" derivative, discussed below) also support the conclusion that p66 is the catalytically active subunit (120).

Structure of HIV-1 RT

Two groups have reported X-ray crystal structures for HIV-1 RT (121–123). In the enzyme-inhibitor (Nevirapine) complex described by Kohlstaedt et al (121), the polymerase regions of p66 and p51 are divided into four subdomains denoted "finger," "palm," "thumb," and "connection." Remarkably, these subdomains are arranged differently in each subunit and thus the heterodimer is asymmetric. In p66 the finger, palm, and thumb domains form a cleft, or "hand," presumed to grasp the template-primer, with the polymerase active-site residues positioned in the "palm" domain (Figure 7*b*). The subdomain arrangement of p66 shows limited homology to the polymerase region of the Klenow fragment of *E. coli* DNA polymerase I and to the bacteriophage T7 RNA polymerase (124) (see chapter by Joyce & Steitz in this volume). The Nevirapine inhibitor is bound in a pocket near the base of the thumb and may act by restricting the movement of this domain. Alternatively, it could affect the nearby catalytic residues in the palm. Nevirapine contacts Y181, which is consistent with the observation that mutations of this residue are associated with resistance to this drug (see below). The major contacts between the p66 and p51 subunits occur within the connection domains. The extended thumb of p51 contacts the RNase H domain of p66, and this interaction may be required for RNase H activity (125). In the p51 subunit, the connection domain is rotated such that it occupies the palm, and buries the active-site residues. This is consistent with the biochemical results discussed above. The fact that the p51 subunit displays a distinct structure (and function) from the p66 subunit appears to be a novel example of economical use of viral coding capacity; a single DNA coding sequence produces polypeptides with the same amino-acid sequence that are structurally and functionally distinct (121). It has been proposed that the p51 subunit may bind the tRNA primer as well as the template (121).

The structure determined by Arnold and coworkers (122) contains a DNA template-primer duplex bound in the cleft region. The cleft is flanked on one end by a Hg-UTP-binding site and by the RNase H active site on the other. The arrangement is consistent with models derived from biochemical data, which suggest that the polymerase active site (inferred by the Hg-UTP-binding site) and the RNase H active site act coordinately during minus-strand DNA synthesis (discussed below, see Figure 7*b*). This group has recently reported a higher-resolution (3.0 Å) structure, which shows the putative polymerase "catalytic triad"—D110, D185, D186 (see below)—positioned near to the primer strand 3′OH terminus (123). In this structure, the bound template-primer shows both A- and B-form regions of DNA duplex separated by a bend. Contacts between the template strand and the palm and thumb were also noted.

Conserved Residues and Mutagenesis of Polymerase

Phylogenetic comparisons of protein sequences frequently reveal conserved amino acids that provide essential functions. Deduced protein sequence alignments from a large number of polymerase families, including both RNA and DNA polymerases, have revealed two motifs that are common to all polymerases (126). Both of these ("A" and "C") are composed of conserved aspartic acid residues embedded in a hydrophobic region. The invariant aspartic acid in motif C is included in the "polymerase signature sequence," tyrosine-X-aspartic acid–aspartic acid (YXDD). In retroviral and retrotransposon RTs, the aspartic acid residues in motifs A and C correspond to HIV-1 D110, D185, and D186 (Figure 7). Possible roles for these residues would be to coordinate the required metal ion, participate in the dXTP-binding site, or function in catalysis. Early (127, 128) and more recent alignments (129) of only the retroviral RTs have revealed several blocks of conservation within the N-terminal polymerase region (Figure 7*a*).

Polymerase activities of bacterially expressed enzymes have been measured in partially purified extracts, in "activity gels," or after purification to homogeneity. Initial studies were devised to identify minimal functional domains for polymerase and RNase H. These studies were guided by protein sequence alignments of Johnson et al (127), which predicted the order: N-polymerase–tether–RNase H-C. Linker insertion-deletion analysis, and independent expression of polymerase and RNase H domains, confirmed this prediction for MLV RT (130). With HIV-1 p66, most insertions in the predicted polymerase domain inactivate polymerase function (131, 132). However, some also inactivate RNase H, suggesting either a global effect, or a disruption of specific interactions between the domains. Similarly, some insertions in the RNase H region of p66 had a significant effect on polymerase activity (131, 132). It is generally observed that there is an interdependence of HIV-1 polymerase and RNase H domains (discussed further below).

Single amino-acid replacement studies have confirmed the requirement for the highly conserved polymerase residues. Conservative substitutions of D110 or D185 with glutamic acid resulted in loss of DNA- and RNA-dependent DNA polymerase activities (133, 134), while RNase H (134) and template-primer binding activity (133) were maintained. D186—which is conserved in most RNA-dependent polymerases, and in all RTs—is also essential for polymerase activity (133, 134).

Functional Domains

In addition to the active sites for polymerase and RNase H activities, RTs likely contain regions that bind the primers and template. The determinants involved in selection of the specific host tRNA primer are not fully understood. Different tRNAs are utilized by different retroviruses, and their identities can be deduced from the complementary PBS sequence (Figure 5); for HIV-1 this corresponds to human $tRNA^{Lys3}$. Analysis of MLV deletion mutants suggested a role for RT in assembling the $tRNA^{Pro}$ primer into the particle (135). Selective binding of RT to the correct tRNA in vitro was observed with ASLV RT–$tRNA^{Trp}$, but not with MLV RT–$tRNA^{Pro}$ (6). HIV-1 RT specifically binds $tRNA^{Lys3}$ (136). At the time of viral assembly, RT is embedded in the Gag-Pol precursor (Figure 1), and thus it may be the RT precursor that specifically recognizes the tRNA.

The events leading to initiation of minus-strand DNA synthesis likely include: (*a*) specific incorporation of the host tRNA primer in the viral particle; (*b*) partial unwinding of the tRNA structure to expose 18 or more nt of the ...CCA3′ end of the tRNA; (*c*) unwinding of the viral RNA template structure; and, (*d*) annealing of the tRNA to the PBS and other neighboring regions (137, 138). Although these events are likely mediated by viral proteins, RT and/or possibly NC, precise details are still unknown. In vitro, HIV-1 RT and the $tRNA^{Lys3}$ primer are sufficient for initiation at the PBS (138, 139). This suggests that RT is indeed able to bind and unwind its tRNA primer, although addition of the NC protein may enhance this process (140). Chemical cross-linking studies indicate that the $tRNA^{Lys3}$ forms contacts with both the p66 and p51 subunits of HIV-1 RT (136) and that the heterodimer form is the most active in $tRNA^{Lys3}$ binding (116).

RTs can use a variety of RNA and DNA primers in vitro. It may be that part of the primer recognition site is generic, recognizing features associated with the 3′OH end of RNA or DNA primers, while another part recognizes the specific features of the tRNA primer. Many in vitro studies that address template-primer recognition utilize short oligonucleotides. Although these model substrates differ considerably from the natural template-primer complexes, they have proven useful. Kinetic studies with the HIV-1 enzyme indicate that template-primer recognition involves a high-affinity interaction with the primer portion (141). Also, cross-linking studies with a $d(T)_{15}$ oligodeoxynucleotide (142) identified amino-acid residues 288–307 as a possible primer recognition site in HIV-1 RT. These residues lie in the thumb portion observed in the crystal structure described above (121).

Inhibitors and Drug Resistance

A comprehensive survey of RT inhibitors is beyond the scope of this review and has been presented elsewhere (143, 144). Here, we focus on several studies in which both biochemical and genetic data have been obtained.

As of this writing, the only approved antiviral treatments for AIDS in the United States are compounds that target HIV-1 RT. These include the nucleoside analogs 3′-azido-2′,3′-dideoxythymidine (Zidovudine, AZT) and 2′,3′dideoxyinosine (ddI). These deoxynucleoside analogs are taken up by cells, become converted to the triphosphate forms, and presumably inhibit DNA synthesis by acting as chain terminators.

Several factors are important in considering the potential effectiveness of anti-RT drugs in AIDS treatment: (*a*) the high mutation rate of retroviruses, which leads to drug resistance; (*b*) the molecular basis of resistance; and, (*c*) possible cross-resistance within or between classes of drugs. AZT-resistant viruses appear during the course of treatment of AIDS patients and in tissue culture (see Ref. 144 for a review). Nucleotide sequencing of highly resistant viruses from patients revealed several amino-acid changes in RT (e.g. D67N, K70R, T215F/Y, K219Q, and M41L) that appear in an ordered fashion with a concomitant increase in resistance. Unexpectedly, the RTs purified from these mutant viruses or bacterially expressed counterparts are not resistant to AZTTP in vitro (145), and similar results have been reported for RTs from AZT-resistant feline immunodeficiency virus (146). Thus, the biochemical basis for AZT resistance is obscure. One possible explanation for these observations is that AZT is further metabolized in vivo to its active form and the genetic changes in RT do not produce resistance to the pro-drug (147). Residues associated with AZT, ddI, and ddC resistance lie in the fingers and palm and probably do not play a direct role in recognition of these nucleoside inhibitors, but may affect template interactions (121).

Prasad and coworkers (148 and references therein) have developed a rapid screening assay to select mutations in bacterially expressed HIV-1 RT that confer resistance to nucleotide analogs. Selection for ddGTP resistance revealed a mutation, E89G, that confers cross-resistance to ddTTP, ddCTP, AZTTP as well as a pyrophosphate analog, foscarnet. The metal preference of the mutant enzyme is also affected. The position of the mutation in the HIV-1 RT structure suggests that E89 may also play a role in template interactions (148).

A group of non-nucleoside inhibitors have recently been identified by intensive screening of pharmacological compounds that were developed for other purposes. The two best characterized are the tetrahydroimidazo[4,5,1-*jk*][1,4]-benzodiazepin-2(1*H*)-one and -thione (TIBO) derivatives and the dipyridodiazepinones (e.g. BI-RG-587, Nevirapine) (reviewed in Ref. 143). The anti-HIV-1 activity of the TIBO compounds was detected in a tissue-culture replication assay (149), while Nevirapine was discovered as an HIV-1 RT inhibitor (150). Photoaffinity labeling with Nevirapine is confined largely to the p66 subunit of the heterodimer (151). Formation of the photoadduct could be inhibited by a TIBO derivative, suggesting common or overlapping binding sites (151). Peptide analysis has indicated that the Nevirapine photoadduct is formed by linkage to nonconserved tyrosine residues Y181 and Y188 (152), which flank the highly conserved YXDD motif in HIV-1 RT. In the co-crystal structure, the side chains of Y181 and Y188 are in contact with Nevirapine (121). Consistent with this, HIV-2 RT, which lacks Y181 and Y188, is resistant to Nevirapine. Furthermore, replacement of residues 176–190 of HIV-2 with those from HIV-1 results in an HIV-2 enzyme that is sensitive to Nevirapine as well as a TIBO compound (153). Passage of HIV-1-infected cells in the presence of Nevirapine results in the appearance of resistant viruses containing a Y181C mutation or a K103N, Y181C double mutation (154, 155). Viruses selected for Nevirapine resistance show dual resistance to a TIBO derivative, which is consistent with the biochemical results (155). A third, unrelated drug (a pyronidone) seems to bind to the same site (155). In summary, these results explain the specificity of these compounds for HIV-1, and reveal a common drug-binding site which, when occupied, may affect the flexibility of the thumb region (121).

The RNase H Domain

RNase H enzymes, by definition, degrade the RNA portion of an RNA-DNA hybrid. Retroviral RNases H (see Ref. 155a for a review) are thought to provide two important functions during reverse transcription (Figure 5): (*a*) removal of the RNA template strand to prepare for plus-strand DNA synthesis and both template jumps, (*b*) specific cleavages involved in formation and removal of RNA primers. In vitro, RT RNases H have been shown to have both endo- and 3′ to 5′ exonuclease activities (156–159), which is consistent with these roles. Current research has focused on (*a*) possible coordination of polymerase and RNase H activities, (*b*) how specificity is determined for formation and removal of RNA primers, and (*c*) the possible role of RNase H in jumping between templates and in recombination.

Mutational analyses established MLV RNase H as a separate RT domain. The situation with HIV-1 is more complex, with a greater interdependence of the polymerase and RNase H (discussed above). The RNase H domain is present on both the p66 subunit and a p15 fragment that is released during PR processing (Figure 6). The p15 fragment (or close facsimiles) appears to be structurally unstable, and several studies indicated that it had little or no RNase H activity. The biologically relevant form of RNase H likely resides on the p66 subunit (125) (discussed below), although a role for the p15 fragment cannot be totally ruled out (160).

Activities and Roles of RNase H

The first role for RNase H is to remove all or part of the 5′ end of the viral RNA, which is hybridized to the minus strong stop DNA (Figure 5, *Step 1*). This may facilitate the first jump to the 3′ end of the RNA, via R region complementarity. The minus strand is elongated after the first jump. Tanese et al (130) showed that virus particles containing mutations in RNase H are competent for initiation and DNA synthesis, but not for subsequent elongation of the minus strong stop DNA. The defect is presumed to be in translocation (Figure 5, *Step 4*).

An attractive model for elongation is one in which the template RNA is degraded as the RNA-DNA hybrid product passes through the RNase H domain (161, 156) (Figure 7*b*). RNase H would thereby function in a processive, or polymerase-coupled, mode. Oyama et al (156) and several other groups (162–167, 167a) have studied the coordination of polymerase and RNase H activities using the following rationale. When the 3′ end of the primer is positioned in the polymerase active site, the position of cleavages on the RNA template by RNase H should define the distance between the two active sites. As polymerization proceeds, the template RNA may be removed in the wake of polymerization (Figure 5, *Step 4;* Figure 7*b*). At the end of the template, a remnant of the RNA should be left whose length may define the distance between polymerase and RNase H active sites. Using such logic, several laboratories have reported separations of about 10–19 nts between the active sites. A separation of 18 or 19 nucleotides for HIV-1 RT (165, 167) is consistent with modeling of an A-form DNA-RNA template-product in the putative binding cleft (121). Although there is considerable support for coupling of polymerase and RNase H activities, there are both biochemical (94, 168) and genetic data (169) that reveal uncoupled activities. One way to rationalize these results is to imagine that RNase H can function in two modes, one dependent and the other independent of polymerization. Genetic studies indicate that the uncoupled mode may be sufficient for low-level viral replication (169). Recent in vitro studies have indicated that during the first jump, the template RNA may be removed in two steps, using the two separate modes (170).

The role of RNase H in formation and removal of RNA primers has stirred particular interest because these events help determine the precise ends of the linear viral DNA, which are then recognized by IN for insertion into the host chromosome (7). The "plus-strand primer" is derived from a polypurine tract (PPT) in the genomic RNA (Figure 5, *Steps 4 and 5*). It is not clear what features of the PPT make it resistant to further hydrolysis by RNase H. The PPT is immediately adjacent to the U3 region, and the first deoxynucleotide added should specify the "U3 end of the viral DNA plus strand (Figure 5, bottom). The primer must be selected and then removed by precise cleavage at the RNA-DNA junction. Although other internal plus-strand primers may be used (Figure 5, *Step 6*), the PPT primer is a unique and consistent element in reverse transcription.

Using a model substrate, Luo et al (171) showed that ASLV RT RNase H activity was able to nick viral RNA sequences generating the expected 12 nt PPT RNA primer. They showed further that MLV or *E. coli* RNase H activities could not make the appropriate nicks in the ASLV RNA-DNA hybrid. However, once formed, the primer could be extended by any polymerase. This suggested a model in which the primer is formed by specific recognition by the RT RNase H, but is then selected over other RNA segments by a general feature of the RNA-DNA hybrid.

Two groups (172, 173) used an HIV-1 RNA-DNA hybrid substrate and HIV-1 RT to show that the plus strand starts with 5′ACTG. . . , consistent with the removal of the AC during integration (see INTEGRASE section below). Studies by Huber & Richardson (172) showed that HIV-1 PPT primer formation, extension, and removal could be uncoupled and that the primer could be removed intact.

The removal of the tRNA primer for minus-strand DNA synthesis plays a role in defining the "U5 end" of viral DNA (Figure 5), but it is not understood how specificity is determined. Also, removal of the tRNA must not occur before its 3′ end is copied in order to re-create a DNA copy of the PBS. Early studies with ASLV showed that RNase H activity of RT could cleave at the RNA-DNA junction, producing the predicted 5′ end of the minus-strand DNA (174). However, with HIV-1 the tRNA is apparently not removed precisely in vivo. The terminal ribonucleotide (rA) is left attached to DNA, and this specifies the "U5 end" of HIV-1 DNA (175, 176). Using model substrates, Smith & Roth (177) have shown that a small protein corresponding to the HIV-1 RNase H domain can select this ribonucleotide cleavage site. Thus, positioning by the polymerase domain is not required for selection of this site. Other studies have indicated that sugar conformations at the RNA-DNA junction may play a role in specifying the cleavage site (178).

Genetic recombination in retroviruses requires the packaging of two genetically distinct genomic RNAs, and takes place during the reverse transcription process (178a). One model for recombination ("forced copy-choice model") proposes template switching during minus-strand DNA synthesis (179). The model predicts that DNA synthesis must stop when RT encounters an interruption in one viral RNA template, and presumably a blunt RNA-DNA hybrid terminus would be formed. RNase H action at this point could help expose a single-strand DNA terminus that might then anneal to a homologous location on the second viral RNA; this would be analogous to the "first jump" in reverse transcription, where RNase H has been shown to play a role. In vitro, RNase H activity may facilitate, but is not essential for, jumping between model templates (163, 98). It is possible that the DNA terminus can be exposed by melting as well as by RNase H.

RNase H Structure

The crystal structure of the RNase H domain of HIV-1 RT has been determined using an RNase H domain produced by PR-mediated cleavage from a larger fusion protein (180) (Figure 8). The general folding is similar to that of *E. coli* RNase H (181, 182), which is quite striking since a (constrained) alignment shows only 24% amino-acid homology between the two proteins (181). Included in this homology are seven residues that are conserved among all retroviral and *E. coli* RNases H (180) (Figure 8). These are clustered in what is believed to be the active site. In the HIV-1 RNase H the four conserved acidic residues (D443, E478, D498, and D549) compose binding sites for two Mn^{2+} ions in the crystal. Site-directed mutations have confirmed that many of the conserved residues are important for activity of both *E. coli* and HIV-1 RNases H (183–185). One major

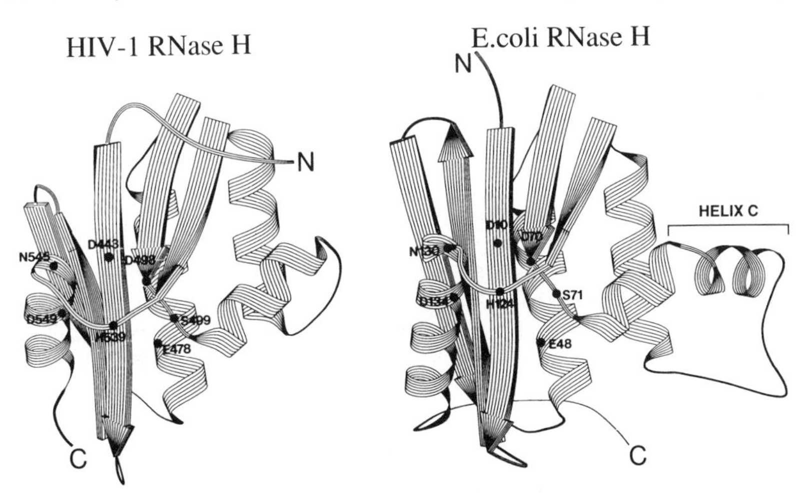

Figure 8 Ribbon diagrams based on the crystal structures of *E. coli* and HIV-1 RNase H. The seven highly conserved residues that likely define the active site are superimposed on the structures. Note that the *E. coli* helix "C" and adjacent loop are absent in HIV-1. Drawing was provided by Zuzana Hostomska, David Mathews, and Zdenek Hostomsky.

difference between the two structures is the absence of helix C and the adjacent loop in the HIV-1 RNase H (Figure 8).

The crystallized form of the HIV-1 RNase H domain is inactive, but activity can be reconstituted by the addition of HIV-1 RT p51 (125). It was proposed that a tryptophan-rich region, nearby in p51, might be important for substrate binding (180). This upstream region could compensate for the missing helix C, which is also tryptophan-rich. However, studies with N-terminal extended versions of HIV-1 RNase H suggest that this is not the case (177). Alternatively, the p51 subunit could stabilize a disordered loop that contains the essential histidine 539 (Figure 8) (180). The HIV-1 p15 RNase H fragment can be activated by various N-terminal extensions (summarized in Ref. 177), but the mechanisms of activation are unknown.

The cleavage site that separates p51 and p15 is buried within the N-terminal β sheet of the conserved RNase H structure (180). This supports a model for proteolytic processing in which the p66 homodimer is structurally asymmetric, with the RNase H domain unfolded in the subunit destined to give rise to p51, exposing the cleavage site. Formation of the p66 homodimer may force the unfolding of the RNase H domain in one subunit (180, 186). This would account for asymmetric cleavage by PR and for the observation that p15 RNase H released from a p66 homodimer is unstable (125). This model is supported by results of proteolysis of p66 with α-chymotrypsin (187) and by studies that show that p66 monomer is an inefficient substrate for processing (188).

Recently an activity that cleaves RNA-RNA hybrids has been found associated with RNase H (189). The biological role of this activity is not yet known.

INTEGRASE

General Features and Roles

Integration is an important step in the life cycle, since it ensures a stable association between viral DNA and the host-cell chromosome (Figures 1 and 9). Also, it is the integrated viral DNA that is transcribed by host RNA polymerase II to produce the viral RNA genome and mRNAs required to complete the replication cycle.

Integration is site-specific with respect to viral DNA (it occurs at the linear DNA ends), but is essentially random with respect to host DNA. The reaction requires the viral protein integrase (IN), which has been defined both genetically and biochemically, and *cis*-acting DNA sequences at the ends of the LTRs. The integration reaction is highly coordinated; two viral linear DNA ends, separated by 7–10 kilobasepairs, must be juxtaposed at

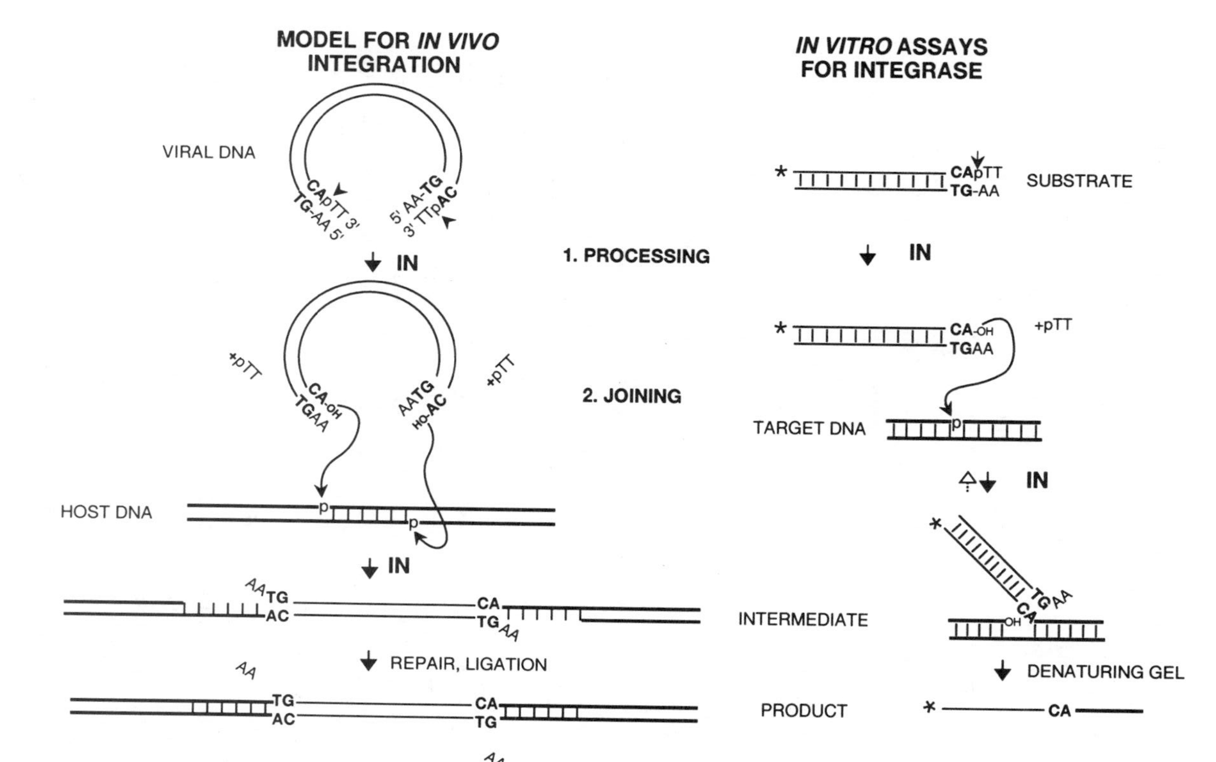

Figure 9 Retroviral DNA Integration. *Left panel:* General model for the in vivo reaction. Sequences shown correspond to the ASLV and MLV DNA ends. See text for details. *Right panel:* in vitro assay for IN activities. An oligodeoxynucleotide duplex (e.g. 15 bp) substrate is shown. The asterisks indicate a radioactive 5'phosphate. IN nicks at the in vivo processing site, and the shortened strand is detected by denaturing gel electrophoresis. IN joins the processed end to a target DNA, forming a strand longer than the substrate. The initial product of joining is a Y intermediate; the joined strands are detected by denaturing gel electrophoresis. Incubation of a Y-intermediate substrate with IN results in regeneration of a processed end and target DNA. This reaction has been termed "disintegration."

the host target site in a pairwise manner. In vitro studies have shown that IN alone can carry out the cutting and joining reactions. As with other retroviral enzymes, IN is a potential target for antiviral therapy, since it provides an essential function. However, to date this target has not been exploited for drug development. In spite of intensive efforts, IN has not yet been crystallized.

IN is encoded at the 3′ end of the *pol* gene and is assembled into the virus particle as part of the Gag-Pol precursor (Figures 1 and 2). IN is produced by PR-mediated cleavage of the precursor during virion maturation. The molecular weight of mature IN ranges from ca. 30,000 to 46,000 depending on the virus of origin. As with the other retroviral enzymes, the assembly process results in localization of IN to the viral core, where, after entry into a host cell, it can act upon the completed DNA product of reverse transcription.

The first insights into the mechanism of the retroviral DNA integration came in the early 1980s when nucleotide sequencing of avian and murine retroviral DNAs and host-viral DNA junctions revealed two characteristic features: the loss of two basepairs (bp) from each viral DNA end during integration, and duplications (4 to 6 bp) of host DNA at the insertion site. Insertion site duplications, as well as the presence of inverted and direct repeats (LTRs) at the ends of retroviral DNA, are reminiscent of bacterial transposable elements. Duplications of target sequences were consistent with models proposed previously for insertion of these elements and suggested that there was a staggered cut of host DNA followed by repair (Figure 9). The fact that the number of host bp duplicated was a characteristic of the retrovirus (MLV, 4; HIV-1, 5; ASLV, 6), rather than the host cell, suggested that a viral function was responsible. The first biochemical evidence that a retroviral protein might be involved in integration came in 1978, when Grandgenett and coworkers (190) identified endonuclease and DNA-binding activities associated with a 32-kDa phosphoprotein (pp32, now denoted IN) isolated from ASLV retroviral cores.

A Model for Integration In Vivo

A generally accepted model for retroviral integration has emerged over the past several years, and is supported by both in vivo and in vitro data (Figure 9). Reverse transcription of the viral RNA occurs within a subviral protein complex in the cytoplasm (191). The final product is a blunt-ended DNA duplex with terminal sequences: 5′-AA**TG**...**CA**TT-3′ for MLV and ASLV; and 5′-AC**TG**...**CA**GT-3′ for HIV-1. The **TG**...**CA** repeat is conserved throughout the retrovirus and retrotransposon families and in many prokaryotic DNA transposable elements as well. Next, the linear duplex is nicked by IN on the 3′ side of the **CA**, to produce new 3′ hydroxyl ends (**CA**OH-3′)

that are recessed by two nucleotides (192–194). This reaction can occur in the cytoplasm (195), presumably within the subviral structure (191). We refer to this site-specific endonuclease activity as the "processing" reaction because this step prepares the viral DNA for integration. It seems likely that IN monomers or multimers bind to each end of viral DNA and that the ends are held together through protein-protein interactions to coordinate processing of both ends (196). After the subviral complex enters the nucleus, processed viral DNA is joined to host target DNA by IN. This "joining" reaction includes a coupled 4–6 bp staggered cleavage of the target host DNA and ligation of processed **CA**OH-3′ viral DNA ends to the 5′phosphate ends of the target DNA (see below). The joining reaction produces a gapped intermediate in which the unpaired 5′phosphate ends of viral DNA are not linked to 3′hydroxyl ends of host DNA (192, 193). The gaps are repaired, producing flanking direct repeats of host DNA with loss of the two 5′ end noncomplementary nucleotides from the viral DNA. Repair may be accomplished by host-cell enzymes; however, a role for RT, and possibly IN, can also be envisioned (197, 198).

Biochemical Activities and Assays

Using purified ASLV IN, Katzman et al (194) showed specific nicking of short oligodeoxynucleotide duplexes that mimic either end of the retroviral DNA. This in vitro system, which reproduced the in vivo processing reaction, provided a simple assay for IN activity (Figure 9). This reaction required only IN, the model DNA substrate, and a metal cofactor, Mg^{2+} or Mn^{2+}. Using similar substrates, it was later shown that IN alone could also carry out the joining reaction in vitro (199, 200). After processing of the model substrate, the new ...**CA**OH-3′ end of the "donor" DNAs can be joined to numerous sites on either strand of other DNA duplexes, which act as surrogate "target DNAs." The product is a "Y intermediate" in which one strand of viral DNA is joined to one strand of target DNA (Figure 9). An appropriately processed end is a prerequisite for efficient joining; this end can be produced during the processing reaction, or an oligodeoxynucleotide duplex can be prepared that mimics a processed end. These studies revealed that the processing and joining steps could be biochemically uncoupled (199, 200), and correlated well with the observation that processing can occur in the cytoplasm (195), while joining takes place in the nucleus. In addition to these "one-sided reactions," ASLV and MLV INs can carry out coordinated insertions of two ends in vitro, where the target site duplication is reproduced (200, 201). Similar in vitro assay systems have been described for HIV-1 IN, but coordinated insertions are less frequent for reasons not yet understood (202).

A "disintegration" activity has recently been described, in which IN can regenerate donor and acceptor molecules from a Y-intermediate model substrate (197; Figure 9). It seems unlikely that this reaction represents a true reversal of joining (203), but it has proven to be a useful parameter in evaluating the fundamental catalytic functions of IN mutants (see below).

DNA Recognition

The retroviral DNA ends are characterized by short, and sometimes degenerate, inverted repeats of 2–13 bp, which always include the conserved 5′-**TG**...**CA**-3′ sequences. The role of the terminal sequences in processing and joining has been investigated in vitro using model DNA substrates. It was first shown with ASLV IN that sequences greater than 8 bp from the end could be substituted without severely affecting the efficiency of processing (194). Similar observations were made with MLV (204) and HIV-1 (198, 205–208). Simultaneous replacement of both the conserved **C** and **A** residues drastically reduces processing efficiency; individual replacements produce somewhat less severe effects (205–208). Adduct interference assays have also shown that only a few base pairs at the termini are important in the joining reaction (209). In contrast to the conserved **CA**, the terminal nucleotides that are removed during processing can be substituted with minimal effect (207). In many cases it appears that the sequence requirements for processing and joining are similar.

The fact that the sequence requirements for IN recognition are quite limited suggests that other structural features of the viral DNA may be important. If viral DNA terminal sequences are positioned internally on a DNA substrate, the efficiency of their processing is dramatically reduced in vitro (208) and in vivo (195). Thus, a DNA end may be important for recognition by IN, and this supports the generally accepted model that linear DNA is the in vivo substrate.

Although some slight DNA-binding preferences have been noted (209a), IN generally does not demonstrate sequence specificity (207, 210). This may reflect the need for IN to bind to host DNA in a sequence-independent manner during the joining step. Mutagenesis experiments have failed to segregate clearly viral and host DNA-binding sites.

Catalytic Mechanism

Based on work with the Mu bacteriophage A protein (reviewed in Ref. 211), Engelman and coworkers (212) tested a "one-step mechanism" for joining where the 3′-hydroxyl oxygen of the processed viral DNA strand directly attacks the host DNA phosphate, resulting in a concerted cleavage-ligation. Stereochemical analysis of the target phosphate revealed that an odd number of bond exchanges take place in both the processing and joining reactions. These results, obtained with HIV-1 IN, are most easily explained by an analogous one-step mechanism for both processing and joining.

In vitro, the target phosphate in processing (**CA**pXXOH) appears to be accessible to attack by many different oxygen nucleophiles, including the 3′ hydroxyl oxygen at the end of the same strand (forming a cyclic product) (212) and the hydroxyl groups of free serine, threonine, and glycerol (213). The phosphates can also be attacked by a hydroxyl group of a serine (or threonine) on IN itself, forming a protein-DNA covalent complex as observed with ASLV IN (214, 214a). These results suggest a mechanism for processing that involves appropriate preparation of the target phosphate, with little discrimination for the attacking group. In a similar way, the host DNA phosphates may also be prepared for nucleophilic attack in the joining reaction. In this case, the 3′-hydroxyl group of viral DNA acts as the nucleophile, but other nucleophiles must be excluded to prevent simple cleavage of the target DNA. The biochemical similarity of the two reactions, as well as mutagenesis studies (see below), suggests that a single catalytic site prepares phosphates for attack in both the processing and joining reactions (211).

Structure-Function Studies

Alignment of deduced IN amino-acid sequences has identified three potential functional domains (127, 214–218) (Figure 10). The N-terminal region is characterized by a HHCC "zinc finger"-like sequence. This homology extends into the retrotransposon integrases, but the function of this region is unknown. Mutagenesis studies have shown that independent replacements of the component His and Cys residues do not inactivate processing or joining activities, although the effects vary depending on the precise substitution (215, 219–221). However, deletions in this region generally produce

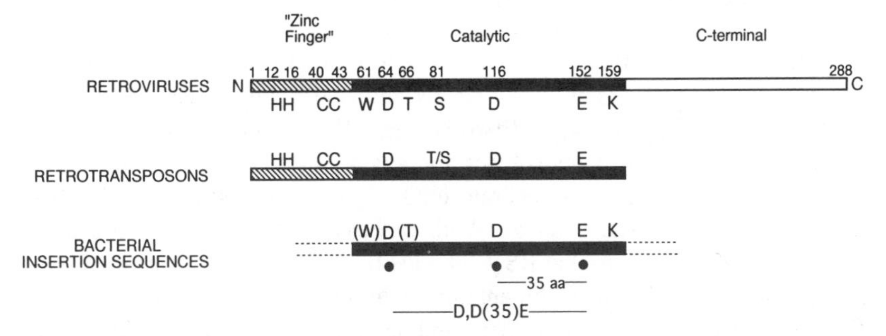

Figure 10 Evolutionary relationships of retroviral IN. Three domains of IN are shown. The most highly conserved amino acids are indicated using the single-letter code. The numbering corresponds to HIV-1 IN. Invariant residues, in the D,D(35)E constellation, are indicated by closed circles.

significant defects (222–224). When expressed independently, the HHCC domain does not bind DNA (215, 225), but it may interact with viral DNA in the context of the whole protein (219). Other studies have indicated that this region can bind zinc (223). A peptide corresponding to the HHCC region of HIV-1 IN also binds zinc, and this binding stabilizes its structure (226). However, the addition of exogenous zinc is not required for IN activity in vitro.

The central region (catalytic domain) is characterized by a D,D(35)E constellation; the component acidic residues have been proposed to be involved in binding the required metal ion(s), Mg^{2+} or Mn^{2+} (216). The D,D(35)E region homology extends to transposases of some bacterial insertion sequences, suggesting a common function in DNA breaking and joining reactions (216). Mutagenesis studies have shown that the conserved acidic residues are essential for both processing and joining activity (216, 220–222, 227). These results support a model in which the central region encodes a single catalytic core that functions in both reactions. Further support for this model comes from recent experiments that show that the isolated central region, although defective for authentic processing and joining reactions, retains cleavage-ligation activity, which is detected using a "disintegration," or "Y-intermediate"-type, substrate in which a hydroxyl group is positioned next to a target phosphate (223, 224; J Kulkosky and AM Skalka, in preparation). Early studies detected some DNA-binding activity in this region of ASLV IN (215), which might be expected if it contains the active site. This central region of HIV-1 is resistant to proteases, suggesting that it is folded into a separate domain (227).

The C-terminal region of IN is not highly conserved and its precise function is unknown. Deletion analysis, as well as independent expression experiments, indicate that the region contains strong sequence-independent DNA-binding activity (215, 224, 225, 228).

There are two general models for how IN may function in a multimeric complex to catalyze the concerted insertion of two ends of viral DNA. In the tetramer model (212), separate IN molecules bind each of the two processed viral DNA ends, and two separate IN molecules bind host DNA to prepare the two host phosphates for attack. In the dimer model (219), each IN monomer contains separate binding sites for host and viral DNA, and a single active site catalyzes both processing and joining. Studies by Jones et al (229) indicate that RSV IN must be at least a dimer for processing and joining. In addition, sedimentation analyses showed a reversible equilibrium between monomer, dimer, and tetramer forms in the absence of DNA (229). Recent complementation studies have provided further support for separate functional domains and for a model in which IN functions as a multimer (230, 231).

Since they function sequentially, it seems possible that RT and IN could act in a complex. IN could be delivered to the ends of the viral DNA upon completion of DNA synthesis by RT. RT could also be involved in repair of the gapped intermediate, with IN playing a role in joining of the 5′ ends of viral DNA to host DNA (197). The ASLV IN domain is present as part of the β chain of RT (Figure 6), and MLV IN appears to be linked to RT in a complex that is dissociated in vitro under denaturing conditions (232). The possible functional relationships between these two enzymes warrant further studies.

SUMMARY

We have reviewed the current state of knowledge concerning the three enzymes common to all retroviruses. It is informative to consider them together, since their activities are interrelated. The enzymatic activities of RT and IN depend on processing of polyprotein precursors by PR. Furthermore, RT produces the viral DNA substrate to be acted upon by IN.

All three of these retroviral enzymes function as multimers, and it is conceivable that specific polyprotein precursor interactions facilitate the multimerization of all of them. The multimeric structures of the enzymes are, however, quite different. PR is a symmetric homodimer whose subunits contribute to formation of a single active site. RT (of HIV, at least) is an asymmetric heterodimer in which one subunit appears to contribute all of the catalytic activity and the second is catalytically inactive, but structurally important. IN also functions minimally as a dimer for processing and joining.

The retroviral enzymes represent important targets for antiviral therapy. Considerable effort continues to be focused on developing PR and RT inhibitors. As more is learned about IN, such efforts can be extended. Since these enzymes are critical at different stages in the retroviral life cycle, one optimistic hope is that a combination of drugs that target all of them may be maximally effective as therapy for AIDS.

ACKNOWLEDGMENTS

We would like to thank many of our colleagues for providing us with reprints, preprints, and comments. Among them, J. Leis, S. Le Grice, J. G. Levin, V. R. Prasad, L. Kohlstaedt, S. Goff, B. Preston, J. Taylor, M. Otto, R. Swanstrom, S. Oroszlan, I. Weber, A. Telesnitsky, R. F. G. Booth, R. Craigie, R. Plasterk, and C. Carter were particularly helpful. We also thank J. Taylor and B. Müller for critical comments on the manuscript. We are grateful to Marie Estes, Mary Williamson, and Julia Chan for their patient and excellent assistance in preparing this manuscript. Work in our laboratory is supported by National Institutes of Health grants

CA47486, CA06927, RR05539, a grant from the Pew Charitable Trust, a grant for infectious disease research from Bristol-Myers Squibb Foundation, and by an appropriation from the Commonwealth of Pennsylvania. The contents of this review are solely the responsibility of the authors and do not necessarily represent the official views of the National Cancer Institute, or any other sponsoring organization.

Literature Cited

1. Weiss RA, Teich N, Varmus HE, Coffin JM, eds. 1982. *RNA Tumor Viruses,* Vol. 1. Cold Spring Harbor, NY: Cold Spring Harbor Lab.
2. Weiss RA, Teich N, Varmus HE, Coffin JM, eds. 1985. *RNA Tumor Viruses,* Vol. 2. Cold Spring Harbor, NY: Cold Spring Harbor Lab.
3. Varmus H, Brown P. 1989. In *Mobile DNA,* ed. DE Berg, MM Howe, pp. 53–108. Washington, DC: Am. Soc. Microbiol.
4. Coffin JM. 1990. In *Virology,* ed. BN Fields, DM Knipe, pp. 1437–1500. New York: Raven
5. Luciw PA, Leung NJ. 1992. In *The Retroviridae,* ed. J Levy, pp. 159–298. New York: Plenum
6. Goff SP. 1990. *J. Acquired Immune Defic. Syndr.* 3:817–31
7. Whitcomb JM, Hughes SH. 1992. *Annu. Rev. Cell. Biol.* 8:275–306
8. Johnson MI, McGowan JJ. 1992. In *AIDS,* ed. VT Devita Jr., S Hellman, SA Rosenberg. Philadelphia: Lippincott. 3rd ed.
9. Debouck C. 1992. *AIDS Res. Hum. Retroviruses* 8:153–64
10. Wlodawer A, Erickson JW. 1993. *Annu. Rev. Biochem.* 62:543–85
11. Kulkosky J, Skalka AM. 1990. *J. AIDS* 3:839–51
12. Brown PO. 1990. *Curr. Top. Microbiol. Immunol.* 157:19–48
13. Goff SP. 1992. *Annu. Rev. Genet.* 26:527–44
13a. Kulkosky J, Skalka AM. 1993. *Pharmacol. Ther.* In press
14. Skalka AM, Goff SP, eds. 1993. *Reverse Transcriptase.* Cold Spring Harbor, NY: Cold Spring Harbor Lab. Press
15. Kaplan AH, Zack JA, Knigge M, Paul DA, Kempf DJ, et al. 1993. *J. Virol.* 67:4050–55
16. Kuzmic P. 1993. *Biochem. Biophys. Res. Commun.* 191:998–1003
17. Grinde B, Cameron CE, Leis J, Weber I, Wlodawer A, et al. 1992. *J. Biol. Chem.* 267:9481–90
18. Sielecki AR, Hayakawa K, Fujinaga M, Murphy MEP, Fraser M, et al. 1989. *Science* 243:1346–51
19. Kotler M, Danho W, Katz RA, Leis J, Skalka AM. 1989. *J. Biol. Chem.* 264:3428–35
20. Cameron CE, Grinde B, Jacques P, Jentoft J, Leis J, et al. 1992. *J. Biol. Chem.* 268:11711–20
21. Tözsér J, Weber IT, Gustchina A, Bláha I, Copeland TD, et al. 1992. *Biochemistry* 31:4793–800
22. Sachdev GP, Fruton JS. 1970. *Biochemistry* 9:4465–70
23. Pettit SC, Simsic J, Loeb DD, Everitt L, Hutchison CA III, Swanstrom R. 1991. *J. Biol. Chem.* 266:14539–47
24. Bu M, Oroszlan S, Luftig RB. 1989. *AIDS Res. Hum. Retroviruses* 5:259–68
25. Kotler M, Katz RA, Danho W, Leis J, Skalka AM. 1988. *Proc. Natl. Acad. Sci. USA* 85:4185–89
26. Roberts NA, Martin JA, Kinchington D, Broadhurst AV, Craig JC, et al. 1990. *Science* 248:358–61
27. Tözsér J, Gustchina A, Weber IT, Blaha I, Wondrak EM, Oroszlan S. 1991. *FEBS Lett.* 279:356–60
28. Leis J, Bizub D, Weber I, Cameron C, Katz R, et al. 1989. In *Current Communications in Molecular Biology: Viral Proteinases as Targets for Chemotherapy,* ed. H Kräusslich, S Oroszlan, E Wimmer, p. 235. Cold Spring Harbor, NY: Cold Spring Harbor Lab. Press

29. Richards AD, Phylip LH, Farmerie WG, Scarborough PE, Alvarez A, et al. 1990. *J. Biol. Chem.* 265:7733–36
30. Matayoshi ED, Wang GT, Krafft GA, Erickson J. 1989. *Science* 247: 954–58
31. Geoghegan KF, Spencer RW, Danley DE, Contillo LG Jr, Andrews GC. 1990. *FEBS Lett.* 262:119–22
32. Tomaszek TA, Magaard VW, Bryan HG, Moore ML, Meek TD. 1990. *Biochem. Biophys. Res. Commun.* 168:274–80
33. Hyland LJ, Dayton BD, Moore ML, Shy AYL, Heys JR, Meek TD. 1990. *Anal. Biochem.* 188:408–15
34. Phylip LH, Richards AD, Kay J, Konvalinka J, Strop P, et al. 1990. *Biochem. Biophys. Res. Commun.* 171: 439–44
35. Tamburini PP, Dreyer RN, Hansen J, Letsinger J, Elting J, et al. 1990. *Anal. Biochem.* 186(2):363–68
36. Billich A, Winkler G. 1990. *Peptide Res.* 3:274–76
37. Rittenhouse J, Turon MC, Helfrich RJ, Albrecht KS, Weigl D, et al. 1990. *Biochem. Biophys. Res. Commun.* 171: 60–66
38. Otto MJ, Reid CD, Garber S, Lam PY-S, Scarnati H et al. 1993. *Antimicrob. Agents Chemother.* In press
39. Lam PY-S, Jadhav PK, Eyermann CJ, Hodge CN, Ru Y, et al. 1993. *Science.* In press
40. Wlodawer A, Miller M, Jaskólski M, Sathyanarayana BK, Baldwin E, et al. 1989. *Science* 245:616–21
41. Weber IT. 1990. *J. Biol. Chem.* 265: 10492–96
42. Schramm HJ, Nakashima H, Schramm W, Wakayama H, Yamamoto N. 1991. *Biochem. Biophys. Res. Commun.* 179: 847–51
43. Zhang Z-Y, Poorman RA, Maggiora LL, Heinrikson RL, Kézdy FJ. 1991. *J. Biol. Chem.* 266:15591–94
44. Babé LM, Rosé J, Craik CS. 1992. *Protein Sci.* 1:1244–53
44a. Miller M, Jaskólski M, Rao JKM, Leis J, Wlodawer A. 1989. *Nature* 337:576–79
44b. Navia MA, Fitzgerald PMD, McKeever BM, Leu CT, Heimbach JC, et al. 1989. *Nature* 337:615–20
45. Rao JKM, Erickson JW, Wlodawer A. 1991. *Biochemistry* 30:4663–71
45a. Weber IT. 1990. *Proteins: Struct. Funct. Genet.* 7:172–84
46. Jaskólski M, Tomasselli AG, Sawyer TK, Staples DG, Heinrikson RL, et al. 1991. *Biochemistry* 30:1600–9
47. Dreyer GB, Boehm JC, Chenera B, DesJarlais RL, Hassell AM, et al. 1993. *Biochemistry* 32:937–47
48. Grinde B, Cameron CE, Leis J, Weber IT, Wlodawer A, et al. 1992. *J. Biol. Chem.* 267:9491–98
49. Kent SBH, Schneider J, Clawson L, Selk L, Delahunty C, Chen Q. 1989. See Ref. 28, p. 223
50. Jaskólski M, Miller M, Rao JKM, Leis J, Wlodawer A. 1990. *Biochemistry* 29:5889–98
51. Hyland LJ, Tomaszek TA, Meek TD. 1991. *Biochemistry* 30:8454–63
52. Hyland LJ, Tomaszek TA, Roberts GD, Carr SA, Magaard VW, et al. 1991. *Biochemistry* 30:8441–53
53. Suguna K, Padlan EA, Smith CW, Carlson WD, Davies DR. 1987. *Proc. Natl. Acad. Sci. USA* 84:7009–13
54. Craven RC, Bennett RP, Wills JW. 1991. *J. Virol.* 65:6205–17
55. Loeb DD, Swanstrom R, Everitt L, Manchester M, Stamper SE, Hutchison CA III. 1989. *Nature* 340:397–400
56. Manchester M, Loeb DD, Everitt L, Moody M, Hutchison CA III, Swanstrom R. 1991. *Adv. Exp. Med. Biol.* 306:493–97
57. Cameron CE, Grinde B, Jacques P, Jentoft J, Leis J, et al. 1993. *J. Biol. Chem.* 268:11711–20
58. Weber IT, Tözsér J, Wu J, Friedman D, Oroszlan S. 1993. *Biochemistry* 32:3354–62
59. Konvalinka J, Horejsi M, Andreánsky M, Novek P, Pichova I, et al. 1992. *EMBO J.* 11:1141–44
60. Craig JC, Whittaker L, Duncan IB, Roberts NA. 1993. *Antiviral Chem. Chemother.* 4:In press
61. Jacobsen H, Ahlborn-Laake L, Gugel R, Mous J. 1992. *J. Virol.* 66:5087–91
62. Otto MJ, Garber S, Stack S, Winslow D. 1993. Reduced sensitivity to HIV PR inhibitors is associated with single amino acid substitutions in HIV-1 PR. *HIV Drug-Resistance 2nd Int. Workshop, Noordwijk, The Netherlands.* 21 pp. (Abstr.)
62a. Otto MJ, Garber S, Winslow DL, Reid CD, Aldrich P, et al. 1993. *Proc. Natl. Acad. Sci. USA* 90:7543–47
63. DiIanni CL, Davis LJ, Holloway MK, Herber WK, Darke PL, et al. 1990. *J. Biol. Chem.* 265:17348–54
63a. Cheng Y-SE, Yin FH, Foundling S, Blomstrom D, Kettner CA. 1990. *Proc. Natl. Acad. Sci. USA* 87:9660–64
64. Bizub D, Weber IT, Cameron CE, Leis JP, Skalka AM. 1991. *J. Biol. Chem.* 266:4951–58
64a. Patterson CE, Seetharam R, Kettner

CA, Cheng Y-SE. 1992. *J. Virol.* 66:1228–31
65. Milton RCD, Milton SCF, Kent SBH. 1992. *Science* 256:1445–48
66. Schnölzer M, Kent SBH. 1992. *Science* 256:221–25
67. Burstein H, Bizub D, Skalka AM. 1991. *J. Virol.* 65:6165–72
68. Jørgensen EC, Kjeldgaard NO, Pedersen FS, Jørgensen P. 1988. *J. Virol.* 62:3217–23
69. Rein A, McClure MR, Rice NR, Luftig RB, Schultz AM. 1986. *Proc. Natl. Acad. Sci. USA* 83:7246–50
70. Rhee SS, Hunter E. 1987. *J. Virol.* 61:1045–53
71. Schultz AM, Rein A. 1989. *J. Virol.* 63:2370–73
72. Wills JW, Craven RC, Achacoso JA. 1989. *J. Virol.* 63:4331–43
73. Luban J, Lee C, Goff SP. 1993. *J. Virol.* 67:3630–34
74. Fäcke M, Janetzko A, Shoeman RL, Kräusslich HG. 1993. *J. Virol.* 67: 4972–80
75. Vogt VM, Burstein H, Skalka AM. 1992. *Virology* 189:771–74
76. Kräusslich HG, Schneider H, Zybarth G, Carter CA, Wimmer E. 1988. *J. Virol.* 62:4393–97
77. Farmerie WG, Loeb DD, Casavant NC, Hutchison CA III, Edgell MH, Swanstrom R. 1987. *Science* 236:305–8
78. Kramer RA, Schaber MD, Skalka AM, Ganguly K, Wong-Staal F, Reddy EP. 1986. *Science* 231:1580–85
79. Gheysen D, Jacobs E, deForesta F, Thiriart C, Francotte M, et al. 1989. *Cell* 59:103–12
80. Karacostas V, Wolffe EJ, Nagashima K, Gonda MA, Moss B. 1993. *Virology* 193:661–71
81. Kaplan AH, Swanstrom R. 1991. *Proc. Natl. Acad. Sci. USA* 88:4528–32
82. Kräusslich HG. 1991. *Proc. Natl. Acad. Sci. USA* 88:3213–17
83. Kräusslich HG. 1992. *J. Virol.* 66:567–72
84. Poorman RA, Tomasselli AG, Heinrikson RL, Kézdy FJ. 1991. *J. Biol. Chem.* 266:14554–61
85. Burstein H, Bizub D, Kotler M, Schatz G, Vogt VM, Skalka AM. 1992. *J. Virol.* 66:1781–85
86. Louis JM, McDonald RA, Nashed NT, Wondrak EM, Jerina DM, et al. 1991. *Eur. J. Biochem.* 199:361–69
87. Zybarth G, Kräusslich HG, Partin K, Carter C. 1994. *J. Virol.* 68:240–50
88. Roberts MM, Oroszlan S. 1989. *Biochem. Biophys. Res. Commun.* 160: 486–94
89. Baboonian C, Dalgleish A, Bountiff L, Gross J, Oroszlan S, et al. 1991. *Biochem. Biophys. Res. Commun.* 179: 17–24
89a. Fu W, Rein A. 1993. *J. Virol.* 67: 5443–49
90. Oertle S, Bowles N, Spahr PF. 1992. *J. Virol.* 66:3873–78
91. Park J, Morrow CD. 1993. *Virology* 194:843–50
92. Telesnitsky A, Goff SP. 1993. *Proc. Natl. Acad. Sci. USA* 90:1276–80
93. Majumdar C, Abbotts J, Broder S, Wilson SH. 1988. *J. Biol. Chem.* 263:15657–65
94. Huber HE, McCoy JM, Seehra JS, Richardson CC. 1989. *J. Biol. Chem.* 264:4669–78
95. Telesnitsky A, Goff SP. 1993. See Ref. 14, pp. 49–83
96. Tanese N, Telesnitsky A, Goff SP. 1991. *J. Virol.* 65:4387–97
97. Telesnitsky A, Blain SW, Goff SP. 1992. *J. Virol.* 66:615–22
98. Buiser RG, DeStefano JJ, Mallaber LM, Fay PJ, Bambara RA. 1991. *J. Biol. Chem.* 266:13103–9
99. Katz RA, Skalka AM. 1990. *Annu. Rev. Genet.* 24:409–45
100. Preston BD, Garvey N. 1992. *Pharm. Tech.* 16:34–51
101. Bebenek K, Kunkel TA. 1993. See Ref. 14, pp. 85–102
102. Varela-Echavarría A, Garvey N, Preston BD, Dougherty JP. 1992. *J. Biol. Chem.* 267:24681–88
103. Perrino FW, Preston BD, Sandell LL, Loeb LA. 1989. *Proc. Natl. Acad. Sci. USA* 86:8343–47
104. Mendelman LV, Petruska J, Goodman MF. 1990. *J. Biol. Chem.* 265:2338–46
105. Bebenek K, Abbotts J, Wilson SH, Kunkel TA. 1992. *J. Biol. Chem.* 268:10324–34
106. Pathak VK, Temin HM. 1990. *Proc. Natl. Acad. Sci. USA* 87:6019–23
107. Temin HM. 1993. *Proc. Natl. Acad. Sci. USA* 90:6900–3
108. Elder JH, Lerner DL, Hasselkus-Light CS, Fontenot DJ, Hunter E, et al. 1992. *J. Virol.* 66:1791–94
109. Threadgill DS, Steagall WK, Flaherty MT, Fuller FJ, Perry ST, et al. 1993. *J. Virol.* 67:2592–600
110. Boone LR, Skalka AM. 1993. See Ref. 14, pp. 119–33
111. Stewart L, Vogt VM. 1991. *J. Virol.* 65:6218–31
112. Peng C, Chang NT, Chang TW. 1991. *J. Virol.* 65:2751–56
113. Veronese FD, Copeland TD, DeVico AL, Rahman R, Oroszlan S, et al. 1986. *Science* 231:1402–5

114. Le Grice SFJ, Grüninger-Leitch F. 1990. *Eur. J. Biochem.* 187:307–14
115. Müller B, Restle T, Weiss S, Gautel M, Sczakiel G, Goody RS. 1989. *J. Biol. Chem.* 264:13975–78
116. Richter-Cook NJ, Howard KJ, Cirino NM, Wöhrl BM, Le Grice SFJ. 1992. *J. Biol. Chem.* 267:15952–57
117. Restle T, Müller B, Goody RS. 1990. *J. Biol. Chem.* 265:8986–88
118. Le Grice SFJ, Naas T, Wohlgensinger B, Schatz O. 1991. *EMBO J.* 10:3905–11
119. Hostomsky Z, Hostomska Z, Fu TB, Taylor J. 1992. *J. Virol.* 66:3179–82
120. Howard KJ, Frank KB, Sim IS, Le Grice SFJ. 1991. *J. Biol. Chem.* 266: 23003–9
121. Kohlstaedt LA, Wang J, Friedman JM, Rice PA, Steitz TA. 1992. *Science* 256:1783–90
122. Arnold E, Jacobo-Molina A, Nanni RG, Williams RL, Lu X, et al. 1992. *Nature* 357:85–89
123. Jacobo-Molina A, Ding J, Nanni RG, Clark AD Jr, Lu X, et al. 1993. *Proc. Natl. Acad. Sci. USA* 90:6320–24
124. Sousa R, Chung YJ, Rose JP, Wang BC. 1993. *Nature* 364:593–99
125. Hostomsky Z, Hostomska Z, Hudson GO, Moomaw EW, Nodes BR. 1991. *Proc. Natl. Acad. Sci. USA* 88:1148–52
126. Delarue M, Poch O, Tordo N, Moras D, Argos P. 1990. *Protein Eng.* 3:461–67
127. Johnson MS, McClure MA, Feng DF, Gray J, Doolittle RF. 1986. *Proc. Natl. Acad. Sci. USA* 83:7648–52
128. Larder BA, Purifoy DJM, Powell KL, Darby G. 1987. *Nature* 327:716–17
129. Jacobo-Molina A, Arnold E. 1991. *Biochemistry* 30:6351–61
130. Tanese N, Goff SP. 1988. *Proc. Natl. Acad. Sci. USA* 85:1777–81
131. Prasad VR, Goff SP. 1989. *Proc. Natl. Acad. Sci. USA* 86:3104–8
132. Hizi A, Hughes SH, Shaharabany M. 1990. *Virology* 175:575–80
133. Lowe DM, Parmar V, Kemp SD, Larder BA. 1991. *FEBS Lett.* 282:231–34
134. Boyer PL, Ferris AL, Hughes SH. 1992. *J. Virol.* 66:1031–39
135. Levin JG, Seidman JG. 1981. *J. Virol.* 38:403–8
136. Barat C, Lullien V, Schatz O, Keith G, Nugeyre MT, et al. 1989. *EMBO J.* 8:3279–85
137. Aiyar A, Cobrinik D, Ge Z, Kung HJ, Leis J. 1992. *J. Virol.* 66:2464–72
138. Kohlstaedt LA, Steitz TA. 1992. *Proc. Natl. Acad. Sci. USA* 89:9652–56
139. Weiss S, König B, Müller H-J, Seidel H, Goody RS. 1992. *Gene* 111:183–97
140. Barat C, Schatz O, Le Grice SFJ, Darlix JL. 1993. *J. Mol. Biol.* 231: 185–90
141. Majumdar C, Stein CA, Cohen JS, Broder S, Wilson SH. 1989. *Biochemistry* 28:1340–46
142. Basu A, Ahluwalia KK, Basu S, Modak MJ. 1992. *Biochemistry* 31: 616–23
143. De Clercq E. 1992. *AIDS Res. Hum. Retroviruses* 8:119–34
144. Larder BA. 1993. See Ref. 14, pp. 205–22
145. Lacey SF, Reardon JE, Furfine ES, Kunkel TA, Bebenek K, et al. 1992. *J. Biol. Chem.* 267:15789–94
146. Remington KM, Chesebro B, Wehrly K, Pedersen NC, North TW. 1991. *J. Virol.* 65:308–12
147. Kedar PS, Abbotts J, Kovács T, Lesiak K, Torrence P, Wilson SH. 1990. *Biochemistry* 29:3603–11
148. Song QG, Yang GZ, Goff SP, Prasad VR. 1992. *J. Virol.* 7568–71
149. Pauwels R, Andries K, Desmyter J, Schols D, Kukla MJ, et al 1990. *Nature* 343:470–74
150. Merluzzi VJ, Hargrave KD, Labadia M, Grozinger K, Skoog M, et al. 1990. *Science* 250:1411–13
151. Wu JC, Warren TC, Adams J, Proudfoot J, Skiles J, et al. 1991. *Biochemistry* 30:2022–26
152. Cohen KA, Hopkins J, Ingraham RH, Pargellis C, Wu JC, et al. 1991. *J. Biol. Chem.* 266:14670–74
153. Shih C-K, Rose JM, Hansen GL, Wu JC, Bacolla A, Griffin JA. 1991. *Proc. Natl. Acad. Sci. USA* 88:9878–82
154. Richman D, Shih C-K, Lowy I, Rose J, Prodanovich P, et al. 1991. *Proc. Natl. Acad. Sci. USA* 88:11241–45
155. Nunberg JH, Schleif WA, Boots EJ, O'Brien JA, Quintero JC, et al. 1991. *J. Virol.* 65:4887–92
155a. Champoux JJ. 1993. See Ref. 14, pp. 103–17
156. Oyama F, Kikuchi R, Crouch RJ, Uchida T. 1989. *J. Biol. Chem.* 264: 18808–17
157. Krug MS, Berger SL. 1989. *Proc. Natl. Acad. Sci. USA* 86:3539–43
158. Schatz O, Mous J, Le Grice SFJ. 1990. *EMBO J.* 9:1171–76
159. DeStefano JJ, Buiser RG, Mallaber LM, Bambara RA, Fay PJ. 1991. *J. Biol. Chem.* 266:24295–301
160. Schulze T, Nawrath M, Moelling K. 1991. *Arch. Virol.* 118:179–88
161. Levin JG, Crouch RJ, Post K, Hu SC,

McKelvin D, et al. 1988. *J. Virol.* 62:4376–80
162. Wöhrl BM, Moelling K. 1990. *Biochemistry* 29:10141–47
163. Luo G, Taylor J. 1990. *J. Virol.* 64:4321–28
164. Furfine ES, Reardon JE. 1991. *J. Biol. Chem.* 266:406–12
165. Kati WM, Johnson KA, Jerva LF, Anderson KS. 1992. *J. Biol. Chem.* 267:25988–97
166. Furfine ES, Reardon JE. 1991. *Biochemistry* 30:7041–46
167. Gopalakrishnan V, Peliska JA, Benkovic SJ. 1992. *Proc. Natl. Acad. Sci. USA* 89:10763–67
167a. Post K, Guo J, Kalman E, Uchida T, Crouch RJ, Levin JG. 1993. *Biochemistry* 32:5508–17
168. DeStefano JJ, Buiser RG, Mallaber LM, Myers TW, Bambara RA, Fay PJ. 1991. *J. Biol. Chem.* 266:7423–31
169. Telesnitsky A, Goff SP. 1993. *EMBO J.* 12:4433–38
170. Peliska JA, Benkovic SJ. 1992. *Science* 258:1112–18
171. Luo G, Sharmeen L, Taylor J. 1990. *J. Virol.* 64:592–97
172. Huber HE, Richardson CC. 1990. *J. Biol. Chem.* 265:10565–73
173. Pullen KA, Champoux JJ. 1990. *J. Virol.* 64:6274–77
174. Omer CA, Faras AJ. 1982. *Cell* 30: 797–805
175. Pullen KA, Ishimoto LK, Champoux JJ. 1992. *J. Virol.* 66:367–73
176. Smith JS, Roth MJ. 1992. *J. Biol. Chem.* 267:15071–79
177. Smith JS, Roth MJ. 1993. *J. Virol.* 67:4037–49
178. Salazar M, Champoux JJ, Reid BR. 1993. *Biochemistry* 32:739–44
178a. Hu W-S, Temin HM. 1990. *Science* 250:1227–33
179. Coffin JM. 1979. *J. Gen. Virol.* 42:1–26
180. Davies JF II, Hostomska Z, Hostomsky Z, Jordan SR, Matthews DA. 1991. *Science* 252:88–95
181. Yang W, Hendrickson WA, Crouch RJ, Satow Y. 1990. *Science* 249:1398–405
182. Katayanagi K, Miyagawa M, Matsushima M, Ishikawa M, Kanaya S, et al. 1990. *Nature* 347:306–9
183. Schatz O, Cromme FV, Grüninger-Leitch F, Le Grice SFJ. 1989. *FEBS Lett.* 257:311–14
184. Kanaya S, Kohara A, Miura Y, Sekiguchi A, Iwai S, et al. 1990. *J. Biol. Chem.* 265:4615–21
185. Mizrahi V, Usdin MT, Harington A, Dudding LR. 1990. *Nucleic Acids Res.* 18:5359–63
186. Hostomska Z, Matthews DA, Davies JF II, Nodes BR, Hostomsky Z. 1991. *J. Biol. Chem.* 266:14697–702
187. Lowe DM, Aitken A, Bradley C, Darby GK, Larder BA, et al. 1988. *Biochemistry* 27:8884–89
188. Deibel MR Jr, McQuade TJ, Brunner DP, Tarpley WG. 1990. *AIDS Res. Hum. Retroviruses* 6:329–40
189. Ben-Artzi H, Zeelon E, Le Grice SFJ, Gorecki M, Panet A. 1992. *Nucleic Acids Res.* 20:5115–18
190. Grandgenett DP, Vora AC, Schiff RD. 1978. *Virology* 89:119–32
191. Bowerman B, Brown PO, Bishop JM, Varmus HE. 1989. *Genes Dev.* 3:469–78
192. Brown PO, Bowerman B, Varmus HE, Bishop JM. 1989. *Proc. Natl. Acad. Sci. USA* 86:2525–29
193. Fujiwara T, Mizuuchi K. 1988. *Cell* 54:497–504
194. Katzman M, Katz RA, Skalka AM, Leis J. 1989. *J. Virol.* 63:5319–27
195. Roth MJ, Schwartzberg PL, Goff SP. 1989. *Cell* 58:47–54
196. Murphy JE, Goff SP. 1992. *J. Virol.* 66:5092–95
197. Chow SA, Vincent KA, Ellison V, Brown PO. 1992. *Science* 255:723–26
198. Sherman PA, Dickson ML, Fyfe JA. 1992. *J. Virol.* 66:3593–601
199. Craigie R, Fujiwara T, Bushman F. 1990. *Cell* 62:829–37
200. Katz RA, Merkel G, Kulkosky J, Leis J, Skalka AM. 1990. *Cell* 63:87–95
201. Fujiwara T, Craigie R. 1989. *Proc. Natl. Acad. Sci. USA* 86:3065–69
202. Bushman FD, Fujiwara T, Craigie R. 1990. *Science* 249:1553–58
203. Jonsson CB, Roth MJ. 1993. *J. Virol.* 67:5562–71
204. Bushman FD, Craigie R. 1990. *J. Virol.* 64:5645–48
205. Bushman FD, Craigie R. 1991. *Proc. Natl. Acad. Sci. USA* 88:1339–43
206. Leavitt AD, Rose RB, Varmus HE. 1992. *J. Virol.* 66:2359–68
207. LaFemina RL, Callahan PL, Cordingley MG. 1991. *J. Virol.* 65:5624–30
208. Vink C, van Gent DC, Elgersma Y, Plasterk RHA. 1991. *J. Virol.* 65: 4636–44
209. Bushman FD, Craigie R. 1992. *Proc. Natl. Acad. Sci. USA* 89:3458–62
209a. Ishimoto LK, Halperin M, Champoux JJ. 1991. *Virology* 180:527–34
210. van Gent DC, Elgersma Y, Bolk MWJ, Vink C, Plasterk RHA. 1991. *Nucleic Acids Res.* 19:3821–27

211. Mizuuchi K. 1992. *J. Biol. Chem.* 267:21273–76
212. Engelman A, Mizuuchi K, Craigie R. 1991. *Cell* 67:1211–21
213. Vink C, Yeheskiely E, van der Marel GA, van Boom JH, Plasterk RHA. 1992. *Nucleic Acids Res.* 19:6691–98
214. Katzman M, Mack JPG, Skalka AM, Leis J. 1991. *Proc. Natl. Acad. Sci. USA* 88:4695–99
214a. Katz RA, Mack JPG, Merkel G, Kulkosky J, Ge Z, et al. 1992. *Proc. Natl. Acad. Sci. USA* 89:6741–45
215. Khan E, Mack JPG, Katz RA, Kulkosky J, Skalka AM. 1991. *Nucleic Acids Res.* 19:851–60
216. Kulkosky J, Jones KS, Katz RA, Mack JPG, Skalka AM. 1992. *Mol. Cell. Biol.* 2:2331–38
217. Rowland SJ, Dyke KGH. 1990. *Mol. Microbiol.* 4:961–75
218. Fayet O, Ramond P, Polard P, Prere MF, Chandler M. 1990. *Mol. Microbiol.* 4:1771–77
219. Vincent KA, Ellison V, Chow SA, Brown PO. 1993. *J. Virol.* 67:425–37
220. van Gent DC, Groeneger AAMO, Plasterk RHA. 1992. *Proc. Natl. Acad. Sci. USA* 89:9598–602
221. Leavitt AD, Shiue L, Varmus HE. 1993. *J. Biol. Chem.* 268:2113–19
222. Drelich M, Wilhelm R, Mous J. 1992. *Virology* 188:459–68
223. Bushman FD, Engelman A, Palmer I, Wingfield P, Craigie R. 1993. *Proc. Natl. Acad. Sci. USA* 90:3428–32
224. Vink C, Oude Groeneger AAM, Plasterk RHA. 1993. *Nucleic Acids Res.* 21:1419–25
225. Mumm SR, Grandgenett DP. 1991. *J. Virol.* 65:1160–67
226. Burke CJ, Sanyal G, Bruner MW, Ryan JA, LaFemina RL, et al. 1992. *J. Biol. Chem.* 267:9639–44
227. Engelman A, Craigie R. 1992. *J. Virol.* 66:6361–69
228. Woerner AM, Marcus-Sekura CJ. 1993. *Nucleic Acids Res.* 21:3507–11
229. Jones KS, Coleman J, Merkel GW, Laue TM, Skalka AM. 1992. *J. Biol. Chem.* 267:16037–40
230. van Gent DC, Vink C, Oude Groeneger AAM, Plasterk RHA. 1993. *EMBO J.* 12:3261–67
231. Engelman A, Bushman FD, Craigie R. 1993. *EMBO J.* 12:3269–75
232. Hu SC, Court DL, Zweig M, Levin JG. 1986. *J. Virol.* 60:267–74

Annu. Rev. Biochem. 1994. 63:175–95

NITRIC OXIDE: A Physiologic Messenger Molecule

D. S. Bredt and S. H. Snyder
Departments of Neuroscience, Pharmacology and Molecular Sciences, Psychiatry and Behavioral Sciences, Johns Hopkins University School of Medicine, 725 North Wolfe Street, Baltimore, Maryland 21205

KEY WORDS: nitric oxide synthase, cyclic GMP, ADP ribosylation, cytochrome P450, calmodulin

---

CONTENTS

INTRODUCTION

Though only recently uncovered as a physiologic messenger, nitric oxide (NO) is increasingly appreciated as a major regulator in the nervous, immune, and cardiovascular systems. Besides mediating normal functions, NO has been implicated in pathophysiologic states as diverse as septic shock, hypertension, stroke, and neurodegenerative diseases.

Biological roles for NO were first established in inflammatory responses and blood vessel reactivity. Studies of nitrosamines as carcinogens demonstrated the existence of endogenous nitrates, since germ-free rats excrete large amounts of nitrates as do humans, whose excretion rises markedly

0066-4154/94/0701-0175$05.00

during infections (1, 2). Urinary nitrates were found to arise from macrophages through oxidation of one of the amidine nitrogens of L-arginine, giving rise to L-citrulline and a reactive substance subsequently shown to be NO. The ability of macrophages to kill tumor cells and fungi depended upon external arginine, whose effects were blocked by arginine derivatives that also blocked the formation of nitrite, leading to identification of NO as the active substance (3, 4, 4a).

Evidence for a physiologic role of NO in blood vessels was preceded by studies implicating NO as the active metabolite of nitroglycerin and other organic nitrates in dilating blood vessels by stimulating cGMP formation through activation of guanylyl cyclase (GC) (5, 6). Furchgott and associates (7) had shown that blood vessel relaxation in response to acetylcholine and other substances requires the endothelial lining, which releases a labile substance that diffuses to the adjacent smooth muscle. The active agent was identified as NO (8, 9).

NO was first implicated in the brain when cerebellar cultures stimulated by excitatory amino acids were found to release a substance with the properties of NO (10). A definitive involvement of NO was demonstrated by the ability of NOS inhibitors, such as L-N^{ω} nitroarginine (L-NNA) and L-N^{ω} methyl-arginine (L-NMA), to block the pronounced stimulation of cGMP in brain slices that is elicited by the excitatory neurotransmitter glutamate acting at *N*-methyl-D-aspartate (NMDA) subtype receptors (11, 12).

NO FORMATION

Biosynthetic regulation is more important for NO than for other neurotransmitters, because NO cannot be stored, released, or inactivated after synaptic release by conventional regulatory mechanisms. Indeed, NOS is one of the most regulated enzymes in biology. Initial efforts to purify the enzyme were unsuccessful because of a rapid loss of enzyme activity upon purification. Observations that calmodulin is required for NOS activity in the brain led to a simple purification of brain NOS (bNOS) to homogeneity (13). Using this approach, other groups purified bNOS (14, 15), macrophage NOS (macNOS) (16–19), and endothelial NOS (eNOS) (20). Molecular cloning of the cDNA for brain (21, 22), endothelial (23–25), macrophage (26–28), and nonmacrophage-inducible (29) forms of NOS has helped elucidate NOS function (Figure 2). The structure of NOS reveals numerous regulatory mechanisms.

NOS oxidizes the guanidine group of L-arginine in a process that consumes five electrons and results in the formation of NO with stoichiometric formation of L-citrulline (Figure 1). L-N^{ω}-substituted arginines function as NOS inhibitors. The inhibition of NOS by these substrate analogs can

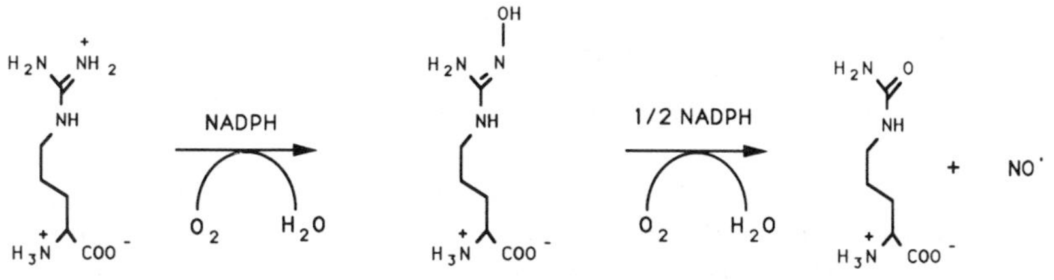

ARGININE HYDROXY-ARGININE CITRULLINE

Figure 1 Biosynthesis of nitric oxide. NOS catalyzes a five-electron oxidation of an amidine nitrogen of L-arginine to generate NO and L-citrulline. L-hydroxyarginine is formed as an intermediate that is tightly bound to the enzyme. Both steps in the reaction are dependent upon calcium and calmodulin and are enhanced by tetrahydrobiopterin.

initially be reversed by simultaneous application of excess arginine, consistent with their competitive blockade of the active site. However, following prolonged exposure, NOS is irreversibly inhibited by some of these agents. The irreversible inactivation of the macrophage enzyme and brain enzymes by L-NMA requires simultaneous incubation with NOS cofactors, suggesting "mechanism-based" inhibition (30, 31). The time-dependent inactivation of the brain enzyme by L-NNA (32) is independent of NOS enzymatic turnover (33).

NOS isoforms display modest differences in their sensitivity to various arginine analogs. L-NNA is the most potent known inhibitor of the brain and endothelial enzymes (K_i = 200–500 nM), and L-N^{ω} aminoarginine (L-NAA) is the most potent blocker of the macrophage enzyme (K_i = 1–5 μM). Clinically useful inhibitors of NOS will likely need to be isoform specific. For instance, potential neuroprotective effects of NOS inhibition might be mitigated by hypertension elicited by inhibition of eNOS. Recent studies suggest that 7-nitroindazole can preferentially influence bNOS functionally, though its K_i values for bNOS and eNOS are similar (34).

NOS Cofactors

Oxidative enzymes generally employ redox-active cofactors. NOS is unprecedented in employing five. The cloning of NOSs (discussed below) indicates their close homology with cytochrome P450 reductase (CPR), including consensus sequences for NADPH, FAD, and FMN binding. While NADPH is a stoichiometric substrate, the two flavins copurify with NOS in a ratio of 1 eq each of FAD and FMN per NOS monomer (18, 35, 36). FAD slowly dissociates from NOS and must be exogenously supplied for maximal activity. The close homology of NOS with CPR suggests that electrons follow the same path through NOS as they do through CPR, that is NADPH initially reduces FAD, which in turn reduces FMN. In fact, NOS and CPR share a domain thought to be involved in this electron transfer (21).

CPR supplies the reducing equivalents from NADPH to the heme-containing cytochrome P450 enzymes. This mechanism is apparently shared by NOS, as both bNOS and macNOS contain 1 eq of iron-protoporphyrin IX per NOS monomer (37–39). Furthermore, NOS displays reduced CO difference spectra typical of cytochrome P450, with a wavelength absorbance maximum at 445 nM indicative of a heme-binding cysteinyl ligand. Purified NOS is inhibited by CO, which is also consistent with the participation of a cytochrome P450-type heme in the reaction. NO itself appears to exert feedback inhibition of NOS (40–42), perhaps by interacting with the enzyme's heme prosthetic group. Optical difference spectroscopy indicates that heme binds to the substrate arginine prior to participating in the oxygenation reactions (43). Heme-substrate binding is also the initial event in catalysis with the various cytochrome P450s. Unlike other mammalian cytochrome P450 enzymes, NOSs are unique because they are not integral membrane proteins and their flavin and heme-containing domains are fused in a single polypeptide. A bacterial fatty acid monooxygenase, $P450_{BM3}$, also has been identified as a soluble, self-contained P450 system (44).

NOS is also regulated by tetrahydrobiopterin (H_4B). While macNOS is absolutely dependent upon H_4B (45, 46), purified bNOS retains substantial activity in the absence of added H_4B (13, 15). This discrepancy is explained by the tight binding of H_4B to bNOS such that H_4B copurifies with the enzyme (47). It was initially assumed that H_4B functions directly in the hydroxylation of arginine by analogy to its role in aromatic amino-acid hydroxylase enzymes. This notion was challenged in experiments by Kaufman and coworkers (48), who proposed that H_4B stabilizes NOS. This conclusion was based on experiments with bNOS that showed that H_4B functions catalytically, is not recycled, and does not affect the initial rate of NOS. Marletta and colleagues (49) also suggested that H_4B stabilizes NOS, based on experiments with pterin analogs used to probe the macNOS reaction.

The conversion of arginine to NO is catalyzed in two independent steps (Figure 1). The first step is a two-electron oxidation of arginine to N^{ω}-hydroxyarginine (NHA) (50). Although this hydroxylated intermediate is tightly bound to NOS, under certain conditions NHA can be isolated as a product (33). This hydroxylation step resembles a classical P450-type monooxygenation reaction utilizing 1 eq of NADPH and 1 eq of O_2 (50).

The hydroxylation reaction is accelerated by H_4B, requires calcium and calmodulin as activators, and is blocked by CO (33, 38, 50).

The second step, i.e. the pathway from NHA to NO and citrulline, is less clear. Any proposed mechanism should account for experiments that find that this oxidation (*a*) utilizes 0.5 eq NADPH, (*b*) requires O_2 and calcium/calmodulin, (*c*) is accelerated by H_4B, and (*d*) is inhibited by CO and arginine analogs with a pharmacology similar to that seen in the initial hydroxylation reaction (33, 38, 50). In one model consistent with these data, NOS would use both its reductase and heme domains for successive independent oxidations of arginine at a common active site, with heme directly functioning in the activation of molecular oxygen. For the first hydroxylation, both reducing equivalents for oxygen activation derive from NADPH. It has been speculated that NHA and NADPH each provide one electron for the second oxidation step (51). This both explains the 0.5 stoichiometry of NADPH utilization and accounts for the unusual five-electron chemistry of NO biosynthesis.

Regulation of NOS

NOS enzymes can be discriminated by their regulation by calcium. In the brain, a stimulus such as glutamate acting at NMDA receptors triggers Ca^{2+} influx, which binds to calmodulin thereby activating NOS. This mode of activation explains the ability of glutamate neurotransmission to stimulate NO formation in a matter of seconds. In blood vessels, acetylcholine acting at muscarinic receptors on endothelial cells activates the phosphoinositide cycle to generate Ca^{2+}, which stimulates NOS. Thus, calcium-regulated NOS accounts for the role of NO in mediating rapid events such as neurotransmission and blood vessel dilatation. The calcium requirement for NOS activity is typical for a calmodulin-activated enzyme (EC_{50} = 200–400 nM). Calmodulin binding regulates the electron transfer and oxygen activation activities of NOS (33, 52). Arginine binding, however, is unaffected by calcium or calmodulin (43). The brain and endothelial enzymes are inhibited by calmodulin antagonists (13), such as trifluoperazine (IC_{50} = 10 μM).

The inducible NOS of macrophage and nonmacrophage sources is neither stimulated by Ca^{2+} nor blocked by calmodulin antagonists. Surprisingly, inducible NOS enzymes possess calmodulin recognition sites (Figure 2). Nathan and colleagues (53) have shown that calmodulin is very tightly bound to inducible NOS, with the binding unaffected by Ca^{2+}, whereas calmodulin cannot bind to neuronal NOS unless Ca^{2+} is present. The fact that calmodulin binds so tightly to inducible NOS that it can be considered an enzyme subunit accounts for the resistance of inducible NOS to Ca^{2+} activation (54).

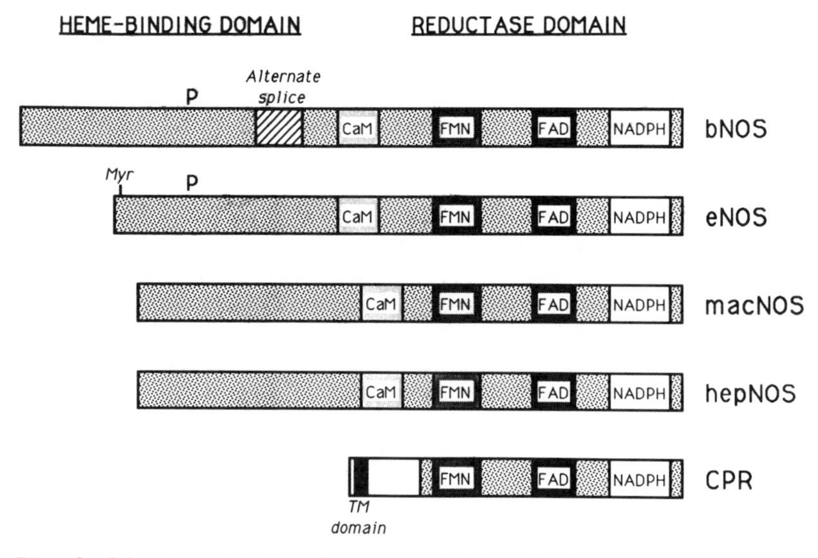

Figure 2 Schematic model of the cofactor recognition sites within NOS enzymes and cytochrome P450 reductase (CPR). Predicted sites for calmodulin binding (CaM), protein kinase A phosphorylation (P), alternative splicing, and myristoyation (Myr) within the NOS sequences and the transmembrane (TM) domain in the CPR sequence are noted.

Under normal circumstances macrophages possess no detectable NOS protein. Stimuli such as interferon-γ and lipopolysaccharide (LPS) elicit new NOS protein synthesis over 2–4 hrs, mediating the NO responses to inflammatory stimuli. It was first thought that macrophages contained the only form of inducible NOS. Following endotoxin treatment, inducible NOS activity has been demonstrated in a great diversity of animal tissues lacking macrophages (54). The hepatocyte-inducible NOS, which has been recently cloned (29), might represent the prototype for nonmacrophage-inducible NOS. Conceivably, the ubiquitous distribution of this form of inducible NOS reflects a primitive sort of immune response. The simplicity of the NO system might have sufficed to repel invading microorganisms early in evolution.

NOS can also be regulated by phosphorylation. Consensus sequences for phosphorylation by cAMP-dependent protein kinase are evident in bNOS and eNOS and hepatic-inducible NOS (Figure 2). These are not as obvious in macNOS. Consensus sites for phosphorylation by other kinases have not been characterized in detail. However, biochemical studies indicate that neuronal NOS can be phosphorylated by cAMP-dependent protein kinase, protein kinase C, cGMP-dependent protein kinase, and Ca^{2+}/calmodulin-dependent protein kinase (35, 55, 56). Phosphorylation by all of these enzymes decreases enzyme catalytic activity (35; JL Dinerman, JP Steiner, TM Dawson, and SH Snyder, in preparation). This provides for multiple levels of enzyme regulation. For instance, Ca^{2+}-calmodulin can directly activate the enzyme, and, by phosphorylation through Ca^{2+}/calmodulin-dependent protein kinase, inhibit enzyme activity. Ca^{2+}, together with lipids, also activates protein kinase C, whose actions would also inhibit NOS. NO stimulates GC to form cGMP, which via cGMP-dependent protein kinase can inhibit NOS.

Phosphorylation of the endothelial enzyme regulates both its enzymatic activity and its subcellular distribution. Unlike bNOS and macNOS, which are largely cytosolic, eNOS is predominately localized to the plasma membrane (20). Michel et al (57) found that eNOS is rapidly phosphorylated in intact endothelial cells in response to bradykinin. Phosphorylation of NOS is associated with translocation of the enzyme from membrane to soluble fractions. Since cytosolic eNOS is catalytically inactive, NO will not be generated within the endothelial cell. Instead, catalytically active, nonphosphorylated NOS is localized to the plasma membrane, where it presumably generates NO that is released into the extracellular environment. While neuronal NOS has been thought to be predominantly soluble, about 50% of NOS activity in brain homogenates is particulate and cannot be solubilized even with strong salt treatment (58). Thus, in neurons as well as blood vessels, the active form of NOS may be the unphosphorylated enzyme localized to the plasma membrane to release NOS to the exterior.

MOLECULAR CLONING

Brain NOS

Isolation of the brain isoform (13) permitted its molecular cloning (21). A two-step PCR cloning strategy was used with oligonucleotide primers based on tryptic peptides sequenced from purified bNOS. The cDNA predicts a polypeptide of 160 kDa and was striking in having 36% identity to CPR in its C-terminal half, the NOS reductase domain, which contains the binding sites of NADPH, FAD, and FMN (Figure 2). This homology to CPR is shared by all NOSs cloned to date and reflects the oxidative mechanism of NO biosynthesis. The sequence of the N-terminal half of NOS, the heme domain, is not similar to that of any cloned gene. Although the classic P450 heme-binding cysteinyl peptide sequence is absent, the amino acids surrounding cysteine 414 show some of the expected homology. Comparison with $P450_{BM3}$, the bacterial heme and flavin-containing P450, shows close alignment of cysteine 675 with the putative heme-binding site in $P450_{BM3}$.

Site-directed mutagenesis of these cysteines may clarify the site of heme binding.

The reductase domain of NOS shares many functional properties with CPR. bNOS catalyzes a rapid NADPH-dependent reduction of cytochrome c. In the absence of arginine, NOS can transfer electrons from NADPH to O_2 and form $O_2^{\dot{-}}$ and H_2O_2. The formation of these reactive oxygen intermediates may contribute to glutamate neurotoxicity and neurodegeneration, as discussed below. Interestingly, L-NNA but not L-NMA blocks the formation of $O_2^{\dot{-}}$ and H_2O_2. Deletional mutagenesis indicates that amino acids 527–1429 fully account for NOS reductase activity (DS Bredt and SH Snyder, in preparation). Presumably, early in evolution CPR donated electrons for NOS and at some point a fusion between CPR and NOS took place. Indeed, when the N-terminal and C-terminal halves are expressed separately and mixed together, one obtains NOS catalytic activity (DS Bredt and SH Snyder, in preparation).

Near the middle of the bNOS cDNA there is an amphipathic α-helix domain, which conforms to the consensus sequence for calmodulin binding. This assignment was confirmed by experiments showing that a peptide corresponding to this region binds calmodulin with low nanomolar affinity in a calcium-dependent manner (59). A consensus sequence for protein kinase A phosphorylation is present at amino acid number 372. Whether this serine is actually phosphorylated by protein kinase A or other enzymes is not yet known.

Recently, bNOS was cloned from human (22) and mouse (60) cerebella. The rat bNOS shares 94% and 98% amino acid identity with the human and mouse, respectively. The human gene was mapped to chromosome 12. Curiously, northern blot analysis reveals a greater abundance of bNOS mRNA in human skeletal muscle than in human brain, while rat skeletal muscle is almost devoid of bNOS mRNA (22). Cloning of the mouse bNOS reveals alternative splicing of the mRNA in brain (60). In the mouse cerebellum, 10% of NOS mRNA has a 415-nucleotide deletion, corresponding to nucleotides 1510–1824. Interestingly, this is in a region of NOS that is highly conserved between the various isoforms. Regulation of this alternative splicing of NOS and functional differences between bNOS expressed from the differentially spliced mRNAs remain important issues.

Endothelial NOS

Endothelial NOS (eNOS) was cloned independently by three laboratories using low-stringency screening strategies based on the DNA sequence of bNOS (23–25). Overall, the predicted sequence shares 60% identity with bNOS. Consensus binding sites for FAD, FMN, NADPH, calmodulin, and protein kinase A phosphorylation are conserved between the brain and endothelial isoforms. A unique feature of the eNOS gene is a consensus sequence for N-terminal myristoylation. This explains the particulate localization of the eNOS protein despite its lack of a membrane-spanning domain. [^{3}H]Myristate is directly incorporated into eNOS, and mutation of the myristoylation sequence renders eNOS soluble (60a). Insertion of the myristoyl group into the plasma membrane presumably accounts for the enzyme's particulate location.

Human eNOS is a large gene, which contains 25 exons spanning 21 kilobases on the 7q35-36 region of chromosome number 7, the same chromosome that contains CPR (61, 61a). The 5′ promoter region of the human gene contains AP-1, AP-2, NF-1, heavy metal, acute-phase response shear stress, and sterol-regulatory elements. These 5′ sequence motifs fit with a recent study showing induction of eNOS in cerebral blood vessels following ischemia (62).

Inducible NOSs

Macrophage NOS-cloning of NOS from macrophages was independently achieved by three laboratories. Two groups used bNOS cDNA as a homologous probe (27, 28), while the third used a macNOS antibody for expression cloning (26). Overall, the amino-acid sequence is 50% identical to bNOS and 51% identical to eNOS. The macNOS cDNA predicts a protein of 133 kDa, with consensus-binding sequences for NADPH, FAD, FMN, and calmodulin. The protein kinase A phosphorylation site conserved between bNOS and eNOS is absent. Putative alternative splicing of macNOS mRNA predicts two isoforms. The shorter version has 22 fewer amino acids at the COOH terminus with 10 terminal amino acids that differ from the longer form (26).

NOS has not yet been cloned from human macrophages. In fact, NOS catalytic activity has only once been identified in human macrophages (63) despite extensive searches in many laboratories. An inducible calcium-independent NOS activity has been well characterized in human hepatocytes following treatment with LPS, interferon-γ, tumor necrosis factor-α, and interleukin-1β. This cDNA was recently cloned and found to share only 82% identity with mouse macNOS, suggesting that it may represent a distinct inducible isoform (29). Independently, an identical human inducible NOS gene was cloned from articular chondrocytes activated with interleukin-1β (64). This human-inducible NOS gene maps to chromosome 17 (65).

For inducible NOS, one would expect the regulatory region of the gene to determine the rate of synthesis of enzyme protein. Characterization of the promoter region of the gene for macrophage-inducible NOS (macNOS) reveals a pattern for complex regulation (66, 67). There appear to be two distinct regulatory regions upstream of the TATA box, which is 30 basepairs upstream of the transcription start site. One of these, region 1, lies about 50–200 basepairs upstream of the start site. Region 1 contains LPS-related response elements such as the binding site for NF-IL6 and the κB binding site for NFκB, indicating that this region regulates the LPS-induced expression of macNOS. Region 2, which is about 900–1000 bases upstream of the start site, does not itself directly regulate NOS expression, but provides a 10-fold increase above the 75-fold increase in NOS expression provided by region 1. Region 2 contains motifs for interferon-γ-related transcription factors and thus is presumably responsible for interferon-γ-mediated regulation. In sum, LPS and interferon-γ-responsive elements occur in two distinct regulatory regions; LPS stimulates macNOS expression directly and interferon-γ acts only in the presence of LPS.

This unique organization of gene enhancers may explain important aspects of inflammation. In sepsis, LPS is released from gram-negative bacterial cell walls and circulates throughout the body to stimulate inflammatory responses. By contrast, interferon-γ is released locally and serves to augment inflammatory responses in specific cell populations close to its release. LPS alone stimulates macrophages only to a limited extent. Interferon-γ elaborated by infiltrating lymphocytes can prime the macrophages for a maximal response to LPS. Thus maximal production of NO is restricted to those cells needed to kill the invader, thereby minimizing damage to adjacent tissue.

NEURONAL FUNCTIONS OF NOS

Neurotransmitter localizations often help clarify function. Following purification of neuronal NOS, antibodies were developed for immunohistochemical staining (68). Throughout the central nervous system, NOS occurs only in neurons. In the cerebral cortex, NOS neurons account for only about 2% of all the cells, disposed in no organized pattern and shaped like medium to large aspiny neurons. In the hippocampus, pyramidal cells lack bNOS, but granule cells of the dentate gyrus have abundant NOS. In the corpus striatum, NOS occurs in both the cell bodies and terminals of the medium aspiny neurons. In the pedunculopontine nucleus and diagonal band of Broca, bNOS occurs in cholinergic neurons. In most areas, NOS appears prominently in neuronal cell bodies, while in the Islands of Callejae NOS staining is confined to a dense fiber bundle.

Unlike the scattered NOS in the cerebral cortex, cerebellar NOS lies in well-defined cell types. NOS is enriched in all granule and basket cells, but not in Purkinje cells. These localizations reveal how glutamate influences cGMP in the cerebellum. GC- and cGMP-dependent protein kinases are selectively concentrated in Purkinje cells, upon which synapse terminals of granule and basket cells. Granule and basket cells possess NMDA receptors, which receive inputs from excitatory mossy fibers. Stimulation of NMDA receptors on basket and granule cells likely triggers formation of NO, which diffuses to Purkinje cells to activate GC.

GC is certainly a target for NO in the cerebellum. In other brain areas, NO may act through other targets. If NO transmission occurred exclusively through GC and if all the GC in the brain were associated with NO transmission, then GC and NOS localizations should be closely similar. However, they differ markedly, indicating that NO may act in other ways than via GC and/or GC may be regulated by other transmitters besides NO.

Virtually all neurons in the brain are thought to utilize more than one neurotransmitter. No consistent colocalizations of NOS with individual transmitter have yet been detected. Thus, in the cerebellum, NOS occurs in the glutamate-containing granule cells as well as the GABA-containing basket cells. Many of the cerebral cortical NOS neurons also contain GABA and neuropeptide transmitters. In the corpus striatum, all NOS neurons stain for somatostatin and neuropeptide Y, but in areas such as the pedunculopontine nucleus of the brain stem, NOS neurons lack somatostatin and neuropeptide Y but stain for choline acetyltransferase (69).

Specific physiologic roles for NOS neurons are not well established. NO certainly is responsible for cGMP generation in some brain regions. Exact functions of cGMP are also unclear. cGMP activates a serine/threonine protein kinase which, in the brain, is selectively expressed in cerebellar Purkinje cells (70). cGMP also regulates ion channels in visual (71) and olfactory tissue (72); however, these channels have not been found in the brain. cGMP can modulate cAMP signalling by regulating the activity of certain cAMP phosphodiesterases. For example, interaction between cAMP and cGMP mediates NO activation of immediate early gene expression in PC12 cells (73). NO appears to influence neurotransmitter release. In several model systems, NOS inhibitors such as nitroarginine block the release of neurotransmitters (74–77). In brain synaptosomes, the release of neurotransmitter evoked by stimulation of NMDA receptors is blocked by nitroarginine (77, 78), while release elicited by potassium depolarization is not affected (78). Presumably glutamate acts at NMDA receptors on NOS terminals to stimulate the formation of NO, which diffuses to adjacent terminals to enhance neurotransmitter release so that blockade of NO formation inhibits release. Potassium depolarization will release transmitter from all terminals, so that any effect of NO would be masked.

PC12 cells, which develop neuronal properties in the presence of nerve growth factor, provide a valuable system linking NO to transmitter release. Rogers and colleagues (79, 80) showed that the release of acetylcholine in response to depolarization is markedly enhanced after eight days of nerve growth factor application. NOS staining and NOS catalytic activity, which are absent in untreated PC12 cells, do not appear until eight days, coincident with marked enhancement of neurotransmitter release. Release of both acetylcholine and dopamine from the cells is blocked by NOS inhibitors and reversed by excess L-arginine (78).

NO may influence differentiating and regenerating neurons. bNOS is transiently expressed throughout the embryonic nervous system. Though absent from dividing cells, NOS is co-expressed together with the earliest markers of the neuronal phenotype. In the rat cerebral cortex, bNOS peaks at gestational day 16 and is nearly absent at birth, which occurs at gestational day 21 (DS Bredt and SH Snyder, unpublished). New bNOS protein is also transiently expressed following neuronal injury. Axotomy induces a marked upregulation of NOS in neurons of the spinal cord and dorsal root ganglion (81). The precise functions of NO in developing and regenerating neurons remain unclear.

Direct evidence for specific neurotransmitter functions of NO comes from studies in the peripheral autonomic nervous system. NOS neurons occur in the myenteric plexus throughout the gastrointestinal pathway (68, 69). Depolarization of myenteric plexus neurons is associated with relaxation of the smooth muscle associated with peristalsis. The blockade of this process by NOS inhibitors indicates that NO is the transmitter (82–85). Recently, homologous recombination techniques have been employed to disrupt the gene for bNOS, resulting in homozygous bNOS "knockout" mice (86). NOS catalytic activity is depleted from the brain, and NOS staining is undetectable in central and peripheral neurons. Yet, in most respects these animals appear normal. Microscopic examination fails to reveal morphologic abnormalities in the brain or most peripheral tissues. The stomachs of bNOS-deficient mice are greatly distended compared to age-matched control mice. Histologic examination reveals circular muscular hypertrophy, especially in the pyloric region, which is likely the result of chronic muscle contraction. The pathology of the stomach resembles that observed in hypertrophic pyloric stenosis, suggesting a role for nitric oxide in this disorder. This conclusion is supported by studies showing a lack of NADPH diaphorase activity in the myenteric neurons of human newborns with pyloric stenosis (87). In these patients, diaphorase staining is normal outside the pyloric region, so that generalized bNOS deficiency is not likely the cause of the disorder.

In blood vessels, besides localizations in the endothelium, NOS occurs in autonomic nerves in the outer, adventitial layers of various large blood vessels (68, 88). In the cerebral circulation and the retina, these neurons derive from cells in the sphenopalatine ganglia at the base of the skull (88, 89). Approximately 40% of the NOS neurons in this ganglia contain the neuropeptide vasoactive intestinal polypeptide (VIP) (88). NOS neurons are prominent in penile tissue—specifically the pelvic plexus and its axonal processes that form the cavernous nerve—as well as in the nerve plexus in the adventitia of the deep cavernosal arteries, and in the sinusoids in the periphery of the corpora cavernosa (90). Electrical stimulation of the cavernous nerve in intact rats produces prominent penile erection, which is blocked by low doses of intravenously administered NOS inhibitors (90). Nerve stimulation–induced relaxation of isolated corpus cavernosum strips is also blocked by NOS inhibitors (91). Accordingly, NO is presumably the transmitter of these nerves, which regulate penile erection.

In the adrenal gland, NOS occurs in discrete ganglion cells and fibers in the medulla (68, 69). Splanchnic nerve stimulation augments both blood flow and catecholamine secretion from the adrenal medulla, with nitroarginine blocking the increased blood flow but not catecholamine secretion (92, 93). In the kidney, NOS is enriched in the macula densa, where it regulates blood flow and glomerular capillary pressure (94). NOS is also prominent in fibers and terminals in the posterior pituitary gland (68, 69), but its relation to function is unclear.

NO has been implicated in long-term potentiation (LTP) in the hippocampus. Nitroarginine application to hippocampal slices blocks LTP formation (95–97). Injection of nitroarginine into pyramidal cells of the hippocampus also inhibits LTP, suggesting that NO might act as a retrograde messenger for LTP, passing from pyramidal cells to Schaffer collateral terminals (98). However, bNOS is not demonstrable in hippocampal pyramidal cells (99).

ROLE OF NO IN NEUROTOXICITY

Although NO participates in normal synaptic transmission, excess levels of NO are neurotoxic. Glutamate released in excess, acting at NMDA receptors, mediates neurotoxicity in the focal ischemia of vascular stroke (100). Glutamate neurotoxicity may also contribute to dysfunction in neurodegenerative diseases such as Alzheimer's and Huntington's diseases. Since glutamate, via NMDA receptors, stimulates NO formation, one might expect excess NMDA receptor stimulation to destroy NOS neurons. Surprisingly, NOS neurons are resistant to NMDA neurotoxicity (101). This conclusion derives from the demonstration that NOS neurons are identical to those that stain for NADPH-diaphorase (69, 102). Diaphorase staining reflects a blue precipitate obtained with tetrazolium dyes in the presence of NADPH (103, 104). Numerous studies have demonstrated that diaphorase-staining neurons are notably resistant to destruction in Huntington's and Alzheimer's diseases, vascular stroke, and NMDA neurotoxicity (105–109). The similarity in localizations of diaphorase and NOS staining neurons in the brain is striking. Transfection of bNOS cDNA into human kidney 293 cells lacking NOS or diaphorase results in staining of the cells for diaphorase and NOS in exactly the same proportions as neurons in the brain (69). Since diaphorase can derive from many NADPH-utilizing enzymes, most diaphorase in brain tissues is unrelated to NOS, but a discrete portion represents NOS (102).

If NMDA stimulates NOS neurons to make NO, but these cells are themselves resistant to neurotoxicity, could the released NO damage other cells? Exposure of cerebral cortical cultures to NMDA kills 60–90% of neurons, with NOS-diaphorase cells being undamaged (101, 105, 106). Treatment with nitroarginine or other NOS inhibitors, removal of arginine from the media, or scavenging NO with hemoglobin block this neurotoxicity (101, 110). The toxicity is also prevented by flavoprotein and calmodulin inhibitors. Superoxide dismutase attenuates neurotoxicity. Since this enzyme removes superoxide, which interacts with NO to form the toxic radical peroxynitrite, NO presumably kills via peroxynitrite. NO has been implicated in NMDA neurotoxicity in a variety of models, including hippocampal slices (111–113), striatal slices (114), and several culture systems (115–119). Others have failed to show that NO is involved in NMDA neurotoxicity (120–122), and NO may be neuroprotective (123). NO may exert both neurodestruction and neuroprotection, depending on its oxidation-reduction status (124, 125), with $NO^{\bullet}$ being neurodestructive and NO^{+} being neuroprotective (124). If NMDA neurotoxicity is responsible for neuronal damage in vascular stroke, then NOS inhibitors should be neuroprotective. Administration of low doses of nitroarginine blocks neural damage following middle cerebral artery occlusion in mice (126), rats (127–129), and cats (130). High doses of NOS inhibitors exacerbate the damage following occlusion of the middle cerebral artery (131, 132), presumably through decreased cerebral blood flow.

NO IN IMMUNE AND CARDIOVASCULAR FUNCTIONS

The first convincingly established role for NO involves its capacity to mediate the bactericidal and tumoricidal actions of macrophages (3, 54). Inflammatory stimuli such as endotoxin enable macrophages to kill tumor cells and bacteria. These stimuli also stimulate the synthesis of very large amounts of new macNOS protein from negligible basal levels. Cytokines that induce macNOS include interferon-γ, factor-α, interleukin-1, and lipopolysaccharide (LPS), a bacterial cell-wall component that elicits symptoms of sepsis. Pharmacologic inhibition of NOS activity in macrophage cultures or deletion of arginine from incubation media blocks tumoricidal and bactericidal activities. Recent studies indicate that besides killing bacteria,

NO can inhibit viral replication (133, 134). Transfection of NOS into cells in culture lowers viral titers (133). In intact rats, NOS inhibitors elevate Coxsackie viral titers and augment mortality from the viral infection (134).

Macrophage NOS may mediate pathologic conditions such as septic shock. In an animal model of sepsis elicited by LPS, the associated hypotension is reversed by NOS inhibitors (135). NOS inhibitors also reverse hypotension in patients with septic shock (136, 137).

NO is likely the major endogenous vasodilator. When a dilator such as acetylcholine or bradykinin acts at receptors on endothelial cells, an influx of calcium binds to calmodulin to activate eNOS. The newly synthesized NO diffuses into adjacent smooth muscle cells where it activates GC. The cGMP provokes muscle relaxation by mechanisms that are not definitely established. Most likely, activation of cGMP-dependent protein kinase with phosphorylation of myosin light chain alters muscle contractility.

NO is implicated in various vascular reflexes. For instance, systemic vasodilatation evoked by hypoxia is prevented by L-NNA (138). Unlike other vascular beds, the pulmonary artery constricts during hypoxia in association with lowered cGMP release, suggesting that hypoxia reduces NO formation in the pulmonary artery (139).

A role for NO in ischemic cardiovascular conditions is controversial. NO donors do reduce myocardial infarct size in animal models (140). Moreover, NO donors diminish mortality in shock associated with ischemia and reperfusion of the splanchnic artery (141).

NO can also regulate the vascular system through its ability to inhibit platelet aggregation (142) and adhesion (143). Molecular mechanisms responsible for influences on platelets have been elusive, but could include effects on GC, phospholipase C (144), or ADP ribosylation (145–148).

Whether NO abnormalities are etiologic in any immune or cardiovascular diseases is unclear. Hypertension and atherosclerosis are attractive candidates, as defects in endothelial-mediated vasodilation are well known in essential hypertension and atherosclerosis (149, 150). For instance, acetylcholine vasodilates the tracheal artery less in hypertensive patients than in controls (150). Moreover, NOS inhibitors reduce brachial artery blood flow less in hypertensive patients than in controls (150). Even if NOS is abnormal in hypertension, one cannot readily ascertain if the disorder is primary or a consequence of the hypertensive process.

TARGETS OF NO ACTION

NO activates GC by binding to iron in the heme, which is at the active site of the enzyme, altering the enzyme's conformation to augment catalysis. NO can bind to nonheme iron in numerous enzymes, such as NADH-ubiquinone oxidoreductase, NADH:succinate oxidoreductase, and *cis*-aconitase, all iron-sulfur enzymes (54, 151). NO can bind to the iron in ferritin, an iron-storage protein, liberating the iron, which could cause lipid peroxidation (152). NO also binds to the nonheme iron of ribonucleotide reductase to inhibit DNA synthesis (153, 154). Its ability to bind iron enables NO to influence iron metabolism. Iron metabolism is regulated posttranscriptionally by specific mRNA-protein interactions between iron regulatory factor (IRF), an iron-sulfur protein, and iron-responsive elements (IRE), which occur in the untranslated regions of the mRNA transcripts for the erythroid form of 5-aminolevulinate synthase, the transferrin receptor, and ferritin (155, 156). NO appears to influence this system. Two laboratories found that NO formed by macrophages stimulates the IRF, which causes translational repression of IRE-containing messenger RNA (157, 157a). Also, NO produced by NMDA receptor activation of cerebellar slices augments the RNA-binding activity of IRF by displacing its iron-sulfur cluster (S Jaffrey, R Klausner, and SH Snyder, in preparation).

NO can stimulate the S-nitrosylation of numerous proteins (124, 158, 159). NO also stimulates the incorporation of NAD into glyceraldehyde-3-phosphate dehydrogenase (145–147, 159a). This modification involves a cysteine at the active site of the enzyme, hence inhibiting its activity and potentially depressing glycolysis.

Which, if any, of these actions is responsible for physiologic or pathologic actions of NO? Recent studies provide evidence that DNA damage is central to NO neurotoxicity (160). NO, like other free radicals, can damage DNA by base deamination (161). DNA damage stimulates the activity of poly (ADP ribose) synthetase (PARS). PARS is a nuclear enzyme that utilizes NAD as a substrate to catalyze the attachment of 50–100 ADP-ribose units to nuclear proteins such as histones and, most prominently, to PARS itself (162). In brain homogenates incubated with [^{32}P]NAD, NO stimulates the poly-ADP-ribosylation of PARS (160). NMDA neurotoxicity in cortical cultures is blocked by a series of PARS inhibitors, with neuroprotective potencies closely paralleling their potencies in inhibiting PARS. These observations implicate the following series of events in NO neurotoxicity (Figure 3). NO damages DNA to activate PARS. Massive activation of PARS depletes the cell of NAD and ATP, as four high-energy phosphate bonds, the equivalent of four molecules of ATP and one of NAD, are consumed in the activation of PARS and the regeneration of NAD, respectively. Considering that PARS is a particularly abundant protein and that catalytic activity involves the addition of up to 100 ADP-ribose units to a single protein molecule, it is not surprising that energy sources are depleted after PARS is activated. The resulting cell death accordingly can be blocked by PARS inhibitors. With lesser degrees of DNA damage, PARS activation is thought to facilitate DNA repair (163).

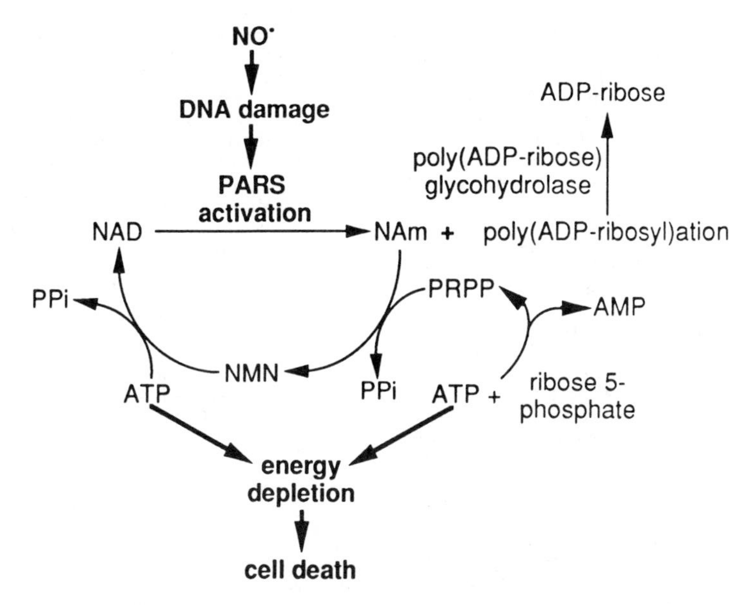

Figure 3 Mechanism of NO-mediated neurotoxicity. DNA damaged by NO activates PARS, which depletes cells of NAD by poly-ADP-ribosylating nuclear proteins. Poly (ADP-ribose) is rapidly degraded by poly (ADP-ribose) glycohydrolase. This futile cycle continues during the prolonged PARS activation. It takes an equivalent of four ATPs to resynthesize NAD from nicotinamide (NAM) via nicotinamide mononucleotide (NMN), a reaction that requires phosphoribosyl pyrophosphate (PRPP) and ATP. The depletion of energy ultimately leads to cell death. Figure is from (160).

Literature Cited

1. Green LC, de Luzuriaga KR, Wagner DA, Rand W, Istfan N., et al 1981. *Proc. Natl. Acad. Sci. USA* 78:7764–68
2. Green LC, Tannenbaum SR, Goldman P. 1981. *Science* 212:56–58
3. Hibbs JB Jr., Taintor RR, Vavrin Z. 1987. *Science* 235:473–76
4. Stuehr DJ, Gross SS, Sukuma I, Levi R, Nathan CF. 1989. *J. Exp. Med.* 169:1011–20

4a. Marletta MA, Yoon PS, Iyengar R, Leaf CD, Wishnok JS. 1988. *Biochemistry* 27:8706–11

5. Arnold WP, Mittal CK, Katsuki S, Murad F. 1977. *Proc. Natl. Acad. Sci. USA* 74:3203–7
6. Ignarro LJ, Lippton H, Edwards JC, Baricos WH, Hyman AL, et al. 1981. *J. Pharmacol. Exp. Ther.* 218:739–49
7. Furchgott RF, Zawadzki JV. 1980. *Nature* 288:373–76
8. Ignarro LJ, Buga GM, Wood KS, Byrns RE, Chaudhuri G. 1987. *Proc. Natl. Acad. Sci. USA* 84:9265–69
9. Palmer RMJ, Ferrige AG, Moncada S. 1987. *Nature* 327:524–26
10. Garthwaite J, Charles SL, Chess-Williams R. 1988. *Nature* 336:385–88

11. Bredt DS, Snyder SH. 1989. *Proc. Natl. Acad. Sci. USA* 86:9030–33
12. Garthwaite J, Garthwaite G, Palmer RMJ, Moncada S. 1989. *Eur. J. Pharmacol.* 172:413–16
13. Bredt DS, Snyder SH. 1990. *Proc. Natl. Acad. Sci. USA* 87:682–85
14. Schmidt HHHW, Pollock JS, Nakane M, Gorsky LD, Forstermann U, Murad F. 1991. *Proc. Natl. Acad. Sci. USA* 88:365–69
15. Mayer B, John M, Bohme E. 1990. *FEBS Lett.* 277:215–19
16. Yui Y, Hattori R, Kosuga K, Eizawa H, Hiki K, Kawai C. 1991. *J. Biol. Chem.* 266:12544–47
17. Stuehr DJ, Cho HJ, Kwon NS, Weise MF, Nathan CF. 1991. *Proc. Natl. Acad. Sci. USA* 88:7773–77
18. Hevel JM, White KA, Marletta MA. 1991. *J. Biol. Chem.* 266:22789–91
19. Evans T, Carpenter A, Cohen J. 1992. *Proc. Natl. Acad. Sci. USA* 89:5361–65
20. Pollock JS, Forstermann U, Mitchell JA, Warner TD, Schmidt HHHW, et al. 1991. *Proc. Natl. Acad. Sci. USA* 88:10480–84
21. Bredt DS, Hwang PM, Glatt C, Lowenstein C, Reed RR, Snyder SH. 1991. *Nature* 351:714–18
22. Nakane M, Schmidt HHHW, Pollock JS, Forstermann U, Murad F. 1993. *FEBS Lett.* 316:175–80
23. Lamas S, Marsden PA, Li GK, Tempst P, Michel T. 1992. *Proc. Natl. Acad. Sci. USA* 89:6348–52
24. Sessa WC, Harrison JK, Barber CM, Zeng D, Durieux ME, et al. 1992. *J. Biol. Chem.* 267:15274–76
25. Janssens SP, Shimouchi A, Quertermous T, Bloch DB, Bloch KD. 1992. *J. Biol. Chem.* 267:14519–22
26. Xie Q-W, Cho HJ, Calaycay J, Mumford RA, Swiderek KM, et al. 1992. *Science* 256:225–28
27. Lyons CR, Orloff GJ, Cunningham JM. 1992. *J. Biol. Chem.* 267:6370–74
28. Lowenstein CJ, Glatt CS, Bredt DS, Snyder SH. 1992. *Proc. Natl. Acad. Sci. USA* 89:6711–15
29. Geller DA, Lowenstein CJ, Shapiro RA, et al. 1993. *Proc. Natl. Acad. Sci. USA* 90:3491–95
30. Pufahl RA, Nanjappan PG, Woodard RW, Marletta MA. 1992. *Biochemistry* 31:6822–28
31. Feldman PL, Griffith OW, Hong H, Stuehr DJ. 1993. *J. Med. Chem.* 36(4): 491–96
32. Dwyer MA, Bredt DS, Snyder SH. 1991. *Biochem. Biophys. Res. Commun.* 176:1136–41
33. Klatt P, Schmidt K, Uray G, Mayer B. 1993. *J. Biol. Chem.* 268:14781–87
34. Moore PK, Babbedge RC, Wallace P, Gaffen ZA, Hart SL. 1993. *Br. J. Pharmacol.* 108:296–97
35. Bredt DS, Ferris CD, Snyder SH. 1992. *J. Biol. Chem.* 267:10976–81
36. Mayer B, John M, Heinzel B, Werner ER, Wachter H, et al. 1991. *FEBS Lett.* 288:187–91
37. McMillan K, Bredt DS, Hirsch DJ, Snyder SH, Clark JE, Masters BSS. 1992. *Proc. Natl. Acad. Sci. USA* 89:11141–45
38. White LA, Marletta MA. 1992. *Biochemistry* 31:6627–31
39. Stuehr DJ, Ikeda-Saito M. 1992. *J. Biol. Chem.* 267:20547–50
40. Rengasamy A, Johns RA. 1993. *Mol. Pharmacol.* 44:124–28
41. Assreuy J, Cunha FQ, Liew FY, Moncada S. 1993. *Br. J. Pharmacol.* 108: 833–37
42. Rogers NE, Ignarro LJ. 1992. *Biochem. Biophys. Res. Commun.* 189: 242–49
43. McMillan K, Masters BSS. 1993. *Biochemistry*. In press
44. Narhi LO, Fulco AJ. 1986. *J. Biol. Chem.* 261:7160–69
45. Tayeh MA, Marletta MA. 1989. *J. Biol. Chem.* 264:19654–58
46. Kwon NS, Nathan CF, Stuehr DJ. 1989. *J. Biol. Chem.* 264:20496–501
47. Schmidt HH, Smith RM, Nakane M, Murad F. 1992. *Biochemistry* 31:3243–49
48. Giovanelli J, Campos KL, Kaufman S. 1991. *Proc. Natl. Acad. Sci. USA* 88:7091–95
49. Hevel JM, Marletta MA. 1992. *Biochemistry* 31:7160–65
50. Stuehr DJ, Kwon NS, Nathan CF, Griffith OW. 1991. *J. Biol. Chem.* 266:6259–63
51. Stuehr DJ, Ikeda SM. 1992. *J. Biol. Chem.* 267:20547–50
52. Pou S, Surichamorn W, Bredt DS, Snyder SH, Rosen GM. 1992. *J. Biol. Chem.* 267:24173–76
53. Cho HJ, Xie Q-W, Calaycay J, Mumford RA, Swiderek KM, et al. 1992. *J. Exp. Med.* 176:599–604
54. Nathan C. 1992. *FASEB J.* 6:3051–64
55. Nakane M, Mitchell J, Forstermann U, Murad F. 1991. *Biochem. Biophys. Res. Commun.* 180:1396–402
56. Brune B, Lapetina EG. 1991. *Biochem. Biophys. Res. Commun.* 181:921–26
57. Michel T, Li GK, Busconi L. 1993. *Proc. Natl. Acad. Sci. USA* 90:6252–56
58. Hiki K, Hattori R, Kawai C, Yui Y. 1992. *J. Biochem.* 111:556–58

59. Vorherr T, Knfel L, Hofmann F, Mollner S, Pfeuffer T, Cafol E. 1993. *Biochemistry* 32:6081–88
60. Ogura T, Yokoyama T, Fujisawa H, Kurshima Y, Esumi H. 1993. *Biochem. Biophys. Res. Commun.* 193(3):1014–22
60a. Busconi L, Michel T. 1993. *J. Biol. Chem.* 268:8410–13
61. Marsden PA, Heng HHQ, Scherer SW, et al. 1993. *J. Biol. Chem.* 268:17478–88
61a. Robinson LI, Weremowicz S, Morton C, Michel T. 1993. *Genomics*. In press
62. Zhang ZG, Chopp M, Zaloga C, Pollock JS, Forstermann U. 1993. *Stroke*. In press
63. Denis M. 1993. *J. Leukocyte Biol.* 49:380–87
64. Charles IG, Palmer MJ, Hickery MS, et al. 1993. *Proc. Natl. Acad. Sci. USA*. In press
65. Xu W, Charles IG, Moncada S, et al. 1993. *Gene*. In press
66. Lowenstein CJ, Alley EW, Raval P, et al. 1993. *Proc. Natl. Acad. Sci. USA*. In press
67. Xie QW, Whisnant R, Nathan C. 1993. *J. Exp. Med.* 177:1779–84
68. Bredt DS, Hwang PM, Snyder SH. 1990. *Nature* 347:76–70
69. Dawson TM, Bredt DS, Fotuhi M, Hwang PM, Snyder SH. 1991. *Proc. Natl. Acad. Sci. USA* 88: 7797–801
70. Lohmann SM, Walter U, Miller PE, Greengard P, De Camilli P. 1981. *Proc. Natl. Acad. Sci. USA* 78:653–57
71. Kaupp UB. 1991. *Trends Neurosci.* 14:150–57
72. Nakamura T, Gold GH. 1987. *Nature* 325:442–44
73. Peunova N, Enlkolopov G. 1993. *Nature* 364:450–53
74. Zhu X-Z, Luo L-G. 1992. *J. Neurochem.* 59:932–35
75. Lonart G, Wang J, Johnson KM. 1992. *Eur. J. Pharmacol.* 220:271–72
76. Dickie BGM, Lewis MJ, Davies JA. 1992. *Neurosci. Lett.* 138:145–48
77. Hanbauer I, Wink D, Osawa Y, Edelman GM, Gally J. 1992. *NeuroReport* 3:409–12
78. Hirsch DB, Steiner JP, Dawson TM, Mammen A, Hayek E, Snyder SH. 1993. *Curr. Biol.* In press
79. Sandberg K, Berry CJ, Eugster E, Rogers TB. 1989. *J. Neurosci.* 9:3946–54
80. Sandberg K, Berry CJ, Rogers TB. 1989. *J. Biol. Chem.* 264:5679–86
81. Verge VMK, Xu Z, Xu X-J, Wiesenfeld-Hallin Z, Hokfelt T. 1992. *Proc. Natl. Acad. Sci. USA* 89:11617–21
82. Bult H, Boeckxstaens GE, Pelckmans PA, Jordaens FH, Van Maercke YM, Herman AG. 1990. *Nature* 345:346–47
83. Boeckxstaens GE, Pelckmans PA, Ruytjens IF, Bult H, Deman JG, et al. 1991. *Br. J. Pharmacol.* 103:1085–91
84. Tottrup A, Svane D, Forman A. 1991. *Am. J. Physiol.* 260:G385-G389
85. Desai KM, Sessa WC, Vane JR. 1991. *Nature* 351:477–79
86. Huang PL, Dawson TM, Bredt DS, Snyder SH, Fishman MC. 1993. *Cell*. In press
87. Vanderwinden J-M, Mailleux P, Schiffmann SN, Vanderhaeghen J-J, DeLaet MH. 1992. *New Engl. J. Med.* 327.8: 511–15
88. Nozaki K, Moskowitz MA, Maynard KI, et al. 1993. *J. Cerebral Blood Flow Metab.* 13:70–79
89. Yamamoto R, Bredt DS, Snyder SH, Stone RA. 1993. *Neuroscience* 54:189–200
90. Burnett AL, Lowenstein CJ, Bredt DS, Chang TSK, Snyder SH. 1992. *Science* 257:401–3
91. Rajfer J, Aronson WJ, Bush PA, Dorey FJ, Ignarro LJ. 1992. *New Engl. J. Med.* 326:90–94
92. Breslow MJ, Jordan DA, Thellman ST, Traystman RJ. 1987. *Am. J. Physiol.* 252:H521-H528
93. Breslow MJ, Tobin JR, Bredt DS, Ferris CD, Snyder SH, Traystman RJ. 1992. *Eur. J. Pharmacol.* 87:682–85
94. Wilcox CS, Welch WJ, Murad F, Gross SS, Taylor G, Levi R. 1992. *Proc. Natl. Acad. Sci. USA* 89:11993–97
95. Bohme GA, Bon C, Stutzmann J-M, Doble A, Blanchard J-C. 1991. *Eur. J. Pharmacol.* 199:379–81
96. O'Dell TJ, Hawkins RD, Kandel ER, Arancio O. 1991. *Proc. Natl. Acad. Sci. USA* 88:11285–89
97. Haley JE, Wilcox GL, Chapman PF. 1992. *Neuron* 8:211–16
98. Schuman EM, Madison DV. 1991. *Science* 254:1503–6
99. Bredt DS, Snyder SH. 1992. *Neuron* 8:3–11
100. Choi DW. 1988. *Neuron* 1:623–34
101. Dawson VL, Dawson TM, Bartley DA, Uhl GR, Snyder SH. 1993. *J. Neurosci.* 13:2651–61
102. Hope BT, Michael GJ, Knigge KM, Vincent SR. 1991. *Proc. Natl. Acad. Sci. USA* 88:2811–14

103. Thomas E, Pearse AGE. 1964. *Acta Neuropathol.* 3:238–49
104. Thomas E, Pearse AGE. 1961. *Histochemistry* 2:266–82
105. Koh J-Y, Peters S, Choi DW. 1986. *Science* 234:73–76
106. Koh J-Y, Choi DW. 1988. *J. Neurosci.* 8:2153–63
107. Ferrante RJ, Kowall NW, Beal MF, Richardson EP Jr, Bird ED, Martin JB. 1985. *Science* 230:561–63
108. Uemura Y, Kowall NW, Beal MF. 1990. *Ann. Neurol.* 27:620–25
109. Hyman BT, Marzloff K, Wenniger JJ, Dawson TM, Bredt DS, Snyder SH. 1992. *Ann. Neurol.* 32:818–20
110. Dawson VL, Dawson TM, London ED, Bredt DS, Snyder SH. 1991. *Proc. Natl. Acad. Sci. USA* 88:6368–71
111. Izumi Y, Benz AM, Clifford DB, Zorumski CF. 1992. *Neurosci. Lett.* 135:227–30
112. Moncada C, Lekieffre D, Arvin B, Meldrum B. 1992. *NeuroReport* 3:530–32
113. Wallis RA, Panizzon K, Wasterlain CG. 1992. *NeuroReport* 3:645–48
114. Kollegger H, McBean GJ, Tipton KF. 1993. *Biochem. Pharmacol.* 45:260–64
115. Lustig HS, von Brauchitsch KL, Chan J, Greenberg DA. 1992. *J. Neurochem.* In press
116. Cazevieille C, Muller A, Meynier F, Bonne C. 1993. *Free Radical Biol. Med.* 14:389–95
117. Corasaniti M, Tartaglia RL, Melino G, Nistico G, Finazzi-Agro A. 1992. *Neurosci. Lett.* 147:221–23
118. Reif DW. 1993. *NeuroReport* 4:566–68
119. Tamura Y, Sato Y, Akaike A, Shiomi H. 1992. *Brain Res.* 592:317–25
120. Demerle-Pallardy C, Lonchampt MO, Chabrier PE, Braquet P. 1991. *Biochem. Biophys. Res. Commun.* 181: 456–64
121. Pauwels PJ, Leysen JE. 1992. *Neurosci. Lett.* 143:27–30
122. Regan RF, Renn KE, Panter SS. 1993. *Neurosci. Lett.* 153:53–56
123. Lei SZ, Pan ZH, Aggarwal SK, et al. 1992. *Neuron* 8:1087–99
124. Lipton SA, Choi YB, Pan Z-H, et al. 1993. *Nature* 364:626–32
125. Snyder SH. 1993. *Nature* 364:577
126. Nowicki JP, Duval D, Poignet H, Scatton B. 1991. *Eur. J. Pharmacol.* 204:339–40
127. Trifiletti RR. 1992. *Eur. J. Pharmacol.* 218:197–98
128. Buisson A, Plotkine M, Boulu RG. 1992. *Br. J. Pharmacol.* 106:766–67
129. Nagafuji T, Matsui T, Koide T, Asano T. 1992. *Neurosci. Lett.* 147:159–62
130. Nishikawa T, Kirsch JR, Koehler RC, Bredt DS, Snyder SH, Traystman RJ. 1993. *Stroke.* In press
131. Yamamoto S, Golanov EV, Berger SB, Reis DJ. 1992. *J. Cerebral Blood Flow Metab.* 12:717–26
132. Dawson DA, Kusumoto K, Graham DI, McCulloch J, Macrae IM. 1992. *Neurosci. Lett.* 142:151–54
133. Karupiah G, Xie Q-W, Buller RML, Nathan C, Duarte C, MacMicking JD. 1993. *Science* 261:1445–48
134. Lowenstein CJ, Herskowitz A, Snyder SH. 1993. *J. Clin. Invest.* In press
135. Kilbourn RG, Jubran A, Gross SS, Griffith OW, Levi R, Lodato RF. 1990. *Biochem. Biophys. Res. Commun.* 172:1132–38
136. Hotchkiss RS, Karl IE, Parker JL, Adams HR. 1992. *Lancet* 339:434–35
137. Petros A, Bennett D, Vallance P. 1991. *Lancet* 338:1557–58
138. Sun MK, Reis DJ. 1992. *Life Sci.* 50:555–65
139. Rodman DM, Yamaguchi T, Hasunuma K, O'Brien RF, McMurtry IF. 1990. *Am. J. Physiol.* 258:L207-L214
140. Siegfried MR, Erhardt J, Rider T, Max L, Lefer AM. 1992. *J. Pharmacol. Exp. Ther.* 260:668–75
141. Carey C, Siegfried MR, Ma XL, Weyrich AS, Lefer AM. 1992. *Circ. Shock* 38:209–16
142. Radomski MW, Palmer RMJ, Moncada S. 1990. *Proc. Natl. Acad. Sci. USA* 87:5193–97
143. Sneddon JM, Vane JR. 1988. *Proc. Natl. Acad. Sci. USA* 85:2800–4
144. Durante W, Kroll MH, Vanhoutte PM, Schafer AI. 1992. *Blood* 79:110–16
145. Zhang J, Snyder SH. 1992. *Proc. Natl. Acad. Sci. USA* 89:9382–85
146. Kots AY, Skurat AV, Sergienko EA, Bulargina TV, Severin ES. 1992. *FEBS Lett.* 300:9–12
147. Dimmeler S, Lottspeich F, Brune B. 1992. *J. Biol. Chem.* 267:16771–74
148. Molina y Vedia L, McDonald B, Reep B, Brune B, DiSilvio M, et al. 1992. *J. Biol. Chem.* 267:4929–32
149. Ludmer PL, Selwin AP, Shook TL, Wayne RR, Mudge GH, et al. 1986. *New Engl. J. Med.* 315:1046–51
150. Panza JA, Quyyumi AA, Brush JE, Epstein SE. 1990. *New Engl. J. Med.* 323:22–27
151. Hibbs JB Jr, Taintor RR, Vavrin V , et al. 1990. In *Nitric Oxide from L-Arginine: A Bioregulatory System,* ed. S Moncada, EA Higgs, pp. 189–223. Amsterdam: Elsevier

152. Reif DW, Simmons RD. 1990. *Arch. Biochem. Biophys.* 283:537–41
153. Lepoivre M, Chenais B, Yapo A, Lemaire G. 1990. *J. Biol. Chem.* 265:14143
154. Kwon NS, Stuehr DJ, Nathan CF. 1991. *J. Exp. Med.* 174:761–68
155. Klausner RD, Rouault TA. 1993. *Mol. Biol. Cell* 4:1–5
156. Munro H. 1993. *Nutr. Rev.* 51:65–73
157. Weiss G, Goossen B, Doppler W, et al. 1993. *EMBO J.* 12:3651–57
157a. Drapier J-C, Hirling H, Wietzerbin J, Kaldy P, Kühn LC. 1993. *EMBO J.* 12:3643–49
158. Stamler JS, Simon DI, Osborne JA, et al. 1992. *Proc. Natl. Acad. Sci. USA* 89:444–48
159. Stamler JS, Singel DJ, Loscalzo J. 1992. *Science* 258:1898–902
159a. McDonald LJL, Moss J. 1993. *Proc. Natl. Acad. Sci. USA* 90:6238–41
160. Zhang J, Dawson VL, Dawson TM, Snyder SH. 1993. *Science*. Submitted
161. Wink DA, Kasprzak KS, Maragos CM, Elespuru RK, Misra M, et al. 1991. *Science* 254:1001–3
162. de Murcia G, Menissier-de Murcia J, Schreiber V. 1991. *BioEssays* 13:455–62
163. Gaal JC, Smith KR, Pearson CK. 1987. *Trends Biol. Sci.* 12:129–30

Annu. Rev. Biochem. 1994. 63:197–234

STRUCTURE, FUNCTION, REGULATION, AND ASSEMBLY OF D-RIBULOSE-1,5-BISPHOSPHATE CARBOXYLASE/OXYGENASE*

Fred C. Hartman and Mark R. Harpel
Biology Division, Oak Ridge National Laboratory, Oak Ridge, Tennessee 37831-8077

KEY WORDS: Rubisco, chaperonins, activase, site-directed mutagenesis, three-dimensional structure

---

CONTENTS

INTRODUCTION

The importance of D-ribulose-1,5-bisphosphate carboxylase/oxygenase (Rubisco, EC 4.1.1.39) would be difficult to exaggerate, because it provides the only quantitatively significant link between the pools of inorganic and organic carbon in the biosphere. As catalyst for the net biosynthesis of carbohydrates from atmospheric carbon dioxide, Rubisco enables plants to meet their total life-sustaining requirements for organic carbon; photosynthetic organisms (whether living now or in the past), in turn, provide food, oxygen, and energy for the animal kingdom.

Apart from the aura associated with the realization that life is underpinned by a single catalyst, Rubisco has attracted enormous attention for a variety of specific reasons: (*a*) The enzyme is schizophrenic, catalyzing the oxidative degradation of D-ribulose-1,5-bisphosphate (RuBP) in competition with the biosynthetic carboxylation of RuBP, and is inherently inefficient (low turnover number) as well. These constraints dictate CO_2 fixation as the rate-limiting step in the overall photosynthetic assimilation of carbon. Such considerations invite questions of whether Rubisco can be redesigned to increase its k_{cat} or to improve its specificity for CO_2 or both. If yes, would a plant transformed with the gene encoding the redesigned enzyme exhibit superior growth and yield characteristics? (*b*) Another notable feature of the enzyme includes its abundance (50% of the total soluble protein in green leaves), which is probably compensatory for its inefficiency. (*c*) The inherent oxygenase activity without intervention of cofactors and the multiple transition states that must be stabilized during the conversion of RuBP to 3-phospho-D-glycerate (PGA) are mechanistically intriguing. (*d*) Distinct from catalytic mechanism per se, a complex hierarchy for light-coupled regulation of Rubisco has evolved with several unique features. (*e*) Rubisco serves as a model for defining the roles of chaperonins in folding and assembly of multimeric proteins. (*f*) Rubisco continues as a paradigm for plant molecular biology, including gene regulation, posttranslational processing, and interorganellar communications.

Given the long-term multidisciplinary assault on Rubisco, the associated literature is truly voluminous. In this chapter, we merely highlight recent advances in mechanism, structure/function relationships, regulation, and assembly. In-depth surveys, combined with historical perspectives, are available for each of these topics (1–14).

SYNTHESIS AND ASSEMBLY

Nature elaborates two architecturally distinct, but functionally analogous, classes of Rubisco: a homodimer (L_2) of 50-kDa subunits as found in purple, nonsulfur bacteria (e.g. *Rhodospirillum rubrum*), and a hexadecamer (L_8S_8) of eight large (L) subunits (53-kDa) and eight small (S) subunits (14-kDa) as present in all other photosynthetic organisms (see Ref. 1). Synthesis and assembly of functional Rubisco in plants and green algae require communication between organelles, because S subunits are encoded by the nuclear genome and synthesized in the cytosol, whereas L subunits are encoded by the chloroplast genome and synthesized on chloroplast ribosomes. Control of the expression of genes for Rubisco occurs both transcriptionally and posttranscriptionally, but apparently differs in the nucleus as compared to the chloroplast (2, 15–20). After translation, newly synthesized S subunits must be translocated across the chloroplast membrane, where an N-terminal signal peptide is proteolyzed prior to assembly with L subunits (21–24). Proteolysis of a small N-terminal region of nascent L subunit is also observed (25, 26). Covalent posttranslational modifications, including acetylation, *N*-methylation, phosphorylation, and possibly transglutamination, have been noted in plant and cyanobacterial Rubiscos, but a correlation of these modifications with Rubisco function or stability has yet to be demonstrated (27–32). Assembly of both L_2 and L_8S_8 Rubisco from photosynthetic prokaryotes and rhodophytic algae is apparently simpler, given that the genes encoding their subunits are found within the same operon and that synthesis and assembly are cytosolic events (2, 33, 34).

Efforts to express functional Rubisco in *Escherichia coli* for structure/function studies have revealed another aspect of Rubisco synthesis: Proper folding of Rubisco is promoted by, and at least in the case of higher-plant L_8S_8 enzymes, dependent on chaperonins. As first defined by the group of Ellis (35), the chaperonins are a class of sequence-related proteins, found in prokaryotes, mitochondria, and plastids, which assist in the correct folding of nascent polypeptide chains without themselves being incorporated into the final structure. Excellent reviews (5, 6, 13, 36–39) and an entire volume (41) are available on the evolving understanding of chaperonin action in protein folding.

To Fold or Not to Fold?

Directed development of a superior Rubisco catalyst is a potential avenue for increasing crop yields. However, this goal requires a genetic system by which systematic alterations of the protein can be rapidly accomplished (i.e. via site-directed mutagenesis) and systematically evaluated. Although the genes encoding the large (L) and small (S) polypeptide chains that make up plant Rubisco are readily expressed in *E. coli,* it has not been possible to synthesize functional enzyme in this foreign host (42). The problem is traced to instability of nascent chains of the higher-plant type Rubisco; plant Rubisco L subunit synthesized in *E. coli* in the absence of S (42, 43) and L subunit extracted from plant L_8S_8 enzyme (44) are either insoluble or improperly folded. However, functional L_2 and L_8S_8 forms of Rubisco from bacteria and cyanobacteria can be synthesized in *E. coli* at levels varying from 0.1% to >10% of total soluble protein (42, 45–59). Stable L_8 core has also been obtained by stripping S subunits from the cyanobacterial L_8S_8 Rubisco (60, 61) or by expression of the *Anacystis nidulans rbc*L gene in *E. coli* in the absence of S subunit (52, 62–65). Therefore, subtle structural differences between the plant and prokaryotic L subunits seem to influence folding stability.

It was subsequently demonstrated that *E. coli* strains cotransformed with plasmids for overexpressing Rubisco genes and the groE operon produced greater yields of both cyanobacterial L_8S_8 Rubisco and *R. rubrum* L_2 Rubisco, without increasing the rate or extent of expression. The groE proteins are heat-response chaperonins first characterized as accessory factors necessary for phage assembly and propagation in *E. coli.* Interestingly, active Rubisco was still produced from the *R. rubrum rbc* gene in a groES$^-$ deletion strain of *E. coli*, but the level of expression was increased fivefold upon introduction of groES (66). Thus, the *E. coli* chaperonins actively promote folding of some polypeptides that can fold spontaneously.

Chaperonins participate during normal cellular function as well as times of stress, such as heat shock (although not all chaperonins are heat-shock proteins). The chaperonins that function in protein folding are classified as cpn60 (also denoted hsp60 or groEL, depending upon source), which forms oligomeric structures composed of two stacked rings of seven subunits (M_r ~60,000) each, and cpn10 (hsp10 or groES), which oligomerizes as a single ring of seven subunits (M_r ~10,000). The cpn60 and cpn10 oligomers act in concert to assist protein folding in an ATP-coupled process by inhibiting incorrect assembly pathways.

The enhanced in vivo expression of functional *R. rubrum* Rubisco elicited by overexpression of the groE operon has led to the development of an in vitro reconstitution system for studying the refolding of denatured Rubisco. A scheme for the requirements and mechanism of chaperonin-assisted protein folding has resulted from these studies (67). Rubisco denatured with either acid or chaotropic agents assumes unique unfolded conformations, which differ in degree of secondary structure (67, 68). These denatured forms are capable of refolding in solution, but only within a restricted set of conditions (68, 69). As expected, the efficiency of the refolding process is greatly promoted by purified groEL/groES chaperonins in the presence of MgATP.

Independent of denaturing method (and therefore initial conformations), chaperonin-assisted refolding proceeds with the same rate constant but to different extents (67). This suggests involvement of a common folding intermediate formed upon dilution of either acid- or chaotrope-denatured Rubisco. The extent of refolding is limited by a partition between irreversible aggregation of the partially folded polypeptide and committed folding to the native form. Demonstration of a binary complex between this intermediate (or a derived equilibrium state) and the groEL oligomer (67, 68) indicates that the chaperonin commits this semi-stable state to complete folding by trapping it from solution.

In the absence of groES, groEL catalyzes uncoupled hydrolysis of ATP. Oligomeric groES blocks the uncoupled hydrolysis reaction through a MgATP- and potassium-dependent binding interaction with oligomeric groEL. This interaction causes the release of stably folded intermediate from the binary complex of Rubisco intermediate and groEL. Whereas discharge of the groEL/Rubisco binary complex by groES clearly requires coupled potassium-dependent ATP hydrolysis, chaperonin-independent spontaneous folding of Rubisco in solution does not show this dependence (69). Potassium enhances the affinity of groEL for ATP (70). The protein released from groEL by groES is monomeric and apparently already committed to assembly of dimeric Rubisco, as demonstrated with a mutant whose dissociation constant greatly favors the monomeric species (71). Nonhydrolyzable analogs of ATP do not support the chaperonin-dependent refolding reaction. Purified chloroplast and yeast cpn60s can functionally replace groEL, but are less effective in assisting the folding reaction (67).

The chaperonin-assisted refolding process for L_2 Rubisco is then defined by several steps: partial spontaneous folding of denatured states to an intermediate recognized by groEL, formation of the groEL/partially folded intermediate binary complex, chaperonin-assisted folding, potassium/MgATP-dependent dissociation of the complex modulated by groES, and chaperonin-independent assembly into stable dimers. In addition to the denatured intermediates discussed above, several other folding intermediates along the Rubisco refolding pathway have been identified kinetically and spectroscopically (68) and shown to resemble intermediates reported in the folding pathways of other proteins (72). In vitro, chaperonins enhance the yield of correctly folded protein by inhibiting unproductive folding pathways. A similar role in vivo is likely.

Cognate proteins have been identified in a number of photosynthetic organisms that may also act as chaperonins in the folding, translocation, and assembly of newly synthesized Rubisco subunits (35, 73–77). Both in vitro and in isolated chloroplasts, the majority of synthesized Rubisco L subunits are associated with another major chloroplast protein, initially termed "Rubisco-binding protein" (78, 79). The binding protein (hereafter referred to as Chcpn60) is composed of two sequence-divergent polypeptides, designated α and β, which associate in a mixed tetradecameric double-toroidal assembly resembling that of the groE oligomer (80–82). Cloning and sequencing of the corresponding cDNAs revealed homology with the *E. coli* groEL chaperonin (35). The amount of L chains bound by this accessory protein decreased upon cessation of protein synthesis in a MgATP-dependent manner as Rubisco activity increased (78, 79, 83, 84), and antibodies against the Chcpn60 protein inhibited appearance of oligomeric L (85). Therefore, Chcpn60 may function analogously to groEL in Rubisco folding in the heterologous system. Chcpn60 also reportedly binds newly imported S subunits (86, 87).

Complete assembly of L into L_8S_8 in the presence of exogenous S subunit, KCl, and MgATP has been demonstrated with an in vitro system of chloroplast extracts (88, 89). A putative L_8 folding intermediate accumulates during protein synthesis in the absence of KCl and S (90). Although ATP is required for formation of the L_8 intermediate, synthesis of L_8S_8 from the intermediate and S is no longer ATP-dependent. Therefore, binding of S appears to be a final, ATP-independent step. Previously, reassociation of the cyanobacterial L_8 core with isolated S subunit from either homologous or heterologous sources to form functional Rubisco also pointed to spontaneous association of S as the final step in assembly of the L_8S_8 Rubiscos (60, 61, 91).

Future Directions

As pointed out by Gatenby & Ellis (6), the next requirement for defining the role of plastid chaperonins in the assembly of Rubisco is the demonstration of a defined in vitro reconstitution system composed of isolated components. Recently, the genes encoding the chloroplast cpn60 α and β subunits have been cloned and functionally analyzed in *E. coli* (92, 93). It was shown that the plant cpn60 subunits were functional in the foreign host, but only the β-type subunits promoted the folding of coexpressed cyanobacterial Rubisco; expression of α alone did not enhance the level of functional Rubisco formed and coexpression of α and β actually suppressed assembly. This suppression was relieved by an increased expression of the groES protein. Therefore, the α-type subunits of Chcpn60 may impose specificity, perhaps for a plastid cpn10. Likewise, *E. coli* cpn10 does not bind the *Rhodobacter sphaeroides* cpn60 with specificity (76). Unfortunately, heterologous expression of plant Rubisco remains elusive. Evidence for a cpn10 in higher plants (94) may supply the missing component necessary for reconstituting an efficient higher-plant Rubisco folding system in vitro and in *E. coli*. Other molecular chaperones may also be required for complete assembly, although such a requirement is not yet documented. For example, the hsp70-type chaperones bind nascent polypeptides and transfer them to the cpn60 chaperonin (13). A chloroplast hsp70 cognate has been identified (77), so such an interaction may also be involved in the folding pathway of plant Rubisco.

Optimization of the chaperonin-assisted folding system is also necessary. Exploiting information gained from studies of *E. coli* chaperonin-assisted folding of Rubisco, Larimer & Soper (59) have increased the yield of functional *Anabaena* L_8S_8 Rubisco in *E. coli* by coordinated expression of the groE operon with the *rbc* operon of *Anabaena* 7120 in combination with a regimen of slow growth and inclusion of KCl in the medium. Optimized expression of all forms of Rubisco in *E. coli* should be attainable through the correct combination of chaperonin and culture conditions.

REGULATION

As catalyst for a rate-limiting, irreversible step in photosynthetic carbon assimilation, Rubisco is highly regulated through diverse, but interrelated, light-coupled processes. Although these processes are not completely understood, several key features have been uncovered and extensively investigated. The equilibrium between the noncarbamylated (inactive) and carbamylated (active) forms of the enzyme may be viewed as the fulcrum for regulation. Numerous experiments with chloroplasts and leaves have shown a greater proportion of the total Rubisco in the active form in response to increased light intensity. Efforts have thus focused on identifying chloroplast factors that alter the ratio of the two forms, and on how these factors are influenced by light.

Spontaneous Activation

As discovered by the groups of Laing (95) and Lorimer (96), catalytic competence of Rubisco, irrespective of species, in both carboxylation and oxygenation is obligatorily dependent on reversible condensation of CO_2 with an internal lysyl residue (Lys191)[1] to form a carbamate, which is stabilized by the catalytically essential Mg^{2+}:

$$\text{Enz-NH}_2 \underset{CO_2 \quad H^+}{\overset{CO_2 \quad H^+}{\rightleftharpoons}} \text{Enz-NHC}(=O)O^{\ominus} \underset{Mg^{2+}}{\overset{Mg^{2+}}{\rightleftharpoons}} \text{Enz-NHC}(O^{\ominus})(O)\cdots Mg^{2+}$$

[1]Residue numbers designate sequence locations relative to the *R. rubrum* enzyme unless noted otherwise.

Kinetic and equilibria studies show that carbamylation is rate limiting in formation of the active, ternary complex (95–97). Activator CO_2, distinct from substrate CO_2, is not incorporated into PGA during catalysis (98, 98a). Based on the pH dependence of activation, the acceptor lysyl side chain displays a pK_a of 8.0 (96). The high acidity, and hence selectivity in carbamylation, has been attributed to the microenvironment characterized by a positive potential that would stabilize the unprotonated ϵ-amino group and by a potential gradient that would assist the movement of the proton liberated during carbamate formation (99).

Despite the spontaneity of in vitro activation, little occurs at 10 μM CO_2 as prevails in chloroplasts. However, at suboptimal concentrations of CO_2, numerous phosphoesters are positive effectors of activation due to preferential binding to the ternary complex (enzyme•CO_2•Mg^{2+}) (100–102). Enhancement of activation state can only be observed provided enzyme is preincubated with effector prior to assaying at substantial dilution for catalytic activity and provided the effector rapidly dissociates from enzyme. These requirements are dictated by the common binding site for both RuBP and effectors; when included in an assay solution, the positive effector behaves as a competitive inhibitor. By contrast, negative effectors of activation invariably exhibit slow, tight-binding properties. Hence, they are not readily displaced by RuBP and appear inhibitory whether included during a preincubation with enzyme or directly in the assay solution. In fact, RuBP, due to preferential interaction with noncarbamylated enzyme, behaves as a negative effector of activation (103).

Activase-Mediated Activation

A major breakthrough in understanding the in vivo regulation of Rubisco was the discovery by Ogren and his colleagues (104) of a nuclear-encoded chloroplast protein, denoted Rubisco activase, that facilitates activation of Rubisco at physiological concentrations of CO_2. During the screening of *Arabidopsis,* grown from seeds mutagenized with ethyl methanesulfonate, a mutant was recovered that was deficient in Rubisco activation even at high light intensity and high levels of CO_2. Surprisingly, the Rubisco purified from the mutant plant was indistinguishable from the wild-type enzyme. Gel electrophoretic analysis of soluble proteins showed that the mutant phenotype correlated with the absence of 41- and 45-kDa polypeptides (105). Subsequently, a 200-kDa protein (composed of similar amounts of the two subunits) was purified from spinach chloroplasts and shown to stimulate activation of Rubisco at physiological concentrations of CO_2 ($\sim$10 μM) (106).

The two polypeptides of activase from both spinach and *Arabidopsis* are encoded by a single nuclear gene. Alternative splicing of the transcript leads to two mRNAs, whose translation products differ by a 37-amino-acid appendage at the C terminus (107). A point mutation (G-to-A transition) at the 5′-splice junction of intron 3 of the activase gene accounts for the deficiency of the *Arabidopsis* mutant (108). The situation is somewhat different in barley, which contains two activase genes in tandem and three mRNAs (109). Activase appears ubiquitous among higher plants and occurs in algae as well (110); activase genes have also been detected in some, but not all, cyanobacteria (111).

cDNAs for both subunits of spinach activase have been cloned, sequenced, and expressed in *E. coli* (112). Both purified, native isoforms (each containing only one type of subunit) facilitate activation of Rubisco, so the significance of subunit heterogeneity of wild-type activase is unclear. As predicted by the sequence of the cDNA and verified by in vitro synthesis of activase as directed by the cDNA, the initial translation product contains a transit peptide of 58 amino acids as necessary for chloroplast import (113). The mature activase does not appear to share sequence similarities with other proteins except for two short segments indicative of nucleotide-binding domains.

The requirements for activase activity were initially defined by use of a reconstituted system with activase partially purified from spinach chloroplasts (114). Necessary components included CO_2, Mg^{2+}, thylakoid membranes (in conjunction with an electron acceptor), non-activated Rubisco, activase, and RuBP. Upon illumination, rapid activation of Rubisco proceeded. Several critical observations were made in this early examination of the activase system: (*a*) In the absence of RuBP, irrespective of the CO_2 concentration, activase had no influence on the kinetics or equilibrium of the spontaneous activation of Rubisco; (*b*) activase did not alter the rate nor extent of deactivation of Rubisco in the dark as promoted by RuBP; (*c*) the K_{act} for CO_2, i.e. the concentration that supports one-half of the maximal extent of activation, was 4 μM (< physiological concentrations) in contrast to 23 μM as observed for spontaneous activation; and (*d*) an absolute requirement for high levels of RuBP (3 mM optimal) even though RuBP strongly inhibits spontaneous activation.

It was soon recognized that the light dependency of activase in the reconstituted system actually represented a requirement for ATP. Furthermore, activase possesses inherent ATPase activity (115). Nonhydrolyzable analogs cannot substitute for ATP, so clearly the energy derived from hydrolysis is utilized in the activase-facilitated activation of Rubisco. The two processes are not tightly coupled as judged by the V_{max} for ATP hydrolysis of 1.5 units/mg as compared to 30–60 microunits/mg for Rubisco activation (115, 116). When taking into account the high concentrations of Rubisco (3 mM in active sites) and activase (0.04 mM) in chloroplasts

(106), this seemingly low V_{max} of activase is quite adequate to account for the in vivo rates of Rubisco activation upon illumination of plants. Lack of tight coupling between activation and ATPase activities of activase has also been observed by site-directed substitution of amino acid residues in the ATP-binding domain, whereby some mutants have an altered ratio of the two activities (117). Replacement of Lys111 in this domain eliminates both activities (112), consistent with the known essentiality of the corresponding residue in other ATP-binding proteins.

Given prior demonstration that RuBP binds avidly and preferentially to noncarbamylated Rubisco (103) (a K_d of 20 nM vs a K_m of 20 μM with the active ternary complex), in conjunction with the activase dependence on RuBP, the idea that activase might accelerate dissociation of the enzyme•RuBP complex was put forward (114). The group of Portis (118) verified this concept directly by the demonstration that activase, in the presence of ATP, enhances the exchange rate of Rubisco-bound [^{3}H]RuBP with unlabeled substrate. This study also showed that RuBP binding is a two-step process of rapid equilibrium followed by a slow conformational change resulting in stable complexation. Activase affects only the second step, suggesting that the free energy of ATP hydrolysis induces a reversal of the original conformational change and hence release of bound RuBP. The interaction between activase and the enzyme•RuBP complex is rather weak, as indicated by a K_m of 2.7 μM (116).

Based on the above, a partial model for in vivo regulation of Rubisco by activase emerges (10, 119). In the dark, deactivated Rubisco is favored by tight complexation with RuBP. During illumination, activase is activated, perhaps in part by an increased level of ATP (additional processes must be operative, however, because of strong inhibition of activase by ADP) (115). Irrespective of its mode of activation, activase destabilizes the enzyme•RuBP complex and thereby promotes release of RuBP. The free enzyme can then undergo carbamylation, and the resulting ternary complex can be stabilized by the binding of RuBP. In short, activase shifts the preferential binding of RuBP from deactivated to activated Rubisco.

With respect to light activation of the Rubisco/activase system, a thylakoid-mediated stimulation by photon flux was observed with lysed chloroplasts even in the presence of high concentrations of ATP (120). It is tempting to envision regulation of activase through light-controlled phosphorylation/dephosphorylation; however, counterindications have been presented (7).

Although knowledge of activase function is derived primarily from in vitro observations, in vivo studies are entirely compatible. Rubisco activation following illumination of plants appears biphasic (119, 121). The initial phase saturates at low photon flux and correlates with activation of activase.

The second phase correlates with Rubisco activation and saturates in parallel with the rate of photosynthesis.

Role of 2-Carboxyarabinitol 1-Phosphate

Evaluation of the activation level of Rubisco in vivo is normally based on measurement of Rubisco activity in freshly prepared leaf extracts before and after preincubation at high [CO_2] as necessary to achieve full activation. The final activity after preincubation should be consistent with the total concentration of Rubisco active sites [quantified with the [^{14}C]-labeled synthetic reaction intermediate analog 2-carboxy-D-arabinitol 1,5-bisphosphate (CABP)] and the known specific activity of fully activated, purified enzyme. In many cases, the final activity obtained with extracts of darkened leaves was considerably lower than forecast by the active-site concentration. If the leaves were illuminated prior to extraction, no such disparity was found (see Ref. 122 and citations therein). This anomaly with extracts of darkened leaves led to the discovery of the nocturnal inhibitor 2-carboxy-D-arabinitol 1-phosphate (CA1P) (123–127), which differs from CABP only in lacking the C5 phosphate group.

CA1P exhibits avid, preferential binding to the carbamylated form of Rubisco (a K_d of 32 nM) (127). Hence, the apparent inability of Rubisco in extracts of darkened leaves to undergo full activation is not a consequence of impaired carbamylation, but rather due to the failure of RuBP to displace bound CA1P from the potentially active ternary complex. Whereas RuBP can be properly viewed as a negative effector of activation, CA1P is actually a positive effector of activation (stabilization of the carbamylated enzyme). However, the net consequence is inhibition of catalysis because the active site is essentially irreversibly blocked.

The strong inhibition of Rubisco by CA1P and the inverse correlation between light intensity and concentration of CA1P in chloroplasts are certainly suggestive that this metabolite plays a role in Rubisco regulation. Although CA1P can be detected in most plants, its concentration is quite variable among species (128). So little CA1P occurs in spinach, wheat, and *Arabidopsis* that regulation of Rubisco in these plants cannot significantly involve this tight-binding inhibitor. *Phaseolus vulgaris* (garden bean) represents the other extreme, in which the concentration of CA1P can exceed the concentration of Rubisco active sites. In such a species, modulation of Rubisco activity may be achieved almost entirely through association/dissociation of CA1P and carbamylated enzyme. By analogy with activase-facilitated dissociation of RuBP from noncarbamylated Rubisco, activase also promotes release of CA1P from carbamylated enzyme (129). Hence, CA1P can only regulate Rubisco activity by coordination with the activase system.

Although light activation of activase may account for the restoration of catalytic competence to CA1P-inhibited Rubisco, additional factors must be responsible for the diminution of CA1P in the dark and its subsequent formation in the light. In vivo, light-coordinated interconversion of CA1P and its hydrolysis product 2-carboxy-D-arabinitol (CA) has been demonstrated (130). These recent data are compatible with the presence of a light-activated phosphatase and a kinase whose activity, relative to phosphatase, would be diminished in light.

CA kinase has not yet been reported, but two laboratories independently have purified a phosphatase from tobacco with stringent substrate selectivity for CA1P (131, 132). The phosphatase is stimulated by fructose 1,6-bisphosphate and inhibited by inorganic phosphate (133). In the absence of exogenous thiols, the enzyme loses activity, thereby implicating a need of free sulfhydryl groups for maximal activity. Interestingly, DTT inhibits the phosphorylation of CA in vivo (130); this could be explained by activation of CA1P phosphatase and concurrent deactivation of CA kinase. Such a scenario is very reminiscent of the light regulation mediated through ferredoxin/thioredoxin, whereby target proteins may be either activated or deactivated by reduction of disulfide bonds (134).

The brevity of this synopsis of Rubisco regulation precludes adequate coverage of the genetic variation of this regulation and the extensive literature on the influence of environmental factors on photosynthesis by leaves and whole plants, which have generally supported the regulatory processes as deduced in vitro. Perhaps, the most convincing validation of in vivo regulation of Rubisco by activase and CA1P acting in concert is provided by transgenic tobacco plants as reported by Andrews and his colleagues (135). Plants transformed with antisense DNA directed against mRNA for activase synthesize twice as much Rubisco as compensation for reduced levels of activase, but are nevertheless impaired in light-stimulated Rubisco carbamylation and in release of CA1P from the enzyme. For further accounts of in vivo regulation of Rubisco, the reader is referred to other sources (119, 136–142) and the citations found therein.

CATALYSIS

Reaction Pathway

As depicted in Figure 1, Rubisco-catalyzed substrate transformations entail sequential partial reactions linked through discrete intermediates (reviewed in Refs. 1, 12, 143). The carboxylation and oxygenation pathways each follow an ordered Theorell-Chance mechanism, in which enzyme-bound RuBP is converted to an enediol(ate), which reacts bimolecularly with CO_2 or O_2. Rubisco does not form Michaelis complexes with either gaseous

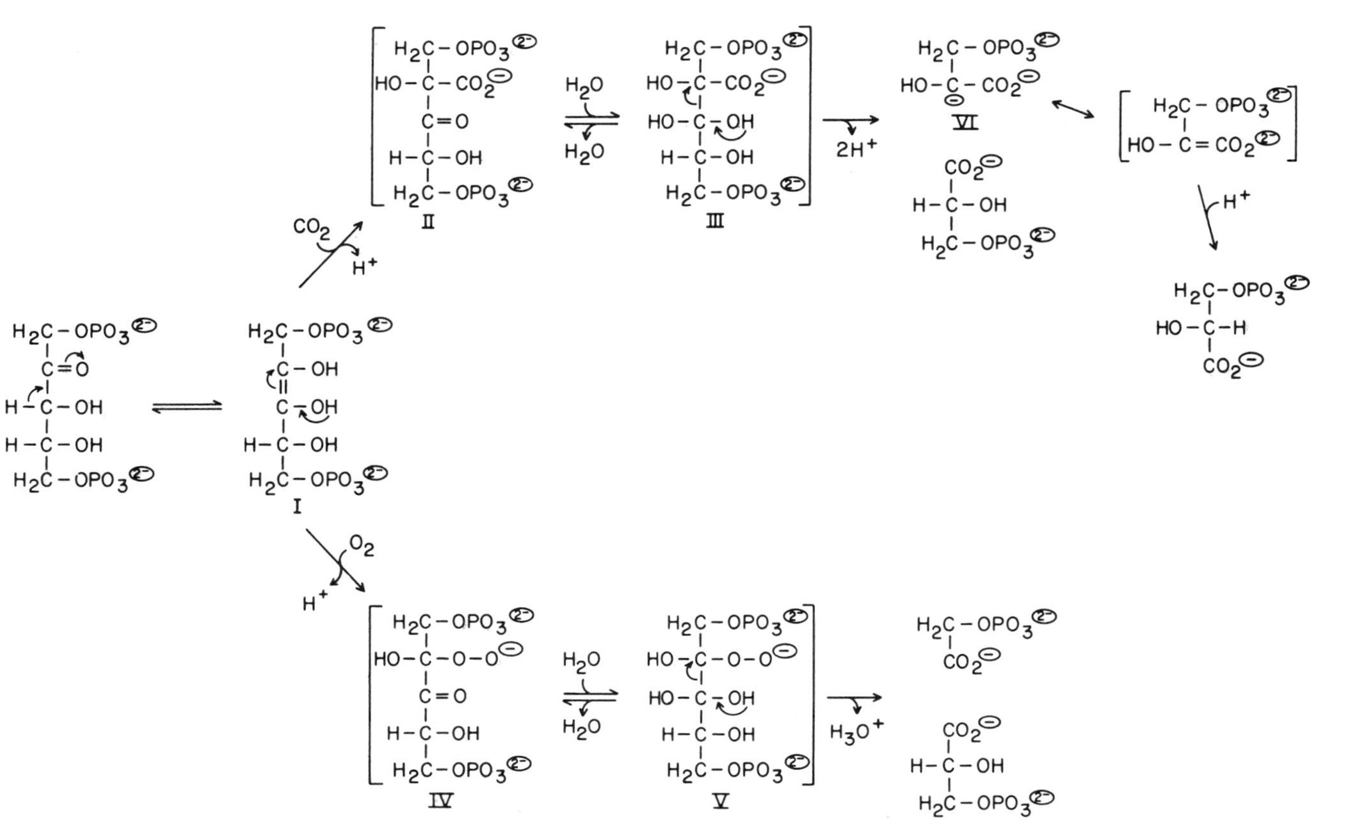

Figure 1 Reaction pathways for the carboxylation and oxygenation of RuBP as catalyzed by Rubisco.

substrate, although each exhibits saturation kinetics with apparent K_m values derived from kinetic rather than equilibrium terms (144). Isotope effects (reviewed extensively in Refs. 12, 143) reveal that rate limitation is distributed throughout multiple steps in the carboxylation pathway.

Deprotonation of C3 of RuBP by Rubisco was demonstrated by the exchange of the C3 proton (labeled with either deuterium or tritium) with solvent and the occurrence of deuterium and tritium isotope effects (145–148). The enediol(ate) intermediate (I) was detected directly in rapid-quench experiments (149, 150). Abstraction of the C3 proton occurs at a rate greater than the overall k_{cat} for carboxylation and is independent of substrate CO_2 (144), although pre-steady-state rapid-quench experiments indicated that tautomerization of the enediol(ate) intermediate to the C2 carbanion is a discrete, partially rate-limiting step in overall catalysis (143). Failure of CO_2 or O_2 to react with the enediol(ate) produced by certain mutant carboxylases demonstrated that this step is enzyme assisted rather than spontaneous (151), as contemplated earlier (152).

Rapid acid-quench of enzyme•RuBP solutions followed by exposure to sodium borohydride detected the 3-keto-2-carboxy intermediate (II) as an epimeric mixture of 2-carboxyarabinitol and 2-carboxylyxitol bisphosphates. Direct quench into borohydride failed to produce the carboxypentitol bisphosphates, presumably due to inaccessibility of the enzyme-bound intermediate and/or its existence as a *gem*-diol. The *R*-configuration at C2 of the isolated carboxypentitol bisphosphates demonstrated that carboxylation occurs at the *si* face of the planar enediol(ate) (153). The crystal structure of a Rubisco•CABP complex (154) supports this interpretation. CABP (chemically synthesized from RuBP) is a slow, tight-binding inhibitor (155, 156), whereas the C2 epimer 2-carboxyribitol bisphosphate behaved merely as a traditional competitive inhibitor (156). Indeed, theoretical consideration of the carboxylation reaction has identified a first-order saddle point in the transition-state energy profile that closely resembles the conformation of the bound CABP molecule modeled crystallographically (157). Inhibition (158) and isotope labeling (159) experiments were consistent with a hydrated 3-keto intermediate (III) bound to the enzyme.

Despite its lability, the 3-keto intermediate (II) can be isolated from acid-quenched reaction mixtures, thereby permitting an examination of its properties as an alternate substrate (160). Rubisco correctly processes the intermediate to the D-isomer of PGA, but at a rate that is only ~3% of overall k_{cat}. Despite this slow rate, the forward commitment factor for processing approaches unity: i.e. decarboxylation, indicative of reversal, is not observed. Based on the premise of a *gem*-diol intermediate (III) as requisite to C2-C3 scission, and the fact that in solution the free ketone form of this intermediate predominates (159), processing of the isolated intermediate at a rate considerably less than k_{cat} can be explained in either of two ways. One is to invoke a rate-limiting conformational change prior to hydration, logically extending the fact that tight binding of CABP entails a slow conformational change (156). Alternatively, hydration and carboxylation of the enediol(ate) (I) may be concerted (161). In this latter situation, the keto form of the six-carbon intermediate (II) would never occur during normal catalytic turnover, and thus would be inefficiently processed. In either case, the hydrolysis rate is limited by an event not on the normal catalytic pathway.

Abstraction of a hydroxyl proton from the *gem*-diol intermediate promotes C2-C3 scission with liberation of PGA (VI) derived from C3, C4, and C5 of RuBP. The resulting *aci*-acid of PGA (derived from C1 and C2 of RuBP and from CO_2) must undergo inversion of configuration at C2 and protonation prior to its release as the D-isomer of PGA. The *aci*-acid intermediate was demonstrated by the observed solvent isotope discrimination against protonation (146) and the discovery of pyruvate as a side product resulting from the β-elimination of phosphate (162) (see below). The frequency of pyruvate formation is invariant with pH and among Rubiscos of differing catalytic efficiencies, suggesting that protonation of the *aci*-acid may be solvent catalyzed.

Oxygenation

Rubisco-catalyzed oxygenation of RuBP, as discovered by the group of Ogren (163) and verified and rigorously characterized by the group of Tolbert (164), occurs in the absence of redox-active metals or organic cofactors such as flavins or pterins; oxygenation reflects inherent reactivity of the enzyme-bound enediol(ate) of RuBP. Carbon-carbon scission of the putative hydroperoxide intermediate (IV) liberates phosphoglycolate directly. Thus, an intermediate analogous to the terminal *aci*-acid of carboxylation does not exist, and one less protonation step is required in the overall pathway. The isotope effect with $^{18}O_2$ is close to the anticipated intrinsic value, suggesting that hydroperoxide formation is the major rate-limiting step (165); scission of the peroxyl O-O bond could also contribute to the isotope effect.

Direct reaction of an unactivated singlet molecule such as RuBP with the O_2 spin triplet is a spin-forbidden process. However, carbanion reactivity towards dioxygen is well known in organic reactions, and so it is not altogether unexpected that the enediol(ate) of RuBP (or a carbanion derived therefrom) can interact with O_2 (for elaboration, see Refs. 1, 166, 167). Accordingly, oxygenation is a far more common, although not universal, reaction of enzyme-bound carbanionic intermediates than was formerly realized (168). The problem of spin inversion can be overcome in several manners. Direct formation of singlet O_2 is energetically unfavorable and unprecedented in biological oxygenation reactions. A more likely mechanism is formation of an intermediate caged radical pair, formally superoxide radical anion and the C2-radical of substrate, which can recombine to produce the hydroperoxy intermediate (164). Because of the distinctive properties of electrophilic CO_2 and the spin triplet O_2, the competition between carboxylation and oxygenation for the enediol(ate) intermediate may involve unique resonance-stabilized tautomers of the enediol(ate). Recently, theoretical treatment of the Rubisco carboxylation and oxygenation reactions based on ab initio energy calculations suggested that geometric perturbation of the enediol(ate) intermediate may be sufficient to decrease the energy gap between the ground-state singlet and the first triplet excited state of the enediol(ate) to allow thermal population of the triplet state (157, 169). This "intersystem crossover" could produce a reactive supermolecule of enediol(ate) and O_2 to form radical intermediates through spin coupling. Thus, Rubisco may act as an oxygenase due to subtle perturbations of the reactive enediol(ate).

Another aspect of oxygenation is stabilization of the hydroperoxy intermediate. Indeed, replacement of the active-site Mg^{2+} of Rubisco with redox-active metals with greater affinities for oxygen or hydroperoxide, such as Mn^{2+} or Co^{2+}, enhances the specificity of Rubisco for oxygenation (170–172). Spectroscopic evidence for the coordination of a putative hydroperoxy intermediate to the active-site metal has been observed with the Cu^{2+} form of this enzyme (173). However, oxygenation may also be enhanced by electronic properties of these transition-metal activator ions not shared by Mg^{2+}. Evidence for such an interaction between the activator metal and carbon intermediates has been obtained in the recently reported chemiluminescence associated only with the Mn^{2+}-substituted enzyme (174). This chemiluminescence, which was oxygen- and RuBP-dependent and suppressed by CABP and CO_2, was initially attributed to singlet oxygen production (dimol emission) during substrate oxygenation. However, the emission phenomenon appears more complex. Chemiluminescence and O_2-uptake correlate reasonably well for wild-type and certain mutant enzymes (174–176), but not for others (177). The oxygenase-related chemiluminescence has since been reinterpreted as arising from a transition of the Mn^{2+} ion, perhaps induced by proximity to the caged radical pair or an excited state of the enediol(ate) (176). An initial burst of luminescence was observed with Rubisco from *R. rubrum*, in contrast to a more linear response for spinach Rubisco, suggesting different rate limitations between the L_2 and L_8S_8 enzymes in the events responsible for chemiluminescence. The potential of this new assay as a probe of the oxygenase reaction is intriguing, but additional characterization and reconciliation of the initial data are needed to exploit this potential fully.

CO_2/O_2 Specificity

Since the discovery (163) and subsequent characterization (164) of the oxygenation reaction catalyzed by Rubisco, a continuing quest has been to obtain a clear definition of the basis for CO_2/O_2 specificity. Phenomenologically, this specificity is not immutable, as observed in the natural variation in the enzyme from species to species (178) and in the altered specificities effected by replacement of the active-site metal (170–172), random and site-directed mutagenesis (177, 179–187), chemical modification (188, 189), and hybridization of heterologous subunits (190). However, the molecular bases of these variations are not entirely clear.

The partition between carboxylation and oxygenation reactions (v_c/v_o) is enumerated by the following relationship: $v_c/v_o = \tau \bullet ([CO_2]/[O_2])$, where τ is a specificity factor defined as the ratio of kinetic properties of the two activities, i.e. V_cK_o/V_oK_c (191). The τ-value can be further analyzed in terms of individual rate constants on the bases that carboxylation and oxygenation are each kinetically ordered reactions and that CO_2 and O_2 react bimolecularly with enediol(ate) without formation of a Michaelis complex. Within this model, the gaseous substrates are also assumed to react irreversibly with the same form of the enediol(ate), although the commitment to forward catalysis has been experimentally proven only for processing of the carboxylated intermediate (160). Given these assumptions, τ can be simplified to the ratio of second-order rate constants for the chemical addition of each gaseous substrate to the enediol(ate) (160, 188). Therefore, specificity can be related to the free energy for the respective transition states as follows (188, 192, 193): $RT\ln\tau = (G^{\ddagger}_{O} - G^{\ddagger}_{C}) - (G^{\ddagger}_{O2} - G^{\ddagger}_{CO2}) = \Delta G^{\ddagger}_{O-C}$, where R is the gas constant, T is the absolute temperature, G_O and G_C are the free energies of the transition states for oxygenation and carboxylation, respectively, G_{O2} and G_{CO2} are the free energies for the gaseous substrates, and ΔG_{O-C} is the free energy difference between transition states.

According to this free energy analysis, alteration of τ must derive from relative changes of the respective transition-state energies as a result of differential stabilization by the enzyme. Consideration of known variations in specificity emphasizes the consequence of even small differences in free energies. For example, the large specificity difference between spinach Rubisco ($\tau = 80$) and the enzyme from *R. rubrum* ($\tau = 10$) equates to a free energy difference of only 1.2 kcal/mol (188), which approximates the energy of a single hydrogen bond. A plant containing the latter enzyme would be incapable of net synthesis of RuBP at air concentrations of CO_2.

The assumptions used in developing the relationships between free energy and specificity may be oversimplistic. In particular, this approach requires that the form of enediol(ate) with which CO_2 or O_2 interacts must be the same. As introduced earlier in this section, the unique requirements of the oxygenation reaction may argue that at least on the microscopic level, this cannot be the case. Indeed, ab initio modeling of these transition states suggests that although the enediol(ate) is deformed and polarized in carboxylation, as is expected as enediol progresses through the C2 carbanion to an sp^3-hybridized adduct with CO_2, further small geometric perturbations of the enediol may be required to attain a low-lying triplet state that couples with O_2 (157, 169). The impact of such a subtle distinction upon the preceding analysis is problematic. However, constraint of enediol(ate) conformations reactive towards O_2 may be one basis for selectivity in favor of carboxylation.

Side Reactions

Decline of carboxylase activity during the course of assay, which is particularly endemic to higher-plant Rubiscos, has been denoted as "fallover" (194). Characterization of the products of RuBP turnover under fallover conditions has shown that this process is a result of enediol(ate)-derived side reactions (195–198). Therefore, chemical promiscuity at the level of the enediol(ate) intermediate is not limited to the oxygenation pathway. Misprotonation at C3, which occurs about once per 400 turnovers (196), gives rise to D-xylulose-1,5-bisphosphate (XuBP), a potent inhibitor of Rubisco (199). The inhibitor accumulates because its utilization as substrate is exceedingly slow. XuBP is carboxylated (with the formation of PGA) at ~0.03% the rate of the normal substrate by the spinach enzyme and displays a very similar K_m (200). In contrast, *R. rubrum* Rubisco, which does not exhibit obvious fallover, carboxylates XuBP at ~0.2% of the RuBP carboxylation rate (187). In theory, turnover of both XuBP and RuBP should utilize a common enediol(ate) intermediate, although this has not been verified experimentally. Abstraction of the C3 proton of XuBP is presumably rate limiting in this reaction, since the active-site base that mediates enolization of RuBP would be pointing towards the C3 hydroxyl.

Another inhibitor is formed in similar amounts to XuBP during RuBP turnover by spinach Rubisco. The chemical properties of this inhibitor are suggestive of 3-keto-arabinitol-1,5-bisphosphate (196, 198), which would result from protonation (rather than carboxylation or oxygenation) at C2 of the enediol(ate) on the same face as carboxylation. Both XuBP and the keto compound were found tightly associated with totally inhibited enzyme (198), specifically linking their binding to the fallover event. Fallover now appears to derive from either the stabilization of non-carbamylated enzyme or the inhibition of catalytically competent enzyme by these two inhibitors, dependent upon reaction conditions (198, 201). One function of activase protein (see section on REGULATION) may be to mitigate fallover in vivo by facilitating the dissociation of Rubisco•inhibitor complexes (10).

Another potential side reaction of the enediol(ate) intermediate is degradation by β-elimination of the C1-phosphate resulting from improper stabilization and/or premature dissociation from the enzyme active site, as was observed for enediol released upon rapid acid quench of Rubisco-RuBP reactions (149). β-Elimination from enediols can occur when significant orbital overlap exists between the π-system of the double bond and the potential leaving group (202). Thus, elimination can be elevated if the phosphate group is not maintained coplanar with the enediol double bond, as suggested by the occasional loss of phosphate from enediol intermediate during triosephosphate isomerase turnover (203) and the observed enhancement of this decomposition following truncation of a crucial stabilizing loop by site-directed mutagenesis (204). Insignificant β-elimination of phosphate is detected in reactions with wild-type *R. rubrum* Rubisco. However, replacement of C1-phosphate ligands by site-directed mutagenesis induces formation of a dicarbonyl compound clearly attributed to elimination of C1 phosphate, demonstrating the required role of these amino acid side chains in stabilizing the enediol(ate) intermediate (see section on SITE-DIRECTED MUTAGENESIS) (205).

As mentioned above, another β-elimination reaction occurs during formation of pyruvate from the terminal *aci*-acid (VI) of phosphoglycerate. Pyruvate formation occurs approximately once per 125 turnovers, and more frequently in the presence of $^{2}H_2O$, as would be expected if protonation and dephosphorylation involve the same intermediate (162). By analogy to studies of the β-elimination of phosphate from the enediol intermediate in the triosephosphate isomerase reaction, it was postulated that pyruvate formation results from imperfect stabilization of the enediol-like *aci*-acid intermediate. Despite an earlier report to the contrary (206), 3-phospho-L-glycerate is not formed during carboxylation, so β-elimination predominates over misprotonation as a side reaction of the *aci*-acid intermediate. β-Elimination may be the price paid during conformational adjustment of the active site and bound catalytic intermediate as necessary to form a stereochemically inverted *aci*-acid intermediate at C2.

THREE-DIMENSIONAL STRUCTURE

Relationships Between L_2 and L_8S_8 Enzymes

Representatives of both classes in various ligand states have been subjected to high-resolution crystallographic analyses by the laboratories of Brändén (8), Schneider (11), and Eisenberg (213) (Table 1). Despite only 30% sequence identity between the two classes, the three-dimensional structure of the L subunit within L_8S_8 holoenzyme is strikingly similar to the L subunit of the L_2 holoenzyme (207, 212). The L subunit consists of two domains illustrated schematically in Figure 2. The small N-terminal domain is built from a five-stranded β-sheet and two α-helices on one side of the sheet. The large C-terminal domain is an eight-stranded β/α-barrel, a folding motif common to numerous proteins exhibiting diverse functions (216). Subunits of the L_2 dimer are aligned antiparallel to form a distorted ellipsoid (45 × 70 × 105 Å). Two major areas of intersubunit interactions are observed: these include barrel-barrel contacts through β-strands and α-helices and contacts between two segments of the N-terminal domain with loops of the barrel domain. These latter interactions are crucial to function, because the active site resides at the interface between the N-terminal domain and the C-terminal barrel domain of the adjacent subunit (217). Thus, a monomeric large subunit cannot elaborate a complete active site, thereby accounting for a homodimer as the smallest functional unit.

Another mechanistically relevant structural feature is the presence of a flexible loop (loop 6), as commonly observed among β/α-barrel proteins. Conformational flexibility of this region was indicated initially by chemical crosslinking studies (218) and then rigorously established by comparing the structures of Rubiscos in different ligand states. In the complex of activated

Table 1 Available three-dimensional structures of Rubisco

Species	Form	Resolution (Å)	Reference
R. rubrum	Nonactivated (ligand-free)	1.7	(207)
R. rubrum	Nonactivated (PGA-bound)	2.9	(208)
R. rubrum	Nonactivated (CABP-bound)	2.6	(209)
R. rubrum	Activated (CO_2- and Mg^{2+}-bound)	2.3	(210)
R. rubrum	Activated (CO_2-, Mg^{2+}-, and RuBP-bound)	2.6	(211)
Spinach	Activated (CO_2-, Mg^{2+}-, and CABP-bound)	2.4	(212)
Tobacco	Nonactivated (ligand-free)	2.0	(213)
Tobacco	Activated (CO_2-, Mg^{2+}-, and CABP-bound)	2.7	(214)
A. nidulans	Activated (CO_2-, Mg^{2+}-, and CABP-bound)	2.2	(215)

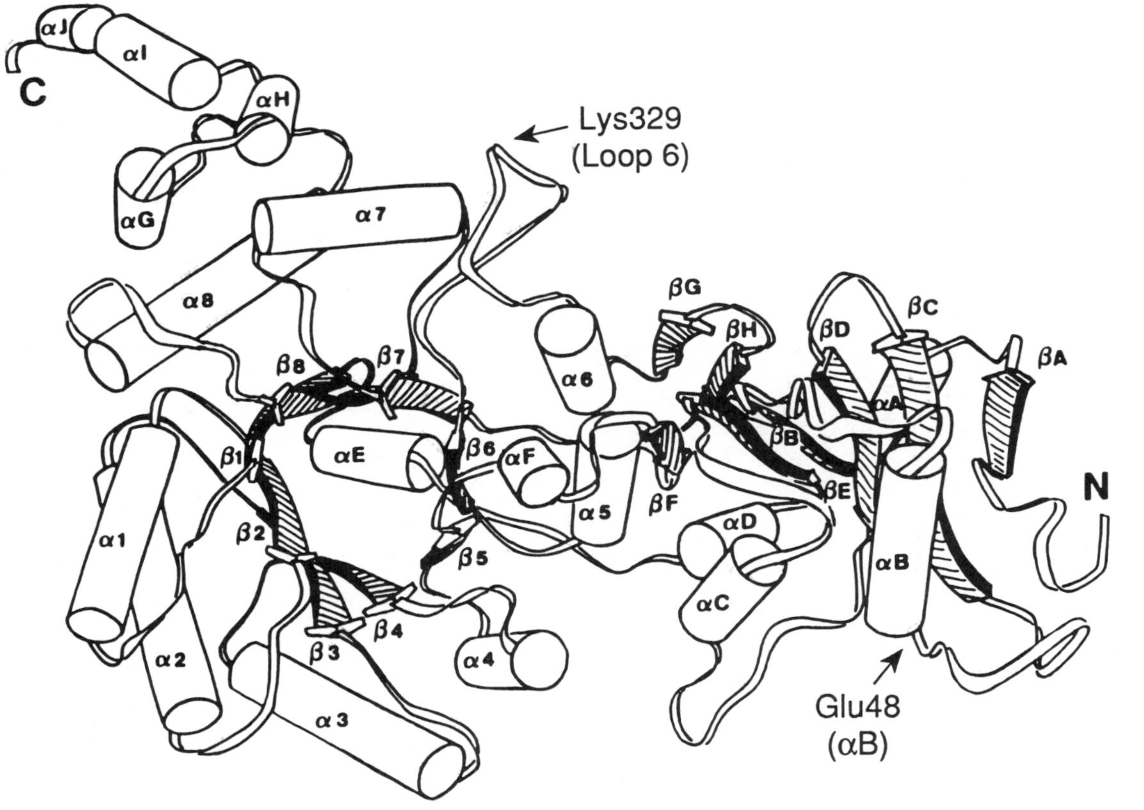

Figure 2 Schematic of an L subunit. Cylinders denote α-helices, and arrows represent β-strands. Reproduced from Ref. 209 with permission of author and publisher.

spinach enzyme with CABP, loop 6 is in the closed conformation and covers the top of the barrel domain, thereby sequestering the bound ligands from external solvent (154, 212, 214). By contrast, loop 6 is observed in the open conformation in the non-activated form of the tobacco enzyme (213, 219) and not observed at all (absence of electron density) in the non-activated *R. rubrum* enzyme (207, 220). These comparisons provide rather compelling arguments for loop movement as a dynamic event during catalytic turnover.

With the recognition that a homodimer is the smallest functional unit, the L_8 core of plant-type Rubisco is best described as a tetramer of L_2. Hexadecameric Rubisco approximates a cube with rounded edges of 110 Å, in which the four L_2 dimers are arranged longitudinally in opposite directions around the side faces of the cube. Dimers are related by a fourfold axis of symmetry. The top and bottom faces of the cube accommodate the two clusters of four S subunits; these extend into crevices between the tips of adjacent dimers. A solvent channel traverses the molecule along its fourfold axis. The small subunit consists primarily of a four-stranded antiparallel β-sheet covered on one side by two α-helices. Despite extensive interactions between S and L subunits, these are weaker collectively than L-L interactions, as concluded by the relative ease by which S subunits can be removed from the cyanobacterial holoenzyme without dissociation of the L_8 core (60).

Structural elucidation of an L_8S_8 Rubisco has settled considerable prior speculation about the function of S subunits. Despite their importance to catalysis (an L_8 core displays only 1% of the activity of holoenzyme) and an unaltered τ-value (62, 64), S subunits are far removed from the active site. Thus, the enhancement of catalytic rate by S subunits can only be mediated through induced conformational changes in L subunits. Schneider et al (221) have defined these conformational changes by detailed comparisons of L_2 and L_8S_8 structures. Functional demonstration of long-distance communication between L and S subunits has been provided by site-directed alterations of S subunits, which lead to significant changes of kinetic parameters (222–226).

Although there are not any large conformational differences between non-activated Rubisco and the activated quaternary complex (enzyme•CO_2 • Mg^{2+}•CABP), the closer approximation of loop 6 to a segment of the N-terminal domain as occurs in the activated complex is indispensable for catalysis. Apart from conformational differences associated with activation, the microenvironment of the active site is obviously altered by the binding of Mg^{2+}.

Negative cooperativity has been observed physically and kinetically in the binding of sugar phosphates to the active site of spinach Rubisco (195, 227, 228). Although the basis of cooperativity is unclear, crystal structures suggest inherent nonequivalence of active sites, but this apparent lack of uniformity could merely reflect crystal-lattice interactions (211). Independently, direct binding and kinetic analyses have provided the basis for invoking a set of regulatory sites, distinct from catalytic sites, capable of binding RuBP or other effectors (229–231). However, such sites are not observed in any of the crystallographic structures.

The Active Site

An active-site schematic of the activated spinach enzyme complexed with CABP is illustrated in Figure 3. Note that Glu48, Thr53, Asn54, and Asn111 are contributed by the N-terminal domain and all of the other residues are located in the β/α-barrel domain of the adjacent subunit. Also note the intersubunit salt bridge between Glu48 and Lys329 of flexible loop 6. Perhaps not surprisingly in view of the strongly anionic character of the substrate, most active-site residues are ionic or polar. Of the nineteen residues within ~5 Å of the bound inhibitor, seven have cationic side chains and four have anionic side chains; only Met 330 is hydrophobic.

The multiple roles of Mg^{2+} in Rubisco function are vividly illuminated by the three-dimensional structures. Coordination of metal ion by both the carbamate oxyanion (derived from activator CO_2) and the CABP carboxylate (corresponding to substrate CO_2) accounts for its essentiality in activation and catalysis. Based on coordination of Mg^{2+} with the carbonyl group of RuBP in the quaternary complex of the *R. rubrum* enzyme (211), the divalent metal ion specifically serves as an electrophile to polarize the substrate carbonyl and thereby promote enolization. Mg^{2+} is also crucial for proper orientation of substrate at the active site, because CABP binds to the non-activated enzyme in inverted fashion (209).

The identity of the phosphate ligands also becomes apparent from direct visualization of the quaternary complex. By close analogy with triose phosphate isomerase (232), the ligands for the P_1 phosphate are polar and include Gly370, Gly393, Gly394, Thr53, and Asn54. The P_2 phosphate group is anchored electrostatically by Arg288 and His321. Functions of other active-site residues are not automatically defined by the structure, but most of these have been probed by site-directed mutagenesis.

SITE-DIRECTED MUTAGENESIS

General Findings

Based on a combination of sequence alignment, chemical modification, and site-directed mutagenesis studies, numerous features of the active site of Rubisco and possible roles of active-site residues had been deduced prior

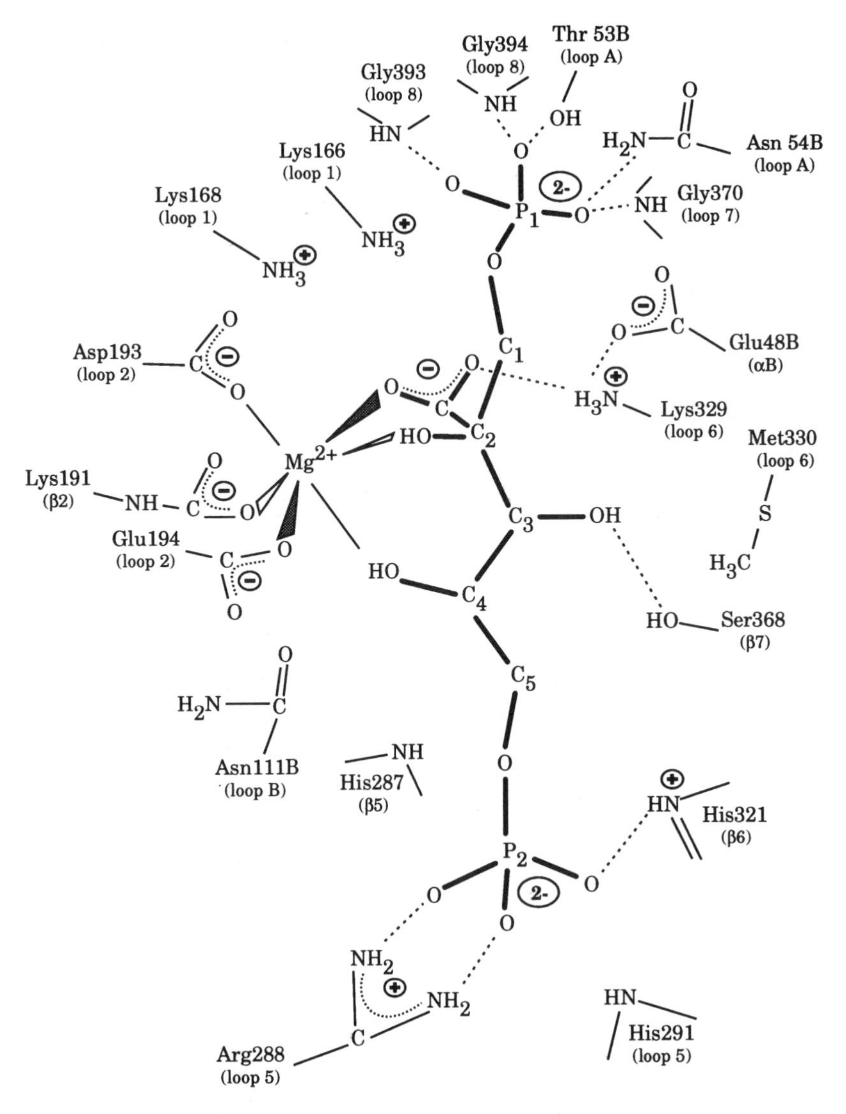

Figure 3 Active-site residues of Rubisco in the immediate vicinity of bound CABP as determined by X-ray crystallography. "B" denotes that the residue is located in the adjacent subunit. Locations of residues in the secondary structure are specified in parentheses. The *trans* conformation is shown relative to hydroxyl groups at C2 and C3; the 2.4 Å resolution of the published structure (212) does not distinguish between *cis* and *trans*. However, recent refinement suggests that the proper conformation may be *cis*, in which the C3 hydroxyl group replaces the C4 hydroxyl as a metal ligand (215). (Adapted from the schematic of the spinach enzyme illustrated in Ref. 154 with permission of author and publisher.)

to the elucidation of the three-dimensional structure of the enzyme (reviewed in Ref. 12). Residues assigned to the active site included Lys191, His291, Lys166, Lys329, and Glu48. Lys191 (the site of carbamylation) was identified by chemically trapping the carbamate via esterification (233), and His291 was assigned to the active site based on selective modification with diethylpyrocarbonate (234). Multiple affinity labels placed Lys166 and Lys329 at the active site, and enhanced acidities and nucleophilicities of the ε-amino groups of these residues suggested key catalytic roles (235). Their intrasubunit proximity, as demonstrated by chemical crosslinking (218), was dependent on the activation state of the enzyme, indicative of conformational flexibility as later visualized directly by crystallography (207, 212). The assignment of Glu48 to the active site was prompted by chemical crosslinking (236) and species invariance of both the residue and adjacent sequence.

Site-directed mutagenesis lent support to the likelihood of catalytic roles for Lys166, Lys329, and Glu48 (237–239). In each case, replacement with any other residue (even conservative substitutions) abolished most carboxylase activity without inducing major structural changes. In contrast, replacement of His291 with Ala only diminished k_{cat} by 2.5-fold, a finding that discounted direct participation in catalysis (240). As the site of carbamylation, Lys191 was known a priori to be required for function. However, mutagenesis was used to ascertain whether a permanently activated Rubisco might result from the replacement of Lys by Glu (241). Despite the sound logic of such a strategy, success was precluded by the distance between the lysyl α-carbon and bound Mg^{2+} (which cannot be spanned by a glutamyl side chain), as subsequently measured in the three-dimensional structure.

Another feature of the active site, discovered prior to availability of a high-resolution structure, is its interfacial location (217). This conclusion followed formation of a catalytically active heterodimer from inactive site-directed mutant proteins. The K166G[2] and E48Q mutants of the *R. rubrum* enzyme are severely impaired catalytically, but when their respective genes were coexpressed in *E. coli*, an active hybrid was formed. The specific activity was half that of wild-type enzyme in accordance with one functional active site per dimeric molecule. Independent of any crystallographic analysis, in vivo hybridization provides a facile, functional test for active sites created by interacting domains of different subunits. The general utility of the approach and variations thereof have been documented (242–245).

[2]The single-letter code is used to describe mutant proteins. The first letter denotes the amino acid present in the wild-type enzyme at the numbered position. The final letter denotes the amino acid present at the corresponding position in the mutant protein.

Strategies for Mutant Characterization

The full potential of site-directed mutagenesis as a structure/function probe cannot be achieved without high-resolution structural analyses of the mutant proteins under investigation. Frequently, this idealized circumstance is lacking. Fortunately, Rubisco offers a number of conveniently monitored properties apart from normal carboxylase/oxygenase activities. Even partial retention of these properties by mutants can allay fears that detrimental consequences of amino acid substitutions at the active site are indirect effects brought about by major conformational changes. These properties can also be invaluable toward pinpointing the role of active-site residues. Proper activation chemistry and integrity of the substrate binding site can be gauged with CABP, which forms an exchange-resistant complex exclusively with the activated form of Rubisco (156, 246). Independent of overall catalytic activity (see Figure 1), enolization of RuBP (146, 151) and turnover of the isolated six-carbon reaction intermediate (160) can be assayed as distinct partial reactions. Enhanced formation of XuBP may serve as an indicator of compromised processing of the enediol(ate) intermediate (196). The status of the final step in overall carboxylation (protonation of the PGA carbanion) should be reflected by the ratio of protonation (PGA formation) to β-elimination (pyruvate formation) (162).

Among the dozens of Rubisco mutants that have been analyzed biochemically, only the inactive D193N mutant of the *R. rubrum* enzyme (in which one metal ion ligand has been replaced by a potentially functional isostere) has been subjected to crystallographic analysis. The sobering finding is that major conformational changes, including alterations in subunit-subunit packing, have occurred (247). This is not terribly surprising given the severe deficiencies of the mutant in activation and CABP binding (248).

Asn111

Crystal structures show that in the activated ternary complex the amide side chain of Asn111 serves as one of the ligands for Mg^{2+} but in the quaternary complex moves away from the metal ion and closer to C3 (<5 Å) of RuBP (210, 211). These structural considerations prompted the suggestion that Asn111 could be the base that abstracts the C3 proton of RuBP and thereby initiates the catalytic pathway. Despite severe impairment of carboxylase activity (<0.01% of wild-type activity), the N111Q mutant retains 35% of the wild-type enolization activity and a normal K_m for RuBP (182). Asn111 must then be involved catalytically at some step beyond enolization. The CO_2/O_2 specificity factor (τ) of both the N111Q and N111G is lowered 10-fold relative to wild-type (182, 183).

Lys329, Lys168, Glu48

Lys329 is located at the apex of flexible loop 6. In its closed conformation as visualized in the activated enzyme•CABP complex (Figure 3) (212), the loop extends over the top of the barrel domain and thereby sequesters the bound inhibitor from solvent. Electrostatic interactions of the ϵ-amino group of Lys329 with both the γ-carboxylate of Glu48 from the adjacent subunit and the carboxylate of bound inhibitor presumably assist in stabilizing the closed-loop conformation. Consistent with this interpretation, mutants with substitutions for Lys329 are unable to form stable complexes with CABP (238). When the loop swings away from the top of the barrel (a 15-Å movement of the ϵ-amino group) as seen in the non-activated enzyme, Glu48 forms an intersubunit salt bridge with Lys168 (209). Thus, loop 6 movement and associated alterations of intersubunit interactions at the active site appear to be dynamic aspects of catalytic turnover.

Based solely on structural grounds, Lys329 would appear to stabilize the transition states leading to the carboxylated or hydroperoxy intermediates. Properties of position 329 mutants of the *R. rubrum* enzyme are strongly supportive of this supposition. The K329G mutant catalyzes both enolization of RuBP (151) and turnover of the carboxylated intermediate to PGA (188), even though it lacks detectable carboxylase activity. Hence, the carboxylation of the enediol(ate) of RuBP is preferentially impaired. Furthermore, the length of the aminoalkyl side chain at position 329 has a large impact on the τ-value. For example, the inactive K329C mutant can be chemically rescued by aminoethylation (249) and aminopropylation (188) which, in effect, displace the ϵ-amino group by 0.4 Å and 1.7 Å, respectively, from the α-carbon. The τ-value of the aminoethylcysteinyl mutant is half that of wild-type, and the τ-value of the aminopropylcysteinyl mutant is fourfold lower than wild-type (188). Another apparent reflection of transition-state perturbation is provided by replacement of the corresponding lysyl residue of the *Anacystis* L_8S_8 enzyme by Arg, in which the mutant retains slight carboxylase activity (0.5% of wild-type) so that the τ-value can be determined; it is lowered from 56 to 0.3 (186).

Failure of the enediol(ate) generated by position 329 mutants to react with CO_2 suggests that Lys329 may play a specific role in activating this intermediate in addition to its role in stabilizing the subsequent transition state (151). Invoking a discrete step of enediol(ate) activation is compatible with the detection of a rate-limiting step between enolization and carboxylation in pre-steady-state kinetic analyses (143).

As in the case of Asn111, Lys168 was suggested on structural grounds as a candidate for the base that facilitates enolization of RuBP (212).

However, this candidate must also be discarded, because the K168Q mutant retains 2% of the wild-type activity in enolization despite a decline in carboxylase activity by 10^4 (250). The precise role of Lys168 has not been determined, but its chief participation must be beyond enolization in the reaction coordinate.

Characterization of the E48Q mutant reveals both similarities and differences in comparison to K329G. While only slight carboxylase activity can be detected (0.6% of wild-type), the mutant retains full activity in processing the carboxylated reaction intermediate and about 10% of the normal activity in enolization (187). Hence, as with Lys329, Glu48 clearly exerts its greatest impact at the stage of reaction of gaseous substrates with the enediol(ate) of RuBP. Whether this is a direct effect or a consequence of increased flexibility of the side chain of Lys329 concomitant with elimination of the intersubunit electrostatic interactions is not known. In either event, the structural change leads to a 35-fold reduction in the τ-value, thereby signifying a shift of the relative reactivity of the enediol(ate) in favor of O_2. Once again the τ-value is sensitive to the length of the pertinent side-chain. Retention of the negative charge at position 48, but with a 1.3 Å displacement as achieved by carboxymethylation of inactive E48C, partially restores carboxylase activity and yields a τ-value that is fivefold less than normal (189). Thus, removal of the negative charge (i.e. E48Q) exerts a greater negative impact on CO_2 selectivity than repositioning the charge more distal from the α-carbon.

Unlike the position 329 mutants, E48Q is very prone to misprotonate the enediol(ate) and consequently form XuBP (187). This epimerization reaction proceeds as rapidly as forward carboxylation of RuBP. The significance of this observation is emphasized by the fact that the wild-type *R. rubrum* enzyme, in contrast to spinach Rubisco, does not exhibit detectable epimerase activity. Upon abolishment of the Glu48—Lys329 salt bridge, the N-terminal segment of the active site may move away from loop 6, allowing water to enter and misprotonation of the protein-bound enediol(ate) to occur.

Loop 6

The crucial catalytic role of Lys329 (Lys334 in plant Rubisco), and the discovery that random mutations in loop 6 influence the τ-value, have stimulated systematic analysis of this region. Chemical mutagenesis of *Chlamydomonas reinhardtii* generated a Rubisco mutant (V331A) with a τ-value that is 37% lower than that of the wild-type enzyme (179). A suppressor mutant with a second amino acid substitution (T342I) partially restores the normal specificity (180). The corresponding residues in *Anacystis* Rubisco have been substituted with several different amino acids by site-directed mutagenesis. With respect to substitutions for Val331, the Gutteridge laboratory (186) observes declines in τ-value consistent with analysis of random mutants of *C. reinhardtii;* however, in a report from another laboratory, the τ-value of the V331A mutant of *Anacystis* Rubisco is described as unaltered (184). This conflict may reflect inherent pitfalls in measurements of oxygenase activities (particularly with impaired mutants) and the lack of standardization among laboratories (for a discussion, see Ref. 251). Enhancements of τ-values were not observed upon single-amino-acid replacements of the suppressor site (Thr342) (186). Although disappointing, this result is not surprising in view of van der Waals contact between the side chains of Val331 and Thr342 (212). Thus, the suppressor mutation (V331A/T342I) may be accounted for by steric complementation (180).

Just beyond the C-terminal side of loop 6, a highly variable four-residue segment is found; DKAS in *Anacystis* Rubisco is replaced by ERDI or EREI in plant Rubiscos. To explore the possibility that this sequence might account for the improved specificity of the plant enzyme, the four-residue substitution was effected in *Anacystis* Rubisco (184, 186). Not only was k_{cat} reduced twofold, but the τ-value was unaltered (186). Although not a part of loop 6, the residue at position 309 was speculated to account for kinetic differences of Rubisco from C_3 and C_4 plants (higher turnover rates and larger K_m for CO_2 in the latter) (252). Since the kinetic parameters of cyanobacterial Rubiscos more closely resemble those of C_4 plants, Met309 (common among Rubiscos of C_4 plants and cyanobacteria) of the *Anacystis* enzyme was replaced with Ile as occurs in higher-plant Rubisco. Significant effects on kinetic parameters were not observed. These two examples do not bode well for prospects of endowing one class of Rubiscos with kinetic parameters displayed by another based on interchanges of short segments of variable sequence.

An important role in stabilizing the closed conformation of loop 6 can be assigned to Phe327 (253). A number of substitutions have little impact on k_{cat} or τ-values, but rather increase the K_m for RuBP by as much as 165-fold. With the realization that RuBP cannot dissociate from the enzyme when loop 6 is in the closed conformation, this drastic increase in the off rate must signify that loop 6 of the mutant spends more time in the open conformation. Hydrophobic interactions (eliminated in the mutants) of Phe327 with Phe379 and Ile447 may favor the closed conformation.

Lys329 is followed in sequence by Met in the L_2 Rubiscos and Leu in the L_8S_8 enzymes. The consequences of interchange of these residues has been examined. The M330L mutant of the *R. rubrum* enzyme has a fivefold decreased k_{cat}, a 10-fold increased K_m for RuBP, and an unaltered τ-value relative to wild-type (254). Similarly, the L330M mutant of *Anacystis* Rubisco is impaired in k_{cat} (threefold decrease); but unlike the L_2 enzyme, the L_8S_8 mutant exhibits a τ-value that is threefold lower than wild-type (177). Once again, alterations in kinetic parameters cannot be predicted from considerations of comparative sequences.

His321, His287, Ser368, Lys166

Each of these residues has been offered as a candidate for the general base that promotes enolization of RuBP. Generally, rate enhancements of 10^6 or greater are exerted by active-site bases that abstract protons from carbon atoms (see discussion in Ref. 12). Thus, the modest 10-fold impairment of enolization by substitutions for His321 and Ser368 would seem to exclude them from further consideration analogously to candidates Asn111 and Lys168 as already discussed (255). Interestingly, the S368A mutant, which retains only 2% of the wild-type carboxylase activity, displays a 60% enhancement of τ (181). In contrast, the corresponding substitution in the *Anacystis* enzyme decreases τ by a substantial (but not quantified) amount (175). His287 facilitates both enolization and overall carboxylation by $>10^3$ and is also required to process the six-carbon reaction intermediate (257). If His287 is the essential base, it must also be crucial to some additional step beyond enolization; otherwise, this function should have been preferentially impaired in the H287N mutant. Multiple catalytic roles for single acid/base groups are certainly possible, so His287 cannot be dismissed as the acceptor of the C3 proton from RuBP.

Prior to the solving of the Rubisco structure, what appeared to be a compelling case for Lys166 as the essential base had been constructed. The side chain of Lys166 is endowed with enhanced nucleophilicity and acidity (235); its pK_a of 7.9 matches that of a catalytic group, which must be unprotonated, observed in the pH-dependence of the deuterium isotope effect with [3-^{2}H]RuBP (148). Furthermore, the K166G mutant lacks detectable carboxylase and enolization activities but retains catalytic competence in turnover of the six-carbon reaction intermediate (237, 258). Finally, the k_{cat} is influenced by the pK_a of the side chain at position 166, as judged by kinetic analysis of the chemically rescued aminoethylated K166C mutant (249).

Despite these seemingly convincing, albeit indirect, data that implicate Lys166 as the proton abstractor, the crystal structures place the ϵ-amino group too far away (6.5 Å) from the C3 proton of RuBP and in the wrong orientation to serve as the base (8, 212). The apparent conflict between structural and functional data would be reconciled if Lys166 actually facilitates enolization through polarization of the substrate carbonyl (general acid catalysis) rather than through proton abstraction (general base catalysis) (212). Lys166 is also observed to be the only group properly positioned to protonate the terminal carbanion of PGA and thereby complete the catalytic cycle (212). Such an assignment, however, contradicts the ability of the K166G mutant to process the six-carbon reaction intermediate.

Unequivocal identification of the key base has been hampered by the absence in the electron-density maps of any proton acceptor within van der Waals contact of the C3 proton of RuBP. Additional possibilities of groups to abstract this proton include the carbamate (212) and the P_1 phosphate of RuBP (211). Chemical rescue of the inactive K191C by aminomethanesulfonate is consistent with some unspecified catalytic role of the carbamate (259). Experimental data concerning the issue of substrate-assisted catalysis has not been presented.

P1-Binding Site

Five amino-acid residues with the potential to serve as hydrogen-bond donors are clustered around the P_1 phosphate oxygens of RuBP, including Gly370, Gly393, and Gly394 from the C-terminal domain of one subunit and Thr53 and Asn54 from the N-terminal domain of the adjacent subunit (Figure 3). The roles of these amino acids in catalysis and intermediate stabilization were evaluated by individual replacement with Ala in the *R. rubrum* Rubisco and analysis for catalytic properties and the propensity to catalyze side reactions (205). The most debilitating effects on catalysis and K_m(RuBP) were with T53A, G370A, and G393A. These same mutants accumulate a dicarbonyl side product, derived from β-elimination of phosphate from enediol(ate), showing their importance in the stabilization of that intermediate. An increase in the occurrence of pyruvate formation due to β-elimination from the *aci*-acid intermediate was found only with T53A, N54A, and G393A. Thus, the ligands for P_1 phosphate appear to be dynamic during turnover, as required to accommodate the distinct conformations of the two planar intermediates (260). Gly370 and Asn54 may preferentially stabilize the enediol(ate) and *aci*-acid, respectively. Clearly, these amino acids are crucial to overall stabilization of intermediates and suppression of undesirable side reactions. Similarly, valinyl replacement of Thr65 (the cognate of Thr53) in *Anacystis* Rubisco also drastically decreased the carboxylase activity and increased the prevalence of side reactions (260).

P2-Binding Site

The only residues that interact electrostatically with the P_2 phosphate group of RuBP are His321 and Arg288 (Figure 3). Replacement of His321 with Ala increases the K_m for RuBP only fivefold, indicating that the imidazolium side chain contributes only modestly to substrate binding (255). However, the k_{cat} of H321A is decreased 30-fold compared to wild-type enzyme, and whereas the CABP of the quaternary complex of the latter exchanges with exogenously added CABP with a $t_{1/2}$ of 35 hr, the bound ligand of the mutant quaternary complex exchanges with a $t_{1/2}$ of 30 min. These results suggest that His321 contributes preferentially to the binding of a transition state, thereby enhancing the catalytic rate.

Catalytic activity cannot be detected in mutants of the L_8S_8 Rubisco of *Anacystis*, in which Arg292 (corresponding to Arg288 of the *R. rubrum* enzyme) is replaced with Lys or Leu (261). These results are compatible with the arginyl residue also interacting with greater affinity with the P_2 phosphate of a transition state as compared to substrate. Interpretation is somewhat tenuous, because neither mutant is able to form a stable quaternary complex with CABP. Thus, structural perturbations that compromise activation chemistry or binding of phosphorylated ligands could be the source of catalytic impairment.

CONCLUDING REMARKS

Despite major advances toward comprehensive understanding of the structure, mechanism, physiology, and molecular biology of Rubisco, the feasibility of designing a more efficient enzyme remains problematic. The recent flurry of observed alterations of τ-values as correlated with structural manipulations is encouraging in the sense of validating external intervention as an avenue for rendering changes in kinetic parameters. Counter to this cautious optimism, most altered τ-values are in the wrong direction (i.e. relative enhancement of oxygenase activity) and are accompanied by impaired turnover rates. Apart from improving the most efficient Rubisco provided by nature, even the more modest goal of converting a low-τ enzyme (bacterial) to a high-τ enzyme (plant) has not been achieved. Fundamental information concerning the structural and electronic requirements for preferential stabilization of the transition state for carboxylation is simply lacking. Witness the 10-fold difference in τ-value between the primitive L_2 enzyme and the more evolved L_8S_8 enzyme, which has not been reconciled in molecular terms even though high-resolution structures have been available for scrutiny for some time.

The catalytic efficiency (k_{cat}/K_m) of Rubisco is several orders of magnitude less than the upper limit imposed by diffusion rates, and the relative substrate specificity of Rubisco pales by comparison with other enzymes that must distinguish closely related substrates. Thus, a credible case can be presented that evolution of Rubisco is unfinished and that if armed with thorough mechanistic information, researchers should be able to accelerate evolution of the enzyme in the laboratory (260). However, the absence of a binding site for CO_2/O_2 and the necessity of stabilizing at least three distinct transition states with energy derived solely from the binding of RuBP may represent inherent barriers to achieving evolutionary perfection. Further attention to these issues is warranted within the dual context of enzyme mechanisms and biomass yields.

With regard to the ultimate goal of engineering plants with superior growth characteristics, recent dissection of processes that govern the expression, assembly, and regulation of Rubisco has provided requisite information and approaches for manipulation and evaluation of the enzyme in vivo. Rubisco represents a prime example in which chemical, physical, biological, and theoretical disciplines have merged in pursuit of a common goal.

ACKNOWLEDGMENTS

Our research was supported by the Office of Health and Environmental Research, US Department of Energy under contract DE-AC05-84OR21400 with Martin Marietta Energy Systems, Inc. We are especially grateful to Ms. Bettye Phifer for expert typing and proofing of the manuscript.

Literature Cited

1. Andrews TJ, Lorimer GH. 1987. *The Biochemistry of Plants*, ed. MD Hatch, NK Boardman, 10:131–218. New York: Academic
2. Tabita FR. 1988. *Microbiol. Rev.* 52: 155–89
3. Woodrow IE, Berry JA. 1988. *Annu. Rev. Plant Physiol. Plant Mol. Biol.* 39:533–94
4. Gutteridge S. 1989. *Biochim. Biophys. Acta* 1015:1–14
5. Roy H. 1989. *The Plant Cell* 1:1035–42
6. Gatenby AA, Ellis RJ. 1990. *Annu. Rev. Cell Biol.* 6:125–49
7. Portis AR Jr. 1990. *Biochim. Biophys. Acta* 1015:15–28
8. Brändén C-I, Lindqvist Y, Schneider G. 1991. *Acta Cryst.* B47:824–35
9. Hartman FC. 1992. *Plant Protein Engineering*, ed. P Shewry, S Gutteridge, pp. 61–92. United Kingdom: Cambridge Univ. Press
10. Portis AR Jr. 1992. *Annu. Rev. Plant Physiol. Plant Mol. Biol.* 43:415–37
11. Schneider G, Lindqvist Y, Brändén C-I. 1992. *Annu. Rev. Biophys. Biomol. Struct.* 21:119–43
12. Hartman FC, Harpel MR. 1993. *Adv. Enzymol.* 67:1–75
13. Hendrick JP, Hartl F-U. 1993. *Annu. Rev. Biochem.* 62:349–84
14. Spreitzer RJ. 1993. *Annu. Rev. Plant Physiol. Plant Mol. Biol.* 44: 411–34
15. Gruissem W. 1989. *Cell* 56:161–70
16. Akada S, Xu YQ, Machii H, Kung SD. 1990. *Gene* 94:195–99
17. Klein RR, Mullet JE. 1990. *J. Biol. Chem.* 265:1895–902
18. Fritz CC, Herget T, Wolter FP, Schell J, Schreier PH. 1991. *Proc. Natl. Acad. Sci. USA* 88:4458–62
19. Nantel AM, Lafleur F, Boivin R, Baszcynski CL, Bellemare G. 1991. *Plant Mol. Biol.* 16:955–66
20. Winder TL, Anderson JC, Spalding MH. 1992. *Plant Physiol.* 98:1409–14
21. Chua NM, Schmidt GW. 1978. *Proc. Natl. Acad. Sci. USA* 75:6110–14
22. Robinson C, Ellis RJ. 1984. *Eur. J. Biochem.* 142:343–46
23. Wasmann CC, Reiss B, Bohnert HJ. 1988. *J. Biol. Chem.* 263:617–19
24. Wasmann CC, Ramage RT, Bohnert HJ, Ostrem JA. 1989. *Proc. Natl. Acad. Sci. USA.* 86:1198–202
25. Langridge P. 1981. *FEBS Lett.* 123:85–89
26. Houtz RL, Stults JT, Mulligan RM,

Tolbert NE. 1989. *Proc. Natl. Acad. Sci. USA* 86:1855–59
27. Mulligan RM, Houtz RL, Tolbert NE. 1988. *Proc. Natl. Acad. Sci. USA* 85:1513–17
28. Houtz RL, Royer M, Salvucci ME. 1991. *Plant Physiol.* 97:913–20
29. Houtz RL, Mulligan RM. 1991. *Plant Physiol.* 96:335–39
30. Houtz RL, Poneleit L, Jones SB, Royer M, Stults JT. 1992. *Plant Physiol.* 98:1170–74
31. Gouitton C, Mache R. 1987. *Eur. J. Biochem.* 166:249–54
32. Margosiak SA, Dharma A, Bruce-Carver MR, Gonzales AP, Louie D, Kuehn GD. 1989. *Plant Physiol.* 92: 86–96
33. Hwang S-R, Tabita FR. 1989. *Plant Mol. Biol.* 13:69–79
34. Hwang S-R, Tabita FR. 1991. *J. Biol. Chem.* 266:6271–79
35. Hemmingsen SM, Woolford C, van der Vies SM, Tilly K, Dennis DT, et al. 1988. *Nature* 333:330–34
36. Roy H, Cannon S, Gilson M. 1988. *Biochim. Biophys. Acta* 957:323–34
37. Ellis RJ, van der Vies SM. 1991. *Annu. Rev. Biochem.* 60:321–47
38. Zeilstra-Ryalls J, Fayet O, Georgopoulos C. 1991. *Annu. Rev. Microbiol.* 45:301–25
39. Georgopoulos C. 1992. *Trends Biochem. Sci.* 17:295–99
40. Deleted in proof
41. Ellis RJ, ed. 1990. *Seminars in Cell Biology,* Vol. 1. Philadelphia: Saunders
42. Bradley D, van der Vies SM, Gatenby AA. 1986. *Philos. Trans. R. Soc. London Ser. B* 313:447–58
43. Gatenby AA. 1984. *Eur. J. Biochem.* 144:361–66
44. Voordouw G, van der Vies SM, Bouwmeister PJ. 1984. *Eur. J. Biochem.* 141:313–18
45. Somerville CR, Somerville SC. 1984. *Mol. Gen. Genet.* 193:214–19
46. Larimer FW, Machanoff R, Hartman FC. 1986. *Gene* 41:113–20
47. Larimer FW, Mural RJ, Soper TS. 1990. *Protein Eng.* 3:227–31
48. Gibson JL, Tabita FR. 1986. *Gene* 44:271–78
49. Gurevitz M, Somerville CR, McIntosh L. 1985. *Proc. Natl. Acad. Sci USA* 82:6546–50
50. Christeller JT, Terzaghi BE, Hill DF, Laing WA. 1985. *Plant Mol. Biol.* 5:257–63
51. Tabita FR, Small CL. 1985. *Proc. Natl. Acad. Sci. USA* 82:6100–3
52. van der Vies S, Bradley D, Gatenby AA. 1986. *EMBO J.* 5:2439–44
53. Kettleborough CA, Parry MAJ, Burton S, Gutteridge S, Keys AJ, Phillips AL. 1987. *Eur. J. Biochem.* 170:335–42
54. Gatenby AA, van der Vies SM, Bradley D. 1985. *Nature* 314:617–20
55. McFadden BA, Small CL. 1988. *Photosynth. Res.* 18:245–60
56. Viale AM, Kobayashi H, Akazawa T. 1990. *J. Biol. Chem.* 265:18386–92
57. Newman J, Gutteridge S. 1990. *J. Biol. Chem.* 265:15154–59
58. Fitchen JH, Knight S, Andersson I, Brändén C-I, McIntosh L. 1990. *Proc. Natl. Acad. Sci. USA* 87:5768–72
59. Larimer FW, Soper TS. 1993. *Gene* 126:85–92
60. Andrews TJ, Ballment B. 1983. *J. Biol. Chem.* 258:7514–18
61. Andrews TJ, Lorimer GH. 1985. *J. Biol. Chem.* 260:4632–36
62. Andrews TJ. 1988. *J. Biol. Chem.* 263:12213–19
63. Lee B, Tabita FR. 1990. *Biochemistry* 29:9352–57
64. Gutteridge S. 1991. *J. Biol. Chem.* 266:7359–62
65. Smrcka AV, Ramage RT, Bohnert HJ, Jensen RG. 1991. *Arch. Biochem. Biophys.* 286:6–13
66. Goloubinoff P, Gatenby AA, Lorimer GH. 1989. *Nature* 337:44–47
67. Goloubinoff P, Christeller JT, Gatenby AA, Lorimer GH. 1989. *Nature* 342: 884–89
68. van der Vies SM, Viitanen PV, Gatenby AA, Lorimer GH, Jaenicke R. 1992. *Biochemistry* 31:3635–44
69. Viitanen PV, Lubben TH, Reed J, Goloubinoff P, O'Keefe DP, Lorimer GH. 1990. *Biochemistry* 29:5665–71
70. Todd MJ, Viitanen PV, Lorimer GH. 1993. *Biochemistry*. 32:8560–67
71. Erijman L, Lorimer GH, Weber G. 1993. *Biochemistry* 32:5187–95
72. Matthews CR. 1993. *Annu. Rev. Biochem.* 62:653–83
73. Torres-Ruiz JA, McFadden B. 1988. *Arch. Biochem. Biophys.* 261:196–204
74. Webb R, Reddy KJ, Sherman LA. 1990. *J. Bacteriol.* 172:5079–88
75. Chitnis PR, Nelson N. 1991. *J. Biol. Chem.* 266:58–65
76. Terlesky KC, Tabita FR. 1991. *Biochemistry* 30:8181–86
77. Wang H, Goffreda M, Leistek T. 1993. *Plant Physiol.* 102:843–50
78. Barraclough R, Ellis RJ. 1980. *Biochim. Biophys. Acta* 608:19–31
79. Roy H, Bloom M, Milos P, Monroe M. 1982. *J. Cell Biol.* 94:20–27
80. Hemmingsen SM, Ellis RJ. 1986. *Plant Physiol.* 80:269–76

81. Musgrove JE, Johnson RA, Ellis RJ. 1987. *Eur. J. Biochem.* 163:529–34
82. Martel R, Cloney LP, Pelcher LE, Hemmingsen SM. 1990. *Gene* 94:181–87
83. Milos P, Roy H. 1984. *J. Cell. Biochem.* 24:153–62
84. Bloom MV, Milos P, Roy H. 1983. *Proc. Natl. Acad. Sci. USA* 80:1013–17
85. Cannon S, Wang P, Roy H. 1986. *J. Cell. Biol.* 103:1327–35
86. Ellis RJ, van der Vies SM. 1988. *Photosynth. Res.* 16:101–15
87. Gatenby AA, Lubben TH, Ahlquist P, Keegstra K. 1988. *EMBO J.* 7:1307–14
88. Hubbs AE, Roy H. 1992. *Plant Physiol.* 100:272–81
89. Hubbs AE, Roy H. 1993. *Plant Physiol.* 101:523–33
90. Hubbs AE, Roy H. 1993. *J. Biol. Chem.* 268:3519–25
91. Andrews TJ, Greenwood DM, Yellowlees D. 1984. *Arch. Biochem. Biophys.* 234:313–17
92. Cloney LP, Wu HB, Hemmingsen SM. 1992. *J. Biol. Chem.* 267:23327–32
93. Cloney LP, Bekkaoui DR, Wood MG, Hemmingsen SM. 1992. *J. Biol. Chem.* 267:23333–36
94. Bertsch U, Soll J, Seetharam R, Viitanen PV. 1992. *Proc. Natl. Acad. Sci. USA* 89:8696–700
95. Laing WA, Christeller JT. 1976. *Biochem. J.*159:563–70
96. Lorimer GH, Badger MR, Andrews JT. 1976. *Biochemistry* 15:529–36
97. Belknap WR, Portis AR. 1986. *Biochemistry* 25:1864–69
98. Miziorko HM. 1979. *J. Biol. Chem.* 254:270–72
98a. Lorimer GH. 1979. *J. Biol. Chem.* 254:5599–601
99. Lu G, Lindqvist Y, Schneider G. 1992. *Proteins: Struct. Funct. Genet.* 12:117–27
100. McCurry SD, Pierce J, Tolbert NE, Orme-Johnson WH. 1981. *J. Biol. Chem.* 256:6623–28
101. Badger MR, Lorimer GH. 1981. *Biochemistry* 20:2219–25
102. Jordan DB, Chollet R, Ogren WL. 1983. *Biochemistry* 22:3410–18
103. Jordan DB, Chollet R. 1983. *J. Biol. Chem.* 258:13752–58
104. Somerville CR, Portis AR Jr, Ogren WL. 1982. *Plant Physiol.* 70:381–87
105. Salvucci ME, Portis AR Jr, Ogren WL. 1985. *Photosynth. Res.* 7:191–203
106. Robinson SP, Streusand VJ, Chatfield JM, Portis AR Jr. 1988. *Plant Physiol.* 88:1008–14
107. Werneke JM, Chatfield JM, Ogren WL. 1989. *Plant Cell* 1:815–25
108. Orozco BM, McClung CR, Werneke JM, Ogren WL. 1993. *Plant Physiol.* 102:227–32
109. Rundle SJ, Zielinski RE. 1991. *J. Biol. Chem.* 266:4677–85
110. Salvucci ME, Werneke JM, Ogren WL, Portis AR Jr. 1987. *Plant Physiol.* 84:930–36
111. Li LA, Gibson JL, Tabita FR. 1993. *Plant Mol. Biol.* 21:753–64
112. Shen JB, Orozco EM Jr, Ogren WL. 1991. *J. Biol. Chem.* 266:8963–68
113. Werneke JM, Zielinski RE, Ogren WL. 1988. *Proc. Natl. Acad. Sci. USA* 85:787–91
114. Portis AR Jr, Salvucci ME, Ogren WL. 1986. *Plant Physiol.* 82:967–71
115. Robinson SP, Portis AR Jr. 1989. *Arch. Biochem. Biophys.* 268:93–99
116. Lan Y, Mott KA. 1991. *Plant Physiol.* 95:604–09
117. Shen JB, Ogren WL. 1992. *Plant Physiol.* 99:1201–07
118. Wang ZY, Portis AR Jr. 1992. *Plant Physiol.* 99:1348–53
119. Mott KA, Woodrow IE. 1993. *Plant Physiol.* 102:859–66
120. Campbell WJ, Ogren WL. 1990. *Plant Physiol.* 92:110–15
121. Woodrow IE, Mott KA. 1992. *Plant Physiol.* 99:298–303
122. Seemann JR, Kobza J, Moore Bd. 1990. *Photosynth. Res.* 23:119–30
123. Seemann JR, Berry JA, Freas SM, Krump MA. 1985. *Proc. Natl. Acad. Sci. USA* 82:8024–28
124. Servaites JC. 1985. *Plant Physiol.* 78:839–43
125. Vu JCV, Allen LH, Bowes G. 1983. *Plant Physiol.* 73:729–34
126. Gutteridge S, Parry MAJ, Burton S, Keys AJ, Mudd A, et al. 1986. *Nature* 324:274–76
127. Berry JA, Lorimer GH, Pierce J, Seemann JR, Meeks J, Freas S. 1987. *Proc. Natl. Acad. Sci. USA* 84:734–38
128. Moore Bd, Kobza J, Seemann JR. 1991. *Plant Physiol.* 96:208–13
129. Robinson SP, Portis AR Jr. 1988. *FEBS Lett.* 233:413–16
130. Moore Bd, Seemann JR. 1992. *Plant Physiol.* 99:1551–55
131. Gutteridge S, Julien B. 1989. *FEBS Lett.* 254:225–30
132. Salvucci ME, Holbrook GP. 1989. *Plant Physiol.* 90:679–85
133. Holbrook GP, Galasinski SC, Salvucci ME. 1991. *Plant Physiol.* 97:894–99
134. Buchanan BB. 1991. *Arch. Biochem. Biophys.* 288:1–9
135. Mate CJ, Hudson GS, von Caemmerer S, Evans JR, Andrews TJ. 1993. *Plant Physiol.* 102:1119–28

136. Perchorowicz JT, Raynes DA, Jensen RG. 1981. *Proc. Natl. Acad. Sci. USA* 78:2985–89
137. Seemann JR, Kobza J. 1988. *Plant Physiol. Biochem.* 26:461–71
138. Salvucci ME. 1989. *Physiol. Plant.* 77:164–71
139. Woodrow IE, Mott KA. 1989. *Aust. J. Plant Physiol.* 16:487–500
140. Sassenrath-Cole GF, Pearcy RW. 1992. *Plant Physiol.* 99:227–34
141. Muthuchelian K. 1992. *Photosynthetica* 26:333–39
142. Sage RF. 1993. *Photosynth. Res.* 35: 219–26
143. Schloss JV. 1990. *The Proceedings of NATO ASI on Enzymatic and Model Carboxylation and Reduction Reactions for Carbon Dioxide Utilization*, ed. M Aresta, JV Schloss, pp. 321–45. Netherlands: Kluwer Academic
144. Pierce J, Lorimer GH, Reddy GS. 1986. *Biochemistry* 25:1635–44
145. Fiedler F, Müllhofer G, Trebst A, Rose IA. 1967. *Eur. J. Biochem.* 1:395–99
146. Saver BG, Knowles JR. 1982. *Biochemistry* 21:5398–403
147. Sue JM, Knowles JR. 1982. *Biochemistry* 21:5404–10
148. VanDyk DE, Schloss JV. 1986. *Biochemistry* 25:5145–56
149. Jaworowski A, Hartman FC, Rose IA. 1984. *J. Biol. Chem.* 259:6783–89
150. Jaworowski A, Rose IA. 1985. *J. Biol. Chem.* 260:944–48
151. Hartman FC, Lee EH. 1989. *J. Biol. Chem.* 264:11784–89
152. Lorimer GH, Andrews TJ. 1973. *Nature* 243:359–60
153. Schloss JV, Lorimer GH. 1982. *J. Biol. Chem.* 257:4691–94
154. Andersson I, Knight S, Schneider G, Lindqvist Y, Lundqvist T, et al. 1989. *Nature* 337:229–34
155. Siegel MK, Lane MD. 1973. *J. Biol. Chem.* 248:5486–98
156. Pierce J, Tolbert NE, Barker R. 1980. *Biochemistry* 19:934–42
157. Tapia O, Andrés J. 1992. *Mol. Eng.* 2:37–41
158. Schloss JV. 1988. *J. Biol. Chem.* 263:4145–50
159. Lorimer GH, Andrews TJ, Pierce J, Schloss JV. 1986. *Philos. Trans. R. Soc. London Ser. B* 313:397–407
160. Pierce J, Andrews TJ, Lorimer GH. 1986. *J. Biol. Chem.* 261:10248–56
161. Cleland WW. 1990. *Biochemistry* 29: 3194–97
162. Andrews TJ, Kane HJ. 1991. *J. Biol. Chem.* 266:9447–52
163. Bowes G, Ogren WL, Hageman RH. 1971. *Biochem. Biophys. Res. Commun.* 45:716–22
164. Lorimer GH, Andrews TJ, Tolbert NE. 1973. *Biochemistry* 12:18–23
165. Kreckl W, Kexel H, Melzel E, Schmidt H-L. 1989. *J. Biol. Chem.* 264:10982–86
166. Lorimer GH. 1981. *Annu. Rev. Plant Physiol.* 32:349–83
167. Miziorko HM, Lorimer GH. 1983. *Annu. Rev. Biochem.* 52:507–35
168. Abell LM, Schloss JV. 1991. *Biochemistry* 30:7883–87
169. Andrés J, Safont VS, Tapia O. 1992. *Chem. Phys. Lett.* 198:515–20
170. Robison PD, Martin MN, Tabita FR. 1979. *Biochemistry* 18:4453–58
171. Christeller JT, Laing WA. 1979. *Biochem. J.* 183:747–50
172. Jordan DB, Ogren WL. 1983. *Arch. Biochem. Biophys.* 227:425–33
173. Brändén R, Nilsson T, Styring S. 1984. *Biochemistry* 23:4378–82
174. Mogel SN, McFadden BA. 1990. *Biochemisty* 29:8333–37
175. Lee GJ, McFadden BA. 1992. *Biochemistry* 31:2304–8
176. Lilley RM, Riesen H, Andrews TJ. 1993. *J. Biol. Chem.* 268:13877–84
177. Lee GJ, McDonald KA, McFadden BA. 1993. *Protein Sci.* 2:1147–54
178. Jordan DB, Ogren WL. 1981. *Nature* 291:513–15
179. Chen Z, Chastain CJ, Al-Abed SR, Chollet R, Spreitzer RJ. 1988. *Proc. Natl. Acad. Sci. USA* 85:4696–99
180. Chen Z, Yu W, Lee J-M, Diao R, Spreitzer RJ. 1991. *Biochemistry* 30: 8846–50
181. Harpel MR, Hartman FC. 1992. *J. Biol. Chem.* 267:6475–78
182. Soper TS, Larimer FW, Mural RJ, Lee EH, Hartman FC. 1992. *J. Biol. Chem.* 267:8452–57
183. Chène P, Day AG, Fersht AR. 1992. *J. Mol. Biol.* 225 891–96
184. Parry MAJ, Madgwick P, Parmar S, Cornelius MJ, Keys AJ. 1992. *Planta* 187:109–12
185. Chen Z, Spreitzer RJ. 1989. *J. Biol. Chem.* 264:3051–53
186. Gutteridge S, Rhoades DF, Herrmann C. 1993. *J. Biol. Chem.* 268:7818–24
187. Lee EH, Harpel MR, Chen Y-R, Hartman FC. 1993. *J. Biol. Chem.* 268: 26583–91
188. Lorimer GH, Chen Y-R, Hartman FC. 1993. *Biochemistry* 32:9018–24
189. Smith HB, Larimer FW, Hartman FC. 1990. *J. Biol. Chem.* 265:1243–45
190. Read BA, Tabita FR. 1992. *Biochemistry* 31:5553–60

191. Laing WA, Ogren WL, Hageman RH. 1974. *Plant Physiol.* 54:678–85
192. Chen Z, Spreitzer RJ. 1991. *Planta* 183:597–603
193. Chen Z, Spreitzer RJ. 1992. *Photosynth. Res.* 31:157–64
194. Edmondson DL, Badger MR, Andrews TJ. 1990. *Plant Physiol.* 93:1376–82
195. Edmondson DL, Badger MR, Andrews TJ. 1990. *Plant Physiol.* 93:1390–97
196. Edmondson DL, Kane HJ, Andrews TJ. 1990. *FEBS Lett.* 260:62–66
197. Zhu G, Jensen RG. 1991. *Plant Physiol.* 97:1348–53
198. Zhu G, Jensen RG. 1991. *Plant Physiol.* 97:1354–58
199. McCurry SD, Tolbert NE. 1977. *J. Biol. Chem.* 252:8344–46
200. Yokota A. 1991. *Plant Cell Physiol.* 32:755–62
201. Edmondson DL, Badger MR, Andrews TJ. 1990. *Plant Physiol.* 93:1383–89
202. Rose IA. 1981. *Philos. Trans. R. Soc. London Ser. B* 293:131–43
203. Richard JP. 1991. *Biochemistry* 30: 4581–85
204. Pompliano DL, Peyman A, Knowles JR. 1990. *Biochemistry* 29:3186–94
205. Larimer FW, Harpel MR, Hartman FC. 1993. *Protein Sci. Suppl. 1* 2:68
206. Brändén R, Nilsson T, Styring S. 1980. *Biochem. Biophys. Res. Commun.* 92: 1297–305
207. Schneider G, Lindqvist Y, Lundqvist T. 1990. *J. Mol. Biol.* 211:989–1008
208. Lundqvist T, Schneider G. 1988. *J. Biol. Chem.* 236:3643–46
209. Lundqvist T, Schneider G. 1989. *J. Biol. Chem.* 264:7078–83
210. Lundqvist T, Schneider G. 1991. *Biochemistry* 30:904–8
211. Lundqvist T, Schneider G. 1991. *J. Biol. Chem.* 266:12604–11
212. Knight S, Andersson I, Brändén C-I. 1990. *J. Mol. Biol.* 215:113–60
213. Curmi PMG, Cascio D, Sweet RM, Eisenberg DS, Schreuder H. 1992. *J. Biol. Chem.* 267:16980–89
214. Schreuder HA, Knight S, Curmi PMG, Andersson I, Cascio D, et al. 1993. *Protein Sci.* 2:1136–46
215. Newman J. 1992. *Structural studies of ribulose 1,5-bisphosphate carboxylase/oxygenase from the cyanobacterium* Synechococcus *PCC301*. PhD thesis. Swedish Univ. Agric. Sci., Uppsala, Sweden
216. Farber GK, Petsko GA. 1990. *Trends Biochem. Sci.* 15:228–34
217. Larimer FW, Lee EH, Mural RJ, Soper TS, Hartman FC. 1987. *J. Biol. Chem.* 262:15327–29
218. Lee EH, Stringer CD, Hartman FC. 1986. *Proc. Natl. Acad. Sci. USA* 83:9383–87
219. Chapman MS, Suh SW, Curmi PMG, Cascio D, Smith WW, Eisenberg DS. 1988. *Science* 241:71–74
220. Schneider G, Lindqvist Y, Brändén C-I, Lorimer GH. 1986. *EMBO J.* 5:3409–15
221. Schneider G, Knight S, Andersson I, Brändén C-I, Lindqvist Y, Lundqvist T. 1990. *EMBO J.* 9:2045–50
222. Voordouw G, de Vries PA, van den Berg WAM, de Clerck EPJ. 1987. *Eur. J. Biochem.* 163:591–98
223. Lee B, Berka RM, Tabita FR. 1991. *J. Biol. Chem.* 266:7417–22
224. Paul K, Morell MK, Andrews TJ. 1991. *Biochemistry* 30:10019–26
225. Paul K, Morell MK, Andrews TJ. 1993. *Plant Physiol.* 102:1129–37
226. Read BA, Tabita FR. 1992. *Biochemistry* 31:519–25
227. Johal S, Partridge BE, Chollet R. 1985. *J. Biol. Chem.* 260:9894–904
228. Zhu G, Jensen RG. 1990. *Plant Physiol.* 93:244–49
229. Yokota A. 1991. *J. Biochem.* 110:246–52
230. Yokota A, Higashioka M, Wadano A. 1991. *J. Biochem.* 110:253–56
231. Yokota A, Higashioka M, Wadano A. 1992. *Eur. J. Biochem.* 208:721–27
232. Knowles JR. 1991. *Nature* 350:121–24
233. Lorimer GH. 1981. *Biochemistry* 20: 1236–40
234. Igarashi Y, McFadden BA, El-Gul T. 1985. *Biochemistry* 24:3957–62
235. Hartman FC, Stringer CD, Milanez S, Lee EH. 1986. *Philos. Trans. R. Soc. London Ser. B* 313:379–95
236. Lee EH, Soper TS, Mural RJ, Stringer CD, Hartman FC. 1987. *Biochemistry* 26:4599–604
237. Hartman FC, Soper TS, Niyogi SK, Mural RJ, Foote RS, et al. 1987. *J. Biol. Chem.* 262:3496–501
238. Soper TS, Mural RJ, Larimer FW, Lee EH, Machanoff R, Hartman FC. 1988. *Protein Eng.* 2:39–44
239. Hartman FC, Larimer FW, Mural RJ, Machanoff R, Soper TS. 1987. *Biochem. Biophys. Res. Commun.* 145: 1158–63
240. Niyogi SK, Foote RS, Mural RJ, Larimer FW, Mitra S, et al. 1986. *J. Biol. Chem.* 261:10087–92
241. Estelle M, Hanks J, McIntosh L, Somerville CR. 1985. *J. Biol. Chem.* 260:9523–26
242. Distefano MD, Moore MJ, Walsh CT. 1990. *Biochemistry* 29:2703–13
243. Scrutton NS, Berry A, Deonarian MP,

Perham RN. 1990. *Proc. R. Soc. London Ser. B* 242:217–24
244. Deonarian MP, Scrutton NS, Perham RN. 1992. *Biochemistry* 31: 1498–504
245. Tobias KE, Kahana C. 1993. *Biochemistry* 32:5842–47
246. Miziorko HM, Sealy RC. 1980. *Biochemistry* 19:1167–71
247. Söderlind E, Schneider G, Gutteridge S. 1992. *Eur. J. Biochem.* 206:729–35
248. Gutteridge S, Lorimer G, Pierce J. 1988. *Plant Physiol. Biochem.* 26:675–82
249. Smith HB, Hartman FC. 1988. *J. Biol. Chem.* 263:4921–25
250. Mural RJ, Soper TS, Larimer FW, Hartman FC. 1990. *J. Biol. Chem.* 265:6501–5
251. Harpel MR, Lee EH, Hartman FC. 1993. *Anal. Biochem.* 209:367–74
252. Morell MK, Kane HJ, Hudson GS, Andrews TJ. 1992. *Arch. Biochem. Biophys.* 299:295–301
253. Day AG, Chène P, Fersht AR. 1993. *Biochemistry* 32:1940–44
254. Terzaghi BE, Laing WA, Christeller JT, Peterson GB, Hill DF. 1986. *Biochem. J.* 235:839–46
255. Harpel MR, Larimer FW, Hartman FC. 1991. *J. Biol. Chem.* 266:24734–40
256. Deleted in proof
257. Harpel MR, Larimer FW, Lorimer GH, Hartman FC. 1993. *Protein Sci. Suppl. 1*. 2:68
258. Lorimer GH, Hartman FC. 1988. *J. Biol. Chem.* 263:6468–71
259. Smith HB, Hartman FC. 1991. *Biochemistry* 30:5172–77
260. Morell MK, Paul K, Kane HJ, Andrews TJ. 1992. *Aust. J. Bot.* 40:431–41
261. Haining RL, McFadden BA. 1990. *J. Biol. Chem.* 265:5434–39

Annu. Rev. Biochem. 1994. 63:235–64

NITROGENASE: A Nucleotide-Dependent Molecular Switch

James B. Howard
Department of Biochemistry, University of Minnesota School of Medicine, Minneapolis, Minnesota 55455

Douglas C. Rees
Division of Chemistry and Chemical Engineering 147-75CH, California Institute of Technology, Pasadena, California 91125

KEY WORDS: nitrogen fixation, nucleotide-binding proteins, ADP/ATP, enzyme mechanism, metal clusters

---

CONTENTS

INTRODUCTION

In the simplest terms, the biological nitrogen cycle is the reduction of atmospheric dinitrogen (N_2) to ammonia with the subsequent reoxidation of ammonia to dinitrogen (1). At the reduction level of ammonia, nitrogen is incorporated into precursors for biological macromolecules such as proteins

0066-4154/94/0701-0235$05.00

and nucleic acids. Reoxidation of ammonia to dinitrogen ("denitrification") by a variety of microbes (by way of nitrite and other oxidation levels of nitrogen) leads to the depletion of the "fixed," biologically usable, nitrogen pool. Besides the relatively small contribution from commercial ammonical fertilizer production, replenishing of the nitrogen pool falls mainly to a limited number of physiologically diverse microbes (e.g. eubacteria and archaebacteria; free-living and symbiotic; aerobic and anaerobic) that contain the nitrogenase enzyme system.

During the past 25 years, steady progress has been made in elucidating the essential elements of the nitrogenase reaction (reviewed in 2–8a), which are summarized in Figure 1. Ammonia synthesis requires eight electrons: six for the reduction of dinitrogen and two for the coupled, obligatory synthesis of H_2 (9). These reactions are catalyzed by the terminal component in the complex, the MoFe-protein,[1] so-designated because it contains iron and molybdenum atoms. Electrons are transferred to the MoFe-protein from the Fe-protein in a process coupled to the hydrolysis of two ATP per electron. Because a minimum of 16 ATP are hydrolyzed for the reduction of one molecule of dinitrogen, the organisms carrying out nitrogen fixation have a vigorous energy metabolism. The Fe-protein is an integral component of the nitrogenase reaction in that other, low-redox-potential electron donors do not support dinitrogen reduction, undoubtedly due to the requirement for coupled ATP hydrolysis. In contrast, a variety of electron donors (ferredoxins and flavodoxins) provide a connection between cellular metabolism and the reduction of the Fe-protein.

In the past, nitrogenase and nitrogen fixation have been of interest primarily to bioinorganic chemists and spectroscopists, and to those working in microbial physiology related to agriculture. However, the similarity of the ATP-dependent electron transfer in nitrogenase to many other nucleotide-dependent energy-transducing systems in higher organisms should make this enzyme of general interest to biochemists. The recent solution of three-dimensional structures for both the nitrogenase proteins (13–20) em-

[1]The nomenclature for the nitrogen fixation enzyme is rich and varied. The molybdenum-iron protein is frequently referred to as MoFe-protein, component 1, or dinitrogenase, while the iron protein is referred to as Fe-protein, component 2, or dinitrogenase reductase. A useful shorthand nomenclature for a component from a specific organism is to abbreviate the species followed by the number of the component, e.g. the MoFe-protein from *Azotobacter vinelandii* is Av1 and the Fe-protein from *Clostridium pasteurianum* is Cp2. We use the shorthand when discussing specific results, and the MoFe-protein and Fe-protein designations when discussing generic conclusions. Amino acid residue numbering for Fe-protein is based upon the protein sequence of Av2 (10), which is one residue shorter than the gene sequence (11). The numberings for MoFe-protein subunits are based upon the gene sequences for Av1 subunits (11), which are one residue longer than for the isolated protein subunits (12). In both cases, the protein apparently is processed by removal of the initiation Met.

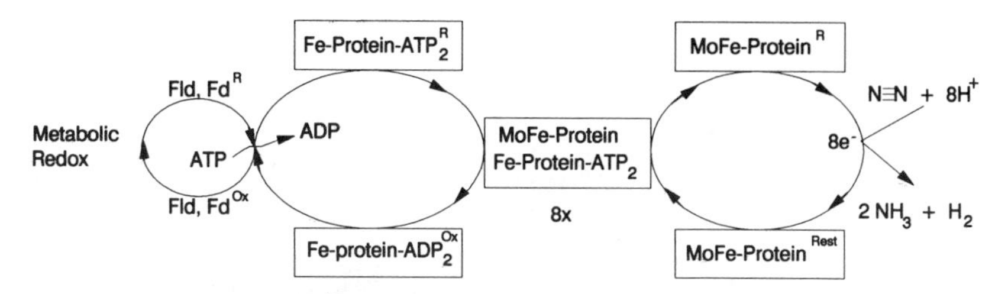

Figure 1 The nitrogenase reaction. The electron transfer proteins ferredoxin (Fd) and flavodoxin (Fld) serve to couple the nitrogenase reaction to metabolically generated reducing equivalents.

phasizes not only the functional relatedness, but also the structural relatedness of nitrogenase to such diverse systems as *ras* p21 (21, 22), membrane-bound transporters (23, 24), muscle contraction (25), the recA protein involved in DNA recombination (26), and elongation factors in protein biosynthesis (27, 28). This review aims to summarize for the general biochemist the status of nitrogenase structure-function studies and to discuss the relevance of nitrogenase chemistry to other systems. For those wanting more detailed analysis of specific questions relating to nitrogenase, there are reports from several recent symposia (29, 30).

PROPERTIES OF THE NITROGENASE PROTEINS

The nitrogenase Fe-protein and MoFe-protein have been sequenced and/or characterized from a variety of nitrogen-fixing organisms. Generally speaking, the structural and functional properties of these proteins are highly conserved among different organisms. For many combinations of Fe-protein and MoFe-protein from different species, substantial activity is obtained from these heterocomplexes; yet for others, little or no activity is observed (31, 32). Consequently, while we emphasize "consensus" features of these proteins, important species-specific variations exist, and are discussed when relevant.

Undoubtedly, no feature more dominates the experimental study of these proteins than their extreme oxygen lability. All manipulations of the proteins must be performed with an atmosphere of <1 ppm oxygen. Because of the destructive effects of even traces of oxygen, it is often difficult to separate legitimate results from artifacts generated by the experimental conditions. For nitrogenase, probably more so than for any other enzyme, the consistency of results between laboratories must be demonstrated before an "observation" should be considered a "fact."

Figure 2 Ribbons diagram of the polypeptide fold of the Fe-protein dimer, with space-filling models for the 4Fe:4S cluster and ADP. Prepared with the program MOLSCRIPT (36a) and reproduced with permission from (8a).

Fe-Protein

The Fe-protein is a ~60,000-dalton dimer of identical subunits bridged by a single 4Fe:4S cluster. The three-dimensional crystal structure of Fe-protein (13) confirmed the hypothesis (33, 34) that the cluster is symmetrically coordinated by Cys97 and Cys132 from each subunit. Significantly, the subunits have the α-helical/β-sheet type of architecture commonly associated with a major class of nucleotide-binding proteins (35) that includes adenylate kinase (36) and the *ras* p21 oncogene protein (21, 22). The large, single domain encompassing the entire Fe-protein subunit consists of an eight-stranded β-sheet flanked by nine α-helixes (Figure 2). The two subunits are related by a molecular two-fold rotation axis that passes through the 4Fe:4S cluster, which is located at one end of the dimer, and the subunit-subunit interface. Besides the cluster serving to crosslink subunits, there are numerous hydrophobic and salt interactions in the interface beneath the cluster that help to stabilize the dimer structure. Indeed, these interactions are sufficiently strong that the cluster can be removed and the dimer structure is still maintained (34).

One of the two principal functional features of the Fe-protein is the binding of the nucleotides, MgATP and MgADP. Two nucleotide-binding

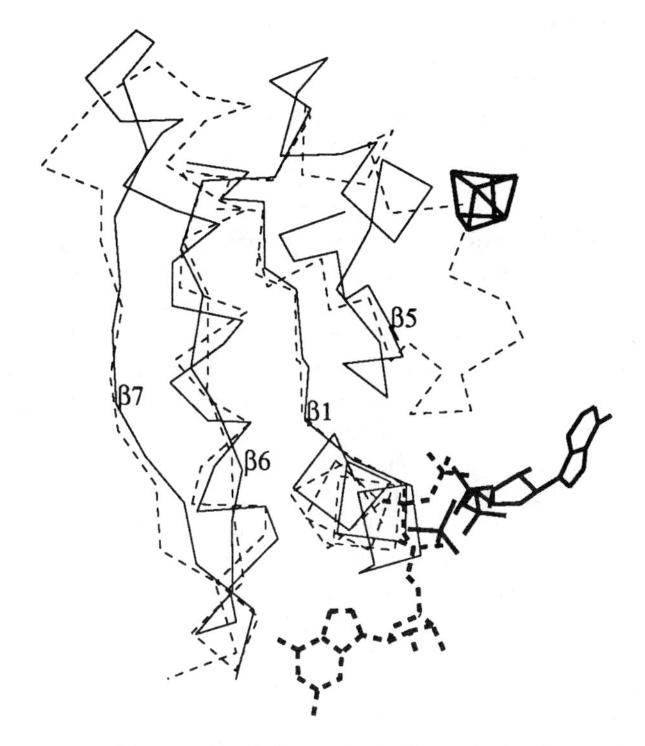

Figure 3 Two potential nucleotide-binding modes in Fe-protein based upon the observed ADP-binding mode in the Av2 crystal structure (13; solid lines) and the GTP-binding mode in *ras* p21 (21; dashed lines). The numbers of the β strands in Av2 are indicated. Reproduced with permission from (13).

sites per dimer have been reported with dissociation constants of ~100 μM, although there is considerable variation in the measured values (37), due in part to the difficulty of determining equilibrium values with oxygen-sensitive proteins. Residues 9–16 near the amino terminus of Fe-protein exhibit the amino acid sequence motif, GXXXXGKS [the Walker motif A (38)], that is characteristic of a major class of nucleotide-binding sites (35). This sequence adopts a β-strand—loop—α-helix conformation that interacts with the β,γ-phosphate groups and the Mg of bound nucleotides. Mutations in Fe-protein leading to substitutions at Lys15 and to the three hydroxy amino acids at residues 16–18 are consistent with this sequence being part of the phosphate-binding site in Av2 (39, 40). In addition, the second Walker motif, DxxG, that in combination with motif A completes the Mg-phosphate protein interactions, is also found in Av2 at residues 125–128. Substitution of glutamic acid for Asp125 confirmed the role of this residue as a ligand to Mg (41).

Although a *ras*-type nucleotide phosphate-binding site is clearly present in Fe-protein, it is less evident how the nucleotide would be oriented with respect to this site. The Walker motif A is located near the molecular two-fold axis of Fe-protein, and the nucleotide-binding sites project into the subunit-subunit interface. In the Av2 crystal structure, one molecule of ADP was observed to be positioned across the subunit-subunit interface, perpendicular to the two-fold axis (Figure 3). With this orientation, the purine ring is bound to one subunit and the phosphates to the other. An alternative and more speculative binding mode can be modeled by analogy to the *ras* protein. In this mode, the nucleotide would lie along the two-fold axis with the purine ring and phosphate groups bound by the same subunit (see Figure 3). There is mutagenesis evidence that may support this orientation of nucleotide binding. If invariant residue Arg213 (comparable to Lys117 in the *ras* structure, which is part of the purine ring–binding domain) is substituted by cysteine, the Fe-protein is inactive (42). Because Arg213 is located on the surface of the protein, at the extreme edge of the intersubunit cleft, substitution here would seem to be innocuous, unless it were part of a specific interaction, such as the extended nucleotide site. As described in more detail below, it is conceivable that both modes of nucleotide binding may be functionally relevant.

The second functional site in Fe-protein is the 4Fe:4S cluster, which undergoes a one-electron redox cycle between the $2Fe^{2+}2Fe^{3+}$ state and the $3Fe^{2+}Fe^{3+}$ state. Both cluster ligands are located near the amino-terminal end of α-helices that are directed towards the cluster. Peptide bonds within these helices form NH-S hydrogen bonds to the cluster that may provide stabilizing electrostatic interactions to this anionic center (43). In contrast to the 4Fe:4S clusters observed in ferredoxin-type proteins, a striking feature of the Fe-protein is the exposure of the 4Fe:4S cluster to solvent, a property anticipated by spectroscopic studies (44). Other than the cysteinyl ligands, there is little contact between the 4Fe:4S cluster and other amino acid side chains. Consequently, the emerging picture of the Fe-protein cluster is that of an exposed, loosely packed redox center that may function as a pivot or hinge to accommodate conformational rearrangements between subunits during the course of the nitrogenase reaction mechanism.

Although Fe-protein has a number of spectroscopic signatures typical of 4Fe:4S clusters, more important to understanding the functional properties of the protein is the response of the cluster to nucleotide binding. Some of these responses indicate that the Fe-protein is intimately involved in the coupling of ATP hydrolysis to electron transfer. For example, the electron paramagnetic resonance (EPR) spectrum becomes more axial (45), and there is a 60–100-mV decrease in the redox potential when AXP[2] is bound to Fe-protein (46). A dramatic effect of nucleotide binding is evidenced by the change in accessibility of the cluster to chelators (47–49). Specifically, the cluster irons can be removed by chelators only in the presence of MgATP; MgADP inhibits chelation, whereas no chelation occurs in the absence of nucleotide. Given that the cluster is already exposed to solvent, it is surprising that chelation is dependent on the presence of MgATP. Furthermore, the chelation process is dependent upon the oxidation state of the cluster (50). The crystal structure clearly shows that the binding site for the terminal nucleotide phosphates (where hydrolysis occurs) is located ~20 Å from the cluster, so that the nucleotide does not interact directly with the cluster. Rather, the location of the cluster and nucleotide-binding sites at the interface between the subunits suggests there is an allosteric coupling mechanism that connects these two functional sites. Different conformational states of the Fe-protein might exist that differ in details of the intersubunit interactions, e.g. changes in number or location of salt bonds. The equilibrium between these conformations must be sensitive to nucleotide binding, oxidation state of the cluster, and complex formation with the MoFe-protein. Hence, ligand-binding and/or redox reactions would shift the equilibrium position between states, thereby coupling spatially distinct processes. As described below, the details of the coupling process are one of the major, unanswered problems in the nitrogenase mechanism.

MoFe-Protein

The MoFe-protein is an $\alpha_2\beta_2$ tetramer with a total molecular weight of ~240,000. For some time, it has been recognized from the spectroscopic properties of the protein that there are two groups of metal centers: the diamagnetic, EPR-silent P-cluster pairs (or P-cluster or P-center), and the unusual $S=3/2$ paramagnetic FeMo-cofactor (or M-center or "cofactor"). Although there has been much speculation as to the arrangement and structure of the 2 Mo, 30 Fe, and ~34 inorganic S that form these clusters, the recent three-dimensional structures of the protein (14–20) revealed how truly unique they are. Each of the two FeMo-cofactor units contains 1 Mo:7 Fe:9 S and one homocitrate molecule, and almost certainly provides the site of substrate binding and reduction. The FeMo-cofactor may be considered to be formed from 4Fe:3S and Mo:3Fe:3S partial cubanes that are bridged by three nonprotein ligands, most probably sulfides (Figure 4). The cofactor is buried ~10 Å beneath the protein surface, in an environment primarily provided by the α subunit. Only two protein ligands, Cys α275 and His α442, coordinate the cofactor to the protein, resulting in the unusual situation

[2]The notation AXP designates either ADP or ATP.

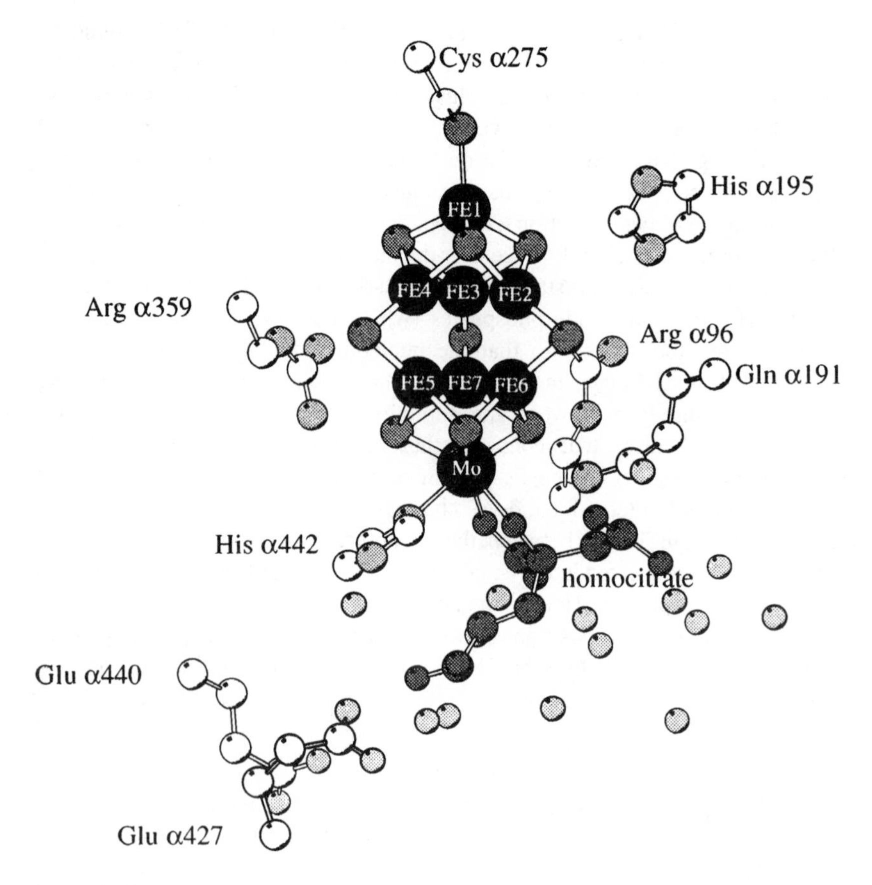

Figure 4 Structure of the FeMo-cofactor with surrounding protein and water molecules indicated. Prepared with the program MOLSCRIPT (36a), and reproduced with permission from (8a).

in which the six Fe atoms bridged by nonprotein ligands are three-coordinate. It is tempting to consider these unsaturated sites as suitable for binding of ligands, including substrates. The octahedral coordination sphere of the Mo is completed by bidentate binding of homocitrate. Hydrogen bonds to sulfurs in the cluster are provided by the side chains of residues Arg α96, His α195, Arg α359, and the NH groups of residues α356 and α358. These hydrogen bonds may provide a mechanism for funneling protons to substrate bound to the FeMo-cofactor. The homocitrate is hydrogen bonded to the side chain of Gln α191, and is also surrounded by a pool of buried waters, which could conceivably function as a proton source for substrate reduction.

The remaining Fe and S are organized into P-cluster pairs, which are

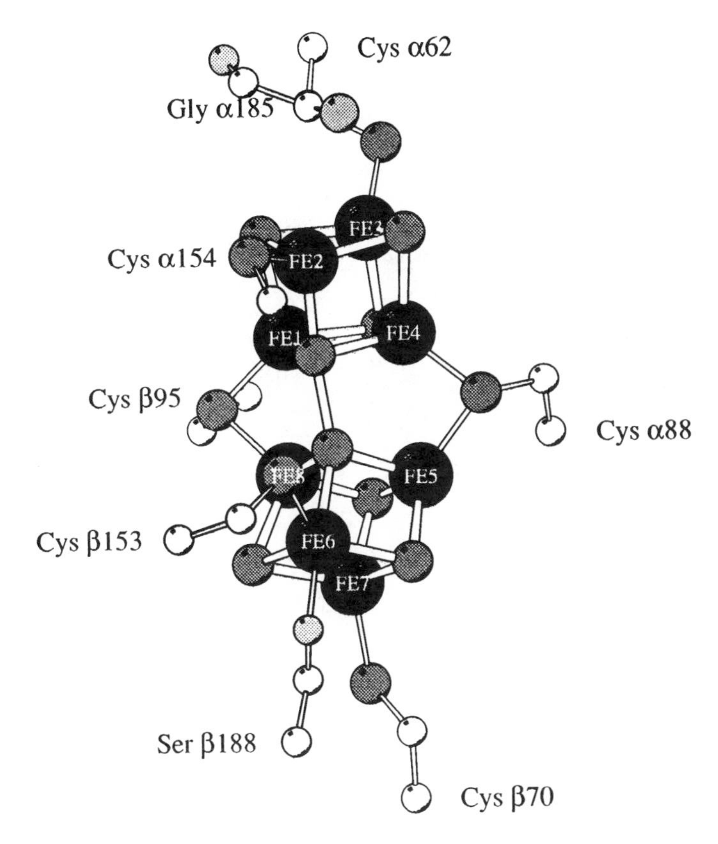

Figure 5 Structure of the P-cluster pair and surrounding protein. Prepared with the program MOLSCRIPT (36a), and reproduced with permission from (8a).

present in two copies per MoFe-protein tetramer. Each P-cluster pair contains two bridged 4Fe:4S clusters, i.e. the P-cluster pair can be considered as a 8Fe:8S cluster. What makes the P-cluster pair an unusual inorganic structure is that the two 4Fe:4S clusters are bridged by the thiol side chains of Cys α88 and β95, and a disulfide bond between two cluster inorganic sulfurs (Figure 5). Although this disulfide bridge is clear in the Av1 structure (17), Bolin's current analysis of the Cp1 structure suggests that perhaps only one sulfur is present in his structure (20). This disulfide bond is located on the side of the P-cluster pair closest to the surface of the protein. Singly coordinating cysteinyl thiols (from residues α62, α154, β70, and β153) ligate the remaining four irons, such that nonbridging cysteines coordinated to a specific 4Fe:4S cluster are from the same subunit. In addition to the

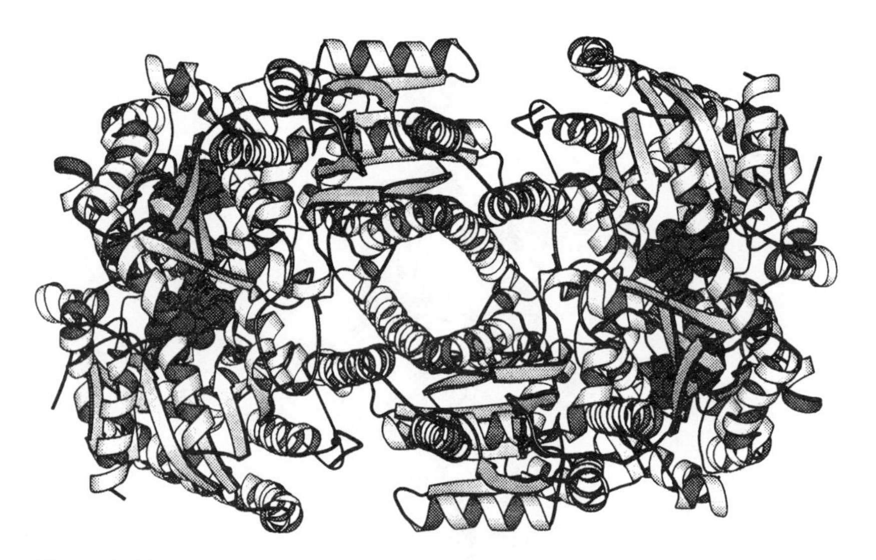

Figure 6 Ribbons diagram of the polypeptide fold of the MoFe-protein tetramer, with space-filling models for the P-cluster pair and the FeMo-cofactor. The view is along the tetramer two-fold axis. Prepared with the program MOLSCRIPT (36a) and reproduced with permission from (8a).

cysteinyl ligands, Ser β188 appears to coordinate the same iron as Cys β153, although in Bolin's analysis of Cp1, this side chain does not appear to serve as an iron ligand, but rather donates a hydrogen bond to a cluster sulfur (20). The presence of the disulfide bridge in the P-cluster pair implies that this center may be able to serve as a two-electron redox group. Since the Fe-protein is generally considered to be a one-electron donor with the P-cluster pair as the immediate acceptor, the P-cluster pair may exhibit both one- and two-electron redox chemistry, analogous to quinones and flavins in other biological systems. Spectroscopic studies indicate that the P-cluster pair is highly reduced, with all eight Fe atoms most likely in the ferrous form (51). Although the all ferrous oxidation state is unprecedented in known biological four-iron clusters, the P-cluster pair cannot be directly compared to these other clusters. Two isolated 4Fe:4S clusters with all Fe in the ferrous state would have a combined net charge of −8 (including the eight thiol ligands), whereas the P-cluster pair, with two bridging ligands and a disulfide, would have a net charge of only −4. That is, the effective cluster charge per iron atom is similar to an oxidized ferredoxin cluster, not to a "super-reduced" cluster.

Although there is minimal amino acid sequence homology between subunits, the α and β subunits of Av1 exhibit similar polypeptide folds, which consist of three domains of the α-helical/β-sheet type with some extra helices (Figure 6). At the interface between the three domains is a wide, shallow cleft; in the α subunit, the FeMo-cofactor occupies the bottom of this cleft. The P-cluster pair is buried at the interface between a pair of α- and β-subunits with a pseudo two-fold rotation axis passing between the two 4Fe:4S halves of the P-cluster pair and relating the two subunits. Coordination of metal centers by ligands at the interface between homologous or identical subunits has been previously described for Fe-protein and bacterial photosynthetic reaction centers (52, 53), and may be a common feature of multisubunit metalloproteins. It should be noted that in some places, the symmetry away from the cluster interface is more apparent than real. For example, some of the symmetry-related secondary elements on the protein surface are, in fact, from different segments of the primary sequence in the individual subunits.

The extensive interaction between α and β subunits in an αβ dimer suggests they form a fundamental functional unit. Intriguingly, there is an open channel of ~8 Å diameter between the two pairs of αβ dimers with the tetramer two-fold axis extending through the center. The tetramer interface is dominated by interactions between helices from the two β subunits, along with a cation-binding site, presumably calcium, that is coordinated by residues from both β subunits.

Species-Specific Variations in Nitrogenase Proteins

A notable exception to heterocomplexes (mixes of Fe-protein and MoFe-protein from different species) having enzymatic activity are both the nitrogenase components from *C. pasteurianum* (Cp1 and Cp2) (31, 32). Structural differences between the nitrogenase proteins from *A. vinelandii* and *C. pasteurianum* have been identified that could be the source of the inability of *C. pasteurianum* proteins to complement other nitrogenase proteins. Cp2 differs from Av2 in the length of the carboxy-terminal residues (Cp2 is 13 residues shorter than Av2) and in the region around residue 65, which exhibits significant sequence variability in different Fe-proteins, including a two-residue deletion in Cp2 relative to Av2. The carboxy-terminal residues of Av2 form intersubunit contacts that would not be possible in Cp2, due to the shorter length of this protein. Incorporation of the Cp2 carboxy-terminal residues into the Av2 sequence results in ~50% reduction in Av2 activity (54), indicating that while these residues are not essential for Fe-protein function in the nitrogenase mechanism, they do contribute to the overall kinetics, perhaps by influencing the relative stabilities of different Fe-protein conformations. The region around residue 65 of Av2 projects from the "top" surface of the protein that has been implicated in the binding of Fe-protein to MoFe-protein. Hence, sequence alterations in this region may perturb the structure or stability of the Fe-protein—MoFe-protein complex.

Cp1 differs from Av1 by a ~50-residue insertion in the α subunit and a ~50-residue deletion in the β subunit. In Av1, the amino-terminal ~50 residues (that are deleted in Cp1) extend from the β subunit, and interact with both an α subunit and the other β subunit. Hence, it is possible that these amino-terminal residues of the β subunit function in stabilization of the quaternary structure of Av1. Interestingly, these amino-terminal residues are also missing from the alternative vanadium-iron protein (55, 56; also, see below). The insertion in the α subunit of Cp1 forms a polypeptide loop that covers the protein surface over the FeMo-cofactor; as this region has been suggested to participate in the interaction between Fe-protein and MoFe-protein, the presence of this additional protein segment could interfere with, or alter, complex formation between these proteins isolated from different sources.

GENETICS AND ASSEMBLY OF THE NITROGENASE PROTEINS

The complexity of the nitrogenase protein structures is reflected in the rather large number of genes in the *nif* regulon. Besides the three genes for the subunits of Fe-protein (*nifH*) and MoFe-protein (*nifD* and *nifK* for the α and β subunits, respectively), at least 17 other genes have been identified and sequenced (reviewed in 57); depending on the species, the number of genes may be considerably greater than 20. Several of the *nif* genes (*nifA* and *nifL*) are part of the regulatory mechanism that allows the expression of nitrogenase only when ammonia is depleted and, in some species, when the organism is growing anaerobically. Also encoded are flavodoxin (*nifF*) and, depending on the species, a pyruvate dehydrogenase (*nifJ*), the so-called "phosphorylclastic" enzyme, that catalyzes pyruvate conversion to acetyl phosphate concomitant with ferredoxin reduction. The phosphorylclastic reaction coupled to ferredoxin reduction provides both the electrons and the ATP required by nitrogenase.

The most important genes from the point of view of the functioning nitrogenase are those encoding processing enzymes. One of the earliest biochemical genetics contributions to nitrogenase chemistry was the recognition that mutations at loci other than the structural genes resulted in inactive enzymes. One of these mutants, UW45, had a MoFe-protein that could be activated by the acid extract of active, wild-type MoFe-protein (58). The purified activating material had the $S=3/2$ EPR signature of the wild-type protein and had a composition equivalent to all the molybdenum and approximately half the iron and inorganic sulfur (59, 60). This material, designated FeMo-cofactor, was subsequently determined to contain homocitrate by Ludden, Shah, and coworkers using elegant, classical biochemical methods (61). To date, five proteins, all encoded by the *nif* regulon, have been identified as essential for the assembly of cofactor (62). Two genes (*nifE, nifN*) encode subunits for a protein with substantial sequence homology to the MoFe-protein, including the ligand Cys-α275 and some of the residues that interact with the cofactor in the MoFe-protein (63). However, there are several critical residues that are not conserved, including three of the P-cluster pair cysteinyl ligands and the cofactor ligand His α442. The working hypothesis is that NIFE,N forms a scaffolding protein on which the cofactor is at least partially assembled before transfer to the des-FeMo-cofactor protein (64). Another required protein is the gene product of *nifV*, which appears to be homocitrate synthetase (61). In the absence of this enzyme, citrate is incorporated in place of homocitrate, with the resulting cofactor able to support reduction of most substrates, except dinitrogen (65). The function of *nifB*, the locus defined in UW45, has yet to be determined. Finally, the gene product of *nifH* is required. Because *nifH* encodes the Fe-protein, one would assume that Fe-protein functions similarly to that in nitrogenase turnover, namely as an ATP-dependent redox donor. This is unlikely to be the case, however. Several mutant strains have been generated with Fe-proteins that are unable to hydrolyze ATP or to support ammonia production, yet which have normal MoFe-protein. Perhaps one clue to the role of Fe-protein in MoFe-protein assembly is the observation that des-FeMo-cofactor protein from *nifH*$^-$ strains is poorly activated by isolated FeMo-cofactor, while des-FeMo-cofactor protein from the other *nif*$^-$ strains can be (66–69). This implies that the des-FeMo-cofactor protein must be acted on by Fe-protein. One possibility for this requirement is the need to phosphorylate the des-FeMo-cofactor protein. This behavior could explain the difference in electrophoretic mobility of the protein, depending upon from what genetic background the des-FeMo-cofactor protein was isolated (68–70).

The insertion of the metal centers into apo-MoFe-protein also provides several perplexing problems. The des-FeMo-cofactor protein contains the P-cluster pairs, yet the analogous des-P-cluster pair protein has not been observed (69). No genetic loci have been specifically ascribed to assembly and/or insertion of the P-cluster pair. The most likely explanation is that the α and β subunits can only assemble around the P-cluster pairs, and that the subunits serve as the scaffolding and final processing center for introducing the unique structural changes to the more conventional 4Fe:4S clusters. From chemical modification studies, it has been suggested that most of the holoenzyme three-dimensional structure is intact in the des-FeMo-cofactor protein (J Magnuson and JB Howard, in preparation), including the Fe-protein docking site (see below for discussion of the docking mechanism). Although the cofactor pocket is deeply buried, the pocket is located at the interface of the three large α subunit domains, and could be exposed by as little as a 15–20° rotation of the carboxy-terminal domain (JB Howard, unpublished). One role of Fe-protein in MoFe-protein assembly might be to stabilize these altered conformations apparently necessary for cofactor insertion.

The Fe-protein also requires processing for full activity. Although apo-Av2 can be reconstituted using inorganic Fe and S^{2-} salts, the resulting activity is highly variable (71). Recently, Dean and coworkers have reported (72) that *nifS* encodes a sulfuryl transferase which, in combination with apo-Av2, regenerates full activity of Fe-protein (D Dean, personal communication). *nifS* is one of several genes that are not absolutely required for biosynthesis of active nitrogenases, presumably because similar enzymes are expressed in cells as part of a more general need to synthesize proteins containing iron-sulfur clusters. In contrast, *nifM* is essential for active Fe-protein (73, 74). The function of NIFM is unknown, except that it is not directly involved in cluster synthesis or insertion. It should be noted that Fe-protein from *nifM*$^-$ strains is still fully active in FeMo-cofactor synthesis (34).

Finally, some prokaryotic species contain other, so-called alternate, forms of nitrogenase in addition to the conventional, molybdenum-containing proteins, which are expressed under various conditions of molybdenum depletion (reviewed in (75, 76)). The amino acid sequences of the alternate nitrogenases are nearly identical to the conventional proteins. The most interesting feature of the alternate enzymes is the lack of molybdenum, which is apparently replaced by either vanadium or iron. Indeed, cofactors extracted from these proteins have only iron or vanadium and iron, yet will substitute for FeMo-cofactor in the UW45 reconstitution assay. Although several genes are common to the alternate and conventional nitrogenase systems, the structural genes are different, and there are additional control elements for regulating expression.

MECHANISM OF NITROGENASE

Often the nitrogenase mechanism is intellectually divided into two parts, the redox cycle between the Fe-protein and the MoFe-protein, and the substrate reduction cycle. Although this framework has been useful for obtaining some of the individual reaction rates, we attempt to integrate the two cycles with the new molecular structures. The salient, experimental observations that must be accounted for by a mechanism may be summarized:

1. Electron transfer proceeds from the Fe-protein to the MoFe-protein. 2. Two ATP are hydrolyzed per electron transferred at maximum efficiency. 3. Although ATP and ADP are bound by the Fe-protein, only the Fe-protein/MoFe-protein complex turns over ATP; Fe-protein-MgATP does not. 4. The rate of substrate reduction is dependent on both the ratio of the two protein components and the absolute protein concentration. 5. Various salts are inhibitors of the substrate reduction activity. 6. There is a burst of ATP hydrolysis before substrate reduction. 7. At a fixed concentration and ratio of protein components, electron flux is independent of which substrate is reduced. 8. Carbon monoxide is an inhibitor of all substrate reduction except for protons. Two additional observations are less convincingly documented yet are generally accepted conditions. 9. The direction of electron transfer in the MoFe-protein is from the P-cluster pair to the FeMo-cofactor. 10. All substrates are reduced at the cofactor [with the possible exception of some reduction of H^+ to H_2 at the P-cluster pair (17)].

Redox Cycle

Hageman & Burris (77–79) first proposed that only one electron-transfer event occurs before dissociation of the Fe-protein/MoFe-protein complex. Because all nitrogenase substrates require multiple electrons, several cycles of complex formation and dissociation must occur before product formation. This concept was expanded by Thorneley & Lowe (80) who, from pre-steady-state and turnover reaction kinetics, determined that at saturating Fe-protein ratios, the overall rate-determining step was the dissociation of Complex II (controlled by the rate constant k_{-3}). An expanded version of the Thorneley-Lowe model is shown in Figure 7. Implicit in this conclusion is that oxidized and reduced Fe-protein$[AXP]_2$ complexes have different affinities for MoFe-protein; the corollary is that the two Fe-protein forms must have different conformations. The cyclic process also accounts for the dependence of the substrate reduction rate on both the total protein concentration and the ratio of the two proteins, since both factors contribute to the concentration of Complex I. The pronounced salt inhibition of substrate reduction has been explained as a simple competitive inhibition of complex formation as indicated Figure 7 (81). Although the inhibition is kinetically equivalent when applied to either protein component, Diets & Howard (81) chose to emphasize salt binding to the Fe-protein, because the strong salt inhibition of ATP-dependent iron chelation from Fe-protein implies a direct effect on this component. A final important facet of the cycle is that ATP hydrolysis appears to precede electron transfer (82). For purposes of this discussion, the single, fast, ATP hydrolysis–coupled electron-transfer step of the Thorneley-Lowe model has been separated into two distinct events: hydrolysis of ATP and electron transfer.

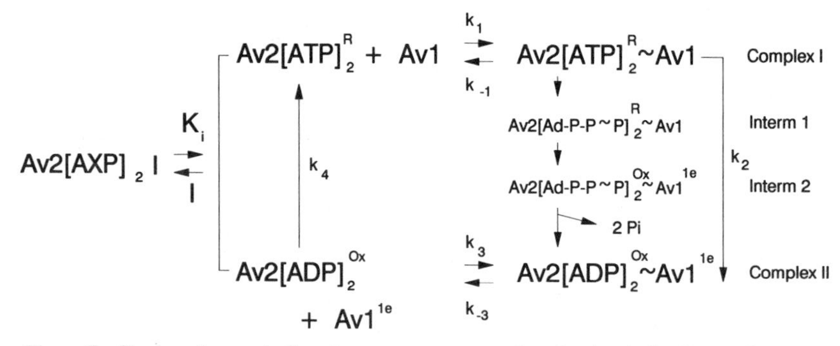

Figure 7 Single redox cycle for nitrogenase turnover using the *A. vinelandii* proteins as an example (77–81). Av1 is depicted as undergoing a one-electron reduction. Intermediates 1 and 2 are hypothetical stages on the path for electron transfer coupled to ATP hydrolysis. It is assumed that hydrolysis of at least one ATP precedes electron transfer. Both intermediates are considered as metastable states in which ATP is hydrolyzed without the release of inorganic phosphate (Pi). Whether both ATP are hydrolyzed at the same time or sequentially is unknown. Salt (I) inhibition is shown as a complex with either oxidation state of Av2, and with either nucleotide or no nucleotide bound (81).

The Hagerman-Burris-Thorneley-Lowe model has made a significant contribution to our concept of the nitrogenase mechanism and predicts many of the observed properties (77, 80). With the availability of altered proteins with site-specific amino acid substitutions, many of the original assumptions about the mechanism can be reconsidered. For example, the Thorneley-Lowe (80) assumption that 50% of the Fe-protein is inactive, yet can compete kinetically with the active form, should be reconciled with the observation that inactive and low-activity mutant proteins fail to compete kinetically, even though they can be chemically crosslinked to the MoFe-protein (34, 39–41, 83–85). Likewise, the assumption that all Fe-protein to MoFe-protein electron transfer rates are the same can be tested, as well as the hypothesis that the rate of proton reduction does change with the redox state of the MoFe-protein. Finally, the effects of salts on the individual first- and second-order rates are largely unknown. The observed overall inhibition of substrate reduction by salts (81) needs to be reconciled with the fact that the rate-determining step is increased by salts (80). The Diets & Howard (81) treatment of salt inhibition may be an oversimplification, and a detailed analysis of the effect of salts on each step in the Thorneley-Lowe model is needed.

Role of ATP Hydrolysis

One of the primary, unanswered questions for understanding the nitrogenase reaction is how MgATP hydrolysis is coupled to electron transfer. It is important to keep in mind that there is no inherent thermodynamic requirement for ATP hydrolysis in substrate reduction, e.g. the redox potential of the Fe-protein at −380 mV is below that for hydrogen evolution at pH < ~7. Likewise, only the first step of nitrogen reduction (breaking the first bond in dinitrogen triple bond to form diimide) has a lower redox potential than does Fe-protein, so that most electron transfer steps along the reaction pathway should be thermodynamically favorable (4). Consequently, it appears that ATP must be required for kinetic reasons, such as controlling a conformational gate, that ensures quasi-unidirectional electron transfer.

Related to the question of how ATP is utilized is the perplexing problem of how the putative electron acceptor in the reaction, the P-cluster pair, can be reduced. First, it appears that all the iron sites in the P-cluster pairs are already in the reduced ferrous state, leaving only the disulfide bridge as an acceptor. Long-distance, outer-sphere electron transfer would seem unlikely for the disulfide reduction because (*a*) the bridging cysteinyl ligands constrain the linked 4Fe:4S segments too closely (by ~1 Å) to accommodate non-bonded contact between reduced cluster sulfurs, and (*b*) disulfide reduction usually proceeds by inner-sphere chemical coupling such as disulfide exchange and in activated carbon adducts. Consistent with this view, it has not been possible, for example, to reduce the P-cluster pair in the isolated MoFe-protein by either low-molecular-weight reductants or by electrochemical methods. In contrast, the P-cluster pairs can be oxidized by dyes, demonstrating that these cofactors are electrochemically accessible. The second problem is that the one electron per FeMo-cofactor reduced state of the MoFe-protein has the electron on the cofactor, and not on the P-cluster pair, i.e. the cofactor becomes EPR silent (87). At least in the generation of this form of MoFe-protein, any proposed reduced P-cluster pair must exist transiently at best.

Thus, the role of ATP in electron transfer may be even more complex than has been generally appreciated. One plausible explanation is that conformational changes associated with ATP binding and hydrolysis might be utilized in two alternate ways: 1. to induce conformational changes in the MoFe-protein required for electron transfer from the P-cluster pair to the cofactor and 2. to induce conformational changes in Fe-protein required for electron transfer to the P-cluster pair. This hypothesis requires that conformational changes occur in the MoFe-protein as well as in the Fe-protein during the complex formation. The critical event in the process is the reorganization at the P-cluster pair such that it can become a better redox acceptor/donor. This is accomplished only by formation of the active complex between MoFe-protein and Fe-protein. The unique and probably essential property of the P-cluster pair is to act as the gate for electron transfer from the Fe-protein to the FeMo-cofactor. As stated above, in the resting state (whose structure presumably has been determined by X-ray diffraction), the cluster disulfide is unlikely to be a good electron acceptor.

A $[8Fe^{2+}:6S^{2-}:1(S\text{-}S)^{2-}:6RS^{1-}]^{4-}$

$\updownarrow$

B $[1Fe^{3+}:7Fe^{2+}:6S^{2-}:1(S\text{-}S)^{3\dot{-}}:6RS^{1-}]^{4-}$

$\updownarrow$

C $[2Fe^{3+}:6Fe^{2+}:8S^{2-}:6RS^{1-}]^{4-}$

Figure 8 Three-state hypothesis for internal redox states of the P-cluster pair.

However, as a consequence of a Fe-protein-induced conformational change, an electron could be transferred from the P-cluster pair to the cofactor, prior to reduction of the former by the Fe-protein. The resulting one-electron oxidized P-cluster pair would now be a good acceptor. In addition, once the physical restraints on the cluster were relaxed, the P-cluster pair could undergo internal redox chemistry as outlined in Figure 8. In this scheme, the disulfide (state A) could be reduced to either a disulfide radical (state B) or to inorganic sulfide (state C) by internal inner-sphere chemistry of the ferrous ligands, without net reduction of the cluster. In either state B or C, electrons could be donated to the FeMo-cofactor or accepted from the Fe-protein by outer-sphere, long-distance transfer mechanisms usually envisioned for two redox-coupled clusters. This model satisfies the observed electronic states of the clusters and the unique requirement for Fe-protein as the reductant, i.e. only the Fe-protein can serve as the key required to unlock the necessary conformational changes in MoFe-protein that allow electron transfer from the P-cluster pair to the FeMo-cofactor. The net result of the scheme is to have the P-cluster pair in the same oxidation state at the beginning and the end of the redox cycle, while the FeMo-cofactor and Fe-protein are reduced and oxidized, respectively, by one electron.

Complex Formation and ATP Hydrolysis

The minimum requirements for formation and turnover of a kinetically competent complex must include the following conditions. 1. Intimate and precise orientation of the two nitrogenase components must precede electron transfer and, as suggested above, at least some differences in Complexes I and II must exist. 2. In Complex I, the MoFe-protein must induce the changes in Fe-protein necessary for ATP hydrolysis. That is, there must be some signal sent from the MoFe-protein to the Fe-protein nucleotide site. 3. During the transition between the ATP and ADP states, the protein components must undergo conformational changes that lead to the correct alignment of the respective donor and acceptor pairs. As discussed above, both proteins may undergo such changes. 4. After electron transfer to the MoFe-protein, the complex must relax sufficiently so that electron transfer back to the oxidized Fe-protein is rare.

The inhibitory effects of salt on enzyme activity (81, 83, 88) and on chemical crosslinking (89, 90) support the idea that ionic interactions contribute to these complexes. The results with several altered Fe-proteins, primarily involving substitutions for Arg100, suggest there are several different classes of ionic interactions, some contributing more than others (83). Chemical crosslinking with the water-soluble carbodiimide is highly specific between Glu112 of Av2 and Lys β400 of Av1. Interestingly, Glu112 is part of an acidic patch of seven carboxylic acid residues at the end of the long helix extending from the cluster in both subunits. At the other end of this helix, next to the cluster, is Arg100. Thus, the two well-characterized ionic regions implicated in complex formation are on the same surface as the cluster (Figure 9).

Potential docking modes for the Av2 and Av1 structures have been presented that superimpose the molecular two-fold axis of Av2 and the pseudo-two-fold axis of Av1 (15). A groove in the "crown" above the P-cluster pair can begin to accommodate the extended α-helices that frame the Av2 4Fe:4S cluster (Figure 10). In this orientation, the side chains of Glu112 and Lys β400 are juxtaposed, but are ~8 Å apart, too far for the rapid and specific crosslinking reaction to proceed. Likewise, Lys α50 and Lys α51 are positioned near Glu112 on the symmetrically related Av2 helix. Formation of a crosslink at this site would appear to be as probable as for Lys β400 on the other side, but experimentally this does not occur. Although crosslinking is strictly dependent on having a native structure in the proteins, it does not require nucleotide. These results suggest a specific, preliminary association in which crosslinking precedes the more symmetrically superimposed two-fold axis state (see Figures 9 and 10; Ref. 86).

In the symmetrical "helix in the groove" orientation, the Fe-protein cluster and the P-cluster pair are separated by a distance >16 Å at van der Waals contact between the two proteins. A large number of ionic interactions between the two proteins are possible, with one notable exception. Namely, the critical residue Arg100 of Fe-protein cannot reach any acidic residue on the MoFe-protein-docking face. The consequence of this is considered further below. In addition to the ionic interactions, there are several potentially important hydrophobic interactions. Phe-α and β125 of the MoFe-protein can interact with the hydrophobic residues 102–109 that are part of the two surface helices on the Fe-protein. At contact, these Phe side chains could fit into pockets created by Ile103 and Thr104 and connected

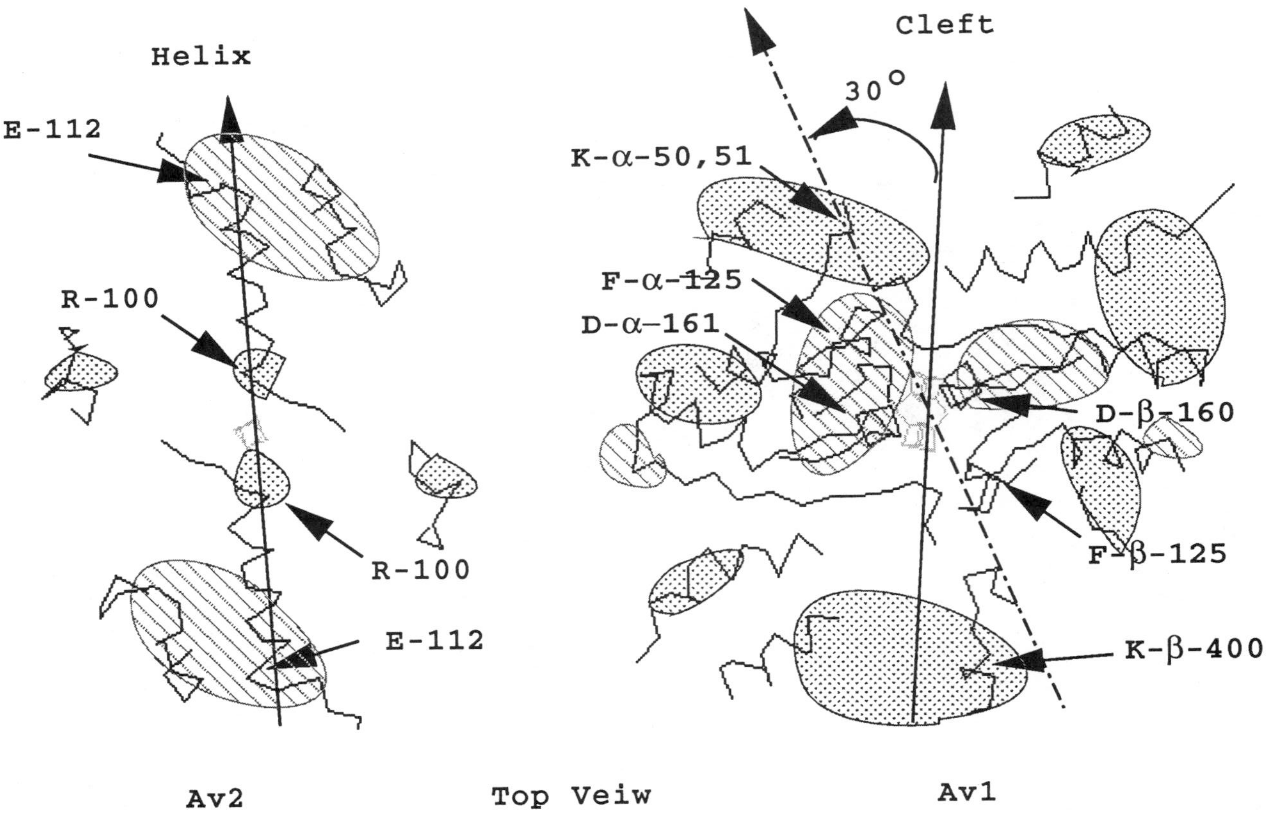

Figure 9 Top surfaces of Av1 and Av2 based on the crystal structures. Patches of positive (stippled) and negative (hatched) charged residues are outlined. Important contact residues are indicated (see text).

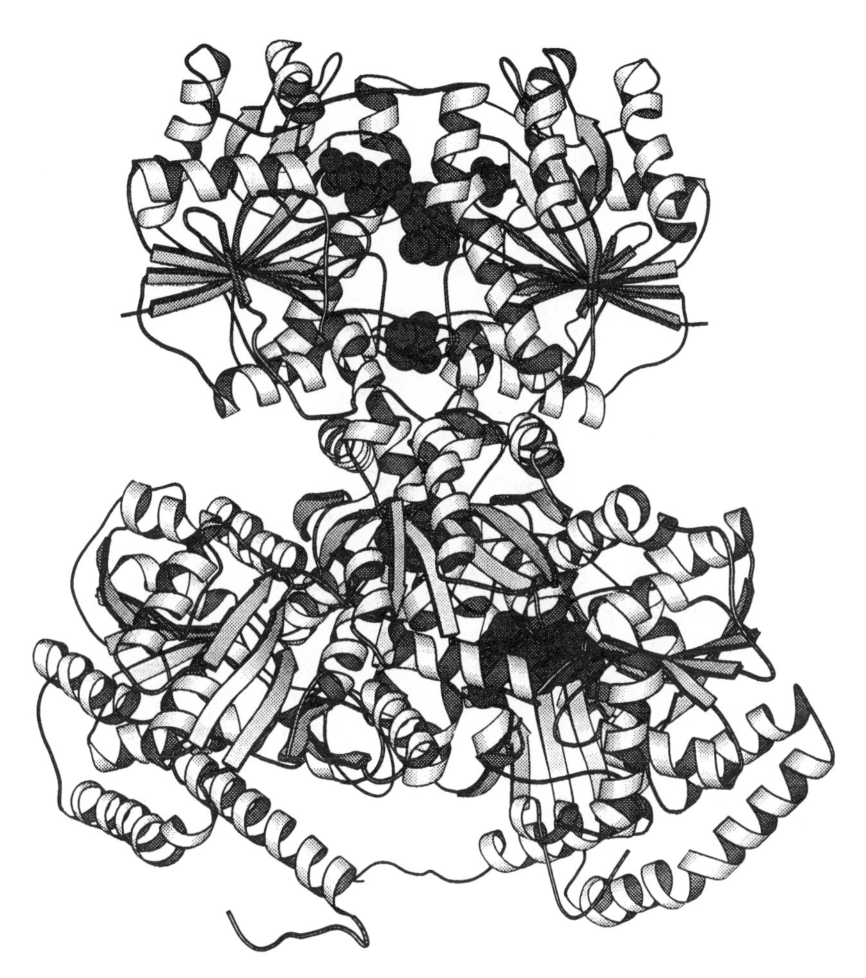

Figure 10 Ribbons diagram of a computer-generated model of the docking complex between Fe-protein (*top molecule*) and an αβ subunit pair of the MoFe-protein (*bottom*). The metal centers and ADP molecule are represented by space-filling models. Prepared with the program MOLSCRIPT (36a) and reproduced with permission from (8a).

to Val102 of Fe-protein which, in turn, interacts with a β-sheet below the Fe-protein surface at Cys38 (see Figure 11). Substitution of Ser for this internal Cys38 has profound effects on the ATP requirement, with a ~50-fold decrease in the efficiency of ATP hydrolyzed per electron transferred (J Magnuson and JB Howard, in preparation). As shown in Figure 11, the amide backbone hydrogen bonds from Cys38 to Asp125 completes a direct

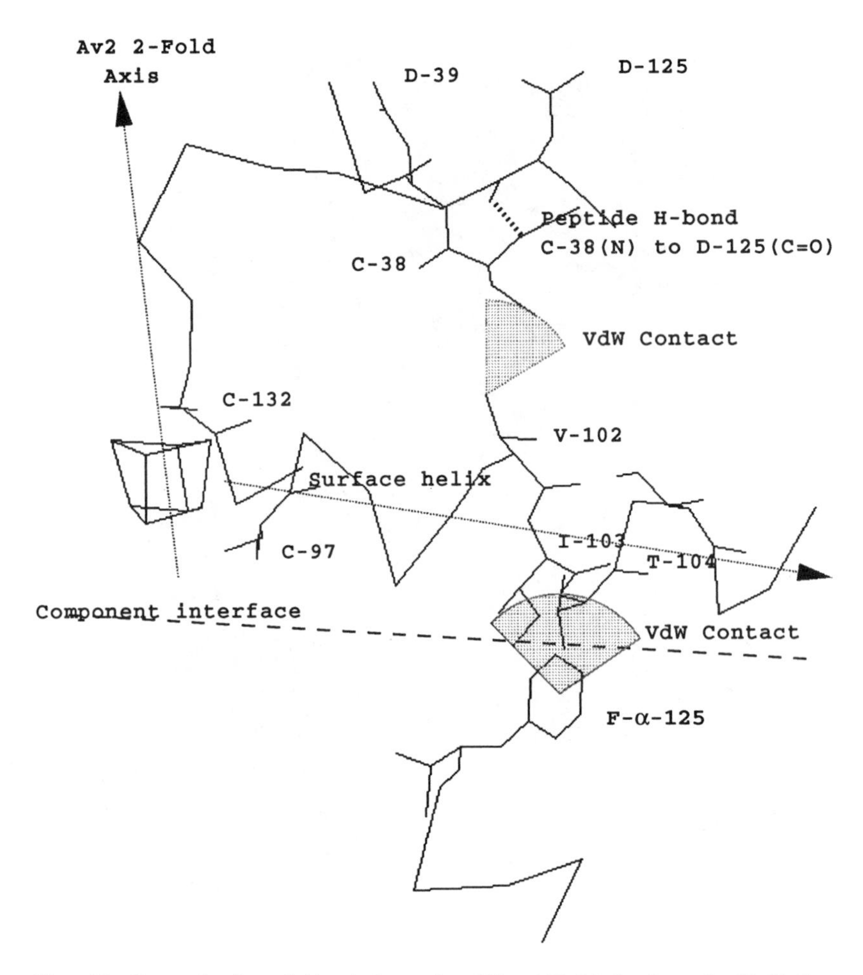

Figure 11 Proposed pathway linking the interaction of Phe α125 of Av1 to the nucleotide-binding site in Av2. A symmetrically related interaction between Phe β125 and the second Av2 subunit is not shown for clarity of presentation. The α-helix from cluster ligand Cys97 to Glu112 on the surface of Av2 is positioned over a second helix (not shown) that includes Cys38 and Asp39, and interacts with that helix through van der Waals contact between Cys38 and Val102. A β-strand containing Asp125 and a bend leading to the cluster ligand, Cys132, are connected to Cys38 by an amide hydrogen bond.

link between the MoFe-protein contact site and the nucleotide- and Mg-binding sites. The potential importance of Phe β125 of the MoFe-protein has been verified by substitution by several hydrophobic residues (91), some of which have considerably lower turnover rates and increased nucleotide affinity in turnover. Hence, the ATP hydrolysis signal from the MoFe-protein may follow this path.

For more intimate contact allowing the 4Fe:4S cluster and P-cluster pair to approach to <15 Å, the groove in the crown must be opened. An approximately 30° rotation of the Fe-protein would force the groove open, allowing new ionic interactions (Figure 9). Importantly, the two Arg100s (from the two symmetrically related Fe-protein subunits) can form salt bridges with Asp α160 and Asp β161. These two invariant acidic residues are located on helices connected to cysteinyl ligands of the P-cluster pair. Thus, in addition to bringing the two clusters closer together, the model building shows that the rotation would affect the environment of the P-cluster pair directly. The rotation in the cleft would be a true "work" step, which could be driven by reorientation of the Fe-protein subunits in response to ATP hydrolysis. A metastable complex similar to Intermediate 1 in Figure 7 can be envisioned, where the ATP is hydrolyzed, but the inorganic phosphate has not been released. After the electron transfer, the phosphate dissociates and the system relaxes when the Fe-protein achieves the MgADP conformation leading to Complex II.

A mechanism for utilizing the hydrolysis of ATP to drive conformational transitions has been proposed, based upon the two potential binding modes for nucleotides in Av2 (41, 86; see Figure 12). In this hypothesis, MgATP is bound in the *ras*-like mode parallel to the two-fold axis separating the subunits. The residue analogous to Asp125 in *ras* p21 (Asp57) is hydrogen bonded to a water in the coordination sphere of the Mg (21, 22). After hydrolysis, the water is displaced and the aspartyl residue is liganded directly. The shift causes a movement along the attached helix-turn which, in the case of Av2, leads directly to Cys132, the bottom ligand of the Fe:S cluster. It is tempting to complete the picture of the conformational changes by suggesting that the purine ring moves to the ADP-binding mode perpendicular to the two-fold axis and the subunits contract. Two observations indicate that this hypothesis is more than just fantasy. First, in the presence of ADP and ATP, the ligands on the 4Fe:4S cluster are clearly in different environments as detected by proton NMR spectroscopy (JB Howard, unpublished), suggesting some reorientation of the subunits occurs. Second, only in the *ras*-like ATP-binding mode is the γ-phosphate correctly positioned for water attack, assisted by Asp39 and/or Asp129, the likely candidates for the general base in hydrolysis (Figure 12). Notice that Asp39 is also likely to sense the MoFe-protein binding by the path involving its neighbor, Cys38 (Figures 11 and 12), while Asp129 could be effected by the polypeptide connection to Asp125. It should be noted that Asp129 in Fe-protein is analogous to Gln61 in *ras*, a candidate general base for GTP hydrolysis.

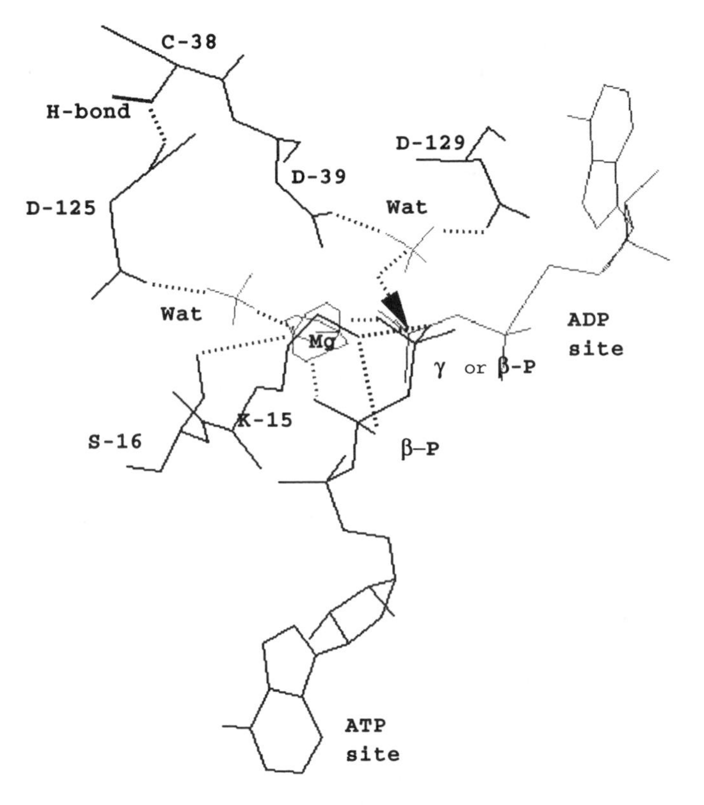

Figure 12 Hypothetical arrangement of ATP, Mg, and two water (Wat) molecules in the Av2 structure based upon comparison to *ras* p21 (21). ADP is shown as observed in the Av2 structure, with superposition of Lys15 of Av2 and Lys16 of *ras* p21. With this superposition, the γ-phosphate of ATP in the "*ras*"-binding mode is only ~0.2 Å displaced from the β-phosphate of ADP. A water molecule hydrogen bonded to Asp125 is included as ligand of Mg, as observed in *ras* p21. A second water molecule is placed, again by comparison to *ras* p21, for attack on the terminal phosphate of ATP, assisted by Asp39 and/or Asp129.

Electron Transfer and Substrate Reduction

The distance of closest approach between metal sites in the P-cluster pair and the FeMo-cofactor is ~14 Å, which suggests that the electron transfer rate between these centers could be faster than the rate of nitrogenase turnover (92, 93). The rate of this electron transfer would determine the lifetime of a reduced or oxidized P-cluster pair: i.e. whether or not such a species would exist transiently or long enough to probe spectroscopically (if this species exists at all). Potential electron-transfer pathways between these centers have been described (15, 19); in particular, the helices α63-α74 and α88-α92 adjacent to the P-cluster pair ligands Cys α62 and Cys α88 are directed towards the FeMo-cofactor. The locations of homocitrate and the Mo on the side of the FeMo-cofactor closest to the P-cluster pair suggest that these groups might also participate in electron transfer between these two redox centers.

The structural details of substrate binding to the FeMo-cofactor and the sequence of electrons and protons transferred to the bound substrate are critical questions. In addition to dinitrogen and protons, nitrogenase stereospecifically reduces a variety of substrates, including acetylene, azide, cyclopropene and 1-butyne, all of which, except for protons, are inhibited by CO. Significantly, the rate of electron flow through the nitrogenase system, under a given set of conditions, is independent of the substrate reduced (94, 95). In the absence of N_2 or in the presence of CO, for example, the entire electron flux can be funneled into reduction of protons to H_2, which occurs at the same rate (per electron) as if N_2 were present or CO absent. Under conditions where the rate of electron flux through nitrogenase is decreased (by lowering the concentration of external reductant or Fe-protein, for example), the proportion of hydrogen and other products requiring fewer electrons increases at the expense of products requiring more electrons. These observations suggest that electron transfer from Fe-protein to MoFe-protein is independent of substrate binding to the FeMo-cofactor.

Since all known substrates of nitrogenase are reduced by an even number of electrons (and, almost always, require an equivalent number of protons), most mechanistic schemes have focused on addition of one (or more) pairs of electrons to dinitrogen, leading to the formal reduction sequence: dinitrogen, diimide, hydrazine, and ammonia. The major barrier in this case should be the two-electron reduction of dinitrogen to diimide. One interesting possibility is that this barrier could be sidestepped by a four-electron reduction process, reducing dinitrogen directly to the hydrazine oxidation level. This could be achieved, for example, by a combination of a two-electron donation from the P-cluster pair (possibly involving sulfide-disulfide conversion), coupled with two electrons stored in the FeMo-cofactor. Funneling of the requisite number of protons into the buried active center is also a critical process during substrate reduction; various ionizable groups in the vicinity of the FeMo-cofactor and the presence of an extensive water network near the homocitrate may participate in this process.

In the absence of any experimental evidence indicating how substrates might bind to the FeMo-cofactor, a range of hypothetical models have been proposed (17, 96, 97). Binding interactions between substrates and one or more of the Fe, Mo, and/or S sites are all conceivable, and at this stage, it would seem prudent not to dismiss any potential candidates. Definitive experimental information concerning the structure(s) of substrate(s) bound to FeMo-cofactor will be essential for formulating more detailed mechanisms concerning the reduction and protonation reactions.

Two basic types of experiments have been conducted to probe the functional significance of groups surrounding the FeMo-cofactor: site-directed mutagenesis, and replacement of homocitrate with various carboxylic acids. Other than replacement of the cofactor ligands (which results in loss of cofactor and activity), His α195 and Gln α191 have been the most mutagenized targets (98, 99). Substitution of His α195, which is hydrogen bonded to one of the bridging sulfurs of the FeMo-cofactor, by Asn, Gln, Thr, Gly, Leu, or Tyr, results in a Nif^- phenotype, i.e. the organism is no longer capable of diazotrophic growth. More varied Nif phenotypes are found with substitution of Gln α191. The side chain of this residue is hydrogen bonded to both a carboxyl group of homocitrate and the NH of residue α61 (adjacent to the P-cluster pair ligand α62), and hence is positioned between the cofactor and the P-cluster pair. While some substitutions, such as Gln to Lys, result in Nif^- phenotype, others, such as Gln to Ala or Pro, support diazotrophic growth rates comparable to wild type. These latter observations are particularly surprising, since these side chains cannot participate in the hydrogen-bonding interactions exhibited by the Gln side chain. Preliminary reports of mutagenesis experiments of residues Arg α96, Arg α359, Phe α381, and Gln α440 in the vicinity of the cofactor have also appeared (99).

The function of homocitrate in nitrogenase is intriguing, given that it is coordinated to the Mo, is surrounded by a number of buried water molecules, and is on the side of the cofactor nearest to the P-cluster pair. To probe the role of homocitrate, Ludden & Shah have pioneered the development of in vitro methods for substituting other organic acids for homocitrate in the FeMo-cofactor (100). In general, the minimal requirements for functional activity by homocitrate substituents include two carboxyls and a hydroxyl group. The specific stereochemistry of the organic acid can also profoundly influence the ability of this group to support reduction of various substrates by nitrogenase. An example of this behavior is provided by the replacement of homocitrate with either *erythro*-fluorohomocitrate or *threo*-fluorohomocitrate (101). These compounds are substituted with a single fluoro group on the single methylene-containing arm of homocitrate, which links the Mo-liganding groups to the carboxyl that hydrogen bonds to Gln α191. While the *erythro*-fluorohomocitrate-substituted FeMo-cofactor has high activities in most nitrogenase activities (including N_2 reduction), the *threo*-isomer has very low N_2 reduction activity, although it has more normal levels of acetylene and proton-reduction activities. Examination of the crystal structure shows that, without any rearrangement, a *threo*-substituent would be directed towards the cofactor and would be in close contact with one of the cluster sulfurs, while an *erythro*-substituent would point away from the cofactor. Apparently, the proximity of the *threo*-fluoro to the cofactor could drive a rearrangement of this region that results in altered activity, while the *erythro*-fluoro group can be accommodated without any mechanistically significant changes. Alternatively, in view of the acidity of these methylene protons in homocitrate, substitution of a hydrogen by fluorine may block stereospecific transfer of one of the hydrogens to the cluster, during a putative step in protonation of bound substrate.

It should be noted that characterization of nitrogenase variants, which is central to understanding the functional roles played by a residue, is far from a trivial matter. The complexity of this problem is illustrated by observations that while certain variants cannot reduce dinitrogen to ammonia, some of them can still reduce protons and/or acetylene (including the production of ethane from acetylene, a reaction that is not catalyzed by the wild-type enzyme), or exhibit altered responses to inhibition by CO and/or other ligands. Mutations in the Fe-protein have exhibited altered coupling between ATP hydrolysis and electron transfer, pronounced changes in nucleotide and metal requirements, substantial increases in salt sensitivity, as well as altered biophysical/spectral properties. It is essential that each variant be characterized as thoroughly and carefully as possible, to obtain the most complete understanding of the functional consequences associated with an amino acid substitution.

RELATIONSHIP OF NITROGENASE TO OTHER SYSTEMS

From a protein structural perspective, nitrogenase provides a tantalizing combination of protein-protein interactions and conformational changes coupled to ATP hydrolysis. The coupling between nucleotide-binding and redox behavior of Fe-protein not only is an important problem for the nitrogenase mechanism, but also is representative of a much broader biochemical phenomenon in which nucleotide hydrolysis is coupled to a second process, such as membrane transport, cellular regulation, or molecular motors (21–26). A common theme that is emerging is that switching between alternate conformational states of a protein, driven by nucleotide binding and hydrolysis at the interface between different subunits or domains, provides a general transducing and timing mechanism for coupling the energy of nucleotide hydrolysis to a variety of biochemical processes. Sequence analyses by Koonin (102) suggest that the Fe-protein may represent an ancestral form of a now widespread assortment of nucleotide-binding proteins, so that the basic structural machinery utilized for transducing the energy of ATP hydrolysis in nitrogenase may have been recruited for a diverse range of other biological functions.

ACKNOWLEDGMENTS

The efforts and contributions of J. Kim, M. Georgiadis, D. Woo, M. K. Chan, M. W. Day, J. Schlessman, H. Komiya, L. Joshua-Tor, M. H. B. Stowell, B. T. Hsu and A. J. Chirino are deeply appreciated, as are discussions, sharing of preprints, etc with our nitrogenase colleagues. Research in the authors' laboratories was supported by grants from NIH and NSF.

Literature Cited

1. Ferguson SJ. 1988. In *The Nitrogen and Sulphur Cycles,* ed. JA Cole, SJ Ferguson, pp. 1–30. Cambridge: Cambridge Univ. Press
2. Burgess BK. 1984. In *Advances in Nitrogen Fixation,* ed. C Veeger, WE Newton, pp. 103–14. Boston: Martinus Nijhoff
3. Orme-Johnson WH. 1985. *Annu. Rev. Biophys. Biophys. Chem.* 14:419–59
4. Stiefel EI, Thomann H, Jin H, Bare RE, Morgan TV, et al. 1988. In *Metal Clusters in Proteins,* ed. L Que, pp. 372–89. Washington, DC: Am. Chem. Soc.
5. Smith BE, Eady RR. 1992. *Eur. J. Biochem.* 205:1–15
6. Burris RH. 1991. *J. Biol. Chem.* 266: 9339–42
7. Rees DC, Chan MK, Kim J. 1993. *Adv. Inorg. Chem.* 40:89–119
8. Stacey G, Burris RH, Evans HJ, eds. 1992. *Biological Nitrogen Fixation.* New York: Chapman & Hall. 943 pp.
8a. Kim J, Rees DC. 1994. *Biochemistry.* In press
9. Simpson FB, Burris RH. 1984. *Science* 224:1095–97
10. Hausinger RP, Howard JB. 1982. *J. Biol. Chem.* 257:2483–90
11. Brigle KE, Newton WE, Dean DR. 1985. *Gene* 37:37–44
12. Lundell D, Howard JB. 1978. *J. Biol. Chem.* 253:3422–26
13. Georgiadis MM, Komiya H, Chakrabarti P, Woo D, Kornuc JJ, Rees DC. 1992. *Science* 257:1653–59
14. Kim J, Rees DC. 1992. *Science* 257: 1677–82
15. Kim J, Rees DC. 1992. *Nature* 360: 553–60
16. Kim J, Woo D, Rees DC. 1993. *Biochemistry* 32:7104–15
17. Chan MK, Kim J, Rees DC. 1993. *Science* 260:792–94
18. Rees DC, Kim J, Georgiadis MM, Komiya H, Chirino AJ, et al. 1993. See Ref. 29, pp. 170–85
19. Bolin JT, Campobasso N, Muchmore SW, Minor W, Morgan TV, Mortenson LE. 1993. See Ref. 30, pp. 89–94
20. Bolin JT, Campobasso N, Muchmore SW, Morgan TV, Mortenson LE. 1993. See Ref. 29, pp. 186–95
21. Pai EF, Krengel U, Petsko GA, Goody RS, Kabsch WK, Wittinghofer A. 1990. *EMBO J.* 9:2351–59
22. Tong L, de Vos AM, Milburn MV, Kim S-H. 1991. *J. Mol. Biol.* 217:503–16
23. Riordan JR, Rommens JM, Kerem BS, Alon N, Rozmahel R, et al. 1989. *Science* 245:1066–72
24. Karkaria CE, Chen CM, Rosen BP. 1990. *J. Biol. Chem.* 265:7832–36
25. Rayment I, Rypniewski WR, Schmidt-Bäse K, Smith R, Tomchick DR, et al. 1993. *Science* 261:50–58
26. Story RM, Steitz TA. 1992. *Nature* 355:374–76
27. Jurnak F. 1985. *Science* 230:32–36
28. Kjeldgaard M, Nyborg J. 1992. *J. Mol. Biol.* 223:721–42
29. Stiefel EI, Coucouvanis D, Newton

WE, eds. 1993. *Molybdenum Enzymes, Cofactors and Model Systems*. Washington, DC: Am. Chem. Soc. 387 pp.
30. Palacios R, Mora J, Newton WE, eds. 1993. *New Horizons in Nitrogen Fixation*. Dordrecht: Kluwer Academic. 788 pp.
31. Emerich DW, Burris RH. 1976. *Proc. Natl. Acad. Sci. USA* 73:4369–73
32. Emerich DW, Hageman RV, Burris RH. 1981. *Adv. Enzymol.* 52:1–22
33. Hausinger RP, Howard JB. 1983. *J. Biol. Chem.* 258:13486–92
34. Howard JB, Davis R, Moldenhauer B, Cash VL, Dean D. 1989. *J. Biol. Chem.* 264:11270–74
35. Schulz GE. 1992. *Curr. Opin. Struct. Biol.* 2:61–67
36. Schulz GE, Elzinga M, Marx F, Schirmer RH. 1974. *Nature* 250:120–23
36a. Kraulis PJ. 1991. *J. Appl. Crystallogr.* 24:946–50
37. Yates MG. 1992. See Ref. 8, pp. 685–735
38. Walker JE, Saraste M, Runswick MJ, Gay NJ. 1982. *EMBO J.* 8:945–81
39. Seefeldt LC, Morgan TV, Dean DR, Mortenson LE. 1992. *J. Biol. Chem.* 267:6680–88
40. Seefeldt LC, Mortenson LE. 1993. *Protein Sci.* 2:93–102
41. Wolle D, Dean DR, Howard JB. 1992. *Science* 258:992–95
42. Chang CL, Davis LC, Rider M, Takemoto DJ. 1988. *J. Bacteriol.* 170: 4015–22
43. Adman ET, Watenpaugh KD, Jensen LH. 1975. *Proc. Natl. Acad. Sci. USA* 72:4854–58
44. Morgan TV, McCracken J, Orme-Johnson WH, Mims WB, Mortenson L, Peisach J. 1990. *Biochemistry* 29:3077–82
45. Zumft W, Mortenson L, Palmer G. 1973. *Biochim. Biophys. Acta* 292:413–21
46. Watt GD, Wang Z-C, Knotts RR. 1986. *Biochemistry* 25:8156–62
47. Walker GA, Mortenson LE. 1974. *Biochemistry* 13:2382–88
48. Ljones T, Burris RH. 1978. *Biochemistry* 17:1866–72
49. Deits TL, Howard JB. 1989. *J. Biol. Chem.* 264:6619–28
50. Anderson GL, Howard JB. 1984. *Biochemistry* 23:2118–22
51. Surerus KK, Hendrich MP, Christie PD, Rottgardt D, Orme-Johnson WH, Munck E. 1992. *J. Am. Chem. Soc.* 114:8579–90
52. Deisenhofer J, Epp O, Miki K, Huber R, Michel H. 1985. *Nature* 318:618–24
53. Allen JP, Feher G, Yeates TO, Komiya H, Rees DC. 1987. *Proc. Natl. Acad. Sci. USA* 84:5730–34
54. Jacobson MR, Cantwell JS, Dean DR. 1990. *J. Biol. Chem.* 265:19429–33
55. Joerger RD, Loveless TM, Pau RN, Mitchenall LA, Simon BH, Bishop PE. 1990. *J. Bacteriol.* 172:3400–8
56. Robson RL, Woodley PR, Pau RN, Eady RR. 1989. *EMBO J.* 8:1217–24
57. Dean DR, Jacobson MR. 1992. See Ref. 8, pp. 763–834
58. Shah VK, Brill W. 1977. *Proc. Natl. Acad. Sci. USA* 74:3249–53
59. Rawlings J, Shah VK, Chisnell RJ, Brill WJ, Zimmermann R, et al. 1978. *J. Biol. Chem.* 253:1001–4
60. Nelson MJ, Levy MA, Orme-Johnson WH. 1983. *Proc. Natl. Acad. Sci. USA* 80:147–50
61. Hoover TR, Robertson AD, Cerny RL, Hayes RN, Imperial J, et al. 1987. *Nature* 329:855–57
62. Hoover TR, Imperial J, Ludden PW, Shah VK. 1988. *Biofactors* 1:199–205
63. Brigle KE, Weiss MC, Newton WE, Dean DR. 1987. *J. Bacteriol.* 169: 1547–53
64. Paustian TD, Shah VK, Roberts GP. 1989. *Proc. Natl. Acad. Sci. USA* 86:6082–86
65. Liang J, Madden M, Shah VK, Burris RH. 1990. *Biochemistry* 29:8577–81
66. Robinson AC, Dean DR, Burgess BK. 1987. *J. Biol. Chem.* 262:14327–32
67. Robinson AC, Chu W, Li J-G, Burgess BK. 1989. *J. Biol. Chem.* 264:10088–95
68. Tal S, Chun TW, Gavini N, Burgess BK. 1991. *J. Biol. Chem.* 266:10654–57
69. Paustian TD, Shah VK, Roberts GP. 1990. *Biochemistry* 29:3515–22
70. White TC, Harris GS, Orme-Johnson WH. 1992. *J. Biol. Chem.* 267:24007–16
71. Howard JB, Anderson G, Deits T. 1985. *In Nitrogen Fixation Research Progress*, ed. HJ Evans, PJ Bottomley, WE Newton, pp. 559–66. Dordrecht: Martinus Nijhoff
72. Zheng L, White RH, Cash VL, Jack RF, Dean DR. 1993. *Proc. Natl. Acad. Sci. USA* 90:2754–58
73. Howard KS, McLean PA, Hansen FB, Lemley PV, Koblan KS, Orme-Johnson WH. 1986. *J. Biol. Chem.* 261:772–78
74. Paul W, Merrick M. 1989. *Eur. J. Biochem.* 178:675–82
75. Eady RR. 1991. *Adv. Inorg. Chem.* 36:77–102
76. Bishop PE, Premakuman R. 1992. See Ref. 8, pp. 736–62

77. Hageman RV, Burris RH. 1978. *Biochemistry* 17:4117–24
78. Hageman RV, Orme-Johnson WH, Burris RH. 1980. *Biochemistry* 19: 2333–42
79. Hageman RV, Burris RH. 1978. *Proc. Natl. Acad. Sci. USA* 75:2699–702
80. Thorneley RNF, Lowe DJ. 1985. In *Molybdenum Enzymes,* ed. TG Spiro, pp. 221–84. New York: Wiley
81. Deits TL, Howard JB. 1990. *J. Biol. Chem.* 265:3859–67
82. Thorneley RNF. 1992. *Philos. Trans. R. Soc. London Ser. B* 336:73–82
83. Wolle D, Kim C, Dean D, Howard JB. 1992. *J. Biol. Chem.* 267:3667–73
84. Gavini N, Burgess BK. 1992. *J. Biol. Chem.* 267:21179–86
85. Lowery RG, Chang CL, Davis LC, McKenna M-C, Stevens PJ, Ludden PW. 1989. *Biochemistry* 28:1206–12
86. Howard JB. 1993. See Ref. 29, pp. 271–89
87. Fisher K, Lowe D, Thorneley R. 1991. *Biochem. J.* 279:81–85
88. Burns A, Watt GD, Wang ZC. 1985. *Biochemistry* 24:3932–36
89. Willing AH, Georgiadis MM, Rees DC, Howard JB. 1989. *J. Biol. Chem.* 264:8499–503
90. Willing A, Howard JB. 1990. *J. Biol. Chem.* 265:6596–99
91. Thorneley RNF, Ashby GA, Fisher K, Lowe DJ. 1993. See Ref. 29, pp. 290–302
92. Wuttke DS, Bjerrum MJ, Winkler JR, Gray HB. 1992. *Science* 256:1007–9
93. Farid RS, Moser CC, Dutton PL. 1993. *Curr. Opin. Struct. Biol.* 3:225–33
94. Hageman RV, Burris RH. 1980. *Biochim. Biophys. Acta* 591:63–75
95. Wherland S, Burgess BK, Stiefel EI, Newton WE. 1981. *Biochemistry* 20: 5132–40
96. Orme-Johnson WH. 1992. *Science* 257: 1639–40
97. Deng H, Hoffmann R. 1993. *Angew. Chem. Int. Ed.* 32:1062–65
98. Scott DJ, Dean DR, Newton WE. 1992. *J. Biol. Chem.* 267:20002–10
99. Newton WE, Dean DR. 1993. See Ref. 29, pp. 216–30
100. Imperial J, Hoover TR, Madden MS, Ludden PW, Shah VK. 1989. *Biochemistry* 28:7796–99
101. Madden MS, Kindon ND, Ludden PW, Shah VK. 1990. *Proc. Natl. Acad. Sci. USA* 87:6517–21
102. Koonin EV. 1993. *J. Mol. Biol.* 229: 1165–74

Annu. Rev. Biochem. 1994. 63:265–97

ROLE OF CHROMATIN STRUCTURE IN THE REGULATION OF TRANSCRIPTION BY RNA POLYMERASE II

Suman M. Paranjape, Rohinton T. Kamakaka, and James T. Kadonaga

Department of Biology and Center for Molecular Genetics, University of California at San Diego, La Jolla, California 92093-0347

KEY WORDS: gene regulation, transcription factors, enhancers, antirepression, nucleosomes

---

CONTENTS

OVERVIEW

In this review, we examine the mechanisms by which gene expression is regulated in the context of the native chromatin template. Over the past

0066-4154/94/0701-0265$05.00

several years, there have been many reviews in the general area of chromatin and transcription (see, for example 1–22), and we gratefully acknowledge their influence on the design of this manuscript. In addition, we apologize in advance for any oversights that were unintended yet inevitable.

WHAT IS THE GROUND STATE OF THE GENOME? This is a semantic issue, but it would be useful to establish a definition of the ground state. In this review, we define the ground state of a gene to be the inactive or uninduced state. A model for chromatin structure and gene activity is thus as follows. An inactive gene in the ground state exists as a transcriptionally repressed chromatin structure that may be folded or compacted and inaccessible to transcription factors. In the activated (or derepressed) state, the chromatin would be decondensed and unfolded, nucleosome structure may be altered, and histones (probably histone H1 rather than core histones) may be depleted. Recent studies have shown that an important function of promoter- and enhancer-binding factors is to relieve chromatin-mediated repression of transcription. This phenomenon is sometimes referred to as "antirepression." In addition, sequence-specific transcription factors possess the ability to facilitate the intrinsic transcription reaction, which is termed "true activation." Therefore, in the course of the activation of a gene, promoter- and enhancer-binding factors may function both to counteract chromatin-mediated repression and to increase the rate of the transcription process. This postulated model for gene activation is depicted in Figure 1. Although this topic is not discussed in the review, it is also important to consider the involvement of nonhistone transcriptional repressors in the analysis of gene regulation. These nonhistone factors include sequence-specific DNA-binding proteins that bind to transcriptional control regions (such as silencer elements) or factors that affect components of the basal transcription machinery (reviewed in Refs. 23–25).

ORGANIZATION OF THE REVIEW We initially describe fundamental aspects of chromatin structure and then examine the relationship between chromatin structure and the regulation of gene activity. It has been difficult to devise a coherent framework for a systematic discussion of the interrelated topics in this field of study. We therefore suggest that the reader survey the Table of Contents of this review, and consider each subsection as an independent, short essay that addresses specific issues. Throughout the review, we discuss questions and problems in the analysis of chromatin and transcription. In virtually all of these instances, the available data do not yet provide definitive answers. Thus, the aim of this review is to present current ideas and theories as a point of reference for further studies.

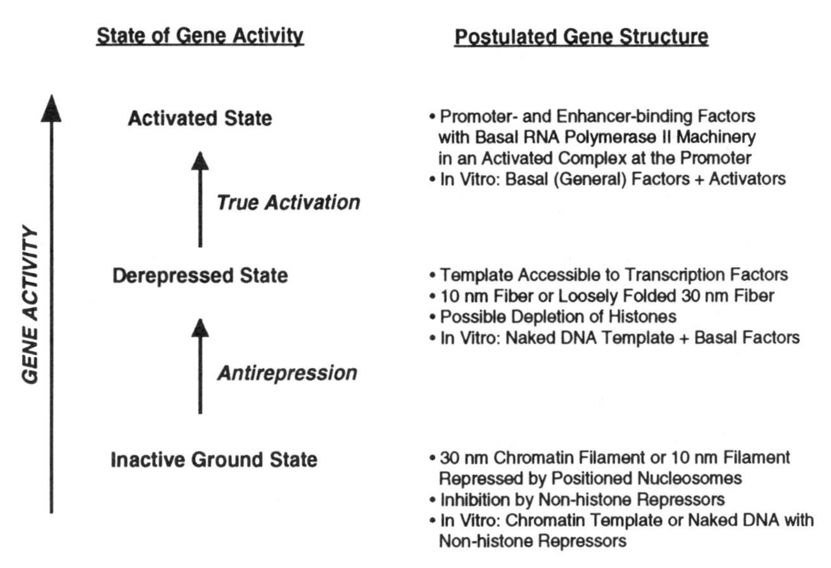

Figure 1 A paradigm for the role of chromatin structure in the regulation of gene expression. In addition to the postulated structures of the inactive, derepressed, and active states in vivo, it is worthwhile to note that a naked DNA template in vitro has no apparent physiological counterpart and corresponds to an unrepressed state in which the transcription of genes is promiscuous and unregulated. In another related issue, template DNA that is introduced into cells by transient transfection is packaged into chromatin (288, 289) that appears to be in a weakly repressed state.

CHROMATIN STRUCTURE

In this section, we summarize several aspects of chromatin structure that are relevant to the analysis of chromatin and transcription. These topics include the nucleosome, the core histone octamer, histone H1, structure of the chromatin filament, acetylation of the core histones, and high mobility group (HMG) proteins. There are also several recent reviews that are particularly informative in their description of chromatin structure (1, 2, 7, 19, 26–30).

In the nucleus, DNA is packaged into a nucleoprotein complex known as chromatin, which consists of roughly a 2:1 mass ratio of protein to DNA and a 1:1 mass ratio of histones to DNA. The packaging of the template into chromatin provides the compaction and organization of the DNA for processes such as transcription, recombination, replication, and mitosis. In the electron microscope, the extended chromatin fiber is a 10-nm diameter filament with an appearance that resembles "beads-on-a-string" (Figure 2,

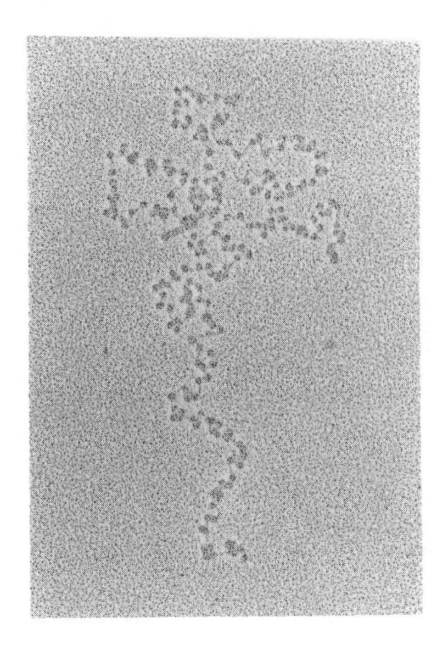

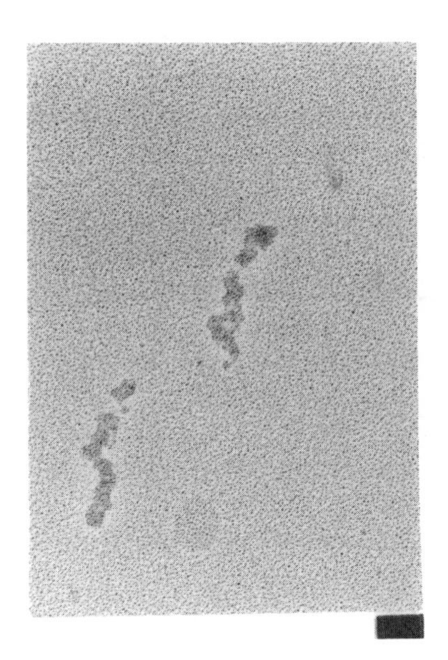

Figure 2 Electron micrographs of native *Drosophila* chromatin. *Left panel:* the extended 10-nm chromatin filament. Note the zig-zag arrangement of the nucleosomes. *Right panel:* the folded 30-nm-diameter chromatin filament. The solid black bar represents a length of 100 nm. Photographs courtesy of M Bulger, University of California, San Diego.

left panel). Each of these "beads" is a nucleosome, the unit repeat of chromatin.

In a nucleosome, the DNA is wrapped twice around a core histone octamer (a structure that consists of two copies each of the core histones, H2A, H2B, H3, and H4) in a left-handed superhelix, and there is one copy of the linker histone H1 (or one of its variants, such as H1° or H5). The nucleosomal repeat length is typically from 180 to 210 bp. When chromatin is lightly digested with micrococcal nuclease, mono- and oligo-nucleosomal fragments are generated. Upon electrophoresis, the resulting DNA from the digested chromatin gives a characteristic "ladder" of fragments that derive from multiple units of the nucleosomal repeat (31). If the chromatin is digested further with micrococcal nuclease, a species referred to as the chromatosome (32) is obtained. The chromatosome contains about 166 bp of DNA wrapped twice around the octamer and one molecule of H1. Finally, if the nuclease digestion proceeds further, the core particle is generated. The core particle contains 146 bp of DNA wrapped 1.8 times around the octamer, and it does not contain H1.

The Core Histone Octamer

The X-ray structure of the core histone octamer was determined at 3.1-Å resolution (33), and the X-ray structure of the core particle was solved to 7-Å resolution (34). The two structures are consistent with each other. In Figure 3, a model of the core histone octamer at 3.1-Å resolution (33) is shown with a 20-Å wide ribbon that depicts the regions of the octamer that might be covered by DNA in a nucleosome. The core histone octamer has been shown to be reversibly disassembled into an H3-H4 tetramer and two H2A-H2B dimers that are probably physiologically significant (35, 36). For example, the process of chromatin assembly in vivo appears to occur by initial deposition of an H3-H4 tetramer followed by the incorporation of two H2A-H2B dimers (37, 38). In the octamer, each of the core histones has a basic, unstructured N-terminal tail and a globular C-terminal domain that is highly α-helical (33). In the X-ray analysis of the core histone octamer (33), the structure of the flexible N-terminal tails could not be

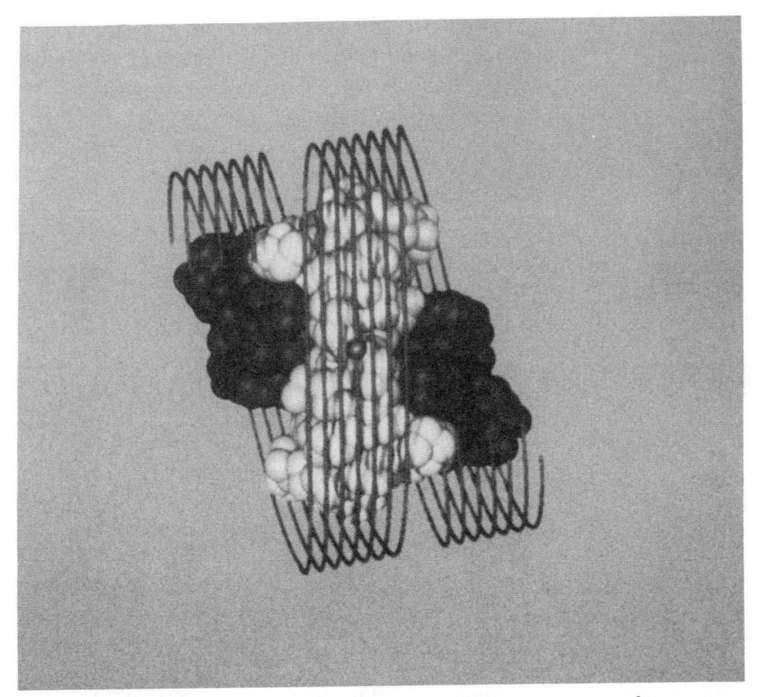

Figure 3 Structure of the globular region of the core histone octamer at 3.1-Å resolution (33). The ribbon has a width of 20 Å, which is roughly equivalent to the diameter of double-stranded DNA. Photograph courtesy of Dr. EN Moudrianakis, Johns Hopkins University.

determined, and the model shown in Figure 3 depicts only the central globular region. It has been hypothesized that the DNA in a nucleosome is wrapped around the globular core of the octamer (as displayed in Figure 3), while the basic core histone tails interact (like the arms of an octopus) with the DNA (for example, see Ref. 9). These N-terminal tails of the core histones are subject to a variety of posttranslational modifications, such as acetylation, and have been studied extensively.

Histone H1

Histone H1 and its variants, such as H1° and H5, are often referred to as linker histones because they are associated with the linker DNA between the nucleosomal cores (for reviews, see Refs. 39–42). There is roughly one molecule of H1 per nucleosome (43). H1 has three distinct domains: a short, basic unstructured N-terminal "tail," a folded globular domain that is highly conserved between species, and a long C-terminal tail. The structure of the globular domain of H5 (an H1 variant from chicken erythrocytes) has been solved to 2.5-Å resolution by X-ray crystallography (44) and also determined by NMR analysis (45). The structure of the globular domain of H5 is similar to the structure of the DNA-binding region of the catabolite gene activator protein CAP (44). Although the exact location of H1 in chromatin is not definitively known, it is generally believed that the H1 tails bind to the linker DNA and that the globular domain interacts with DNA in the vicinity of the nucleosomal pseudo-dyad [the face of the core particle where the DNA enters and exits (46–47a)]. Binding of H1 to naked DNA has been shown to be cooperative (48–50), whereas binding of H1/H5 to chromatin, which occurs in an asymmetric head-to-tail fashion (51), has not been experimentally demonstrated to be cooperative. Although H1-deficient nucleosomal arrays condense into faster-sedimenting species (52, 53) that appear as amorphous, disordered clumps by electron microscopy (54), H1 facilitates the folding of the 10-nm filament into the 30-nm chromatin fiber and stabilizes the compacted structure (53–56). There is considerable evidence that H1 participates in transcriptional repression, which is discussed later. Therefore, H1 may have both structural and regulatory functions, which may be interrelated.

IS THERE HISTONE H1 IN YEAST? It is commonly asked if there exists a version of histone H1 in the budding yeast *Saccharomyces cerevisiae,* an organism that has been widely used in the study of gene expression (15). Several laboratories have attempted unsuccessfully to identify a yeast homolog of H1. These experiments have typically resulted in the purification of a mitochondrial protein termed ABF2 (also known as HM and mtTF1; 57, 58). ABF2 has an "HMG1 box" motif (59, 60; see later) and is functionally interchangeable in vivo with the *Escherichia coli* histonelike protein HU (58). Given the available data, it is likely that there does not exist a yeast version of H1. In addition, the properties of yeast chromatin resemble those of H1-deficient chromatin. For instance, the repeat length of yeast chromatin is about 165 bp (61), and thus, the nucleosomes are closely packed and there is no linker DNA. Moreover, there is a paucity of condensed 30-nm diameter chromatin in yeast. It is possible that histone H1 provides an additional level of structure and regulation of chromatin that is required in higher eukaryotes, but not in budding yeast. In addition, it should be noted that a similar hypothesis has been considered for the existence of CpG methylation in vertebrates and plants, but not in other eukaryotes such as yeast and *Drosophila*.

The Chromatin Filament

The bulk of the chromatin in an interphase nucleus is present as a 30-nm diameter filament rather than the extended 10-nm fiber ("beads-on-a-string"), which is commonly shown in electron micrographs of chromatin (Figure 2) (for reviews of the 30-nm chromatin fiber, see Refs. 26–28). [To unfold the 30-nm fiber into the 10-nm fiber, it is necessary to subject the chromatin to nonphysiologically low salt concentrations (54).] The detailed structure of the 30-nm chromatin fiber is not known, and several models have been proposed as follows. First, in the solenoid model, the 10-nm chromatin filament is coiled like a spring into a left-handed superhelix with about six nucleosomes per turn, and histone H1 is located in the central, lengthwise axis of the fiber (54). Second, the supercoiled spacer model is similar to the solenoid model, except that histone H1 could be located in either the interior or the periphery of the fiber as a function of the linker length (62). Third, in the crossed-linker model, the 10-nm filament folds like an accordion into a zig-zag structure that twists along the lengthwise axis of the fiber into a double-superhelical structure (63). (Imagine a double helix in which nucleosomes at the periphery of the fiber are connected by linker DNAs that are perpendicular to the lengthwise axis of fiber and define the fiber diameter.) Fourth, in the twisted ribbon model, the 10-nm fiber folds into the zig-zag structure/ribbon (as in the crossed-linker model), and this ribbon is coiled like a spring into a superhelical structure (64, 65; imagine a ribbon wrapped around a thin rod). (The pictures in Refs. 27, 63 are particularly helpful for visualization of these structures.) At present, there does not exist evidence that definitively proves or disproves any of these proposed models.

It is important to consider the structure of the 30-nm filament, because it is the native form of the bulk of chromatin at interphase. There is roughly a sevenfold compaction of the DNA upon packaging into the 10-nm filament, and another six- to sevenfold compaction of the 10-nm chromatin fiber occurs upon folding into the 30-nm fiber. Hence, there is about 40–50-fold compaction of DNA upon packaging and condensation into the 30-nm chromatin filament. The 30-nm fiber is not, however, a homogeneous rod, but instead, is a somewhat irregular and uneven structure (Figure 2). In addition, the chromatin filament appears to be a dynamic structure that either folds or unfolds depending on the state of transcriptional activity. Studies of actively transcribed polytene chromosomes of *Chironomus tentans* (termed Balbiani rings) by electron microscopy have revealed that the transcribed chromatin fiber (about 5-nm diameter) condenses into a thick (20–30-nm diameter) chromatin fiber if transcription is inhibited with DRB (a purine nucleotide analog), and that transcription of this thick fiber could be resumed upon removal of the inhibitor (30, 66, 67). Thus, the 30-nm chromatin filament should be viewed as a potentially flexible and dynamic structure.

Acetylation of the Core Histones

The core histones, particularly H3 and H4, are reversibly acetylated at the ϵ-amino groups of lysines in the N-terminal tails. It has been postulated that core histone acetylation may be involved in the process of transcriptional activation because acetylation of the lysine residues neutralizes their positive charge and would presumably reduce the interaction of the core histone tails with DNA (for a recent review, see Ref. 68). Consistent with this hypothesis, immunofractionation of chromatin with antibodies that recognize acetylated H4 resulted in significant (15–30-fold) enrichment for active, but not inactive, gene sequences (69). In addition, it was found that chromatin preparations of CpG islands, which appear to represent promoter regions of active genes, contained hyperacetylated H3 and H4 (70). Hyperacetylation (as well as many other effects) can be induced in vivo by treatment of cells with sodium butyrate, which inhibits histone deacetylases. In general, butyrate treatment of cells causes increased expression of genes by mechanisms that may involve histone hyperacetylation, but there are also instances where gene induction has been found to be inhibited upon treatment with butyrate (71, 72).

The use of antibody preparations that recognize specifically acetylated residues in histone H4 (at positions 5, 8, 12, or 16) has provided useful information regarding the distribution of acetylated histones in the cell (68, 73, 74). For instance, in *Drosophila* polytene chromosomes, H4 histones acetylated at positions 5 or 8 were found to be distributed at overlapping, but not identical regions in euchromatin; H4 acetylated at lysine 12 was enriched in β-heterochromatin; and H4 acetylated at lysine 16 was localized to sites in the male X chromosome, but not in the male autosomes or any of the female chromosomes (note that heterochromatin is the term for regions of chromatin that are always highly condensed in which gene activity is generally repressed, whereas the remainder of the genome is designated as euchromatin) (73). In addition, similar experiments with the inactive X chromosome of mouse and human cells derived from females revealed they were deficient in acetylated H4 (74). In these studies, the accessibility of the antibodies to the histones in the folded chromatin filament is also an important consideration. Notwithstanding, the nonrandom distribution of specifically acetylated H4 suggests a functional role for histone acetylation.

An important question in the study of histone acetylation is whether it is actively involved in the process of gene activation or if it occurs as a consequence of gene activation (i.e.: Is acetylation a cause or effect?). To address this issue, the binding of the RNA polymerase III transcription factor TFIIIA to nucleosomal DNA was examined with core particles that were reconstituted with hyperacetylated histones, underacetylated histones, or trypsin-treated core histones (from which the tails were removed) (75). It was found that TFIIIA was able to bind to DNA packaged into core particles containing either the hyperacetylated or the trypsinized core histones, but not to DNA packaged into core particles prepared from the underacetylated core histones. These results were taken to indicate that acetylation of the core histone tails renders DNA packaged into a nucleosome more accessible to DNA-binding transcription factors. It is not known, however, whether hyperacetylation of the core histones affects the ability of TFIIIA to activate transcription. Nevertheless, these data, along with genetic studies in yeast (76–79; discussed later), suggest that core histone acetylation has an important role in potentiation of the chromatin template prior to gene activation.

The High Mobility Group (HMG) Proteins

The high mobility group (HMG) proteins are a set of nonhistone chromosomal proteins (for reviews, see Refs. 59, 60, 80–82). The HMG proteins are traditionally (and somewhat arbitrarily) defined to be those proteins (with the exception of ubiquitin) that can be extracted from nuclei with 0.35 M NaCl and are soluble in 2% trichloroacetic acid (80). There are three subclasses of HMG proteins: the HMG1/2 proteins, the HMG14/17 proteins, and the HMG-I/Y proteins. HMG proteins from different subclasses are not related to each other, and have been commonly termed "HMG proteins" only because of their similar extraction and solubility properties.

HMG1/2 HMG1 and HMG2 are related proteins (25–30 kDa) that have a distinct three-domain structure. From the N terminus to the C terminus of the polypeptide, these domains are named A, B, and C (83, 84). Domains

A and B are homologous, and domain C is an acidic C-terminal tail. The A and B domains correspond to the protein sequence that is now popularly referred to as the "HMG1 box" (59, 60, 85), a motif that has been found in a variety of proteins, such as the RNA polymerase I transcription factor UBF (86), the protein encoded by the sex-determining region Y (SRY) gene (87, 88), and many others (for a recent list, see Ref. 60). A consensus amino-acid sequence, or "signature," for the HMG1 box has been deduced from statistical analysis of a variety of protein sequences (60), and it was found that several proteins that have been reported to have HMG1 boxes, such as CCG1, SIN1, and TFC3, do not possess this "signature" (59, 60).

The biological function of HMG1/2 is not yet known, but some of its biochemical properties have been characterized. The structure of the B-domain of rat HMG1 has been determined by NMR spectroscopy (89). The domain is L-shaped and contains three α-helices that compose about 75% of the molecule. HMG1/2 binds preferentially to cruciform DNA (90), single-stranded DNA (91, 92), or DNA modified by the drug cisplatin (93, 94) relative to double-stranded DNA. Based on these results, it has been suggested that HMG1/2 may participate in recombination, replication, or DNA repair. In addition, either intact HMG1 or the B-domain of HMG1 can bend DNA efficiently (95). In vitro transcription experiments have suggested that HMG1/2 activates transcription by RNA polymerases II and III (96–98) and stimulates DNA binding by USF/MLTF (99). Since those studies were performed, however, in vitro transcription systems have advanced considerably, and it would be worthwhile to re-examine the effects of HMG1/2 upon transcription in vitro with reconstituted chromatin templates. The abundance of HMG1 and HMG2 has been estimated to be from 0.1 to 0.5 molecules (of HMG1 + HMG2) per nucleosome (100). In contrast to the situation with histone H1, homologs of HMG1/2, termed NHP6A and NHP6B, have been identified in *S. cerevisiae* (101). Given the abundance of HMG1/2 in chromatin and the conservation of the protein among humans, *Drosophila* (102), and yeast, it is likely that HMG1/2 performs an important function in the nucleus.

HMG14/17 HMG14 and HMG17 are small (~10-kDa), highly charged proteins that contain a preponderance of basic amino-acid residues in the N-terminal portion of the protein and acidic amino-acid residues in the C-terminal region. The abundance of HMG14/17 has been estimated to be from 0.1 to 0.5 molecules (of HMG14 + HMG17) per nucleosome (100). HMG14 and HMG17 bind to core particles with a stoichiometry of two HMG proteins per core particle (103–105). In chromatin reconstitution reactions, phosphorylated but not unphosphorylated HMG14/17 can increase the nucleosome repeat length from 145 bp to 160–165 bp (106, 107). In addition, like HMG1/2, HMG14/17 binds preferentially to single-stranded DNA relative to double-stranded DNA (108). There has been some debate over whether or not HMG14/17 is specifically associated with active chromatin. By immunofractionation of oligonucleosomes containing HMG17, it was suggested that HMG17 is specifically located downstream, but not upstream of the RNA start sites of active genes (109). In addition, UV crosslinking experiments have shown that HMG14/17 is modestly enriched (1.5–2.5-fold) in an actively transcribed gene relative to inactive genes (110). Further studies are necessary to clarify the biological function of HMG14/17.

HMG-I/Y HMG-I and HMG-Y are small proteins (~11 kDa) that differ only by an internal deletion of an 11-amino-acid segment in HMG-Y relative to HMG-I (111, 112). These proteins are also known as α-protein because of their ability to bind to A/T-rich α-satellite DNA (113, 114). HMG-I was found to be localized to A/T-rich repetitive sequences, centromeres, and telomeres in mammalian metaphase chromosomes (115). It thus appears that HMG-I/Y binds to A/T-rich sequences in heterochromatin. In addition, HMG-I/Y has been observed to interact with A/T-rich sequences in promoters and enhancers of protein-coding genes (116, 117). For instance, HMG-I binds to an A/T-rich upstream activating sequence of the lymphotoxin (also known as tumor necrosis factor-β) gene, but it was not determined if the binding of HMG-I affects transcription of the gene (116). In studies of the human interferon-β gene, HMG-I was found to interact with the PRDII promoter element and to stimulate the simultaneous binding of NF-κB to this element (117). Furthermore, viral induction of the interferon-β gene was decreased by the presence of antisense HMG-I/Y RNA. These results provide evidence that HMG-I/Y may participate in the regulation of gene expression, but it is not yet known whether it functions in a structural capacity (i.e. contributes to the promoter architecture) or if it affects the activity of the transcriptional machinery in a manner similar to promoter- and enhancer-binding factors.

NUCLEOSOME STRUCTURE AND GENE ACTIVITY

Modification of Core Histones In Vivo in Yeast

Genetic studies with the yeast *S. cerevisiae* have provided some of the most convincing evidence of a functional relationship between chromatin structure and transcriptional regulation (for reviews, see Refs. 5, 7, 9, 10, 15, 16, 21–23). The efficacy of these experiments has derived from a combination of the ease of genetic manipulation of the organism and the presence of only two copies of each of the genes encoding the core histones per haploid genome (by comparison, in *Drosophila melanogaster,* there are an estimated 110 copies of each of the histone genes). This propitious situation has enabled modulation of the levels of the core histones in vivo, substitution of the wild-type core histones with specifically modified mutant variants, and the use of genetic techniques to identify factors that affect chromatin structure and gene activity.

Studies of transcriptional regulation in yeast have included experiments in which genes encoding the core histones were identified as suppressors of defective promoters or enhancers, which are typically termed "upstream activating sequences" (abbreviated as UAS) in yeast. For instance, the yeast histone locus *HTA1-HTB1,* which encodes histones H2A and H2B, was identified as a high-copy-number suppressor of a transcription unit that was rendered defective by a solo δ insertion mutation (118). By variation of the in vivo expression of either H2A-H2B or H3-H4, it was found that the suppression of the transcription defect was due to an imbalance in the synthesis of H2A-H2B relative to H3-H4. These results suggested that the stoichiometry of H2A-H2B relative to H3-H4 affected transcription. In addition, modulation of the expression of H2A-H2B in yeast was observed to alter the chromatin structure of different genes to varying extents (119). More recently, the *BUR* genes were isolated as suppressors that bypass a requirement for a UAS (Bypass UAS Requirement = *BUR*; Ref. 120). *BUR1* encodes a CDC28-related putative protein kinase, whereas *BUR5* is identical to *HHT1,* one of the two yeast genes that encodes histone H3. The *bur5-1* mutant allele was sequenced and found to possess a G-to-A missense mutation that results in a Thr-to-Ile amino acid substitution at residue 119 of histone H3. In other studies, it was found that *SIN2*—which was identified as a suppressor of mutations in *SWI1, SWI2/SNF2,* and *SWI3,* which encode putative transcriptional activators (121)—also encodes yeast histone H3 (122). These findings further suggest an important relationship between chromatin structure and transcriptional activity.

To examine specifically the function of histone H4 in vivo, a yeast strain was constructed in which the two endogenous H4 genes were deleted and an exogenous H4 gene was located on a plasmid (9, 15, 16, 77, 123–125). The plasmid-borne histone H4 gene was placed under the transcriptional control of the *GAL1* promoter, which is induced by galactose and repressed by glucose. Thus, in the presence of glucose, histone H4 was depleted. Under such conditions, the chromatin structure was altered (for simplicity, this was termed "nucleosome loss"), and transcription of the *PHO5, CYC1,* and *GAL1* promoters was increased in the absence of inducers (123–125). Depletion of H4 also resulted in activation of the *CUP1* and *HIS3* promoters when they were present as fusions with the bacterial *lacZ* gene (77). Similar experiments with histone H3 revealed that replacement of the wild-type protein with mutant variants containing deletions or amino-acid substitutions at potential acetylation sites in the N-terminal tail resulted in "hyperactivation," in which activation occurred to higher levels than with the wild-type H3, of several *GAL4*-regulated genes, but not the *PHO5* gene (126). These experiments indicated that alteration of chromatin structure leads to activation (or perhaps more correctly, derepression) of genes that probably occurs by relief of nucleosome-mediated repression.

Next, a systematic set of deletion and substitution mutants of histone H4 was generated and characterized in vivo (76, 127). Deletions in the hydrophobic core of H4 were lethal, whereas a strain termed del(4–28) containing a large deletion of the N-terminal tail of H4 (residues 4–28, inclusively) was viable (127). Furthermore, in the del(4-28) strain, the silent mating type loci, *HMLα* and *HMRa,* were derepressed. These and other results suggested that silencing of the mating type loci involved genetic interactions between histone H4 and SIR3, a protein that is required for silencing (23), and that acetylation of Lysine 16 of H4 caused derepression of the *HMLα* and *HMRa* loci (78, 79). [Note, however, that the viability of the H4 del(4-28) strain implies that acetylation of the H4 tail may not be an essential cellular function.] Further analysis of mutant variants of H4 revealed that full activation of the *GAL1* (20-fold reduction in expression) and *PHO5* (5-fold reduction in expression) promoters was not achieved with a variety of deletions or substitutions at potential acetylation sites in the N-terminal tail (76). These findings collectively demonstrate that the histone H4 N-terminal tail functions in both gene activation and repression.

In the context of the paradigm presented in Figure 1, these results could be interpreted as follows. In instances where histone depletion or alteration resulted in an increase in transcriptional activity under noninducing conditions, the genes may have undergone a transition from the inactive ground state to a derepressed state ("antirepression"). In the derepressed state, the chromatin structure is unfolded, the template is accessible to transcription factors, and a basal level of transcription is attained. In the experiments in which mutant variants of H4 inhibited full induction of the *GAL1* and *PHO5* genes, the altered H4 proteins may have interfered with "true activation," in which sequence-specific transcription factors such as GAL4 facilitate the assembly of transcription initiation complexes at the promoter. Alternatively, the wild-type, but not mutant H4 protein may facilitate the true activation process. As we discuss later, the results of genetic studies in yeast are consistent with data obtained in biochemical experiments that further support the antirepression and true activation model of gene activation.

IS THERE PROMOTER-SPECIFICITY IN CORE HISTONE FUNCTION? As a corollary to the studies discussed above, it would be useful to address the issue of whether or not histones can act in a promoter-specific manner. Thus far, biochemical studies (discussed later) have demonstrated a general repression of transcription upon packaging of the template DNA into chromatin. On the other hand, the genetic studies in yeast provide examples of gene specificity of histone function. For instance, in the study of mutant variants of H3, both deletion and substitution mutants were observed to induce "hyperactivation" of the *GAL1*, but not the *PHO5* promoter, where "hyperactivation" is a higher level of transcription at lower concentrations of galactose, which induces *GAL1* (126). A variety of mutations in the N-terminal tail of H4 inhibited activation of *GAL1, GAL4, GAL7, GAL10,* and *PHO5,* but not constitutive expression of *PRC1* (76). The mechanistic basis of this promoter specificity of histone function is not yet known, but may reflect aspects of the chromatin structure at the different genes, such as nucleosome positioning, or specific interactions between histones and transcription factors or regulatory proteins, such as the postulated SIR3-H4 interaction (78).

Nucleosome Positioning

Several facets of the structure of nucleosomal arrays should be considered in the analysis of chromatin and transcription. Properties of bulk chromatin include the protein composition (core histones, histone H1, HMG proteins, etc), the regularity of nucleosomal spacing (which includes the nucleosome repeat length), posttranslational modification of histones, and the higher-order folding of the chromatin filament. In the analysis of a specific gene or DNA sequence, another important consideration is nucleosome positioning (for recent reviews, see Refs. 21, 128).

TRANSLATIONAL AND ROTATIONAL POSITIONING OF NUCLEOSOMES Nucleosome positioning can be subcategorized into "translational" and "rotational" positioning. Rotational positioning refers to the orientation of the DNA relative to the surface of the core histone octamer (for instance, at a specific base pair, whether the minor groove of the DNA is facing toward or away from the surface of the octamer). The rotational positioning of nucleosomes is usually characterized by DNase I (or hydroxyl radical) digestion of chromatin. When chromatin is treated with DNase I, the nuclease digests the DNA at 10-bp intervals (i.e. once every turn of the helix, where the minor groove of the DNA is facing away from the octamer and is accessible to the enzyme) and yields a repeated 10-bp ladder (129, 129a; but also see 130). Nucleosomes exhibit a thermodynamic preference for positioning that allows the minor groove of A/T-rich DNA to face inward toward the octamer, and for the minor groove of G/C-rich DNA to face away from the octamer

(131, 132; but see also 133). This observation has led to the design of synthetic, curved DNA sequences containing alternating A/T- and G/C-rich stretches of DNA (with the period of a helical turn of DNA on the octamer) that have been found to exhibit strong rotational positioning when reconstituted into nucleosomes (see, for example Refs. 134–136).

Translational positioning of nucleosomes refers to the location of a nucleosome along a gene. For example, there may be a core histone octamer translationally positioned from −300 bp to −144 bp relative to the transcription start site of a gene. Translational positioning of nucleosomes can be deduced by a method known as indirect end-labeling (137–140), which is outlined in Figure 4. More recently, the development of ligation-mediated polymerase chain reaction methodology (141, 142) has advanced the resolution and sensitivity of the analysis of nucleosome positioning (143). These techniques are generally useful for the analysis of chromatin structure, which also includes identification of DNase I–hypersensitive sites (Figure 4) and the binding of factors to transcriptional control regions in vivo (see, for example Refs. 141–143). It is important to note, however, that indirect end-labeling techniques do not prove conclusively that nucleosomes are located between sites of micrococcal digestion, but rather they suggest, based on deduction, the positioning of putative nucleosomes. In addition, the lack of a distinct pattern of micrococcal nuclease cleavage sites in an indirect-labeling experiment does not necessarily indicate the absence of nucleosomes, but instead suggests that there may not be preferentially positioned nucleosomes in the region. Finally, it should be noted that these experiments generally do not discern whether 80% or more of the putative nucleosomes are translationally positioned, and the remaining 20% uncertainty can affect the interpretation of these data. It is therefore judicious to analyze data on nucleosome positioning somewhat cautiously.

NUCLEOSOME POSITIONING AND GENE ACTIVITY The location of nucleosomes in vivo is typically at specific positions along the DNA (21). In general, alterations in the positioning of nucleosomes have been observed to correlate with the conversion of genes from a repressed state to a transcriptionally competent state. A few examples of preferential nucleosome positioning in vivo include the yeast TRP1ARS1 plasmid (144), the mouse β-major globin gene (145), the mouse mammary tumor virus promoter (146), the *Drosophila hsp26* gene (147), and the yeast *PHO5*/*PHO3* locus (148, 149). It has also been possible to reconstitute nucleosomes in vitro with short DNA fragments (150–450 bp) to give translationally and rotationally positioned mono- or di-nucleosomes (see, for instance, Refs. 134–136, 150–155).

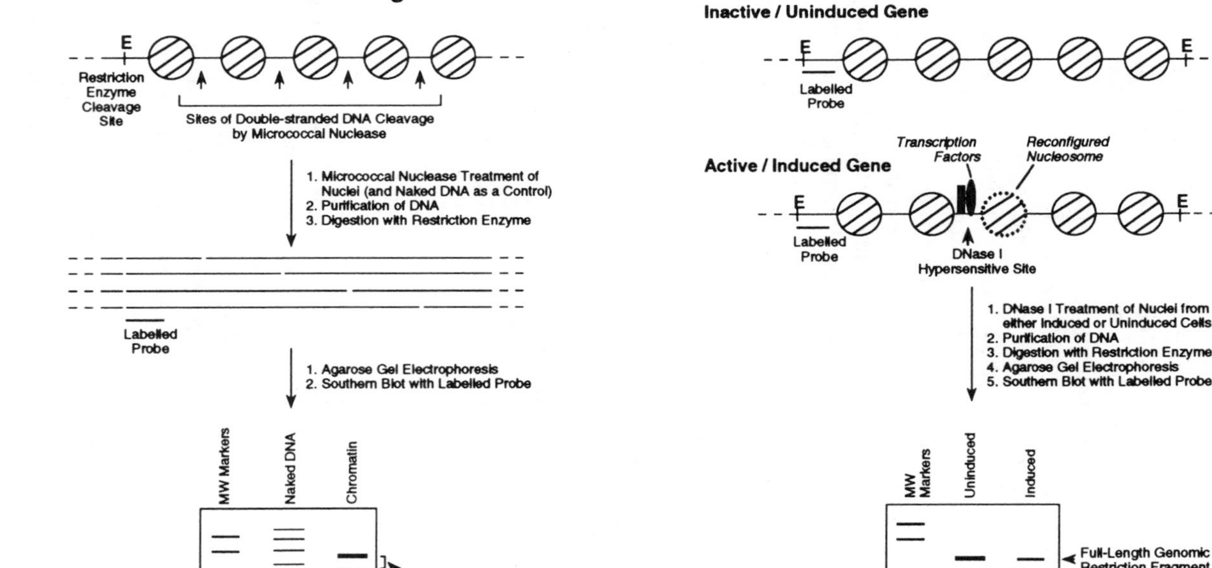

Figure 4 The use of indirect end-labeling techniques in the analysis of nucleosome positioning and DNase I hypersensitivity. A commonly employed variation of the DNase I hypersensitivity assay involves the use of restriction enzymes instead of DNase I to examine the accessibility of the chromatin at specific locations.

The mechanistic basis for nucleosome positioning may involve interactions between the core histone octamer and DNA and/or the influence of factors such as sequence-specific DNA-binding proteins. As discussed above, there is a preference for nucleosomes to be located on curved DNA (131, 132; but also see 133). There is not, however, a good correlation between the positioning of nucleosomes in vitro with short DNA fragments and the positioning of nucleosomes integrated into the genome in vivo. For example, an artificially designed curved DNA sequence at which nucleosome positioning occurs in vitro with short DNA fragments (less than 450 bp) (134, 135) does not mediate nucleosome positioning in vivo when integrated into the genome of yeast (144) or *Drosophila* (Q Lu and SCR Elgin, personal communication). These and other (156, 157) results collectively suggest the following: (*a*) the positioning of nucleosomes in vitro with short DNA fragments may be affected by the boundary constraints imposed by the ends of the fragment; and (*b*) the DNA sequence alone may not be sufficient for determination of nucleosome positioning in vivo.

In yeast, two sequence-specific DNA-binding factors, α2 repressor (158–161) and GRF2 (also known as factor Y; Refs. 162–164), appear to be involved in the positioning of nucleosomes. In studies of the α2 repressor, which inhibits transcription of **a** cell–specific genes in α cells but not **a** cells, it has been shown that a nucleosome is positioned next to the α2 operator site and that α2 repressor is required for this nucleosome positioning to occur. In α cells but not **a** cells, α2 repressor was found to mediate the positioning of a nucleosome over the TATA box of the **a** cell–specific genes, *STE6* and *BAR1*. Moreover, in yeast strains containing deletion or substitution mutations in histone H4, the α2 repressor–mediated positioning of nucleosomes in the *STE6* gene was altered, and this apparent destabilization of nucleosome positioning correlated with partial derepression (to about 5–15% of full expression) of transcription from the promoter of the **a** cell–specific gene *MF**a**2*. These results have led to the hypothesis that α2 repressor functions to inhibit transcription of **a** cell–specific genes by positioning a nucleosome over the TATA box of each of the genes. Although there is convincing evidence for α2 repressor–mediated positioning of nucleosomes, the exact function of α2 repressor in the inhibition of transcription remains a point of debate (165). It is possible that α2 repressor blocks transcription by nucleosome positioning and/or by interaction with components of the transcriptional machinery. Nevertheless, the available data suggest that nucleosome positioning may be an important component of transcriptional regulation.

Biochemical Analysis of Transcriptional Regulation with Chromatin Templates

The biochemical analysis of transcription by RNA polymerase II has been carried out predominantly with naked DNA templates rather than with chromatin templates, which may more accurately reflect the natural state of genes in vivo. The use of purified DNA templates has been simple and expedient, but recent experiments with reconstituted chromatin templates suggest that it may be more appropriate to employ chromatin rather than naked DNA templates. The packaging of purified plasmid DNA templates into chromatin has been observed to cause a general repression of transcription that can be relieved by sequence-specific DNA-binding factors and/or basal transcription factors, such as the TATA box–binding factor TFIID, when these factors are bound to the DNA template prior to the reconstitution of chromatin (166–186). These studies have led to the hypothesis that sequence-specific transcription factors function, at least in part, to counteract chromatin-mediated repression.

In the context of the paradigm presented in Figure 1, chromatin templates are a model for the transcriptionally repressed ground state of inactive or uninduced genes, naked DNA templates correspond to the derepressed state of a gene, and templates (either chromatin or naked DNA) with transcriptional activators are in the activated state. This conclusion is based on the following observations. First, a broad range of genes are promiscuously transcribed in an unregulated manner by the basal transcriptional machinery with naked DNA templates. For example, in the absence of histones, *Drosophila* embryo nuclear extracts will indiscriminately transcribe genes with comparable efficiency whether or not they are normally expressed in the embryo (JT Kadonaga laboratory, unpublished observations). Second, the magnitude of transcriptional activation by sequence-specific DNA-binding factors that is observed in vivo can generally be recreated in vitro with transcriptionally repressed chromatin, but not with histone-free template DNAs (see, for example Refs. 166, 167). Third, phenomena such as long-distance (> 1 kbp) activation of transcription by sequence-specific transcription factors can be recreated in vitro with chromatin templates (169–171). These experiments suggested that enhancers may function to counteract chromatin-mediated repression rather than to increase the rate of transcription, and that the packaging of DNA into chromatin may bring distally bound enhancer-binding factors into the proximity of the TATA box and RNA start site (see also Refs. 147, 187). [Note, however, that the requirement for a chromatin template to achieve enhancerlike function in vitro is not absolutely certain, because long-range activation of transcription has been observed in the apparent absence of nucleosomes (188).] Therefore, the biochemical studies have led to the hypothesis that an important function of promoter- and enhancer-binding factors is to relieve chromatin-mediated inhibition of transcription (antirepression). Hence, sequence-specific transcription factors appear to function both to counteract chromatin-mediated repression and to facilitate the intrinsic transcription process in the apparent absence of histones (including H1) when the factors are bound at the proximal promoter (166, 168). These findings, along with the results from genetic studies in yeast, support the antirepression and true activation model presented in Figure 1.

Histone H1 and Transcriptional Regulation

The linker histone, H1, appears to repress transcription by all three eukaryotic RNA polymerases. Crosslinking studies performed in vivo have demonstrated that H1 is depleted in active or competent genes relative to inactive genes (110, 189–191). Biochemical experiments have shown that H1 is a potent repressor of transcription by RNA polymerases I (192), II (166, 168–170, 181, 193–198), and III (199–201). The presence of sequence-specific transcription factors and/or basal transcription factors can counteract histone H1–mediated repression with the template as either H1-DNA complexes (168, 181, 192–198) or H1-containing chromatin (166, 169, 170, 181). [The abundance of H1 in chromatin of higher eukaryotes is about 1 molecule per nucleosome (43). Therefore, in biochemical studies, H1-containing chromatin is a more accurate model for the natural state of the template in higher eukaryotes than H1-deficient chromatin.] These data suggest that histone H1 participates along with nucleosomal cores and other components of chromatin to repress gene activity. It is important to note, however, that in the crosslinking experiments, H1 was partially rather than completely depleted upon transcriptional activation (110, 189–191). Consistent with these findings, it was observed by immunoelectron microscopy of Balbiani ring genes (30) that H1 was localized to both the fully extended, actively transcribed genes and the repressed 30-nm chromatin fiber (202). Thus, the data indicate that partial rather than complete depletion of histone H1 occurs during gene activation. It is not known, however, whether the incomplete depletion of H1 in the promoter region of active genes is due to partial loss of H1 over the entire region or to complete loss of H1 at a specific location.

There has been some controversy regarding the biochemical experiments with H1-DNA complexes (186). In particular, it was questionable whether H1-DNA complexes (i.e. a mixture of purified histone H1 and plasmid DNA) could be used as a model for H1-containing chromatin. To address this issue, H1-DNA complexes and H1-containing chromatin were directly compared, and it was found that they possessed similar, though not identical, transcriptional properties that were quite distinct from those of naked DNA templates (169, 181). Hence, although H1-DNA complexes are aesthetically unattractive as a model for chromatin, they may be useful as a simple but imperfect version of transcriptionally repressed template DNA. In this context, it is also pertinent to mention that nuclear extracts that have been commonly used for in vitro transcription experiments (such as a 0.42 M NaCl extract from HeLa cells; 203) contain high levels of histone H1 (168, 193). Therefore, studies that have been carried out with such extracts have actually employed transcriptionally repressed H1-DNA complexes rather than naked DNA templates. In the future, it will be important to determine whether or not histone H1 is present in the reaction medium of transcription experiments. It may be useful to note that a highly active low-salt nuclear extract, termed the soluble nuclear fraction, can be used as a source of basal transcription factors that is deficient in histone H1 (Ref. 204; except that 0.1 M KCl is used in the nuclear extraction instead of 0.4 M potassium glutamate).

Nucleosome Reconfiguration During Gene Activation

It is well established that changes in chromatin structure accompany gene activation or potentiation to a pre-activated state (18, 19, 205), but whether nucleosomes are lost or reconfigured during this process has yet to be clarified (for recent reviews, see Refs. 3, 12, 14). The data clearly demonstrate that chromatin structure is altered upon conversion of genes from the repressed to the active state, during which increased nuclease accessibility to the promoter region and disappearance of canonical nucleosome structure is commonly observed (see, for example Refs. 70, 146, 149, 206). Yet, in many instances where nuclease digestion studies have revealed hypersensitive sites and/or the generation of subnucleosomal particles upon gene activation, it is possible that nucleosomes may have been "reconfigured" or "remodeled" rather than "removed" or "lost." This issue may, in part, be semantic. In this review, the term "nucleosome displacement" denotes the loss of core histones from the template DNA, whereas the term "nucleosome reconfiguration" indicates that nucleosome structure has been altered but core histones have not been removed.

ARE NUCLEOSOMES DISPLACED UPON GENE ACTIVATION? To address this question, the interaction of histones with either active or inactive genes has been examined. First, two separate studies involving chemical crosslinking of histones to the *Drosophila hsp70* genes indicated that the core histones remain bound to the genes upon heat activation (189, 207). Second, UV-crosslinking experiments with the mouse mammary tumor virus promoter revealed that histone H1, but not histone H2B, is depleted in the region of the promoter (termed nucleosome B) that displays increased accessibility to nucleases upon transcriptional induction with hormone (191). Third, a study of the yeast *GAL1-GAL10* genes (Southern blot analysis with mono- and di-nucleosomes that were subjected to native gel electrophoresis) demonstrated that the promoter region is associated with intact but detectably altered [as determined by either nuclease or methidiumpropyl EDTA-Fe(II) digestion experiments] nucleosomes upon gene induction (162). These experiments indicate that chromatin structure is altered but core histones are not displaced upon transcriptional activation, and are therefore consistent with nucleosome reconfiguration rather than nucleosome displacement.

These findings lead to the following question: If nucleosomes are not displaced upon gene activation, then what is the structure of the reconfigured chromatin that exhibits properties such as nuclease hypersensitivity? Several models, which are not mutually exclusive, are as follows. (*a*) Depletion of histone H1 (70, 110, 189–191) may increase the accessibility of factors to the DNA template and also allow greater mobilization or sliding of nucleosomes. (*b*) The binding of transcription factors to DNA wrapped around a nucleosome may induce nuclease sensitivity. For instance, it has been found that the glucocorticoid receptor (153–155), yeast GAL4 protein (208, 209), or heat-shock factor along with TATA box–binding protein (208) bind to reconstituted mono- or di-nucleosomes. In addition, the ability of different promoter- and enhancer-binding factors to interact with nucleosomal cores appears to be variable (208). (*c*) The core nucleosomal structure may be altered in a manner that has been described as "unfolded," "open," or "split." For example, mercury [Hg(II)] affinity chromatography has been employed to isolate "unfolded" nucleosomes (from nuclei treated with micrococcal nuclease) in which the thiol group of the cysteine residue at position 110 of histone H3 (which is located in the center of the globular region of the core histone octamer; Ref. 33) has become accessible to the affinity resin (210–216). These experiments have revealed that there is a good correlation between the accessibility of the sulfhydryl group to the affinity resin and transcriptional activity. The structure of unfolded nucleosomes is not known, but they contain hyperacetylated core histones with the four core histones present in roughly equivalent stoichiometry. The possible existence of unfolded nucleosomes has also been suggested by nuclease digestion studies in which "atypical" or "split" nucleosomes were observed in actively transcribed genes (217, 218). (*d*) A variety of other factors, such as histone modification and binding of HMG proteins, may be involved in the reconfiguration of chromatin structure. It should be noted, however, that these studies are at an early stage, and the structural and functional consequences of alterations in chromatin that occur upon gene activation have yet to be elucidated.

Mechanisms for Alteration of Chromatin Structure

Whereas in the previous section we had discussed the structure of active chromatin, we now consider the events and mechanisms that lead to the formation or establishment of the transcriptionally active state. In a simple sense, this process can be subdivided into replication-dependent and replication-independent pathways (Figure 5). In the DNA replication–dependent mechanism, transcription factors bind to the DNA template immediately after passage of the replication fork and prior to the assembly of the newly replicated DNA into chromatin (see, for instance Refs. 20, 170, 219–224). Thus, in this situation, the disruption of chromatin structure that occurs upon DNA replication allows the binding of factors for potentiation of gene activity. In the DNA replication–independent model, transcription factors may act to reconfigure chromatin structure as a key step in gene activation. For example, with the yeast *PHO5* and mouse mammary tumor virus promoters, factors bind to the template, chromatin structure is altered, and transcriptional activation occurs in the absence of DNA replication (146, 225).

In either the occurrence or the absence of DNA replication, the establishment of the active state and concomitant reconfiguration of chromatin structure may involve promoter- and enhancer-binding factors as well as other proteins that are known to affect transcriptional activity, such as those

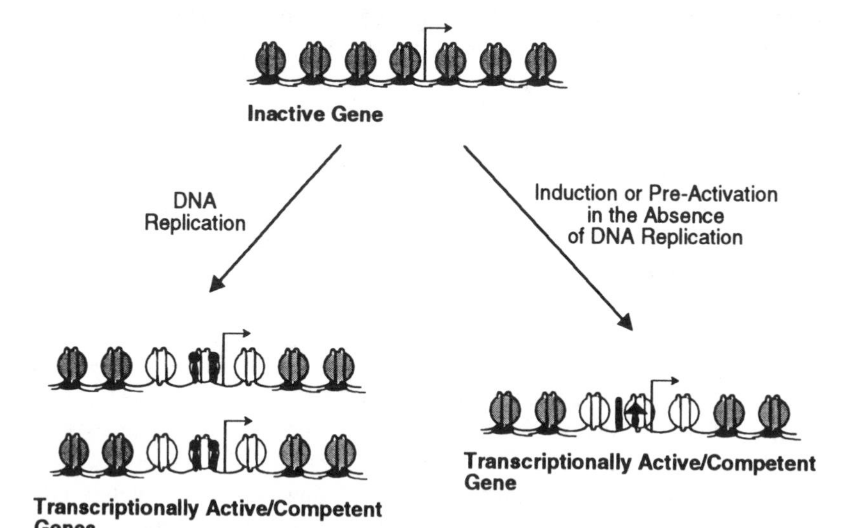

Figure 5 DNA replication-dependent and -independent models for reconfiguration of chromatin structure and potentiation of gene activity. The unshaded nucleosomes represent a reconfigured chromatin structure in which histone H1 is depleted and canonical nucleosomal cores are altered.

encoded by the *SWI* and *SNF* genes in yeast (10, 226). Moreover, the binding of factors to the template need not lead to immediate activation of the gene. It is possible that the transcription factors may bind to the template to set up a pre-activated (competent or potentiated) state from which the transcriptionally poised gene could be rapidly induced in the absence of replication. Studies of the *Drosophila* heat-shock genes have revealed such a mechanism, in which RNA polymerase II is transcriptionally engaged but paused approximately 25 nucleotides downstream of the RNA start site in the absence of heat shock (227, 228). The binding of a ubiquitously expressed sequence-specific DNA factor known as the GAGA factor (229) or CT-binding protein (230) appears to be required for the establishment of the pre-activated state of the *Drosophila* heat-shock genes in vivo (231–233). Consistent with these findings, biochemical studies of the GAGA factor have revealed that it is able to counteract histone H1–mediated inhibition of transcription (168, 197). Whether or not the GAGA factor binds to chromatin in a replication-dependent manner is not known, however.

Other Related Topics

Many subjects are relevant to the analysis of chromatin structure and transcription, and we have been able to include only a subset of these areas in this review. A few topics are discussed below, and we refer the interested reader to the cited papers and reviews as a starting point for further investigation.

METHODOLOGY FOR THE ANALYSIS OF CHROMATIN RECONSTITUTED IN VITRO

The biochemical characterization of transcription with chromatin templates is not straightforward, and a critical aspect of these experiments is the properties of the reconstituted chromatin. Although it may be a bit tedious, it is useful from a practical standpoint to list some criteria that can be applied to assess the quality of chromatin that is prepared in vitro. (*a*) A simple and highly informative assay for determination of the quality of reconstituted chromatin is the micrococcal nuclease digestion assay (31), which indicates the regularity of nucleosomal spacing as well as the repeat length of the chromatin. If the chromatin displays a micrococcal digestion ladder that reveals mono-, di-, tri-, tetra-, and penta-nucleosomes (or more), it is of reasonably good quality. (*b*) The average nucleosome density of the reconstituted chromatin can be estimated by DNA supercoiling analysis when circular plasmid DNA is employed for chromatin reconstitution. The deposition of one nucleosome causes a change in the linking number of –1 (234, 235). Therefore, the number of nucleosomes on a circular template that has been relaxed by topoisomerase I can be deduced from the linking number, which can be analyzed by two-dimensional agarose gel electrophoresis (166, 170, 236). In the interpretation of this assay, it is important to note that subnucleosomal particles such as H3-H4 tetramers will also supercoil DNA (237). (*c*) It is thus necessary to perform the third assay, which is to determine the protein composition of the purified chromatin by gel electrophoresis. For instance, polyacrylamide-SDS gel electrophoresis confirms the purity of the chromatin and ensures that the four core histones are present in equimolar amounts. In addition, a Triton-acid-urea gel system detects acetylation of the core histones (238). (*d*) DNase I treatment of DNA wrapped around a nucleosomal core yields a repeated 10-bp digestion ladder, as mentioned previously in the section on *Nucleosome Positioning* (129, 129a). This assay is useful for the determination of the rotational positioning of nucleosomes, but it is important to consider that complexes of individual core histones with DNA also yield a repeated 10-bp ladder upon treatment with DNase I (130). Hence, the 10-bp DNase I ladder is not a unique feature of intact nucleosomes. (*e*) The proper incorporation of histone H1 can be monitored by native gel electrophoresis of chromatin that has been digested to mono- and di-nucleosomes with micrococcal nuclease (46, 239, 240). (*f*) Nucleosome positioning, if it occurs, can be detected by indirect end-labeling (Figure 4) (137–140) or ligation-mediated polymerase chain reaction (141–143) techniques. (*g*) Finally, electron microscopy can be used to observe features of the chromatin filament such as the zig-zag nature of histone H1–containing chromatin or folding into the 30-nm diameter filament (Figure 2) (54, 170, 199).

TRANSCRIPTIONAL ELONGATION WITH CHROMATIN TEMPLATES (for reviews, see Refs. 241–243) Biochemical experiments have shown that nucleosomes appear to impede, but not block transcription in vitro by either RNA polymerase II (174, 244–246) or various prokaryotic RNA polymerases (174, 244, 247–249). It has been suggested that the transient positive supercoiling that is generated downstream of elongating polymerases (250) destabilizes the interaction of histone octamers with the DNA (248, 251). Some models for elongation suggest dissolution of histone octamers into H3-H4 tetramers and H2A-H2B dimers during transcription, but transcriptional studies with nucleosomal templates prepared from octamers in which the core histones have been chemically crosslinked together suggest that such disassembly of the octamer is not necessary for transcriptional elongation through DNA packaged into nucleosomes (252).

LOCUS BOUNDARY (INSULATOR) ELEMENTS It has been hypothesized that genetic loci are, in some instances, separated by boundary elements that may correspond to sites of attachment of the DNA to the nuclear scaffold or matrix (253–258). These boundary elements are thought to be intrinsically neutral with regard to transcription, as they do not appear either to activate or to repress gene activity, and it has been postulated that these elements function to insulate the regulatory effects of neighboring genes from each other. Experiments carried out either with stably transformed genes in tissue culture cells (259, 260) or with transgenic *Drosophila* (260–262) or mice (263, 264) have characterized the transcriptional insulating effects of putative locus boundary elements in vivo. In addition, the region of the gypsy retrotransposon that is bound by the protein encoded by the *suppressor of Hairy-wing* gene in *Drosophila* appears to function as a locus boundary element (265–269). Studies of the nuclear scaffold/matrix have led to the identification of topoisomerase II as a major component of the scaffold/matrix (270–272). In addition, factors have been identified that interact with DNA elements known as scaffold-associated regions (SARs) or matrix-association regions (MARs) (273–275), which may correspond to sites of attachment of DNA to the nuclear scaffold/matrix. It has also been postulated that DNA unwinding occurs at sites of attachment to the scaffold/matrix (276). Furthermore, genetic studies in yeast have suggested that calmodulin may be involved in MAR function in vivo (277). The relationship between scaffold- or matrix-associated regions and locus boundary elements is not clear, however, and there is considerable controversy regarding the physiological relevance of biochemical preparations of the nuclear scaffold or matrix (see, for example, Refs. 255, 278, 279).

LOCUS CONTROL REGIONS Studies of the expression of the human β-globin gene cluster, which spans 70 kbp and contains five genes (in the order 5′ ε, Gγ, Aγ, δ, and β 3′), have led to the identification of a *cis*-acting transcriptional regulatory element known as the locus control region (LCR; for reviews, see Refs. 8, 280–284), which has been formerly referred to as either the dominant control region (DCR; Ref. 285) or the locus activating region (LAR; Ref. 286). The LCR is required for the tissue- and developmental-specificity of the transcription of the genes in the β-globin cluster. In stable transformation assays, the LCR confers position-independent (integration site–independent) and copy number–dependent expression (in many instances) to genes with which it is linked. The β-globin LCR corresponds to four or five DNase I hypersensitive sites that are distributed throughout a 15-kbp region that is located upstream (5′ side) of the ε-globin gene. The mechanism by which the LCR functions remains to be clarified. The LCR contains a plethora of recognition sites for sequence-specific DNA-binding transcription factors, such as GATA-1, NF-E2/AP-1, and Sp1 (or an Sp1-like factor). The simultaneous binding of these transcription factors may create a potent activating domain and significantly alter the structure of the chromatin template. It has been hypothesized that the LCR exerts a global influence upon the β-globin locus to allow increased access of transcription factors to the genes, although no direct evidence supports this theory. It is also possible that the LCR acts as an hyperactive version of a conventional transcriptional enhancer element. The human β-globin LCR has become the model with which the properties of an LCR are being defined. There are, however, LCR-like elements that have been identified in other genetic loci, and it will be important to illuminate the generality and mechanism of LCR function.

SUMMARY AND PERSPECTIVES

In conclusion, it is important to ask critically whether the alterations in chromatin structure that occur upon gene potentiation are required for transcriptional activation. This question reflects the central issue of this review—the role of chromatin structure in the regulation of gene activity. We attempt here to address this point from both "pro-chromatin" and "anti-chromatin" perspectives.

PRO-CHROMATIN Genetic studies in yeast have shown that modification of the core histones in vivo can affect transcriptional activity. In addition, numerous studies have shown that changes in chromatin structure either precede or accompany gene activation.

ANTI-CHROMATIN It is possible that under normal conditions with the wild-type histones, gene activation occurs without a requirement for alteration of chromatin structure. The mutant variants of the histones may simply interfere with the normal regulation of transcription. The observed changes in chromatin structure, such as nuclease hypersensitivity, may be due to the binding of transcription factors to DNA, and any consequent alterations in chromatin structure may not be intrinsically necessary for transcriptional activation.

PRO-CHROMATIN Studies performed in vivo in yeast have shown that alterations in chromatin structure that normally accompany gene activation also occur with transcriptionally inactive promoters that contain mutations in the TATA box motif, but require the action of proteins such as SNF5 or GAL4 (226, 287). Hence, active transcription is not required for changes in chromatin structure.

ANTI-CHROMATIN The observation that chromatin structure can be altered in the absence of transcription does not prove that the alterations are required for gene activation.

PRO-CHROMATIN Biochemical experiments have shown that promoter- and enhancer-binding factors can function to counteract chromatin-mediated repression. This effect requires the presence of a transcriptional activation region in the sequence-specific factor (167, 181). Moreover, binding of the sequence-specific activator GAL4-VP16 to the chromatin template is required but not sufficient to allow access of the basal transcriptional apparatus to the chromatin (180, 181), which indicates that the antirepression process is more complex than a simple removal of histones by the transcriptional activator.

ANTI-CHROMATIN The biochemical experiments do not provide an accurate model for the native state of chromatin in vivo. Aside from histones, many agents (proteins or chemicals) can inhibit transcription. The observed transcriptional repression is nonspecific, and has not been clearly proven to be relevant to the natural process of transcription in the nucleus.

This debate can be carried on indefinitely. The aim of this review is not to argue in favor of the postulate that chromatin structure is an important component of gene regulation, but rather to present current ideas and concepts regarding the relationship between chromatin structure and gene activity. Perhaps the most compelling reason for the consideration of chromatin structure in the analysis of gene regulation is the fact that chromatin is the natural state of DNA in vivo. The study of chromatin and transcription is presently at an early stage, and the next several years should be a fruitful period in which many of the strategies and mechanisms of gene regulation will be elucidated in the context of the native chromatin template.

ACKNOWLEDGMENTS

We thank EN Moudriankis for kindly providing the figure of the core histone octamer; M Bulger for the electron micrographs of chromatin; our many colleagues who have generously provided us with information prior to publication (with apologies to those whose work was not included due to space and subject limitations); M Grunstein, SY Roth, J Gottesfeld, Q Lu, SCR Elgin, AD Johnson, and C Wu for helpful suggestions; and EP Geiduschek, D Reinberg, B Zimm, EM Blackwood, T Burke, LA Kerrigan, M Bulger, M Pazin, and CP George for critical reading of the manuscript. We are also very grateful to D Frank and S Alignay for their invaluable help in organizing the information for this review. JTK is a Lucille P. Markey Scholar in the Biomedical Sciences and a Presidential Faculty Fellow. Studies in the Kadonaga laboratory have been supported by grants from the National Institutes of Health, the National Science Foundation, the Council for Tobacco Research, and the Lucille P. Markey Charitable Trust.

Literature Cited

1. van Holde KE. 1989. *Chromatin*. New York: Springer-Verlag. 497 pp.
2. Wolffe AP. *Chromatin: Structure and Function*. San Diego, CA: Academic. 213 pp.
3. Croston GE, Kadonaga JT. 1993. *Curr. Opin. Cell Biol.* 5:417–23
4. Adams CC, Workman JL. 1993. *Cell* 72:305–8
5. Svaren J, Hörz W. 1993. *Curr. Opin. Genet. Dev.* 3:219–25
6. Workman JL, Buchman AR. 1993. *Trends Biochem. Sci.* 18:90–95
7. Kornberg RD, Lorch Y. 1992. *Annu. Rev. Cell Biol.* 8:563–87
8. Felsenfeld G. 1992. *Nature* 355:219–23
9. Grunstein M. 1992. *Sci. Am.* 267:68–74B
10. Winston F, Carlson M. 1992. *Trends Genet.* 8:387–91
11. Wolffe AP. 1992. *FASEB J.* 6:3354–61
12. Hayes JJ, Wolffe AP. 1992. *BioEssays* 14:597–603
13. Wolffe AP. 1991. *Trends Cell Biol.* 1:61–66
14. Thoma F. 1991. *Trends Genet.* 7:175–77
15. Grunstein M. 1990. *Annu. Rev. Cell Biol.* 6:643–78
16. Grunstein M. 1990. *Trends Genet.* 6:395–400
17. Elgin SCR. 1990. *Curr. Opin. Cell Biol.* 2:437–45
18. Elgin SCR. 1988. *J. Biol. Chem.* 263:19259–62
19. Gross DS, Garrard WT. 1988. *Annu. Rev. Biochem.* 57:159–97
20. Weintraub H. 1985. *Cell* 42:705–11
21. Simpson RT. 1991. *Prog. Nucleic Acid Res. Mol. Biol.* 40:143–84
22. Fedor MJ. 1992. *Curr. Opin. Cell Biol.* 4:436–43
23. Laurenson P, Rine J. 1992. *Microbiol. Rev.* 56:543–60
24. Levine M, Manley JL. 1989. *Cell* 59:405–8
25. Drapkin R, Merino A, Reinberg D. 1993. *Curr. Opin. Cell Biol.* 5:469–76
26. Felsenfeld G, McGhee JD. 1986. *Cell* 44:375–77
27. Pederson DS, Thoma F, Simpson RT. 1986. *Annu. Rev. Cell Biol.* 2:117–47
28. Widom J. 1989. *Annu. Rev. Biophys. Biophys. Chem.* 18:365–95
29. Svaren J, Chalkley R. 1990. *Trends Genet.* 6:52–56
30. Daneholt B. 1992. *Cell Biol. Int. Rep.* 16:709–15
31. Noll M, Kornberg RD. 1977. *J. Mol. Biol.* 109:393–404
32. Simpson RT. 1978. *Biochemistry* 17:5524–31
33. Arents G, Burlingame RW, Wang BC, Love WE, Moudrianakis EN. 1991. *Proc. Natl. Acad. Sci. USA* 88:10148–52
34. Richmond TJ, Finch JT, Rushton B, Rhodes D, Klug A. 1984. *Nature* 311:532–37
35. Ruiz-Carrillo A, Jorcano JL. 1979. *Biochemistry* 18:760–68
36. Godfrey JE, Eickbush TH, Moudrianakis EN. 1980. *Biochemistry* 19:1339–46
37. Worcel A, Han S, Wong ML. 1978. *Cell* 15:969–77
38. Jackson V. 1987. *Biochemistry* 26:2315–25
39. Cole RD. 1987. *Int. J. Peptide Protein Res.* 30:433–49
40. Garrard WT. 1991. *BioEssays* 13:87–88
41. Zlatanova J. 1990. *Trends Biochem. Sci.* 15:273–76
42. Zlatanova J, van Holde KE. 1992. *J. Cell Sci.* 103:889–95
43. Bates DL, Thomas JO. 1981. *Nucleic Acids Res.* 9:5883–94
44. Ramakrishnan V, Finch JT, Graziano V, Lee PL, Sweet RM. 1993. *Nature* 362:219–23
45. Clore GM, Gronenborn AM, Nilges M, Sukumaran DK, Zarbock J. 1987. *EMBO J.* 6:1833–42
46. Allan J, Hartman PG, Crane-Robinson C, Aviles FX. 1980. *Nature* 288:675–79
47. Staynov DZ, Crane-Robinson C. 1988. *EMBO J.* 7:3685–91

47a. Hayes JJ, Wolffe AP. 1993. *Proc. Natl. Acad. Sci. USA* 90:6415–19

48. Renz M, Day LA. 1976. *Biochemistry* 15:3220–28
49. Clark DJ, Thomas JO. 1986. *J. Mol. Biol.* 187:569–80
50. De Bernardin W, Losa R, Koller T. 1986. *J. Mol. Biol.* 189:503–17
51. Lennard AC, Thomas JO. 1985. *EMBO J.* 4:3455–62
52. Hansen JC, Ausio J, Stanik VH, van Holde KE. 1989. *Biochemistry* 28:9129–36
53. Butler PJG, Thomas JO. 1980. *J. Mol. Biol.* 140:505–29
54. Thoma F, Koller T, Klug A. 1979. *J. Cell Biol.* 83:403–27
55. Finch JT, Klug A. 1976. *Proc. Natl. Acad. Sci. USA* 73:1897–1901
56. Renz M, Nehls P, Hozier J. 1977. *Proc. Natl. Acad. Sci. USA* 74:1879–83

57. Diffley JFX, Stillman B. 1992. *J. Biol. Chem.* 267:3368–74
58. Megraw TL, Chae CB. 1993. *J. Biol. Chem.* 268:12758–63
59. Landsman D. 1993. *Nature* 363:590
60. Landsman D, Bustin M. 1993. *BioEssays* 15:539–46
61. Thomas JO, Furber V. 1976. *FEBS Lett.* 66:274–80
62. McGhee JD, Nickol JM, Felsenfeld G, Rau DC. 1983. *Cell* 33:831–41
63. Williams SP, Athey BD, Muglia LJ, Schappe RS, Gough AH, Langmore JP. 1986. *Biophys. J.* 49:233–48
64. Worcel A, Strogatz S, Riley D. 1981. *Proc. Natl. Acad. Sci. USA* 78:1461–65
65. Woodcock CLF, Frado LLY, Rattner JB. 1984. *J. Cell Biol.* 99:42–52
66. Andersson K, Mähr R, Björkroth B, Daneholt B. 1982. *Chromosoma* 87:33–48
67. Andersson K, Björkroth B, Daneholt B. 1984. *J. Cell Biol.* 98:1296–1303
68. Turner BM. 1991. *J. Cell Sci.* 99:13–20
69. Hebbes TR, Thorne AW, Crane-Robinson C. 1988. *EMBO J.* 7:1395–1402
70. Tazi J, Bird A. 1990. *Cell* 60:909–20
71. Bresnick EH, John S, Berard DS, LeFebvre P, Hager GL. 1990. *Proc. Natl. Acad. Sci. USA* 87:3977–81
72. Bresnick EH, John S, Hager GL. 1991. *Biochemistry* 30:3490–97
73. Turner BM, Birley AJ, Lavender J. 1992. *Cell* 69:375–84
74. Jeppesen P, Turner BM. 1993. *Cell* 74:281–89
75. Lee DY, Hayes JJ, Pruss D, Wolffe AP. 1993. *Cell* 72:73–84
76. Durrin LK, Mann RK, Kayne PS, Grunstein M. 1991. *Cell* 65:1023–31
77. Durrin LK, Mann RK, Grunstein M. 1992. *Mol. Cell. Biol.* 12:1621–29
78. Johnson LM, Kayne PS, Kahn ES, Grunstein M. 1990. *Proc. Natl. Acad. Sci. USA* 87:6286–90
79. Johnson LM, Fisher-Adams G, Grunstein M. 1992. *EMBO J.* 11:2201–9
80. Johns EW, ed. 1982. *The HMG Chromosomal Proteins.* New York: Academic. 251 pp.
81. Bustin M, Lehn DA, Landsman D. 1990. *Biochim. Biophys. Acta* 1049:231–43
82. Lilley DMJ. 1992. *Nature* 357:282–83
83. Reeck GR, Isackson PJ, Teller DC. 1982. *Nature* 300:76–78
84. Cary PD, Turner CH, Mayes E, Crane-Robinson C. 1983. *Eur. J. Biochem.* 131:367–74
85. Kolodrubetz D. 1990. *Nucleic Acids Res.* 18:5565
86. Jantzen HM, Admon A, Bell SP, Tjian R. 1990. *Nature* 344:830–36
87. Sinclair AH, Berta P, Palmer MS, Hawkins JR, Griffiths BL, et al. 1990. *Nature* 346:240–44
88. Gubbay J, Collignon J, Koopman P, Capel B, Economou A, Münsterberg A, Vivian N, Goodfellow P, Lovell-Badge R. 1990. *Nature* 346:245–50
89. Weir HM, Kraulis PJ, Hill CS, Raine ARC, Laue ED, Thomas JO. 1993. *EMBO J.* 12:1311–19
90. Bianchi ME, Beltrame M, Paonessa G. 1989. *Science* 243:1056–59
91. Isackson PJ, Fishback JL, Bidney DL, Reeck GR. 1979. *J. Biol. Chem.* 254:5569–72
92. Bonne C, Sautiere P, Duguet M, de Recondo AM. 1982. *J. Biol. Chem.* 257:2722–25
93. Pil PM, Lippard SJ. 1992. *Science* 256:234–37
94. Hughes EN, Engelsberg BN, Billings PC. 1992. *J. Biol. Chem.* 267:13520–27
95. Paull TT, Haykinson MJ, Johnson RC. 1993. *Genes Dev.* 7:1521–34
96. Stoute JA, Marzluff WF. 1982. *Biochem. Biophys. Res. Commun.* 107:1279–84
97. Tremethick DJ, Molloy PL. 1986. *J. Biol. Chem.* 261:6986–92
98. Tremethick DJ, Molloy PL. 1988. *Nucleic Acids Res.* 16:11107–23
99. Watt F, Molloy PL. 1988. *Nucleic Acids Res.* 16:1471–86
100. Kuehl L, Salmond B, Tran L. 1984. *J. Cell Biol.* 99:648–54
101. Kolodrubetz D, Burgum A. 1990. *J. Biol. Chem.* 265:3234–39
102. Wagner CR, Hamana K, Elgin SCR. 1992. *Mol. Cell. Biol.* 12:1915–23
103. Crippa MP, Alfonso PJ, Bustin M. 1992. *J. Mol. Biol.* 228:442–49
104. Sandeen G, Wood WI, Felsenfeld G. 1980. *Nucleic Acids Res.* 8:3757–78
105. Mardian JKW, Paton AE, Bunick GJ, Olins DE. 1980. *Science* 209:1534–36
106. Tremethick DJ, Drew HR. 1993. *J. Biol. Chem.* 268:11389–93
107. Drew HR. 1993. *J. Mol. Biol.* 230:824–36
108. Isackson PJ, Reeck GR. 1981. *Nucleic Acids Res.* 9:3779–91
109. Dorbic T, Wittig B. 1987. *EMBO J.* 6:2393–99
110. Postnikov YV, Shick VV, Belyavsky AV, Khrapko KR, Brodolin KL, et al. 1991. *Nucleic Acids Res.* 19:717–25
111. Johnson KR, Lehn DA, Elton TS, Barr PJ, Reeves R. 1988. *J. Biol. Chem.* 263:18338–42

112. Johnson KR, Lehn DA, Reeves R. 1989. *Mol. Cell. Biol.* 9:2114–23
113. Strauss F, Varshavsky A. 1984. *Cell* 37:889–901
114. Solomon MJ, Strauss F, Varshavsky A. 1986. *Proc. Natl. Acad. Sci. USA* 83:1276–80
115. Disney JE, Johnson KR, Magnuson NS, Sylvester SR, Reeves R. 1989. *J. Cell Biol.* 109:1975–82
116. Fashena SJ, Reeves R, Ruddle NH. 1992. *Mol. Cell. Biol.* 12:894–903
117. Thanos D, Maniatis T. 1992. *Cell* 71:777–89
118. Clark-Adams CD, Norris D, Osley MA, Fassler JS, Winston F. 1988. *Genes Dev.* 2:150–59
119. Norris D, Dunn B, Osley MA. 1988. *Science* 242:759–61
120. Prelich G, Winston F. 1993. *Genetics* 135:665–76
121. Sternberg PW, Stern MJ, Clark I, Herskowitz I. 1987. *Cell* 48:567–77
122. Kruger W, Herskowitz I. 1991. *Mol. Cell. Biol.* 11:4135–46
123. Kim UJ, Han M, Kayne P, Grunstein M. 1988. *EMBO J.* 7:2211–19
124. Han M, Kim UJ, Kayne P, Grunstein M. 1988. *EMBO J.* 7:2221–28
125. Han M, Grunstein M. 1988. *Cell* 55:1137–45
126. Mann RK, Grunstein M. 1992. *EMBO J.* 11:3297–3306
127. Kayne PS, Kim UJ, Han M, Mullen JR, Yoshizaki F, Grunstein M. 1988. *Cell* 55:27–39
128. Thoma F. 1992. *Biochim. Biophys. Acta* 1130:1–19
129. Noll M. 1974. *Nucleic Acids Res.* 1:1573–78
129a. Lutter LC. 1989. *Methods Enzymol.* 170:264–69
130. Kerrigan LA, Kadonaga JT. 1992. *Nucleic Acids Res.* 20:6673–80
131. Travers AA. 1987. *Trends Biochem. Sci.* 12:108–12
132. Satchwell SC, Drew HR, Travers AA. 1986. *J. Mol. Biol.* 191:659–75
133. Satchwell SC, Travers AA. 1989. *EMBO J.* 8:229–38
134. Shrader TE, Crothers DM. 1989. *Proc. Natl. Acad. Sci. USA* 86:7418–22
135. Shrader TE, Crothers DM. 1990. *J. Mol. Biol.* 216:69–84
136. Wolffe AP, Drew H. 1989. *Proc. Natl. Acad. Sci. USA* 86:9817–21
137. Wu C. 1980. *Nature* 286:854–60
138. Nedospasov SA, Georgiev GP. 1980. *Biochem. Biophys. Res. Commun.* 92: 532–39
139. Wu C. 1989. *Methods Enzymol.* 170: 269–89
140. Nedospasov SA, Shakhov AN, Georgiev GP. 1989. *Methods Enzymol.* 170:408–20
141. Mueller PR, Wold B. 1989. *Science* 246:780–86
142. Garrity PA, Wold B. 1992. *Proc. Natl. Acad. Sci. USA* 89:1021–25
143. Pfeifer GP, Riggs AD. 1991. *Genes Dev.* 5:1102–13
144. Thoma F, Bergman LW, Simpson RT. 1984. *J. Mol. Biol.* 177:715–33
145. Benezra R, Cantor CR, Axel R. 1986. *Cell* 44:697–704
146. Richard-Foy H, Hager GL. 1987. *EMBO J.* 6:2321–28
147. Thomas GH, Elgin SCR. 1988. *EMBO J.* 7:2191–2201
148. Almer A, Hörz W. 1986. *EMBO J.* 5:2681–87
149. Almer A, Rudolph H, Hinnen A, Hörz W. 1986. *EMBO J.* 5:2689–96
150. Simpson RT, Stafford DW. 1983. *Proc. Natl. Acad. Sci. USA* 80:51–55
151. Chao MV, Gralla JD, Martinson HG. 1979. *Biochemistry* 18:1068–74
152. Ramsay N, Felsenfeld G, Rushton BM, McGhee JD. 1984. *EMBO J.* 3:2605–11
153. Perlmann T, Wrange Ö. 1988. *EMBO J.* 7:3073–79
154. Piña B, Brüggemeier U, Beato M. 1990. *Cell* 60:719–31
155. Archer TK, Cordingley MG, Wolford RG, Hager GL. 1991. *Mol. Cell. Biol.* 11:688–98
156. Thoma F, Simpson RT. 1985. *Nature* 315:250–52
157. Thoma F, Zatchej M. 1988. *Cell* 55: 945–53
158. Simpson RT. 1990. *Nature* 343:387–89
159. Roth SY, Dean A, Simpson RT. 1990. *Mol. Cell. Biol.* 10:2247–60
160. Roth SY, Shimizu M, Johnson L, Grunstein M, Simpson RT. 1992. *Genes Dev.* 6:411–25
161. Shimizu M, Roth SY, Szent-Gyorgyi C, Simpson RT. 1991. *EMBO J.* 10: 3033–41
162. Fedor MJ, Kornberg RD. 1989. *Mol. Cell. Biol.* 9:1721–32
163. Fedor MJ, Lue NF, Kornberg RD. 1988. *J. Mol. Biol.* 204:109–27
164. Chasman DI, Lue NF, Buchman AR, LaPointe JW, Lorch Y, Kornberg RD. 1990. *Genes Dev.* 4:503–14
165. Herschbach BM, Johnson AD. 1993. *Mol. Cell. Biol.* 13:4029–38
166. Laybourn PJ, Kadonaga JT. 1991. *Science* 254:238–45
167. Workman JL, Taylor ICA, Kingston RE. 1991. *Cell* 64:533–44
168. Croston GE, Kerrigan LA, Lira L, Marshak DR, Kadonaga JT. 1991. *Science* 251:643–49

169. Laybourn PJ, Kadonaga JT. 1992. *Science* 257:1682–85
170. Kamakaka RT, Bulger M, Kadonaga JT. 1993. *Genes Dev.* 7:1779–95
171. Schild C, Claret FX, Wahli W, Wolffe AP. 1993. *EMBO J.* 12:423–33
172. Knezetic JA, Luse DS. 1986. *Cell* 45:95–104
173. Matsui T. 1987. *Mol. Cell. Biol.* 7:1401–8
174. Lorch Y, LaPointe JW, Kornberg RD. 1987. *Cell* 49:203–10
175. Workman JL, Roeder RG. 1987. *Cell* 51:613–22
176. Workman JL, Abmayr SM, Cromlish WA, Roeder RG. 1988. *Cell* 55:211–19
177. Workman JL, Roeder RG, Kingston RE. 1990. *EMBO J.* 9:1299–1308
178. Meisterernst M, Horikoshi M, Roeder RG. 1990. *Proc. Natl. Acad. Sci. USA* 87:9153–57
179. Corthésy B, Léonnard P, Wahli W. 1990. *Mol. Cell. Biol.* 10:3926–33
180. Lorch Y, LaPointe JW, Kornberg RD. 1992. *Genes Dev.* 6:2282–87
181. Croston GE, Laybourn PJ, Paranjape SM, Kadonaga JT. 1992. *Genes Dev.* 6:2270–81
182. Becker PB, Rabindran SK, Wu C. 1991. *Proc. Natl. Acad. Sci. USA* 88:4109–13
183. Barton MC, Madani N, Emerson BM. 1993. *Genes Dev.* 7:1796–1809
184. Batson SC, Sundseth R, Heath CV, Samuels M, Hansen U. 1992. *Mol. Cell. Biol.* 12:1639–51
185. Batson SC, Rimsky S, Sundseth R, Hansen U. 1993. *Nucleic Acids Res.* 21:3459–68
186. Kamakaka RT, Kadonaga JT. 1993. *Cold Spring Harbor Symp. Quant. Biol.* 58:In press
187. Majumder S, Miranda M, DePamphilis ML. 1993. *EMBO J.* 12:1131–40
188. Carey M, Leatherwood J, Ptashne M. 1990. *Science* 247:710–12
189. Nacheva BA, Guschin DY, Preobrazhenskaya OV, Karpov VL, Ebralidse KK, Mirzabekov AD. 1989. *Cell* 58:27–36
190. Kamakaka RT, Thomas JO. 1990. *EMBO J.* 9:3997–4006
191. Bresnick EH, Bustin M, Marsaud V, Richard-Foy H, Hager GL. 1992. *Nucleic Acids Res.* 20:273–78
192. Kuhn A, Grummt I. 1992. *Proc. Natl. Acad. Sci. USA* 89:7340–44
193. Croston GE, Lira LM, Kadonaga JT. 1991. *Protein Expr. Purif.* 2:162–69
194. Dusserre Y, Mermod N. 1992. *Mol. Cell. Biol.* 12:5228–37
195. Pan D, Huang JD, Courey AJ. 1991. *Genes Dev.* 5:1892–1901
196. Eriksson P, Wrange Ö. 1993. *Eur. J. Biochem.* 215:505–11
197. Kerrigan LA, Croston GE, Lira LM, Kadonaga JT. 1991. *J. Biol. Chem.* 266:574–82
198. Klucher KM, Sommer M, Kadonaga JT, Spector DH. 1993. *Mol. Cell. Biol.* 13:1238–50
199. Shimamura A, Sapp M, Rodriguez-Campos A, Worcel A. 1989. *Mol. Cell. Biol.* 9:5573–84
200. Schlissel MS, Brown DD. 1984. *Cell* 37:903–13
201. Wolffe AP. 1989. *EMBO J.* 8:527–37
202. Ericsson C, Grossbach U, Björkroth B, Daneholt B. 1990. *Cell* 60:73–83
203. Dignam JD, Lebovitz RM, Roeder RG. 1983. *Nucleic Acids Res.* 11:1475–89
204. Kamakaka RT, Tyree CM, Kadonaga JT. 1991. *Proc. Natl. Acad. Sci. USA* 88:1024–28
205. Weintraub H, Groudine M. 1976. *Science* 193:848–56
206. McGhee JD, Wood WI, Dolan M, Engel JD, Felsenfeld G. 1981. *Cell* 27:45–55
207. Solomon MJ, Larsen PL, Varshavsky A. 1988. *Cell* 53:937–47
208. Taylor ICA, Workman JL, Schuetz TJ, Kingston RE. 1991. *Genes Dev.* 5:1285–98
209. Workman JL, Kingston RE. 1992. *Science* 258:1780–84
210. Allfrey VG, Chen TA. 1991. *Methods Cell Biol.* 35:315–35
211. Allegra P, Sterner R, Clayton DF, Allfrey VG. 1987. *J. Mol. Biol.* 196:379–88
212. Chen TA, Allfrey VG. 1987. *Proc. Natl. Acad. Sci. USA* 84:5252–56
213. Chen TA, Sterner R, Cozzolino A, Allfrey VG. 1990. *J. Mol. Biol.* 212:481–93
214. Walker J, Chen TA, Sterner R, Berger M, Winston F, Allfrey VG. 1990. *J. Biol. Chem.* 265:5736–46
215. Chen TA, Smith MM, Le S, Sternglanz R, Allfrey VG. 1991. *J. Biol. Chem.* 266:6489–98
216. Chen-Cleland TA, Smith MM, Le S, Sternglanz R, Allfrey VG. 1993. *J. Biol. Chem.* 268:1118–24
217. Sun YL, Xu YZ, Bellard M, Chambon P. 1986. *EMBO J.* 5:293–300
218. Lee MS, Garrard WT. 1991. *EMBO J.* 10:607–15
219. Weintraub H. 1979. *Nucleic Acids Res.* 7:781–92
220. Brown DD. 1984. *Cell* 37:359–65
221. Wolffe AP. 1991. *J. Cell Sci.* 99:201–6

222. Almouzni G, Wolffe AP. 1993. *Exp. Cell Res.* 205:1–15
223. Solomon MJ, Varshavsky A. 1987. *Mol. Cell. Biol.* 7:3822–25
224. Almouzni G, Méchali M, Wolffe AP. 1990. *EMBO J.* 9:573–82
225. Schmid A, Fascher KD, Hörz W. 1992. *Cell* 71:853–64
226. Hirschhorn JN, Brown SA, Clark CD, Winston F. 1992. *Genes Dev.* 6:2288–98
227. Lis J, Wu C. 1993. *Cell* 73:1–4
228. Rougvie AE, Lis JT. 1988. *Cell* 54:795–804
229. Biggin MD, Tjian R. 1988. *Cell* 53:699–711
230. Gilmour DS, Thomas GH, Elgin SCR. 1989. *Science* 245:1487–90
231. Lu Q, Wallrath LL, Allan BD, Glaser RL, Lis JT, Elgin SCR. 1992. *J. Mol. Biol.* 225:985–98
232. Lu Q, Wallrath LL, Granok H, Elgin SCR. 1993. *Mol. Cell. Biol.* 13:2802–14
233. Lee H, Kraus KW, Wolfner MF, Lis JT. 1992. *Genes Dev.* 6:284–95
234. Germond JE, Hirt B, Oudet P, Gross-Bellard M, Chambon P. 1975. *Proc. Natl. Acad. Sci. USA* 72:1843–47
235. Simpson RT, Thoma F, Brubaker JM. 1985. *Cell* 42:799–808
236. Peck LJ, Wang JC. 1983. *Proc. Natl. Acad. Sci. USA* 80:6206–10
237. Camerini-Otero RD, Felsenfeld G. 1977. *Nucleic Acids Res.* 4:1159–81
238. Zweidler A. 1978. *Methods Cell Biol.* 17:223–33
239. Varshavsky AJ, Bakaye VV, Georgiev GP. 1976. *Nucleic Acids Res.* 3:477–92
240. Nelson PP, Albright SC, Wiseman JM, Garrard WT. 1979. *J. Biol. Chem.* 254:11751–60
241. Spencer CA, Groudine M. 1990. *Oncogene* 5:777–85
242. Kornberg RD, Lorch Y. 1991. *Cell* 67:833–36
243. van Holde KE, Lohr D, Robert C. 1992. *J. Biol. Chem.* 267:2837–40
244. Lorch Y, LaPointe JW, Kornberg RD. 1988. *Cell* 55:743–44
245. Izban MG, Luse DS. 1991. *Genes Dev.* 5:683–96
246. Izban MG, Luse DS. 1992. *J. Biol. Chem.* 267:13647–55
247. Losa R, Brown DD. 1987. *Cell* 50:801–8
248. Pfaffle P, Gerlach V, Bunzel L, Jackson V. 1990. *J. Biol. Chem.* 265:16830–40
249. O'Neill TE, Roberge M, Bradbury EM. 1992. *J. Mol. Biol.* 223:67–78
250. Liu LF, Wang JC. 1987. *Proc. Natl. Acad. Sci. USA* 84:7024–27
251. Clark DJ, Felsenfeld G. 1992. *Cell* 71:11–22
252. O'Neill TE, Smith JG, Bradbury EM. 1993. *Proc. Natl. Acad. Sci. USA* 90:6203–7
253. Gasser SM, Laemmli UK. 1987. *Trends Genet.* 3:16–22
254. Laemmli UK, Käs E, Poljak L, Adachi Y. 1992. *Curr. Opin. Genet. Dev.* 2:275–85
255. Jackson DA, Dolle A, Robertson G, Cook PR. 1992. *Cell Biol. Int. Rep.* 16:687–96
256. Mirkovitch J, Mirault ME, Laemmli UK. 1984. *Cell* 39:223–32
257. Gasser SM, Laemmli UK. 1986. *Cell* 46:521–30
258. Cockerill PN, Garrard WT. 1986. *Cell* 44:273–82
259. Stief A, Winter DM, Strätling WH, Sippel AE. 1989. *Nature* 341:343–45
260. Chung JH, Whiteley M, Felsenfeld G. 1993. *Cell* 74:505–14
261. Kellum R, Schedl P. 1991. *Cell* 64:941–50
262. Kellum R, Schedl P. 1992. *Mol. Cell. Biol.* 12:2424–31
263. Bonifer C, Vidal M, Grosveld F, Sippel AE. 1990. *EMBO J.* 9:2843–48
264. McKnight RA, Shamay A, Sankaran L, Wall RJ, Hennighausen L. 1992. *Proc. Natl. Acad. Sci. USA* 89:6943–47
265. Corces VG, Geyer PK. 1991. *Trends Genet.* 7:86–90
266. Geyer PK, Corces VG. 1992. *Genes Dev.* 6:1865–73
267. Roseman RR, Pirrotta V, Geyer PK. 1993. *EMBO J.* 12:435–42
268. Jack J, Dorsett D, Delotto Y, Liu S. 1991. *Development* 113:735–47
269. Dorsett D. 1993. *Genetics* 134:1135–44
270. Earnshaw WC, Heck MMS. 1985. *J. Cell Biol.* 100:1716–25
271. Gasser SM, Laroche T, Falquet J, Boy de la Tour E, Laemmli UK. 1986. *J. Mol. Biol.* 188:613–29
272. Berrios M, Osheroff N, Fisher PA. 1985. *Proc. Natl. Acad. Sci. USA* 82:4142–46
273. von Kries JP, Buhrmester H, Strätling WH. 1991. *Cell* 64:123–35
274. Tsutsui Ken, Tsutsui Kimiko, Okada S, Watarai S, Seki S, Yasuda T, Shohmori T. 1993. *J. Biol. Chem.* 268:12886–94
275. Dickinson LA, Joh T, Kohwi Y, Kohwi-Shigematsu T. 1992. *Cell* 70:631–45
276. Bode J, Kohwi Y, Dickinson L, Joh T, Klehr D, et al. 1992. *Science* 255:195–97
277. Fishel BR, Sperry AO, Garrard WT.

1993. *Proc. Natl. Acad. Sci. USA* 90:5623–27
278. Eggert H, Jack RS. 1991. *EMBO J.* 10:1237–43
279. Jackson DA, Dickinson P, Cook PR. 1990. *Nucleic Acids Res.* 18:4385–93
280. Orkin SH. 1990. *Cell* 63:665–72
281. Townes TM, Behringer RR. 1990. *Trends Genet.* 6:219–23
282. Stamatoyannopoulos G. 1991. *Science* 252:383
283. Grosveld F, Antoniou M, Berry M, De Boer E, Dillon N, et al. 1993. *Philos. Trans. R. Soc. London Ser. B* 339:183–91
284. Crossley M, Orkin SH. 1993. *Curr. Opin. Genet. Dev.* 3:232–37
285. Grosveld F, van Assendelft GB, Greaves DR, Kollias G. 1987. *Cell* 51:975–85
286. Forrester WC, Takegawa S, Papayannopoulou T, Stamatoyannopoulos G, Groudine M. 1987. *Nucleic Acids Res.* 15:10159–77
287. Axelrod JD, Reagan MS, Majors J. 1993. *Genes Dev.* 7:857–69
288. Cereghini S, Yaniv M. 1984. *EMBO J.* 3:1243–53
289. Reeves R. 1985. *Nucleic Acids Res.* 13:3599–615

Annu. Rev. Biochem. 1994. 63:299–344

QUINOENZYMES IN BIOLOGY[1]

Judith P. Klinman and David Mu
Department of Chemistry, University of California, Berkeley, California 94720

KEY WORDS: quinone cofactors, pyrroloquinoline quinone, topa quinone, tryptophan tryptophylquinone, copper amine oxidases

---

CONTENTS

[1]Abbreviations used: AO, amine oxidase; SAO, serum amine oxidase; GDH, glucose dehydrogenase; MDH, methanol dehydrogenase; MADH, methylamine dehydrogenase; HPLC, high-pressure liquid chromatography; NMR, nuclear magnetic resonance; EPR, electron paramagnetic resonance; EXAFS, extended X-ray absorption fine structure; ENDOR, electron nuclear double resonance; ESEEM, electron spin echo envelope modulation; PQQ, pyrroloquinoline quinone; TPQ, topa quinone; TTQ, tryptophan tryptophylquinone; amu, atomic mass units; NOESY, nuclear Overhauser and exchange spectroscopy; MCD, magnetic circular dichroism; bpy, 2,2′-bipyridine; SCE, saturated calomel electrode.

0066-4154/94/0701-0299$05.00

1. INTRODUCTION

Although quinones have been recognized as important electron-transfer agents in biology, they have not been viewed as important redox cofactors. As we describe in this chapter, the past three years has seen the emergence of a new field, referred to as quinoenzymes. The story begins in the early 1980s, with the description of pyrroloquinoline quinone as a dissociable redox cofactor in gram-negative bacteria. This was followed by the erroneous attribution of a role for a covalently bound form of pyrroloquinoline quinone in eukaryotes. The intimation of a wider role for quino-cofactors was, however, correct: As reviewed herein, two new cofactors—tryptophan tryptophylquinone and topa quinone—have been recently demonstrated in prokaryotic and eukaryotic enzymes. An excellent book, *Principles and Applications of Quinoproteins,* has been published (1). The purpose of this review is to clarify the status of each of these new quino-cofactors, as well as to summarize the recent exciting findings regarding their structure and function.

2. CLARIFICATION OF THE STATUS OF PYRROLOQUINOLINE QUINONE (PQQ) AS A REDOX COFACTOR

PQQ Does Not Function as a Covalently Bound Redox Cofactor in Eukaryotic Enzymes

The discovery of new cofactors in eukaryotes is a rare occurrence in modern biochemistry. The initial claim of a role for PQQ (Figure 1) in the mammalian copper amine oxidases was therefore greeted with a great deal of excitement and interest. Though under investigation for several decades, the structure of the active-site redox cofactor in these proteins had completely eluded detection. The ability of copper amine oxidases (AOs) to undergo derivatization with carbonyl reagents such as phenylhydrazine had led early on to the conjecture of a role for pyridoxal phosphate (2–4). However, unlike other pyridoxal phosphate–requiring enzymes, the cofactor in copper AOs is not released upon protein denaturation, and further, the chemistry

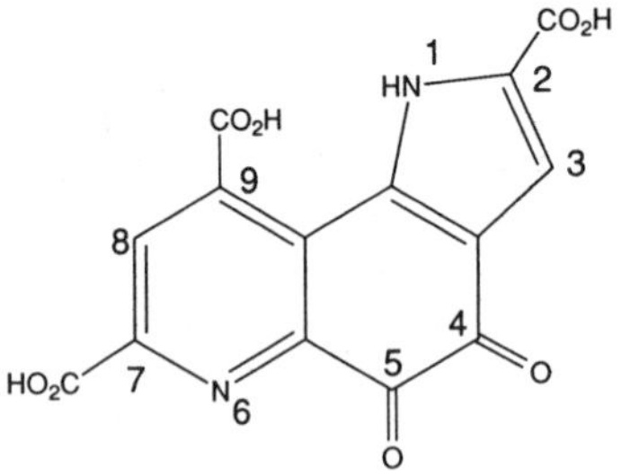

Figure 1 Structure for pyrroloquinoline quinone (PQQ).

of these proteins fails to conform to properties predicted from simple pyridoxal models. Prior to 1984, a number of alternatives to pyridoxal phosphate in copper AOs had been proposed, including a modified flavin (5) and a redox-active cysteine connected to an unknown residue (6, 7).

The announcement in 1984 of covalently bound PQQ in bovine serum amine oxidase (SAO) appeared both to resolve a decades-long controversy and to provide a structural basis for the chemical and spectroscopic properties of the copper AOs (8, 9). It should be emphasized that the evidence in support of covalently bound PQQ was, from the outset, quite weak, involving either a fluorescence characterization of intact protein (9) or the isolation of a dinitrophenylhydrazone derivative from proteolyzed protein, which displayed the same retention time on HPLC as an authentic derivative of PQQ (8). Resonance Raman spectroscopy subsequently showed that the active-site cofactor in copper AOs had properties much closer to PQQ than pyridoxal phosphate (10, 11), while reductive trapping experiments could be rationalized from a quino-structure (12). As is described in greater detail below, we now know that the active-site cofactor in the copper AOs is not PQQ, but rather a trihydroxyphenylalanine contained within the protein polypeptide chain. It is of considerable interest that trihydroxyphenylalanine is a redox-active compound undergoing oxidation in air to a hydroxy-quinone. This provides a retrospective explanation for the interpretations of resonance Raman studies (10) and reductive trapping experiments (12) in terms of a quino-structure.

Subsequent to the initial claim of PQQ in bovine SAO (8, 9), a large number of eukaryotic proteins were erroneously reported to contain covalently bound PQQ (Table 1). The current status of each of the proteins in Table 1 is as follows. All of the copper AOs (which includes the serum AOs, diamine oxidases, and the plant AOs) contain an alternate quino-cofactor, within the protein backbone and designated topa quinone (TPQ, see below). Lysyl oxidase, while it is formally assigned to the class of copper AOs (24) and shows properties of a quinoprotein (25), is characterized by significant difference in mass, sequence (26, 27), and spectroscopic properties (17) relative to other proteins of this class. Although it is tempting to speculate that as a copper AO, lysyl oxidase contains TPQ, the identity of its protein-bound cofactor must await the isolation and characterization of an active-site, cofactor-containing peptide. X-ray diffraction studies have led to a recent structure for galactose oxidase (28), showing a unique organic structure, consisting of a tyrosyl radical (29) covalently linked to cysteine at C-3 of the tyrosyl ring; this structure is in close proximity to the active-site copper and has been ascribed a role in catalysis (30). The crystal structure of lipoxygenase has been determined independently by two groups (31, 32): The single-iron center is reported to be tetra-coordinated in one case (31) and penta-coordinated in another (32); there is no evidence for unusual amino-acid side chains or bound cofactor. Although an

Table 1 Eukaryotic proteins claimed to contain covalently bound PQQ

Protein	Source	Evidence	Ref(s).
Serum amine oxidase	Bovine	HPLC retention time of dinitrophenylhydrazone.	8
		Resonance Raman.	10, 11
		Fluorescence spectrum and stimulation of bacterial growth by acid hydrolysate.	9
	Porcine	Resonance Raman spectrum.	11
Diamine oxidase	Porcine kidney	HPLC retention time and proton NMR spectrum of dinitrophenylhydrazone adduct isolated at low yield.	13
		Fluorescence spectrum.	9
Seedling amine oxidase	Pea	HPLC retention time of native cofactor from enzyme hydrolysate and reactivity with carbonyl reagents. Fluorescence spectrum.	14
	Lentil	Immunologic detection with antibody raised against PQQ.	15
Lysyl oxidase	Bovine aorta	HPLC retention time of dinitrophenylhydrazone.	16
		Resonance Raman.	17
Dopamine β-monooxygenase	Bovine adrenal gland	HPLC retention time of phenylhydrazone and optical spectrum.	18
Galactose oxidase	Fungal	HPLC analysis of the dihexyl ketal derivative extracted with hexanol in conc HCl and optical spectrum.	19
Laccase	Fungal	As above.	20
Dopa decarboxylase	Porcine kidney	As above.	21
Lipoxygenase	Soybean	Optical absorption of enzyme phenylhydrazone.	22
Choline dehydrogenase	Dog kidney	Chromophore from direct acid hydrolysate activates apo-glucose dehydrogenase.	23

Table 2 Summary of evidence against covalently bound PQQ in eukaryotic proteins

Protein	Source	Evidence	Refs.
Serum amine oxidases	Bovine	Mass spectrometric studies, NMR, optical absorbance properties.	39
		Resonance Raman.	40
		Resonance Raman and absorbance of nitrophenylhydrazone.	41
Diamine oxidase	Porcine kidney	Resonance Raman and absorbance of nitrophenylhydrazone.	41
Seedling amine oxidases	Pea	Same as above.	41
	Lentil	Sequence homology with copper amine oxidases.	42
		EPR and absorbance spectra.	43
Dopamine β-monooxygenase	Bovine adrenal gland	Absorbance properties, amino-acid sequencing.	36, 37, 37a
		Negative redox-cycling assay.	44
Galactose oxidase	Fungal	X-ray crystallography.	28
		Negative redox-cycling assay.	45
Laccase	Fungal	Sequence homology to ascorbate oxidase.	33
		Negative redox-cycling assay.	45
		MCD and EPR studies.	46
Dopa decarboxylase	Porcine kidney	Absorbance properties.	38
		Ability to reconstitute with pyridoxal phosphate.	38
Lipoxygenase	Soybean	X-ray crystallography.	31, 32
		Negative redox-cycling assay.	44

X-ray structure is not available for laccase, sequence homology to ascorbate oxidase (33), a protein whose structure has been refined to 1.9 Å (34), argues that the cofactors in fungal laccase are limited to the three copper centers. Dopamine β-monooxygenase is known to be inhibited by phenylhydrazine (35); however, this inhibition occurs at the copper center and does not lead to the expected absorbance spectrum of a PQQ phenylhydrazone derivative (36). Additionally, amino-acid sequencing of this large protein fails to provide any evidence for posttranslational processing to generate a covalently bound organic cofactor (37, 37a). Finally, for dopa decarboxylase, absorbance spectra argue against the presence of a PQQ-like structure, as does the ability to restore full activity to apo-enzyme with pyridoxal phosphate (38). Consistent with these properties, expression of recombinant rat liver dopa decarboxylase in a strain of *Escherichia coli* lacking PQQ gave rise to protein with the expected spectral and kinetic properties (38). Significantly, none of the proteins in Table

1 can now be concluded to contain PQQ as an active-site cofactor. The evidence against PQQ in each case is summarized in Table 2.

In light of the failure to confirm a role for PQQ in a single eukaryotic protein, the question arises: What was wrong with the assay conditions used to demonstrate covalently bound PQQ? One important method used to identify PQQ involved derivatization of native protein with the appropriate phenylhydrazine, followed by extensive proteolysis and elution of a product from HPLC with the same retention time as an authentic adduct of PQQ and phenylhydrazine. Often, the yields of product were reported to be high, close to stoichiometric with protein subunits. As discussed in Janes et al (39), it is possible to write a reaction scheme for the production of a PQQ-like molecule from the phenylhydrazone of TPQ. This reaction scheme requires first, proteolytic release of the TPQ derivative as a free amino acid, second, its cyclization to form a bicyclic structure, and lastly, a condensation with glutamate to form a PQQ analog. In the case of other copper- and iron-containing proteins that lack TPQ (e.g. galactose oxidase and lipoxygenase), it would also be necessary to postulate a production of topa from a tyrosine precursor in the presence of an active-site redox-active metal center. Given the large number of improbable steps, it appears difficult (if not impossible) to rationalize high yields for PQQ phenylhydrazone-like products in this manner. Another explanation, advanced at a conference on quinoproteins, is the possible contamination of the HPLC system with PQQ phenylhydrazone adducts injected repeatedly as standards (JA Jongejan, personal communication). Residual PQQ phenylhydrazone might be expected to co-elute with protein digests, leading to the artifactual assignment of PQQ in a wide variety of protein systems.

Is There a Role for PQQ in Eukaryotes?

Despite the failure to implicate PQQ as a covalently bound redox cofactor in enzymes of eukaryotic origin, Rucker and coworkers have repeatedly found evidence for a "failure to thrive" syndrome in germ-free rodents fed chemically defined diets devoid of PQQ (47, 47a). The observed abnormalities can be reversed by the addition of PQQ to the diets. Breeding of mice deprived of PQQ leads to significant reduction in litter sizes and cannibalization of pups (47). In an early experiment (48), it was observed that PQQ deprivation reduced lysyl oxidase levels and had a marked negative effect on animal posture and skin quality. Since it appears highly unlikely (although still unproven) that lysyl oxidase contains PQQ, these effects are now attributed to indirect effects.

Gallop and coworkers have pursued the detection of PQQ in tissues of mammalian origin using a redox-cycling assay (49) (see section on DETECTION OF QUINONE COFACTORS AND QUINOPROTEINS). Using this technique, evidence has been presented for the presence of PQQ in milk, plasma, red cell hemolysates, and brain homogenates (49). There has been some controversy regarding the specificity of the redox-cycling assay when applied to cell extracts (50, 51, 51a), and a more specific assay was subsequently developed that involves a multichannel electrocoulometric detection of PQQ as it is eluted from HPLC (52). This new assay system, depending on a compound having not only the correct retention time on HPLC but also the redox potential of PQQ, has been used to confirm the presence of PQQ in milk at a level of 500 pmol/ml (52). Another laboratory also reports PQQ from mammalian sources, but at significantly reduced levels (53). At this juncture, two issues require resolution. First, the presence of PQQ or a derivative of PQQ needs to be proven definitively by structural analysis (through, for example, mass spectrometry). Second, the discrepancy in yields of PQQ needs to be resolved. Gallop has proposed that PQQ (a trianionic compound) binds to cations in a manner that reduces its redox capability (52). It is also likely that isolation of PQQ under oxidizing conditions in the presence of high levels of protein will lead to significant reaction of the PQQ quinone with protein side chains. Both of these properties could lead to highly variable yields for PQQ.

If PQQ is, indeed, present at significant levels in mammalian tissue, what role is PQQ likely to play in cellular physiology? Two recent studies propose that PQQ may play a role in mitochondria and neutrophils. In the mitochondrion, PQQ has been reported to reverse the effects of rotenone inhibition of NADH:CoQ reductase, leading to the suggestion that PQQ acts as a reversibly bound cofactor in this enzyme system (54). In stimulated neutrophils, trapping of PQQ prevents the production of superoxide, and a catalytic cycle that places PQQ between NADPH and O_2^- has been postulated (55). An interesting study of the *N*-methyl-D-aspartic acid receptor has shown that PQQ has a neuroprotective effect, possibly via a modulation of the redox state of the receptor (56). A number of laboratories have reported protective effects of administered PQQ, for example, the prevention of cataract formation and hepatic pigmentation in chick embryos exposed to hydrocortisone, the prevention of carrageenin-induced rat paw edema, and the prevention of liver damage upon exposure to hepatotoxins such as carbon tetrachloride (57). Paz et al (55) have written a provocative essay, summarizing many of these effects and proposing multiple routes whereby PQQ could exert its biological effects. Quite recently, PQQ has been shown to serve as a substrate for an NADPH-dependent methemoglobin reductase via the production of $PQQH_2$ (58). It has also been found that PQQ protects against oxygen damage in reperfusion experiments, attributed to a $PQQH_2$-mediated attenuation in the level of damaging oxygen species (58, 59). With regard to plants, Weber & Bach have recently reported a 5–10-fold stimulation of the enzyme catalyzing the conversion of acetyl-CoA to HMG-CoA in the presence of Fe^{2+} and PQQ (60).

As the above-documented studies indicate, a possible role for PQQ in plants and animals should not be ignored. While many of the functions currently being ascribed to PQQ must be viewed with considerable caution, the confluence of multiple lines of inquiry (e.g. the detection of PQQ in animal tissues, the well-documented nutritional effects, and the demonstration of direct effects of PQQ in several enzyme systems) indicates that PQQ may, after all, perform important physiologic roles in eukaryotes.

PQQ in Prokaryotes

In contrast to the findings on eukaryotes, the presence of PQQ as an enzymatic cofactor and growth regulator has been well established in gram-negative bacteria since ca. 1980 (61). The presence of PQQ-dependent enzymes in the periplasmic space and membrane surface of gram-negative bacteria provides a mechanism whereby oxidation of alcohol and sugar substrates can be efficiently linked to respiration via a series of electron-transport proteins. The enzymes for which PQQ has been unambiguously demonstrated as an essential cofactor include glucose dehydrogenase (GDH, from a range of gram-negative bacteria) (62, 63), methanol dehydrogenase (MDH, from methanotrophs as well as other gram-negative bacteria) (61, 64, 65), and a series of bacterial alcohol and aldehyde (hydrate) dehydrogenases (66–68). As the term dehydrogenase indicates, each of these enzymes undergoes recycling from a reduced to oxidized form by oxidants other than dioxygen: The physiologic electron acceptors vary from ubiquinone in the case of membranous GDH to a cytochrome c in the case of MDHs. The electron-transfer aspect of PQQ-dependent enzymes has been extensively discussed in the current literature (69–71) and is not repeated in this review. There is a recent example in which a MDH from *Paraccocus* has been shown to react with oxygen, generating superoxide ion, which can be trapped by superoxide dismutase (72); however, this process is only observed in the presence of an inefficient electron acceptor and is unlikely to play a role under physiologic conditions. In a recent study, lupanine hydroxylase from *Methylobacterium extorquens* has been implicated as a PQQ-requiring enzyme system (73); from a mechanistic perspective this enzyme has been proposed to function as a dehydrogenase, forming an intermediate imine that is subsequently hydrolyzed to form the hydroxylated product. Thus, lupanine hydroxylase fits within the rubric of PQQ-dependent dehydrogenases.

In virtually all cases (possible exceptions discussed below), PQQ is bound noncovalently to protein, becoming dissociated under variable reaction conditions, which are protein dependent. Although early studies had suggested that glutamate decarboxylase from *E. coli* (74) and tryptophan decarboxylase from *Catharanthus roseus* (75) contained covalently bound PQQ, it is now accepted that these enzymes depend solely on the organic cofactor pyridoxal phosphate for enzymatic activity (76). Additionally, the iron-sulfur-containing nitrile hydratase (77) had been erroneously deduced to contain covalently bound PQQ (78). In certain organisms (e.g. *E. coli* or *Salmonella*), the soluble form of GDH is synthesized in an apo-form, requiring the addition of exogenous PQQ for the reconstitution of activity; this property has provided a long-standing assay for free PQQ (63). From a biologic perspective, the production of an apo-form of enzyme is puzzling and not yet understood. Recent studies suggest that PQQ may be covalently attached to protein in the cases of amine dehydrogenases from *Citrobacter freundii* (78a) and *Pseudomonas* sp. (78b) and an aldehyde dehydrogenase from *Methylobacillus glycogenes* (79); in each instance, protein hydrolysis in 6N HCl was required for the release of PQQ. This observation may not be sufficient to prove covalently attached PQQ, especially if cofactor is noncovalently attached in a particularly tight manner. Given the dismaying history of the misidentity of covalently bound PQQ in eukaryotes, it would appear particularly important to isolate an active-site peptide from these dehydrogenases and to demonstrate unambiguously the presence of PQQ. We note that earlier studies by Tabor and coworkers had indicated the presence of an FAD cofactor in a spermidine dehydrogenase from *Serratia marcescens* (80). Resolution of the nature of the cofactor with the amine dehydrogenases is especially pertinent, since with the exception of lupanine hydroxylase (73), the role of PQQ appears restricted to the oxidation of alcohol (or sugar) functional groups. As is described below, another bacterial amine dehydrogenase (methylamine dehydrogenase, MADH) contains a covalently bound cofactor; however, in this case the cofactor structure is a quinone derived from tryptophan, not PQQ.

In addition to the "simple" alcohol dehydrogenases, which contain only PQQ as cofactor, a number of PQQ-dependent alcohol dehydrogenases have been described that also contain heme. These are referred to as quinohemoproteins, and encompass both soluble and membrane-bound enzyme forms. In the former case, heme is covalently bound, whereas PQQ is released upon protein denaturation; this class of enzyme consists of a single subunit of ca. 70 kDa (81). The membrane-associated alcohol dehydrogenase is more complex, being composed of three subunits, one of which contains PQQ and heme; a second subunit contains heme; and a third, considerably smaller subunit performs an as-yet-undefined function (82, 83).

From a structure/function perspective, the two best-studied systems are GDH and MDH. Both immunoblotting (84) and genetic studies (85) of GDH support the view that the soluble and membranous forms of this protein are unique gene products with different subunit structures (α_2 of 50 kDa each for the soluble enzyme, and α of ca. 85 kDa for the membranous form). A gene sequence is available for the soluble enzyme from *Acinetobacter calcoaceticus* (86), while the membranous form of enzyme has been cloned from a number of sources [*A. calcoaceticus* (87), *E. coli* (88), and

Gluconobacter oxydans (89)]. Both protein forms require divalent metal ion to bind PQQ (67). The biologic role of the soluble form of GDH is unclear, in light of the observation that a mutant strain of *A. calcoaceticus* containing the soluble but not the membranous form of enzyme is incapable of utilizing glucose as a substrate (85). A topological analysis of the membranous GDH from the inner membrane of *E. coli* suggests five membrane-spanning segments, with the N terminus of the protein residing in the cytoplasm and the C terminus in the periplasm. The amino acids predicted to interact with both PQQ and its electron acceptor ubiquinone are assigned to the periplasmic side of the membrane (90). A model has been suggested in which the generation of an electrochemical potential results from the interaction of reduced ubiquinol with a ubiquinol oxidase located on the periplasmic face of the inner membrane (90). This view differs from an earlier one that invoked reduction of ubiquinone at the cytosolic face of GDH to yield an electrochemical gradient directly (91).

Among studies of PQQ-containing enzymes, studies of MDHs are structurally the most advanced. These enzymes are of the $\alpha_2\beta_2$-type, with a large (α) subunit of ca. 65–70 kDa and a small (β) subunit of ca. 8–10 kDa (92). Additionally, a MDH from *Acetobacter methanolicus* has been recently isolated that contains a third subunit (93), designated the *mox J* product since it maps to a gene cluster encoding the α and β subunits (Mox F and Mox I) as well as cytochrome c_L (Mox G) (94–96). PQQ has been demonstrated to bind to the α subunits with a stoichiometry of one per subunit (97). A single Ca^{2+} ion [which can be replaced by other divalent metal ions such as Sr^{2+} (98, 99)] is required for tight binding of PQQ and enzyme activity (99). Genetic studies of MDH production indicate a very large number of gene products (at least 20) that are essential for functional enzyme; among these the *mox A, mox K,* and *mox L* gene products have been concluded to play a role in the insertion of the essential metal ion, either through the maintenance of a high calcium level or through a direct effect on protein folding and metal insertion (cf Refs. 100, 101 for reviews). Enzyme isolated from strains lacking the *mox A, mox K,* and *mox L* gene products can be reconstituted in a time-dependent manner upon incubation with Ca^{2+} (102).

Amino-acid sequences have been available for a number of MDHs, e.g. from *M. extorquens* (103), *M. organophilum* (104), and *Paracoccus denitrificans* (105). Although a preliminary crystallographic study of enzyme from *M. extorquens* had been reported, full (2.6-Å) structures for MDHs from *Methylophilus methylotrophus* and W3A1 became available in 1992 (65). These structures define the unusual topology of the heavy subunit, which consists of eight antiparallel β-sheets, each composed of four strands shaped like a "W," as well as the relationship of the light to the heavy

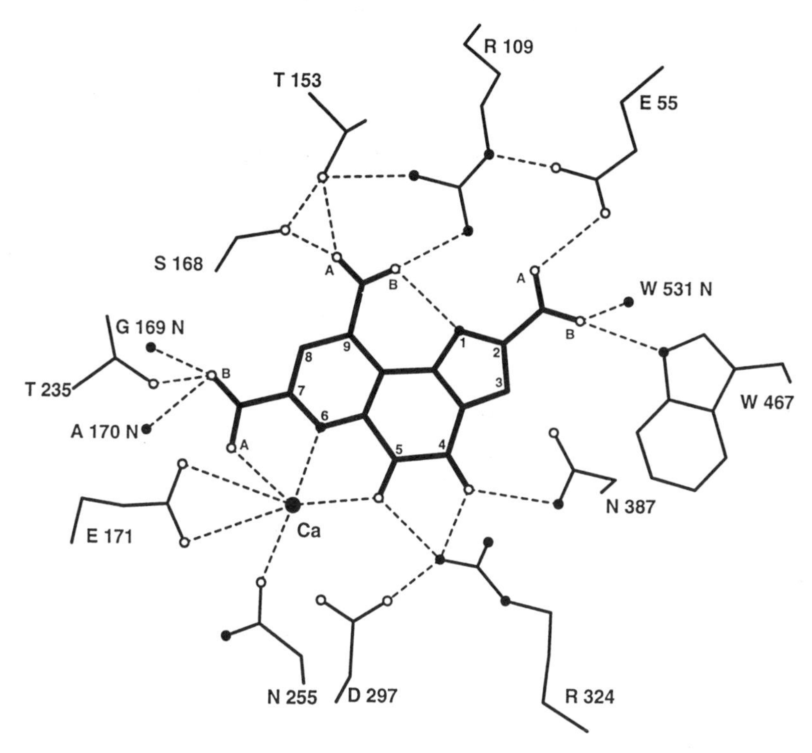

Figure 2 Schematic illustrating binding interactions with PQQ and Ca^{2+} at the active site of MDH. Reprinted with permission from Ref. 105a. Copyright 1993 American Chemical Society.

protein chains. The light subunit lies away from the PQQ-binding site, as well as the region of the heavy chain deduced to interact with cytochrome c. Its function is unknown.

In the absence of primary sequence data for either *M. methylotrophus* or W3A1, it was not possible to define the PQQ-binding site with much precision (65). This situation has been corrected with the recent availability of the gene sequence for the heavy subunit of enzyme from bacterium W3A1, permitting a refinement of the initial diffraction map (105a). The residues interacting with the bound PQQ are shown in Figure 2. Of particular note is the proximity of the Ca^{2+} ion, which is shown to be six coordinate, with three of its ligands coming from the PQQ itself. The binding pocket for the PQQ has considerable hydrophobic character, with the cofactor being sandwiched between a tryptophan 237 and a disulfide bond (C103-C104). The role of this disulfide bond was unexpected. The bond may function to shuttle electrons to cytochrome c or possibly simply to anchor the cofactor into the enzyme active site. Although a putative PQQ-binding site had been previously identified from sequence homology among PQQ-dependent dehydrogenases (88, 106), this is not supported by the recent crystallographic refinement.

The mechanism of substrate oxidation by enzyme-bound PQQ remains unproven. Two likely pathways involve either a direct hydride transfer from substrate to cofactor or the formation of a hemiketal between an hydroxyl group of substrate and the C-5 carbonyl of PQQ (107). Support for hemiketal formation comes from the demonstrated property of PQQ in solution to form adducts with nucleophiles (108); a crystal structure has been published for a complex between acetone and the C-5 carbonyl of PQQ (61). If adduct formation precedes substrate oxidation, the mode of C-H bond cleavage is likely to be proton abstraction, accompanied by electron delocalization into the cofactor ring. As described in the recent crystallographic refinement by Mathews and coworkers (105a), only two residues can be identified with the potential to perform the function of base catalysis: Asp297 and Glu171. The coordination of Ca^{2+} to the cofactor is particularly interesting in that one of the ligands to the metal ion is the oxygen of the C-5 carbonyl. This type of coordination is expected to provide electrophilic catalysis either for a direct hydride transfer or covalent adduct formation. As is discussed in the context of TPQ enzymes, Cu^{2+} is an essential cofactor, playing an important role in the cycling of electrons from reduced cofactor to dioxygen. This metal is also believed to be bound in fairly close proximity to the active-site organic cofactor and may play the role of electrophilic catalyst in the initial oxidation of substrate.

An area that has been particularly active in recent years concerns the mode of PQQ biogenesis. Stable isotope labeling has been used to demonstrate that the phenol ring of tyrosine yields the o-quinone structure of PQQ, with an internal cyclization of the ethylamine side chain yielding the pyrrole-2-carboxylate portion of the cofactor (109, 110). Although the pathway for PQQ synthesis has been proposed to involve a condensation of dopaquinone with glutamate (111), efforts to demonstrate direct incorporation of labeled glutamate into the final PQQ product have thus far been unsuccessful. Genetic studies have identified a 5-kb fragment from the *A. calcoaceticus* genome capable of restoring PQQ expression to PQQ^- mutants of *A. lwoffi* and *E. coli* (112). Within this 5-kb fragment, three gene products have been identified, corresponding to proteins of 10.8, 29.7, and 43.6 kDa. Additionally, a reading frame encoding a 24-amino-acid peptide was identified. Two particularly curious features result from this study. First, both tyrosine and glutamate are found in the 24-amino-acid peptide, leading to speculation that this peptide provides the amino-acid building blocks for

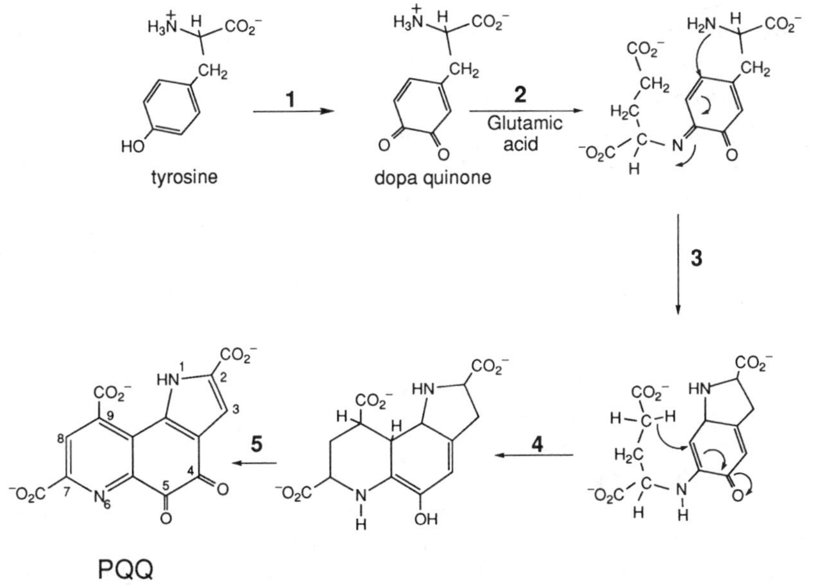

Figure 3 Proposed pathway for PQQ biogenesis from tyrosine and glutamic acid. Step 1 shows the hydroxylation of Tyr to dopa and its oxidation to dopa quinone. Step 2 shows the addition of glutamic acid, followed by an internal cyclization of the ethylamine side chain of tyrosine (step 3). The γ-carbon of glutamate is shown condensing in step 4, and a final, multi-electron oxidation in step 5 (cf Ref. 113).

PQQ (112). Second, the presence of only three protein products may be inadequate to generate PQQ, given the extensive chemistry required for this process (cf Figure 3). As noted by Unkefer, the proposed pathway from tyrosine and glutamate requires the loss of a total of 12 electrons (113)!

Analogous to the demonstration of PQQ production in *E. coli* following transformation with a 5-kb fragment of DNA from *Acinetobacter,* a 6.7-kb fragment from *Klebsiella pneumoniae* has been shown to generate PQQ in *E. coli* (114). Both of these studies follow from the conventional view that *E. coli* lacks the genes for production of PQQ. However, this view has been recently challenged by studies from Gasser and coworkers, who have demonstrated production of PQQ following growth of a PTS^- strain of *E. coli* in glucose minimal media (115). The authors argue that the PTS^- strain, impaired in its ability to phosphorylate and transport glucose through the cytoplasmic membrane, can generate mutants capable of production and excretion of PQQ to the periplasmic space (where PQQ reconstitutes the apo-form of membranous GDH). It has been suggested that wild-type *E.*

coli lacks an active promoter for the expression of PQQ. In light of these very interesting results, transformation studies in *E. coli* (112, 114) should be interpreted with some caution. Further, studies of PQQ production in *M. organophilum* implicate six genes (116), rather than the four genes reported in *A. calcoaceticus* (112). It would appear that the total number of genes involved in PQQ biogenesis is "soft," and must await further experimentation. Elaboration of the enzymatic properties of each of these gene products is also expected to yield a great deal of fascinating chemistry.

3. DISCOVERY OF TOPA QUINONE (TPQ) IN A MAMMALIAN ENZYME

The surprising discovery of covalently bound quinone cofactors in mammalian enzymes resulted from efforts to confirm the presence of PQQ. As noted above, bovine SAO was the first eukaryotic protein implicated as PQQ-containing. In light of the indirect evidence in support of PQQ, it was clear that an active-site, cofactor-containing peptide would have to be isolated and characterized. This proved considerably more difficult than anticipated. Proteolysis of underivatized protein routinely failed to yield a chromophoric peptide in reasonable amount, attributed to the inherent chemical reactivity of quinones toward amino acids and their side chains. The facile reaction of SAOs with phenylhydrazines had been known for years to lead to a stable form of protein with a characteristic absorbance at ca. 450 nm (118). Although there was some discrepancy regarding the stoichiometry of phenylhydrazone adduct to protein subunit (see section on PROPERTIES OF TPQ ENZYMES below), Janes & Klinman (119) had shown that active preparations of bovine SAO incorporated close to one mole of phenylhydrazine per enzyme subunit. Thus, the phenylhydrazone derivative appeared a suitable enzyme derivative for structural elucidation.

In principle, proteolysis of derivatized protein should have been a straightforward matter. However, it was known that proteolysis of a porcine kidney diamine oxidase derivative with trypsin led to peptide in unacceptably low yields (0.1%) (120). Labeling of bovine SAO with [^{14}C]-phenylhydrazine, reductive alkylation of SH groups, and proteolysis with a range of standard proteases (e.g. trypsin, chymotrypsin, V8, and thrombin) produced a hydrophobic smear of radiolabeled peptides (121). Ultimately, a somewhat unorthodox approach involving proteolysis of native protein in 2M urea with thermolysin was found to produce an active-site peptide in approximately 40% yield (39). Since no precursor peptides were observed to accumulate, Janes et al suggested that the active cofactor may lie close to the surface, providing accessibility to thermolysin. More recent experiments have shown that reductive alkylation of bovine SAO prior to proteolysis with thermolysin is also a suitable protocol for the preparation of active-site peptides (X Wang, JP Klinman, unpublished results). Two features of the thermolytic digests from bovine SAO were fortuitous: First, the radiolabeled peptide was found to elute from reverse-phase HPLC at the end of the peptide profile, greatly facilitating its purification; and second, the peptide was quite small and easily analyzed by amino-acid sequencing (providing the sequence Leu, Asn, Unknown, Asp, Tyr) (39).

The major challenge was the identification of the unknown residue. High-resolution mass spectrometry of the pentapeptide from bovine SAO provided a mass to seven significant figures; subtraction of masses for the four known amino acids led to an assignment of 283.0967 amu to the unknown. Subsequent computer analysis of this mass indicated that, within 5 ppm, a single empirical formula could be written: $C_{15}H_{13}N_3O_3$. At this juncture it was clear that PQQ could not be the missing cofactor, and two structures were proposed that were consistent both with the mass of the unknown and biochemical intuition. These are shown as structures **A** and **B** in Figure 4, where **A** is an adduct between the hydroxyl group of a serine and the oxidized form of catechol derivatized with phenylhydrazine; **B** represents the oxidized form of a trihydroxyphenylalanine, also derivatized as a phenylhydrazone (39).

Distinction between structures **A** and **B** required the synthesis of authentic analogs, followed by a comparison of physical properties for the synthetic and protein-derived compounds. Initial attention was focused on an analog of **B**, since the mass spectrometric fragmentation pattern observed with the protein-derived pentapeptide indicated cleavage between an α- and β-carbon to form a benzylic radical (expected from **B**, whereas **A** was expected to cleave between the β-carbon and oxygen to form a phenolic radical). Using a synthetic analog of **B** in which the peptide moiety was tied back into a cyclic, hydantoin structure, comparative NMR (39) and subsequent reso-

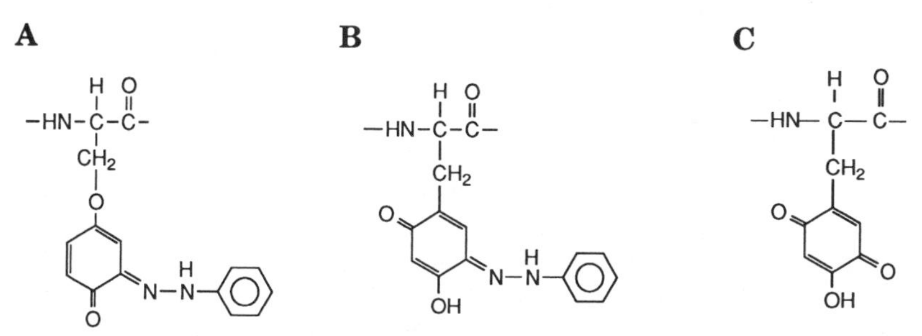

Figure 4 **A** and **B** represent structures proposed to explain the mass spectrometric data for the active-site peptide derived from bovine serum amine oxidase (39). **C** is the structure of the underivatized cofactor, topa quinone (TPQ).

nance Raman studies (40) established its identity to the unknown amino acid in the bovine SAO-derived peptide. In this manner, a trihydroxy-phenylalanine (in its oxidized, quino-form) was demonstrated to be present in the polypeptide backbone of a mammalian protein (**C** in Figure 4).

This observation is somewhat less surprising when placed in the context of the pioneering work of Waite and coworkers on quinoprotein production in marine invertebrates. A protein produced by the marine mussel (*Mytilus edulis*) has been shown to contain a high content of the amino-acid dopa (in its quino-form) (122). This protein, which is secreted by a gland located in the foot of the mussel, serves the function of a glue, allowing the organism to remain tightly affixed to a variety of supports (123). Structural analysis of an *M. edulis* adhesive protein has indicated the presence of ca. 10% dopa residues contained within a repeating decapeptide: Ala, Lys, Pro/Hyp, Ser, Tyr/Dopa, Hyp, Hyp, Thr, Dopa/Tyr, Lys (as indicated, several positions show mixtures of either Hyp/Pro or Dopa/Tyr) (122). Although the exact mode of formation of dopa from tyrosine is not known, the possible involvement of a tyrosinase-like enzyme has been considered (124). A dopa-rich protein has also been shown to constitute a class of eggshell precursor proteins in the liver fluke *Fasciola hepatica* (125). Once again dopa is found to constitute ca. 10% of the total protein, although the sequence of the prominent dopa peptide differs from that in the muscle: Gly, Gly, Gly, Dopa, Gly, Gly, Dopa, Gly, Lys. It is remarkable that murine and parasitic organisms have used a similar strategy to construct adhesive and sclerotizing agents, respectively. The strength of dopa-containing proteins almost certainly derives from the crosslinking properties of dopa quinone, through reaction of one of its quinone carbonyls with protein-bound nucleophilic side chains or by Michael attack on the quinone ring directly (123). Given the ability of invertebrates to construct structural quinoproteins through a posttranslational modification of tyrosine to dihydroxyphenylalanine, the evolution of trihydroxyphenylalanine as an active-site cofactor in mammals appears to be a reasonable "next step."

It should be noted that topa had been known to neurophysiologists for several decades, following the demonstration of the extreme neuro- and cytotoxic property of this compound as its free amino acid (referred to as 6-hydroxydopa in the earlier literature) (126). In recent studies it has been suggested that the toxicity of topa results from its ability to act as an agonist at the non-*N*-methyl-D-aspartic acid glutamatergic receptor (127). The fact that toxicity can be prevented by glutathione has further led to the proposal that topa is only toxic in its oxidized quino-form (128). As described below (see section on PROPERTIES OF TPQ ENZYMES), TPQ exists as a dianion at physiologic pH, due to the acidity of its ring hydroxyl, leading to an unexpected structural homology between TPQ and a naturally occurring agonist, glutamate. Although free topa may arise spontaneously in vivo, from either tyrosine or dopa, it is unlikely to accumulate to any great extent from quinoprotein precursors containing a single topa residue per polypeptide chain. A more serious complication may, however, occur under conditions where large concentrations of dopa are administered therapeutically, as is the case in the treatment of Parkinson's disease (129).

4. DISCOVERY OF TRYPTOPHAN TRYPTOPHYLQUINONE (TTQ) IN A BACTERIAL ENZYME

Methylamine dehydrogenase had been isolated from the periplasm of several methylotrophic bacteria and implicated to contain a quinone-type cofactor from the analyses of the EPR and ENDOR spectroscopy (130). The native enzyme is devoid of metal ion cofactors and composed of two large (α) subunits of molecular weight of 40,000–50,000 and two quinone cofactor-containing small (β) subunits of molecular weight of 13,000–16,000 (for a recent review, cf Ref. 131). The entire amino-acid sequence of the small cofactor-containing subunit of MADH from *M. extorquens* AM1 was obtained from sequencing overlapped peptide fragments (132). From the peptide-sequencing data, it was shown that the organic cofactor was attached to the protein backbone through two unknown residues at positions 55 and 106 (132). PQQ had been concluded to be the quinone cofactor of MADH based on the optical similarities of the isolated phenylhydrazine and hexanol adducts to the corresponding model compounds prepared from PQQ (133, 134). However, when McIntire and coworkers applied mass spectrometry to the peptidyl semicarbazide-cofactor adduct of MADH from bacterium W3A1, PQQ would not explain the observed mass (without invoking a structural modification, e.g. removing the three carboxyl groups) (135). Approximately three years later, the first quinoprotein crystal structure was determined at a resolution of 2.25 Å for a MADH isolated from *Thiobacillus versutus* (136). Due to the lack of sequence information of the *T. versutus* MADH at that time, the identity of each side chain was assigned solely on the analysis of electron density. Surprisingly, only the quinone indole moiety of PQQ could be found in the electron-density map; the remaining electron density in the cofactor region was not coplanar with that of the quinone indole group. In order to fit the observed electron density in the context of PQQ, an uncyclized form of PQQ, termed "pro-PQQ," was proposed to be the cofactor (136, 137). A hypothesized pathway leading to the formation of PQQ from the pro-PQQ liberated during proteolysis was also put forward (137). This presents another example in which a structural modification of

PQQ was needed to explain the experimental data obtained from direct structural analysis.

A major advancement came from the cloning and sequencing of the DNA of the small subunit of MADH isolated from *M. extorquens* AM1 in Lidstrom's group (138). The DNA-derived protein sequence revealed that Trp55 and Trp106 were the two unknown residues that had eluded identification in the previous peptide-sequencing experiment. In analogy to the tyrosine-derived TPQ cofactor, Lidstrom and coworkers proposed a posttranslational modification of the two tryptophan residues to form a novel reactive center. Since the active-site peptide sequence is quite conserved between the W3A1 and the *M. extorquens* AM1 enzymes, it was assumed that two tryptophans were involved in the active site of W3A1 enzyme as well (139).

Given the above important piece of information, McIntire et al (139) isolated a total of 3.8 mg of the peptidyl semicarbazone from the W3A1 MADH and subjected it to rigorous NMR and mass spectrometric analyses. An exact mass of 940.3262 was obtained for the parent ion of this peptidyl cofactor-semicarbazone by mass spectrometry. Only one ($C_{41}H_{53}N_{11}O_{13}S_1$) out of the 29 possible empirical formulas made chemical sense and fit the results of ^{13}C NMR. A chemical formula of $C_{16}H_8N_2O_2$ was assigned to the cofactor, after subtraction of the known portions of the peptide chains and the semicarbazide inhibitor. A benzenoid ring with four neighboring protonated positions, two isolated unsaturated C-H groups, and two heterocyclic ring N-H groups was deduced from one-dimensional NMR and double-quantum filtered correlated spectroscopy. Two carbonyls and eight other unsaturated carbons were shown in the ^{13}C NMR spectra. Furthermore, NOESY connectivity experiments identified the first tryptophan residue (Trp55) as the location of the indole-6,7-dione moiety. Based on all the above experimental data combined with the results of gene sequencing, McIntire et al (139) proposed a novel structure, tryptophan tryptophyquinone (TTQ), as the MADH cofactor (Figure 5). Although several structures could

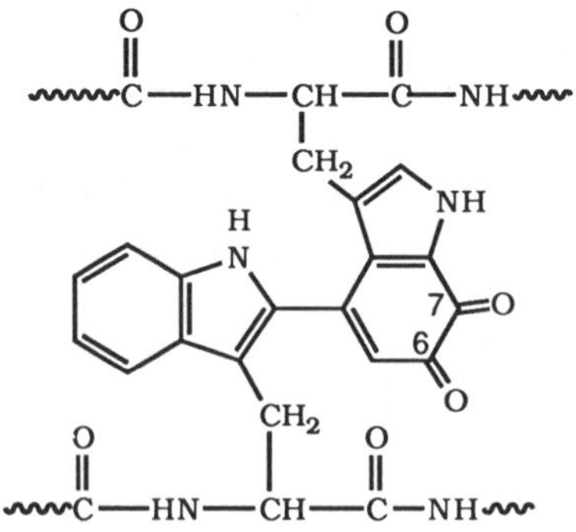

Figure 5 Structure for TTQ.

be deduced from the above information, TTQ was the only one that would explain all the nuclear Overhauser effect experiments.

Immediately following the proposal of TTQ by McIntire et al, the structure of TTQ was placed in the active site of the crystal structures of the MADHs from *T. versutus* and *P. denitrificans* (140). The introduction of TTQ into the active site was found to improve the fitting of the electron density map dramatically without any of the geometrical strain that was seen with PQQ in the active site, providing confirmation for TTQ as the cofactor. Moreover, the crystallographic studies established the generality of TTQ dependency within the class of bacterial MADHs from bacterium W3A1, *P. denitrificans*, and *T. versutus* (140).

5. DETECTION OF QUINONE-COFACTORS AND QUINOPROTEINS

Following the discovery of the two new cofactors, TPQ and TTQ, quino-cofactor detection in other enzyme systems was actively pursued. Although a combination of physicochemical methods (e.g. mass spectrometry and NMR spectroscopy) provides a powerful tool for structural elucidation, the commonly encountered limited availability of low-molecular-weight, cofactor-containing peptides made it important to pursue alternative methods. Three techniques have become accepted for the detection of quinone cofactors and proteins: resonance Raman spectroscopy (for a recent review, cf ref. 141), a pH-dependent λ_{max} shift of TPQ-*p*-nitrophenylhydrazone (41), and a redox-cycling assay system (142, 44). The specific biological assay for PQQ, which is based on the reconstitution of apo-GDH, has been reviewed elsewhere (63, 143) and is not discussed here.

Resonance Raman Spectroscopy

In spite of initial reports that had used resonance Raman spectroscopy to implicate erroneously the role of PQQ in eukaryotic enzymes (10, 11), resonance Raman spectra can provide rich structural information regarding the nature and the environment of quinone cofactors in enzymes (141). Due to the low extinction coefficient of quinone cofactors, chromophores are normally generated by labeling the quinone cofactor with phenylhydrazine or other substituted hydrazines to enhance the Raman signals. Considerable differences can be seen from a comparison of reported resonance Raman spectra for the substituted hydrazine-derivatives of PQQ and topa hydantoin (a TPQ model compound) (40). Additionally, many peaks present in the spectrum of MADH are absent from those of PQQ and topa hydantoin (144). As a result of these differences, resonance Raman spectroscopy can be used as a diagnostic tool to characterize the nature of a cofactor in a new quinoenzyme. One potential complication is the appearance of frequency shifts and intensity variations for spectra of native proteins relative to isolated active-site peptides or model compounds. However, when the resonance Raman spectrum of derivatized bovine SAO was compared to that of a model TPQ compound, only minor differences were observed (40). Resonance Raman studies have been extended to include labeled cofactor-containing peptides isolated from yeast (145), pea seedling, porcine kidney, and porcine SAOs (41), and labeled proteins isolated from *Arthrobacter* P1 (146) and *E. coli* K-12 (147), establishing the presence of TPQ in each case. As pointed out earlier, lysyl oxidase remains a special case among copper AOs; characterization of a phenylhydrazine-derivatized 5.5-kDa peptide by resonance Raman shows significant shifts relative to other TPQ enzymes. These differences may indicate either a different cofactor or possibly, unusual interactions between the cofactor and its active-site surroundings.

Resonance Raman spectra of native MADH have been determined for enzymes from bacterium W3A1 (144), *P. denitrificans*, and *T. versutus* (148). The close similarity among these spectra supports the existence of the same TTQ cofactor (148), consistent with physiochemical studies on the first enzyme (139) and crystallographic studies on the second and third (140). Although the resonance Raman spectrum of a TTQ model compound was reported to be similar to those of native MADHs (148a), a detailed spectrum has not yet been published.

*pH-Dependent λ_{max} Shift of TPQ-*p*-Nitrophenylhydrazone*

Janes et al (41) observed a pH-dependent shift of λ_{max} for the *p*-nitrophenylhydrazone of native copper AOs, with λ_{max} increasing from 457–463 nm at neutral pH (pH 7.2), to 575–587 nm in basic solution (1–2 M KOH).

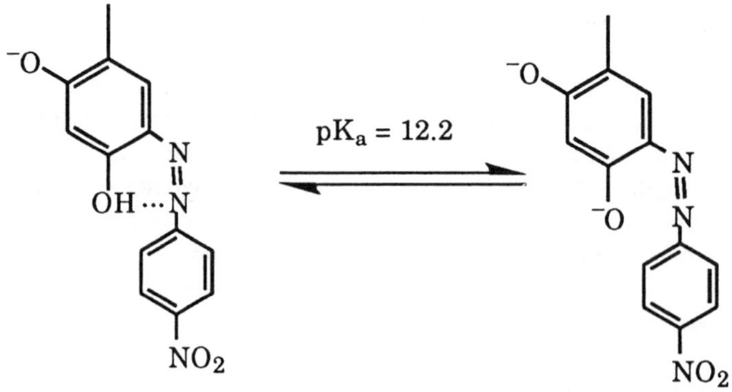

$$pK_a = 12.2 \qquad (1.)$$

This behavior was found to be identical to that observed with active-site peptides and a TPQ model compound, and appears to be unique to TPQ. An apparent pK_a of 12.2 was observed from the pH titration of the *p*-nitrophenylhydrazone of a TPQ model compound. The properties of the unique pH-dependent absorbance shift have been attributed to the ionization of the azo-form of the derivatized TPQ.

The unsubstituted phenylhydrazone derivatives of both TPQ-containing peptides and enzymes indicate an incomplete pH-induced red shift, attributed to a higher pK_a for the phenylhydrazone than *p*-nitrophenylhydrazone derivatives. A major advantage of this assay is its simplicity, since it can be performed with a desktop spectrophotometer. Using this new detection method, two AOs from sheep plasma and chick pea seedling have been added to the list of TPQ-containing enzymes (41).

Redox Cycling Assay

Quinones are well known to undergo redox reactions. In the presence of nitroblue tetrazolium, dioxygen, and excess glycine at alkaline pH (pH 10), quinones and quinonoid-type compounds can serve as an electron shuttle, reducing nitroblue tetrazolium to formazan, which absorbs at 530 nm (142). These properties constitute a redox-cycling detection for free quinones in solution. Under these assay conditions, it was shown (*a*) that PQQ could be detected at nM levels and (*b*) that it is one to two orders of magnitude more reactive than trihydroxyphenylalanine (52). Flückiger et al have summarized the detection sensitivity towards several types of quinone or quinonoid compounds as well as inhibitory effects from divalent cations and possible side reactions (52). As already noted, there has been considerable discussion in the literature regarding the suitability of redox-cycling assays for the detection of PQQ in crude cell extracts (cf 49–51a).

When a quinone is protein-bound, side reactions such as polymerization are eliminated. It has been shown that quinone redox cycling can be performed on electroblotted nitrocellulose paper following gel electrophoresis of quinoprotein-containing samples (44). With this novel staining method, several quinoproteins have shown positive identification. These include bovine SAO, porcine kidney diamine oxidase, lysyl oxidase, *Arthrobacter* P1 methylamine oxidase, MADH β subunit, and dopa-containing vitelline B and C (44), as well as two isozymes of TPQ-dependent AOs from yeast (D Cai, JP Klinman, unpublished results). Nonquinoproteins such as dopamine β-monooxygenase (44), soybean lipoxygenase (44), galactose oxidase (45), and laccase (45) did not stain with the method. The overall results obtained by redox cycling are consistent with our current knowledge of quinoproteins.

6. PROPERTIES OF TPQ ENZYMES

Ubiquity and Function

Following the initial demonstration of TPQ in the AO from bovine serum, it was logical to assume that a similar cofactor would be found at the active site of other members of the copper AO class. As discussed above (see section on DETECTION OF QUINONE COFACTORS AND QUINOPROTEINS), evidence now exists for TPQ in bovine, porcine, and sheep SAOs, porcine kidney diamine oxidase, pea seedling and chick pea seedling AO, yeast AOs, and bacterial AOs from both gram-positive and gram-negative bacteria. At this juncture it can be concluded that TPQ is a ubiquitous cofactor, occurring in enzymes from bacteria, yeast, plants, and mammals. In the case of PQQ and TTQ, well-established roles for these alternate quino-cofactors still appear restricted to bacteria. However, given the recent investigations of possible roles for PQQ in eukaryotes, this quinone may also prove to be ubiquitous.

Despite the widespread occurrence of TPQ-containing enzymes, their biologic roles often remain poorly defined. The reaction catalyzed by these enzymes is shown in equation 2, illustrating the oxidative release of aldehyde, ammonia, and hydrogen peroxide:

$$RCH_2NH_2 + O_2 + H_2O \rightarrow RCHO + NH_3 + H_2O_2 \qquad 2.$$

In the case of bacteria (e.g. 149) and yeast (e.g. 150), this reaction permits growth on a range of amines as a nitrogen and, in some instances, carbon source. Turning to plants, a major function of the AOs has been proposed to center on hydrogen peroxide production, which is used for wound healing via cell wall and lignin formation (151). Additionally, the AOs in plants are likely to play a role in cell growth through the regulation of polyamine levels (151). It is in the area of eukaryotes that the precise function of the AOs becomes most diverse and elusive. The best-defined role for a mammalian AO occurs with lysyl oxidase (151a), an enzyme that catalyzes connective tissue maturation by way of crosslinking of elastin and collagen. However, as already discussed, the active-site cofactor in lysyl oxidase has not yet been elucidated and may not be TPQ.

The full range of mammalian AOs is still unknown (151), although recent cloning and sequencing results (see below) indicate the presence of at least two distinct enzyme forms, one of which is released to the plasma from liver and a second of which is produced as a soluble, intracellular form in kidney (152). Additionally, numerous investigators have demonstrated the presence of a membrane-associated protein from vascular smooth muscle cells, often referred to (for historical reasons) as the semicarbazide-sensitive AOs (153, 154). Speculative, although unproven functions for each of these enzymes vary from the oxidative removal of biogenic amines from plasma, to the intracellular regulation of polyamines as growth factors. Additionally, a number of investigators have speculated that the hydrogen peroxide produced by AOs could play a role in cell regulation (155, 156). The finding (157) that the gene for lysyl oxidase is 98% homologous to a tumor suppresser gene (*rrg*) has led to the hypothesis that high levels of expression of lysyl oxidase may correlate negatively with uncontrolled cell growth.

Protein and DNA Sequencing

As increasing numbers of active-site peptides from copper AOs became available for cofactor identification, comparative sequence studies quickly followed. Alignment of thermolytic peptides from five proteins has indicated a conserved sequence surrounding topa: Asn, Topa, Asp/Glu; additionally, a hydrophobic residue is always found toward the N terminus of Asn, and Tyr is preferentially located toward the C terminus of Asp. Using extended active-site peptides (of 11–23 amino acids), it could be further established that the degree of amino acid identity was much greater for proteins from a common tissue (e.g. 89% between SAOs from bovine and porcine) than between intra- and extracellular proteins from a common organism (e.g. only 56% identity between porcine serum and porcine kidney AOs) (41).

Full-length DNA sequences are currently available for five copper AOs: yeast AO from *Hansenula polymorpha* (158), lentil seedling AO (42), bacterial methylamine oxidase (158a),[2] bovine SAO (152), and human kidney diamine oxidase (152). The latter two examples, representing the first complete sequences for mammalian proteins, have only recently become available. In the case of the human kidney diamine oxidase sequence, this was initially ascribed to a human amiloride-binding protein believed to function as a sodium channel (159). However, screening of the Protein Identification Resource database (160) with the cofactor containing extended tryptic peptide from bovine SAO indicated a high degree of homology to the human amiloride-binding protein; it was later shown that the sequence of an extended active-site peptide isolated from porcine kidney diamine oxidase was identical to that of human amiloride-binding protein, and human

[2]A cloned bacterial tyramine oxidase from *Klebsiella aerogenes* (158b) was proposed to be a TPQ/copper amine oxidase, based on its protein sequence homology with the yeast and *Arthrobacter* P1 amine oxidase (158a).

amiloride-binding protein has now been redesignated human kidney diamine oxidase (152).

The question of the amino acid precursor to topa could be easily addressed, once protein and DNA sequences were available from a common source. This comparison was first conducted with the yeast AO from *H. polymorpha,* demonstrating tyrosine as the precursor for topa (145); although there was a less than perfect match between protein and DNA sequences, this can now be ascribed to the presence of multiple forms of yeast AO in *H. polymorpha* [such that the cloned and sequenced DNA encoded an aliphatic amine-specific enzyme (158), whereas the isolated and sequenced protein was specific for aromatic amines (D Cai, JP Klinman, unpublished results)]. Confirmation of the involvement of tyrosine as the precursor to topa has come from comparative studies of DNA and protein from lentil seedling (42), human and porcine kidney (152), and bovine serum (152).

As the name implies, the copper AOs contain tightly bound copper in addition to TPQ. In an earlier study, comparison of DNA-derived protein sequences for a yeast and a plant AO had led to the identity of three conserved histidines as ligands for the active-site copper [positions 8, 246, and 357 in the lentil seedling enzyme (42)]. With the availability of gene sequences for five TPQ-containing enzymes, a more rigorous analysis of structural homologies has become possible (152). In Figure 6, partial sequences for pairwise alignments of DNA-derived sequences are presented. One His-X-His motif, 40–50 residues toward the C terminus from the cofactor, is conserved in all alignments. This most likely contains two of the histidine ligands to copper, based on the observation that histidines in His-X-His motifs have been identified as ligands to type II or III copper in the crystal structures of several copper proteins (cf Ref. 161 for a review). It is interesting to note that the His-X-His motif in yeast AO is contained in a stretch of sequence HNHQH. Although HQH is the conserved motif when yeast is aligned with bovine serum, lentil seedling, or bacterial AOs, HNH appears when yeast is aligned with the human kidney diamine oxidase. Thus, it is not yet possible to discern which motif constitutes ligands to copper in the yeast enzyme (152). In addition to the histidines found in the His-X-His motif, a second histidine 20–30 residues toward the N terminus from the cofactor is also conserved in all sequences (152). A fourth histidine near the N terminus has been reported to be conserved when the DNA-derived protein sequences of AOs from yeast and lentil seedling were aligned with that of the AO from *Arthrobacter* P1 (162, 158a); however, this does not hold true when mammalian enzymes are included in the alignment (152). Despite variations in length of the polypeptide chains (587 residues for lentil seedling enzyme, 692 for the yeast enzyme, 713 for human kidney diamine

```
BSAO   436 GLPLRRH-HSDFLS--HYFGGVAQTVLVFRSVSTMLNYDYVWDMVFYPNGAIEVKLHATGYISSAFLFGAARRYGNQVGEHTLGPVHTHS-AHYKVD 528
HKDAO  422 GVPLRRHFNSNFKGGFNFYAGLKGQVLVLRTTSTVYNYDYIWDFIFYPNGVMEAKMHATGYVHATF-YTPEG-CARHSPAHPPDWQHTHSLVHYRVD 517

BSAO   436 GLPLRRHHSDFLSHYFGGVA--QTVLVFRSVSTMLNYDYVWDMVFYPNGAIEVKLHATGYISSAFLFGA---ARRYGNQVGEHTLGPVHTHSAHYKVD 528
YAO    371 GLLFK--HSDFRDNFATSLVTRATKLVVSQIFTAANYEYCLYWVFMQDGAIRLDIRLTG-ILNTYILGDDEEAGPWGTRVYPNVNAHNHQHLFSLRID 465

HKDAO  422 GVPLRRHFNSNFKGGFNFYAGLKGQVLVLRTTSTVYNYDYIWDFIFYPNGVMEAKMHATG--YVHATFYTPEGCARHSPAHPPDWQHTH-SLVHYRVD 517
YAO    371 G-LLFKH--SDFRDNFATSLVTRATKLVVSQIFTAANYEYCLYWVFMQDGAIRLDIRLTGILNTYILGDDEEAGPWGTRVYPNVNAHNHQHLFSLRID 465

YAO    371 GLLFKHSDFRDNFATSLVTRA-TKLVVSQIFTAANYEYCLYWVFMQDGAIRLDIRLTGILN---TYILGDDEEAGPWGTR-VYPNVNAHNHQHLFSLRID 465
LNSAO  370 NIMWRHTETGIPNESIEESRTEVDLAIRTVVTVGNYDNVLDWEFKTSGWMKPSIALSGILEIKGTNIKHKDEIKEEIHGKLVSANSIGIYHDHFYIYYLD 469

YAO    371 GLLFKHSDFRDNFATSLVTRA-TKLVVSQIFTAANYEYCLYWVFMQDGAIRLDIRLTGILNTYILGDDEEAGPWGTR-VYPN-VNAHNHQHLFSLRID 465
MAO    353 SILWKHFDFRE--GTAETRRS-RKLVISFIATVANYEYAFYWHLFLDGSIEFLVKATGILST--AGQLPGEKNPYGQSLNNDGLYAPIHQHMFNVRMD 445
```

Figure 6 Pairwise alignments for DNA-derived protein sequences for bovine serum amine oxidase (BSAO), human kidney diamine oxidase (HKDAO), yeast amine oxidase (YAO), lentil seedling amine oxidase (LNSAO) (152), and *Arthrobacter* P1 methylamine oxidase (MAO).

oxidase, and 762 for bovine SAO), the positions of conserved histidines appear to be maintained relative to the cofactor consensus sequence. Overall, conservation of structure is considerably higher in the C-terminal region of these proteins, leading to the suggestion that an N-terminal domain has diverged in order to accommodate the range of substrate specificities demonstrated by the copper AOs (152).

Once again, lysyl oxidase must be considered in a category separate from the other AOs. Analysis of the cloned gene for lysyl oxidase fails to reveal the expected consensus sequence observed in other topa-containing enzymes. Additionally, no obvious conservation of copper-binding ligands can be inferred by comparison of the lysyl oxidase gene to genes for other AOs (26, 27). With regard to overall structure, lysyl oxidase is also considerably smaller than any of the other characterized AOs, which exist as dimers of subunit size in the range of 70–90 kDa: Kagan and coworkers have demonstrated that this protein undergoes processing from a 50-kDa precursor to a mature 32-kDa form (163).

Studies of Enzyme Mechanism

Detailed information regarding the kinetic properties of the AOs from numerous sources predates their characterization as TPQ-containing enzymes. The overall kinetic reaction has been shown to consist of two half reactions, involving enzyme reduction by substrate (equation 3), followed by enzyme reoxidation by molecular oxygen (equation 4).

$$E_{ox} + RCH_2NH_2 \rightarrow E_{red} + RCHO \qquad 3.$$

$$E_{red} + O_2 \rightarrow E_{ox} + H_2O_2 \qquad 4.$$

The reduction of enzyme proceeds in a catalytic fashion in the absence of O_2, and the enzymes are normally categorized kinetically as "ping pong" (reviewed in 164, 151). Depending on the enzyme source, either the reductive or oxidative half reaction has been found to be predominately rate limiting (165, 166).

REDUCTIVE HALF REACTION The first mechanistic evidence for a quino-cofactor came from reductive trapping experiments in which it was shown that substrate could be reduced onto enzyme during catalytic turnover in the presence of the mild reductant sodium cyanoborohydride. Using alternatively C-14-labeled substrate or tritiated cyanoborohydride, it was shown that reductively inactivated enzyme retained C-14 but not tritium (12). This was rationalized by the presence of a dicarbonyl (quino-) structure, which tautomerized subsequent to reduction, releasing tritium to solvent. Although this phenomenon was first seen in bovine SAO, it was later reproduced with the bacterial AO from *Arthrobacter* P1 (167).

An important aspect of the reductive inactivation experiments was the ability to trap substrate in a covalent bond to enzyme, providing support for a Schiff base complex between substrate and cofactor. In a key early study, Knowles and coworkers (168) had demonstrated that the rate of ammonia loss to solvent with porcine SAO correlated with enzyme reoxidation; this result indicated that the amino group of the substrate was retained on the enzyme during the reductive half of the reaction, implicating an aminotransferase type of reaction mechanism. Retention of ammonia in the absence of oxygen was later confirmed with bovine SAO, where use of strict anaerobic conditions and benzylamine as substrate yielded benzaldehyde as the sole product; only upon addition of O_2 to these reaction mixtures could free ammonia be detected (119).

The failure to see any incorporation of tritium in reductively inactivated protein samples (12, 167) indicated that product Schiff base complexes were not accumulating to any significant extent, since reduction of these complexes would be expected to lead to an irreversible incorporation of tritium. The conclusion of a very short lifetime for the product Schiff base complex was consistent with earlier kinetic studies of the effects of pH and isotopic substitution on steady-state and pre-steady-state rates (166, 169). Of particular note is the very large size of deuterium isotope effects seen in the bovine SAO-catalyzed oxidation of benzylamines [with isotope effects on V_{max}/K_m in the range of 7–16 (166)]. The latter observation indicates rate limitation by C-H bond cleavage from the substrate Schiff base complex and hence, significant accumulation of this species during turnover conditions. It has been noted that the size of deuterium isotope effects in the bovine SAO reaction lies outside the range expected from semiclassical considerations. Detailed studies of tritium isotope effects as a function of temperature have led to the conclusion of both protio- and deuterio-tunneling in the course of substrate oxidation (170).

Formation of covalent adducts between substrate and TPQ in the course of catalysis has important implications for the mode of substrate activation. By analogy to the majority of pyridoxal phosphate-dependent systems, where Schiff base complexes serve to delocalize negative charge following substrate activation, it is expected that C-H bond cleavage in the copper AOs will occur by a proton abstraction mechanism. Support for this pathway comes from early investigations by Abeles and coworkers on substrate analogs; among these studies one of the more informative was the demonstration of an anaerobic elimination of HCl from β-chlorophenethylamine (171). Numerous copper AOs use aromatic amines as substrates, making structure reactivity correlations with ring-substituted benzylamines a possibility. This has been pursued in both the porcine (172) and bovine SAOs (173). The latter study combined structure reactivity correlations with isotope effects, leading to a clear separation of rate constants for individual steps. In this manner, rate constants for the C-H bond cleavage step could be shown to increase with electron-releasing substituents, $\rho = 1.47 \pm 0.27$ (173).

The involvement of a proton abstraction mechanism raises questions regarding the active-site base responsible for this process. From pH-dependent studies of partial reactions in the bovine SAO reaction, a residue of pK_a of ca. 5 (169) has been implicated. Despite efforts to derivatize the active-site base and establish its identity, virtually all mechanism-based inhibitors have been found to give at least partial recovery of activity; the reversibility of inactivation suggests a carboxylate yielding unstable ester or anhydride intermediates (SM Janes, J Plastino, JP Klinman). In this context, it is worth noting that crystallographic studies of both MDH (105a) and MADH (174) suggest the participation of carboxylate side chains in catalysis. In light of the presence of a hydroxyl group in the TPQ, it is reasonable to question whether this functional group on the cofactor could catalyze substrate oxidation. However, this seems unlikely, given a very low pK_a for the hydroxyl group of enzyme-bound TPQ (see section on *Chemical Properties* below). Further, the orientation of the C-4 functional group would require C-H bond cleavage from substrate to occur in the plane of the ring formed by TPQ; it is expected that C-H cleavage perpendicular to the cofactor ring would generate optimal orbital overlap and delocalization of incipient negative charge on the substrate.

The stereochemistry of proton abstraction in the copper AOs has been studied in some detail. Early investigations of bovine SAOs demonstrated a curious phenomenon in which benzylamine showed absolute stereochemistry (175, 176), but phenethylamines displayed random processing of substrates (177). The latter observation was proposed to be due to "mirror image" binding of phenethylamine derivatives, with each binding mode giving rise to absolute but opposite stereochemistry. This work has been extended by Yu (178, 179) and Palcic (180–182), yielding the conclusions summarized in Table 3. As indicated, enzymes can be divided into two categories, depending on whether they catalyze an exchange at C-2 of substrate (186), in addition to oxidation at C-1 (180). These data, which show all possible stereochemical paths at C-1, indicate that the copper AOs differ from the majority of enzyme systems, which show conservation of stereochemistry within a reaction class (181). Given the uncertainty regarding the number of distinct TPQ-containing enzymes and their physiologic roles, MM Palcic (personal communication) has proposed that stereochemical distinctions can be used to implicate distinct gene products with unique functions.

Table 3 Summary of stereochemical trends among TPQ enzymes with phenethylamines as substrates

Protein source	C-1 abstraction	C-2 exchange	Ref.
Serum			
Bovine	random	yes	177, 179, 180
Sheep	random	yes	181
Rabbit	random	yes	181
Porcine	pro-R	yes	180
Seedling			
Pea	pro-S	no	183, 184, 180
Chick pea	pro-S	no	181
Soybean	pro-S	no	181
Porcine kidney	pro-S	no	179, 181
Aorta			
Rat	pro-S	unknown	179
Porcine	pro-S	yes	182
Bovine	pro-S	yes	182
Lysyl oxidase	pro-S	yes	185

As a result of the detailed mechanistic investigations of TPQ enzymes, a working mechanism has been proposed for the reductive half reaction (illustrated in Figure 7). Among the species shown, direct evidence exists for the substrate Schiff base, **1** [reductive trapping experiments (12, 167)], a carbanionic intermediate, **2** [substrate analogs (171) and structure reactivity correlations (173)], and an aminoquinol, **4** [transfer of ammonia from substrate to cofactor (119)]. The most elusive of the species in Figure 7 has been the product Schiff base, **3**, due to its rapid rate of hydrolysis; however, recent rapid-scanning stopped-flow studies (187) present evidence for a new spectral intermediate between 400 and 500 nm using benzylamines with electron-releasing substituents, attributed to a tautomerization of the product Schiff base to form a quinonoid intermediate. Thus, evidence now exists for each of the species postulated to arise in the course of TPQ interaction with substrate.

The role of the enzyme-bound copper in copper AOs [one Cu(II) per subunit] is less well understood than that of the organic cofactor. The spectroscopic properties of copper have been investigated by a variety of techniques [EXAFS (188–190), EPR and ESEEM (191), and ENDOR (192)], leading to a description of the copper-binding site that includes three equatorial histidine ligands, together with an equatorial and axial water molecule. The evidence in support of this structure and its properties are documented in an excellent forthcoming review by Knowles & Dooley

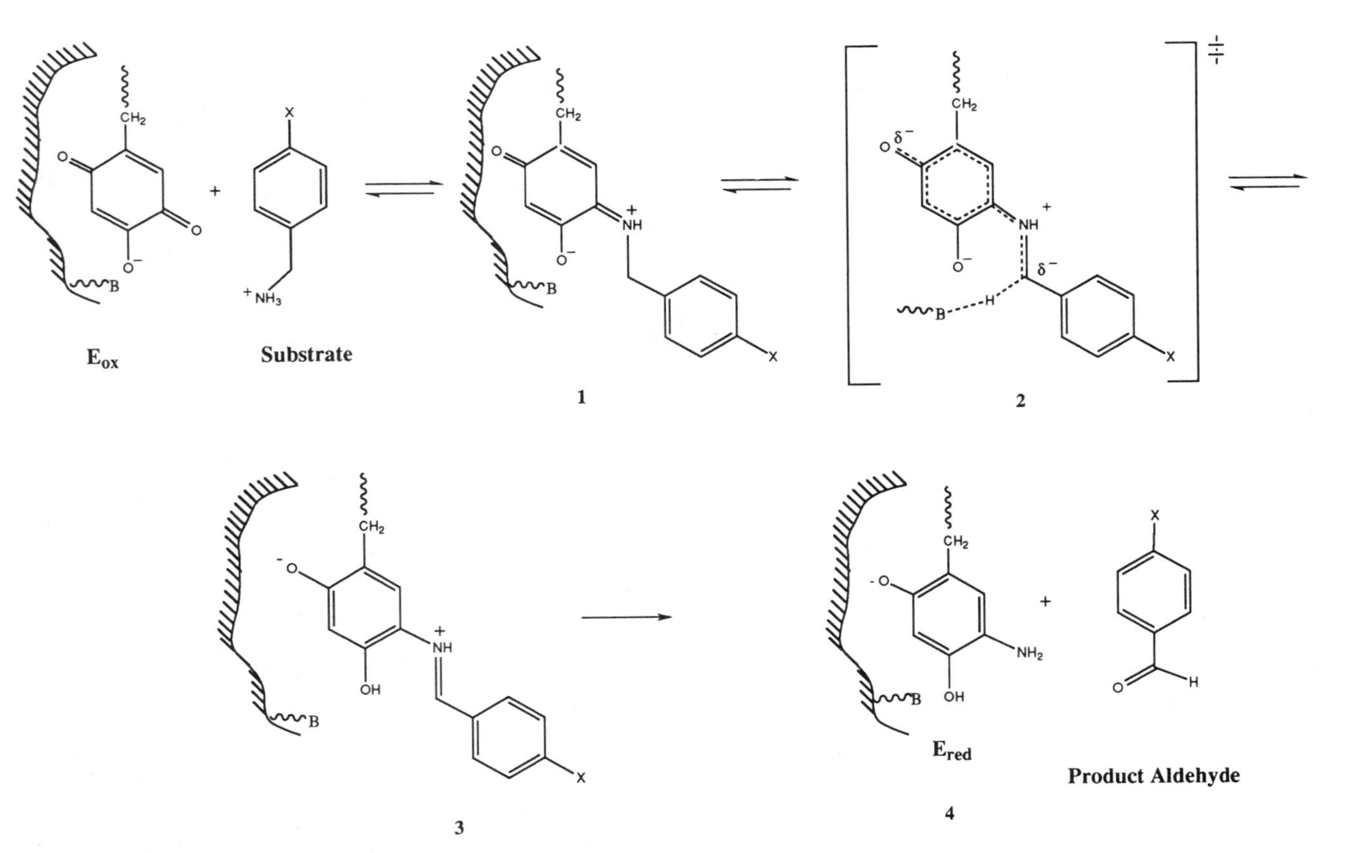

Figure 7 Pathway for the reductive half reaction of TPQ with amine substrates. Species **1** is the substrate Schiff base, **2** is the carbanionic transition state, **3** is product Schiff base, and **4** is aminoquinol (187).

(192a). The distance between the copper and TPQ has been the subject of a number of investigations using reagents that covalently bind to the cofactor and can be characterized by NMR (193), EPR (194), or fluorescent energy transfer (195). In all cases, the copper is concluded to be reasonably far from the reactive carbonyl of the cofactor, which, at the time these experiments were performed, was still believed to be PQQ. Reevaluation of these data with the structure for topa leads to an estimate of the distance between the C-2 oxygen of TPQ and active-site copper of approximately 3.0 Å (196).

As discussed under the section on DISCOVERY OF TOPA QUINONE IN A MAMMALIAN ENZYME, the initial characterization of TPQ was carried out with protein, which was determined to contain close to two functional cofactors per enzyme dimer. The stoichiometry of bound cofactor has, in fact, been quite controversial over the years, with numerous reports of half of the sites reactivity (cf Ref. 151), but only one well-documented example, porcine SAO (117). In general, as purer or more active preparations of enzyme have become available, the stoichiometry of bound cofactor has risen from one toward a value of two per enzyme dimer. This phenomenon has been documented recently with enzymes from lentil seedling (197) and bovine serum (119). As discussed by Mondavi and coworkers, the correct determination of cofactor stoichiometry relies ultimately on accurate values for both the protein subunit molecular weight and its extinction coefficient; using somewhat different values from Janes & Klinman, these authors continue to conclude that bovine SAO contains only a single reactive cofactor per dimer (198).

Throughout this review, the status of the quino-cofactor in lysyl oxidase has been treated with caution. As summarized in Table 3, the stereochemical properties of lysyl oxidase are identical to those for AOs derived from the aorta; however, none of the aortal enzymes have yet been demonstrated to contain TPQ. From a mechanistic perspective, lysyl oxidase displays all of the characteristics expected for TPQ, in particular, the formation of carbanion intermediates (199) in the course of substrate oxidation and the ability to undergo inactivation with the dicarbonyl reagents, vicinal diamines (200). The latter reaction, which leads to very effective inactivation of enzyme, was originally attributed to PQQ. In more recent experiments, TPQ in either bovine SAO (X Wang, JP Klinman, unpublished results) or model compounds (LM Sayre, personal communication) has been shown to be inhibited with vicinal diamines, providing further mechanistic support for a TPQ-like structure in lysyl oxidase.

OXIDATIVE HALF REACTION The interaction of molecular oxygen with copper AOs leads to an overall two-electron reduction of O_2 to hydrogen peroxide, concomitant with the reoxidation of the aminoquinol form of reduced cofactor. The nature of intermediates in this process, together with the role of copper, has been under investigation for a number of years. In early experiments, Finazzi-Agro and coworkers (201) showed that the addition of cyanide ion to substrate-reduced AOs from *Euphobia latex* or lentil seedlings led to the production of an organic radical. Although the interaction of enzyme with cyanide may occur at cofactor as well as copper (FT Greenaway, personal communication), a reasonable explanation advanced for the observed behavior was binding of cyanide ion to the Cu(I) form of enzyme, thereby trapping the semiquinone form of cofactor (201). In more recent experiments, Dooley and coworkers (202, 203) have detected cofactor-derived semiquinone species directly by varying the temperature at which EPR spectra were collected. It appears that an equilibrium between Cu(II) and reduced cofactor to form Cu(I) and semiquinone shifts toward semiquinone as the temperature is elevated (202). Kinetic studies indicate intramolecular electron-transfer rates between protein-bound copper and cofactor that are sufficiently rapid to be catalytically competent (203a). In retrospect, an earlier failure to detect semiquinone formation in the absence of cyanide can be attributed to the low temperatures routinely employed in EPR studies.

The involvement of copper in cofactor oxidation is supported by a reduction in the Cu(II) signal that accompanies semiquinone formation. Quantitation of signals indicates that the diminution of the Cu(II) signal actually exceeds the increase in amplitude due to semiquinone formation, likely due to disproportionation of the semiquinone to produce a mixture of reduced and oxidized cofactor (202). As a result of the above-described experiments, it is reasonable to assume that reduced TPQ undergoes reoxidation in two one-electron steps. Presumably, Cu(I)-semiquinone intermediate can be trapped by dioxygen, leading to the (transient) formation of superoxide. In a series of important experiments, the unpaired electron in the semiquinone species was shown to couple to at least one nitrogen nucleus; labeling experiments with either [^{14}N]- or [^{15}N]amine substrate indicate that this nitrogen nucleus is derived from substrate (203). It is therefore expected that two-electron oxidation of reduced TPQ will lead to an iminoquinone, which may either undergo hydrolysis with water or possibly, a direct transimination with substrate. Overall, the available experimental results for the oxidative half reaction are in full accord with the proposal of a transamination process during the reductive half reaction. For the future, X-ray crystallographic studies are expected to elucidate the spatial relationship between the active-site cofactor and copper [cf a recent preliminary report of X-ray diffraction quality crystals for pea seedling AO (204)].

Chemical Properties

Given the relative novelty of TPQ, the extent of chemical characterization of this cofactor has thus far been fairly limited. Work completed to date has focused primarily on the physical characterization of topa model compounds in which the reactive amino group of the free amino acid has either been tied back into a cyclic structure (topa hydantoin) or eliminated altogether (205). The absorbance spectrum of topa hydantoin indicates a prominent absorbance at 480 nm, giving rise to the characteristic red color of both the model compound and enzyme. Reduction of TPQ to topa quinol leads to a loss of the 480-nm band, together with the appearance of a new band at around 300 nm.

Spectroscopic titrations of both the oxidized and reduced forms of topa hydantoin indicate very different pK_a values: The oxidized form of cofactor shows a single pK_a at 4.13, whereas the reduced cofactor reveals three pK_a's with values of 9.17, 11.66, and >12. Comparison of the pH-dependent absorbance properties of protein-bound TPQ to those of the model compound indicates that the protein environment reduces its pK_a from 4.1 to 3.0. Electrochemical redox titrations of model compounds indicate similar pK_a values to spectroscopic studies, with an estimate for the third pK_a of reduced cofactor of 13.0. When the two-electron redox potential for topa hydantoin was compared to the dopa analog, topa was found to be more reducing by ca. 300 mV (e.g. the values of E_m for topa and dopa have been measured at pH 7.2 as -0.187 and 0.116 V, respectively, vs SCE). Of particular interest is the position of the cofactor undergoing attack by nucleophiles such as amines and hydrazines. Detailed analysis of the reaction of *p*-nitrophenylhydrazine with topa hydantoin by ^1H and ^{13}C-NMR demonstrates that the C-5 carbonyl of cofactor is the center where hydrazone formation occurs (205). Since the product of reaction of hydrazines with either model compounds or protein-bound cofactor is identical by resonance Raman (40), this result implies reaction of nucleophiles at the C-5 carbonyl of cofactor within the enzyme active site.

Given the anticipated formation of an aminoquinol product upon cofactor reduction by amine substrates, the properties of aminoquinol have been compared to quinol. The presence of the amino group has very little effect on redox potentials, but does lead to a small blue shift in absorbance for reduced cofactor from ca. 300 to 310 nm at neutral pH (consistent with absorbance spectra for reduced forms of AOs). pK_a determinations for the aminoquinol indicate values of 5.88, 9.59, and 11.62, with the first pK_a being assigned to the amino group (205).

These model studies provide some insight into the mechanistic role of TPQ in enzyme-catalyzed oxidation of substrate and reduction of dioxygen. First, the low pK_a of the oxidized form of cofactor has been postulated to provide an electrostatic stabilization of the initially formed substrate Schiff base (cf Figure 7). The large increase in pK_a observed upon cofactor reduction has been further postulated to lead to protonation of the C-4 oxyanion, thereby eliminating an electrostatic stabilization from the product Schiff base complex and leading to its rapid hydrolysis. The presence of the C-4 hydroxyl group also has a pronounced effect on the driving force for cofactor reoxidation, making this process more favorable by 0.3 V relative to a simple quinone such as dopa quinone. The unique ionization and electrochemical properties of topa may explain why this amino acid, rather than the naturally occurring amino acid dopa, has evolved as a redox active-site cofactor (205). It is expected that mechanistic studies contrasting the behavior of TPQ with other model compounds will provide further insight into the relationship between structure and function for this new cofactor (M Mure, JP Klinman, unpublished results; Sayre, personal communication).

In order to investigate the interaction of copper with topa, Suzuki and coworkers have constructed a topa-containing copper complex (206). X-ray studies of $[Cu(DL\text{-topa})bpy(H_2O)]BF_4 \cdot 3H_2O$ indicate complexation of both the amino and carboxylate functional groups of topa to the metal center. Since these functional groups have been incorporated into the polypeptide chain in protein-bound topa, the mode of complexation (if any) between copper and cofactor in protein will be different. It is of interest that the copper-topa complex leads to a ca. 14-fold rate acceleration of benzylamine oxidation relative to free topa (206). As noted by the authors, the distance between the C-2 carbonyl oxygen of topa and the copper center is 4.073 Å in the model compound (206), compared to an estimated value of ca. 3.0 Å in the protein (196).

7. PROPERTIES OF TTQ ENZYMES

Studies of Enzyme Mechanism

Periplasmic MADHs, the only known TTQ-containing enzymes,[3] exist in several obligate and facultative methylotrophic bacteria. The net reaction catalyzed by MADHs is shown in equation 5:

$$CH_3NH_2 + 2H_2O \xrightarrow{2(1e^- \text{ acceptor})} CH_2O + NH_4^+ + H_3O^+ \qquad 5.$$

Steady-state kinetic analyses of the MADH-catalyzed reactions have been performed on enzymes isolated from *M. extorquens* AM1 (206a), *Meth-*

[3]An aromatic amine dehydrogenase from *Alcaligenes faecalis*, has also been claimed to contain TTQ from resonance Raman studies (VL Davidson, personal communication).

ylomonas sp. J (207), *P. denitrificans* (208), and bacterium W3A1 (209); in each case, a ping-pong type of mechanism has been implicated. Thus, the working model of the enzymatic mechanism can be viewed as a reductive half reaction leading to cofactor reduction by methylamine, followed by an oxidative half reaction involving a sequential transfer of two electrons from reduced cofactor to a separate copper protein. This is reminiscent of the mechanism of TPQ (equations 3 and 4), with the exception that TTQ reoxidation requires an exogenous copper protein acceptor (210, 211).

THE REDUCTIVE HALF REACTION A deuterium kinetic isotope effect of 3.0 has been observed on V_{max} under steady-state studies of methylamine oxidation catalyzed by *P. denitrificans* (208), suggesting that the abstraction of a methyl proton is partially rate limiting. Stopped-flow kinetic studies of cofactor reduction have been pursued with MADHs from W3A1 (212) and *P. denitrificans* (213). In the case of the W3A1 enzyme, two transients were observed: The first, relatively fast transition was attributed to quinone reduction, while the second, slow transition was attributed to the release of the product aldehyde. Primary deuterium kinetic isotope effects of 20 and 17 have been assigned to the cofactor reduction step directly using enzymes from W3A1 (212) and *P. denitrificans* (213). The origin of these large isotope effects is unknown. However, a similar large kinetic isotope effect has been analyzed in detail in the TPQ-dependent bovine SAO (170), and attributed to the phenomenon of quantum-mechanical tunneling.

The mechanism of C-H activation leading to quinone reduction was probed by structure-reactivity correlations using substituted benzylamines. Although aromatic amines do not support catalytic turnover with MADHs, benzylamines have been observed to reduce the TTQ cofactor of MADH stoichiometrically (214). As a result of such studies, the reductive half reaction of MADHs has been suggested to involve carbanionic species and a transamination mechanism, analogous to TPQ-dependent enzymes (cf Figure 7). From the close examination of the immediate environment of the TTQ cofactor in the crystal structure of the MADH from *T. versutus,* Huizinga et al (174) hypothesized that Asp76 may perform the function of base catalyses in substrate oxidation.

The position of the reactive carbonyl group of TTQ undergoing nucleophilic attack has been addressed by X-ray crystallography studies of two complexes. An enzyme-methylhydrazone complex was obtained by soaking the enzyme crystal in the inhibitor solution, while a (2,2,2-trifluoroethyl)hydrazone-enzyme complex was prepared by crystallizing the inhibited enzyme directly. In the electron density maps, the C-6 carbonyl of cofactor was found to possess extra electron density corresponding to the two nitrogen atoms of the hydrazone. This implicates the C-6 carbonyl of the ortho-quinone moiety as the active one toward substrates. The poor electron density of the methyl or trifluoroethyl groups of the bound inhibitors was attributed to incomplete inhibition or disorder in the crystal structures (174).

THE OXIDATIVE HALF REACTION In an early experiment, a low-molecular-weight copper protein, amicyanin, was shown to replace the artificial electron acceptor phenazine methosulfate in the assay of MADH from *M. extorquens*. While this enzyme was not able to reduce c-type cytochrome directly, reduction occurred in the presence of amicyanin. Consequently, Tobari et al (210) postulated the sequential electron transfer pathway of equation 6.

$$\text{methylamine} \rightarrow \text{MADH} \rightarrow \text{amicyanin} \rightarrow \text{c-type cytochrome} \qquad 6.$$

More recently, it has been shown that disruption of the amicyanin gene abolishes the ability of certain methylotrophic bacteria to utilize methylamine (215).

Resonance Raman (215a), and X-ray crystallographic studies (216) of amicyanin indicate that this protein is similar to other small, type I copper proteins such as azurin and plastocyanin (217). The crystal structure of the binary complex of MADH and amicyanin, both isolated from *P. denitrificans,* has been solved at a resolution of 2.5 Å (218), with a composition of $\alpha_2\beta_2$Amicyanin$_2$. The location of the copper atom followed from a comparison of copper-free apo-amicyanin to the holo-protein. Each amicyanin molecule is in contact with both the large and small subunits of MADH, but the interaction with the TTQ-bearing small subunit is more extensive. The nonquinone portion (Trp107) of TTQ and the copper atom of amicyanin are separated by a distance of 9.3 Å. One of the four copper ligands of amicyanin, His95, was found on the surface of amicyanin and located halfway between TTQ and copper. Therefore, it was speculated that His95 may play a role in mediating the electron transfer from TTQ to copper; the closest distance between His95 and TTQ is 5.4 Å. The ortho-quinone-containing portion (Trp57) of TTQ is pointing in a direction opposite to that for the putative electron-transfer triad, [(Trp107)—(His95)—(Copper)]. This indicates that the turnover of substrate is taking place at a distance from the electron-transfer site, eliminating any possible interference between the two crucial functions (218). The reduction of amicyanin by methylamine and MADH was found to decrease as the solution ionic strength was raised, while the reduction of an artificial electron acceptor (phenazine ethosulfate) was not affected by ionic strength. To interpret these results, electrostatic interactions were proposed to be the major force in the interaction of MADH with amicyanin (219). However, the binary complex crystal structure indicates a predominance of hydrophobic interactions at the interface of these two proteins (218). Since a high salt condition (2.3–2.6 M of sodium/potassium phosphate) was used to obtain the crystals of the protein complex, it is possible that this structure differs from the catalytic complex formed under enzyme turnover.

A preliminary crystal structure of the ternary complex of MADH, amicyanin, and cytochrome, isolated from *P. denitrificans,* has been reported by Mathews and coworkers (220). Two types of ternary complexes were prepared from apo- or holo-amicyanin, at resolutions of 2.4 Å and 2.75 Å, respectively. In this preliminary structure, the iron-heme of cytochrome and the copper redox center of amicyanin could be unambiguously located, even though protein portions were not processed at high resolution. The crystal composition, $\alpha_2\beta_2$Amicyanin$_2$Cytochrome$_2$, corroborates the stoichiometry previously advanced (221, 131). The positions of the three redox centers (TTQ-copper-iron) fall approximately on a line, confirming the long-believed sequential relationship of electron transfer (equation 6). The ternary complex structure demonstrates that the iron of the cytochrome heme group is 24.8 Å away from the copper of amicyanin. The structure of the interface between MADH and amicyanin in the ternary complex is similar to what was observed in the crystal structure of MADH-amicyanin binary complex (218), suggesting little structural influence from the binding of cytochrome c (220). Very recently, FS Mathews and coworkers completed a refinement of the crystal structure of the ternary complex (personal communication). This allows a theoretical estimation of the electron-transfer efficiency between each pair of donor and acceptor using most probable electron-transfer routes. The calculated results predict that the coupling of TTQ cofactor and copper is 1000-fold more efficient than that of the copper and heme-iron group. This result contrasts with stopped-flow studies, which suggest similar rate constants for electron transfer from MADH to amicyanin and from amicyanin to cytochrome c (222, 223). The inconsistency between theory and experiment may reflect the involvement of additional barriers such as solvent or protein reorganization (FS Mathews, personal communication).

The redox midpoint potentials (E_m vs Ag/AgCl) of free amicyanin was measured to be +294 mV (224), which is thermodynamically unfavorable with regard to its electron acceptor, cytochrome c_{551} [E_m = +190 mV (224)]. However, the E_m of amicyanin in the MADH binary complex is shifted to +221 mV (219). This decrease of E_m by 73 mV may arise from a charge neutralization of residues near the redox center, as suggested from the recently refined crystal structure for ternary complex (FS Mathews, personal communication).

The semiquinone species of TTQ has been actively investigated. MADH from bacterium W3A1 was initially shown to go through a biphasic reduction by dithionite, ascribed to the appearance and disappearance of TTQ semiquinone (225). For a second MADH from *P. denitrificans,* the reductive titration of 50% of the enzyme-bound cofactor by methylamine at pH 7.5 was, on the contrary, monophasic and no semiquinone species could be detected; however, elevation of the pH to 9.0 led to a rapid appearance of the TTQ semiquinone from the 50% reduced enzyme (226). These results have been attributed to a pH-dependent disproportionation of the oxidized and reduced forms of TTQ. The protein concentration dependence of this reaction supports an intermolecular rather than an intramolecular process. The mechanistic importance of semiquinone intermediate in the oxidative half reaction is apparent from the one-to-one stoichiometry of TTQ cofactor to copper atom, as seen in the crystal structure of the MADH-amicyanin binary complex (218). Warncke et al (227) designed an important experiment to understand further the status of TTQ semiquinone in the reductive half reaction. Following the mixing of fully oxidized MADH with an amount of either [^{14}N]- or [^{15}N]-methylamine equal to 50% of enzyme-bound cofactor, TTQ semiquinone was quantitatively generated by electron disproportionation. Studies of this semiquinone species using ESEEM or pulsed-EPR spectroscopy indicate that the substrate-derived nitrogen atom is covalently attached to the semiquinone. This provides direct evidence for the involvement of substrate preceding the formation of semiquinone and suggests that release of ammonia takes place either after or in concert with the oxidation of TTQ semiquinone (227).

Chemical Properties

The crosslinked di-tryptophanyl structure of TTQ presents a synthetic challenge. Within a year of the discovery of TTQ, Itoh et al (148a) succeeded in the chemical synthesis of a model compound for the TTQ cofactor, **5**:

5

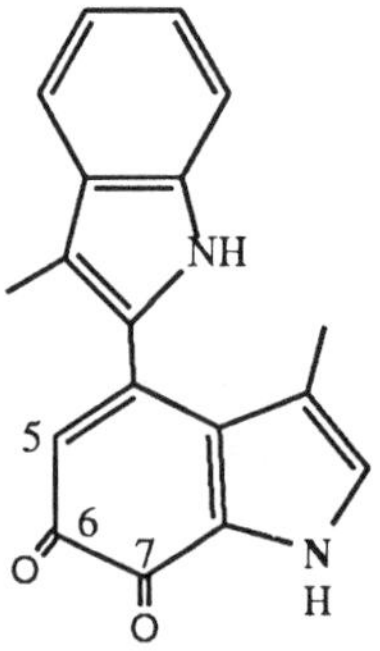

The optical absorption properties, redox potential, and resonance Raman spectrum of compound **5** are not very different from those of TTQ in the native MADH. A dihedral angle of 54.7° is suggested for the two indole rings of **5** using either computational method or a nuclear Overhauser effect experiment. Preliminary studies on the catalytic ability of the TTQ model compound have shown that **5** is very efficient in catalyzing the oxidation of benzylamine into N-benzylidenebenzylamine, under aerobic conditions in organic solvents such as methanol and acetonitrile. However, no detailed account of this reaction has been provided thus far (148a).

Genetic Studies

The name "*mau*" has been assigned to the genes involved in methylamine utilization (138). The structural gene for the TTQ-containing MADH small subunit had been cloned from *M. extorquens* AM1 (138), and subsequently from *P. denitrificans* (215), and *T. versutus* (228). The *M. extorquens* AM1 system has been pursued in detail by Lidstrom and coworkers and is reviewed as follows.

A 5.2-kb BamI-HindIII chromosomal DNA fragment was isolated from *M. extorquens* AM1; following the expression of different deletion mutants, five genes were located on the 5.2-kb fragment in the following order (229):

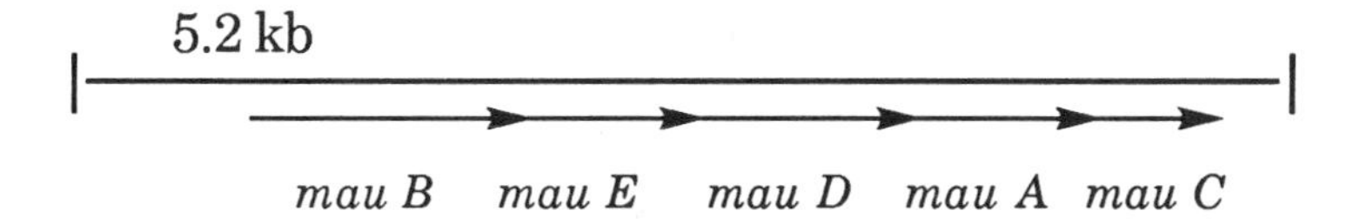

The large and small subunits of MADH correspond to the products of *mau* B and A, while *mau* C encodes amicyanin. A mutant containing disrupted *mau* D exhibits a negative methylamine phenotype, indicating that *mau* D is important for methylamine utilization. Since *mau* E is in the stack of methylamine utilization genes, it is not unreasonable to suggest that it also belongs to the *mau* family. However, the enigmatic functions of *mau* E and D gene products remain to be clarified. Little is known about the transcriptional regulation of *mau* genes. Putative hairpin structures, which may terminate transcription, have been identified at both the 5′ and 3′ ends of *mau* C. Additionally, gene fusion studies in *E. coli* indicate at least two other functional transcriptional terminators located on both sides of *mau* B. As a consequence of these studies, it has been suggested that *mau* genes may be transcribed separately (229).

The nucleotide sequence of the TTQ-containing small subunit revealed an especially long leader sequence (57 residues) with an uncommon positively charged region preceding the cleavage site. Gene fusion experiments have indicated that this leader sequence is unique to *M. extorquens* AM1 and cannot be recognized by *E. coli*. It has been suggested that the presence of this leader peptide may reflect an unusual secretory mechanism that is relevant to TTQ production (230).

The genes of the MADH large and small subunits were recently sequenced and mapped onto a 5.6-kb chromosomal fragment from *P. denitrificans* (231). This is the first complete DNA-derived amino acid sequence of MADH that includes both the large and small subunits. The DNA-deduced protein sequence was used to refine the crystal structure of *P. denitrificans* MADH to a resolution of 2.5 Å. In addition to the five *mau* genes (B, E, D, A, and C) identified in *M. extorquens* AM1, a sixth gene (designated *mau* F, with unknown function) was found upstream of *mau* B on the 5.6-kb DNA fragment (231). In light of this sixth gene (*mau* F) from *P. denitrificans* and the presence of an open reading frame downstream of *mau* C in *M. extorquens* AM1 (229), it is likely that more genes are involved in methylamine utilization. In terms of TTQ biogenesis, it would be very informative to investigate the integrity of TTQ on the MADH small subunit purified from the different deletion mutants generated by Lidstrom and coworkers (229). In contrast to the possible self-catalyzed TPQ biogenesis of copper AOs (145), the TTQ biogenesis of metal-free MADH is likely to be carried out by a separate class of enzymes. The elucidation of the *mau* D, E, and F genes should provide a promising beginning toward studies of cofactor biogenesis, secretion, and transcriptional regulation.

8. PERSPECTIVE AND FUTURE DIRECTIONS

It is expected that investigations during the next several years will lead to numerous advances in our understanding of the role of quino-cofactors in biology. Of particular urgency is the resolution of the role played by PQQ in eukaryotes. Can agreement be reached regarding the actual levels of PQQ present in mammalian tissues and further, whether this cofactor functions as a general antioxidant? It will also be interesting to see if a role emerges for PQQ as a reversibly bound cofactor in select redox enzymes in eukaryotes. In prokaryotes, what is the number of gene products and enzyme chemistry leading to PQQ production from tyrosine and glutamate? The full scope of TPQ in eukaryotes is not yet clear. Is this protein only present in the copper AOs, or can it be demonstrated in additional proteins with functions other than amine oxidation? What is the nature of the cofactor in lysyl oxidase, i.e. does this unique protein possess the same cofactor as the other copper AOs? The biogenesis of TPQ is a particularly interesting area, given the potential for "self-processing" of an active-site tyrosine to topa by the active-site copper center in the copper AOs. From a structural perspective, it is expected that the preliminary report of diffractable crystals for a plant copper AO will lead to the first three-dimensional structure for a TPQ-containing enzyme. Turning to TTQ, is this unusual cofactor restricted to prokaryotes, or can it be shown to have a more ubiquitous role in nature? How does a dimer of tryptophan arise in an enzyme active site and what is the nature of the enzymes leading to tryptophan oxidation and dimerization? As this partial list indicates, many provocative and engaging questions remain to be answered in the field of quinoenzymes.

Literature Cited

1. Davidson VL, ed. 1993. *Principles and Applications of Quinoproteins*. New York: Dekker
2. Von Werle E, Von Rechman E. 1949. *Justus Liebigs Ann. Chem.* 562:44–60
3. Yamada H, Yasunobu KT. 1962. *J. Biol. Chem.* 237:3077–82
4. Pettersson G. 1985. In *Structure and Functions of Amine Oxidases,* ed. B Mondovi, pp. 105–20. Boca Raton, Fla: CRC Press
5. Hamilton GA. 1981. In *Copper Proteins,* ed. TG Spiro, pp. 205–18. New York: Wiley-Interscience
6. Suva RH, Abeles RH. 1978. *Biochemistry* 17:3538–45
7. Berg KA, Abeles RH. 1980. *Biochemistry* 19:3186–89
8. Lobenstein-Verbeek CL, Jongejan JA, Frank J, Duine JA. 1984. *FEBS Lett.* 170:305–9
9. Ameyama M, Hayashi U, Matsushita K, Shinagawa E, Adachi O. 1984. *Agric. Biol. Chem.* 48:561–65
10. Moog RS, McGuirl MA, Cote CE, Dooley DM. 1986. *Proc. Natl. Acad. Sci. USA* 83:8435–39
11. Knowles PF, Pandeya KB, Rius FX, Spencer CM, Moog RS, et al. 1987. *Biochem. J.* 241:603–8
12. Hartmann C, Klinman JP. 1987. *J. Biol. Chem.* 262:962–65
13. van der Meer RA, Jongejan JA, Frank J, Duine JA. 1986. *FEBS Lett.* 206: 111–14
14. Glatz Z, Kovar J, Macholan L, Pec P. 1987. *Biochem. J.* 242:603–6
15. Citro G, Verdina A, Galati R, Floris G, Sabatini S, Finazzi- Agro A. 1989. *FEBS Lett.* 247:201–4
16. van der Meer RA, Duine JA. 1986. *Biochem. J.* 239:789–91
17. Williamson PR, Moog RS, Dooley DM, Kagan HM. 1986. *J. Biol. Chem.* 261:16302–5
18. van der Meer RA, Jongejan JA, Duine JA. 1988. *FEBS Lett.* 231:303–7
19. van der Meer RA, Jongejan JA, Duine JA. 1989. *J. Biol. Chem.* 264:7792–94
20. Karhunen E, Niku-Paavola M-L, Viikara L, Haltia T, van der Meer RA, Duine JA. 1990. *FEBS Lett.* 267:6–8
21. Groen BW, van der Meer RA, Duine JA. 1988. *FEBS Lett.* 237:98–102
22. van der Meer RA, Duine JA. 1988. *FEBS Lett.* 235:194–200
23. Ameyama M, Shinagawa E, Matsushita K, Takimoto K, Nakashima K, Adachi O. 1985. *Agric. Biol. Chem.* 49:3623–26
24. Gacheru SN, Trackman PC, Shah MA, O'Gara CY, Spacciapoli P, et al. 1990. *J. Biol. Chem.* 265:19002–27
25. Williamson PR, Kittler JM, Thanassi JW, Kagan HM. 1986. *Biochem. J.* 235:597–605
26. Trackman PC, Pratt AM, Wolanski A, Tang S-S, Offner GD, et al. 1990. *Biochemistry* 29:4863–70
27. Trackman PC, Pratt AM, Wolanski A, Tang S-S, Offner GD, et al. 1991. *Biochemistry* 30:8282
28. Ito N, Phillips SEV, Stevens C, Ogel ZB, McPherson MJ, et al. 1991. *Nature* 350:87–90
29. Whittaker MM, Whittaker JW. 1990. *J. Biol. Chem.* 265:9610–13
30. Branchaud BP, Montague-Smith MP, Kosman DJ, McLaren FR. 1993. *J. Am. Chem. Soc.* 115:798–800

31. Boyington JC, Gaffney BJ, Amzel LM. 1993. *Science* 260:1482–86
32. Minor W, Steczko J, Bolin JT, Otwinowski Z, Axelrod B. 1993. *Biochemistry* 32:6320–23
33. Messerschmidt A, Huber R. 1990. *Eur. J. Biochem.* 187:341–52
34. Messerschmidt A, Ladenstein R, Huber R, Bolognesi M, Abigliano L, et al. 1992. *J. Mol. Biol.* 224:179–205
35. Fitzpatrick PE, Villafranca JJ. 1986. *J. Biol. Chem.* 261:4510–18
36. Robertson JG, Kumar A, Mancewicz JA, Villafranca JJ. 1989. *J. Biol. Chem.* 264:19916–21
37. Farrington GK, Kumar A, Villafranca JJ. 1990. *J. Biol. Chem.* 265:1036–40
37a. Wang N, Southan C, DeWolf WE, Wells TNC, Kruse LI, et al. 1990. *Biochemistry* 29:6466–74
38. Takagishi T, Tanaka T, Imamura I, Mizuguchi H, Kuroda M, et al. 1990. In *Enzymes Dependent on Pyridoxal Phosphate and Other Carbonyl Compounds as Cofactors,* ed. T Fukui, H Kagamiyama, H Soda, H Wada, pp. 487–88. New York: Pergamon
39. Janes SM, Mu D, Wemmer D, Smith AJ, Kaur S, et al. 1990. *Science* 248:981–87
40. Brown DE, McGuirl MA, Dooley DM, Janes SM, Mu D, Klinman JP. 1991. *J. Biol. Chem.* 266:4049–51
41. Janes SM, Palcic MM, Scaman CH, Smith AJ, Brown DE, et al. 1992. *Biochemistry* 31:12147–54
42. Rossi A, Petruzzelli R, Finazzi-Agro A. 1992. *FEBS Lett.* 301:253–57
43. Pederson JZ, El-Sherbini S, Finazzi-Agro A, Rotilio G. 1992. *Biochemistry* 31:8–12
44. Paz MA, Flückiger R, Boak A, Kagan HM, Gallop PM. 1991. *J. Biol. Chem.* 266:689–92
45. Maccarrone M, Veldink GA, Vliegenthart JFG. 1991. *J. Biol. Chem.* 266:21014–17
46. Solomon EI, Hemming BL, Root DE. 1993. In *Bioinorganic Chemistry of Copper,* ed. KD Karlin, Z Tyeklár, pp. 3–20. New York: Chapman & Hall
47. Steinberg FM, Smidt C, Kilgore J, Romero-Chapman N, Tran D, et al. 1993. See Ref. 1, pp. 367–79
47a. Smidt CR, Steinberg FM, Rucker RB. 1991. *Proc. Soc. Exp. Biol. Med.* 197:19–26
48. Killgore J, Smidt C, Duich L, Romero-Chapman N, Tinker D, et al. 1989. *Science* 245:850–52
49. Paz MA, Flückiger R, Henson E, Gallop PM. 1988. In *PQQ and Quinoproteins,* ed. JA Jongejan, JA Duine, pp. 131–43. Dordrecht, The Netherlands: Kluwer
50. van der Meer RA, Groen BW, Jongejan JA, Duine JA. 1990. *FEBS Lett.* 261: 131–34
51. Paz MA, Flückiger R, Gallop PM. 1990. *FEBS Lett.* 264:283–84
51a van der Meer RA, Groen BW, Jongejan JA, Duine JA. 1990. *FEBS Lett.* 264: 284
52. Flückiger R, Paz MA, Henson E, Gallop PM. 1993. See Ref. 1, pp. 331–41
53. Kumazawa T, Seno H, Urakami T, Matsumoto T, Suzuki O. 1992. *Biochim. Biophys. Acta* 1156:62–66
54. Paz MA, Mah J, Flückiger R, Stang PJ, Gallop PM. 1993. *Cofactor PQQ in mitochondrial complex I is the target rotenone, MPP^+ and iodonium compounds.* Presented at the Annu. Meet. Am. Soc. Biochem. Mol. Biol., San Diego, California (Abstr.)
55. Paz MA, Flückiger R, Gallop PM. 1993. See Ref. 1, pp. 381–93
56. Aizenman E, Hartnett KA, Zhong C, Gallop PM, Rosenberg PA. 1992. *J. Neurosci.* 12:2362–69
57. Watanabe A, Tsuchida T, Nishigori H, Urakami T, Hobara N. 1993. See Ref. 1, pp. 395–408
58. Xu F, Mack CP, Quandt KS, Shlafer M, Massey V, Hultquist DE. 1993. *Biochem. Biophys. Res. Commun.* 193: 434–39
59. Hultquist DE, Xu F, Quandt KS, Shlafer M, Mack CP, et al. 1993. *Am. J. Hemat.* 42:13–18
60. Webber T, Bach TJ. 1992. In *Metabolism, Structure and Utilization of Plant Lipids,* ed. O. Cherif, D Ben-Miled-Daoud, B Marzouk, A Smaoui, M Zarrouk, pp. 243–46. Tunis, Tunisia: Centre Natl. Pedagogigue
61. Salisbury SA, Forrest HS, Cruse WBT, Kennard O. 1979. *Nature* 280: 843–44
62. Duine JA, Frank J, van Zeeland JK. 1979. *FEBS Lett.* 108:443–46
63. Ameyama M, Nonobe M, Shinagawa E, Matsushita K, Adachi O. 1985. *Anal. Biochem.* 151:263–67
64. Duine JA, Frank J, Verwiel PEJ. 1980. *Eur. J. Biochem.* 108:187–92
65. Xia Z-X, Dai W-W, Xiong J-P, Hao Z-P, Davidson VL, et al. 1992. *J. Biol. Chem.* 267:22289–97
66. Hommel R, Kleber H-P. 1990. *J. Gen. Microbiol.* 136:1705–11
67. Matsushita K, Adachi O. 1993. See Ref. 1, pp. 47–63
68. Matsushita K, Adachi O. 1993. See Ref. 1, pp. 65–71

69. Anthony C. 1993. See Ref. 1, pp. 223–44
70. Anthony C, Chen HTC, Cox JM, Richardson IW. 1992. In *Microbial Growth on C1 Compounds,* ed. JC Murrel, DP Kelly, pp. 221–33. Hampshire, UK: Intercept
71. Anthony C. 1992. *Biochim. Biophys. Acta* 1099:1–15
72. Davidson VL. 1992. *Biochemistry* 31: 1504–8
73. Hopper DJ, Rogozinski J, Toczko M. 1991. *Biochem. J.* 279:105–9
74. van der Meer RA, Groen BW, Duine JA. 1989. *FEBS Lett.* 246:109–12
75. Pennings EJM, Groen BW, Duine JA, Verpoorte R. 1989. *FEBS Lett.* 255:97–100
76. De Biase D, Maras B, John RA. 1991. *FEBS Lett.* 278:120–22
77. Jin HY, Turner IM, Nelson MJ, Gurbiel RJ, Doan PE, Hoffman BM. 1993. *J. Am. Chem. Soc.* 115:5290–91
78. Nagasawa T, Yamada H. 1987. *Biochem. Biophys. Res. Commun.* 147: 701–9
78a. Hsiano T, Murata K, Kimura A, Matsushita K, Adachi O. 1992. *Biosci. Biotech. Biochem.* 56:311–14
78b. Shinagawa E, Matsushita K, Nakashima K, Adachi O, Ameyama M. 1988. *Agric. Biol. Chem.* 52: 2255–63
79. Adachi O, Matsushita K, Shinagawa E, Ameyama M. 1990. *Agric. Biol. Chem.* 54:3123–29
80. Tabor CW, Kellogg PD. 1970. *J. Biol. Chem.* 245:5424–33
81. Groen BW, van Kleef MAG, Duine JA. 1986. *Biochem. J.* 234:611–15
82. Ameyama M, Adachi O. 1982. *Methods Enzymol.* 89:450–57
83. Matsushita K, Takaki Y, Shinagawa E, Ameyama M, Adachi O. 1992. *Biosci. Biotech. Biochem.* 56:304–10
84. Matsushita K, Shinigawa E, Inoue T, Adachi O, Ameyama M. 1986. *FEMS Microbiol. Lett.* 37:141–44
85. Cleton-Jansen A-M, Goosen N, Wenzel TJ, van de Putte P. 1988. *J. Bacteriol.* 170:2121–25
86. Cleton-Jansen A-M, Goosen N, Vink K, van de Putte P. 1989. *Mol. Gen. Genet.* 217:430–36
87. Cleton-Jansen A-M, Goosen N, Odle G, van de Putte P. 1988. *Nucleic Acids Res.* 16:6228
88. Cleton-Jansen A-M, Goosen N, Fayet O, van de Putte P. 1990. *J. Bacteriol.* 172:6308–15
89. Cleton-Jansen A-M, Dekker S, van de Putte P, Goosen N. 1991. *Mol. Gen. Genet.* 229:206–12
90. Yamada M, Sumi K, Matsushita K, Adachi O, Yamada Y. 1993. *J. Biol. Chem.* 268:12812–17
91. Friedrich T, Strohdeicher M, Hofhaus G, Preis D, Sahm H, Weiss H. 1990. *FEBS Lett.* 265:37–40
92. Nunn DN, Day D, Anthony C. 1989. *Biochem. J.* 260:857–62
93. Matsushita K, Takahashi K, Adachi O. 1993. *Biochemistry* 32:5576–82
94. Anderson DJ, Lidstrom ME. 1988. *J. Bacteriol.* 170:2254–61
95. Nunn DN, Lidstrom ME. 1986. *J. Bacteriol.* 166:581–90
96. Nunn DN, Lidstrom ME. 1986. *J. Bacteriol.* 166:591–97
97. Duine JA, Frank J, Verwiel PEJ. 1981. *Eur. J. Biochem.* 118:395–99
98. Mutzel A, Görisch H. 1991. *Agric. Biol. Chem.* 55:1721–26
99. Adachi O, Matsushita K, Shinagawa E, Ameyama M. 1990. *Agric. Biol. Chem.* 54:2833–37
100. Laufer K, Lidstrom ME. 1993. See Ref. 1, pp. 193–221
101. Lidstrom ME. 1990. *FEMS Microbiol. Rev.* 87:431–36
102. Richardson IW, Anthony C. 1992. *Biochem. J.* 287:709–15
103. Anderson DJ, Morris CJ, Nunn DN, Anthony C, Lidstrom ME. 1990. *Gene* 90:173–76
104. Machlin SM, Hanson RS. 1988. *J. Bacteriol.* 170:4739–47
105. Harms N, de Vries GE, Maurer K, Hoogendijk J, Stouthamer AH. 1987. *J. Bacteriol.* 169:3969–75
105a. White S, Boyd G, Mathews FS, Xia Z-X, Dai W-W, et al. 1993. *Biochemistry*. In press
106. Inoue T, Sunagawa M, Mori A, Imai C, Fukuda M, et al. 1990. *J. Ferment. Bioeng.* 70:58–60
107. Anthony C. 1993. See Ref. 1, pp. 17–45
108. Frank J, van Krimpen SH, Verwiel PEJ, Jongejan JA, Mulder AC, Duine JA. 1989. *Eur. J. Biochem.* 184:187–95
109. Houck DR, Hanners JL, Unkefer CJ. 1991. *J. Am. Chem. Soc.* 113:3162–66
110. van Kleef MAG, Duine JA. 1988. *FEBS Lett.* 237:91–97
111. Houck DR, Hanners JL, Unkefer CJ, van Kleef MAG, Duine JA. 1989. *Antonie van Leeuwenhoek* 56:93–101
112. Goosen N, Horsman HPA, Huinen RGM, de Groot A, van de Putte P. 1989. *Antonie van Leeuwenhoek* 56:85–91
113. Unkefer CJ. 1993. See Ref. 1, pp. 343–53
114. Meulenberg JJM, Sellink E, Loenen

WAM, Riegman NH, van Kleef M, Postma PW. 1990. *FEMS Microbiol. Lett.* 71:337–44
115. Biville F, Turlin E, Gasser F. 1991. *J. Gen. Microbiol.* 137:1775–82
116. Biville F, Turlin E, Gasser F. 1989. *J. Gen. Microbiol.* 135:2917–29
117. Collison D, Knowles PF, Mabbs FE, Rius FX, Singh I, et al. 1989. *Biochem. J.* 264:663–69
118. Yamada H, Yasunobu KT. 1963. *J. Biol. Chem.* 238:2669–75
119. Janes SM, Klinman JP. 1991. *Biochemistry* 30:4599–605
120. van der Meer RA, van Wassenaar PD, van Brouwershaven JH, Duine JA. 1989. *Biochem. Biophys. Res. Commun.* 159:726–33
121. Janes SM. 1990. *Mechanistic and structural characterization of bovine serum amine oxidase.* PhD thesis. Univ. Calif., Berkeley
122. Waite JH. 1983. *J. Biol. Chem.* 258: 2911–15
123. Waite JH. 1987. *ChemTech.* 17:692–97
124. Marumo K, Waite JH. 1986. *Biochim. Biophys. Acta* 872:98–103
125. Waite JH, Rice-Ficht AC. 1987. *Biochemistry* 26:7819–25
126. Graham DG, Tiffany SM, Bell WR, Gutknecht WF. 1978. *Mol. Pharmacol.* 14:644–53
127. Rosenberg PA, Loring R, Xie Y, Zaleskas V, Aizenman E. 1991. *Proc. Natl. Acad. Sci. USA* 88:4865–69
128. Aizeman E, Boeckman FA, Rosenberg PA. 1992. *Neurosci. Lett.* 144:233–36
129. Duvoisin RC. 1984. In *Parkinson's Disease: A Guide for Patient and Family,* pp. 81–104. New York: Raven
130. De Beer R, Duine JA, Frank J, Large PJ. 1980. *Biochim. Biophys. Acta* 622: 370–74
131. Davidson VL. 1993. See Ref. 1, pp. 73–95
132. Ishii Y, Hase T, Fukumori Y, Matsubara H, Tabori J. 1983. *J. Biochem.* 93:107–19
133. van der Meer RA, Jongejan JA, Duine JA. 1987. *FEBS Lett.* 221:299–304
134. van der Meer RA, Mulder AC, Jongejan JA, Duine JA. 1989. *FEBS Lett.* 254:99–105
135. McIntire WS, Stults JT. 1986. *Biochem. Biophys. Res. Commun.* 141: 562–69
136. Vellieux FMD, Huitema F, Groendijk H, Kalk KH, Jzn JF, et al. 1989. *EMBO J.* 8:2171–78
137. Vellieux FMD, Hol WGJ. 1989. *FEBS Lett.* 255:460–64
138. Christoserdov AY, Tsygankov YD, Lidstrom ME. 1990. *Biochem. Biophys. Res. Commun.* 172:211–16
139. McIntire WS, Wemmer DE, Christoserdov A, Lidstrom ME. 1991. *Science* 252:817–24
140. Chen L, Mathews FS, Davidson VL, Huizinga EG, Vellieux FMD, et al. 1991. *FEBS Lett.* 287:163–66
141. Dooley DM, Brown DE. 1993. See Ref. 1, pp. 275–305
142. Flückiger R, Woodtli T, Gallop PM. 1988. *Biochem. Biophys. Res. Commun.* 153:353–58
143. Geiger O, Dörisch H. 1987. *Anal. Biochem.* 164:418–23
144. McIntire WS, Bates JL, Brown DE, Dooley DM. 1991. *Biochemistry* 30: 125–33
145. Mu D, Janes SM, Smith AJ, Brown DE, Dooley DM, Klinman JP. 1992. *J. Biol. Chem.* 267:7979–82
146. Dooley DM, McIntire WS, McGuirl MA, Cote CE, Bates JL. 1990. *J. Am. Chem. Soc.* 112:2782–89
147. Cooper RA, Knowles PF, Brown DE, McGuirl MA, Dooley DM. 1992. *Biochem. J.* 288:337–40
148. Backes G, Davidson VL, Huitema F, Duine JA, Sanders-Loehr J. 1991. *Biochemistry* 30:9201–10
148a. Itoh S, Ogino M, Komatsu M, Ohshiro Y. 1992. *J. Am. Chem. Soc.* 114:7294–95
149. Parrott S, Jones S, Cooper RA. 1987. *J. Gen. Microbiol.* 133:347–51
150. Zwart KB, Harder W. 1983. *J. Gen. Microbiol.* 129:3157–69
151. McIntire WS, Hartmann C. 1993. See Ref. 1, pp. 97–171
151a. Kagan HM, Trackman PC. 1993. See Ref. 1, pp. 173–89
152. Mu D, Medzihradszky KF, Adams GW, Meyer P, Hines WM, et al. 1993. *J. Biol. Chem.* Submitted
153. Callingham BA, Holt A, Elliott J. 1990. *J. Neural Transm.(Suppl.)* 32: 279–90
154. Callingham BA, Holt A, Elliott J. 1991. *Biochem. Soc. Trans.* 19:228–33
155. Burke TM, Wolin MS. 1987. *Am. J. Physiol.* 252:H721–32
156. Henle KJ, Moss AJ, Nagle WA. 1986. *Cancer Res.* 46:175–82
157. Kenyon K, Contente S, Trackman PC, Tang J, Kagan HM, et al. 1991. *Science* 253:802
158. Bruinenberg PG, Evers M, Waterham HR, Kuipers J, Arnberg AC, Ab G. 1989. *Biochim. Biophys. Acta* 1008: 157–67
158a. Zhang X, Fuller JH, McIntire WS. 1993. *J. Bacteriol.* 175:5617–27

158b. Sugino H, Sasak M, Azakami H, Yamashita M, Murooka Y. 1992. *J. Bacteriol.* 174:2484–92
159. Barbry P, Champe M, Chassande O, Munemitsu S, Champigny G, et al. 1990. *Proc. Natl. Acad. Sci. USA* 87:7347–51
160. George DG, Barker WC, Hunt LT. 1986. *Nucleic Acids Res.* 14:11–15
161. Adman ET. 1991. *Adv. Protein Chem.* 42:144–97
162. Dooley DM, Brown DE, Clague AW, Kemsley JN, McCahon CD, et al. 1993. See Ref. 46, pp. 459–70
163. Trackman PC, Bedell-Hogan D, Tang J, Kagan HM. 1992. *J. Biol. Chem.* 267:8666–71
164. Knowles PF, Yadav DS. 1984. In *Copper Proteins and Copper Enzymes,* ed. R Lontie, pp. 103–29. Boca Raton, Fla: CRC Press
165. Taylor CE, Taylor RS, Rasmussen C, Knowles PF. 1972. *Biochem. J.* 130: 713–28
166. Palcic MM, Klinman JP. 1983. *Biochemistry* 22:5957–66
167. Hartmann C, Klinman JP. 1990. *FEBS Lett.* 261:441–44
168. Rius FX, Knowles PF, Pettersson G. 1984. *Biochem. J.* 220:767–72
169. Farnum M, Palcic M, Klinman JP. 1986. *Biochemistry* 25:1898–904
170. Grant KL, Klinman JP. 1989. *Biochemistry* 28:6597–605
171. Neumann R, Hevey RC, Abeles RH. 1975. *J. Biol. Chem.* 250: 6362–67
172. Lindström A, Olsson B, Olsson J, Pettersson G. 1976. *Eur. J. Biochem.* 64:321–26
173. Hartmann C, Klinman JP. 1991. *Biochemistry* 30:4605–11
174. Huizinga EG, van Zanten BAM, Duine JA, Jongejan JA, Huitema F, et al. 1992. *Biochemistry* 31:9789–95
175. Battersby AR, Staunton J, Klinman JP, Summers MC. 1979. *FEBS Lett.* 99:297–98
176. Suva RH, Abeles RH. 1978. *Biochemistry* 17:3538–45
177. Summers MC, Markovic R, Klinman JP. 1979. *Biochemistry* 18: 1969–79
178. Yu PH, Davis BA. 1988. *Int. J. Biochem.* 20:1197–201
179. Yu PH. 1987. *Biochem. Cell Biol.* 66:853–61
180. Coleman AA, Hindsgaul O, Palcic MM. 1989. *J. Biol. Chem.* 264:19500–5
181. Coleman AA, Scaman CH, Kang YJ, Palcic MM. 1991. *J. Biol. Chem.* 266:6795–800
182. Scaman CH, Palcic MM. 1992. *Biochemistry* 31:6829–41
183. Battersby AR, Staunton J, Summers MC. 1976. *J. Chem. Soc. Perkin Trans. I* 1976:1052–56
184. Battersby AR, Summers MC, Southgate R. 1979. *J. Chem. Soc. Perkin Trans. I* 1979:45–52
185. Shah MA, Scaman CH, Palcic MM, Kagan HM. 1993. *J. Biol. Chem.* 268:11573–79
186. Lovenberg W, Beaven MA. 1971. *Biochim. Biophys. Acta* 250:452–55
187. Hartmann C, Brzovic P, Klinman JP. 1993. *Biochemistry* 32:2234–41
188. Scott RA, Dooley DM. 1985. *J. Am. Chem. Soc.* 107:4348–50
189. Scott RA, Coté CE, Dooley DM. 1988. *Inorg. Chem.* 27:3859–61
190. Knowles PF, Strange RW, Blackburn NJ, Hasnain SS. 1989. *J. Am. Chem. Soc.* 111:102–7
191. McCracken J, Peisach J, Dooley DM. 1987. *J. Am. Chem. Soc.* 109:4064–72
192. Baker GJ, Knowles PF, Pandeya KB, Rayner JB. 1986. *Biochem J.* 237:609–12
192a. Knowles PF, Dooley DM. 1993 In *Metal Ions in Biological Systems,* ed. H. Seigel. In press
193. Williams TJ, Falk MC. 1986. *J. Biol. Chem.* 261:15949–54
194. Greenaway FT, O'Gara CY, Marchena JM, Poku JW, Urtiaga JG, Zou Y. 1991. *Arch. Biochem. Biophys.* 285: 291–96
195. Lamkin MS, Williams TJ, Falk MC. 1988. *Arch. Biochem. Biophys.* 261: 72–79
196. McGuirl MA, Brown DE, McCahon CD, Turawski PN, Dooley DM. 1991. *J. Inorg. Biochem.* 43:186
197. Padiglia A, Medda R, Floris G. 1992. *Biochem. Int.* 28:1097–107
198. Morpurgo L, Agostinelli E, Mondovi B, Avigliano L, Silvestri R, et al. 1992. *Biochemistry* 31:2615–21
199. Williamson PR, Kagan HM. 1987. *J. Biol. Chem.* 262:8196–201
200. Gacheru SN, Trackman PC, Calaman SD, Greenaway FT, Kagan HM. 1989. *J. Biol. Chem.* 264:12963–69
201. Finazzi-Agro A, Rinaldi A, Floris G, Rotilio G. 1984. *FEBS Lett.* 176:378–80
202. Dooley DM, McGuirl MA, Brown DE, Turowski PN, McIntire WS, Knowles PF. 1991. *Nature* 349:262–64
203. McCracken J, Peisach J, Cote CE, McGuirl MA, Dooley DM. 1992. *J. Am. Chem. Soc.* 114:3715–20
203a. Turowski PN, McGuirl MA, Dooley

DM. 1993. *J. Biol. Chem.* 268:17680–82
204. Vignevich V, Dooley DM, Guss JM, Harvey I, McGuirl MA, Freeman HC. 1993. *J. Mol. Biol.* 229:243–45
205. Mure M, Klinman JP. 1993. *J. Am. Chem. Soc.* 115:7117–27
206. Nakamura N, Kohzuma T, Kuma H, Suzuki S. 1992. *J. Am. Chem. Soc.* 114:6550–52
206a. Eady RR, Large PJ. 1971. *Biochem. J.* 123:757–71
207. Matsumoto T. 1978. *Biochim. Biophys. Acta* 522:291–302
208. Davidson VL. 1989. *Biochem. J.* 261: 107–11
209. McIntire WS. 1987. *J. Biol. Chem.* 262:11012–19
210. Tobari J, Harada Y. 1981. *Biochem. Biophys. Res. Commun.* 101:502–8
211. Ohta S, Tobari J. 1981. *J. Biochem.* 90:215–24
212. McWhirter RB, Klapper MH. 1988. In *PQQ and Quinoproteins,* ed. JA Jongejan, JA Duine, pp. 259–68. Dordrecht, The Netherlands: Kluwer
213. Brooks HB, Jones LH, Davidson VL. 1993. *Biochemistry* 32:2725–29
214. Davidson VL, Jones LH, Graichen ME. 1992. *Biochemistry* 31:3385–90
215. Van Spanning R, Wansell C, Reijnders W, Oltmann L, Stouthamer A. 1990. *FEBS Lett.* 275:217–20
215a. Sharma KD, Loehr TM, Sanders-Loehr J, Husain M, Davidson VL. 1988. *J. Biol. Chem.* 263:3303–6
216. Durley R, Chen L, Lim LW, Mathews FS, Davidson VL. 1993. *Protein Sci.* 2:739–52
217. Chapman SK. 1991. In *Perspectives on Bioinorganic Chemistry,* ed. RW Hay, JR Dilworth, KB Nolan, 1:95–140. London: JAI
218. Chen L, Durley R, Poliks BJ, Hamada K, Chen Z, et al. 1992. *Biochemistry* 31:4959–64
219. Gray KA, Davidson VL, Knaff DB. 1988. *J. Biol. Chem.* 263: 13987–90
220. Chen L, Mathews FS, Davidson VL, Tegoni M, Rivetti C, Rossi GL. 1993. *Protein Sci.* 2:147–54
221. Chen L, Mathews FS, Davidson VL, Huizinga EG, Vellieux FMD, et al. 1992. *Proteins* 14:288–99
222. Brooks HB, Davidson VL. 1993. *Biochem. J.* 294:211–13
223. Davidson VL, Jones LH. 1991. *Anal. Chim. Acta* 249:235–40
224. Gray KA, Knaff DB, Husain M, Davidson VL. 1986. *FEBS Lett.* 207:239–42
225. Kenny WC, McIntire W. 1983. *Biochemistry* 22:3858–68
226. Davidson VL, Jones LH, Kumar MA. 1990. *Biochemistry* 29:10786–91
227. Warncke K, Brooks HB, Babcock GT, Davidson VL, McCracken J. 1993. *J. Am. Chem. Soc.* 115:6464–65
228. Ubbink M, van Kleef MAG, Kleinjan D, Hoitink CWG, Huitema F, et al. 1991. *Eur. J. Biochem.* 202:1003–12
229. Christoserdov AY, Tsygankov YD, Lidstrom ME. 1991. *J. Bacteriol.* 173: 5901–8
230. Christoserdov AY, Lidstrom ME. 1991. *J. Bacteriol.* 173:5909–13
231. Christoserdov AY, Boyd J, Mathews FS, Lidstrom ME. 1992. *Biochem. Biophys. Res. Commun.* 184:1226–34

Annu. Rev. Biochem. 1994. 63:345–82

INTERMEDIATE FILAMENTS: Structure, Dynamics, Function, and Disease

Elaine Fuchs[1]
Howard Hughes Medical Institute and Department of Molecular Genetics and Cell Biology, The University of Chicago, Chicago, Illinois 60637

Klaus Weber
Department of Biochemistry, Max Planck Institute for Biophysical Chemistry, D-37018, Göttingen, Germany

KEY WORDS: cytoskeleton, protein structure/function, genetic disease, macromolecular assembly, multigene family

---

CONTENTS

[1]To whom correspondence should be addressed: Howard Hughes Medical Institute, Dept. of Molecular Genetics and Cell Biology, The University of Chicago, 5841 S. Maryland Avenue, Room N314, Chicago, IL 60637

0066-4154/94/0701-0345$05.00

I. INTRODUCTION

Although it is only relatively recently that intermediate filaments (IFs) have been recognized as a superfamily of 10-nm fibers ubiquitous in multicellular eukaryotes, their existence has been known for nearly a century, since the time when cytoskeletal fibrillar structures—or neurofilaments—were revealed upon silver staining of neurons. The first X-ray diffraction patterns of IFs were obtained from wool fibers, revealing that keratins, which were subsequently found to be members of this family, are richly α-helical polypeptides that intertwine in a coiled-coil fashion to form the subunit structure of 10-nm filaments (1–3). When electronmicroscopy was more extensively used on vertebrate cells, the widespread occurrence of IFs began to be recognized.

IFs derive their name from their diameter (10 nm). Originally used to emphasize their intermediate nature between thin actin filaments and thick myosin filaments (4), the name later focused on width differences between actin filaments (6 nm), IFs, and microtubules (23 nm). Together, these three filamentous networks constitute the basic features of the higher eukaryotic cytoskeleton. In contrast to actins and tubulins, which are highly evolutionarily conserved, IF proteins, which include the nuclear lamins, sometimes share as little as 20% sequence identity. Members of the IF superfamily exhibit cell-type-specific and often complex patterns of expression. In various combinations in vitro, they have the intrinsic ability to self-assemble into 10-nm filaments containing several 10,000s of individual IF polypeptide chains. Their filament-forming capacity thus seems to be critical for some common function; their diversity in sequence and expression suggests that in addition, IFs perform specialized cellular functions.

In this review, we summarize studies that probe the structure and dynamics of IFs. Discoveries in recent years have projected new and exciting insights into the functions of IFs and their involvement in human disease.

II. THE SUPERFAMILY OF IF PROTEINS

Although 10-nm filaments have now been intensively studied for more than 30 years, it only became clear in the 1980s that IFs are composed of proteins that share common sequence and structural features. Initially, IF-assembling polypeptides were grouped into five categories: keratins, desmin, vimentin, neurofilament (NF) proteins, and glial fibrillary acidic protein (GFAP) (5). Early studies revealed biochemical, immunological, and structural similarities among IF proteins (e.g. 6–8). Subsequent analyses of partial amino-acid sequences of fragments from four porcine IF proteins left no question that desmin, vimentin, an NF protein, and GFAP were evolutionarily and structurally related (9–11).

When the first cytoskeletal keratin cDNAs were sequenced (12–14), it became clear that prior partial amino-acid sequences from chemically derived "type I" and "type II" wool fragments (15, 16) arose from two distinct proteins. These analyses also established a relation between epidermal and hair (wool) keratins, admitting them into the IF superfamily. The findings explained earlier hybridization results demonstrating the existence of two distinct classes of keratin mRNAs and genes (17). In keeping with the nomenclature of the two wool fragments, keratins were divided into type I and type II sequences (13). Among keratins of a single type, the α-helical domains share 50–99% sequence identity, while keratins of opposite type display only ~30% homology in these regions (12–14).

As additional sequences accumulated, IF proteins were grouped relative to homologies within their α-helical domains. Keratins remained as the sole members of the type I and type II classes. Vimentin (57 kDa), desmin (53–54 kDa), GFAP (50 kDa), and peripherin (57 kDa) were designated as type III IF proteins, sharing >70% sequence identity among themselves, but only 25–30% and 35–40% identities with type I and II keratins, respectively (18–20). Sharing ~50% sequence identity with their closest type III relative, neurofilament proteins and later α-internexin were grouped separately as type IV. Finally, in 1986, electronmicroscopy unveiled a meshwork of 10-nm filaments beneath the nuclear envelope (21), and sequence analyses disclosed the relation of these proteins, the nuclear lamins, to the IF superfamily (22, 23). Lamins now constitute the type V group.

Presently, there are nearly 50 known human IF proteins that are differentially expressed in different tissues (for a review of IF gene regulation, see 24). Most sequences fall into the five established types. A few are unique and will likely be assigned new types. Based on continuing discoveries of new IF proteins, it is predicted that the IF superfamily will continue to foster new members.

A. Type I and Type II IF Proteins: The Keratins

The largest and most complex group of IF proteins are keratins. There are at least 30 keratins ranging in size from 40 to 67 kDa (25). Type II keratins are basic (pKi = 6–8) and include eight epithelial proteins, K1–K8, and four hair keratins, Hb1–Hb4. Type I keratins are more acidic (pKi = 4–6) and include eleven epithelial keratins, K9–K20, and four hair keratins, Ha1–Ha4. Pairs of type II and type I keratins are expressed differentially in most epithelial cells at various stages of development and differentiation (26–28). The major keratin pair broadly expressed in most stratified squamous epithelia is K5 and K14 (29). K8 and K18 is the pair produced by nearly all simple epithelia (27). Differentiating epithelial cells synthesize other keratins in a pairwise fashion (28), with the epidermal appendages displaying the most elaborate programs (30, 31). Surprisingly, despite the comprehensive studies of the 1980s, a new keratin, K20, has recently been cloned (32).

Keratins assemble in vitro as obligatory heteropolymers (6), in virtually a 1:1 ratio of any combination of at least one type I and one type II keratin (33, 34). However, filaments generated from different keratins have distinct physical properties, suggesting that the differential expression of specific keratin pairs in vivo may tailor IF networks to suit tissue-specific structural requirements of tensile strength, flexibility, and dynamics.

B. Type III Intermediate Filament Proteins

Of the type III IF proteins, vimentin is the most widely expressed, produced by mesenchymal cell types and by a variety of transformed cell lines and tumors (35). Desmin is more restricted, displayed in smooth muscle and in skeletal and cardiac muscle cells, where it concentrates at Z discs (36). GFAP is expressed in glial cells and astrocytes, and peripherin is found in the peripheral nervous system, where it localizes to neurons of the dorsal root ganglion, sympathetic ganglia, cranial nerves, and ventral motor neurons (37).

Unlike keratin IFs, type III IFs can exist as homopolymers (38). In vitro and in vivo, they can also heteropolymerize with other type III proteins (38, 39) or with the neurofilament protein, NF-L (40). Type III IFs cannot, however, coassemble with keratins: in cells expressing both vimentin and keratin, two distinct IF networks are formed (41). This ability of keratins and type III IF proteins to segregate into distinct networks stems from two separate domains within the α-helical segments of the IF polypeptides (42; see also 34).

C. Type IV IF Proteins: The Neurofilament and α-Internexin Filament Constituents

Three type IV IF proteins are coexpressed in axons, dendrites, and perikarya. Human NF-L (light), NF-M (medium), and NF-H (heavy) have predicted sizes of 62, 102, and 110 kDa, respectively, manifested by dramatic differences in the lengths of their glutamic and lysine-rich carboxy termini

(11, 43–46). An additional type IV protein, α-internexin (66–70 kDa), is also expressed in neurons, where it seems to play a more prominent role in embryonic development than do the NF proteins (47, 48).

α-Internexin may be the only type IV IF protein able to form homopolymers of 10-nm filaments (49). On their own, NF-M and NF-H appear to be assembly incompetent, and NF-L requires special conditions for in vitro filament assembly (50, and references therein). NF-L seems to form the filament backbone, while NF-M and NF-H associate more peripherally and form extensive interfilament crossbridges (51). As judged by transient transfection assays in cultured cells, the type III IF protein, vimentin, can substitute for NF-L in providing a backbone for NF-M (and presumably NF-H) assembly (40). Surprisingly, tail-less NF-M and NF-H as well as their wild-type counterparts are also able to complement NF-L for assembly (50, 52). Thus, NFs are obligate heteropolymers.

D. Type V IF Proteins: Proteins Composing the Nuclear Lamina

Probably universally expressed in higher eukaryotes, lamins form a fibrous meshwork, the nuclear lamina, on the inner surface of the nuclear membrane (for review, see 53). This structure seems to provide a framework for the nucleus, and it may facilitate chromatin organization. In a few cell types, the lamina can be visualized as a net of orthogonally arranged IFs that appear to interconnect nuclear pore complexes (21).

Early vertebrate embryos possess only B-type lamins (lamins B_1-B_2, 63–68 kDa; Refs. 54, 55). The *Xenopus* oocyte possesses a unique B-type lamin, L_{III} (56, 57 and references therein), while mouse spermatocytes have an unusual lamin B_3 with an entirely distinct N-terminal third of the molecule (58). The B_3 mRNA seems to arise by differential splicing from the normal lamin B_2 gene.

In contrast, somatic cells synthesize A-type lamins (lamin A, 70 kDa, and the "truncated" mammalian splice variant lamin C, 60 kDa) in addition to B-type lamins (22, 23, 55, 56, 59, 60). The A-type genes probably arose in evolution from an ancestral B-type gene by exon shuffling (56). Expression of A-type lamins is often coincident with major changes in tissue differentiation (59, 60), although ectopic expression of lamin A in embryonal cells does not alter their normal program of differentiation (61). Nevertheless, A-type lamins may still play a specialized role in cellular differentiation or chromosome organization, perhaps in concert with other as-yet-unidentified proteins.

All lamins possess sequences that signal their dispatch to the nucleus (62). With the exception of lamin C, lamins also possess C-terminal CaaX box sequences that target them for modification by isoprenylation and carboxymethylation (62–65). Despite these similarities between A- and B-type lamins, a difference can be observed during mitosis when the nuclear lamina transiently breaks down: A-type lamins disperse as soluble oligomers while B-type lamins remain associated with remnants of the nuclear membrane (63, 66, 67). Two factors contribute to this difference: (*a*) A-type lamins are subject to an additional processing event that removes the C-terminal cysteine, likely releasing the protein from an otherwise membrane-associated fate, and (*b*) the interaction between the nuclear membrane and lamin B may be strengthened by specific inner-nuclear-membrane proteins, probably acting as lamin receptors (68–70).

The nuclear lamina associates with chromatin, although electronmicroscopy suggests that only a fraction of the chromatin may be involved (71, 72). The molecular nature(s) of these interactions has only recently been explored, and several lamin-associated proteins (LAPs) have been implicated (70). In addition, A-type lamins associate generally with chromatin in vitro (73–75), while B lamins may bind to specific regions of chromatin (76). Candidate DNA sequences for specific associations with lamins include nuclear matrix attachment sites (76, 77 and references therein), DNA replication sites (78), and telomeres (72). In this regard, it is interesting that a *Xenopus* oocyte extract depleted of its single lamin (L_{III}) can encapsulate chromatin and assemble nuclear pores, but the nuclei fail to synthesize DNA (79, 80; for somewhat different results, see 81, 82). Thus, lamins might function not only to impart structural integrity to the nuclear envelope, but also to provide a foundation for DNA replication, either by acting as a solid substrate for formation of replication complexes, or by regulating processes such as DNA decondensation and nuclear scaffold-DNA loop organization.

Interestingly and conversely, the assembly of lamin IFs seems to be influenced by chromatin: Under physiological salt conditions, A-type lamins can polymerize on chromosome surfaces (74, 75). Whether B-type lamins require chromatin for assembly and whether they can coassemble with A-type lamins are presently unresolved issues.

E. IF Proteins that Do Not Classify as One of the Five Major Types

Several recently discovered proteins are clearly members of the IF superfamily based on sequence and structure comparisons, but do not fall into one of the five major types. The first of this class is nestin, a protein expressed in proliferating stem cells of the developing mammalian central nervous system and to a lesser extent (and only transiently) in developing skeletal muscle (83). Nestin's predicted α-helical sequences can be aligned with other cytoplasmic IF types without insertions. Within the α-helical domains, nestin's closest relatives are type III IF proteins (33% sequence identity with vimentin). However, nestin's sequence predicts a 200-kDa protein, with an enormous, nonhelical carboxy tail, and in this respect it resembles the type IV NF proteins. Its gene structure is also that of a type IV IF protein (84).

Filensin is a protein expressed during differentiation of the vertebrate lens epithelial cell (85 and references therein). In conjunction with a 47-kDa protein, filensin forms heteropolymers that constitute the beaded-chain filaments in these cells. Analyses of cloned cDNA sequences reveal that filensin is a member of the IF superfamily, sharing the greatest similarities with nestin, type III, and type IV IF proteins. This said, the sequence identities shared between filensin and any of these IF proteins are considerably less than 50%.

III. IF SECONDARY STRUCTURE

IF proteins are predicted to share a common secondary structure (Figure 1). All IF proteins have a central α-helical domain, the rod, which is flanked by nonhelical head (amino-end) and tail (carboxy-end) domains. The rods of two polypeptide chains intertwine in a coiled-coil fashion, a feature originally predicted from X-ray diffraction data (1–3) and later substantiated by sequence data and model building (91, 92). Throughout the α-helical sequences are repeats of hydrophobic amino acids, such that the first and fourth of every seven residues are frequently apolar. This provides a hydrophobic seal on the helical surface, enabling the coiling between two IF polypeptides.

The IF α-helical rod is subdivided by three short nonhelical linker segments, which often contain proline or multiple glycine residues (12, 13, 18). The rod segments are referred to as helices 1A, 1B, 2A, and 2B, while the linker segments are referred to as L1, L1-2, and L2 (91). The rods of type I-IV IF polypeptides are ~310 amino acids (aa) in length (~46 nm); for the lamins, the rods are 356 aa (~53 nm), due to an insertion of 42 aa (6 heptads) in helical domain 1B (see Figure 1); for filensin, the rod is ~281 aa due to a 29-aa deletion extending from helix 2A to 2B. The sequence identity among all IF proteins is particularly high at the start of helix 1A and near the end of helix 2B, and these sequences are among the most diagnostic for an IF protein.

In addition to the heptad repeat pattern, a nonrandom periodic distribution of acidic and basic residues in the rod has been noted. A series of zones of alternating acidic and basic residues seem to repeat at ~9.5-residue intervals along the molecule, or 3 times every 28 residues (93, 94). It has been suggested that this charge periodicity may reflect an importance of

```
                 AMINO TERMINUS (NON-HELICAL HEAD DOMAIN)

hu K14 :   1     --MTTCSRQFTSSSSMKGSCGIGGGIGAGSSRISSVLAGGSCRAPNTYGGGLSVSSSRFSSGGAYGLGGGYGGGFSSSSSSFGSGFGGGYGGGLGAGLGGGFGGGFAGG
hu K5  :   1     ----------------------------------------------------------MSRQSSVSSGAGGSPSFSTASAITPSVSRTSFTSVSRSGGGGGGGFGRVSL
                 AGACGVGGYGSRSLYNLGGSKRISISTSGGSFRNRFGAGAGGGYGFGGGAGSGFGFGGGAGGGFGLGGGAGFGGGFGGPGFPVCPPGGIQEVTVNQSLLTPLNLQIDPS
hu Vim :   1     ----------------STRSVSSSSYRRMFGGPGTASRPSSSRSYVTTSTRTYSLGNLRPSTSRSLYASSPGGVYVTRSSAVRLRSSVPGVRLLQDSVDFSLADAINTE
ha Des :   1     ------------SQAYSSSQRVSSYRRTFGGAPSFSLGSPLSSPVFPRAGFGTKGSSSSVTSRVYQVSRTSGGAGGLGSLRASRLGSTRAPSYGAGELLDFSLADAVNQE
mu NFL :   1     ---------------------------MSSFASDPIFSTSYKRRYVETPRVHISSVRSGYSTARSAYSSYSAPVSSSLSVRRSYSSSSGSLKPSLENLDVSQVAAISND
hu Lam :   1     --------------------------------------------------------------------------------------ETPSQRRATRSGAQASSTPLSPT
as IFB :   1     ------------------------------------SLKQSQESSEYEIAYRSTIQPRTAVRTQSRQSGAYSTGAVSGGGGRVLKMVTEMGSASIGGISPALSANAAKS


                 HELICAL DOMAIN 1A (ROD DOMAIN)

hu K14 : 108     DGSSVGSEKVTMQNLNDRLASYLDKVRALEEANADLEVKIRDW-YQRQRPAEIKDYSP-Y
hu K5  : 161     IQRVRTEEREQIKTLNNKFASFIDKVRFLEQQNKVLETKWTLLQEQGTKTVRQNL-EPLF
hu Vim :  94     FKNTRTNEKVELQELNDRFANYIDKVRFLEQQNKILLAELEQLKGQG-KS-RLGD---LY
ha Des :  98     FLATRTNEKVELQELNDRFANYIEKVRFLEQQNAALAAEVNRLKGREP-T-RVAE---LY
mu NFL :  83     LKSIRIQEKAQLQDLNDRFASFIERVHELEQQNKVLEAGLLVLRQKH--SGPSRFRA-LY
hu Lam :  24     RITTRLQEKEDLQELNDRLAVYIDRVRSLETENAGLRLRITESEEVVSREV-SGIKA-AY
as IFB :  74     FLEATDKEKKEMQGLNDRLGNYIDRVKKLEEQNRKLVADLDELRGRWGKDT-SEIKIQ-Y


                 HELICAL DOMAIN 1B

hu K14 : 166     FKTIEDLRNKILTATVDNANVLLQIDNARLAADDFRTKYETELNLRMSVEADINGLRRVLDELTLARADLEMQIESLKEELAYLKKNHEEEMNAL
hu K5  : 220     EQYINNLRRQLDSIVGERGRLDSELRNMQDLVEDFKNKYEDEINKRTTAKNEFVMLKKDVDAAYMNKVELEAKVDALMDEINFMKMFFDAELSQM
hu Vim : 149     EEEMRELRRQVDQLTNDKARVEVERDNLAEDIMRLREKLQEEMLQREEAENTLQSFRQDVDNASLARLDLERKVESLQEEIAFLKKLHEEEIQEL
ha Des : 154     EEEMRELRRQVEVLTNQRARVDVERDNLIDDLQRLKAKLQEEIQLREEAENNLAAFRADVDAATLARIDLERRIESLNEEIAFLKKVHEEEIREL
mu NFL : 141     EQEIRDLRLAAEDATNEKQALEGEREGLEETLRNLQARYEEEVLSREDAEGRLMEARKGADEAALARAELEKRIDSLMDEIAFLKKVHEEEIAEL
hu Lam :  82     EAELGDARKTLDSVAKERARLQLELSKVREEFKELKA LHDLRGQVAKLEAALGEAKKQLQDEMLRRVDAENRLQTMKEELDFQKNIYSEELRET

                           RNTKKEGDLIAAQARLKDLEALLNSKEAALSTALSEKRTLEGE
                           RYEDVQHRRESDREKINQWQHAIEDAQSELEMLRARWRQLTEE

as IFB : 132     SDSLRDARKEIDDGARRKAEIDVKVARLRDDLAELRN EKRLNGDNARIWEELQKARNDLDEETLGRIDFQNQVQTLMEELEFLRRVHEQEVKEL
```

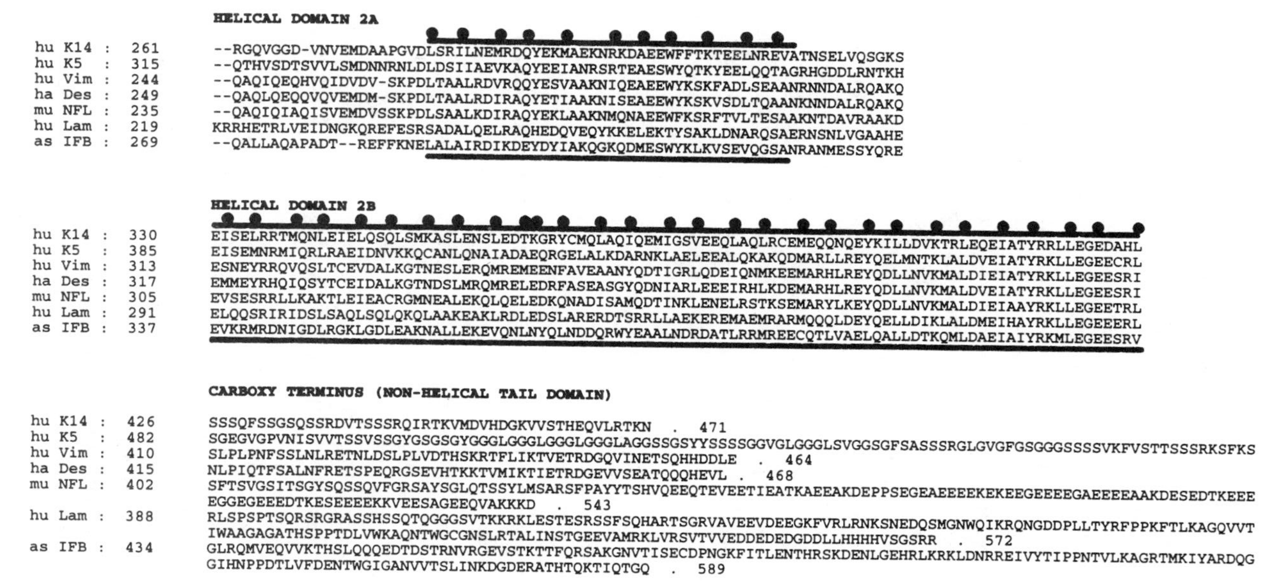

HELICAL DOMAIN 2A
hu K14 : 261 --RGQVGGD-VNVEMDAAPGVDLSRILNEMRDQYEKMAEKNRKDAEEWFFTKTEELNREVATNSELVQSGKS
hu K5 : 315 --QTHVSDTSVVLSMDNNRNLDLDSIIAEVKAQYEEIANRSRTEAESWYQTKYEELQQTAGRHGDDLRNTKH
hu Vim : 244 --QAQIQEQHVQIDVDV-SKPDLTAALRDVRQQYESVAAKNIQEAEEWYKSKFADLSEAANRNNDALRQAKQ
ha Des : 249 --QAQLQEQQVQVEMDM-SKPDLTAALRDIRAQYETIAAKNISEAEEWYKSKVSDLTQAANKNNDALRQAKQ
mu NFL : 235 --QAQIQIAQISVEMDVSSKPDLSAALKDIRAQYEKLAAKNMQNAEEWFKSRFTVLTESAAKNTDAVRAAKD
hu Lam : 219 KRRHETRLVEIDNGKQREFESRSADALQELRAQHEDQVEQYKKELEKTYSAKLDNARQSAERNSNLVGAAHE
as IFB : 269 --QALLAQAPADT--REFFKNELALAIRDIKDEYDYIAKQGKQDMESWYKLKVSEVQGSANRANMESSYQRE
HELICAL DOMAIN 2B
hu K14 : 330 EISELRRTMQNLEIELQSQLSMKASLENSLEDTKGRYCMQLAQIQEMIGSVEEQLAQLRCEMEQQNQEYKILLDVKTRLEQEIATYRRLLEGEDAHL
hu K5 : 385 EISEMNRMIQRLRAEIDNVKKQCANLQNAIADAEQRGELALKDARNKLAELEEALQKAKQDMARLLREYQELMNTKLALDVEIATYRKLLEGEECRL
hu Vim : 313 ESNEYRRQVQSLTCEVDALKGTNESLERQMREMEENFAVEAANYQDTIGRLQDEIQNMKEEMARHLREYQDLLNVKMALDIEIATYRKLLEGEESRI
ha Des : 317 EMMEYRHQIQSYTCEIDALKGTNDSLMRQMRELEDRFASEASGYQDNIARLEEEIRHLKDEMARHLREYQDLLNVKMALDVEIATYRKLLEGEESRI
mu NFL : 305 EVSESRRLLKAKTLEIEACRGMNEALEKQLQELEDKQNADISAMQDTINKLENELRSTKSEMARYLKEYQDLLNVKMALDIEIAAYRKLLEGEETRL
hu Lam : 291 ELQQSRIRIDSLSAQLSQLQKQLAAKEAKLRDLEDSLARERDTSRRLLAEKEREMAEMRARMQQQLDEYQELLDIKLALDMEIHAYRKLLEGEEERL
as IFB : 337 EVKRMRDNIGDLRGKLGDLEAKNALLEKEVQNLNYQLNDDQRWYEAALNDRDATLRRMREECQTLVAELQALLDTKQMLDAEIAIYRKMLEGEESRV
CARBOXY TERMINUS (NON-HELICAL TAIL DOMAIN)
hu K14 : 426 SSSQFSSGSQSSRDVTSSSRQIRTKVMDVHDGKVVSTHEQVLRTKN . 471
hu K5 : 482 SGEGVGPVNISVVTSSVSSGYGSGSGYGGGLGGGLGGGLGGGLAGGSSGSYYSSSSGGVGLGGGLSVGGSGFSASSSRGLGVGFGSGGGSSSSVKFVSTTSSSRKSFKS . 590
hu Vim : 410 SLPLPNFSSLNLRETNLDSLPLVDTHSKRTFLIKTVETRDGQVINETSQHHDDLE . 464
ha Des : 415 NLPIQTFSALNFRETSPEQRGSEVHTKKTVMIKTIETRDGEVVSEATQQQHEVL . 468
mu NFL : 402 SFTSVGSITSGYSQSSQVFGRSAYSGLQTSSYLMSARSFPAYYTSHVQEEQTEVEETIEATKAEEAKDEPPSEGEAEEEEKEKEEGEEEEGAEEEEAAKDESEDTKEEE
EGGEGEEEDTKESEEEEKKVEESAGEEQVAKKKD . 543
hu Lam : 388 RLSPSPTSQRSRGRASSHSSQTQGGGSVTKKRKLESTESRSSFSQHARTSGRVAVEEVDEEGKFVRLRNKSNEDQSMGNWQIKRQNGDDPLLTYRFPPKFTLKAGQVVT
IWAAGAGATHSPPTDLVWKAQNTWGCGNSLRTALINSTGEEVAMRKLVRSVTVVEDDEDEDGDDLLHHHHVSGSRR . 572
as IFB : 434 GLRQMVEQVVKTHSLQQQEDTDSTRNVRGEVSTKTTFQRSAKGNVTISECDPNGKFITLENTHRSKDENLGEHRLKRKLDNRREIVYTIPPNTVLKAGRTMKIYARDQG
GIHNPPDTLVFDENTWGIGANVVTSLINKDGDERATHTQKTIQTGQ . 589

Figure 1 *Comparison of selected IF sequences aligned for optimal homology*. Shown are amino-acid (aa) sequences of K14 (hu K14) (86), K5 (hu K5) (87), vimentin (hu Vim) (88), desmin (ha Des) (89), NF-L (mu NFL) (43), lamin C (hu Lam) (22, 23), and nematode as IF-B protein (90). Hu, human; ha, hamster; mu, murine; as, *Ascaris lumbricoides*. Bars denote α-helical domains. Dots indicate heptad repeats of hydrophobic residues (for helix 1B insert, these aas are in bold). Note: While the sequence in helix 2A is predicted to be α-helical, it does not conform to a good coiled-coil; hence, there exist some ambiguities regarding the positioning of the end of helix 2A and the beginning of helix 2B (for a different prediction, see 91). Note also the heptad reversal in helix 2B.

electrostatic interactions in stabilizing associations between coiled-coil dimers or higher-ordered structures (91, 93, 95). While this may be true, it is also notable that many of these acidic and basic residues are spaced 4 aa apart (95a; A Letai, E Fuchs, unpublished observations), and such spacing is optimal for formation of ionic salt bridges, which can stabilize α-helices (96 and references therein). This type of periodicity is reflective of intrachain, rather than interchain, interactions.

It may be relevant that studies on myosin, another coiled-coil protein, indicate that charge interactions are unlikely to be the driving force behind tetramer formation (97). However, the nature of ionic interactions in IFs may be very different from those in thick filaments, and clearly further studies are necessary to determine experimentally the extent to which either of the periodicities described above are important in IF assembly. One additional possibility worth considering is that IF assembly may be in part facilitated through the switching of acidic and basic interactions from helix-stabilizing intrachain salt bridges to interchain ionic associations.

The nonhelical head and tail segments of IF proteins vary in length and amino-acid composition. Length variations are greatest in the tail, where predicted sizes range from 9 residues in keratin K19 (98) to 1491 residues in nestin (83). The functional significance of these differences and the role of head and tail domains in IF assembly are discussed in Section IV.D.

IV. ASSEMBLY AND STRUCTURE OF INTERMEDIATE FILAMENTS

A. Early Steps in Filament Assembly: Formation of Dimers and Tetramers

Cytoplasmic IFs can assemble in vitro in the absence of auxiliary proteins or factors, suggesting that the information necessary to form a 10-nm filament is intrinsic, contained within the primary sequence of the IF polypeptide. Given the remarkable sequence heterogeneity among IF proteins, it seems most likely that the ability of proteins to assemble into IFs relies upon common secondary and tertiary structures.

For all IFs, the first step in assembly is the formation of parallel and in-register dimers, dependent upon the right-handed α-helical rods of two polypeptide chains coiling around a common axis in a left-handed fashion (2, 3, 18, 99–101). Based on a study of many coiled-coil proteins, the pitch of the coil is likely ~14 nm (102 and references therein). Rotary shadowing and electronmicroscopy of A- and B-type lamins have provided a unique glimpse of the type V IF dimer, a 53-nm, double-stranded α-helical rod with separate globular end domains. Upon assembly, lamin dimers align in a head-to-tail fashion to form linear polymers, which are not seen with cytoplasmic IFs (21, 103).

Most IF proteins can form functional homodimers (100 and references therein). In contrast, while keratin homodimers have been described (33, 34, 104, 105), they do not appear to be competent for filament assembly (104, 105). Rather, keratin IFs appear unique in that their elemental subunit for assembly is a heterodimer of one type I and one type II polypeptide (104–106). This accounts for the heteropolymeric nature of keratin IFs. In addition, it resolves a key issue in keratin IF structure, and is in agreement with the subunit structure predicted by model building (91, 92).

Dimers of cytoplasmic IF proteins associate readily to form stable tetramers (99, 107). Whether dimer units in the tetramer are in register or

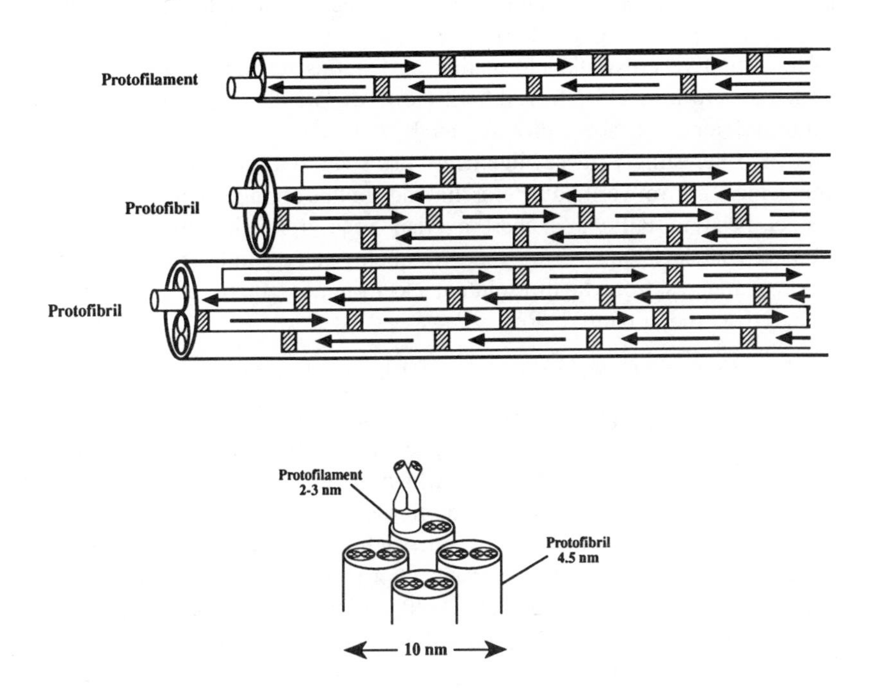

Figure 2 Model of the Alignments of Subunits in IFs. Model is adapted from those described recently (113–115; in particular 50). Box denotes rod segment of a dimer; arrow indicates direction of polypeptides, from base (N terminus) to tip (C terminus). Hatched boxes denote regions of putative rod overlap between helix 2B of one dimer and helix 1A of another (see 115, 50). Note: In this model, unstaggered antiparallel alignments of dimers occur at the level of protofibril-protofibril associations (50).

staggered has been controversial, and both unstaggered (106, 108, 109) and staggered (51, 106, 109–112) forms have been described. The presence of different modes of dimer association is explained by nearest-neighbor analyses of recent crosslinking data on (*a*) protofilaments, filaments, and the tetrameric rod of desmin (113), and (*b*) K1 and K10 oligomers that form below the critical concentration for polymerization (114, 115). Three distinct antiparallel arrangements of dimers are described: (*a*) Near half-staggered dimers bringing coil 1B segments into approximate alignment. This form prevails in desmin crosslinks (114) and is also found in keratin (115); the arrangement resembles that found in a smaller wool keratin particle (99) and in paracrystalline GFAP rods (112; for lamins, see 135). (*b*) Near half-staggered dimers bringing the coil 2B segments into approximate alignment. This form is documented in keratin crosslinks (114, 115) and also accounts for a minor crosslinked species in desmin filaments (113). (*c*) Dimers in approximate register, without stagger. This form is prevalent in keratin crosslinks (114, 115).

The extensive crosslink information on keratins refines estimates of the rod length (46 nm; 115). In addition, the three lateral alignments predict a two-dimensional surface lattice for an IF (115), which supersedes earlier versions. Interestingly, these alignments also suggest a fourth mode of dimer arrangement not yet verified by crosslinking. In this arrangement, two dimers interact head to tail via a short overlap zone, reflecting the conserved sequences at each end of the rod. A topologically similar model for IF packing emerges from high-resolution electronmicrographs of paracrystals of the NF-L rod (Figure 2; Ref. 50).

B. Packing of IF Subunits into 10-nm Filaments

Electron microscopy provided several early insights into the higher-ordered interactions involved in IF structure. Each 10-nm filament is composed of smaller protofibrils, seemingly intertwined in a left-handed fashion. IFs appear to have approximately four protofibrils per unit width (116), although this number can vary from two to six, depending upon IF type and assembly conditions (U Aebi, personal communication). Scanning-transmission electronmicroscopy has indicated that 24–40, commonly 32, polypeptides (8 tetramers) contribute to the overall width of an IF (117, 118).

It has been suggested that each protofibril might consist of two smaller protofilaments, and that the protofilaments consist of tetramers, linked end to end (116). If the head-to-tail arrangement visualized for the lamin dimer (21, 103, 141) is maintained for cytoplasmic IFs, this would impose certain constraints on tetramer packing. However, in contrast to lamins, linear arrays of wild-type cytoplasmic IF dimers have never been generated in vitro, nor have protofilaments or protofibrils assembled from these proteins been visualized as distinct entities. Based on such observations, it seems that under most conditions, the free energies for lateral associations of cytoplasmic IF dimers are greater than those for end-to-end associations, while for tetramers, the free energies for lateral and end-to-end associations are similar. An example in point is that in 6M urea, stable tetramers of K5 and K14 form, and when the urea concentration is reduced from 6M to 4M, lateral and end-to-end interactions take place simultaneously, giving rise to a mixture of higher-ordered structures (106; see also 119). This said, under some circumstances, hexamers of keratins have been detected, unveiling an even greater level of complexity in the nature of higher-ordered interactions among IF dimers (114, 115).

C. The Conserved Ends of the Rod: A Special Role in IF Assembly

Cells engineered to express genes encoding truncated or point-mutated IF proteins brought to light the concept that the highly conserved rod ends play a special role in IF network formation (62, 120–127). In vitro assembly studies provided experimental evidence that even conservative amino-acid substitutions in the amino end of helix 1A or the carboxy end of 2B can elicit marked perturbations in 10-nm filament structure (125, 127–130). Point substitutions more central within the rod or in the linker regions often have less severe effects on IF assembly (62, 127). More recently, synthetic peptides have been used to identify critical sequences in the process. A peptide encompassing the highly conserved motif at the carboxy end of the rod completely disassembles IF when present at 50 molar excess (131, 132). Similarly, peptides from the conserved amino end of the rod and certain other conserved regions also lead to IF disassembly (114, 115). It seems that the peptides most active at promoting disassembly are those that make multiple contact points in the IF surface lattice (115).

Why are the rod ends so important for packing IF subunits into higher-ordered structures? Some insights can be gained by considering the axial pitch of 21 to 22 nm, visible in rotary-shadowed IFs (50, 51, 133, 134) and deduced from X-ray diffraction patterns of wool keratins (for review, see 95). This "beading" of IFs has been interpreted to represent the periodicity in protofibrillar coiling. Another explanation is that the repeat is influenced by the few-nm overlap between tetramers due to a head-to-tail interaction between the beginning of helix 1A and end of helix 2B (see above; 50, 115). This could also explain why the beading is slightly smaller than the value for half the rod length. Although this interaction is not yet verified by crosslinking, it arises from the lattice described above, and it underscores the importance of the rod ends, first realized in mutagenesis studies.

A different view emphasizing a critical role for the rod ends stems from earlier studies of paracrystals formed by a GFAP derivative (112). The paracrystals are based on only one type of dimer arrangement, namely antiparallel, and staggered such that helix 1B segments come into approximate alignment. While this arrangement is consistent with one of the alignments found by crosslinking (see above), the tetramers have only a short overlap involving only the carboxy ends of the rod, a model that does not fit the recent crosslinking data. Since NF-L paracrystals are distinct from GFAP paracrystals and follow the structural parameters predicted from the crosslinking data (50), it seems that the packing in GFAP paracrystals may reflect a particular arrangement of the GFAP rod tetramer favored when divalent cations are present.

D. The Role of IF Heads and Tails in IF Structure

Although most sequences within the nonhelical head and tail domains of IF proteins are not conserved, all IF proteins assemble to form 10-nm structures that are morphologically similar. Thus based on sequence alone, it seems doubtful that the head and tail domains play a role in structural features shared among IFs. For most IF tails, this notion may be correct. Proteolytic removal of a desmin tail segment (136), or deletion of the complete vimentin tail, has only very minor effects on IF assembly in vitro (130, 137). Similarly, both type I and type II keratin tails can be removed in vitro without a major impact on assembly (138, 139; see also 128). This is consistent with the findings that (*a*) keratin filaments sheared of their end domains by mild proteolysis appear indistinguishable from wild-type filaments (14), and (*b*) K19, the naturally occurring "tail-less" IF protein, assembles into bona fide IFs with a type II partner (140). Finally, tail-less lamins retain the ability to form head-to-tail polymers in vitro (141), and under some circumstances, removal of the long NF-L tail does not impede IF formation (142; but see 50). Ironically, for both type III and IV filaments, removal of the tail may actually accelerate IF assembly (137, 142).

The apparent dispensability of the tail to 10-nm filament structure does not imply lack of function. One role of IF tails may be to control lateral associations at the protofilament and protofibril level, thereby influencing filament diameter. For example, tail-less type III and type IV IFs laterally aggregate under certain assembly conditions (50, 110, 136, 143). Moreover, wild-type type III IFs aggregate when a peptide conserved among type III IF tails is added to assembly mixtures (144, 145). Interestingly, the converse holds for keratins: Removal of tails from either K8 and K18 (138) or K5 and K14 (139) seems to lessen aggregation of IFs under near physiological conditions. Finally, for both type III IF proteins (130) and keratins (138, 139), filaments formed from tail-less proteins have a tendency to unravel under conditions where wild-type IFs do not, suggesting a general role for the tails in stabilizing protofibrillar interactions and filament diameter.

IF head domains seem to have more prominent and diverse roles in filament structure than do IF tails. Thus, although amino-terminally truncated desmin and vimentin can each copolymerize into IFs when combined with wild-type counterparts, they cannot form IFs or proper IF networks on their own (126, 136, 146, 147). Similarly, NF-L and NF-M mutants missing >30% or >70%, respectively, of their head domains fail to incorporate into IF networks in vivo (123, 124). Under standard assembly buffers, headless NF-Ls only form linear arrays of tetramers (50). Upon addition of 1–2 mM calcium, IFs assemble from headless NF-Ls, but scanning transmission electronmicroscopy indicates that these IFs possess fewer protofibrils than normal (50). These observations infer that the NF head might promote lateral associations of protofilaments and protofibrils. In contrast, headless lamins cannot form a linear array of dimers typical of tail-less and wild-type lamins (141), suggesting that the lamin head might function in tetramer elongation.

The only IF head thus far scoring as dispensable for in vitro filament assembly is the K14 head (128). While part of the explanation for the functionality of headless K14 most certainly resides in the obligatory heteropolymeric nature of keratins, it is worth considering that conversely, headless K5 does not assemble into filaments with wild-type K14 (139). This suggests a dichotomy of the type II and type I head domains, a feature also implied by computer modelling (91). This said, for the simple epithelial keratins, both K8 and K18 head domains seem to play a role in IF network formation at least in vivo (122, 140). Thus even within an IF type, there appears to be some variability in the extent to which the head domains contribute to various aspects of IF assembly.

Mutagenesis studies have begun to identify sequences within the head domains that are involved in IF assembly. One candidate is a motif, T/S S Y R R X F/Y, where X is variable, which occurs in the heads of desmin, vimentin, peripherin, internexin, and NF-L (for GFAP, see Ref. 146). Deletion of this sequence blocks vimentin's ability to assemble into homopolymers (148). Point-mutagenesis studies reveal that the two arginines and aromatic residues of the motif are indispensable for assembly. How does this motif function? While the precise role of IF head domains in 10-nm filament structure is not yet understood, one suggestion is that the type III head interacts with the carboxy end of the rod domain, and that this interaction may serve to position dimers appropriately (149).

In summary, the tail and head domains of IF proteins seem to enhance favorable thermodynamic interactions between subunits. For most IF proteins, the head seems to govern both end-to-end and lateral associations, while the tail seems primarily involved in controlling lateral interactions.

E. The Dynamics and Regulation of IF Assembly

1. PHOSPHORYLATION-MEDIATED REGULATION: THE ROLE OF HEAD AND TAIL SEQUENCES The notion that IFs in cells are dynamic has emerged from early discoveries that IFs reorganize in response to cell-cycle-specific or differentiation-specific cues. In contrast to cytoplasmic microfilaments and microtubules, which are depolymerized during the cell cycle and subsequently recruited for mitotic-specific functions, IFs exhibit more variable behaviors. Some types of cells undergo a major reorganization of their IF networks during mitosis, while others show milder effects resembling a severing of filament bundles at the cleavage furrow of the dividing cell (150 and references therein). Such observations led to the view that cell-cycle-mediated as well as regional regulation of IF assembly and disassembly must occur in cells. From the beginning, phosphorylation-mediated modifications in IF proteins was offered as an attractive mechanism to govern these dynamic processes by transiently modulating thermodynamic interactions between subunits.

Some of the earliest studies invoking phosphorylation as a regulator of IF assembly state involved kinase treatment of purified type III IF proteins in vitro. Cyclic AMP–dependent and Ca^{2+}-activated phospholipid-dependent kinases can each phosphorylate desmin IFs in vitro, resulting in their disassembly (151, 152). Conversely, phosphatase treatment of phosphorylated subunits restores their competence for assembly. This reversible process is controlled at kinase target sites, which include residues 29, 35, and 50 of the head domain of desmin (151). Vimentin's 12-kDa head domain is also a substrate for kinases, including protein kinase A (PKA), protein kinase C (PKC), and cAMP-dependent kinase (153 and references therein).

Temporal phosphorylation can also orchestrate IF assembly and disassembly events in vivo (for review, see Ref. 154). For example, mitosis-mediated reorganization of vimentin filaments is associated with a transient increase in IF protein phosphorylation. In addition, cell-cycle-dependent increase in phosphorylation of the lamins coincides with nuclear lamina disassembly. Both of these in vivo processes are controlled by a cdc2 kinase-mediated phosphorylation cascade. Mitotic-stage lamin A is specifically phosphorylated at serines 22 and 392, flanking the central rod domain (155, 156). Elimination of these serines by mutagenesis leads to inhibition of lamina breakdown during mitosis (156). Conversely, in vitro phosphorylation of lamin B_2 by cdc2 kinase blocks head-to-tail alignment (157). Phosphorylation of vimentin IFs by purified cdc2 kinase also causes vimentin disassembly in vitro (158), and serine residue 55 in the head domain of vimentin appears to be the crucial regulatory site (159). In contrast to the case of lamins, the degree of vimentin reorganization during mitosis varies among cell types, and consequently the functional significance of vimentin reorganization is difficult to assess.

Lamins are phosphorylated not only at mitosis but also at interphase, a phenomenon whose underlying significance has only recently been explored. PKA and PKC each phosphorylate lamins when cells are treated with agonists that stimulate their activity. At interphase, lamin B_2 is specifically phosphorylated by PKC at sites located in the tail, and in contrast to cdc2 kinase, PKC-mediated phosphorylation does not interfere with lamin polymer disassembly (160). Rather, such phosphorylation seems to inhibit nuclear transport of lamin B_2. What might be the functional consequences of a temporal reduction in the nuclear levels of lamin B_2? As pointed out by Henekes et al (160), newly synthesized lamins exchange with polymeric ones, suggesting that a reduced nuclear pool size coupled with a change in phosphorylation patterns might induce lamin filament rearrangements and possibly changes in lamin-chromatin interactions.

Non-cdc2 kinase phosphorylation has also been implicated in the partial disassembly of IFs that occurs regionally at sites such as the cleavage furrow. As detected by monoclonal antibodies specific for phosphorylated peptides of GFAP, a mitosis-specific, phosphorylated form of GFAP is confined to the cleavage furrow of mitotic astroglial cells (161). At mitosis, there is both an increase in phosphorylation of GFAP, and a transient change in the pattern of serines that are phosphorylated. Thus, cdc2 kinase acts as a glial filament kinase only at the G2-M transition, whereas other GFAP kinases are activated at the cleavage furrow (162). In this manner, kinases acting at specific sites and times in the cell may allow efficient separation of IF networks to daughter cells.

The regulatory functions controlled by phosphorylation extend beyond cell-cycle-mediated changes in IF networks. For example, NF-M and NF-H tails are heavily phosphorylated by various kinases, yielding up to 50 moles of phosphate per mole of protein, predominantly at serine residues within the Lys-Ser-Pro-(Val) repeats (163–165). This process occurs as NFs move from cell body to axon, and does not appear to regulate filament assembly and disassembly. Rather, it may modulate surface properties of the filaments, which in turn could affect functional interactions. There is some evidence that this comes about through kinase-regulated dissociation of the binding of NF-H to microtubules (166). Phosphorylation has also been implicated in reducing spacing between NFs, thereby shrinking axonal caliber, and possibly slowing axonal transport (167).

Although the operational significance of many kinase-mediated modifications remains to be elucidated, most if not all IF networks are modified by phosphorylation, and this extends to the keratins. A number of head- and tail-domain sites of phosphorylation have been identified for K1 (168) and also for K8 and K18 (169). Interestingly, hyperphosphorylation of K8 during meiotic maturation of *Xenopus* oocytes correlates with keratin filament severing and disappearance of the cortical cytoskeleton (169). It has been proposed by Klymkowsky and coworkers that since the organization of keratin filament systems in the oocyte differs substantially from that of the early embryo, this kinase-mediated fragmentation of the oocyte IF network may serve to facilitate the reorganization of IF architectures that occurs during development.

2. INTERMEDIATE FILAMENT DYNAMICS: NOT JUST AN INTERNAL STEEL INFRASTRUCTURE Although it has generally been presumed that IFs are the most stable component of the mammalian cytoskeleton, several lines of evidence suggest that IFs are dynamic structures, not only with respect to their ability to reorganize within a cell, but also with regard to the assembly of their networks. Kinetic analyses demonstrate that in vivo, vimentin IFs rapidly assemble from a small soluble precursor pool apparently composed of tetrameric subunits (170). Experiments involving microinjection of epidermal mRNAs (171), transient expression of epidermal keratin transgenes (120), rapid disassembly of IFs upon addition of certain peptides (115, 131, 132), and microinjection of biotinylated IF proteins into cells (172) all indicate that foreign IF proteins (peptides) can rapidly integrate throughout an existing IF network. That IF networks are not simply static, but undergo exchange or incorporation of subunits, is further suggested by the finding that an existing IF network can be disrupted by a transiently expressed truncated keratin (120; see also 173). Finally, when a vimentin transgene is hormonally induced, newly synthesized subunits incorporate along the length of existing IFs (174, 175). Such studies leave little doubt that IF networks in cells are dynamic, and that subunits integrate in a nonpolarized fashion at many sites along IFs.

A variety of biophysical methods have been implemented to examine more precisely the dynamics of IF subunit exchange. Angelides and coworkers used fluorescently labeled NF proteins coupled with a fluorescence resonance energy transfer assay to discover that subunit exchange is preceded by dissociation of subunits from NFs and generation of a kinetically active pool of soluble subunits (176). Similar studies on desmin indicate that a significant level of unpolymerized desmin exists in steady-state equilibrium with intact IFs (177). More recent studies have used photobleaching of Xrhodamine fluorescently labeled vimentin protein in order to track the kinetics of fluorescence recovery subsequently (178). Since X rhodamine does not recover its fluorescence after photobleaching, all fluorescence recovery is due to incorporation of unbleached vimentin. Fluorescence recovery of photobleached IF networks occurs relatively slowly, in ~30-40 minutes, and with no apparent polarity. In similar studies, photobleaching of fluorescently labeled NF-L protein followed by electronmicroscopy enabled visualization of NF dynamics in living neurons (179). Interestingly, recovery of fluorescence is more rapid in growing axons than in quiescent ones, indicating a growth-dependent regulation of the turnover rate. In this case, incorporation of biotinylated NF-L subunits seems to occur at discrete sites, which grow and become continuous after ~24 hr (179). Collectively, the fluorescence-bleaching studies demonstrate that IFs can accept and exchange subunits along their entire surface.

V. ASSOCIATIONS BETWEEN IFs AND OTHER PROTEINS AND STRUCTURES: FORMING A CYTOPLASMIC IF NETWORK

Cytoplasmic IFs associate with the nuclear and plasma membranes (66, 180–182). IF-nuclear envelope interactions have been documented for peripherin, vimentin, and desmin, and appear to be mediated by B lamins. Binding to the nuclear envelope is cooperative and saturable, and in addition, it seems to support IF polymerization. In contrast, binding to the plasma membrane fraction is not saturable and cannot support filament polymerization. Furthermore, binding assays using NH_2- and COOH-terminally truncated proteins reveal a polarity to the binding: IF polypeptides containing the tail domain bind selectively to the nuclear envelope fraction, while polypeptides containing the head domain bind selectively to the plasma-membrane fraction.

These studies support the hypothesis by Eckert et al (183) that initiation of IF assembly may occur preferentially at nuclear envelope–associated sites. Also relevant to these findings is the observation that following a seemingly complete disruption of a cytoplasmic IF network by a wave of transiently expressed keratin mutants, subsequent de novo assembly of wild-type keratins, now in the absence of interference by the mutants, results in a cage of normal IFs around the nucleus (120). In vitro assembly and microinjection studies clearly demonstrate that IFs do not require a nuclear envelope for de novo IF assembly, and dynamic exchange experiments show that subunits integrate throughout the IF network without apparent polarity. This said, the nucleus still might provide preferential sites for de novo IF network formation, a notion that is presently controversial but unresolved.

The associations of IFs with plasma membranes vary with cell type. In erythrocytes, vimentin binds to the plasma membrane through ankyrin, a constituent of the plasma-membrane skeleton (180). Stratified squamous epithelial cells produce two kinds of specialized plaques in the plasma membrane, which serve as attachment sites for keratin filaments. Desmosomes allow cells to adhere to each other, and hemidesmosomes, which are restricted to basal cells, allow them to affix to the basement membrane (for review, see 184; 185). Human meningiomal cells, myocardium, and dendritic follicular cells also have desmosomes, which share many features with epithelial desmosomes, except that they associate with type III IFs rather than keratin IFs. In all cases, IFs appear to loop through the matrix of desmosomal and hemidesmosomal plaques rather than terminate within them.

Although the precise interactions between IFs and desmosomes or hemidesmosomes remain unknown, a growing family of membrane-associated proteins, the desmoplakinlike proteins, have been implicated (186). These proteins share similar predicted secondary structures, including a long coiled-coil rod domain and a variable number of globular carboxy-terminal domains made of tandem 38-residue repeats (187–189). Recent studies show that desmoplakin (DP), a major component of most if not all desmosomes, is corecruited with IFs to the sites of artificial gap junctions produced by transfected cells expressing a hybrid desmocollin-connexin protein (190). This finding suggests that either indirectly or directly, desmoplakin and IFs associate with one another. Deletion mutagenesis studies on desmoplakin suggest that the union between desmosomes and IFs might be mediated through the tail portion of the desmoplakin molecule (191).

Plectin, another member of the DP family, was initially identified as a prominent component of vimentin IF preparations (for review see 192). It is broadly expressed in different cell types, binds to most if not all IF proteins, and localizes to the IF cytoskeleton as well as desmosomes and hemidesmosomes. Based on deletion mutagenesis, the six repeats in plectin's tail may be involved in the interactions with IFs (193). A third member of the DP family, bullous pemphigoid antigen (230 kDa; BP230), is also a component of hemidesmosomes, but in contrast to plectin, BP230 is not found in desmosomes, nor does it decorate or colocalize with the IF cytoskeleton (189). Although BP230 has not yet been subjected to mutagenesis, it seems likely that the repeats within its tail domain will participate in linking hemidesmosomes to IFs.

How does the cytoskeletal architecture of a cell change with growth and differentiation state? While an exact answer to this question is not yet at hand, it is interesting that desmosomes and hemidesmosomes assemble in a calcium-dependent fashion, suggesting that intracellular fluxes in calcium might regulate these contacts. This could impart dynamic and flexible properties to the interactions between adjacent cells as well as between a cell and its basement membrane (for review, see 184). Interestingly, the plectin-IF alliance appears to be governed by kinases, and PKA, PKC, and cdc2 kinase may all be involved (194; G Wiche, personal communication). Although the role of phosphorylation in controlling desmosomal and hemidesmosomal connections with IFs has not yet been explored, phosphorylation-mediated events may offer additional versatility to IF cytoskeletons.

The cytoarchitecture of IFs is likely to be influenced by interactions with other IF-associated proteins (IFAPs) and with other cytoskeletal elements, particularly microtubules (for review, see 192). It is well known that selective depolymerization of microtubules causes collapse of type III IFs into a perinuclear cap (4). Recent microinjection studies indicate that this interaction involves the microtubule motor kinesin (195). Of the few additional IFAPs that have been characterized thus far, most seem to be expressed in a tissue-specific fashion, and their functions remain largely unknown. Among the best-studied IFAPs are those expressed in the epidermis and its appendages. They include (*a*) filaggrin, a calcium-regulated protein that is expressed late in epidermal terminal differentiation and that can laterally crosslink IFs into bundles (196, 197); (*b*) trichohyalin, a large (>200-kDa) α-helical, calcium-regulated IFAP produced and retained in the cells of the hardened inner-root sheath and medulla of the hair follicle (198); and (*c*) hair-specific IFAPs, which embed IFs in a matrix of globular glycine-rich and tyrosine-rich proteins (199) and sulfur-rich proteins (200).

The reader is referred to other reviews for more comprehensive descriptions of IFAPs (192). Although the field of IFAPs has not yet been extensively explored, it seems likely that these proteins play roles in establishing and maintaining IF networks, and perhaps in governing dynamic changes that occur in IF cytoarchitectures during cellular growth and differentiation. In addition, because expression of most IFAPs is cell-type specific, unions between filaments and IFAPs may provide a means to specify the properties of IF cytoskeletons further, and in so doing may tailor the IF network to suit the distinctive needs of each higher-eukaryotic cell type.

VI. EVOLUTIONARY ORIGIN OF INTERMEDIATE FILAMENTS

The mammalian IF superfamily presently consists of ~50 different genes. The striking homologies in IF amino-acid sequences and immunological properties across vertebrates suggest that their genes are derived from a common ancestor (18, 201). Moreover, with the exception of the mammalian type IV gene family (202) and nestin (84), conservation of intron positioning reflects a mutual evolutionary origin for IF genes (86, 89). Thus, the non-neuronal mammalian genes have five to seven nearly identically positioned introns that are located within sequences encoding the rod domain. They have an additional one to four variably positioned introns located in the hypervariable sequences encoding the carboxy-tail segment. In contrast, mammalian type IV genes and the nestin gene have only two or three introns within rod sequences and one or none in tail segments. Moreover, despite conservation among these genes, two of their rod introns are positioned differently from those in other IF genes.

Given the universal importance of nuclear structure in the eukaryotic kingdom, it has long been postulated that all eukaryotes might have lamins and that the putative primordial IF gene encoded a lamin. Despite the widespread acceptance of such a notion, repeated attempts to identify a bona fide IF protein in single-cell eukaryotes have been largely unsuccessful. The best evidence to date in support of a yeast IF protein stems from the characterization of the *Saccharomyces cerevisiae* mutant, MDM1 (203). This MDM1 protein plays a role in orienting mitotic spindle and in transferring mitochondria into developing daughter buds. The MDM1 gene encodes a 51-kDa protein that can assemble in vitro into filaments of approximately 10-nm diameter. Moreover, affinity-purified antibodies against MDM1 recognize IF networks in a variety of animal cells. This said, the sequence of MDM1 bears little recognizable relation to the IF superfamily of proteins. Most notably, the heptad repeats of hydrophobic residues are missing in parts of the α-helical domains, and there is no discernable relation of the rod ends to the highly conserved IF consensus motifs. Thus, while certain characteristics of MDM1 resemble those of IF proteins, others do not.

To date, multicellular invertebrates are the earliest eukaryotes known to possess proteins that maintain the overall structural principles that apply to vertebrate IF proteins, i.e. the presence of hydrophobic heptad repeats, IF consensus sequences at the ends of the rod, and linker segments spaced within the rod (90, 204, 205). *Drosophila* has nuclear lamins analogous to vertebrate lamins (204, 206), and indeed it was recently discovered that ablation of the *Drosophila* lamin Dmo gene is lethal, thereby demonstrating that this gene is essential (Y Gruenbaum, personal communication). In addition, at least as far down the evolutionary scale as the nematodes and mollusks, invertebrates have cytoplasmic IF proteins that, intriguingly, share many structural features with lamins (90, 205). These features include (*a*) the 42-residue insert in helix 1B, (*b*) differential splicing of transcripts to produce proteins that differ only in their carboxy termini, and (*c*) a moderate homology at the carboxy termini, providing that alignment of residues is optimized by deleting a 15–20-residue stretch corresponding to the nuclear localization signal. In contrast to the lamins, these invertebrate IF proteins lack a CaaX box and nuclear localization signal, thereby accounting for their residence in the cytoplasm.

Overall, the cytoplasmic invertebrate IF proteins have a significantly lower complexity than the mammalian IF superfamily. For example, the mollusk

Helix aspersa has a single non-neuronal IF gene, which is broadly expressed (207). The gene produces a transcript that is differentially spliced to generate two mRNAs encoding proteins that differ in their carboxy-tail sequences. Interestingly, both polypeptides are expressed in the epithelia of the snail, although these "keratins" do not follow the obligatory heteropolymeric principles of vertebrate keratins (205). The single IF gene has six introns residing within sequences encoding the rod segment, and they are identically positioned in vertebrate type III IF genes. Moreover, three of its four introns in sequences encoding the tail segment are identically positioned in the vertebrate lamin genes (54, 207).

A clue to the evolution of the type IV IF family comes from recent analysis of the neurofilament gene of the squid, another member of the mollusks (208). In contrast to mammalian NF proteins, which arise from individual genes, squid NF proteins stem from splice variants of a single gene. This gives rise to 60-kDa, 70-kDa, and 220-kDa NF proteins, differing in their carboxy-tail segments. The 220-kDa NF protein contains an acidic domain and 21 Lys-Ser-Pro repeats, similar to those in mammalian NF-M and NF-H, but distinct from non-neuronal IF proteins. In striking contrast to mammalian NF genes but analogous to non-neuronal IF genes of gastropods and nematodes, the squid NF gene encodes a 42-aa laminlike insert in helix 1B and displays the type III IF pattern for rod introns. These data establish an evolutionary link between the mammalian type IV IF genes and the rest of the IF family. Moreover, they suggest that the mammalian NF genes may have evolved either from a reverse transcription–mediated mechanism of an invertebrate NF mRNA (202), or from a loss of introns occurring somewhere after the evolution of higher invertebrates.

The shortened helix 1B domain of vertebrate cytoplasmic IF proteins has been found in a non-neuronal IF gene of *Branchiostoma lanceolatum,* a member of the cephalochordates, which are a sister group of the vertebrates in the phylum of chordates. The six introns of the *Branchiostoma* gene are identically placed in the vertebrate type III IF genes in sequences encoding the rod domain (209).

In summary, these results imply that the archetypal IF gene had a laminlike structure. With increasing evolutionary pressure, the primordial gene underwent duplication and divergence. One early progeny lost its CaaX box and nuclear localization signal to become a cytoplasmic IF gene with a laminlike rod domain (54, 207). Subsequently, evolving cytoplasmic IF genes lost sequences encoding the 42-residue insert in helix 1B. Although the precise timing of this event is unknown, it may have coincided with the origin of the chordates or occurred earlier, during the evolution of deuterostomes (209).

Given the sharp cutoff in immunological and sequence homologies between the lower-vertebrate keratins and invertebrate IF proteins and the prominence of keratins in the epidermis and its appendages, it seems possible that type I and type II keratin genes emerged during the transition from organisms bearing an exoskeleton to those with an endoskeleton (90, 201, 205). Given the substantial sequence variation across IF types, with higher sequence identities within types, much of the intratype gene duplications probably occurred relatively recently.

VII. INTERMEDIATE FILAMENTS AND DISEASE

Given that genetically engineered mutations in IF genes can act in a dominant negative fashion to perturb endogenous cytoskeletal architectures, it seems likely that naturally occurring mutations in IF genes might give rise to genetic disease. In the past few years, the first three genetic diseases of IF genes have been discovered. All of these are blistering skin diseases involving mutations in keratin genes.

A. *Epidermolysis Bullosa Simplex*

Epidermolysis Bullosa Simplex (EBS) is a rare genetic skin disease affecting 1 in 50,000 in the human population. It is typified by mechanical stress-induced intraepidermal blistering due to cytolysis within the basal layer (for review, see 210). In most cases, the disease is autosomal dominant, although recessiveness has been reported, and clinical manifestations are often present at birth. EBS has been subdivided into three major subtypes. Weber-Cockayne (W-C) EBS is the mildest form, where blisters occur predominantly on the hands and feet. Basal keratin filaments appear nearly normal, and as in all EBS subtypes, suprabasal layers are unperturbed, indicative of a normal differentiation process. Koebner (K) EBS is intermediate in severity, and Dowling-Meara (D-M) EBS is the severest form. The major ultrastructural feature distinguishing D-M from other EBS subtypes is the occurrence of clumps of keratin within the basal cell cytoplasm (211).

Compelling evidence has mounted to indicate that structural defects in keratin underlie the EBS disease. Tonofilament clumping precedes cell cytolysis, an indication that it is an early event in the blistering process (211). Cultured D-M EBS keratinocytes exhibit a perturbed keratin network, similar in appearance to that of normal keratinocytes transfected with mutant K14 genes (212, 213). Finally, when engineered to express a mutant human K14 that severely disrupts IF assembly, transgenic mice produce nearly all the symptoms of D-M EBS (214). Transgenic mice expressing a mildly disrupting K14 mutant display features typical of W-C EBS (215). Thus, structural defects in K14 and K5 genes can generate an EBS phenotype,

LOCATION OF EBS AND EH MUTATIONS
RELATIVE TO KERATIN SECONDARY STRUCTURE

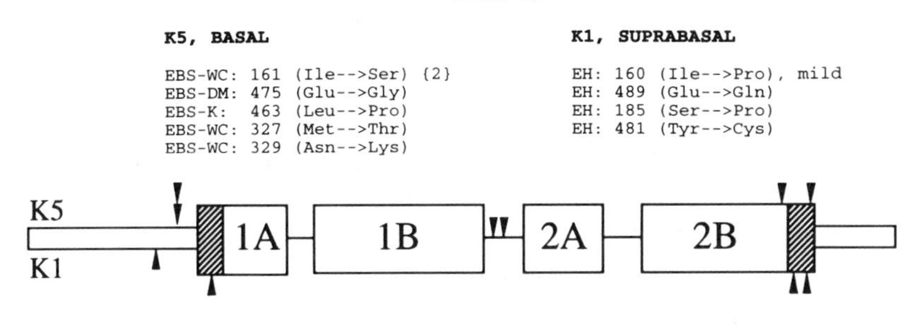

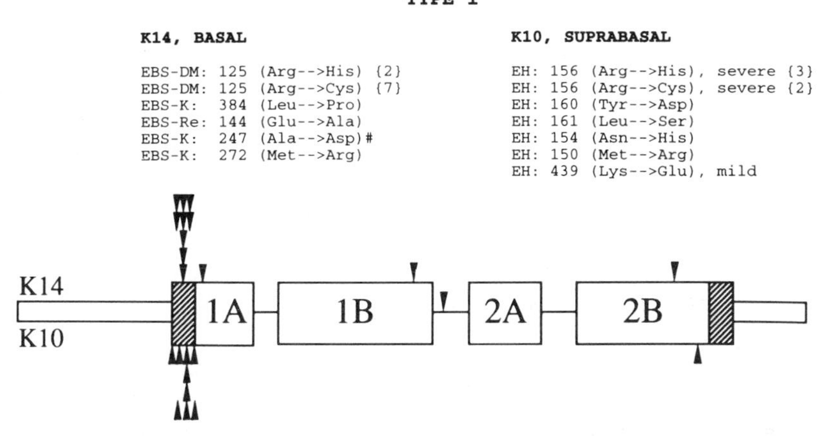

#Hutton and Fuchs, unpublished.
EBS-Re, recessive EBS.

Figure 3 Correlation between mutation location and disease severity in genetic disorders of keratin. Stick figures depict secondary structures of human type II and type I keratins (12, 13). Large boxes encompass the α-helical rod domain, interrupted by the short nonhelical linker segments. Hatched boxes denote conserved ends of the rod. Open bars denote nonhelical head and tail domains. Arrowheads denote positions of mutations found in EBS (K5 and K14) and EH (K1 and K10) (for review, see Ref. 222).

and the degree to which a specific K14 or K5 mutant perturbs filament assembly correlates with corresponding severity of the EBS phenotype.

The genetic defects of several EBS families map to human chromosomes 17 or 12 (216–218, 218a), at locations corresponding to the loci for epidermal type I and type II keratin gene clusters, respectively (219). In addition, patients with EBS have point mutations in their K14 or K5 genes (213, 216, 218, 218a, 220, 221, 221a, 221b). Most D-M EBS cases analyzed have a single amino-acid substitution, 125 Arg→Cys/His, in the conserved amino end of the K14 α-helical rod (213, 221, 222). Affected members of another D-M EBS family have a 475 Glu→Gly mutation in the conserved carboxy end of the K5 rod (220). In contrast, K-EBS mutations have been found more centrally in the α-helical segments of the rod (216, 222), while six W-C mutations have been localized to nonhelical segments, either the head domain of K5 (218) or the linker segments (218a, 221b). All of these mutations correlate with affected individuals, and they have not been found in unaffected family members or the normal population. The locations of these mutations are illustrated in Figure 3.

Functional evidence suggests that the keratin point mutations identified in these EBS cases are in fact responsible for generating the EBS phenotype (213, 223). When expressed in transfected human epidermal cells, the D-M EBS point mutants cause perturbations in the IF networks similar to those detected in cultured D-M EBS keratinocytes. When combined with their wild-type partner in vitro, bacterially expressed D-M point mutants cause shortening of filaments in a fashion similar to the altered filaments formed with keratins isolated from D-M EBS keratinocytes (213, 223). The D-M mutations are particularly severe, presumably because they occur in domains that are involved in multiple contacts within IFs (see Figure 2). When these assays are performed on milder cases of EBS, the perturbations are significantly reduced, underscoring the correlation between the degree to which IF assembly is perturbed and the disease severity (218a, 223).

B. Epidermolytic Hyperkeratosis and Epidermolytic Palmoplantar Keratoderma

Epidermolytic Hyperkeratosis (EH) is also an autosomal-dominant disease, with clinical manifestations usually present at birth. The histopathology of EH is typified by a basal epidermal layer with normal morphology, and cytolysis in suprabasal layers, beginning in the lower spinous layers and increasing during terminal differentiation (for review, see 211). In addition, tonofilament clumping and perinuclear shells of tonofilament aggregates occur in suprabasal cells, and there is a markedly thickened granular layer and stratum corneum in the upper layers of EH skin.

Clues to the genetic basis of EH stem from (*a*) the striking similarity between tonofilament clumping in suprabasal cells of EH skin and in basal cells of EBS skin, (*b*) the realization that EBS is a disorder of K5 and K14, and (*c*) the knowledge that as epidermal cells commit to differentiation, they downregulate expression of K5 and K14 and switch on expression of K1 and K10. Indeed, when expressed in transgenic mice, a truncated human keratin 10 gene generates the pathobiological and biochemical characteristics of EH (for review, see 222). In addition, genetic mapping of EH families reveals linkage to the keratin clusters on chromosomes 12 and 17 (224). Point mutations in the K1 and K10 genes of patients with human EH provide even stronger evidence that EH is a K1-K10 disorder (225–227).

Epidermolytic Palmoplantar Keratoderma (PPK) resembles EH except that its clinical manifestations are confined to palmoplantar skin. Recently, patients with PPK were found to have point mutations in the gene encoding K9, a keratin specifically epxressed in the suprabasal layers of this type of skin (228).

Remarkably, one residue mutated in K10 of EH (156 Arg→His/Cys) and in K9 of PPK (162 Arg→Trp/Gln) is equivalent to the arginine residue that when mutated in K14 gives rise to EBS (225, 226, 228). Thus it appears that the same mutation in a highly conserved residue of three genes can lead to three distinct genetic diseases by virtue of the differential expression of the genes. The high frequency with which this residue is mutated in EBS, EH, and PPK is due to two factors: (*a*) This residue is critical for IF assembly (156, 213, 227), and (*b*) this is a hot-spot for C-to-T mutagenesis by CpG methylation and deamination (229).

C. What Does Our Knowledge of IFs Tell Us About These Diseases?

The cell biology of keratins provides an understanding of many previously unexplained aspects of human EBS and EH. The autosomal dominance arises because mutant keratins can recognize their obligatory heterotypic partners, and act in a dominant-negative fashion to perturb keratin network formation. The manifestation of the diseases in either the basal (EBS) or suprabasal (EH) layers of the epidermis is due to the differential expression of the K5-K14 and K1-K10 keratin gene pairs. In severe cases of EBS, clinical manifestations can occur in other stratified squamous epithelia, including nail, oral mucosa, cornea, and esophagus, because these cells also express K5 and K14. In contrast, EH is predominantly manifested in the skin, because K1 and K10 have a more restricted pattern of expression.

D. What Do These Diseases Tell Us About the Functions of IFs?

A major question uncovered from these studies is: How do mutant keratin genes cause cytolysis? Several findings point to the hypothesis that without a proper IF network, cells become fragile and prone to breakage upon mechanical stress. Thus, in mice and in men, EBS cells often rupture in a defined zone, beneath the nucleus and above the hemidesmosomes. This is the longest portion of the columnar basal cell, and the zone expected to be most fragile when mutations arise that compromise filament elongation, as do many of the EBS and EH mutations. Thus, an important function of keratin filaments seems to be to impart a mechanical framework to an epidermal cell.

Do other IFs also play a role as mechanical integrators of space, as originally suggested by Lazarides (5)? Several lines of evidence suggest that they do. Rheologic studies show that wild-type type III IFs harden and resist breakage under stresses that cause other cytoskeletal networks to rupture (230). Newport and colleagues (79) discovered that a *Xenopus* oocyte extract depleted of its single lamin (L_{III}) can encapsulate chromatin by fused membrane vesicles and assemble nuclear pores, but the resulting nuclei are fragile. Transgenic mice overexpressing a wild-type hair keratin gene have fragile cortical cells, leading to hair brittleness and breakage (231). Finally, when the keratin 8 gene, normally expressed in liver, is ablated in mice by embryonic stem-cell technology, developing mouse embryos often do not survive past a critical stage in development when the liver becomes vascularized, i.e. when potentially substantial mechanical stress is exerted on this tissue (232, 233).

E. Are There Other Diseases of IF Genes?

If most IFs function in part to impart mechanical integrity to cells, then it seems likely that genetic defects in other IF genes might lead to cell fragility and degeneration in other tissues. Given the nature of IF assembly, if such diseases exist, they might also be expected to be autosomal dominant and involve abnormalities in their IF organization.

Several possible candidate hair diseases share these features (for review, see 211), and in some cases, abnormalities in two-dimensional gel patterns and amino-acid compositions of hair keratins from patients with hair diseases have been reported (see for example 234). It may also be relevant that several autosomal-dominant mouse mutants, Re, Bsk, and Re^{den}, map in close proximity to the type I epidermal keratin genes (235 and references therein): Rex mice have curly whiskers and bent hair shafts, and denuded mutant mice and bareskin mice undergo hair loss after completion of the first hair cycle, i.e. at ~4–5 weeks post-birth.

Another class of candidate IF disorders are those that might involve NF genes. Although not present in all neurons, NFs are preferentially located in the axon, where they are thought to provide mechanical strength. NF defects have been suspected in the pathogenesis of several types of neurodegenerative diseases, such as amyotrophic lateral sclerosis (ALS) and infantile spinal muscular atrophy (236 and references therein). These motor neuron diseases are clinically characterized by progressive muscle wasting, leading to total muscle paralysis, with the primary lesion site in the spinal and cortical motor neurons. Neuronal abnormalities include massive accumulation of neurofilaments in cell bodies and proximal axons. Interestingly, transgenic mice that overexpress wild-type NFs or produce mutant NFs exhibit the hallmarks of motor neuron disease, including axonal degeneration and skeletal muscle atrophy (237, 238). Thus, primary changes in the cytoskeleton, specifically in NFs, are sufficient to trigger a neurodegenerative process, and may be a key intermediate in the pathway of pathogenesis leading to neuronal loss of the type that is seen in ALS and other motor-neuron diseases.

Another class of diseases are familial cardiomyopathies. Many of these disorders are myosin defects (239, 240). However, some are not, and in several of these cases, abnormalities in desmin filament networks have been observed electronmicroscopically (241). Thus, although at present no functional evidence exists to indicate that desmin mutations in mice or in man can cause familial cardiomyopathies, it is possible that some of these cases may stem from desmin mutations.

Finally, it should be mentioned that overexpression of the chicken vimentin gene in transgenic mice leads to cataract formation in the lens, apparently as a consequence of impaired fiber cell denucleation and elongation (242). Cataract formation is confined to animals expressing high (>3× the) levels of endogenous vimentin. This has also been noted for transgenic mice expressing NFs in the lens (243), and in the lens of transgenic animals expressing a hybrid vimentin-desmin gene (244). Interestingly, a number of similarities between the transgenic lens and the early stages of congenital cataract formation can be found, including ultrastructural changes in the integrity of the plasma membrane, alteration in fiber shape and orientation, and the restriction of these changes to defined regions of the lens. It may be, therefore, that alterations in the control of IF expression in the lens is a factor in cataract formation.

VIII. SPECIALIZED FUNCTIONS OF CYTOPLASMIC INTERMEDIATE FILAMENTS

Given the heterogeneity among cytoplasmic IF proteins and their sometimes remarkable abundance, it seems extraordinary that specialized functions, like general functions, of IFs have remained elusive since their discovery in the early 1960s. Recent studies have provided provocative insights to such functions, bringing answers to this question closer than ever before.

In contrast to lamins, cytoplasmic IFs cannot be essential for housekeeping roles, since several cultured cell lines exist that lack these filaments altogether (245, 246). Given recent links between IFs and cell degenerative diseases, it could be argued that cells flattened over a large surface might not require the same mechanical strength as cells in a tissue. Can cytoplasmic IFs be dispensed with in vivo? Recent studies suggest that in some instances, they can. A mutant vimentin causing disruption of endogenous vimentin assembly is without effect when present during early embryogenesis in the frog (247). Even more surprising is that some mice lacking keratin K8 altogether can escape hemorrhaging in the liver and develop to adulthood without any apparent complications (233). Despite the seemingly ominous implications these studies might have for cytoplasmic IF function, it could be that cytoplasmic IFs are not needed in certain cell types, species, or stages of development, but are critical in others. This might explain why depletion of K8 and K18 is without apparent effect in early-stage mouse embryos (232), but in frog embryos, it results in a loss of the compacted epithelial surface of the blastula, an inability to close the wounded surface, and defective gastrulation (248).

In some tissues, a particular IF protein may be essential even when multiple IF proteins are coexpressed. A case in point is the glial cell line U251, which constitutively expresses both GFAP and vimentin, forming a single 10-nm filament network. These cells respond to neurons as astrocytes do in vivo: They withdraw from the cell cycle, extend complex processes, and promote neural survival and outgrowth. When GFAP expression is suppressed by antisense GFAP RNA, the ability of U251 cells to respond to neurons by extending glial processes is inhibited, even though glial proliferation is still arrested and neurite outgrowth is still supported. This inhibition occurs in spite of continued vimentin expression and IF network formation, suggesting a role for GFAP in formation of stable astrocytic processes in response to neurons (249).

Mounting evidence has strengthened the view that many functions for IFs are tissue specific. Although NFs are not abundant during neurite extension, their density increases greatly at a time when neurites expand radially, leading to the hypothesis that these IFs may be involved in establishing neuronal caliber (250 and references therein). Moreover, when axons are severed, NF synthesis is downregulated, with a subsequent decrease in axonal NFs and a shrinkage in neuronal caliber. In addition, when NF-H, but not NF-L, is overexpressed in transgenic mice, radial growth of the axons is increased (237, 238, 243). Finally, a nonsense mutation in the quail NF-L gene leads to an absence of NFs and a concomitant decrease in axonal calibers (251). Collectively, these data demonstrate convincingly that NFs are intrinsic determinants of axonal caliber.

While variation in IF density can influence axon diameter, overexpression of up to 10× the physiological levels of wild-type vimentin leads to no detectable morphological abnormalities during the early stages of *Xenopus* embryogenesis, suggesting that cells in early development can tolerate substantial changes in IF density (247). Similarly, the creation of a dense

NF network in transgenic mouse kidney (normally NF^-) is without apparent consequence (243). This said, too many cytoplasmic IFs can be deleterious: Overexpression of type III or type IV IFs in lens epithelium ($vimentin^+$) of transgenic mice leads to cataract formation, as mentioned above.

The localization of desmin at Z discs has precipitated the hypothesis that desmin might be involved in either linking Z-bands of adjacent myofibrils to one another and to the sarcolemma laterally, or in connecting successive Z-bands longitudinally in elongating myofibrils. However, when a truncated desmin is expressed transiently in postmitotic myoblasts and multinucleated myotubes, the preexisting IF networks are completely dismantled, and yet striated myofibrils are still assembled and aligned laterally, and contraction proceeds apparently unaffected (252). Thus, the specialized role of desmin in muscle cells is at present unclear.

Perhaps more than any other tissue, the epidermis and its appendages have evolved to utilize cytoplasmic IFs as their major structural components. Here, a function of keratin IFs may be to survive, since most other cytoplasmic proteins and organelles are lost during terminal differentiation. Indeed, epidermal keratins form some of the most stable protein-protein interactions known in nature. Coupled with this remarkable stability, epidermal keratin IFs are progressively packaged into larger bundles as cells differentiate, and this may further enhance their chances of survival. Hair keratins are even more stable than epidermal keratins: They contain cysteine-rich sequences, possibly resulting in complex disulfide linkages not only among keratins, but also with hair-specific, cysteine-rich IFAPs. In contrast, keratins of simple epithelial cell types differ from keratins of epidermis with regard to filament solubility, tensile strength, and flexibility. These differences may affect filament network structure and interactions with other proteins and cytoskeletal elements. In addition, cell-type-specific IFAPs might contribute to IF network differences.

In summary, it is clear that in addition to their potential role in providing mechanical integrity to cells, IFs serve more specialized roles, ranging from stabilizers of DNA replication complexes as manifested by lamin IFs, to tough, resilient survivors of terminal differentiation, as manifested by keratin IFs. As additional studies are conducted, roles of IFs in determining the intricate architecture and physiology of different cell types and tissues should become increasingly apparent.

IX. CONCLUSIONS AND PERSPECTIVES

IF proteins are a complex and still-expanding family of proteins that share a remarkable conservation in structure. Although functions are still to be defined for many IFs, the developmental, differentiation, and tissue-specific expression of these proteins provide strong evidence that their 10-nm networks are crucial not only to mechanical integrity, but also to specialized roles of cells and tissues. Combined approaches involving DNA recombinant technology, biochemistry, and biology have led to significant recent progress in our understanding of the structure, evolution, expression, and functions of IFs. X-ray diffraction and electronmicroscopy studies, coupled with chemical crosslinking analyses, have provided major insights into the structure of 10-nm filaments. Additionally, the ability to manipulate IF gene sequences and express them or ablate them in cultured cells, in bacteria, and in mice have provided powerful experimental systems in which to dissect complex issues of IF structure, function, and disease. Microinjection and gene transfection studies have revealed that IFs form a dynamic network with rapid subunit exchange, rendering it susceptible to disruption upon incorporation of abnormal subunits. From such studies, it has been possible to identify specific sequences in IF polypeptides that are critical for filament assembly and network formation.

The realization that EBS and EH are genetic disorders of keratins has enabled predictions as to what other types of genetic diseases might be disorders of IF genes. Present studies on the relation between NF aberrancies and motor-neuron diseases are promising, and future studies should reveal if such human diseases involve defects in NF genes. In addition, similar studies on other IF genes should yield a greater understanding of the extent to which defects in the ~50 human IF genes might give rise to a whole family of IF disorders.

In the past 10 years, it has become increasingly clear that IF networks exhibit close morphological associations with membranes and other cell structures. Although the genes for some IFAPs have been cloned and sequenced, most known IFAPs are not well characterized and many more IFAPs remain to be identified. Understanding the biological nature of these interactions is an important prerequisite to elucidating the underlying basis of the supramolecular structure and function of the IF network within cells.

ACKNOWLEDGMENTS

We thank our many scientific colleagues for sharing unpublished information and for stimulating discussions. Their exciting contributions to this field have unraveled many of the mysteries underlying the field of intermediate filaments, and have made it a pleasure for us to write this review. We also thank the members of our laboratory and collaborators with whom we were privileged to conduct research in this area. We regret that due to space constraints, we have not been able to cite all relevant literature on this topic. The research that led up to this review was provided by grants from the National Institutes of Health (EF), from the Howard Hughes Medical

Institute (EF) and the Max Planck Society (KW) and the Deutsche Forschungsgemeinschaft (KW).

Literature Cited

1. Astbury WT, Street A. 1931. *Gen. Philos. Trans. R. Soc. London Ser. A* 230:75–78
2. Pauling L, Corey RB. 1953. *Nature* 171:59–61
3. Crick FHC. 1953. *Acta Cryst.* 6:689–97
4. Ishikawa H, Bischoff R, Holtzer H. 1968. *J. Cell Biol.* 38:538–55
5. Lazarides E. 1980. *Nature* 283:249–56
6. Steinert PM, Idler WW, Zimmerman SB. 1976. *J. Mol. Biol.* 108:547–67
7. Pruss RM, Mirsky R, Raff MC, Thorpe R, Dowding AJ, Anderton BH. 1981. *Cell* 27:419–28
8. Franke WW, Schmid E, Osborn M, Weber K. 1978. *Proc. Natl. Acad. Sci. USA* 75:5034–38
9. Geisler N, Weber K. 1981. *Proc. Natl. Acad. Sci. USA* 78:4120–23
10. Geisler N, Weber K. 1983. *EMBO J.* 2:2059–63
11. Geisler N, Plessmann U, Weber K. 1982. *Nature* 296:448–50
12. Hanukoglu I, Fuchs E. 1982. *Cell* 31:243–52
13. Hanukoglu I, Fuchs E. 1983. *Cell* 33:915–24
14. Steinert PM, Rice RH, Roop DR, Trus BL, Steven AC. 1983. *Nature* 302: 794–800
15. Crewther WG, Inglis AS, McKern NM. 1978. *Biochem. J.* 173:365–71
16. Gough KH, Inglis AS, Crewther WG. 1978. *Biochem. J.* 173:373–85
17. Fuchs E, Coppock S, Green H, Cleveland D. 1981. *Cell* 27:75–84
18. Geisler N, Weber K. 1982. *EMBO J.* 1:1649–56
19. Quax-Jeuken YEF, Quax WJ, Bloemendal H. 1983. *Proc. Natl. Acad. Sci. USA* 80:3548–52
20. Thompson MA, Ziff EB. 1989. *Neuron* 2:1043–53
21. Aebi U, Cohn JB, Gerace LL. 1986. *Nature* 323:560–64
22. McKeon FM, Kirschner MW, Caput D. 1986. *Nature* 319:463–68
23. Fisher DZ, Chaudhary N, Blobel G. 1986. *Proc. Natl. Acad. Sci. USA* 83:6450–54
24. Oshima RG. 1992. *Curr. Opin. Cell Biol.* 4:110–16
25. Moll R, Franke WW, Schiller DL, Geiger B, Krepler R. 1982. *Cell* 31:11–24
26. Fuchs E, Green H. 1980. *Cell* 19:1033–42
27. Wu YJ, Parker LM, Binder NE, Beckett MA, Sinard JH, et al. 1982. *Cell* 31:693–703
28. Sun TT, Eichner R, Schermer A, Cooper D, Nelson WG, Weiss RA. 1984. In *The Cancer Cell*, ed. A Levine, W Topp, G van de Woude, JD Watson, 1:169–76. New York: Cold Spring Harbor Lab.
29. Nelson W, Sun TT. 1983. *J. Cell Biol.* 97:244–51
30. Lynch MH, O'Guin WM, Hardy C, Mak L, Sun TT. 1986. *J. Cell Biol.* 103:2593–606
31. Heid HW, Werner E, Franke WW. 1986. *Differentiation* 32:101–19
32. Moll R, Zimbelmann R, Goldschmidt MD, Keith M, Laufer J. et al. 1993. *Differentiation* 53:75–93
33. Hatzfeld M, Franke WW. 1985. *J. Cell Biol.* 101:1826–41
34. Hatzfeld M, Maier G, Franke WW. 1987. *J. Mol. Biol.* 197:237–55
35. Osborn M. 1983. *J. Invest. Dermatol.* 81:s104-7
36. Lazarides E. 1982. *Annu. Rev. Biochem.* 51:219–50
37. Portier MM, de Nechaud B, Gros F. 1983. *Dev. Neurosci.* 6:335–44
38. Steinert PM, Idler WW, Cabral F, Gottesman MM, Goldman RD. 1981. *Proc. Natl. Acad. Sci. USA* 78:3692–96
39. Quinlan RA, Franke WW. 1982. *Proc. Natl. Acad. Sci. USA* 79:3452–56
40. Monteiro MJ, Cleveland DW. 1989. *J. Cell Biol.* 108:579–93
41. Osborn M, Franke W, Weber K. 1980. *Exp. Cell Res.* 125:37–46
42. McCormick MB, Coulombe P, Fuchs E. 1991. *J. Cell Biol.* 113:1111–24

43. Lewis SA, Cowan NJ. 1985. *J. Cell Biol.* 100:843–50
44. Levy E, Liem RKH, D'Eustachio P, Cowan N. 1987. *Eur. J. Biochem.* 166:71–77
45. Myers MW, Lazzarini RA, Lee VMY, Schlaepfer WW, Nelson DL. 1987. *EMBO J.* 6:1617–26
46. Lees JF, Schneidman PS, Skuntz SF, Carden MJ, Lazzarini RA. 1988. *EMBO J.* 7:1947–55
47. Pachter JS, Liem RKH. 1985. *J. Cell Biol.* 101:1316–22
48. Napolitano EW, Pachter JS, Chin SSM, Liem RKH. 1985. *J. Cell Biol.* 101: 1323–31
49. Ching G, Liem R. 1993. *J. Cell Biol.* 122:1323–35
50. Heins S, Wong PC, Muller S, Goldie K, Cleveland DW, et al. 1993. *J. Cell Biol.* 123:1517–33
51. Hisanaga S, Hirokawa N. 1988. *J. Mol. Biol.* 202:297–305
52. Lee MK, Xu Z, Wong PC, Cleveland DW. 1993. *J. Cell Biol.* 122:1337–50
53. Gerace L, Burke B. 1988. *Annu. Rev. Cell Biol.* 4:335–74
54. Vorburger K, Lehner CF, Kitten GT, Eppenberger HM, Nigg EA. 1989. *J. Mol. Biol.* 208:405–15
55. Peter M, Kitten GT, Lehner CF, Vorburger K, Bailer SM, et al. 1989. *J. Mol. Biol.* 208:393–404
56. Stick R. 1992. *Chromosoma* 101:566–74
57. Doering V, Stick R. 1990. *EMBO J.* 9:4073–81
58. Furukawa K, Hotta Y. 1993. *EMBO J.* 12:97–106
59. Roeber RA, Weber K, Osborn M. 1989. *Development* 105:365–78
60. Lehner CF, Stick R, Eppenberger HM, Nigg EA. 1987. *J. Cell Biol.* 105:577–87
61. Peter M, Nigg EA. 1991. *J. Cell Sci.* 100:589–98
62. Loewinger L, McKeon F. 1988. *EMBO J.* 7:2301–9
63. Holtz D, Tanaka RA, Hartwig J, McKeon F. 1989. *Cell* 59:969–77
64. Kitten GT, Nigg EA. 1991. *J. Cell Biol.* 113:13–23
65. Krohne G, Waizenegger I, Hoeger TH. 1989. *J. Cell Biol.* 109:2003–11
66. Georgatos SD, Blobel G. 1987. *J. Cell Biol.* 105:117–25
67. Stick R, Angres B, Lehner CF, Nigg EA. 1988. *J. Cell Biol.* 107: 397–406
68. Worman HJ, Evans CD, Blobel G. 1990. *J. Cell Biol.* 111:1535–42
69. Simos G, Georgatos SD. 1992. *EMBO J.* 11:4027–36
70. Foisner R, Gerace L. 1993. *Cell* 73: 1267–79
71. Paddy MR, Belmont AS, Saumweber H, Agard DA, Sedat JW. 1990. *Cell* 62:89–106
72. de Lange T. 1992. *EMBO J.* 11:717–24
73. Burke B. 1990. *Exp. Cell Res.* 186: 169–76
74. Glass JR, Gerace L. 1990. *J. Cell Biol.* 111:1047–57
75. Yuan J, Simos G, Blobel G, Georgatos SD. 1991. *J. Biol. Chem.* 266: 9211–15
76. Luderus MEE, de Graaf A, Mattia E, den Blaauwen JL, Grande MA. et al. 1992. *Cell* 70:949–59
77. Romig H, Fackelmayer FO, Renz A, Ramsperger U, Richter A. 1992. *EMBO J.* 11:3431–40
78. Hozak P, Hassan AB, Jackson DA, Cook PR. 1993. *Cell* 73:361–73
79. Newport JW, Wilson KL, Dunphy WG. 1990. *J. Cell Biol.* 111:2247–59
80. Meier J, Campbell KHS, Ford CC, Stick R, Hutchison CJ. 1991. *J. Cell Sci.* 98:271–79
81. Dabauvalle MC, Loos K, Merkert H, Scheer U. 1991. *J. Cell Biol.* 112: 1073–82
82. Ulitzur N, Harel A, Feinstein N, Gruenbaum Y. 1992. *J. Cell Biol.* 119:17–25
83. Lendahl U, Zimmerman LB, McKay RDG. 1990. *Cell* 60:585–95
84. Dahlstrand J, Zimmerman LB, McKay RDG, Lendahl U. 1992. *J. Cell Sci.* 103:589–97
85. Gounari F, Merdes A, Quinlan R, Hess J, FitzGerald PG. et al. 1993. *J. Cell Biol.* 121:847–53
86. Marchuk D, McCrohon S, Fuchs E. 1984. *Cell* 39:491–98
87. Lersch R, Stellmach V, Stocks C, Giudice G, Fuchs E. 1989. *Mol. Cell. Biol.* 9:3685–97
88. Ferrari S, Battini R, Kaczmarek L, Rittling S, Calabretta B, et al. 1986. *Mol. Cell Biochem.* 6:3614–20
89. Quax W, Egberts WV, Hendriks W, Quax-Jeuken Y, Bloemendal H. 1983. *Cell* 35:215–23
90. Weber K, Plessmann U, Ulrich W. 1989. *EMBO J.* 8:3221–27
91. Conway JF, Parry DAD. 1990. *Int. J. Biol. Macromol.* 12:328–34
92. McLachlan AD. 1978. *J. Mol. Biol.* 124:297–304
93. Parry DAD, Crewther WG, Fraser RD, MacRae TP. 1977. *J. Mol. Biol.* 113: 449–54
94. McLachlan AD, Stewart M. 1982. *J. Mol. Biol.* 162:693–98
95. Fraser RDB, MacRae TP, Suzuki E,

Parry DAD, Trajstmann AC, Lucas I. 1985. *Int. J. Biol. Macromol.* 7:258–74
95a. Letai A. 1993. PhD thesis. The University of Chicago
96. Huyghues-Despointes BM, Scholtz JM, Baldwin RL. 1993. *Protein Sci.* 2:80–85
97. Atkinson SJ, Stewart M. 1992. *J. Mol. Biol.* 226:7–13
98. Bader BL, Magin TM, Hatzfeld M, Franke WW. 1986. *EMBO J.* 5:1865–75
99. Woods EF, Inglis AS. 1984. *Int. J. Biol. Macromol.* 6:277–83
100. Quinlan RA, Hatzfeld M, Franke WW, Lustig A, Schulthess T, Engel J. 1986. *J. Mol. Biol.* 192:337–49
101. Parry DAD, Steven AC, Steinert PM. 1985. *Biochem. Biophys. Res. Commun.* 127:1012–18
102. Phillips GN. 1992. *Proteins* 14:425–29
103. Heitlinger E, Peter M, Haner M, Lustig A, Aebi U, Nigg EA. 1991. *J. Cell Biol.* 113:485–95
104. Hatzfeld M, Weber K. 1990. *J. Cell Biol.* 110:1199–10
105. Steinert PM. 1990. *J. Biol. Chem.* 265:8766–74
106. Coulombe P, Fuchs E. 1990. *J. Cell Biol.* 111:153–69
107. Quinlan RA, Cohlberg JA, Schiller DL, Hatzfeld M, Franke WW. 1984. *J. Mol. Biol.* 178:365–88
108. Geisler N, Kaufmann E, Weber K. 1985. *J. Mol. Biol.* 182:173–77
109. Steinert PM. 1991. *J. Struct. Biol.* 107:157–74
110. Ip W, Hartzer MK, Pang YYS, Robson RM. 1985. *J. Mol. Biol.* 183:365–75
111. Potschka M, Nave R, Weber K, Geisler N. 1990. *Eur. J. Biochem.* 190:503–8
112. Stewart M, Quinlan RA, Moir RD. 1989. *J. Cell Biol.* 109:225–34
113. Geisler N, Schuenemann J, Weber K. 1992. *Eur. J. Biochem.* 206:841–52
114. Steinert PM, Parry DAD. 1993. *J. Biol. Chem.* 268:2878–87
115. Steinert PM, Marekov LN, Fraser RDB, Parry DAD. 1993. *J. Mol. Biol.* 230:436–52
116. Aebi U, Fowler WE, Rew P, Sun TT. 1983. *J. Cell Biol.* 97:1131–43
117. Steven AC, Hainfeld JF, Trus BL, Wall JS, Steinert PM. 1983. *J. Biol. Chem.* 258:8323–29
118. Engel A, Eichner R, Aebi U. 1985. *J. Ultrastruc. Res.* 90:323–35
119. Steinert PM. 1991. *J. Struct. Biol.* 107:175–88
120. Albers K, Fuchs E. 1987. *J. Cell Biol.* 105:791–806
121. Albers K, Fuchs E. 1989. *J. Cell Biol.* 108:1477–93
122. Lu X, Lane EB. 1990. *Cell* 62:681–96
123. Wong PC, Cleveland DW. 1990. *J. Cell Biol.* 111:1987–203
124. Gill SR, Wong PC, Monteiro MJ, Cleveland DW. 1990. *J. Cell Biol.* 111:2005–19
125. Raats JMH, Henderik JBJ, Verdijk M, van Oort FLG, Gerards WLH, et al. 1991. *Eur. J. Cell Biol.* 56:84–103
126. Raats JMH, Pieper FR, Egberts WTM, Verrijp KN, Ramaekers FCS, Bloemendal H. 1990. *J. Cell Biol.* 111:1971–85
127. Letai A, Coulombe P, Fuchs E. 1992. *J. Cell Biol.* 116:1181–95
128. Coulombe P, Chan YM, Albers K, Fuchs E. 1990. *J. Cell Biol.* 111:3049–64
129. Hatzfeld M, Weber K. 1991. *J. Cell Sci.* 99:351–62
130. McCormick MB, Kouklis P, Syder A, Fuchs E. 1993. *J. Cell Biol.* 122:395–407
131. Hatzfeld M, Weber K. 1992. *J. Cell Biol.* 116:157–66
132. Kouklis PD, Traub P, Georgatos SD. 1992. *J. Cell Sci.* 102:31–41
133. Henderson D, Geisler N, Weber K. 1982. *J. Mol. Biol.* 155:173–76
134. Milam L, Erickson HP. 1982. *J. Cell Biol.* 94:592–96
135. Moir RD, Donaldson AD, Stewart M. 1991. *J. Cell Sci.* 99:363–72
136. Kaufmann E, Weber K, Geisler N. 1985. *J. Mol. Biol.* 185:733–42
137. Eckelt A, Herrmann H, Franke WW. 1992. *Eur. J. Cell Biol.* 58:319–30
138. Hatzfeld M, Weber K. 1990. *J. Cell Sci.* 97:317–24
139. Wilson AK, Coulombe PA, Fuchs E. 1992. *J. Cell Biol.* 119:401–14
140. Bader BL, Magin TM, Freundenmann M, Stumpp S, Franke WW. 1991. *J. Cell Biol.* 115:1293–307
141. Heitlinger E, Peter M, Lustig A, Villiger W, Nigg EA, Aebi U. 1992. *J. Struct. Biol.* 108:74–91
142. Nakamura Y, Takeda M, Aimoto S, Hariguchi S, Kitajima S, Nishimura T. 1993. *J. Biochem.* 212:565–71
143. Shoeman RL, Mothes E, Kesselmeier C, Traub P. 1990. *Cell Biol. Int. Rep.* 14:583–94
144. Birkenberger L, Ip W. 1990. *J. Cell Biol.* 111:2063–75
145. Kouklis PD, Papamarcaki T, Merdes A, Georgatos SD. 1991. *J. Cell Biol.* 114:773–86
146. Hatzfeld M, Dodemont H, Plessmann U, Weber K. 1992. *FEBS Lett.* 302: 239–42
147. Traub P, Vorgias CE. 1983. *J. Cell Sci.* 63:43–67

148. Herrmann H, Hofmann I, Franke WW. 1992. *J. Mol. Biol.* 223:637–50
149. Traub P, Scherbarth A, Wiegers W, Shoeman RL. 1992. *J. Cell Sci.* 101: 363–81
150. Franke WW, Schmid E, Grund C, Geiger B. 1982. *Cell* 30:103–13
151. Geisler N, Weber K. 1988. *EMBO J.* 7:15–20
152. Inagaki M, Gonda Y, Matsuyama M, Nishizawa K, Nishi Y, Sato C. 1988. *J. Biol. Chem.* 263:5970–78
153. Geisler N, Hatzfeld M, Weber K. 1989. *Eur. J. Biochem.* 183:441–47
154. Nigg EA. 1992. *Curr. Opin. Cell Biol.* 4:105–9
155. Peter M, Nakagawa J, Doree M, Labbe JC, Nigg EA. 1990. *Cell* 61:591–602
156. Heald R, McKeon F. 1990. *Cell* 61: 579–89
157. Peter M, Heitlinger E, Haner M, Aebi U, Nigg EA. 1991. *EMBO J.* 10:1535–44
158. Chou YH, Bischoff JR, Beach D, Goldman RD. 1990. *Cell* 62:1063–71
159. Chou YH, Ngai KL, Goldman R. 1991. *J. Biol. Chem.* 266:7325–28
160. Hennekes H, Peter M, Weber K, Nigg EA. 1993. *J. Cell Biol.* 120:1293–1304
161. Nishizawa K, Yano T, Shibata M, Ando S, Saga S, et al. 1991. *J. Biol. Chem.* 266:3074–79
162. Matsuoka Y, Nishizawa K, Yano T, Shibata M, Ando S, et al. 1992. *EMBO J.* 11:2895–902
163. Julien JP, Mushynski WE. 1983. *J. Biol. Chem.* 258:4019–25
164. Geisler N, Vandekerckhove J, Weber K. 1987. *FEBS Lett.* 221:403–7
165. Lee VMY, Otvos LJ, Carden MJ, Hollosi M, Dietzchold B, Lazzarini RA. 1988. *Proc. Natl. Acad. Sci. USA* 85:1998–2002
166. Hisanaga SI, Kusubata M, Okumura E, Kishimoto T. 1991. *J. Biol. Chem.* 266:21798–903
167. de Waegh SM, Lee VMY, Brady ST. 1992. *Cell* 68:451–63
168. Steinert PM. 1988. *J. Biol. Chem.* 263:13333–39
169. Klymkowsky MW, Maynell LA, Nislow C. 1991. *J. Cell Biol.* 114:787–97
170. Soellner P, Quinlan RA, Franke WW. 1985. *Proc. Natl. Acad. Sci. USA* 82:7929–33
171. Franke WW, Schmid E, Mittnacht S, Grund C, Jorcano JL. 1984. *Cell* 36: 813–25
172. Vikstrom KL, Borisy GG, Goldman RD. 1989. *Proc. Natl. Acad. Sci. USA* 86:549–53
173. Klymkowsky MW, Miller RH, Lane EB. 1983. *J. Cell Biol.* 96:494–509
174. Ngai J, Coleman TR, Lazarides E. 1990. *Cell* 60:415–27
175. Sarria AJ, Nordeen SK, Evans RM. 1990. *J. Cell Biol.* 111:553–65
176. Angelides KJ, Smith KE, Takeda M. 1989. *J. Cell Biol.* 108:1495–606
177. Ip W, Fellows ME. 1990. *Anal. Biochem.* 185:10–16
178. Vikstrom KL, Lim SS, Goldman RD, Borisy GG. 1992. *J. Cell Biol.* 118: 121–29
179. Okabe S, Miyasaka H, Hirokawa N. 1993. *J. Cell Biol.* 121:375–86
180. Georgatos SD, Weaver DC, Marchesi VT. 1985. *J. Cell Biol.* 100:1962–67
181. Georgatos SD, Weber K, Geisler N, Blobel G. 1987. *Proc. Natl. Acad. Sci. USA* 84:6780–84
182. Djabali K, Portier MM, Gros F, Blobel G, Georgatos S. 1991. *Cell* 64:109–21
183. Eckert BS, Daley RA, Parysek LM. 1982. *J. Cell Biol.* 92:575–78
184. Garrod DR. 1993. *Curr. Opin. Cell Biol.* 5:30–40
185. Green KJ, Jones JCR. 1990. In *Cellular and Molecular Biology of Intermediate Filaments,* ed. RD Goldman, PM Steinert, pp. 147–74. New York: Plenum
186. Green KJ, Virata MLA, Elgart GW, Stanley JR, Parry DAD. 1992. *Int. J. Biol. Macromol.* 14:145–53
187. Green KJ, Stappenbeck TS, Parry DA, Virata ML. 1992. *J. Dermatol.* 19:765–69
188. Wiche G, Becker B, Luber K, Weitzer G, Castanon MJ, et al. 1991. *J. Cell Biol.* 114:83–99
189. Tanaka T, Parry DAD, Klaus-Kovtun V, Steinert PM, Stanley JR. 1991. *J. Biol. Chem.* 266:12555–59
190. Troyanovsky SM, Eshkind LG, Troyanovsky RB, Leube RE, Franke WW. 1993. *Cell* 72:561–74
191. Stappenbeck TS, Bornslaeger EA, Corcoran CM, Luu HH, Virata MLA, et al. 1993. *J. Cell Biol.* 123:691–706
192. Foisner R, Wiche G. 1991. *Curr. Opin. Cell Biol.* 3:75–81
193. Wiche G, Gromov D, Donovan A, Castanon MJ, Fuchs E. 1993. *J. Cell Biol.* 121:607–19
194. Foisner R, Traub P, Wiche G. 1991. *Proc. Natl. Acad. Sci. USA* 88:3812–16
195. Gyoeva FK, Gelfand VL. 1991. *Nature* 353:445–48
196. Presland RB, Haydock PV, Fleckman P, Nirunsuksiri W, Dale BA. 1992. *J. Biol. Chem.* 267:23772–81
197. Markova NG, Marekov LN, Chipev CC, Gan SQ, Idler WW, Steinert PM. 1993. *Mol. Cell Biol.* 13:613–25
198. Fietz MJ, McLaughlan CJ, Campbell

MT, Rogers GE. 1993. *J. Cell Biol.* 121:855–65
199. Fratini A, Powell BC, Rogers GE. 1993. *J. Biol. Chem.* 268:4511–18
200. MacKinnon PJ, Powell BC, Rogers GE. 1990. *J. Cell Biol.* 111:2587–600
201. Fuchs E, Marchuk D. 1983. *Proc. Natl. Acad. Sci. USA* 80:5857–61
202. Lewis SA, Cowan NJ. 1986. *Mol. Cell. Biol.* 6:1529–34
203. McConnell SJ, Yaffe MP. 1993. *Science* 260:687–89
204. Gruenbaum Y, Landesman Y, Drees B, Bare JW, Saumweber H. et al. 1988. *J. Cell Biol.* 106:585–96
205. Weber K, Plessmann U, Dodemont H, Kossmagk-Stephan K. 1988. *EMBO J.* 7:2995–3001
206. Bossie CA, Sanders MM. 1993. *J. Cell Sci.* 104:1263–72
207. Dodemont H, Riemer D, Weber K. 1990. *EMBO J.* 9:4083–94
208. Way J, Hellmich MR, Jaffe H, Szaro B, Pant HC, et al. 1992. *Proc. Natl. Acad. Sci. USA* 89:6963–67
209. Riemer D, Dodemont H, Weber K. 1992. *Eur. J. Cell Biol.* 58:128–35
210. Fine JD, Bauer EA, Briggaman RA, Carter DM, Eady RAJ, et al. 1991. *Am. Acad. Dermatol.* 24:119–35
211. Anton-Lamprecht I. 1983. *J. Invest. Dermatol.* 81:s149-56
212. Kitajima Y, Inoue S, Yaoita H. 1989. *Arch. Dermatol. Res.* 281:5–10
213. Coulombe PA, Hutton ME, Letai A, Paller AS, Fuchs E. 1991. *Cell* 66: 1301–11
214. Vassar R, Coulombe PA, Degenstein L, Albers K, Fuchs E. 1991. *Cell* 64:365–80
215. Coulombe PA, Hutton ME, Vassar R, Fuchs E. 1991. *J. Cell Biol.* 115:1661–74
216. Bonifas JM, Rothman AL, Epstein EH. 1991. *Science* 254:1202–5
217. Ryynanen M, Knowlton RG, Uitto J. 1991. *Am. J. Hum. Genet.* 49:978–84
218. Chan YM, Yu QC, Fine JD, Fuchs E. 1993. *Proc. Natl. Acad. Sci. USA* 90:7414–18
218a. Chan Y-M, Yu Q-C, LeBlanc-Strasecki J, Kucherlapati R, Uitto J, et al. 1994. *J. Cell Sci.* In press
219. Rosenberg M, Fuchs E, Le Beau MM, Eddy R, Shows TB. 1991. *Cell Cytogen.* 57:33–38
220. Lane EB, Rugg EL, Navsaria H, Leigh IM, Heagerty AHM, et al. 1992. *Nature* 356:244–46
221. Stephens K, Sybert VP, Wijsman EM, Ehrlich P, Spencer A. 1993. *J. Invest. Dermatol.* 101:240–43
221a. Hovnanian A, Pollack E, Hilal L, Rochat A, Prost C, et al. 1993. *Nature Genet.* 3:327–32
221b. Rugg El, Morley SM, Smith FJD, Boxer M, Tidman MJ, et al. 1993. *Nature Genet.* 5:294–300
222. Fuchs E, Coulombe P, Cheng J, Chan YM, et al. 1994. *J. Invest. Dermatol.* In press
223. Letai A, Coulombe PA, McCormick MB, Yu QC, Hutton E, Fuchs E. 1993. *Proc. Natl. Acad. Sci. USA* 90:3197–201
224. Compton JG, DiGiovanna JJ, Santucci SK, Kearns KS, Amos CI, et al. 1992. *Nat. Genet.* 1:301–5
225. Rothnagel JA, Dominey AM, Dempsey LD, Longley MA, Greenhalgh DA, et al. 1992. *Science* 257:1128–30
226. Cheng J, Syder AJ, Yu QC, Letai A, Paller AS, Fuchs E. 1992. *Cell* 70:811–19
227. Chipev CC, Korge BP, Markova N, Bale SJ, DiGiovanna JJ, et al. 1992. *Cell* 70:821–28
228. Reis A, Hennies HC, Langbein L, Digweed M, Mischke D, et al. 1994. *Nature Genet.* In press
229. Cooper DN, Youssoufian H. 1988. *Hum. Genet.* 78:151–55
230. Janmey PA, Euteneuer U, Traub P, Schliwa M. 1991. *J. Cell Biol.* 113: 155–60
231. Powell BC, Rogers GE. 1990. *EMBO J.* 9:1485–93
232. Baribault H, Oshima RG. 1991. *J. Cell Biol.* 115:1675–84
233. Baribault H, Price J, Miyai K, Oshima RG. 1993. *Genes Dev.* 7:1191–201
234. Gillespie JM, Marshall RC. 1989. In *The Biology of Wool and Hair,* ed. GE Roger, PJ Reis, KA Ward, RC Marshall, pp. 257–85. London: Chapman & Hall
235. Nadeau JH, Berger FG, Cox DR, Crosby JL, Davisson MT, et al. 1989. *Genomics* 5:454–62
236. Wiley CA, Love S, Skoglund RR, Lampert PW. 1987. *Acta Neuro-Pathol.* 72:369–76
237. Xu Z, Cork LC, Griffin JW, Cleveland DW. 1993. *Cell* 73:23–33
238. Cote F, Collard JF, Julien JP. 1993. *Cell* 73:35–46
239. Geisterfer-Lowrance AA, Kass S, Tanigawa G, Vosberg HP, McKenna W, et al. 1990. *Cell* 62:999–1006
240. Tanigawa G, Jarcho JA, Kass S, Solomon SD, Vosberg HP, et al. 1990. *Cell* 62:991–98
241. Pellissier JF, Pouget J, Charpin C, Figarella D. 1989. *J. Neurol. Sci.* 89:49–61

242. Capetanaki YG, Starnes S, Smith S. 1989. *Proc. Natl. Acad. Sci. USA* 86:4882–86
243. Monteiro MJ, Hoffman PN, Gearhart JD, Cleveland DW. 1990. *J. Cell Biol.* 111:1543–57
244. Dunia I, Pieper F, Manenti S, Vandekamp A, Devilliers G, et al. 1990. *Eur. J. Cell Biol.* 53:59–74
245. Venetianer A, Schiller DL, Magin T, Franke WW. 1983. *Nature* 305:730–33
246. Hedberg KK, Chen LB. 1986. *Exp. Cell Res.* 163:509–17
247. Christian JL, Edelstein NG, Moon RT. 1990. *New Biol.* 2:700–11
248. Torpey N, Wylie CC, Heasman J. 1992. *Nature* 357:413–15
249. Weinstein DE, Shelanski ML, Liem RKH. 1991. *J. Cell Biol.* 112:1205–13
250. Hoffman PN, Cleveland DW, Griffin JW, Landes PW, Cowan NJ, Price DL. 1987. *Proc. Natl. Acad. Sci. USA* 84:3472–76
251. O'Hara O, Gahara Y, Miyake T, Teraoka H, Kitamura T. 1993. *J. Cell Biol.* 121:387–95
252. Schultheiss T, Lin ZX, Ishikawa H, Zamir I, Stoeckert CJ, Holtzer H. 1991. *J. Cell Biol.* 114:953–66

Annu. Rev. Biochem. 1994. 63:383–417

5-LIPOXYGENASE

A. W. Ford-Hutchinson, M. Gresser, and R. N. Young
Merck Frosst Centre for Therapeutic Research, Kirkland, Quebec, Canada

KEY WORDS: 5-lipoxygenase, leukotrienes, arachidonic acid, 5-lipoxygenase activating protein, hydroxyeicosatetraenoic acids

---

CONTENTS

INTRODUCTION—BIOLOGICAL IMPLICATIONS OF LEUKOTRIENE SYNTHESIS AND INHIBITION

5-Lipoxygenase is a member of a family of lipoxygenases that includes 12- and 15-lipoxygenase. 5-Lipoxygenase catalyzes both the first step in the oxygenation of arachidonic acid to produce 5(S)-hydroperoxy-6,8,11,14-(E,Z,Z,Z)-eicosatetraenoic acid (5-HPETE) and, the second, the dehydration of hydroperoxide intermediate, to produce the epoxide, 5(S),6(S)-oxido-7,9,11,15(E,E,Z,Z)-eicosatetraenoic acid (leukotriene A_4) (Figure 1). There are two subsequent metabolic routes from leukotriene A_4 that lead to the production of biologically active compounds. The first involves the enzymatic, stereochemical hydrolysis of leukotriene A_4 by the enzyme leukotriene A_4 hydrolase, which leads to the production of leukotriene B_4 [5(S),12(R)-dihydroxy-6,8,10,14(Z,E,E,Z)-eicosatetraenoic acid]. Leukotriene B_4 interacts with high-affinity, structurally specific, G protein–coupled receptors on the surfaces of leukocytes and lymphocytes. Activation of these receptors stimulates a number of leukocyte and lymphocyte functions, including the chemotaxis, chemokinesis, and aggregation of polymorphonuclear leukocytes (1, 2). Because of this, leukotriene B_4 has been implicated as a potential mediator of inflammation (2).

The second enzymic route of metabolism of leukotriene A_4 involves conjugation with glutathione by the enzyme leukotriene C_4 synthase to produce 5(S)-hydroxy-6(R)-glutathionyl-7,9,11,14(E,E,Z,Z)-eicosatetraenoic acid (leukotriene C_4) (3). Leukotriene C_4 can then be metabolized through

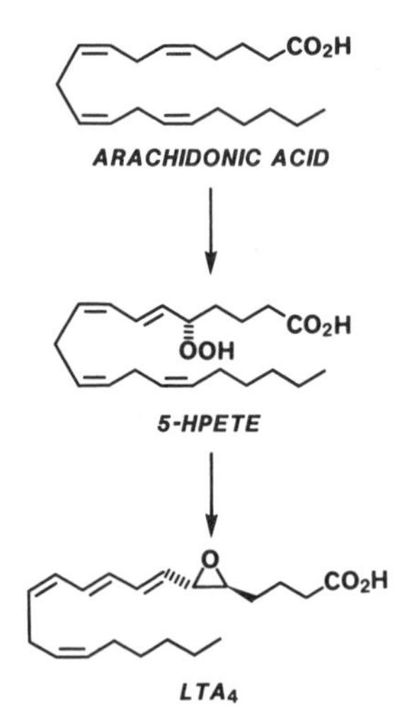

Figure 1 Metabolism of arachidonic acid by 5-lipoxygenase

cleavage of glutamic acid by a membrane-bound γ-glutamyl transferase to produce leukotriene D_4. Leukotriene D_4 in turn can be converted to leukotriene E_4, the cysteinyl conjugate, by a specific membrane-bound peptidase with loss of glycine. Leukotrienes C_4, D_4, and E_4 are known as the sulfidopeptide leukotrienes and collectively account for a biological activity originally described as slow-reacting substance of anaphylaxis (4). These leukotrienes also interact with high-affinity, structurally specific, G protein–coupled receptors to produce contractile responses, which can lead, for example, to bronchoconstriction and changes in vascular permeability in the lung (4, 5). The sulfidopeptide leukotrienes have been implicated as potentially important mediators of immediate hypersensitivity reactions and allergic conditions, and leukotriene D_4 receptor antagonists have shown promise in the treatment of human bronchial asthma (6).

PURIFICATION, SEQUENCING, AND CLONING OF 5-LIPOXYGENASE

All the known lipoxygenases, including 5-lipoxygenase, catalyze the addition of molecular oxygen to a 1,4-*cis,cis*-pentadiene moiety to produce a 1-hydroperoxy-2,4-*trans,cis*-pentadiene unit (7). 5-lipoxygenase causes the removal of a pro-S hydrogen atom from C_7 of arachidonic acid by a redox mechanism that involves reduction of iron at the active site of the enzyme, formation of a radical intermediate, and subsequent reoxidation of the iron and formation of the hydroperoxide intermediate following oxygenation of the substrate as outlined below. The enzyme then catalyzes a second redox cycle, which results in the conversion of 5-HPETE to leukotriene A_4 following the stereospecific removal of the pro-R hydrogen from position C_{10}.

5-lipoxygenase has been characterized from a number of mammalian sources, including pig (8), human (9, 10), rat (11), and guinea pig (12). Human 5-lipoxygenase is a 78-kDa protein with a requirement for Ca^{2+} and ATP. Both human and rat 5-lipoxygenase have been cloned and expressed in a number of expression systems, and the expressed enzyme has all the characteristics of the originally purified enzyme from either the human or rat leukocyte (13–16).

The genomic sequences for 5-lipoxygenase have been studied, and the human 5-lipoxygenase gene is estimated to be <82 kb and consists of 14 exons divided by 13 introns (17). Exons 1–7, which encode the NH_2-terminal half of the protein, stretch over 65 kb, whereas the carboxy-terminal half of the protein (exons 8–14) is contained in a smaller (~6 kb) section of DNA. A major site of transcription or initiation in leukocytes was mapped to a thymidine residue 65 base pairs upstream of the ATG initiation codon by nuclease S_1 protection and primer extension experiments. The putative promoter region in the 5′ end of the human 5-lipoxygenase gene is unusual in that it does not contain a characteristic "TATA" or "CCAAT" box and thus has some of the characteristics of a housekeeping gene. This region, however, contains multiple GC boxes within a (G+C)-rich region, as does the immediate 5′ region of the first intron. Despite the absence of obvious regulatory sequences in the 5-lipoxygenase gene, it is clear that 5-lipoxygenase must be regulated in a highly cell-specific fashion as substantial production of 5-lipoxygenase is only well documented for the cells of myeloid lineage and not for many other cells. The guinea pig 5-lipoxygenase gene has also been cloned, and nucleotide sequence of its promoter determined (18). Analysis of the guinea-pig promoter region has suggested that this gene is probably also regulated by *cis*-acting nucleotide sequences. In particular, putative nucleotide sequences that may bind factors such as Sp-1, Ab-2, NF-κb, and C-Ha-*ras* were identified. One interesting aspect of the regulation of 5-lipoxygenase derived from the finding that a heat-stable, but protease-sensitive, component in human serum induced an increased 5-lipoxygenase activity in maturing HL-60 pro-myelocytic leukemia cells (19). This factor was subsequently identified as transforming growth factor β, which was shown to regulate 5-lipoxygenase activity without affecting 5-lipoxygenase mRNA expression (20).

Comparative Structure with Other Lipoxygenases

Lipoxygenases are a family of enzymes, there being an overall 60% sequence similarity among human 5-, 12- and 15-lipoxygenase, and each human lipoxygenase being roughly 25% identical to plant lipoxygenases (7). There are two regions of striking conservation within the lipoxygenase family. The first such region runs from position 350 to 421 in 5-lipoxygenase, where there is considerable sequence homology between 5-lipoxygenase, 12-lipoxygenase, and 15-lipoxygenase. This region contains the hydrophobic center of the mammalian lipoxygenases and sequence homology to lipases, suggesting that it might be a site for either substrate binding or membrane association. This residue also contains five histidine residues that are highly conserved in all lipoxygenase sequences to date, and as shown below, three of these histidine residues are known to be coordination sites for iron at the active site of the enzyme. The second region of homology occurs between residues 545 and 579 of 5-lipoxygenase. A sixth histidine that is conserved in all the lipoxygenases is found within this region. The function of this region remains to be determined.

STRUCTURE AND MECHANISM

Detailed structural and functional studies of human leukocyte 5-lipoxygenase were very difficult for many years due to the low natural abundance, low yields upon purification, and the instability of this enzyme. The availability of high-level expression systems, including recombinant baculovirus and *Escherichia coli* (21, 22), made it possible to obtain the pure enzyme in milligram quantities (23), and to discover conditions under which it is reasonably stable (24). The recombinant enzyme, in terms of its catalytic activities, immunoreactivity, and sensitivity to inhibitors, appears to be identical to that isolated from human leukocytes (23, 25). On the basis of these advances, the presence of nonheme iron, which is essential for activity, was confirmed (26), and methods were developed for obtaining the pure enzyme with specific activities about 10-fold higher than had been previously reported (24, 27). Using highly active pure enzyme, studies of the requirements and behavior of 5-lipoxygenase have been repeated and extended. In some cases, results have been found to depend on the particular enzyme preparation used, and not to reflect the intrinsic nature of 5-lipoxygenase itself. In this review, the major focus is on recent advances from studies of recombinant human 5-lipoxygenase. The earlier literature on this enzyme has been thoroughly and repeatedly reviewed elsewhere (25, 28–32).

ACTIVATION

By ATP

Stimulation by ATP of 5-lipoxygenase from various sources has been reported many times (28). The extent of stimulation has usually been reported to be on the order of twofold, although rate increases of around several hundredfold have been reported (33). Since there is no ATP-binding consensus sequence apparent in the 5-lipoxygenase amino-acid sequence, chemical modification studies and elucidation of the three-dimensional structure of 5-lipoxygenase are expected to reveal a novel ATP-binding domain. Other lipoxygenases are not activated by ATP. Activation by ATP is dependent upon the presence of Ca^{2+}, and lower levels of activation have been obtained with other adenine nucleotides and nucleoside triphosphates (34). Concentrations of ATP used to achieve maximum stimulation typically are about one millimolar. In this laboratory, the optimum conditions obtained for use with highly active, pure recombinant human 5-lipoxygenase include 0.1 mM ATP (24).

By Leukocyte Factors

Although the binding of ATP and its activation of 5-lipoxygenase are poorly understood, this phenomenon has been put to good use in the purification of the enzyme. The use of ATP-agarose affinity chromatography to purify 5-lipoxygenase was first reported for the rat enzyme (35). This procedure was later modified and scaled up to provide a one-step purification of the recombinant human enzyme to greater than 95% homogeneity as assessed by SDS-PAGE. The pure enzyme was used in an investigation of the role of leukocyte stimulatory factors. It had been found that addition of these uncharacterized factors to 5-lipoxygenase purified from human leukocytes was necessary in order to achieve maximum activity (10, 36). Without the addition of these factors, human 5-lipoxygenase had very low levels of activity, making the study of the fully active enzyme impossible under defined conditions. Using the affinity-purified recombinant enzyme, it was found that the leukocyte factors protected the enzyme against inactivation during pre-incubation in the standard assay protocol of Rouzer & Samuelsson (10). However, they were not found to enhance the activity of the enzyme when phosphatidylcholine was present in the assay solution (see below). Use of a different protocol that did not stress the enzyme as much during pre-incubation eliminated the need for leukocyte factors (23). The effect of the new protocol on 5-lipoxygenase purified from human leukocytes was not investigated.

It was observed that 3–5-fold stimulation of activity of affinity-purified 5-lipoxygenase could be obtained by adding back aliquots of fractions from the ATP-agarose column that did not contain 5-lipoxygenase activity (26). Since stimulation by these fractions was obviously due to different causes than previously reported for the leukocyte factors, this phenomenon was further investigated. It was discovered that the column flow-through aliquots could be substituted by ethylenediaminetetraacetic acid (EDTA), dithiothreitol (DTT), and a small amount of γ-globulin (24). The γ-globulin most likely serves simply as a carrier protein, to minimize loss of the very small amounts of highly active 5-lipoxygenase that were used in these experiments. The EDTA probably chelates inhibitory metal ions or prevents metal-catalyzed generation of reduced oxygen species, and the small additional activating effect of DTT in the assay is not understood, but may be due to scavenging of free radical species that are generated during turnover. As discussed below, it was found that it was best to avoid DTT in the buffer during storage of purified 5-lipoxygenase (24). These advances made it possible to study highly active pure human 5-lipoxygenase under defined conditions, which greatly facilitated mechanistic studies.

By Ca^{2+}

Although other lipoxygenases do not require Ca^{2+} for activity, crude and purified 5-lipoxygenase from various sources is very strongly activated by Ca^{2+} (28). As discussed above, the 5-lipoxygenase amino-acid sequence does not contain an obvious Ca^{2+}-binding domain, although weak homologies are present with a sequence common to a group of calcium-dependent membrane-binding proteins, such as lipocortin (13, 15). Calcium concentrations in the millimolar range are normally used in assays, and in this laboratory the standard assay conditions include 0.3 mM $CaCl_2$ (24). When the dependency of both rate and total product formed on free Ca^{2+} concentration was studied, it was found that free Ca^{2+} concentrations in the one micromolar range are sufficient to give half-maximal product formation (24). Somewhat higher Ca^{2+} concentration is necessary to give half-maximal rates. This is because the rate of turnover-dependent inactivation decreases considerably as the Ca^{2+} concentration is decreased into the low micromolar range. The mechanism by which Ca^{2+} activates 5-lipoxygenase is very likely by causing the enzyme to attach itself to phospholipid membranes in a manner similar to that reported for protein kinase C (37), cytosolic phospholipase A_2 (38), and other calcium-dependent enzymes whose substrates are found in membranes, micelles, or other aggregated structures. Ca^{2+} may have additional activating effects as well, beyond simply driving translocation. In cell-free systems it has been observed that 5-lipoxygenase associates with the membrane fraction in the presence of Ca^{2+}, and that the association is reversed upon removal of Ca^{2+} (39). Calcium ionophore has also been shown to cause 5-lipoxygenase in various types of whole cells to translocate to the membrane, although in whole cells this translocation is largely dependent upon the presence of active 5-lipoxygenase-activating protein (FLAP), as discussed below. It was found that with purified 5-lipoxygenase, Mn^{2+}, Ba^{2+}, and Sr^{2+} can stimulate activity in the absence of Ca^{2+}, although to lesser extents than Ca^{2+}. The enzyme is strongly inhibited by Zn^{2+}, Co^{2+}, and Cu^{2+} in a manner that is apparently not competitive with Ca^{2+}. The inhibition by Co^{2+} and Zn^{2+} is reversible by EDTA, while inhibition by Cu^{2+} is not (24).

By Phosphatidylcholine

Phosphatidylcholine has been observed to stabilize the rat 5-lipoxygenase during purification and to activate the rat (40) and porcine enzymes (41). Binding of the rat enzyme to phosphatidylserine vesicles has been reported (42), and Ca^{2+}-dependent binding of human 5-lipoxygenase to phosphatidylcholine vesicles has been observed in this laboratory (MD Percival and K Neden, unpublished results). In the absence of phospholipid or detergent, rates of 5-lipoxygenase-catalyzed oxygenation of arachidonic acid by the pure enzyme are very slow (43). It is apparent that monomeric arachidonic acid is a very poor substrate for 5-lipoxygenase and that catalysis occurs at an interface between the aqueous solution and a membrane or particle surface, depending on the constituents of the reaction mixture. Whereas the influence of the nature of the interface has been extensively investigated for some interfacial enzymes, such as the phospholipases A_2 (44–46), this area has not yet been systematically investigated for 5-lipoxygenase. In one recent study, it was shown clearly that it is the mole fraction of arachidonic acid in phosphatidylcholine vesicles, rather than just the total arachidonic acid concentration in the solution, that determines the point at which maximum reaction rate is reached, as well as when substrate inhibition becomes apparent (43). Considerably more work of this nature will have to be done before the role of the nature of the interface in 5-lipoxygenase catalysis and inhibition begins to become clearer.

CATALYTIC MECHANISM

A proposed catalytic mechanism for lipoxygenases is shown in Figure 2. Indicated as well are proposed ways in which redox inhibitors such as

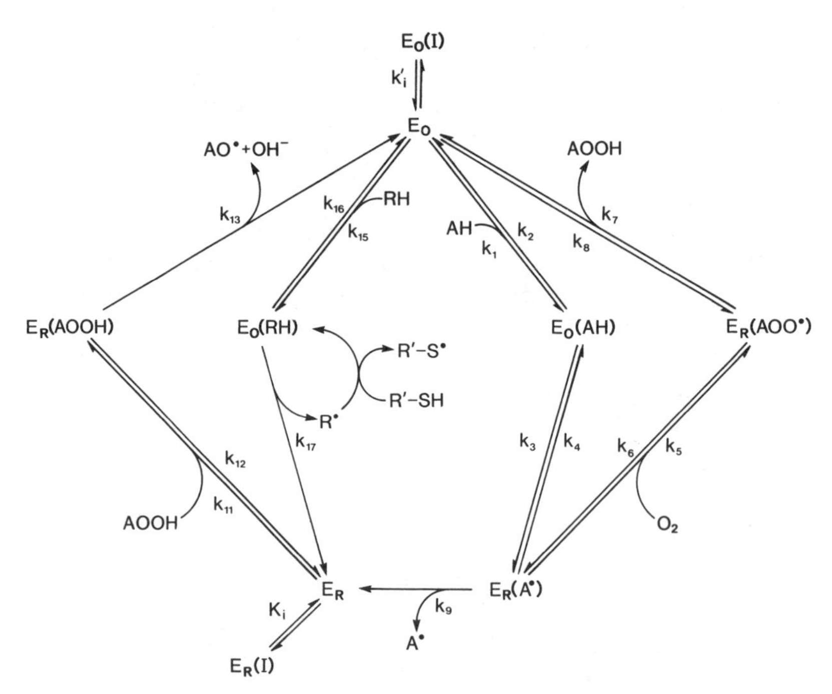

Figure 2 A kinetic mechanism for the main activities of lipoxygenases. Oxygenase activity (*righthand side*) results from the reaction of Fe(III) enzyme (E_O) with fatty acid substrate (AH) to yield the Fe(II) enzyme (E_R) with bound pentadienyl radical ($A^{\bullet}$), followed by binding of O_2 and reaction to yield hydroperoxyl radical ($AOO^{\bullet}$) bound to E_R, followed by reaction to regenerate E_O and release of fatty acid hydroperoxide product (AOOH). Free E_R can result either from dissociation of $A^{\bullet}$ from E_R ($A^{\bullet}$), or from reduction of E_O by various reducing agents (RH). The product of one-electron oxidation of RH, $R^{\bullet}$, can be reduced by thiols, R′-SH, to regenerate RH. Reduced enzyme, E_R, can be reoxidized by fatty acid hydroperoxide, AOOH, to yield hydroxide ion and alkoxyl radical ($AO^{\bullet}$). Nonredox, reversible dead-end inhibitors (I) can bind to E_R, E_O, and possibly other states of the enzyme as well.

N-hydroxyureas and phenols, as well as reversible dead-end inhibitors, are considered to inhibit lipoxygenases. The definitions of the symbols in the figure are given in the figure legend. We consider that this mechanism accommodates all of the relevant observations reported thus far on 5-lipoxygenase, although it is based largely on detailed kinetic studies that have been done over the years with 15-lipoxygenases from reticulocytes and from soybeans (47, 51, 52). Such studies have not been practical with human 5-lipoxygenase because of the unavailability of pure enzyme that could be studied under defined conditions. Even now, detailed kinetic studies with 5-lipoxygenase are difficult to interpret quantitatively, because of the presence of both lags and rapid turnover inactivation with half-times <1 min with the fully activated enzyme. It is difficult to find a range of conditions over which one can be confident that one is observing true steady-state rates. Thus, mechanistic studies of 5-lipoxygenase have of necessity been of a more qualitative nature than the most detailed studies of the 15-lipoxygenases.

Nonheme Iron, the Redox Center

Although the presence of nonheme iron had been shown for several other lipoxygenases (28), it was only recently definitely shown to be present in 5-lipoxygenase and required for activity (26). It was found that the specific activity of purified 5-lipoxygenase from different preparations correlated linearly with the iron to protein stoichiometry (26, 27), and electron paramagnetic resonance (EPR) studies showed that one molar equivalent of fatty acid hydroperoxide per mole of Fe(II) in the native enzyme could convert all of the Fe(II) to Fe(III). The Fe(III) could be completely reduced to Fe(II) by one equivalent of an N-hydroxyurea (48). The iron redox titrations could also be monitored by UV absorbance and by fluorescence. The interconversion of Fe(II) and Fe(III) forms was also monitored by taking advantage of the irreversible inhibition of the Fe(III) form, but not the Fe(II) form, of 5-lipoxygenase by 4-nitrocatechol (48). This effect of 4-nitrocatechol had been observed previously for the soybean lipoxygenase-1 (49). One molar equivalent of fatty acid hydroperoxide per mole of iron was found to be sufficient to expose all of the activity of the native Fe(II) enzyme to irreversible inactivation by 4-nitrocatechol, and one equivalent of an N-hydroxyurea was sufficient to reverse this effect of fatty acid hydroperoxide. These observations established that all of the activity of 5-lipoxygenase is due to protein that contains nonheme iron, which can be interconverted between Fe(II) and Fe(III) oxidation states. These are thus the E_R and E_o states, respectively, as shown in Figure 2.

The native enzyme, E_R, is inactive as a catalyst of oxygenation of arachidonic acid, and must be oxidized to E_o before turnover can begin.

There has been considerable uncertainty over the years, based on studies with other lipoxygenases, concerning whether oxidation of the Fe(II) enzyme to the Fe(III) form was sufficient to activate the enzyme fully. For example, it was found that while one equivalent of fatty acid hydroperoxide was sufficient to convert native soybean lipoxygenase-1 to the Fe(III) form, a large molar excess of fatty acid hydroperoxide was necessary in assay solutions for the enzyme to be fully active upon addition of substrate (50). The apparent inconsistencies with the Fe(III) enzyme being fully active can be rationalized in terms of the mechanism shown in Figure 2 if the pentadienyl radical dissociates from E_R via the k_9 step once every few turnovers to yield the inactive free E_R. Fatty acid hydroperoxide must be present in the solution at a concentration sufficient to rapidly convert E_R to E_o via the k_{11}, k_{13} pathway and thus keep E_R from accumulating. This rationalization has been supported by a series of detailed kinetic experiments with soybean lipoxygenase-1 (51, 52). The kinetics of the system are further complicated by the binding of the substrate fatty acid to E_R as a reversible dead-end inhibitor. Thus, at low initial concentrations of fatty acid hydroperoxide, initial rates actually decrease with increasing fatty acid substrate concentration. This behavior pattern has been observed in this laboratory for 5-lipoxygenase as well (KI Skorey, unpublished results).

It has been shown that addition of glutathione peroxidase and glutathione can completely inhibit 5-lipoxygenase and 15-lipoxygenase activities in human leukocyte homogenates (53), and that conditions that reduce glutathione levels in intact human leukocytes result in increased levels of leukotriene biosynthesis (54). Other studies have shown the importance of selenium-dependent glutathione peroxidase activity in intact leukocytes in regulating 5-lipoxygenase activity (55). Thus, it appears that the Fe(III) form of the lipoxygenases is fully active and that redox cycling of the nonheme iron as shown in Figure 2 is the basis for catalysis. It should be mentioned, however, that acceptance of a mechanism involving obligatory redox cycling of the nonheme iron at each turnover is not universal (56).

REACTIONS CATALYZED BY 5-LIPOXYGENASE

Dioxygenation

The reaction for which lipoxygenases are named, the oxygenation of unsaturated fatty acids, can be viewed in terms of the mechanism shown in Figure 2 as proceeding via the sequence of steps corresponding to k_1, k_3, k_5, and k_7. The product is a fatty acid hydroperoxide. The fatty acid substrates of 5-lipoxygenase contain one or more 1,4-pentadiene moieties, and can be partially oxidized fatty acids, as in the cases of 12- or 15-HPETE

(8, 28). The natural substrate is arachidonic acid, and the product is 5-HPETE. The reaction involves stereospecific removal of the pro-S C-7 hydrogen atom of arachidonic acid and addition of O_2 to the opposite side of the pentadiene system (57, 58). The evidence is consistent with the reaction proceeding via one-electron transfer reactions. For example, in experiments with 15-lipoxygenases, dimers, presumably formed from radicals that dissociated from E_R via the k_9 step, were detected (59), and free-radical intermediates have been detected by spin-trapping experiments (60, 61). Formation of these free radical–derived species was favored under anaerobic or nearly anaerobic conditions, consistent with trapping by oxygen of the radical bound to E_R as shown in Figure 2. In terms of the mechanism, the $E_R(A^\bullet)$ species partitions between the k_9 and k_5 steps in a ratio that is dependent on oxygen concentration. Enzyme-bound alkyl and peroxy radicals have been detected by EPR experiments with soybean lipoxygenase-1 (62, 63). The observations were consistent with the signals being due to pentadienyl and peroxy radicals of the structures expected to be intermediates in the reaction, although it was not possible to assign structures conclusively to the enzyme-bound radicals. The various mechanisms that have been proposed for the chemistry occurring at the catalytic site during the oxygenase reaction have been summarized and discussed elsewhere (58). No results have been reported for analogous experiments with 5-lipoxygenase.

Leukotriene A_4 Synthase

The formation of leukotriene A_4 from arachidonic acid, presumably via 5-HPETE, or from exogenous 5-HPETE, proceeds via stereospecific removal of the pro-R C-10 hydrogen atom. The mechanism could involve removal of a proton and an electron from C-10 of 5-HPETE, followed by allylic shifts and donation of an electron and a proton to the leaving hydroxyl of the hydroperoxyl moiety, to generate H_2O and the epoxide, leukotriene A_4.

In cells capable of making leukotrienes, very high ratios of products derived from leukotriene A_4 to those derived from 5-HPETE, mainly 5-HETE, are observed. However, under most cell-free assay conditions for 5-lipoxygenase, only about 20% of the total products is leukotriene A_4, with 5-HPETE accounting for the remainder. It was observed from isotope-labelling experiments that under normal assay conditions about 20% of the arachidonate consumed by the rat 5-lipoxygenase is converted directly to leukotriene A_4 without mixing with exogenous 5-HPETE (64). In the same study, it was observed that arachidonate and 5-HPETE behaved as competitive substrates for 5-lipoxygenase, suggesting that the oxygenase and leukotriene A_4 synthase reactions both occur at the same catalytic site. More recently it was shown that when sufficient purified recombinant human 5-lipoxygenase was used to ensure that all arachidonate was consumed before the enzyme had all become irreversibly inactivated, more than 90% of the substrate was converted to leukotriene A_4 (43). This behavior was observed both when arachidonate was added as a substrate to phosphatidylcholine vesicles, and when it was generated in situ by the action of phospholipase A_2 on arachidonyl phospholipids. The ratio of leukotriene A_4 to 5-HPETE produced is dependent on the relative concentrations of arachidonate and 5-HPETE, and the rate and efficiency of conversion of 5-HPETE to leukotriene A_4 may depend on the partitioning of 5-HPETE between the interface to which the enzyme is adsorbed, and the aqueous solution where there is little enzyme when Ca^{2+} is present. This latter factor could explain why the rate of the 5-lipoxygenase-catalyzed leukotriene A_4 synthase reaction with exogenous 5-HPETE as substrate is only 2–5% the rate of the oxygenase reaction with arachidonate as substrate (64).

Anaerobic, Arachidonate-Dependent Hydroperoxidase

Under anaerobic conditions, soybean lipoxygenase and the reticulocyte 15-lipoxygenase catalyze the linoleic acid–dependent decomposition of fatty acid hydroperoxides, such as 13-HPODE (50, 59, 65–68). This type of activity of lipoxygenase is often referred to as a pseudoperoxidase activity, since the reaction involves a one-electron reduction of the fatty acid hydroperoxide, rather than a two-electron reduction as with peroxidases. The products of the one-electron reduction have been proposed to be hydroxide ion and an alkoxide radical (69), whose decomposition is considered to give rise to the oxodienoic acids and pentane, which have been observed products of the soybean and reticulocyte lipoxygenases (references cited above). The analogous process has been found to occur with recombinant human 5-lipoxygenase with the substrates arachidonate and 13-HPODE (27). In terms of the proposed mechanism, this reaction can be viewed as proceeding via the sequence of steps corresponding to k_1, k_3, k_9, k_{11}, and k_{13}. The mechanism is consistent with the strong inhibition of this process by oxygen. As mentioned earlier, the free radicals that are generated in this reaction catalyzed by the soybean lipoxygenase-1 have been observed to dimerize (59). A detailed kinetic study of this reaction catalyzed by 5-lipoxygenase has not been done; however, its rate does appear to be considerably slower than that of the oxygenase reaction (27). In a steady-state kinetic study of the soybean lipoxygenase-1–catalyzed reaction, k_{cat} for this reaction was found to be about 13-fold smaller than k_{cat} for the oxygenase reaction, consistent with k_9 being the rate-limiting step (70).

Anaerobic, Decomposition of Fatty Acid Hydroperoxide

The soybean lipoxygenase-1 has been observed to catalyze the decomposition of 13-HPODE in the absence of linoleic acid or any other reducing agent

(71). This reaction is strongly inhibited by oxygen, and was reported to be about 1500-fold slower than the anaerobic, linoleic acid–dependent hydroperoxidase activity discussed above (71). In terms of the proposed mechanism, this reaction is considered to proceed via the sequence of steps corresponding to k_8, k_6, k_9, k_{11}, and k_{13}. This activity has not been reported for 5-lipoxygenase, in spite of efforts to observe it in this laboratory. It is mentioned here for the sake of completeness, and to indicate that the analogy between 5-lipoxygenase and other lipoxygenases is not perfect.

Reducing Agent–Dependent Hydroperoxidase

Soybean lipoxygenase-1 has been observed to catalyze the oxidation of various reducing agents, which are capable of reducing the Fe(III) enzyme to the Fe(II) form, by fatty acid hydroperoxides (72–74). This reaction is not significantly inhibited by oxygen, and in terms of the proposed mechanism is considered to proceed via the sequence of steps corresponding to k_{15}, k_{17}, k_{11}, and k_{13}. This reaction has also been observed to be catalyzed by porcine 5-lipoxygenase (75), and by human 5-lipoxygenase (76–79). The types of reducing agents that have been found to be 5-lipoxygenase substrates for this activity include N-hydroxyureas, hydroxybenzofurans, hydroxamic acids, hydroxylamines, and catechols. Compounds of these classes are inhibitors of lipoxygenases, and those that are better substrates for the pseudoperoxidase activity of 5-lipoxygenase are in general more potent inhibitors (75). Thus these compounds derive at least a part of their potency from their action as alternate substrates for 5-lipoxygenase, and reduce the Fe(III) enzyme to the inactive Fe(II) form. It is difficult to determine quantitatively to what extent the inhibitory effect on the oxygenase activity of compounds that are substrates for the pseudoperoxidase activity of 5-lipoxygenase is due to their action as reducing agents. From a detailed steady-state kinetic study of the soybean lipoxygenase-1, it has been concluded that inhibition of the oxygenase activity by N-(4-chlorophenyl)-N-hydroxy-N′-(3-chlorophenyl) urea (CPHU) can be entirely attributed to its action as a substrate in the pseudoperoxidase activity (70). No such detailed studies have been reported for 5-lipoxygenase.

In the case of reducing agents whose one-electron oxidation products, $R^\bullet$ in Figure 2, are fairly reactive, it is likely that $R^\bullet$ reacts nonselectively with proteins and other species that are present. This has been shown to be the case for the 5-lipoxygenase inhibitor 2-[4′-methoxyphenyl)methyl]-3-methyl-4-hydroxy-5-propyl-7-chlorobenzofuran (76). It was found that this compound became covalently attached to 5-lipoxygenase and to other proteins present in the solution when both active 5-lipoxygenase and 13-HPODE were present. Degradation of reducing substrates as well as 13-HPODE during the 5-lipoxygenase-catalyzed pseudoperoxidase reaction has been shown for several reducing agents (75, 76). Dithiothreitol has been shown to regenerate the reducing agent, RH, from $R^\bullet$, and in the case of reducing agents that give relatively stable free radicals, such as N-hydroxyureas, it has been found that there was essentially no net consumption of the reducing substrate during the pseudoperoxidase reaction (78).

It appears that all of the inhibitors of 12- and 15-lipoxygenases thus far reported are reducing agents that inhibit by virtue of their activity as pseudoperoxidase substrates. Until the report by the ICI/Zeneca group of the methoxyalkyl thiazole class of inhibitors (80, 81), and of the thiopyranoindole inhibitors by the Merck group (82), this was true of 5-lipoxygenase inhibitors as well. The new classes of inhibitors do not function as pseudoperoxidase substrates, and in fact inhibit this activity as well as the oxygenase activity of 5-lipoxygenase (79). These inhibitors also differ from redox inhibitors in that they considerably slow the turnover-dependent irreversible inactivation of the enzyme, which is discussed below (79). The nonredox inhibitors are considered to act as reversible dead-end inhibitors, indicated as I in Figure 2. They may act by binding to E_o, E_R, and/or other states of the enzyme at the catalytic site. No detailed studies that might yield further insights concerning the mechanism of action of these nonredox inhibitors have yet been reported.

Irreversible Inactivation

The irreversible inactivation of 5-lipoxygenase, at least the turnover-dependent inactivation, must be considered as an activity of the enzyme. Neither the mechanism of the turnover-dependent inactivation nor the nature of the structural modification of the enzyme that results in this loss of function is known. It was reported that the turnover-dependent inactivation of the reticulocyte 15-lipoxygenase was accompanied by oxidation of a single methionine residue, presumably to methionine sulfoxide (83). However, more recent work by the same group failed to confirm this earlier observation (84). It has been reported that inactivation of the soybean and reticulocyte lipoxygenases either by turnover or by exposure to H_2O_2 is accompanied by loss of iron, and it has been proposed that this may be due to oxidation of histidine residues (84).

Turnover-dependent inactivation of soybean and reticulocyte lipoxygenases is a relatively slow process, but with fully active 5-lipoxygenase it proceeds with half-times on the order of one minute and is kinetically approximately first order (79). The half-times for this process are considerably longer in the presence of low Ca^{2+} concentrations (24) or some classes of nonredox inhibitors (79), both of which conditions slow the oxygenase rate considerably as well. Not surprisingly, redox inhibitors, which act as alternate substrates, do not increase the half-times for turnover-dependent inactivation (79).

The irreversible inactivation of human 5-lipoxygenase that occurs when the enzyme is dissolved in aerobic buffer has been shown to be due primarily to hydrogen peroxide, since the inactivation rate can be considerably reduced by addition of catalase (24). Since the Fe(II) enzyme is much more labile than the Fe(III) enzyme toward hydrogen peroxide, it was proposed that the Fe(II) enzyme itself likely is oxidized by hydrogen peroxide to the Fe(III) enzyme in a Fenton-type reaction. The resulting hydroxyl radical then oxidizes residues in the catalytic site (24). Although the structural change in the enzyme is not known, the inactivation appears to be at least a two-step process, the first of which does not involve the loss of iron. This conclusion was reached because a number of different batches of 5-lipoxygenase prepared by two different procedures gave two different linear correlations of specific activity with iron content (24). The highest specific activity was obtained if the recombinant enzyme was prepared in the absence of dithiothreitol or other sulfhydryl compounds. Hydrogen peroxide is a product of the nonenzymic reaction of oxygen with dithiothreitol. Since dithiothreitol had been reported to stabilize the activity of 5-lipoxygenase during its purification from leukocytes (29, 85), it was not obvious that eliminating it during purification of the recombinant enzyme would be advantageous.

THE IRON LIGANDS, SITE-DIRECTED MUTAGENESIS

It has been concluded on the basis of X-ray absorption spectroscopy studies that the nonheme iron of 15-lipoxygenase is coordinated by 6 ± 1 nitrogen and/or oxygen ligands, four of which are proposed to be imidazoles (86), one of which dissociates upon oxidation of the Fe(II) enzyme to the Fe(III) form (87). Two crystal structures of the soybean 15-lipoxygenase show that the iron atom in the Fe(II) enzyme is coordinated by three histidines and the carboxylate of the C-terminal isoleucine (88, 89). One of the structures also shows an asparagine ligand (89). The six histidine residues which, as mentioned earlier, are conserved in all known lipoxygenase sequences include, in human 5-lipoxygenase, His368, His373, and His550, which have been shown by site-directed mutagenesis studies to be essential for activity (90–92). His373 and His550 are also essential for iron binding (22, 27). The three histidines of the soybean lipoxygenase that correspond to these three essential 5-lipoxygenase histidines are the histidine iron ligands that appear in the crystal structures (88, 89), and site-directed mutagenesis studies have shown them to be essential for activity of the soybean enzyme (93). It has been found by site-directed mutagenesis that, although His368 is essential for oxygenase activity of 5-lipoxygenase, the Ser368 mutant retains most of its pseudoperoxidase activity (27).

Additional studies with 5-lipoxygenase have shown that the conserved residues His363, His391, His400, and Met436 are not essential for 5-lipoxygenase activity (22, 90). Also found to be inessential were Tyr384 and Phe394 (90). Mutations in the 416–418 region of the human 15-lipoxygenase have been found to alter the positional specificity of this enzyme (94).

DISCOVERY AND DEVELOPMENT OF DIRECT 5-LIPOXYGENASE INHIBITORS

Screening strategies to find direct inhibitors of 5-lipoxygenase have involved testing of natural products and synthetic compounds for inhibition of leukotriene biosynthesis in inflammatory cells (such as macrophage cell lines, and polymorphonuclear leukocytes), and then followup of active structures in broken cell preparations, semi-purified or purified 5-lipoxygenase prepararations or, more recently, recombinant 5-lipoxygenase, to demonstrate direct activity on the enzyme. Compounds have also been evaluated for inhibitory potency in human (and other species) whole blood and for ability to inhibit systemic leukotriene biosynthesis of leukotrienes as measured by urinary excretion of leukotriene E_4 and metabolites.

Redox Inhibitors

A wide variety of compounds has been reported with direct inhibitory activity of 5-lipoxygenase (95). The structures of a variety of 5-lipoxygenase inhibitors are shown in Figure 3. Many are compounds that by virtue of their redox potentials can act as alternative substrates for 5-lipoxygenase either by 5-lipoxygenase-induced oxidation of the native drug or by oxidation of products resulting from metabolic reduction of a prodrug. Although many of these compounds exhibit potent and specific inhibition of 5-lipoxygenase, they have frequently been attended with a variety of adverse effects, either due to direct interference with other biological processes (e.g. methemoglobin formation), or possibly as a result of the production of reactive radical species produced as by-products of 5-lipoxygenase inhibition.

Screening efforts at Merck identified a class of potent inhibitors of leukotriene synthesis in polymorphonuclear leukocytes (phenothiazines) exemplified by L-615,919. In broken cell preparations, activity was dependent on addition of NADH. The compound was also a potent inducer of methemoglobin formation in dog blood. Correlation of redox potential with methemoglobinemia—but not with 5-lipoxygenase inhibitory potency—was observed (P Bélanger, unpublished results), and the series evolved to identify

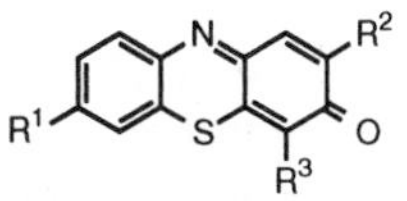

L-615919 ; $R^1=R^2=H$, $R^3=Cl$
L-651392 ; $R^1=R^2=CH_3$, $R^3=Br$

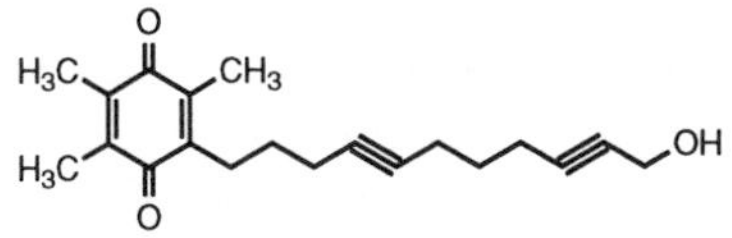

AA-861

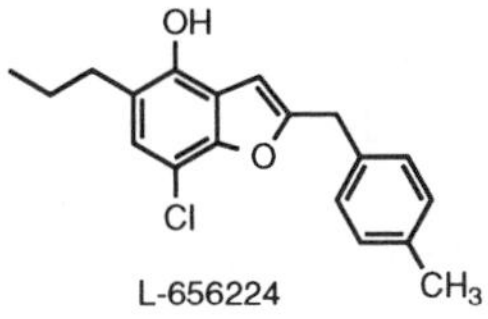

L-656224

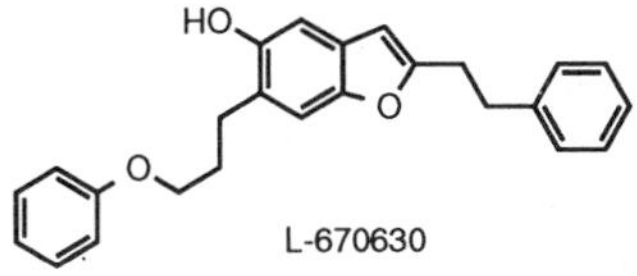

L-670630

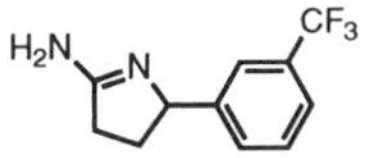

BW-755C

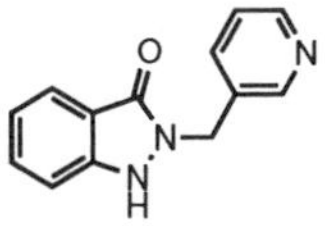

ICI-207968

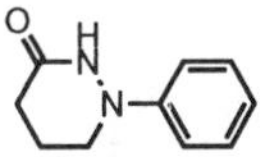

A-53612

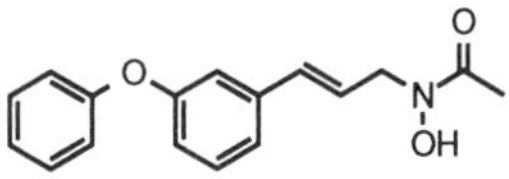

BW-A4C

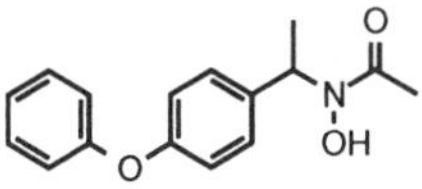

A-63162

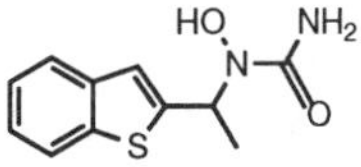

A-64077 (ZILEUTON)

Figure 3 Structures of 5-lipoxygenase inhibitors

L-651,392 (96, 97). Although not toxic in dog blood, this compound was poorly soluble and not developed further.

Quinones and hydroquinones (or prodrugs thereof) have also been reported to inhibit 5-lipoxygenase. The most widely studied compound, AA-861, is reported to be in clinical development (98, 99). This lipophilic quinone is notable for its structural similarity with natural quinones such as coenzyme Q. Close analogs of AA-861, which by virtue of their structure cannot enter into a redox cycle, are inactive, thus attesting to the redox mechanism of action of these type of compounds (100, 101).

Screening of natural products has revealed a wide variety of phenolic compounds that inhibit lipoxygenases through redox mechanisms. Observation of activity in compounds such as caffeic acid (102, 103), NDGA (104–106), and vignafuran (107) prompted efforts by pharmaceutical research groups to identify analogs with the desired specificity and potency suitable for development. Such studies resulted in the identification of potent specific and orally active phenol 5-lipoxygenase inhibitors such as L-656,224 (108, 109) and L-670,630 (110). Again in these series, some compounds caused methemoglobin formation in canine blood, but no correlation of this effect with 5-lipoxygenase inhibitory potency was observed (110). Although the optimal compounds were devoid of this activity, subsequent studies have shown that phenols such as L-656,224 serve to reduce the oxidized form of the 5-lipoxygenase to an inactive form, and the drug is itself oxidized and rapidly degraded to unstable by-products (76). These observations and other factors (such as rapid metabolism to glucuronides) have impeded the further development of such compounds. However, a number of phenolic 5-lipoxygenase inhibitors have entered development and clinical trials, such as TMK-688 (111), DUP 654 (112), and R68151 (113). It remains to be seen if these compounds will be developed to market.

A variety of redox-active heterocycles have also been identified as inhibitors of 5-lipoxygenase. The observation of nonspecific inhibitory activity in pyrazolidinones such as phenidone prompted concerted efforts to derive safe and selective compounds based on this structure. Studies indicated phenidone and analogs were also prone to induction of methemoglobin in blood, and that this toxicity did not correlate with 5-lipoxygenase activity or with redox potential (114). The further development of compounds such as ICI-207968 (114), BW755C (115), and homologous structures such as A-53612 (116) has been precluded by this type of observation.

Another class of redox 5-lipoxygenase inhibitors that has received much attention is the hydroxamic acids, and related N-hydroxyureas. Based on the expectation of inhibitory activity of compounds incorporating functional groups that might chelate iron, leukotriene analogs incorporating hydroxamic acids (117) were prepared. Subsequently, researchers from Burroughs-Wellcome and Abbott laboratories identified N-acetyl hydroxylamine compounds BW-A4C (118) and A-63162 (119) as potent and selective 5-lipoxygenase inhibitors with in vivo activity superior to that of the analogous N-methylhydroxamic acids. However, BW-A4C is rapidly metabolized in humans (120), and has been shown to be oxidized by lipoxygenase to form nitroxide radicals (121). The Abbott group continued studies in this area, concentrating on the heterologous series of N-hydroxyureas, which exhibited hydrolytic stability, reduced glucuronidation, and therefore superior in vivo properties. The product of these efforts, A-64077 (zileuton), has undergone extensive clinical evaluation (122–126), and is now reported to be in Phase III trials for asthma and inflammatory bowel disease (127). However, zileuton is also metabolized by lipoxygenase to form nitroxides and as such functions as a reducing substrate for 5-lipoxygenase (79).

Competitive Inhibitors of 5-Lipoxygenase

The search for nonredox, competitive inhibitors for 5-lipoxygenase has in some senses been inhibited by the ease of finding inhibitors that act by redox mechanisms. Only as a better understanding of the mechanism of the enzyme has evolved from kinetic and structural studies has it been possible to develop reliable criteria for the identification of nonredox inhibitors (79).

The identification of a series of methoxyalkylthiazoles (81) and methoxytetrahydropyrans (128) by the ICI (Zeneca) as potent and in some cases enantioselective inhibitors of 5-lipoxygenase suggested that they might act in a nonredox fashion. Structurally they also seemed unlikely to enter into redox reactions. Studies have subsequently shown that compounds in these series (such as ICI-211965) (*a*) do not act as reducing substrates in the 5-lipoxygenase-catalyzed decomposition of lipid hydroperoxides, (*b*) inhibit the 5-lipoxygenase-catalyzed reaction of reducing agents with lipid hydroperoxides, and (*c*) strongly inhibit the turnover-dependent inactivation of 5-lipoxygenase, and therefore can be considered to be true nonredox inhibitors of 5-lipoxygenase (79). Optimization of these compounds from the initially poorly bioavailable lead structures has resulted in the discovery of D-2138 (Figure 4) (128), a specific and orally bioavailable 5-lipoxygenase inhibitor currently undergoing clinical evaluation.

Screening of synthetic and natural compounds by the Merck Frosst group identified a class of lignans derived from a natural product, Justicidin E (129, 130), as moderately potent nonredox inhibitors of 5-lipoxygenase (131). Observation of structural similarity of these compounds with compounds such as D-2138 suggested the synthesis of hybrid molecules such as L-697,198, which were not only markedly more potent than the original lignans but, by virtue of the possibility of dosing as the prodrug hydroxy acid form, show excellent bioavailability and oral activity (132).

Identification of another class of nonredox inhibitors of 5-lipoxygenase,

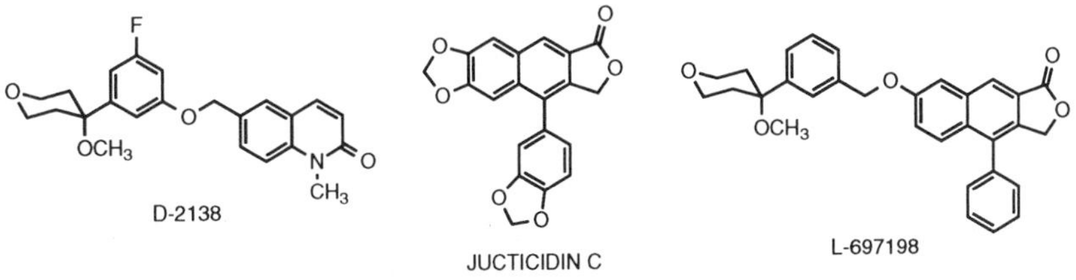

Figure 4 Structures of 5-lipoxygenase inhibitors

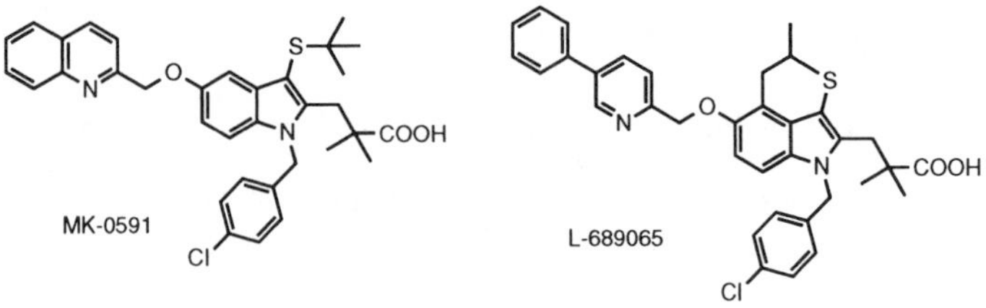

Figure 5 Structures of 5-lipoxygenase inhibitors

the thiopyranoindoles, derived from observations of moderate direct 5-lipoxygenase inhibitory activity in analogs of the indirect (or FLAP) inhibitor MK-0591. Addition of the thiopyrano ring and substitution of the quinoline moiety in MK-0591 for a 5-phenylpyridinyl unit derived potent and selective nonredox inhibitors of 5-lipoxygenase, exemplified by L-689,065 (Figure 5) (82), which have essentially no activity as FLAP inhibitors. It is indeed striking how such subtle structural changes can derive specific inhibitors for such apparently diverse biochemical targets, although similarities may rest in the interaction of both of these targets with the common fatty acid substrate, arachidonic acid.

5-LIPOXYGENASE ACTIVATION

Addition of arachidonic acid to cells containing either cyclooxygenase-1 or cyclooxygenase-2 results in the synthesis of prostaglandins, implying that the key event in the production of these eicosanoids is the activation of phospholipase A_2 and release of substrate. This is not the case with 5-lipoxygenase, and leukotriene synthesis only occurs in intact cells following exposure to certain stimuli that induce a significant rise in intracellular calcium. This is consistent with the suggestion that there is a defined regulatory process for the activation of 5-lipoxygenase within the cell. Initial studies have suggested that the activity of 5-lipoxygenase was dependent on a number of factors, including calcium, ATP, and undefined protein or membrane fractions (10, 133, 134). In particular, it was suggested that membranes had a potential role in the activation process because, first, 5-lipoxygenase was stimulated by either phosphatidylcholine vesicles or membrane fractions. Secondly, the substrate, arachidonic acid, is derived from membrane phospholipids through the action of a specific high-molecular-weight phospholipase A_2, which undergoes translocation to the membrane (38). Finally, sequence homologies in the 5-lipoxygenase have been described with the interfacial binding domains of lipases, which are indicative of membrane interactions (13).

The first evidence that 5-lipoxygenase activation was associated with the translocation of the enzyme from the cytosol to a membrane fraction was provided by Rouzer & Kargman (135). They showed that following stimulation of human polymorphonuclear leukocytes with ionophore A23187, there was a loss of 5-lipoxygenase activity from the supernatant, loss of 5-lipoxygenase protein from the cytosol, and the appearance of inactive 5-lipoxygenase protein in the 100,000 × g membrane. This suggested that following activation of the cell, 5-lipoxygenase translocated to a membrane site, where it became activated, synthesized leukotrienes, and underwent suicide inactivation. In support of this, it was observed that the cytosolic 5-lipoxygenase retained its enzymatic activity, indicating that this pool was not used for leukotriene synthesis. Other evidence was the facts that varying the ionophore concentration resulted in varying amounts of leukotriene synthesis, and varying degrees of enzyme translocation and termination of leukotriene synthesis during ongoing cell activation by addition of EDTA resulted in termination of enzyme translocation.

5-LIPOXYGENASE ACTIVATING PROTEIN (FLAP)

Discovery of FLAP

The key to the understanding of the mechanisms of 5-lipoxygenase activation came through mechanistic studies on the leukotriene biosynthesis inhibitor, MK-886 (L-663,536: 3-[1-(4-chlorobenzyl-3-*t*-butyl-thio-5-isopropyl-2-yl)-2,2-dimethyl-propanoic acid] (136) (Figure 6). This compound was shown to be a potent inhibitor of leukotriene biosynthesis in a variety of intact cell preparations, but had no effect on the production of either cyclooxygenase or 15- or 12-lipoxygenase-derived products of arachidonic acid metabolism (136). These potent in vitro effects on leukotriene biosynthesis were also observed in in vivo studies in which MK-886 was shown to inhibit leukotriene levels in inflammatory exudates, in rat bile, or in functional studies where the compound was shown to inhibit antigen-induced

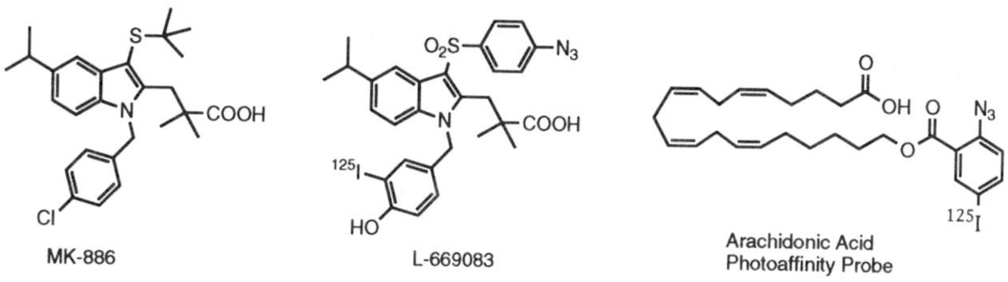

MK-886

L-669083

Arachidonic Acid Photoaffinity Probe

Figure 6 Structures of MK-886 and photoaffinity probes

bronchoconstriction in the rat and the squirrel monkey (136). Surprisingly, despite the fact that this compound was a potent inhibitor of leukotriene biosynthesis in intact cells, MK-886 had no effect on the 5-lipoxygenase enzyme itself, either in broken cell preparations or in purified enzyme preparations. In addition, MK-886 had no effect on phospholipase A_2, and did not inhibit arachidonic acid release (Merck Frosst, unpublished results).

An obvious explanation for this apparent paradox was that MK-886 might affect the activation of 5-lipoxygenase. The first evidence for this was obtained in studies using intact human polymorphonuclear leukocytes. In these experiments, MK-886 was shown to cause a concentration-dependent inhibition of the translocation of 5-lipoxygenase to the membrane that could be correlated with inhibition of leukotriene biosynthesis (137). Other pieces of evidence were, first, that a range of structural analogs of MK-886 were shown to inhibit the membrane translocation of 5-lipoxygenase with an apparent rank order of potency that correlated with their potencies as inhibitors of leukotriene biosynthesis. Secondly, addition of MK-886 to cells already activated with ionophore A23187, and thus after enzyme translocation had occurred, resulted in the release back into the cytosol of inactivated, translocated enzyme (137). This argued against the possibility that translocation was unrelated to cell activation, and simply reflected deposition of inactive enzyme on the membrane after leukotriene biosynthesis had occurred.

Of particular interest was the high intrinsic potency of MK-886, which suggested attachment to a target protein with high affinity. The original hypothesis was that such a target protein could be a membrane "docking" protein or 5-lipoxygenase-activating protein (FLAP). In order to look for this putative activating protein, a number of experimental approaches were utilized. The first involved the preparation of an analog of MK-886 labelled with ^{125}I and containing a photoactivatable azido group (^{125}I-L-669,083) (Figure 6). Following incubation of this photoaffinity probe with neutrophils, the cells were irradiated with ultraviolet light and the labelling of a number of proteins was observed in both the cytosolic and the membrane fractions (100,000 $\times$ g supernatant and pellet) (138). However, when these incubations were carried out in the presence of excess, cold MK-886, only a single 18-kDa membrane protein was specifically labelled. This protein appeared to be present only in the membranes of leukocytes, and was not present in a variety of other cells lines not known to have 5-lipoxygenase (138).

In order to isolate and purify this 18-kDa protein, a series of affinity columns, to which were coupled analogs of MK-886, were prepared. The 18-kDa protein was solubilized from neutrophil membranes with the detergent CHAPS, bound to affinity columns, and selectively eluted with MK-886

(138). The rat protein was then purified to homogeneity by chromatography on superose-12 and TSK3000 columns, and sequence information from the native protein as well as cyanogen bromide and tryptic cleavage products was obtained (123). Sequence information obtained on the rat protein was used to construct oligonucleotide probes, which were used to screen libraries of rat neutrophil and basophil leukemia cells to obtain the rat cDNA (139). The rat cDNA was then used to screen an HL-60 cell library to obtain the corresponding human cDNA. Both cDNA clones were shown to encode proteins of 101 amino acid residues, which were 92% identical and contained all the sequence data obtained from purified rat protein. The structure of these novel proteins, which showed no significant sequence homologies with other known proteins, were proposed to consist of three transmembrane-spanning regions consisting of two hydrophilic loops, with the C and N termini on opposite sides of the membrane (139). This protein has been termed 5-lipoxygenase-activating protein (FLAP). Immunoprecipitation experiments using a polyclonal rabbit antibody to the internal sequence of the rat 18-kDa protein demonstrated that the cloned protein was the same protein that was selectively labelled by the photoaffinity probe (138).

The first evidence that FLAP was required for cellular leukotriene biosynthesis came from a series of transfection experiments carried out in human osteosarcoma cells (139). When these cells were transfected with the gene for either 5-lipoxygenase or FLAP alone and treated with ionophore A23187, no arachidonic acid metabolites were produced. However, when both proteins were co-expressed in the same cell, cellular leukotriene biosynthesis was observed following ionophore challenge, and this synthesis was inhibited in the presence of MK-886 (124). In addition, in cells transfected with both 5-lipoxygenase and FLAP, translocation of 5-lipoxygenase to the membrane following exposure to ionophore A23187 was observed (140). This translocation, together with the accompanying leukotriene biosynthesis, was inhibited by MK-886. Since the original discovery of FLAP, a number of studies have demonstrated that both FLAP and 5-lipoxygenase are necessary for cellular leukotriene synthesis. In the human monocytic cell line, U937, the expression of FLAP, but not 5-lipoxygenase, was observed and the cells failed to produce leukotrienes upon challenge with ionophore A23187 (141). Upon differentiation of these cells with dimethylsulfoxide (DMSO) towards a more mature monocyte-macrophage lineage, induction of FLAP but not 5-lipoxygenase was seen. These induced cells also lacked the ability to produce leukotrienes. When these cells were infected with DNA coding for 5-lipoxygenase and then differentiated with DMSO, leukotriene synthesis could be observed following challenge with ionophore A23187; this synthesis was inhibited by MK-886 (141). Similar results have been obtained in sf9 insect cells infected with recombinant baculoviruses for both human 5-lipoxygenase and FLAP, in which both proteins were required for cellular leukotriene synthesis, which is inhibited by MK-886 (142).

Biological Function of FLAP

The initial hypothesis for the involvement of FLAP in leukotriene biosynthesis was that cellular activation through a receptor-mediated event induced a significant rise in intracellular calcium. This rise in intracellular calcium resulted in the activation and translocation of 5-lipoxygenase, as well as high-molecular-weight phospholipase A_2, with 5-lipoxygenase moving to the "docking" protein, FLAP. This model would require a stable complex to be formed at the membrane between active 5-lipoxygenase and FLAP, as well as possibly other components of the leukotriene biosynthetic pathway, such as cytosolic phospholipase A_2 and leukotriene A_4 hydrolase. Formation of this complex could regulate the interaction of the enzyme with its substrate, arachidonic acid, resulting in efficient leukotriene biosynthesis. The 5-lipoxygenase enzyme would then undergo turnover associated–suicide inactivation, resulting in dead enzyme remaining bound to the membrane protein. These studies assumed that drugs such as MK-886 were binding to a site on FLAP that mediated the interaction of FLAP with 5-lipoxygenase and hence were blocking a protein-protein interaction. If this hypothesis were correct, then there should be an excellent correlation between inhibition of leukotriene biosynthesis by various indirect leukotriene biosynthesis inhibitors and inhibition of 5-lipoxygenase translocation. Initial results using polymorphonuclear leukocytes challenged with ionophore A23187 supported this hypothesis (137). Subsequently, translocation of 5-lipoxygenase was also observed in the human myeloid cell line, HL-60, in response to both calcium ionophore A23187 and the receptor-mediated stimulus, *N*-formylmethionyl-leucyl-phenylalanine (143). In this study, using both MK-886 and a quinoline leukotriene synthesis inhibitor, a correlation was observed between inhibition of leukotriene production and translocation. Similarly, leukotriene production, membrane translocation of 5-lipoxygenase, and increases in intracellular calcium were shown to be tightly coupled in rat basophilic leukemia cells (144). In these experiments, translocation was observed in response to ionomycin, thapsigargin, and crosslinking of high-affinity IgE receptors, and translocation was inhibited by MK-886. Other authors have shown translocation of 5-lipoxygenase to occur in human alveolar macrophages challenged with ionophore A23187; this translocation and leukotriene synthesis are also inhibited by MK-886 (145).

Coffey et al were the first authors to suggest that the correlation between leukotriene biosynthesis and translocation was not universal (146). They carried out a series of studies on rat alveolar and peritoneal macrophages and compared them with human polymorphonuclear leukocytes. The peritoneal macrophages resembled polymorphonuclear leukocytes in that most of their cell-free 5-lipoxygenase activity, as well as 5-lipoxygenase protein content, was localized in the cytosol fraction. In contrast, resting alveolar macrophages contained most of their activity and almost half of their immunoreactive protein in the crude membrane fractions. In the latter cells, MK-886 was unable to reverse this membrane association, suggesting that either FLAP was not the site of binding of 5-lipoxygenase in the resting cell or that MK-886 was not binding to a site mediating the interaction of FLAP with 5-lipoxygenase. Despite the fact that this drug did not reverse the membrane association of 5-lipoxygenase, leukotriene biosynthesis in these cells was completely inhibited by MK-886, indicating that FLAP was essential for cellular leukotriene biosynthesis (146). FLAP-independent translocation of 5-lipoxygenase following stimulation with ionophore A23187 has been seen in osteosarcoma cells infected with the 5-lipoxygenase gene alone; this translocation is unaffected by MK-886 (140). Further evidence that the correlation between inhibition of translocation and inhibition of leukotriene biosynthesis did not exist was obtained in a series of studies on analogs of BAY X1005, a quinoline-based leukotriene biosynthesis inhibitor (147, 148). In these studies, the binding of a series of compounds to FLAP was compared with their ability to inhibit leukotriene biosynthesis and their ability to inhibit translocation of 5-lipoxygenase. As observed previously, there was a good correlation between the binding to FLAP and inhibition of leukotriene biosynthesis, but the correlation broke down when the effects on translocation were studied.

The above results suggest that the MK-886 binding site on FLAP does not directly mediate an interaction with 5-lipoxygenase. The current hypothesis is that this binding site is an arachidonic acid–binding site on FLAP, and that FLAP facilitates the transfer of arachidonic acid to 5-lipoxygenase, allowing the 5-lipoxygenase reaction to occur in a more efficient fashion. Three studies support this hypothesis. The first has shown, through the use of a novel photoaffinity analog of arachidonic acid (Figure 6), that FLAP is an arachidonic acid–binding protein (149). This photoaffinity probe was shown to label human FLAP expressed in sf9 insect cells infected with recombinant baculovirus specifically, and the binding of this photoaffinity probe was inhibited by both arachidonic acid and MK-886. Secondly, Hill et al obtained active 5-lipoxygenase on the membranes of human polymorphonuclear leukocytes by stimulating the cells with ionophore A23187 in the presence of a direct 5-lipoxygenase inhibitor, zileuton (150). Under these conditions, accumulation of active membrane-associated 5-lipoxygenase was inhibited and reversed by MK-886. More interestingly, the membrane-associated 5-lipoxygenase was two times more efficient in the production of leukotriene A_4 from arachidonic acid–derived 5-HPETE than was the cytosolic enzyme. In addition, unlike the cytosolic enzyme, membrane-associated 5-lipoxygenase could metabolize 12(*S*)- and 15(*S*)-HETE to 5(*S*),12(*S*)- and 5(*S*),15(*S*)-dihydroxyeicosatetraenoic acid, respectively. This ability to metabolize hydroxy fatty acids was dependent on FLAP and was lost if 5-lipoxygenase was eluted from the membrane by MK-886. This indicates that association with FLAP can alter the substrate specificity of 5-lipoxygenase. Thirdly, in sf9 insect cells infected with recombinant baculoviruses to express either human 5-lipoxygenase alone or human 5-lipoxygenase and human FLAP together, FLAP was shown to stimulate the utilization of arachidonic acid by 5-lipoxygenase and increases the efficiency with which 5-lipoxygenase converts 5-HPETE to leukotriene A_4 (142).

FLAP Gene Structure

The human gene for FLAP has been isolated from two different genomic libraries and cloned (151). The gene spans >31 kb and consists of five small exons and four large introns. The presence of a single FLAP gene per haploid genome has been suggested through southern blot analysis of human genomic DNA. The transcription initiation site was located at an adenine residue, 74 basepairs upstream of the ATG initiation codon. The presence of a possible TATA box (TGTAAT) 22 basepairs upstream and potential AP-2 and glucocorticoid receptor–binding sites were indicated by an examination of the sequence of the gene 5′ to the mRNA start site. Functional analysis of the FLAP gene promoter was assessed using transient transfections of mouse P388D_1 cells (macrophagelike) and human HepG2 (hepatoma) cells with 5′ flanking sequences of the FLAP gene fused upstream to a chloramphenicol acetyltransferase reporter gene. Only a minimal level of activity was observed in the hepatoma cell line, but expression in the mouse macrophage cell line of various FLAP gene promoter constructs revealed both tissue specificity and enhancerlike activities. The human FLAP gene is localized on chromosome 13, as opposed to 5-lipoxygenase, which is found on chromosome 10.

Regulation of FLAP Production

As both FLAP and 5-lipoxygenase are required for cellular leukotriene biosynthesis, it would be obvious to speculate that both gene products might be regulated together. Initial studies with the human promyelocytic cell line, HL-60, supported this hypothesis. Thus, when these cells were differentiated towards granulocytes by exposure to DMSO, the concurrent induction of FLAP and 5-lipoxygenase, and an increased capacity to synthesize leukotriene, were observed (152). These results were confirmed by Bennett et al, but these authors also showed a discordant regulation of 5-lipoxygenase and FLAP mRNA following differentiation of HL-60 cells to monocytes following treatment with 1,25-dihydroxyvitamin D_3 and phorbol ester (153). Thus, treatment with the former agent resulted in a sixfold increase in 5-lipoxygenase mRNA and a 1.3-fold increase in FLAP mRNA, and treatment with phorbol ester failed to induce 5-lipoxygenase mRNA but increased FLAP mRNA twofold. These authors also further examined the mechanism of upregulation of mRNA and showed that the transcriptional rate of the 5-lipoxygenase and FLAP genes did not change upon differentiation, suggesting that the increase in the mRNA coding for these proteins was not due to transcriptional activation of their respective genes. In addition, the mRNA half-life for 5-lipoxygenase did not change significantly upon treatment, whereas the FLAP mRNA half-life only increased from approximately 3.5 to 4.5 h upon differentiation. These data have been interpreted to indicate that expression of 5-lipoxygenase and FLAP is controlled by a posttranscriptional event other than stabilization of the mRNA (153). Further evidence for discordant regulation of the two genes has been obtained in various cell lines. Thus, the human monocytic cell line, U937, expresses only FLAP without 5-lipoxygenase (141). Differentiation of these cells with DMSO increased the expression of FLAP fourfold with no induction of 5-lipoxygenase. Similarly, five lymphoblastoid T cell lines have been investigated and found to express the FLAP gene but not the 5-lipoxygenase gene (154). Differentiation of human peripheral blood monocytes by 1,25-dihydroxyvitamin D_3 upregulates 5-lipoxygenase metabolism and increases the expression of FLAP but not 5-lipoxygenase (155). In addition, "de-differentiation" of alveolar macrophages obtained from 1,25-dihydroxyvitamin D_3-deficient rats resulted in a reduced 5-lipoxygenase metabolic capacity, which was associated with decreased FLAP but not 5-lipoxygenase expression. To date, no cell types have been described that express 5-lipoxygenase in the absence of FLAP.

The above results show that while both the 5-lipoxygenase and FLAP genes show a high degree of tissue selectivity, both genes being predominantly expressed in myeloid cells, they are also clearly regulated in different ways as observed in differentiation experiments with various cell lines. This is consistent with the differences observed in the 5′ upstream regulatory sequences for both gene products. For example, the 5-lipoxygenase gene lacks a TATA box and has many of the characteristics of a housekeeping gene, whereas the FLAP gene has an identifiable TATA box and other regulatory motives not common to the 5-lipoxygenase gene (151).

Tissue Distribution and Subcellular Localization

Results published to date indicate that FLAP is expressed in a number of cells of myeloid origin, including polymorphonuclear leukocytes, monocytes, macrophages, eosinophils, and basophils, as well as lymphoblastoid cells, including various B and T cell lines. FLAP is not present in the platelet, in normal human T cells, or in various murine T cell lines (152). Unpublished data indicates that FLAP is not present in a variety of tissues, and is probably largely confined to cells of the myeloid origin.

Of particular interest is the subcellular localization of FLAP. Woods et al have used immuno-electronmicroscopic labelling of ultrathin frozen sections to visualize the intracellular location of 5-lipoxygenase and FLAP in resting and ionophore-activated human leukocytes (156). In activated leukocytes, both FLAP and 5-lipoxygenase were localized in the lumen of the nuclear envelope and endoplasmic reticulum. Neither protein could be detected in any other cell compartment or along the plasma membrane. In contrast, in resting cells the distribution of FLAP was identical to that observed in activated cells, but 5-lipoxygenase was not present on the nuclear envelope and, except for weak labelling of the euchromatin region, could not be readily detected in any cellular compartment. These results were the first direct demonstration that in unstimulated cells FLAP is present in a membrane, and that following stimulation 5-lipoxygenase moves to the same membrane site as FLAP. Subcellular fractionation of polymorphonuclear leukocytes was carried out in parallel with the electron microscopy studies (156). The presence of FLAP in various fractions was analyzed by immunoblot and a FLAP-binding assay. These results also showed that >83% of the immunoblot-detectable FLAP protein and approximately 64% of the FLAP ligand-binding activity was found in a nuclear membrane fraction.

INHIBITORS OF LEUKOTRIENE BIOSYNTHESIS THAT BIND TO FLAP

As described above, the discovery of FLAP was driven through an elucidation of the mechanism of action of the leukotriene biosynthesis inhibitor, MK-886, an indole structure. This compound was shown to be an effective inhibitor of leukotriene biosynthesis in a variety of intact cell preparations (136), and entered clinical trials in humans, where it significantly inhibited leukotriene production as measured through inhibition of ex vivo challenge of whole blood with ionophore A23187 and inhibition of excretion of leukotriene E_4 in the urine (157). The compound was also shown to have significant effects on the antigen-induced bronchoconstriction in asthmatic subjects (158). It was rapidly realized that in addition to indole structures, a second class of chemical structures, termed quinolines, also bound to FLAP and inhibited leukotriene biosynthesis in intact cells (159). A member of this class, Bay X1005 (Figure 7), is also being evaluated in clinical

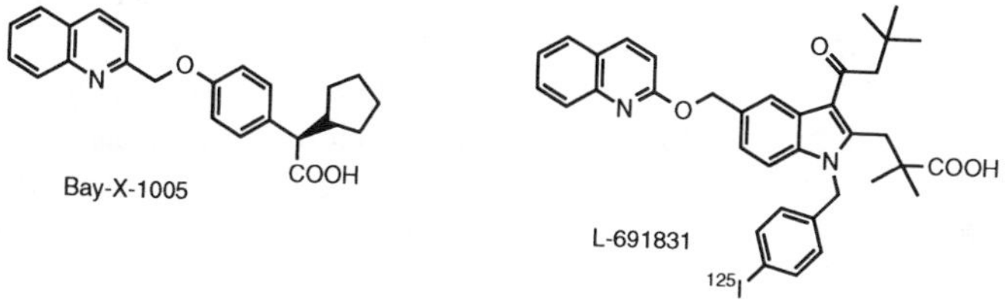

Figure 7 Structures of Bay X1005 and radioiodinated FLAP ligand L-691,831

trials. In order to assess these various compounds as potential inhibitors of leukotriene biosynthesis, a FLAP-binding assay has been devised using human leukocyte membrane as a source of FLAP, and using a radioiodinated leukotriene synthesis inhibitor, ^{125}I-L-691,831, as ligand (160). Using this assay, an excellent correlation was shown between the affinity for FLAP in the binding assay and inhibition of leukotriene synthesis in polymorphonuclear leukocytes for compounds from the two structurally distinct classes, indoles and quinolines. With the recognition that two separate classes of drugs bound to FLAP, a novel class of hybrid structures based on the indole and quinoline classes of inhibitors and termed quindoles was developed (161). This led to the development of MK-0591, an extremely potent inhibitor of leukotriene biosynthesis in a variety of intact cells (162). This compound is also a potent inhibitor of leukotriene biosynthesis in vivo and is currently in clinical trials.

A number of studies have been done on the binding site for these drugs on FLAP. In order to initially define this, a cross-species comparison of FLAP was done, and binding assays were carried out in 10 mammalian species (163). Using the polymerase chain reaction, cDNA clones for FLAP from six species (monkey, horse, pig, sheep, rabbit, and mouse) were isolated and sequenced. The deduced sequences of FLAP showed a high degree of identity to each other and to the published sequence for human and rat FLAP. Two regions of the protein are almost totally conserved among all the species analyzed, and this suggested that these regions may have functional significance and could be involved in the binding of inhibitors. In order to further pursue this, the binding of photoaffinity analogs of two classes of potent leukotriene biosynthesis inhibitors was studied (164). As human FLAP contains only a single tryptophan residue and two internal methionine residues, it is possible to use reagents that specifically cleave at these residues in conjunction with antipeptide antisera to localize the site of attachment of the photoaffinity ligands. Thus inhibitors were shown to bind to FLAP amino terminal to Trp72. To further localize the site, a series of site-specific mutagenesis and deletion mutant studies were carried out (164). The results indicated that a negative charge associated with residue 62 (Asp62) was critical for inhibitor binding and that mutants containing deletions in a highly conserved region of the protein (residues 42–61) do not bind inhibitors. These results have localized the binding site to the hydrophilic loop between the proposed first and second transmembrane regions.

CONCLUSIONS

Understanding of the structure and mechanism of 5-lipoxygenase is still at a very primitive stage. However, with the availability of methods to obtain pure recombinant enzyme with high specific activity, which can be studied under defined conditions, much more rapid progress should occur in the future. The availability of a number of potent reversible dead-end inhibitors with varied structures will also be of considerable help in efforts to understand how this enzyme works. The most intriguing question regarding this enzyme is why it does not work in cells unless uninhibited FLAP is also present. Although, as summarized above and below, a number of ingenious experiments have been carried out and interpreted in terms of hypotheses regarding the role of FLAP, it is important to recall that in cell-free experiments, 5-lipoxygenase has exhibited FLAP-independent activities corresponding to essentially all of its known cellular activities. It might be very fruitful to focus attention on understanding what keeps the enzyme from functioning significantly in the cell, even when Ca^{2+} and arachidonate are available, unless active FLAP is also available. It is apparent that FLAP does not simply activate 5-lipoxygenase in the cell, it also counteracts some sort of inhibition or constraint whose nature no apparent effort has yet been made to discover.

While purified 5-lipoxygenase can function effectively as an oxygenase to convert arachidonic acid to leukotriene A_4, in the test tube it is clear that in the intact cell, arachidonic acid metabolism through the 5-lipoxygenase pathway occurs only in the presence of FLAP. The activation of the leukotriene biosynthetic pathway requires a rise in intracellular calcium, which causes 5-lipoxygenase to translocate, presumably from the cytosol, to the nuclear envelope, where FLAP acts as an arachidonic acid transfer protein, allowing 5-lipoxygenase to utilize the substrate more efficiently (Figure 8*a*). Drugs, such as MK-886, bind to the arachidonic acid–binding site on FLAP and may indirectly prevent an association of 5-lipoxygenase with FLAP (Figure 8*b*). It has been postulated that FLAP acts as a putative "docking" for 5-lipoxygenase in the membrane, although there is no direct evidence for this at the present time. This unique mechanism appears to

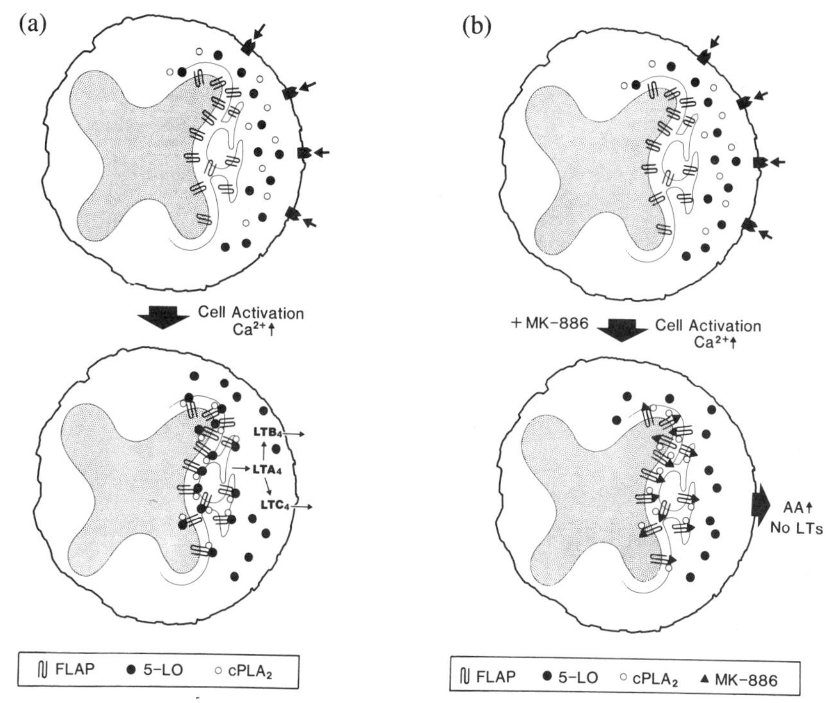

Figure 8 *a*. Proposed mechanism for leukotriene biosynthesis in intact cells showing translocation of 5-lipoxygenase (5-LO) and cytosolic phospholipase A_2 ($cPLA_2$) to FLAP. *b*. Inhibition of leukotriene biosynthesis in the presence of FLAP inhibitors such as MK-886.

distinguish 5-lipoxygenase from other arachidonic acid oxygenases, such as cycloxygenases, 12-lipoxygenase, and 15-lipoxygenase.

Two therapeutic approaches have been developed for blocking the production of 5-lipoxygenase products. These are direct 5-lipoxygenase inhibitors, which can act through various mechanisms on the enzyme, and indirect inhibitors, which bind to the putative arachidonic acid–binding site on FLAP. It appears that both therapeutic approaches result in effective inhibition of leukotriene biosynthesis in vivo, and drugs such as these may have therapeutic uses in the treatment of asthma and allergic diseases as well as inflammatory conditions, such as inflammatory bowel disease.

Acknowledgements

The authors wish to thank Louise Charlton, Carolyn Green, and Mary Wacasey for their help in the preparation of this manuscript.

Literature Cited

1. Ford-Hutchinson AW, Bray MA, Doig MV, Shipley ME, Smith MJH. 1980. *Nature* 286:264–65
2. Ford-Hutchinson AW. 1990. *Crit. Rev. Immunol.* 10:1–12
3. Nicholson DW, Ali A, Vaillancourt JP, Calaycay JR, Mumford RA, et al. 1993. *Proc. Natl. Acad. Sci. USA* 90:2015–19
4. Piper PJ. 1985. *Int. Arch. Allergy Appl. Immunol.* 76(Suppl. 1):43–48
5. Snyder DW, Fleisch JH. 1989. *Annu. Rev. Pharmacol. Toxicol.* 29:123–43
6. Ford-Hutchinson AW. 1993. *Springer Semin. Immunopathol.* 15:37–50
7. Sigal E. 1991. *Am. J. Physiol.* 260: L13–28
8. Ueda N, Kaneko S, Yoshimoto T, Yamamoto S. 1986. *J. Biol. Chem.* 261:7982–88
9. Rouzer CA, Matsumoto T, Samuelsson B. 1986. *Proc. Natl. Acad. Sci. USA* 83:857–61
10. Rouzer CA, Samuelsson B. 1985. *Proc. Natl. Acad. Sci. USA* 82:6040–44
11. Skoog MT, Nichols JS, Wiseman JS. 1986. *Prostaglandins* 31:561–76
12. Aharony D, Stein RL. 1986. *J. Biol. Chem.* 261:11512–19
13. Dixon RAF, Jones RE, Diehl RE, Bennett CD, Kargman S, Rouzer CA. 1988. *Proc. Natl. Acad. Sci. USA* 85:416–20
14. Matsumoto T, Funk CD, Rådmark O, Höög J-O, Jörnvall H, Samuelsson B. 1988. *Proc. Natl. Acad. Sci. USA* 85:26–30
15. Balcarek JM, Theisen TW, Cook MN, Varrichio A, Hwang S-M, et al. 1988. *J. Biol. Chem.* 263:13937–41
16. Rouzer CA, Rands E, Kargman S, Jones RE, Register RB, Dixon RAF. 1988. *J. Biol. Chem.* 263:10135–40
17. Funk CD, Hoshiko S, Matsumoto T, Rådmark O, Samuelsson B. 1989. *Proc. Natl. Acad. Sci. USA* 86:2587–91
18. Chopra A, Ferreira-Alves DL, Sirois P, Thirion JP. 1992. *Biochem. Biophys. Res. Commun.* 185:489–95
19. Steinhilber D, Hoshiko S, Grunewald J, Rådmark O, Samuelsson B. 1993. *Biochim. Biophys. Acta* 1178:1–8
20. Steinhilber D, Rådmark O, Samuelsson B. 1993. *Proc. Natl. Acad. Sci. USA* 90:5984–88
21. Funk CD, Gunne H, Steiner H, Izumi T, Samuelsson B. 1989. *Proc. Natl. Acad. Sci. USA* 86:2592–96
22. Zhang Y-Y, Lind B, Rådmark O, Samuelsson B. 1993. *J. Biol. Chem.* 268:2535–41
23. Denis D, Falgueyret J-P, Riendeau D, Abramovitz M. 1991. *J. Biol. Chem.* 266:5072–79
24. Percival MD, Denis D, Riendeau D, Gresser MJ. 1992. *Eur. J. Biochem.* 210:109–17
25. Samuelsson B, Funk CD. 1989. *J. Biol. Chem.* 264:19469–72
26. Percival MD. 1991. *J. Biol. Chem.* 266:10058–61
27. Percival MD, Ouellet M. 1992. *Biochem. Biophys. Res. Commun.* 186:-1265–70
28. Yamamoto S. 1992. *Biochim. Biophys. Acta* 1128:117–31
29. DeWolf WE Jr. 1991. In *Lipoxygenases and Their Products*, ed. ST Crooke, A Wong, pp. 105–35. San Diego, CA: Academic
30. Yamamoto S. 1989. *Prostaglandins Leukot. Essent. Fatty Acids* 35:219–29
31. McMillan RM, Walker ER. 1992. *Trends Pharmacol. Sci.* 13:323–30
32. Musser JH, Kreft AF. 1992. *J. Med. Chem.* 35:2501–24
33. Hogaboom GK, Cook M, Newton JF, Varrichio A, Shorr RGL, et al. 1986. *Mol. Pharmacol.* 30:510–19
34. Ochi K, Yoshimoto T, Yamamoto S, Taniguchi K, Miyamoto T. 1983. *J. Biol. Chem.* 258:5754–58
35. Wiseman JS. 1989. *Biochemistry* 28: 2106–11
36. Rouzer CA, Samuelsson B. 1990. *Methods Enzymol.* 187:312–19
37. Stabel S, Parker PJ. 1991. *Pharmacol. Ther.* 51:71–95
38. Clark JD, Lin L-L, Kriz RW, Ramesha CS, Sultzman LA, et al. 1991. *Cell* 65:1043–51
39. Rouzer CA, Samuelsson B. 1987. *Proc. Natl. Acad. Sci. USA* 84:7393–97
40. Goetze AM, Fayer L, Bouska J, Bornemeier D, Carter GW. 1985. *Prostaglandins* 29:689–701
41. Riendeau D, Falgueyret J-P, Nathaniel

DJ, Rokach J, Ueda N, Yamamoto S. 1989. *Biochem. Pharmacol.* 38:2313–21
42. Wong A. Crooke ST. 1991. *Lipoxygenases and Their Products,* ed. ST Crooke, A Wong, pp. 67–87. San Diego, CA: Academic
43. Riendeau D, Falgueyret J-P, Meisner D, Sherman MM, Laliberté F, Street IP. 1993. *J. Lipid Mediat.* 6:23–30
44. Jain MK, Berg OG. 1989. *Biochim. Biophys. Acta* 1002:127–56
45. Jain MK, Gelb MH. 1991. *Methods Enzymol.* 197:112–25
46. Ghomashchi F, Schüttel S, Jain MK, Gelb MH. 1992. *Biochemistry* 31: 3814–24
47. Schewe T, Rapoport SM, Kühn H. 1986. *Adv. Enzymol.* 58:191–272
48. Chasteen ND, Grady JK, Skorey KI, Neden KJ, Riendeau D, Percival MD. 1993. *Biochemistry* 32:9763–71
49. Galpin JR, Tielens LGM, Veldink GA, Vliegenthart JFG, Boldingh J. 1976. *FEBS Lett.* 69:179–82
50. de Groot JJMC, Garssen GJ, Veldink GA, Vliegenthart JFG, Boldingh J. 1975. *FEBS Lett.* 56:50–54
51. Schilstra MJ, Veldink GA, Verhagen J, Vliegenthart JFG. 1992. *Biochemistry* 31:7692–99
52. Schilstra MJ, Veldink GA, Vliegenthart JFG. 1993. *Biochemistry* 32:7686–91
53. Hatzelmann A, Schatz M, Ullrich V. 1989. *Eur. J. Biochem.* 180:527–33
54. Hatzelmann A, Ullrich V. 1987. *Eur. J. Biochem.* 169:175–84
55. Weitzel F, Wendel A. 1993. *J. Biol. Chem.* 268:6288–92
56. Wang Z-X, Killilea SD, Srivastava DK. 1993. *Biochemistry* 32:1500–9
57. Corey EJ, Lansbury PT Jr. 1983. *J. Am. Chem. Soc.* 105:4093–94
58. Stubbe J. 1989. *Annu. Rev. Biochem.* 58:257–85
59. Garssen GJ, Vliegenthart JFG, Boldingh J. 1972. *Biochem. J.* 130: 435–42
60. Connor HD, Fischer V, Mason RP. 1986. *Biochem. Biophys. Res. Commun.* 141:614–21
61. Iwahashi H, Parker CE, Mason RP, Tomer KB. 1991. *Biochem. J.* 276: 447–53
62. Nelson MJ, Seitz SP, Cowling RA. 1990. *Biochemistry* 29:6897–903
63. Nelson MJ, Cowling RA. 1990. *J. Am. Chem. Soc.* 112:2820–21
64. Wiseman JS, Skoog MT, Nichols JS, Harrison BL. 1987. *Biochemistry* 26: 5684–89
65. Garssen GJ, Vliegenthart JFG, Boldingh J. 1971. *Biochem. J.* 122:327–32
66. Verhagen J, Veldink GA, Egmond MR, Vliegenthart JFG, Boldingh J, van der Star J. 1978. *Biochim. Biophys. Acta* 529:369–79
67. Härtel B, Ludwig P, Schewe T, Rapoport SM. 1982. *Eur. J. Biochem.* 126:353–57
68. Salzmann U, Kühn H, Schewe T, Rapoport SM. 1984. *Biochim. Biophys. Acta* 795:535–42
69. de Groot JJMC, Veldink GA, Vliegenthart JFG, Boldingh J, Wever R, van Gelder BF. 1975. *Biochim. Biophys. Acta* 377:71–79
70. Desmarais S, Riendeau D, Gresser MJ. 1994. *Biochemistry.* In press
71. Verhagen J, Bouman AA, Vliegenthart JFG, Boldingh J. 1977. *Biochim. Biophys. Acta* 486:114–20
72. Clapp CH, Banerjee A, Rotenberg SA. 1985. *Biochemistry* 24:1826–30
73. Kemal C, Louis-Flamberg P, Krupinski-Olsen R, Shorter AL. 1987. *Biochemistry* 26:7064–72
74. Mansuy D, Cucurou C, Biatry B, Battioni JP. 1988. *Biochem. Biophys. Res. Commun.* 151:339–46
75. Riendeau D, Falgueyret J-P, Guay J, Ueda N, Yamamoto S. 1991. *Biochem. J.* 274:287–392
76. Rouzer CA, Riendeau D, Falgueyret J-P, Lau CK, Gresser MJ. 1991. *Biochem. Pharmacol.* 41:1354–73
77. Riendeau D, Denis D, Falgueyret J-P, Percival D, Gresser MJ. 1991. In *Prostaglandins, Leukotrienes, Lipoxins and PAF,* ed. JM Bailey, pp. 31–38. New York: Plenum
78. Falgueyret J-P, Desmarais S, Roy, PJ, Riendeau D. 1992. *Biochem. Cell Biol.* 70:228–36
79. Falgueyret J-P, Hutchinson JH, Riendeau D. 1993. *Biochem. Pharmacol.* 45:978–81
80. McMillan RM, Girodeau J-M, Foster SJ. 1990. *Br. J. Pharmacol.* 101:501–3
81. Bird TGC, Bruneau P, Crawley GC, Edwards MP, Foster SJ, et al. 1991. *J. Med. Chem.* 34:2176–86
82. Hutchinson JH, Prasit P, Choo LY, Riendeau D, Charleson S, et al. 1992. *Bioorg. Med. Chem. Lett.* 2: 1699–702
83. Rapoport SM, Härtel B, Hausdorf G. 1984. *Eur. J. Biochem.* 139:573–76
84. Höhne WE, Kojima N, Thiele B, Rapoport SM. 1991. *Biomed. Biochem. Acta* 50:125–38
85. Skoog MT, Nichols JS, Harrison BL, Wiseman JS. 1988. *Prostaglandins* 36: 373–84
86. Navaratnam S, Feiters MC, Al-Hakim M, Allen JC, Veldink GA, Vliegenthart

JFG. 1988. *Biochim. Biophys. Acta* 956:70–76
87. Van der Heijdt LM, Feiters MC, Navaratnam S, Nolting H-F, Hermes C, et al. 1992. *Eur. J. Biochem.* 207:793–802
88. Boyington JC, Gaffney BJ, Amzel LM. 1993. *Science* 260:1482–86
89. Minor W, Steczko J, Bolin JT, Otwinowski Z, Axelrod B. 1993. *Biochemistry* 32:6320–23
90. Nguyen T, Falgueyret J-P, Abramovitz M, Riendeau D. 1991. *J. Biol. Chem.* 266:22057–62
91. Zhang Y, Rådmark O, Samuelsson B. 1992. *Proc. Natl. Acad. Sci. USA* 89:485–89
92. Ishii S, Noguchi M, Miyano M, Matsumoto T, Noma M. 1992. *Biochem. Biophys. Res. Commun.* 182: 1482–90
93. Steczko J, Muchmore CR, Smith JL, Axelrod B. 1990. *J. Biol. Chem.* 265:11352–54
94. Sloane DL, Leung R, Craik CS, Sigal E. 1991. *Nature* 354:149–52
95. Batt DG. 1992. *Prog. Med. Chem.* 29:1–63
96. Guindon Y, Girard Y, Maycock A, Ford-Hutchinson AW, Atkinson JG, et al. 1987. *Adv. Prostaglandin Thromboxane Leukot. Res.* 17:554–57
97. Guindon Y, Fortin R, Lau CK, Rokach J, Yoakim C. 1984. *Eur. Patent Appl. EP 115,394.*
98. Ancill RJ, Takahashi Y, Kibune Y, Campbell R, Smith JR. 1990. *J. Int. Med. Res.* 18:75–88
99. Fujimura M, Sasaki F, Nakatsumi Y, Takahashi Y, Hifumi S, et al. 1986. *Thorax* 41:955–59
100. Yamamoto S, Yoshimoto T, Furukawa M, Horie T, Watanabe-Kohno S. 1984. *J. Allergy Clin. Immunol.* 74: 349–52
101. Yoshimoto T, Yokoyama C, Ochi K, Yamamoto S, Maki Y, et al. 1982. *Biochim. Biophys. Acta* 713:470–73
102. Sekiya K, Okuda H, Arichi S. 1982. *Biochim. Biophys. Acta* 713:68–72
103. Koshihara Y, Neichi T, Murota S, Lao A, Fujimoto Y, Tatsuno T. 1984. *Biochim. Biophys. Acta* 792:92–97
104. Walenga RW, Showell HJ, Feinstein MB, Becker EL. 1980. *Life Sci.* 27: 1047–53
105. Bokoch GM, Reed PW. 1981. *J. Biol. Chem.* 256:4156–59
106. Levine L. 1983. *Biochem. Pharmacol.* 32:3023–26
107. Miller DK, Sadowski S, Han GQ, Joshua H. 1989. *Prostaglandins Leukot. Essent. Fatty Acids* 38:137–43
108. Bélanger P, Maycock A, Guindon Y, Baca T, Dollob AL, et al. 1987. *Can. J. Physiol. Pharmacol.* 65:2441–48
109. Lau CK, Bélanger PC, Scheigetz J, Dufresne C, Williams HWR, et al. 1989. *J. Med. Chem.* 32:1190–97
110. Lau CK, Bélanger PC, Dufresne C, Scheigetz J, Thérien M, et al. 1992. *J. Med. Chem. 35:1299–1318*
111. *Scrip*. 1988. No. 1306, p. 28
112. Harris RR, Batt DG, Galbraith W, Ackerman NR. 1989. *Agents Actions* 27:297–99
113. Marien K, Morren M, Degreef H, De Doncker P, Rooman RP, Gauwenbergh G. 1992. *Arch. Dermatol.* 128:993–94
114. Bruneau P, Delvare C. 1991. *J. Med. Chem.* 34:1028–36
115. Fort FL, Pratt MC, Carter GW, Lewkowski JP, Heyman IA, et al. 1984. *Fundam. Appl. Toxicol.* 4:216–20
116. Albert DH, Machinist J, Young PR, Dyer R, Bouska J, et al. 1989. *FASEB J.* 3:A735
117. Corey EJ, Cashman JR, Kantner SS, Wright SW. 1984. *J. Am. Chem. Soc.* 106:1503–4
118. Tateson JE, Randall RW, Reynolds CH, Jackson WP, Bhattacherjee P, et al. 1988. *Br. J. Pharmacol.* 94:528–39
119. Summers JB, Gunn BP, Martin JG, Martin MB, Mazdiyasni H, et al. 1988. *J. Med. Chem.* 31:1960–64
120. Payne AN, Jackson WP, Salmon JA, Nicholls A, Yeadon M, Garland LG. 1991. *Agents Actions Suppl.* 34:189–99
121. Chamulitrat W, Mason RP, Riendeau D, 1992. *J. Biol. Chem.* 267:9574–79
122. Knapp HR. 1990. *New Engl. J. Med.* 323:1745–48
123. Israel E, Dermarkarian R, Rosenberg M, Sperling R, Taylor G, et al. 1990. *New Engl. J. Med.* 323:1740–44
124. Collawn C, Rubin P, Perez N, Reyes E, Bobadilla J, et al. 1989. *Am. J. Gastroenterol.* 84:1178 (Abstr. 158)
125. Laursen LS, Naesdal J, Bukhave K, Lauritsen K, Rask-Madsen J. 1990. *Lancet* 335:683–85
126. Weinblatt M, Kremer J, Helfgott S, Coblyn J, Maier A, et al. 1990. *Arthritis Rheum.* 33:(Suppl. 9)S152
127. *Pharmaprojects*. 1993. 14:M10
128. Crawley GC, Dowell RI, Edwards PN, Foster SJ, McMillan RM, et al. 1992. *J. Med. Chem.* 35:2600–9
129. Wada K, Munakata R. 1970. *Tetrahedron Lett.* 23:2017–19
130. Stevenson R, Weber JV. 1989. *J. Nat. Prod.* 52:367–75
131. Thérien M, Fitzsimmons BJ, Scheigetz J, Macdonald D, Choo LY, et al. 1993. *Bioorg. Med. Chem. Lett.* 3: 2063–66

132. Ducharme Y, Brideau C, Chan C, Dubé D, Falgueyret JP, et al. 1993. *ACS Natl. Meet., 206th, Chicago*. Div. Med. Chem. (Abstr. 38)
133. Rouzer CA, Samuelsson B. 1986. *FEBS Lett*. 204:293–96
134. Rouzer CA, Shimizu T, Samuelsson B. 1985. *Proc. Natl. Acad. Sci. USA* 82:7505–9
135. Rouzer CA, Kargman S. 1988. *J. Biol. Chem*. 263:10980–88
136. Gillard J, Ford-Hutchinson AW, Chan C, Charleson S, Denis D, et al. 1989. *Can. J. Physiol. Pharmacol*. 67:456–64
137. Rouzer CA, Ford-Hutchinson AW, Morton HE, Gillard JW. 1990. *J. Biol. Chem*. 265:1436–42
138. Miller DK, Gillard JW, Vickers PJ, Sadowski S, Léveillé C, et al. 1990. *Nature* 343:278–81
139. Dixon RAF, Diehl RE, Opas E, Rands E, Vickers PJ, et al. 1990. *Nature* 343:282–84
140. Kargman S, Vickers PJ, Evans JF. 1992. *J. Cell. Biol*. 119:1701–9
141. Kargman S, Rousseau P, Reid GK, Rouzer CA, Mancini JA, et al. 1993. *J. Lipid Mediat*. 7:31–45
142. Abramovitz M, Wong E, Cox ME, Richardson CD, Li C, Vickers PJ. 1993. *Eur. J. Biochem*. 215:105–11
143. Kargman S, Prasit P, Evans JF. 1991. *J. Biol. Chem*. 266:23745–52
144. Wong A, Cook MN, Hwang SM, Sarau HM, Foley JJ, Crooke ST. 1992. *Biochemistry* 31:4046–53
145. Pueringer RJ, Bahns CC, Monick MM, Hunninghake GW. 1992. *Am. J. Physiol*. 262:L454–58
146. Coffey M, Peters-Golden M, Fantone JC III, Sporn PHS. 1992. *J. Biol. Chem*. 267:570–76
147. Hatzelmann A, Fruchtmann R, Mohrs KH, Raddatz S, Müller-Peddinghaus R. 1993. *Biochem. Pharmacol*. 45: 101–11
148. Hatzelmann A, Fruchtmann R, Mohrs KH, Matzke M, Pleiss U, et al. 1994. *Advances in Prostaglandin, Thromboxane and Leukotriene Research*, ed. B Samuelsson, SE Dahlén, J Fritsh, P Hedqvist. New York: Raven. In press
149. Mancini JA, Abramovitz M, Cox ME, Wong E, Charleson S, et al. 1993. *FEBS Lett*. 318:277–81
150. Hill E, Maclouf J, Murphy RC, Henson PM. 1992. *J. Biol. Chem*. 267:22048–53
151. Kennedy BP, Diehl RE, Boie Y, Adam M, Dixon RAF. 1991. *J. Biol. Chem*. 266:8511–16
152. Reid GK, Kargman S, Vickers PJ, Mancini JA, Léveillé C, et al. 1990. *J. Biol. Chem*. 265:19818–23
153. Bennett CF, Chiang M-Y, Monia BP, Crooke ST. 1993. *Biochem. J*. 289:33–39
154. Jakobsson P-J, Steinhilber D, Odlander B, Rådmark O, Claesson H-E, Samuelsson B. 1992. *Proc. Natl. Acad. Sci. USA* 89:3521–25
155. Coffey MJ, Gyetko M, Peters-Golden M. 1993. *J. Lipid Mediat*. 6:43–51
156. Woods JW, Evans JF, Ethier D, Scott S, Vickers PJ, et al. 1993. *J. Exp. Med*. 178:1935–46
157. Depre M, Friedman B, Tanaka W, Van Hecken A, Buntinx A, DeSchepper PJ. 1993. *Clin. Pharmacol. Ther*. 53:602–7
158. Friedman BS, Bel EH, Buntinx A, Tanaka W, Han Y-HR, et al. 1993. *Am. Rev. Respir. Dis*. 147:839–44
159. Evans JF, Léveillé C, Mancini JA, Prasit P, Thérien M, et al. 1991. *Mol. Pharmacol*. 40:22–27
160. Charleson S, Prasit P, Léger S, Gillard JW, Vickers PJ, et al. 1992. *Mol. Pharmacol*. 41:873–79
161. Mancini JA, Prasit P, Cappolino MG, Charleson P, Leger S, et al. 1992. *Mol. Pharmacol*. 41:267–72
162. Brideau C, Chan C, Charleson S, Denis D, Evans JF, et al. 1992. *Can. J. Physiol. Pharmacol*. 70:799–807
163. Vickers PJ, O'Neill GP, Mancini JA, Charleson S, Abramovitz M. 1992. *Mol. Pharmacol*. 42:1014–19
164. Vickers PJ, Adam M, Charleson S, Coppolino MG, Evans JF, Mancini JA. 1992. *Mol. Pharmacol*. 42:94–102

Annu. Rev. Biochem. 1994. 63:419–50

THE EXPRESSION OF ASYMMETRY DURING CAULOBACTER CELL DIFFERENTIATION

Yves V. Brun
Department of Biology, Indiana University, Bloomington, Indiana 47405

Greg Marczynski and Lucille Shapiro
Department of Developmental Biology, Stanford University School of Medicine, Stanford, California 94305-5427

KEY WORDS: cell division, flagellum, polarity, replication, cell cycle

---

CONTENTS

At the top of each stalk was a cell, and, as the fluid currents shifted, the cells and stalks waved much as tall grass in the wind (1),

0066-4154/94/0701-0419$05.00

INTRODUCTION

Asymmetric cell divisions are critical events in the differentiation and development of a large number of organisms. This is the case in organisms as different as *Caulobacter crescentus, Bacillus subtilis,* yeast, *Drosophila, Caenorhabditis elegans,* and mammals. The question of how cellular asymmetry is generated is thus central to our understanding of the development of both prokaryotic and eukaryotic biological systems.

The bacterium *Caulobacter crescentus* dedicates the major portion of its life cycle to the expression of asymmetry in its predivisional cell. Each cell division produces progeny cells that differ in structure, transcriptional properties, and chromosome replicative ability. The predivisional cell has distinct structures at each pole: a flagellum and several pili at one pole and a stalk at the other (Figure 1). The progeny swarmer cell has a single polar

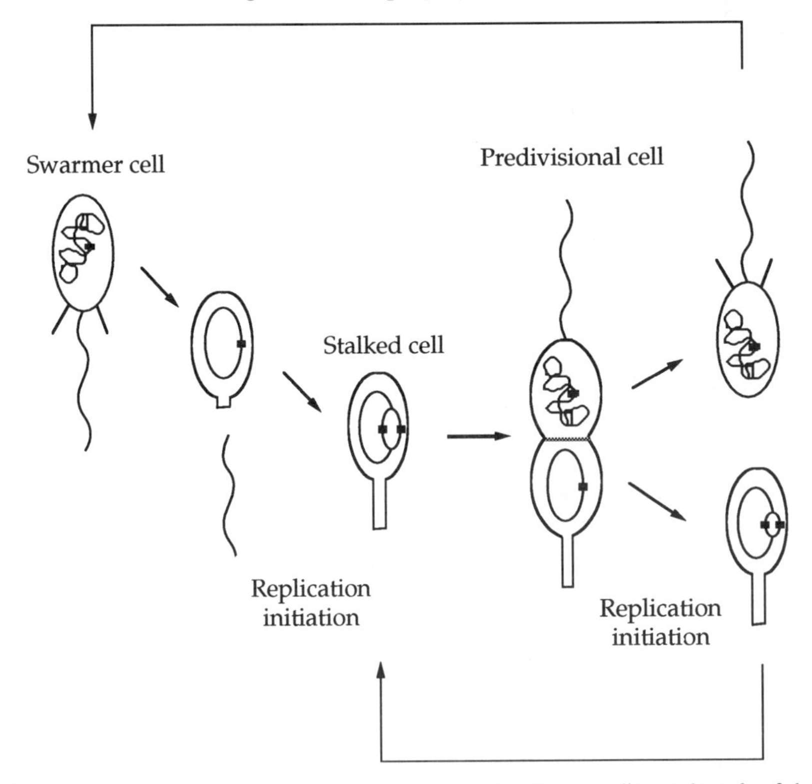

Figure 1 The *Caulobacter crescentus* cell division cycle. The wavy line at the pole of the swarmer cell represents the single polar flagellum and the adjacent straight lines represent polar pili. The stalk grows at the site previously occupied by the flagellum. The structures within cells represent the condensed and open chromosomes. The solid box on the chromosome indicates the origin of DNA replication.

flagellum and is chemotactically competent, whereas the stalked cell is sessile. *C. crescentus* progeny cells have different proliferative capacities: The stalked cell initiates chromosomal replication immediately upon cell division, whereas the swarmer cell delays chromosomal replication until it differentiates into a stalked cell later in the cell cycle (2, 3). What are the molecular events in the predivisional cell that cause cellular polarization, yielding different progeny cells?

During the swarmer-to-stalked cell transition, the flagellum is released, pili are lost, and a stalk is formed at the site previously occupied by the flagellum. At the same time, the structure of the chromosome changes and chromosome replication is initiated. The stalked cell effectively acts as a stem cell, expressing asymmetry so as to generate a swarmer cell from the opposite cell pole. Because cultures of *C. crescentus* can be easily synchronized, the mechanisms that control polar morphogenesis can be examined in individual cell types throughout the cell cycle.

In this review we consider the mechanisms used to express asymmetry. These involve protein targeting, temporal and spatial regulation of gene expression, polar organelle formation, differential initiation of DNA replication, and DNA methylation.

PROTEIN LOCALIZATION

The de novo assembly of the flagellum and the chemosensory apparatus at the pole opposite the stalk is a clear demonstration of protein localization. In the case of the flagellum, large numbers of proteins are targeted to the cell pole and assembled into a complex structure. A simpler model for protein localization is presented by the chemosensory proteins that are positioned at the incipient swarmer pole of the predivisional cell. The *C. crescentus* chemosensory proteins and their spatial disposition during the cell cycle represent a prototype system for the localization of a functionally related group of proteins.

Chemoreceptor Polar Targeting

The *C. crescentus* chemotaxis apparatus, as is the case in *Escherichia coli,* consists of chemoreceptor proteins (MCPs) and proteins involved in the transduction of a signal to the flagellum. The chemotaxis proteins are synthesized in the predivisional cell (4–6) but appear only in the progeny swarmer cell upon division (6, 7). This was first demonstrated by enzymatic assays of methyl-accepting, carboxymethyltransferase, and carboxymethylesterase activity in reconstituted fractions from synchronized cells. These experiments established that the chemotaxis machinery is present only in the progeny cell type that is capable of responding to chemotactic signals (the swarmer).

The transcription of a large operon containing a gene encoding the McpA chemoreceptors, as well as the phosphoproteins involved in signal transduction to the flagellar motor, occurs only in the predivisional cell (5). Chemotaxis proteins pulse-labeled in the predivisional cell are selectively localized in the progeny swarmer cell (6, 7). Experiments with polar membrane vesicles isolated from predivisional cells gave the first indication that the chemoreceptors might be localized to the incipient swarmer pole prior to division (8). Immunoaffinity chromatography using antiflagellin antibody (9) allowed a separation of flagellated and nonflagellated vesicles, and nascent chemoreceptors were found to be preferentially located in the flagellated vesicles (8). However, the precise localization of the chemoreceptors was only possible once the gene encoding McpA was cloned and antibodies obtained to the McpA protein (6). Immunogold electron microscopy of sectioned cells revealed that the chemoreceptors are present only at the flagellated pole of the predivisional cell and, upon division, at the flagellated pole of the swarmer cell (6, 10). The chemoreceptors form a small cap at the cell pole bearing the flagellum. The polar location of chemoreceptors is not unique to *C. crescentus* or polarly flagellated bacteria; chemoreceptors are similarly clustered at the cell poles in *E. coli,* though flagella are distributed over the entire surface (11).

Information about the mechanism of chemoreceptor targeting in the bacterial cell has come, in part, from *E. coli,* where the structure and function of the chemotaxis sensory transduction pathway has been elegantly described (12–14). A ternary complex of the membrane chemoreceptor and two cytoplasmic proteins, the CheA histidine phosphokinase and CheW, has been shown to occur in vitro with purified components (15–18), and in vivo using genetic analysis of allele-specific suppressors of chemoreceptor mutants (19). All three members of this ternary complex were shown by immunogold electron microscopy and indirect immunofluorescence light microscopy to be preferentially located at the pole of the *E. coli* cell (11). It was shown that it is the complex of membrane and cytoplasmic proteins that is required for polar localization. The sequence of the region on the chemoreceptor protein that interacts with the CheA and CheW proteins is conserved among all chemoreceptors that have been analyzed, including the *C. crescentus* McpA (6). When this highly conserved domain (HCD) is deleted from McpA, the chemoreceptors are not localized to the pole of the cell (6, 10; JR Maddock, MRK Alley, L Shapiro, unpublished). Possible models for targeting the chemoreceptors to the pole are (*a*) that they are initially inserted in the membrane only at the cell pole via a pole-specific secretory apparatus, and (*b*) that they are initially inserted randomly over the cell surface and then migrate to the cell pole where they are retained. In either case, complex formation with CheA and CheW could function to retain the chemoreceptors at the cell pole. A polar structure termed the "periseptal annulus" has been identified in *E. coli* by Rothfield and coworkers (20, 21). This structure might contribute to polar constraints of the chemoreceptor complex.

Cell Type–Specific Chemoreceptor Turnover

The chemoreceptors are lost when the swarmer cell differentiates into the stalked cell. This loss coincides with the release of the flagellum, suggesting that both events may be caused by a regulated proteolytic process (10). When an *E. coli* chemoreceptor gene, *tsr,* is expressed in *C. crescentus,* its transcription is temporally controlled in parallel with the endogenous *mcpA* gene (22). However, unlike the *C. crescentus* McpA chemoreceptor, Tsr is not turned over during the swarmer-to-stalked cell transition (6; JR Maddock, MRK Alley, L Shapiro, unpublished). This observation suggests that the *C. crescentus* chemoreceptor is a special substrate for a regulated proteolytic event. Comparison of the predicted protein sequences of the *C. crescentus* and *E. coli* chemoreceptors (6, 10) revealed a unique C-terminal region in the *C. crescentus* McpA chemoreceptor. A small deletion of 14 carboxy-terminal amino acids from McpA yielded a chemoreceptor that was not degraded during the swarmer-to-stalked cell transition, indicating that a sequence at the C terminus of McpA is essential for cell type–specific degradation. The terminally deleted McpA chemoreceptor was nevertheless targeted to the *C. crescentus* cell pole, arguing that proteolysis is not required for polar localization. Thus, the appearance of the chemoreceptors at the cell pole is not due to random insertion into the cytoplasmic membrane followed by degradation of chemoreceptors at the sides of the cell. These results also suggest that the stalked cell pole is able to retain chemoreceptors if a signal for turnover has been eliminated. Overexpression of the wild-type chemoreceptor results in saturation of the swarmer pole in the predivisional cell and the appearance of chemoreceptors at the stalked pole of these cells. Upon division, the chemoreceptors left at the stalked pole are selectively degraded, arguing that a cell type–specific protein degradation mechanism helps maintain asymmetry in the *C. crescentus* cell. Either the C-terminal domain of the McpA protein is selectively modified at the stalked cell pole, thus making it a substrate for a ubiquitous protease, or a protease that recognizes the McpA C-terminal sequence is present or is activated solely at the stalked cell pole. Additional proteins that are specifically turned over during the swarmer-to-stalked cell transition include the carboxymethyltransferase and the carboxymethylesterase (7). It may well be that other morphogenetic events that occur during this transition, such as the release of the flagellum and the marked change in sedimentation coefficient of the nucleoid (23–26), are also initiated by spatially constrained and targeted proteolytic events.

ASYMMETRY OF POLAR ORGANELLE BIOGENESIS

The two distinct polar organelles on *C. crescentus* cells, the flagellum and the stalk, are not synthesized de novo in the same cell (Figure 1). The flagellum is synthesized and assembled in the predivisional cell at the pole opposite that bearing an existing stalk. It is not known what marks this pole as the site of flagellum assembly, but the site is invariant. The progeny swarmer cell sheds its flagellum later in the cell cycle, and it is at the site vacated by the flagellum that the new stalk is assembled. The asymmetric distribution of proteins used to build these structures is controlled in part by the temporally and spatially regulated transcription of the genes encoding these proteins. In turn, the regulated expression of these genes is a function of key cell-cycle events, including chromosome replication and cell division, that help establish the timing of stalk and flagellum biosynthesis.

The Flagellum

STRUCTURE The *C. crescentus* flagellum is a highly conserved structure that is composed of three subassemblies. These include a motor (basal body) that traverses the cell envelope (27), a universal joint (hook) (28), and a helical filament (propeller) (29). The basal body, made up of several rings threaded on a rod (Figure 2), is the first substructure to be assembled (30). The M ring embeds the motor in the cytoplasmic membrane, while the P and L rings are located in the vicinity of the peptidoglycan and the outer membrane, respectively. Although electron micrographs of the basal body reveal a ring (the S ring) directly above and in apposition to the M ring (Figure 2), in *Salmonella typhimurium* this structure is apparently part of the M ring and the protein subunit of both rings is encoded by the *fliF* (formerly called *flaO*) gene (31). The genes encoding the *C. crescentus* homologs of the protein subunits of these rings, *fliF* (M ring), *flgI* (formerly called *flaP*, encoding the P ring), and *flgH* (formerly called *flbN*, encoding the L ring), have all been cloned and sequenced (G Ramakrishnan, JL Zhao, A Newton, personal communication; 32, 33). In addition to these rings, the *Caulobacter* basal body has a ring between the M+S and P rings, which is referred to as the E ring (27). The rod upon which these rings are mounted has two parts, a proximal rod and a distal rod, and the genes encoding the *C. crescentus* homologs of the protein subunits of these rings, *flgF* (formerly called *flaG*) and *flgG* (formerly called *flaB*), have been isolated (34).

The external components of the flagellum are assembled next. The hook

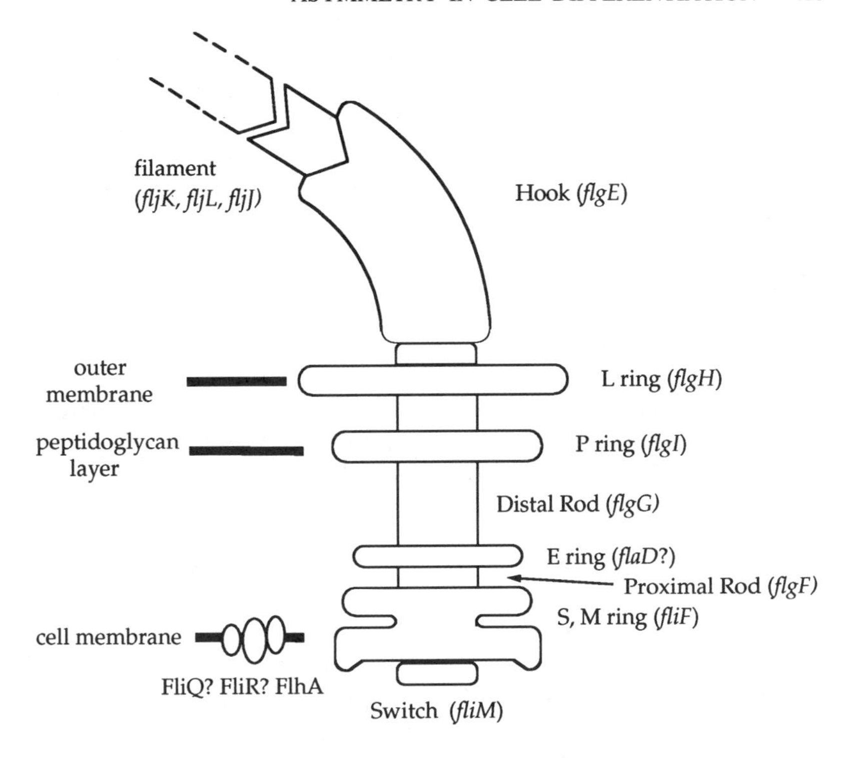

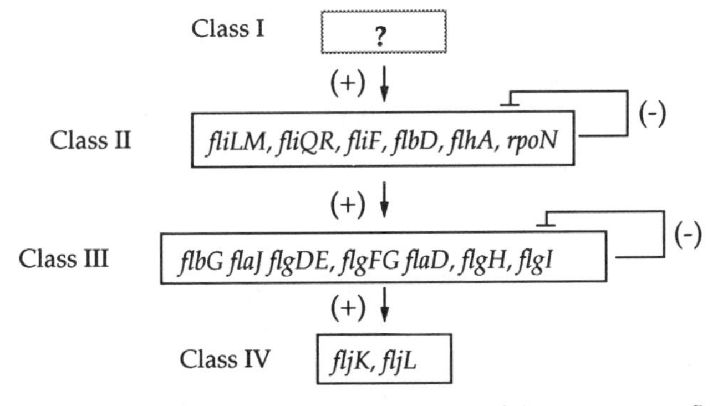

Figure 2 The top diagram is a schematic representation of the *C. crescentus* flagellar basal body. The name for each structure is accompanied by the gene designation, shown in parentheses. The bold lines represent the layers of the cell envelope and their location with respect to the basal body (34). A schematic of the flagellar regulatory hierarchy is shown below. Arrows indicate a positive epistatic interaction between the classes of flagellar genes. Several gene names have been changed to correlate with the nomenclature of *E. coli* and *S. typhimurium* flagellar genes (G Ramakrishnan, JL Zhao, A Newton, personal communication).

is assembled by adding 70-kDa subunits of FlgE (formerly called FlaK) (35–37) through a central pore in the rod. It is thought that the subunits traverse the pore of the growing hook and add on to the distal portion of the hook. Although a hook-filament junction has been identified in *E. coli* and *S. typhimurium* flagella, the hook-associated proteins (HAPs) presumed to assemble in this region have only been identified in *C. crescentus* by sequence analysis of cloned genes. The *C. crescentus* flagellar filament, which assembles next, differs from the filament in *E. coli* and *S. typhimurium* in that each filament is composed of three distinct flagellin subunits (38–42). Immunoelectron microscopy has revealed the order of the flagellin subunits and the length of filament occupied by each subunit (43).

How is such a complex structure, requiring the sequential assembly of at least 15 proteins from the innermost ring to the flagellins, synthesized specifically at one pole of the cell? How is expression of the flagellar genes regulated such that their products are available at the right time and at the right place to build the flagellum? The next sections review the current knowledge of the temporal and spatial regulation of flagellar gene expression.

TRANSCRIPTIONAL REGULATION OF FLAGELLAR GENES Approximately 50 genes are required for the assembly of the flagellum (44). As summarized in Figures 2 and 3, epistasis experiments have established that most flagellar genes can be grouped into four classes (45–49; for reviews see Refs. 50–52). The *C. crescentus* flagellar regulatory hierarchy is similar in concept to the flagellar hierarchies of *E. coli* and *S. typhimurium* in that the expression of genes in an earlier class is required for the expression of the genes in the subsequent classes (Figure 2). Class I is still hypothetical and consists of genes that would be required for the expression of Class II genes (53). Class II genes are expressed relatively early in the cell cycle, shortly after the swarmer-to-stalked cell transition, and they are required for the expression of Class III and Class IV genes. For example, Ohta et al (54) showed that delaying expression of the Class II *fliF* operon delays the expression of the Class III *flbG* operon. The flagellin genes compose the lowest level of the regulatory hierarchy, and their expression is dependent on Class III genes. Studies have established that the resulting periodic expression of flagellar genes is controlled mostly at the transcriptional level (26, 52). Furthermore, the genes that encode the structural components of the flagellum are transcribed in the order that their gene products are assembled into the nascent structure. The promoter elements that are responsible for temporally controlled transcription of these flagellar genes have been identified. In addition, some of the genes encoding flagellar regulatory proteins have been identified.

Comparison of flagellar gene promoters has revealed common sequence

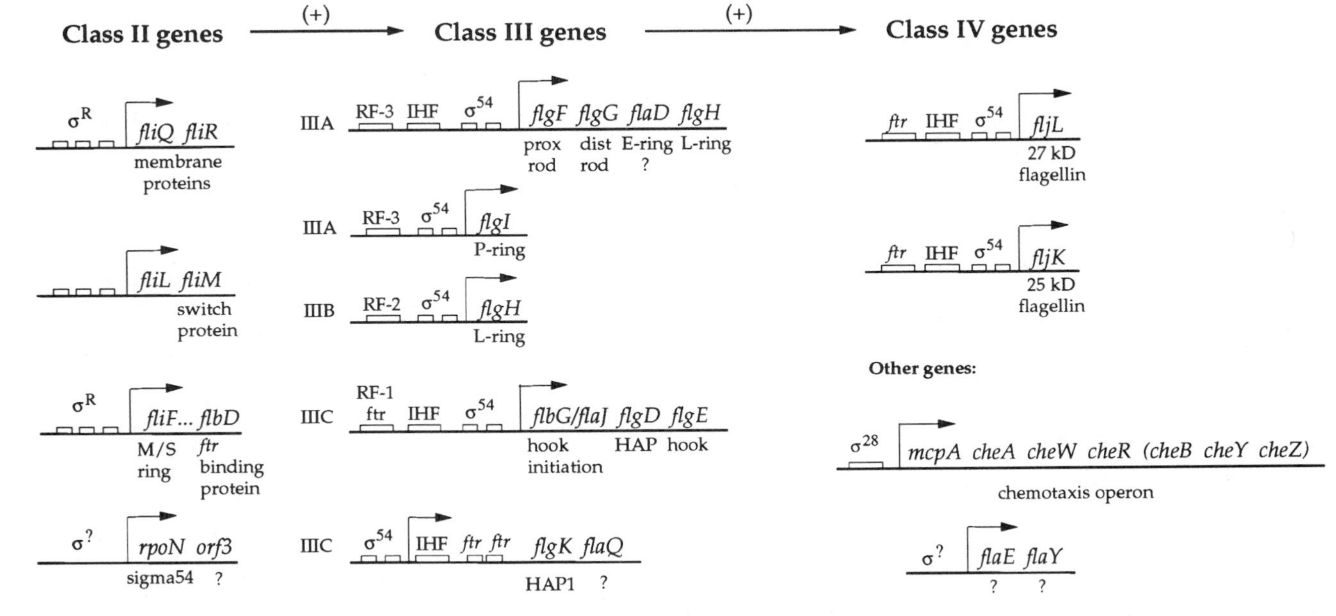

Figure 3 A summary of the genes within each class of the flagellar regulatory hierarchy. The organization of the promoter region is shown for each gene or operon.

motifs in each class. Almost all the genes in classes III and IV possess a sequence similar to the consensus for promoters requiring the alternative sigma factor, σ^{54}, for their transcription (32–34, 55–59). These promoters can be recognized by *E. coli* σ^{54} in vitro (60) and in vivo (33, 61), and are dependent on the presence of σ^{54} for their expression in *C. crescentus* (62; see next section). A characteristic of all known bacterial σ^{54}-dependent promoters is that the σ^{54}-RNA polymerase holoenzyme, even though it can bind to the promoter, is unable to catalyze the isomerization of the closed transcription complex to the open transcription complex. To do so, the holoenzyme must interact with a transcriptional activator that usually binds to an enhancer sequence approximately 100–150 bp upstream of the transcription start site (63). Three different enhancer sequences have been identified in *C. crescentus* σ^{54}-dependent flagellar promoters (Figure 3). One such conserved sequence element, called *ftr* for "flagellar transcription regulation" (57), is located at approximately 100 bp upstream of the promoters of the hook operon and of the flagellin genes *fljK* and *fljL* (formerly called *flgK* and *flgL*), and downstream of the *flgK* (formerly called *flaN*) promoter. Mutational analysis established that the *ftr* elements are required for promoter activity (56–59, 64). The promoter for the *flgH* gene, encoding the L ring, possesses a different activator site (RF-2) (33). Finally, the *flgI* (P ring) and *flgF* operon (rods) promoters have yet another conserved sequence element, approximately 100 bp upstream of the transcription start site, which has been hypothesized to be an activator-binding sequence (32, 34). The genes that share a common enhancer sequence are transcribed at approximately the same time in the cell cycle. This suggests that the precise timing of the transcription of Class III and Class IV flagellar genes is accomplished by the utilization of different transcriptional activators (34).

Using gel retardation experiments, two different *ftr*-binding proteins have been detected, a 95-kDa protein, RF-1, and a 55-kDa protein (26, 59). The binding activity of these two proteins is temporally controlled and is maximal during the period in the cell cycle when flagellar genes are expressed at their highest level (26, 59). The identity of the genes encoding these two proteins is unknown. Another *ftr*-binding protein is encoded by the *flbD* gene. This gene is required for the expression of Class III and Class IV genes (see Figure 2; 49), and its product can activate transcription of both *E. coli* and *C. crescentus* σ^{54}-dependent promoters in *E. coli* (61). *flbD* encodes a protein with a predicted size of 52 kDa that is homologous to transcriptional activators of σ^{54}-dependent promoters of different bacteria (61). A purified FlbD fusion protein has been shown to bind to all known *ftr* sequences (65). Although both the transcription of *flbD* (66) and the synthesis of the FlbD protein (65) vary in a cell cycle–dependent manner, the level of FlbD varies only slightly during the cell cycle. This indicates that it is not the level of FlbD that is responsible for the temporal expression of *ftr* promoters. The enhancer-binding transcriptional activators of σ^{54}-dependent promoters in other bacteria often require phosphorylation by another protein in order to be able to activate transcription (63, 67). It is therefore possible that it is the activated FlbD and not the absolute amount of protein that controls the expression of *ftr*-containing flagellar genes. The activity of FlbD would be dependent on its phosphorylation by a sensor protein that transduces cell cycle–dependent cues. Indeed, ^{32}P labelled FlbD can be immunoprecipitated from extracts of *C. crescentus* cells grown in the presence of $^{32}PO_4$, and cell extracts can catalyze the phosphorylation of purified FlbD in vitro (65). Furthermore, the kinase activity is expressed in a cell cycle–dependent manner and is maximal when the *ftr*-containing promoters are expressed (65). Thus, the phosphorylation of FlbD is probably involved in both the temporal control of flagellar genes and in the control of the spatial transcription of late flagellar genes (see later section). However, the complex regulation of flagellar genes is also subject to other forms of control (see next paragraph on IHF and next section on the role of *rpoN*).

As is the case for other bacterial σ^{54}-dependent promoters (68, 69), *C. crescentus* σ^{54}-dependent flagellar promoters can possess a binding site for IHF between the promoter and the enhancer sequence (70). IHF is a small heterodimeric DNA-bending protein initially identified as a host requirement for bacteriophage lambda integration (reviewed in 71). Mutagenesis of the putative IHF-binding sites of several flagellar promoters has shown that they are required for maximal gene expression, and gel retardation and footprinting experiments have shown that IHF from *E. coli* binds specifically to these sequences (70). Western blot analysis using antibody raised against *E. coli* IHF indicates that *C. crescentus* possesses two cross-rcacting proteins of similar size to those of the *E. coli* IHF subunits (59). These two proteins were partially purified and were shown to substitute for *E. coli* IHF in an in vitro lambda recombination assay (59). The level of the putative IHF proteins was also shown to be under cell-cycle control, being maximal in predivisional cells when Class III and IV flagellar genes are expressed (59).

The *flbD* (encoding the *ftr*-binding transcriptional activator) and *rpoN* (encoding σ^{54}) genes have been placed among the flagellar Class II genes on the basis of epistasis experiments (Figure 2; 48, 49, 62). Class II genes are not all regulatory genes. Most nonregulatory Class II genes identified to date encode proteins that are required early in flagellar biosynthesis. *fliM* encodes a homolog of a *S. typhimurium* switch protein (72) shown to assemble in a "C-ring complex" residing at the cytoplasmic face of the M ring (73–75). *fliF* (the first gene of the operon that also encodes *flbD*) encodes the first basal body ring to be assembled, the M ring (G Ramakrishnan, JL Zhao, and A Newton, unpublished), *fliQR* encodes integral membrane proteins required for the initiation of flagellar assembly (76; W Zhuang, L Shapiro, in preparation), and FlhA (formerly called FlbF) is an integral cytoplasmic membrane protein (W Zhang, D Mullin, personal communication) homologous to a *Yersinia enterolytica* virulence protein, LcrD, which is a putative Ca^{2+}-binding protein (77–80). Thus, except for the two known transcription factor genes, *flbD* and *rpoN,* the requirement of the Class II genes for the expression of genes lower in the regulatory hierarchy is probably due to checkpoints for assembly generated by the early steps in flagellum biogenesis.

Since Class II genes initiate the known flagellar regulatory hierarchy, their promoters are being extensively studied in order to identify sequence elements important for their cell cycle–dependent regulation. This information will allow access to the proposed Class I regulatory proteins and signals involved in initiating the flagellar hierarchy. The sequences directly upstream of the transcription start sites of the *fliQR, fliLM,* and *fliF* operons show extensive conservation (53, 72, 76, 81). This conserved sequence element has no obvious similarity to previously identified bacterial promoters. Deletion analysis of the *fliLM* promoter has established that sequences between −43 and +12 are sufficient for wild-type promoter activity and temporal regulation (81). Mutational analysis has shown that most of the conserved bases are functionally important for the *fliLM* promoter (81) and the *fliQR* promoter (W Zhuang and L Shapiro, in preparation). Finally, the negative effects of point mutations and deletions in the *fliF* promoter indicate that the equivalent region is necessary for normal promoter activity (53). In addition to these conserved sequences, the *fliF* promoter contains a negative regulatory element called *ftr4,* adjacent to the transcription start site. The *ftr4* element could be recognized by FlbD and thus turn off the *fliF* operon at the correct time in the cell cycle (53). Part of the conserved Class II regulatory sequence, from −22 to −29 bp, has also been found in the region directly upstream of the *rpoN* gene, although it is not yet known if it is required for its transcription (YV Brun, L Shapiro, unpublished results). In contrast to other Class II promoter regions, that of *flhA* exhibits only limited similarity to this sequence (53).

Although the order of flagellar gene expression reflects the order that their gene products are assembled into the flagellum, it is not known if the temporal regulation of transcription is essential for normal flagellum biogenesis. The transcription of the Class II flagellar genes is coupled to cell-cycle events (72, 76, 81). Completion of DNA replication and cell division may be required in order to ensure that the polar site destined to accept the flagellum is ready to do so. In fact, analysis of cell-cycle mutants by Huguenel & Newton (82) suggests that early steps in cell division must be completed in order for formation of new flagellum assembly sites. Since the Class II genes are among the earliest known flagellar genes to be expressed during the cell cycle and encode the first proteins to be assembled, they are the genes most likely to be intimately coupled to the cell cycle. Indeed, mutations in almost all known Class II genes result in aberrant cell division, and the activity of at least three Class II promoters, for the *fliQR*, *fliLM*, and the *fliF* operons, is sensitive to an interruption in DNA replication (76, 81). Thus, the observed inhibition of Class III and Class IV flagellar protein synthesis by an interruption of DNA replication (40, 83) is probably due to its effect on the transcription of Class II genes. Extensive screens are in progress to identify "Class I" genes that would be required for the expression of Class II genes. It is possible that such genes are involved in coordinating cell-cycle events and morphogenesis, and thus might be essential for viability.

As seen in Figure 2, the *rpoN* gene has been included in the Class II flagellar genes by virtue of its requirement for the expression of Class III and IV flagellar genes. However, the *rpoN* gene differs from other Class II genes in the scope of cellular events dependent on its expression (62). In the next section, we examine the role of this sigma factor gene in the control of polar morphogenesis during cell division.

THE ROLE OF *rpoN* IN CELLULAR MORPHOGENESIS A common feature of most Class III and Class IV flagellar genes is that their promoters require σ^{54} for recognition by RNA polymerase. Does the variation of σ^{54} levels contribute to the temporal regulation of flagellar gene expression? Is this sigma factor required for other cellular functions, perhaps also related to the expression of asymmetry? The *rpoN* gene, encoding σ^{54}, was cloned (62), and mutants in *rpoN* were shown to be nonmotile and to have morphological defects in addition to the absence of a flagellum: They lack stalks and they are filamentous (62, 84). The expression of the σ^{54}-dependent Class III and Class IV flagellar genes is absolutely dependent on a wild-type *rpoN* gene, indicating that *rpoN* is the only σ^{54} gene in *C. crescentus* (62). σ^{54} is not required for the expression of the Class II flagellar genes, nor for the expression of the *mcpA* operon predicted to use σ^{28} (5, 22). Although σ^{54} is used for the expression of genes involved in numerous metabolic functions in other bacteria (67), it is not required for general metabolic functions in *C. crescentus* (62).

The stalkless phenotype of *rpoN* mutants is partially suppressed by starvation for phosphate. Phosphate limitation has a dramatic effect on stalk biosynthesis (85). Stalks become progressively longer as phosphate is diluted, and can attain lengths of 20 μm under conditions of phosphate starvation (85). This effect of phosphate concentration on stalk biosynthesis occurs at the stalk elongation stage and does not seem to affect stalk initiation, since cells limited for phosphate still undergo the proper series of developmental steps (85). Phosphate starvation seems to trigger similar regulatory mechanisms in *C. crescentus* as it does in other gram-negative bacteria. The *oprP* gene encoding a phosphate-starvation-inducible porin from *Pseudomonas aeruginosa* responds normally to phosphate starvation when introduced into *C. crescentus* (86). The OprP protein was synthesized when *C. crescentus* cells were starved for phosphate, and its induction was coincident with the induction of stalk elongation, suggesting that *C. crescentus* possesses the transcriptional regulatory elements necessary for the recognition of the "pho box" of the *oprP* promoter (86). Thus, *C. crescentus* probably contains a pho regulatory system similar to that of other bacteria, and it is possible that the induction of stalk elongation by phosphate starvation is controlled by this system. Partial suppression of the stalkless phenotype of *rpoN* mutants by phosphate starvation implies that transcription of genes encoding enzymes directly involved in stalk elongation cannot be completely dependent on *rpoN*.

The finding that the transcription of *rpoN* is under cell-cycle control provides some insight into possible global regulatory mechanisms that mediate differentiation in *C. crescentus*. Transcription of *rpoN* is low in swarmer cells and increases 10-fold commensurate with stalk biosynthesis and just prior to the onset of flagellum biosynthesis (62). Its transcription is maximal during the period when most flagellar genes are transcribed, and then decreases towards the end of the cell cycle, prior to cell division. This transcription pattern may result in a parallel cell-cycle variation of the level of the σ^{54} protein, thus restricting the expression of σ^{54}-dependent genes to the corresponding period of the cell cycle. For example, in a *flbD* mutant in which the FlbD transcription factor should constitutively activate transcription independent of its phosphorylation state, the transcription of the σ^{54}-dependent *fljK* flagellin promoter is still under cell-cycle control, although it is somewhat relaxed (65). This residual *fljK* control could result from the temporal control of *rpoN* transcription and also the cell-cycle variation of IHF protein levels (59).

Thus, in *C. crescentus,* σ^{54} contributes to the control of genes involved in building the polar structures. It is, however, the transcriptional activators that are responsible for the fine tuning of the expression of flagellar genes as described in the previous section, and probably for the cell type–specific transcription of certain flagellar genes described in the next section.

SPATIAL REGULATION OF LATE FLAGELLAR GENE EXPRESSION Whereas Class II (early) genes are expressed from the replicating chromosome, a subset of Class III and Class IV (late) genes are expressed after chromosome replication is completed. These genes, including the hook operon (*flbG*),

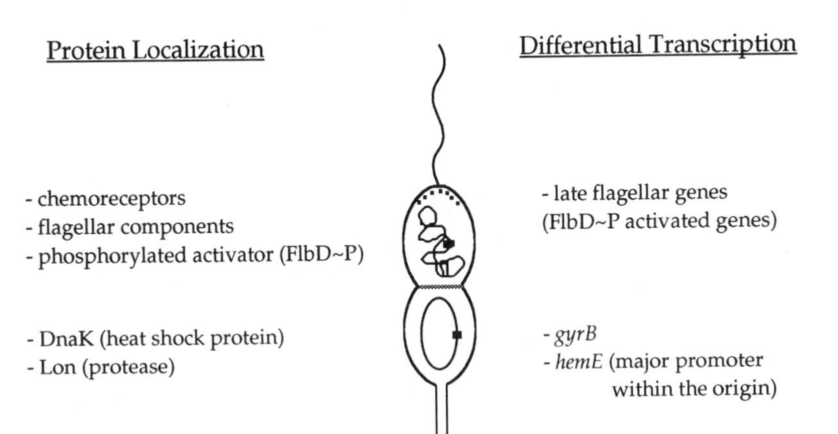

Figure 4 The proteins that are localized to a specific portion of the predivisional cell are shown adjacent to either the flagellated or stalked part of the cell (*left*). The genes that are transcribed from the chromosome in either the flagellated or stalked part of the predivisional cell are shown at the appropriate site to the right of the cell.

the 25-kDa flagellin gene (*fljK*), and the *flgKflaQ* (*flgK* was formerly called *flaN*) operon, appear to be transcribed only from the chromosome residing in the swarmer portion of the predivisional cell (59, 65, 87; Figure 4).

Investigation of polarized transcription began in 1983 when Milhausen & Agabian (88) reported that flagellin mRNA was synthesized in the predivisional cell and then selectively partitioned to the progeny swarmer cell. Since the halflife of these mRNAs is relatively long (~15 min), selective transcription from the chromosome in the incipient swarmer pole of the predivisional cell would provide a mechanism to target flagellar protein synthesis to the correct cell pole. Protein targeting, as described earlier for the localization of the chemoreceptors, and pole-specific transcription need not be mutually exclusive events. However, pole-specific transcription requires compartment barriers to prevent diffusion of the mRNA to the other half of the cell. Recently, electron micrographs have revealed an ordered region at the incipient division plane that excludes cytoplasmic proteins marked by colloidal gold–conjugated antibodies, and suggests that a barrier exists between the two halves of the predivisional cell well before division is completed (JR Maddock, L Shapiro, unpublished). This "barrier" may physically separate the newly replicated chromosomes.

There are at least three possible basic mechanisms to direct mRNA compartmentalization in the predivisional cells: (*a*) *mRNA targeting model*: Flagellar mRNA could be sequestered to a target at the swarmer pole, ensuring localized translation. (*b*) *mRNA stability model*: The mRNA, transcribed in both poles, could be selectively degraded in the stalked pole compartment. (*c*) *Localized transcription model*: The mRNA could be preferentially transcribed from the chromosome in the swarmer pole compartment of the predivisional cell. In the last case, signals in the promoter region of the gene could be recognized by spatially constrained transcription factors. Using constructs that delete the entire mRNA region containing the flagellar gene sequences, but that fuse Class III or Class IV flagellar promoter regions directly to an intact reporter gene, it was demonstrated that the reporter proteins, pulse-labeled in synchronized predivisional cells, were preferentially localized in the progeny swarmer cells (87). The first two mechanisms, which require the presence of flagellar mRNA, could not account for the selective appearance of the reporter gene product in the swarmer cell. Selective localization in the swarmer cell depends solely on the upstream regulatory region of these genes (59, 87). In other words, the presence of the reporter gene protein in the swarmer cell, but not in the stalked cell progeny, is a biochemical "snapshot" of transcription in the predivisional cell (when labeling occurred), suggesting that the chimeric gene is specifically transcribed from the chromosome residing in the "swarmer" compartment of the predivisional cell. As is described later in this review, the *hemE* and *gyrB* genes, which are associated with DNA replication, are preferentially transcribed from the chromosome residing in the stalked portion of the predivisional cell (89, 90).

Differential transcription of the chromosomes in the swarmer and stalked cell compartments of the predivisional cell implies the presence of compartmentalized transcription factors. These might include sigma factors, transcription activators [as has been observed in forespore and mother cells during *Bacillus subtilis* sporulation (91)], and unique chromosomal structures inferred in *C. crescentus* by the significant differences in the sedimentation coefficients of the two chromosomes. For example, differences in the tertiary structure of the chromosomes might make some genes more accessible to transcription factors. The promoter organization of the late Class III and IV genes suggests that all three factors may, in fact, combine to direct the swarmer pole–specific transcription of these genes. All of these genes have a σ^{54} promoter and RF-1 upstream enhancer elements separated by an integration host factor (IHF)–binding site (26, 59, 70). As discussed earlier, the transcriptional activator of these genes, the product of the *flbD* gene, is present in both the swarmer and stalked cell compartments. However, only the phosphorylated form of the FlbD protein is active, and it is only present in the swarmer pole of the predivisional cell (65). A mutation in the *flbD* gene that causes persistent activation alters flagellar transcription so that flagellins appear in both swarmer and stalked cell progeny. Therefore, at least for σ^{54}-controlled flagellar genes that require phosphorylated FlbD for transcription activation, localized transcription is due to cell cycle–controlled kinase activity.

The Stalk

STRUCTURE AND LOCALIZED BIOSYNTHESIS OF THE STALK The structure that gave *Caulobacter* its name ("caulo" means stalk in Latin) is the polar stalk. This thin cylindrical extension of the cell surface is synthesized at a specific pole of the cell, the one previously occupied by the flagellum. The biogenesis of the stalk is initiated at a specific time in the cell cycle, the swarmer-to-stalked-cell transition.

Electron microscopic examination of stalks has shown that the cell-surface layers of the stalk are continuous with those of the cell body (92). The core of the stalk contains cytoplasmic material that seems to be continuous with the cell's cytoplasm but that is devoid of ribosomes and DNA (92). The stalk is traversed at intervals by dense rings known as crossbands that are composed, at least in part, of peptidoglycan (93–96), which may provide rigidity to the stalk by attaching the inner and the outer membranes (93). It has been proposed that the crossbands are an indication of cell age, with one crossband being synthesized during each cell cycle (97, 98; and JS Poindexter, personal communication). The structure of isolated crossbands (99) is indistinguishable from basal plates found to encircle the flagellar motor in *Aquaspirilum serpens* (100, 101). Perhaps the origin of the stalk crossband is its displacement from the site previously occupied by the flagellum as the stalk is elongated.

Another polar structure intimately associated with the stalk is the holdfast, which is found at the tip of the stalk. This adhesive organelle, composed of a complex polysaccharide (102–104), is synthesized at the base of the flagellum (105, 106) and solely mediates the attachment of *C. crescentus* to surfaces or to other bacteria. A cluster of four genes (*hfaABDC*) involved in the attachment of the holdfast to the cell was recently identified (107, 108; J Smit, personal communication). The similarity of the C terminus of HfaA to the PapG and SmfG adhesins suggests that HfaA is a bridging protein between the holdfast and the holdfast attachment site (108). The *hfaB* gene is similar to bacterial transcriptional activator genes, suggesting that it is involved in the regulation of holdfast biosynthesis (108). The sequence of HfaD suggests that it possesses transmembrane domains, while HfaC has similarity to ATP-binding proteins (J Smit, personal communication). The precise role of these genes in holdfast synthesis and attachment is still unknown.

STALK FORMATION AS A POLAR CELL DIVISION EVENT In practical terms, the biosynthesis of the stalk is a formidable problem for the cell. The cell has to (*a*) redirect the biosynthesis of the cell surface in a perpendicular direction at a specific pole, (*b*) initiate stalk biosynthesis at the proper time in the cell cycle (coordinate with flagellum release and DNA synthesis initiation), (*c*) control stalk elongation in subsequent cell cycles in response to environmental conditions, and (*d*) provide stabilizing rings (the cross-bands) for this structure. It is likely that some of the proteins involved in stalk biosynthesis are also involved in cell division (discussed below). In both stalk biosynthesis and cell division, those proteins have to be localized to specific regions of the cell. Stalk biosynthesis is localized to a relatively confined area at its base (85). The periodic surface array, a hexagonally packed periodic surface layer composed of three major proteins that covers the entire surface of the cell (for a review see 109), is inserted at the stalk–cell body junction, during both stalk initiation and elongation (110). Surface array biogenesis also occurs at the site of cell division during the late stages of division, another common point between cell division and stalk biosynthesis.

One of the most obvious aspects of stalk biosynthesis in common with cell division is the biosynthesis of the cell wall. Penicillin-binding proteins (PBPs) are likely to play an important role during both stalk biosynthesis and cell division. *C. crescentus* possesses between 15 and 20 different PBPs (111–117). Given the morphological difference between stalks and the cell proper, stalk biosynthesis might require the action of specific PBPs. When stalks shed from a stalk abscission mutant (118) or sheared from a wild-type strain were analyzed, it was found that their complement of PBPs differred from those in the cell envelope (114). Stalks lacked PBP 1A and PBP 3, and two other PBPs, PBP X (M_r 93,000) and PBP Y (M_r 85,000) were greatly enriched in stalk preparations. Since the body of the stalk does not contain areas of active peptidoglycan biosynthesis (85), it is likely that the PBPs are localized at the base of the stalk, where stalk biosynthesis occurs. A spontaneous mecillinam-resistant mutant that is unable to synthesize elongated stalks was found to possess normal levels of the major PBPs but to have a greatly reduced level of PBP X and PBP Y (114). This suggests that PBP X and PBP Y are involved in stalk biosynthesis. Recently, a mutant that is unable to undergo the swarmer-to-stalked-cell transition and with an increased cell diameter at the nonpermissive temperature was isolated (PJ Kang, L Shapiro, in preparation). The mutated gene encodes a protein with a high degree of similarity to PBP 2 of *E. coli* (32% identity, 54% similarity). The predicted size of this PBP 2 homolog agrees well with the observed size of PBP Y. In *E. coli,* PBP 2 is a specific target of mecillinam, a β-lactam antibiotic that causes an increase in cell diameter. In *C.*

crescentus, mecillinam not only causes an increase in cell diameter, but also inhibits stalk biosynthesis (114, 117). These results strongly suggest that the *C. crescentus* homolog of *E. coli* PBP 2 is PBP Y and is required for stalk biosynthesis. This mutant is not viable at the restrictive temperature, and it is thus possible that PBP Y is also required for general cell wall biosynthesis in *C. crescentus.*

The results of experiments reviewed here imply that stalk biosynthesis and cell division are dependent on similar processes (114, 119, 120). It may be that part of the machinery required for stalk biosynthesis is laid down at the site of cell division, which will form the new poles of the cell (82). A subset of the cell-division proteins could play a dual role in stalk biosynthesis and cell division. The FtsZ protein is a good candidate. It has been shown in *E. coli* and *B. subtilis* that FtsZ, an essential GTP-binding protein required for the initiation of cell division, is assembled as an annular ring at the future site of cell division (for a review, see 121). Contrary to the case in *E. coli* and *B. subtilis,* the level of FtsZ is under cell-cycle control in *C. crescentus* (Y Brun and L Shapiro, unpublished results). FtsZ does not appear to be present in swarmer cells. FtsZ first appears at the time of initiation of stalk biosynthesis, and its level is highest during cell division. When the *ftsZ* gene is expressed from a heterologous constitutive promoter, stalk defects appear: The stalks are longer than normal and they are often bifurcated with a flagellum still attached. These results support the notion that *ftsZ* is involved in stalk biosynthesis. Once the relationship between cell division and stalk biosynthesis is established, stalk biosynthesis could become a useful model for the study of some aspects of cell division.

REGULATORY GENES INVOLVED IN STALK BIOSYNTHESIS AND CELL DIVISION

In their study of *C. crescentus* DNA phage ϕCbK-resistant mutants, Fukuda et al (122, 123) and subsequently, Ely et al (124) and Sommer & Newton (125) obtained a pleiotropic mutant that was resistant to phage ϕCbK, lacked stalks and pili, and possessed inactive flagella. All motile revertants of these mutants simultaneously regained phage sensitivity, stalks, and pili, suggesting that a single mutation was responsible for the pleiotropic phenotype. Most of these mutations mapped to the *pleC* gene. Like *rpoN* mutants, *pleC* mutants can synthesize stalks under conditions of phosphate limitation (84, 125).

The predicted product of the *pleC* locus has a high degree of similarity to the C-terminal region of histidine protein kinases of two-component regulatory systems (84). PleC also possesses two highly hydrophobic regions that could be membrane-spanning domains, which are common in histidine protein kinase sensors of two-component regulatory systems. Another common property of these protein kinases is that they are capable of autophosphorylation. Analysis of a PleC fusion protein showed that the C-terminal region of PleC is capable of autophosphorylation. These phosphate groups are removed by acid but not by base treatment, suggesting that phosphorylation occurs at a histidine residue (84). Two independent PleC-β-galactosidase fusions each partially complement a *pleC* mutant even though they lack the putative membrane-spanning domains of PleC. Thus, although not essential for PleC function, the membrane-spanning domains seem to be required for maximum efficiency. Given the phenotype of *pleC* mutants and the homology of PleC to sensor proteins, it is possible that PleC could play a role in sensing a cell cycle–dependent signal involved in the regulation of polar development and transducing this signal to one or more response regulators.

Pseudoreversion analysis of a *pleC* thermosensitive mutant identified a number of suppressors that map to three different genes, *divJ, divK,* and *divL* (126). Mutations in these genes confer a filamentous phenotype, strengthening the concept that PleC and other proteins with which it interacts act as part of a system connecting polar differentiation and cell division. *divJ* was found to encode another homolog of the histidine protein kinases, with 48% amino acid identity to the kinase domain of PleC (127). DivJ possesses the characteristic putative membrane-spanning domains of this family of proteins. The transcription of *divJ* is temporally controlled and is highest during the swarmer-to-stalked cell transition, suggesting an early involvement in cell division, consistent with the phenotype of some of the *divJ* alleles. It is not known whether *divJ* is essential or if it is indirectly involved in cell division, since a filamentous phenotype is caused by a variety of mutations, including mutations in Class II flagellar genes (see earlier section).

If PleC and DivJ are sensor proteins, what are their cognate response regulators? The predicted product of *divK,* identified as a suppressor of a *pleC* mutant, is homologous to the N terminus of bacterial response regulators (G Hecht, A Newton, personal communication). It has been suggested that PleC and DivJ are sensor kinases involved in modulating the activity of the DivK response regulator. These three proteins, and possibly others, could be part of a putative protein phosphorylation signal transduction system involved in coupling polar differentiation and cell division (127). Further characterization of this putative signal transduction system should clarify its role in linking asymmetric polar differentiation and the cell cycle.

CELL TYPE–SPECIFIC DNA REPLICATION

A fundamental example of asymmetry in *C. crescentus* is the differential replicative ability of the chromosomes of the two progeny cells. Upon cell division, the progeny stalked cell immediately initiates DNA replication, whereas the chromosome that partitions to the progeny swarmer cell is unable to initiate DNA replication until later in the cell cycle when it differentiates into a stalked cell. What are the molecular events that cause this marked difference in chromosome behavior?

Nucleoid Polarization and Chromosome Partitioning

The bacterial nucleoid is an aggregate of cell components that specifically cosediment with the genomic DNA in a sucrose gradient (128). *C. crescentus* nucleoids can be isolated gently, such that they retain many subtle structures, including active RNA polymerase molecules and polysomes, and interesting "looping domains" and "beaded" chromatinlike features as seen in the electron microscope (25, 129). Several studies of *C. crescentus* nucleoids suggest that swarmer cells and stalked cells have distinctive higher-order chromosome structures (24, 25, 59). Swarmer cells have fast-sedimenting (FS) nucleoids (>6000 S), while stalked cells have slow-sedimenting (SS) nucleoids (3000 S). This is contrary to expectations based solely on DNA content, since the stalked cell genome is partially replicated. In synchronized cultures, the transition from FS to SS nucleoid type is abrupt, and it coincides with the swarmer-to-stalked cell transition (24). When nucleoids are prepared from asynchronous cultures, representing all transitional cell types, only two distinct nucleoid peaks are observed (25, 129), rather than the gradual nucleoid changes predicted if the sedimentation difference was due to DNA replication. The simplest interpretation of these data is that the swarmer nucleoid has a more compact conformation that is generated and reversed in a stage-specific "remodeling" process.

Both FS and SS nucleoids are present in predivisional cells (23, 24). It would be of interest if *C. crescentus* nucleoid states were found to be analogous to euchromatin and heterochromatin states that coinhabit the nucleus of animal cells with properties beyond simple DNA condensation state. It is not clear whether or not the two distinct *C. crescentus* nucleoids are contained within the same cytoplasmic compartment of the predivisional cell. As discussed earlier, a barrier may form between the daughter cells well before cell division takes place, and before the nucleoids differentiate.

Following DNA replication, chromosomes must be partitioned to opposite cell poles in order to prevent production of anucleate cells. There is evidence in *E. coli* that nucleoids are moved by an active process (130), and it is likely that similar mechanisms operate in *C. crescentus*. In addition to the problem of chromosome mobility, the differentially sedimenting nucleoids must be targeted to the correct cell poles. Nucleoid differentiation could take place after segregation with signals provided by each pole. In this case, immediately following DNA replication, the nucleoids might be identical (both slow sedimenting), and then become restructured by specific polar induction signals only after they have been moved into place. Alternatively, it may be that the nucleoids are marked prior to segregation in order to ensure their proper polar positioning. However, it has been demonstrated that old and new DNA strands are randomly segregated (3, 131). This conclusion was reached using significantly different methods for assaying the fate of new DNA strands that were pulse labeled with [3-H]deoxyguanosine. Osley & Newton (131) employed the plate elution technique for cell synchrony, whereas Marczynski et al (3) employed the ludox synchrony technique to obtain homogenous populations of recombination-deficient stalked cells. The absence of a chromosome partitioning bias suggests that chromosomes are developmentally equal following DNA replication, and that they are chosen at random to become swarmer- or stalked-cell specific. These results imply that special inductive signals for nucleoid differentiation emanate from the cell poles.

The C. crescentus *Chromosomal Origin of Replication*

One early indication that the *C. crescentus* origin may have unique features was the observation that the *Klebsiella pneumoniae* origin, which replicates in *E. coli,* could not replicate in *C. crescentus* (132). Subsequently, it was demonstrated that the *C. crescentus* origin cannot replicate in *E. coli* (G Marczynski, L Shapiro unpublished results). The *C. crescentus* chromosomal origin of replication (Cori) was cloned (89), and sequence analysis revealed that it contains significant structural differences as well as similarities with the *E. coli* origin of replication (Figure 5). Like other bacterial origins, the *C. crescentus* origin contains potential binding sites for the DnaA protein ("DnaA boxes") (for a review see 133) as well as an 85% A+T-rich stretch of DNA, exceptional considering that *C. crescentus* averages 67% G+C overall (89). Biochemical analyses of the *E. coli* origin established that the DnaA molecules (bound to four DnaA boxes) unwind and open the *E. coli* origin at its adjacent "A+T region," prior to the assembly of the replication apparatus (134). Although the *C. crescentus* origin contains five potential DnaA boxes, none of these motifs agree perfectly with the eubacterial consensus, and three of the DnaA boxes (above the dotted line in Figure 5) are dispensable for autonomous replication. However, point mutations in one *C. crescentus* DnaA box (marked by a star in Figure 5) completely abolish autonomous replication, implying that the *C. crescentus* DnaA protein serves an essential role at this origin as well.

These observations are reminiscent of the *E. coli* P1 plasmid origin, where DnaA is essential for replication, but plays a subordinate role to the plasmid-encoded RepA protein, which binds to iterated target sites (135, 136). By analogy, it is possible that the *C. crescentus* origin utilizes auxiliary

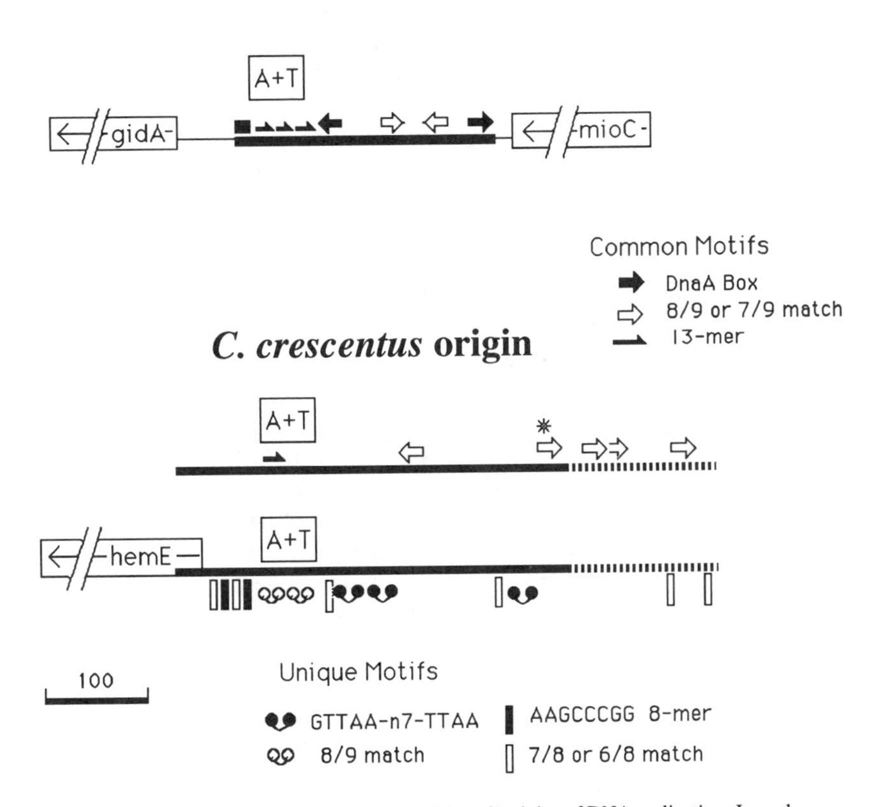

Figure 5 Comparison of the *C. crescentus* and *E. coli* origins of DNA replication. In each case, the minimal region of the origin required for origin-driven (autonomous) plasmid replication is indicated by a thick black line. Protein-coding sequences surrounding the origins are indicated by open boxes. Sequence motifs that are shared by both origins are indicated by symbols above the line, and motifs unique to the *C. crescentus* origin are indicated below the line. Portions of this figure are from (89). Details shown are from (89, 158, 159).

proteins that bind iterated target sites, in addition to DnaA. This hypothesis is supported by the existence of iterons that are unique to the *C. crescentus* origin of replication (Figure 5). In particular, three perfect matches to an essential 9-bp motif GTTAA-n7-TTAA are contained within the minimal origin, and two near matches (both 8/9) span the A+T-rich region. The integrity of this motif is absolutely essential for replication. Both linker-insertion and site-directed mutagenesis experiments demonstrated that all three perfect motifs are required for autonomous replication (G Marczynski, K Lentine, L Shapiro, submitted for publication).

DIFFERENTIAL TRANSCRIPTION FROM A PROMOTER WITHIN THE ORIGIN OF REPLICATION RNA polymerase has been implicated in the earliest steps of *E. coli* chromosomal replication (137, 138). In the *C. crescentus* origin, the minimal sequence requirements for autonomous replication cannot be separated from the promoters of an overlapping heme biosynthetic gene (G Marczynski, K Lentine, L Shapiro, submitted for publication). Two promoters have been identified upstream of the *hemE* gene: a weak promoter overlapping a cluster of four 8-mer motifs (AAGCCCGG), and an upstream strong promoter, overlapping the A+T region and two imperfect 9-mer GTTAA-n7-TTAA motifs (Figure 5). Deletions in either promoter region abolish autonomous plasmid replication. The weak promoter has significant similarity to other *C. crescentus* "housekeeping" promoters (presumably analogous to the *E. coli* sigma-70 promoters) (J Malakooti, SP Wang, and B Ely, personal communication). The strong promoter, also similar to σ^{70} promoters, shows significant similarity to the -10 region of *Rhizobium meliloti* heme promoters (139).

Transcription from the strong *hemE* promoter correlates with the replicative abilities of the swarmer and stalked cells. The strong promoter, but not the weak promoter, is specifically inactivated at the swarmer cell pole 20 min before cell separation. It remains inactive in the nascent swarmer cells, but it is again activated in the nascent stalked cells. This differential transcriptional control of the strong *hemE* promoter may be mediated by the two 9-mer GTTAA-n7-TTAA motifs that overlap the -10 and -35 regions of this promoter. Differential transcription in the predivisional cell is not observed for most *C. crescentus* genes. However, as described in an earlier section, swarmer pole–specific transcription is seen among several flagellar genes, and recently, stalked pole-specific transcription of the *gyrB* has been observed (90).

Cell-Cycle Plasmid Replication in C. crescentus

Apart from their utility as vectors, plasmids can serve as interesting experimental systems to probe cellular replication functions. The fact that different plasmids have different requirements for host-encoded replication proteins suggested that plasmids may act as sensors for fluctuating *C. crescentus* replication factors. incQ plasmids supply their own origin recognition (DNA-binding), DNA helicase, and primase (plasmid RepA, B, C) functions. In contrast, incP-1 minimal replicons have an absolute requirement for all host replication proteins (including DnaA) because they only encode one small origin-binding protein (TrfA). Despite these major differences between incP-1 and incQ replication strategies, both of these broad-host-range replicons were found to replicate with the same cell-cycle pattern: All plasmids can initiate replication at any point in the cell cycle, but there is a 5–10-fold higher probability of initiation in stalked cells as compared to swarmer cells (3). This observation demonstrates that swarmer cells are competent for DNA replication, and that the absence of chromosomal replication in swarmer cells is not simply due to the absence of a replication protein. Also, the uniform replication behavior of both replicon types suggests that replication proteins involved in the earliest steps of initiation (up to the formation of an open and primed replication complex) are not rate limiting in the swarmer cells. If this were the case, incQ plasmids would replicate more readily in the swarmer cells than do the incP-1 plasmids. It is not known why plasmids tend to replicate more frequently in stalked cells. Perhaps they are responding to the factors that unfold the nucleoids (as discussed above), or to the availability of additional replication proteins that act at later stages of the replication process, and whose abundance or activity may be rate limiting in the swarmer cells. Some candidate proteins are suggested below.

DNA Replication Genes

Genes that affect DNA replication in *C. crescentus* have been identified by screening conditional temperature-sensitive mutations that block progression of the cell cycle (120). Osley & Newton found several temperature-sensitive DNA synthesis mutants, and grouped them into those that block DNA chain elongation ("DNAe" mutants) or those that block the initiation of DNA synthesis ("DNAi" mutants) (120). One DNAi mutant (PC2076) is especially interesting, because careful synchrony and temperature-shift experiments suggested that its gene product(s) are required in predivisional cells in order for the progeny stalked cells to initiate chromosomal replication (140). These experiments support the concept that replication potential is established and polarized in the predivisional cell.

Newton and coworkers have also ordered DNAi, DNAe, cell-division mutants (*div* genes), and developmental "landmarks" (e.g. stalk, pillin, flagellin synthesis) within the cell cycle as a series of dependent and interdependent steps (82, 83, 140–142). One inference from this analysis is that the initiation of chromosomal DNA replication requires the completion of chromosomal replication as well as cell division in the previous generation (140, 142). This stage dependency in *C. crescentus* contrasts with *E. coli* and *B. subtilis,* where chromosomal DNA replication can reinitiate before the termination of the previous replication cycle (143).

The homologs of several well-characterized *E. coli* replication genes, including *dnaA* and *gyrB,* have been identified in *C. crescentus*. The *dnaA* homolog was cloned from the same cosmid that contained the *C. crescentus* origin of replication by virtue of its homology with the *Pseudomonas putida dnaA* gene (143a). The *C. crescentus* gene shares significant sequence homology with all other bacterial *dnaA* genes. It is most abundantly transcribed as swarmers differentiate into stalked cells, but it is clearly expressed at other times during the cell cycle as well. Inappropriate expression of the *C. crescentus dnaA* from a constitutive promoter appears to cause additional DNA synthesis, as judged by fluorescent flow cytometry experiments (143a). These observations, and the presence of an essential DnaA box inside the *C. crescentus* origin of replication (as discussed above), suggest that the DnaA protein is a key component of the *C. crescentus* replication process. The *gyrB* gene is maximally transcribed in the stalked cell, commensurate with the initiation of DNA replication (90). In addition, a temperature-sensitive DNA chain elongation mutant (PC2179, renamed *dnaC303,* but unrelated to the *E. coli dnaC* gene), has been cloned by complementing its temperature-sensitive phenotype (144). The transcription of this gene peaks at the start of S-phase, implying that DnaC may be rate limiting for chromosomal and perhaps also plasmid DNA synthesis in the swarmer cells (144).

Cell-Cycle Control of DNA Methylation

In *E. coli, dam*-directed methylation is required to mediate the attachment of the hemimethylated origin and the membrane, a process that is required for the proper timing of chromosomal DNA replication (137, 145, 146). Like most bacteria, *C. crescentus* does not have a *dam* sequence (GATC) methylase (147). Nonetheless, alternative methyltransferases must be present, because approximately 1% of all adenine residues occur as N^6-methyladenine (6MA) and approximately 2% of all cytosine residues occur as 5-methylcytosine (5MC) within the *C. crescentus* genome (148). Crude protein extracts contain both 6MA and 5MC methyltransferase activities that appear to be severalfold higher when the extracts are prepared from stalked and predivisional cells rather than from swarmer cells (148). *C. crescentus* contains at least one site-specific endonuclease (*CcrI*) (149), an isoschizomer of *XhoI*. It is very likely that *C. crescentus* also possesses a methyltransferase that recognizes the sequence CTCGAG.

Recently, *C. crescentus* was found to possess a site-specific DNA methyltransferase (*ccrM*) that methylates the adenine of the *HinfI* recognition sequence, GANTC (150). Consequently, *C. crescentus* DNA is resistant to digestion by *HinfI* as well as some enzymes with recognition sequences overlapping GANTC. The *ccrM* gene was cloned, and DNA sequence analysis revealed that the predicted amino-acid sequence has 49% identity to the *Haemophilus influenzae* methyltransferase HinfM.

The CcrM DNA methyltransferase may have an important role in the expression of asymmetry in *C. crescentus,* because the expression and function of the *ccrM* gene is restricted to the *C. crescentus* predivisional cells. When three specific genomic GANTC sites were assayed for their methylation states (by sensitivity to restriction endonucleases with overlapping sites), it was discovered that swarmer cells contain GANTC sites fully methylated on both strands. These sites become hemimethylated by DNA replication in the stalked cells, and they persist in the hemimethylated state until they are remethylated in the late predivisional cells. Consequently, GANTC sites close to the origin of DNA replication can remain hemimethylated for as long as 100 min out of a 180 min cell-division cycle. This is in contrast to the *E. coli dam* GATC sites, which are remethylated within a few min after replication, except at the *E. coli* origin of replication and the *dnaA* gene, where hemimethylated GATC sites persist, apparently due to a specific binding reaction with membrane proteins (146, 151, 152). One consequence of the delayed remethylation by CcrM is that the two nascent *C. crescentus* chromosomes become chemically distinct molecules, since the top strand remains methylated in one chromosome while the bottom strand remains methylated in the other. While asymmetric methylation could theoretically serve to mark the developmental fates of the chromosomes, random partitioning of these chromosomes (as discussed above) indicates that, at least for this process, differential methylation is not relevant.

The transcription of the *ccrM* gene is temporally controlled, with the promoter active in only predivisional cells. When *ccrM* is driven by a constitutive promoter, the chromosomal DNA becomes persistently fully methylated, and cells exhibit abnormal morphologies. They become elongated, pinch irregularly, are often stalkless, and they have tapered cell poles. Fluorescence flow cytometry experiments showed that many of these cells have three chromosomes. Such defects may reflect a combination of abnormal chromosome replication, chromosome partitioning, and cell division, suggesting that the CcrM-mediated functions are important for key cell-cycle events.

CONCLUDING REMARKS

In the early 1960s, Jeanne Poindexter and Roger Stanier published an article characterizing the biology of *Caulobacter* (153). They showed that the *C. crescentus* cell cycle included a series of differentiation events that culminated in an asymmetric predivisional cell bearing different structures. They also showed that cells can be easily synchronized by differential centrifugation. With the more recent ability to manipulate this organism genetically at hand (for a review see 154), both the spatial and the temporal control of development can be fully explored in this relatively simple organism.

The biogenesis of external polar structures, such as the flagellum and the stalk, is the most visible characteristic of the asymmetry that leads to the generation of unique progeny. However, numerous internal processes, such as the localization of proteins, cell type–specific DNA replication, establishment of chromosome structure, temporally controlled DNA methylation, and localized transcription, are some of the underlying events that yield asymmetry in the predivisional cell. An easily accessible aspect of asymmetry is the polar targeting of chemotaxis and flagellar proteins, and their recent study has provided new insight into the mechanisms by which the differentiating cell generates asymmetry. Yet to be resolved is how the cell directs surface biosynthesis at a specific pole to make a stalk. It is certain that here too, localized proteins will be involved.

The mechanisms that are used to control the initiation of DNA replication are of fundamental importance in all organisms. The problem is particularly amenable to study in *Caulobacter,* because replication initiation is cell-type specific. Why do the two newly replicated chromosomes in the predivisional cell differ in their replication potential? Are localized factors responsible for polarized replication potential? There is evidence that differential transcription within the chromosomal origin may trigger replication initiation. Although only approximately 10% of cellular proteins are periodically expressed during the *C. crescentus* cell cycle, most of the genes involved in cell differentiation have been shown to be temporally controlled. It is likely that proper temporal regulation of specific genes is important for cell differentiation to occur normally. Work in the past few years has shown that not only genes encoding structural proteins but also regulatory genes are temporally regulated. How is the regulation of these genes integrated into the cell cycle? What is the signal that initiates the cascade of events that leads to the differentiation of the swarmer cell into a stalked cell and to the establishment of asymmetry?

Numerous genes involved in *C. crescentus* differentiation have been identified, bringing us closer to the answers to many of these questions. The study of the different aspects of the generation of asymmetry in *Caulobacter* might help to solve this immensely complex aspect of cellular differentiation, yielding results applicable to both prokaryotes and eukaryotes (155–157).

ACKNOWLEDGMENTS

We thank Austin Newton, Bert Ely, John Smit, David Mullin, Jeanne Poindexter, and Jim Gober for communicating results prior to publication. We are particularly grateful to Rick Roberts and Alfred Spormann for their extremely helpful comments on the form and contents of the manuscript. We thank Austin Newton, David Mullin, Bert Ely, John Smit, Janine Maddock, Craig Stephens, Gary Zweiger, the members of the Gober laboratory, and Ellen Quardokus for critical reading of the manuscript, and

Lara London for invaluable editorial assistance. Portions of the work described in this review was supported by US Public Health Service Grant GM 32506 from the NIH, and grant NP-938B from the American Cancer Society. Yves Brun was a postdoctoral fellow of the Natural Science and Engineering Research Council (NSERC) and of the Medical Research Council (MRC) of Canada, and Greg Marczynski of the American Cancer Society during part of this work.

Literature Cited

1. Bowers LE, Weaver RH, Grula EA, Edwards OF. 1954. *J. Bacteriol.* 68: 194
2. Degnen ST, Newton A. 1972. *J. Mol. Biol.* 64:671–80
3. Marczynski GT, Dingwall A, Shapiro L. 1990. *J. Mol. Biol.* 212:709–22
4. Shaw P, Gomes SL, Sweeny K, Ely B, Shapiro L. 1983. *Proc. Natl. Acad. Sci. USA* 80:5261–65
5. Alley MRK, Gomes SL, Alexander W, Shapiro L. 1991. *Genetics* 129: 333–42
6. Alley MRK, Maddock JR, Shapiro L. 1992. *Genes Dev.* 6:825–36
7. Gomes SL, Shapiro L. 1984. *J. Mol. Biol.* 177:551–68
8. Nathan P, Gomes SL, Hahnenberger K, Newton A, Shapiro L. 1986. *J. Mol. Biol.* 191:433–40
9. Huguenel E, Newton A. 1984. *Proc. Natl. Acad. Sci. USA* 81:3409–13
10. Alley MRK, Maddock JR, Shapiro L. 1993. *Science* 259:1754–57
11. Maddock JR, Shapiro L. 1993. *Science* 259:1717–23
12. Parkinson JS. 1993. *Cell* 73:857–71
13. Lukat GS, Stock JB. 1993. *J. Cell. Biochem.* 51:41–46
14. Bray D, Bourret RB, Simon MI. 1993. *Mol. Biol. Cell.* 4:469–82
15. Gegner JA, Graham DR, Roth AF, Dahlquist FW 1992. *Cell* 70:975–82
16. Borkovich KA, Kaplan N, Hess JF, Simon MI. 1989. *Proc. Natl. Acad. Sci. USA* 86:1208–12
17. Borkovich KA, Simon MI. 1990. *Cell* 63:1339–48
18. Ninfa EG, Stock A, Mowbray J, Stock J. 1991. *J. Biol. Chem.* 266:9764–70
19. Lui J, Parkinson JS. 1991. *J. Bacteriol.* 173:4941–51
20. MacAlister TJ, MacDonald B, Rothfield LI. 1983. *Proc. Natl. Acad. Sci. USA* 80:1372–76
21. Cook WR, Kepes F, Joseleau-Petit D, MacAlister TJ, Rothfield LI. 1987. *Proc. Natl. Acad. Sci. USA* 84:7144–48
22. Frederikse PH, Shapiro L. 1989. *Proc. Natl. Acad. Sci. USA* 86:4061–65
23. Evinger M, Agabian N. 1977. *J. Bacteriol.* 132:294–301
24. Evinger M, Agabian N. 1979. *Proc. Natl. Acad. Sci. USA* 76:175–78
25. Swoboda UK, Dow CS, Vitkovic L. 1982. *J. Gen. Microbiol.* 128:279–89
26. Gober JW, Shapiro L. 1991. *BioEssays* 13:277–83
27. Stallmeyer MJB, Hahnenberger K, Sosinsky GE, Shapiro L, DeRosier D. 1989. *J. Mol. Biol.* 205:511–18
28. Wagenknect T, DeRosier D, Shapiro L, Weissborn A. 1981. *J. Mol. Biol.* 151:439–65
29. Trachtenberg S, DeRosier DJ. 1988. *J. Mol. Biol.* 202:787–808
30. Hahnenberger K, Shapiro L. 1987. *J. Mol. Biol.* 194:91–103
31. Ueno T, Oosawa K, Aizawa S. 1992. *J. Mol. Biol.* 227:672–77
32. Khambaty FM, Ely B. 1992. *J. Bacteriol.* 174:4101–9
33. Dingwall A, Gober JW, Shapiro L. 1990. *J. Bacteriol.* 172:6066–76
34. Dingwall A, Garman D, Shapiro L. 1992. J. Mol. Biol. 228:1147–62
35. Lagenaur C, DeMartini L, Agabian N. 1978. *J. Bacteriol.* 136:795–98
36. Johnson RC, Walsh JP, Ely B, Shapiro L. 1979. *J. Bacteriol.* 138:984–89
37. Sheffery M, Newton A. 1979. *J. Bacteriol.* 138:575–83
38. Lagenaur C, Agabian N. 1976. *J. Bacteriol.* 128:435–44

39. Lagenauer C, Agabian N. 1978. *J. Bacteriol.* 135:1062–69
40. Osley MA, Sheffery M, Newton A. 1977. *Cell* 12:393–400
41. Fukuda A, Koyasu S, Okada Y. 1978. *FEBS Lett.* 95:70–75
42. Weissborn A, Steinman HM, Shapiro L. 1982. *J. Biol. Chem.* 257:2066–74
43. Driks A, Bryan R, Shapiro L, DeRosier D. 1989. *J. Mol. Biol.* 206:627–36
44. Ely B, Ely TW. 1989. *Genetics* 123: 649–54
45. Bryan R, Purucker M, Gomes SL, Alexander W, Shapiro L. 1984. *Proc. Natl. Acad. Sci. USA* 81:1341–45
46. Ohta N, Chen LS, Swanson E, Newton A. 1985. *J. Mol. Biol.* 186:107–15
47. Champer R, Dingwall A, Shapiro L. 1987. *J. Mol. Biol.* 194:71–81
48. Xu H, Dingwall A, Shapiro L. 1989. *Proc. Natl. Acad. Sci. USA* 86:6656–60
49. Newton A, Ohta N, Ramakrishnan G, Mullin D, Raymond G. 1989. *Proc. Natl. Acad. Sci. USA* 86:6651–55
50. Bryan R, Glaser D, Shapiro L. 1990. *Adv. Genet.* 27:1–31
51. Dingwall A, Shapiro L, Ely B. 1990. In *Biology of the Chemotactic Response. Soc. Gen. Microbiol. Symp.* 46:155–76. Cambridge: Cambridge Univ. Press,
52. Newton A, Ohta N. 1990. *Annu. Rev. Microbiol.* 44:689–719
53. Van Way SM, Newton A, Mullin AH, Mullin DA. 1993. *J. Bacteriol.* 175: 367–76
54. Ohta N, Chen LS, Mullin DA, Newton A. 1991. *J. Bacteriol.* 173:1514–22
55. Minnich SA, Newton A. 1987. *Proc. Natl. Acad. Sci. USA* 84:1142–46
56. Mullin D, Minnich S, Chen LS, Newton A. 1987. *J. Mol. Biol.* 195:939–43
57. Mullin DA, Newton A. 1989. *J. Bacteriol.* 171:3218–27
58. Gober JW, Xu H, Dingwall A, Shapiro L. 1991. *J. Mol. Biol.* 217:247–57
59. Gober JW, Shapiro L. 1992. *Mol. Biol. Cell* 3:913–26
60. Ninfa AJ, Mullin DA, Ramakrishnan G, Newton A. 1989. *J. Bacteriol.* 171:383–91
61. Ramakrishnan G, Newton A. 1990. *Proc. Natl. Acad. Sci.* 87:2369–73
62. Brun YV, Shapiro L. 1992. *Genes Dev.* 6:2395–408
63. Kustu S, North AK, Weiss DS. 1991. *Trends Biochem. Sci.* 16:397–402
64. Mullin DA, Newton A. 1993. *J. Bacteriol.* 175:2067–76
65. Wingrove JA, Mangan EK, Gober JW. 1993. *Genes Dev.* 7:1979–92
66. Newton A, Ohta N, Ramakrishnan G, Mullin DA, Raymond G. 1989. *Proc. Natl. Acad. Sci. USA* 86:6651–55
67. Kustu S, Santero E, Keener J, Popham D, Weiss D. 1989. *Microbiol. Rev.* 53:367–76
68. Santero E, Hoover T, Keener J, Kustu S. 1989. *Proc. Natl. Acad. Sci. USA* 86:7346–50
69. Hoover TR, Santero E, Porter S, Kustu S. 1990. *Cell* 63:11–21
70. Gober JW, Shapiro L. 1990. *Genes Dev.* 4:1494–504
71. Landy A. 1989. *Annu. Rev. Biochem.* 58:913–49
72. Yu J, Shapiro L. 1992. *J. Bacteriol.* 174:3327–38
73. Francis NR, Sosinsky GE, Thomas D, DeRosier DJ. 1994. *J. Mol. Biol.* 235:1261–70
74. Magariyama Y, Yamaguchi S, Aizawa SI. 1990. *J. Bacteriol.* 172:4359–69
75. Sockett H, Yamaguchi S, Kihara M, Irikura VM, Macnab RM. 1992. *J. Bacteriol.* 174:793–806
76. Dingwall A, Zhuang WY, Quon K, Shapiro L. 1992. *J. Bacteriol.* 174: 1760–68
77. Ramakrishnan G, Zhao JL, Newton A. 1991. *J. Bacteriol.* 173:7283–92
78. Sanders LA, Van Way S, Mullin DA. 1992. *J. Bacteriol.* 174:857–66
79. Viitanen AM, Toivanen P, Skurnik M. 1990. *J. Bacteriol.* 172:3152–62
80. Plano GV, Barve SS, Straley SC. 1991. *J. Bacteriol.* 173:7293–303
81. Stephens CM, Shapiro L. 1993. *Mol. Microbiol.* 9:1169–79
82. Huguenel ED, Newton A. 1982. *Differentiation* 21:71–78
83. Sheffery M, Newton A. 1981. *Cell* 24:49–57
84. Wang SP, Sharma P, Schoenlein PV, Ely B. 1993. *Proc. Natl. Acad. Sci. USA* 90:630–34
85. Schmidt JM, Stanier RY. 1966. *J. Cell Biol.* 28:423–36
86. Walker SG, Hancock REW, Smit J. 1991. *FEMS Microbiol. Lett.* 77:217–22
87. Gober JW, Champer R, Reuter S, Shapiro L. 1991. *Cell* 64:381–91
88. Milhausen M, Agabian N. 1983. *Nature* 302:630–32
89. Marczynski GT, Shapiro L. 1992. *J. Mol. Biol.* 226:959–77
90. Rizzo MF, Shapiro L, Gober J. 1993. *J. Bacteriol.* 175:6970–81
91. Losick R, Stragier P. 1992. *Nature* 355:601–4
92. Poindexter JS, Cohen-Bazire G. 1964. *J. Cell Biol.* 23:587–607
93. Pate JL, Ordal EJ. 1965. *J. Cell Biol.* 27:133–50

94. Schmidt JM. 1973. *Arch. Mikrobiol.* 89:33–40
95. Jones HC, Schmidt JM. 1973. *J. Bacteriol.* 116:466–70
96. Schmidt JM, Swafford JR. 1975. *J. Bacteriol.* 124:1601–1603
97. Stanley JT, Jordan TL. 1973. *Nature* 246:155–56
98. Poindexter JS. 1991. *Abstr. Am. Soc. Microbiol. Meet.* p. 96 (I-34)
99. Driks A, Schonlein PV, DeRosier DJ, Shapiro L, Ely B. 1990. *J. Bacteriol.* 172:2113–23
100. Coulton JW, Murray RGE. 1977. *Biochim. Biophys. Acta* 465:290–310
101. Coulton JW, Murray RGE. 1978. *J. Bacteriol.* 136:1037–49
102. Merker RI, Smit J. 1988. *Appl. Environ. Microbiol.* 54:2078–85
103. Ong C, Wong MLY, Smit J. 1990. *J. Bacteriol.* 172:1448–56
104. Poindexter JS. 1981. *Microbiol. Rev.* 45:123–79
105. Poindexter JS. 1964. *Bacteriol. Rev.* 28:231–95
106. Umbreit TH, Pate JL. 1978. *Arch. Microbiol.* 118:157–68
107. Mitchell D, Smit J. 1990. *J. Bacteriol.* 172:5425–31
108. Kurtz HD, Smit J. 1992. *J. Bacteriol.* 174:687–94
109. Bingle WH, Walker SG, Smit J. 1993. *Advances in Bacterial Paracrystalline Surface Layers*. New York: Plenum
110. Smit J, Agabian N. 1982. *J. Cell Biol.* 95:41–49
111. Koyasu S, Fukuda A, Okada Y. 1980. *J. Biochem.* 87:363–66
112. Koyasu S, Fukuda A, Okada Y. 1981. *J. Gen. Microbiol.* 126:111–21
113. Koyasu S, Fukuda A, Okada Y. 1982. *J. Gen. Microbiol.* 128:1117–24
114. Koyasu S, Fukuda A, Okada Y, Poindexter JS. 1983. *J. Gen. Microbiol.* 129:2789–99
115. Koyasu S, Fukuda A, Okada Y. 1984. *J. Biochem.* 95:593–95
116. Markiewicz Z, Sniezek Z. 1983. *Acta Microbiol. Polonica* 32:207–12
117. Nathan P, Newton A. 1988. *J. Bacteriol.* 170:2319–27
118. Poindexter JS. 1978. *J. Bacteriol.* 135: 1141–45
119. Terrana B, Newton A. 1976. *J. Bacteriol.* 128:456–62
120. Osley MA, Newton A. 1977. *Proc. Natl. Acad. Sci. USA* 74:124–28
121. Lutkenhaus J. 1993. *Mol. Microbiol.* 9:403–9
122. Fukuda A, Iba H, Okada Y. 1977. *J. Bacteriol.* 131:280–87
123. Fukuda A, Asada M, Koyasu S, Yoshida H, Yaginuma K, Okada Y. 1981. *J. Bacteriol.* 145:559–72
124. Ely B, Croft RH, Gerardot CJ. 1984. *Genetics* 108:523–32
125. Sommer JM, Newton A. 1989. *J. Bacteriol.* 171:392–401
126. Sommer JM, Newton A. 1991. *Genetics* 129:623–30
127. Ohta N, Lane T, Ninfa E, Sommer JM, Newton A. 1992. *Proc. Natl. Acad. Sci. USA* 89:10297–301
128. Drlica K. 1987. In *Escherichia coli and Salmonella typhimurium Cellular and Molecular Biology,* ed. FC Neidhardt, JL Ingraham, KB Low, B Magasanik, M Schaechter, HE Umbarger, 1:91–101. Washington, DC: Am. Soc. Microbiol.
129. Swoboda UK, Dow CS, Vitkovic L. 1982. *J. Gen. Microbiol.* 128: 291–301
130. Niki H, Jaffe A, Imamura R, Ogura T, Hiraga S. 1991. *EMBO J.* 10:183–93
131. Osley MA, Newton A. 1974. *J. Mol. Biol.* 90:359–70
132. O'Neill EA, Bender RA. 1988. *J. Bacteriol.* 170:3774–77
133. Yoshikawa H, Ogasawara N. 1991. *Mol. Microbiol.* 5:2589–697
134. Kornberg A, Baker TA. 1992. *DNA Replication*. New York: Freeman. 2nd ed.
135. Abeles AL, Reaves LD, Austin SJ. 1990. *J. Bacteriol.* 172:4386–91
136. Wickner S, Hoskins J, Chattoraj D, McKenny K. 1990. *J. Biol. Chem.* 265:11622–27
137. Lark KG. 1972. *J. Mol. Biol.* 64:47–60
138. Messer W. 1972. *J. Bacteriol.* 112:7–12
139. Leong SA, Williams PH, Ditta GS. 1985. *Nucleic Acids Res.* 13:5965–76
140. Nathan P, Osley MA, Newton A. 1982. *J. Bacteriol.* 151:503–6
141. Osley MA, Newton A. 1978. *J. Bacteriol.* 135:10–17
142. Osley MA, Newton A. 1980. *J. Mol. Biol.* 138:109–28
143. Donachie WD, Begg KJ, Sullivan NF. 1984. In *Microbial Development,* ed. R Losick, L Shapiro, pp. 27–62. Cold Spring Harbor, NY: Cold Spring Harbor Lab.
143a. Zweiger G, Shapiro L. 1994. *J. Bacteriol.* 176:401–8
144. Ohta N, Masurekar M, Newton A. 1990. *J. Bacteriol.* 172:7027–34
145. Bakker A, Smith DW. 1989. *J. Bacteriol.* 171:5738–42
146. Ogden GB, Pratt MJ, Schaechter M. 1988. *Cell* 54:127–35

147. Barbeyron T, Kean K, Forterre P. 1984. *J. Bacteriol.* 160:586–90
148. Degnen ST, Morris NR. 1973. *J. Bacteriol.* 116:48–53
149. Syddall R, Stachow C. 1985. *Biochim. Biophys. Acta* 825:236–43
150. Zweiger G, Marczynski G, Shapiro L. 1994. *J. Mol. Biol.* 235:472–85
151. Campbell JL, Kleckner N. 1990. *Cell* 62:967–79
152. Chakraborti A, Gunji S, Shakibai N, Cubeddu J, Rothfield L. 1992. *J. Bacteriol.* 174:7202–6
153. Stove JL, Stanier RY. 1962. *Nature* 196:1189–92
154. Ely B, Shapiro L. 1989. *Genetics* 123:427–29
155. Horvitz HR, Herskowitz I. 1992. *Cell* 68:237–55
156. Nelson WJ. 1992. *Science* 258:948–55
157. Shapiro L. 1993. *Cell* 73:841–55
158. Oka A, Sugimoto K, Takanami M. 1980. *Mol. Gen. Genet.* 178:9–20
159. Asai T, Takanami M, Imai M. 1990. *EMBO J.* 9:4065–72

Annu. Rev. Biochem. 1994. 63:451–86

MOLECULAR MECHANISMS OF ACTION OF STEROID/THYROID RECEPTOR SUPERFAMILY MEMBERS

Ming-Jer Tsai and Bert W. O'Malley
Baylor College of Medicine, Houston, Texas 77030

KEY WORDS: transcription, hormone, retinoic acid, vitamin

CONTENTS

INTRODUCTION

Classical steroid hormones, such as estrogen, progesterone, androgens, glucocorticoids, and mineralocorticoids, are synthesized and secreted by

0066-4154/94/0701-0451$05.00

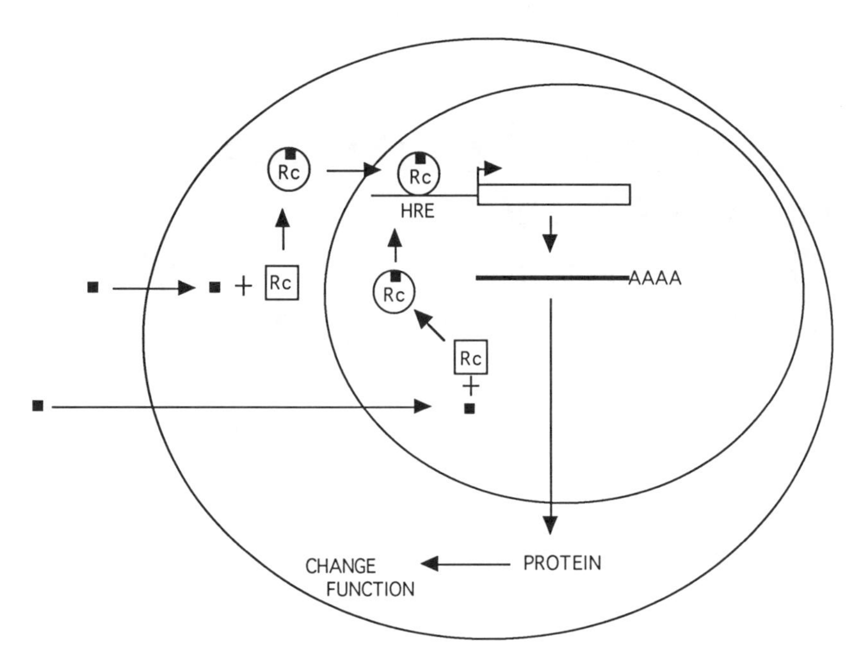

Figure 1 A simplified model of steroid hormone action. See text for description. n, steroid hormone; Rc, receptor; and HRE, hormone response element.

endocrine cells. They travel via the blood stream to their target cells, enter these cells by simple or facilitated diffusion, and then bind to specific receptors (Figure 1). The steroid hormone receptors are intracellular transcription factors that exist in inactive apoprotein forms either in the cytoplasm or nucleus. Upon binding their respective hormonal ligands, the receptors undergo an activation or "transformation" step. The activated receptor can bind effectively to a DNA element (hormone response element, HRE) and activate transcription of a *cis*-linked gene (Figure 1). In addition to regulating transcription, steroid hormones occasionally regulate gene expression by affecting mRNA stability and translational efficiency.

Although the overall pathway was proposed in the late 1960s (1–3), a detailed mechanism of action remained unclear until more recently. Since the mid-1980s, when cDNAs encoding various steroid hormone receptors were cloned, a large amount of structural information regarding steroid hormone receptors has accumulated (4). Also, the development of a cell-free system that mimicked in vivo receptor activity has made it possible to dissect potential molecular mechanisms of action (5–8). For discussions of the identification and characterization of steroid receptors, cDNA cloning, definitions of functional domains, and identification of phosphorylation sites, readers should refer to reviews by Evans, Yamamoto, Chambon, Beato, O'Malley, Gronemeyer, and Moudgil (4, 9–15). In the present review, we focus on recent data concerning receptor-DNA interactions, the role of receptors in target gene activation or repression, the role of agonists or antagonists in receptor activation, and dimeric interactions of receptors with one another or with other transcription factors that activate or silence responsive genes.

STEROID RECEPTOR SUPERFAMILY

The steroid hormone receptor superfamily consists of a surprisingly large number of genes (4, 16–22), and represents the largest known family of transcription factors in eukaryotes. It includes receptors for the steroids, estrogen (ER), progesterone (PR), glucocorticoid (GR), mineralocorticoid (MR), and androgen (AR). In addition, it includes receptors for thyroid hormone (TR), vitamin D (VDR), retinoic acid (RAR) and 9-*cis* retinoic acid (RXR), and ecdysone (EcR). Furthermore, cloning by various means has identified a large number of previously unknown genes having sequence homology to the steroid hormone receptor superfamily (22–38). Since the ligands for these genes are not known, they have been termed "orphan receptors." Finally, an unexpected variety of isoforms of TR, RAR, RXR, ER, PR, and EcR have been identified (20, 21, 39–48). These isoforms may be expressed in distinct cell types and developmental stages, suggesting that they play a variety of physiological roles.

Amino acid sequence analysis and mutational dissection of intracellular receptors indicate that they can be subdivided into several domains as indicated in Figure 2. The N-terminal A/B domain is highly variable in sequence and in length. Usually, this domain contains a transactivation function (AF), which activates target genes presumably by interacting with components of the core transcriptional machinery, coactivators, or other transactivators (49–53). This region may be important also for determining target gene specificity for receptor isoforms, which recognize the same response element (54, 55). The C region contains two type II zinc (Zn) fingers, which are responsible for DNA recognition and dimerization (4, 56, 57, and references therein). A more detailed description of sequence-specific recognition by the steroid receptor DNA-binding domain (DBD) and its structure is given below.

Downstream of the C region, a variable hinge region exists (D region). This region may allow the protein to bend or alter conformation, and often contains a nuclear localization domain (GR, PR) and/or transactivation domain (TR, GR) (49, 50, 53, 58, 59). The ligand-binding domain (LBD

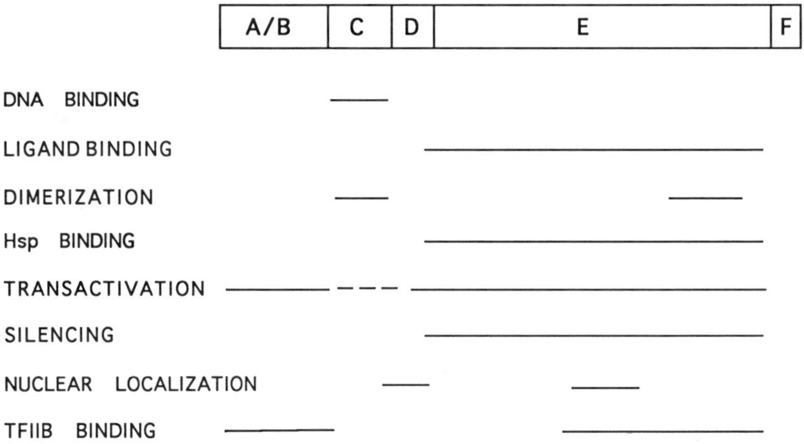

A/B | C | D | E | F

DNA BINDING

LIGAND BINDING

DIMERIZATION

Hsp BINDING

TRANSACTIVATION

SILENCING

NUCLEAR LOCALIZATION

TFIIB BINDING

Figure 2 Functional domains of steroid receptors. The structure of steroid receptors can be divided into six domains: A, B, C, D, E, and F. The function of each domain is indicated by solid lines.

or E region) is located carboxy-terminal to the D region. The E region is relatively large (~250 aa) and is functionally complex. It usually contains regions important for heat-shock protein association, dimerization, nuclear localization, transactivation, intermolecular silencing (TR, RAR, COUP-TF, etc), intramolecular repression (PR) and, most importantly, ligand binding (4, 51–53, 58–70 and references therein). Although most of these functions require only small stretches of amino acid sequence, ligand binding appears to involve a majority of the E region, since most of the mutations identified in the LBD compromise the ability of the altered receptor to bind hormones. The major dimerization domain of receptors has been localized in the C-terminal half of the LBD (62, 63). This region contains leucine-rich sequences that may form coil-coil interactions as the receptor dimerizes. Finally, located at the C-terminal end of certain receptors is the variable F region, for which no specific function has been identified. For example, deletion of the F region in ER does not affect a known ER function (51).

Genomic genes for most of the steroid/thyroid hormone receptors have been cloned. For PR, GR, AR, ER, MR, TR, and RAR, the exon-intron organization is quite complex. The genes usually encompass ~ 60 kb and are interrupted by numerous introns. The promoter region resembles that of a housekeeping gene and is often embedded in a GC-rich island (32, 71–80). Multiple sites for initiation of transcription are the rule. In contrast, the structural genes for two orphan members of the superfamily, COUP-TF I and COUP-TF II, are more simple (81). The entire genes for these two members cover ~4–6 kb and contain only two introns. This simple organizational pattern for COUP-TF I and COUP-TF II genes may correlate with the notion that they are extreme ancestral members of the superfamily.

PROTEIN-DNA INTERACTIONS

DNA sequences responsive to steroid hormones have been termed hormone response elements (HREs). This concept was substantiated when an HRE for glucocorticoids (GRE) was first identified by mutational analysis of a target gene (MMTV) (9 and references therein). A deletion or mutation of that sequence eliminated hormonal responsiveness. Later it was demonstrated that a short oligonucleotide corresponding to the identified sequence conferred glucocorticoid binding and responsiveness to a heterologous promoter (82). The GRE consisted of two short, imperfect inverted repeats separated by three nucleotides. Later, response elements for progesterone, mineralocorticoid, and androgen receptors were shown to be similar to that of GR (78, 82–87). Estrogen response elements also have been identified. Their sequence is similar to the GRE and comprises an inverted palindrome (88). Unlike GR, PR, MR, and AR, the ER does not bind to or act upon a GRE, nor does the estrogen response element (ERE) respond to GR, AR, PR, or MR. Interestingly, Klock et al (29) converted an ERE into a GRE by substituting one or two bases at the homologous position in the palindrome. Similar results were obtained by Martinez et al (89).

By the same token, conservation among HREs suggests that the receptor amino acids responsible for DNA binding must be conserved, so that only minor change(s) in the amino acid sequence of the DBD might change the specificity of DNA binding. Indeed, Mader et al (97) have identified three amino acids in the so-called P-box (EGckA) that are essential for ERE recognition (Figure 3). Mutation of these amino acids to GSckV changes ERE recognition to GRE recognition. Similar observations were made by Danielsen et al (91) and Umesono & Evans (92), who generated a version of GR that bound functionally to an ERE/TRE by changing amino acids in the P-box. Using the P-box sequence as an indicator of DNA recognition specificity, Forman & Samuels (63) classified receptors into several groups (Figure 3). It is clear that the P-box sequence cGSckV recognizes the TGTTCT half site, while cEGckG, cEGckS, cEAckA, cESckG, cDGckG, or cEGckA P-boxes recognize the AGGTCA half site. Since receptors prefer to bind to DNA as dimers, it may be expected that the recognition sequence is a repeat, either direct or inverted, depending on receptor origin and on whether the receptor binds as a homo- or heterodimer (see below).

The structure of the receptor DBD was first solved by NMR and later

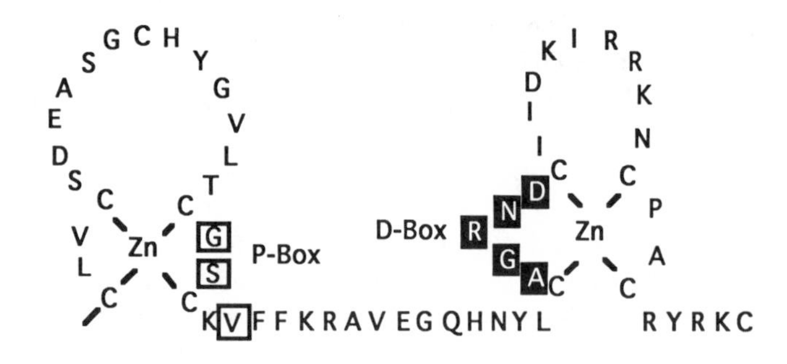

RECEPTORS	P-Box	HALF SITE
TR, RAR,VDR,RXR,PPAR, NGF1-B, NURR-1,TR2,TR2R1 EAR1,REVERBa	cEGcKG	GGTCA
HNF4	cDGcKG	GGTCA
EAR2	cEGcKS	GGTCA
SF1	cEScKG	GGTCA
ERR1	cEAcKA	GGTCA
ER	cEGcKA	GGTCA
GR,MR,PR,AR	cGScKV	TGTTCT

Figure 3 Sequence-specific recognition by steroid receptors. Amino acid sequences of glucocorticoid receptor DNA-binding domain (C) are shown. Amino acids in P-box and D-box that are important for the DNA recognition and dimerization, respectively, are highlighted.

refined and confirmed by X-ray crystallographic analysis (56, 93–96). The DBD of the GR is a globular structure that can be subdivided into two modules. Each module consists of a Zn coordination center and an amphipathic α helix. The first module contains the first Zn finger. It starts with a short segment of antiparallel β sheet and ends with an α-helical structure between the second pair of Zn coordinating cysteines. The β sheet helps to orient the residues that contact the phosphate backbone of DNA. The helical structure (P-box and downstream amino acids, Figure 3) provides important deoxynucleotide contacts and fits into the major groove of the DNA helix. The second module is more important for phosphate contacts and for dimerization (D-box between A476 and D481) of the two DBD molecules. The two modules form a globular structure through the interaction of aromatic side chains of conserved amino acids in the amphipathic helices (F463, F464, Y497, Y452, and Y474, Figure 3).

A similar structure was obtained for the DBD of ER (94, 95) except that the ER structure is at somewhat higher resolution and allows one to address the question of how ER and GR discriminate between their binding sites. While the GR-specific amino acid Val29 contacts the GRE-specific thymidine base of TG(T)TCT, the ER-specific amino acid Glu25 interacts with the cytidine of TGA(C)C. Surprisingly, the conserved amino acid Lys32, which plays no role in GR interaction with its GRE, interacts with central G-C and T-A base pairs that are specific for an ERE. To some degree, these results confirmed earlier results obtained by biochemical and mutational analyses (91, 92, 97), which suggested that amino acids in the P-box were responsible for base recognition and discrimination. Also, prior studies had indicated that the DNA-binding domain of GR binds to a GRE preferentially as a dimer (98, 99), suggesting the existence of a dimerization function in the DNA-binding domain.

While GR, PR, ER, AR, and MR bind to DNA as homodimers and recognize a palindromic response element, other receptors such as TR, RAR, VDR, RXR, PRAR, ultraspiracle, and COUP-TF can recognize direct repeat response elements (100–102). This latter group of receptors can form heterodimers with each other. More importantly, TR, RAR, VDR, and PPAR bind to their cognate DNA responsive elements with higher affinity as heterodimers with RXR than as homodimers (69, 103–111). Consequently, it has been predicted that the heterodimer is the major functional complex for this group of receptors. Similarly, the ecdysone receptor functions by heterodimerizing with ultraspiracle, a functional homolog of RXR from *Drosophila* cells (112).

Since TR, RAR, VDR, COUP-TF, PPAR, and RXR all bind to direct repeats of AGGTCA, a question could be raised as to whether any binding-site discrimination occurs among these receptors. Using AGGTCA direct repeats with various spacers, Umesono et al (100) and Naar et al (101) have derived the so-called "3,4,5 rule." DR-1 (AGGTCA direct repeat with one nucleotide spacer) can bind and act as a response element for the RXR homodimer, COUP-TF homodimer, RXR-COUP-TF heterodimer, and PPAR homodimer or heterodimer with RXR. DR-3, DR-4, and DR-5 serve as response elements for VDR, TR, and RAR, respectively. Also, DR-2 has been shown to bind to TR. Although the "3,4,5 rule" appears generally correct, receptors in this category may bind to and activate each other's response elements, albeit at lower efficiency.

Promiscuous binding and gene regulation by receptor family members may be more frequent than expected. For example, COUP-TF can be considered the "plastic man" of this superfamily. It binds as a homodimer with reasonable affinity to the GGTCA direct repeat with a spacer of anywhere from 1 to 10 nucleotides (102); it can bind also to inverted repeats with different spacers. This promiscuous binding indicates that the two DBDs of the COUP-TF dimer are not fixed in the dimeric structure. Instead, they are flexible and can be induced by the DNA to swivel and accommodate different spacers and orientations between the two half sites on the DNA target element. This conclusion is supported by studies showing that COUP-TF binds across various response elements differently (113). It wraps around the COUP sequence of the ovalbumin gene, but only contacts one face of the DNA response element (RIPE-1) of the insulin gene (113). The functional consequence of the promiscuous nature of COUP-TF binding to GGTCA repeat sequences is reflected in its ability to bind a variety of response elements and negatively affect the activity of many other receptors (69, 102, 111, 114).

A three-dimensional structure for this subgroup of receptors has been determined only for the RXR DBD. In addition to the two amphipathic α helices of the GR and ER DBDs mentioned above, RXR contains another α-helix immediately after the second Zn finger (96). Mutational analysis indicates that this helix is necessary for both DNA recognition and homodimerization. The requirement for an additional helix for dimerization is not fully understood. However, since RXR binds to direct repeat elements, this additional helix may be required to hold two RXR DBDs in a head-to-tail configuration. If that is the case, COUP-TF and TR, which can bind to both direct and inverted repeat response elements (102), also may have differential requirements for a third helix. Additional questions regarding the mechanism by which RXR forms heterodimers with TR, RAR, PPAR, and VDR remain to be resolved. Considering the rapid progress in X-ray crystallographic studies of the receptor DBDs, the structure of a heterodimer should be forthcoming.

In addition to receptor dimer binding to DNA repeat elements, several orphan receptors appear to recognize their response elements as monomers. RevTRα (EAR1) and NGFI-β recognize the AGGTCA sequence without repetition (115, 116). However, additional AT-rich sequences upstream of the AGGTCA sequence are also important for this recognition (116).

While most receptors can bind in vitro to their cognate DNA response elements in the apparent absence of a ligand, steroid hormone receptors usually do not bind to DNA in the absence of ligand in vivo. Furthermore, PR and ER can be demonstrated to bind to DNA in a hormone-dependent manner in vitro (117–120). Similarly, induction of homodimer binding to DNA by the RXR ligand (9-*cis*-retinoic acid) has been reported (121).

The mechanism by which ligands induce steroid receptors to bind DNA is of interest. It has been proposed that the ligand-binding domain blocks the DBD in the absence of a ligand; subsequently, ligand binding changes the structure and exposes the DBD. Also, since this group of aporeceptors is thought to exist in a complex with heat-shock proteins (hsp90, hsp70, and hsp56) prior to exposure to ligand, it has been proposed that these heat-shock proteins interfere with DNA binding (60, 122). Binding of ligand is thought to disperse heat-shock proteins, relieving the inhibition. Nevertheless, these two hypotheses are not consistent with the observation that the ligand-binding domain of ER or GR can be fused to heterologous transcription factors in different positions and can render these transcription factors ligand-dependent (122–125). One would predict that if the ligand-binding domain or heat-shock proteins simply block DNA binding, the ligand-binding domain located at different distances or positions from the DBD should have quantitatively different effects on receptor-DNA interaction. Furthermore, removal of heat-shock proteins by biochemical means does not make human PR bind DNA constitutively (126); ligand is still required for inducing DNA binding by the progesterone receptor free of heat-shock protein. These results suggest that ligands must play an additional role to induce steroid receptors to bind to DNA.

Recently, Allan et al took another approach to answer this question. This approach took advantage of the observation that GR (also PR and ER) binds to DNA as a dimer (98, 127, 128). The monomeric form of the receptor (DBD only) binds to DNA with sufficiently lower affinity that one must use purified receptor and low levels of nonspecific competitor DNA in binding assays in order to detect binding by gel mobility shift assays. Since it is possible that the lack of detection of DNA binding in the absence of ligand is due to an inability of receptor to form a dimer, Allan et al used an antibody as the dimerization agent (one divalent antibody forces two receptor molecules to bind to DNA at adjacent sites) in the absence of ligand (129). Indeed, divalent monoclonal antibodies that recognize human PR induced the receptor to bind to its response elements, while control antibody did not. Thus, the role of ligand in DNA binding is likely to be induction of a conformational change in the ligand-binding domain (discussed below), which exposes the major dimerization function present in this region. The dimeric receptor can then bind to its HRE with high affinity. This conclusion is consistent with the ability of other nuclear receptors such as TR, RAR, COUP-TF, and VDR to dimerize in the absence of ligand and thereby bind to DNA constitutively (100, 101).

ROLE OF RECEPTOR IN GENE ACTIVATION AND SILENCING

Gene Activation

Upon binding hormone, receptors bind to DNA and activate target gene expression. Progress in understanding the mechanism of transactivation was hampered by the lack of appropriate in vitro systems. Several years ago, a

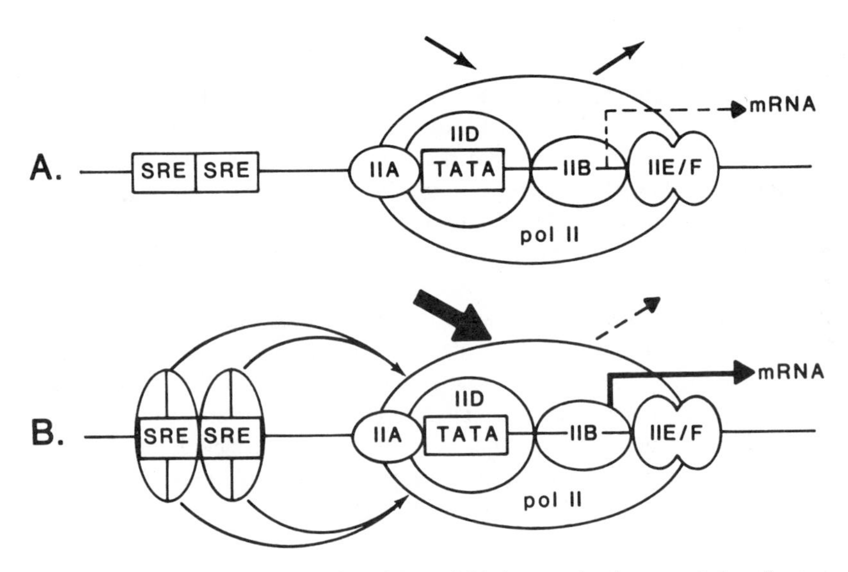

Figure 4 Steroid receptor stabilization of the preinitiation complex for transcription. See text for description.

cell-free transcription system that allowed one to demonstrate activation of target genes by exogenous purified steroid receptor was developed (5–8). Similarly, baculovirus-expressed GR, PR, and ER were shown to activate target genes (118, 130–132). Activation was observed with minimal as well as naturally occurring promoter constructs (133). The ability of steroid receptors to activate minimal promoter constructs suggested that activation might occur via direct interaction of the receptor with the core (TATA) transcriptional machinery. Indeed, using template commitment assays in vitro, we demonstrated that cPR, mER, and hGR activate transcription by enhancing the formation of stable preinitiation complexes at their target promoters (6, 130, 132).

A simplified illustration of receptor induction of transcription at a typical target gene is summarized in Figure 4. This scheme proposes that the receptor stimulates the formation of a preinitiation complex by increasing its rate of formation and/or by stabilizing a preformed preinitiation complex. The action of receptors to stimulate preinitiation complex formation could be either by direct interaction with components of the transcriptional machinery or through an intermediate factor such as a coactivator. Formation of the preinitiation complex at a core promoter is a sequential process (Figure 5). First, TFIID recognizes the TATA box; for a TATA-less promoter, binding of an initiator protein may be required. Next, TFIIA and

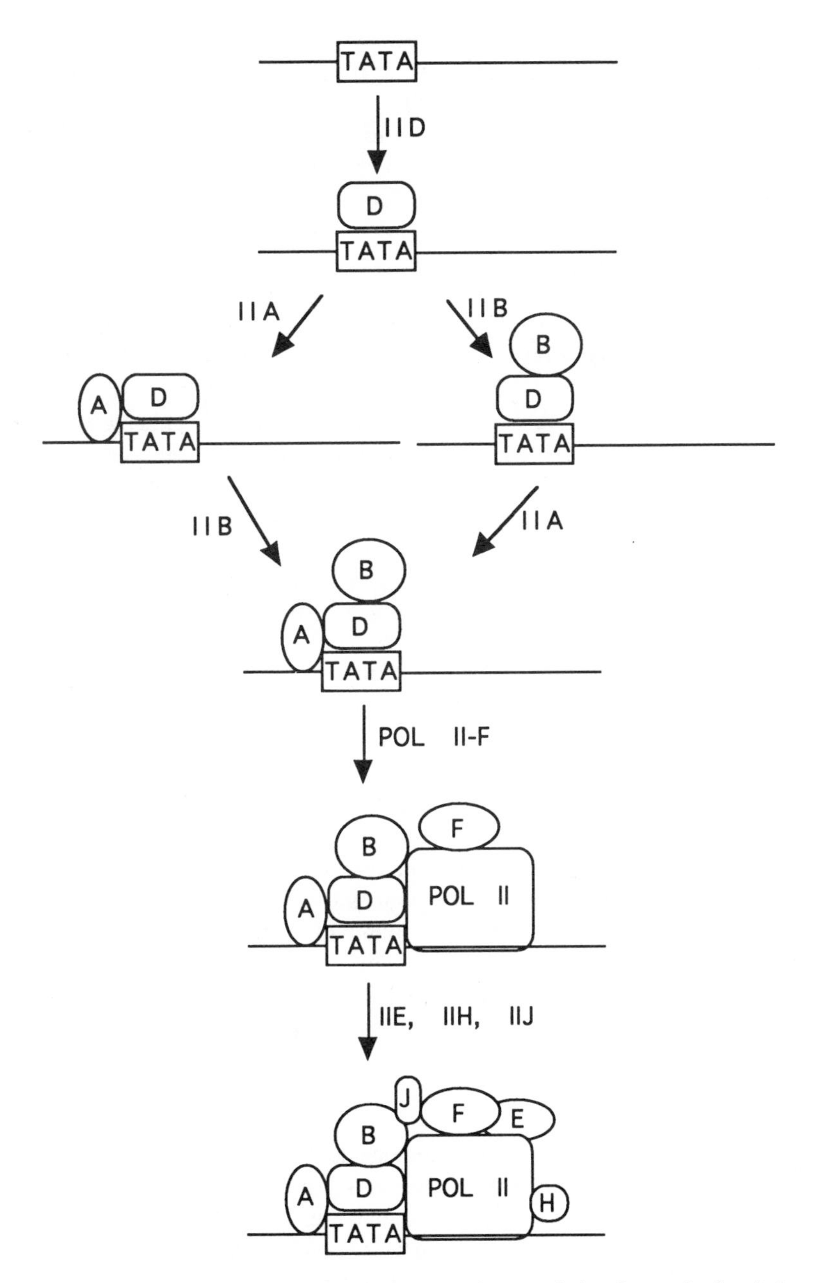

Figure 5 Steps in assembly of the preinitiation complex for transcription. See text for description.

TFIIB bind independently to the TFIID-DNA complex. Importantly, when TFIIB enters the complex, the TFIIF-RNA polymerase complex can then enter to form the ABDF-polymerase-DNA complex. Lastly, the remaining general transcription factors assemble to form the complete preinitiation complex (for review see Ref. 134). It is possible that a receptor could act on any of these steps to enhance formation of the stable preinitiation complex.

A hint at which step the receptor enhances gene transcription came from earlier studies that indicated that COUP-TF, a member of the steroid/thyroid hormone receptor superfamily, interacted specifically with a transcription factor that we called S300-II (135). After cloning of the S300-II factor, it was found to be TFIIB (136). This interaction probably does not require a helper protein, since renaturated COUP-TF from a single gel band was able to bind specifically to TFIIB. Similar observations were made for PR, ER, TR, and RAR (136, 137; XH Leng, SY Tsai, M-J Tsai, BW O'Malley, unpublished results). These results are consistent with results obtained using other activators, such as VP16 (138) and Ftz protein (139). VP16, a viral transactivator, has been shown most clearly to activate target genes by interacting with TFIIB. Mutation of amino acids necessary for the VP16-dependent activity of TFIIB but not for its basal activity destroys the specific interaction between TFIIB and VP16 (140).

Using TR as a model, Baniahmad et al (137) have found that the N-terminal end of the receptor interacts specifically with TFIIB. Like VP16, this interaction requires the second half of TFIIB, which contains a direct repeat. Since binding of TFIIB to the TFIID-DNA complex is one of the rate-limiting steps in preinitiation complex formation, it is plausible that through this interaction TR enhances its formation. It should be emphasized, however, that the receptor may also interact with other components of the transcriptional machinery. TBP (a subunit of TFIID) has been shown to interact with TR and other transactivators (141, 142; A Baniahmad, SY Tsai, M-J Tsai, BW O'Malley, unpublished results). However, since TBP is a relatively sticky protein, the significance of this interaction remains to be determined.

Gene Silencing

Classical steroid receptors have been considered to function in vivo only in the presence of their hormonal ligands. In the absence of ligands these receptors cannot bind to DNA, and thus remain functionally silent. In contrast, the subgroup of smaller nuclear receptors including TR can bind to DNA in the absence of ligand (see below). Although in most cases they cannot activate target genes, they are not neutral in regulating target gene expression. In the absence of hormones, they frequently silence basal promoter activity (64, 66–69, 143). The region important for the silencing activity is localized within the C-terminal hormone-binding domain. Within this domain, both the proximal and distal C-terminal regions are required (67). Consequently, the LBD can be subdivided into two halves, each of which has no silencing activity by itself. In contrast, cotransfection of both halves into cells restores silencing activity (A Baniahmad, SY Tsai, M-J Tsai, BW O'Malley, unpublished data). Since silencing activity requires a DNA-binding site, inhibition of basal promoter activity is unlikely to be due to cellular squelching of a coactivator. Furthermore, silencing activity can be detected with a simple promoter construct containing only a TATA box and an HRE (137), suggesting that silencing is a result of direct interaction of the receptor with the core transcriptional machinery.

Recently, Fondell et al (144) have duplicated silencing activity in vitro using TR expressed in *Escherichia coli*. They have demonstrated that the silencing activity of TR is due to its ability to inhibit the formation of a preinitiation complex. They demonstrated that the receptor can interact with TFIIB, but in their in vitro system the ligand (T3) was unable to convert the receptor from a silencer to an activator as happens in vivo. Recent evidence to support the notion that this interaction with TFIIB is important for silencing activity came from work carried out by Baniahmad et al (137). It was demonstrated that TR interacts with TFIIB specifically in two distinct regions. The N-terminal domain of the receptor and the C-terminal half of the ligand-binding domain can interact separately with TFIIB. The N-terminal region of TR interacts with the C-terminal half of TFIIB and, as discussed earlier, this interaction may be important for transactivation activity. In contrast, the C-terminal domain of TR interacts with the N-terminal Zn finger of TFIIB, which is important for basal activity. Importantly, this interaction is sensitive to hormone. In the presence of physiological concentrations of T3, binding is decreased significantly. Since this region of TR is important for silencing and since the interaction with TFIIB is sensitive to ligand, it was proposed that the interaction is functional for silencing in the absence of ligand. Since the Zn finger region of TFIIB is important for basal activity due to its interaction with the RNA polymerase–TFIIF complex (via RAP30 subunit of TFIIF) (145), it is tempting to speculate that competition for binding to the putative Zn finger of TFIIB between the TR ligand-binding domain and the TFIIF-RNA polymerase complex is the underlying mechanism for silencing by TR. Finally, similar results have been obtained with RAR; addition of retinoic acid decreases the strength of specific binding between RAR and TFIIB (XH Leng, SY Tsai, BW O'Malley, M-J Tsai, unpublished observation).

Recent evidence has provided a strong argument for the existence of negative HREs, at least in the case of the glucocorticoid receptor (146,

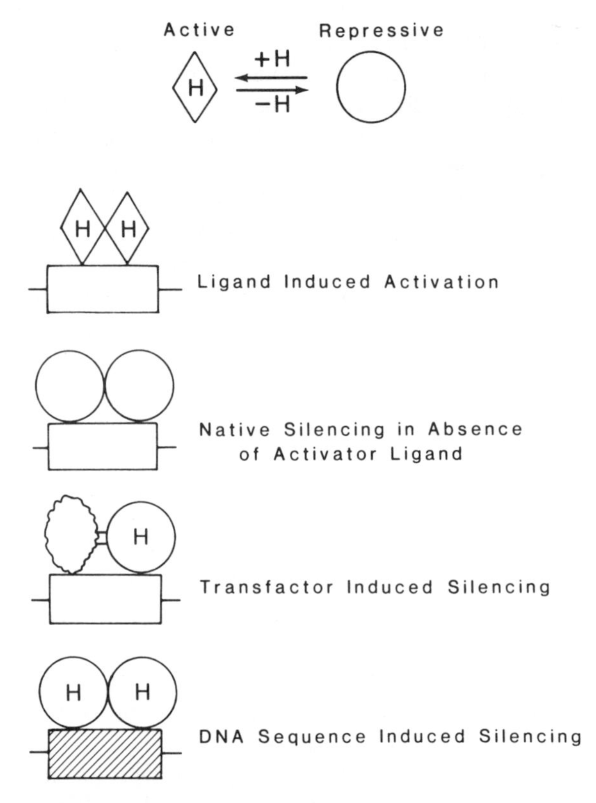

Figure 6 Conformational states of steroid/thyroid receptors. Models depict different modes of action by steroid hormone receptors to activate or repress target gene expression. See text for description.

147). These are specific DNA sequences that allow high-affinity binding of receptor but that appear to force it into a conformation that exposes peptide domains that silence transcription. It had been thought previously that silencing activity was likely to be induced solely by adjacent *cis* elements occupied by *trans* factors, which exert their negative combinatorial effects via protein-protein interactions with receptors (Figure 6). This appears not to be the case, since glucocorticoid receptor silencing activity is seen with unrelated combinations of positive activators and appears to require only the core transcriptional machinery located at the TATA promoter for its negative impact. A summary of some of the known mechanisms by which receptors may silence or repress target genes is shown in Figure 6.

SYNERGISM BETWEEN DIFFERENT *CIS*-ACTING ELEMENTS

Many eukaryotic genes are under the control of multiple hormones and environmental cues. Consequently, steroid hormone response elements are usually found in multiple copies or clustered with other *cis*-acting elements. The tryptophan oxygenase gene, the MMTV LTR, and the vitellogenin genes contain multiple HREs (148–150). In addition, the MMTV LTR, rat tryptophan oxygenase, and phosphoenolpyruvate carboxykinase gene contain other *cis*-acting elements in close conjunction with the HREs (149, 151, 152). In these cases, when either one of the HREs or the adjacent *cis*-acting elements is mutated, promoter activity is drastically decreased. Thus, HREs often interact synergistically with other HREs or with unrelated *cis*-acting elements. Using artificial reporter genes in transfected cells, one can demonstrate synergistic activation when multiple HREs or an HRE together with other *cis*-acting elements are inserted in front of a target gene (150, 153–157).

As depicted in Figure 7, Ptashne has proposed two models to explain synergism that are relevant to receptor superfamily members (158). First, binding of transcription factor dimers to two or more DNA response elements can be cooperative. Binding of one receptor complex facilitates the binding of a second (Figure 7, *top*). This synergistic interaction allows both complexes to bind with greater affinity and consequently in greater occupancy of the *cis*-acting elements, thus promoting more transcription. This type of synergism would be observed only if the transcription factors are in limited concentration, and indeed, cooperative binding of receptors to multiple response elements has been observed with PR, GR, and ER (154, 159, 160). It is likely that protein-protein interactions between the two dimers

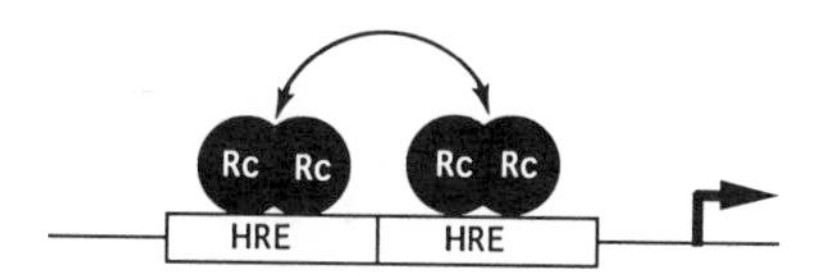

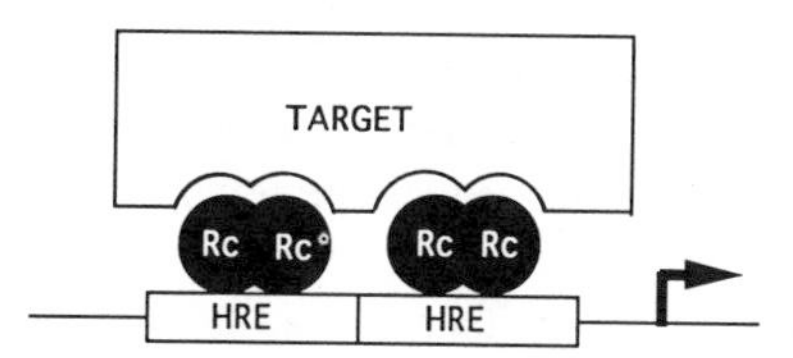

Figure 7 Synergism between two HREs. Model depicts how two HREs could synergize in a target gene activation. Models that depend upon cooperative binding of receptor dimers to DNA (*top*) and synergistic interaction of receptor dimers with a target on the transcription machinery (*bottom*) are shown.

facilitate cooperativity. Tsai et al found that two molecules of the *E. coli*-expressed DBD of GR were able to bind to a single response element in a cooperative manner (98). However, two dimers of GR DBDs were not able to bind cooperatively to two adjacent GRE/PREs, but the full-length receptor could do so (154; SY Tsai, M-J Tsai, BW O'Malley, unpublished observation). Since both the DBD and the full-length receptor can bind DNA, cooperative binding between two receptors appears to require protein-protein interactions outside the DBD domains. By deletion analysis (155), the region required for cooperative binding was localized to position 250–417 of cPR, which includes the DNA-binding domain (amino acids 280–369).

Although cooperative binding is an attractive model, it is unlikely to account for the total level of transcriptional synergism observed at target genes, especially when different HREs are present. For example, a PRE and an ERE can activate transcription of a linked target gene synergistically, but very little cooperative binding between PR and ER can be observed (155). The second mechanism proposed in Figure 7 (*lower model*) suggests that cooperative interactions of receptors or transcription factors with multiple target sites (presumably components of the aggregate transcriptional machinery) may play a role in synergism. Attempts to define the regions important for this second type of cooperativity have proven to be unsuccessful. Mutation of a variety of regions in PR decreases its ability to synergize with ER (155). Similarly, synergistic interaction has been observed between GR and other transcription factors, and the structure important for this interaction has been shown to be complex (156).

ROLE OF LIGAND IN RECEPTOR TRANSFORMATION AND ACTIVATION

Role of Ligand

Intracellular hormone receptors have either neutral or silencing activity in the absence of ligand. Upon binding their cognate ligand, they usually become positive regulators, although certain genes can be repressed if the *cis*-elements are arranged appropriately. The receptor activation process is called transformation. Members of the receptor superfamily can be roughly divided into two subgroups according to their functional properties. Group A consists of the larger steroid receptors GR, AR, PR, MR, and ER, which have longer A/B domains. In the absence of hormone they have been shown to exist as 8–10S complexes associated with heat-shock proteins (hsp90, hsp70, and hsp56) (161–169). For the most part, unliganded group A receptors do not bind to DNA and thus have neither transcriptional nor silencing activity. Upon binding hormone, heat-shock proteins dissociate, the receptors sediment as a 4S complex, and they are able to dimerize, bind to DNA, and transactivate a target gene. In contrast, Group B receptors such as TR, RAR, VDR, RXR, PPAR, and most (if not all) of the orphan receptors have short A/B domains. They are nuclear receptors and have not been shown to be associated with heat-shock proteins. They appear to be able to bind to DNA in the absence of ligand. Upon binding hormone, all receptors appear to undergo transformation to activators. Although there are distinguishing features between the two groups of intracellular receptors, the high degree of conservation in both the DNA- and ligand-binding domains suggests that the underlying mechanism of hormonal regulation is similar. In this section we discuss hormonal control of these groups of receptors separately, then we offer a unified model to resolve their distinguishing features.

Due to the observation that ligand dissociates heat-shock proteins from a Group A receptor coincident with activation, it was considered that the role of the ligand is simply to remove the inhibitory heat-shock proteins, thus allowing DNA binding and transcription (60, 122). In this scenario, the receptor should be constitutively active after heat-shock protein removal. Picard et al, using a yeast genetic system to reduce the cellular level of hsp90, found that the response of GR to its ligand had been altered (170). However, receptor activity remained dependent on exogenous ligand. These results argue that removal of hsp90 itself is not sufficient for receptor transformation. It is not clear from these experiments, however, whether other receptor-associated heat-shock proteins, including hsp70 and hsp56 or a residual amount of hsp90, had any effect on the hormone dependency.

Direct evidence that removing heat-shock proteins is not sufficient to convert a receptor from a hormone-dependent to a constitutively active molecule came from in vitro studies (126). Human PR from T47D cells was used to develop a hormone-dependent DNA-binding and transcription system (5, 117). Using this hormone-dependent, cell-free system and purified receptor free of hsp90, hsp70, and hsp56, it was demonstrated that the heat-shock protein–free receptor still binds to DNA and transactivates a target gene in a hormone-dependent manner. These results argue strongly that the steroid hormone plays an additional role(s) beyond removal of heat-shock proteins.

One can envision two possible roles that hormones may have in receptor activation. First, the receptor may require covalent modification (e.g. phosphorylation) for DNA binding and transactivation. Indeed, ligand-dependent phosphorylation has been observed with PR (171–173). However, as is discussed below, a great deal of phosphorylation occurs after the receptor binds to DNA (174–176). Furthermore, removal of receptor phosphates by alkaline phosphatase treatment did not prevent hormone-dependent binding to an HRE (G Allan, SY Tsai, M-J Tsai, BW O'Malley, unpublished observation). Thus, while it may be important for maximal gene transactivation activity, phosphorylation of receptor is not absolutely required for ligand-dependent activation of receptor.

The second possibility is that a conformational change is induced by hormones. It has been observed in a number of laboratories that upon binding hormone, certain steroid receptor-DNA complexes migrate to a different position from the ligand-free receptor in a gel mobility shift electrophoretic assay (118, 177–179). In addition, Fritsch et al showed that hormone-bound estrogen receptor partitioned in PEG-palmitate differently from the unliganded receptor (180). These results suggested that the receptor may have a different conformation after binding ligand. In order to show more directly that this is the case, Allan et al used protease digestion and monoclonal antibody mapping as probes to detect ligand-dependent conformational changes (181). It was reasoned that if ligand induced a different conformation in the receptor, protease digestion sites may become more or less sensitive to digestion, depending on their location. Indeed, it was found that PR is generally quite sensitive to protease digestion in the absence of ligands. Upon binding progesterone or other agonists, a portion of the receptor becomes very resistant to protease digestion and a ~30-kDa protease-resistant fragment appears. Resistance to protease digestion appears due to a major structural change, since resistant bands of about the same size were observed regardless of what proteases were used. Thus, progesterone induces a conformational change converting virtually the entire ligand-binding domain (E region, Figure 2) to a very compact structure. The conformational change was confirmed also by monoclonal epitope mapping in this region (70, 182).

The protease-resistant conformational change occurs before heat-shock proteins are released from the PR (181). In fact, it is likely that this conformational change induces their dissociation from the 8–10S receptor complex. In the absence of heat-shock proteins and with the new conformation in the ligand-binding domain, the receptor is able to dimerize and bind to target DNA response elements.

The observation that protease-resistant conformations are induced in receptors by hormones has been extended to estrogen, glucocorticoid, and androgen receptors (118). Importantly, experimental conditions that induce the protease-resistant conformation in the absence of ligand also create a receptor that is active in transcription. Thus, it is speculated that this conformation is essential for a receptor to carry out transactivational regulation.

Substantial new evidence on the mechanism of action of hormone antag-

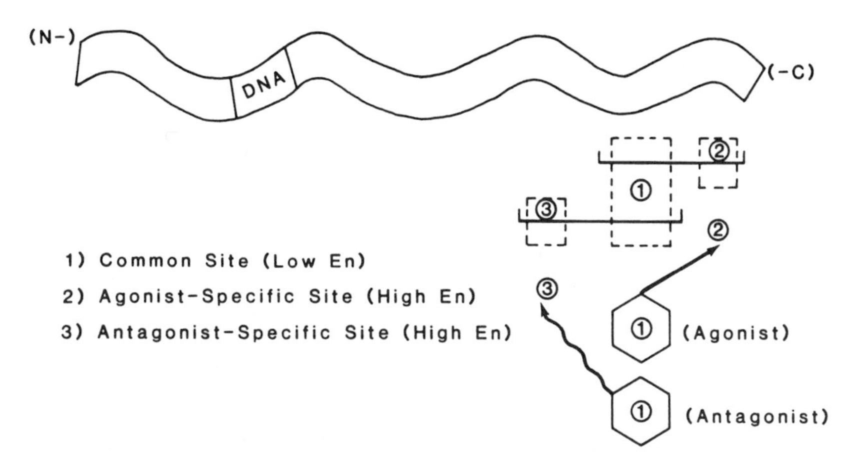

Figure 8 Two-step binding of ligand to progesterone receptor. This model argues that agonists and antagonists of the progesterone receptor recognize distinct regions of the ligand-binding domain. It is suggested that each may interact initially with a low-affinity site and then make high-energy (high-En) contacts with different specific binding sites, consequently inducing different C-terminal structures.

onists comes from recent studies (181). Previously, it was believed that hormone antagonists are compounds that can bind competitively with authentic agonists at the identical site on their cognate receptors. Recent evidence has changed our appreciation of the sites at which agonistic and antagonistic ligands bind to the LBD of intracellular receptors. Mutational analyses have clearly established that the agonist- and antagonist-binding sites on the human PR are distinct (70, 183). Rather than agonists and antagonists competing for the identical sites, they appear to compete for agonist- or antagonist-induced structures. For example, agonists bind most tightly to the extreme C-terminal tail of the LBD, while antagonists make their high-affinity contacts at a more proximal region, as indicated in Figure 8. Variations on this theme may apply also to the estrogen receptor (184) and glucocorticoid receptor (183).

Most antagonists are able to stimulate receptors to bind to DNA and to undergo phosphorylation (117, 175, 179). Nevertheless, the antagonist-bound receptor is not active in turning on target genes, although some antagonists have partial agonist activity. A question can be raised as to why an antagonist is able to promote most of the events observed with an agonist, yet the DNA-bound receptor is transcriptionally inactive. From the above discussion, we would predict that an antagonist must also cause conformational change to expose the dimerization domain, allowing the receptor to bind to DNA with high affinity. Nevertheless, this conformational change must differ functionally from that induced by the agonist. This hypothesis is consistent with the observation that an antagonist-bound receptor-DNA complex migrates differently electrophoretically from that of an agonist-receptor-DNA complex (118, 177–179). More direct evidence came from Allan et al (181). It was observed that like the agonist-bound receptor, the antagonist-bound receptor is resistant to protease digestion, but the resistant fragment is slightly shorter than that of the agonist-bound receptor (27 vs 30 kDa). The results suggest that antagonist binding induces a conformational change in a receptor that is different from that induced by an agonist. A similar conclusion has been drawn from antibody studies (70, 182). It was found that an antibody that recognizes the C-terminal tail of PR recognizes the receptor when it is ligand-free or antagonist-bound. However, it cannot recognize the epitope when the receptor is bound to an agonist.

The difference between agonist- and antagonist-induced conformations has been localized to the extreme C terminus of the LBD. The protease digestion site 3 kDa upstream of the C-terminal end is available to the protease when the receptor is complexed with an antagonist, but the same site in the agonist-bound receptor is hidden from the protease. Therefore, we reasoned that the sequences within the final 3 kDa of the PR must play a role in distinguishing between agonist and antagonist activity. Functional evidence for the role of the C-terminal end comes from studies by Vegeto et al, who used yeast as a system to generate mutants that have altered ligand specificity (70). Using this approach, a PR mutant termed UP-1 was generated. This mutant was no longer responsive to progesterone agonists but was activated by progesterone antagonists such as RU486. Sequence analysis indicated that the UP-1 mutation results from a frameshift replacing the last 54 amino acids of the receptor. More detailed studies showed that the last 23 amino acids are required for progesterone binding but that the last 54 amino acids are not needed for RU486 binding (G Allan et al, unpublished observation). The results indicate that the C-terminal end of PR contains a repressor function that downregulates the activity of aporeceptor or receptor bound to an antagonist. Inactivation of this repressor function by deletion creates a receptor that is now able to activate a target gene in the presence of antagonists.

These results are consistent with the conclusion drawn from protease digestion studies, which indicated that the C-terminal end of PR plays an important role in determining antagonist activity. Similar differential protease digestion patterns were observed with ER and GR, suggesting that a related repressor function might also exist in these receptors. However, with these receptors the story may be somewhat more complicated, since a simple deletion of C-terminal amino acids results in similar decreases in receptor activity for both agonist and antagonist in each case (185; G Allan et al, unpublished observation). Nevertheless, the hypothesis is consistent with work described by Danielian et al (186), who also demonstrated the existence of a transactivation domain in the C-terminal region of ER. Strong evidence for a repressor function in GR comes from a report by Lanz et al (185), who showed that point mutations of two hydrophobic amino acids (770 and 771) and deletion of another pair of conserved amino acids (780 and 781) present in the C terminus resulted in a receptor responsive to the antagonist, RU486, but not to the agonist, dexamethasone. Similarly, an androgen receptor isolated from LNCap cells, which has a mutation at amino acids 868 (Thr to Ala) is activated now by anti-androgens (187). Taken together, these results support the general existence of repressor functions within the C terminus of the steroid receptor superfamily members.

The identification of an intramolecular repressor function in Group A receptors raises the question of whether the repressor function of the Group A receptor is related to the silencer function of Group B receptors. It is interesting to note that both of these two functions are localized within the C-terminal end of the molecule. Exchange of the C-terminal sequence between these two groups of receptors could answer this question.

Model

The hypothetical model presented in Figure 9 illustrates our current hypothesis for activation of a Group A steroid receptor by a hormonal agonist (H) or antagonist (AH). In the absence of ligands, the receptor exists in a conformation such that the DBD could be available for DNA binding. However, the dimerization domain is not available, so the receptor cannot bind to DNA with high affinity. Furthermore, the activation domain(s) (AF) is under the control of the intramolecular repressor (R), and therefore, is unable to activate transcription. Upon binding agonist (H), the receptor alters conformation, resulting in the exposure of the dimerization domain; the receptor dimerizes then and binds to DNA. In parallel, the R domain is sequestered from the surface of the receptor by the hormone (since the R region is required for agonist binding) into a compacted structure that is not available to protease or antibody and thus is not able to continue repressing the AF(s). Without inhibition by the R region, the AF can interact with its target, presumably components of the transcriptional machinery, to activate gene transcription. Since there are multiple AFs in a given receptor (49–53, 64–66), it is possible that only the C-terminal AF in the hormone-binding domain is under the R regulation. This possibility may explain why some antagonists have partial agonist activity, presumably by utilizing other AFs in the N terminus. It should be noted also that inhibition of AF by the R region is not necessarily as depicted in Figure 9. It could either block

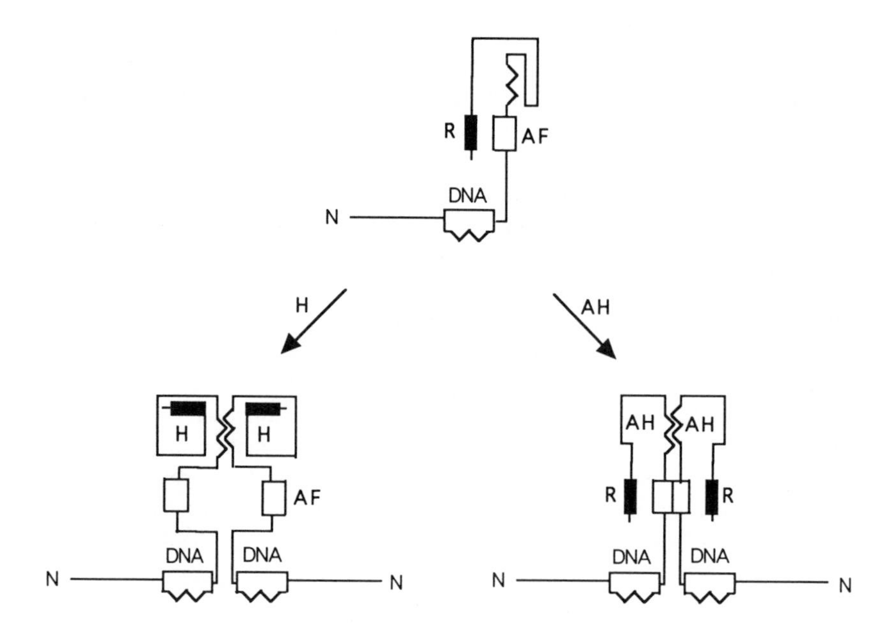

Figure 9 Mechanism of action of agonist and antagonist ligands. A key point is that an antihormone (AH) does not induce a structure that removes a surface repressor function. See text for description. H, hormone; AH, antihormone; AF, activation function; R, repressor function; DNA, DNA-binding domain; ⁓, dimerization domain; N, N terminus.

the AF by physical means, compete for the target component, or serve as a binding site for a distinct repressor protein.

In contrast, when the receptor binds to the antagonist (AH), there is an incomplete conformational change, also resulting in exposure of the dimerization domain. Thus, the receptor can dimerize and bind to DNA. In contrast, the C-terminal R region is still available for protease and antibody recognition and to interfere with the AF. Thus, while the antagonist-occupied receptor can bind to DNA, it is not able to activate gene transcription.

The Group B receptors have certain distinguishing features. They are able to bind to DNA in the absence of ligand. The lack of heat-shock protein association with this group of receptors may contribute to their ability to bind DNA without a ligand. Also, certain receptors in this group (e.g. TR) may bind to DNA as monomers. However, it is more likely that their ability to form a homodimer or heterodimer (with RXR) in the absence of hormone is the key feature that allows them to bind to DNA with high affinity (69, 100, 101, 103–111). The fact that they cannot activate target genes even though they are bound to DNA may again relate to conformational studies obtained by protease digestion. Leng et al (188) found that in the absence of ligand, TR and RAR are sensitive to protease digestion. Ligands also make the ligand-binding domains of these receptors protease resistant. Thus, the story with this group of receptors is similar to that of Group A: The protease-resistant conformation is required for receptor to be able to activate transcription. Recently, Toney et al (189) demonstrated a ligand-dependent conformation change in the TR ligand-binding domain using circular dichroism.

Figure 10 summarizes the overall pathway for steroid/thyroid hormone receptor activation of target genes. For Group A receptors, the conformation changes upon hormone binding, and results in the dissociation of hsps, dimerization, and binding to target HREs. After phosphorylation by a DNA-dependent protein kinase (discussed below), the receptor is able to facilitate the formation of the preinitiation complex and induce gene transcription. Group B receptors, which are bound to DNA as dimers, await conformational activation by ligand in situ. Transcriptional effects follow. Further studies are needed to determine if the silencing function of Group B receptors is identical to the repressor function of Group A receptors and to substantiate the apparent ligand-induced conformational changes by structural analyses of crystals.

FACTORS INFLUENCING RECEPTOR ACTIVITY

Phosphorylation

A major additional process that can enhance ligand-dependent activation of receptors is phosphorylation. Steroid hormone receptors are highly phosphorylated proteins (190), and several of the phosphorylation sites have been identified on selected receptors (171–173). Certain receptor residues are phosphorylated in a ligand-dependent fashion and some are ligand-independent (171, 173, 191–193). Importantly, a significant level of phosphorylation occurs only after the activated receptor binds to DNA (174–176). Weigel et al demonstrated that this phosphorylation was carried out by a nuclear DNA–dependent protein kinase (174).

Although we know that receptors are phosphoproteins, we do not understand the precise consequences of this phosphorylation. From functional correlations, it has been proposed that receptor phosphorylation is important for gene transactivation capacity (175). Such studies have shown hormone-induced phosphorylation that precedes gene activation, both in cells (194–196), and in cell-free conditions (175). The latter appears inconsistent with one theory that predicted that phosphorylation of receptor leads to inactivation. Additional evidence for a positive role for phosphorylation arises from studies on the activation of ligand-dependent receptor activity by agents

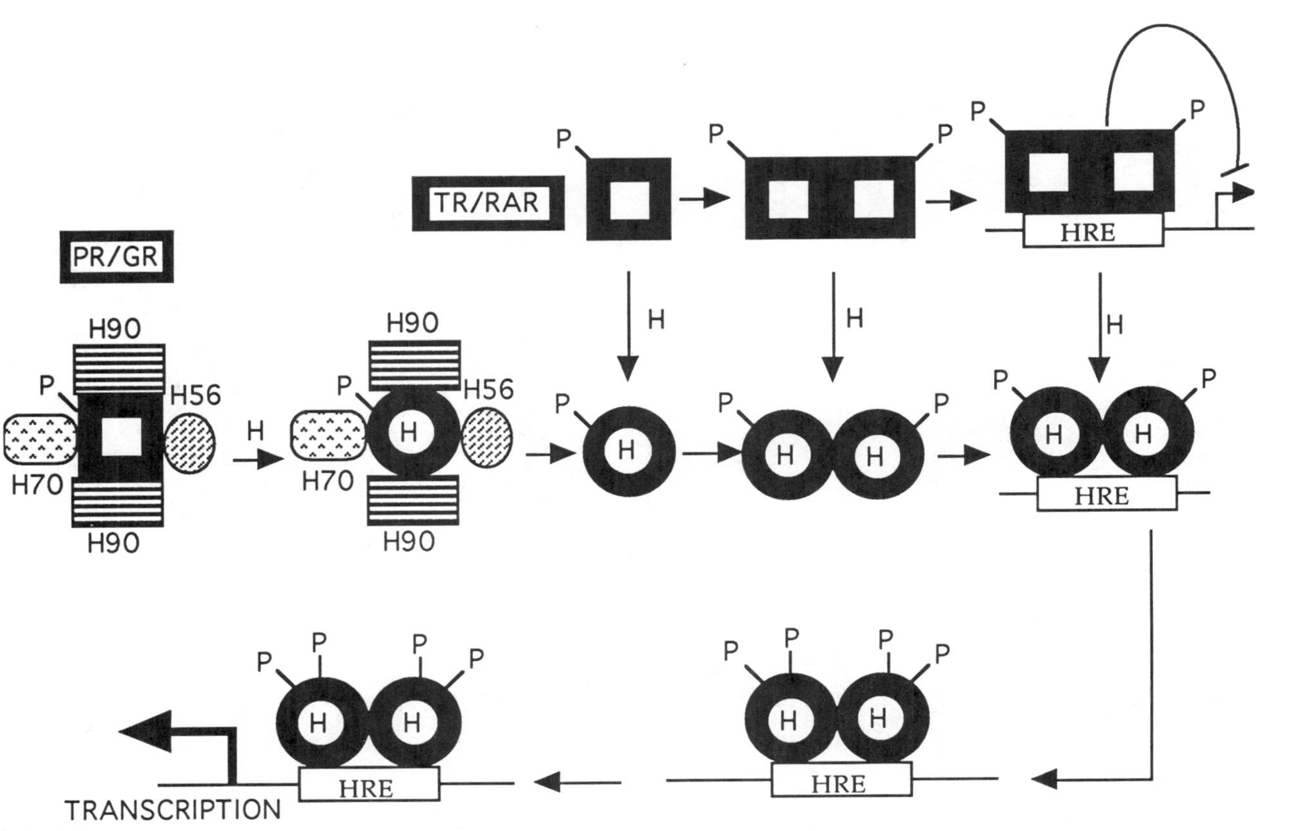

Figure 10 Hormone-dependent activation of the steroid/thyroid receptor family. Group A receptors represented by PR/GR and Group B by TR/RAR. See text for description. H90, heat-shock protein 90; H70, heat-shock protein 70; H56, heat-shock protein 56; and P, phosphorylation (does not represent precise number of sites).

that induce phosphorylation or inhibit cellular phosphatases (194, 196–198). The ability of H8, a kinase inhibitor, to dampen receptor activity in cell culture systems further supports the importance of phosphorylation in gene activation (194, 198). To date, mutational analysis has shown that a point mutation of any one of the phosphorylation sites of progesterone receptors has only a marginal effect on its transactivation potential (196). Perhaps the most dramatic effect has been demonstrated for the estrogen receptor, where a mutation in one of the N-terminal amino acids caused a significant decrease in transactivation capacity (199). It is possible that functional redundancy among phosphorylation sites makes it impossible to see a drastic effect by mutating only one of the many sites. Appropriate combinations of mutations may need to be generated to address this question. It is quite possible that still-unidentified phosphorylation sites are contributing to function. We suggest that it may be more accurate to consider intracellular liganded receptors as inherently active transcription factors whose potential simply is enhanced by phosphorylation in a cell and promoter context manner, rather than to consider phosphorylation as a "switch" for activation.

In addition to a transcriptional role, phosphorylation has been proposed to be important for receptor intracellular trafficking. Nuclear retention of GR is inhibited by okadaic acid, a strong phosphatase inhibitor of types 1 and 2A (200). Thus, phosphorylation may be important for proper shuttling of receptor between the cytoplasm and the nucleus. Finally, phosphorylation has been proposed to be involved in receptor transformation and processing. However, RU486 has been shown to induce phosphorylation without an increase in receptor processing (201). Thus, it is unlikely that phosphorylation plays its major role in this event.

Ligand-Independent Activation

In addition to the putative synergistic effect of phosphorylation on ligand-dependent activation of receptor (197), several recent reports indicate that agents that stimulate intracellular phosphorylation pathways can also activate receptors in the "absence" of ligand. It was first demonstrated by Denner et al that 8-Br-cAMP and okadaic acid can activate chicken PR in the complete absence of progesterone (197). Shortly thereafter, dopamine was shown to be a biologic activator of certain steroid receptors (202, 203); also, cAMP inducers such as cholera toxin plus isobutylmethylxanthine were reported to activate native steroid receptors in a ligand-independent manner (204, 205).

To our great surprise, dopamine was shown to activate certain receptors (e.g. cPR and hER) in the absence of ligand and by events initiated at its cell membrane D1 receptors. Glucocorticoid and mineralocorticoid receptor subtypes were not activated (203). Pure antagonist ligands (e.g. ICI164,384) prevent this crosstalk from the membrane by driving the steroid receptor into an unresponsive form (203, 205a). Finally, recent studies indicate that dopamine requires an intact PR for its effect on sexual behavior in the ventral medial nucleus of the intact rat (S Mani, J Clark, BW O'Malley, unpublished observation).

Additional recent evidence indicates that growth factors are another class of hormone that appears able to activate certain steroid receptors, particularly the human ER. In cultured cells, IGF-1, EGF, and TGF-α have been demonstrated to stimulate specific transcription of estrogen target genes by activating hER in a ligand-independent fashion (204, 206, 206a). There is a notable in vivo corollary with these experiments. EGF, acting via its membrane receptor, has been shown to cause uterine growth in ovariectomized mice that is indistinguishable from that caused by estrogen itself (205). In fact, EGF has been shown to act via its membrane receptor to drive unoccupied ER to its nuclear form and turn on estrogen target genes in mice; this effect of EGF can be blocked also by administration of a pure antiestrogen (206, 206b).

This newly discovered crosstalk between membrane receptors and intracellular receptor pathways has exciting biological implications. Most certainly, it raises the question as to whether we must expand our concept of the role of intracellular receptors. As shown in Figure 11, perhaps we should expand our model for activation of intracellular receptors to include activation from cell membrane sites. This revision may be more palatable if we consider these receptors to be transcription factors under pleiotropic regulation. Nevertheless, before we can accept the broad physiologic importance of this alternate and ligand-independent pathway for activation of receptors, we need more in vivo substantiation. Also, we do not understand the mechanism by which hormones whose receptors are located at the cell membrane can activate intracellular receptors; it is presumed that it might involve either phosphorylation of the receptor itself or a specific coactivator required by receptor. The fact, however, that select membrane receptor activation (e.g. growth factors, neurotransmitters, etc) can play a synergistic or additive role along with intracellular ligands to activate genes is gaining rapid acceptance. Setting aside the concept of complete ligand-independence, it appears almost certain that "cooperation" between intracellular and plasma membrane signaling pathways plays a role in the cellular physiology of certain of these receptor superfamily members.

Nuclear Transcription Factors

Another way to modulate receptor activity is via protein-protein interactions at the level of the nucleus. Interactions of certain steroid receptors with components of AP1 (Fos and Jun) have been shown to downregulate a

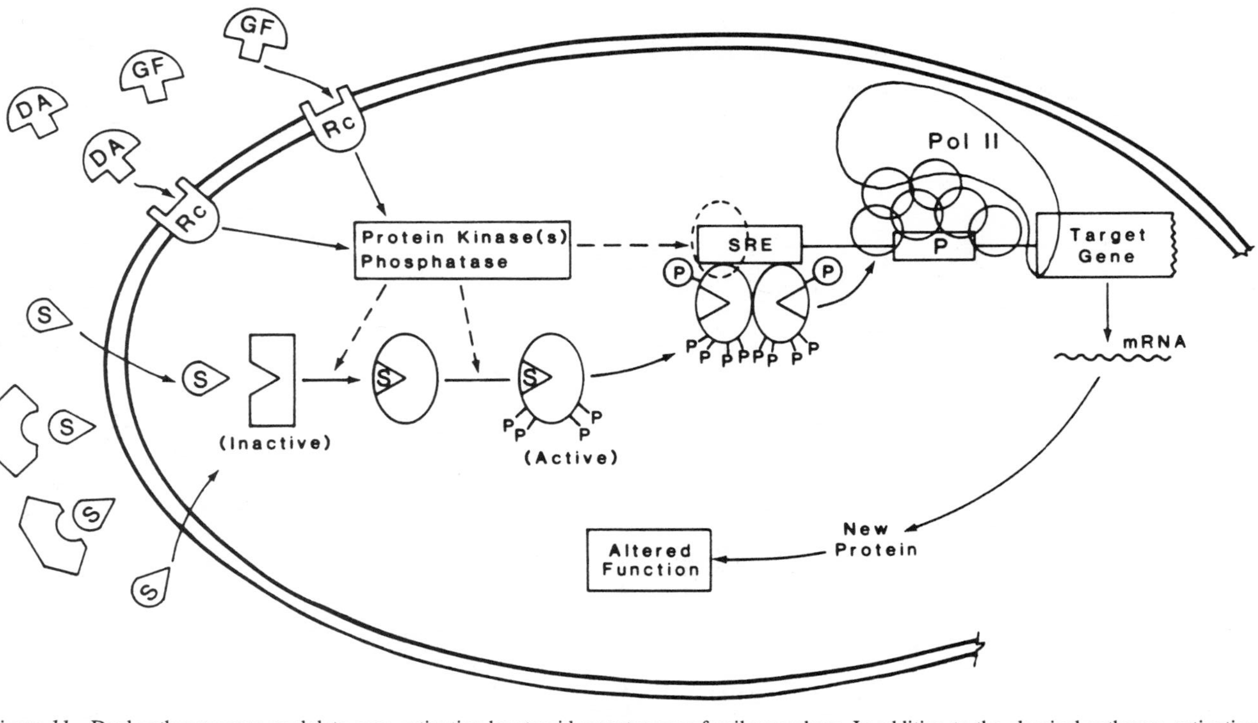

Figure 11 Dual pathways may modulate gene activation by steroid receptor superfamily members. In addition to the classical pathway, activation may be initiated at cell-surface receptors (e.g. dopamine, growth factors, etc), which results in either a presumptive phosphorylation cascade that phosphorylates receptor directly or in some "specific" cofactor needed for receptor function. See text for description. S, steroid; DA, dopamine; GF, growth factor.

target gene (147, 207–209). The mechanism of this modulation remains controversial. Some studies showed that interaction of the glucocorticoid receptor with AP1 results in a complex that is unable to bind to DNA (209); another report indicated that the interaction takes place while the components remain bound to DNA (210). This latter mechanism appears to be gaining favor. Others have proposed that, depending on the relative compositions of Jun and Fos, GR receptor activity either can be enhanced or inhibited (147). The confusion may derive from the use of different target genes, levels of receptors, AP1 types, and cell types. In any event, there appears to be some validity to the existence of this additional form of crosstalk between ligand-activated and phosphorylation-regulated pathways at the "nuclear" level. Again, in vivo studies are needed to assess the physiological importance of this phenomenon.

Finally, logic argues that intracellular receptor activity is dependent also on the cellular environment and on the local concentrations of attendant transcription factors. It is clear that overexpression of one receptor can result in the squelching of its own or another receptor's activity (211–213). Furthermore, it has been shown that the functional activity of various transactivation domains of ER varies in the context of different cell types (214). This latter result suggests that the targets for various activation domains may be distinct. Since the basic transcriptional machinery is similar for all genes, it is likely that factors other than general (TATA) transcription factors, such as coactivators, play a role in receptor activation. Using an in vitro reconstituted system, Shemshedini et al have identified a fraction (called TIF for transcription intermediary factor) that can eliminate the self-squelching effect of AF-1 but has no influence on basal transcription (214). Yamamoto and we have used a yeast-based genetic system to demonstrate that several factors, including SWI1, SWI2, SWI3, SIN3, and SPT6, have an effect on the transcriptional activity of different receptors (215–217). These coactivators may act to bridge receptors to the core transcription machinery or to alter chromatin structure. Overall, these reports suggest that multiple helper factors are involved in receptor transactivational activity.

McDonnell et al have used a genetic selection system to identify yeast SSN6 protein as a transcriptional repressor protein for receptors, since mutation of SSN6 results in a drastic increase in receptor activity (218). A mammalian homolog has not been identified yet. Consistent with these data, however, the silencing activity of TR can be competed by cellular overexpression of an internal fragment of the TR ligand-binding domain, indicating the existence of a soluble corepressor (A Baniahmad, SY Tsai, M-J Tsai, BW O'Malley, unpublished results). Therefore, in addition to coactivators, we suggest that corepressors exist within cells. The mechanisms by which these cofactors enhance or reduce receptor activity remain to be determined, but are a subject of intense interest.

CHROMATIN STRUCTURE AND RECEPTOR ACTION

Chromatin structure must play an important modulatory role in gene transcription in vivo (for reviews see 219, 220). The interaction of transcription factors with *cis*-acting elements is altered by chromatin structure, depending on whether that particular gene is in an active or repressed state. Genes located in heterochromatin are unable to bind to transacting factors and thus remain inactive. In a general sense, heterochromatin has a positive effect on transcription since it prevents the loss of limited transfactors bound in unproductive complexes at inactive genes. The expression of genes in euchromatin regions also depends on the chromatin structure and the availability of *trans*-acting factors necessary for their expression. When an inducible *trans*-acting factor binds to a promoter or enhancer region of a gene, either between nucleosomes or within a nucleosome, changes are likely to be created in the local chromatin structure; general transcription factors can assemble then to form a preintiation complex at the promoter (for review see 220). Overall, nucleosomal structure appears to act as a general repressor to restrict the availability of DNA regulatory elements.

The effect of chromatin structure on the action of steroid hormone receptors has been studied by a number of investigators (221–224). The MMTV promoter is the best-studied example of how hormone receptors alter chromatin structure to allow the expression of a target gene. In the absence of hormone, the mouse MMTV promoter contains six phased nucleosomes that prevent NF-1 and other transcription factors from binding to their *cis*-acting elements, thus preventing the expression of viral RNA (221, 222, 225, 226). The initial nucleosomal structure does not prevent activated receptors from binding to their response elements. Rather, in the presence of glucocorticoid, the GR can interact with the enhancer region of the MMTV promoter, resulting in the creation of DNase I hypersensitive sites around the GRE and in other factor-binding sites. This change is presumably caused by the removal or the modification of a nucleosome(s) within this region. Thus, a GR-induced change in the chromatin structure allows transcription factors NF-1 and others to interact with their *cis*-acting element, leading to transcription. This conclusion is supported further by Pina et al (223), who used an in vitro reconstitution system to demonstrate that GR can bind to in vitro reconstituted chromatin with an affinity similar to its binding to protein-free DNA. In contrast, NF-1, which is able to bind to protein-free DNA with high affinity, is unable to interact with the NF-1-binding site within reconstituted chromatin. Upon binding GR or PR, the reconstituted nucleosome now becomes more accessible to exonuclease digestion, and presumably to transcription factors binding in vivo.

In addition to a general inhibitory role in factor binding, nucleosomes may promote receptor induction of transcription more directly. Schild et al (224), using the *Xenopus* vitellogenin B1 gene promoter and an estrogen-dependent in vitro transcription system, demonstrated that the formation of a nucleosome between −300 and −140 is necessary for the vitellogenin gene to be transcribed efficiently. Formation of the nucleosome is thought to create a static loop, which facilitates the interaction of an ERE located at −300 with the promoter transcriptional machinery located downstream of the nucleosome. Thus, receptor regulation of gene activity may be influenced by chromatin structure in three possible ways. It can pack unwanted genes into heterochromatin, providing a higher concentration of attendant transcription factors for induced genes. It can inhibit gene transcription by preventing binding of other required *trans*-acting factors. Finally, it can enhance receptor-factor interactions to facilitate the transcription process.

SUMMARY AND PERSPECTIVES

In recent years, the availability of receptor cDNAs has allowed rapid progress in understanding receptor functional domains. The development of in vitro systems for assessing DNA binding and transactivation has further hastened the pace of our understanding of the detailed mechanisms by which receptors activate their target genes and of the role hormones play in this process. In this review, we have concentrated our discussion on more recent molecular advances. As is usually the case for biologic processes, however, many questions remain to be answered. An incomplete list of needed tasks and/or information includes: (*a*) purification of mammalian coactivators and corepressors and determination of their mechanism of action; (*b*) understanding of the physiological role of ligand-independent receptor activation by neurotransmitters and growth factors that regulate intracellular phosphorylation; (*c*) knowledge of the precise functional relevance of specific interactions of receptors with general transcription factors; (*d*) understanding of the relationship between silencing of the core promoter (e.g. TR) and intramolecular repression of aporeceptor (e.g. PR); (*e*) determination of the location and mechanism of action of positive versus negative transactivational functions within the receptor; (*f*) a more detailed understanding of receptor-induced changes in chromatin structure; (*g*) knowledge of the generality of negative HREs; (*h*) information on gene activational selectivity with distinct combinations of heterodimers; (*i*) more knowlege of receptor phosphorylation and gene regulatory capacity; and (*j*) determination of the precise structure of the LBD when it is coupled to ligand (agonist or antagonist).

In addition to the receptors with known ligands, there is a vast group of orphan receptors whose function and ligand are still unknown. A detailed discussion of these molecules, which represent the largest division of the superfamily, is beyond the scope of this review. At present, we are not certain if these orphans have a ligand. From their developmental distribution and timing of expression, however, it is accepted that these orphans must play important roles in tissue development and cellular physiology (227, 228). Because of the large number (>40) of orphan receptors identified to date, a great deal of experimentation is required to sort out this complex group of transcription factors. Finally, many different isoforms of both classical receptors and orphan receptors have been identified. Several isoforms of ER and a large number of forms of RAR have been shown to exist to date, most arising via alternative splicing of mRNA precursors (20, 21, 39–48, 229, 230). Variant forms of ER and PR have been demonstrated to have differential gene regulatory activities in cells in culture (41, 55, 214, 231, 232). Some isoforms of orphans are expressed in different cells, but many have overlapping expression patterns. Why do so many different receptor isoforms exist? Since this and the other questions listed above are complex, it is likely that a continuing series of breakthroughs in technology is required to achieve a complete understanding of receptor transactivation of genes. In any event, these unanswered questions should provide a good foundation for additional experimentation in this explosive field of study and will almost certainly serve as a basis for future reviews on the steroid/thyroid receptor superfamily.

ACKNOWLEDGMENTS

We thank Drs. George Allan and Sophia Y. Tsai for their critical comments. Also, we extend our appreciation to all members of our laboratories, both past and present, for their significant experimental contributions. This work was supported in part by NIH grants to Drs. Bert W. O'Malley and Ming-Jer Tsai.

Literature Cited

1. Jensen EV, Suzuki T, Kawashima T, Stumpf WE, Jungblut PW, Desombre ER. 1968. *Proc. Natl. Acad. Sci. USA* 59:632–38
2. O'Malley BW, McGuire WL, Kohler PO, Koreman SG. 1969. *Recent Prog. Horm. Res.* 25:105–60
3. Jensen EV, Jacobson HI. 1962. *Recent Prog. Horm. Res.* 18:387–414
4. Evans RM. 1988. *Science* 240:889–95
5. Bagchi MK, Tsai SY, Tsai M-J, O'Malley BW. 1990. *Nature* 345:547–50
6. Klein-Hitpass L, Tsai SY, Weigel NL,

Allan GF, Riley D, et al. 1990. *Cell* 60:247–57
7. Freedman LP, Yoshinaga SK, Vanderbilt JN, Yamamoto KR. 1989. *Science* 245:298–301
8. Corthesy B, Hipskind R, Theulaz I, Wahli W. 1988. *Science* 239:1137–39
9. Yamamoto KR. 1985. *Annu. Rev. Genet.* 19:209–52
10. Beato M. 1989. *Cell* 56:335–44
11. Beato M. 1991. *FASEB J.* 5:2044–51
12. O'Malley BW, Tsai SY, Bagchi MK, Weigel NL, Schrader WT, Tsai M-J. 1991. *Recent Prog. Horm. Res.* 47:1–26
13. O'Malley BW. 1990. *Mol. Endocrinol.* 4:363–69
14. Moudgil VK. 1990. *Biochem. Biophys. Acta* 1055:243–58
15. Gronemeyer H, Green S, Kumar V, Jeltsch JM, Chambon P. 1988. *Steroid Receptors and Disease: Cancer, Autoimmune, Bone and Circulatory Disorders,* ed. PJ Sheridan, K Blum, MC Trachtenberg, pp. 153–87. New York: Basel
16. Chang CS, Kokontis J, Liao SS. 1988. *Proc. Natl. Acad. Sci. USA* 85:7211–15
17. Lubahn DB, Joseph DR, Sullivan PM, Willard HF, French FS, Wilson EM. 1988. *Science* 340:327–30
18. McDonnell DP, Mangelsdorf DJ, Pike WJ, Haussler MR, O'Malley BW. 1987. *Science* 235:1214–17
19. Hamada K, Gleason SL, Levi BZ, Hirschfeld S, Appella E, Ozato K. 1989. *Proc. Natl. Acad. Sci. USA* 26:8289–93
20. Mangelsdorf DJ, Borgmeyer U, Heyman RA, Zhou JY, Ong ES, et al. 1992. *Genes Dev.* 6:329–44
21. Koelle MR, Talbot WS, Segraves WA, Bender MT, Cherbas P, Hogness DS. 1991. *Cell* 67:59–77
22. Becker-Andre M, Andre E, Delamarter JF. 1993. *Biochem. Biophys. Res. Commun.* 194:1371–79
23. Wang LH, Tsai SY, Cook RG, Beattie WG, Tsai M-J, O'Malley BW. 1989. *Nature* 340:163–66
24. Wang LH, Ing NH, Tsai SY, O'Malley BW, Tsai M-J. 1991. *Gene Express.* 1:207–16
25. Ladias JAA, Karathanasis SK. 1991. *Science* 251:561–65
26. Miyajima N, Kadowaki Y, Fukushige SI, Shimizu SI, Semba K, et al. 1988. *Nucleic Acids Res.* 16:11057–66
27. Issemann I, Green S. 1990. *Nature* 347:645–50
28. Schmidt A, Endo N, Rutledge SJ, Vogel R, Shinar D, Rodan GA. 1992. *Mol. Endocrinol.* 6:1634–41
29. Klock G, Strahle U, Schutz G. 1987. *Nature* 329:734–36
30. Wheeler C, Komm BS, Lyttle CR. 1987. *Endocrinology* 120:919–23
31. Miyajima N, Horiuchi R, Shibuya Y, Fukushige SI, Matsubara KI, et al. 1989. *Cell* 57:31–39
32. Lazar MA, Hodin RA, Darling DS, Chin WW. 1989. *Mol. Cell. Biol.* 9:1128–36
33. Milbrandt J. 1988. *Neuron* 1:183–88
34. Hazel TG, Misra R, Davis IJ, Greenberg ME, Lau LF. 1991. *Mol. Cell. Biol.* 11:3239–46
35. Ryseck RP, MacDonald-Bravo H, Mattei MG, Ruppert S, Bravo R. 1989. *EMBO J.* 8:3327–35
36. Lala DS, Rice DA, Parker KL. 1992. *Mol. Endocrinol.* 6:1249–58
37. Sladek FM, Zhong WM, Lai E, Darnell JE Jr. 1990. *Genes Dev.* 4:2353–65
38. Law SW, Conneely OM, DeMayo FJ, O'Malley BW. 1992. *Mol. Endocrinol.* 6:2129–35
39. Zelent A, Mendelsohn C, Kastner P. 1991. *EMBO J.* 10:71–81
40. Leroy P, Krust A, Zelent A, Mendelsohn C, Garnier JM, et al. 1991. *EMBO J.* 10:59–69
41. Kastner P, Krust A, Mendelsohn C, Garnier JM, Zelent A, et al. 1990. *Proc. Natl. Acad. Sci. USA* 87:2700–4
42. Conneely OM, Maxwell BL, Toft DO, Schrader WT, O'Malley BW. 1987. *Biochem. Biophys. Res. Commun.* 149: 493–501
43. Sap J, Munoz A, Damm K, Goldberg Y, Ghysdael J, et al. 1986. *Nature* 324:635–40
44. Weinberger C, Thompson CC, Ong ES, Lebo R, Gruol DJ, Evans RM. 1986. *Nature* 324:641–46
45. Benbrook D, Pfahl M. 1987. *Science* 238:788–91
46. Thompson CC, Weinberger C, Lebo R, Evans RM. 1987. *Science* 237: 1610–14
47. Koenig RJ, Warne RL, Brent GA, Harney JW, Larsen PR, Moore DD. 1988. *Proc. Natl. Acad. Sci. USA* 85:5031–35
48. Giguere V, Ong ES, Segui P, Evans RM. 1987. *Nature* 330:624–29
49. Giguere V, Hollenberg SM, Rosenfeld MG, Evans RM. 1986. *Cell* 46:645–52
50. Godowski PJ, Picard D, Yamamoto KR. 1988. *Science* 241:812–16
51. Kumar V, Green S, Stack G, Berry M, Jin J, Chambon P. 1987. *Cell* 51:941–51
52. Webster NJG, Green S, Jin J, Chambon P. 1988. *Cell* 54:199–207

53. Hollenberg SM, Evans RM. 1988. *Cell* 55:899–906
54. Tora L, Gronemeyer H, Turcotte B, Gaub M, Chambon P. 1988. *Nature* 333:185–88
55. Bocquel MT, Kumar V, Stricker C, Chambon P, Gronemeyer H. 1989. *Nucleic Acids Res.* 17:2581–94
56. Luisi BF, Xu WX, Otwinowski Z, Freedman LP, Yamamoto KR, Sigler PB. 1991. *Nature* 352:497–505
57. Freedman LP. 1992. *Endocr. Rev.* 13:129–45
58. Picard D, Yamamoto KR. 1987. *EMBO J.* 6:3333–40
59. Guiochon-Mantel A, Loosfelt H, Lescop P, Sar S, Atger M, et al. 1989. *Cell* 57:1147–54
60. Pratt WB, Jolly DJ, Pratt DV, Hollenberg SM, Giguere V, et al. 1988. *J. Biol. Chem.* 263:267–73
61. Chambraud B, Berry M, Redeuilh G, Chambon P, Baulieu E. 1990. *J. Biol. Chem.* 265:20686–91
62. Fawell SE, Lees JA, White R, Parker MG. 1990. *Cell* 60:953–62
63. Forman BM, Samuels HH. 1990. *Mol. Endocrinol.* 4:1293–301
64. Damm K, Thompson CC, Evans RM. 1989. *Nature* 339:593–97
65. Sap J, Munoz A, Schmitt J, Stunnenberg H, Vennstrom B. 1989. *Nature* 340:242–44
66. Graupner G, Wills KN, Tzukerman M, Zhang XK, Pfahl M. 1989. *Nature* 340:653–56
67. Baniahmad A, Kohne AC, Renkawitz R. 1992. *EMBO J.* 11:1015–23
68. Baniahmad A, Tsai SY, O'Malley BW, Tsai M-J. 1992. *Proc. Natl. Acad. Sci. USA* 89:10633–37
69. Cooney AJ, Leng X, Tsai SY, O'Malley BW, Tsai M-J. 1993. *J. Biol. Chem.* 268:4152–60
70. Vegeto E, Allan GF, Schrader WT, Tsai M-J, McDonnell DP, O'Malley BW. 1992. *Cell* 69:703–13
71. Huckaby CS, Conneely OM, Beattie WG, Dobson ADW, Tsai M-J, O'Malley BW. 1987. *Proc. Natl. Acad. Sci. USA* 84:8380–84
72. Ponglikitmongkol M, Green S, Chambon P. 1988. *EMBO J.* 7:3385–88
73. Watson MA, Milbrandt J. 1989. *Mol. Cell. Biol.* 9:4213–19
74. Liu Q, Linney E. 1993. *Mol. Endocrinol.* 7:651–58
75. Lehmann JM, Hoffmann B, Pfahl M. 1991. *Nucleic Acids Res.* 19:573–78
76. Van der Leede BM, Folkers GE, Kruyt FAE, van der Saag PT. 1992. *Biochem. Biophys. Res. Commun.* 188:695–702
77. Encio IJ, Detera-Wadleigh SD. 1991. *J. Biol. Chem.* 266:7182–88
78. Arriza JL, Weinberger C, Cerelli G, Glaser TM, Handelin BL, et al. 1987. *Science* 237:268–75
79. Zahraoui A, Cuny G. 1987. *Eur. J. Biochem.* 166:63–69
80. Marcelli M, Tilley WD, Wilson CM, Griffin JE, Wilson JD, McPhaul MJ. 1990. *Mol. Endocrinol.* 4:1105–16
81. Ritchie HH, Wang LH, Tsai SY, O'Malley BW, Tsai M-J. 1990. *Nucleic Acids Res.* 18:6857–62
82. Strahle U, Klock G, Schutz G. 1987. *Proc. Natl. Acad. Sci. USA* 84:7871–75
83. Ham J, Thomson A, Needham M, Webb P, Parker M. 1988. *Nucleic Acids Res.* 16:5263–77
84. Darbre P, Page M, King RJB. 1986. *Mol. Cell. Biol.* 6:2847–54
85. Cato ACB, Miksicek R, Schutz G, Arnemann J, Beato M. 1986. *EMBO J.* 5:2237–40
86. Cato ACB, Weinmann J. 1988. *J. Cell Biol.* 106:2119–25
87. Cato ACB, Henderson D, Ponta H. 1987. *EMBO J.* 6:363–68
88. Klein-Hitpass L, Schorpp M, Wagner U, Ryffel GU. 1986. *Cell* 46:1053–61
89. Martinez E, Givel F, Wahli W. 1987. *EMBO J.* 6:3719–27
90. Deleted in proof
91. Danielsen M, Hinck L, Ringold GM. 1989. *Cell* 57:1131–32
92. Umesono K, Evans RM. 1989. *Cell* 57:1139–46
93. Hard T, Kellenbach E, Boelens R, Maler BA, Dahlman K, et al. 1990. *Science* 249:157–60
94. Schwabe JWR, Chapman L, Finch JT, Rhodes D. 1993. *Cell* 75:567–78
95. Schwabe JWR, Neuhaus D, Rhodes D. 1990. *Nature* 340:458–61
96. Lee MS, Kliewer SA, Provencal J, Wright PE, Evans RM. 1993. *Science* 260:1117–21
97. Mader S, Kumar V, deVerneuil H, Chambon P. 1989. *Nature* 338:271–74
98. Tsai SY, Carlstedt-Duke J, Weigel NL, Dahlman K, Gustafsson JA, et al. 1988. *Cell* 55:361–69
99. Hard T, Dahlman K, Carlstedt-Duke J, Gustafsson JA, Rigler R. 1990. *Biochemistry* 29:5358–64
100. Umesono K, Murakami KK, Thompson CC, Evans RM. 1991. *Cell* 65:1255–66
101. Naar AM, Boutin JM, Lipkin SM, Yu VC, Holloway JM, et al. 1991. *Cell* 657:1267–79
102. Cooney AJ, Tsai SY, O'Malley BW, Tsai M-J. 1992. *Mol. Cell. Biol.* 12: 4153–63

103. Yu VC, Delsert C, Andersen B, Holloway JM, Devary OV, et al. 1991. *Cell* 657:1251–66
104. Leid M, Kastner P, Lyons R, Nakshatri H, Saunders M, et al. 1992. *Cell* 68:377–97
105. Marks MS, Hallenbeck PL, Nagata T, Segars JH, Appella E, et al. 1992. *EMBO J.* 11:1419–35
106. Zhang XK, Hoffmann B, Tran PBV, Graupner G, Pfahl M. 1992. *Nature* 355:441–46
107. Kliewer SA, Umesono K, Mangelsdorf DJ, Evans RM. 1992. *Nature* 355:446–49
108. Bugge TH, Pohl J, Lonnoy O, Stunnenberg HG. 1992. *EMBO J.* 11: 1409–18
109. Kliewer SA, Umesono K, Noonan DJ, Heyman RA, Evans RM. 1992. *Nature* 358:771–74
110. Kliewer SA, Umesono K, Heyman RA, Mangelsdorf DJ, Dyck JA, Evans RM. 1992. *Proc. Natl. Acad. Sci. USA* 89:1448–52
111. Widom RL, Rhee M, Karathanasis SK. 1992. *Mol. Cell. Biol.* 12:3380–89
112. Yao TP, Segraves WA, Oro AE, McKeown M, Evans RM. 1992. *Cell* 71:63–72
113. Hwung YP, Wang LH, Tsai SY, Tsai M-J. 1988. *J. Biol. Chem.* 263:13470–74
114. Tran PB, Zhang XK, Salbert G, Hermann T, Lehmann JM, Pfahl M. 1992. *Mol. Cell. Biol.* 12:4666–76
115. Wilson TE, Fahrner TJ, Johnston M, Milbrandt J. 1991. *Science* 252:1296–300
116. Harding HP, Lazar MA. 1993. *Proc. 75th Endocrinol. Soc. Meet.*, p. 1621a
117. Bagchi MK, Elliston JF, Tsai SY, Edwards DP, Tsai M-J, O'Malley BW. 1988. *Mol. Endocrinol.* 2:1221–29
118. Beekman JM, Allan GF, Tsai SY, Tsai M-J, O'Malley BW. 1993. *Mol. Endocrinol.* 1:1266–74
119. Brown M, Sharp PA. 1990. *J. Biol. Chem.* 265:11238–43
120. Edwards DP, Kuhnel B, Estes PA, Nordeen SK. 1989. *Mol. Endocrinol.* 3:381–91
121. Zhang X, Lehmann JM, Hoffmann B, Dawson MI, Cameron J, et al. 1992. *Nature* 358:587–91
122. Picard D, Salser SJ, Yamamoto KR. 1988. *Cell* 54:1073–80
123. Becker DM, Hollenberg SM, Ricciardi R. 1989. *Mol. Cell. Biol.* 9:3878–87
124. Jackson SP, Baltimore D, Picard D. 1993. *EMBO J.* 12:2809–19
125. Burk O, Klemprauer KH. 1991. *EMBO J.* 10:3713–19
126. Bagchi MK, Tsai SY, Tsai M-J, O'Malley BW. 1991. *Mol. Cell. Biol.* 11:4998–5004
127. Kumar V, Chambon P. 1988. *Cell* 55:145–56
128. Rodriguez R, Weigel NL, O'Malley BW, Schrader WT. 1990. *Mol. Endocrinol.* 4:1782–90
129. Allan GF, Tsai SY, Tsai M-J, O'Malley BW. 1992. *Proc. Natl. Acad. Sci. USA* 89:11750–54
130. Tsai SY, Srinivasan G, Allan GF, Thompson EB, O'Malley BW, Tsai M-J. 1990. *J. Biol. Chem.* 265:17055–61
131. Elliston JF, Beekman JM, Tsai SY, O'Malley BW, Tsai M-J. 1992. *J. Biol. Chem.* 267:5193–98
132. Elliston JF, Fawell SE, Klein-Hitpass L, Tsai SY, Tsai M-J, et al. 1990. *Mol. Cell. Biol.* 10:6607–12
133. Allan GF, Ing NH, Tsai SY, Srinivasan G, Weigel NL, et al. 1991. *J. Biol. Chem.* 266:5905–10
134. Zawel L, Reinberg D. 1993. *Nucleic Acid Res. Mol. Biol.* 44:69–108
135. Tsai SY, Sagami I, Wang LH, Tsai M-J, O'Malley BW. 1987. *Cell* 50: 701–9
136. Ing NH, Beekman JM, Tsai SY, Tsai M-J, O'Malley BW. 1992. *J. Biol. Chem.* 267:17617–23
137. Baniahmad A, Ha I, Reinberg D, Tsai SY, Tsai M-J, O'Malley BW. 1993. *Proc. Natl. Acad. Sci. USA* 90:8832–36
138. Lin YS, Ha I, Maldonado E, Reinberg D, Green MR. 1991. *Nature* 353:569–71
139. Colgan J, Wampler S, Manley JL. 1993. *Nature* 362:549–53
140. Roberts SGE, Ha I, Maldonado E, Reinberg D, Green MR. 1993. *Nature* 363:741
141. Ingles CJ, Shales M, Cress WD, Triezenberg SJ, Greenblatt J. 1991. *Nature* 351:588–90
142. Lieberman PM, Berk AJ. 1991. *Genes Dev.* 5:2441–54
143. Baniahmad A, Steiner C, Kohne AC, Renkawitz R. 1990. *Cell* 61:505–14
144. Fondell JD, Roy AL, Roeder RG. 1993. *Genes Dev.* 7:1400–10
145. Ha I, Roberts S, Maldonado E, Sun X, Kim LU, et al. 1993. *Genes Dev.* 7:1021
146. Sakai D, Helms S, Carlstedt-Duke J, Gustafsson JA, Rottman FM, Yamamoto KR. 1988. *Genes Dev.* 2:1144–54
147. Diamond MI, Miner JN, Yoshinaga SK, Yamamoto KR. 1990. *Science* 249:1266–72
148. Danesch U, Gloss B, Schmid W,

Schultz GS, Schule R, Renkawitz R. 1987. *EMBO J.* 6:625–30
149. Miksicek R, Hebner A, Schmid W, Danesch U, Posseckert G, et al. 1986. *Cell* 46:283–90
150. Klein-Hitpass L, Kalins M, Ryffel GU. 1988. *J. Mol. Biol.* 201:537–44
151. Schule R, Muller E, Otsuka-Murakami H, Renkawitz R. 1988. *Nature* 332:87–90
152. Granner DK, O'Brien R, Imai E, Forest C, Mitchell J, Lucas P. 1991. *Recent Prog. Horm. Res.* 47: 319–46
153. Schule R, Muller E, Kaltschmidt C, Renkawitz R. 1988. *Science* 242:1418–20
154. Tsai SY, Tsai M-J, O'Malley BW. 1989. *Cell* 57:443–48
155. Bradshaw MS, Tsai SY, Leng XH, Dobson ADW, Conneely OM, et al. 1991. *J. Biol. Chem.* 266:16684–90
156. Baniahmad C, Muller M, Altschmied J, Renkawitz R. 1991. *J. Mol. Biol.* 222:155–65
157. Ponglikitmongkol M, White JH, Chambon P. 1990. *EMBO J.* 9:2221–31
158. Ptashne M. 1988. *Nature* 335:683–89
159. Schmid W, Strahle U, Schutz G, Schmidt TJ, Stunnenberg H. 1989. *EMBO J.* 8:2257–63
160. Martinez E, Wahli W. 1989. *EMBO J.* 8:3781–91
161. Denis M, Poellinger L, Wikstrom A, Gustafsson JA. 1988. *Nature* 333:686–88
162. Mendel DB, Bodwell JE, Gametchu B, Harrison RW, Munck A. 1986. *J. Biol. Chem.* 261:2758–63
163. Pratt WB. 1987. *J. Cell Biochem.* 35:51–68
164. Sanchez ER, Faber LE, Henzel WJ, Pratt WB. 1990. *Biochemistry* 29: 5145–52
165. Sanchez ER. 1990. *J. Biol. Chem.* 265:22067–70
166. Denis M, Wikstrom A, Gustafsson JA. 1987. *J. Biol. Chem.* 262:11803–6
167. Joab I, Radanyi C, Renoir M, Buchou T, Catelli MG, et al. 1984. *Nature* 308:850–53
168. Kost SL, Smith DF, Sullivan WP, Welch WJ, Toft DO. 1989. *Mol. Cell. Biol.* 9:3829–32
169. Tai PKK, Maeda Y, Nakao K, Wakim NG, Duhring JL, Faber LE. 1986. *Biochem. J.* 25:5269–75
170. Picard D, Khursheed B, Garabedian MJ, Fortin MG, Lindquist S, Yamamoto KR. 1990. *Nature* 348:166–68
171. Denner LA, Schrader WT, O'Malley BW, Weigel NL. 1990. *J. Biol. Chem.* 265:16548–55
172. Poletti A, Weigel NL. 1993. *Mol. Endocrinol.* 7:241–46
173. Sheridan PL, Evans RM, Horwitz KB. 1989. *J. Biol. Chem.* 264:6520–28
174. Weigel NL, Carter TH, Schrader WT, O'Malley BW. 1992. *Mol. Endocrinol.* 6:8–14
175. Bagchi MK, Tsai SY, Tsai M-J, O'Malley BW. 1992. *Proc. Natl. Acad. Sci. USA* 89:2664–68
176. Takimoto GS, Tasset DM, Eppert AC, Horwitz KB. 1992. *Proc. Natl. Acad. Sci. USA* 89:3050–54
177. Lees JA, Fawell SE, Parker MG. 1989. *Nucleic Acids Res.* 17:5477–88
178. Sabbah M, Gouilleux F, Sola B, Redeuilh G, Baulieu EE. 1991. *Proc. Natl. Acad. Sci. USA* 88:390–94
179. El-Ashry-Stowers D, Onate SA, Nordeen SK, Edwards DP. 1989. *Mol. Endocrinol.* 3:1545–58
180. Fritsch M, Leary CM, Furlow JD, Ahrens H, Schuh TJ, et al. 1992. *Biochemistry* 31:5303–11
181. Allan GF, Leng XH, Tsai SY, Weigel NL, Edwards DP, et al. 1992. *J. Biol. Chem.* 267:19513–20
182. Weigel NL, Beck CA, Estes PA, Prendergast P, Altmann M, et al. 1992. *Mol. Endocrinol.* 6:1585–97
183. Benhamou B, Garcia T, Lerouge T, Vergezac A, Gofflo D, et al. 1992. *Science* 255:206–9
184. Pakdel F, Katzenellenbogen BS. 1992. *J. Biol. Chem.* 267:3429–37
185. Lanz R, Wieland S, Rusconi S. 1993. *J. Cell Biol.* 17A:163
186. Danielian PS, White R, Lees JA, Parker MG. 1992. *EMBO J.* 11:1025–33
187. Veldscholte J, Ris-Stalpers C, Kuiper GGJM, Jenster G, Berrevoets C, et al. 1990. *Biochem. Biophys. Res. Commun.* 173:534–40
188. Leng XH, Tsai SY, O'Malley BW, Tsai M-J. 1993. *J. Steroid Biochem. Mol. Biol.* 46:643–61
189. Toney JH, Wu L, Summerfield AE, Sanyal G, Forman BM, Zhu J, Samuels HH. 1993. *Biochemistry* 32:2–6
190. Orti E, Bodwell JE, Munck A. 1992. *Endocr. Rev.* 13:105–28
191. Orti E, Mendel DB, Smith LI, Munck A. 1989. *J. Biol. Chem.* 264:9728–31
192. Sullivan WP, Madden B, McCormick DJ, Toft DO. 1988. *J. Biol. Chem.* 263:14717–23
193. Logeat F, LeCunff M, Pamphile R, Milgrom E. 1985. *Biochem. Biophys. Res. Commun.* 131:421–27
194. Beck CA, Weigel NL, Edwards DP. 1992. *Mol. Endocrinol.* 6:607–20
195. Schroeder C, Raynoschek C, Fuhrmann

U, Damm K, Vennstrom B, Beug H. 1990. *Oncogene* 5:1445–53
196. Weigel NL, Denner LA, Poletti A, Beck CA, Edwards DP, Zhang Y. 1993. *Adv. Prot. Phosphatases* 7:237–69
197. Denner LA, Weigel NL, Maxwell BL, Schrader WT, O'Malley BW. 1990. *Science* 250:1740–43
198. Cho H, Katzenellenbogen BS. 1993. *Mol. Endocrinol.* 7:441–42
199. Ali S, Metzger D, Bornert JM, Chambon P. 1993. *EMBO J.* 12:1153–60
200. DeFranco DB, Qi M, Borror KC, Garabedian MJ, Brautigan DL. 1991. *Mol. Endocrinol.* 5:1215–28
201. Sheridan PL, Krett NL, Gordon JA, Horwitz KB. 1988. *Mol. Endocrinol.* 2:1329–42
202. Power RF, Lydon JP, Conneely OM, O'Malley BW. 1991. *Science* 252: 1546–8
203. Power RF, Mani SK, Codina J, Conneely OM, O'Malley BW. 1991. *Science* 254:1636–39
204. Aronica SM, Katzenellenbogen BS. 1993. *Mol. Endocrinol.* 7:743–52
205. Ignar-Trowbridge D, Nelson KG, Bidwell MC, Curtis SW, Washburn TF, et al. 1992. *Proc. Natl. Acad. Sci. USA* 89:4658–62
205a. Smith CL, Conneely OM, O'Malley BW. 1993. *Proc. Natl. Acad. Sci. USA* 90:6120–24
206. Smith CL, Conneely OM, O'Malley BW. 1993. *Steroid Receptors in Health and Disease*, ed. VK Moudgil, pp. 333–56. Boston/Basel/Berlin: Birkhauser
206a. Ignar-Trowbridge DM, Teng CT, Ross KA, Parker MG, Korach KS, McLachlan JA. 1993. *Mol. Endocrinol.* 7:992–98
206b. Nelson KG, Takahashi T, Bossert NL, Walmer DK, McLachlan JA. 1991. *Proc. Natl. Acad. Sci. USA* 88:21–25
207. Jonat C, Rahmsdorf HJ, Park KK, Cato ACB, Gebel S, et al. 1990. *Cell* 62:1189–204
208. Schule R, Rangarajan PN, Kliewer SA, Ransone LJ, Bolado J, et al. 1990. *Cell* 62:1217–26
209. Yang-Yen H, Chambard J, Sun Y, Smeal T, Schmidt TJ, et al. 1990. *Cell* 62:1205–15
210. Konig H, Ponta H, Rahmsdorf HJ, Herrlich P. 1992. *EMBO J.* 11:2241–46
211. Meyer ME, Gronemeyer H, Turcotte B, Bocquel MT, Tasset D, Chambon P. 1989. *Cell* 57:433–42
212. Conneely OM, Kettelberger DM, Tsai M-J, O'Malley BW. 1989. *Gene Regulation by Steroid Hormones IV,* ed. AK Roy, J Clark, pp. 220–23. New York/Berlin/Heidelberg/London/Paris/Tokyo: Springer-Verlag
213. Webb P, Lopez GN, Greene GL, Baxter JD, Kushner PJ. 1992. *Mol. Endocrinol.* 6:157–67
214. Shemshedini L, Ji JW, Brou C, Chambon P, Gronemeyer H. 1992. *J. Biol. Chem.* 261:1834–39
215. Yoshinaga SK, Peterson CL, Herskowitz I, Yamamoto KR. 1992. *Science* 258:1598–604
216. Baniahmad C, Baniahmad A, Tsai M-J, O'Malley BW. 1994. *Mol. Cell. Biol.* Submitted
217. Nawaz Z, Baniahmad C, Stillman DJ, O'Malley BW, Tsai M-J. 1994. *Mol. Cell. Biol.* Submitted
218. McDonnell DP, Vegeto E, O'Malley BW. 1992. *Proc. Natl. Acad. Sci. USA* 89:10563–67
219. Telsenfeld G. 1992. *Nature* 355:219–24
220. Hayes JL, Wolffe AP. 1992. *BioEssays* 14:597–603
221. Archer TK, Lefebvre P, Wolfcrd RG, Hager GL. 1992. *Science* 255:1573–76
222. Archer TK, Cordingley MG, Wolford RG, Hager GL. 1991. *Mol. Cell. Biol.* 11:688–98
223. Piña B, Brüggemeier U, Beato M. 1990. *Cell* 60:719–31
224. Schild C, Claret FX, Wahli W, Wolffe AP. 1993. *EMBO J.* 12:423–33
225. Cordingley MG, Riegel AT, Hager GL. 1987. *Cell* 48:261–70
226. Richard-Foy H, Hager GL. 1987. *EMBO J.* 6:2321–28
227. Qiu YH, Cooney AJ, Kuratani S, DeMayo FJ, Tsai SY, Tsai M-J. 1994. *Proc. Natl. Acad. Sci. USA*. In press
228. Lutz B, Kuratani S, Cooney AJ, Wawersik S, Tsai SY, et al. 1994. *Development*. In press
229. McGuire WL, Chamness GC, Fuqua SAW. 1991. *Mol. Endocrinol.* 5:1571–77
230. Skipper JK, Young LJ, Bergeron JM, Tetzlaff MT, Osborn CT, Crews D. 1993. *Proc. Natl. Acad. Sci. USA* 90:7172–75
231. Vegeto E, Shahbaz MM, Wen DX, Goldman ME, O'Malley BW, McDonnell DP. 1993. *Mol. Endocrinol.* 7:1244–55
232. Wei LL, Gonzalez-Aller C, Wood WM, Miller LA, Horwitz KB. 1990. *Mol. Endocrinol.* 4:1833–40

Annu. Rev. Biochem. 1994. 63:487–526

HOMEODOMAIN PROTEINS

Walter J. Gehring, Markus Affolter, and Thomas Bürglin[1]
Biozentrum, University of Basel, Klingelbergstr. 70, CH-4056 Basel, Switzerland

KEY WORDS: homeobox, homeotic genes, NMR, DNA binding, transcription regulation

---

CONTENTS

INTRODUCTION

Homeotic genes are master control genes that specify the body plan and regulate development of higher organisms. They share a common sequence element of 180 bp, the homeobox, which was first discovered in *Drosophila* (1–3). Subsequently, the homeobox was shown to occur in all metazoa that have been analyzed, ranging from sponges (4) to vertebrates (5) and humans (6), and also in plants (7, 8) and in fungi (9, 10). The homeobox encodes the 60-amino-acid homeodomain, which represents the DNA-binding domain

[1]Dept. of Genetics, Harvard Medical School, Massachussetts General Hospital, Boston, MA 02114

0066-4154/94/0701-0487$05.00

of the much larger homeodomain proteins. The homeodomain allows the sequence-specific recognition of sets of target genes by the homeodomain proteins, which primarily serve a gene regulatory function and act as transcription factors that regulate target genes in a precise spatial and temporal pattern. The homeotic genes, which specify segmental identity and positional information along the antero-posterior axis, are organized into gene complexes and are aligned along the chromosome in the same order as they are expressed along the antero-posterior axis. This colinearity evolved prior to the separation of vertebrates and invertebrates, and it has been largely conserved in both major branches of the evolutionary tree. In this review, our current knowledge about the structure and function of the homeodomain is summarized, and the functional role of homeodomain proteins in DNA binding, transcriptional regulation, and developmental control is discussed.

HOMEODOMAIN SEQUENCE ANALYSIS

Prototype and Consensus Sequence

The organization of the *Antennapedia (Antp)* gene and its protein, and the amino acid sequence of the *Antp* homeodomain with some of its structural features are shown as a prototype in Figure 1*a*. Its sequence differs at nine positions only from that of a consensus sequence derived from a compilation of 346 homeodomain sequences (11) (Figure 1*b*) listed in the appendix figure. There are seven positions in the consensus sequence (black triangles) that are occupied by the same amino acid in more than 95% of the 346 sequences. These include L16, F20, W48, and F49, which belong to the hydrophobic core amino acids, and R5, N51, and R53, which are involved directly in DNA binding. Another 10 amino acids are conserved in more than 80% of the sequences (open triangles), and at 12 additional positions only two different amino acids are found in more than 80% of the sequences, e.g. R or K at position 2. These highly conserved amino acids, forming the highest pillars in the frequency diagram (Figure 1*b*), define the homeodomain. A very similar frequency diagram is obtained when only the sequences from a single species are plotted; the diagrams of *Drosophila*, mouse, and human are very similar (12), indicating a high degree of evolutionary conservation.

Homeobox Gene Complexes

Homeodomain sequences can be subdivided into at least 20 different classes (11, 13) on the basis of several criteria, including sequence identity, sequence similarity in the flanking regions, organization into gene clusters, association

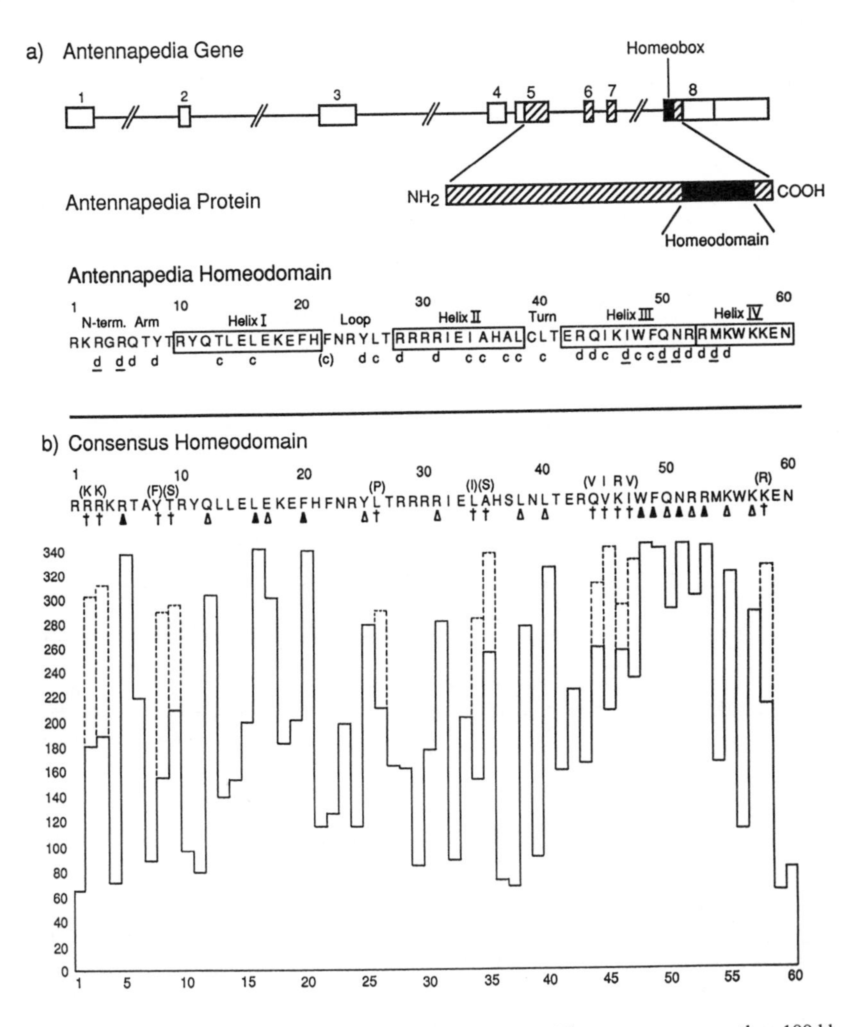

Figure 1 (*a*) Organization of the *Antennapedia (Antp)* gene. The gene spans more than 100 kb and consists of 8 exons indicated by numbered blocks. The protein-coding region (hatched blocks) contains the homeobox (black bar) encoding the homeodomain of the *Antp* protein. The amino acid sequence of the *Antp* homeodomain (1–60) is given in the single-letter code. The main structural features are indicated: c, core amino acids; d, amino acids involved in DNA binding; d, amino acids contacting the DNA bases. (*b*) Consensus sequence derived from 346 homeodomain sequences. In the frequency diagram below, the highest amino acid frequencies for positions 1–60 are given. For those positions in which two amino acids are found very frequently, the frequency of the second amino acid is indicated by a dashed line and the respective amino acid shown above the consensus in parentheses.

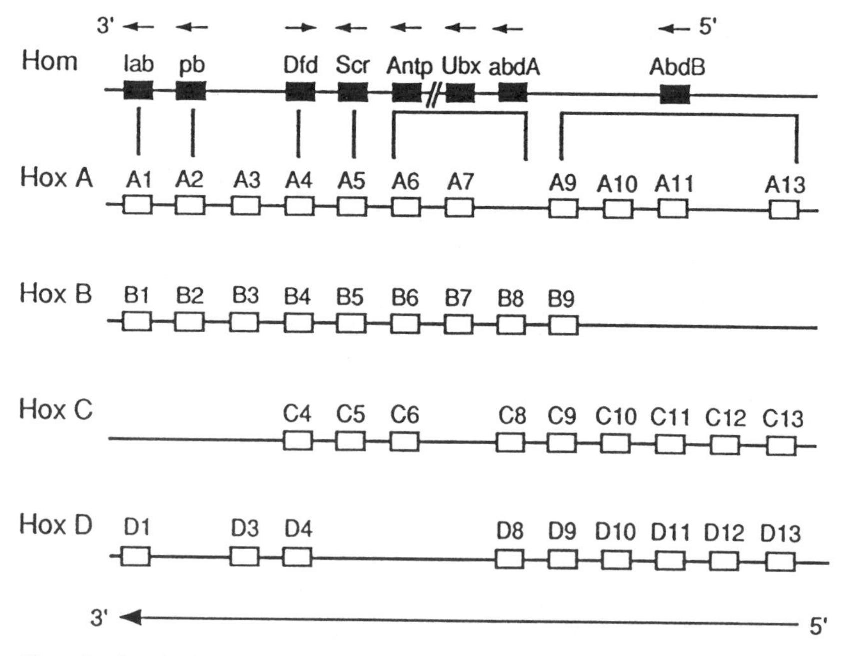

Figure 2 Organization of the homeotic gene complexes in *Drosophila* (*Hom*) and mammals (*Hox A–D*) (143). The arrows indicate the direction of transcription (5′→3′). Orthologous relationships between the *Drosophila* genes and mammalian cognates are indicated by vertical lines. Abbreviations—see text.

with other sequence motifs, and position of introns. These criteria allow us in some cases to distinguish between orthologous genes, which are derived from a speciation event, and paralogous genes, which are derived from a gene duplication event. This distinction is particularly important for the analysis of the homeotic gene complexes (Figure 2).

As first demonstrated in *Drosophila,* homeotic genes are arranged in closely linked clusters or complexes, and the genes are arranged in the same order along the chromosome as they are expressed along the antero-posterior body axis (14, 15). This phenomenon is designated as the colinearity rule, and was subsequently found to apply also to the *Hox* gene clusters of mouse and human (16–20). In *Drosophila* there are two clusters, the Bithorax and the Antennapedia Complexes (14, 15), which are thought to have originated from a single primordial cluster as found in the beetle *Tribolium* (21), a more primitive insect. In mammals there are four such *Hox* clusters (*Hox A* to *Hox D,* Figure 2), presumably generated by duplications of an entire primordial gene cluster (22–26). In support of this idea, there is evidence for one *Hox* cluster in the acorn worm, a hemichordate; two *Hox* clusters in amphioxus, a cephalochordate; and three (or possibly four) in the lamprey, a primitive vertebrate (27). Therefore, the corresponding genes in the four mammalian clusters are considered to be paralogous. As shown in Figure 2, the mammalian genes can be assigned to 13 paralog groups (*Hox 1–13*), which contain two to four paralogous genes.

In both insects and mammals, the homeotic genes specify segmental identity and the body plan along the antero-posterior axis. Based upon sequence identity of their homeodomains, some of their flanking sequences, associated sequence motifs, colinear clustering, and conservation of function, the homeotic genes of *Drosophila* have to be considered as homologs (orthologs) of the mammalian *Hox* genes. The orthology between certain *Drosophila* homeotic genes and their mouse counterparts is strongly supported by functional studies in which the mouse genes are tested in *Drosophila* by gene transfer (28–30).

Little is known about the *origin of the primordial gene* cluster, which presumably arose by gene duplication. The nematode *Caenorhabditis* (11, 31, 32) shows some clustering of homeobox genes (*ceh-13, lin-39, mab-5, egl-5,* and *ceh-23*), which are also roughly arranged according to the colinearity rule, but the homeodomain sequences have diverged considerably from those of insects and mammals, indicating that nematodes are on a separate branch of the evolutionary tree. In order to reconstruct the evolution of the primordial gene cluster, other more primitive metazoa must be examined in detail. We must find out not only how the colinear arrangement evolved in the first place, but also why it has been conserved over millions of years, while the members of other gene families have largely been dispersed in the genome. A hint of how the cluster might have evolved is provided by the finding that *Antp* and its cognate *Hox* genes (*Hox 6* and *7*), located in the center of the cluster, deviate the least from the consensus sequence, and that the sequences progressively diverge as we proceed to more anterior and posterior genes, respectively (Figure 3*a*). Finally, the terminal genes *labial (lab)* (*Hox 1*) and *Abdominal-B (Abd-B)* (*Hox 13*) contain the most divergent homeodomains. This relationship is also apparent in a distance matrix based on pairwise sequence comparisons in all possible combinations to quantify sequence similarities (22a), and in evolutionary trees rooted on the *Antp* ortholog, which show the same *Hox 1* to *Hox 13* arrangement as indicated in Figure 3*a* (A Adoutte, personal communication). These observations, and the fact that all the genes in the *Hox* clusters are transcribed in the same direction, suggest that the primordial gene cluster arose by a series of consecutive tandem duplications due to unequal crossingover. The first step presumably led to the generation of the two terminal genes, which consequently evolved into an anterior and a posterior gene and had the longest time to diverge. Subsequently, a series of consecutive

a)

```
Consensus    RRRKRTAYTRYQLLELEKEFHFNRYLTRRRRIELAHSLNLTERQVKIWFQNRRMKWKKEN   Δ

lab/ Hox1    PNAI··NF·TK··T·········K····A··V·I·AT·Q·N·T············Q··RE   21
pb/  Hox2    S··L·····NT············K··C·P··V·I·AL·D·······V········H·RQT   17
     Hox3    SK·A·····SA··V············C·P··V·M·NL·······I··········Y··DQ   16
Dfd/ Hox4    PK·S······Q·V········Y···········I··T·C·S···I·············DH   13
Scr/ Hox5    GK·A········T····················I··A·C·S···I·············D·   10
     Hox6    ···G·QT·····T····················I·NA·C·····I···············    9
Antp         ·K·G·QT·····T····················I··A·C·····I···············    9
     Hox7    ·K·G·QT·····T····················I··A·C·····I··············H   10
Ubx/abdA     ···G·QT···F·T··········H·········I··A·C·····I··········L···L   12
     Hox8    ···G·QT·S···T·······L··P····K····VS·A·G·····················   12
AbdB/Hox9    T·K··CP··K··T·······L··M····D··Y·V·RL··················M··M·   15
     Hox10   G·K··CP··KH·T·······L··M····E··L·ISK·V···D·············L··M·   18
     Hox11   T·K··CP··K··IR···R··F··V·INKEK·LQ·SRM····D·············E··L·   23
     Hox12   S·K··KP··KQ·IA···N··LV·EFIN·QK·K··SNR···SDQ············K·RVV   28
     Hox13   G·K··VP··KL··K···N·YAI·KFINKDK·RRISATT··S····T·······V·E··VV   31
```

b)

```
Consensus    RRRKRTAYTRYGLLELEKEFHFNRYLTRRRRIELAHSLNLTERQVKIWFQNRRMKWKKEN   Δ

eve          V··Y···F··E·IAR·····YREN·VS·P··C···AA···P·TTI·V········D·RQR   26
ems          PK·I···FSPS···R··HA·EK·H·VVGAE·KQ··Q··S·S·T···V······T·H·RQK   30
Dll          I·KP··I·SSL··QA·NRR·QKTQ··ALPE·A···A··G··QT··········S·F··LM   29
cad          KDKY·VV··DH·R········YS··I·I··KA···AN·G··············A·ER·V·   22
Hlx          ·SWS·AVFSNL·RKG···R·QIQK·I·KPD·KK··AM·G··DA···V·········RHSK   33
TCL/NEC      KKKP··SFS·S·VF····R·ERKK··SSAE·AA··KA·HM··T··········T···RQT   30
msh          N·KP··PF·TS···A··RK·RQKQ··SIAE·A·FSS······T··········A·A·RLQ   28
NK-2         K·KR·VLFSQA·VY···RR·RQQ···SAPE·EH··SLIR··PT········H·Y·M·RQA   34
NK-1         P··A···F·YE··VA··NK·KTT···SVCE·LN··L··S···T··········T····Q·   23
en           EK·P···FSAE··AR·KA··QE·····EQ··QS··QE·G·N·S·I······K·A·I··AS   28
prd          Q··S··TF·AE··EA··RA·ERTH·PDVYT·E···QRTD···ARIQV··S···AR·R·QE   35
prd-like     Q··H··TF·SE··E···NL·QETH·PDIFM·E···RKI····SR·QV······A··RRQE   31
cut          LKKP·VVLAPEEKEA·KRAYQQKP·PSPKTIE···TQ···KTST·IN··H·Y·SRIRR·L   45
LIM          T··P··TIKAK··ET·KNA·AA·PKPD·HV·EQ··QETG·SM·VIQV······A·ERRLK   38
ZF           KK·L··RI·DE··KV·KEF·DL·SSPS·EMIE·ISNK·G·PK·V·QV····T·A·ER·RY   38
Pou          K·K···SIEVNVKGA··SH·LKCPKPSAQEITS··D··Q·EKEV·RV··C···Q·E·RMT   40
```

Figure 3 Consensus homeodomain sequences of the homeodomain families and paralog groups of the Complex Superclass (*a*) and the Dispersed Superclass (*b*). Amino acids shared by adjacent families are boxed. Δ indicates the number of amino acids that differ from the general consensus sequence listed on the top.

unequal crossingovers probably generated the interior genes. Due to these multiple recombination events, the internal genes are mosaics combining segments of their ancestral genes. Since the interior genes were generated later in evolution, they have diverged less from the primordial gene than the external genes. This hypothesis is supported by the finding that the external genes *lab, proboscipedia (pb),* and *Abd-B* share an intron at the same position in the homeodomain (amino acid 44/45). However, the successive duplication model provides only a partial explanation for the colinearity rule, and we need more information about intermediary evolutionary stages.

With respect to the maintenance of the chromosomal order, several genetic mechanisms may be involved. For certain neighboring genes in both *Drosophila* and mouse, evidence for shared *cis*-regulatory sequences has been found [(33, 34) and R. Krumlauf, personal communication]. Therefore, such genes are functionally linked, since translocation or inversion events separate one or the other gene from its *cis*-regulatory control region and therefore, are detrimental. Nevertheless, the gene cluster may occasionally be split, as in *Drosophila,* and a gene (as for example *Deformed*) can be inverted (see Figure 2). Another stabilizing factor may be exon sharing between transcripts of different genes by differential splicing as, for example, in the human *Hox* cluster (23). Higher-order chromatin structure has also been invoked in maintaining the chromosomal order of the homeotic genes (35), but the colinear gene arrangement remains largely enigmatic.

Classification of Homeodomain Sequences

On the basis of sequence similarity and chromosomal clustering, which reflect evolutionary relationships, we can classify the available sequences into two superclasses, the *Complex Superclass,* which comprises the homeodomains of all the genes clustered in the known homeotic gene complexes (36), and the *Dispersed Superclass,* whose genes have been largely dispersed in the genome, even though in some cases close linkage is still detectable and smaller clusters may still be discovered.

In the appendix figure, the Complex Superclass has been divided into six classes, named after the corresponding *Drosophila* genes, and subdivided into the cognate paralog groups of vertebrates. These classes are relatively homogeneous in sequence. Their consensus sequences arranged according to the colinearity rule are shown in Figure 3*a,* in which the characteristic amino acids at a given position that differ from the general consensus sequence are boxed. The sequence similarities, as reflected by the continuity of most boxes, indicate that neighboring consensus sequences are most closely related in evolution.

The genes of the *labial (lab)* class are located near the 3′ end of the homeotic gene clusters. *lab* is most closely related to the *Hox 1* group (80–85% identity). The consensus sequence of this class differs at 21 positions from the general consensus of all 346 known homeodomain sequences. Interestingly, all of the basic amino acids at positions 1–4 in the general consensus sequence are substituted by neutral amino acids.

The *proboscipedia (pb)* class shares most sequence identity with the *Hox 2* group and is most closely related on the one hand to the *lab/Hox 1* class and on the other hand to the *Hox 3* group, for which there is no other *Drosophila* ortholog.

The *Deformed (Dfd)* class corresponds to the *Hox 4* cognate group, which is supported by sequence similarities both inside the homeodomain and outside in the flanking sequences. The mouse gene *Hox D4,* when tested in *Drosophila,* behaves in the manner of *Dfd* (29).

The *Sex combs reduced (Scr)* class includes the *Hox 5* group. The mouse gene *Hox A5,* when tested in *Drosophila,* confers an *Scr* phenotype (30).

The relationships within the *Antennapedia (Antp)* class, which also includes the *Drosophila* genes *Ultrabithorax (Ubx)* and *abdominal-A (abd-A),* is less clear. The *Antp* homeodomain is 98% identical with the mammalian homeodomains of the *Hox 7* group, but also closely related to *Hox 6. Ubx* and *abd-A* are also very similar to *Hox 8.*

The *Abdominal-B (Abd-B)* class includes only one *Drosophila* gene, but five *Hox* paralog groups, suggesting that there has been an expansion of the complex in the vertebrates in the posterior (5′) direction. The homeodomains of *Hox 9* to *13* progressively diverge from the general consensus (Figure 3*a*). Interestingly, the homeoboxes of *Abd-B, lab,* and *pb* all have an intron at the same position.

As a secondary criterion for classifying homeodomains and tracing their evolutionary origin, we can use conserved sequence motifs outside of the homeodomain. With the exception of the *Abd-B* class, all homeobox genes within the Complex Superclass share a *hexapeptide motif* of the consensus sequence IYPWMK, which is located N-terminal of the homeodomain (11). This motif is also found in the *caudal (cad), empty spiracles (ems),* and *TCL* classes, outside of the homeotic gene complexes. Therefore, it is possible that these genes were originally located inside the clusters and that they were subsequently translocated to other chromosomal positions. This hypothesis is supported by the fact that some members of the *cad* and *ems* classes have an intron at the same position (amino acid 44/45) as *lab, pb,* and *Abd-B.*

In *Drosophila,* the *Antennapedia* Complex contains additional homeobox genes with different functions: *fushi tarazu (ftz),* a segmentation gene; *bicoid (bcd),* a maternal-effect gene specifying anterior-posterior polarity; and the two *zerknüllt* genes (*zen* 1 and 2), which are involved in determining dorso-ventral positional information. *Ftz* belongs to the *Antp* class, and the two *zen* genes are related to the *pb* class, whereas *bcd* is the most highly diverged gene in the *Antennapedia* Complex. In mammals, two *even-skipped*-like genes, Evx1 and 2, are closely linked to the *Hox A* and the *Hox D* cluster, respectively, but transcribed in the opposite direction (23), whereas in *Drosophila, even-skipped (eve)* functions as a segmentation gene and is located on a different chromosome from the homeotic complexes.

The *Dispersed Superclass* can be subdivided into more than 16 classes, whose consensus sequences are listed in Figure 3*b*. They differ at more than 22 positions from the general consensus sequence, and in the case of *cut* as many as 45 positions are occupied by different amino acids. In addition, there are a number of unique homeodomain sequences, which are listed under *curios* in the appendix.

Within the Dispersed Superclass, the *cad, ems,* and *TCL/NEC* classes share the hexapeptide motif with the Complex Superclass, and some members of these classes contain an intron at position 44/45 as do *lab, pb,* and *Abd-B,* suggesting a close evolutionary relationship to the genes in the homeotic complexes.

Some members of the *Distalless (Dll), HLX, msh, NK-1,* and *NK-2* classes also retain an intron at position 44/45, but they all lack the hexapeptide motif.

The *even-skipped (eve)* class is more distantly related to the Complex Superclass, even though two of its members are located close to the mammalian *Hox* clusters. The *eve* class lacks the hexapeptide motif, and some of its members have an intron at a different position (amino acid 46/47) from that of the Complex Superclass (amino acid 44/45). An *eve* class gene has been isolated from corals (37), indicating that this class is very ancient. Since *eveC* in the coral *Acropora* is closely linked to an *Antp*-like gene, the association with the *Hox* cluster may represent the ancestral gene arrangement (37a).

The remaining classes listed in Figure 3*b* are characterized by their association with other sequence motifs besides the homeodomain.

The *engrailed (en)* class is highly conserved in evolution and characterized by four highly conserved protein segments (EH in Figure 4) outside of the homeodomain, which supports the notion that the insect and vertebrate genes are orthologous (38). The *en* genes of *Drosophila* and *Bombyx* both have an intron at position 17/18, which is absent in the honeybee and in vertebrates.

The *paired (prd)* class most clearly illustrates "evolutionary tinkering" (39–42). In this case the homeodomain is associated with a second DNA-binding domain of 128 amino acids designated as the *paired domain* (Figure 4). Some proteins contain only a paired domain, and some contain both a paired domain and a homeodomain. The presence of two DNA-binding domains in the same protein increases the DNA-binding specificity strongly by adding a second binding site. In addition, the spacing between the two binding sites is important. The presence of two DNA-binding domains in the same protein suggests that those proteins that contain only a homeodomain may in some cases associate with other DNA-binding proteins. The finding that the *ftz* protein containing a deletion in the homeodomain

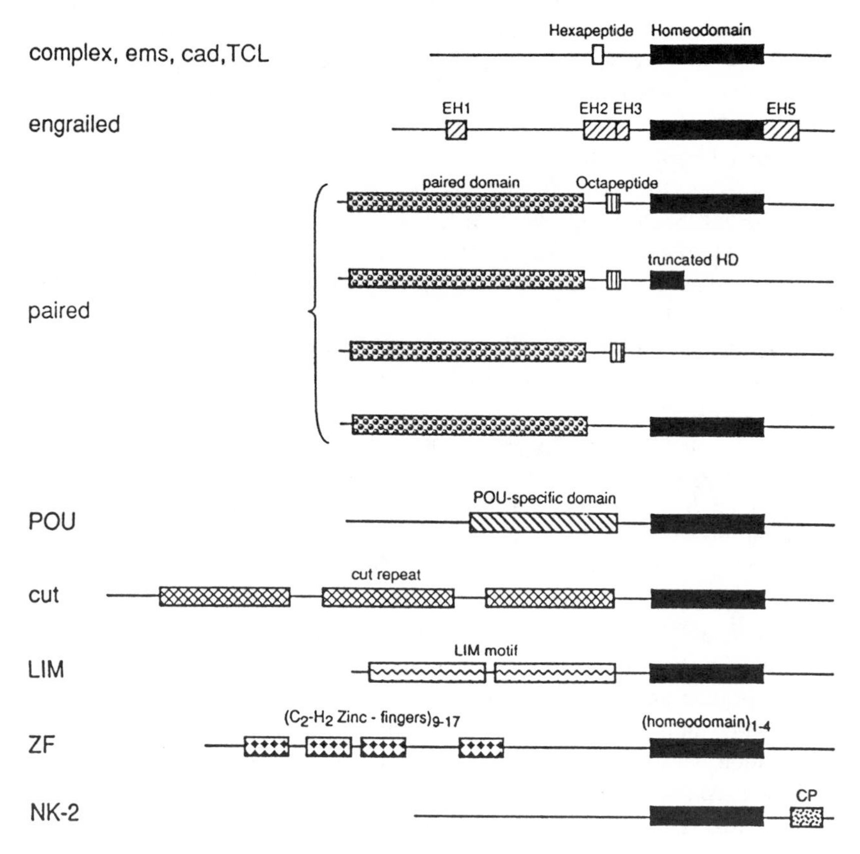

Figure 4 Schematic representation of the conserved protein domains associated with the homeodomain in various classes of homeodomain proteins. Explanation in the text.

is still partially active (43) points in the same direction. In addition, an *octapeptide motif* is located between the paired domain and the homeodomain, in some but not all *prd* class proteins. Interestingly, some proteins contain a truncated homeodomain comprising amino acids 7 to 29, including α-helix I, which may still be functionally important. Therefore, the *prd* class proteins provide a perfect example of evolutionary tinkering in which various motifs and domains are recombined in various ways to form proteins of different function. Since all *prd* homeodomains are uniquely characterized by a serine residue at position 50 of the recognition helix, the association of the paired and homeodomain probably occurred very early in evolution, prior to the extensive diversification of the homeodomain.

The *paired-like (prd-like)* class of homeodomains includes sequences that are similar to the *prd* class homeodomains, although they lack the characteristic serine residue at position 50 and they are not associated with a paired domain. Recently, the mammalian homologs of the *Drosophila* gene *orthodenticle (otd)* have been isolated and shown to be expressed in the forebrain, similar to *ems* and its homologs (44–46). Since *otd* and *ems* are also expressed in the head region of *Drosophila,* their expression pattern has apparently been conserved in evolution, even though those genes are not part of the homeotic gene clusters.

The *POU* class was originally defined on the basis of four independently isolated genes Pit-1, Oct-1 and 2, and unc-86 (47). The POU proteins were first isolated biochemically as transcription factors and only later shown to contain a homeodomain, which provided the most convincing evidence that homeobox genes encode transcription factors (48–50). The POU proteins also contain two DNA-binding domains, the 80-amino-acid POU-specific domain and the homeodomain (Figure 4), but in contrast to the *prd* class, the POU-specific domain has so far always been found in association with the homeodomain, and never alone. The homeodomains of all 35 known *POU* genes ranging from nematodes to human contain a characteristic cysteine residue at position 50, indicating that the association of the POU-specific domain and the homeodomain is very ancient.

The *cut* class is relatively small, but clearly defined by the presence of three copies of an 80-amino-acid *cut*-repeat (51) upstream of the homeodomain (Figure 4). The *cut* homeodomain is characterized by a histidine residue at position 50.

The *LIM* class contains a repeated motif of about 60 amino acids, the *LIM* motif, containing conserved cysteine and histidine residues. The *LIM* class homeodomains are much better conserved than the *LIM* motifs (52, 53), which are quite variable.

The *Zinc-finger (ZF)* class contains some of the most unusual proteins combining both zinc-finger motifs and homeodomains; *zfh-2* contains no less than 16 zinc fingers and three homeodomains (54), whereas its putative human homolog ATBF-1 contains 17 zinc fingers and four homeodomains (55). The zinc fingers are of the C_2-H_2 type and most closely related to the gap genes *Krüppel* and *hunchback* of *Drosophila*. The *ZF* class is likely to be due to a recombination event between two developmental control genes, one containing zinc-finger motifs and another with one or more homeodomains.

Finally, the *NK-2* class is a complex collection of vertebrate and fly genes, which contain a "conserved peptide" (CP) downstream of the homeodomain (56). The first vertebrate gene isolated from this class was the thyroid transcription factor (TTF-1), which shows interesting DNA-binding properties (see below). Besides these rather well-defined classes, there are a number of unique homeodomains and "*curios*" (see appendix figure).

In addition, some *atypical homeodomains* characterized by insertions or deletions have been described. Interestingly, the overall three-dimensional structure of these variants shows little change. The sequence similarity between the yeast mating type genes *MAT a1* and *MAT* α*2* and the homeodomain was first detected soon after the discovery of the homeodomain (9), and provided the first hint at the function of the homeodomain. As more sequence data were available, it became apparent that *MAT a1* has a typical homeodomain, whereas for *MAT* α*2* a better alignment of the helix 1 region could be achieved by looping out three amino acids (57). The X-ray crystallographic analysis of the *MAT* α*2* DNA complex later demonstrated that indeed three amino acids are looping out between helices I and II. In the meantime, human, *Drosophila,* and nematode genes have been found, which also contain three extra amino acids at this position (58–61). The human gene (*Prl*) is involved in leukemias (58, 59). Since several other *Hox* genes can mutate to oncogenes and induce transformation in vitro, homeobox genes can be considered as a family of nuclear protooncogenes (62).

The rat liver transcription factor LFB1 (=HNF1) is even more extreme, in that it contains 81 amino acids in its atypical homeodomain rather than 60 amino acids (63, 64). As shown by NMR spectroscopy, two amino acids in the turn of the helix-turn-helix motif (see below) are replaced by a 23-residue linker region between the two helices, accommodating the extra amino acids without any major effects on the overall conformation of the molecule (65).

Typical and atypical homeodomains have been found in fungi other than yeast (66–68), and also in plants (7, 69). Interestingly, several *Arabidopsis* genes contain a putative leucine-zipper region downstream of the homeodomain, allowing these proteins to dimerize (70).

HOMEODOMAIN STRUCTURE DETERMINATION BY NMR SPECTROSCOPY

The solution structure of a 68-amino-acid recombinant polypeptide corresponding to the *Antp* homeodomain was determined at high resolution by NMR spectroscopy (71–73). A nearly identical structure was obtained for the *Antp* (C39→S) homeodomain (74), in which cysteine 39 is replaced by serine, a residue that is normally found at this position in several homeodomains of the *Antp* class, e.g. in *ftz*. This analog has the advantage that it is not subject to oxidative dimerization so that no reducing agents need to be added for the structure determination. The well-defined core of the

Antp homeodomain structure consists of helix I from residues 10–21, a connecting loop of residues 22–27 leading to helix II (residues 28–38), a tight turn of three residues (39–41), a third helical region including helix III (residues 42–52), and a more disordered and flexible helix IV consisting of residues 53–59. The two ends of the polypeptide chain (residues 0–6 and 60–67) are, in contrast, not well defined by the NMR spectra, presumably because they are flexible in solution. Helices I and II are connected by a hexapeptide and arranged in an antiparallel fashion relative to each other, whereas helix III and its direct extension, helix IV, are aligned essentially perpendicularly to the first two helices.

Helices II and III are connected by a tight turn and form a helix-turn-helix motif as previously described for prokaryotic gene regulatory proteins (75). A comparison of the backbone structure from five prokaryotic gene regulatory proteins with that of the corresponding residues 31–50 of the *Antp* homeodomain reveals that these structures are readily superimposable (76). This represents a remarkable evolutionary conservation with respect to the three-dimensional structure of this motif from *Escherichia coli* to *Drosophila* and human, even though very little amino acid sequence homology has been retained. Only three amino acids are to some extent conserved in the prokaryotic gene regulatory proteins: glycine in the turn, alanine in the first helix, and isoleucine or valine in the second helix of the helix-turn-helix motif corresponding to residues 39, 35, and 45 in the *Antp* homeodomain, respectively (75). The glycine in the first position of the turn forms a left-handed α-helical conformation, which is energetically favorable for glycine only. Nevertheless, in the *Antp* homeodomain, a cysteine residue is located at this position and assumes an energetically unfavorable conformation. Therefore, the presence of glycine in the turn is not obligatory.

The molecular architecture of the homeodomain is held together by a core of 11 amino acids (see Figure 1*a*), which are tightly packed, with a displacement of the side-chain heavy atoms of less than 2.0 Å. All but one of these residues are hydrophobic, and six (L16, L38, L40, I/V45, W48, F49) are very highly conserved or invariant (Figure 1). Since the hydrophobic core of a protein plays a key role in determining its structure, and since many of the core amino acids are highly conserved, one can predict that different homeodomains fold into similar three-dimensional structures. This prediction is supported by more recent experimental data. The homeodomain structures of *en* and *MAT* α2 in the crystal structures of their DNA complexes (77, 78) clearly have similar molecular architectures as determined for the free *Antp* homeodomain in solution. A more recent NMR structure determination of the free *ftz* homeodomain in solution revealed the same molecular fold for residues 8–53 (79). However, the more flexible helix IV comprising residues 53–59 is completely disordered in *ftz*.

Since the flexibly disordered N-terminal arm of the homeodomain is important for DNA binding and functional specificity (see below), the structures of an N-terminally truncated *Antp* homeodomain polypeptide lacking residues 1–6 was determined (80). The truncated and the intact homeodomain have identical molecular architectures. The removal of the N-terminal arm does not noticeably affect helix I from residues 10–21. Furthermore, an N-terminally extended *Antp* homeodomain polypeptide containing residues −13 to +67 including the hexapeptide (L)YPWM(R) motif was analyzed by NMR spectroscopy (81). Despite the hydrophobic nature of the hexapeptide motif, the N-terminal arm consisting of residues −14 to +6 is flexibly disordered, and the well-defined part of the homeodomain structure with residues 7–59 is indistinguishable from that of the intact 68-amino-acid polypeptide. These studies indicate that the homeodomain is connected through a flexible linker arm to the main body of the *Antp* protein.

HOMEODOMAIN-DNA COMPLEXES

NMR Structure of an Antennapedia Homeodomain-DNA Complex

The global mode of binding of the *Antp* homeodomain monomer to a 14-bp DNA duplex was determined by NMR spectroscopy (82, 83). As expected, the recognition helix (III/IV) lies in the major groove of the DNA. However, the recognition helix is considerably longer than that in prokaryotic repressors, and the turn of the helix-turn-helix motif is shifted away from the DNA by about 7 Å relative to complexes formed between prokaryotic repressor proteins and DNA. Furthermore, the N-terminal flexible arm makes additional specific contacts to bases in the minor groove, and the loop between helices I and II interacts with the DNA backbone on the other side of the major groove. This mode of binding differs in several respects from that of prokaryotic repressors and shows more similarities to the one proposed for Hin recombinase (84, 85), but the structure of the Hin recombinase-DNA complex remains to be determined. Using uniformly ^{13}C-labelled homeodomain in the complex, the solution structure of an *Antp* (C39→S) homeodomain-DNA 14-mer complex was refined (86, 87), and the individual contacts between the protein, the DNA, and the hydration water located at the protein-DNA interface were analyzed. The homeodomain interacts with both the major groove and the minor groove of the DNA, and establishes numerous contacts to the DNA backbone (Figure 5).

DNA CONTACTS OF THE N-TERMINAL ARM OF THE HOMEODOMAIN Arginine residues at positions 3 and 5 contact the DNA in the minor groove. While

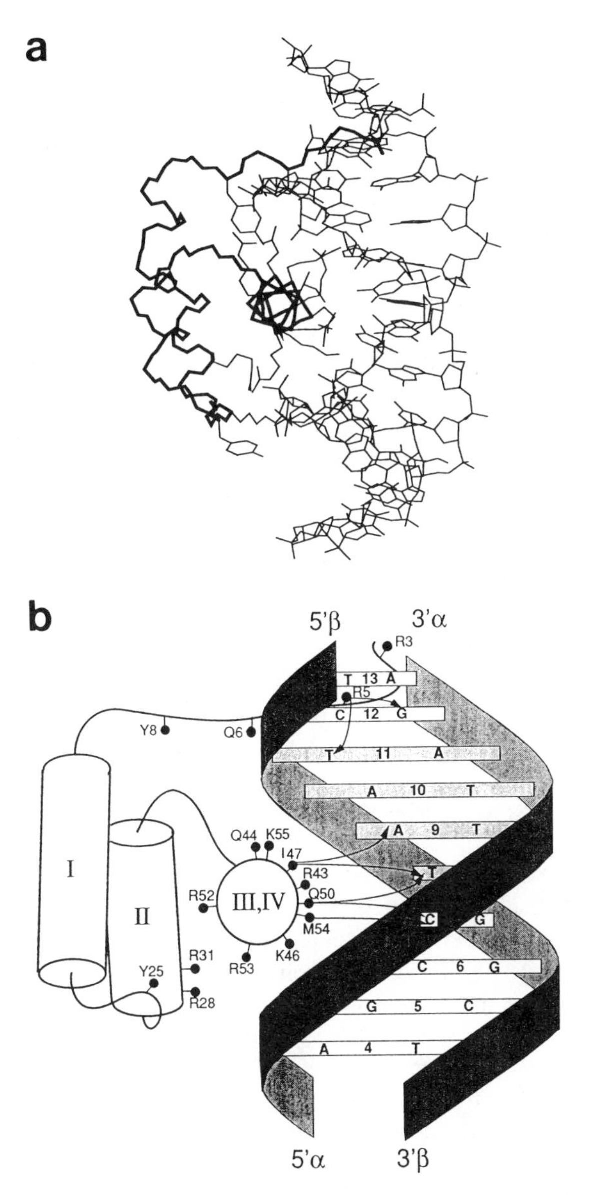

Figure 5 Schematic drawings of the *Antp (C39S)* homeodomain-DNA (BS2) complex. The view is along the axis of helix III from the C to the N terminus. (*a*) The drawing shows the backbone (dark line) of the homeodomain residues 3–55 and all side chains (thin lines) that contact the DNA (thin lines). (*b*) Schematic representation of the homeodomain-DNA contacts. The helices of the homeodomain are represented by cylinders and the amino acid residues contacting the DNA are shown by filled circles. Arrows indicate specific contacts with the DNA bases. The bars representing the core basepairs (TAAT/ATTA) are shaded. Based on data from Ref. 87.

Arg3 forms a salt bridge to the phosphate group of G12, Arg5 makes hydrophobic contacts to the sugar moieties of G12 and A13, and further short contacts to the bases T11 and G12. The Arg5-T11 contact might be a hydrogen bond, but the local precision of the structure in this region does not allow resolution of this question. Gln6 and Tyr8, which are also located in the polypeptide segment preceding helix I, contact the backbone groups of A10 in the β-strand of the DNA (Figure 5).

DNA CONTACTS OF THE RECOGNITION HELIX Numerous contacts are observed between the recognition helix III/IV and both strands of the DNA (Figure 5). Salt bridges from the side chains of Arg43, Arg52, and Lys55 to phosphate groups on the β-strand, and from Arg53 to the α-strand, connect the recognition helix to both sides of the major groove. Additional backbone contacts include hydrophobic interactions from Arg43 and Ile47 to the β-strand, and from Met54 and Lys46 to the α-strand. These are supplemented by contacts from Arg43, Gln44, and Met54 to phosphate groups. Sequence-specific recognition is provided by amino acid side chain contacts to DNA bases: from Ile47 to T8 and A9, from Gln50 to C7 and T8, and from Met54 to C7.

DNA CONTACTS FROM AMINO ACID RESIDUES NEAR THE START OF HELIX II The loop between helices I and II and the start of helix II approach the α-strand of the DNA. The side chain of Tyr25 contacts the sugar moiety of G5 and the phosphate of C6, while the side chains of Arg28 and Arg31 form salt bridges to the phosphate groups of A4 and G5, respectively.

HYDRATION AND INTERNAL MOBILITY OF THE HOMEODOMAIN DNA-COMPLEX Not withstanding the large number of direct protein-DNA contacts, the *Antp* (C39→S) homeodomain-DNA complex contains a well-defined cavity at the interface between the recognition helix and the DNA duplex, which was shown to contain several molecules of hydration water (88). The lifetimes of the water molecules in the cavity with respect to the exchange with the bulk water were estimated to be about 2 ns to 20 ms (89). This timescale overlaps with that for the slow-motional averaging for the side chain of Asn51, which indicates that this side chain jumps back and forth between two or more contact sites on the DNA. A molecular dynamics simulation of the complex in a water bath (87) implies the presence of up to five water molecules in the cavity. The water molecules can exchange positions with some amino acid side chains. In particular, the functionally important side chains of Gln50 and Asn51 occupy, on average, the same space as some of the interior water molecules. If we adopt the hypothesis that these structural variations represent conformational fluctuations in time, these findings are best rationalized as a fluctuating network of short-lived, weak-bonding contacts between Gln50 and Asn51, and the DNA bases A9 and A10 as well as hydration water molecules. This implies that the water molecules mediate specific hydrogen bonds between Gln50 and Asn51 on the one hand, and the polar groups of the DNA on the other, which otherwise could not be formed since the atomic distances across the interfacial cavity are too long.

X-ray Crystallographic Analysis of the engrailed *and* MAT *α2 Homeodomain-DNA Complexes*

The cocrystal structures of a *Drosophila engrailed* homeodomain-DNA complex (77) and a yeast *MAT α2* homeodomain-operator DNA complex (78) have been determined at 2.8 and 2.7 Å resolution, respectively. For *MAT α2*, a naturally occurring operator DNA sequence was studied, while it is not known whether the *engrailed (en)* binding site that was used is functional in vivo. The X-ray crystallographic data are in good agreement with the NMR data obtained for *Antp*. The overall structures of the polypeptides are quite similar, despite of the fact that the *MAT α2* homeodomain shares only 27% sequence identity with that of *en*. In addition, *MAT α2* has an insertion of three amino acids in the loop between helices I and II, but this is accommodated without any major changes in the overall folding of the protein. This clearly shows that *MAT α2* indeed encodes a homeodomain. The only minor difference of the *Antp* structure was found in the recognition helix, which is not subdivided into helices III and IV, but rather forms a continuous α-helix. However, as mentioned above, the *ftz* homeodomain lacks helix IV in its solution structure, indicating that a distinction between these two helices has to be made. Also, the two homeodomains of *en* and *MAT α2* dock against the DNA in very similar ways as *Antp*. In all three complexes, the N-terminal arm fits into the minor groove, where Arg3 and Arg5 of *en* and Arg7 of *MAT α2* make contacts to the bases in the DNA. Phe8 (Tyr8 in *Antp*), Tyr25 in the loop between helices I and II, and Lys55 in helix IV establish identical contacts in all three complexes to phosphates in the DNA backbone, which locate the recognition helix in the major groove. In the *en* complex, direct contacts to the bases in the major groove are made by Ile47, Gln50, and Asn51 in helix III, whereas the side chain of residue 54 (Ala) is too short to reach the bases. In both the *en* and *MAT α2* complexes, the invariant Asn51 makes a pair of hydrogen bonds to an adenine in the TAAT core binding site, which in *Antp* are thought to be mediated by hydration water molecules. Finally, Arg31 and Arg53 contact the DNA backbone in slightly different positions in the three different complexes.

As mentioned above, the "recognition helix" of the homeodomain is much longer than those found in prokaryotic helix-turn-helix motifs and elongated by a fourth helical segment. Comparing the three homeodomain DNA complexes, it is interesting to note that the recognition helix actually can be subdivided into two parts: Arg/Lys residues in the C-terminal part (mainly helix IV) contact phosphate groups, whereas the N-terminal part (mainly helix III) interacts with the DNA bases (see Ref. 90). In contrast to prokaryotic helix-turn-helix units, the elongated recognition helix of the eukaryotic homeodomain is thus intimately involved in not only sequence-specific recognition of bases, but also fixing the reading head in the major groove. It has been suggested that such bifunctional recognition helices might occur in several other DNA-binding domains of multigene families of transcription factors (90).

Is the three-dimensional structure of homeodomains unique to eukaryotic proteins? The DNA-binding domain of the *c-myb* protooncogene product consists of three imperfect tandem repeats of 51–52 amino acids. The primary amino acid sequences of these repeats match well with the sequences of different homeodomains (91). The three-dimensional structure of the third repeat, which is essential for sequence-specific recognition of DNA, has recently been determined by NMR (92). This third repeat consists of three well-characterized helices defining a hydrophobic core and helices II and III forming a helix-turn-helix motif. A comparison with *en* homeodomain shows (92) that the overall structure of the third uncomplexed *myb* repeat is significantly different from that of the homeodomain as judged from the size of the root mean square deviation (RMSD) values. Despite the differences presented by Ogata and coworkers, the helix-turn-helix motifs of the *myb* DNA-binding domain and the *Antp* homeodomain seem to be embedded in a similar global architecture.

DNA-BINDING SPECIFICITY OF HOMEODOMAINS

The binding specificity of a large number of homeodomain polypeptides has been analyzed in vitro and the results are summarized in Figure 6. Sequences recognized by polypeptides containing additional DNA-binding domains besides the homeodomain have been omitted. Many of the sequences presented have been selected from random oligonucleotide libraries [e.g. *ftz-*, *otd-*, *prd*(K50)-, and *Oct*1 (POU) homeodomain-binding sites], and therefore represent short DNA ligands that interact optimally with the contact surface of a given homeodomain. Other sequences represent in vivo targets recognized by specific homeodomain proteins (i.e. α2, *ftz* UAS1, and *bcd* sites), and thus reflect sequences that are selectively recognized by the given homeodomain protein in the extremely complex and dense mixture of molecules in a nucleus. In those cases where both in vitro and in vivo

a) Protein					major groove Q50 M54	I47 N51			minor groove R3,5			Origin	Ref. #
Antp HD	α	A	G	C	C	A	T	T	A	G	A	Regions of contact between BS2 and amino acid side chains of the *Antp* HD	(82,86,87)
	β	T	C	G	G	T	A	A	T	C	T		
b)													
MAT α2 HD[1]		[3] A	C	A	T	G	T	A	A	T	T	MAT α2 operator sequence; used in X-ray studies of the MAT α2·DNA complex	(78)
en HD[1]		G	G	T	A	A	T	T	A	C	A	MAT α2 operator sequence; used in X-ray studies of the *en*·DNA complex	(77,144,145)
			T	C	A	A	T	T	A	A	A	Consensus sequence obtained by footprint analysis	
Antp HD[1]		A	G	C	C	A	T	T	A	G	A	BS2; used in NMR structural analysis of the *Antp*·DNA complex	(146,147)
ftz HD		A	G	C	A	A	T	T	A	A	G	Consensus obtained through the analysis of synthetic DNA binding sites	(108)
ftz UAS1		A	G	C	A	A	T	T	A	C	A	Two high affinity sites of the *ftz* autoregulatory enhancer identified by footprint analysis	(106)
UAS2		G	G	C	A	A	T	T	A	A	G		
Hox A5(Scr)		C	Y	Y	N	A	T	T	A	T/G	Y	Consensus sequence derived from comparing band shift activity of various oligonucleotides	(148)
Ubx HD(*Ubx*1b)		G	G	C>T	C>A	A	T	T	A	A		Obtained via selection of binding site oligonucleotides. The same consensus binding site has been obtained using full-length *Ubx*1b for selection.	(133)
Dfd				T	C	A	T	T	A			Identified by footprint analysis	(149)
Hox A4(*Dfd*)				T	C	A	T	T	A	A		Identified by footprint analysis	(133)
Dfd HD		G	T	T>C	C>A	A	T	T	A	G		Obtained via selection of binding site oligonucleotides	(150)
bcd			G	G	G	A	T	T	A	G	A	High affinity site consensus identified for *bcd* binding sites	(110)
otd HD			G/C	G	G	A	T	T	A	G		Obtained via selection of binding site oligonucleotides	Wilson & Desplan (pers. comm.)
prd (K50) HD				G	G	A	T	T	A			Obtained via selection of binding site oligonucleotides	
ftz (K50) HD			G	G	G	A	T	T	A	G	A	Obtained via analysis of base pair substitutions in BS2	(97)
*Oct*1 HD				T	N	A	T	T	A	Y		Obtained via binding site selection with the *Oct*1 Pou HD	(103)
cad					C	A	T	A	A	A[2]		Binding site consensus derived from sites identified in the *ftz* zebra element	(151)
*TTF*1 HD				C	A	C	T	C	A	A	G[2]	Obtained by footprint analysis; aligned via the A and T residues that occur at conserved positions in the three HD·DNA complexes described above	(100)

[1]Sequences aligned through conserved contacts in the HD·DNA complex structures

[2]Temptative alignment of *cad* and *TTF*1 binding sites

[3]The ATTA (TAAT) core motif is shaded in grey

Figure 6 Binding specificities of homeodomain polypeptides.

data are available, it seems that the in vivo binding sites closely resemble sequences selected in vitro, validating the importance of a detailed analysis of homeodomain DNA interactions.

Recognition of High-Affinity Binding Sites

Surprisingly, most of the DNA sequences that interact favorably with homeodomain molecules contain a tetranucleotide ATTA, the so-called core motif (TAAT on the complementary strand). This core motif is selectively bound by homeodomains encoded by distantly related genes, such as *ftz, otd,* or *Oct*1. In both the *en* and the *Antp* homeodomain-DNA complexes, bases of the ATTA (TAAT) core interact with the highly conserved side chains of Ile47 and Asn51 in the major groove (TAAT in the *Antp* and TAAT in the *en* complex) and with Arg5 and eventually Arg3 of the N-terminal arm in the minor groove (TAAT in the *Antp,* TAAT in the *en* complex) (77, 80) (Figure 6). In support of these structural data, mutational analysis confirmed that each conserved basepair in the ATTA core contributes to the high binding affinity (see Ref. 93 for a comprehensive review). The extreme conservation of the four amino acid residues Arg3, Arg5, Ile47, and Asn51 across different classes of homeobox genes (Figure 1), as well as the conserved docking arrangement of divergent homeodomains against DNA (see above) are thus reflected in a concomitant conservation of the ATTA core motif in the high-affinity binding sites. In contrast, *MAT* $\alpha 2$, which differs from *Antp* and *en* in the residues contacting the minor groove at the N-terminal arm as well as in the amino acid side chain in position 47 (Asn instead of a highly conserved Ile), recognizes the sequence GTAA, which diverges from the ATTA core motif at two positions.

An additional remarkable feature of the sequences summarized in Figure 6*b* is the strict preference of a GG dinucleotide preceding the ATTA core in those sites that bind with high affinity to homeodomains containing a Lys residue at position 50 in the recognition helix [*bcd, otd, prd* (K50), and *ftz* (K50)]. Although structural data on a Lys50-containing homeodomain-DNA complex are still elusive, the results obtained from a genetic dissection of the DNA-binding specificity of the *bcd* gene product in yeast support the view that Lys hydrogen bonds with the guanine base immediately 5′ to the ATTA core motif (94). Consistent with this model, the corresponding nucleotides in the *Antp* and *en* homeodomain-DNA complexes (CC and TA, respectively) are contacted by the Gln side chain at position 50 in the recognition helix [*Antp* (87), *en* (77)] and by the Met54 in helix IV [*Antp* (87)].

Although the amino acid at position 50 is only one of several amino acids that interacts with the DNA ligand in a sequence-specific manner, it is the only single residue that has clearly been shown to be involved in the discriminative recognition of distinct classes of DNA sequences. This is best illustrated by the complete change of DNA specificity of *bcd* by replacing the lysine with a glutamine as found in the *Hox* gene families (95). Reciprocal swaps in which serine or glutamine is exchanged for lysine [*prd* (K50), *ftz* (K50)] also result in a preferred recognition of sites containing GG preceding the ATTA core and a concomitant low affinity for non-G-containing sites (96, 97). More recently, it has been demonstrated in an elegant series of experiments that position 50 is indeed a major determinant for the in vivo DNA-binding specificity in *Drosophila* embryos (98). By mutating the putative *ftz*-binding sites in the autoregulatory enhancer of *ftz* from CAATTA (or related sequences) to GGATTA, *ftz*'s transcriptional activity in vivo is reduced. This downregulatory effect is specifically suppressed in vivo by the DNA-binding specificity mutant *ftz* Q50K, which recognizes the mutated (GGATTA) motif. This experiment definitively proves that *ftz* is a transcriptional activator, which interacts directly with its binding sites in the autoregulatory enhancer. Consistent with structural studies, the binding specificity of the ftz protein in vivo also depends on the two nucleotides preceding the ATTA motif (98).

Besides residue 50 (and most likely residue 54, see above), little is known about the role of other particular amino acid side chains in the discrimination of binding sequences among homeodomains of different gene products. In three independent studies, the preferred binding sites for *Dfd* homeodomain–containing peptides or proteins revealed the presence or selection of a TC dinucleotide upstream of the ATTA core (Figure 6). The structural basis for this apparent selectivity is not understood, and all the side chains that contact DNA in the *Antp* complex are identical in the *Dfd* homeodomain (the only exception being Gln6, which is changed to Thr in *Dfd*). It is possible that the base preference at these two positions is modulated by differences in the positioning of the Gln50 and Met54 side chains of the recognition helix in the major groove.

Recent studies on the function of the helix-turn-helix motif of the *ftz* gene product revealed the importance of a set of class-specific DNA backbone–contacting residues in the helix-turn-helix, particularly Arg28 and Arg43 (99). These residues, which are distinct from the highly conserved DNA backbone contacting residues because they vary considerably between members of different classes, are required for efficient target site recognition of *ftz* both in vitro and in vivo. However, both residues, Arg28 as well as Arg43, are not present in the *en* gene product, which can bind virtually identical DNA sequences as those bound by the *ftz* and *Antp* gene products and forms a DNA complex similar to that of the *Antp* homeodomain. These studies suggest that slight differences exist in the requirement for DNA backbone contacts in different homeodomain-DNA complexes. Such minor differences might translate into a modification of the positioning of the recognition helix in the major groove and/or the N-terminal arm in the minor groove of the DNA, resulting in a slightly different readout mechanism. Minor differences are actually observed for certain identical amino acids in the *Antp* and the *en* complexes, in which both the N-terminal arm as well as Arg31 and Arg53 contact the DNA backbone in a slightly different way. Although little experimental evidence is available to demonstrate that such altered homeodomain-DNA complexes might change the sequence specificity, it has been found that the TTF1 gene product recognizes a divergent binding site characterized by the absence of an ATTA core (100). This is surprising, since the TTF1 homeodomain is identical to *Antp* in all residues that interact with the core motif. The absence of the class-specific contacts Arg28 and Arg43 as well as the presence of a Tyr residue at position 54 (i.e. in a position that makes base-specific contacts in the *MAT* $\alpha 2$ and the *Antp* complexes) might result in a different packing of the TTF-1 homeodomain and DNA, and result in a different readout. Alternatively, it is conceivable that despite a similar folding of the protein backbone, amino acid side chains that contact the DNA surface are held in positions in the TTF1 protein that differ from those observed in other homeodomains through interactions with non-DNA-contacting residues (see Ref. 75), and that this results in the recognition of a different high-affinity binding site.

The apparent (and rather unexpected) similarity between the DNA sequences recognized by various homeodomains (Figure 6) might suggest that the in vitro DNA-binding properties of bacterially produced, isolated homeodomains may be somewhat relaxed and that the selected sequences may not reflect sequences as they occur in in vivo target elements. This is certainly the case for those proteins that contain additional DNA-binding domains (e.g. POU- or *prd*-domain containing proteins), in which the additional domains influence the DNA-binding behavior of the linked homeodomain (101–105). However, the close match between two *ftz* target sequences in the *ftz* upstream enhancer (106, 98, 107) and the *ftz* consensus sequence as derived from binding site selection and mutational studies (108; see Figure 6) argues that isolated homeodomains do recognize sites in vitro that match sequence elements that are found to be functional in vivo. This is strongly supported for the *ftz* gene product by the demonstration that in vitro measured affinity differences between *ftz*-binding sites show a good correlation to the transcriptional activity of complex regulatory elements containing these sites as measured in transgenic flies (98, 109). A correlation between the in vitro affinity and the in vivo activity of binding sites has previously also been reported in the case of *bcd* (110).

The extreme similarity between sites recognized by a large number of homeodomain proteins is, thus, in sharp contrast to the distinctly different biological effects exerted by these proteins in vivo. Therefore, the question of how the different homeodomain proteins achieve their functional specificity in the developing organism has to be investigated in great detail if one aims at an understanding of the function of homeodomain proteins at the molecular level.

Recognition of Low-Affinity Binding Sites

In Figure 6, only those DNA sequences that bind to a given homeodomain with highest affinity have been compared. However, recent studies on a minimal upstream enhancer element that represents a direct target for the *ftz* gene product in the *Drosophila* embryo have revealed that deletion of the single high-affinity binding site does not detectably affect the function of a minimal upstream element, and that multiple medium-affinity binding sites are sufficient to mediate *ftz* DNA binding and transcriptional regulation of this particular regulatory element (106, 98, 107). Strikingly, most of the medium-affinity sites are highly conserved in *Drosophila virilis* (P Baumgartner, WJ Gehring, unpublished) and *Drosophila hydei* (111), which have presumably diverged from *Drosophila melanogaster* some 60 million years ago. The strong sequence conservation of these medium-affinity binding sites (which differ from the high-affinity consensus sequence shown in Figure 6*b* and often do not contain an ATTA core motif) suggests the intriguing possibility that the precise DNA sequence of medium-affinity sites might help to discriminate against the recognition of these sites by other homeodomain proteins expressed in the same cells at the same time. It is thus possible that different closely related homeodomain proteins do recognize the same sequence element as a highest-affinity binding site in vitro, but differ in their binding behavior towards non-optimal sites. Such differences might contribute to the functional specificity of homeodomain proteins in vivo.

The importance of medium-affinity binding sites is also highlighted by studies on the morphogen *bcd* and its potential to activate reporter genes carrying sites of high or medium affinity in the *Drosophila* embryo. These studies revealed that the level of *bcd* gene product required to activate artificial reporter gene constructs in vivo depends on the affinity of the binding sites as determined in vitro (110). However, it remains to be shown that DNA regulatory elements that respond to different levels of *bcd* in vivo indeed harbor binding sites of correspondingly different affinities.

FUNCTIONAL SPECIFICITIES OF HOMEODOMAIN PROTEINS

Consistent with the presence of a DNA-binding homeodomain motif, it has been shown that homeodomain proteins function as DNA-dependent transcriptional regulators in vitro and in vivo (reviewed in 112–115). Since distantly related homeodomains can recognize the same sequence elements in vitro, major efforts have been concentrated on the question of how in vivo target and regulatory specificity is achieved among different homeobox genes, many of which are expressed in the same cells and contain very similar homeodomains. Recently, sequences within or flanking the homeodomain have been shown to be major determinants for in vivo target specificity, and in certain cases, biochemical explanations have been put forward to account for the specificity.

The most detailed studies concerning molecular aspects of gene regulation via homeodomain proteins have been done in yeast and investigate the function of the MAT α2 homeodomain protein. MAT α2 interacts with two distinct operator sequences controlling two different sets of downstream target genes (reviewed in 116, 117). Although the sequence elements recognized by MAT α2 in each type of operator are CATGTAA (118; C Goutte, A Johnson, personal communication), MAT α2 is recruited to the two operators via the selective interaction with two different DNA-binding proteins, MCM1 and MAT a1. Analysis of MAT α2 deletion mutants and hybrid proteins revealed that the interaction between MCM1 and MAT α2 is mediated by a short, disordered protein region immediately preceding the homeodomain of MAT α2; in contrast, interaction between the two homeodomain-containing gene products MAT a1 and MAT α2 requires an unstructured 20-amino-acid tail that extends from the carboxy-terminal side of the MAT α2 homeodomain (119, 117). In both cases, the relevant protein-protein interactions result in high-affinity binding of the regulatory protein complexes and determine the orientation and the spacing between the binding sites for the respective proteins (see Ref. 120). Thus, MAT α2 recognizes two different types of operators by using peptide regions immediately adjacent to either the N or C terminus of its homeodomain to interact with partner proteins that bind adjacent to MAT α2. It is conceivable that in higher eukaryotes different proteins that contain similar homeodomain sequences but divergent N- and C-terminal sequences do interact with different transcription factors. Such homeodomain proteins differ in their ability to recognize operator sequences in vivo, despite the fact that they have identical in vitro binding properties.

Sequences within the homeodomain have also been shown to be responsible for the discrimination of proteins carrying similar homeodomains. This has been revealed by studying the POU domain proteins Oct-1 and Oct-2 and their different behavior with respect to the recruitment of the transcriptional activator, VP-16, to form a multicomponent transcription complex (121–124). Extensive domain-swap and substitution mutagenesis experiments revealed that the unique property of Oct-1 (as compared to Oct-2) to interact with VP-16 mainly resides in the presence of a Glu residue at position 22 at the end of the first helix in the POU homeodomain. Replacement of the corresponding Ala residue in the Oct-2 protein with a Glu confers to it the ability to associate with VP-16 and mediate VP-16-induced positive control in vivo. Thus, the different regulatory activity of these two proteins can be determined by a single amino acid residing on the protein surface facing away from the DNA. Interestingly, the two flanking amino acids at the C terminus of helix 1 (residues 21 and 23) are also involved in determining protein-protein interactions in other POU proteins (125).

The lack of knowledge on specific in vivo DNA target sites for most homeodomain proteins in higher eukaryotes has thus far hindered a detailed biochemical description of the regulatory specificity of these proteins. However, a large number of studies have aimed to learn more about the protein portions that might contribute to the functional specificity of *Drosophila* homeodomain proteins. Due to the important function of the homeotic genes with respect to the morphological specification of different body segments and due to the astonishing degree of conservation of the homeotic gene complexes during evolution, these studies have mainly concentrated on the identification of sequences that distinguish the functional specificities of various homeotic gene products.

Functional Specificity of Homeotic Gene Products

The dramatic demonstration of the specific functions inherent to the different homeotic genes of *Drosophila,* when assayed in ectopic expression experiments using heat-inducible promoters, has stimulated the analysis of a large number of mutant and chimeric proteins. These studies have shown that most (though not all) of the developmental and regulatory specificity differences observed among the homeotic proteins are encoded by the homeobox, the most conserved region among the different homeotic genes (126–129). More recently, it has been shown that the N terminus of the homeodomain is critical in determining the specific effects of *Antp, Scr,* and *Dfd*/*Ubx* chimeras observed upon ubiquitous expression (130–132).

At present, the molecular basis that explains the dramatic biological effects observed upon the exchange of the N-terminal arm between different homeotic proteins is not understood. Because this flexible arm of the homeodomain makes sequence-specific interactions in the minor groove in both the NMR and the X-ray crystal structures of homeodomain-DNA complexes, it is possible that the differences in N-terminal sequences between various homeotic gene products translate into slightly different binding preferences, which in turn contribute to different biological functions. Preliminary in vitro DNA-binding studies, with a purified chimeric homeodomain that introduces the *Scr* N-terminal four amino acids into the sequence of the *Antp* homeodomain, suggest that changing these four amino acids results in similar but distinct affinities for a series of synthetic binding sites (G Halder and WJ Gehring, personal communication). In line with these results, Ekker et al (133) have demonstrated that the exchange of amino-terminal sequences between *Ubx* and *Dfd* homeodomains results in a slightly different sequence preference 3′ to the ATTA core. Interestingly, different full-length Hox proteins produced in tissue-culture cells have also been shown to bind differentially to a number of DNA-binding sites depending on their N-terminal arm sequences (G Urier and D Duboule, personal communication).

Alternatively, the N-terminal arm of homeotic gene products might be involved in selective protein-protein interactions (similar to those described above for *MAT* $\alpha 2$ or Oct-1), and in this way contribute to the exquisite functional specificity of homeotic proteins. Such interactions have also been postulated to occur between the homeodomain proteins of *extradenticle (exd)* and *Ultrabithorax (Ubx),* which may form heterodimers with a DNA-binding specificity different from that of the respective monomers or homodimers (60). Very little is known at present about the sequence composition of cis-acting DNA elements, which mediate the gene regulatory effects of homeotic gene products, and factors that interact with homeotic proteins have not been characterized yet. Several elements that are targets of homeotic gene control have recently been identified (reviewed in Ref. 134), but it remains to be shown whether these putative target sites interact directly with the respective homeotic protein. The use of DNA-binding specificity mutants and their introduction into living organisms via reverse genetics (98) allows the definitive identification of target genes (135) and opens the way for biochemical and genetic study of potential cofactors that might interact with homeotic proteins.

CONCLUSIONS

During the past 10 years, more than 300 homeobox genes from taxonomic groups ranging from yeast to human have been isolated. A vast amount of sequence data on homeodomains has been accumulated, which provides useful and important information about the evolution of the homeobox gene family and the phylogeny of eukaryotic organisms. The sequence similarity and the colinear arrangement of the homeobox genes in the homeotic gene complexes indicates that these genes are true homologs (orthologs) in insects and mammals. This conclusion is strongly supported by functional studies. The patterns of *Hox* gene expression, the effects of gain-of-function (136) and loss-of-function (137) mutations in transgenic mice, and embryological studies (138) indicate that the vertebrate cognate genes have similar homeotic effects as the homeotic genes of *Drosophila,* and that they also specify the body plan and provide positional information along the antero-posterior axis (20). A striking demonstration of the orthology between the mammalian *Hox* genes and the homeotic *Drosophila* genes is provided by gene transfer experiments of *Hox* genes into *Drosophila* (28–30).

The detailed knowledge of the three-dimensional structure of homeodomain and homeodomain-DNA complexes has provided the framework for understanding the functional aspects of homeodomain proteins. The NMR studies emphasize the importance of the more flexible parts of the homeodomain in intermolecular interactions with the DNA. In particular, the readout of the DNA sequence by flexible amino acid side chains in the recognition helix and their interaction with hydration water represents a novel finding. Also, the flexible N-terminal arm, and the more disordered helix IV are important in DNA binding. Knowledge of the specific DNA contacts made it possible to construct a mutant homeodomain with an altered DNA-binding specificity, which is capable of suppressing the corresponding mutations in the DNA-binding site (98). This crucial experiment shows that homeodomain proteins interact directly with their target sites in vivo and regulate transcription.

This same method can now be used to identify direct target genes (135) and to elucidate the network of interactions controlling development. Putative target genes that have been identified include other transcription factors (139, 140), growth factor genes (135), and genes encoding cell-adhesion molecules (141, 142, 142a). In the future, the cascade of gene action from the nucleus to the final morphological differentiation will have to be elucidated.

ACKNOWLEDGMENTS

This paper is in memory of Hansruedi Widmer.

We wish to thank our colleagues for providing manuscripts and information prior to publication. We are grateful to Kurt Wüthrich and Martin Billeter for providing figures and Georg Halder for reading the manuscript. Also, we are indebted to Magrit Jaeggi and Verena Grieder for their assistance in preparing the illustrations, and to Erika Marquardt-Wenger for typing and editing this manuscript. This work was supported by the Kantons Basel and the Swiss National Science Foundation.

Due to space limitations, we regret that we have not been able to cite all relevant literature on the topic.

APPENDIX

```
....|...10....|...20....|...30....|...40....|...50....|...60
RRRKRTAYTRYQLLELEKEFHFNRYLTRRRRIELAHSLNLTERQVKIWFQNRRMKWKKEN   Consensus    Class
                                                                            (Paralog group)

NNSG..NF.NK..T..............A....I.NT.Q.N.T............Q..RV   lab          lab
PNAV..NF.TK..T.........K....A..V.I.A..Q.N.T............Q..RE   m HoxA1
PNAV..NF.TK..T.........K....A..V.I.A..Q.N.T............Q..RE   h HoxA1
PNTA..NF.TK..T.........K....A..V.I.AA.Q.N.T............Q..RE   x HoxA1
PGGL..NF.TR..T.........K..S.A..V.I.AT.E.N.T............Q..RE   m HoxB1
PSGL..NF.TR..T.........K..S.A..V.I.AT.E.N.T............Q..RE   h HoxB1
PNTI..NF.TK..T.........K....A..V.I.AT.E.N.T............Q..RE   c HoxB1      (Hox1)
QNSI..NF.TK..S..............A..V.I.AT.E.N.T............Q..RE   ax HoxB1
PSAI..NFSTK..T.........K....A....I.NC.Q.NDT............Q..RE   m HoxD1
SSAI..NFSTK..T.........K....A....I.NC.H.NDT............Q..RE   h HoxD1
PCNV..NF.TK..T.........K....A....I.N..Q.NDT............Q..RE   x HoxD1
AFSL..SFSTR..T........S...S.A..L.V.R..R.RDA...V........Q..RE   CHbx-1
NGTN..NF.TH..T.......TAK.VN.T..T.I.SN.K.Q.A............E..RE   ceh-13

P..L.....NT............K..C.P....I.A..D.......V........H.RQT   pb           pb
S..L.....NT............K..C.P..V.I.AL.D.......V........H.RQT   h HoxA2
S..L.....NT............K..C.P..V.I.AL.D.......V........H.RQT   m HoxA2
S..L.....NT............K..C.P..V.I.AL.D.......V........H.RQT   m HoxB2      (Hox2)
S..L.....NT............K..C.P..V.I.AL.D.......V........H.RQT   Nv HoxB2
A..L.....NT............K..C.P..V.I.AL.D.......V........H.RQT   h HoxB2
PG.L.....NT............K..C.P..V.I.AL.D.......V........H.RQT   Ss pS6

SK.A.....SA..V............V.P..V.M.NL.......I..........Y..DQ   h HoxA3
SK.G......P..V............M.P..V.M.NL.......I..........Y..DQ   m HoxA3
SK.A.....SA..V............C.P..V.M.NL...S...I..........Y..DQ   m HoxB3
SK.A.....SA..V............C.P..V.M.NL...S...I..........Y..DQ   h HoxB3
SK.A.....SA..V............C.P..V.M.NL...S...I..........Y..DQ   x HoxB3
SK.A.....SA..V............C.P..V.M.NL...S...I..........Y..DE   c HoxB3      (Hox3)
SK.AA....SA..V............C.P..V.M.NL.S.....I..........Y..DQ   Nv HoxB3
SK.V.....SA..V............C.P..V.M.NL.......I..........Y..DQ   h HoxD3
SK.V.....SA..V............C....V.M.NL.......I..........Y..DQ   r HoxD3
SKSV..T..SA..V............C.P..V.M.NL.......I..........Y..DQ   m HoxD3
SK.A.V.F.SS...........SA..C.N..L.M.EL.K..D..I..........Y..DH   zf-114
GK.A.....SA..V............C.P..V.M.AM.......I..........Y...Q   AmphiHox3

LK.S...F.SV..V...N..KS.M..Y.T....I.QR.S.C..............F..DI   zen1         Diverse
SK.S...FSSL..I...R...L.K..A.T....ISQR.A................L..ST   zen2
SK.M...F.ST......R..AS.M..S.L....I.TY...S.K..........V.H..G    Gsh-1
GK.M...F.ST......R..SS.M..S.L....I.TY...S.K..........V.H..G    Gsh-2
NK.T....S.S..F........DK.IS.P..V...S.......HI.............ME   Htr-A2
NK.T......A.........L..K.IS.P..V...VM......HI..............E   Xlhbox8

PK.Q......H.I........Y...........I..T.V.S...I.............D.   Dfd          Dfd
PK.S......Q.V....................I..T.C.S.................DH   c HoxA4
PK.S......Q.V....................I..T.C.S.................DH   m HoxA4
PK.S......Q.V....................I..T.C.S.................DH   h HoxA4
PK.S......Q.V........Y.........V.I..A.C.S...I.............DH   m HoxB4
PK.S......Q.V........Y.........V.I..A.C.S...I.............DH   h HoxB4
PK.S......Q.V........Y.........V.I....C.S...I.............DH   c HoxB4
AK.S......Q.V........Y.........V.I..T.R.S...I.............DH   x HoxB4      (Hox4)
PK.S......Q.V........Y.........V.I..T.C.S...I.............DH   zf-13
PK.S.A....Q.V........Y...........I....C.S...I.............DH   h HoxC4
PK.S......Q.V........Y...........I....C.S...I.............DH   m HoxC4
GPARGV.NC.Q.V........Y...........I....C.S...I.............DH   r HoxC4
PK.S......Q.V....................I..T.C.S...I.............DH   c HoxD4
PK.S......Q.V....................I..T.C.S...I.............DH   h HoxD4
PK.S......Q.V....................I..T.C.P...I.............DH   m HoxD4
```

```
....|...10....|...20....|...30....|...40....|...50....|...60
RRRKRTAYTRYQLLELEKEFHFNRYLTRRRRIELAHSLNLTERQVKIWFQNRRMKWKKEN  Consensus      Class
                                                                       (Paralog group)

TK.Q..S.....T....................I..A.C.....I..........L...H  Scr            Scr
VK.Q..S.....T....................I..A.C.....I..............H  Am H55

GK.A........T....................I..A.C.S...I.............D.  m HoxA5
GK.A........T....................I..A.C.S...I.............D.  h HoxA5
GK.A........T....................I..A.C.S...I.............D.  c HoxB5
GK.A........T....................I..A.C.S...I.............D.  m HoxB5
GK.A........T....................I..A.C.S...I.............D.  h HoxB5
GK.A........T....................I..A.C.S...I.............D.  Ov HoxB5
GK.A........T....................I..T.C.S...I.............D.  x HoxB5
GK.A........T....................I..A.C.S...I.............D.  zf-21          (Hox5)
EN.A........A....................IR.A.C.S...I.............D.  zf-54
G.GRGQT.....T....................M..A.C.S...I.............D.  Ss pS12-B
GK.S..S.....T....................I.NN.C.N...I.............DS  m HoxC5
GK.S..S.....T....................I.NN.C.N...I.............DS  h HoxC5
GK.S..S.....T....................I.NN.C.N...I.............DS  Nv HoxC5
GK.S..S.....T................D...I.NN.C.N...I.............DT  x HoxC5
GK.S..S.....T....................I.NN.C.N...I.............DS  zf-25

GK.G.QT...Q.T.........S..V.....F.I.Q..G.S...I............R.H  TgHbox3
DK.A..S.S...T.............NG.....I....G.....I.............D.  Hr Lox6        Diverse
EK.Q......N.V........THK....K....V....M................H....  lin-39

.K.G.QT.....T....................I..A.C.....I...............  Antp           Antp
.K.G.QT.....T........Y...........I..A.C.....I...............  Am H90
...G.QT...F.T..........H.........I..A.C.....I..........L...L  abd-A
...G.QT...F.T..........H.........I..A.C.....I..........L...L  Aa abd-A
...G.QT...F.T..........H.........I..A.C.....I..........L...L  Bm abd-A
...G.QT...F.T..........H.........I..A.C.....I..........L...L  Ms abd-A
...G.QT...F.T..........H.........I..A.C.....I..........L...L  Sg abd-A
...G.QT...F.T........Y.H.........I..A.C.....I..........L...L  Am H15
...G.QT.....T........T.H.........M..A.C.....I..........L...I  Ubx
...G.QT.....T........T.H.........M..A.C.....I..........L...I  Bm Ubx
...G.QT.....T.......K.............S.T.Y.....I..........E...V  Hm Lox2
...G.QT.....T.......K.............S.T.Y.....I..........E...V  Hr Lox2
 K.T.QT.....T.......YS...........I....A.S...I...............  Hr Lox5
SK.T.QT.....T............I......DI.NA.S.S...I..........S..DR  ftz

G..G.QT.....T....................I.NA.C.....I...............  m HoxA6
G..G.QT.....T....................I.NA.C.....I...............  h HoxA6
G..G.QT.....T....................I..A.C.....I...............  zf-22
G..G.QT.....T........Y...........I..A.C.....I..............S  m HoxB6
G..G.QT.....T........Y...........I..A.C.....I..............S  h HoxB6
A..G.QT.....T....................I....C.....I...............  c HoxB6
...G.QI.S...T....................I.NA.C.....I..............S  m HoxC6        (Hox6)
...G.QI.S...T....................I.NA.C.....I..............S  h HoxC6
...G.QI.S...T....................I.NA.C.....I..............S  x HoxC6
...G.QI.S...T....................I.NASC.....I..............S  Nv HoxC6
...G.QI.S...T....................I.NA.C.....I..............T  zf-61
.K.G.QT...A.T........Y......K....I.QAVC.S...I..............R  PaHbox6
.K.G.QT...A.T........Y......K....I.QAVC.S...I..............R  TgHbox6
```

		Class (Paralog group)
....|...10....|...20....|...30....|...40....|...50....|...60		
RRRKRTAYTRYQLLELEKEFHFNRYLTRRRRIELAHSLNLTERQVKIWFQNRRMKWKKEN	Consensus	Class (Paralog group)
.K.G.QT.....T....................I..A.C.....I..............H	c HoxA7	Antp
.K.G.QT.....T....................I..A.C.....I..............H	m HoxA7	
.K.G.QT.....T....................I..A.C.....I..............H	h HoxA7	
.K.G.QT.....T....................I..A.C.....I..............H	x HoxA7	
.K.G.QT.....T....................Y..A.C.....I..............H	qu HoxA7	
GK.G.QT.....T.................AV.I..A.C.....I..............H	r HoxA7	
.K.G.QT.....T........Y...........I..T.C.....I...............	m HoxB7	(Hox7)
.K.G.QT.....T........Y...........I..A.C.....I...............	h HoxB7	
.K.G.QT.....T....................I..T.C.....I...............	x HoxB7β	
.K.G.QT.....T....................I..V.C.....I...............	x HoxB7α	
.K.G.QT.....T........Y...........I..GVC.....I...............	r HoxB7	
.K.GSQT.....T..................V.I..V.C.....I.............DH	Ss pS12-A	
...G.QT.S...T.......L..P....K....VS.A.G.....................	m HoxB8	
...G.QT.S...T.......L..P....K....VS.A.G.....................	h HoxB8	
...G.QT.S...T.......L..P....K....VS.A.G.....................	r HoxB8	
...G.QT.S...T.......L..P....K....VS.A.G.....................	x HoxB8	
..SG.QT.S...T.......L..P....K....VS.A.G.....................	cat HoxC8	
..SG.QT.S...T.......L..P....K....VS.A.G.....................	m HoxC8	
..SG.QT.S...T.......L..P....K....VS.A.G.....................	h HoxC8	(Hox8)
..SG.QT.S...T.......L..P....K....VS.A.G.....................	r HoxC8	
...G.QT.S.F.T.......L..P....K....VS.A.G.....................	c HoxD8	
...G.QT.S.F.T.......L..P....K....VS.A.A.....................	h HoxD8	
...G.QT.S.F.T.......L..P....K....VS.T.A.....................	m HoxD8	
H..GLQTCS...T.......QC.P...CK.W..VS.A.G.....I.............RE	pig Hbx24	
.K.C.QT.....T.....................S.L.G.....I..........Y...S	TgHbox1	
QK.T.QT.....T..........K.........I..T.T.....I.......	smox-1	
KK.G.QT.S.H.T........Y.K..........NR..G.S...I..........K...D	Alhb-1	
SK.T.QT.S.S.T........YHK....K..Q.ISET.H................H...A	mab-5	
SK.I.....SI.........QN....S.L...QI.AI.D...K..........V....DK	Cv cnox2	
SK.I.....SI.........QN....S.L...QI.AI.D...K..........V....DK	Hv cnox2	
CSFGHRKII...R..KY......D..L.F.RN.D.S.S.I.V........Q...Q	Cv cnox1	
V.K..KP.SKF.T.......L..A.VSKQK.W...RN.Q................N..NS	Abd-B	Abd-B
V.K..KP.SKF.T.......L..A.VSKQK.W...RN..................N..NS	Bm Abd-B	
T.K..CP..KH.T.......L..M....D..Y.V.RL..................M..I.	c HoxA9	
T.K..CP..KH.T.......L..M....D..Y.V.RL..................M..I.	Ca HoxA9	
T.K..CP..KH.T.......L..M....D..Y.V.RL..................M..I.	m HoxA9	
T.K..CP..KH.T.......L..M....D..Y.V.RL..................M..I.	h HoxA9	
T.K..CP..KH.T.......L..M....D..Y.V.RL..................M..I.	Xb HoxA9	
S.K..CP..K..T.......L..M....D..H.V.RL...S..............M..M.	m HoxB9	
S.K..CP..K..T.......L..M....D..H.V.RL...S..............M..M.	h HoxB9	(Hox9)
S.K..CP.SK..T.......L..M....D..H.V.RL...S..............M..L.	x HoxB9	
T.K..CP..K..T.......L..M....D..Y.V.RV..................M..M.	m HoxC9	
T.K..CP..K..T.......L..M....D..Y.V.RV..................M..M.	h HoxC9	
T.K..CP..K..T.......L..M....D..Y.V.RI..................M..MS	c HoxD9	
T.K..CP..K..T.......L..M....D..Y.V.RI..................M..MS	m HoxD9	
T.K..CP..K..T.......L..M....D..Y.V.RI..................M..MS	h HoxD9	
G.K..CP..KF.T.......L..M....D..L.I.RL.S................M..Q.	TgHbox4	

....|...10....|...20....|...30....|...40....|...50....|...60		Class (Paralog group)
RRRKRTAYTRYQLLELEKEFHFNRYLTRRRRIELAHSLNLTERQVKIWFQNRRMKWKKEN	Consensus	
G.K..CP..KH.T.......L..M....E..L.ISR.VH..D.............L..M.	m HoxA10	Abd-B
G.K..CP..KH.T.......L..M....E..L.ISR.VH..D.............L..M.	h HoxA10	
G.K..CP..KH.T.......L..M....E..L.ISKTI...D.............L..M.	m HoxC10	
G.K..CP..KH.T.......L..M....E..L.ISKTI...D.............L..M.	h HoxC10	
G.K..CP..KH.T.......L..M....E..L.ISK.I...D.............L..M.	NvHoxC10	(Hox10)
G.K..CP..KH.T.......L..M....E..L.ISK.V...D.............L..MS	c HoxD10	
G.K..CP..KH.T.......L..M....E..L.ISK.V...D.............L..MS	h HoxD10	
G.E..CP..KH.T.......L..M....E..L.ISK.V...D.............L..MS	m HoxD10	
G.K..CP..KH.T.......L..M....E..L.ISK.V...D.............L..MS	Nv HoxD10	
T.K..CP..K..IR...R..F.SV.INKEK.LQ.SRM....D.............E..I.	c HoxA11	
T.K..CP..K..IR...R..F.SV.INKEK.LQ.SRM....D.............E..I.	m HoxA11	
T.K..CP..K..IR...R..F.SV.INKEK.LQ.SRM....D.............E..I.	h HoxA11	
T.K..CP.SKF.IR...R..F..V.INKEK.LQ.SRM....D.............E..LS	m HoxC11	
T.K..CP.SKF.IR...R..F..V.INKEK.LQ.SRM....D.............E..LS	h HoxC11	(Hox11)
S.K..CP..K..IR...R..F..V.INKEK.LQ.SRM....D.............E..L.	c HoxD11	
S.K..CP..K..IR...R..F..V.INKEK.LQ.SRM....D.............E..L.	m HoxD11	
S.K..CP..K..IR...R..F..V.INKEK.LQ.SRM....D.............E..L.	h HoxD11	
S.K..CP..K..IR...R..F..V.INKEK.LQ.SRM....D.............E..L.	Nv HoxD11	
S.K..KP.SKL..A...G..LV.EFI..Q..R..SDR...SDQ............K.RLL	h HoxC12	
A.K..KP..KQ.IA...N..LV.EFIN.QK.K..SNR...SDQ............K.RVV	m HoxD12	
A.K..KP..KQ.IA...N..LV.EFIN.QK.K..SNR...SDQ............K.RVV	h HoxD12	(Hox12)
S.K..KP..KQ.IA...N..LL.EFIN.QK.K..SNR...SDQ............K.RVV	c HoxD12	
G.K..VP..KV..K...R.YAT.KFI.KDK.RRISATT..S....T.......V.E..VI	h HoxA13	
G.K..VP..KV..K.....YAASKFI.KEK.RRISATT..S....T.......V.E..VV	h HoxC13	
G.K..VP..KL..K...N.YAI.KFINKDK.RRISAAT..S....T.......V.D..IV	c HoxD13	(Hox13)
G.K..VP..KL..K...N.YAI.KFINKDK.RRISAAT..S....T.......V.D..IV	h HoxD13	
SKKG.QT.Q...TSV..AK.QQSS.VSKKQ.E..RLQTQ..D..I..........A...K	egl-5	
V..Y...F..D..GR.....YKEN.VS.P..C...AQ...P.STI.V........D.RQR	eve	eve
T..Y...F..E..SR.....LREN.VS.T..S...SM...S.TTI..........A.RRR	Af eveC	
M..Y...F..E.IAR.....YREN.VS.P..C...AA...P.TTI.V........D.RQR	m Evx-1	
M..Y...F..E.IAR.....YREN.VS.P..C...AA...P.TTI.V........D.RQR	x Evx-1	
V..Y...F..E.IAR.....YREN.VS.P..C...AA...P.TTI.V........D.RQR	m Evx-2	
V..Y...F..E.IAR.....YREN.VS.P..C...AA...P.TTI.V........D.RQR	h EVX2	
PK.I...FSPS...K..HA.ES.Q.VVGAE.KA..QN...S.T...V......T.H.RMQ	ems	ems
PK.I...FSPS...R..RA.EK.H.VVGAE.KQ..G..S.S.T...V......T.Y.RQK	m Emx1	
PK.I...FSPS...R..RA.EK.H.VVGAE.KQ..G..S.S.T...V......T.Y.RQK	h EMX1	
PK.I...FSPS...R..HA.EK.H.VVGAE.KQ.....S...T...V......T.F.RQK	m Emx2	
PK.I...FSPS...R..HA.EK.H.VVGAE.KQ.....S...T...V......T.F.RQK	h EMX2	
PK.V...FSPT...K..HA.EG.H.VVGAE.KQ..QG.S...T...V......T.H.RMQ	E5	
H.KA..I.GTT.TQQ..DM.KGQM.VVGAE.EN..QR.G.SPS..R.......S.HRRKQ	ceh-23	
M.KP..I.SSL..QQ.NRR.QRTQ..ALPE.A...A..G..QT..........S.Y..MM	Dll	Dll
V.KP..I.SSF..AA.QRR.QKTQ..ALPE.A...A..G..QT..........S.F..MW	Tes-1	
I.KP..I.SS...AA.QRR.QKAQ..ALPE.A...AQ.G..QT..........S.F..LY	zf dlx-3	
I.KP..I.SSL..QA.NRR.QQTQ..ALPE.A...A..G..QT........K.S.F..LM	m dlx	
I.KP..I.SSL..QA.NHR.QQTQ..ALPE.A...A..GV.QT........K.S.Y..LI	Xdll	

```
....|...10....|...20....|...30....|...40....|...50....|...60
RRRKRTAYTRYQLLELEKEFHFNRYLTRRRRIELAHSLNLTERQVKIWFQNRRMKWKKEN  Consensus     Class
                                                                         (Paralog group)

KDKY.VV..DF.R......YCTS..I.I..KS...QT.S.S............A.ERTS.  cad           cad
KDKY.VV..DH.R........YS..I.I..KA...AA.G..............A.ER.V.  c cdxA
KDKY.VV..DH.R........YS..I.I..KA...AA.G..............A.ER.V.  Xcad2
KDKS.VV..DH.R........YS..I.I..KS...AN.G..............A.ER.V.  m cdx-1
KDKY.VV..DH.R........YS..I.I..KA...AT.G.S............A.ER.IN  sh cdx-3
KDKY.VV..DQ.R........YS..I.I..KA...VN.G.S.T..........A.ER.I.  Xcad1
KEKY.VV..DH.R............I.I..KS...VN.G.S............A.ER.LI  zf cad1
ADKY.MV.SD..R........TSPFI.SD.KSQ.STM.S.....I........A.DRRDK  pal-1

.SWS.AVFSNL.RKG..IQ.QQQK.I.KPD.RK..AR....DA...V.........RHTR  H2.0          HLX
.SWS.AVFSNL.RKG...R.EIQK.V.KPD.KQ..AM.G..DA...V.........RHSK  m Hlx
.SWS.AVFSNL.RKG...R.EIQK.V.KPD.KQ..AM.G..DA...V.........RHSK  HB24
.KWN.AVFRLM.RRG...S.QSQK.VAKPE.RK..DA.S..DA.............RQ.I  Hal-hlx
GMLR.AVFSDV.RKA...T.QKQK.ISKPD.KK..SK.G.KDS.............RNSK  m Dbx
GILR.AVFSED.RKA...M.QKQK.ISKTD.KK..IN.G.K.S.............RNSK  Chlx-A

.KKA..TFSGK.VF....Q.EAKK..SSSD.S...KR.DV..T..........T....IE  ceh-9         NEC
KKKT..VFS.S.VFQ..ST.EVK...SSSE.AG..AN.H...T..........N...RQM  TgHbox5
KKKT..IFSKS.VFQ..ST.DVK...SSAE.AG..AA.H...T..........N.L.RQL  soho

KKKP..SF..L.IC....R..RQK..ASAE.AA..KA.KM.DA...T......T..RRQT  TCL-3         TCL
.KKP..SFS.S.V....RR.LRQK..ASAE.AA..KA.RM.DA...T......T..RRQT  MUR10F

N.KP..PF.TQ...S...K.REKQ..SIAE.A.FSS..R...T..........A.A.RLQ  msh           msh
N.KP..PF.TQ...S...K.REKQ...IAE.A.FSS..H...T..........A.A.RLQ  Am H17
N.KP..PF.TA...A..RK.RQKQ..SIAE.A.FSS..S...T..........A.A.RLQ  zf msh-A
N.KP..PF.TA...A..RK.RQKQ..SIAE.A.FSS..S...T..........A.A.RLQ  c msx-1
N.KP..PF.TA...A..RK.RQKQ..SIAE.A.FSS..S...T..........A.A.RLQ  m msx-1
N.KP..PF.TA...A..RK.RQKQ..SIAE.A.FSS..S...T..........A.A.RLQ  h msx-1
N.KP..PF.TS...A..RK.RQKQ..SIAE.A.FSS......T..........A.A.RLQ  x msx-1
N.KP..PF.TA...A..RK..QKQ..SIAE.A.FSS..S...T..........A.A.RLQ  m msh-3
N.KP..PF.TS...A..RK.RQKQ..SIAE.A.FSS......T..........A.A.RLQ  qu msx-2
N.KP..PF.TS...A..RK.RQKQ..SIAE.A.FSS......T..........A.A.RLQ  c msx-2
N.KP..PF.TS...A..RK.RQKQ..SIAE.A.FSS......T..........A.A.RLQ  m msx-2
N.KP..PF.TS...A..RK.RQKQ..SIAE.A.FSS......T..........A.A.RLQ  x msx-2
N.KP..PF.TS...A..RK.RQKQ..SIAE.A.FSN......T..........A.A.RLQ  zf msh-C
N.KP..PF.TS...A..RK.RQKQ..SIAE.A.FSS..T...T..........A.A.RLQ  zf msh-D
N.KP..PFSTS...S..RK.RQKQ..SIAE.A.FSN......T..........A.A.RLQ  zf msh-B
N.KP..PF.TQ..MS...K.REKQ..SIAE.A.FSN..S...T..........A.S.RLQ  Ci msh
N.KP..PFSVN...T..QK.KRKQ..SISE.A..SEL.R...T.I........A.Q.RSK  Cv cnmsh

K.KR.VLF.KA.TY...RR.RQQ...SAPE.EH..SLIR..PT........H.Y.T.RAQ  NK-2          NK-2
K.KR.VLFSKA.TY...RR.RQQ...SAPE.EH..SLIR..PT........H.Y.M.RAR  Nkx-2.2
K.KR.VLFSKA.TY...RR.RQQ...SAPE.EH..SLIR..PT........H.Y.M.RAR  x XeNK-2
..KR.VLFSQA.VY...RR.KQQK..SAPE.EH..SMIH..PT........H.Y.M.RQA  r TTF-1
..KR.VLFSQA.VY...RR.KQQK..SAPE.EH..SMIH..PT........H.Y.M.RQA  m TTF-1
   R.VLFSQA.VY...RR.KQQK..SAPE.EH..SMIH..PT........H.Y.M.RQA  Nkx-2.4
..KR.ILFSQA.IY...RR.KQQK..SAPE.EH..NLI...PT........H.Y.C.RSQ  Dth-2
..KP.VLFSQA.VF...RR.KQQ...SAPE.EH..S..K..ST..........Y.C.RQR  Nkx-2.3
K.KR.VLFSKK.I....RH.RQKK..SAPE.EH..NLIG.SPT........H.Y.M.RAH  Dth-1
KK.S.A.FSHA.VF...RR.AQQ...SGPE.S.M.K..R...T..........Y.T.RKQ  bap(NK-3)
K.KP.VLFSQA.V....CR.RLKK...GAE.EII.QK...SAT..........Y.S.RGD  tin (NK-4)
QSKR.VLFNKF.ISQ...R.RKQ....AQE.Q....TIG..PT........HAY.M.RLF  EgHbx3
```

....|...10....|...20....|...30....|...40....|...50....|...60		
RRRKRTAYTRYQLLELEKEFHFNRYLTRRRRIELAHSLNLTERQVKIWFQNRRMKWKKEN	Consensus	Class (Paralog group)
P..T..TF.SS.IA...QH.LQG....AP.LAD.SAK.A.GTA......K...RRH.IQS	bcd	curios
P..T..TF.SS.IA...QH.LQG....AP.LAD.SAK.A.GTA......K...RRH.IQS	Dps bcd	
Q.KA...F.DH..QT...S.ERQK..SVQE.Q....K.D.SDC...T.Y....T...RQT	BarH1	
Q.KA...F.DH..QT...S.ERQK..SVQE.Q..S.K.D.SDC...T.Y....T...RQT	Da BarH1	
Q.KA...F.DH..QT...S.ERQK..SVQD.M...NK.E.SDC...T.Y....T...RQT	BarH2	
C.KP..VFSDL..MV..R..NNRK..STPQ.TN..DR.G.NQT...T.Y.........T	Cv cnox3	
..KA..VFSDP..SG...R.EGQ...SPPE.V...TA.G.S.T...T........H..QL	bsh	
E.KP.Q..SAR..DR..T..QTDK..SVNK..Q.SQT.....T.I.T......T....QL	ceh-19	
QK.A.VSFSSS.VHV..ER.DRQK..SSAE.A.MSRD.G.S.T..........Y.T..RA	EgHbx2	
.KGGQVRFSNE.TI....K.ETQK..SPPE.KR..KL.Q.S.....T......A..RRLK	c Prh	
.KGGQVRFSND.TI....K.ETQK..SPPE.KR..KM.Q.S.....T......A..RRLK	h Prh	
S..R...F.SE..........CKK..SLTE.SQI..A.K.S.V..........A...RIK	c Ovx1	
SF.N...F.D...IC..R..SHIQ..S.ID..H..QN.....K..........VR.R.R.	smox-4	
A.KE...F.KE..R...A..AHHN....L..Y.I.VN.D.S.....V..........RVK	mox-1	
P.KE...F.KE.IR...A..AHHN....L..Y.I.VN.D.......V..........RVK	mox-2	
IP.R..TF.VE..YL..MY.AQSQ.VGCDE.ER..RI.S.D.Y..........IRMRR.A	ceh-7	
MK.I..VF.PE..EK.....LKQQ.MVGTE.VD..ST.....T...V......I..R.QS	x Xnot	
KKHS.PTFSGQ.IFA...T.EQTK..AGPE.AR..Y..GM..S...V......T..R.RH	Gtx	
PK.P..VF.DE..EK..ES.NTSG..SGST.AK..E..G.SDN...V......T.Q..ID	ceh-5	
..KT..TFSNC..N...NN.NRQ....PTD.DRI.KH.G..NT..IT......A.L.R.A	smox-5	
GKCR.SRTAFTSQQL..L..K..K..S.PK.F.V.T..M...T..............RSK	HB9	
Q..Q..TFSTE.T.R..V...R.E.IS.S..F...ET.R...T.I........A.D.RIE	ro	
Q..Q..TFSTE.T.R..V...R.E.IS.S..F...ET.R.S.T.I........A.D.RIE	Dv ro	
EK.P...FSSE..AR.KR..NE.....E...QQ.SSE.G.N.A.I......K.A.I..ST	en	en
EK.P...FSSE..AR.KR..NE.....E...QQ.SSE.G.N.A.I......K.A.I..ST	Dv en	
DK.P...FSGT..AR.KH..NE.....EK..QQ.SGE.G.N.A.I......K.A.L..SS	inv	
EK.P...FSAE..AR.KR..AE.....E...QQ.SRD.G...A.I......K.A.I..AS	Am E30	
EK.P...FSGE..AR.KR..AE.....E...QQ.SRD.G.N.A.I......K.A.I..AS	Am E60	
EK.P...FSGA..AR.KH..AE.....E...QS..AE.G.A.A.I......K.A.I..AS	Bm en	
EK.P...FSGP..AR.KH..AE.....E...QS..AE.G.A.A.I......K.A.I..AS	Bm inv	
EK.P...FSGE..AR.KH..TE.....E...Q...RE.G.N.A.I......K.A.I..AS	Sa en	
DK.P...F.AE..QR.KA..QT.....EQ..QS..QE.S.N.S.I......K.A.I..AT	h En-2	
DK.P...F.AE..QR.KA..QT.....EQ..QS..QE.S.N.S.I......K.A.I..AT	m En-2	
DK.P...F.AD..QR.KA..QT.....EQ..QS..QE.S.N.S.I......K.A.I..AT	x En2α	
DK.P...F.AE..QR.KA..QT.....EQ..QS..QE.G.N.S.I......K.A.I..ST	x En2β	
DK.P...F.AE..QR.KA..QA...I.EQ..QT..QE.S.N.S.I......K.A.I..AT	m En-1	
DK.P...F.AE..QR.KA..QT.....EQ.AQS..QE.G.N.S.I......K.A.I..AS	zf en	
EK.P...FSAS..QR.KQ..QQSN...EQ..RS..KE.T.S.S.I......K.A.I..AS	TgHbox-en	
EK.P...F.GD..AR.KR..SE.K...EQ..TC..KE...N.S.I......K.A.M..AS	Htr en	
EK.P...F.GD..DR.KT..RES....EK..Q....E.G.N.S.I......K.A.L..ST	ceh-16	
LK.P..SF.VP..KR.SQ..EK....DEL..KK..TE.D.R.S........K.A.T..AS	smox-2	
P..A...F.YE..VS..NK.KTT...SVCE.LN..L..S...T..........T....Q.	S59/NK-1	NK-1
A..A...F.YE..VA..NK.KTT...SVCE.LN..L..S...T..........T....Q.	Am H40	
P..A...F.YE..VA..NK.RAT...SVCE.LN..L..S...T..........T....QH	c Sax1	
M..A...F.YE..VA..NK.KTS...SVVE.LN..IQ.Q.S.T..........T....H.	ceh-1	
...A...F.YE..VT..NK.QST...SVYE.LN..L......T..........T....Q.	EgHbx1	
Q..C..TFSAS..D...RA.ERTQ.PDIYT.E...QRT....ARIQV..S...ARLR.QH	prd	prd
Q..S..TF.AE..EA..RA.SRTQ.PDVYT.E...QTTA...ARIQV..S...ARLR.HS	gsb-p	
Q..S..TFSND.IDA..RI.ARTQ.PDVYT.E...Q.TG...AR.QV..S...ARLR.QL	gsb-d	
LQ.N..SF.QE.IEA.....ERTH.PDVFA.ER..AKID.P.ARIQV..S...A..RR.E	pax[zf-a]	
LQ.N..SF.QE.IEA.....ERTH.PDVFA.ER..AKID.P.ARIQV..S...A..RR.E	AN	
LQ.N..SF.QE.IEA.....ERTH.PDVFA.ER..AKID.P.ARIQV..S...A..RR.E	Pax-6	
Q..S..TF.AE..E...RA.ERTH.PDIYT.E...QRAK...AR.QV..S...AR.R.QA	Pax-3	
Q..S..TF.AE..E....A.ERTH.PDIYT.E...QRTK...ARFQV..S...AR.R.QA	Pax-7	

....|...10....|...20....|...30....|...40....|...50....|...60		
RRRKRTAYTRYQLLELEKEFHFNRYLTRRRRIELAHSLNLTERQVKIWFQNRRMKWKKEN	Consensus	Class (Paralog group)
Q..Y..TF.SF..E....A.SRTH.PDVFT.E...MKIG...ARIQV......A..R.QE	al	prd-like
Q..N..TFNSS..QA..RV.ERTH.PDAFV.E...RRV..S.AR.QV......A.FRRNE	m S8	
Q..N..TFNSS..QA..RV.ERTH.PDAFV.ED..RRV....AR.QV......A.FRRNE	m MHbx/K-2	
Q..N..TFNSS..QA..RV.ERTH.PDAFV.ED..RRV....AR.QV......A.FRRNE	h PHBX1	
Q..I..TF.SL..K...RA.QETH.PDIYT.ED..LRID...AR.QV......A.FR.TE	smox-3	
K..H..IF.Q..ID....A.QDSH.PDIYA.EV..GKTE.Q.DRIQV......A..R.TE	ceh-10	
NEAH.PDVYA.EM..MKTE.P.DRIQV	Ca G10	
...T..NFSGW..E...SA.EASH.PDVFM.EA..MR.D.L.SR.QV......A..R.RE	unc-4	
QK.H..RF.PA..N...RC.SKTH.PDIFM.E.I.MRIG...SR.QV......A....RK	W26	
Q.....FF.QA..DI..QF.QT.M.PDIHH.E...RHIYIP.SRIQV......A.VRRQG	Mix.1	
K..H..IF.DE..EA..NL.QETK.PDVGT.EQ..RKVH.R.EK.EV..K...A..RRQK	m gsc	
K..H..IF.DE..EA..NL.QETK.PDVGT.EQ..RRVH.R.EK.EV..K...A..RRQK	x gsc	
K..H..IF.DE..EA..NL.QETK.PDVGT.EQ..RKVH.R.EK.EV..K...A..RRQK	zf zgsc	
S..E..I..PE..EAM.EV.GV...PDVSM.E...SR.GIN.SKIQV..K...A.LRNLE	EgHbx4	
G..P...F..S.IEI..NV.RV.S.PGIDV.E...SK.A.D.DRIQ.......A.L.RSH	XANF-1	
G..P...F.QN.VEV..NV.RV.C.PGIDI.ED..QK...E.DRIQ.......A.M.RSR	m HES-1	
Q..E..TF..A..DV..AL.GKT..PDIFM.E.V.LKI..P.SR.QV..K...A.CRQQL	otd	
Q..E..TF..S..DV..AL.AKT..PDIFM.E.V.LKI..P.SR.QV..K...A.CRQQQ	m Otx1	
Q..E..TF..A..DV..AL.AKT..PDIFM.E.V.LKI..P.SR.QV..K...A.CRQQQ	m Otx2	
SKKQ.VLFSEE.KEA.RLA.ALDP.PNVGTIEF..NE.G.AT.TITN..H.H..RL.QQV	cut	cut
LKKP.VVLAPEEKEA.KRAYQQKP.PSPKTIED..TQ...KTST.IN..H.Y.SRIRR.L	CDP	
LKKP.VVLAPEEKEA.KRAYQQKP.PSPKTIE...TQ...KTST.IN..H.Y.SRIRR.L	Clox	
TT.V..VLNEK..HT.RTCYAA.PRPDALMKEQ.VEMTG.SP.VIRV....K.C.D..RS	r Isl-1	LIM
TT.V..VLNEK..HT.RTCYNA.PRPDALMKEQ.VEMTG.SP.VIRV....K.C.D..RT	Ot Isl-1	
PK.P..IL.TQ.RRAFKAS.EVSSKPC.KV.ET..AETG.SV.V.QV....Q.A.M..LA	sh lmx-1	
TK.M..SFKHH..RTMKSY.AI.HNPDAKDLKQ.SQKTG.PK.VLQV....A.A..RRMM	ap	
TK.M..SFKHH..RTMKSY.AI.HNPDAKDLKQ..QKTG..K.VLQV....A.A.FRRNL	r LH-2	
..GP..TIKQN..DV.NEM.SNTPKPSKHA.AK..LETG.SM.VIQV......S.ERRLK	mec-3	
..GP..TIKQN..DV.NEM.SNTPKPSKHA.AKK.LETG.SM.VIQV......S.ERRLK	Cvmec-3	
..GP..TIKAK..ET.KNA.AATPKP..HI.EQ..AETG.NM.VIQV......S.ERRMK	lin-11	
..GP..TIKAK..ET.KAA.AATPKP..HI.EQ..QETG.NM.VIQV......S.ERRMK	Xlim1	
NK.P..TISAKS.ET.KQAYQTSSKPA.HV.EQ..SETG.DM.V.QV......A.E.RLK	ceh-14	
.AK..ET.KNAYDNSPKPA.HV.EQ.SSETG.DM.V.QV	Xlim3	
DK.L..TI.PE..EI.YQKYLLDSNP..KMLDHI..EVG.KK.V.QV....T.ARER.GQ	ATBF1 HD3	ZF
NK.L..TILPE..NF.YECYQSESNPS.KMLE.ISKKV..KK.V.QV....S.A.D..SR	zfh-2 HD3	
QK.F..QM.NL..KV.KSC.NDY.TP.MLECEV.GNDIG.PK.V.QV....A.A.E..SK	ATBF1 HD4	
KV.V...INEE.QQQ.KQHYSL.ARPS.DEFRMI.AR.Q.DP.V.QV....N.SRER.MQ	zfh-1	
NK.P..RI.DD..RV.RQY.DI.NSPSEEQIK.M.DKSG.PQKVI.H..R.TLF.ERQR.	ATBF1 HD1	
QK.A..RI.DD..KI.RAH.DI.NSPSEESIM.MSQKA..PMKV..H..R.TLF.ERQR.	zfh-2 HD1	
K.SS..RF.D...RV.QDF.DA.A.PKDDEFEQ.SNL...PT.VIVV....A.Q.AR.NY	ATBF1 HD2	
K.AN..RF.D..IKV.QEF.EN.S.PKDSDLEY.SKL.L.SP.VIVV....A.Q.QR.IY	zfh-2 HD2	
K.KR..TISIAAKDA..RH.GEQNKPSSQEILRM.EE...EKEV.RV..C...QRE.RVK	Bt Pit-1	POU
K.KR..TISIAAKDA..RH.GEQNKPSSQEIMRM.EE...EKEV.RV..C...QRE.RVK	h Pit-1	
K.KR..TISVAAKDA..RH.GEHSKPSSQEIMRM.EE...EKEV.RV..C...QRE.RVK	m Pit-1	
K.KR..TISIAAKDA..RH.GEHSKPSSQEIMRM.EE...EKEV.RV..C...QRE.RVK	r Pit-1	
..K...SIETNVRFA...S.LA.QKP.SEEILLI.EQ.HMEKEVIRV..C...Q.E.RI.	h Oct-2	
..K...SIETNVRFA...S.LA.QKP.SEEILLI.EQ.HMEKEVIRV..C...Q.E.RI.	m Oct-2	
..K...SIETNIRVA...S.LE.QKP.SEEITMI.DQ..MEKEVIRV..C...Q.E.RI.	c Oct-1	
..K...SIETNIRVA...S.LE.QKP.SEEITMI.DQ..MEKEVIRV..C...Q.E.RI.	h Oct-1	
..K...SIETNIRVA...S.LE.QKP.SEEITMI.DQ..MEKEVIRV..C...Q.E.RI.	x Oct-1	
..K...SIETNIRVA...S.ME.QKP.SEDITLI.EQ..MEKEVIRV..C...Q.E.RI.	m Oct-1	
..K...SIETTIRGA...A.LA.QKP.SEEITQ..DR.SMEKEV.RV..C...Q.E.RI.	pdm-1	
..K...SIETTVRTT...A.LM.CKP.SEEISQ.SER..MDKEVIRV..C...Q.E.RI.	pdm-2	

```
....|...10....|...20....|...30....|...40....|...50....|...60
```

RRRKRTAYTRYQLLELEKEFHFNRYLTRRRRIELAHSLNLTERQVKIWFQNRRMKWKKEN	Consensus	Class (Paralog group)
K.KH..SIEDNVRHT..NY.MQCSKPSAQEIAQI.RE..MEKDV.RV..C...Q.G.RQV	x Oct-79	POU
K.KH..SIENNVKCT..NY.MQCSKPSAQEIAQI.RE..MEKDV.RV..C...Q.G.RQV	x Oct-91	
K.KR..NIENIVKGT..SY.MKCPKPGAQEMVQI.KE..MDKDV.RV..C...Q.G.RQG	x Oct-25	
K.K...SIEVSVKGA..QH..KQPKPSAQEITS..D..Q.EKEV.RV..C...Q.E.RMT	Bm POU-M1	
K.K...SIEVSVKGA..QH..KQPKPSAQEITS..D..Q.EKEV.RV..C...Q.E.RMT	Cf1a	
K.K...SIEVSVKGV..TH.LKCPKPAAQEISS..D..Q.EKEV.RV..C...Q.E.RMT	m Brn-4	
K.K...SIEVSVKGV..TH.LKCPKPAAQEISS..D..Q.EKEV.RV..C...Q.E.RMT	r Brn-4	
K.K...SIEVSVKGV..TH.LKCPKPAALEITS..D..Q.EKEV.RV..C...Q.E.RMT	XlPOU2	
K.K...SIEVSVKGA..SH.LKCPKPSAQEITN..D..Q.EKEV.RV..C...Q.E.RMT	m Brn-1	
K.K...SIEVSVKGA..SH.LKCPKPSAQEITS..D..Q.EKEV.RV..C...Q.E.RMT	m Brn-2	
K.K...SIEVGVKGA..SH.LKCPKPSAHEITG..D..Q.EKEV.RV..C...Q.E.RMT	r SCIP	
K.K...SIEVGVKGA..SH.LKCPKPSAHEITG..D..Q.EKEV.RV..C...Q.E.RMT	m SCIP	
K.K...SIEVGVKGA..NH.LKCPKPSAHEITS..D..Q.EKEV.RV..C...Q.E.RMT	XLPOU1	
K.K...SIEVSVKGA..SH.LKCPKPSAPEITS..D..Q.EKEV.RV.CC...Q.E.RMT	XLPOU3	
K.K...SIEVSVKGA..SH.LKCPKPSAQEITS..D..Q.EKEV.RV..C...Q.E.R	h Brn-2	
K.K...SIEVSVKGA..SH.LKCPKPSSQEITN..D..Q.EKEV.RV..C...Q.E.R	h Brn-1	
K.K...SIEVNVKSR..FH.QS.QKPNAQEITQV.MF.Q.EKEV.RV..C...Q.E.RIA	ceh-6	
K.K...SIEANVKSI..SS.MKLSKPSAQDISS..EK.S.EKEV.RV..C...Q.E.RIT	Djpou1	
.K....SIENRVRWS..TM.LKCPKPSLQQITHI.NQ.G.EKDV.RV..C...Q.G.RSS	m Oct-3/4	
K.KM..CFDTVLKGQ..GH.MC.QKPGA.ELT.I.KE.S.EKDV.RV..C...Q.E.SKF	x Oct-60	
KK....SIAAPEKRS..AY.AVQPRPSSEKIAAI.EK.D.KKNV.RV..C.Q.Q.Q.R	r Brn-3	
KK....SIAAPEKRS..AY.AVQPRPSGEKIAAI.EK.D.KKNV.RV..C.Q.Q.Q.RIV	tI-POU	
KK....SIAAPEKR...QF.KQQPRPSGE.IASI.DR.D.KKNV.RV..C.Q.Q.Q.RDF	unc-86	
Q.P...RAKGEA.DV.KRK.EI.PTPSLVE.KKISDLIGMP.KN.R.......A.LR.KQ	PHO2	Fungi & Plants
SKKP.PKFHSEYTPL..LY.R..A.P.YAD.RV..EKTGMLT..ITV....H.RRA.GPL	Sc AαY1	
GK.S.PKFHSEYTPV..LY....A.P.YAD.RI..EKTGMLT..ITV....H.RRA.GPL	Sc AαY4	
YKKP.PKFHSEYTPL..LY....A.P.FAD.RM..EKTGMQT..ITV....H.RRA.GPL	Sc AαY3	
PPS.KQ.FNVHYIPV...Y.EY.A.P.AQD.AL..RKSMMSA..IEV....H.ARAR..G	Cc Aα2-1	
LEV...PFNSEYTPL...Y.EY.A.PSA.D.EW..RKTMMSV..IEV....H.RRAR..G	Cc Aβ2-1	
PLKTGRGHDSEAVRI..QA.KHSPNI.PAEKFR.SEVTG.KPK..T.......NRKG.K.	Um bW1	
PLKTGRGHDSEAVRI..QA.KHSPNI.PAEKFR.SEVTG.KPK..T.......NRKG.K.	Um bW2	
PLKTGRGHDSEAVRI..QA.KHSPNI.PAEKFR.SEVTG.KPK..T.......NRKG.K.	Um bW4	
PLKTGRGHDSEAVQI..EA.KHSPNI.PAEKFR.SEVTG.KPK..T.......NRKG.K.	Um bW3	
LPE.KRRL.TE.VHL...S.ETENK.EPE.KTQ..KK.G.QP...AV......AR..TKQ	Athb-1	
LGE.KKRLNLE.VRA...S.ELGNK.EPE.KMQ..KA.G.QP..IA.......AR..TKQ	Athb-3	
NS..KLRLSKD.SAI..ET.KDHST.NPKQKQA..KQ.G.RA...EV......ART.LKQ	Athb-2	
STARKGHFGPVINQK.HEH.KTQP.PS.SVKES..EE.G..F...NK..ET..HSARVAS	Zmhbx1a	
IKDRKGHFGPVISQK.HEH.KTQP.PS.SLKES..EE.G..FH..NR..E...HFARLAS	Zmhbx1b	
KKP..QMK.PF..ET..RVYAMET.PSEAI.A..SEK.G..D..LQM..CH..L.D.NTS	Lp hbx7	
SPKGKSSISPQARAF..QV.RRKQS.NSKEKE.V.KKCGI.PL..RV..I.K..RS.*	MATa1	

Species nomenclature used for figures

h	human (Homo sapiens)
m	mouse (Mus musculus)
r	rat (Rattus norvegicus)
pig	Sus suum
cat	Felis catus
Ca	Cavia sp. (guinea pig)
Ov	Ovis sp. (sheep)
Bt	Bos taurus (cow)
sh	Syrian hamster
c	chicken (Gallus gallus)
qu	quail (Coturnix coturnix)
x	Xenopus laevis (clawed frog)
Xb	Xenopus borealis
ax	axolotl (Ambystoma mexicanum)
Nv	Notophthalmus viridescens (newt)
zf	zebra fish (Brachydanio rerio)
Ot	Oncorhynchus tschawytscha (Salmon)
Ss	Salmo salar (Atlantic salmon)
Lp	Lampetra planeri (brook lamprey)
Mg	Myxine glutinosa (Atlantic hagfish)
Ci	Ciona intestinalis (ascidia)
Hal	Halocynthia roretzi (ascidia)
Tr	Terebratulina retusa (sea urchin)
Tg	Tripneustes gratilla (sea urchin)
Pa	Parechinus angulosus (sea urchin)
Htr	Helobdella triserialis (leech)
Hr	Helobdella robusta (leech)
Hm	Hirudo medicinalis (leech)
Dt	Dugesia tigrina (flatworm)
Dj	Dugesia japonica (flatworm)
Eg	Echinococcus granulosus (parasitic flatworm)
Cv	Caenorhabditis vulgarensis (roundworm)
ce	Caenorhabditis elegans (roundworm)
Al	Ascaris lumbricoides (parasitic roundworm)
Hv	Hydra vulgaris (hydra) (in conjunction with cnox)
Cv	Chlorohydra viridissima (hydra) (in conjunction with cnox)
Af	Acropora formosa (coral)
Bm	Bombyx mori (silk worm)
Ms	Manduca sexta (moth)
Sa	Schistocerca americana (grasshopper)
Sg	Schistocerca gregaria (grasshopper)
Aa	Aedes aegypti (mosquito)
Am	Apis mellis (honeybee)
d	Drosophila melanogaster
Dv	Drosophila virilis
Dps	Drosophila pseudoobscura
Da	Drosophila ananassae
Df	Drosophila funebris
Md	Musca domestica (house fly)
At	Arabidopsis thaliana (plant)
Lp	Lycopersicon peruvianum (plant)
Um	Ustilago maydis (smut fungus)
Cc	Coprinus cinereus (fungus)
Sc	Schizophyllum commune (bracket fungus)
Sp	Schizosaccharomyces pombe (fission yeast)

Literature Cited

1. McGinnis W, Levine MS, Hafen E, Kuroiwa A, Gehring WJ. 1984. *Nature* 308:428–33
2. McGinnis W, Garber RL, Wirz J, Kuroiwa A, Gehring WJ. 1984. *Cell* 37:403–8
3. Scott MP, Weiner AJ. 1984. *Proc. Natl. Acad. Sci. USA* 81:4115–19
4. Seimiya M, Ishiguro H, Miura K, Watanabe Y, Kurosawa Y. 1993. *Eur. J. Biochem.* In press
5. Carrasco AE, McGinnis W, Gehring WJ, De Robertis EM. 1984. *Cell* 37:409–14
6. Levine M, Rubin GM, Tjian R. 1984. *Cell* 38:667–74
7. Vollbrecht E, Veit B, Sinha N, Hake S. 1991. *Nature* 350:241–43
8. Ruberti I, Sessa G, Lucchetti S, Morelli G. 1991. *EMBO J.* 10:1787–91
9. Shepherd JCW, McGinnis W, Carrasco AE, De Robertis EM, Gehring WJ. 1984. *Nature* 310:70–71
10. Schulz B, Banuett F, Dahl M, Schlesinger R, Schäfer W, et al. 1990. *Cell* 60:295–306
11. Bürglin T. 1994. *The Homeobox Guidebook,* ed D Duboule. New York: Oxford Univ. Press. In press
12. Wüthrich K, Gehring W. 1992. *Transcriptional Regulation,* ed. SL McKnight, KR Yamamoto, pp. 535–77. Cold Spring Harbor, NY: Cold Spring Harbor Lab. Press
13. Scott MP, Tamkun JW, Hartzell GWI. 1989. *Biochim. Biophys. Acta* 989:25–48
14. Lewis EB. 1978. *Nature* 276:565–70
15. Kaufman TC, Seeger MA, Olsen G. 1990. *Adv. Genet.* 27:309–62
16. Gaunt SJ, Sharpe PT, Duboule D. 1988. *Development* 104(Suppl.):169–79
17. Duboule D, Dollé P. 1989. *EMBO J.* 8:1497–505
18. Graham A, Papalopulu N, Krumlauf R. 1989. *Cell* 57:367–78
19. Boncinelli E. 1988. *Hum. Reprod.* 3:880–86
20. McGinnis W, Krumlauf R. 1992. *Cell* 68:283–302
21. Beeman RW, Brown SJ, Stuart JJ, Denell R. 1993. *Evolutionary Conservation of Developmental Mechanisms,* ed. A Spradling, pp. 71–84. New York: Wiley
22. Kappen C, Schughart K, Ruddle FH. 1989. *Proc. Natl. Acad. Sci. USA* 86:5459–63

22a. Kappen C, Schughart K, Ruddle FH. 1993. *Genomics* 18:54–70

23. Boncinelli E, Simeone A, Acampora D, Mavilio F. 1991. *Trends Genet.* 7:329–34
24. Krumlauf R. 1992. *BioEssays* 14:245–52
25. Holland P. 1992. *BioEssays* 14:267–73
26. Schubert FR, Nieselt-Struwe K, Gruss P. 1993. *Proc. Natl. Acad. Sci. USA* 90:143–47
27. Pendelton JW, Nagai BK, Murtha MT, Ruddle FH. 1993. *Proc. Natl. Acad. Sci. USA* 90:6300–4
28. Malicki J, Schughart K, McGinnis W. 1990. *Cell* 63:961–67
29. McGinnis N, Kuziora MA, McGinnis W. 1990. *Cell* 63:969–76
30. Zhao JJ, Lazzarini RA, Pick L. 1993. *Genes Dev.* 7:343–54
31. Bürglin TR, Ruvkun G. 1993. *Curr. Opin. Genet. Dev.* 3:615–20
32. Wang B, Müller-Immerglück M, Austin J, Robinson N, Chisholm A, et al. 1993. *Cell* 74:29–42
33. Peifer M, Karch F, Bender W. 1987. *Genes Dev.* 1:891–98
34. Simon J, Peifer M, Bender W, O'Connor M. 1990. *EMBO J.* 9:3945–56
35. Gyurkovics H, Gausz J, Kummer J, Karch F. 1990. *EMBO J.* 9:2579–85
36. Akam M. 1989. *Cell* 57:347–49
37. Miles A, Miller DJ. 1992. *Proc. R. Soc. London Ser. B* 248:159–61

37a. Miller DJ, Miles A. 1993. *Nature* 365:215–16

38. Joyner AL, Hanks M. 1991. *Semin. Dev. Biol.* 2:435–45
39. Jacob F. 1977. *Science* 196:1161–66
40. Noll M. 1993. *Curr. Opin. Genet. Dev.* 3:595–605
41. Deutsch U, Gruss P. 1991. *Semin. Dev. Biol.* 2:413–24
42. Walther C, Guenet J-L, Simon D, Deutsch U, Jostes B, et al. 1991. *Genomics* 11:424–34
43. Fitzpatrick VD, Percival-Smith A, Ingles CJ, Krause HM. 1992. *Nature* 356:610–12

44. Simeone A, Acampora D, Mallamaci A, Stornaiuolo A, Boncinelli E. 1993. *EMBO J.* 12:2735–47
45. Simeone A, Gulisano M, Acampora D, Stornaiuolo A, Rambaldi M, Boncinelli E. 1992. *EMBO J.* 11:2541–50
46. Simeone A, Acampora D, Gulisano M, Stornaiuolo A, Boncinelli E. 1992. *Nature* 358:687–90
47. Herr W, Sturm RA, Clerc RG, Corcoran LM, Baltimore D, et al. 1988. *Genes Dev.* 2:1513–16
48. Ruvkun G, Finney M. 1991. *Cell* 64:475–78
49. Schöler HR. 1991. *Trends Genet.* 7: 323–29
50. Rosenfeld MG. 1991. *Genes Dev.* 5: 897–907
51. Blochlinger KR, Bodmer R, Jack J, Jan LY, Yan YN. 1988. *Nature* 333: 629–35
52. Freyd G, Kim SK, Horvitz HR. 1990. *Nature* 344:876–82
53. Karlsson O, Thor S, Norberg T, Ohlsson H, Edlund T. 1990. *Nature* 344: 879–82
54. Fortini ME, Lai Z, Rubin GM. 1991. *Mech. Dev.* 34:113–22
55. Morinaga K, Galili N, Nourse J, Saltman D, Cleary ML. 1991. *Mol. Cell. Biol.* 11:6041–49
56. Price M, Lazzaro D, Pohl T, Mattei M-G, Rüther U, et al. 1992. *Neuron* 8:241–55
57. Hall MN, Johnson AD. 1987. *Science* 237:1007–12
58. Nourse J, Mellentin JD, Galili N, Wilkinson J, Stanbridge E, et al. 1990. *Cell* 60:535–45
59. Camps MP, Murre C, Sun X-h, Baltimore D. 1990. *Cell* 60:547–55
60. Rauskolb C, Peifer M, Wieschaus E. 1993. *Cell* 74:1101–12
61. Bürglin TR, Ruvkun G. 1992. *Nature Genet.* 1:319–20
62. Maulbecker CC, Gruss P. 1993. *Cell Growth Differ.* 4:431–41
63. De Simeone V, Cortese R. 1991. *Curr. Opin. Cell Biol.* 3:950–65
64. Mendel DB, Crabtree GR. 1991. *J. Biol. Chem.* 266:677–80
65. Leiting B, De Francesco R, Tomei L, Cortese R, Otting G, Wüthrich K. 1993. *EMBO J.* 12:1797–803
66. Gillissen B, Bergemann J, Sandmann C, Schroeer B, Bölker M, Kahmann R. 1992. *Cell* 68:647–57
67. Kües U, Richardson WVJ, Tymon AM, Mutasa ES, Göttgens B, et al. 1992. *Genes Dev.* 6:568–77
68. Tymon AM, Kües U, Richardson WVJ, Casselton LA. 1992. *EMBO J.* 11: 1805–13
69. Bellmann R, Werr W. 1992. *EMBO J.* 11:3367–74
70. Sessa G, Morelli G, Ruberti I. 1993. *EMBO J.* 12:3507–17
71. Otting G, Qian YQ, Müller M, Affolter M, Gehring W, Wüthrich K. 1988. *EMBO J.* 7:4305–9
72. Qian YQ, Billeter M, Otting G, Müller M, Gehring WJ, Wüthrich K. 1989. *Cell* 59:573–80
73. Billeter M, Qian YQ, Otting G, Müller M, Gehring WJ, Wüthrich K. 1990. *J. Mol. Biol.* 214:183–97
74. Güntert P, Qian YQ, Otting G, Müller M, Gehring W, Wüthrich K. 1991. *J. Mol. Biol.* 217:531–40
75. Pabo CO, Sauer RT. 1992. *Annu. Rev. Biochem.* 61:1053–95
76. Gehring WJ, Müller M, Affolter M, Percival-Smith A, Billeter M, et al. 1990. *Trends Genet.* 6:323–29
77. Kissinger CR, Liu B, Martin-Blanco E, Kornberg TB, Pabo CO. 1990. *Cell* 63:579–90
78. Wolberger C, Vershon AK, Liu B, Johnson AD, Pabo CO. 1991. *Cell* 67:517–28
79. Qian YQ, Furukubo-Tokunaga K, Müller M, Resendez-Perez D, Gehring WJ, Wüthrich K. 1994. *J. Mol. Biol.* In press
80. Qian YQ, Resendez-Perez D, Gehring WJ, Wüthrich K. 1994. *Proc. Natl. Acad. Sci. USA.* In press
81. Qian YQ, Otting G, Furukubo-Tokunaga K, Affolter M, Gehring WJ, Wüthrich K. 1992. *Proc. Natl. Acad. Sci. USA* 89:10738–42
82. Otting G, Qian YQ, Billeter M, Müller M, Affolter M, Gehring WJ, Wüthrich K. 1990. *EMBO J.* 9:3085–92
83. Wüthrich K, Otting G, Qian YQ, Billeter M, Gehring W. 1992. *Molecular Structure of Life,* ed. Y Kyogoku, Y Nishimura, pp. 115–27. Tokyo: Jpn. Soc. Press
84. Sluka JP, Horvath SJ, Bruist MF, Simon MI, Dervan PB. 1987. *Science* 238:1129–32
85. Affolter M, Percival-Smith A, Müller M, Billeter M, Qian YQ, et al. 1991. *Cell* 64:879–80
86. Qian YQ, Otting G, Billeter M, Müller M, Gehring W, Wüthrich K. 1993. *J. Mol. Biol.* 234:1070–83
87. Billeter M, Qian YQ, Otting G, Müller M, Gehring W, Wüthrich K. 1993. *J. Mol. Biol.* 234:1084–97
88. Qian YQ, Otting G, Wüthrich K. 1993. *J. Am. Chem. Soc.* 115:1189–90
89. Otting G, Liepinsh E, Wüthrich K. 1991. *J. Am. Chem. Soc.* 113:4363–64
90. Suzuki M. 1993. *EMBO J.* 12:3221–26

91. Frampton J, Leutz A, Gibson TJ, Graf T. 1989. *Nature* 342:134
92. Ogata K, Hojo H, Aimoto S, Nakai T, Nakamura H, et al. 1992. *Proc. Natl. Acad. Sci. USA* 89:6428–32
93. Laughon A. 1991. *Biochemistry* 30: 11357–67
94. Hanes SD, Brent R. 1991. *Science* 251:426–30
95. Hanes SD, Brent R. 1989. *Cell* 57: 1275–83
96. Treisman J, Gönczy P, Vashishtha M, Harris E, Desplan C. 1989. *Cell* 59: 553–62
97. Percival-Smith A, Müller M, Affolter M, Gehring WJ. 1990. *EMBO J.* 9:3967–74
98. Schier AF, Gehring WJ. 1992. *Nature* 356:804–7
99. Furukubo-Tokunaga K, Müller M, Affolter M, Pick L, Kloter U, Gehring WJ. 1992. *Genes Dev.* 6:1082–96
100. Damante G, Di Lauro R. 1991. *Proc. Natl. Acad. Sci. USA* 88:5388–92
101. Ingraham HA, Flynn SE, Voss JW, Albert VR, Kapiloff MS, et al. 1990. *Cell* 61:1021–33
102. Aurora R, Herr W. 1992. *Mol. Cell. Biol.* 12:455–67
103. Verrijzer CP, Alkema MJ, van Weperen WW, Van Leeuwen HC, Strating MJJ, van der Vliet PC. 1992. *EMBO J.* 11:4993–5003
104. Treisman J, Harris E, Desplan C. 1991. *Genes Dev.* 5:594–604
105. Brugnera E, Xu L, Schaffner W, Arnosti DN. 1992. *FEBS Lett.* 314: 361–65
106. Pick L, Schier A, Affolter M, Schmidt-Glenewinkel T, Gehring WJ. 1990. *Genes Dev.* 4:1224–39
107. Schier AF, Gehring WJ. 1993. *EMBO J.* 12:1111–19
108. Florence B, Hondron R, Laughon A. 1991. *Mol. Cell. Biol.* 11:3613–23
109. Schier A, Gehring WJ. 1993. *Proc. Natl. Acad. Sci. USA* 90:1450–54
110. Driever W. 1992. *Transcriptional Regulation,* ed. SL McKnight, KR Yamamoto, pp. 1221–50. Cold Spring Harbor, NY: Cold Spring Harbor Lab.
111. Maier D, Preiss A, Powell JR. 1990. *EMBO J.* 9:3957–66
112. Levine M, Hoey T. 1988. *Cell* 55:537–40
113. Affolter M, Schier A, Gehring WJ. 1990. *Curr. Opin. Cell Biol.* 2:485–95
114. Hayashi S, Scott MP. 1990. *Cell* 63: 883–94
115. Treisman J, Harris E, Wilson D, Desplan C. 1992. *BioEssays* 14:145–50
116. Herskowitz I. 1989. *Nature* 342:749–57
117. Mak A, Johnson AD. 1993. *Genes Dev.* 7:1862–70
118. Johnson AD, Herskowitz I. 1985. *Cell* 42:237–47
119. Vershon AK, Johnson AD. 1993. *Cell* 72:1–20
120. Smith DL, Johnson AD. 1992. *Cell* 68:133–42
121. Gerster T, Roeder RG. 1988. *Proc. Natl. Acad. Sci. USA* 85:6347–51
122. Stern S, Herr W. 1991. *Genes Dev.* 5:2555–66
123. Lai J-S, Cleary MA, Herr W. 1992. *Genes Dev.* 6:2058–65
124. Pomerantz JL, Kristie TM, Sharp PA. 1992. *Genes Dev.* 6:2047–57
125. Treacy MN, Neilson LI, Turner EE, He X, Rosenfeld MG. 1992. *Cell* 68:491–505
126. Kuziora MA, McGinnis W. 1989. *Cell* 59:563–71
127. Mann RS, Hogness DS. 1990. *Cell* 60:597–610
128. Gibson G, Schier A, LeMotte P, Gehring WJ. 1990. *Cell* 62:1087–103
129. Kuziora MA, McGinnis W. 1991. *Mech. Dev.* 33:83–94
130. Lin L, McGinnis W. 1992. *Genes Dev.* 6:1071–81
131. Zeng W, Andrew DJ, Mathies LD, Horner MA, Scott MP. 1993. *Development* 118:339–52
132. Furukubo-Tokunaga K, Flister S, Gehring WJ. 1993. *Proc. Natl. Acad. Sci. USA* 90:6360–64
133. Ekker SC, von Kessler DP, Beachy PA. 1992. *EMBO J.* 11:4059–72
134. Andrew DJ, Scott MP. 1992. *New Biol.* 4:5–15
135. Capovilla M, Brandt M, Botas J. 1994. *Cell.* In press
136. Kessel M, Gruss P. 1991. *Cell* 67:89–104
137. Condie BG, Capecchi MR. 1993. *Development.* 119:579–95
138. Cho KWY, Blumberg B, Steinbeisser H, De Robertis EM. 1991. *Cell* 67: 1111–20
139. Röder L, Vola C, Kerridge S. 1992. *Development* 115:1017–33
140. Wagner-Bernholz JT, Wilson C, Gibson G, Schuh R, Gehring WJ. 1991. *Genes Dev.* 5:2467–80
141. Gould AP, White RAH. 1992. *Development* 116:1163–74
142. Jones FS, Prediger EA, Bittner DA, De Robertis EM, Edelman GM. 1992. *Proc. Natl. Acad. Sci. USA* 89:2086–90
142a. Tomotsune D, Shoji H, Wakamatsu Y, Kondoh H, Takahashi N. 1993. *Nature* 365:69–72
143. Scott MP. 1992. *Cell* 71:551–53

144. Desplan C, Theis J, O'Farrell PH. 1988. *Cell* 54:1081–90
145. Hoey T, Levine M. 1988. *Nature* 332:858–61
146. Affolter M, Percival-Smith A, Müller M, Leupin W, Gehring WJ. 1990. *Proc. Natl. Acad. Sci. USA* 87:4093–97
147. Müller M, Affolter M, Leupin W, Otting G, Wüthrich K, Gehring WJ. 1988. *EMBO J.* 7:4299–304
148. Odenwald WF, Garbern J, Arnheiter H, Tournier-Lasserve E, Lazzarini RA. 1989. *Genes Dev.* 3:158–72
149. Regulski M, Dessain S, McGinnis N, McGinnis W. 1991. *Genes Dev.* 5:278–86
150. Sasaki H, Yokoyama E, Kuroiwa A. 1990. *Nucleic Acids Res.* 18:1739–47
151. Dearolf CR, Topol J, Parker CS. 1989. *Nature* 341:340–43

Annu. Rev. Biochem. 1994. 63:527–70

ESCHERICHIA COLI SINGLE-STRANDED DNA-BINDING PROTEIN: Multiple DNA-Binding Modes and Cooperativities

Timothy M. Lohman and Marilyn E. Ferrari
Department of Biochemistry and Molecular Biophysics, Washington University School of Medicine, Box 8231, 660 S. Euclid Avenue, St. Louis, Missouri 63110

KEY WORDS: single-stranded DNA-binding proteins, SSB proteins, cooperativity, DNA replication, DNA recombination and repair

---

CONTENTS

0066-4154/94/0701-0527$05.00

INTRODUCTION

Single-stranded DNA-binding proteins (SSB proteins, or SSBs) bind with high affinity to single-stranded (ss) DNA and play essential roles as accessory proteins in DNA replication, recombination, and repair. SSBs have been referred to by a number of different names since their discovery in 1970 (1), including DNA-unwinding proteins (2, 3), DNA-melting proteins, and helix-destabilizing proteins (4, 5); however, the terms single-stranded DNA-binding protein, SSB protein (or SSB), or helix-destabilizing protein are used most commonly. SSBs are essential for DNA replication in phage T4 (6), phage T7 (6a), *Escherichia coli* (7–10), and yeast (11–13), and it is assumed that such proteins are essential in all organisms. As discussed by Chase & Williams (14), both biochemical and genetic criteria should be used to identify true SSB proteins involved in DNA replication. In fact, without genetic information, it is difficult to classify such proteins unambiguously, since they do not possess enzymatic activity and too few SSB proteins have been characterized in detail to establish a single set of biochemical criteria. The proteins discussed in this review bind preferentially and with high affinity to ssDNA and also satisfy one or more of the following criteria: (*a*) a genetic requirement for DNA replication, (*b*) specific stimulation of its cognate DNA polymerase, (*c*) interaction with or stimulation of other replication enzymes (e.g. helicases), and (*d*) a requirement for stoichiometric amounts of protein with respect to the ssDNA template (14). On the other hand, it is not clear that SSBs that function in recombination and repair should also be expected to satisfy these criteria.

The first example of this class of proteins, the bacteriophage T4 gene 32 protein, was discovered and characterized by Alberts & Frey (1); soon after, the *E. coli* SSB protein (*Eco* SSB) was identified (2, 15). Since then, other prokaryotic SSBs that are known to function in DNA metabolism have been identified in bacteria [*Serratia marcescens* SSB (16)], conjugative bacterial plasmids [F factor (17–19), CoIIb-P9 (20), IP231a (17, 21), IP71a (17, 21), R64 (17, 21)], and bacteriophages [ϕ29 gene 5 (22, 23), N4 SSB (24), T7 gene 2.5 (25–28)]. Functionally analogous SSBs have also recently been identified in eukaryotes. Heterotrimeric nuclear SSBs (RP-A or RF-A) have been found in human (29–31), yeast (*Saccharomyces cerevisiae*) (32), *Drosophila* (33), and calf thymus (34). These proteins are essential for replication (11, 29–31, 35–37), are involved in recombination (12, 38, 39) and repair (40, 41), and also interact specifically with other proteins involved in DNA metabolism (33, 34, 42–48). Mitochondrial SSBs that share extensive similarity with *Eco* SSB have been found in *S. cerevisiae* (RIM1) (13), rat (49–51), human (51, 52), and *Xenopus laevis* (53, 54). SSBs encoded by animal viruses [Herpes ICP8 (55–59) and Adenovirus SSB (59–64)] have also been characterized. However, with only a few exceptions, these proteins show little sequence similarity and display very different ssDNA-binding properties. It is even unclear whether highly cooperative binding to ssDNA is functionally important for all SSBs, even though this has been generally thought to be a hallmark of SSB proteins.

Only two SSBs that stimulate DNA replication,[1] the T4 gene 32 protein and *Eco* SSB, have undergone extensive biochemical and biophysical characterization. Even though these two proteins show some similarities in their ssDNA-binding properties, major differences are also apparent (65), including the fact that the T4 gene 32 protein is monomeric, whereas the *Eco* SSB is a homotetramer. Most general perceptions of how SSBs bind to ssDNA are based on the properties of the phage T4 gene 32 protein, since this was the first SSB to undergo extensive biochemical and biophysical characterization and still serves as a paradigm for studies of SSBs [for reviews see Kowalczykowski et al (66) and Karpel (67)]. The T4 gene 32 protein binds to ssDNA with high cooperativity of the "unlimited" type that allows the formation of continuous protein clusters that can readily saturate the ssDNA. As a result, this property has generally been assumed to be important for the function of SSBs in general. In fact, this view was supported by early studies that showed that the *Eco* SSB (2, 15, 68) as well as the phage f1 gene V protein (69, 70) also bind cooperatively to

[1]Studies of the f1 gene V protein have contributed substantially to our understanding of protein interactions with ssDNA. Even though this protein binds cooperatively to ssDNA and is often grouped with the T4 gene 32 protein and the *E. coli* SSB, as suggested by Chase & Williams (14), we do not consider the gene V protein as a member of the class of SSBs involved in DNA metabolism since it does not function at the replication fork to facilitate replication, but rather binds to the replicated ssDNA product in order to inhibit its further replication (217–219). For this reason, it may be misleading to consider the general properties of the phage f1 gene V protein in attempts to form a general view of how SSBs function in DNA replication, recombination, and repair.

ssDNA to form long protein clusters. However, recent studies indicate that the DNA-binding properties of *Eco* SSB differ significantly from those of the T4 gene 32 protein under many conditions in vitro, hence the "consensus" view of SSBs based on the T4 gene 32 protein is not generally applicable to the *Eco* SSB, at least in vitro (65).

There has been much confusion regarding the ssDNA-binding properties of the *Eco* SSB protein. Therefore, rather than present a general discussion of all SSBs, this review focuses on *Eco* SSB, emphasizing recent biochemical and biophysical studies that have led to a clearer picture of its interactions with ssDNA and how its properties differ from those of the T4 gene 32 protein. We also review other SSBs from both prokaryotes and eukaryotes that share extensive sequence similarities with *Eco* SSB and appear to bind similarly to ssDNA. These include SSBs from conjugative plasmids (17, 19–21, 71) and from *S. marcescens* (16), as well as several eukaryotic mitochondrial SSBs (13, 49–54, 72, 73). By focusing on *Eco* SSB and how it differs from T4 gene 32 protein, we hope to facilitate future comparisons with other SSBs as these become better characterized in order to develop a better understanding of this class of DNA-binding proteins.

Excellent reviews have discussed the many biological roles of *Eco* SSB in DNA replication, recombination, and repair (14, 74), hence these aspects are not covered in detail here. However, we note that *Eco* SSB is also required for late transcription in phage N4 and thus also seems to serve as a transcription factor in that system (75). The subject of single-stranded DNA-binding proteins was last reviewed in *Annual Review of Biochemistry* in 1986 by Chase & Williams (14); however, the reader is also referred to previous general reviews on prokaryotic SSBs (5, 66, 76) and eukaryotic SSBs (59), as well as reviews specifically on the T4 gene 32 protein (67) and *Eco* SSB (74, 77, 78). The kinetics of *Eco* SSB–ssDNA interactions have also been reviewed recently (77), and thus are not discussed here.

E. COLI SSB PROTEIN STRUCTURE

Structure and Physical Properties

Details of the structure and physical properties of *Eco* SSB have been reviewed in detail previously (14, 77, 78). The *Eco ssb* gene has been cloned (79) and sequenced (80), and the protein has been overexpressed (81–84). The monomer contains 177 amino acids (M_r=18,843) (80, 82) with an isoelectric point of 6.0 (15, 80). In the absence of DNA, the protein assembles to a homotetramer (3, 15), which is stable at concentrations as low as 30 nM (tetramer) at 25°C (85) over a wide range of solution conditions (82, 85–88). The sedimentation coefficient of the tetramer is $s_{20,w}$ =

4.4–4.9S (15, 87, 88) and its diffusion coefficient has been estimated as $D_{20,w}$ = 5.6 × 10^{-7} cm^2sec^{-1}, corresponding to a Stokes radius of 38Å (15). Single crystals of tetramers have been obtained in which ~40–60 amino acids had been removed from the C terminus of some subunits; preliminary X-ray analysis of these crystals indicates a tetramer with D_2 symmetry (89–92). However, no high-resolution structural information is yet available.

N-Terminal Domain and Amino Acids Implicated in DNA Binding

Proteolysis studies (93) indicate that the ssDNA-binding site of *Eco* SSB is contained within the first 115 N-terminal amino acids. The N-terminal "core" polypeptides, SSB^*_T and SSB^*_C, formed by trypsin cleavage after Arg115 or chymotrypsin cleavage after Trp135, respectively, also form tetramers; however, these are not as stable as wild-type tetramers, suggesting that the C-terminal region facilitates tetramerization (93). Both SSB^*_T and SSB^*_C lower the melting temperature of poly[d(A-T)] more than does intact SSB, suggesting that the C terminus modulates the ability of *Eco* SSB to melt duplex DNA (93). The sensitivity of the C-terminal domain to proteolysis increases upon binding poly(dT), although not $(dT)_{16}$, hence it was suggested that cooperative interactions between tetramers on poly(dT) facilitate this conformational change (93). However, since $(dT)_{16}$ is now known to interact with only a single subunit (protomer) of the SSB tetramer (94–96), the poly(dT)-induced conformational change may occur with any nucleic acid that is long enough to span two or more SSB subunits and may not be associated with cooperative interactions between tetramers.

Chemical modification studies suggest that lysine and tryptophan residues are important for ssDNA binding, whereas surface arginine and tyrosine residues do not seem to play a major role (97, 98). There are four tyrosine and four tryptophan residues per SSB monomer (80); however, the fluorescence emission spectrum of the protein shows contributions only from tryptophan (52, 98). The tryptophan fluorescence is partially quenched upon binding ss nucleic acids (88, 98–100), with the majority of the effect resulting from nearly complete quenching of Trp54 and Trp88 fluorescence (52). Analysis of a series of site-directed mutants indicates that Trp40 and Trp54 are required for high-affinity binding to ssDNA, whereas Trp135 is not (101, 102). Further spectroscopic evidence suggests that both Trp40 and Trp54, although not Trp88, are involved in stacking interactions with the nucleic acid bases (103); however, other mutagenesis studies indicate that Trp88 does influence SSB binding in vitro, at least in mutants lacking Trp54 (52). The importance of Trp54 and Trp88 for SSB function in vivo is clearly evident, since a Δ*ssb* strain containing either *ssb*W54S or *ssb*W88T shows increased sensitivity to UV; however, no increased UV sensitivity is observed with *ssb*W40T (104). As discussed below, Trp54 and Trp88 also influence the transitions among the SSB-polynucleotide-binding modes (52).

Phe60 has been implicated in DNA binding since it is crosslinked to $d(pT)_8$ upon UV treatment (105). Site-directed mutagenesis of Phe60 to Ala also results in a decrease in affinity of SSB for ssDNA (106). Phe60 has also been replaced by Tyr, His, Val, Leu, Trp, and Ser (84). Although substitution of Tyr, His, Val, and Leu did not appear to affect the protein structure as detected by NMR, the Trp and Ser substitutions showed considerable changes in protein conformation. The binding of the mutant proteins to poly(dT) were examined by fluorescence stopped-flow kinetics (0.3 M NaCl, 0.5 M $MgCl_2$, pH 7.8, unspecified temperature), and the following relative affinities were inferred: Trp > Phe (wild-type) > Tyr > Leu > His > Val > Ser. Based on the observation that an aromatic residue at position 60 seems to increase SSB affinity for ssDNA, Phe60 may be involved in a hydrophobic type of interaction with poly(dT) (84). However, based on their proximity to His55, which is important for tetramerization (82), mutations at residues 54 or 60 may affect the stability of the tetramer in addition to ssDNA binding. These data also suggest that ssDNA binding may occur near the interface between SSB subunits, although since isolated SSB-1 monomers can bind with reasonable affinity to the oligonucleotide, $dT(pT)_{15}$, tetramer formation is clearly not required for ssDNA binding even though it does enhance affinity (96).

C-Terminal Acidic Tail

Five of the last 12 amino acids of *Eco* SSB (residues 166–177) are negatively charged, making the C terminus highly acidic; this region also contains the site of the *ssb-113* mutation (Pro176 to Ser) (107). This region has been implicated in interactions with other proteins involved in DNA metabolism (93, 107–109), as first suggested by the observations that the SSB-113 protein shows no defects in ssDNA binding (107) and that the *ssb-113* mutation inhibits binding of a monoclonal antibody that recognizes the C terminus (108). Although direct interactions between the C terminus and other proteins have yet to be demonstrated, this possibility is supported by other lines of evidence. For instance, Williams et al (93) have noted that the C termini of *Eco* SSB, T4 gene 32 protein (110), and T7 gene *2.5* protein (111), although not the fd gene V protein, are each highly acidic and become more exposed to proteolysis upon binding ssDNA, with the resulting N-terminal polypeptides possessing increased helix-destabilizing activity (76, 112–116). Furthermore, there is good evidence that the acidic C terminus of T4 gene 32 protein interacts with other phage T4 proteins involved in DNA metabolism (93, 107, 117, 118). The functional importance of the C-terminal acidic region of *Eco* SSB is also suggested by the fact that nearly identical acidic tails are found in the SSBs encoded by three conjugative plasmids (20, 21), by the *E. coli* F episome (encoding F-SSB) (18, 19), and the *S. marcescens* chromosome (16). In fact, even though most of the C-terminal domains of the plasmid SSBs show little similarity with *Eco* SSB, six of the last seven C-terminal amino acids are identical (19, 21).

STRUCTURAL ASPECTS OF *Eco* SSB PROTEIN–SINGLE-STRANDED DNA COMPLEXES

SSB Protein Forms Multiple Binding Modes on ss-Polynucleotides

A characteristic of any protein–nucleic acid complex is its "site size," i.e. the average number of nucleotides occluded by the bound protein. In addition to providing gross structural information, knowledge of the site size is required for quantitative analyses of the nonspecific binding of a protein to linear DNA (119–121). Early estimates of the site size of *Eco* SSB on ssDNA ranged from 30 to 73 nucleotides per tetramer (2, 15, 93, 97, 122, 123), which caused some confusion until it was recognized that the SSB tetramer can bind to ss-polynucleotides in several distinct modes that differ in site size (100, 124–126). However, some of this site size variation was also due to the use of SSB extinction coefficients that differed from the value of $\epsilon_{280} = 1.13 \times 10^5\ M^{-1}$ (tetramer) cm^{-1}, which has now been determined by several groups (68, 82, 93, 100, 127, 128), as well as the use of natural ssDNA (rather than ss-homopolynucleotides), which can result in overestimates of the site size (77, 100).

Evidence that *Eco* SSB can form different complexes or binding modes on ssDNA was obtained independently by Griffith et al (126), who observed a protein concentration–dependent transition between two morphologies of SSB complexes on ss-fd DNA by electron microscopy (EM), and Lohman & Overman (100), who observed a [NaCl]-dependent transition between two SSB-poly(dT) complexes possessing site sizes of 35 ± 3 and 65 ± 5 nucleotides per tetramer using titration methods that monitored quenching of intrinsic SSB fluorescence. The different binding modes observed with ss-polynucleotides have been designated $(SSB)_n$, where n is the average site size per tetramer in that mode (see Figure 1). The different site sizes in the $(SSB)_{35}$- and $(SSB)_{65}$-binding modes reflect at least two effects. Firstly, ssDNA interacts with only two SSB subunits in the $(SSB)_{35}$ complex, but with all four SSB subunits in the $(SSB)_{65}$ complex (86, 94, 95, 100, 124, 129), resulting in different extents of ssDNA compaction in each complex

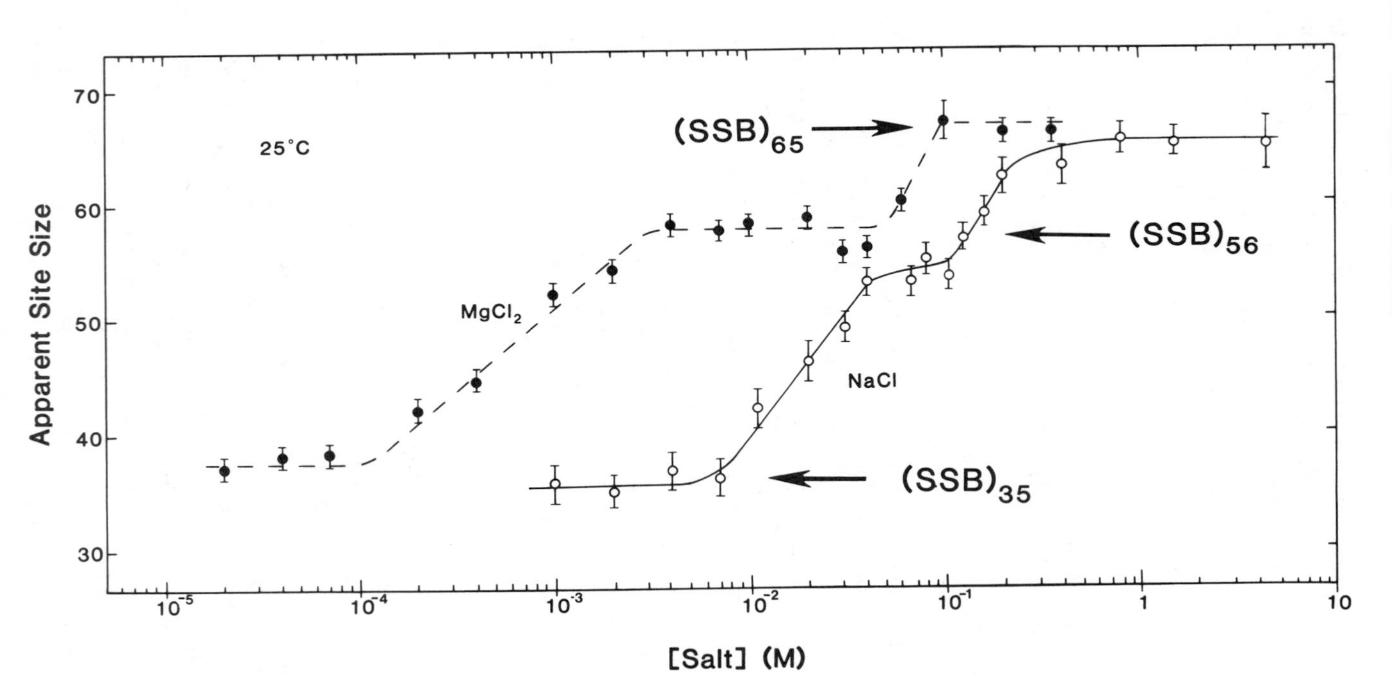

Figure 1 The apparent site size of the SSB tetramer-poly(dT) complex (nucleotides per SSB tetramer) as a function of NaCl and $MgCl_2$ concentrations (pH 8.1, 25°C) as measured by monitoring the quenching of the SSB tryptophan fluorescence upon addition of poly(dT) (124). Reprinted with permission from Ref. 124. Copyright 1986 American Chemical Society.

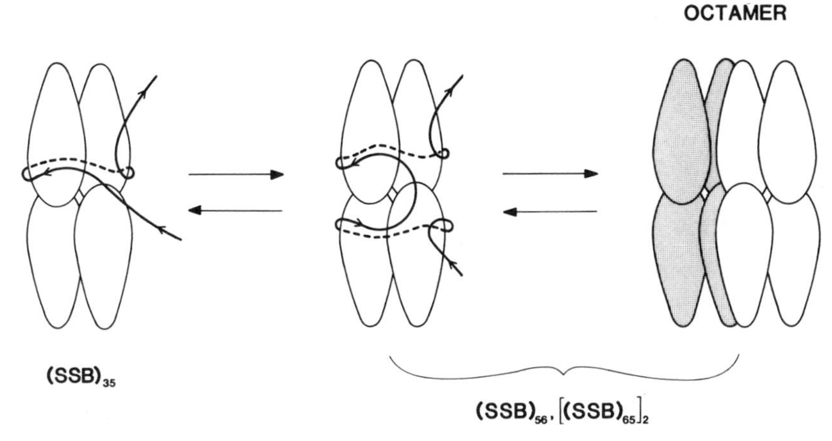

Figure 2 A cartoon depicting that in the $(SSB)_{35}$-binding mode, ssDNA interacts with only two subunits of the SSB tetramer, whereas in both the $(SSB)_{56}$- and $(SSB)_{65}$-binding modes, ssDNA interacts with all four SSB subunits (65, 86). In the $(SSB)_{65}$-binding mode (and possibly the $(SSB)_{56}$), bound tetramers are in equilibrium with octamers. The precise topology by which the ssDNA is wrapped around the SSB in these complexes is unknown. Although not indicated in the cartoon, the ssDNA is believed to interact with all subunits of the octamer. Reprinted with permission from *The Biology of Nonspecific DNA-Protein Interactions,* ed. A. Revzin, 1990, CRC Press. Copyright CRC Press, Boca Raton, FL.

(see Figure 2) (122, 125, 126). Secondly, in at least two of these binding modes, SSB binds to ssDNA with two different types of intertetramer cooperativity (86, 130) (see Figure 3). The different SSB-binding modes have also been observed by others, using a variety of techniques, including circular dichroism (131), electron spin resonance (132), and SSB fluorescence quenching (52).

The stability of these modes is influenced by the concentration of monovalent salt (both anion and cation effects), divalent cations, protein-binding density, pH, temperature, and multivalent cations such as polyamines (124–126, 133). Figure 1 shows the salt-dependent transitions among the $(SSB)_{35}$-, $(SSB)_{56}$-, and $(SSB)_{65}$-binding modes formed with poly(dT) at 25°C (pH 8.1) in either NaCl or $MgCl_2$ (124). The $(SSB)_{35}$ mode is stable at [NaCl] $\leq$ 10 mM, whereas increasing the [NaCl] induces a transition to the $(SSB)_{56}$ mode, followed by a second transition to the $(SSB)_{65}$ mode, which is stable at [NaCl] $\geq$ 0.2 M (100, 124). The same binding modes are observed in the presence of $MgCl_2$; however, the transitions occur at much lower [$MgCl_2$] (see Figure 1) (124, 125). A progressive compaction of the SSB-poly(dT) complex occurs throughout the transition from the $(SSB)_{35}$- to the $(SSB)_{65}$-binding mode, as seen by sedimentation analysis

A

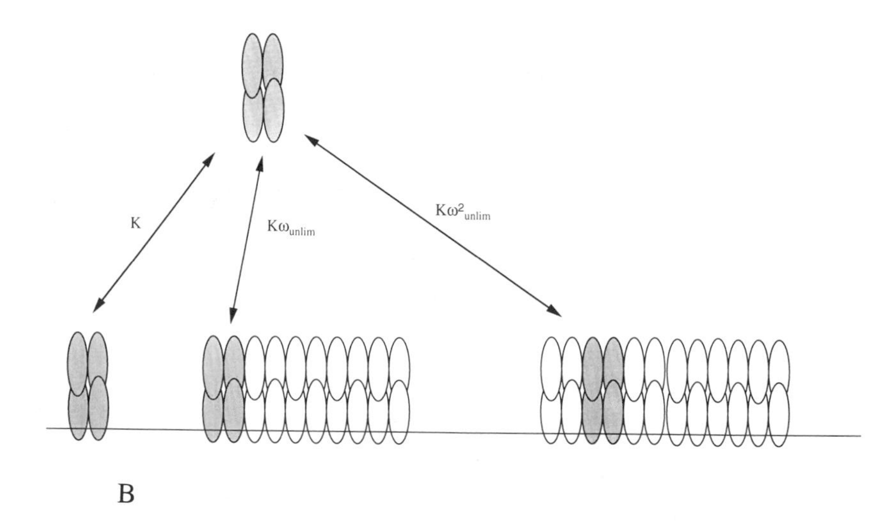

B

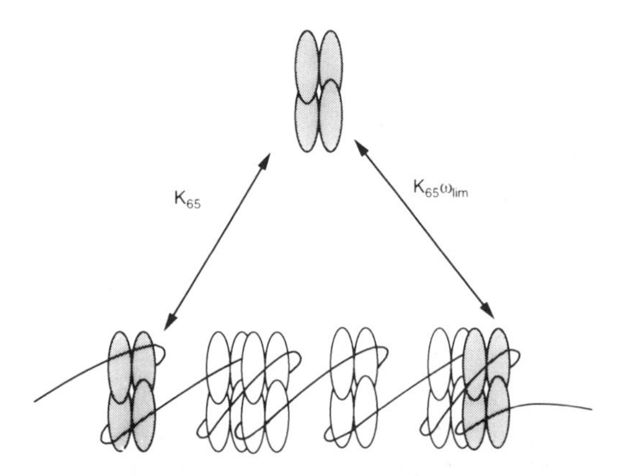

Figure 3 Schematic view of the two types of nearest-neighbor cooperative binding models used to describe nonspecific binding of *Eco* SSB tetramers to linear DNA. *A*. "Unlimited" cooperativity, in which long protein clusters can form due to the nearest-neighbor interactions occurring on both sides of a bound protein. This type of cooperativity seems to be associated with the $(SSB)_{35}$-binding mode. The three shaded SSB tetramers represent binding in an isolated manner (with equilibrium constant K), a singly contiguous manner (with equilibrium constant $K\omega$), and a doubly contiguous manner (with equilibrium constant $K\omega^2$). In the limit of very large ω ($\rightarrow \infty$), a single protein cluster saturates the ssDNA. "Unlimited" cooperativity also describes the binding of the T4 gene 32 protein to ssDNA, although the binding unit is a gene 32 protein monomer, rather than an SSB tetramer. *B* "Limited" cooperativity, typified by the *Eco* SSB tetramer binding to ssDNA in its $(SSB)_{65}$-binding mode. In this case, all four SSB subunits interact with ssDNA and clustering is limited to the formation of dimers of tetramers (octamers), such that there are no doubly contiguously bound tetramers. With this type of cooperativity, even in the limit of very large ω_{lim}, only octamers form along the ssDNA, hence complete saturation of the ssDNA is difficult.

(125). At 37°C, a fourth binding mode with a site size of 42±2 nucleotides per tetramer is also observed for the SSB-poly(dT) interaction(124).

In complexes with poly(dT), the SSB tryptophan fluorescence is quenched by 89±2% in both the $(SSB)_{56}$- and $(SSB)_{65}$-binding modes, whereas in the $(SSB)_{35}$-binding mode only 53±3% quenching is observed (100, 124, 129). Comparisons of SSB binding to poly(dT) vs the oligodeoxynucleotides, $dT(pT)_{15}$, $dT(pT)_{34}$, and $dT(pT)_{69}$, indicate that only two subunits of the tetramer interact with DNA in the $(SSB)_{35}$-binding mode, whereas all four subunits interact with DNA in both the $(SSB)_{56}$- and the $(SSB)_{65}$-binding modes (94, 129). The cartoon in Figure 2 shows the differences in the number of SSB subunits that interact with ssDNA in the $(SSB)_{35}$ vs the $(SSB)_{56}/(SSB)_{65}$ modes, although the details by which ssDNA wraps around the tetramer in each mode are not known. In Figure 2 the $(SSB)_{56}$ and $(SSB)_{65}$ modes are grouped together since the molecular differences between them are unknown. However, one possibility is that the $(SSB)_{56}$ and $(SSB)_{65}$ modes differ in their higher-order structure on the DNA (e.g. tetramer vs octamer binding as discussed below). Although the ssDNA undergoes considerable compaction due to its wrapping around the SSB tetramer, CD experiments with polyA indicate that the local base-base separation is increased and the bases are slightly tilted in both the $(SSB)_{35}$ and $(SSB)_{65}$ binding modes (131).

The effects of salt concentration on the binding mode transitions are complex; however, there is a general requirement for the uptake of cations in progressing from the lower to the higher site-size modes, although anion-binding effects are also indicated (100, 125). Multivalent cations, such as the polyamines, spermidine^{3+}, and spermine^{4+}, can induce the transition from the $(SSB)_{35}$- to the $(SSB)_{56}$-binding mode at micromolar concentrations (pH 8.1, 25°C) (133), which is significantly lower than the 10 mM and 100 mM concentrations required to induce the same transition using divalent and monovalent cations, respectively. This reflects the fact that cation binding induces the transitions and that the more highly charged cations have increased affinity for the SSB-DNA complexes (124, 133). In fact, the $(SSB)_{35}$ to $(SSB)_{56}$ transition is accompanied by the cooperative binding of ~10–12 spermine ions to the SSB-poly(dT) complex (133). However, although cation charge is a major factor that influences the SSB-binding mode transition, it is not the only determining factor, since $Co(NH_3)_6^{3+}$ is more effective than either spermine^{4+} or spermidine^{3+} (133).

Griffith and colleagues have demonstrated two distinct morphologies of *Eco* SSB-ss fd DNA complexes by electron microscopy (122, 126). A "beaded" complex is observed at low SSB-to-DNA ratios and higher salt concentrations, whereas a "smooth-contoured" complex, with a twofold increase in apparent contour length, is observed at higher protein-to-DNA ratios and low salt concentrations in the absence of divalent cations. The "beaded" morphology is observed under conditions that favor the formation of the $(SSB)_{56}$- and/or the $(SSB)_{65}$-binding modes (100, 124). Each "bead" is composed of two SSB tetramers (octamer) with $\sim$145$\pm$20 nucleotides of ssDNA ($\sim$73$\pm$10 nucleotides per tetramer) (122). [Since the EM studies were performed before the discovery of the intermediate $(SSB)_{56}$-binding mode, it is not yet known whether the $(SSB)_{56}$- and the $(SSB)_{65}$-binding modes can be distinguished by EM.] The apparent contour length of the smooth-contoured complexes is $\sim$twofold longer than that of the "beaded" form (126), consistent with the proposal that SSB is bound in the $(SSB)_{35}$ mode in this complex (100); however, the structure and composition of the smooth-contoured complexes have not been examined in detail. At this point, the direct site-size measurements and the EM studies seem to provide a consistent (low-resolution) picture of the SSB–ss-polynucleotide-binding modes.

For a protein that can bind to a linear DNA in two modes differing in site size, it is expected that the mode with the smaller site size will be favored at high protein-binding densities (protein-to-DNA ratios), since more protein can bind to the DNA in the low site-size mode (134). In fact, the SSB-binding mode transitions display this behavior, since EM experiments (126) show that the smooth-contoured complex is favored at high protein-binding densities and fluorescence titrations show that the $(SSB)_{35}$ mode is favored at high SSB-binding densities (125, 135). Therefore, although the transition between the $(SSB)_{35}$- and $(SSB)_{56}$/$(SSB)_{65}$-binding modes is reversible at high binding densities (100, 135), at low binding densities, the $(SSB)_{56}$/$(SSB)_{65}$ modes are preferred, since these interact with a larger region of DNA and thus the $(SSB)_{35}$ mode is not stable at low binding densities (86, 125).

The majority of the site-size studies have been performed with poly(dT), since the affinity of *Eco* SSB for poly(dT) is so high; however, the different SSB–ss-polynucleotide-binding modes also form on poly(dA), poly(U), and poly(A) (100, 131), although the salt concentration required to stabilize each mode varies slightly with each polynucleotide (W Bujalowski, T Lohman, unpublished). However, a major exception is that the $(SSB)_{35}$ mode is not observed to form on poly(dC) at low [NaCl], even at high protein-to-DNA ratios; rather only the $(SSB)_{56}$ and $(SSB)_{65}$ modes are observed (W Bujalowski, T Lohman, unpublished). The absence of the $(SSB)_{35}$ mode on poly(dC) correlates with a much lower degree of negative cooperativity for binding of $dC(pC)_{34}$ to the subunits of the SSB tetramer (see below); the presence of this negative cooperativity is partially responsible for stabilizing the $(SSB)_{35}$–ss-polynucleotide mode (94, 95, 129). Although a range of SSB site sizes are also observable in studies with natural ssDNA (e.g. ss M13 DNA), it is difficult to interpret these due to the competitive effects of intramolecular basepairing (100, 136), which is why it is best to use homopolynucleotides for such studies.

Wrapping of ssDNA around the SSB Tetramer

A number of studies based on EM (2, 68, 122, 126, 137), nuclease digestion (122, 138), and oligonucleotide-binding stoichiometry (94, 95, 123) indicate that ssDNA is wrapped around the SSB tetramer in at least the high site-size SSB-ssDNA complexes.

ELECTRON MICROSCOPY EM studies have shown that ssDNA undergoes a significant compaction when complexed with *Eco* SSB (2, 68, 122, 126, 137). Sigal et al (2) first noted that ssDNA complexed with *Eco* SSB displayed a "beaded" morphology with a shortened contour length (1.8 Å per nucleotide), indicating that the DNA must either follow a broad helix or be folded regularly. At high [NaCl] and low SSB-binding densities or in the presence of $MgCl_2$, Griffith et al (122, 126) also observed "beads," which they interpret as "octamers" of SSB, whereas at high SSB-binding densities a "smooth-contoured" complex was observed. The apparent contour lengths differed for the "beaded" vs the "smooth" complexes, although both were significantly shorter than expected for uncomplexed ssDNA. The "beaded" complexes were compacted by $78 \pm 2\%$, whereas the "smooth" complexes were compacted by only $60 \pm 2\%$ (126), consistent with the latter complex having less DNA interacting per SSB tetramer (100). "Beaded" structures have also been observed in complexes with poly(dT) (0.3 M NaCl), although each bead appears to represent an individual tetramer rather than an "octamer" (137). This apparent discrepancy may reflect the fact that these two studies were performed under different solution conditions, which can influence the distribution of DNA-bound tetramers vs octamers within the $(SSB)_{65}$ mode (88, 139, 140). Systematic EM studies of the SSB-poly(dT) complexes as a function of solution conditions and binding density would provide valuable insight into these questions and allow a better correlation between the different binding modes and their morphologies in the EM.

NUCLEASE DIGESTION Nuclease digestion studies also indicate that ssDNA is wrapped around SSB in the "beaded" complexes observed by Chrysogelos & Griffith (122). Micrococcal nuclease treatment and gel electrophoretic analysis of "beaded" complexes at 37°C and three [NaCl] (20, 50, and 170 mM) (pH 7.5, 5 mM $CaCl_2$) showed multiple bands of 160 ± 25 nucleotides, with each "bead" composed of eight SSB monomers and ~145 nucleotides of ssDNA. Hence it was concluded that *Eco* SSB organizes ssDNA into "nucleosomelike" units, in which the "core" is an octamer (presumably two tetramers) of SSB monomers associated with 145 nucleotides and these octamers are separated by an average of ~30 nucleotides (122). Further treatment of the 145-nucleotide–SSB "core" complexes with DNase I showed a ladder of ~9 bands each differing by ~15 nucleotides, suggesting that DNase I can cut between SSB monomers within the octamer and that ssDNA is wrapped around the protein (122). This value of ~15 nucleotides per monomer or ~60 nucleotides per tetramer is very close to the high site-size value of 65±5 nucleotides per tetramer determined by fluorescence titrations (100, 124), suggesting that the $(SSB)_{65}$ mode and the "beaded" complexes are equivalent. These results are also consistent with equilibrium binding studies showing that saturation of an SSB tetramer with $dT(pT)_{15}$ occurs with one $dT(pT)_{15}$ bound per SSB subunit (94, 95).

Similar studies using nuclease P1 were also carried out on SSB-poly(dT) complexes, although under different solution conditions (20°C, 30 mM NaCl, 1 mM $MgCl_2$, 1 mM $ZnSO_4$, pH 6.2) (138). However, in these studies, protected fragments of ~75±10, 170±30, 235±35, 285±35, and 335±35 were observed, with the 75-nucleotide fragment being by far the most intense. This differs from the micrococcal nuclease results (122), in which only a 145-nucleotide fragment was observed at 0.17 M NaCl (5 mM $CaCl_2$), although at 20 mM NaCl, smaller DNA fragments were also apparent but were not characterized. The discrepancies between these two studies may be explained by the different conditions (pH 6.2 vs 7.5; 20°C vs 37°C, 30 mM vs 170 mM NaCl), DNA samples, and nuclease probes that were used [see Lohman & Bujalowski (77) for further discussion]. It is also interesting that P1 nuclease digestion of T4 gene 32 protein–poly(dT) complexes resulted in a protected fragment of ~80 nucleotides as well as a minor band at ~140–180 nucleotides, suggesting that T4 gene 32 protein may also form higher-order complexes with poly(dT) (138); however, this observation has not been pursued further.

STOICHIOMETRY OF OLIGODEOXYNUCLEOTIDE BINDING The stoichiometries for binding ss oligodeoxynucleotides to the SSB tetramer support the basic features of the different SSB polynucleotide-binding modes and provide additional evidence that ssDNA wraps around an SSB tetramer. The proposal that ssDNA wraps around an SSB tetramer was made first by Krauss et al (123) based on stoichiometry estimates of >3, 2, and 1 molecules per SSB tetramer for $d(pT)_8$, $d(pT)_{16}$, and $d(pT)_{30-40}$, respectively, which suggested that a 30–40-nucleotide-long ssDNA can interact with all four SSB subunits. However, these stoichiometry estimates are too low by a factor of two, as shown by Bujalowski & Lohman (94, 95), who measured stoichiometries of 4, 2, and 1 for saturation of the SSB tetramer by $dT(pT)_{15}$, $dT(pT)_{34}$, and $dT(pT)_{69}$, respectively. Thus, an ssDNA at least 64 nucleotides long is in fact needed to occupy all of the sites on the SSB tetramer simultaneously, whereas a 35-nucleotide-long ssDNA can only occupy half of the sites on the SSB tetramer (94, 95). These studies are consistent with the site sizes measured for SSB-poly(dT) complexes and demonstrate that only two SSB subunits interact with ssDNA in the $(SSB)_{35}$ polynucleotide-binding mode, whereas all four subunits interact with ssDNA in both the $(SSB)_{56}$ and $(SSB)_{65}$ modes (94, 95, 129). The discrepancy between these stoichiometry measurements reflects the fact that only half of the sites on the SSB tetramer were filled with oligodeoxynucleotides in the studies of Krauss et al (123), due to negative cooperativity, which decreases the affinity of ssDNA for the third and fourth subunits of the *Eco* SSB tetramer (94, 95, 129).

INFLUENCE OF SALT CONCENTRATION ON PROTEIN-DNA EQUILIBRIA

In the discussions that follow, it is important to recognize that changes in solution conditions generally produce significant effects on the observed equilibrium constants for protein-DNA interactions. This is true for changes in temperature and pH as is generally appreciated; however, it is especially true for changes in salt concentration and type (including ion valence). The bases of such salt effects have been discussed in detail (141–145); we briefly mention them here due to their central importance for SSB-ssDNA interactions.

Equilibrium binding constants, K_{obs}, for most protein–nucleic acid interactions are extremely sensitive to the concentration and type of salt in solution, with K_{obs} generally decreasing with increasing salt concentration (141, 142). This effect stems primarily from the fact that linear nucleic acids are highly charged polyanions and as a result sequester cations (e.g. K^+, Na^+, Mg^{2+}) in their vicinity (141, 142, 146). Therefore, when a protein binds to a nucleic acid, cations are released from the DNA. The release of these bound cations into the bulk salt solution provides a favorable entropic contribution to the free energy of binding. In fact, at low salt concentrations, counterion release from the nucleic acid provides the major driving force for the binding to linear nucleic acids of simple oligocations (141, 147, 148), as well as some nonspecific binding proteins (149, 150).

More complicated effects of salt concentration are generally observed for protein-DNA interactions, due to the fact that cations and anions (as well as protons and water) also interact with most proteins, hence binding or release of ions from the protein can occur in addition to the release of cations and/or water from the nucleic acid (145, 151). The dependence of K_{obs} on monovalent salt concentration (MX) at constant temperature, pressure, and pH for this case can be expressed as in equation 1 (neglecting preferential hydration),

$$\frac{\partial \log K_{obs}}{\partial \log [MX]} = \Delta c + \Delta a \quad 1.$$

where Δc and Δa are the net number of cations and anions, respectively, released or bound in a thermodynamic sense upon formation of the protein-DNA complex [a negative value of ($\Delta c + \Delta a$) indicates a net release of ions]. For most protein–nucleic acid interactions, cation release from the nucleic acid ($\Delta c < 0$) is a dominant component of Δc and ($\Delta c + \Delta a$) is negative, indicating that ions are released upon formation of the protein-DNA complex. However, both ion release and ion uptake can contribute to the net value of ($\Delta c + \Delta a$), as is the case for SSB binding in its $(SSB)_{65}$ mode (88). One consequence of these effects is that the precise value of K_{obs} in vitro can easily be manipulated simply by changing the salt concentration and/or type. As a result, values of K_{obs} can only be meaningfully compared under identical solution conditions, and are of little use if the conditions of the experiment are not specified.

Eco SSB BINDING TO SINGLE-STRANDED OLIGODEOXYNUCLEOTIDES

Equilibrium binding of ss-oligodeoxynucleotides to the SSB tetramer has been investigated by a number of laboratories (87, 93–95, 123, 129). Early equilibrium (87, 93, 123) and kinetic (123) studies, performed before the different SSB-binding modes were recognized, have been reviewed (14, 77), hence we discuss mainly recent studies for which binding stoichiometries and affinities have been determined (94, 95, 129).

Negative Cooperativity among DNA-Binding Sites within Individual SSB Tetramers

The observation of negative cooperativity for binding ssDNA among the four subunits of the *Eco* SSB tetramer (94, 95, 129) has provided some insight into the molecular basis for at least one of the SSB polynucleotide-binding mode transitions. Negative cooperativity is exhibited as a reduction in the affinity of ss-oligodeoxynucleotides for the third and fourth subunits of the SSB tetramer and exists under all solution conditions tested; however, the degree of negative cooperativity increases dramatically with decreasing salt concentration and is also influenced by cation charge. This negative cooperativity was identified based on equilibrium binding studies of $dT(pT)_{15}$, $dT(pT)_{27}$, and $dT(pT)_{34}$ (94, 95, 129). In the case of $dT(pT)_{34}$, which binds to two sites on the SSB tetramer, the affinity of the second

$dT(pT)_{34}$ is significantly lower than that of the first (129). This negative cooperativity is not due to steric effects, since it is also observed with $dT(pT)_{27}$, which binds to two sites per tetramer, as well as $dT(pT)_{15}$, which binds to four sites per tetramer (94, 95). Negative cooperativity is observed for $dT(pT)_{15}$ binding to the SSB-1 tetramer, although not for $dT(pT)_{15}$ binding to an SSB-1 monomer, hence negative cooperativity is a property of the SSB tetramer (96).

Interestingly, the quantitative degree of negative cooperativity increases as the salt concentration decreases, making it more difficult to bind ssDNA to the third and fourth SSB subunits. For example, at 1.5 mM NaCl (25°C, pH 8.1), one molecule of $dT(pT)_{34}$ binds with high affinity, yet a second $dT(pT)_{34}$ molecule does not bind even at 10-fold higher DNA concentrations, whereas at 200 mM NaCl, two molecules of $dT(pT)_{34}$ bind nearly stoichiometrically (129). However, upon raising the [NaCl] above 0.2 M, the macroscopic affinities for the binding of $dT(pT)_{34}$ to both the first and second sites decrease, as expected for a protein-DNA interaction (141, 145). These salt-dependent effects on negative cooperativity are also observed with changes in [$MgCl_2$]; however $MgCl_2$ is effective at concentrations that are 100-fold lower than observed with NaCl. In fact, the range of [NaCl] and [$MgCl_2$] that most affects the quantitative degree of negative cooperativity also induces the transition between the $(SSB)_{35}$ and the $(SSB)_{56}$ polynucleotide-binding modes (129). Thus, it appears that the $(SSB)_{35}$-binding mode is stabilized, at least in part, by the negative cooperativity for ssDNA binding that exists at low salt concentrations, which is an intrinsic property of the SSB tetramer (129). Relief of the negative cooperativity at higher salt concentrations results in ssDNA binding to the third and fourth SSB subunits to form the $(SSB)_{56}$-binding mode (129).

Part of the molecular basis for this negative cooperativity is electrostatic, since it is relieved upon raising the salt concentration. However, the extent of negative cooperativity is also influenced by DNA base composition. At the same salt concentration, negative cooperativity is greatest for $dA(pA)_{34}$ and weakest for $dC(pC)_{34}$, with $dT(pT)_{34}$ being intermediate between these (W Bujalowski, T Lohman, unpublished). These observations are consistent with the fact that poly(dC) is the only ss-polynucleotide that does not form the $(SSB)_{35}$ mode at $[NaCl] \leq 10$ mM, and further support the proposal that the $(SSB)_{35}$-binding mode is stabilized by this high degree of negative cooperativity (129).

Thermodynamics of Oligodeoxynucleotide Binding

Equilibrium binding of oligodeoxynucleotides to *Eco* SSB has been examined as a function of DNA length, salt concentration, and temperature, and analyzed using a statistical thermodynamic "square" model that accounts for the negative cooperativity (94, 95). The "square" model is described by two constants: the intrinsic equilibrium binding constant, K_N, and a negative cooperativity constant, σ_N (where N refers to the length of the oligodeoxynucleotide). Negative cooperativity exists when $\sigma_N < 1$, whereas binding is noncooperative for $\sigma_N = 1$. The basic premise of the model, as applied to the binding of $dT(pT)_{15}$, is that $dT(pT)_{15}$ can bind to any of the four initially equivalent subunits of the SSB tetramer with intrinsic affinity K_{16}. However, a subunit that has $dT(pT)_{15}$ bound exerts a negative cooperative effect on only two of its neighboring subunits, resulting in a decrease in affinity of $dT(pT)_{15}$ for these two subunits ($\sigma_{16} < 1$). The other subunit, however, can still bind $dT(pT)_{15}$ with higher affinity ($\sigma_{16} = 1$). The biphasic isotherms that result for $dT(pT)_{15}$ binding to the SSB tetramer at low salt can be fully explained by this model, on the basis of only these two interaction constants. The equilibrium binding of $dT(pT)_{34}$, which interacts with two SSB subunits, is also well described by this model (94, 95).

The intrinsic affinity, K_{16}, decreases, whereas the cooperativity constant, σ_{16}, increases with increasing NaCl concentration. Furthermore, K_{16} is slightly dependent on the type of anion (Cl vs Br), whereas σ_{16} is not. This indicates that the intrinsic binding interaction results mainly in a net release of cations, whereas negative cooperativity is accompanied by the uptake of approximately four sodium ions upon saturation of the tetramer with $dT(pT)_{15}$. However, although σ_{16} increases with increasing NaCl concentration, it always remains less than one; i.e. negative cooperativity is not eliminated at high monovalent salt concentration (95). The salt dependence and the lack of a temperature dependence for σ_{16} suggest that the molecular basis of the negative cooperativity is due in part to the electrostatic repulsion between segments of ssDNA when they are bound simultaneously to both "halves" of the SSB tetramer, where each half represents two SSB subunits. An uptake of cations might be necessary to partially neutralize the repulsion between these ssDNA segments if the binding sites are in close proximity (94, 95, 100, 124).

The intrinsic binding constant, K_N, for $dT(pT)_{N-1}$ binding to an SSB tetramer increases with increasing N for N = 16, 35, or 70 nucleotides (1M NaBr, pH 8.1, 25°C) (77, 94, 95). However, increasing the oligonucleotide length by a factor of two from 16 to 35 nucleotides or from 35 to 70 does not double the free energy of complex formation; in fact the free energy change for binding one $dT(pT)_{34}$ is considerably less than the free energy change for binding two $dT(pT)_{15}$, (i.e. $K^2_{16} >> K_{35}$). The free energy changes for saturating the SSB tetramer with one molecule of $dT(pT)_{69}$, two molecules of $dT(pT)_{34}$, or four molecules of $dT(pT)_{15}$, are −15, −20, and −29 kcal, respectively (after correction for the cratic entropy terms due to the different oligonucleotide stoichiometries) (95). This indicates that additional unfavorable interactions accompany the binding of the longer oligodeoxynucleotides. The unfavorable free energy contribution due to DNA that crosses two subunits has been estimated to be $\sim 4.7 \pm 0.3$ kcal. Interestingly, the SSB tetramer binds with considerably higher affinity to $dT(pT)_{69}$ than to poly(dT) in its isolated complex in the $(SSB)_{65}$ mode (77, 95).

The equilibrium binding constants for oligodeoxynucleotide binding (94, 95) as well as for ss-polynucleotide binding (88, 139) are quite sensitive to the salt concentration and type in solution. These observations indicate that electrostatic interactions contribute significantly to SSB-ssDNA binding. This salt sensitivity is qualitatively similar to the behavior observed for T4 gene 32 protein binding to ssDNA (152–155) as well as the majority of protein–nucleic acid interactions. A number of incorrect claims have been made that *Eco* SSB-ss DNA interactions are not salt-dependent (14, 87, 123, 156); however, these claims were based on early studies of oligonucleotide binding (87), as well as on studies with poly(dT) that were performed under conditions such that only the same minimum estimate of the binding constant was obtained at each salt concentration (93). In fact, K_N decreases with increasing salt concentration for all oligonucleotides, with $\partial \log K_N / \partial \log [NaBr] = -2.7 \pm 0.3$, -4.6 ± 0.5, and -7.1 ± 0.6 for the binding of one mole of $dT(pT)_{N-1}$ with N = 16, 35, and 70, respectively (95) (see below for salt effects on ss-polynucleotide binding). Major effects of anions are also observed when the oligonucleotide is long enough to interact with more than one SSB subunit. Whereas σ_{16} is not very sensitive to temperature, K_{16} is quite dependent on temperature with $\Delta H^\circ = -26 \pm 3$ kcal/mole $dT(pT)_{15}$, and binding is enthalpically driven (95).

COOPERATIVE BINDING TO SINGLE-STRANDED POLYNUCLEOTIDES

Although it appears that many SSB proteins bind with some degree of positive cooperativity to ssDNA, the magnitude and in some cases the type of cooperativity (e.g. limited vs unlimited, see below) varies considerably among different SSBs. For this reason, it is still not clear whether the ability to form long protein clusters along ssDNA (unlimited cooperativity) is a necessary feature for SSB function in DNA metabolism. Positive cooperativity reflects the fact that a protein binds with higher affinity to a DNA molecule to which protein is already bound than to an isolated (unliganded) DNA molecule. This can result from direct protein-protein interactions between nearest neighbors, as seems to be the case for the T4 gene 32 protein (76, 157, 158); however, cooperativity can also result from protein-induced distortions of adjacent DNA. Furthermore, cooperative interactions can have contributions from non-nearest-neighbor interactions, as seems to be true for the f1 gene V protein dimer, which interacts with two ssDNA segments that are separated along the DNA contour length (69, 70, 159).

Quantitative descriptions of cooperative, nonspecific binding of proteins to linear DNA require statistical mechanical models that necessarily approximate the actual physical situation in order to limit the parameters of the model to a number that can be determined unambiguously from the experimental data. However, one should always use a physically appropriate model and define the interaction constants (binding constant, cooperativity parameter, etc) in terms of the important molecular features of the system. Otherwise, the "interaction constants" are no more than fitting parameters, which cannot be related to free energy changes for the interactions. This is particularly important for cooperativity parameters, ω (see below), since the quantitative value of ω obtained from analysis of experimental data depends on how cooperativity is defined in the model. As a result, it is always important to consider the molecular basis for the model as well as the resulting value of the cooperativity parameter itself. Of course, in cases where there is insufficient information to discriminate among models, it is still appropriate to use a model; however, one must consider that the parameters obtained from such a model may have little relationship to actual molecular processes. Accurate quantitative estimates of cooperativity parameters can also be difficult due to experimental considerations. For example, the use of natural ssDNA, which can form intramolecular basepairs that compete with protein binding and thus even affect protein cluster distributions, makes meaningful estimates of cooperativity impossible. For this reason it is best to use ss-homopolynucleotides that cannot form intramolecular basepairs.

A wide range of estimates of nearest-neighbor cooperativity parameters (from 50 to 10^5) have been reported for *Eco* SSB binding to ss-polynucleotides, resulting in much confusion concerning this important quantity [for a discussion see Lohman & Bujalowski (77)]. Most of these estimates were made before the different SSB binding modes were recognized and were generally based on experiments performed under conditions that support a mixture of binding modes. Furthermore, it is now known that SSB can bind to ssDNA with at least two qualitatively different types of cooperativity that are associated with two different binding modes. Therefore, even if a particular binding mode can be studied in isolation, use of a model that correctly describes the type of cooperativity in that mode is essential for obtaining meaningful cooperativity estimates.

At least two types of cooperative binding to ssDNA, referred to as "limited" and "unlimited," have been observed for *Eco* SSB (65, 86), and these yield complexes with quite different properties in vitro (86, 130, 135) (see Figure 3). Although many details remain to be determined, it is clear that the two different types of cooperativities reflect the binding properties of at least two of the *Eco* SSB-binding modes. The $(SSB)_{65}$-binding mode displays a "limited" type of cooperativity (88, 130, 139) in which SSB clusters appear to be limited to the formation of dimers of SSB tetramers (octamers), which are seen as "beads" in the EM (122); long clusters of SSB tetramers do not form along ssDNA in this mode (86). In the "unlimited" cooperativity mode, which appears to be correlated with the $(SSB)_{35}$-binding mode (86, 135), long clusters of SSB can form (2, 68, 86), as observed for T4 gene 32 protein binding (1). Studies of ssDNA binding in the $(SSB)_{40}$ and $(SSB)_{56}$ modes (124, 125) have not been carried out, hence information on the cooperativity in these modes is not available.

Statistical Thermodynamic Models for Nonspecific, Cooperative Binding of SSB to ss-Polynucleotides

The "unlimited" cooperativity model is the most common model used to describe nonspecific, cooperative binding of proteins to infinite (119, 120, 160, 161) as well as finite linear nucleic acids (162, 163). This model provides an excellent description of the cooperative binding of the T4 gene 32 protein to ssDNA (1, 66, 152–155, 164, 165), and appears to be applicable to *Eco* SSB binding in its $(SSB)_{35}$ mode (135). The "limited" cooperativity model was developed to describe the unique cooperative aspects of *Eco* SSB binding to ssDNA in its $(SSB)_{65}$ mode (130, 166). Three parameters are needed to describe either model (see Figure 3): n—the number of nucleotides occluded by the bound protein (site size), which is not the same as the number of nucleotides contacted (167, 168); K—the intrinsic equilibrium constant for the binding of the protein to an isolated site on the DNA lattice; ω—the nearest-neighbor cooperativity parameter, which is a unitless equilibrium constant for the process of moving two proteins, bound to DNA in isolated modes, so that they are contiguous. Both models account for the statistical effect of "overlap" of potential protein-binding sites on the DNA; i.e. that each nucleotide of a linear nucleic acid represents the start of a potential nonspecific binding site for a protein, but a protein occludes more than one nucleotide upon binding the nucleic acid [see McGhee & von Hippel (119, 120) for an excellent discussion].

"UNLIMITED" NEAREST-NEIGHBOR COOPERATIVITY In this model, nearest-neighbor cooperative interactions can occur on either side of a bound SSB tetramer, thus enabling a continuous protein cluster to form along the entire length of the nucleic acid. Detailed descriptions of this model have been given (119, 120, 160, 161, 166), and the cartoon in Figure 3*A* defines the binding parameters for this model. (For clarity, we designate cooperativity parameters based on the "unlimited" model as ω_{unlim}.) In this model, a protein can bind to an infinite, linear, homogeneous nucleic acid lattice in any of three modes: an isolated mode, with intrinsic equilibrium constant, K; a singly contiguous mode (one nearest-neighbor protein), with equilibrium constant $K\omega$; a doubly contiguous mode (two nearest-neighbor proteins), with equilibrium constant, $K\omega^2$. The cooperativity parameter, ω, is a unitless equilibrium constant for the process of moving two bound isolated proteins so that they are bound in a singly contiguous manner, and is related to the standard free energy change for this process by $\Delta G°_{coop} = -RT \ln\omega$. Positive cooperativity is reflected by $\omega > 1$ ($\Delta G°_{coop} < 0$), whereas noncooperative interactions are described by $\omega = 1$ ($\Delta G°_{coop} = 0$). Useful average quantities, such as the average protein cluster size on the DNA, can be calculated from the known values of n, K, and ω (119, 160, 166, 169). For $\omega > 1$, proteins form clusters of variable length along the nucleic acid, where the average cluster length increases with increasing protein-binding density, n (for given values of n, K, and ω), and with increasing ω at a given protein-binding density (119, 120, 160, 164, 170).

"LIMITED" NEAREST-NEIGHBOR COOPERATIVITY The "limited" cooperativity model was developed to describe the equilibrium binding of the *Eco* SSB to ssDNA in its $(SSB)_{65}$ mode and provides a better description for cooperativity in this binding mode than does the "unlimited" cooperativity model (130). It accounts for the observation that in the "beaded" $(SSB)_{65}$ mode, the SSB cluster size is "limited" to clusters of two SSB tetramers (122). Figure 3*B* defines the interaction constants, K and ω_{lim}, for this model of SSB tetramers binding in the $(SSB)_{65}$ mode. (We designate the cooperativity parameter based on the "limited" model as ω_{lim}, although it has also been designated $\omega_{T/O}$, where the T/O refers to tetramer/octamer.) Note the absence of doubly contiguously bound proteins in this model. Average quantities, such as the fraction of tetramers bound as octamers, can be calculated using this model and the known values of n, K, and ω_{lim}. The "limited" cooperativity model predicts that only tetramers and dimers of tetramers (octamers) occur on the DNA so that long protein clusters are never formed (130).

"Limited" Cooperativity in the (SSB)65 Mode

The $(SSB)_{65}$ mode, in which ssDNA interacts with all four SSB subunits, is the best-studied of the SSB-binding modes since it can be studied in isolation at [NaCl] > 0.2 M (100, 124, 125). Cooperative binding in this mode has been examined as a function of salt concentration and type, pH, and temperature (86, 88, 139, 140). From analysis of equilibrium isotherms for SSB tetramer binding to ss-homopolynucleotides, a value of $\omega_{65,lim} = 420 \pm 80$ was determined at 25°C, pH 8.1, independent of monovalent salt

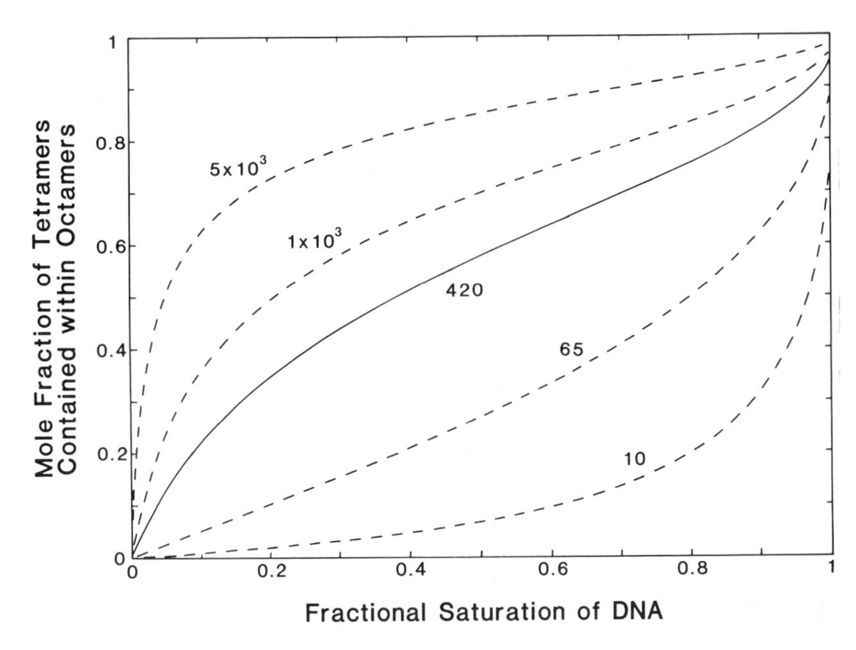

Figure 4 Theoretical simulations of the mole fraction of SSB tetramers contained within SSB octamers as a function of the fractional saturation of ss nucleic acid, for different values of the "limited" cooperativity parameter, ω_{lim}. The plots are based on the "limited" cooperativity model (130) with n = 65 nucleotides per tetramer and the value of ω_{lim} as indicated for each curve. The continuous line (ω_{lim} = 420) represents the predicted distribution for *Eco* SSB tetramers binding to ssDNA in the $(SSB)_{65}$ mode at 25°C, pH 8.1. Reprinted with permission from the *Journal of Molecular Biology* (130).

concentration and type, polynucleotide lattice (88, 130, 171), and pH (140). However, $\omega_{65,lim}$ does depend on temperature, increasing from ~170±130 at 10°C to ~1000±400 at 37°C (pH 8.1) (140). Figure 4 shows the distribution of SSB tetramers and octamers (clusters of two tetramers) as a function of fractional saturation of the ssDNA, predicted by the "limited" cooperativity model for values of $\omega_{65,lim}$ = 10, 65, 420, 10^3, and 5×10^3 (130). Note that the "limited" cooperativity model does not predict that DNA-bound SSB tetramers are found exclusively as part of octamers, as might be inferred from EM studies (122). In fact, for $\omega_{65,lim}$ = 420 (25°C, pH 8.1), SSB tetramers in the $(SSB)_{65}$ mode are predicted to be distributed equally between isolated tetramers and octamers at 50% saturation of the ssDNA (130).

Before the "limited" cooperativity model was developed, the "unlimited" cooperativity model was used to analyze SSB binding in the $(SSB)_{65}$ mode and a value of ω_{unlim} = 50±10 (25°C, pH 8.1) was reported (86). However, since the SSB tetramer does not form unlimited clusters when bound in the $(SSB)_{65}$-binding mode (86, 130), this value of ω_{unlim} = 50±10 has no defined physical meaning and therefore should not be used for quantitative descriptions of the cooperative behavior in the $(SSB)_{65}$ mode. Reanalysis of the same equilibrium binding data using the more realistic "limited" cooperativity model yields $\omega_{65,lim}$ = 420±80 (130).

"Unlimited" Cooperative Binding Correlates with the (SSB)35 Mode

The first EM studies of SSB complexes with ssDNA showed evidence of high "unlimited" cooperativity (2, 68). In fact, at low protein-binding densities, ssDNA molecules nearly saturated with protein were observed in the same population with free DNA. This type of non-uniform (biphasic) distribution of DNA-bound SSB was also observed in gel electrophoresis studies of SSB/ssDNA complexes formed by direct mixing at low [NaCl] and low protein-binding densities (86). However, this "unlimited" cooperative behavior was metastable at low SSB-binding density, even at low [NaCl], and a more uniform distribution pattern was eventually formed, identical to that observed for complexes formed at higher [NaCl] ≥0.2 M (86). Although this observation is not completely understood, it seems that the highly cooperative complexes observed by EM (2) are the same as the metastable complexes observed by gel electrophoresis, since both were observed after direct mixing of SSB with ssDNA at low [NaCl] (77, 86). Since the metastable "unlimited" cooperativity forms under conditions that favor the $(SSB)_{35}$ mode, in which ssDNA interacts with only two subunits of the SSB tetramer, it was proposed that the $(SSB)_{35}$ mode may play some role in the transient stabilization of this "unlimited" cooperativity (65, 86).

Ruyechan & Wetmur (68) estimated an "unlimited" nearest-neighbor cooperativity parameter (ω_{unlim}) of ~3 × 10^5 based on an EM analysis of the distribution of SSB cluster lengths on heat-denatured phage λ DNA. [A reanalysis of these data (135) accounting for "overlap" of potential protein-binding sites (119, 120) increases this estimate to ~2 × 10^6.] However, since the EM studies were performed in conditions (0.15 M NaCl, 4°C) that should favor a mixture of SSB-binding modes, it is impossible to determine from these data whether a single SSB-binding mode is responsible for this high "unlimited" cooperativity behavior (and if so, which one?).

In order to estimate the nearest-neighbor cooperativity parameter of the $(SSB)_{35}$ mode, Ferrari et al (135) examined the equilibrium binding of SSB tetramers to $dA(pA)_{69}$, under conditions that permit the binding of two tetramers to $dA(pA)_{69}$. In these complexes only two subunits of each SSB tetramer interact with the DNA, hence this should mimic the $(SSB)_{35}$ mode

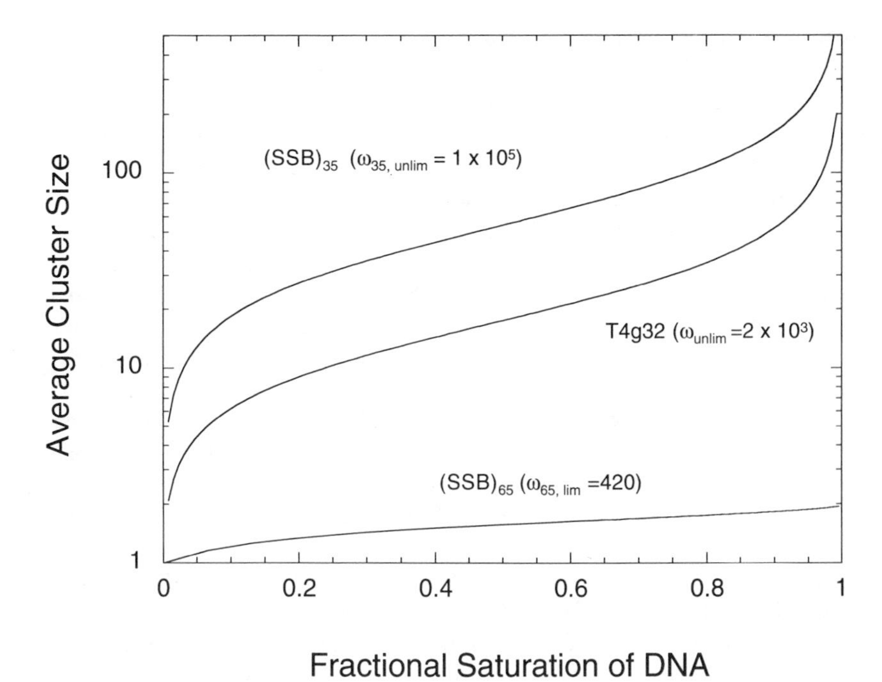

Figure 5 The average lengths of protein clusters predicted for SSB tetramers bound in the $(SSB)_{35}$ mode with "unlimited" cooperativity are compared with those predicted for SSB tetramers bound in the $(SSB)_{65}$ mode with "limited" cooperativity. The average cluster lengths for T4 gene 32 (T4 g32p) monomers bound with "unlimited" cooperativity are also shown for comparison. The cluster distributions for $(SSB)_{35}$ ($n = 35$, $\omega_{35,unlim} = 10^5$) and T4 g32p ($n = 7$, $\omega_{unlim} = 2 \times 10^3$) were calculated using the "unlimited" nearest-neighbor cooperativity model (119, 120), whereas the distribution for the $(SSB)_{65}$ mode ($n = 65$, $\omega_{65,lim} = 420$) was calculated using the "limited" cooperativity model (130). All calculations assume an infinite ssDNA lattice.

and allow an estimate of cooperativity (if end effects are absent). These studies conclude that a large positive cooperativity exists between SSB tetramers in this mode, although only a minimum estimate of $\omega_{35} \geq 10^5$ (0.125 M NaCl, pH 8.1, 25°C) could be obtained (135). However, this minimum estimate is consistent with the SSB cluster distributions observed in the EM (68) and gel electrophoresis studies (86) at low [NaCl]. Taken together, these studies support the view that the $(SSB)_{35}$ mode binds with high "unlimited" cooperativity, and that the EM studies of Ruyechan & Wetmur (68) at 0.15 M NaCl, which showed "unlimited" cooperative behavior, were probing primarily the $(SSB)_{35}$-binding mode.

Upon increasing the [NaCl] from 0.15 to 0.3 M, Ruyechan & Wetmur (68) observed a decrease in the average length of SSB–ssDNA complexes, which was interpreted as a decrease in cooperativity. However, since only protein-bound DNA complexes could be detected, Ruyechan & Wetmur (68) were mainly detecting the twofold reduction of the apparent contour length of the SSB-ssDNA complexes accompanying the [NaCl]-dependent transition from the $(SSB)_{35}$ mode to the $(SSB)_{56}/(SSB)_{65}$ mode (100).

To illustrate the dramatic differences between "unlimited" cooperativity in the $(SSB)_{35}$ mode ($n = 35$, $\omega_{35,unlim} = 10^5$) and "limited" cooperativity in the $(SSB)_{65}$ mode ($n = 65$, $\omega_{65,lim} = 420$) (88, 130, 140), Figure 5 shows the average cluster size of SSB tetramers as a function of the fractional saturation of ssDNA, assuming exclusive binding in each mode. For comparison, Figure 5 also shows the average cluster sizes predicted for the phage T4 gene 32 protein monomers on ssDNA, using an "unlimited" cooperativity model with $\omega_{unlim} = 2 \times 10^3$ (152). The average cluster lengths for SSB binding in the $(SSB)_{35}$ mode are considerably longer than those for the T4 gene 32 protein. However, both of these are considerably larger than the average cluster size for SSB tetramers binding in the $(SSB)_{65}$ mode. Of course, actual cluster distributions for *Eco* SSB tetramers on ssDNA will differ due to the fact that both the $(SSB)_{35}$ and $(SSB)_{65}$ modes can coexist on a long ssDNA(135).

The observation that "unlimited" cooperative binding of SSB is metastable at low SSB-binding densities, reverting to the "limited" cooperativity SSB-binding mode at equilibrium, indicates that the $(SSB)_{35}$ mode and the resulting protein clusters must form more rapidly than the $(SSB)_{65}$ mode even at low binding densities. It is likely that the $(SSB)_{65}$ mode forms more slowly, since in this mode ssDNA must wrap around the tetramer fully, interacting with all four subunits, whereas only two subunits bind to ssDNA in the $(SSB)_{35}$ mode. Furthermore, the $(SSB)_{35}$ mode is likely to be an intermediate along the pathway to formation of the $(SSB)_{65}$ mode (86, 100, 172). The pathway(s) by which SSB tetramers, when bound in the $(SSB)_{35}$ mode, can rapidly form long protein clusters is not yet known, although evidence for rolling (173) and direct transfer (174) has been presented, and sliding is also possible (175, 176). In fact, the $(SSB)_{35}$ mode is well suited for a direct transfer mechanism, since this would require two DNA-binding sites on the SSB tetramer (77). In any event, the observations that the "unlimited" cooperative binding of SSB is metastable at low SSB-binding densities, but still possesses an equilibrium value of $\omega_{35,unlim} \sim 10^5$, are not necessarily inconsistent.

EQUILIBRIUM BINDING IN THE $(SSB)_{65}$ MODE

Early studies of *Eco* SSB binding to ss-polynucleotides were generally performed under conditions that are now known to support SSB binding in a mixture of modes. Such experiments are difficult to interpret quantitatively, since resolution of the equilibrium binding and cooperativity parameters for each individual mode is nearly an intractable problem. The simplest approach is to use conditions that populate only a single binding mode. To date, only the $(SSB)_{65}$-binding mode has been studied in isolation on ss polynucleotides, since it exists as the sole binding mode at high salt conditions ([NaCl] $\geq$ 0.2 M) (86, 88, 130, 139, 140). Equilibrium binding in the $(SSB)_{65}$ mode has been analyzed using the "limited" cooperativity model to obtain the equilibrium constant for isolated binding of an SSB tetramer, K^{65}, and $\omega_{65,lim}$ (also designated $\omega_{T/O}$) (130). In the following sections, we discuss the effects of solution variables and polynucleotide base composition on K^{65} (the effects on $\omega_{65,lim}$ have been discussed above).

Most studies of SSB tetramer–ss-polynucleotide binding have been performed by monitoring the tryptophan fluorescence quenching of SSB upon complexation. It is important to recognize that the use of such indirect optical methods can result in incorrect binding isotherms unless the analyses used are not based on assumptions about the relationship between the optical signal and the fraction of bound protein. Such methods of analysis have been reviewed in detail (121, 171, 177).

Base Specificity

Although *Eco* SSB is a nonspecific DNA-binding protein, it displays a high degree of base specificity in its binding to ss-nucleic acids. As a result, *Eco* SSB likely displays some nucleotide sequence specificity in its binding; however, this has not been examined and is not thought to be relevant functionally. Base specificity was first demonstrated by filter binding and gel filtration studies (15) and fluorescence quenching experiments (99), which showed that *Eco* SSB affinity is greatest for poly(dT), although SSB was likely bound in a mixture of binding modes in these early studies.

Base specificity in the $(SSB)_{65}$-binding mode has been examined using equilibrium titrations in which quenching of the Trp fluorescence of SSB was monitored upon binding a series of ss-homopolynucleotides (86, 88); in 0.20 M NaCl (25.0°C, pH 8.1): $K^{65}{}_{obs}$ [poly(dT)] > $K^{65}{}_{obs}$ (dC) >> $K^{65}{}_{obs}$ (ss M13 DNA) > $K^{65}{}_{obs}$ (rI) > $K^{65}{}_{obs}$ (rU) = 8 $K^{65}{}_{obs}$ (dA) = 87 $K^{65}{}_{obs}$ (rA) >> $K^{65}{}_{obs}$ (rC). Mainly qualitative rankings are indicated, since equilibrium binding constants for poly(dT) could not be measured accurately under these conditions. In fact, *Eco* SSB affinity for poly(dT) is so high that it cannot be measured accurately by fluorescence titrations even in buffers containing 5 M NaCl at 25°C (100). However, in 1 M NaBr (25°C, pH 8.1), where the affinity for poly(dT) can be measured accurately, $K^{65}{}_{obs}$ is a factor of $\sim 10^8$ greater for poly(dT) than for poly(U) (88). Furthermore, since the relative affinity of SSB for poly(C) is $\sim 10^7$–10^8 lower than for poly(U) (15 mM NaCl, 25°C, pH 8.1) (TM Lohman, unpublished experiments), the largest affinity ratio for binding of an isolated SSB tetramer in the $(SSB)_{65}$ mode at constant solution conditions is estimated to be a factor of 10^{15}–10^{16} [for poly(dT) vs poly(C)].

Similar base specificity rankings have been reported for T4 gene 32 protein (153, 178) and fd gene 5 protein (179–182), including the fact that these proteins also bind with highest affinity to poly(dT) and lowest to poly(C). However, the affinity ratio for isolated binding to poly(dT) vs poly(C) is only $\sim 10^5$–10^6 for T4 gene 32 protein (153) and $\sim 10^4$ for fd gene 5 protein (179). The much larger specificity for isolated binding of the SSB tetramer in the $(SSB)_{65}$-binding mode may be due to the fact that all four protomers interact with the ssDNA in the $(SSB)_{65}$ mode, whereas T4 gene 32 protein interacts as a monomer and gene 5 protein interacts as a dimer. Interestingly, although oligolysines containing Trp do not discriminate among poly(dT), poly(A), poly(dU), and poly(U), they do bind with significantly lower affinity to poly(C) (183). Therefore, although the interaction of aromatic amino acids may be the source of some of this base specificity, it cannot explain all of it.

Bobst et al (132) have estimated relative affinities of SSB tetramers for ss-polynucleotides at low binding densities in 0.125 M NaCl (pH 8.1, 10% glycerol, temperature unspecified) by competition methods using EPR and found: K_{app} (dT) = 40 K_{app} (dA) = 200 K_{app} (rA). Although the qualitative rankings agree with those observed at 0.2 M NaCl where the $(SSB)_{65}$ mode is solely populated (88), the relative affinities differ considerably, especially for poly(dT) vs poly(dA). These quantitative differences may reflect the fact that the EPR studies were performed at lower salt concentrations (0.125 M NaCl), so that the SSB tetramer may be bound as a mixture of the $(SSB)_{56}$ and $(SSB)_{65}$ modes. We also note that the binding constant reported in the EPR studies, K_{app} (132), is not the same as the intrinsic binding constant for isolated SSB tetramer binding (K_{obs} as defined above), rather K_{app} is a complex function of K_{obs} and the cooperativity for any binding mode that is populated, which may also contribute to the different results.

Salt Dependence

Although it has been stated that the *Eco* SSB–ss nucleic acid interactions are not salt dependent and that electrostatic interactions are not important in binding (14, 87, 156), this conclusion is not correct. In fact, the equilibrium constant is quite sensitive to salt concentration and type for SSB tetramer binding to ss-polynucleotides in the $(SSB)_{65}$-binding mode (88, 139) as well as to oligodeoxynucleotides (94, 95), with affinity decreasing dramatically with increasing salt concentration. However, the effects of salt on *Eco* SSB–ss polynucleotide binding are complex, even in a single binding mode, since SSB binding to ssDNA is accompanied by both cation and anion uptake and release (88, 100, 124, 125), and these are differentially affected by changes in pH (139) and temperature (140).

Equilibrium constants for SSB tetramer binding in the $(SSB)_{65}$-binding mode have been measured with poly(dA), poly(dT), poly(A), and poly(U) over a range of monovalent salt concentrations and types (86, 88). K^{65}_{obs} decreases steeply with increasing salt concentration, with $(\partial \log K^{65}_{obs}/\partial \log [NaCl]) = -7.4 \pm 0.5$ for poly(U), -6.1 ± 0.6 for poly(dA), and -6.2 ± 0.3 for poly(A) (25.0°C, pH 8.1). Although SSB-poly(dT) affinity is too high to measure in buffers containing NaCl (even 5 M), it can be measured in NaBr (1.8–2.5 M), where $(\partial \log K^{65}_{obs}/\partial \log [NaBr]) = -5.7 \pm 0.7$, although a slightly lower value of $\omega_{65,lim} = 130 \pm 70$ is measured (compared to $\omega_{65,lim} = 420 \pm 100$ for the other polynucleotides in NaCl). Both the salt dependence and magnitude of K^{65}_{obs} are also affected by anion type; for poly(U): $(\partial \log K^{65}_{obs}/\partial \log [MX]) = -5.7 \pm 0.4$, -4.3 ± 0.4, -6.5 ± 0.3, and -6.7 ± 0.6 in KGlu, NaF, $NaCH_3CO_2$, and NaBr, respectively. Affinity is highest in glutamate, followed by F^-, acetate, Cl^-, and Br^- (88). These data indicate that significant electrostatic interactions occur in the $(SSB)_{65}$ complexes formed with all ss-polynucleotides, resulting in a net release of both cations and anions, although there are also contributions due to cation and anion uptake.

Temperature and pH Dependence

There is a strong linkage between the effects of [NaCl] and pH for *Eco* SSB binding to poly(U) in the $(SSB)_{65}$ mode, such that $(\partial \log K_{obs}/\partial \log [NaCl])$ ranges from -12.0 ± 0.8 at pH 5.5, to -6.0 ± 0.5 at pH 9.0 (25.0°C), whereas pH has little effect on the [NaF] dependence $(\partial \log K_{obs}/\partial \log [NaF] = -4.5 \pm 1)$ (139). This appears to result from an increase in Cl^- binding to SSB at low pH due to a requirement for protonation of Cl^--binding sites on SSB, whereas F^- has only low affinity for these sites. The larger [NaCl] dependence at lower pH is due to an increased release of Cl^- upon formation of the $(SSB)_{65}$ complex at low pH. The observed linked effects of ion concentration and type and pH on K^{65}_{obs} seem to reflect: (*a*) cation release from the polynucleotide; (*b*) release of Cl^- from sites on the SSB tetramer that require protonation to bind Cl^-; (*c*) binding of cations to sites on the SSB tetramer that require deprotonation for cation binding, and (*d*) required binding of 2–3 protons by the SSB tetramer in order to form the SSB-poly(U) complex (139).

The dependence of K^{65}_{obs} on [NaCl] $(\partial \log K^{65}_{obs}/\partial \log [NaCl])$ is also dependent on temperature (140) (M Ferrari, T Lohman, unpublished data), and equivalently the observed van't Hoff ΔH° is dependent upon [NaCl]. For *Eco* SSB binding to poly(U) in the $(SSB)_{65}$ mode (pH 8.1), the van't Hoff ΔH° increases from -67 ± 2 kcal/mol at 0.12 M NaCl to -31 ± 3

kcal/mol at 0.48 M NaCl (140). In addition, $\partial \log K^{65}_{obs}/\partial \log[NaCl]$ varies with increasing temperature, from -9.3 ± 0.3 at 10°C to -5.1 ± 0.4 at 37°C (pH 8.1) (140), which also appears to be due to binding of anions and/or cations to protonatable sites on the SSB tetramer. Although similar anion effects have been observed for the T4 gene 32 protein binding to ssDNA (152), the influences of pH and temperature have not been examined for this protein. The observation of such strong linkages makes it impossible to draw firm conclusions about the relative forces involved in protein-DNA interactions based on studies carried out under a single set of solution conditions. Even conclusions concerning the effects of base composition on relative affinities can potentially be changed by solution conditions.

The effects of temperature on the equilibrium formation of 1:1 SSB tetramer/$dN(pN)_{69}$ complexes (N = C, T, or A), have also been examined by van't Hoff analysis (M Ferrari, T Lohman, unpublished). The same base hierarchy is observed as with ss-polynucleotides (88); however, for $dC(pC)_{69}$ and $dT(pT)_{69}$, $\Delta H°$ is large and negative, -68 ± 7 and -82 ± 17 kcal/mol, respectively, whereas for $dA(pA)_{69}$, the van't Hoff plot displays extensive curvature, with K_{obs} possessing a maximum value of 5.0×10^7 M^{-1} at 25°C (0.17 M NaCl, pH 8.1). These results suggest that adenine bases become unstacked upon SSB tetramer binding, which is consistent with interpretations from electric field–induced birefringence studies of SSB-poly(A) complexes (131).

SSB MUTANTS

Only a few natural *ssb* mutants, *ssb-1, ssb-2, ssb-3, ssb-113,* and *ssb-114,* have been isolated and characterized. Mutations in the *ssb* gene are pleiotropic, resulting in a variety of defects in either replication, recombination, or repair (14) [see Meyer & Laine (74) for a more detailed discussion]. Only the *ssb-1* and *ssb-113* mutants have been characterized in terms of both their effects on *E. coli* physiology and their biochemical properties. The *ssb-1* mutation [His55 to Tyr (82)] provided the first evidence of the essential nature of *Eco ssb* (7, 8, 184); *ssb-1* confers temperature-sensitive lethality due to rapid cessation of replication (7, 8) and also displays defects in recombination (8). The biochemical defect is a destabilization of the SSB-1 tetramer with respect to monomers, which reduces its apparent affinity for ssDNA (82, 85). In fact, overproduction of SSB-1 protein in vivo eliminates the temperature-sensitive lethality of this mutation, presumably due to the formation of functional SSB-1 tetramers at higher protein concentration (185). The *ssb-113* mutation [Pro176 to Ser (107)] (formerly *lexC113*) (9, 186) exhibits a temperature-sensitive defect in replication (187); however, most of the *ssb-113* defects are apparent even at the permissive temperature (9, 187). The *ssb-114* mutation possesses the same Pro176-to-Ser substitution as in the *ssb-113* mutation, yet an *ssb-114* strain displays physiological differences distinct from *ssb-113,* hence *ssb-114* likely contains additional sites of mutation that have not been identified. The *ssb-3* mutation (Gly15 to Asp) renders the cell extremely UV sensitive (much more than either *ssb-1* or *ssb-113*) at the permissive temperature ($\leq$43°C), and shows a 75% reduction in recombination frequency as measured by Hfr mating (188) with only a slight effect on replication (74). The molecular basis of the *ssb-2* mutation (189) is not known.

ssb-1

The SSB-1 mutation (His55 to Tyr) (82) results in a destabilization of the SSB-1 tetramer with respect to monomers (82, 85, 96, 128). At 25°C (pH 8.1), the wild-type protein is fully tetrameric at a total monomer concentration as low as 120 nM; however, only half of the SSB-1 protein is tetrameric at a total monomer concentration of 3 μM under these conditions (85). Williams et al (82) showed by small-zone gel filtration techniques that the SSB-1 tetramer is unstable below ~10 μM (monomer), whereas the wild-type SSB remains tetrameric at these concentrations (25°C, pH 8.0, 0.5 M NaCl). Under the same conditions at 45°C, no change was observed in the SSB-1 monomer-tetramer equilibrium, although the wild-type SSB tetramer was destabilized below ~0.5 μM (monomer). At 25°C (pH 7.0, 50 mM Na_2HPO_4), stoichiometric binding of both SSB and SSB-1 proteins to poly(dT) was observed; however, at 45°C (pH 7.0, 50 mM Na_2HPO_4, 0.5 M NaCl) SSB-1 showed only slight binding to poly(dT) at 0.24 μM (monomer), whereas at a 13-fold higher concentration, SSB-1 protein and wild-type SSB binding to poly(dT) were quite similar. Williams et al (82) proposed that the basis for the temperature-dependent *ssb-1* defect is that SSB-1 tetramers dissociate to monomers to the same extent at both 25°C and 45°C, but that SSB-1 monomers denature at the nonpermissive temperature (>42°C), whereas SSB-1 tetramers remain native at the higher temperature.

The SSB-1 monomer-tetramer association has been examined quantitatively by monitoring the increase in fluorescence anisotropy upon formation of the tetramer (85). Although many of the qualitative features of the SSB-1 monomer/tetramer equilibrium are in agreement with the results of Williams et al (82), some important differences were found (85). The equilibrium formation of tetramers from monomers is essentially a one-step process (i.e. the fraction of dimers within the transition is negligible) with an equilibrium constant of $K_T = 3 \pm 1 \times 10^{18}$ M^{-3} (pH 8.1, 25°C); K_T is independent of [NaCl] (from 50 mM to 1 M), although it is reduced in the presence of $MgCl_2$. Most importantly and in contrast to previous conclusions (82), K_T

was found to be quite temperature dependent, such that SSB-1 tetramers are less stable at higher temperatures (85). The $\Delta H°$ for tetramer formation is pH dependent, with $\Delta H° = -51\pm7$ kcal/mole tetramer at pH 8.1 vs -37 ± 5 kcal/mole tetramer at pH 6.9 (85). This large negative $\Delta H°$ for tetramer formation, coupled with the significantly lower ssDNA-binding affinity of the SSB-1 monomer, is sufficient to explain the temperature sensitivity of the *ssb-1* mutation in vivo (85, 96), hence it is not necessary to invoke the selective denaturation of SSB-1 monomers at the nonpermissive temperature (82), although this is still a possibility.

Isolated SSB-1 monomers can bind $dT(pT)_{15}$, hence the formation of tetramers is not required for ssDNA binding (96). At 50 mM NaCl (pH 8.1, 25°C), the affinity of $dT(pT)_{15}$ for SSB-1 monomers is reduced compared to protomers within the SSB-1 tetramer ($7\pm3 \times 10^5$ vs $8\pm2 \times 10^6$ M^{-1}), although at 0.2 M NaCl, the affinities are comparable ($1.2\pm0.5 \times 10^5$ M^{-1}) (96). SSB-1 tetramers can bind four molecules of $dT(pT)_{15}$, as does wild-type (94, 95), but with lower affinity than wild-type tetramer ($K_{16} = 1.2\pm0.4 \times 10^5$ for SSB-1 vs $1.2\pm0.2 \times 10^6$ M^{-1} at 0.2 M NaCl). Negative cooperativity for ssDNA is also observed for the SSB-1 tetramer, with a negative cooperativity constant, σ_{16}, that is comparable to wild-type; however, negative cooperativity is not observed with SSB-1 monomers, indicating that it is solely a characteristic of the tetramer (96).

His55 has also been replaced by Tyr (Y) (*ssb-1*), Glu (E), Lys (K), Phe (F), or Ile (I) by site-directed mutagenesis, and the assembly state and poly(dT)-binding properties of the mutant proteins have been examined (128). H55F and H55I behaved identically to wild-type; however, the H55K tetramer is even less stable than the H55Y (SSB-1) tetramer (128) and shows even greater UV sensitivity than *ssb-1* (H55Y) in vivo (104). As with SSB-1 protein, H55K protein forms tetramers when bound to poly(dT) (128).

Suppression of the *ssb-1* mutation is observed if cells are grown in media containing high [NaCl] or glucose (190). This suppression has been proposed to result from the decrease in accessible cell volume that accompanies growth of *E. coli* at high osmolarity (191), rather than to be a direct effect of ion binding to the protein, which is consistent with the lack of a dependence of the SSB-1 monomer-tetramer equilibrium on [NaCl] in vitro (85, 96). The decrease in cell volume increases the effective concentration of the SSB-1 monomers, and thus shifts the equilibrium in favor of tetramer formation.

ssb-113

The *ssb-113* mutation (Pro176 to Ser) occurs at the penultimate amino acid within the acidic C-terminal region that is conserved among a number of

SSBs that share similarities with *Eco* SSB [e.g. the F-plasmid SSB and other conjugative plasmids (19, 21) as well as *S. marcescens* (16)]. Studies of ssDNA-binding properties and tetramer stability in vitro show only minor differences between SSB-113 and wild-type SSB, such as a slightly greater ability of SSB-113 to lower the melting temperature of poly[dA-T] (107). The SSB-113 mutant and wild-type SSB undergo the same salt-dependent transition between the $(SSB)_{35}$- and $(SSB)_{56}$-binding modes on poly(dT) (W Bujalowski, T Lohman, unpublished). Equilibrium binding studies with poly(U) indicate similar values of K^{65} and $\omega_{65,\mathrm{lim}}$ for SSB-113 and wild-type SSB binding in the $(SSB)_{65}$ mode (L Overman, T Lohman, unpublished). Therefore, the *ssb-113* mutation does not appear to affect ssDNA binding, although the cooperativity in the $(SSB)_{35}$ mode has not been examined. Based on this, it has been proposed that the *ssb-113* mutation causes a defect in the interactions of the SSB-113 protein with other proteins involved in DNA metabolism (107, 109).

Eco *SSB Mutants that Affect the SSB-Binding Modes*

Curth et al (52) have recently reported that mutations at Trp54 influence the relative stability of the $(SSB)_{35}$-binding mode in vitro, and Carlini et al (104) have shown that one such mutant, *ssbW54S,* is defective in both replication and UV repair in vivo. SSBW54S protein binds poly(dT), although with lower affinity than wild-type; however, the most dramatic effect of this mutation is the increased relative stability of the $(SSB)_{35}$-binding mode (52). Whereas for wild-type SSB, the $(SSB)_{35}$ mode is observed in reverse titrations [addition of poly(dT) to SSB] only below 10 mM NaCl (25°C, pH 8.1) (100, 124, 125), a site size of 35 nucleotides is observed for SSBW54S at NaCl concentrations as high as 0.3 M (52). Therefore, the W54S mutation either stabilizes the $(SSB)_{35}$ mode or destabilizes the higher site-size binding modes. Although Curth et al (52) suggest that all four subunits, rather than only two, interact with poly(dT) in the SSBW54S mode with a site size of 35, conclusions based solely on the extent of fluorescence quenching in the complex can be misleading, and it is more likely that only two subunits of the tetramer interact with poly(dT) in the $(SSB)_{35}$ mode for the SSBW54S mutant, as is the case for the wild-type SSB tetramer (94, 95, 129). However, these biochemical and physiological studies support the suggestion that the different SSB-binding modes may be used preferentially in different metabolic processes (65, 86), and the identification of such mutants will be important for determining functional differences among the SSB-binding modes.

Curth et al (52) have also reported that at low [NaCl] ($<$ 10 mM), SSB mutants lacking Trp54 but containing Trp88 can form an additional binding mode on poly(dT) with a site size of $\sim$27$\pm$3 nucleotides per tetramer. The n = 27 mode is converted to the n = 33 mode upon further addition of poly(dT) (i.e. upon lowering the SSB density). The n = 27 mode is not observed with wild-type SSB (52, 100, 124, 125).

INTERACTIONS OF *Eco* SSB PROTEINS WITH OTHER PROTEINS

There is increasing evidence, although much of it indirect, that *Eco* SSB also functions through direct interactions with other proteins involved in DNA metabolism (107–109). *Eco* SSB stimulates the polymerase activity of both *E. coli* DNA polymerase II and pol III holoenzyme (2, 15, 192, 193). The T4 gene 32 protein does not stimulate these polymerases, although it does stimulate specifically the T4 DNA polymerase (2). Preliminary biochemical evidence also suggests interactions between *Eco* SSB and DNA polymerase II, exo I (192, 194), and the n protein (195). Recently, a folded chromosome-associated protein, as well as two other *E. coli* polypeptides, have been shown to bind to an SSB affinity column (196); however, these proteins have not yet been identified. Genetic evidence suggests potential interactions between *Eco* SSB and the Rep protein (190), Rho protein (197), and the GroEL proteins (198) from *E. coli*. The fact that the SSB-113 mutant protein shows no apparent change in its ssDNA-binding properties has also prompted the suggestion that the carboxy-terminal region is important for interactions with other proteins (107, 109). No direct interactions have been observed between the *Eco* SSB and RecA proteins, although it has been suggested that some interaction may occur when both are bound to ssDNA (199). In this context, we note that if an interaction between SSB and another protein is dependent upon ssDNA binding, then the binding mode used by the SSB protein may influence this interaction (65).

INFLUENCE OF THE *Eco* SSB MODES ON DNA METABOLISM

Since the ssDNA-binding properties of *Eco* SSB vary dramatically with its binding mode, one must consider whether only one particular binding mode is used for all DNA metabolic processes in *E. coli* (and if so, which one?) or whether different SSB modes might be used selectively in replication, recombination, or repair (65, 77, 86). Based on their biochemical properties in vitro, it is unlikely that each SSB-binding mode functions equivalently in all metabolic processes. For example, if SSB binding to ssDNA with "limited" cooperativity facilitates a particular repair process, then an SSB mode that binds with "unlimited" cooperativity would not likely substitute. However, if a major function of SSBs during replication is to saturate regions of ssDNA that are formed transiently, then the "unlimited" cooperativity $(SSB)_{35}$ mode would be best suited for this purpose (65, 135). On the other hand, it remains a possibility that "unlimited" cooperative binding is not necessary for the function of an SSB during replication. Although it has not been shown unambiguously that a particular SSB mode is used exclusively in any metabolic process, there is evidence that the ability of SSB to function in a particular process in vitro is influenced by its binding mode (104, 126, 200, 201). Clearly, systematic studies of the effects of the different SSB/ssDNA-binding modes on DNA replication, recombination, and repair in vitro will be needed to address this question. Such studies have not yet been performed with DNA replication or repair; however, a few such studies have addressed the role of the SSB modes in RecA-catalyzed reactions as discussed below.

Effect of SSB-Binding Modes on RecA Function

The RecA protein is the principal enzyme involved in homologous recombination in *E. coli,* and *Eco* SSB facilitates a number of processes known to be mediated by RecA (for reviews see Refs. 202–206). At least one important function of *Eco* SSB is its ability to stimulate the DNA strand exchange activity of RecA; however, its precise role in this process is not fully understood. Under solution conditions that are optimal for strand exchange (10–12 mM $MgCl_2$), *Eco* SSB facilitates RecA binding to ssDNA (126, 207). It has been presumed that this is due to the ability of SSB to destabilize the intramolecular DNA basepairing that forms within ssDNA at these high $MgCl_2$ concentrations, since RecA can bind ssDNA in the absence of SSB at lower (1 mM) $MgCl_2$ (207–209). This conclusion is supported by the qualitative observation that other SSBs, such as the T4 gene 32 protein, can also stimulate RecA activities, although the quantitative effect of T4 gene 32 protein is clearly different than that of *Eco* SSB (207, 210, 211), suggesting that other effects may also be involved. SSB also functions by inhibiting reinitiation of strand invasion by RecA, although this is believed to be an effect of SSB on the heteroduplex DNA rather than on the displaced ssDNA (212). There have been reports that RecA protein and SSB interact transiently in presynaptic complexes (199, 201), although the fact that RecA protein can completely displace SSB from ssDNA and vice versa suggests that the major effect of SSB results from its interaction with ssDNA (213).

Only a few studies have directly considered that the mode of SSB binding to ssDNA might influence the ability of SSB to stimulate RecA activity. These studies suggest that the "beaded," $(SSB)_{56}/(SSB)_{65}$ "limited" cooperativity mode stimulates RecA protein activity, whereas the smooth-contoured, "unlimited" cooperativity mode is inhibitory. Griffith et al (126) have shown that RecA protein filaments form more rapidly on ssDNA that is precoated with SSB in its "beaded," limited cooperativity [$(SSB)_{56}/(SSB)_{65}$] mode, compared to ssDNA that is precoated with SSB in its "smooth-contoured" [$(SSB)_{35}$] mode. In addition, the "beaded" SSB-ssDNA complex facilitates RecA-stimulation of LexA cleavage (126). Morrical & Cox (200) have also shown that SSB enhances RecA protein binding to ssDNA in 10 mM $MgCl_2$, conditions that favor SSB binding in its $(SSB)_{56}$ mode, whereas SSB inhibits RecA binding to ssDNA in 1 mM $MgCl_2$, conditions that favor SSB binding in its $(SSB)_{35}$ mode. In addition, the "beaded" $(SSB)_{56}/(SSB)_{65}$ modes formed in 12 mM $MgCl_2$ facilitate RecA-catalyzed strand exchange, whereas the "smooth-contoured," $(SSB)_{35}$ mode, formed in 1 mM $MgCl_2$, inhibits strand exchange (201). Therefore, it appears that the "beaded" $(SSB)_{56}/(SSB)_{65}$-binding mode facilitates RecA function in vitro; thus it may also function in this capacity in vivo.

Since any of the SSB-binding modes should be able to "melt out" DNA secondary structure by virtue of their higher affinity for ssDNA, it is possible that a particular SSB-binding mode may be required to allow RecA access to the ssDNA (65, 77, 122, 126, 130). In fact, the proposal has been made that SSB, when bound in its "beaded" $(SSB)_{56}/(SSB)_{65}$ mode, serves as an assembly factor for RecA binding to ssDNA by leaving gaps of protein-free ssDNA to which RecA can bind (126), possibly in addition to its role in destabilizing secondary structure in the ssDNA. This would be expected from the fact that in its "limited" cooperativity $(SSB)_{65}$ mode, regions of ssDNA are left accessible for RecA binding since SSB does not fully saturate the ssDNA even at high protein-binding densities (126, 130). On the other hand, when bound in the $(SSB)_{35}$ mode with "unlimited" cooperativity, SSB would be expected to inhibit RecA binding to ssDNA by fully occluding the ssDNA, even though the $(SSB)_{35}$ mode presumably is also able to remove secondary structure from the ssDNA. Therefore, in the absence of allowing RecA access to the DNA, the simple removal of DNA secondary structure by SSB may not be sufficient to facilitate RecA binding. In this context, it is also important to recognize that the ability of *Eco* SSB to compete with RecA for sites on ssDNA depends not only on the relative interaction constants and affinities of the two proteins (e.g. n, K, and ω), but also on the SSB-binding mode, since the latter influences not only n, K, and ω, but also the type of cooperativity used by SSB (i.e. limited vs unlimited).

Effects of SSB-Binding Modes on Replication and Repair

Recent studies of a site-directed SSB mutant in vivo suggest that at least one of the high site-size SSB-binding modes [$(SSB_{56}$ or $(SSB)_{65}$] functions in DNA metabolism. The SSBW54S mutant protein exhibits a strong preference for binding in the $(SSB)_{35}$ mode in vitro, even at 0.3 M NaCl (52), where the wild-type SSB tetramer binds to poly(dT) in the $(SSB)_{65}$ mode (100, 124). Carlini et al (104) have shown that although the SSBW54S mutant does complement an *ssb* deletion when *ssbW54S* is carried on a low-copy-number plasmid in vivo, slow growth rates and extreme UV-sensitivity result (104). Although it is premature to draw firm conclusions, such defects may be linked to a reduced ability of SSB to bind in the $(SSB)_{56}$- or $(SSB)_{65}$-binding modes. Further studies of the effects of SSB mutants that are defective in forming the different binding modes on the individual stages of replication, recombination, and repair in vitro and in vivo will be required to sort out the roles of the different SSB-binding modes in these processes.

OTHER SSB PROTEINS SHARING SIMILARITIES WITH *Eco* SSB

F-SSB and Conjugative Plasmid-Encoded SSB Proteins

Many conjugative plasmids, including the *E. coli* F episome (F sex factor) (17–19) and plasmids pIP231a, pIP71a, R64 (17, 21), and CoIIb-P9 (20), encode SSBs that are highly similar structurally to *Eco* SSB. The gene encoding F-SSB (*ssf*) partially complements a chromosomal *ssb-1* mutation (18) and complements a full deletion of the *ssb* gene when present on a high-copy-number plasmid (71); however, the growth properties of these cells suggest that F-SSB is partially defective in replication (71). The polypeptides encoded by these plasmids are very similar in size, ranging from 174 to 178 amino acids, and share extensive sequence similarities within the N-terminal DNA-binding region (e.g. 87 of the first 115 amino acids of F-SSB are identical to *Eco* SSB). Furthermore, the C-terminal tails of these proteins are all highly acidic, with six of the last seven residues being identical (Phe-X-Asp-Asp-Ile-Pro-Phe), including the penultimate proline that is changed to Ser in the *ssb-113* mutation (19, 109). This is particularly striking since the remainder of the C-terminal regions of the plasmid-encoded SSBs differ considerably from *Eco* SSB. Residues of known importance for *Eco* SSB function and/or DNA binding, such as Gly15, His55, Pro176, Trp40, Trp54, Trp88, and Phe60, are identical in all of the plasmid SSBs, along with numerous other residues within the N-terminal domain. The F-SSB is a stable tetramer in the absence of DNA at concentrations as low as 0.19 μM (tetramer) (M Ferrari, unpublished experiments), hence it is likely that the other plasmid SSBs also form tetramers, although this has not yet been demonstrated.

The plasmid SSBs are much less soluble than wild-type *Eco* SSB at low salt concentrations, hence ssDNA-binding studies have been limited to [NaCl] > 0.40 M (21, 101). However, under these high salt conditions, all of the plasmid proteins (21) and F-SSB (R Porter, J Fang, M Ferrari, T Lohman, unpublished) bind to poly(dT) with site sizes near 65 nucleotides per tetramer as observed with *Eco* SSB (100). F-SSB also displays negative cooperativity in its binding of $dT(pT)_{34}$ (J Fang, M Ferrari, TM Lohman, unpublished experiments), as observed for *Eco* SSB (94, 95, 129), which suggests that F-SSB is likely to form the $(SSB)_{35}$ mode at lower [NaCl]. Although the negative cooperativity parameter for $dT(pT)_{34}$ binding is the same for F-SSB and *Eco* SSB (σ_{35} = 0.50 at 0.6 M NaCl), the intrinsic binding constant K_{35} is a factor $\sim 10^3$ lower for F-SSB than for *Eco* SSB under the same conditions ($K_{35} = 1.2 \times 10^6$ M^{-1} for F-SSB). The pIP231a, pIP71a, and R64 SSBs seem to bind ssDNA with comparable affinities and more tightly than does F-SSB (21, 101). Interestingly, the IP231a, IP71a, and R64 SSBs are more effective at lowering the T_m of poly[d(A-T)] than wild-type SSB, even though their affinities for poly(dA) do not appear to be significantly different (21), hence these differences may reflect kinetic effects. R64, pIP231a, and pIP71a SSBs also stimulate DNA pol III holoenzyme elongation rates, whereas F-SSB does not (21).

Mitochondrial SSB Proteins

Mitochondrial (mt) SSBs isolated from several eukaryotes, including rat liver (p16 protein) (49, 51), human (214), *X. laevis* (72, 73), and yeast (*S. cerevisiae RIM1*) (13) also show striking similarities to the N-terminal DNA-binding regions of *Eco* SSB. The polypeptide lengths, as predicted from the gene sequences, are 132 and 135 for rat and human, respectively (51), 135 for yeast (13), and 129 for *X. laevis* (53); however, the mature protein isolated from yeast is only 118 amino acids, indicating that processing occurs (13). The *S. cerevisiae RIM1* gene encoding the mtSSB is essential for mtDNA replication (13).

Most of the N-terminal amino acids that are conserved among *Eco* SSB, F-SSB, and the plasmid-encoded SSBs are also conserved in the mtSSBs. Residues corresponding to Trp40, Trp54, His55, and Phe60 of *Eco* SSB are conserved in the rat, human, and *X. laevis* mt proteins (51); in fact, these proteins show greater sequence similarity to *Eco* SSB than to the yeast RIM1 protein (13, 51). For instance, in the RIM1 protein, the residue corresponding to Trp40 in *Eco* SSB is replaced by Arg (Arg36 in RIM1) and the residue corresponding to His55 in *Eco* SSB is replaced by Tyr (Tyr44 in RIM1). Recall that in *Eco* SSB, His55 is important for tetramer stability, being replaced by Tyr in *Eco* SSB-1 (85, 96, 128, 185). Interestingly, only the yeast RIM1 protein has a highly acidic C terminus similar to those in *Eco* SSB, F-SSB, and the plasmid-encoded SSBs, although

RIM1 is significantly smaller than *Eco* SSB (118 vs 177 amino acids) (13). Based on site-directed mutagenesis studies of the human mtSSB, Trp84 (corresponding to Trp54 in *Eco* SSB) is important for binding to poly(dT) (214).

Although the mtSSBs are smaller than *Eco* SSB (177 residues), preliminary studies indicate that they also form tetramers (13, 72, 214). The mt proteins bind tightly and preferentially to ssDNA, requiring ≥1 M NaCl for elution from ssDNA columns (13, 73), whereas both the rat (49) and *X. laevis* (73) mtSSBs have negligible affinity for duplex DNA. *X. laevis* mtSSB also binds preferentially to ss-homopolynucleotides containing pyrimidines, with highest affinity for poly(dT) (73), and in this respect is similar to most SSBs, including *Eco* SSB (15, 88, 132, 194). At high salt (0.1 to 1 M NaCl), a site size of 60±5 nucleotides per tetramer was measured for human mtSSB binding to poly(dT) using fluorescence titrations (214), which is similar to that found for *Eco* SSB under these conditions (100), although low solubility of the human mt protein precluded studies below 50 mM [NaCl]. On the other hand, a site size of 32–36 nucleotides per tetramer was reported for the rat mtSSB at lower salt (50 mM sodium phosphate, pH 7.7) (216), suggesting that the mtSSBs may undergo salt-dependent binding mode transitions similar to those observed with *Eco* SSB (100, 124, 125); however, this needs to be examined further.

SUMMARY

There are now several well-documented SSBs from both prokaryotes and eukaryotes that function in replication, recombination, and repair; however, no "consensus" view of their interactions with ssDNA has emerged. Although these proteins all bind preferentially and with high affinity to ssDNA, their modes of binding to ssDNA in vitro, including whether they bind with cooperativity, often differ dramatically. This point is most clear upon comparing the properties of the phage T4 gene 32 protein and the *E. coli* SSB protein. Depending on the solution conditions, *Eco* SSB can bind ssDNA in several different modes, which display quite different properties, including cooperativity. The wide range of interactions with ssDNA observed for *Eco* SSB is due principally to its tetrameric structure and the fact that each SSB protomer (subunit) can bind ssDNA. This reflects a major difference between *Eco* SSB and the T4 gene 32 protein, which binds DNA as a monomer and displays "unlimited" positive cooperativity in its binding to ssDNA. The *Eco* SSB tetramer can bind ssDNA with at least two different types of nearest-neighbor positive cooperativity ("limited" and "unlimited"), as well as negative cooperativity among the subunits within an individual tetramer. In fact, this latter property, which is dependent upon salt concentration and nucleotide base composition, is a major factor influencing whether ssDNA interacts with all four or only two SSB subunits, which in turn determines the type of intertetramer positive cooperativity. Hence, it is clear that the interactions of *Eco* SSB with ssDNA are quite different from those of T4 gene 32 protein, and the idea that all SSBs bind to ssDNA as does the T4 gene 32 protein must be amended. Although it is not yet known which of the *Eco* SSB-binding modes is functionally important in vivo, it is possible that some of the modes are used preferentially in different DNA metabolic processes. In any event, the vastly different properties of the *Eco* SSB-binding modes must be considered in studies of DNA replication, recombination, and repair in vitro. Since eukaryotic mitochondrial SSBs as well as SSBs encoded by prokaryotic conjugative plasmids are highly similar to *Eco* SSB, these proteins are likely to show similar complexities. However, based on their heterotrimeric subunit composition, the eukaryotic nuclear SSBs (RP-A proteins) are significantly different from either *Eco* SSB or T4 gene 32 proteins. Further subclassification of these proteins must await more detailed biochemical and biophysical studies.

ACKNOWLEDGMENTS

The preparation of this review was supported in part by NIH grant GM30498 to T.M.L. We thank Ron Porter for providing his manuscript prior to publication, Anna Goffinet for assistance with preparation of this article, and Lisa Lohman for assistance with some of the figures.

Literature Cited

1. Alberts BM, Frey L. 1970. *Nature* 227:1313–18
2. Sigal N, Delius H, Kornberg T, Gefter ML, Alberts BM. 1972. *Proc. Natl. Acad. Sci. USA* 69:3537–41
3. Molineux IJ, Friedman S, Gefter ML. 1974. *J. Biol. Chem.* 249:6090–98
4. Alberts B, Sternglanz R. 1977. *Nature* 269:655–61
5. Coleman JE, Oakley JL. 1980. *CRC Crit. Rev. Biochem.* 7:247–89
6. Gold L, O'Farrell PZ, Russel M. 1976. *J. Biol. Chem.* 251:7251–62

6a. Kim YT, Richardson CC. 1993. *Proc. Natl. Acad. Sci. USA* 90:10173–77

7. Meyer RR, Glassberg J, Kornberg A. 1979. *Proc. Natl. Acad. Sci. USA* 76:1702–5
8. Glassberg J, Meyer RR, Kornberg A. 1979. *J. Bacteriol.* 140:14–19
9. Meyer RR, Rein DC, Glassberg J. 1982. *J. Bacteriol.* 150:433–35
10. Porter RD, Black S, Pannuri S, Carlson A. 1990. *Bio/Technology* 8:47–51
11. Brill SJ, Stillman B. 1991. *Genes Dev.* 5:1589–600
12. Heyer W-D, Kolodner RD. 1989. *Biochemistry* 28:2856–62
13. Van Dyck E, Foury F, Stillman B, Brill SJ. 1992. *EMBO J.* 11:3421–30
14. Chase JW, Williams KR. 1986. *Annu. Rev. Biochem.* 55:103–36
15. Weiner JH, Bertsch LL, Kornberg A. 1975. *J. Biol. Chem.* 250:1972–80
16. de Vries J, Wackernagel W. 1993. *Gene* 127:39–45

17. Golub EI, Low KB. 1985. *J. Bacteriol.* 162:235–41
18. Kolodkin AL, Capage MA, Golub EI, Low KB. 1983. *Proc. Natl. Acad. Sci. USA* 80:4422–26
19. Chase JW, Merrill BM, Williams KR. 1983. *Proc. Natl. Acad. Sci. USA* 80:5480–84
20. Howland CJ, Rees CE, Barth PT, Wilkins BM. 1989. *J. Bacteriol.* 171: 2466–73
21. Ruvolo PP, Keating KM, Williams KR, Chase JW. 1991. *Proteins: Struct., Funct., Genet.* 9:120–34
22. Martin G, Salas M. 1988. *Gene* 67: 193–201
23. Martin G, Lazaro JM, Mendez E, Salas M. 1989. *Nucleic Acids Res.* 17:3663–72
24. Lindberg G, Kowalczykowski SC, Rist JK, Sugino A, Rothman-Denes LB. 1989. *J. Biol. Chem.* 264:12700–8
25. Scherzinger E, Litfin F, Jost E. 1973. *Mol. Gen. Genet.* 123:247–62
26. Reuben RC, Gefter ML. 1973. *Proc. Natl. Acad. Sci. USA* 70:1846–50
27. Araki H, Ogawa H. 1981. *Mol. Gen. Genet.* 183:66–73
28. Reuben RC, Gefter ML. 1974. *J. Biol. Chem.* 249:3843–50
29. Fairman MP, Stillman B. 1988. *EMBO J.* 7:1211–18
30. Wobbe CR, Weissbach L, Borowiec JA, Dean FB, Murakami Y, et al. 1987. *Proc. Natl. Acad. Sci. USA* 84:1834–38
31. Wold MS, Kelly T. 1988. *Proc. Natl. Acad. Sci. USA* 85:2523–27
32. Brill SJ, Stillman B. 1989. *Nature* 342:92–95
33. Mitsis PG, Kowalczykowski SC, Lehman IR. 1993. *Biochemistry* 32:5257–66
34. Atrazhev A, Zhang S, Grosse F. 1992. *Eur. J. Biochem.* 210:855–65
35. Heyer W-D, Rao MRS, Erdile LF, Kelly TJ, Kolodner RD. 1990. *EMBO J.* 9:2321–29
36. Hurwitz J, Dean FB, Kwong AD, Lee SH. 1990. *J. Biol. Chem.* 265:18043–46
37. Stillman B. 1992. *DNA Replicaton and the Cell Cycle,* pp. 127–43. Berlin: Springer-Verlag
38. Moore SP, Erdile L, Kelly T, Fishel R. 1991. *Proc. Natl. Acad. Sci. USA* 88:9067–71
39. Alani E, Thresher R, Griffith JD, Kolodner RD. 1992. *J. Mol. Biol.* 227:54–71
40. Coverley D, Kenny MK, Munn M, Rupp WD, Lane DP, Wood RD. 1991. *Nature* 349:538–41
41. Coverley D, Kenny MK, Lane DP, Wood RD. 1992. *Nucleic Acids Res.* 20:3873–80
42. Dornreiter I, Erdile LF, Gilbert IU, von Winkler D, Kelly TJ, Fanning E. 1992. *EMBO J.* 11:769–76
43. Melendy T, Stillman B. 1993. *J. Biol. Chem.* 268:3389–95
44. Erdile LF, Heyer W-D, Kolodner R, Kelly TJ. 1991. *J. Biol. Chem.* 266: 12090–98
45. Kenny MK, Lee S-H, Hurwitz J. 1989. *Proc. Natl. Acad. Sci. USA* 86:9757–61
46. Tsurimoto T, Stillman B. 1989. *EMBO J.* 8:3883–89
47. Tsuji M, van der Vliet PC, Kitchingman GR. 1991. *J. Biol. Chem.* 266: 16178–87
48. Matsumoto T, Eki T, Hurwitz J. 1990. *Proc. Natl. Acad. Sci. USA* 87:9712–16
49. Pavco PA, Van Tuyle GC. 1985. *J. Cell. Biol.* 100:258–64
50. Van Tuyle GC, Pavco PA. 1985. *J. Cell. Biol.* 100:251–57
51. Tiranti V, Rocchi M, DiDonato S, Zeviani M. 1993. *Gene* 126:219–25
52. Curth U, Greipel J, Urbanke C, Maass G. 1993. *Biochemistry* 32:2585–91
53. Tiranti V, Barat-Gueride M, Bijl J, DiDonato S, Zeviani M. 1991. *Nucleic Acids Res.* 19:4291
54. Ghrir R, Lecaer J-P, Dufresne C, Gueride M. 1991. *Arch. Biochem. Biophys.* 291:395–400
55. Bayliss GJ, Marsden HS, Hay J. 1975. *Virology* 68:124–34
56. Schaffer PA, Bone DR, Courtney RJ. 1976. *J. Virol.* 17:1043–48
57. Gao M, Knipe DM. 1993. *J. Virol.* 67:876–85
58. Bortner C, Hernandez TR, Lehman IR, Griffith J. 1993. *J. Mol. Biol.* 231:241–50
59. Williams KR, Chase JW. 1990. *The Biology of Nonspecific DNA-Protein Interactions,* ed. A Revzin, pp. 197–227. Boca Raton, Fla: CRC Press
60. Challberg MD. 1986. *Proc. Natl. Acad. Sci. USA* 83:9094–98
61. Challberg MD, Kelly TJ. 1989. *Annu. Rev. Biochem.* 58:671–717
62. van der Vliet PC, Keegstra W, Jansz HS. 1978. *Eur. J. Biochem.* 86:389–98
63. Ensinger MJ, Ginsberg HS. 1972. *J. Virol.* 10:328–39
64. Lindenbaum JO, Field J, Hurwitz J. 1986. *J. Biol. Chem.* 261:10218–27
65. Lohman TM, Bujalowski W, Overman LB. 1988. *Trends Biochem. Sci.* 13: 250–55
66. Kowalczykowski SC, Bear DG, von Hippel PH. 1981. *The Enzymes* 14: 373–444

67. Karpel RL. 1990. See Ref. 59, pp. 103–30
68. Ruyechan WT, Wetmur JG. 1975. *Biochemistry* 14:5529–34
69. Alberts B, Frey L, Delius H. 1972. *J. Mol. Biol.* 68:139–52
70. Cavalieri SJ, Neet KE, Goldthwait DA. 1976. *J. Mol. Biol.* 102:697–711
71. Porter RD, Black S. 1991. *J. Bacteriol.* 173:2720–23
72. Barat M, Mignotte B. 1981. *Chromosoma* 82:583–93
73. Mignotte B, Barat M, Mounolou J-C. 1985. *Nucleic Acids Res.* 13:1703–16
74. Meyer RR, Laine PS. 1990. *Microbiol. Rev.* 54:342–80
75. Markiewicz P, Malone C, Chase JW, Rothman-Denes LB. 1992. *Genes Dev.* 6:2010–19
76. Williams KR, Konigsberg W. 1981. *Gene Amplification and Analysis,* ed. T Chirikjian, T Papas, pp. 475–508. North Holland/New York: Elsevier
77. Lohman TM, Bujalowski W. 1990. See Ref. 59, pp. 131–68
78. Greipel J, Urbanke C, Maass G. 1989. *Protein-Nucleic Acid Interaction,* ed. W Saenger, U Heinemann, pp. 61–86. Boca Raton, Fla: CRC Press
79. Sancar A, Rupp WD. 1979. *Biochem. Biophys. Res. Commun.* 90:123–29
80. Sancar A, Williams KR, Chase JW, Rupp WD. 1981. *Proc. Natl. Acad. Sci. USA* 78:4274–78
81. Chase JW, Whittier RF, Auerbach J, Sancar A, Rupp WD. 1980. *Nucleic Acids Res.* 8:3215–27
82. Williams KR, Murphy JB, Chase JW. 1984. *J. Biol. Chem.* 259:11804–11
83. Lohman TM, Green JM, Beyer S. 1986. *Biochemistry* 25:21–25
84. Bayer I, Fliess A, Greipel J, Urbanke C, Maass G. 1989. *Eur. J. Biochem.*179:399–404
85. Bujalowski W, Lohman TM. 1991. *J. Biol. Chem.* 266:1616–26
86. Lohman TM, Overman LB, Datta S. 1986. *J. Mol. Biol.* 187:603–15
87. Ruyechan WT, Wetmur JG. 1976. *Biochemistry* 15:5057–64
88. Overman LB, Bujalowski W, Lohman TM. 1988. *Biochemistry* 27:456–71
89. Ollis D, Brick P, Abdel-Meguid SS, Murthy K, Chase JW, Steitz TA. 1983. *J. Mol. Biol.* 170:797–800
90. Monzingo AF, Christiansen C. 1983. *J. Mol. Biol.* 170:801
91. Hilgenfeld R, Saenger W, Schomburg U, Krauss G. 1984. *FEBS Lett.* 170:-143–46
92. Ng JD, McPherson A. 1989. *J. Biomol. Struct. Dyn.* 6:1071–76
93. Williams KR, Spicer EK, LoPresti MB, Guggenheimer RA, Chase JW. 1983. *J. Biol. Chem.* 258:3346–55
94. Bujalowski W, Lohman TM. 1989. *J. Mol. Biol.* 207:249–68
95. Bujalowski W, Lohman TM. 1989. *J. Mol. Biol.* 207:269–88
96. Bujalowski W, Lohman TM. 1991. *J. Mol. Biol.* 217:63–74
97. Anderson RA, Coleman JE. 1975. *Biochemistry* 14:5485–91
98. Bandyopadhyay PK, Wu C-W. 1978. *Biochemistry* 17:4078–85
99. Molineux IJ, Pauli A, Gefter ML. 1975. *Nucleic Acids Res.* 2:1821–37
100. Lohman TM, Overman LB. 1985. *J. Biol. Chem.* 260:3594–603
101. Casas-Finet JR, Khamis MI, Maki AH, Ruvolo PP, Chase JW. 1987. *J. Biol. Chem.* 262:8574–83
102. Khamis MI, Casas-Finet JR, Maki AH, Murphy JB, Chase JW. 1987. *J. Biol. Chem.* 262:10938–45
103. Khamis MI, Casas-Finet JR, Maki AH, Ruvolo PP, Chase JW. 1987. *Biochemistry* 26:3347–54
104. Carlini LE, Porter RD, Curth U, Urbanke C. 1993. *Mol. Microbiol.* In press
105. Merrill BM, Williams KR, Chase JW, Konigsberg WH. 1984. *J. Biol. Chem.* 259:10850–56
106. Casas-Finet JR, Khamis MI, Maki AH, Chase JW. 1987. *FEBS Lett.* 220:347–52
107. Chase JW, L'Italien JJ, Murphy JB, Spicer EK, Williams KR. 1984. *J. Biol. Chem.* 259:805–14
108. Chase JW, Flory J, Ruddle NH, Murphy JB, Williams KR. 1985. *J. Biol. Chem.* 260:7214–18
109. Chase JW. 1984. *BioEssays* 1:218–22
110. Williams KR, LoPresti MB, Setoguchi M, Konigsberg WH. 1980. *Proc. Natl. Acad. Sci. USA* 77:4614–17
111. Dunn J, Studier F. 1981. *J. Mol. Biol.* 148:303–30
112. Moise H, Hosoda J. 1976. *Nature* 259:455–58
113. Hosoda J, Takacs B, Brack C. 1974. *FEBS Lett.* 47:338–42
114. Greve J, Maestre MF, Moise H, Hosoda J. 1978. *Biochemistry* 17:893–98
115. Hosoda J, Moise H. 1978. *J. Biol. Chem.* 253:7547–55
116. Araki H, Ogawa H. 1981. *Mol. Gen. Genet.* 183:66–73
117. Hurley JM, Chervitz SA, Jarvis TC, Singer BS, Gold L. 1993. *J. Mol. Biol.* 229:398–418
118. Jiang H, Giedroc D, Kodadek T. 1993. *J. Biol. Chem.* 268:7904–11

119. McGhee JD, von Hippel PH. 1974. *J. Mol. Biol.* 86:469–89
120. McGhee JD, von Hippel PH. 1976. *J. Mol. Biol.* 103:679
121. Lohman TM, Mascotti DP. 1992. *Methods Enzymol.* 212:424–58
122. Chrysogelos S, Griffith J. 1982. *Proc. Natl. Acad. Sci. USA* 79:5803–7
123. Krauss G, Sindermann H, Schomburg U, Maass G. 1981. *Biochemistry* 20: 5346–52
124. Bujalowski W, Lohman TM. 1986. *Biochemistry* 25:7799–802
125. Bujalowski W, Overman LB, Lohman TM. 1988. *J. Biol. Chem.* 263:4629–40
126. Griffith JD, Harris LD, Register J III. 1984. *Cold Spring Harbor Symp. Quant. Biol.* 49:553–59
127. Shimamoto N, Ikushima N, Utiyama H, Tachibana H, Horie K. 1987. *Nucleic Acids Res.* 15:5241–50
128. Curth U, Bayer I, Greipel J, Mayer F, Urbanke C, Maass G. 1991. *Eur. J. Biochem.* 196:87–93
129. Lohman TM, Bujalowski W. 1988. *Biochemistry* 27:2260–65
130. Bujalowski W, Lohman TM. 1987. *J. Mol. Biol.* 195:897–907
131. Kuil ME, Holmlund K, Vlaanderen CA, van Grondelle R. 1990. *Biochemistry* 29:8184–89
132. Bobst EV, Perrino FW, Meyer RR, Bobst AM. 1991. *Biochim. Biophys. Acta* 1078:199–207
133. Wei T-F, Bujalowski W, Lohman TM. 1992. *Biochemistry* 31:6166–74
134. Schwarz G, Stankowski S. 1979. *Biophys. Chem.* 10:173–81
135. Ferrari ME, Bujalowski W, Lohman TM. 1994. *J. Mol. Biol.* 235:In press
136. Schaper A, Urbanke C, Maass G. 1991. *J. Biomol. Struct. Dyn.* 8:1211–32
137. Greipel J, Maass G, Mayer F. 1987. *Biophys. Chem.* 26:149–61
138. Boidot-Forget M, Saison-Behmoaras T, Toulme J, Helene C. 1986. *Biochimie* 68:1129–34
139. Overman LB, Lohman TM. 1994. *J. Mol. Biol.* 235:In press
140. Overman LB. 1989. *Thermodynamic characterization of E. coli single strand binding protein—single strand polynucleotide interactions*. PhD thesis. Texas A&M Univ.
141. Record MT, Lohman TM, de Haseth PL. 1976. *J. Mol. Biol.* 107:145–58
142. Record MT, Anderson CF, Lohman TM. 1978. *Q. Rev. Biophys.* 11:103–78
143. Record MT Jr., Ha J-H, Fisher MA. 1991. *Methods Enzymol.* 208:291–343
144. Lohman TM. 1986. *CRC Crit. Rev. Biochem.* 19:191–245
145. Lohman TM, Mascotti DP 1992. *Methods Enzymol.* 212:400–24
146. Manning GS. 1978. *Q. Rev. Biophys.* 11:179–246
147. Mascotti DP, Lohman TM. 1990. *Proc. Natl. Acad. Sci. USA* 87:3142–46
148. Mascotti DP, Lohman TM. 1992. *Biochemistry* 31:8932–46
149. de Haseth PL, Lohman TM, Record MT. 1977. *Biochemistry* 16:4783–90
150. Record MT, de Haseth PL, Lohman TM. 1977. *Biochemistry* 16:4791–96
151. Leirmo SL, Harrison C, Cayley S, Burgess RR, Record MT Jr. 1987. *Biochemistry* 26:2095–101
152. Kowalczykowski SC, Lonberg N, Newport JW, von Hippel PH. 1981. *J. Mol. Biol.* 145:75–104
153. Newport JW, Lonberg N, Kowalczykowski SC, von Hippel PH. 1981. *J. Mol. Biol.* 145:105–21
154. Lohman TM. 1984. *Biochemistry* 23: 4656–65
155. Lohman TM. 1984. *Biochemistry* 23: 4665–75
156. Clore GM, Gronenborn AM, Greipel J, Maass G. 1986. *J. Mol. Biol.* 187:119–24
157. Lonberg N, Kowalczykowski SC, Paul L, von Hippel PH. 1981. *J. Mol. Biol.* 145:123–38
158. Giedroc DP, Khan R, Barnhart K. 1990. *J. Biol. Chem.* 265:1444–55
159. Gray CW. 1989. *J. Mol. Biol.* 208:57–64
160. Schellman JA. 1974. *Isr. J. Chem.* 12:219–38
161. Zasedatelev AS, Gurskii GV, Volkenshtein MV. 1971. *Mol. Biol.* 5:194–98
162. Epstein IR. 1978. *Biophys. Chem.* 8:327–39
163. Schellman JA. 1980. *Molecular Structure and Dynamics,* ed. M Balaban, pp. 245–65. Rehovot, Israel: Int. Sci. Serv.
164. Kowalczykowski SC, Paul LS, Lonberg N, Newport JW, McSwiggen JA, von Hippel PH. 1986. *Biochemistry* 25:1226–40
165. Kowalczykowski SC, Paul LS, Lonberg N, Newport JW, McSwiggen JA, von Hippel PH. 1986. *Biochemistry* 25:8473
166. Bujalowski W, Lohman TM, Anderson CF. 1989. *Biopolymers* 28:1637–43
167. Kelly RC, von Hippel PH. 1976. *J. Biol. Chem.* 251:7229–39
168. Draper DE, von Hippel PH. 1978. *J. Mol. Biol.* 122:339–59
169. Lohman TM. 1983. *Biopolymers* 22: 1697–713

170. Schwarz G, Watanabe F. 1983. *J. Mol. Biol.* 163:467–84
171. Bujalowski W, Lohman TM. 1987. *Biochemistry* 26:3099–106
172. Urbanke C, Schaper A. 1990. *Biochemistry* 29:1744–49
173. Romer R, Schomburg U, Krauss G, Maass G. 1984. *Biochemistry* 23:6132–37
174. Schneider RJ, Wetmur JG. 1982. *Biochemistry* 21:608–15
175. Richter PH, Eigen M. 1974. *Biophys. Chem.* 2:255–63
176. Berg OG, Winter RB, von Hippel PH. 1981. *Biochemistry* 20:6929–48
177. Lohman TM, Bujalowski W. 1991. *Methods Enzymol.* 208:258–90
178. Bobst AM, Langemeier PW, Warwick-Koochaki PE, Bobst EV, Ireland JC. 1982. *J. Biol. Chem.* 257:6184–93
179. Porschke D, Rau H. 1983. *Biochemistry* 22:4737–45
180. Bobst A, Ireland J, Bobst E. 1984. *J. Biol. Chem.* 259:2130–34
181. Bulsink H, Harmsen BJM, Hilbers CW. 1985. *J. Biomol. Struct. Dyn.* 3:227–47
182. Sang B-C, Gray DM. 1989. *J. Biomol. Struct. Dyn.* 7:693–706
183. Mascotti DP, Lohman TM. 1993. *Biochemistry* 32:10568–79
184. Meyer RR, Glassberg J, Scott JV, Kornberg A. 1980. *J. Biol. Chem.* 255:2897–901
185. Chase JW, Murphy JB, Whittier RF, Lorensen E, Sninsky JJ. 1983. *J. Mol. Biol.* 164:193–211
186. Johnson BF. 1977. *Mol. Gen. Genet.* 157:91–97
187. Vales LD, Chase JW, Murphy JB. 1980. *J. Bacteriol.* 143:887–96
188. Schmellik-Sandage CS, Tessman ES. 1990. *J. Bacteriol.* 172:4378–85
189. Auerbach JI, Howard-Flanders P. 1981. *J. Bacteriol.* 146:713–17
190. Tessman ES, Peterson PK. 1982. *J. Bacteriol.* 152:572–83
191. Cayley S, Lewis BA, Guttman HJ, Record MT Jr. 1991. *J. Mol. Biol.* 222:281–300
192. Molineux IJ, Gefter ML. 1974. *Proc. Natl. Acad. Sci. USA* 71:3858–62
193. Fay PJ, Johanson KO, McHenry CS, Bambara RA. 1982. *J. Biol. Chem.* 257:5692–99
194. Molineux IJ, Gefter ML. 1975. *J. Mol. Biol.* 98:811–25
195. Low RL, Shlomai J, Kornberg A. 1982. *J. Biol. Chem.* 257:6242–50
196. Perrino FW, Meyer RR, Bobst AM, Rein DC. 1988. *J. Biol. Chem.* 263: 11833–983
197. Fassler JS, Tessman I, Tessman ES. 1985. *J. Bacteriol.* 161:609–14
198. Ruben SM, VanDenBrink-Webb SE, Rein DC, Meyer RR. 1988. *Proc. Natl. Acad. Sci. USA* 85:3767–71
199. Morrical SW, Lee J, Cox MM. 1986. *Biochemistry* 25:1482–94
200. Morrical SW, Cox MM. 1990. *Biochemistry* 29:837–43
201. Muniyappa K, Williams K, Chase JW, Radding CM. 1990. *Nucleic Acids Res.* 18:3967–73
202. Cox MM, Lehman IR. 1987. *Annu. Rev. Biochem.* 56:229–62
203. Radding CM. 1989. *Biochim. Biophys. Acta* 1008:131–45
204. Kowalczykowski SC. 1991. *Annu. Rev. Biophys. Chem.* 20:539–75
205. Roca AI, Cox MM. 1990. *CRC Crit. Rev. Biochem. Mol. Biol.* 25: 415–56
206. Griffith JD, Harris LD. 1988. *CRC Crit. Rev. Biochem.* 23:S43-S86
207. Kowalczykowski SC, Krupp RA. 1987. *J. Mol. Biol.* 193:97–113
208. Muniyappa K, Shaner SL, Tsang SS, Radding CM. 1984. *Proc. Natl. Acad. Sci. USA* 81:2757–61
209. Tsang SS, Muniyappa K, Azhderian E, Gonda DK, Radding CM, Flory J. 1985. *J. Mol. Biol.* 185:295–309
210. Shibata T, DasGupta C, Cunningham RP, Radding CM. 1980. *Proc. Natl. Acad. Sci. USA* 77:2606–10
211. Egner C, Azhderian E, Tsang SS, Radding CM, Chase JW. 1987. *J. Bacteriol.* 169:3422–3428
212. Chow SA, Rao BJ, Radding CM. 1988. *J. Biol. Chem.* 263:200–9
213. Kowalczykowski SC, Clow J, Somani R, Varghese A. 1987. *J. Mol. Biol.* 193:81–95
214. Curth U, Maass G, Urbanke C, Greipel J, Gerberding H, et al. 1993. *J. Biomol. Struct. Dyn.* 10:a035
215. Deleted in proof
216. Hoke GD, Pavco PA, Ledwith BJ, Van Tuyle GC. 1990. *Arch. Biochem. Biophys.* 282:116–24
217. Salstrom JS, Pratt D. 1971. *J. Mol. Biol.* 61:489–501
218. Mazur BZ, Model P. 1973. *J. Mol. Biol.* 78:285–300
219. Kornberg A, Baker TA. 1992. *DNA Replication*. New York: Freeman

Annu. Rev. Biochem. 1994. 63:571–600

THE BIOCHEMISTRY OF SYNAPTIC REGULATION IN THE CENTRAL NERVOUS SYSTEM

Mary B. Kennedy
Division of Biology 216-76, California Institute of Technology, Pasadena, California 91125

KEY WORDS: long-term potentiation, synaptic release, neuronal receptors, protein phosphorylation, nitric oxide

---

CONTENTS

WHY IS SYNAPTIC REGULATION IMPORTANT?

Human thought processes are exquisitely sensitive to past experiences. In 1949, D. O. Hebb hypothesized that electrical activity produced in our brains by experiences changes the properties of the active neurons so that their connections and patterns of firing are altered (1). The basic idea that

0066-4154/94/0701-0571$05.00

neural plasticity underlies processing and storage of information in the brain continues to motivate much of current research on adult brain function. The best physiological evidence available today suggests that two major kinds of alterations are responsible for neural plasticity produced by prior electrical activity: changes in the intrinsic membrane properties of neurons and changes in the strength of synaptic transmission between neurons. Changes in intrinsic membrane properties usually reflect subtle modifications of the numbers and kinetic behavior of ion channels in the neuronal membrane (2). Changes in synaptic strength occur in two major ways. The presynaptic release machinery can be modified to change the amount of neurotransmitter released from presynaptic terminals by each electrical impulse (3, 4). Additionally, postsynaptic receptors or voltage-dependent ion channels can be modified to alter the sensitivity of the receiving neuron to released transmitter (5, 6). Evidence is growing that transmitter release and receptor function can both be modulated by a wide variety of distinct biochemical mechanisms. Furthermore, the mammalian central nervous system contains a large, as-yet-undefined number of different neuronal cell types. The extent to which different types of neurons possess distinct mechanisms for resetting their synapses is unknown. A major challenge for biochemists is to unravel the nature and cellular distribution of the many molecular mechanisms that are responsible for synaptic modulation.

In this review, I introduce some of the major issues in this burgeoning field, then I focus primarily on molecular aspects of recent studies of a form of synaptic regulation called associative long-term potentiation, or LTP. LTP is an enhancement of synaptic efficacy lasting hours to weeks that is induced experimentally by relatively brief (one second or less) high-frequency synaptic stimulation (7). LTP has attracted considerable attention because it may underlie early stages of memory formation in the hippocampus, a mammalian brain structure that plays a critical role in formation of new memories. The possible relationship of LTP to memory has fueled a vast effort to understand its molecular mechanism. Several excellent reviews of pharmacological and physiological studies of LTP have appeared recently (4, 8–11).

WHAT WE KNOW AND WHAT WE DON'T KNOW

Classes of Synapses in the Central Nervous System

EXCITATORY AND INHIBITORY SYNAPSES One major way that CNS synapses are classified is according to the identity and action of their principal chemical transmitter. Excitatory synapses are those at which the chemical transmitter activates receptors that produce depolarization of the membrane potential, usually by opening ion channels that permit the flow of the positively charged cations K^+, Na^+, or Ca^{2+}. Depolarization increases the probability that the neuron will fire an electrical impulse. It is now generally agreed that the amino acid glutamate is the principal excitatory neurotransmitter in the central nervous system; synapses at which glutamate is the transmitter are referred to as glutamatergic. Inhibitory synapses are those at which the chemical transmitter activates receptors that produce hyperpolarization of the membrane potential or make the membrane more difficult to depolarize, usually by opening ion channels that permit the flow of Cl^- ion across the membrane. Hyperpolarization decreases the probability that the neuron will fire an electrical impulse. The principal inhibitory transmitter in the CNS is γ-amino butyric acid, or GABA, which is produced by decarboxylation of glutamic acid. Synapses at which GABA is the transmitter are referred to as GABAergic. The CNS contains additional excitatory (e.g. acetylcholine) and inhibitory (e.g. glycine) transmitters. However, the major relay synapses that transmit and process sensory information and motor output are most often glutamatergic or GABAergic.

MODULATORY SYNAPSES A large number of brain neurotransmitters, including the family of peptide neurohormones, and the biogenic amines—noradrenaline, dopamine, and serotonin—initiate biochemical events that modulate synaptic transmission or membrane excitability rather than excite or inhibit postsynaptic neurons directly. Modulatory influences are usually mediated by the family of "metabotropic" receptors, a group of seven-transmembrane-domain receptors that are coupled to the activation of trimeric G-proteins (12). Activation of particular G-proteins by these receptors can enhance synthesis of cAMP, IP_3, diacylglycerol, and the release of internal Ca^{2+} stores (13). These second messengers in turn stimulate biochemical alterations of synaptic and other neuronal proteins. In some cases, subunits of the activated G-proteins bind directly to channels and regulate their function (14). The remarkable complexity and subtlety of synaptic function in the central nervous system is exemplified by the fact that, although they are usually packaged and released from different vesicle populations, modulatory transmitters are often coreleased from the same presynaptic terminals that release excitatory or inhibitory transmitters. Furthermore, glutamate, GABA, and acetylcholine sometimes themselves act as modulatory transmitters depending upon the nature of their postsynaptic receptors (15).

Over the past five years, genes encoding many of the molecular components of the presynaptic release machinery (16, 17) as well as the subunits of postsynaptic receptors (18–21) have been cloned. Present efforts are directed at identifying additional molecular components of synapses, and understanding how these components interact to mediate synaptic function, and how these functions are regulated. Ultimately, it will be important to understand how distinct modulatory influences interact to control synaptic strength at each different class of CNS synapse.

Forms of Usage-Dependent Synaptic Regulation

SHORT-TERM PLASTICITY Mechanisms of synaptic regulation are usually classified as short-term if they produce effects that last for less than about 20 minutes, and long-term if they last as long as about 20 minutes to days or weeks. Many modulatory transmitters act locally at specific synapses to produce short-term "heterosynaptic" changes in synaptic strength (reviewed recently in Ref. 4). In addition, at most synapses, electrical activity itself results in transient changes in synaptic strength. Three commonly studied forms of short-term, usage-dependent regulation are paired-pulse facilitation, post-tetanic potentiation, and depression (22). Paired-pulse facilitation is an augmentation of the second synaptic potential that occurs when a synapse is activated by a pair of pulses spaced no longer than about half a second apart. Post-tetanic potentiation refers to a more pronounced enhancement of the synaptic potential that is induced by a brief high-frequency train of impulses (a tetanus) and lasts for two to five minutes thereafter (23). Both paired-pulse facilitation and post-tetanic potentiation are believed to be caused by a buildup of residual calcium in the presynaptic terminal that permits more transmitter release per impulse (24, 25). The slower decay of post-tetanic potentiation may reflect a slow phase of calcium removal. In some synapses, post-tetanic potentiation may be caused, at least in part, by a different mechanism, involving reversible phosphorylation of synapsin I, a synaptic vesicle-associated protein (26). Short-term depression is a decline in synaptic strength produced by repetitive stimulation and lasting less than a minute. It is believed to reflect progressive depletion of releasable transmitter stores and predominates over facilitation in some terminals (22).

LONG-TERM PLASTICITY Long-term readjustments of synaptic strength are usually initiated by specific patterns of synaptic activity. Both long-term potentiation (LTP) and long-term depression (LTD) have been observed. They occur in a restricted number of synapses, and appear to have distinct mechanisms of initiation and maintenance in different synapses. Cerebellar synapses between parallel fibers and Purkinje neurons (27), hippocampal synapses in the dentate gyrus and area CA1 (28–35), and neurons in the cerebral cortex (36, 37) display pronounced long-term depression that is highly dependent on the patterns of presynaptic and postsynaptic activity. The mechanism of LTD has only recently been studied intensively (35, 38–40). Long-term potentiation (LTP, defined above) has been documented at the crayfish neuromuscular junction (41), at synapses within sympathetic ganglia (42), at glutamatergic synapses in the hippocampus (7, 9), and in neocortex (43). Although it was once considered a single phenomenon, it now seems clear that LTP encompasses a set of mechanisms that differ both in the molecular events that initiate potentiation and in those that maintain it. Attempts to determine the individual mechanisms of "sub-forms" of LTP are the subject of this review.

A Biochemist's Introduction to Quantal Analysis

Because synaptic transmission can be regulated at both presynaptic and postsynaptic sites, an important first step in the study of a modulatory mechanism is to determine whether its effect is in the presynaptic or postsynaptic compartment, or in both. A method of statistical analysis termed "quantal analysis" is often used to make this determination. The method was invented by Katz and his coworkers, who used it in their electrophysiological studies of transmission at the neuromuscular junction (44); it has been elaborated and refined since then by a variety of investigators (45, 46).

QUANTAL RELEASE OF NEUROTRANSMITTER It is now widely accepted that transmitter release from presynaptic terminals occurs when small membrane-bounded vesicles loaded with transmitter fuse with specialized regions of the presynaptic terminal called release sites and discharge their contents into the extracellular space immediately adjacent to postsynaptic receptor proteins. The fusion event is triggered by the influx of Ca^{2+} that occurs when voltage-gated Ca^{2+} channels in the terminals are opened in response to depolarization of the terminal by an action potential in the presynaptic axon. Quantal analysis derives from a statistical model of this process based upon physiological and anatomical studies of neuromuscular junctions.

Individual motor neuron terminals contain a large number of vesicles, some of which are "docked" at release sites at the presynaptic membrane opposite clusters of postsynaptic receptors. According to the model, when the terminal is not electrically active the probability that any vesicle will fuse with the presynaptic membrane and release its contents is extremely small. However, when the terminal is depolarized by an action potential, the resulting influx of Ca^{2+} produces an increase in the probability of vesicle fusion at release sites. One packet of acetylcholine, released by one vesicle, causes a small transient current of relatively uniform size through the postsynaptic acetylcholine receptors. These "quanta" of current can be observed and measured easily by reducing the concentration of extracellular Ca^{2+}, thus lowering the average probability of electrically evoked fusion so that only a few vesicles release their contents with each nerve impulse.

As the concentration of extracellular Ca^{2+} is increased, the probability of electrically evoked vesicle fusion increases in parallel. A statistical analysis of the amplitudes of these larger composite postsynaptic currents shows that they are always an integral multiple of the single quantal currents, when small variations in the sizes of the quantal currents are taken into account. The mean postsynaptic current is equal to qNp, where q is the mean amplitude of one quantal current produced by fusion of a single synaptic vesicle, N is the number of release sites, and p is the average probability of fusion at release sites after each electrical impulse. This model predicts that fluctuations of postsynaptic current amplitudes around the mean will follow binomial statistics; the prediction is borne out at the neuromuscular junction (44).

PRESYNAPTIC OR POSTSYNAPTIC? In addition to providing a concrete model of synaptic transmission, quantal analysis provides a way to study mechanisms underlying changes in synaptic efficacy. By measuring the statistical properties of changes in endplate currents that accompany a particular change in synaptic strength, it is possible to deduce whether the principal change is in the parameter q, N, p, or some combination of these (45). A change in q reflects a change in the quantal current produced by release of transmitter from a single vesicle. This change could be produced presynaptically by altering the amount of transmitter loaded into each vesicle, or postsynaptically by altering the current produced by binding of one quantum of transmitter to receptors. In practice, measured changes in q have usually been accounted for by alterations in the response of postsynaptic receptors or channels. Therefore, a change in q has been assumed to indicate a postsynaptic mechanism of synaptic plasticity. In contrast, a change in the parameters N or p necessarily reflects changes in the presynaptic release mechanism; either the creation of new release sites (N), or an alteration in the probability of release produced by synaptic depolarization (p) (45, 47). Quantal analysis has been used to establish that paired-pulse facilitation and post-tetanic potentiation at the neuromuscular junction are primarily presynaptic phenomena (22).

Differences Between the Neuromuscular Junction (the Traditional Model) and CNS Synapses

The validity of quantal analysis as a method for determining the locus of a change in synaptic efficacy depends upon a number of assumptions that are true for the neuromuscular junction and for some inhibitory synapses, but that may not hold for most synapses in the mammalian central nervous system (45–48). One of the most important of these assumptions is that q, the quantal size, is determined by the fixed amount of transmitter released by a single vesicle. This condition holds only if three other assumptions are true. Individual vesicles must contain approximately the same amount of transmitter and they must fuse with the presynaptic membrane independently of each other. In addition, postsynaptic receptors must not be saturated by the contents of a single vesicle. The validity of these assumptions at some central nervous system synapses has been questioned.

QUANTAL VARIANCE One important difference between the neuromuscular junction and CNS synapses is that in the CNS the sizes of single quanta of postsynaptic current produced are small, involving 10 to 100 receptor channels, in contrast to several thousand channels at the neuromuscular junction (49–53). In addition, the range of sizes of apparently single quantal currents (the "variance") is higher in the CNS than at the neuromuscular junction (45–47). There is considerable controversy about the source and meaning of this variance in quantal size. Some investigators now suspect that fundamental differences between the structure of CNS synapses and that of the typical neuromuscular junction may render the traditional model underlying quantal analysis invalid in the CNS, requiring that quantal analyses be interpreted in a fundamentally different way (47).

MICROSTRUCTURE OF PRESYNAPTIC AND POSTSYNAPTIC SITES To explain the statistical behavior of quantal variance in the CNS, two groups of investigators have recently suggested that the quantal postsynaptic current is not determined by the amount of transmitter contained in a single vesicle, as it is at the neuromuscular junction. Rather, it is determined by the small number of postsynaptic receptors, which are saturated by the amount of transmitter released by a single vesicle (47, 48, 54). These investigators suggest further that several individual release sites, spatially isolated from each other, may originate from a single axon. In this model, individual "patches" of postsynaptic receptors associated with each release site would also be spatially isolated from each other. Thus, some axons may end in one release site with one receptor patch; others might end in two release sites with two receptor patches, each of which is exposed only to the transmitter released from its own release site; etc. This proposed organization of presynaptic terminals, release sites, and receptors is quite different from the organization at the neuromuscular junction, where several release sites contained within one presynaptic terminal impinge upon a large plaque of acetylcholine receptors that are not fully saturated even by the amount of acetylcholine released by many vesicles. The revised model of the structure of CNS synapses is supported by a recent study of the fine structure of synapses from single axons onto dendrites of individual pyramidal neurons in the hippocampus (55). Twenty percent of axons examined in this study made more than one close apposition with the dendrites of a single pyramidal neuron. Furthermore, a study of serial electron micrographs revealed that 24% of individual synaptic boutons made synapses with more than one dendritic spine, arising from either the same or different dendritic segments. These multiple, spatially isolated release sites emerging from single axons may account for the high quantal variance and unusual transmission statistics displayed by CNS synapses.

Because of the uncertainty about an appropriate quantitative model of synaptic transmission in the CNS, a consensus has not yet been reached about the location of synaptic changes underlying LTP and other forms of CNS synaptic plasticity. Some recent quantal analyses have been interpreted as demonstrating a presynaptic locus (56–60), others a postsynaptic locus (61, 62); still others seem to demonstrate changes in both locations (63, 64). Furthermore, Bliss et al have found evidence for an increase in glutamate release following induction of LTP (65–67), whereas others have found evidence for a selective increase in synaptic currents mediated by AMPA receptors, suggesting a receptor-specific postsynaptic change (68–70; but see Ref. 71). Thus, neither traditional pharmacological studies nor quantal analyses have yet resolved the location of mechanisms underlying long-term potentiation. A more fruitful approach may involve development of cell culture systems, such as organotypic cultures (72), in which investigators could deliver pharmacological agents to both presynaptic and postsynaptic compartments of neurons during induction and maintenance of LTP.

The uncertainty about the "primary" location of the process that maintains LTP continues to cloud investigations of possible molecular mechanisms. Nevertheless, the growing evidence for multiple mechanisms suggests that alterations in synaptic machinery are likely to occur both presynaptically and postsynaptically, perhaps under different conditions (64) or with different time courses (70).

LONG-TERM POTENTIATION AT HIPPOCAMPAL SYNAPSES

Associative and Non-Associative LTP

Despite recent conflicting results concerning the location of mechanisms underlying LTP, several basic aspects of its mechanism have been established and are accepted by most investigators in the field. Much of what is now known comes from work on defined pathways in the hippocampus. Hippocampal neurons are aligned in two tightly organized single layers in which synaptic and cell body regions are well separated (73, 74). It is possible to stimulate groups of incoming axons and record synchronized waves of synaptic responses with extracellular electrodes both in the living animal and in slices of brain tissue maintained in vitro. This accessibility has facilitated pharmacological dissection of some of the early events in the induction of LTP. Such studies have established that in the hippocampus there are two distinct types of LTP that differ fundamentally in their mechanism of induction and probably in the sites at which potentiation is maintained. These two types are found in different sets of synapses that are spatially segregated from one another. The group of large "mossy fiber" synapses made onto CA3 pyramidal neurons displays a form of LTP called homosynaptic, or non-associative, LTP (75–77). It can be induced only by high-frequency stimulation of the presynaptic terminals; the strength of stimulation needed for induction is not influenced by depolarization of the postsynaptic membrane. Two other sets of synapses, those made by axons from the entorhinal cortex synapsing onto neurons called dentate granule cells and those made by axons from the CA3 neurons onto a group of neurons called CA1 pyramidal cells, display an intriguing form of LTP now called associative LTP (78, 79). In these synapses, potentiation can be induced both by high-frequency presynaptic stimulation, and by weak stimulation coupled with depolarization of the postsynaptic membrane (80–83). Thus, stimulation of one set of synapses increases the probability that neighboring synapses on the same neuron will be strengthened if they are activated at the same time or within a few msec of each other. Associative LTP is a particularly intriguing form of synaptic regulation because it provides a possible neural mechanism for encoding associations between different events or sensations that happen concurrently.

Role of the NMDA Receptor

It has been known for some time that ionotropic glutamate receptors, the receptors that produce postsynaptic current in response to release of glutamate, can be divided into at least three distinct pharmacological classes based on their specific activation by the agonists α-amino-3-hydroxy-5-methyl-4-isoxazole propionic acid (AMPA), kainate, or *N*-methyl-D-aspartate (NMDA) (84–86). Activation of one of these, the NMDA-type glutamate receptors, is required to induce associative LTP at hippocampal synapses (87). Blockade of NMDA receptors by specific antagonists prevents induction of LTP in area CA1, but does not appreciably affect the basal depolarizing synaptic potentials. Thus, activation of NMDA receptors is not required for normal synaptic transmission. In contrast, blockade of AMPA receptors with a specific antagonist eliminates the depolarizing synaptic potential, indicating that AMPA receptors generate most of the postsynaptic current during normal excitatory synaptic transmission in these neurons.

MAGNESIUM BLOCKADE Two properties of NMDA receptors that are not shared by AMPA receptors define their role in induction of LTP. First, the NMDA receptor channel is blocked by Mg^{2+}, but only at negative resting membrane potentials. This means that synaptically released glutamate cannot initiate ion flow through the receptor until the membrane is sufficiently depolarized to relieve the Mg^{2+} block (88, 89). Postsynaptic depolarization sufficient to activate NMDA receptors can be accomplished either by high-frequency stimulation of presynaptic afferents or by injection of depolarizing current into the postsynaptic neuron. The second critical property of the NMDA receptor is that its channel is highly permeable to Ca^{2+} ion (90, 91), whereas the AMPA and kainate-type receptor channels are selective for Na^+ and K^+ ions (90, 92, 93). The postsynaptic Ca^{2+} influx initiated by activation of NMDA receptors is critical for induction of LTP, as is discussed in more detail in the next section.

COINCIDENCE DETECTION The voltage-dependent Mg^{2+} block, together with a relatively slow rate of dissociation of glutamate from its binding site (52), makes the NMDA receptor a powerful coincidence detector (21, 94). Although the postsynaptic potential produced by AMPA receptors decays within about 10 msec, glutamate remains bound to NMDA receptors for hundreds of msec. Therefore, strong depolarization of a dendrite by stimulation of several neighboring synapses can, in theory, activate NMDA receptors at all nearby synapses that were active within the previous several hundred msec. Electrophysiological studies of NMDA receptors reconstituted from cloned subunits has revealed that different combinations of subunits produce channels with distinct response decay constants ranging from 100 to 400 msec, presumably reflecting different glutamate dissociation constants (95). Thus, expression of distinct combinations of NMDA receptor subunits may control the effective coincidence detection window in different neuronal types (21).

NON-ASSOCIATIVE LTP NMDA receptor activation does not play a role in induction of homosynaptic, non-associative LTP in area CA3 of the hippocampus (76). Thus, it is not surprising that NMDA receptors are sparse in dendrites in area CA3 where axons from the "mossy fiber" pathway form synapses that display non-associative LTP, whereas they are highly enriched in dendrites in area CA1 and in the dentate gyrus where synapses that show associative LTP are concentrated (96).

Postsynaptic Calcium

It is now generally agreed that entry of Ca^{2+} into the postsynaptic cell is the primary event initiated by activation of NMDA receptors that is required for induction of associative LTP. High-frequency stimuli that induce associative LTP in area CA1 produce a large transient increase in Ca^{2+} concentration confined to dendritic areas near the activated synapses. The increase is prevented by blockade of NMDA receptors (97, 98).

CALCIUM IS NECESSARY At these same synapses, blockade of the rise in postsynaptic Ca^{2+} prevents induction of associative LTP. This blockade was demonstrated by injection of either ethylene glycol-*bis* (β-aminoethyl ether) *N,N,N′,N′*-tetraacetic acid (EGTA) (99) or the light-sensitive Ca^{2+} chelator nitr-5 (100) into CA1 neurons. The injections prevented the development of LTP after tetanic synaptic stimulation. In a converse experiment, Ca^{2+}-loaded nitr-5 was injected into the neurons, then photolysed to release Ca^{2+} into the cytoplasm. Photolysis produced a long-lasting potentiation of the postsynaptic response (100). Using the photoactivatable Ca^{2+} buffer, diazo-4, Malenka et al (101) determined that a time window of about 2.5 seconds of increased Ca^{2+} is necessary to induce LTP. Reversal of the rise in Ca^{2+} by photolysis, 1 sec after the start of a tetanus, blocked induction of LTP. However, a rise in postsynaptic Ca^{2+} lasting just 2.5 seconds after the start of a tetanus was sufficient to produce LTP. Attenuation of the Ca^{2+} rise at 1.5 to 2 seconds after the start of tetanus often produced a short-lasting potentiation, suggesting that sequential activation of a series of Ca^{2+}-stimulated biochemical events may be necessary for induction of LTP.

IS CALCIUM SUFFICIENT? Although an increase in postsynaptic Ca^{2+} is necessary to induce LTP, it is not yet clear that it is sufficient. Several experiments suggest that another factor or process provided by presynaptic stimulation is required for induction of LTP, in addition to the Ca^{2+} influx provided by activation of NMDA receptors. For example, under standard physiological conditions, single iontophoretic applications of glutamate or NMDA onto CA1 neurons in the absence of presynaptic stimulation produce only a relatively short-term facilitation, termed STP, lasting about 15 minutes (102, 103). Thus, activation of NMDA receptors alone is not sufficient to induce LTP. On the other hand, iontophoresis of NMDA in the presence of elevated external Ca^{2+} (6 mM) produces typical LTP (103). Furthermore, a recent report indicates that, in the absence of presynaptic action potentials, five pulses of glutamate applied iontophoretically for 10 sec and spaced 60 sec apart can induce a form of LTP that shares many of the characteristics of tetanus-induced LTP (104). Many of these observations can be reconciled by a hypothesis suggested by Malenka (103) that a threshold concentration of cytosolic Ca^{2+} must be reached for a critical length of time to trigger LTP. A single pulse of activation of NMDA receptors alone may not provide sufficient cytosolic Ca^{2+}. Additional events that enhance the postsynaptic increase in Ca^{2+} may be required, including activation of voltage-dependent Ca^{2+} channels (105) or, as discussed below, activation of metabotropic glutamate receptors. It has not been ruled out that events in addition to the postsynaptic rise in Ca^{2+} concentration, perhaps associated with stimulation of presynaptic afferents, are also necessary to establish "true" LTP.

Role of the Metabotropic Glutamate Receptor

In the past few years, a third class of glutamate receptor, the metabotropic glutamate receptor (mGluR), has been implicated in the induction process. A specific mGluR agonist, aminocyclopentate dicarboxylate (ACPD), augments tetanus-induced LTP (106). Furthermore, in the septal nucleus (107) and hippocampus (108), application of ACPD itself to synapses has been reported to induce a slow-onset, long-lasting potentiation. Finally, a new specific antagonist of the mGluR, (*RS*)-α-methyl-4-carboxyphenylglycine (MCPG) (109), reportedly blocks not only induction of associative, but also induction of non-associative LTP in the hippocampus (110). Thus, the mGluR may participate along with the NMDA receptor in induction of associative LTP and may be a principal trigger for induction of non-associative LTP. To establish whether activation of the mGluR is the postulated additional induction "factor," it will be important to determine whether activation of NMDA receptors together with mGlu receptors is sufficient to induce associative LTP in hippocampal synapses.

In the hippocampus, the metabotropic glutamate receptor is linked via a G-protein to hydrolysis of phosphatidyl inositol (111), which leads to activation of protein kinase C and mobilization of intracellular Ca^{2+} stores. When it is expressed in heterologous cells, activation of the mGluR also stimulates synthesis of cAMP and production of arachidonic acid (112). A recent report suggests that coapplication of NMDA and 8-bromo-cAMP to hippocampal slices produces a long-lasting potentiation of area CA1 synapses (113). It has also been proposed that the mGluR could aid in induction of LTP by augmenting the activation of NMDA receptors through the action of protein kinase C (114, 115). Thus, activation of the mGluR can potentially mobilize a wide variety of signal transduction pathways and activate various protein kinases that have been implicated in induction of LTP (see below).

BIOCHEMICAL ISSUES AND OPPORTUNITIES

A Retrograde Messenger

It is now widely accepted that normal induction of associative LTP requires activation of postsynaptic NMDA receptors and the consequent transient rise in postsynaptic Ca^{2+} concentration. It is also widely accepted that, at least in part, expression of LTP involves an increase in the amount of glutamate released by each presynaptic impulse (8, 9, 11). To reconcile postsynaptic induction and presynaptic expression of LTP, it is necessary to postulate the existence of a signal that travels from the postsynaptic site back to the presynaptic terminal to initiate the modification of presynaptic proteins that must underlie enhanced release. There is no precedent for such a "retrograde signal" at the neuromuscular junction or at synapses in the peripheral nervous system. Furthermore, studies of CNS synapses by electron microscopy have not revealed evidence for a vesicular release apparatus at the postsynaptic membrane. Therefore, investigators have examined the possibility that paracrine messengers that can diffuse across cellular membranes might be responsible for the retrograde signal. The roles in the induction of LTP of two such messengers—arachidonic acid and its metabolites, and nitric oxide (NO)—have been extensively examined. A third potential messenger, carbon monoxide (CO), has received attention recently.

ARACHIDONIC ACID Arachidonic acid is released from membrane lipids along with other fatty acids by phospholipase A2, many forms of which are dependent on Ca^{2+} for activity. Thus the phospholipase is a potential target for the postsynaptic increase in Ca^{2+} that initiates LTP. Arachidonic acid is the common metabolic precursor of two related families of paracrine regulators: the prostaglandins, which are products of the enzyme cyclo-oxygenase; and 5- and 12-hydroxy- and hydroperoxy-eicosatetraenoic acid (5- and 12-HETE and -HPETE), which are products of the lipoxygenase enzymes (116). These regulators have a variety of local actions in adipose tissue, smooth muscle, and in cells mediating the inflammatory response. In the invertebrate *Aplysia,* lipoxygenase metabolites released in response to the peptide hormone FMRFamide inhibit transmitter release by stimulating the opening of a K^+ channel that accelerates repolarization of the presynaptic membrane after an action potential, thus reducing Ca^{2+} influx through voltage-activated Ca^{2+} channels (117).

Early experiments provided support for arachidonic acid as a retrograde messenger in LTP. Activation of NMDA receptors stimulates release of arachidonate and its metabolites into the surrounding medium from striatal (118), cerebellar (119), and hippocampal neurons (120), as well as from hippocampal slices during induction of LTP (121). Furthermore, putative inhibitors of phospholipase A2—nordihydroguaiaretic acid (NDGA), and *p*-bromophenacyl bromide (BPB)—were reported to inhibit induction of LTP (122–124), whereas indomethacin, an inhibitor of cyclo-oxygenase, had no effect on LTP (123).

Examination of the effects of arachidonic acid on glutamatergic synapses has weakened the case that it is the principal retrograde signal, however.

Application of arachidonic acid to hippocampal slices can produce a long-lasting potentiation of synaptic transmission, but only when it is coupled with weak stimulation of presynaptic afferents (125). The effect develops slowly, rising gradually to a maximum over about an hour; whereas LTP is apparent within the first 20 minutes after tetanization. Furthermore, even the gradual potentiation is not observed when NMDA receptor activation is blocked (126). Thus, arachidonic acid is not the sole agent acting downstream of NMDA receptors.

Neither arachidonic acid nor its metabolites have been shown to produce a direct stimulation of transmitter release in hippocampal neurons. However, it has been reported that application of arachidonic acid to glial cells in culture produces a rapid, dose-dependent 20–80% inhibition of Na^+-dependent glutamate uptake that is only partially reversed after removal of arachidonic acid from the medium (127). Therefore, arachidonic acid may increase the effectiveness of glutamate at the synapse by inhibiting uptake of glutamate into glial cells, a process believed to be an important mode of inactivation of the excitatory action of glutamate. This effect of arachidonic acid is not inhibited by NDGA, and so the reported suppression of induction of LTP by NDGA must be accounted for by a different mechanism.

More selective and specific inhibitors of phospholipase A2, lipoxygenase, and cyclooxygenase would be useful in resolving the role of arachidonic acid in LTP, as would structural and regulatory information about the phospholipase A2 isozyme that is activated in response to NMDA. Additional potential targets of arachidonate in the presynaptic terminal are discussed in the section entitled *Possible Control Mechanisms at Presynaptic Sites*.

NITRIC OXIDE The small molecule nitric oxide (NO) has emerged as a strong candidate for a retrograde messenger (11, 128). It was first recognized as a paracrine regulator in 1987 when the elusive Endothelium-Derived Relaxing Factor (EDFR) was identified as NO (129). NO is synthesized from arginine in endothelial cells of blood vessels in response to activation of the metabotropic acetylcholine receptor, and diffuses into adjacent smooth muscle cells, where it activates soluble guanylate cyclase, ultimately resulting in muscle relaxation and vessel dilation (130). Its significance in the brain was recognized with the discovery that activation of NMDA receptors on granule neurons in the cerebellum results in synthesis of NO in those neurons and activation of guanylate cyclase in nearby glial cells (131). It is now clear that NO may be a paracrine messenger in additional tissues, including macrophages in the immune system (132).

Recently, four groups of investigators reported that the arginine derivatives ω-N-nitro-L-arginine or ω-N-methyl-L-arginine, which are specific inhibitors of nitric oxide synthase, severely attenuate induction of associative LTP in

CA1 pyramidal neurons in hippocampal slices (126, 133–135). Two of these groups showed that intracellular application of the inhibitors to the postsynaptic neuron through the recording pipette was sufficient to inhibit induction (126, 134). This result demonstrates that NO is produced in the postsynaptic neurons and thus implicates NO as a diffusible retrograde messenger. In all four studies, the inhibition of induction was reversed by addition of excess L-arginine to compete with the NO synthase inhibitors. Application of ω-N-methyl-D-arginine, a related arginine derivative that does not inhibit NO synthesis, did not inhibit induction of LTP (126, 134, 135). Addition of hemoglobin to the slices also blocked induction (126, 133–135). The heme moiety of hemoglobin binds NO; thus the inhibition by extracellular hemoglobin is consistent with the notion that NO travels from the postsynaptic site to presynaptic terminals during induction of LTP. Methemoglobin, which has a considerably lower affinity for NO, does not inhibit induction.

NO is a short-lived, highly reactive free radical. Perhaps for this reason, only two groups have succeeded in demonstrating that NO or artificial NO donors can modulate synaptic transmission in hippocampal neurons. Application of two different NO donors, hydroxylamine and sodium nitroprusside, was reported to potentiate synaptic transmission in hippocampal slices (133, 136). Furthermore, this potentiation prevented subsequent induction of associative LTP in the neurons, suggesting that the LTP mechanism had been saturated. In another study, direct application of NO gas to cultured hippocampal neurons increased the frequency of spontaneous quantal currents ("miniature postsynaptic potentials"), suggesting that NO can regulate the release apparatus, perhaps by increasing the probability of release (126). The same laboratory also found that application of NO gas to hippocampal slices increased the size of postsynaptic potentials, but only if the application coincided with weak tetanic stimulation (137).

Despite clear and self-consistent pharmacological results from several laboratories implicating NO as a retrograde messenger during induction of LTP, there is considerable controversy about the appropriate interpretation of these experiments, because a number of other laboratories see no effect of NO synthase inhibitors on induction under certain experimental conditions (138; reviewed in Ref. 128). It seems likely that these discrepant results have arisen because LTP can be produced by redundant mechanisms that are brought into play in different proportions by the various LTP induction protocols used by different laboratories. This interpretation is supported by recent reports that at ambient temperature (20–24° C) NO synthase inhibitors reliably inhibit induction of LTP by strong tetanic stimuli, whereas at physiological temperature (30–32° C) they do not inhibit induction by strong tetanic stimuli, but reliably inhibit induction by relatively weak tetanic stimuli (128, 139–141).

NO is synthesized from arginine by NO synthase, an enzyme that is highly homologous to cytochrome P-450 reductase (142). The activity of the brain isozyme of NO synthase purified from the cerebellum depends on Ca^{2+}/calmodulin (143), unlike the inducible isozyme purified from macrophages, the activity of which is unregulated. Its dependence on Ca^{2+} makes NO synthase a possible target for the postsynaptic increase in Ca^{2+} concentration that initiates LTP.

Additional controversy surrounding the hypothesis that NO is a retrograde messenger concerns the localization of NO synthase in the hippocampus. It was first reported that both the mRNA encoding the isozyme originally purified from the cerebellum (144) and its protein product detected by immunocytochemistry (144–147) are located in dentate granule neurons and in a few sparsely scattered multipolar neurons in area CA1 of the hippocampus, but they are not detected in pyramidal neurons in area CA1, where it has been proposed that NO is generated during induction of LTP (126, 134). In contrast, a recent report indicates that one of the antibodies raised against the cerebellar isozyme stains CA1 pyramidal neurons and their dendrites when the tissue is fixed under conditions milder than those in the earlier studies (148). These discrepancies suggest that the CA1 pyramidal neurons may express an NO synthase isozyme distinct from the cerbellar isozyme; or that they contain smaller quantities of the cerebellar isozyme than are present in the intensely staining hippocampal multipolar neurons. However, it has still not been rigorously demonstrated that the CA1 neurons contain any NO synthase.

Cerebellar NO synthase contains consensus phosphorylation sequences for several protein kinases and is phosphorylated on serine residues in vitro by the cAMP-dependent protein kinase (149, 150), Ca^{2+}/calmodulin-dependent protein kinase (149, 151, 152), and protein kinase C (149, 151, 152). The regulatory effects of phosphorylation by each kinase appear to depend upon the specific assay conditions. Phosphorylation by CaM kinase II has been reported to decrease enzyme activity (151) or to have no effect (149); phosphorylation by cAMP-dependent protein kinase has no effect on activity (149, 150); and, finally, phosphorylation by protein kinase C has been reported to increase (151) or decrease (149) activity. To determine whether these protein kinases modulate production of NO, it will be important to measure phosphorylation of NO synthase in living neurons and to determine how phosphorylation alters enzyme activity under appropriate physiological conditions.

CARBON MONOXIDE Carbon monoxide (CO) was initially proposed as a third potential retrograde messenger because heme-oxygenase-2, an enzyme that can catalyze production of CO in the brain, is expressed in CA1 neurons of the hippocampus (153). Inhibitors of heme oxygenase, including zinc-protoporphyrin IX (ZnPP) (137, 154) and zinc deuteroporphyrin IX 2,4-*bis* glycol (ZnBg) (154), were tested for their ability to block LTP and they do so. Interestingly, application of ZnPP was also reported to reverse established LTP (154). Furthermore, application to hippocampal slices of CO, like that of NO, together with weak tetanic stimulation, produces long-lasting enhancement of synaptic potentials (137).

Heme oxygenase-2 is not regulated by Ca^{2+} ion in any known way, therefore it is not clear how its activity would be stimulated by activation of NMDA receptors during induction of LTP. The case for CO as a retrograde messenger is further weakened by the fact that ZnPP and ZnBg have also been reported to inhibit soluble guanylate cyclase, one of the potential targets of NO (155).

Possible Control Mechanisms at Presynaptic Sites

LTP differs from all other known forms of presynaptic regulation of synaptic efficacy in an interesting way: Induction of LTP in a synaptic terminal does not alter the magnitude of subsequent paired-pulse facilitation in that terminal (156–159). It is generally agreed that paired-pulse facilitation and post-tetanic potentiation are presynaptic phenomena caused by a change in regulation of Ca^{2+} concentration resulting in increased free Ca^{2+} in the terminal during each electrical impulse (22). At the peak of post-tetanic potentiation, paired-pulse facilitation disappears, reflecting the fact that the two phenomena share a common mechanism, although they have different time courses of decay. It is therefore said that post-tetanic potentiation "occludes" paired-pulse facilitation. The absence of a change in the magnitude of paired-pulse facilitation following induction of LTP indicates that the effects of LTP on transmitter release do not involve a change in the concentration of Ca^{2+} in the terminal produced by each impulse. Rather LTP must involve regulation of a step downstream from the Ca^{2+} increase, such as an increase in the number of release sites that can be activated by Ca^{2+}. Thus, it seems likely that presynaptic expression of LTP involves covalent modifications of the release machinery itself.

Although identification of the molecular components of the release apparatus is proceeding rapidly (16, 17, 160, 161), the location of the Ca^{2+}-sensitive mechanism that triggers fusion is not yet known. Similarly, we have only small clues about the potential targets of candidate retrograde messengers such as NO and arachidonic acid. The identification of these targets and their relationship to the release mechanism can be expected to be an active area of research in the near future.

POTENTIAL TARGETS OF ARACHIDONIC ACID Among the potential targets of arachidonic acid in the presynaptic terminal are K^+ channels (117) and the α and β isoforms of protein kinase C (163). Protein kinase C is activated during induction of LTP (164, 165), and inhibitors of its activity can block induction (166–168). Phosphorylation of at least one presynaptic protein, B-50 [also known as GAP-43 (169), F-1 (170), or neuromodulin (171)], by protein kinase C is significantly enhanced in hippocampal slices following induction of LTP by tetanic stimulation (172). Its phosphorylation is maximal 10 minutes after delivery of a tetanus and remains high 50 minutes later, suggesting a prolonged presynaptic activation of protein kinase C. A second line of evidence also supports the idea that prolonged protein kinase activation in the presynaptic terminal accompanies LTP. Injection of the general protein kinase inhibitor H-7 into a postsynaptic neuron before application of tetanic stimulation blocks LTP (168, 173). However, if the postsynaptic neuron is impaled with an electrode filled with H-7 after induction of LTP, established LTP is not affected (168). Nevertheless, bath application of H-7 to the same slice depresses the established LTP in the impaled neuron. The simplest explanation for this result is that persistent protein kinase activity in the presynaptic terminal is necessary for expression of LTP under these circumstances. It will be interesting to learn whether inhibitors of arachidonic acid production can block the prolonged presynaptic phosphorylation of B-50 that accompanies induction of LTP.

B-50, (F1, GAP-43, neuromodulin) binds calmodulin in the absence of calcium; but phosphorylation by protein kinase C decreases its affinity for calmodulin (171). Thus, the protein may function to concentrate calmodulin at specific sites, releasing calmodulin locally in response to phosphorylation by protein kinase C (174). It was recently reported that introduction of antibodies against B-50 into synaptosomes inhibited vesicular release of noradrenaline, suggesting that B-50 may interact with the release mechanism (175). However, there is no known direct interaction of B-50 with proteins of the release apparatus. Therefore, it is still premature to conclude that phosphorylation of B-50 (GAP-43) itself underlies presynaptic expression of LTP.

It has been proposed that release of calmodulin from B-50 following its phosphorylation by protein kinase C might affect the release apparatus indirectly by increasing phosphorylation of the vesicle-associated protein synapsin I by Ca^{2+}/calmodulin-dependent protein kinase II (CaM kinase II) (172). One current model for regulation of transmitter release holds that in nerve terminals, small synaptic vesicles are bound within a meshwork of actin filaments by synapsin I, which associates both with actin filaments and with proteins at the surface of synaptic vesicles (3, 176). The associations of synapsin I with vesicles and with actin are modulated by phosphorylation by three different protein kinases, including CaM kinase II. Phosphorylation by CaM kinase II decreases the affinity of synapsin I for vesicles, thus releasing them from the actin meshwork. Therefore, enhanced phosphorylation of synapsin I would, in theory, free vesicles from association with actin filaments and increase the number of vesicles available to dock at release sites, thus increasing transmitter release per impulse. If phosphorylation of B-50 releases calmodulin and produces a small, persistent enhancement of the steady-state activity of CaM kinase II, this could result in a long-lasting increase in transmitter release.

Synthesis of B-50 (GAP-43, F-1, neuromodulin) is up-regulated following axonal injury, and the newly synthesized protein is specifically transported into axons and growth cones, where it is proposed to play a role in axonal growth (169). Thus, phosphorylation of B-50 after LTP induction could simply reflect a preparation for growth or sprouting of the potentiated terminal.

Although arachidonic acid metabolites inhibit release by facilitating the opening of a K^+ channel in *Aplysia* neurons (117), no direct action of arachidonic acid on a mammalian K^+ channel has yet been reported. However, Herrero et al (177) found evidence in mammalian synaptosomes that the presence of arachidonic acid facilitates inhibition of the opening of a presynaptic K^+ channel through phosphorylation by protein kinase C. They constructed a hypothesis for a positive feedback mechanism in which activation of the presynaptic metabotropic glutamate receptor stimulates protein kinase C activity via the phospholipase C pathway. Protein kinase C phosphorylates and thereby inhibits the K^+ channel, but only in the presence of low levels of arachidonic acid produced, for example, by the postsynaptic neuron. Inhibition of the K^+ channel prolongs Ca^{2+} influx through voltage-dependent Ca^{2+} channels, thus increasing transmitter release. Although this mechanism may come into play in glutamatergic terminals during tetanic stimulation, it is unlikely that it accounts for presynaptic LTP, because such a mechanism would be expected to occlude paired-pulse facilitation. However, as discussed above, the mechanism of LTP does not occlude paired-pulse facilitation.

In summary, although arachidonic acid is an attractive candidate for a retrograde messenger, its action on the release mechanism during LTP has yet to be demonstrated. Future research should clarify whether protein kinase C regulates the release machinery and whether such regulation is critically involved in induction or maintenance of LTP.

POTENTIAL TARGETS OF NITRIC OXIDE In non-neural systems, two potential targets for regulation by NO have been identified: a heme-containing guanylate cyclase activated by NO in endothelial cells (178), and an ADP-ribosyl transferase activated by NO in platelets (179).

An indication that guanylate cyclase might be a target for NO in presynaptic terminals was reported by Haley et al (135), who found that application of dibutyryl cGMP to hippocampal slices partially reversed the inhibition of LTP induction produced by inhibitors of NO synthase. In addition, tetanic stimulation caused a large increase in cGMP in hippocampal slices that can be blocked by NO synthase inhibitors (180). In contrast, however, Schuman et al (181) reported that extracellular application of cGMP analogs to hippocampal slices in conjunction with low-frequency presynaptic stimulation had no effect on synaptic efficacy. Furthermore, application of the cGMP analogs followed by high-frequency stimulation in the presence of NMDA receptor blockers produced a transient depression. Thus, it is still unclear whether activation of guanylate cyclase mediates effects of NO on the presynaptic release apparatus.

Two recent reports suggest that an ADP ribosyl transferase might be an NO target in the hippocampus. Three different inhibitors of ADP ribosyl transferase were found to be effective inhibitors of LTP induction (181). When one of the inhibitors was applied intracellularly through an electrode into a postsynaptic neuron, it did not block LTP induction in synapses on that neuron. Thus the externally applied inhibitors may be acting on an ADP-ribosyl transferase in the presynaptic terminal. Duman et al (182) found a marked reduction in the amount of NO-stimulated ADP-ribosyl-transferase substrate proteins that could be labeled with radioactive ADP in homogenates prepared from hippocampal slices in which LTP had been induced. This result suggests that induction of LTP causes increased endogenous ADP-ribosylation. It is intriguing that B-50 (GAP-43) was recently identified as a prominent target of a neural ADP-ribosyl transferase (183). The roles of NO-stimulated guanylate cyclase and ADP-ribosyl transferase in induction and maintenance of LTP promise to be active areas of research over the next few years.

Possible Control Mechanisms at Postsynaptic Sites

As discussed earlier, the evidence is compelling that a postsynaptic increase in Ca^{2+} is required for induction of LTP. One effect of the postsynaptic Ca^{2+} is believed to be activation of enzymes that produce a retrograde messenger to initiate potentiation of presynaptic transmitter release. However, the mechanism of LTP may also include enhancement of the response of postsynaptic receptors. Two groups have reported a long-lasting and selective increase in synaptic currents mediated by AMPA receptors following induction of LTP (68, 69). Another group observed a slowly developing increase in postsynaptic sensitivity to iontophoretic application of AMPA following induction (70). These findings are controversial because others have observed long-term potentiation of NMDA receptor currents as well as of AMPA receptor currents (71, 184–186), suggesting that in their experiments the potentiation could be accounted for simply by an increase in transmitter release. More recently, however, additional investigators have reported that, although both NMDA and AMPA components of the post-synaptic current show long-term potentiation, enhancement of the non-NMDA component is considerably greater, again suggesting specific modification of the AMPA-type glutamate receptors (187–189). The discrepancies in the observations made by different laboratories may once again reflect the diversity of mechanisms for generating LTP.

The leading candidates for Ca^{2+} targets that could directly modify post-synaptic receptors are CaM kinase II and protein kinase C. Application of inhibitors of both of these protein kinases to hippocampal slices can block induction of LTP (167, 168, 173, 185). In one of the most convincing of these experiments, highly specific "pseudosubstrate" peptide inhibitors of each of the two kinases (190, 191) were applied to the postsynaptic neuron through the recording pipette (168). Induction of LTP was blocked in the neurons that received either inhibiting peptide, but not in surrounding neurons that had not received a peptide or in neurons that had been injected with an inactive control peptide. Thus, activity of both CaM kinase II and protein kinase C in the postsynaptic neuron appears to be required for induction.

It has also been reported that general inhibitors of tyrosine kinase activity, lavendustin A and genistein, block induction of LTP when applied to hippocampal slices (192). Furthermore, activation of NMDA receptors in hippocampal cultures can lead by an unknown transduction pathway to activation of MAP kinase, which is usually activated through a protein kinase cascade that includes tyrosine kinases (193). Thus, protein tyrosine kinases may also play a role under some circumstances in postsynaptic induction of LTP.

The requirement for any of these kinase activities could be "permissive" if, for example, continuous basal phosphorylation is required to maintain the integrity of the induction apparatus; or it could be "direct" if NMDA-receptor stimulated phosphorylation of postsynaptic receptors or ion channels contributes to the mechanism of LTP.

CALCIUM/CALMODULIN-DEPENDENT PROTEIN KINASE II CaM kinase II is expressed at very high levels in hippocampal and cortical neurons, accounting for approximately 2% of total hippocampal protein and about 1% of cortical protein (194), suggesting that it plays a specialized role in forebrain neurons. It is considered a likely target for the Ca^{2+} flowing through NMDA receptor channels because it is concentrated in the postsynaptic density fraction (195–197). The postsynaptic density is a specialization of the submembranous cytoskeleton that adheres to the cytoplasmic face of the postsynaptic membrane immediately adjacent to presynaptic active zones (198, 199; reviewed in Ref. 200). It has been postulated to contain concentrated signal transduction molecules that may participate in receptor clustering and regulation, or in control of adhesion between presynaptic terminals and the postsynaptic membrane. In the forebrain, about half of the CaM kinase holoenzymes are soluble and distributed throughout the cytosol (201). The other half sediment with the particulate fraction. One of the particulate structures that contains CaM kinase is the postsynaptic density fraction, where the bound CaM kinase holoenzymes constitute about 17% of the total protein (201).

The CaM kinase holoenzyme is regulated in a complex way by autophosphorylation (202), and this has fueled considerable speculation about its role in LTP and in learning and memory in general (203). The holoenzyme is composed of clusters of 12 catalytic subunits (204). Each subunit contains a regulatory calmodulin-binding domain (205–208). When kinase activity is activated in the presence of Ca^{2+} and calmodulin, a threonine residue adjacent to the calmodulin-binding domain becomes autophosphorylated (209–211) by an intramolecular mechanism (212, 213). The autophosphorylated residue maintains the kinase in an activated state after the Ca^{2+} concentration has returned to basal levels (209, 212, 214–217). Autophosphorylation of less than three of the subunits in a holoenzyme is sufficient to induce maximum activation of the holoenzyme (212, 215, 218), indicating that the activation process is cooperative. Furthermore, the activated kinase holoenzyme continues to autophosphorylate in a Ca^{2+}-independent manner (212, 218, 219).

Taken together, these properties suggest that the CaM kinase can exist in at least two distinct states. In state I, the kinase activity is completely dependent on Ca^{2+} and calmodulin. When autophosphorylation of two to three threonine-286 residues per holoenzyme has occurred, the kinase changes to state II, in which it has substantial Ca^{2+}-independent kinase activity with exogenous substrates, and continues to autophosphorylate in the absence of Ca^{2+}. Meyer et al reported recently that in state II, the CaM kinase also retains a high affinity for calmodulin in the absence of Ca^{2+} and may serve to sequester calmodulin molecules (220). To be returned to state I, in which its activity is Ca^{2+}-dependent, the kinase must be dephosphorylated by cellular phosphatases. Because Ca^{2+}-independent autophosphorylation can oppose the rate of dephosphorylation by cellular phosphatases, kinase activity may be prolonged for quite some time beyond an initial activating Ca^{2+} signal. Thus, it has been suggested that the kinase may be able to retain information in vivo about previous electrical activity of the neuron (212). This model of kinase regulation was formalized by Lisman & Goldring (203), who speculated that information encoded by autophosphorylated kinase might be retained even during turnover of individual protein kinase molecules.

Autophosphorylation of threonine-286 occurs in situ in cerebellar granule neurons (221), dissociated hippocampal neurons (222), acute hippocampal slices (223), and organotypic hippocampal cultures (224). In dissociated cultures, the proportion of CaM kinase in the autophosphorylated state is low at basal Ca^{2+} (2–5%) and can be transiently increased to 9–12% by depolarization (221) or by activation of NMDA receptors (222). However, in acute hippocampal slices or organotypic cultures, in which the neurons are more mature and CaM kinase is expressed at higher levels, the proportion of CaM kinase in the autophosphorylated state is relatively high even at basal Ca^{2+} levels (10–30%) and does not appear to change significantly upon activation of NMDA receptors (223, 224). The basally autophosphorylated CaM kinase is distributed throughout the cytosol of most neurons in organotypic cultures as determined by immunocytochemistry utilizing a phosphokinase-specific antibody (225). Furthermore, the proportion of Ca^{2+}-independent kinase in these cultures is reduced to less than 5% 20 minutes after removal of Ca^{2+} from the medium (224). These findings do not support the model for memory storage as proposed by Lisman & Goldring (203), although it has not been strictly ruled out that the neurons contain a small pool of CaM kinase that is sensitive to regulation by NMDA receptors and behaves in a switch-like fashion as predicted by their model. The results suggest rather that activation by autophosphorylation may serve to keep a substantial portion of the CaM kinase molecules active at low Ca^{2+} concentrations (224). In the absence of activation by autophosphorylation, kinetic equations predict that the level of basal CaM kinase activity in the cytosol of hippocampal neurons would be less than 1%. A mechanism in which the level of active CaM kinase is determined by a continuous equilibrium between Ca^{2+}-stimulated autophosphorylation and dephosphorylation by cellular phosphatases would permit rapid and perhaps highly localized regulation of kinase activity either upward through an increase in Ca^{2+} concentration or downward through a decrease in phosphatase activity. Thus, the kinase may function more as an analog "dimmer" switch than as a digital "on/off" switch.

Additional support for the general notion that CaM kinase II is important for synaptic regulation has come from recent experiments employing molecular genetic techniques. CaM kinase II can phosphorylate heterologously expressed GluR1 subunits of the AMPA glutamate receptor, as well as a protein from the postsynaptic density fraction that is precipitated by anti-GluR1 antibodies (226). Perfusion of hippocampal neurons with a high concentration of recombinant, activated CaM kinase II was reported to increase kainate-induced inward currents in the perfused neurons, suggesting that phosphorylation by CaM kinase II can regulate a kainate-activated receptor ion channel (226).

Finally, transgenic mice in which the most abundant brain subunit of CaM kinase II, the α-subunit, was disrupted by homologous recombination show behavioral deficits in a spatial learning task (227). Furthermore, long-term potentiation in slices prepared from the hippocampi of these mice is severely impaired (228). Perhaps the most interesting effect of this mutation, however, is that the homozygous mutant mice are far more susceptible to induction of severe seizures than their wild-type litter mates (229). The latter phenotype suggests that mutant mice have lost an important molecular "control element," rendering their brains less able to maintain stability in the face of environmental perturbations than those of wild-type mice. It is difficult to assess from the mutant phenotype which molecular actions of CaM kinase II are most important for maintaining stability in wild-type mice. Indeed, the molecular effects of such a mutation are likely to be complex and pleiotropic.

To establish the precise regulatory roles of CaM kinase II in synapses, more research is needed to establish the circumstances under which it is activated in situ, and the identities of its postsynaptic substrates.

PROTEIN KINASE C Although recent speculation has centered on a possible role for presynaptic protein kinase C in LTP (see earlier discussion), experiments with inhibitory peptides have established that postsynaptic protein kinase C activity is also necessary for induction of LTP (168). During induction, protein kinase C could be activated synergistically by the combination of diacylglycerol, released after activation of phospholipase C by the metabotropic glutamate receptor, and Ca^{2+} ion flowing in through the NMDA receptor channel or released from internal stores (230). Application of phorbol esters that activate protein kinase C to hippocampal slices (231), as well as injection of protein kinase C into postsynaptic neurons (232), results in synaptic potentiation. However, potentiation by phorbol ester does not occlude induction of LTP by electrical stimulation (233, 234), indicating that the two forms of potentiation are mediated by distinct, additive mechanisms. Thus, it appears that activation of protein kinase C is necessary but not sufficient for induction of LTP.

At this point, there is only one clearly identified postsynaptic target for protein kinase C that could be involved in the mechanism of LTP. The NR1 subunit of the NMDA receptor has been shown to be phosphorylated by protein kinase C after heterologous expression in 293 cells and in cortical neurons in culture (235). Furthermore, in the trigeminal nucleus, opiate-mediated phosphorylation by the C-kinase has been reported to potentiate

NMDA receptor responses by reducing the receptor's Mg^{2+} block (236, 237). If this mechanism operates in hippocampal neurons, phosphorylation by protein kinase C may participate in long-term potentiation of NMDA-receptor responses. Although AMPA-type glutamate receptors have potential consensus phosphorylation sites for protein kinase C (238), only one group has observed C-kinase phosphorylation of an AMPA-receptor subunit directly (226), and the functional significance of this phosphorylation has not been determined. Thus, as for CaM kinase II, the identification of additional postsynaptic substrates for protein kinase C that are necessary for induction of LTP is an important future goal.

DIRECTIONS FOR THE FUTURE

This review has focused on biochemical mechanisms that may underlie associative LTP of hippocampal synapses in the first half-hour during which it is established. It is evident that, although many individual molecules involved in induction and maintenance of LTP have been identified, we are still unable to construct a coherent molecular model for LTP. Progress seems likely to come from pharmacological and physiological experiments designed to pinpoint key processes, from experiments with new methodologies that permit measurement of the time course of posttranslational modifications at defined locations within neurons, and from structural work on regulatory components of the major presynaptic and postsynaptic organelles of CNS synapses. In addition, mathematical simulations of interacting biochemical regulatory networks are likely to play an increasingly important role in organizing our thinking about molecular processes underlying LTP.

LTP is only one form of long-term synaptic plasticity exhibited by central nervous system neurons. Associative long-term depression (LTD) is now recognized as an equally important and ubiquitous form of modulation (35, 239). Studies of the interplay between these two processes, which may share some common mechanisms (32), will provide opportunities to identify the most critical molecular control points. They are also likely to illuminate how the strength of synaptic transmission is balanced to maintain homeostasis during the processing and storage of information in complex neural pathways.

Most investigators now believe that structural changes in existing synapses as well as growth of new synapses may be involved in LTP that lasts for days or weeks in the animal. The gradual transition from reversible posttranslational modifications during the first minutes of LTP to firmly established structural changes will be a subject of increasing study. Two recent reviews cover initial attempts to describe this transition (240, 241).

Literature Cited

1. Hebb DO. 1949. *The Organization of Behavior*. New York: Wiley
2. Kaczmarek LK, Levitan IB, eds. 1987. *Neuromodulation: The Biochemical Control of Neuronal Excitability*. Oxford: Oxford Univ. Press. 286 pp.
3. De Camilli P, Benfenati F, Valtorta F, Greengard P. 1990. *Annu. Rev. Cell Biol.* 6:433–60
4. Hawkins RD, Kandel ER, Siegelbaum SA. 1993. *Annu. Rev. Neurosci.* 16: 625–65
5. Swope SL, Moss SJ, Blackstone CD, Huganir RL. 1992. *FASEB J.* 6:2514–23
6. Sommer B, Seeburg PH. 1992. *Trends Pharmacol. Sci.* 13:291–96
7. Bliss TVP, Lomo T. 1973. *J. Physiol.* 232:331–56
8. Madison DV, Malenka RC, Nicoll RA. 1991. *Annu. Rev. Neurosci.* 14:379–97
9. Bliss TVP, Collingridge GL. 1993. *Nature* 361:31–39
10. McNaughton BL. 1993. *Annu. Rev. Physiol.* 55:375–96
11. Schuman EM, Madison DV. 1994. *Annu. Rev. Neurosci.* 17:153–83
12. O'Dowd BF, Lefkowitz RJ, Caron MG. 1989. *Annu. Rev. Neurosci.* 12: 67–83
13. Bourne HR, Sanders DA, McCormick F. 1990. *Nature* 348:125–32
14. Lester HA, Dascal N. 1993. *Nature* 364:758–59
15. Masu M, Tanabe Y, Tsuchida K, Shigemoto R, Nakanishi S. 1991. *Nature* 349:760–65
16. Sudhof TC, Jahn R. 1991. *Neuron* 6:665–77
17. Bennett MK, Scheller RH. 1994. *Annu. Rev. Biochem.* 63:63–100
18. Barnard EA, Darlison MG, Seeburg PH. 1987. *Trends Neurosci.* 10:502–9
19. Nakanishi S. 1992. *Science* 258:597–603
20. Gasic GP, Hollmann M. 1992. *Annu. Rev. Physiol.* 54:507–36
21. Seeburg PH. 1993. *Trends Neurosci.* 16:359–65
22. Zucker RS. 1989. *Annu. Rev. Neurosci.* 12:13–31
23. Magleby KL, Zengel JE. 1975. *J. Physiol.* 245:183–208
24. Katz B, Miledi R. 1968. *J. Physiol.* 195:481–92
25. Delaney KR, Zucker RS, Tank DW. 1989. *J. Neurosci.* 9:3558–67
26. Llinas R, McGuinness TL, Leonard CS, Sugimori M, Greengard P. 1985. *Proc. Natl. Acad. Sci. USA* 82:3035–39
27. Ito M. 1989. *Annu. Rev. Neurosci.* 12:85–102
28. Levy WB, Steward O. 1983. *Neuroscience* 8:791–97
29. Bramham CR, Srebro B. 1987. *Brain Res.* 405:100–7
30. Stanton PK, Sejnowski TJ. 1989. *Nature* 339:215–18
31. Dudek SM, Bear MF. 1992. *Proc. Natl. Acad. Sci. USA* 89:4363–67
32. Dudek SM, Bear MF. 1993. *J. Neurosci.* 13:2910–18
33. Mulkey RM, Malenka RC. 1992. *Neuron* 9:967–75
34. Christofi G, Nowicky AV, Bolsover SR, Bindman LJ. 1993. *J. Neurophysiol.* 69:219–29
35. Malenka RC. 1993. *Proc. Natl. Acad. Sci. USA* 90:3121–23
36. Artola A, Brocher S, Singer W. 1990. *Nature* 347:69–72
37. Kirkwood A, Dudek SM, Gold JT, Aizenman CD, Bear MF. 1993. *Science* 260:1518–21
38. Linden DJ, Dickinson MH, Smeyne M, Connor JA. 1991. *Neuron* 7:81–89
39. Linden DJ, Connor JA. 1991. *Science* 254:1656–59
40. Mulkey RM, Herron CE, Malenka RC. 1993. *Science* 261:1051–55
41. Baxter DA, Bittner GD, Brown TH. 1985. *Proc. Natl. Acad. Sci. USA* 82:5978–82
42. Brown TH, McAfee DA. 1982. *Science* 215:1411–13
43. Bindman LJ, Murphy KPSJ, Pockett S. 1988. *J. Neurophysiol.* 60:1053–65
44. Katz B. 1969. *The Release of Neural Transmitter Substances*. Liverpool: Liverpool Univ. Press
45. Korn H, Faber DS. 1991. *Trends Neurosci.* 14:439–45
46. Stevens CF. 1993. *Cell* 72:55–63
47. Edwards F. 1991. *Nature* 350:271–72
48. Edwards FA, Konnerth A, Sakmann B. 1990. *J. Physiol.* 430:213–49
49. Rang HP. 1981. *J. Physiol.* 311:23–55

50. Collingridge GL, Gage PW, Robertson B. 1984. *J. Physiol.* 356:551–64
51. Bekkers JM, Richerson GB, Stevens CF. 1990. *Proc. Natl. Acad. Sci. USA* 87:5359–62
52. Stern P, Edwards FA, Sakmann B. 1992. *J. Physiol.* 449:247–78
53. Ropert N, Miles R, Korn H. 1990. *J. Physiol.* 428:707–22
54. Larkman A, Stratford K, Jack J. 1991. *Nature* 350:344–47
55. Sorra KE, Harris KM. 1993. *J. Neurosci.* 13:3736–48
56. Malinow R, Tsien RW. 1990. *Nature* 346:177–80
57. Bekkers JM, Stevens CF. 1990. *Nature* 346:724–29
58. Baskys A, Carlen PL, Wojtowicz JM. 1991. *Neurosci. Lett.* 127:169–72
59. Malinow R. 1991. *Science* 252:722–24
60. Malgaroli A, Tsien RW. 1992. *Nature* 357:134–39
61. Foster TC, McNaughton BL. 1991. *Hippocampus* 1:79–91
62. Manabe T, Renner P, Nicoll RA. 1992. *Nature* 355:50–55
63. Kullmann DM, Nicoll RA. 1992. *Nature* 357:240–44
64. Larkman A, Hannay T, Stratford K, Jack J. 1992. *Nature* 360:70–73
65. Dolphin AC, Errington ML, Bliss TVP. 1982. *Nature* 297:496–98
66. Bliss TVP, Douglas RM, Errington ML, Lynch MA. 1986. *J. Physiol.* 377:391–408
67. Errington ML, Lynch MA, Bliss TVP. 1987. *Neuroscience* 20:279–84
68. Kauer JA, Malenka RC, Nicoll RA. 1988. *Neuron* 1:911–17
69. Muller D, Joly M, Lynch G. 1988. *Science* 242:1694–97
70. Davies SN, Lester RAJ, Reymann KG, Collingridge GL. 1989. *Nature* 338: 500–3
71. Bashir ZI, Alford S, Davies SN, Randall AD, Collingridge GL. 1991. *Nature* 349:156–58
72. Gahwiler BH. 1988. *Trends Neurosci.* 11:484–89
73. Nicoll RA, Kauer JA, Malenka RC. 1988. *Neuron* 1:97–103
74. Kennedy MB. 1989. *Cell* 59:777–87
75. Johnston D, Brown TH. 1984. In *Brain Slices,* ed. R Dingledine, pp. 51–86. New York: Plenum
76. Harris EW, Cotman CW. 1986. *Neurosci. Lett.* 70:132–37
77. Johnston D, Williams S, Jaffe D, Gray R. 1992. *Annu. Rev. Physiol.* 54:489–505
78. Barrionuevo G, Brown TH. 1983. *Proc. Natl. Acad. Sci. USA* 80:7347–51
79. Levy WB, Steward O. 1979. *Brain Res.* 175:233–45
80. Kelso SR, Ganong AH, Brown TH. 1986. *Proc. Natl. Acad. Sci. USA* 83:5326–30
81. Malinow R, Miller JP. 1986. *Nature* 320:529–30
82. Sastry BR, Goh JW, Auyeung A. 1986. *Science* 232:988–90
83. Wigstrom H, Gustafsson B, Huang Y-Y, Abraham WC. 1986. *Acta Physiol. Scand.* 126:317–19
84. Monaghan DT, Bridges RJ, Cotman CW. 1989. *Annu. Rev. Pharmacol. Toxicol.* 29:365–402
85. Collingridge GL, Lester RAJ. 1989. *Pharmacol. Rev.* 41:143–210
86. Watkins JC, Krogsgaard-Larsen P, Honore T. 1990. *Trends Pharmacol. Sci.* 11:25–33
87. Collingridge GL, Kehl SJ, McLennan H. 1983. *J. Physiol.* 334:33–46
88. Nowak L, Bregestovski P, Ascher P, Herbet A, Prochiantz A. 1984. *Nature* 307:462–65
89. Mayer ML, Westbrook GL, Guthrie PB. 1984. *Nature* 309:261–63
90. Jahr CE, Stevens CF. 1987. *Nature* 325:522–25
91. Mayer ML, Westbrook GL. 1987. *J. Physiol.* 394:501–27
92. Mayer ML, Westbrook GL. 1985. *J. Physiol.* 361:65–90
93. MacDermott AB, Mayer ML, Westbrook GL, Smith SJ, Barker JL. 1986. *Nature* 321:519–22
94. Bourne HR, Nicoll R. 1993. *Cell* 72:65–75
95. Monyer H, Sprengel R, Schoepfer R, Herb A, Higuchi M, et al. 1992. *Science* 256:1217–21
96. Monaghan DT, Cotman CW. 1985. *J. Neurosci.* 5:2909–19
97. Regehr WG, Tank DW. 1990. *Nature* 345:807–10
98. Regehr WG, Tank DW. 1992. *J. Neurosci.* 12:4202–23
99. Lynch G, Larson J, Kelso S, Barrionuevo G, Schottler F. 1983. *Nature* 305:719–21
100. Malenka RC, Kauer JA, Zucker RS, Nicoll RA. 1988. *Science* 242:81–84
101. Malenka RC, Lancaster B, Zucker RS. 1992. *Neuron* 9:121–28
102. Kauer JA, Malenka RC, Nicoll RA. 1988. *Nature* 334:250–52
103. Malenka RC. 1991. *Neuron* 6:53–60
104. Cormier RJ, Mauk MD, Kelly PT. 1993. *Neuron* 10:907–19
105. Kullmann DM, Perkel DJ, Manabe T, Nicoll RA. 1992. *Neuron* 9:1175–83
106. McGuinness N, Anwyl R, Rowan M. 1991. *Eur. J. Pharmacol.* 197:231–32

107. Zheng F, Gallagher JP. 1992. *Neuron* 9:163–72
108. Bortolotto ZA, Collingridge GL. 1993. *Neuropharmacology* 32:1–9
109. Eaton SA, Jane DE, Jones PLS, Porter RHP, Pook PCK. 1993. *Eur. J. Pharmacol.* 244:195–97
110. Bashir ZI, Bortolotto ZA, Davies CH, Berretta N, Irving AJ, et al. 1993. *Nature* 363:347–50
111. Nicoletti F, Meek JL, Iadarola MJ, Chuang DM, Roth BL, et al. 1986. *J. Neurochem.* 46:40–46
112. Aramori I, Nakanishi S. 1992. *Neuron* 8:757–65
113. Musgrave MA, Ballyk BA, Goh JW. 1993. *NeuroReport* 4:171–74
114. Aniksztejn L, Otani S, Ben-ari Y. 1992. *Eur. J. Neurosci.* 4:500–5
115. Otani S, Ben-ari Y, Roisinlallemand MP. 1993. *Brain. Res.* 613:1–9
116. Piomelli D, Greengard P. 1990. *Trends Pharmacol. Sci.* 11:367–73
117. Piomelli D, Volterra A, Dale N, Siegelbaum SA, Kandel ER, et al. 1987. *Nature* 328:38–43
118. Dumuis A, Sebben M, Haynes L, Pin J-P, Bockaert J. 1988. *Nature* 336:68–70
119. Lazarewicz JW, Wroblewski JT, Costa E. 1990. *J. Neurochem.* 55:1875–81
120. Sanfeliu C, Hunt A, Patel AJ. 1990. *Brain Res.* 526:241–48
121. Lynch MA, Clements MP, Voss KL, Bramham CR, Bliss TVP. 1991. *Biochem. Soc. Trans.* 19:391–96
122. Okada D, Yamagishi S, Sugiyama H. 1989. *Neurosci. Lett.* 100:141–46
123. Williams JH, Bliss TVP. 1989. *Neurosci. Lett.* 107:301–6
124. Massicotte G, Oliver MW, Lynch G, Baudry M. 1990. *Brain Res.* 537:49–53
125. Williams JH, Errington ML, Lynch MA, Bliss TVP. 1989. *Nature* 341:739–42
126. O'Dell TJ, Hawkins RD, Kandel ER, Arancio O. 1991. *Proc. Natl. Acad. Sci. USA* 88:1285–89
127. Barbour B, Szatkowski M, Ingledew N, Attwell D. 1989. *Nature* 342:918–20
128. Haley JE, Schuman EM. 1994. *Semin. Neurosci.* In press
129. Palmer RMJ, Ferrige AG, Moncada S. 1987. *Nature* 327:524–26
130. Ignarro LJ. 1989. *FASEB J.* 3:31–36
131. Garthwaite J, Charles SL, Chess-Williams R. 1988. *Nature* 336:385–88
132. Marletta MA. 1989. *Trends Biol. Sci.* 14:488–92
133. Bohme GA, Bon C, Stutzmann J-M, Doble A, Blanchard J-C. 1991. *Eur. J. Pharmacol.* 199:379–81
134. Schuman EM, Madison DV. 1991. *Science* 254:1503–6
135. Haley JE, Wilcox GL, Chapman PF. 1992. *Neuron* 8:211–16
136. Bon C, Bohme GA, Doble A, Stutzmann JM, Blanchard JC. 1992. *Eur. J. Neurosci.* 4:420–24
137. Zhuo M, Small SA, Kandel ER, Hawkins RD. 1993. *Science* 260:1946–50
138. Nicola SM, Cummings JA, Malenka RC. 1993. *Soc. Neurosci. Abstr.* 19:1328
139. Williams JH, Li Y-G, Nayak A, Errington ML, Murphy KPSJ, et al. 1993. *Neuron.* 11:877–84
140. Chetkovich DM, Klann E, Sweatt JD. 1993. *NeuroReport* 4:919–22
141. Haley JE, Malen PL, Chapman PF. 1993. *Neurosci. Lett.* 160:85–88
142. Bredt DS, Hwang PM, Glatt CE, Lowenstein C, Reed RR, et al. 1991. *Nature* 351:714–18
143. Bredt DS, Snyder SH. 1990. *Proc. Natl. Acad. Sci. USA* 87:682–85
144. Bredt DS, Glatt CE, Hwang PM, Fotuhi M, Dawson TM, et al. 1991. *Neuron* 7:615–24
145. Bredt DS, Hwang PM, Snyder SH. 1990. *Nature* 347:768–70
146. Vincent SR, Kimura H. 1992. *Neuroscience* 46:755–84
147. Valtschanoff JG, Weinberg RJ, Kharazia VN, Nakane M, Schmidt HHHW. 1993. *J. Comp. Neurol.* 331:111–21
148. Schweizer FE, Wendland B, Ryan TA, Nakane M, Murad F, et al. 1993. *Proc. Natl. Acad. Sci. USA.* In press
149. Bredt DS, Ferris CD, Snyder SH. 1992. *J. Biol. Chem.* 267:976–81
150. Brune B, Lapetina EG. 1991. *Biochem. Biophys. Res. Commun.* 181:921–26
151. Nakane M, Mitchell J, Forstermann U, Murad F. 1991. *Biochem. Biophys. Res. Commun.* 180:1396–402
152. Schmidt HHHW, Pollock JS, Nakane M, Forstermann U, Murad F. 1992. *Cell Calcium* 13:427–34
153. Verma ADJ, Glatt CE, Ronnett GV, Snyder SH. 1993. *Science* 259:381–84
154. Stevens CF, Wang Y. 1993. *Nature* 364:147–49
155. Ignarro LJ, Ballot B, Wood KS. 1984. *J. Biol. Chem.* 259:6201–7
156. McNaughton BL. 1982. *J. Physiol.* 324:249–62
157. Muller D, Lynch G. 1989. *Brain Res.* 479:290–99
158. Zalutsky RA, Nicoll RA. 1990. *Science* 248:1619–24
159. Manabe T, Wyllie DJA, Perkel DJ, Nicoll RA. 1993. *J. Neurophysiol.* 70:1451–59
160. Söllner T, Whitehart SW, Brunner M,

Erdjument-Bromage H, Geromanos S, et al. 1993. *Nature* 362:318–24
161. DeBello WM, Betz H, Augustine GJ. 1993. *Cell* 74:947–50
162. Deleted in proof
163. Shearman MS, Shinomura T, Oda T, Nishizuka Y. 1991. *FEBS Lett.* 279: 261–64
164. Akers RF, Lovinger DM, Colley PA, Linden DJ, Routtenberg A. 1986. *Science* 231:587–89
165. Reymann KG, Brodemann R, Kase H, Matthies H. 1988. *Brain Res.* 461:388–92
166. Lovinger DM, Wong KL, Murakami K, Routtenberg A. 1987. *Brain Res.* 436:177–83
167. Malinow R, Madison DV, Tsien RW. 1988. *Nature* 335:820–24
168. Malinow R, Schulman H, Tsien RW. 1989. *Science* 245:862–66
169. Skene JHP. 1989. *Annu. Rev. Neurosci.* 12:127–56
170. Benowitz LI, Routtenberg A. 1987. *Trends Neurosci.* 10:527–32
171. Andreasen TJ, Luetje CW, Heideman W, Storm DR. 1983. *Biochemistry* 22:4615–18
172. Gianotti C, Nunzi MG, Gispen WH, Corradetti R. 1992. *Neuron* 8:843–48
173. Malenka RC, Kauer JA, Perkel DJ, Mauk MD, Kelly PT, et al. 1989. *Nature* 340:554–57
174. Alexander KA, Cimler BM, Meier KE, Storm DR. 1987. *J. Biol. Chem.* 262: 6108–13
175. Dekker LV, De Graan PNE, Oestreicher AB, Versteeg DHG, Gispen WH. 1989. *Nature* 342:74–76
176. Llinas R, Gruner JA, Sugimori M, McGuinness TL, Greengard P. 1991. *J. Physiol.* 436:257–82
177. Herrero I, Miras-Portugal MT, Sanchez-Prieto J. 1992. *Nature* 360:163–66
178. Ignarro LJ, Harbison RG, Wood KS, Kadowitz PJ. 1986. *J. Pharmacol. Exp. Ther.* 237:893–900
179. Brune B, Lapetina EG. 1989. *J. Biol. Chem.* 264:8455–58
180. Chetkovich DM, Sweatt JD. 1992. *Soc. Neurosci. Abstr.* 18:761
181. Schuman EM, Meffert MK, Schulman H, Madison DV. 1992. *Soc. Neurosci. Abstr.* 18:761
182. Duman RS, Terwilliger RZ, Nestler EJ. 1993. *J. Neurochem.* 61:1542–45
183. Coggins PJ, McLean K, Nagy A, Zwiers H. 1993. *J. Neurochem.* 60: 368–71
184. Berretta N, Berton F, Bianchi R, Brunelli M, Capogna M, et al. 1991. *Eur. J. Neurosci.* 3:850–54
185. Tsien RW, Malinow R. 1990. *Cold Spring Harbor Symp. Quant. Biol.* 55:147–59
186. Xie XP, Berger TW, Barrionuevo G. 1992. *J. Neurophysiol.* 67:1009–13
187. Asztely F, Wigstrom H, Gustafsson B. 1992. *Eur. J. Neurosci.* 4:681–90
188. Muller D, Arai A, Lynch G. 1992. *Hippocampus* 2:29–38
189. Perkel DJ, Nicoll RA. 1993. *J. Physiol.* In press
190. House CE, Kemp BE. 1987. *Science* 238:1726–28
191. Hardie G. 1988. *Nature* 335:592–93
192. O'Dell TJ, Kandel ER, Grant SGN. 1991. *Nature* 353:558–60
193. Bading H, Greenberg ME. 1991. *Science* 253:912–14
194. Erondu NE, Kennedy MB. 1985. *J. Neurosci.* 5:3270–77
195. Kennedy MB, Bennett MK, Erondu NE. 1983. *Proc. Natl. Acad. Sci. USA* 80:7357–61
196. Kelly PT, McGuinness TL, Greengard P. 1984. *Proc. Natl. Acad. Sci. USA* 81:945–49
197. Goldenring JR, McGuire JS, DeLorenzo RJ. 1984. *J. Neurochem.* 42: 1077–84
198. Cotman CW, Banker B, Churchill L, Taylor D. 1974. *J. Cell Biol.* 63:441–55
199. Cohen RS, Blomberg F, Berzins K, Siekevitz P. 1977. *J. Cell Biol.* 74:181–203
200. Kennedy MB. 1993. *Curr. Opin. Neurobiol.* 3:732–37
201. Miller SG, Kennedy MB. 1985. *J. Biol. Chem.* 260:9039–46
202. Kennedy MB, Bennett MK, Bulleit RF, Erondu NE, Jennings VR, et al. 1990. *Cold Spring Harbor Spring Quant. Biol.* 55:101–10
203. Lisman JE, Goldring MA. 1988. *Proc. Natl. Acad. Sci. USA* 85:5320–24
204. Bennett MK, Erondu NE, Kennedy MB. 1983. *J. Biol. Chem.* 258:12735–44
205. Bennett MK, Kennedy MB. 1987. *Proc. Natl. Acad. Sci. USA* 84:1794–98
206. Hanley RM, Means AR, Ono T, Kemp BE, Burgin KE, et al. 1987. *Science* 237:293–97
207. Lin CR, Kapiloff MS, Durgerian S, Tatemoto K, Russo AF, et al. 1987. *Proc. Natl. Acad. Sci. USA* 84:5962–66
208. Hanley RM, Means AR, Kemp BE, Shenolikar S. 1988. *Biochem. Biophys. Res. Commun.* 152:122–28
209. Miller SG, Patton BL, Kennedy MB. 1988. *Neuron* 1:593–604
210. Schworer CM, Colbran RJ, Keefer JR, Soderling TR. 1988. *J. Biol. Chem.* 263:13486–89

211. Thiel G, Czernik AJ, Gorelick F, Nairn AC, Greengard P. 1988. *Proc. Natl. Acad. Sci. USA* 85:6337–41
212. Miller SG, Kennedy MB. 1986. *Cell* 44:861–70
213. Kwiatkowski AP, Shell DJ, King MM. 1988. *J. Biol. Chem.* 263:6484–86
214. Miller SG, Kennedy MB. 1985. *Soc. Neurosci. Abstr.* 11:645
215. Lai Y, Nairn AC, Greengard P. 1986. *Proc. Natl. Acad. Sci. USA* 83:4253–57
216. Hanson PI, Kapiloff MS, Lou LL, Rosenfeld MG, Schulman H. 1989. *Neuron* 3:59–70
217. Fong YL, Taylor WL, Means AR, Soderling TR. 1989. *J. Biol. Chem.* 264:6759–63
218. Lickteig R, Shenolikar S, Denner L, Kelly PT. 1988. *J. Biol. Chem.* 263: 19232–39
219. Hashimoto Y, Schworer CM, Colbran RJ, Soderling TR. 1987. *J. Biol. Chem.* 262:8051–55
220. Meyer T, Hanson PI, Stryer L, Schulman H. 1992. *Science* 256:1199–1202
221. Fukunaga K, Rich DP, Soderling TR. 1989. *J. Biol. Chem.* 264:21830–36
222. Fukunaga K, Soderling TR, Miyamoto E. 1992. *J. Biol. Chem.* 267:22527–33
223. Ocorr KA, Schulman H. 1991. *Neuron* 6:907–14
224. Molloy SS, Kennedy MB. 1991. *Proc. Natl. Acad. Sci. USA* 88:4756–60
225. Patton BL, Molloy SS, Kennedy MB. 1993. *Mol. Biol. Cell* 4:159–72
226. McGlade-McCulloh E, Yamamoto H, Tan S-E, Brickey DA, Soderling TR. 1993. *Nature* 362:640–42
227. Silva AJ, Paylor R, Wehner JM, Tonegawa S. 1992. *Science* 257:206–11
228. Silva AJ, Stevens CF, Tonegawa S, Wang Y. 1992. *Science* 257:201–6
229. Butler L, Silva A, Tonegawa S, McNamara JO. 1993. *Soc. Neurosci. Abstr.* 19:1030
230. Huang K-P. 1989. *Trends Neurosci.* 12:425–32
231. Malenka RC, Madison DV, Nicoll RA. 1986. *Nature* 321:175–77
232. Hu GY, Hvalby O, Walaas SI, Albert KA, Skjeklo P, et al. 1987. *Nature* 328:426–29
233. Gustafsson B, Huang Y-Y, Wigstrom H. 1988. *Neurosci. Lett.* 85:77–81
234. Muller D, Turnbull J, Baudry M, Lynch G. 1988. *Proc. Natl. Acad. Sci. USA* 85:6997–7000
235. Tingley WG, Roche KW, Thompson AK, Huganir RL. 1993. *Nature* 364: 70–73
236. Chen L, Huang L-YM. 1991. *Neuron* 7:319–26
237. Chen L, Huang L-YM. 1992. *Nature* 356:521–23
238. Keinanen K, Wisden W, Sommer B, Werner P, Herb A, et al. 1990. *Science* 249:556–60
239. Tsumoto T. 1993. *Neurosci. Res.* 16: 263–70
240. Wallace CS, Hawrylak N, Greenough WT. 1991. In *Long-Term Potentiation: A Debate of Current Issues,* ed. M Baudry, JL Davis, pp. 189–232. Cambridge: MIT Press
241. Bailey CH, Kandel ER. 1993. *Annu. Rev. Physiol.* 55:397–426

Annu. Rev. Biochem. 1994. 63:601–37

STRUCTURES AND FUNCTIONS OF MULTILIGAND LIPOPROTEIN RECEPTORS: Macrophage Scavenger Receptors and LDL Receptor-Related Protein (LRP)

Monty Krieger
Department of Biology, Massachusetts Institute of Technology, Cambridge, Massachusetts 02139

Joachim Herz
Department of Molecular Genetics, University of Texas Southwestern Medical Center, Dallas, Texas 75235

KEY WORDS: receptors, lipoproteins, multiligand binding, multidomain proteins, proligands

---

CONTENTS

0066-4154/94/0701-0601$05.00

I. INTRODUCTION

The intercellular transport of lipids through the aqueous circulatory system requires the packaging of these hydrophobic molecules into water-soluble carriers, called lipoproteins, and the regulated targeting of these lipoproteins to appropriate tissues by receptor-mediated endocytic pathways. The major classes of mammalian lipoproteins include low-density lipoprotein (LDL, the principal cholesteryl-ester transporter in human plasma), intermediate-density lipoprotein (IDL), high-density lipoprotein (HDL), very low-density lipoprotein (VLDL, a triglyceride-rich carrier synthesized by the liver), and chylomicrons (dietary triglyceride-rich carriers synthesized in the intestines). These lipoproteins have different protein (called apolipoprotein) and lipid compositions, and varying physiological activities. The best-characterized lipoprotein receptor is the LDL receptor. It recognizes apolipoproteins B-100 (apoB-100) and E (apoE), which are constituents of LDL, VLDL, IDL, and catabolized chylomicrons (chylomicron remnants). The LDL receptor plays a central role in lipid metabolism, and the analysis of its structure and function has laid the groundwork for our current understanding of receptor-mediated endocytosis (reviewed in 1–3).

LDL receptors and most other mammalian cell-surface receptors that mediate endocytosis, adhesion, or signaling exhibit two common ligand-binding characteristics: high affinity and narrow specificity (only one ligand or one class of closely related ligands). The ligand-binding properties of two more recently characterized lipoprotein receptors, macrophage scavenger receptors (types I and II) and LDL receptor-related protein (LRP), do not conform to this "one receptor–one ligand" view, which has dominated the analysis of cell-surface receptor biology. Ligand binding to scavenger receptors and LRP is characterized by high affinity and broad specificity, or "one-receptor–many ligands." As a consequence, these receptors recognize both lipoprotein and nonlipoprotein ligands and appear to participate in a wide variety of biological processes. Here we review the structures, binding properties, and diverse biological functions of these two multiligand lipoprotein receptors.

II. TYPE I AND TYPE II MACROPHAGE SCAVENGER RECEPTORS

Identification of Scavenger Receptors

Macrophage scavenger receptor activity was discovered in the laboratory of M Brown and J Goldstein in Dallas (4). These investigators and their colleagues had previously discovered LDL receptors (2), and were examining potential mechanisms for the deposition of LDL-cholesterol in artery walls during the formation of atherosclerotic plaques. One of the early features of developing atherosclerotic plaques is the accumulation of cholesterol by subendothelial macrophages (5). The observation that elevated plasma LDL levels are correlated with increased risk for atherosclerosis suggested that LDL receptors might mediate the uptake of LDL cholesterol into macrophages. However, in vitro and in vivo studies indicated that the LDL receptor pathway is not required for, and probably not involved in, massive accumulation of cholesterol by macrophages during atherogenesis (2, 6–8). Thus, at least one additional, presumably receptor-mediated, mechanism is required for lipoprotein-cholesterol uptake by macrophages. Currently, an attractive model for macrophage accumulation of lipoprotein-cholesterol is the macrophage scavenger receptor model (2, 6, 7, 9, 10). The Dallas group first showed that macrophage scavenger receptor activity can mediate the endocytosis of large amounts of chemically modified LDL (4, 11). This results in the conversion of cultured macrophages into cholesteryl ester droplet-filled cells, whose morphologies are strikingly similar to those of macrophages in atherosclerotic plaques. Effective chemical modifications that convert LDL into a ligand of scavenger receptors include acetylation or oxidation (6, 7, 12, 13). These receptors, originally named acetyl LDL (AcLDL) receptors, are now called macrophage scavenger receptors because of their broad ligand-binding specificity (see below).

Binding Properties

Initial studies of mouse peritoneal macrophages showed that scavenger receptor activity could mediate the high-affinity binding of ^{125}I-labeled AcLDL and its subsequent internalization and lysosomal degradation (4). Remarkably, this binding could be inhibited by a wide variety of compounds in a simple competition assay (4, 14, 15). Using this assay, as well as direct binding studies, investigators have shown that scavenger receptor ligands include certain types of (*a*) chemically modified proteins, such as AcLDL, oxidized LDL (OxLDL), and maleylated bovine serum albumin (M-BSA), but not their unmodified counterparts (6, 16, 17), (*b*) polyribonucleotides, including poly I and poly G, (*c*) polysaccharides, such as

Table 1 Inhibitors of macrophage scavenger receptors[a]

Effective competitors (all polyanions)	Ineffective competitors (polyanions and polycations)
Modified proteins	Native and modified proteins
AcLDL, OxLDL, maleylated LDL (M-LDL), M-HDL, M-albumin	Poly (D-glutamate), phosvitin, thyroglobulin, orosomucoid fetuin, asialoorosomucoid, lysozyme, acetylated proteins, including albumin, γ-globulin, α-1-antitrypsin, transferrin, ovalbumin, histones, ovomucoid, α-1-acid glycoprotein, and HDL, and methylated LDL
Four-stranded nucleic acids	Nucleic acids (not four-stranded)
Polyinosinic acid (poly I), poly G, poly G:I, polyxanthinylic acid, telomere models [$d(G_4T_4)_5$]	Poly A, poly C, poly U, single- and double-stranded DNA
Polysaccharides	Polysaccharides
Dextran sulfate, fucoidin, carragheenan	Heparin, chondroitin sulfate A and C, colominic acid (polysialic acid), yeast mannan
Phospholipids	Phospholipids
Phosphatidylserine	Phosphatidylcholine
Others	Others
Bovine sulfatides, polyvinyl sulfate, endotoxin, lipoteichoic acid, crocidolite asbestos	Polyphosphates (n = 65)

[a] Data compiled from studies of cultured macrophages and type I and type II receptors expressed by transfected cells. Modified from (6) and reprinted from (39).

dextran sulfate, (*d*) anionic phospholipids, including phosphatidylserine (19, 20), and (*e*) other molecules, such as polyvinyl sulfate. Some of the ligands, e.g. AcLDL and polyvinyl sulfate, are not naturally occurring and thus, not physiologically relevant. They presumably share with natural ligands structural features required for high-affinity (nM dissociation constants) binding.

All known macrophage scavenger receptor ligands are polyanionic molecules or macromolecular complexes; however, many polyanions are not ligands (see Table 1). The exact determinants that confer binding ability upon ligands have not yet been identified. Nevertheless, some progress has been made in defining the structures underlying specificity for one class of ligands, oligodeoxyribonucleotides and polyribonucleotides (21). To bind tightly to scavenger receptors, these nucleic acids must form a base-quartet-stabilized four-stranded helix (quadruplex). This conformational requirement accounts for the polyribonucleotide-binding specificity of scavenger receptors: Poly G, poly I, and other polynucleotides that form stable quadruplexes, bind, while poly A, double-stranded DNA, and others that do not form such structures, do not bind. The spatial distribution of the negatively charged phosphates in polynucleotide quadruplexes presumably forms a charged surface complementary to that of the receptor's ligand-binding domain (see below). Oligonucleotides with telomeric sequences can fold into well-defined quadruplex structures, and at least some of these molecules can bind to scavenger receptors (21). Thus, scavenger receptors may be helpful for the study of telomeric structures in natural nucleic acids.

Isolation and Cloning

After the initial partial purification and characterization of 200–260-kDa ligand-binding proteins from cultured cells and tissues (22–24), a two-step ligand-affinity and immunoaffinity chromatographic procedure was used to purify detergent-solubilized scavenger receptors approximately 240,000-fold from bovine lung membranes (25). These bovine scavenger receptors are composed of related N-glycosylated proteins: ~220-kDa trimers, ~150-kDa dimers, and ~77-kDa monomers. The dimers and trimers are converted into the monomers by reduction. Using a ligand-blotting technique, several groups have reported ligand binding to the trimeric form, but not a monomeric form generated by extensive reduction (25, 26). It has also been reported that ligands can bind to monomers after mild reduction (26). Partial amino-acid sequence of the apparently N-terminally blocked, purified bovine scavenger receptor was used to clone the corresponding cDNAs (27, 28). The murine (29, 30), human (31), and rabbit (32) homologs of the bovine cDNAs have also been cloned (see 30 for detailed sequence comparisons). In addition, genomic DNAs for the bovine, murine, and human receptors have been isolated, their partial or complete patterns of intron/exon organization have been determined, and their promoter sequences and the mechanisms regulating receptor expression are under study (30, 33, 34).

The two scavenger receptor isoforms, called type I and type II, are generated by alternative splicing of a message encoded by a single gene (29, 33). This gene is located on chromosome 8 in both mice (29) and humans (31). The cDNAs for type I and type II receptors have been transfected into COS and Chinese hamster ovary (CHO) cells, and these cells have been used to examine the receptor's binding properties and to study their biosynthesis, trimerization, posttranslational processing, and structures (see below) (9, 21, 27, 28, 30, 35–39). Expression of both types of scavenger receptor has been detected in macrophages in vitro (27, 30, 34) and in vivo (13, 31, 40). In addition, their expression is induced in cultured, phorbol ester–treated smooth muscle cells and fibroblasts (32, 41). The major sites of expression in vivo are tissue macrophages (40; S Gordon, personal communication).

Predicted Quaternary Structure

The cDNA sequences of the bovine, murine, human, and rabbit receptors (27–32, 42) in combination with biochemical (25, 36) and biophysical (38; D Resnick, K Schwartz, H Slayter, and M Krieger, unpublished) methods have been used to study scavenger receptor structure. These methods include chemical crosslinking, electrophoretic analysis, hydrodynamic studies, and electron microscopy of either secreted forms or membrane-bound forms of scavenger receptors. These studies suggest that the type I macrophage scavenger receptors are elongated homotrimeric integral membrane proteins comprising 451–454-amino-acid-long monomers (Figure 1; see also 15, 30). The predicted structure is composed of six distinctive domains (bovine receptor's numbering): I. the N-terminal cytoplasmic domain [amino-acid residues (aa) 1–50], II. a single transmembrane domain per chain (aa 51–76), and four extracellular domains—spacer domain III (aa 77–150), and domains IV (aa 151–271), V (aa 272–343), and VI (aa 344–453). Sequence features shared by all four mammalian type I receptors include eight conserved cysteine residues (one each in domains I and III and six in domain VI), a proline in the middle of the transmembrane domain, which might influence its structure or packing, and six potential asparagine-linked (N-linked) glycosylation sites in domains III and IV. Type II scavenger receptors differ from type I receptors only in that the type I receptor's cysteine-rich 110-residue C-terminal domain, called the SRCR domain (see below), is replaced by a short (6–17-amino-acid) C terminus in the type II molecules (28, 30) (Figure 1). Despite its truncated C terminus, the type II receptor exhibits high-affinity, broad-specificity ligand binding, which is similar to that of the type I receptor (28, 35, 38; however see 30 for key exceptions). Thus, the SRCR domain is not required for the receptor's characteristic broad ligand-binding specificity (see below for further discussion).

The most striking features of the type I receptor's predicted structure are the coiled-coil domains IV and V. Domain IV is composed of a series of as many as 16 seven-amino-acid repeats called heptads. The side chains of the residues in the first (position "*a*") and fourth (position "*d*") sites of the heptads are frequently aliphatic (leucine, isoleucine, valine). In other proteins, heptad repeat–containing domains have been shown to fold into right-handed amphipathic α-helices with 3.5 residues per turn (43). These then assemble into two- or three-stranded parallel bundles, called α-helical coiled-coils, in which the helices wrap around each other to form a left-handed superhelical fiber stabilized by hydrophobic interactions between the aliphatic residues at the "*a*" and "*d*" positions of the adjacent chains. The heptad repeats in α-helical coiled-coil domain IV are interrupted by a skip residue (Asn or Ser) in position *d* (residue 203, bovine sequence). The

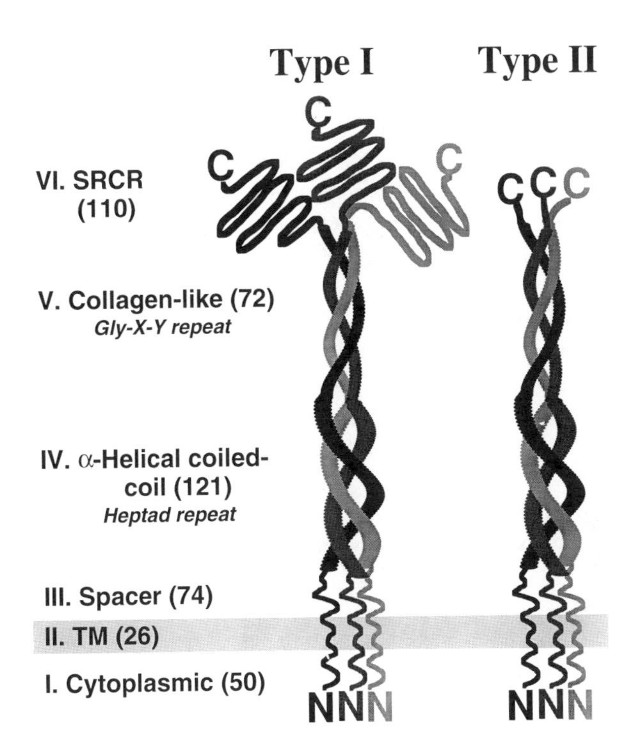

Figure 1 Models of the predicted quaternary structures of the type I and type II macrophage scavenger receptors. The type I and type II scavenger receptors each contain six domains (27, 28, 30). In both types of receptor, the first five domains are identical. They are the N-terminal cytoplasmic, transmembrane (TM), spacer, α-helical coiled coil, and collagen-like domains. The number of amino acids in each domain of the bovine type I receptor is indicated in parentheses. The C-terminal sixth domain of the type I receptor is composed of an eight-residue hinge followed by a 102-amino-acid cysteine-rich domain (SRCR), while in the type II receptor it is a short oligopeptide (6–17 residues depending on the species). This model assumes that the α-helical coiled-coil of domain IV forms a single triple-stranded left-handed superhelix, which merges with the right-handed collagenlike triple helix of domain V to form a single long fibrous stalk. Here the coiled-coil domains IV and V are overwound to emphasize their triple-helical structures. Reprinted in modified form from (15).

disruption of the heptad repeats by this skip might change the superhelical pitch or bend the superhelical axis (43, 44). The N-terminal heptads along with the disrupting residues are designated subdomain IVa, and the remaining C-terminal heptads subdomain IVb. The intron-exon structure of the receptor's genes is consistent with this subdivision (30, 33). It appears likely that the α-helical coiled-coil domain folds into an approximately 160 Å long three-stranded fiber. Imperfections in the heptads (27) might either shorten or insert discontinuities in the superhelix. Similar coiled-coil domains have subsequently been proposed for other cell-surface proteins (45).

The second coiled-coil domain, domain V, contains 23 or 24 uninterrupted Gly-X-Y triplet repeats, with prolines or lysines frequently present in the Y position (Figure 1). An additional triplet in the murine receptor could extend this domain to 25 triplets. All (murine and bovine) or all but one (human and rabbit) of the Gly-X-Y triplets are neutral, zwitterionic, or positively charged at physiological pH. The presence of triplet repeats strongly suggests that domain V forms a classic, right-handed, collagenous triple helix approximately 200 Å long. The macrophage scavenger receptors were the first integral membrane proteins containing a collagenous domain to be identified. Subsequently, the gene encoding the bullous pemphigoid autoantigen BP180 was sequenced. This strongly suggested that BP180 is also an integral membrane protein with C-terminal extracellular collagenous domains (46). Most of the proteins known to have collagenous triple-helical domains are secreted molecules, including the extracellular matrix collagens (47) with very long stretches of Gly-X-Y repeats (e.g. approximately 300 triplets for type I collagen) and a set of smaller mammalian proteins with 19–61 triplets per chain, including complement factor C1q (48), pulmonary surfactant apoproteins A and D (49–52), mannose-binding protein (53–55), conglutinin, and bovine serum lectin (56) (see below and 27, 55, 57 for additional citations). The asymmetric form of acetylcholineesterase (58) and the bacterial enzyme pullulanase (59) also contain collagenous domains.

A distinctive feature of mammalian collagens is that certain proline and lysine side chains are hydroxylated posttranslationally in the endoplasmic reticulum (ER) prior to triple helix formation (see 36 for additional citations). Proline hydroxylation stabilizes the triple helix, and inhibition of hydroxylation interferes with intracellular transport. In addition, mutations that disrupt the Gly-X-Y repeats can produce dominant-negative mutants because an abnormal chain can interfere with the trimerization of normal chains. Such mutations can cause osteogenesis imperfecta, an inherited bone disease (60). Scavenger receptors expressed in CHO cells are trimeric, some of their proline and lysine residues are hydroxylated, their efficient ER-to-Golgi transport requires hydroxylation, and an expression vector encoding a scavenger receptor with a truncated collagenous domain acts as a dominant-negative mutant (36, 37; D Resnick, K Schwartz, and M Krieger, unpublished data; and see below). Thus, experimental evidence strongly supports the sequence-based prediction that scavenger receptors contain a collagenous domain whose superhelical twist, right-handed, is predicted to be opposite to that of the left-handed superhelical twist of the α-helical coiled-coil domain IV (Figure 1). Deletion of all of C-terminal domain VI and most or all of collagenous domain V has established that the collagenous domain is not required for the trimerization and cell-surface expression of the remaining portions of the receptor (domains I–IV) (37, 42). The role of the collagenous domain in ligand binding is considered below.

The C-terminal domain VI of the type I receptors comprises an 8-residue hinge (subdomain VIa), which links the collagenous domain to the 102-residue scavenger receptor cysteine-rich (SRCR) domain (subdomain VIb). The SRCR domain helped in defining a previously unrecognized, ancient, and highly conserved family of cysteine-rich protein domains (9, 29, 60a). To date, eight different cell-surface or secreted proteins, which contain 1–11 SRCR domains per polypeptide chain, have been described (60a). The 47 SRCR domain sequences currently analyzed define two subgroups in the SRCR superfamily (60a). The consensus sequences for these are: Group A: X_{25}-^{2}C-X_{12}-^{3}C-X_{30}-^{5}C-X_{9}-^{6}C-X_{9}-^{7}C-X_{9}-^{8}C-X_{1}; Group B: X_{9}-^{1}C-X_{15}-^{2}C-X_{12}-^{3}C-X_{4}-^{4}C-X_{25}-^{5}C-X_{9}-^{6}C-X_{9}-^{7}C-X_{9}-^{8}C-X_{1}, where X_n represent n residues and ^{1}C, ^{2}C etc represent individual, conserved cysteines. It seems likely that the conserved cysteines within each SRCR domain form three (Group A) or four (Group B) intradomain disulfide bonds and that the domains fold into similar, probably globular, structures. Apparently many, if not most, of the SRCR domains participate in binding extracellular ligands, although most of these ligands, including those for the scavenger receptor, have not yet been identified. Many of the mammalian proteins containing SRCR domains are associated directly or indirectly with the immune system and host defense (60a), a finding reminiscent of the frequent appearance of immunoglobulinlike domains in diverse proteins of the immune system (61, 62).

The SRCR superfamily joins a growing list of conserved cysteine-containing domains found in extracellular portions of membrane proteins or in secreted proteins. Other families or superfamilies of such domains include the EGF-like repeat, the LDL receptor-binding (or complement C9) repeat, immunoglobulin superfamily domains, and clotting factor Kringle domains (60a, 63, 64). These disulfide-crosslinked domains appear to provide stable structures, which (*a*) are able to withstand the rigors of the extracellular environment and endocytic cycling, (*b*) are well suited for diverse biochemical functions, often involving binding, and (*c*) are readily merged with other types of domains to permit the construction of complex mosaic proteins (29, 63, 65), which presumably arose as a consequence of exon shuffling.

Biosynthesis

The pathway of the biosynthesis of bovine and murine scavenger receptors expressed in CHO cells has been examined in detail (9, 30, 36). Newly synthesized receptor subunits are inserted into the membrane of the rough endoplasmic reticulum (ER). In the lumen of the ER, a precursor form of the receptors is generated when their extracellular domains undergo a series of four covalent and conformational modifications. These include N-linked glycosylation [six high-mannose N-linked chains are added to the bovine receptors (38)]; hydroxylation of prolines and lysines; oligomerization of monomers into trimers; and disulfide bond crosslinking of some of the chains via cysteine residues in the spacer domain (domain III). Some or all of the subunits in the precursor can, but need not, be covalently crosslinked by disulfide bonds (30, 36). For example, in bovine scavenger receptors, each trimer is normally composed of a monomer and a disulfide-linked dimer formed by Cys83 residues (36). Conversion of Cys83 to Gly83 prevents interchain disulfide bond formation, but does not interfere with trimerization, intracellular transport, or endocytic activity (36). Because it is relatively simple to use scavenger receptors to measure ER-associated folding (oligomerization), disulfide bond formation, hydroxylation, and N-glycosylation, these receptors provide a useful tool for studies of ER function and the secretory pathway (L Hobbie et al, 1994, submitted). After ER processing, the precursor trimers are transported to and through the Golgi apparatus and trans Golgi network en route to the cell surface. During this transport, they are converted to mature forms by the processing of their N-linked oligosaccharides into complex, endoglycosidase H–resistant chains. From the cell surface, the receptors can participate in the endocytosis of ligands, presumably via coated pits and vesicles (40). The efficiency of the intracellular processing and transport of scavenger receptors may be cell-type dependent. Naito et al (40) have reported that most immunoreactive type I bovine scavenger receptor expressed by transfected COS cells is associated with the nuclear envelope, ER, and Golgi apparatus, but not the cell surface. In contrast, these receptors were observed mainly on the cell surface or in endocytic vesicles in transfected, macrophagelike, phorbol ester–treated HEL cells.

Regulation

The mechanisms regulating scavenger receptor expression have only recently come under investigation. In vitro studies with freshly isolated monocytes and monocytelike cell lines established that these cells express substantial levels of scavenger receptor activity only after differentiation into macrophages (25, 27, 66, 67). In two murine macrophagelike cell lines, P388D1 and RAW264, the type II receptor is the predominant isoform (30, 68), and most of the scavenger receptor activity in P388D1 cells can be attributed to the expression of type II receptors (30). Differences in the levels of expression of type I and type II receptors may be functionally significant because their binding properties, while similar, are not identical (30; see below for examples). Immunohistochemical and in situ hybridization studies suggest that the primary, if not exclusive, sites of type I and type II macrophage scavenger receptor expression in vivo are macrophages in many different tissues (13, 40; S Gordon personal communication). Moulton et al (34) and Emi et al (33) have described sequences in the promoter regions of the bovine and human type I and type II receptor genes. The control elements responsible for the primarily macrophage-specific expression of these receptors have not yet been reported.

A number of reagents have been reported to affect scavenger receptor activity in cultured systems. These include phorbol esters (25, 27, 32, 41, 67, 69, 70), endotoxin (71), poly[I:C] (72), macrophage colony–stimulating factor (M-CSF) (72a, 73), platelet secretory products (74), conditioned medium from stimulated lymphocytes (75), TGF-β1 (76), interferon-γ (77; however, see 72), retinoic acid and dexamethasone (34, 78), prostacyclin agonists (79), and 1,25-dihydroxyvitamin D3 (80). It is noteworthy that M-CSF stimulates macrophage scavenger receptor expression and also lowers plasma cholesterol levels. It is not clear if these activities are causally related. Unlike LDL receptors, scavenger receptor activity is not suppressed by increased levels of cellular cholesterol (4). Thus, unlike LDL receptors, scavenger receptors can mediate the massive accumulation of cholesterol in macrophages.

It is possible that by binding to scavenger receptors, ligands may transmit a signal and regulate cellular metabolism. Scavenger receptor ligands can affect several macrophage activities, including urokinase and interleukin-1 production (81, 82); however, a direct role of scavenger receptors in signaling has not been shown. In the case of endotoxin activation of macrophages, in vitro experiments suggest that the activation mechanism is independent of endotoxin binding to scavenger receptors (20; and see below).

Complex Polyanionic Ligand Binding to the Receptor's Collagenous Domain

Ligand binding to type I and type II macrophage scavenger receptors exhibits a number of complex characteristics. In addition to its broad specificity, (*a*) binding is temperature- and pH-dependent (4, 30, 38, 40, 42) and (*b*) there are both high- and low-affinity binding of AcLDL (30). (*c*) The lipoprotein ligands OxLDL and AcLDL exhibit nonreciprocal cross-competition (30, 35, 83–89). Nonreciprocal cross-competition (NRCC) occurs when one ligand efficiently competes for the binding of a second ligand while the second ligand does not compete, or only partially competes, for the binding of the first (35; for additional examples see 24, 30, 89–91). The NRCC phenomenon was first observed for modified LDL binding to cultured macrophages, and it was suggested that NRCC was a consequence of binding to two or more different kinds of receptor, each of which had different specificities for AcLDL and OxLDL. Therefore, the observation of NRCC in CHO cells expressing only the type I or only the type II scavenger receptors was unexpected (30, 35). Most NRCC experiments have been performed with cells under non-equilibrium conditions; therefore, some aspects of NRCC may be due to complex kinetic phenomena under these conditions. Possible explanations for NRCC include the presence of multiple binding sites on a single receptor or the existence of multiple receptor conformations with differing binding properties (see 35 for further discussion). The relationship, if any, between the nonreciprocal cross-competition and the two classes of AcLDL-binding sites referred to above is not yet clear. NRCC in macrophages may also arise because of the expression of classes of scavenger receptors other than type I and type II (see below). (*d*) There are receptor-species-dependent (bovine vs murine) and receptor-type-dependent (type I vs type II, 30) differences in ligand binding. For example, the ReLPS form of endotoxin inhibits the high-affinity, but not the low-affinity, binding of AcLDL to bovine type II and murine type I receptors, but does not inhibit binding to bovine type I receptors. These differences may have pathophysiological consequences (see 30).

Sequence analysis, mutagenesis experiments, and studies with analogs strongly suggest that the positively charged collagenous domain in scavenger receptors serves as the binding site for polyanionic ligands and plays a key role in conferring on scavenger receptors their distinctive broad, yet circumscribed, specificity (28, 30, 37, 42). Truncation of the receptor's collagenous domain prevents ligand binding (37, 42). Doi, Kodama, and colleagues have also shown that certain lysine residues in the C-terminal portion of the receptor are especially important for binding. They prepared a series of Lys-to-Ala mutants in the collagenous domain of the bovine scavenger receptor and found that mutation of Lys337 alone or in combination with mutations of lysines 327, 334, and 340 could reduce AcLDL and OxLDL binding and endocytosis (42). These and other clusters of positively charged residues in the collagenous domain are highly conserved in the four mammalian species sequenced to date (30, 42). These mutagenesis experiments established that the collagenous domain is necessary for ligand binding; recent binding studies using the soluble collagenous tails of complement subcomponent C1q suggest that the collagenous domain is sufficient to confer broad specificity (37). Although there is substantial sequence divergence between the short collagenous domains of C1q and the scavenger receptor, both contain clusters of positive charges (30, 42, 48). Polyanions bind to the collagenous tails of C1q with a broad specificity that is remarkably similar, but not identical, to that of scavenger receptors (37).

Clusters of positively charged residues apparently provide sticky surfaces that function as a kind of molecular flypaper for the high-affinity binding of specific polyanions (9). We also suspect that at least some of the conserved negatively charged residues (30) in this domain play a key role in differentiating between polyanions that can bind and those that cannot (15). The conserved negatively charged residues may serve to inhibit the binding of many polyanions by charge repulsion. Thus, polyanions will be high-affinity ligands only when their negative charges can interact with the positively charged residues on the receptor without being repelled by the receptor's negatively charged residues. The other extracellular domains of the receptor may also directly or indirectly affect polyanionic ligand binding (e.g. 30).

Other Modified LDL Receptors

The complex binding of OxLDL and AcLDL to macrophages and endothelial cells has led to the suggestion that there may be at least three different classes of modified LDL receptors, including receptors that recognize both AcLDL and OxLDL, or only one or the other of these modified lipoproteins (83, 84, 86, 87). Type I and II scavenger receptors bind both AcLDL and OxLDL, but the binding is complex (see 35; and above). Expression cloning has recently revealed that two previously identified cell-surface molecules, FcγRII-B2 (an Fc receptor, 92) and CD36 (a collagen receptor, 93) can bind OxLDL, and that this binding is not inhibited by AcLDL. The significance of the binding of OxLDL to FcγRII-B2 in transfected COS cells is uncertain because FcγRII-B2 in macrophages apparently does not play a significant role in binding OxLDL. In contrast, antibodies to CD36, which is expressed on macrophages and several other types of cells, can partially inhibit OxLDL uptake by cultured macrophages. Thus, CD36 may play a quantitatively significant role in macrophage catabolism of OxLDL. There may be other, as yet uncharacterized, modified LDL receptors expressed by macrophages.

There are also receptors for modified LDL on endothelial cells (85–87, 94–96). The endothelial cell receptor activity is not mediated by the type I and type II class of receptors described here (32, 40, 86, 87, 96a). In addition, using in vitro binding studies, Ottnad et al (89) and Schnitzer et al (97) have observed scavenger receptor–like modified LDL binding to relatively small (15–86-kDa) membrane-associated proteins. Finally, a mutant Chinese hamster ovary cell line, Var-261, expresses an AcLDL and broad polyanionic ligand-binding scavenger receptor, which is distinct from the type I and type II macrophage scavenger receptors. This type III scavenger receptor's relationship to other modified LDL receptors is currently under investigation (S Acton, M Penman, J Ashkenas, and M Krieger, unpublished data).

Other surface molecules with multiple ligand-binding properties are modified albumin receptors (98–100), LRP (see below), and the β-2 integrins. Integrins are heterodimeric integral membrane proteins, many of which bind one or more extracellular matrix or plasma components: fibronectin, laminin, collagen, vitronectin, fibrinogen, von Willebrand factor, thrombospondin, and osteopontin (101). Recently, the MAC-1 and p150,95, but not the LFA-1, β-2 integrins have been reported to bind to denatured proteins (102). Further analysis of the distinctive binding properties of these multiligand receptors may provide insight into other systems in which broad binding specificity is important, e.g. chaperonins, heat-shock proteins, and multidrug resistance.

Macrophage Scavenger Receptor Functions

The physiologic and pathophysiologic functions of the type I and type II macrophage scavenger receptors have not been established with certainty. However, because of their broad ligand-binding specificities, they have been proposed to participate in several macrophage-associated processes, including atherosclerosis, adhesion, and host defense (9, 15).

ATHEROSCLEROSIS It has been proposed, but not yet been proven, that macrophage scavenger receptors may be involved in cholesterol accumulation during atherogenesis (see above) (6, 7, 9). Several lines of evidence support this proposal. (*a*) Expression of scavenger receptors in cultured macrophages or transfected CHO cells can lead to their conversion to lipid-laden cells similar to those in plaques when the cells are incubated with modified LDL (11, 35). (*b*) Scavenger receptor mRNA and protein have been detected in atherosclerotic plaques (13, 31). (*c*) The ligand OxLDL is present in plaques (12, 13, 103). (*d*) The antioxidant probucol can inhibit plaque formation in an animal model of atherosclerosis (7, 104, 105). Additional genetic and/or pharmacologic studies will be required to establish definitively the role of these or analogous receptors in atherosclerosis.

ADHESION The presence of the fibrous and SRCR domains in scavenger receptors raises the possibility that they may be involved in cell-cell or cell-extracellular matrix interactions (9). This might account for the observation that scavenger receptors are highly expressed on macrophages but not circulating monocytes. Recent experiments of Gordon and colleagues have shown that antimurine macrophage scavenger receptor monoclonal antibodies can inhibit cation-independent adhesion of macrophages to an artificial substrate (68). Thus, macrophages may exploit the adhesive as well as the endocytic capacity of scavenger receptors.

HOST DEFENSE A hallmark of macrophage function is the recognition, internalization, and often destruction, of a variety of foreign and endogenous substances. These include pathogenic organisms, such as bacteria and protozoans (106). Macrophages recognize some pathogens as foreign because they have been coated by antibodies or complement. Neutrophils, macrophages, and other cytotoxic cells also release numerous enzymatic and chemical agents (e.g. oxidants during the oxidative burst), which can modify pathogens, and macrophages may be able to recognize such modified substances. Alternatively, macrophages can directly bind unmodified pathogens, and such direct recognition of foreign substances ("nonself") is a crucial mechanism in host defense (106–109). Receptors with broad binding specificity might be used by macrophages to discriminate directly between self and nonself, or modified and unmodified, and such receptors may have arisen early in the evolution of host defense systems (107). The type I and type II macrophage scavenger receptors are attractive candidates to play such a role in addition to their putative role in atherosclerosis (9, 20, 29).

Recent experiments support the possibility that these receptors may provide a general mechanism for macrophage recognition and internalization of pathogens and their cell-surface components. For example, certain forms of bacterial endotoxin bind to type I and type II scavenger receptors (lipid IV_A, ReLPS) and are cleared from the circulation in vivo by scavenger receptors (20, 30). Endotoxin (lipopolysaccharide) is a crucial component of the outer surface of the outer membrane of gram-negative bacteria and is responsible for inducing toxic shock (110). Thus, scavenger receptors presumably help protect the body from endotoxic shock during gram-negative bacterial sepsis. Scavenger receptors also bind to purified lipoteichoic acid (LTA), an anionic polymer expressed on the surface of many gram-positive bacteria, and to intact gram-positive bacteria, including *Streptococcus pyogenes, Staphylococcus aureus,* and *Listeria monocytogenes* (110a). LTA presumably mediates receptor binding to the intact bacteria. LTAs, for which no other host cell receptors have been identified,[1] are analogous to endotoxin in that they are implicated in the pathogenesis of septic shock due to gram-positive bacteria. Thus, scavenger receptors may participate in host defense by clearing LTA and/or intact bacteria from tissues and the circulation during gram-positive sepsis.

The broad specificity of scavenger receptors might occasionally be deleterious rather than protective. For example, these receptors can specifically bind to crocidolite asbestos, raising the possibility that they participate in the pathologic interactions of inhaled asbestos with alveolar macrophages (38). Also, scavenger receptor recognition of OxLDL may normally be protective because oxidized lipids in OxLDL can kill dividing cells (111); however, in the face of hypercholesterolemia, the massive uptake of cho-

[1]See note added in proof, p. 637

lesterol from OxLDL via scavenger receptors may overwhelm the normal mechanisms for handling excess cholesterol and contribute to atherosclerosis (see above). In addition, many pathogens have evolved to use host cell-surface molecules as receptors. It would not be surprising to find that some intracellular pathogens of macrophages use scavenger receptors to gain entry into the cells.

The unusual binding properties of short collagenous domains may contribute not only to the potential host-defense functions of scavenger receptors, but also to those of other proteins with such domains, including complement subcomponent C1q (57), the mannose-binding protein (53–55), conglutinin, bovine serum lectin (56), and lung surfactant apoproteins A (49) and D (50–52). All of these molecules are thought to participate in clearing the extracellular space of debris and pathogenic material. Proteins with specialized short collagenous domains compose one of at least two broad functional categories of proteins containing collagenous domains: those implicated in host defense [D (defense)-collagens] and those in the extracellular matrix [S (structural)-collagens] (15, 37).

D-collagens may have evolved to perform fundamental host-defense functions, possibly before the humoral immune system developed. In this regard, macrophage scavenger receptor activity is expressed in *Drosophila melanogaster* embryonic macrophages in vivo and in vitro (112). In primary embryonic cell cultures, uptake of fluorescent AcLDL is macrophage specific and exhibits the broad specificity characteristic of mammalian scavenger receptors. Furthermore, Schneider L2 cells, a cultured *Drosophila* cell line, exhibit a scavenger receptor–mediated endocytic pathway, which is almost identical to that of mammalian macrophages. The L2 cell receptors exhibit characteristic scavenger receptor–like broad polyanion-binding specificity, and mediate high-affinity and saturable binding, uptake, and degradation of AcLDL. In L2 cells, the kinetics of intracellular ligand degradation after binding and uptake shows a lag phase, intracellular ligand degradation is chloroquine sensitive, and endocytosis is temperature dependent.

The structure and function of the *Drosophila* scavenger receptors and their relationships to mammalian scavenger receptors are currently under investigation (A Pearson, A Lux, M Krieger, unpublished). The presence of scavenger receptors on both mammalian and *Drosophila* macrophages suggests that they may mediate critical, well-conserved functions, possibly pathogen recognition, and raises the possibility that they may have appeared early in the evolution of host defense systems (9, 29, 112). The application of the genetic techniques available in *Drosophila* should provide a powerful new approach for studying scavenger receptor function. It is noteworthy that postembryonic macrophagelike hemocytes in *Drosophila* participate in wound healing, encapsulation of pathogens, and phagocytosis (113). Furthermore, macrophages play an important role in the recognition of apoptotic or senescent cells during the course of development, normal cell turnover (e.g. red blood cell catabolism), and aging (106, 112, 114, 115). It remains to be determined if scavenger receptors are also involved in these important developmental processes.

III. LDL RECEPTOR-RELATED PROTEIN (LRP)

Identification and Cloning

Two key lipoprotein particles, which transport triglycerides and other lipids from the intestines or the liver to muscle, fat, and other tissues, are chylomicrons and very low-density lipoproteins (VLDL) (115a). During the course of their transport in the circulatory system, these lipoproteins are modified both by hydrolases (e.g. lipoprotein lipase) and by the association and dissociation of proteins (apolipoproteins). The modified forms of these lipoproteins, called remnants, are rapidly cleared from the circulation by receptors on hepatic parenchymal cells (115b). The remnant lipoproteins bind to hepatic receptors via multiple apolipoprotein E (apoE) moieties, which are present on their surfaces. Although genetic analysis showed that the "remnant" receptor was distinct from the well-defined LDL receptor (116–118), both types of receptors can recognize lipoproteins containing apoE.

In 1988, a cDNA-encoding a candidate "remnant" receptor was cloned by a homology screening approach (119). The predicted protein sequence consists of a series of distinctive domains or structural motifs, which are also found in the LDL receptor (reviewed in 1, 2). Thus, this receptor was called the LDL receptor–related protein or LRP. Both the mRNA for LRP and the protein are found in many tissues and cell types (119, 120). LRP is primarily expressed in the liver, the brain, and the placenta. The parenchymal cells synthesize most of the hepatic LRP, although the protein is also present in Kupffer cells. In the brain, LRP is expressed by neuronal cells (121), and in the placenta it is abundant in the syncytiotrophoblast (122, 123). LRP also has been observed in the lesions of patients with Alzheimer's syndrome (124), and has been defined as a monocyte differentiation antigen (125). Studies reviewed below provide strong evidence that LRP is a multiligand receptor for lipoprotein remnants and other physiologically important ligands.

Predicted Domain Structure

The primary sequence of LRP contains 4525 residues, which compose a series of protein domains (see Figure 2). The extracellular, N-terminal

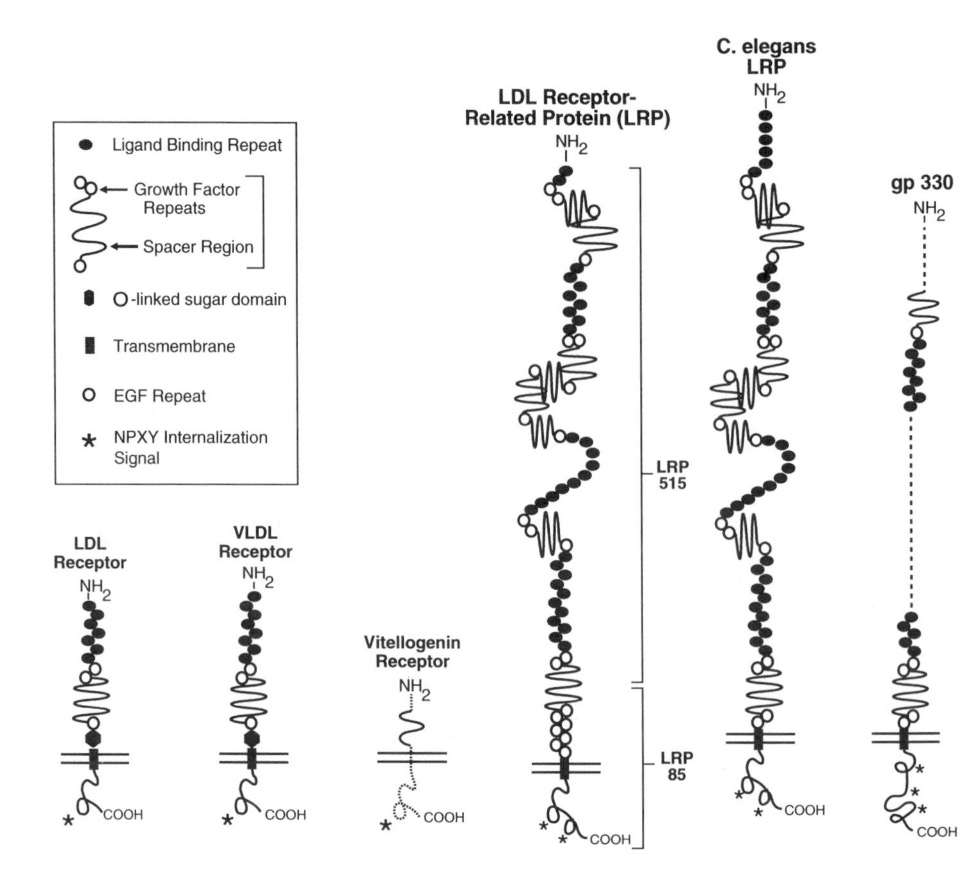

Figure 2 LDL receptor gene family. All members of the LDL receptor gene family consist of the same basic structural motifs. Ligand-binding (complement-type) cysteine-rich repeats of approximately 40 amino acids are arranged in clusters (ligand-binding domains) that contain between 2 and 11 repeats. Ligand-binding domains are always followed by EGF-precursor homologous domains. In these domains, two EGF-like repeats are separated from a third EGF-repeat by a spacer region containing the YWTD motif. In LRP and gp330, EGF-precursor homologous domains are either followed by another ligand-binding domain or by a spacer region. The EGF-precursor homology domain, which precedes the plasma membrane, is separated from the single membrane-spanning segment either by an O-linked sugar domain (in the LDL receptor and VLDL receptor) or by one (in *C. elegans* LRP and gp330) or six EGF-repeats (in LRP). The cytoplasmic tails contain between one and three "NPXY" internalization signals required for clustering of the receptors in coated pits. In a late compartment of the secretory pathway, LRP is cleaved within the eighth EGF-precursor homology domain. The two subunits LRP-515 and LRP-85 (indicated by the brackets) remain tightly and noncovalently associated. Only partial amino acid sequences of the vitellogenin receptor and of gp330 are available.

portion of LRP is composed of multiple copies of three of the four structural motifs found in the extracellular domain of the LDL receptor, including the cysteine-rich repeats, which form the LDL-binding domain ("ligand-binding repeats"), the cysteine-rich epidermal growth factor (EGF) repeat, and an EGF precursor homology domain consisting of two EGF-type cysteine-rich repeats (126), a spacer region, and a third EGF-repeat (see Figure 2). The juxtamembrane O-linked sugar domain in the LDL receptor is not present in LRP.

Seven adjacent copies of the approximately 40-residue-long cysteine-rich LDL receptor ligand-binding repeat are present in the LDL receptor (127, 128; Figure 2). Each of these repeats is characterized by a highly conserved spacing pattern of six cysteine residues that form three intramolecular disulfide bonds (3) and several negatively charged amino acids between the fourth and sixth cysteines, including the highly conserved Ser (S)—Asp (D)—Glu (E) sequence. The SDE-motif is important for the high-affinity binding of corresponding positively charged sequences in the LDL receptor's two ligands, apolipoprotein B-100 (apoB-100) and apoE (129–131). Homologous repeats have also been found in the terminal complement components (132–135), and are therefore sometimes synonymously referred to as complement-type repeats. In LRP, 31 copies of these similar, but not identical, repeats are arranged in four clusters containing 2, 8, 10, and 11 repeats per cluster.

In both the LDL receptor and LRP, an EGF-precursor homology domain is present at the C terminus of each cluster of ligand-binding repeats. Two of the four EGF-precursor homology domains in LRP are adjacent to one or three tandem repeats of a truncated form of this domain containing a spacer region and only one, rather than three, cysteine-rich EGF repeats (see Figure 2). The most C-terminal EGF-precursor homology domain is separated from the protein's single transmembrane domain by six additional EGF repeats. Each of the eight EGF-precursor homology spacer domains in LRP and the single spacer domain in the LDL receptor contain five repeats of a motif whose consensus sequence is Tyr (Y)—Trp (W)—Thr (T)—Asp (D). These are located at approximately 50-amino-acid intervals within the spacer domain. The EGF-precursor homology domain of the LDL receptor is required for the acid-dependent dissociation of ligands from the receptor (136). The function of the YWTD-repeats has not been determined; however, they may play a role in detecting and responding to the low pH in intracellular compartments.

LRP is a type I membrane protein (extracellular N terminus, cytoplasmic C terminus; 137) and is anchored in the plasma membrane by a single transmembrane segment. The C-terminal cytoplasmic domain is 100 amino acids long, twice the length of the LDL receptor's cytoplasmic domain.

The cytoplasmic domain of LRP contains two copies of the NPXY-motif, which has been shown to be an internalization signal used by many endocytic receptors for rapid coated pit–mediated endocytosis (119, 138–140).

Biosynthesis

LRP is synthesized as a single polypeptide chain precursor in the endoplasmic reticulum (119). During its passage through the secretory pathway, it is modified by N-linked glycosylation and by a proteolytic processing event that takes place in a late Golgi compartment (141). There, LRP is cleaved between amino acids 3924 and 3925 within the spacer region of the most C-terminal EGF-precursor homology domain. This tetrabasic cleavage site conforms to the consensus recognition sequence of furin, a resident protein precursor processing hydrolase in the secretory pathway (142, 143). Proteolysis generates a large N-terminal subunit, LRP-515, which contains most of the extracellular portion of the molecule and which remains tightly and noncovalently associated with the smaller transmembrane and cytoplasmic domain containing C-terminal LRP-85 subunit (Figure 2). The significance of this cleavage for the function of LRP has not yet been elucidated. It is possible that cleavage late in the secretory pathway is required to activate LRP and prevent the binding of LRP to potential ligands, which the molecule might encounter during its passage through the endoplasmic reticulum and the Golgi apparatus.

After cleavage, LRP is transported to the cell surface (119, 144). At steady state, the subcellular distribution of LRP is similar to that of the LDL receptor (141). It is predominantly present in endosomes and in recycling vesicles (145).

Functions

LRP IS AN ENDOCYTIC RECEPTOR The primary structure of LRP, its tissue distribution, its location on the cell surface, and especially the presence of two potential endocytosis signals in its cytoplasmic domain suggest that LRP is an endocytic hepatic receptor for circulating ligands. This was established by use of monoclonal antibodies in a series of experiments in cultured cells and in vivo (137). One antibody, designated anti-LRP 515, recognizes the LRP-515 portion of the molecule and has the striking property of pH-dependent binding to LRP. It binds at normal physiologic pH ($\sim$7.4) and is released at low pH ($\sim$5). When this antibody is added to the medium of cultured cells at 37° C, it binds to LRP on the cell surface, and is then internalized, released from LRP (presumably in the acidic lumen of endosomes), and degraded in lysosomes. Thus, the fate of LRP-bound anti-LRP-515 is similar to the fates of the ligands of many endocytic receptors (binding, internalization, dissociation in endosomes, lysosomal degradation; 3).

In contrast, another antibody, anti-LRP 85, binds to an extracellular epitope on the smaller transmembrane LRP-85 subunit, which does not have an intact EGF-precursor homology domain, and is not released from LRP at pH 5. Anti-LRP 85 is endocytosed with the receptor but does not undergo lysosomal degradation. Instead, anti-LRP 85 recycles back to the plasma membrane with the receptor. The pH-dependent release of anti-LRP 515, but not anti-LRP 85, raises the possibility that, as is the case for the LDL receptor, the EGF-precursor homology domains in LRP may play a role as pH-sensors.

When ^{125}I-labeled anti-LRP 515 and anti-LRP 85 are injected intravenously into rabbits, they are rapidly removed from the circulation by the liver. This demonstrates that LRP can function in vivo as an hepatic endocytic clearance receptor.

LIPOPROTEIN LIGANDS The ability of LRP to mediate cellular uptake of apoE-containing, remnant-like lipoproteins, which had been predicted from its primary structure, has been confirmed by several investigators (146, 147). For example, Kowal et al (146) used polyclonal anti-LRP antibodies to show that LRP mediates the apoE-dependent uptake of β-VLDL by LDL receptor-deficient cultured fibroblasts. β-VLDL is a cholesterol-rich lipoprotein fraction isolated from the plasma of cholesterol-fed rabbits, and consists predominantly of chylomicron and VLDL remnants. β-VLDL is an LDL receptor ligand, and, when enriched with apoE, can be recognized by LRP. Independently, Beisiegel et al (147) demonstrated in binding and crosslinking studies that apoE-containing liposomes interact with LRP on the surface of cultured cells.

Ligand blot experiments (148) also demonstrated the dependence of lipoprotein binding to LRP upon enrichment of the particles with exogenous apoE (149). In these experiments, partially purified LRP and LDL receptors were separated by nonreducing SDS-gel electrophoresis, transferred to nitrocellulose, and incubated with biotin-labeled β-VLDL in the absence or presence of exogenous apoE. Binding of the lipoprotein to the immobilized receptors was detected by incubation of the filters with ^{125}I-labeled streptavidin. As had been observed in the cultured cells, β-VLDL bound to the LDL receptor but not to LRP in the absence of exogenous apoE. Addition of apoE to the incubation mixture greatly stimulated the binding of the lipoprotein to LRP, but did not significantly alter the binding of β-VLDL to the LDL receptor. Ligand binding to LRP requires Ca^{2+} (147, 149, 150).

The binding of apoE-containing lipoproteins to purified LRP or to LRP in cultured cells requires enrichment with additional apoE. This may be relevant in vivo. Mahley et al (151) and Yamada and colleagues (152) observed a significant rapid reduction of plasma cholesterol levels following the injection of apoE into cholesterol-fed rabbits and into Watanabe heritable hyperlipidemic rabbits, which lack functional LDL receptors. In the cholesterol-fed rabbit, LDL receptor activity is both suppressed and saturated by the high concentration of circulating β-VLDL (153). The rapid drop in plasma cholesterol levels following apoE injection suggests that the increase in plasma apoE-concentration leads to apoE enrichment of lipoproteins, and thus stimulates a clearance pathway that is independent from the LDL receptor, possibly through LRP.

Additional support for the role of LRP in remnant clearance comes from experiments using another class of apolipoproteins, the C-apolipoproteins (apoCI, CII, CIII). ApoC has been shown to reduce apoE-dependent chylomicron remnant removal by the liver (154–157), and Kowal et al (149) demonstrated an opposing effect of the C and E apolipoproteins on the cellular uptake of β-VLDL by LDL receptor-deficient human fibroblasts and on binding to partially purified LRP using ligand blotting. The inhibitory effect of apoC on apoE-dependent remnant binding and uptake by LRP may be due to apoC competition for apoE-binding sites on the surface of the β-VLDL particles (158). Alternatively, a direct protein interaction of apoCs with apoE might interfere with the interaction of apoE with LRP (159).

Simultaneous binding of apoE molecules to the multiple ligand-binding domains of LRP might be required for high-affinity binding of the lipoprotein particle. Such a mechanism would ensure that only completely processed chylomicron remnants, which have accumulated sufficient apoE on their surface, are substrates for LRP-dependent uptake into the liver. Interestingly, apoE is abundantly expressed in tissues and cell types that either themselves make LRP or that are in close proximity to cells expressing this receptor. ApoE is produced by and found in high concentrations on the surface of hepatocytes (160). These findings form the basis for a secretion/recapture model of apoE-dependent remnant uptake by hepatic parenchymal cells (159). ApoE is also synthesized by glial cells in the brain (161–164) and in the yolk sac during early embryonic development (165). This coordinated expression pattern supports the hypothesis that apoE and LRP work synergistically in the hepatic uptake of lipoproteins and possibly in the transport of lipids between cells.

In a recent study, Beisiegel and colleagues (166) reported the presence of an independent binding site for lipoprotein lipase (LPL) on LRP. LPL is bound to the surface of endothelial cells in peripheral capillaries, predominantly muscle and adipose tissue. This enzyme is responsible for the hydrolysis of triglycerides in chylomicrons and VLDL, a key step in remnant formation. LRP is required for the endocytosis and degradation of LPL by cultured cells (167). LPL can substitute for apoE, and can directly mediate the binding of lipoproteins to the receptor (166–168). LPL might therefore play an active role in the hepatic uptake of chylomicrons and VLDL remnants by LRP. However, LPL association cannot be sufficient for chylomicron remnant removal in vivo, because apoE is also required. Humans (169) and mice (170, 171) genetically deficient in apoE display a profound clearance defect of chylomicron remnants, although they express normal amounts of LPL.

Further indirect evidence for the postulated role of LRP in chylomicron remnant removal by the liver stems from experiments using another LRP ligand, lactoferrin. Willnow et al (168) have shown that very low concentrations of lactoferrin (10 μg/ml) are highly effective in competing for apoE-dependent lipoprotein uptake by LRP in cultured cells, and Huettinger et al (172, 173) found that intravenous injection of bovine lactoferrin inhibited hepatic chylomicron remnant removal in rabbits.

In addition to its role in normal lipoprotein metabolism, LRP may participate in the uptake of lipoprotein cholesterol into macrophages during atherogenesis. In response to the initial accumulation of cholesterol through other pathways, for example via scavenger receptors (see above), cultured macrophages synthesize and secrete large amounts of apoE (174). This apoE would be available to associate with lipoproteins in the subendothelial space of a developing atheromatous plaque. This could initiate an autocatalytic cycle in which LRP, which is expressed on the surface of monocytes, mediates the internalization of these apoE-enriched lipoproteins via a secretion/recapture mechanism similar to that proposed for the hepatic uptake of chylomicron remnants (159). In addition, the expression of LRP, as well as that of scavenger receptors, is stimulated by M-CSF, one of many cytokines present in atherosclerotic lesions (175). Additional experiments are required to explore this possibility.

PROTEASE/INHIBITOR LIGANDS

LRP is the α_2-macroglobulin receptor The discovery by Strickland et al (176) and Kristensen et al (177) that LRP is identical to the receptor for α_2-macroglobulin demonstrated that LRP plays a complex physiological role as a multifunctional receptor for ligands involved in diverse biological processes. α_2-Macroglobulin is a major tetrameric plasma protein, which serves as a circulating inhibitor for all known classes of endoproteinases (reviewed in 178). Binding of an endoproteinase to a "bait region" within the α_2-macroglobulin molecule induces a conformational change in the inhibitor that leads to the exposure of a receptor recognition site. The activated α_2-macroglobulin/proteinase complex is then rapidly removed from the plasma by high-affinity receptors in the liver (179, 180). Moestrup et al (181) used ligand blotting to map the binding site for α_2-macroglobulin/protease complexes to a region in LRP that comprises amino acids 776–1399. This region includes the fourth EGF repeat from the N terminus, the adjacent cluster of eight LDL receptor ligand-binding repeats, and approximately one-half of the EGF precursor homology domain that follows that cluster.

Treatment of α_2-macroglobulin with methylamine results in cleavage of an intramolecular thiol ester bond. This mimics the conformational change induced by endoproteinase attack and generates activated, receptor-binding competent, α_2-macroglobulin (α_2M^*) (182). Hussain et al (183) and Choi & Cooper (184) observed cross-competition of α_2M^* with chylomicron remnants in cultured cells, indicating that both ligands bind to LRP. However, investigators have observed little (183) or no (185, 186) effect of α_2M^* on the in vivo clearance of chylomicrons in normal animals. These in vivo results are probably due to two independent mechanisms that play a role in chylomicron uptake into the liver. First, the LDL receptor can bind apoE-containing chylomicron remnants independent of LRP, and this binding is not inhibitable by α_2M^*. Second, although α_2M^* and chylomicron remnants cross-compete for binding to LRP, α_2M^* may not interfere with chylomicron remnant binding to proteoglycans, and hence might not dramatically reduce in vivo clearance (160, 187).

LRP is a plasminogen activator/inhibitor receptor Another protease/inhibitor complex has been shown to bind to and be internalized by LRP. Complexes of the plasminogen activators tPA (tissue-type plasminogen activator) and uPA (urokinase-type plasminogen activator) with their specific inhibitor, plasminogen activator inhibitor-1 (PAI-1), are recognized by LRP with high affinity. Binding to LRP occurs via the PAI-1 moiety (188, 189). The active protease tPA itself can also be recognized by LRP independently of PAI-1 (190).

Plasminogen activators are serine proteases with a narrow substrate specificity. They convert the inactive zymogen plasminogen to plasmin, a highly active protease with broad substrate specificity (reviewed by 191). Plasmin participates in a variety of biological processes, including tissue restructuring, wound healing, thrombolysis, and metastasis (192–195). Plasminogen activators are used therapeutically to treat the acute occlusion of coronary blood vessels in myocardial infarction. tPA is cleared rapidly from the circulation into the liver following intravenous injection (196). Warshawsky et al (197) have now shown that LRP participates in the clearance of tPA from the circulation. This is another example of how LRP is a highly effective hepatic receptor for several circulating ligands (137, 180).

A separate clearance pathway for tPA involves the mannose receptor, which recognizes high mannose groups on tPA (198).

In contrast to tPA, which is secreted from several cell types as an active protease that is further stimulated by the presence of fibrin, uPA is thought to function primarily while bound to a specific uPA receptor on the cell surface (194, 199, 200). The primary structure of this glycosyl-phosphatidyl inositol–anchored uPA receptor (201) has been elucidated by cDNA cloning (202). Receptor-bound uPA remains on the cell surface for long periods of time. However, inactivation of the receptor-bound protease with PAI-1 results in the rapid removal of the complex from the cell surface by endocytosis and subsequent lysosomal degradation (195, 203, 204). Because the uPA-receptor does not have a cytoplasmic tail containing an endocytosis signal and because only uPA/PAI-1 complex, but not the uncomplexed uPA, is rapidly cleared from the cell surface, it is likely that LRP participates in this process. It has been proposed that the clearance of uPA/PAI-1 complex from the cell surface proceeds after the formation of a tetrameric complex containing uPA-receptor, uPA/PAI-1 complex, and LRP (188, 189, 205).

TOXIC LIGANDS Thompson et al (206) and Kounnas et al (207) have shown that LRP is the long-sought receptor for the exotoxin A from *Pseudomonas aeruginosa*. This toxin consists of three functionally distinct domains (208): a receptor-binding domain, a fusogenic domain, which is required for the translocation of the toxin across the endosomal membrane, and an enzymatic domain, which mediates ADP-ribosylation of elongation factor-2, and thereby blocks the protein synthesis machinery of the cell (209). Endocytosis of the toxin through LRP is necessary for the toxic effect. Embryonic fibroblasts that are homozygous for a disruption of the LRP gene (205), but not heterozygous or wild-type cells, are resistant to this toxin (209a).

LRP may also indirectly contribute to the removal of endotoxins from the systemic circulation. Harris et al (210) found that chylomicrons protected rats from a lethal dose of *E. coli* endotoxin by shunting the toxin to hepatocytes and thereby decreasing cytokine secretion.

Sequential Receptor Binding—Implications for Removal of Chylomicron Remnants and Proteases from the Cell Surface

LRP has been implicated in the cellular uptake of at least three ligands, which are initially bound to other sites on the cell surface (Figure 3). These ligands are apoE and LPL, which are associated with extracellular proteoglycans (160, 187, 210a, 211), and the uPA/PAI-1 complex, which is bound to the uPA-receptor (188, 205). ApoE or LPL is required for or facilitates the uptake of remnant lipoproteins into cells, and treatment of cultured cells with heparinase dramatically reduces LRP-dependent uptake

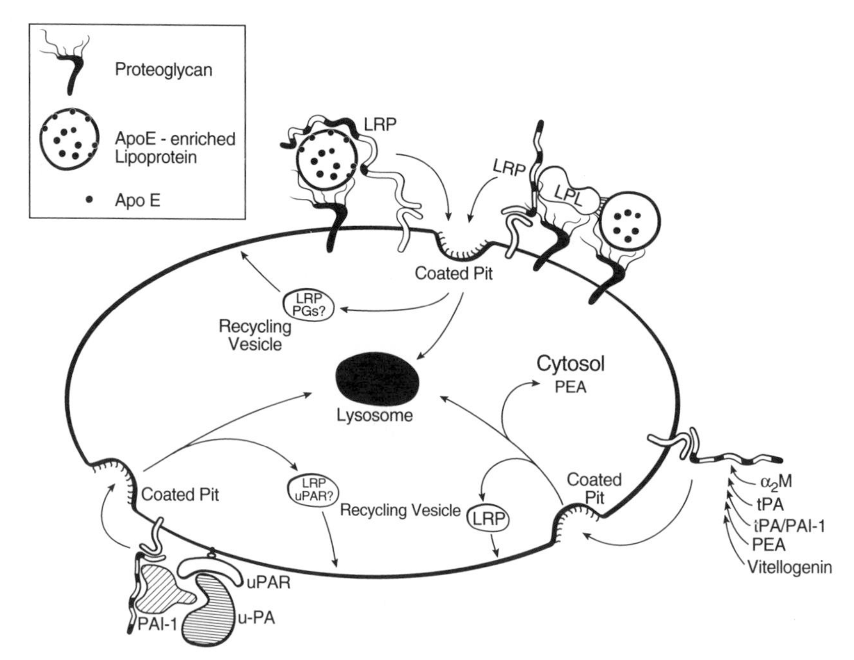

Figure 3 Ligand uptake by LRP. LRP internalizes ligands from the extracellular space by at least three distinct mechanisms: (*a*) uPA/PAI-1 complexes are recognized by LRP probably while they are bound to a uPA-specific receptor. The uPA-receptor does not contain an endocytosis signal, and uPA/PAI-1 complexes are not endocytosed when LRP is blocked (188, 205). It is currently not known if the uPA-receptor is internalized with LRP and the uPA/PAI-1 complex. (*b*) LRP-mediated uptake of apoE-enriched lipoproteins and of lipoprotein lipase (LPL) is enhanced by their prior binding to cell-surface proteoglycans (187, 211). Proteoglycan-bound LPL can be recognized by LRP in the absence of lipoprotein (in contrast to apoE), or it can substitute for apoE and mediate lipoprotein uptake (166–168). (*c*) Soluble ligands bind directly to LRP and are internalized apparently without intermediate binding to other surface molecules. These ligands include α_2-macroglobulin/protease complexes, tPA, tPA/PAI-1 complexes, *Pseudomonas* exotoxin A (PEA), and vitellogenin. In contrast to the other ligands, PEA escapes lysosomal degradation by translocating across intracellular membranes into the cytoplasm, where it enzymatically inactivates elongation factor-2 (208).

of lipoproteins or of LPL (187, 211). Likewise, competition of the binding of uPA/PAI-1 complex to the uPA-receptor with the amino-terminal fragment of uPA reduces uptake and degradation of the uPA/PAI-1 complex by human monocytes (188). These findings suggest that localization of these ligands at the cell surface, where they are either further modified by the action of lipolytic enzymes, LPL, and hepatic lipase (212), or inactivated by inhibitors, is an important step that precedes internalization through LRP (Figure 3). It is not known if the primary binding sites (proteoglycans, uPA-receptor) of these ligands are endocytosed along with the ligand or if they dissociate prior to LRP-mediated internalization. If they are internalized, they may be degraded in lysosomes or they may dissociate from their ligands and recycle to the cell surface. In contrast, sequestration on the cell surface by binding sites other than LRP is apparently not required for the cellular uptake of α_2M^* by LRP (213).

There are precedents for dual receptor systems in which cell-surface proteoglycans are required to present a ligand to its receptor. For example, signal transduction via the basic fibroblast growth factor (bFGF) receptor requires the binding and presentation of bFGF by extracellular proteoglycans (214–216). A similar mechanism appears to be used by the TGFβ receptor system (217, 218).

Regulation

The regulation of LRP gene expression differs considerably from that of the LDL receptor. Although both receptors mediate the cellular uptake of cholesterol-rich lipoproteins, LDL receptor expression is suppressed in response to cholesterol loading of the cell (219, 220), while that of LRP is not significantly affected (146, 221). The molecular basis for this sterol-dependent repression is a small sterol-regulatory response element in the LDL receptor promoter (222), which is absent in the LRP promoter (221). LRP transcription and expression have been shown to be down-regulated in macrophages by lipopolysaccharide (LPS), interferon-γ, and estradiol (223, 224) and increased by colony stimulating factor-1 (225). These factors exert their regulatory function on LRP on the transcriptional level, but the physiological significance of these regulatory mechanisms and the response elements in the LRP promoter have not yet been identified.

The cellular uptake of α_2-macroglobulin and of apoE-enriched β-VLDL by freshly isolated adipocytes is stimulated by insulin. This insulin-stimulated uptake is mediated by LRP. There is a rapid redistribution of LRP from intracellular compartments to the cell surface within seconds of the addition of the hormone (226). Thus, the mechanism of insulin stimulation of LRP activity resembles that of insulin-stimulated glucose transport (227, 227a). Insulin has also been shown to increase the cell-surface expression of several endocytic receptors, e.g. the transferrin receptor (228–230) and the IGF-II receptor (231, 232), as well as the α_2-macroglobulin receptor (233). This insulin sensitivity might be physiologically relevant in the postprandial state, when insulin levels increase with the appearance of chylomicrons in the circulation.

Malignant cells generally appear to bind α_2-macroglobulin/protease complex less efficiently than do normal untransformed cells (234), but the mechanism by which the downregulation of LRP might contribute to the transformed phenotype is unknown.

A receptor-associated protein (RAP) of 323 amino acids, which was originally discovered by Strickland and colleagues (176) and Kristensen and coworkers (177) during the purification of LRP, effectively competes for the binding of all known ligands to the receptor (168, 213, 235). RAP is synthesized in the ER with an atypical signal sequence at the NH_2-terminus. It enters the secretory pathway but is not secreted from the cell. Although it does not contain a membrane-spanning segment (144, 236), RAP is membrane associated (213), mainly inside the endoplasmic reticulum (237). It also interacts with gp330, another member of the LDL receptor gene family that binds several of the same ligands that bind to LRP (168, 238). The inhibition of the binding of all known ligands to both LRP and gp330 by RAP, and its localization in intracellular compartments distinct from those containing LRP, raise the possibility that RAP could serve as a fast-acting physiological modulator of LRP (213, 235, 239) by associating with LRP in recycling vesicles in response to an extracellular signal. Alternatively, RAP might play a role in the synthesis and intracellular transport of large cysteine-rich receptors such as LRP and gp330.

Requirement for LRP During Early Development

Mouse embryos in which both copies of the LRP gene have been disrupted by homologous recombination are nonviable and die early during gestation (before day 13 of development; 205, 240). No one particular function of LRP has so far been shown to be responsible for this developmental defect, and the multifunctional nature of LRP may be the reason why it is an essential gene. Perhaps the disturbance of regulated extracellular proteolysis, which probably occurs as a result of the clearance defect due to the absence of LRP, can be considered as the most likely cause for the developmental arrest. Deficiencies of apoE and of LPL in humans have been described and are not lethal (169, 241).

Another possible explanation of homozygous lethality stems from the proposed role of α_2-macroglobulin as a modulator of cytokine activity (for review see 242). Several different cytokines have been shown to bind to α_2-macroglobulin, which in some instances can enhance and in others dampen the biological response of cells exposed to these complexes. The absence of LRP as a clearance receptor for α_2-macroglobulin/endoproteinase/cytokine complexes could thus conceivably influence a multitude of cellular signalling pathways. Another previously described multiligand endocytic receptor that is also involved in the modulation of cytokine activity is the cation-independent mannose 6-phosphate/insulinlike growth factor-II (IGF-II) receptor (243).

Gene Family

LRP is a member of the LDL receptor gene family, which at this time comprises six known cell-surface receptors (Figure 2). In all members of this gene family, four structural motifs are present:

1. LDL receptor ligand-binding repeats (complement-type repeats),
2. EGF-repeats and EGF-precursor homology domains,
3. a single membrane-spanning segment, and
4. at least one copy of the "NPXY" internalization signal.

The gene family consists of:

1. the LDL receptor, which is responsible for cellular uptake of apoB-100- and apoE-containing lipoproteins (1, 2).
2. the VLDL receptor, whose domain structure differs from that of the LDL receptor only by the presence of an eighth ligand-binding repeat at the N terminus. It binds apoE-containing lipoproteins, but the physiological function of this receptor is not yet defined (244). It is not present in the liver.
3. the vitellogenin receptor, which is abundant in chicken oocytes. Only partial amino-acid sequence is available, which suggests that it is a member of this gene family (245).
4. LRP. The ligands of LRP identified to date include apoE-enriched remnant lipoproteins, lipoprotein lipase, α_2-macroglobulin/endoproteinase complexes, plasminogen activator/inhibitor complexes, tPA, vitellogenin, exotoxin A from *Pseudomonas aeruginosa,* lactoferrin, and RAP. LRP has been proposed to participate in vitellogenesis. However, it probably plays a minor role, if any, because the amount of LRP present in chicken oocytes is very small compared to the amount of the major vitellogenin receptor (246).
5. an LRP-like molecule has been observed in the nematode *Caenorhabditis elegans* (247). Its functions are unknown.
6. gp330, also known as the major Heymann nephritis antigen. gp330 is a major kidney membrane protein. Autoantibodies against it produce a glomerulonephritis-like disease in rats. A partial protein sequence of gp330 has been published, and defines it as a member of the LDL receptor gene family (248). gp330 has binding properties similar to those of LRP; however, its tissue distribution is very different. In particular, it is not expressed in liver. Although gp330 binds apoE-containing lipoproteins (168), it appears to play no role in chylomicron remnant metabolism. Other ligands for gp330 include plasminogen activator/inhibitor complexes (168), lipoprotein lipase (168, 249), lactoferrin (168),

and plasminogen (250). Binding of RAP to gp330 and RAP inhibition of ligand binding has also been demonstrated (168, 237, 238, 251).

The multifunctional nature of several members of this gene family, the partly overlapping functions, and the characteristic tissue distribution of the individual receptors suggest that these genes are involved in specialized functions depending on the site of expression.

IV. COMPARISON OF THE MULTILIGAND-BINDING PROPERTIES OF SCAVENGER RECEPTORS AND LRP

How do scavenger receptors and LRP achieve ligand binding, which is characterized by both high-affinity and broad specificity? The extended array of structural motifs throughout the long, 4400-residue extracellular portion of LRP (31 LDL receptor ligand-binding repeats, 8 EGF precursor spacer regions, and 22 EGF repeats) appears ideally suited to generate multiple, independent binding sites for a variety of ligands (see Figure 2). For some LRP ligands there is only a single high-affinity binding site on the receptor (e.g. α_2-macroglobulin/endoproteinase complexes, 235); two or more different ligands with different binding sites might be able to bind to LRP simultaneously (168). In contrast, the extracellular portions of type I and II scavenger receptors are much smaller than that of LRP (Figures 1 and 2). The relatively short 72-residue positively charged collagenous domain is responsible for most of the scavenger receptor's polyanionic ligand-binding specificity. Both the presence of multiple positively charged residues in each chain of the collagenous binding domain and its trimeric structure raise the possibilities that the scavenger receptor's high affinity might be a consequence of polyvalent ligand binding, and that there may be distinct, independent, or overlapping binding sites for different ligands. LRP clearly shares this potential for polyvalent ligand binding. Some LRP ligands may be able to bind to multiple sites along LRP's polypeptide chain (e.g. RAP and lipoprotein remnants enriched with multiple apoEs). Such multiple binding sites might permit the high-affinity binding of ligands that would otherwise not bind tightly if only a single site on LRP were available. This might account for the requirement that β-VLDL be enriched with apoE molecules before it can bind to LRP. The oligomerization state of LRP on the cell surface has not been established. If LRP forms dimers or higher-order oligomers, additional opportunities for polyvalent binding would be generated. Although the detailed molecular interactions between LRP and its ligands have not yet been defined, it seems likely that the binding of some positively charged ligands, such as apoE, will be mediated by the conserved negatively charged sequences on the LDL receptor ligand-binding repeats.

Some of the ligands of both scavenger receptors and LRP can bind to these receptors only after a precursor, or proligand, form is modified. For example, LDL must be chemically altered (e.g. oxidized), and α_2-macroglobulin must be activated (see above). This requirement for proligand conversion to a binding-competent form suggests that both receptors play a physiologically important role in the efficient clearance of these ligands from the circulation and/or extracellular spaces. This clearance function could in some cases provide a simple scavenging mechanism to remove from the body and destroy inactive or deleterious material. In other cases, the clearance could be integrated into more complex intercellular transport systems (e.g. lipoprotein remnant targeting to hepatocytes via LRP).

Macrophage scavenger receptors (type I and type II) and LRP each bind a wide variety of ligands and appear to play important roles in diverse physiologic and pathophysiologic systems. These include lipoprotein transport, hepatic clearance of inactivated endogenous proteins, removal of pathogenic substances for host defense, and embryonic development. As more receptor-ligand systems are characterized at the molecular level, the multiligand binding exhibited by scavenger receptors and LRP may no longer be regarded as exceptional.

ACKNOWLEDGMENTS

We thank our many colleagues for their contributions to the research reviewed here. The work from our laboratories was supported by grants from the National Institutes of Health (HL41484 to MK) and Arris Pharmaceutical Corporation (MK) and from the Lucille P. Markey Charitable Trust, the Syntex Scholars Program, and the Perot Family Foundation (JH). We also thank K Sweeney for her assistance in preparing the manuscript.

Literature Cited

1. Brown MS, Goldstein JL. 1986. *Science* 232:34–47
2. Goldstein JL, Brown MS. 1989. In *The Metabolic Basis of Inherited Disease,* ed. CR Scriver, AL Beaudet, WS Sly, D Valle, pp. 1215–50. New York: McGraw-Hill. 6th ed.
3. Goldstein JL, Brown MS, Anderson RGW, Russell DW, Schneider WJ. 1985. *Annu. Rev. Cell Biol.* 1:1–39
4. Goldstein JL, Ho YK, Basu SK, Brown MS. 1979. *Proc. Natl. Acad. Sci. USA* 76:333–37
5. Gerrity RG. 1981. *Am. J. Pathol.* 103:181–90
6. Brown MS, Goldstein JL. 1983. *Annu. Rev. Biochem.* 52:223–61
7. Steinberg D, Parthasarathy S, Carew TE, Khoo JC, Witztum JL. 1989. *New Engl. J. Med.* 320:915–24
8. Krieger M. 1994. In *Molecular Cardiovascular Medicine*, ed. E Haber. New York: Sci. Am. Med. In press

9. Krieger M. 1992. *Trends Biochem. Sci.* 17:141–46
10. Fogelman AM, VanLenten BJ, Warden C, Haberland ME, Edwards PA. 1988. *J. Cell Sci. Suppl.* 9:135–49
11. Brown MS, Goldstein JL, Krieger M, Ho YK, Anderson RG. 1979. *J. Cell Biol.* 82:597–613
12. Haberland ME, Fong D, Cheng L. 1988. *Science* 241:215–18
13. Yla-Herttuala S, Rosenfeld ME, Parthasarathy S, Sigal E, Sarkioja T, et al. 1991. *J. Clin. Invest.* 87:1146–52
14. Brown MS, Basu SK, Falck JR, Ho YK, Goldstein JL. 1980. *J. Supramol. Struct.* 13:67–81
15. Krieger M, Acton S, Ashkenas J, Pearson A, Penman M, Resnick D. 1993. *J. Biol. Chem.* 268:4569–72
16. Shechter I, Fogelman AM, Haberland ME, Seager J, Hokom M, Edwards PA. 1981. *J. Lipid Res.* 22:63–71
17. Zhang H, Yang Y, Steinbrecher UP. 1993. *J. Biol. Chem.* 268:5535–42
18. Deleted in proof
19. Nishikawa K, Arai H, Inoue K. 1990. *J. Biol. Chem.* 265:5226–31
20. Hampton RY, Golenbock DT, Penman M, Krieger M, Raetz CR. 1991. *Nature* 353:342–44
21. Pearson AM, Rich A, Krieger M. 1993. *J. Biol. Chem.* 268:3546–54
22. Wong H, Fogelman AM, Haberland ME, Edwards PA. 1983. *Arteriosclerosis* 3:A475
23. Via DP, Dresel HA, Cheng SL, Gotto AM Jr. 1985. *J. Biol. Chem.* 260:7379–86
24. Dresel HA, Friedrich E, Via DP, Sinn H, Ziegler R, et al. 1987. *EMBO J.* 6:319–26
25. Kodama T, Reddy P, Kishimoto C, Krieger M. 1988. *Proc. Natl. Acad. Sci. USA* 85:9238–42
26. Via DP, Kempner ES, Pons L, Fanslow AE, Vignale S, et al. 1992. *Proc. Natl. Acad. Sci. USA* 89:6780–84
27. Kodama T, Freeman M, Rohrer L, Zabrecky J, Matsudaira P, Krieger M. 1990. *Nature* 343:531–35
28. Rohrer L, Freeman M, Kodama T, Penman M, Krieger M. 1990. *Nature* 343:570–72
29. Freeman M, Ashkenas J, Rees DJ, Kingsley DM, Copeland NG, et al. 1990. *Proc. Natl. Acad. Sci. USA* 87:8810–14
30. Ashkenas J, Penman M, Vasile E, Freeman M, Krieger M. 1993. *J. Lipid Res.* 34:983–1000
31. Matsumoto A, Naito M, Itakura H, Ikemoto S, Asaoka H, et al. 1990. *Proc. Natl. Acad. Sci. USA* 87:9133–37
32. Bickel PE, Freeman MW. 1992. *J. Clin. Invest.* 90:1450–57
33. Emi M, Asaoka H, Matsumoto A, Itakura H, Kurihara Y, et al. 1993. *J. Biol. Chem.* 268:2120–25
34. Moulton KS, Wu H, Barnett J, Parthasarathy S, Glass CK. 1992. *Proc. Natl. Acad. Sci. USA* 89:8102–6
35. Freeman M, Ekkel Y, Rohrer L, Penman M, Freedman NJ, et al. 1991. *Proc. Natl. Acad. Sci. USA* 88:4931–35
36. Penman M, Lux A, Freedman NJ, Rohrer L, Ekkel Y, et al. 1991. *J. Biol. Chem.* 266:23985–93
37. Acton S, Resnick D, Freeman M, Ekkel Y, Ashkenas J, Krieger M. 1993. *J. Biol. Chem.* 268:3530–37
38. Resnick D, Freedman NJ, Xu S, Krieger M. 1993. *J. Biol. Chem.* 268:3538–45
39. Krieger M. 1994. In *Lipoproteins in Health and Disease,* ed. DR Illingworth. In press
40. Naito M, Kodama T, Matsumoto A, Doi T, Takahashi K. 1991. *Am. J. Pathol.* 139:1411–23
41. Pitas RE. 1990. *J. Biol. Chem.* 265:12722–27
42. Doi T, Higashino K, Kurihara Y, Wada Y, Miyazaki T, et al. 1993. *J. Biol. Chem.* 268:2126–33
43. Cohen C, Parry DAD. 1990. *Proteins* 7:1–15
44. McLachlan AD, Karn J. 1982. *Nature* 299:226–31
45. Beavil AJ, Edmeades RL, Gould HJ, Sutton BJ. 1992. *Proc. Natl. Acad. Sci. USA* 89:753–57
46. Giudice GJ, Emery DJ, Diaz LA. 1992. *J. Invest. Dermatol.* 99:243–50
47. Miller EJ, Gay S. 1987. *Methods Enzymol.* 144:3–41
48. Reid KB. 1979. *Biochem. J.* 179:367–71
49. Voss T, Eistetter H, Schafer KP, Engel J. 1988. *J. Mol. Biol.* 201:219–27
50. Shimizu H, Fisher JH, Papst P, Benson B, Lau K, et al. 1992. *J. Biol. Chem.* 267:1853–57
51. Lim BL, Lu J, Reid KB. 1993. *Immunology* 78:159–65
52. Crouch E, Rust K, Veile RK, Donis KH, Grosso L. 1993. *J. Biol. Chem.* 268:2976–83
53. Drickamer K, Dordal MS, Reynolds L. 1986. *J. Biol. Chem.* 261:6878–87
54. Kuhlman M, Joiner K, Ezekowitz RA. 1989. *J. Exp. Med.* 169:1733–45
55. Sastry K, Ezekowitz RA. 1993. *Curr. Opin. Immunol.* 5:59–66
56. Holmskov U, Teisner B, Willis AC, Reid KB, Jensenius JC. 1993. *J. Biol. Chem.* 268:10120–25

57. Thiel S, Reid KBM. 1989. *FEBS Lett.* 250:79–84
58. Rosenberry TL, Richardson JM. 1977. *Biochemistry* 16:3550–58
59. Charalambous BM, Keen JN, McPherson MJ. 1988. *EMBO J.* 7:2903–9
60. Byers PH. 1989. See Ref. 2, pp. 2805–42
60a. Resnick D, Pearson A, Krieger M. 1994. *Trends Biochem. Sci.* 19:5–8
61. Hunkapiller T, Hood L. 1989. *Adv. Immunol.* 44:1–63
62. Williams AF, Barclay AN. 1988. *Annu. Rev. Immunol.* 6:381–405
63. Doolittle RF. 1985. *Trends Biochem. Sci.* 10:233–37
64. Krieger M. 1986. In *Molecular Structures of Receptors,* ed. PW Rossow, AD Strosberg, pp. 210–31. Chichester: Ellis Horwood
65. Sudhof TC, Goldstein JL, Brown MS, Russell DW. 1985. *Science* 228:815–22
66. Fogelman AM, Haberland ME, Seager J, Hokom M, Edwards PA. 1981. *J. Lipid Res.* 22:1131–41
67. Via DP, Pons L, Dennison DK, Fanslow AE, Bernini F. 1989. *J. Lipid Res.* 30:1515–24
68. Fraser I, Hughes D, Gordon S. 1993. *Nature* 364:343–46
69. Hara H, Tanishima H, Yokoyama S, Tajima S, Yamamoto A. 1987. *Biochem. Biophys. Res. Commun.* 146: 802–8
70. Pitas RE, Friera A, McGuire J, Dejager S. 1992. *Arterioscler. Thromb.* 12: 1235–44
71. Van Lenten BJ, Fogelman AM, Seager J, Ribi E, Haberland ME, Edwards PA. 1985. *J. Immunol.* 134: 3718–21
72. DeWhalley CV, Riches DW. 1991. *Exp. Cell Res.* 192:460–68
72a. Ishibashi, S, Inaba, T, Shimano, H, Harada, K, Inoue I, et al. 1990. *J. Biol. Chem.* 265:14109–17
73. Clinton SK, Underwood RL, Hayes L, Sherman ML, Kufe DW, Libby P. 1992. *Am. J. Pathol.* 140:301–16
74. Aviram M. 1989. *Metabolism* 38:425–30
75. Fogelman AM, Seager J, Haberland ME, Hokom M, Tanaka R, Edwards PA. 1982. *Proc. Natl. Acad. Sci. USA* 79:922–26
76. Bottalico LA, Wager RE, Agellon LB, Assoian RK, Tabas I. 1991. *J. Biol. Chem.* 266:22866–71
77. Geng YJ, Hansson GK. 1992. *J. Clin. Invest.* 89:1322–30
78. Hirsch LJ, Mazzone T. 1986. *J. Clin. Invest.* 77:485–90
79. Kowala MC, Mazzucco CE, Hartl KS, Seiler SM, Warr GA, et al. 1993. *Arterioscler. Thromb.* 13:435–44
80. Jouni ZE, McNamara DJ. 1991. *Arterioscler. Thromb.* 11:995–1006
81. Falcone DJ, McCaffrey TA, Vergilio JA. 1991. *J. Biol. Chem.* 266:22726–32
82. Palkama T. 1991. *Immunology* 74:432–38
83. Sparrow CP, Parthasarathy S, Steinberg D. 1989. *J. Biol. Chem.* 264:2599–604
84. Arai H, Kita T, Yokode M, Narumiya S, Kawai C. 1989. *Biochem. Biophys. Res. Commun.* 159:1375–82
85. Kamps JA, Kruijt JK, Kuiper J, van Berkel TJ. 1992. *Arterioscler. Thromb.* 12:1079–87
86. Nagelkerke JF, Barto KP, van Berkel TJ. 1983. *J. Biol. Chem.* 258:12221–27
87. van Berkel TJ, De Rijke YB, Kruijt JK. 1991. *J. Biol. Chem.* 266:2282–89
88. Dejager S, Mietus SM, Pitas RE. 1993. *Arterioscler. Thromb.* 13:371–78
89. Ottnad E, Via DP, Frubis J, Sinn H, Friedrich E, et al. 1992. *Biochem. J.* 281:745–51
90. Ottnad E, Via DP, Sinn H, Friedrich E, Ziegler R, Dresel HA. 1988. *Biochem. J.* 253:835–38
91. Ottnad E, Via DP, Sinn H, Friedrich E, Ziegler R, Dresel HA. 1990. *Biochem. J.* 265:689–98
92. Stanton LW, White RT, Bryant CM, Protter A, Endemann G. 1992. *J. Biol. Chem.* 267:22446–51
93. Endemann G, Stanton LW, Madden KS, Bryant CM, White RT, et al. 1993. *J. Biol. Chem.* 268:11811–16
94. Mahley RW, Innerarity TL, Weisgraber KH. 1980. *Ann. NY Acad. Sci.* 348: 265–80
95. Pitas RE, Boyles J, Mahley RW, Bissell DM. 1985. *J. Cell. Biol.* 100: 103–17
96. Eskild W, Kindberg GM, Smedsrod B, Blomhoff R, Norum KR, Berg T. 1989. *Biochem. J.* 258:511–20
96a. Via DP, Fanslow A, Dresel HA, Dennison DK, Levine E, et al. 1992. *FASEB J.* 6:A371
97. Schnitzer JE, Sung A, Horvat R, Bravo J. 1992. *J. Biol. Chem.* 267: 24544–53
98. Horiuchi S, Murakami M, Takata K, Morino Y. 1986. *J. Biol. Chem.* 261: 4962–66
99. Barnes JL, Reznicek MJ, Radnik RA, Venkatachalam MA. 1988. *Kidney Int.* 34:156–63
100. Haberland ME, Rasmussen RR, Olch CL, Fogelman AM. 1986. *J. Clin. Invest.* 77:681–89
101. Hynes RO. 1992. *Cell* 69:11–25

102. Davis GG. 1992. *Exp. Cell. Res.* 200:242–52
103. Hoff HF, O'Neil J, Pepin JM, Cole TB. 1990. *Eur. Heart J.* 11(Suppl. E):105–15
104. Carew TE, Schwenke DC, Steinberg D. 1987. *Proc. Natl. Acad. Sci. USA* 84:7725–29
105. Kita T, Nagano Y, Yokode M, Ishii K, Kume N, et al. 1987. *Proc. Natl. Acad. Sci. USA* 84:5928–31
106. Gordon S, Perry VH, Rabinowitz S, Chung LP, Rosen H. 1988. *J. Cell Sci. Suppl.* 9:1–26
107. Janeway CA. 1992. *Immunol. Today* 13:11–16
108. Ezekowitz RA, Williams DJ, Koziel H, Armstrong MY, Warner A, et al. 1991. *Nature* 351:155–58
109. Taylor ME, Bezouska K, Drickamer K. 1992. *J. Biol. Chem.* 267:1719–26
110. Raetz CRH. 1990. *Annu. Rev. Biochem.* 59:129–70
110a. Dunne DW, Resnick D, Greenberg J, Krieger M, Joiner KA. 1994. *Proc. Natl. Acad. Sci. USA* 91:1863–67
111. Jurgens G, Hoff HF, Chisolm GM 3rd, Esterbauer H. 1987. *Chem. Phys. Lipids* 45:315–36
112. Abrams JM, Lux A, Steller H, Krieger M. 1992. *Proc. Natl. Acad. Sci. USA* 89:10375–79
113. Brehelin M. 1982. *Cell Tissue Res.* 221:607–15
114. Campos-Ortega JA, Hartenstein V. 1985. *The Embryonic Development of Drosophila Melanogaster,* p. 49. Berlin/Heidelberg: Springer-Verlag
115. Abrams JM, White K, Fessler LI, Steller H. 1993. *Development* 117:29–44
115a. Havel RJ, Kane JP. 1989. See Ref. 2, pp. 1129–38
115b. Havel RJ, Hamilton RL. 1988. *Hepatology* 8:1689–1704
116. Kita T, Goldstein JL, Brown MS, Watanabe Y, Hornick CA, Havel RJ. 1982. *Proc. Natl. Acad. Sci. USA* 79:3623–27
117. Rubinsztein DC, Cohen JC, Berger GM, van der Westhuyzen DR, Coetzee GA, Gevers W. 1990. *J. Clin. Invest.* 86:1306–12
118. Ishibashi S, Brown MS, Goldstein JL, Gerard RD, Hammer RE, Herz J. 1993. *J. Clin. Invest.* 92:883–93
119. Herz J, Hamann U, Rogne S, Myklebost O, Gausepohl H, Stanley KK. 1988. *EMBO J.* 7:4119–27
120. Moestrup SK, Gliemann J, Pallesen G. 1992. *Cell Tissue Res.* 269:375–82
121. Wolf BB, Lopes MBS, VandenBerg SR, Gonias SL. 1992. *Am. J. Pathol.* 141:37–42
122. Jensen PH, Moestrup SK, Sottrup-Jensen L, Petersen CM, Gliemann J. 1988. *Placenta* 9:463–77
123. Gåfvels ME, Coukos G, Sayegh R, Coutifaris C, Strickland DK, Strauss JF. 1992. *J. Biol. Chem.* 267:21230–34
124. Tooyama I, Kawamata T, Akiyama H, Moestrup SK, Gliemann J, McGeer PL. 1993. *Mol. Chem. Neuropathol.* 18:153–60
125. Moestrup SK, Kaltoft K, Petersen CM, Pedersen S, Gliemann J, Christensen EI. 1990. *Exp. Cell Res.* 190:195–203
126. Gray A, Dull TJ, Ullrich A. 1983. *Nature* 303:722–25
127. Yamamoto T, Davis CG, Brown MS, Schneider WJ, Casey ML, Goldstein JL, et al. 1984. *Cell* 39:27–38
128. Russell DW, Brown MS, Goldstein JL. 1989. *J. Biol. Chem.* 264:21682–88
129. Mahley RW. 1988. *Science* 240:622–30
130. Chan L. 1992. *J. Biol. Chem.* 267: 25621–24
131. Scott J. 1989. *Mol. Biol. Med.* 6:65–80
132. Stanley KK, Kocher H-P, Luzio JP, Jackson P, Tschopp J, Dickson J. 1985. *EMBO J.* 4:375–82
133. Haefliger J-A, Tschopp J, Nardelli D, Wahli W, Kocher H-P, Tosi M, et al. 1987. *Biochemistry* 26:3551–56
134. Howard OMZ, Rao AG, Sodetz JM. 1987. *Biochemistry* 26:3565–70
135. DiScipio RG, Hugli TE. 1989. *J. Biol. Chem.* 264:16197–206
136. Davis CG, Goldstein JL, Südhof TC, Anderson RGW, Russell DW, Brown MS. 1987. *Nature* 326:760–65
137. Herz J, Kowal RC, Ho YK, Brown MS, Goldstein JL. 1990. *J. Biol. Chem.* 265:21355–62
138. Chen W-J, Goldstein JL, Brown MS. 1990. *J. Biol. Chem.* 265:3116–23
139. Davis CG, van Driel IR, Russell DW, Brown MS, Goldstein JL. 1987. *J. Biol. Chem.* 262:4075–82
140. Bansal A, Gierasch LM. 1991. *Cell* 67:1195–1201
141. Herz J, Kowal RC, Goldstein JL, Brown MS. 1990. *EMBO J.* 9:1769–76
142. Steiner DF, Smeekens SP, Ohagi S, Chan SJ. 1992. *J. Biol. Chem.* 267: 23435–38
143. Wise RJ, Barr PJ, Wong PA, Kiefer MC, Brake AJ, Kaufman RJ. 1990. *Proc. Natl. Acad. Sci. USA* 87:9378–82
144. Strickland DK, Ashcom JD, Williams S, Battey F, Behre E, McTigue K, et al. 1991. *J. Biol. Chem.* 266:13364–69
145. Lund H, Takahashi K, Hamilton RL, Havel RJ. 1989. *Proc. Natl. Acad. Sci. USA* 86:9318–22

146. Kowal RC, Herz J, Goldstein JL, Esser V, Brown MS. 1989. *Proc. Natl. Acad. Sci. USA* 86:5810–14
147. Beisiegel U, Weber W, Ihrke G, Herz J, Stanley KK. 1989. *Nature* 341:162–64
148. Daniel TO, Schneider WJ, Goldstein JL, Brown MS. 1983. *J. Biol. Chem.* 258:4606–11
149. Kowal RC, Herz J, Weisgraber KH, Mahley RW, Brown MS, Goldstein JL. 1990. *J. Biol. Chem.* 265:10771–79
150. Moestrup SK, Kaltoft K, Sottrup-Jensen L, Gliemann J. 1990. *J. Biol. Chem.* 265:12623–28
151. Mahley RW, Weisgraber KH, Hussain MM, Greenman B, Fisher M, et al. 1989. *J. Clin. Invest.* 83:2125–30
152. Yamada N, Shimano H, Mokuno H, Ishibashi S, Gotohda T, et al. 1989. *Proc. Natl. Acad. Sci. USA* 86:665–69
153. Kovanen PT, Brown MS, Basu SK, Bilheimer DW, Goldstein JL. 1981. *Proc. Natl. Acad. Sci. USA* 78:1396-1400
154. Mjos OD, Faergeman O, Hamilton RL, Havel RJ. 1975. *J. Clin. Invest.* 56:603–15
155. Shelburne F, Hanks J, Meyers W, Quarfordt S. 1980. *J. Clin. Invest.* 65:652–58
156. Windler E, Chao Y-S, Havel RJ. 1980. *J. Biol. Chem.* 255:5475–80
157. Windler E, Havel RJ. 1985. *J. Lipid Res.* 26:556–65
158. Weisgraber KH, Mahley RW, Kowal RC, Herz J, Goldstein JL, Brown MS. 1990. *J. Biol. Chem.* 265:22453–59
159. Brown MS, Herz J, Kowal RC, Goldstein JL. 1991. *Curr. Opin. Lipidol.* 2:65–72
160. Hamilton RL, Wong JS, Guo LSS, Krisans S, Havel RJ. 1990. *J. Lipid Res.* 31:1589–1603
161. Boyles JK, Pitas RE, Wilson E, Mahley RW, Taylor JM. 1985. *J. Clin. Invest.* 76:1501–13
162. Boyles JK, Pitas RE, Weisgraber KH, Mahley RW, Gebicke-Haerter PJ, et al. 1986. *Circulation* 74(Suppl.):II-195
163. Boyles JK, Zoeliner CD, Anderson LJ, Kosik LM, Pitas RE, et al. 1989. *J. Clin. Invest.* 83:1015–31
164. Elshourbagy NA, Liao WS, Mahley RW, Taylor JM. 1985. *Proc. Natl. Acad. Sci. USA* 82:203–7
165. Shi WK, Heath JK. 1984. *J. Embryol. Exp. Morphol.* 81:143–52
166. Beisiegel U, Weber W, Bengtsson-Olivecrona G. 1991. *Proc. Natl. Acad. Sci. USA* 88:8342–46
167. Chappell DA, Fry GL, Waknitz MA, Iverius P-H, Williams SE, Strickland DK. 1992. *J. Biol. Chem.* 267:25764–67
168. Willnow TE, Goldstein JL, Orth K, Brown MS, Herz J. 1992. *J. Biol. Chem.* 267:26172–80
169. Schaefer EJ, Gregg RE, Ghiselli G, Forte TM, Ordovas JM, et al. 1986. *J. Clin. Invest.* 78:1206–19
170. Zhang SH, Reddick RL, Piedrahita JA, Maeda N. 1992. *Science* 258:468–71
171. Plump AS, Smith JD, Hayek T, Aalto-Setälä K, Walsh A, et al. 1992. *Cell* 71:343–53
172. Huettinger M, Retzek H, Eder M, Goldenberg H. 1988. *Clin. Biochem.* 21:87–92
173. Huettinger M, Retzek H, Hermann M, Goldenberg H. 1992. *J. Biol. Chem.* 267:18551–57
174. Basu SK, Goldstein JL, Brown MS. 1983. *Science* 219:871–73
175. Ross R. 1993. *Nature* 362:801–9
176. Strickland DK, Ashcom JD, Williams S, Burgess WH, Migliorini M, Argraves WS. 1990. *J. Biol. Chem.* 265:17401–4
177. Kristensen T, Moestrup SK, Gliemann J, Bendtsen L, Sand O, Sottrup-Jensen L. 1990. *FEBS Lett.* 276:151–55
178. Sottrup-Jensen L. 1989. *J. Biol. Chem.* 264:11539–42
179. Gliemann J, Larsen TR, Sottrup-Jensen L. 1983. *Biochim. Biophys. Acta* 756: 230–37
180. Davidsen O, Christensen EI, Gliemann J. 1985. *Biochim. Biophys. Acta* 846: 85–92
181. Moestrup SK, Holtet TL, Etzerodt M, Thøgersen HC, Nykjaer A, et al. 1993. *J. Biol. Chem.* 268:13691–96
182. Strickland DK, Bhattacharya P, Olson S. 1984. *Biochemistry* 23:3115–23
183. Hussain MM, Maxfield FR, Más-Oliva J, Tabas I, Ji Z-S, et al. 1991. *J. Biol. Chem.* 226:13936–40
184. Choi SY, Cooper AD. 1993. *J. Biol. Chem.* 268:15804–11
185. van Dijk MCM, Ziere GJ, Boers W, Linthorst C, Bijsterbosch MK, van Berkel TJC. 1991. *Biochem. J.* 279: 863–70
186. Jäckle S, Huber C, Moestrup S, Gliemann J, Beisiegel U. 1993. *J. Lipid. Res.* 34:309–15
187. Ji Z-S, Brecht WJ, Miranda RD, Hussain MM, Innerarity TL, Mahley RW. 1993. *J. Biol. Chem.* 268:10160–67
188. Nykjaer A, Petersen CM, Moller B, Jensen PA, Moestrup SK, et al. 1992. *J. Biol. Chem.* 267:14543–46
189. Orth K, Madison EL, Gething M-J,

Sambrook JF, Herz J. 1992. *Proc. Natl. Acad. Sci. USA* 89:7422–26
190. Bu G, Williams S, Strickland DK, Schwartz AL. 1992. *Proc. Natl. Acad. Sci. USA* 89:7427–31
191. Lijnen HR, Collen D. 1991. *Thromb. Haemost.* 66:88–110
192. Saksela O, Rifkin DB. 1988. *Annu. Rev. Cell Biol.* 4:93–126
193. Sappino A-P, Huarte J, Belin D, Vassalli J-D. 1989. *J. Cell Biol.* 109:2471–79
194. Stephens RW, Pöllanen J, Tapiovaara H, Leung K-C, Sim P-S, et al. 1989. *J. Cell Biol.* 108:1987–95
195. Estreicher A, Mühlhauser J, Carpentier J-L, Orci L, Vassalli J-D. 1990. *J. Cell Biol.* 111:783–92
196. Wing LR, Hawksworth GM, Bennett B, Booth NA. 1991. *J. Lab. Clin. Med.* 117:109–14
197. Warshawsky I, Bu G, Schwartz AL. 1993. *J. Clin. Invest.* 92:937–44
198. Otter M, Barrett-Bergshoeff MM, Rijken DC. 1991. *J. Biol. Chem.* 266:13931–35
199. Cubellis MV, Andreasen PA, Ragno P, Mayer M, Danø K, Blasi F. 1989. *Proc. Natl. Acad. Sci. USA* 86:4828–32
200. Ellis V, Scully MF, Kakkar VV. 1989. *J. Biol. Chem.* 264:2185–88
201. Ploug M, Rønne E, Behrendt N, Jensen AL, Blasi F, Danø K. 1991. *J. Biol. Chem.* 266:1926–33
202. Roldan AL, Cubellis MV, Masucci MT, Behrendt N, Lund LR, et al. 1990. *EMBO J.* 9:467–74
203. Cubellis MV, Wun T-C, Blasi F. 1990. *EMBO J.* 9:1079–85
204. Jensen PH, Christensen EI, Ebbesen P, Gliemann J, Andreasen PA. 1990. *Cell Regul.* 1:1043–56
205. Herz J, Clouthier DE, Hammer RE. 1992. *Cell* 71:411–21
206. Thompson MR, Forristal J, Kauffmann P, Madden T, Kozak K, et al. 1991. *J. Biol. Chem.* 266:2390–96
207. Kounnas MZ, Morris RE, Thompson MR, FitzGerald DJ, Strickland DK, Saelinger CB. 1992. *J. Biol. Chem.* 267:12420–23
208. Hwang J, FitzGerald DJ, Adhya S, Pastan I. 1987. *Cell* 48:129–36
209. Iglewski BH, Liu PV, Kabat D. 1977. *Infect. Immun.* 15:138–44
209a. Willnow TE, Herz J. 1994. *J. Cell Sci.* In press
210. Harris HW, Grunfeld C, Feingold KR, Read TE, Kane JP, et al. 1993. *J. Clin. Invest.* 91:1028–34
210a. Saxena U, Klein MG, Goldberg IJ. 1990. *J. Biol. Chem.* 265:12880–86
211. Chappell DA, Fry GL, Waknitz MA, Muhonen LE, Pladet MW, et al. 1993. *J. Biol. Chem.* 268:14168–75
212. Borensztajn J, Getz GS, Kotlar TJ. 1988. *J. Lipid Res.* 29:1087–96
213. Herz J, Goldstein JL, Strickland DK, Ho YK, Brown MS. 1991. *J. Biol. Chem.* 266:21232–38
214. Klagsbrun M, Baird A. 1991. *Cell* 67:229–31
215. Rapraeger AC, Krufka A, Olwin BB. 1991. *Science* 252:1705–8
216. Yayon A, Klagsbrun M, Esko JD, Leder P, Ornitz DM. 1991. *Cell* 64:841–48
217. Lopez-Casillas F, Wrana JL, Massague J. 1993. *Cell* 73:1435–44
218. Lin HY, Lodish HF. 1993. *Trends Cell Biol.* 3:14–19
219. Brown MS, Goldstein JL. 1975. *Cell* 6:307–16
220. Russell DW, Yamamoto T, Schneider WJ, Slaughter CJ, Brown MS, Goldstein JL. 1983. *Proc. Natl. Acad. Sci. USA* 80:7501–5
221. Kütt H, Herz J, Stanley KK. 1989. *Biochim. Biophys. Acta* 1009:229–36
222. Südhof TC, Russell DW, Brown MS, Goldstein JL. 1987. *Cell* 48:1061–69
223. Szanto A, Balasubramaniam S, Roach PD, Nestel PJ. 1992. *Biochem. J.* 288:791–94
224. LaMarre J, Wolf BB, Kittler ELW, Quesenberry PJ, Gonias SL. 1993. *J. Clin. Invest.* 91:1219–24
225. Hussaini IM, Srikumar K, Quesenberry PJ, Gonias SL. 1990. *J. Biol. Chem.* 265:19441–46
226. Descamps O, Bilheimer D, Herz J. 1993. *J. Biol. Chem.* 268:974–81
227. Cushman SW, Wardzala LJ. 1980. *J. Biol. Chem. 255:4758–62*
227a. Suzuki K, Kono, T. 1980. *Proc. Natl. Acad. Sci. USA* 77:2542–45
228. Davis RJ, Corvera S, Czech MP. 1986. *J. Biol. Chem.* 261:8708–11
229. Davis RJ, Faucher M, Racaniello LK, Carruthers A, Czech MP. 1987. *J. Biol. Chem.* 262:13126–34
230. Tanner LI, Lienhard GE. 1987. *J. Biol. Chem.* 262:8975–80
231. Oka Y, Mottola C, Oppenheimer CL, Czech MP. 1984. *J. Biol. Chem.* 81:4028–32
232. Wardzala LJ, Simpson IA, Rechler MM, Cushman SW. 1984. *J. Biol. Chem.* 259:8378–83
233. Corvera S, Graver DF, Smith RM. 1989. *J. Biol. Chem.* 264:10133–38

234. Jensen PH, Ebbesen P, Gliemann J. 1989. *In Vivo* 3:7–9
235. Moestrup SK, Gliemann J. 1991. *J. Biol. Chem.* 266:14011–17
236. Furukawa T, Ozawa M, Huang R-P, Muramatsu T. 1990. *J. Biochem.* 108: 297–302
237. Orlando RA, Kerjaschki D, Kurihura H, Biemesderfer D, Farquhar MG. 1992. *Proc. Natl. Acad. Sci. USA* 89: 6698–702
238. Kounnas MZ, Argraves WS, Strickland DK. 1992. *J. Biol. Chem.* 267:21162–66
239. Williams SE, Ashcom JD, Argraves WS, Strickland DK. 1992. *J. Biol. Chem.* 267:9035–40
240. Herz J, Clouthier DE, Hammer RE. 1993. *Cell* 73:428
241. Lalouel J-M, Wilson DE, Iverius PH. 1992. *Curr. Opin. Lipidol.* 3:86–95
242. Borth W. 1992. *FASEB J.* 6:3345–53
243. Nissley P, Kiess W. 1991. In *Molecular Biology and Physiology of Insulin and Insulin-Like Growth Factors,* ed. MK Raizada, D LeRoith, pp. 311–24. New York: Plenum
244. Takahashi S, Kawarabayasi Y, Nakai T, Sakai J, Yamamoto T. 1992. *Proc. Natl. Acad. Sci. USA* 89:9252–56
245. Barber DL, Sanders EJ, Aebersold R, Schneider WJ. 1991. *J. Biol. Chem.* 266:18761–70
246. Stifani S, Barber DL, Aebersold R, Steyrer E, Shen X, et al. 1991. *J. Biol. Chem.* 266:19079–87
247. Yochem J, Greenwald I. 1993. *Proc. Natl. Acad. Sci. USA* 90:4572–76
248. Raychowdhury R, Niles JL, McCluskey RT, Smith JA. 1989. *Science* 244: 1163–66
249. Kounnas MZ, Chappell DA, Strickland DK, Argraves WS. 1993. *J. Biol. Chem.* 268:14176–81
250. Kanalas JJ, Makker SP. 1991. *J. Biol. Chem.* 266:10825–28
251. Christensen EI, Gliemann J, Moestrup SK. 1992. *J. Histochem. Cytochem.* 40:1481–90
252. Dziarski R, Gupta D. 1994. *J. Biol. Chem.* 269:2100–110
253. Green SJ, Chen T-Y, Crawford RM, Nacy CA, Morrison DC, Meltzer MS. 1992. *J. Immunol.* 149:2069–75

NOTE ADDED IN PROOF: Recently, Dziarski & Gupta (252) have reported the binding of LTAs and other components of bacterial cell walls to a 70-kDa receptor on lymphocytes, which has also been characterized by Morrison and colleagues (253).

Annu. Rev. Biochem. 1994. 63:639–74

THE CENTROSOME AND CELLULAR ORGANIZATION

Douglas R. Kellogg, Michelle Moritz, and Bruce M. Alberts[1]
University of California, San Francisco, California 94143-0448

KEY WORDS: centriole, microtubules, spindle pole body, basal body, cytoskeleton

---

CONTENTS

[1]Present address: National Academy of Sciences, 2101 Constitution Avenue NW, Washington, DC 20418

0066-4154/94/0701-0639$05.00

INTRODUCTION

Microtubules are highly dynamic polymers that form a major filament system of the cytoskeleton in eukaryotic cells. The centrosome nucleates the growth of microtubules and thereby determines the number, and to a large extent, the distribution of the hundreds of microtubules in an animal cell. The centrosome and its associated microtubules direct the events of mitosis, and they play a central role in the organization of the interior of the cell during interphase.

The importance of the centrosome as a central organizer of cellular events was recognized as long ago as the 19th century (1). Our present challenge is to understand the molecular events that underlie the observed functions of the centrosome. For example, we would like to know which proteins form the centrosome matrix (or "pericentriolar material"), how they are responsible for microtubule nucleation, and how their activity is regulated during the cell cycle. We would also like to know how the centrosome duplicates during each cell cycle so that only one new centrosome is produced for each daughter cell. Finally, how do the microtubules emanating from the centrosome interact with other cytoplasmic components to organize cellular events?

As will be described, the centrosome and its associated microtubules impart shape, polarity, and internal order to the cell. An important unanswered question is how many of these properties depend on the centrosome itself, as distinct from its attached microtubules. Put another way, suppose that we had available a synthetic bead that would grow an astral array of microtubules, in the same number and with the same polarity as those that grow from a centrosome. If this bead were injected into an interphase cell whose normal centrosome had been destroyed, how many of the functions that depend on microtubules would remain? Recent history shows that scientists have repeatedly underestimated the complexity and sophistication of cells, and there are reasons to suspect that the centrosome may contribute more to cell biology than is commonly recognized.

In this review, we discuss recent progress towards a molecular understanding of the centrosome, with an emphasis on new and promising approaches. Due to the vast literature on the centrosome, readers are referred to a number of previous reviews that cover the older literature and specific topics in greater detail (2–5). In particular, two recent reviews list all known protein components of the centrosome (6, 7), and we therefore do not provide an extensive discussion of specific centrosomal proteins here, except in cases where there are hints as to their function.

THE STRUCTURE OF THE CENTROSOME

More than 100 years ago, cytologists were able to observe a densely staining pair of dots associated with the interphase nucleus of animal cells (1). The dots were surrounded by a cloud of amorphous material that appeared to be the source of a system of fibers reaching throughout the cell. The dots were eventually given the name of *centrioles,* while the centrioles together with their surrounding amorphous material were called the *centrosome.* Subsequent studies using electron microscopy revealed that the centrioles are composed of nine triplet microtubules arranged to form a short cylinder, and are surrounded by a dense amorphous material (the *pericentriolar material,* or *centrosome matrix*) from which microtubules emanate. In an interphase cell, the centrosome lies next to the nucleus and nucleates an array of hundreds of microtubules oriented with their fast-growing plus ends

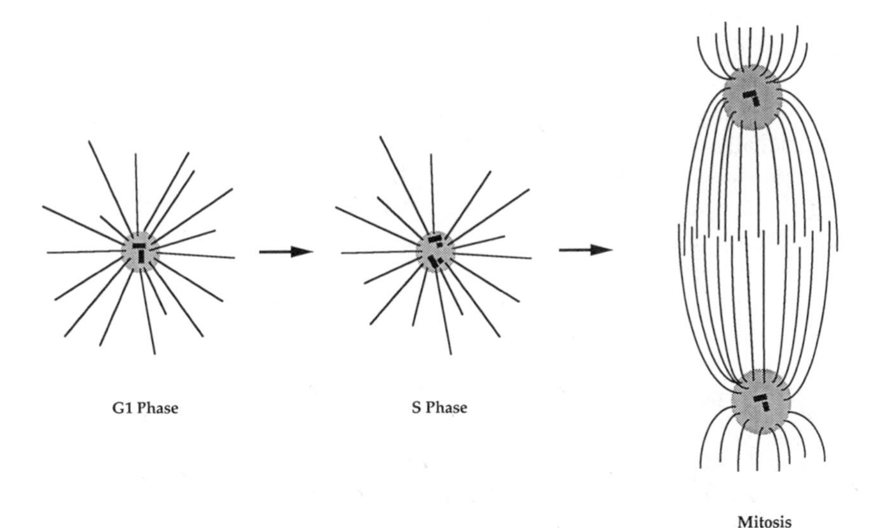

Figure 1 The centrosome undergoes characteristic changes during the cell cycle. Microtubules are oriented with their fast-growing "plus" ends away from the centrosome. Daughter centrioles begin to form during S phase, and the amount of pericentriolar material increases during mitosis, as does the number of microtubules nucleated by the centrosome. The centrosome divides and migrates around the nucleus to form the mitotic spindle poles.

reaching into the cytoplasm, and their minus ends embedded in the centrosome matrix (Figure 1). During mitosis, the centrosome duplicates to form the spindle poles. As is discussed below, many eukaryotic cells do not contain the classic centrosome structure found in animal cells, although they contain functionally analogous organelles. In the budding yeast *Saccharomyces cerevisiae,* for instance, microtubules are organized by a structure called the *spindle pole body,* which is embedded in the nuclear envelope and contains no centrioles. In this review we use the term "microtubule organizing center" to encompass the centrosome and its functional equivalents found in various eukaryotic species.

The Centriole Is a Complicated Structure of Unknown Function

The centrioles found at the center of the animal cell centrosome are beautifully symmetric, with a fine structure that abounds in intricate details (for examples, see Refs. 8–10). Outside the cylinder, at one end of a centriole, there is usually a series of projections, one associated with each of the triplet microtubules. Inside the cylinder, a complicated series of symmetric spokes and bridges often reaches between the triplet microtubules into the interior, particularly in newly forming centrioles. Such morphological details differ from species to species, and can differ in different cell types of the same species.

The centriole bears a structural resemblance to the basal body, which is found at the base of cilia and flagella. These two structures are also functionally related; in many species the basal body of the sperm flagella is incorporated into the centrosome in the fertilized egg, and in the unicellular alga *Chlamydomonas* the flagellar basal bodies are incorporated into the centrosome at each mitosis, and return to flagellar basal bodies during interphase (reviewed in Ref. 11).

The function of the centrioles remains unclear. If mammalian tissue-culture cells are treated with drugs such as nocodazole that prolong mitosis, the centrosome breaks into fragments, giving rise to multiple spindle poles. Many of the spindle poles thus formed do not contain centrioles, but appear to function normally, suggesting that centrioles do not play an essential role in the microtubule-organizing capacity of the centrosome (12). Likewise, the first few divisions of the mouse egg utilize centrosomes that contain no centrioles (13), and a cultured cell line derived from *Drosophila* lacks centrioles (14, 15). Centrioles have not been found in plants (16), apparently being found only in species that have at least some cells with cilia or flagella, where the centriole functions as a basal body.

The Centrosome Undergoes Characteristic Structural Changes during the Cell Cycle

The structure of the centrosome in animal cells changes continually during the cell cycle (10, 17, 18) (Figure 1). When cells are in interphase the centrosome contains two centrioles that lie perpendicular to each other, and microtubules are associated primarily with only the older of these. The microtubules appear to be nucleated both from appendages protruding from the side of the centriole and from dense aggregates in the centrosome matrix. When duplication of the centrioles begins during S phase, the two centrioles move apart from each other slightly and daughter centrioles begin to form next to each of the original centrioles, with the new cylinder at a right angle to the old one. Elongation of the daughter centrioles occurs gradually and is not completed until sometime in mitosis (10, 17). During prophase, each centriole pair begins to migrate to opposite sides of the nucleus and the centriolar appendages disappear, being replaced by an increased amount of centrosome matrix material that surrounds the older centriole in each pair. This increase in the amount of matrix is correlated with a dramatic increase in the number of microtubules nucleated by the centrosome [from 25 to 140 in one in vitro study (19)]. At the end of mitosis, each cell inherits a pair of centrioles and the amount of centrosome matrix material is greatly reduced. The centriolar appendages then reappear on the oldest of the two centrioles (reviewed in Ref. 18).

The Animal Cell Centrosome and the Spindle Pole Body of Yeast Are Structurally Distinct, but Functionally Similar

The spindle pole body found in yeasts and other fungi is the functional equivalent of the animal cell centrosome, although the two organelles bear little structural resemblance. The spindle pole body of the budding yeast *S. cerevisiae* has been studied in the most detail, and is a disclike structure that remains embedded in the nuclear envelope throughout the cell cycle (20–22). When viewed in cross section, the spindle pole body has several distinct layers, including a densely staining layer in the plane of the nuclear envelope, and a more lightly staining layer facing to the outside of the nuclear envelope that is thought to nucleate the microtubules that emanate into the cytoplasm. A second lightly staining layer faces to the inside of the nuclear envelope and appears to be the source of the intranuclear spindle microtubules. At one end of the spindle pole body, there is a short projection called the half bridge that lies in the same plane as the nuclear envelope (Figure 2).

The first morphological manifestation of spindle pole body duplication is a small amorphous mass called the satellite that appears on the cytoplasmic

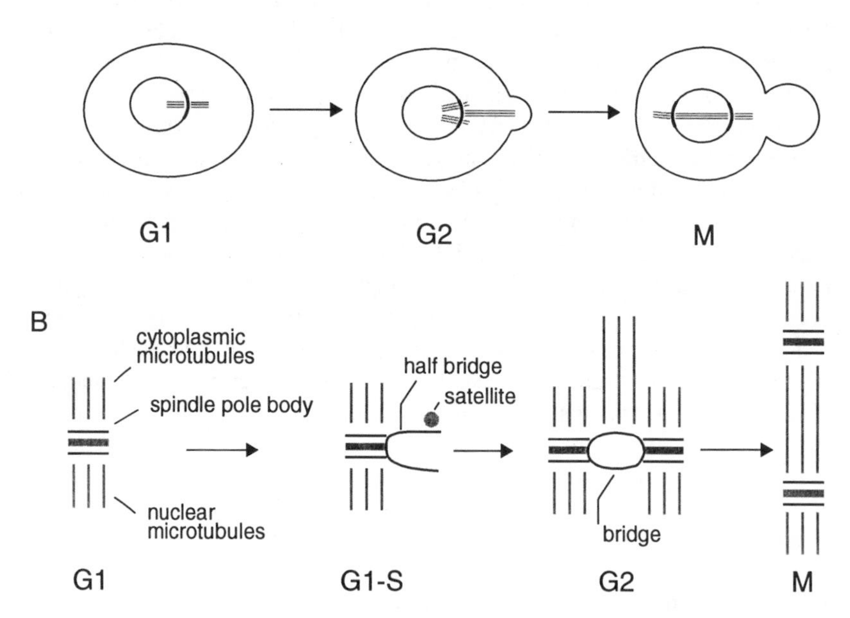

Figure 2 The spindle pole body cycle of *S. cerevisiae*. *A*. Low-resolution view of the spindle pole body cycle in coordination with the cell cycle. By the time the spindle pole body duplicates, a bud begins to grow from the mother cell (G2). The new spindle pole body migrates around the nucleus, and the spindle is oriented so that one end points into the growing bud (M). The nuclear envelope remains intact throughout the cell cycle. *B*. Higher-resolution view of the spindle pole body cycle. Duplication begins in G1-S with the formation of a "satellite" on the cytoplasmic side of the "half bridge." The "satellite" enlarges and becomes the new spindle pole body. During G2, microtubules grow from the "bridge" into the neck of the bud (G2). The mitotic spindle forms between the two spindle pole bodies as they separate and migrate around the nucleus (M).

side of the half bridge during the G1 stage of the cell cycle (Figure 2). By the time the daughter cell begins budding from the mother, there are two complete spindle pole bodies lying adjacent to each other, connected by their half-bridge structures. The duplicated spindle pole bodies are positioned near the site at which bud growth is taking place, with the cytoplasmic microtubules that emanate from the bridge structure reaching into the growing bud neck. At the onset of mitosis, the spindle pole bodies separate and migrate to opposite sides of the nucleus as the two sets of intranuclear microtubules assemble into the mitotic spindle. The ends of microtubules are associated with both faces of the spindle pole body throughout the cell cycle, although immunofluorescent staining experiments suggest that the intranuclear microtubules seen by electron microscopy during interphase are very short and elongate only during mitosis (23).

The contrast between the structure of the yeast spindle pole body and the animal cell centrosome underscores the remarkable diversity seen in the structure of microtubule organizing centers from different species. Indeed, the structure of the microtubule organizing center seems to be at least slightly different in every species examined. At the ultrastructural level, the only common feature of microtubule organizing centers from diverse organisms is that microtubules are nucleated from a material that has an amorphous appearance. It is likely that microtubule organizing centers arose early in the evolution of eukaryotic cells, and that they are much more similar to each other on a molecular level than their outward appearance would suggest. But a more complete understanding of the relationships between the structure and function of microtubule organizing centers will require a molecular characterization of these organelles.

BIOCHEMICAL AND GENETIC STUDIES OF THE CENTROSOME

In theory, the isolation of centrosomes should facilitate a molecular understanding of the centrosome. However, the purification of centrosomes in biochemical quantities has proven to be difficult, both because centrosomes constitute such a small proportion of the total cell protein (there are usually only one or two centrosomes per cell), and because they are usually tightly associated with the nuclear envelope. While the development of protocols for centrosome isolation has not been as fruitful as one might have hoped, it has at least allowed the characterization of some of the basic properties of the centrosome, and it has led to the identification of a number of centrosomal proteins. Proteins of the centrosome have also been identified using autoimmune sera, microtubule affinity chromatography, and genetic approaches—all of which are discussed in the following sections.

Partial Purification of Centrosomes and Spindle Pole Bodies

Early work on microtubule organizing centers of mammalian and yeast cells showed that it is possible to lyse and extract cells, add back purified bovine brain tubulin, and observe the regrowth of microtubules from centrosomes and spindle pole bodies (24–27). More recently, different laboratories have developed protocols for the isolation of functional microtubule organizing centers from mammalian cells, *S. cerevisiae,* oocytes of the surf clam *Spisula solidissima,* and *Drosophila* embryos. Briefly, each of these protocols takes advantage of the large size of the microtubule organizing center relative to most other cellular organelles, allowing a relatively simple

Table 1 Summary of centrosome isolation protocols

Source	Yield	Purification	Protein (μg)	References
Mammalian tissue culture cells,[a] calf thymus	30–50%	3000-fold	1–10	28–32
Saccharomyces cerevisiae	45%	600-fold	4000	39
Spisula solidissima	—	1000-fold	150	42, 43, and R Palazzo, personal communication
Drosophila melanogaster	10%	800-fold	300	M Moritz, BM Alberts, unpublished

[a] N115 neuroblastoma, CHO, lymphoblastoid KE37

isolation and concentration of them by sucrose gradient velocity sedimentation. In all cases, the isolated microtubule organizing centers nucleate microtubules upon incubation with purified tubulin subunits, and they can be frozen and stored indefinitely at −80°C with little loss of activity. The yields and estimated purities of the microtubule organizing centers isolated using each of these protocols are summarized in Table 1. The protein profiles of these partially purified microtubule organizing centers on polyacrylamide gels are complex, suggesting that there could be as many as 100 different proteins associated with the microtubule organizing center.

The isolation of mammalian centrosomes allowed Mitchison & Kirschner (28, 33) to discover important and unexpected properties of microtubules. They found that the centrosome can nucleate microtubules at tubulin concentrations where free microtubules are unstable. This property indicates that microtubule assembly off of the centrosome is preferred, explaining why the centrosome is the source of all microtubules in the cell. In the same experiments, they discovered that microtubules coexist in growing and shrinking populations that rarely interconvert—a phenomenon they called *dynamic instability*. Subsequent work has shown that dynamic instability is an inherent property of microtubules, whether or not they form from a centrosome (33). In addition, the dynamic instability of microtubules has been directly observed in living cells (34–38).

Rout & Kilmartin (39) have raised monoclonal antibodies to spindle pole bodies isolated from *S. cerevisiae*. One of these antibodies has been used to clone the gene encoding a 110-kDa component of the spindle pole body (40). The gene was named *SPC110* and is identical to *NUF1*, a gene previously identified as encoding a potential component of the nucleoskeleton (41). The Spc110/Nuf1 protein is essential for viability and is predicted to contain a large coiled-coil domain (40, 41). Electron microscopy shows that a Spc110/ Nuf1 fusion protein expressed in bacteria forms rods of the predicted length (40). The protein has been localized within the spindle pole body to the space between the central plaque and the origins of the nuclear microtubules near the inner plaque (39), where rodlike structures are visible (40). If the wild-type Spc110/Nuf1 protein is replaced with truncated proteins bearing deletions in the coiled-coil domain, the gap between the central plaque and the nuclear microtubules shortens in proportion to the size of the deletion, strongly suggesting that these rodlike structures consist of the Spc110/Nuf1 protein (40).

In the surf clam, centrioles, centrosomes, and microtubule asters will arise de novo in oocytes that have been activated by KCl treatment (42, 43). Palazzo and colleagues are using this system to identify proteins that assemble onto centrosomes during the maturation stages and to study centriole assembly (43; R Palazzo, personal communication).

Antisera Against Centrosomal Proteins

One approach to the characterization of centrosomal components begins with antibodies that recognize the centrosome. In some cases, these antibodies have been obtained using isolated nuclei or spindles as antigen. In other cases, human autoimmune sera or rabbit preimmune sera were screened for reactivity with the centrosome. Since this information was recently reviewed (6, 7), details are not given here. Although a number of centrosomal proteins have been identified by these means (the reviews list a total of about 10 proteins, from a wide range of organisms), in most cases the functions of these proteins in the centrosome are unknown. Pericentrin, a centrosomal protein identified using human autoimmune serum, is probably the best characterized of the proteins discovered in this way, and it is described later.

Identification of Centrosomal Proteins by Microtubule Affinity Chromatography

Since the primary function of the centrosome is to organize microtubules, one might expect that some proteins of the centrosome could be identified on the basis of their ability to interact with microtubules, and indeed this is the case. Affinity chromatography experiments, in which a *Drosophila* embryo extract is passed over a column containing immobilized microtubules, have identified more than 100 different proteins that interact specifically with the microtubule column (44). To characterize these proteins, polyclonal antisera were raised against 25 of them. Of these antisera, 23 stain some kind of microtubule structure when used in immunofluorescence experiments, including five antibodies that primarily stain the centrosome. These antibodies have been used to clone cDNAs encoding proteins that interact with microtubules, and are localized to the centrosome. One of these proteins has been given the name DMAP190, to indicate that it is a Drosophila microtubule associated protein with an apparent mass on SDS-polyacrylamide gels of 190 kDa (45). This protein is identical to the protein recognized by the Bx63 monoclonal antibody, which was originally identified in a collection of monoclonal antibodies derived from mice immunized with partially purified nuclei (46, 47). The DMAP190/Bx63 protein is a component of a complex of centrosomal proteins (45) (see below); it localizes to the centrosome during mitosis, and to the nucleus during interphase. The significance of the nuclear localization is unknown.

The gene encoding another protein that interacts with microtubules and is localized to the centrosome has been cloned in this way and given the name LK6. On western blots, LK6 antibodies recognize a 180-kDa protein in crude extracts, and an approximately 200-kDa protein in the fraction of proteins that bind to a microtubule affinity column. It appears that these are different forms of the same protein. The sequence of the LK6 cDNA indicates that it encodes a protein kinase, and the LK6 protein has been shown to phosphorylate itself in vitro (J Raff and BM Alberts, unpublished).

Genetic Identification of Centrosome and Spindle Pole Body Proteins

Genetic approaches to understanding the centrosome have proceeded slowly, perhaps because it is difficult to predict the phenotype of mutations in centrosomal proteins. Recently, however, a number of centrosomal proteins have been identified by genetic means.

A screen for extragenic suppressors of a β-tubulin mutation in *Aspergillus nidulans* resulted in the identification of a tubulin variant called γ-tubulin, which is 35% identical to α- and β-tubulin (48). Antibodies raised against γ-tubulin specifically recognize the spindle pole body in *A. nidulans* (49). In addition, homologs of γ-tubulin have been found in species as diverse as humans, plants, and fission yeast. In each case, the γ-tubulin is localized to the centrosome or spindle pole body [e.g. (50–53)]. The only organism in which γ-tubulin has been sought and not found is the yeast *S. cerevisiae* (Y Zheng and B Oakley; T Stearns and M Kirschner, personal communications). These results indicate that γ-tubulin is a highly conserved component of microtubule organizing centers.

A null mutation in the *A. nidulans* γ-tubulin gene is a recessive lethal that strongly inhibits nuclear division and weakly inhibits nuclear migration. The null mutation also causes a reduction in the number and length of cytoplasmic microtubules and a complete loss of the mitotic spindle (49). A null mutation in the *Schizosaccharomyces pombe* gene is also a recessive lethal (50, 52). These and other observations have led to the speculation that γ-tubulin may be the protein that nucleates the growth of microtubules from the centrosome (49–52, 54). This possibility is discussed in greater detail below.

Drosophila is potentially a good system in which to identify mutants defective in centrosome function, because the nuclei in early embryos undergo a series of very rapid divisions without cytokinesis, during which the centrosome plays a central role in cell organization (55–57). Since the mother supplies the early embryo with centrosome components needed for the earliest divisions, mutants with defective centrosomes could be maternal-effect lethals, late larval lethals, or early pupal lethals, depending on when the embryo requires the particular centrosome component that is specified by the mutant gene (58).

There are a number of mutants in *Drosophila* that show defects in mitotic or meiotic divisions suggestive of abnormal centrosomes. For example, the *daughterless-abo-like* maternal-effect mutation causes a defect in centrosome separation during the early nuclear divisions of the embryo (59). Several other *Drosophila* mutants that may have centrosomal defects have recently been reviewed (60). (See also section on *The Centrosome and Cortical Events* below.)

Mutations in a gene called *vfl2* in *Chlamydomonas* cause cells to have variable numbers of flagella (61–63). Recently, it was found that the *vfl2* gene encodes centrin (also known as caltractin) (64), a protein originally identified as an abundant component of the basal body apparatus in the alga *Tetraselmis,* but later found to be a component of the centrosome in all organisms in which it has been sought, including mammals (reviewed in Refs. 65, 65a). Centrin is a member of a family of EF-hand-Ca^{2+}-binding proteins that includes calmodulin and Cdc31 (66, 67). In the green alga *Chlamydomonas,* centrin is found in three different fibrous structures within the basal body–containing nucleo-flagellar apparatus, and all of these structures display Ca^{2+}-mediated contractions (reviewed in Ref. 65). Cells carrying the *vfl2* mutation have defects in all of these structures when viewed by electron microscopy, and they appear to be defective in the localization and segregation of the basal bodies (64). In addition to the *vfl2* mutations, other mutations that affect flagella and basal bodies have been isolated in *Chlamydomonas,* and these may help identify additional components of the centrosome (J Jarvik, personal communication; Ref. 68).

Genes involved in yeast spindle pole body duplication have been identified by genetic approaches, and these are discussed in detail below. Yeast spindle pole body components have also been identified in screens for genes involved in nuclear fusion and chromosome maintenance (69, 70).

Centrosomal Proteins Are Components of Multiprotein Complexes

Many cellular processes are carried out by multiprotein complexes, and the identification of such complexes from the centrosome would be a major step towards unraveling its structure and function. One such complex has been identified by immunoaffinity chromatography (45). In these experiments, an immunoaffinity column is constructed with antibodies that recognize the *Drosophila* centrosomal protein DMAP190. When a *Drosophila* embryo extract is passed over this column, DMAP190 is retained, as well as approximately 10 additional proteins. Characterization of these proteins supports the idea that they interact with DMAP190 to form a complex within the cell. Antibodies that recognize a 60-kDa protein that is retained on the immunoaffinity column reveal that it is colocalized with DMAP190 at the centrosome and binds to a microtubule affinity column in a manner identical to DMAP190. An 85-kDa protein that is retained on the anti-DMAP190 immunoaffinity columns has similar properties (J Raff, BM Alberts, unpublished). These two proteins have been given the names DMAP60 and DMAP85, respectively. The cDNA that encodes DMAP60 has been cloned and sequenced, and has no homology to any known proteins, although it contains six consensus sites for phosphorylation by cyclin-dependent kinases (DR Kellogg and BM Alberts, unpublished).

As discussed in the previous section, γ-tubulin is a highly conserved component of microtubule organizing centers from diverse species. γ-tubulin is also specifically retained on the anti-DMAP190 column, and several lines of evidence suggest that DMAP60 and γ-tubulin interact with each other (71). For instance, both γ-tubulin and DMAP60 in embryo extracts bind to microtubules and, if microtubule-binding proteins are sedimented on a sucrose gradient, γ-tubulin and DMAP60 comigrate with a sedimentation coefficient of 8S. Moreover, if microtubule-associated proteins are passed over an anti-DMAP60 immunoaffinity column, most of the γ-tubulin is retained on the column. This is true even when the experiment is carried out in the presence of 0.5M KCl, indicating that the two proteins interact tightly with each other. Because these experiments are performed in high salt, it is not clear how this 8S complex might relate to a 25S γ-tubulin-containing complex identified in *Xenopus,* described next.

A soluble complex that contains γ-tubulin has been identified in eggs and tissue-culture cells from *Xenopus,* as well as in human 293 cells. This complex has a sedimentation coefficient of 25S on sucrose gradients, and it contains most, if not all, of the γ-tubulin that is not in the centrosome itself (a 25S spherical particle would have a mass of about a million daltons). Surprisingly, about half of the γ-tubulin in actively growing tissue-culture cells is in the form of this soluble complex, the rest being in the centrosome. The complex cosediments with microtubules in spin-down experiments. The stoichiometry of binding suggests that the γ-tubulin-containing complex may bind to one or both ends of a microtubule (71a).

CENTROSOME ASSEMBLY

Images obtained by electron microscopy give the impression that the centrosome is a relatively static structure that simply nucleates microtubules. However, it is becoming increasingly clear that the centrosome is a highly dynamic organelle that is constantly changing during the cell cycle. In the following sections, we discuss how components of the centrosome can be found both at the centrosome and in the cytoplasm, and how a new centrosome can be assembled entirely from cytoplasmic components.

Formation of a Centrosome Occurs de novo in Some Situations

Formation of a new centrosome normally occurs in association with a preexisting centrosome. However, in some situations formation of a new centrosome or a centrosomelike structure can occur independently of a preexisting centrosome. For instance, it has been known for many years that treatment of unfertilized sea urchin eggs with hypertonic solutions can induce the formation of centrosomes that contain centrioles (1, 72–74). Many centrosomes can be induced to form within a single egg, and it has been shown that these centrosomes are able to undergo division in a normal 1—2—4 fashion (G Sluder, personal communication). Since these eggs contain no centrosome or centrioles prior to the treatment, their formation must occur de novo. The de novo appearance of centrioles also occurs normally during the development of the mouse embryo. The first few divisions of the early mouse embryo take place without any centrioles. Instead, the spindle poles are formed by aggregation of microtubule-nucleating material that is initially scattered about the cytoplasm. Normal centrosomes that contain centrioles appear after several divisions (17).

More recently, the de novo formation of microtubule organizing centers has been studied using micromanipulation techniques to generate cell fragments that contain no centrosome. For example, portions of large, flat melanophore cells from fish scales can be cut away with a microneedle, leaving the centrosome and nucleus behind. These centrosome-free cell fragments initially have an unfocused microtubule array, but over the course of 4 hours the microtubules reorganize to form a radial array with their minus ends focused at the cell center (75). Similarly, a cultured mammalian cell (BSC1) in interphase can be manipulated with a microneedle to form two cell fragments, one containing the nucleus and the other the centrosome. The microtubules in the cell fragment lacking the centrosome are initially disorganized, but after 20–30 hours these microtubules reorganize into a radial array centered next to the nucleus (76). As judged by immunofluorescence, the array that forms appears nearly indistinguishable from the arrays seen in normal cells, but the centrosomelike structure that forms does not contain centrioles and does not duplicate.

Assembly of Centrosomes in vitro from Soluble Components

A number of systems have been developed that allow aspects of centrosome assembly to be studied in cell-free extracts. This represents a powerful approach for studying centrosome assembly, because antibodies can be used either to deplete specific centrosomal proteins from the extract, or to block their function in it. In addition, in vitro systems may provide assays that allow biochemical purification of new centrosomal components based on their functional properties.

Radial arrays of microtubules that resemble the kind of arrays nucleated by the centrosome can be induced to form in cell-free extracts from mitotic *Xenopus* eggs by addition of taxol, a drug that stabilizes microtubules. The centers of these asters are composed of a dense material that stains with a number of antibodies that recognize centrosomal components, suggesting that a true centrosome matrix has formed (77). These results suggest that centrosomal subunits dispersed in the extract are able to self-associate to assemble a new microtubule organizing center. In theory, this self-association would be greatly facilitated if the dispersed centrosomal proteins were able to bind anywhere on a microtubule and then be transported to the minus end. This idea is supported by the finding that formation of the taxol-induced asters in *Xenopus* extracts is dependent on cytoplasmic dynein, a minus end–directed microtubule motor protein (77).

Cell-free extracts made from *Xenopus* eggs can also be used to study the assembly of microtubule-nucleating activity onto inactivated centrosomes (78). Partially purified centrosomes from human lymphoid cells that have been extracted with 2–3 M urea lose the ability to nucleate microtubules in the presence of purified tubulin, although this treatment causes only a slight disruption of the morphology of the centrosome as determined by electron microscopy. When these inactivated centrosomes are added to a *Xenopus* egg extract, they regain the ability to nucleate the growth of microtubules, indicating that factors required for microtubule nucleation are present in the extract and can assemble onto the inactivated centrosome template. The factor in the egg extract is present in soluble form, since centrifugation of the extract at 250,000 × g for two hours does not affect the ability of the extract to reconstitute the inactivated centrosomes. The fact that microtubule-nucleating components exist in a soluble form may make it possible to purify and characterize them (78).

A similar assay for centrosome assembly is based on the addition of sperm basal bodies to mitotic *Xenopus* egg extracts (71a, 71b, 78a). The *Xenopus* sperm nucleus has an associated pair of basal bodies that lacks the ability to nucleate microtubules. However, when sperm nuclei are added to a mitotic egg extract, a functional centrosome that nucleates microtubules assembles around the basal bodies after a delay of 4–10 minutes. Immunofluorescence experiments reveal that the highly conserved centrosomal protein, γ-tubulin, is initially absent from the nascent centrosome but always appears coincidentally with microtubule nucleation. The assembly of a functional centrosome in these experiments does not require microtubules, since it will occur in the presence of the microtubule-depolymerizing drug nocodazole (71a, 71b).

Experiments carried out in this system have provided evidence that a protein called pericentrin may be required for centrosome assembly. Pericentrin is a 220-kDa centrosomal protein that was identified using a human autoimmune serum from a scleroderma patient [historically known as 5051 serum (13, 79)]. Pericentrin is found at microtubule organizing centers (with or without centrioles) in organisms as diverse as *Drosophila, Tetrahymena* (a ciliate), *Naegleria* (an amoeba), and mammals. Immunoelectron microscopy localizes pericentrin to the centrosomal matrix surrounding the centrioles in animal cells. In the *Xenopus* extract system described above, antibodies that recognize pericentrin block the ability of the sperm basal bodies to assemble a centrosome that can nucleate microtubule arrays. These antibodies also inhibit mitotic spindle formation when injected into one- or two-cell *Xenopus* embryos, and they inhibit meiotic spindle formation when injected into developing mouse oocytes. It is unclear whether the anti-pericentrin antibodies act by blocking assembly of the centrosome or by blocking microtubule nucleation, and it is possible that the antibody prevents centrosome assembly in a nonspecific manner by sterically hindering assembly of other centrosome components. The anti-pericentrin serum does not disrupt microtubule nucleation by partially purified mature centrosomes. The sequence of mouse pericentrin is not significantly homologous to any known proteins, but it contains a very large coiled-coil domain that could be as long as 285 nm, suggesting that it may be a component of a structural scaffold within the centrosome on which microtubule-nucleating centers are organized (78a).

A cell-free system that supports the assembly of centrosomes that contain centrioles has also been developed (43). Oocytes from the surf clam *Spisula* can be induced to form centrosomes by activation with hypertonic solutions. A crude lysate made just after activation also supports centrosome formation;

but if the lysate is first centrifuged at 29,000 × g for 15 minutes, the ability to form centrosomes is lost. If the centrifuged lysate is mixed with crude lysate from unactivated eggs, centrosomes are formed and increase in number with time. Examination of the centrosomes by electron microscopy reveals that they each contain a single centriole after one minute and a pair of centrioles after 15 minutes (43).

The Centrosome Matrix May Consist of Many Individual Subunits

The experiments described above tell us that proteins present in the cytoplasm can assemble to form a new microtubule organizing center, and in some cases this can occur without requiring a preexisting centrosome to act as a template. It has been suggested that the centrosome matrix is formed by the assembly of many identical subunits (5, 6). In this view, each microtubule would be nucleated by one such complex, and the centrosome matrix would be formed, at least in part, by the linking together of many such identical complexes by fibrous structural proteins. Additional complexes might carry out such functions as the anchoring and release of microtubules (see below). This view of the centrosome implies that there is a much higher degree of organization in the centrosome matrix than is presently recognized by electron microscopy.

CENTROSOME DUPLICATION

The centrosome must duplicate once and only once in order to assemble a bipolar spindle and to segregate the chromosomes accurately. The mechanisms by which centrosome duplication occurs remain unknown, although genetic approaches have recently identified a number of genes that are likely to be involved. Experiments with animal cells have also helped define some of the general requirements for centrosome duplication.

Formation of a New Centrosome Usually Occurs in Association with Preexisting Centrioles

A centrosome that has the capacity to duplicate itself is usually formed only by the duplication of a preexisting centrosome. For instance, the centrosome of a BSC1 cell growing in culture can be removed by microsurgery, leaving the nucleus behind (76). Although a new centrosomelike structure is regenerated in the cell, the structure never duplicates, suggesting that there is some factor associated with the preexisting centrosome that allows duplication, and the cell is not able to replace it.

Experiments using sea urchin embryos also provide evidence for the existence of some kind of duplication potential that usually arises only from a preexisting centrosome. For unknown reasons, treatments that prolong the first mitosis of the sea urchin embryo cause the centrosome at each pole of the mitotic spindle to split and thereby form a tetrapolar spindle (80). Examination of the four centrosomes thus formed shows that each contains one centriole, rather than the normal two (81). Thus, this treatment causes centrosome splitting to occur without accompanying centriole duplication. In the ensuing cytokinesis, each of the four centrosomes is segregated into a separate cell. When the next mitosis arrives, the centrosomes that contain only a single centriole are not competent to divide and a monopolar spindle is formed. Thus, some aspect of the centrosome that allows continued division to occur is lost when the centrosome is induced to divide during mitosis.

The requirements for duplication of a centrosome have been studied to some extent in *Xenopus* eggs. The unfertilized *Xenopus* egg contains no centrosome but can be induced to undergo cell division and develop parthenogenetically if injected with fractions containing partially purified centrosomes or basal bodies (82–85). The ability of the injected centrosomes to support cell division is unaffected by extracting them with 2 M KCl or 2 M urea, even though these treatments completely destroy the ability of the centrosome to nucleate microtubules in vitro (86). Extraction of the centrosomes with 4 M KCl largely destroys their ability to support cell division, but has only minor effects on centrosome or centriole structure, as determined by electron microscopy. These results suggest that something tightly associated with the centrosome is responsible for determining its ability to function in a reproductive capacity. Whether this is a regulatory factor, or some kind of nucleating structure, is unknown. It apparently has not yet been possible to solubilize a factor that complements the 4M KCl–extracted centrosome in the experiment described above.

In normal situations, what aspect of a preexisting centrosome directs formation of a new centrosome that has a full reproductive capacity? The existence of an autonomous centrosome-associated DNA that controls centrosome duplication has been proposed, but there has been no convincing evidence to support this idea (reviewed in Ref. 87). Since the reproductive capacity of the centrosome is often correlated with the presence of centrioles, it has been proposed that the centriole pair, or something very tightly associated with it, is responsible for determining the reproductive capacity of the centrosome. However, it seems unlikely that the centriole structure itself is the critical determinant that allows duplication of the centrosome. As discussed previously, there are numerous examples of cells that lack centrioles and still undergo normal cell division, and the mechanisms that control the accurate duplication of microtubule organizing centers are likely to have been conserved during evolution. Perhaps the best hope for understanding centrosome duplication lies in the identification of mutations that affect this process in organisms such as yeast, *Chlamydomonas*, and *Drosophila*. Some candidates for such mutants are discussed in the following section.

Mutations in Budding Yeast Define Genes Required for Spindle Pole Body Duplication

In budding yeast, mutations that block spindle pole body duplication can be recognized at the level of immunofluorescence because they cause cells to form monopolar spindles. Mutations in the *CDC31*, *KAR1*, *NDC1*, *MPS1*, and *MPS2* genes can all confer this phenotype. Careful examination of the phenotypes of these mutants reveals that each has unique effects on spindle pole body duplication and the cell cycle (reviewed in Refs. 88–90).

Examination of *cdc31* cells shifted to the nonpermissive temperature reveals that they arrest with a large bud and a single unreplicated spindle pole body that is somewhat larger than normal (91). The mutation affects both mitotic and meiotic divisions, and the phenotype suggests that the defect is at an early step in spindle pole body duplication. Recently, the Cdc31 protein has been localized by immunoelectron microscopy to the cytoplasmic sides of the half-bridge and bridge structures of the spindle pole body (92), which is an expected location for a protein that plays an early role in spindle pole body duplication (see Figure 2).

Sequence analysis indicates that the Cdc31 protein is a member of a family of Ca^{2+}-binding proteins that includes centrin and calmodulin (67, 93). The Cdc31 protein is most similar (50% identical) to centrin, a highly conserved centrosomal protein. Centrin mutations in *Chlamydomonas* appear to cause defects in the localization and segregation of the basal bodies (64), but it is unclear whether they block centrosome duplication in the same way that *cdc31* mutations block spindle pole body duplication. It may be that Cdc31 and centrin are different members of a family of calcium-binding regulatory proteins, and that each carries out a different function.

The *KAR1* gene was initially identified by mutations that prevent nuclear fusion during mating of yeast cells (*karyogamy*) (94–96). Isolation of conditional alleles of the *KAR1* gene demonstrated that it is also required for spindle pole body duplication and for the formation of normal cytoplasmic microtubule arrays (96). At the restrictive temperature, *kar1* mutant cells arrest with a single unduplicated spindle pole body in a manner similar to *cdc31* cells. Overproduction of the Kar1 protein also prevents spindle pole body duplication. Expression of Kar1 fusion proteins in yeast has demonstrated that a 72-amino-acid internal fragment directs localization to the newly forming spindle pole body (70). Localization of the endogenous Kar1 protein has not been possible because it is a very low-abundance protein whose overexpression is toxic to the cell.

A 17-amino-acid deletion in the spindle pole body localization domain of the Kar1 protein was recently found to confer a temperature-sensitive phenotype on cells. Interestingly, alleles of the *CDC31* gene were isolated as spontaneous suppressors of this *kar1* allele, providing genetic evidence for an interaction between the Cdc31 and Kar1 proteins. In a gel overlay system, the Cdc31 protein was found to interact with the spindle pole body localization domain of the Kar1 protein. These results suggest that an interaction between the Kar1 and Cdc31 proteins plays a role in spindle pole body duplication (MD Rose, personal communication).

The *mps1* mutant causes yeast cells to arrest with a single unreplicated spindle pole body that has an unusually large half-bridge structure (97). The sequence of the *MPS1* gene suggests that it encodes a protein kinase (M Winey, personal communication). The *mps2* mutation leads to formation of a nonfunctional and morphologically aberrant spindle pole body lying near a normal one (97). The spindle pole body that is formed in the *mps2* mutant is not inserted into the nuclear envelope and only nucleates microtubules from its cytoplasmic face. It also lacks a layer of electron dense material that is normally found on the nuclear side of the spindle pole body.

Another yeast gene that has been implicated in spindle pole body duplication is *NDC1*. It was originally identified as a conditional-lethal mutant (*ndc1-1*) that arrests with a monopolar spindle (98). Examination of this mutant by electron microscopy reveals that a new spindle pole body forms, but fails to insert into the nuclear envelope (99). This phenotype is indistinguishable from that of the *mps2* mutant (M Winey, personal communication). The *NDC1* gene is essential and encodes a protein with six to seven membrane-spanning domains. Immunofluorescence localizes the Ndc1 protein to the nuclear envelope. One therefore suspects that this protein may mediate the insertion of the newly formed spindle pole body into the nuclear envelope (99).

The *ESP1* gene (extra spindle pole) was originally thought to be a negative regulator of spindle pole body duplication because mutations in this gene cause cells to accumulate many spindle pole bodies (66). More recent work, however, has shown that the *esp1* mutation actually causes a defect in spindle assembly that prevents normal segregation of the spindle pole bodies to the two daughter cells (100).

Coordination of Centrosome Duplication with Other Cell-Cycle Events

The duplication of the centrosome must be coordinated with other events of the cell cycle. A number of experiments, however, have demonstrated that centrosome duplication and other cell-cycle events can be uncoupled. The nucleus of the fertilized sea urchin embryo can be removed and the remaining centrosome will continue to divide in a normal 1—2—4—8 fashion, although with a slightly longer cycle time (101). A similar experiment can be carried out by injecting the embryo with aphidicolon at an early stage to block DNA synthesis (102). In this case, the centrosome becomes dissociated from the blocked nucleus and continues to divide in a normal fashion. Similar findings have been made in embryos from starfish and *Drosophila* (103, 104). These experiments tell us that neither DNA replication nor a nucleus is required for continued centrosome duplication.

Centrosome duplication will also continue in embryos from frogs or sea urchins that have been injected with doses of protein synthesis inhibitors that reduce protein synthesis by 97% (105, 106). This result is particularly striking because inhibiting protein synthesis in these embryos is known to block all other events of the cell cycle, such as DNA synthesis, nuclear envelope breakdown, spindle assembly, activation of H1 kinase, etc. The fact that centrosome duplication continues suggests that the centrosome cycle is autonomous from the rest of the cell cycle in these embryos. Alternatively, the protein synthesis inhibitors might arrest the cell cycle at a stage where the signal to initiate centrosome duplication is constitutively turned on. In contrast to frog and sea urchin embryos, when protein synthesis is blocked in embryos from *Drosophila* or starfish, the centrosome cycle arrests (103; J Raff, W Sullivan, personal communication).

The coordination between the cell cycle and the duplication of microtubule organizing centers can also be studied in budding yeast. A critical event in the yeast cell cycle occurs between G1 and S phase and is called "Start" (reviewed in Refs. 107–110). Once cells have passed through Start, they are committed to completing DNA replication and mitosis. Before cells have passed through Start they can be arrested in the cell cycle by mating pheromones and induced to enter the mating pathway. Passage through Start requires the activity of a kinase complex formed by the association of a catalytic subunit encoded by the *CDC28* gene with a member of a group of related proteins called G1 cyclins. Yeast cells can therefore be arrested before Start either by treatment with mating pheromone or by shifting cells that carry temperature-sensitive alleles of *CDC28* to the restrictive temperature. When cells are arrested before Start they have an unduplicated spindle pole body as expected, since duplication is known to occur during S phase (21). However, the unduplicated spindle pole body has a satellite on its bridge structure, which is the first morphological manifestation of spindle pole body duplication (see Figure 2). This result suggests that spindle pole body duplication has already been initiated before Start, but requires a signal generated by the Cdc28 kinase complex to continue. The nature of the signals required for initiating and continuing spindle pole body duplication remains unknown.

As already discussed, yeast cells carrying *cdc31*, *kar1*, and *mps2* mutations are defective in spindle pole body duplication and arrest with large buds after completing DNA replication, showing that DNA replication and bud formation can continue in the absence of spindle pole body duplication (66, 96, 97, 111). Other cell-cycle mutants can complete spindle pole body duplication in the absence of DNA synthesis and bud formation, indicating that the converse is also true (see for example Refs. 112, 113).

The fact that the spindle pole body duplication mutants arrest before mitosis is most likely due to the presence of feedback controls that prevent cells from entering mitosis until they have assembled a normal spindle (114). Interestingly, cells carrying the *mps1-1* mutation cannot undergo spindle pole body duplication, but nevertheless go through multiple rounds of DNA replication, and undergo bud formation to produce anucleate cells, suggesting that the cell cycle continues in the absence of spindle pole body duplication (97). A possible explanation for this result is that *MPS1* also functions in a feedback control mechanism that detects unduplicated spindle pole bodies. The idea that the protein machinery of a particular process might be involved in activating a feedback control has also been proposed in the context of DNA replication (115).

Duplication of Microtubule Organizing Centers Occurs by a Conservative Mechanism

At least part of the centrosome duplicates conservatively. If tissue-culture cells are injected with biotin-labeled tubulin, only the newly forming centriole in each pair incorporates the labelled tubulin. At the end of mitosis, each cell therefore inherits a pair of centrioles that contains one old centriole and one entirely new centriole, indicating that centriole duplication occurs conservatively (116).

The duplication of the spindle pole body in yeast also appears to occur by a conservative mechanism. This conclusion is based on the observation that mutations that disrupt spindle pole body duplication only affect the newly forming one (97). In addition, a Kar1 fusion protein is only incorporated into the newly forming spindle pole body (70).

MICROTUBULE NUCLEATION BY THE CENTROSOME

The nucleation of microtubules at tubulin concentrations below those required for spontaneous microtubule assembly is perhaps the most basic function of the centrosome. In general, nucleation-dependent protein polymerization requires the formation of a seed containing a small number of protein subunits. Formation of this seed is a thermodynamically unfavorable process, but once the seed has formed, further addition of monomers is thermodynamically favorable (reviewed in Ref. 117). Presumably, the centrosome catalyzes the formation of small microtubule seeds. In all cells, microtubules appear to be nucleated from a material that has an amorphous appearance when examined by electron microscopy. This is true even in specialized cells where microtubules are nucleated from noncentrosomal sites. In the case of the centrosome itself, this material is called the centrosome matrix or the pericentriolar material. Ultimately, we would like to be able to reconstitute microtubule nucleation from completely purified, well-characterized components.

The Centrosome Appears to Have Distinct Microtubule-Nucleating Sites that Are Regulated during the Cell Cycle

If permeabilized cells are incubated with increasing amounts of tubulin, the number of microtubules nucleated by each centrosome reaches a plateau (118). A similar result is obtained if partially purified centrosomes or spindle pole bodies are used (22, 29, 119). The existence of a saturable number of microtubule-nucleating sites in the centrosome matrix suggests that the centrosome nucleates microtubules at defined sites, rather than by creating a local environment that favors microtubule nucleation. In addition, most of the microtubules that are nucleated by the centrosome have 13 protofilaments, whereas most microtubules formed by spontaneous nucleation in vitro have 14 protofilaments (120–123). This observation supports the idea that the centrosome has defined sites, which, moreover, serve as precise templates for microtubule nucleation.

Entry into mitosis in animal cells is accompanied by a five- to tenfold increase in the number of microtubules nucleated by the centrosome (19, 27). In vertebrate cells, the increase in the number of microtubules begins at about the time of prophase, and is accompanied by a dramatic increase in both the amount of pericentriolar material and the level of phosphorylation of centrosomal proteins (10, 17, 124). Since cyclin-dependent protein kinase complexes are responsible for inducing the events of mitosis, they are likely to be responsible for causing the increased phosphorylation of centrosomal proteins and the increased ability of the centrosome to nucleate microtubules. It is unknown whether these cyclin-dependent kinases directly phosphorylate centrosomal proteins or instead initiate a cascade of other events that results in the observed changes in the centrosome. However, immunofluorescence experiments have shown that cyclin A is localized to the centrosome during prophase, and cyclin B and the cyclin-dependent protein kinase $p34^{cdc2}$ are localized either at or near the centrosome during G2 and throughout mitosis

(125–129). These observations suggest that cyclin-dependent kinase complexes may directly phosphorylate centrosomal proteins.

An increase in the number of microtubules nucleated by the centrosome during mitosis can also be seen in cell-free systems. As described previously, partially purified centrosomes that have been extracted with 2M urea lose the ability to nucleate microtubules in vitro, and regain nucleating activity when added to a *Xenopus* egg extract. If the extracted centrosomes are added to an extract made from mitotic eggs they regain the ability to nucleate about twice as many microtubules as centrosomes added to an extract made from interphase cells. This increase appears to be ATP dependent, while assembly of a basal number of microtubule-nucleating sites onto an extracted centrosome in interphase extracts is not (78).

Similar results have been obtained using the fission yeast *Schizosaccharomyces pombe*. The spindle pole body of *S. pombe* nucleates no microtubules during interphase, and must somehow be activated to nucleate microtubules during mitosis to assemble a mitotic spindle. The ability of the interphase spindle pole body to nucleate microtubules can be activated if *S. pombe* cells are permeabilized and then added to a mitotic *Xenopus* egg extract, but not if the cells are added to an interphase extract (130).

γ-tubulin and Microtubule Nucleation

A complete understanding of microtubule nucleation requires the identification of the proteins that are directly involved. Currently, the best candidate for such a protein is γ-tubulin, a new member of the tubulin superfamily that was introduced previously. A review of the evidence illustrates some of the difficulties in unambiguously assigning specific functions to centrosomal proteins.

Several lines of evidence support the idea that γ-tubulin is involved in microtubule nucleation. First, γ-tubulin was identified as a suppressor of a β-tubulin mutation, suggesting that the two proteins may interact directly, as expected for a protein that must nucleate microtubules. The fact that γ-tubulin is 35% identical to α- and β-tubulin, which bind tightly to each other, suggests that γ-tubulin might also bind directly to one of these tubulins. A null mutation in the γ-tubulin gene causes a reduction in the number of interphase microtubules and a complete loss of the mitotic spindle in *A. nidulans* (49). A similar phenotype is observed in γ-tubulin null mutants of *S. pombe* (52). A cytoplasmic protein complex that contains γ-tubulin interacts with microtubules in crude extracts, as one might expect if γ-tubulin directly nucleates the growth of microtubules (71, 71a). Injection of cells before or during mitosis with antibodies that recognize γ-tubulin blocks the formation of a mitotic spindle, and the injected antibodies also block formation of an interphase microtubule array in cells recovering from treatments that depolymerize microtubules (e.g. nocadazole or cold) (131). Similarly, antibodies that recognize γ-tubulin block the formation of a microtubule aster on sperm basal bodies that have been added to a *Xenopus* egg extract (71a, 71b). Immunodepletion of γ-tubulin from a *Xenopus* egg extract also prevents formation of a centrosome that can nucleate microtubules (71b). Immunolabeling experiments in many different organisms have demonstrated that γ-tubulin is localized exclusively to the centrosome matrix from which microtubules are nucleated (49–51, 131–135). Finally, γ-tubulin synthesized in reticulocyte lysates appears to bind to one end of microtubules (H Joshi, personal communication).

These results are supportive of a role for γ-tubulin in microtubule nucleation. However, one should keep in mind the following caveats. It may be that γ-tubulin is involved in assembly of the centrosome, and antibodies that recognize γ-tubulin block nucleation of microtubules indirectly by preventing assembly of centrosomal structures. Antibodies that recognize γ-tubulin could also cause nonspecific effects simply by sterically blocking the assembly of other complexes that are directly involved in microtubule nucleation. The finding that γ-tubulin binds to microtubules in crude extracts is intriguing, but we do not know whether γ-tubulin is binding directly to the microtubules, or through secondary interactions with other proteins that interact directly with microtubules. Similar concerns apply to experiments that have provided evidence for the involvement of other proteins in microtubule nucleation (e.g. Refs. 136, 137).

By what criteria should we judge that a protein is responsible for nucleating microtubules? As with other cellular processes, it seems that ultimately one must reconstitute function in vitro with purified components. Since γ-tubulin exists in a soluble form as a component of a large complex, it should be possible to purify and characterize this complex. In addition, the recent development of assays for the assembly of nucleating factors onto centrosomes may provide a systematic approach for identifying and characterizing the proteins involved in microtubule nucleation.

THE CENTROSOME AND THE MITOTIC SPINDLE POLE

How does the centrosome contribute to the assembly of the mitotic spindle? At one extreme, one might imagine that the centrosome serves only to nucleate microtubules in all directions, and other mechanisms are used to capture and utilize microtubules selectively in the formation of the spindle. However, this simple view of the centrosome cannot be entirely correct.

As is discussed below, recent work has demonstrated that the mitotic centrosome can somehow allow depolymerization to occur at the minus ends of kinetochore microtubules, and that it is likely that the centrosome actively releases microtubules during mitosis. These examples demonstrate that the centrosome is likely to play a more active role in spindle assembly.

Nevertheless, it remains unclear how the centrosome contributes to the organization of the mitotic spindle poles. In newt lung cells, the centrosome can occasionally become dissociated from the spindle. In these rare cases, the microtubules of the spindle remain focused at a defined spindle pole, and chromosome segregation is unaffected (138). These observations suggest that at some point during mitosis the centrosome may no longer be needed for maintenance of a focused spindle pole. There are also some cases in which the assembly of a bipolar spindle has been observed to occur in the absence of any kind of conventional centrosome. For example, the poles of the meiotic spindle in *Drosophila* are not recognized by a variety of antibodies that recognize known centrosomal components, including γ-tubulin (138a; W Theurkauf, personal communication). In this unusual case, it appears that the chromatin and kinesinlike proteins may play an important role in the organization of the spindle (reviewed in Refs. 139, 140).

Depolymerization of the Minus Ends of Microtubules at the Centrosome

Experiments that follow the behavior of the microtubules that connect the centrosome to the kinetochore reveal that the minus ends of these microtubules can depolymerize at the centrosome (138, 141, 142). In these experiments, mitotic cells are injected with tubulin that has been labeled with a photoactivatable probe that becomes fluorescent only after illumination with light of a specific wavelength. After incorporation of the injected tubulin into the mitotic spindle, a laser is used to illuminate a small region of the spindle next to the chromosomes, creating a fluorescently labeled region on the kinetochore microtubules. This region is observed to move towards the spindle pole at a rate of 0.4 microns per minute, showing that kinetochore microtubules must be constantly adding tubulin subunits at the kinetochore, and losing subunits at the centrosome. Therefore, the centrosome must somehow allow controlled depolymerization to occur at the minus ends of kinetochore microtubules, indicating that the centrosome is able to do more than simply anchor and stabilize these microtubule ends. This constant treadmilling of kinetochore microtubules is thought to be involved in maintaining a balance of forces in the spindle. Whether subunits are also constantly being lost from the ends of microtubules in the interphase centrosome is not known.

Microtubule Release by the Centrosome

When the cell passes from interphase into mitosis it must rapidly disassemble the interphase microtubule network and replace it with the mitotic spindle, and similarly, when the cell leaves mitosis it must rapidly disassemble the spindle. Since the minus ends of microtubules depolymerize much more rapidly than do the plus ends, the disassembly of microtubule networks would be facilitated if the centrosome were to release microtubules. There is some evidence to support this idea. When cells from the slime mold *Dictyostelium* enter prophase it appears that all of the interphase microtubules are released from the microtubule organizing center, and they depolymerize shortly thereafter (143). There is no evidence for such a mechanism in vertebrate cells during the transition from interphase to mitosis (T Mitchison, personal communication), although it may be difficult to observe a release of microtubules in cells with such complicated microtubule arrays.

During mitosis, some microtubules appear to be released from the centrosome in vertebrate cells. In mitotic *Xenopus* egg extracts, microtubules nucleated from the centrosome are occasionally observed to be released, and they then move away from the centrosome, apparently due to the action of microtubule-based motor proteins bound to the glass coverslip. In contrast to the *Dictyostelium* example, the minus ends of the released microtubules remain stable (144). Electron microscopic reconstruction of mitotic spindles from PTK cells has revealed that the minus ends of nonkinetochore microtubules are often quite distant from the centrosome during metaphase and anaphase, whereas the minus ends of kinetochore microtubules are closer to the centrosome (144a). This observation supports the idea that microtubules nucleated by the centrosome can be released.

The minus ends of microtubules are unstable in vitro, and it is thought that they must be stabilized to prevent their depolymerization within the cell. The mechanism by which this stabilization occurs is unknown. The simplest hypothesis is that once a microtubule has been nucleated, it remains bound to the nucleation site and is thereby anchored and stabilized. But the examples described above suggest that nucleation, stabilization, and anchoring can be independent events.

THE CENTROSOME AND CELLULAR ORGANIZATION

The centrosome and its associated microtubules direct the events of mitosis and determine the localization of the Golgi apparatus, the endoplasmic reticulum, the nucleus, and the site of cleavage furrow formation during cytokinesis. Examples of how the centrosome organizes the interior of the cell are discussed in the following sections. The centrosome can be viewed as an apparatus for positioning the contents of the cell appropriately, although little is known about how it functions in this way.

The Centrosome Can Be Positioned at the Cell Center

As described previously, the microtubules in cell fragments that lack a centrosome reorganize around a new microtubule organizing center (75, 76, 145, 146). Strikingly, the newly formed microtubule organizing center is positioned at the center of the cell, even in cells of quite irregular shape (75). This suggests that the centrosome and its array of microtubules can be viewed as a "surveying device," whose ability to find the cell center has important consequences for the organization of the cell interior. Although the final position of the centrosome is different in different cells, its centering behavior presumably sets the framework on which other forces act to determine the final position of cell contents.

How does the centrosome find the center of a cell? A plausible mechanism is based on the dynamic instability of microtubules, which could cause a constant pushing of the growing plus ends of microtubules against the cell cortex (146a). Because the number of microtubules that reach the cortex decreases as the distance of this cortex from the centrosome increases, the net effect of the pushing would be to propel the centrosome to a central location.

The Centrosome Determines the Position of Membrane-Bounded Organelles

Virtually all membrane-bounded organelles, including the endoplasmic reticulum, the Golgi apparatus, endosomes, lysosomes, and mitochondria, have a close association with microtubules (reviewed in Refs. 147–152). Because most microtubules have their minus ends embedded in the centrosome matrix, the centrosome ultimately determines the position of these organelles. Presumably, membrane-bounded organelles are decorated with receptors for microtubule motor proteins, and it is through the action of these motor proteins that the organelles are specifically positioned within the cell (147–150, 153, 156). At least one minus-end-directed motor, cytoplasmic dynein, has been shown to be involved in the clustering of the Golgi apparatus around the centrosome during interphase (154, 155).

Branches of the endoplasmic reticulum extend out into the cytosol from their perinuclear origin along centrosomal microtubules, suggesting that they achieve this position via plus-end-directed microtubule motors, such as kinesin. If cells are treated with microtubule-depolymerizing drugs, both the Golgi apparatus and the endoplasmic reticulum lose their positions within the cell: The Golgi complex vesiculates and disperses throughout the cytoplasm (157, 158), and the endoplasmic reticulum collapses inward toward the nucleus and centrosome (159–161).

Interactions of Microtubules with the Cell Cortex Can Position the Centrosome

A number of experiments have suggested that interactions between the microtubules attached to the centrosome and the cell cortex can alter the position of the centrosome within cells. For example, at the two-cell stage of the *C. elegans* embryo, one of the cells undergoes an unequal division (162). In this cell, the duplicated centrosomes migrate to opposite sides of the nucleus during prophase, and then, together with the nucleus, they rotate through 90 degrees to form a new axis of division (163). This rotation

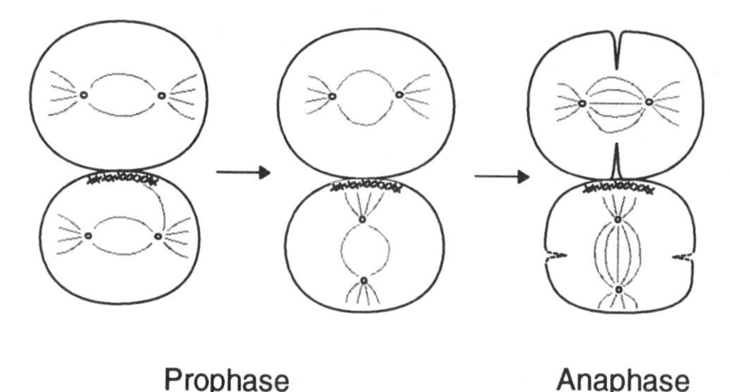

Figure 3 Interaction of microtubules with a specialized region of the actin-rich cell cortex (cross-hatching) repositions the centrosome and nucleus during prophase in *C. elegans* embryos at the two-cell stage (*left*). Thus, the ensuing cleavage furrow forms in a new plane (*right*).

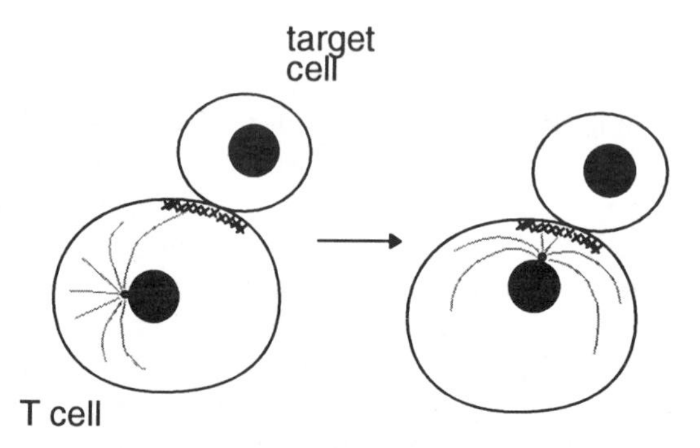

Figure 4 Interaction of microtubules with the cell cortex reorients the killer T cell centrosome towards the target cell. An actin-rich cortical site (cross-hatching) is contacted by microtubules that mediate the repositioning of the centrosome. After reorientation, secretory vesicles are presumed to be released from the T cell and the target cell is killed.

appears to be the result of an interaction of centrosomal microtubules with a specialized site at the cell cortex (Figure 3). Microtubules reach from the rotating centrosome to this site. The rotation is sensitive to microtubule-depolymerizing drugs, and irradiation of one of the centrosomes with UV light blocks its ability to rotate towards the cortical site, as does irradiation of the cortical site itself, or the region between the site and the centrosome (164). During rotation the plasma membrane invaginates near the former midbody, indicating that a force acts on the cortex.

Interestingly, disruption of actin filaments with cytochalasin D blocks the rotation, suggesting that maintenance of the cortical site requires the actin filaments there (164). The ability of a specific cortical site to position the meiotic spindle by pulling on the microtubules emanating from one spindle pole has also been observed in the egg of the marine worm *Chaetopterus* (165, 166).

Similar mechanisms may be involved in positioning the mitotic spindle in the budding yeast *S. cerevisiae*. During cell division in yeast, the mitotic spindle must be positioned so that it points into the growing bud. Studies with conditional tubulin mutants and microtubule-disrupting drugs demonstrate that astral microtubules are required both for migration of the nucleus and proper spindle orientation (167, 168). Immunofluorescence shows that microtubules extend from one of the spindle pole bodies into the bud (169). Cytoplasmic dynein is apparently involved in spindle positioning (170, 170a). Conditional actin mutants show that actin is also required for proper spindle orientation in yeast (171). These observations suggest that microtubules from the spindle pole body interact with a special cortical actin site in the growing bud, or bud neck, to orient the spindle. Interestingly, it is always the newly formed spindle pole body that migrates into the bud neck (70).

A final example of centrosome positioning is found in the polarization of a cytotoxic T lymphocyte cell towards its target cell (reviewed in Refs. 172–174). When a killer T cell binds to its target cell, the centrosome and nucleus rotate until the centrosome faces the target cell, and killing of the target cell soon follows (Figure 4). This localization of the centrosome positions the Golgi complex near the site of contact with the target cell, allowing transport of materials needed for cytotoxicity from the Golgi along centrosomal microtubules to the target site, where they are secreted. In situations where a killer T cell is bound to more than one target cell, it kills the cells one at a time, orienting the centrosome and Golgi apparatus first in each case. The reorientation of the centrosome in killer T cells is blocked by drugs that inhibit microtubule polymerization. There is evidence to suggest that centrosomal microtubules interact with a cortical site at the region of cell contact to bring about this reorientation (172–174).

The examples discussed above suggest a general model for how microtubule organizing centers can be positioned within cells. Centrosomal microtubules that contact a specific cortical site could become stabilized, and microtubule-based motors (acting at the centrosome, along the length of the microtubules, or at the cortical site) could then move the centrosome towards the cortical site. In most cases, the proteins responsible for positioning microtubule organizing centers are presently unknown. They might be found by screening for mutants that affect spindle positioning in *C. elegans* or *S. cerevisiae*; alternatively, they may exist in collections of microtubule motor protein genes cloned by DNA sequence homology.

The Centrosome and Cortical Events

When animal cells divide, the cleavage furrow always forms midway between the two spindle poles, thus ensuring equal partitioning of the cell contents at the end of mitosis. The sea urchin embryo can be manipulated so that a single cell containing two separate mitotic spindles is produced. If a centrosome from one spindle pole lies close enough to a centrosome from the other, a cleavage furrow forms between them, even though there is no mitotic spindle or chromosomes between them (reviewed in Ref. 175). This strongly suggests that a critical signal for formation of the cleavage furrow at the cell cortex comes from the spindle poles.

The ability of the centrosome and its associated microtubules to affect the cell cortex is especially clear in the early *Drosophila* embryo. The first 14 nuclear divisions of the *Drosophila* embryo take place in a syncytial cytoplasm. Nuclei initially divide in the interior of the embryo, but subsequently migrate to the embryo cortex where they form an even monolayer. The nuclei in the monolayer divide several times and then cellularize synchronously. Before nuclear migration, actin and a number of actin-binding proteins appear to be evenly distributed at the embryo cortex. When nuclei and their associated centrosomes arrive at the cortex, they induce a major reorganization of the actin cytoskeleton. During interphase, an "actin cap" forms between each nucleus and the plasma membrane. During mitosis, the actin caps spread laterally into plasma membrane furrows that invaginate down around each mitotic spindle. Nuclei reach the posterior end of the embryo one cycle earlier than they reach the cortex. These nuclei also organize cortical actin caps, but they cellularize earlier than other nuclei, forming the "pole cells" that are the future germ line of the embryo (reviewed in Ref. 55).

If *Drosophila* embryos are injected with the DNA synthesis inhibitor aphidicolon prior to nuclear migration, the nuclear cycle is arrested and the centrosomes become dissociated from their nuclei. Remarkably, these dissociated centrosomes continue to divide synchronously and migrate to the embryo cortex where they organize actin caps and membrane invaginations that are similar to those in normal embryos. In addition, the centrosomes that migrate to the posterior tip of the embryo induce the formation of "pole cells" that lack nuclei (56). Further evidence that centrosomal microtubules organize cortical actin comes from experiments in which antibodies against tyrosinated α-tubulin are injected into living *Drosophila* embryos. The antibodies cause surface nuclei to lose their cortical position and fall into the interior of the embryo. The actin caps that had been associated with such nuclei then disintegrate (176). These results make it clear that the centrosome and its associated microtubules play a central role in the organization of actin filaments and the cell cortex in *Drosophila* embryos.

The mechanisms by which the centrosome directs cortical events remain largely unknown. During cleavage furrow formation, centrosomal microtubules may induce contraction of the actin/myosin network to form the furrow (177). Alternatively, a diffusible signal, such as calcium released from vesicles centered around the centrosome, may be responsible. A number of mutations have been identified in *Drosophila* that affect the cortical reorganization induced by centrosomes (178, 179). Genetic studies such as these should provide inroads into this difficult but important problem.

FUTURE DIRECTIONS

In spite of the many difficulties inherent in studying the centrosome, many new components of the centrosome (and spindle pole body) have recently been identified. Moreover, with modern biochemical and genetic techniques, the identification of each new protein provides a starting point for the identification of additional interacting proteins, which should allow all of the components of the centrosome to be obtained.

A current limitation is that different centrosomal proteins have been identified in different species, making it difficult to build up a complete set of components for any one centrosome. It will be important to identify the homologs of centrosomal proteins in diverse species, since this not only will reveal those that are evolutionarily conserved, but also will allow their functions to be characterized in whatever organism is most experimentally accessible.

It seems likely that the functions of many centrosomal proteins will be difficult to study in purified systems. For example, it is not possible to determine the function of a protein that is a structural component of the centrosome by studying its behavior in isolation. Similarly, it is likely to be very difficult to obtain purified systems that carry out such events as centrosome duplication or centrosome separation. We therefore need to develop approaches that allow the functions of centrosomal proteins to be studied in vivo or in crude extract systems. The study of mutant phenotypes in yeast and other genetically tractable organisms is one means of getting clues to function. Here, specific mutagenesis can be used to study the effect of altering functional domains and phosphorylation sites. The recent demonstration in yeast that internal deletions of the Spc110/Nuf1 protein can dramatically affect spindle pole body structure in vivo provides a striking example of how one can use such approaches to study the function of proteins that would be very difficult to study in vitro (40).

Protein function can also be disrupted in vivo with antibodies, although one would prefer to use antibodies that block localization of a protein to the centrosome, so as to minimize the concern that the antibodies are disrupting centrosome function in a nonspecific manner. Antibodies can also be used to deplete soluble components of the centrosome from crude extract systems that carry out centrosome assembly. Recently developed extracts that support centrosome assembly and microtubule nucleation should provide useful systems in which to carry out these kinds of experiments (43, 71a, 71b, 78).

Another promising means of disrupting centrosome function is through the use of dominant negative mutations (180). One could start by mapping a domain on a centrosomal protein that is responsible for localization to the centrosome, and protein fragments that bear this domain might be used to block the localization of the endogenous protein.

A traditional means of understanding a cellular function is to reconstitute it from purified components. This is a somewhat daunting task for the centrosome, which is complex and difficult to purify. Perhaps the best approach is to try to break the problem down into smaller parts. The recent identification of soluble complexes of centrosomal proteins is perhaps a step in the right direction, since these soluble multiprotein complexes should be considerably easier to purify and characterize than the centrosome itself. As a next step, one needs to use these complexes to nucleate microtubules, or to add them back to inactivated centrosomes to reconstitute function.

Until recently, a molecular understanding of the centrosome seemed beyond reach, but as more centrosomal proteins are identified, and new powerful tools to characterize them become available, we can expect a renaissance of interest in this long-neglected field.

ACKNOWLEDGMENTS

We thank K. Bloom, S. Doxsey, M. Felix, H. Joshi, I. Gibbons, J. Kilmartin, M. Kirschner, J. R. McIntosh, R. Palazzo, M. Rose, E. Scheibel, G. Sluder, T. Stearns, and M. Winey for sharing data before publication, and we thank R. Kellum, T. Mitchison, K. Oegema, J. Raff, K. Sawin,

K. Schneider, T. Stearns, A. Straight, M. Winey, and Y. Zheng for helpful discussions and critical reading of the manuscript. We also thank F. Dependahl for help in preparing the manuscript, and M. L. Wong for preparing the figures. Supported by NIH grant GM-23928.

Literature Cited

1. Wilson EB. 1925. *The Cell in Development and Heredity*. New York: Macmillan. 1232 pp.
2. Kalnins VI, ed. 1992. *The Centrosome*. San Diego: Academic
3. Vandre DD, Borisy GG. 1989. In *Mitosis: Molecules and Mechanisms,* ed. JS Hyams, BR Brinkley, pp. 39–75. San Diego: Academic
4. Vorobjev IA, Nadezhdina ES. 1987. *Int. Rev. Cytol.* 106:227–93
5. Mazia D. 1987. *Int. Rev. Cytol.* 100: 49–92
6. Kimble M, Kuriyama R. 1992. *Int. Rev. Cytol.* 136:1–51
7. Kalt A, Schliwa M. 1993. *Trends Cell Biol.* 3:119–28
8. Baron AT, Salisbury JL. 1988. *J. Cell Biol.* 107:2669–78
9. Paintrand M, Moudjou M, Delacroix H, Bornens M. 1992. *J. Struct. Biol.* 108:107–28
10. Vorobjev IA, Chentsov YS. 1982. *J. Cell Biol.* 93:938–49
11. Levine RD, Ebersold WA. 1960. *Annu. Rev. Microbiol.* 14:197–216
12. Keryer G, Ris H, Borisy GG. 1984. *J. Cell Biol.* 98:2222–29
13. Calarco-Gillam PD, Siebert MC, Hubble R, Mitchison T, Kirschner M. 1983. *Cell* 35:621–29
14. Debec A, Szollosi A, Szollosi D. 1982. *Biol. Cell* 44:133–38
15. Debec A, Abbadie C. 1989. *Biol. Cell* 67:307–11
16. Clayton L, Black CM, Lloyd CW. 1985. *Cell* 35:621–29
17. Rieder CL, Borisy GG. 1982. *Biol. Cell* 44:117–32
18. Rattner JB. 1992. See Ref. 2, pp. 51–65
19. Kuriyama R, Borisy GG. 1981. *J. Cell Biol.* 91:822–26
20. Peterson JB, Ris H. 1976. *J. Cell Sci.* 22:219–42
21. Byers B, Goetsch L. 1975. *J. Bacteriol.* 124:511–23
22. Byers B, Shriver K, Goetsch L. 1978. *J. Cell Sci.* 30:331–52
23. Kilmartin JV, Adams AEM. 1984. *J. Cell Biol.* 98:922–33
24. Gould RR, Borisy GG. 1977. *J. Cell. Biol.* 73:601–15
25. McGill M, Brinkley BR. 1975. *J. Cell Biol.* 67:189–99
26. Hyams JS, Borisy GG. 1978. *J. Cell Sci.* 3:235–53
27. Snyder JA, McIntosh JR. 1975. *J. Cell Biol.* 67:744–60
28. Mitchison TJ, Kirschner MW. 1984. *Nature* 312:232–37
29. Mitchison TJ, Kirschner MW. 1986. *Methods Enzymol.* 134:261–68
30. Bornens M, Paintrand M, Berges J, Marty MC, Karsenti E. 1987. *Cell Motil. Cytoskel.* 8:238–49
31. Gosti-Testu F, Marty MC, Berges J, Maunoury R, Bornens M. 1986. *EMBO J.* 5:2545–50
32. Komesli S, Tournier F, Paintrand M, Margolis RL, Job D, Bornens M. 1989. *J. Cell Biol.* 109:2869–78
33. Mitchison TJ, Kirschner MW. 1984. *Nature* 312:237–42
34. Cassimeris L, Pryer NK, Salmon ED. 1988. *J. Cell Biol.* 107:2223–31
35. Sammak PJ, Gorbsky GJ, Borisy GG. 1987. *J. Cell Biol.* 104:395–405
36. Sammak PJ, Borisy GG. 1988. *Nature* 332:724–26
37. Schulze E, Kirschner M. 1988. *Nature* 334:356–59
38. Sheldon E, Wadsworth P. 1993. *J. Cell Biol.* 120:935–45
39. Rout MP, Kilmartin JV. 1990. *J. Cell Biol.* 111:1913–17
40. Kilmartin JV, Dyos SL, Kershaw D, Finch JT. 1993. *J. Cell Biol.* 1994. 123:1175–84

41. Mirzayan C, Copeland CS, Snyder M. 1992. *J. Cell Biol.* 123:1175–84
42. Palazzo RE, Brawley JB, Rebhun LI. 1988. *Zool. Sci.* 5:603–11
43. Palazzo RE, Vaisberg E, Cole RW, Rieder CL. 1992. *Science* 256:219–21
44. Kellogg DR, Field CM, Alberts BM. 1989. *J. Cell Biol.* 109:2977–91
45. Kellogg DR, Alberts BM. 1992. *Mol. Biol. Cell* 3:1–11
46. Frasch M, Glover D, Saumweber H. 1986. *J. Cell Sci.* 82:155–72
47. Whitfield WG, Millar SE, Saumweber H, Frasch M, Glover DM. 1988. *J. Cell Sci.* 89:467–80
48. Oakley CE, Oakley BR. 1989. *Nature* 338:662–64
49. Oakley BR, Oakley CE, Yoon Y, Jung MK. 1990. *Cell* 61:1289–301
50. Stearns T, Evans L, Kirschner M. 1991. *Cell* 65:825–36
51. Zheng Y, Jung MK, Oakley BR. 1991. *Cell* 65:817–23
52. Horio T, Uzawa S, Jung MK, Oakley BR, Tanaka K, Yanagida M. 1991. *J. Cell Sci.* 99:693–700
53. Liu B, Marc J, Joshi HC, Palevitz BA. 1993. *J. Cell Sci.* 104:1217–28
54. Oakley BR. 1992. *Trends Cell Biol.* 2:1–5
55. Schejter ED, Wieschaus E. 1993. *Annu. Rev. Cell Biol.* 9:67–99
56. Raff JW, Glover DM. 1989. *Cell* 57: 611–19
57. Foe VE. 1989. *Development* 107:1–22
58. Gatti M, Baker BS. 1989. *Genes Dev.* 3:438–53
59. Sullivan W, Minden JS, Alberts BM. 1990. *Development* 110:311–23
60. Glover DM. 1992. See Ref. 2, pp. 219–34
61. Kuchka MR, Jarvik JW. 1982. *J. Cell Biol.* 92:170–75
62. Wright RL, Salisbury J, Jarvik JW. 1985. *J. Cell Biol.* 101:1903–12
63. Wright RL, Adler SA, Spanier JG, Jarvik JW. 1989. *Cell Motil. Cytoskel.* 14:516–26
64. Taillon BE, Adler SA, Suhan JP, Jarvik JW. 1992. *J. Cell Biol.* 119:1613–24
65. Baron AT, Salisbury JL. 1992. See Ref. 2, pp. 167–95
65a. Lee VD, Huang B. 1993. *Proc. Natl. Acad. Sci. USA* 90:11039–43
66. Baum P, Yip C, Goetsch L, Byers B. 1988. *Mol. Cell Biol.* 8:5386–97
67. Huang B, Mengersen A, Lee VD. 1988. *J. Cell Biol.* 107:133–40
68. Dutcher SK, Lux FG. 1989. *Cell Motil. Cytoskel.* 14:104–17
69. Page BD, Snyder M. 1992. *Genes Dev.* 6:1414–29
70. Vallen EA, Scherson TY, Roberts T, van Zee K, Rose MD. 1992. *Cell* 69: 505–15
71. Raff JW, Kellogg DR, Alberts BM. 1993. *J. Cell Biol.* 121:823–35
71a. Stearns T, Kirschner MW. 1994. *Cell* 76:In press
71b. Felix M, Antony C, Wright M, Maro B. 1994. *J. Cell Biol.* 124:19–31
72. Dirksen ER. 1961. *J. Cell Biol.* 11: 211–17
73. Kallenbach RJ. 1985. *J. Cell Sci.* 73: 261–78
74. Kallenbach RJ, Mazia D. 1982. *Eur. J. Cell Biol.* 28:68–76
75. McNiven MA, Porter KR. 1988. *J. Cell Biol.* 106:1593–605
76. Maniotis A, Schliwa M. 1991. *Cell* 67:495–504
77. Verde F, Berrez J-M, Antony C, Karsenti E. 1991. *J. Cell Biol.* 112: 1177–87
78. Buendia B, Draetta G, Karsenti E. 1992. *J. Cell Biol.* 116:1431–42
78a. Doxsey SJ, Stein P, Evans LM, Calarco PD, Kirschner MW. 1994. *Cell* 76:In press
79. Tuffanelli DL, McKeon F, Kleinsmith DM, Burnham TK, Kirschner MW. 1983. *Arch. Dermatol.* 119:560–66
80. Mazia D, Harris PJ, Bibring T. 1960. *J. Biophys. Biochem. Cytol.* 7:1–20
81. Sluder G, Rieder CL. 1985. *J. Cell Biol.* 100:887–96
82. Heidemann SR, Kirschner MW. 1975. *J. Cell Biol.* 67:105–17
83. Tournier F, Karsenti E, Bornens M. 1989. *Dev. Biol.* 136:321–29
84. Maller JD, Poccia C, Nishioka D, Kidd P, Gerhart J, Hartman H. 1976. *Exp. Cell Res.* 99:285–94
85. Karsenti E, Newport J, Hubble R, Kirschner MW. 1984. *J. Cell Biol.* 98: 1730–45
86. Klotz C, Dabauvalle M-C, Paintrand M, Weber T, Bornens M, Karsenti E. 1990. *J. Cell Biol.* 110:405–15
87. Johnson KA, Rosenbaum JL. 1991. *Trends Cell Biol.* 1:145–49
88. Winey M, Byers B. 1993. *Trends Cell Biol.* 9:300–4
89. Winey M, Byers B. 1992. See Ref. 2, pp. 197–218
90. Rose M, Biggins S, Satterwhite L. 1993. *Curr. Opin. Cell Biol.* 5:105–15
91. Byers B. 1981. In *Molecular Genetics in Yeast,* ed. D von Wettstein, A Stenderup, M Kielland-Brandt, J Friis, pp. 119–33. Copenhagen: Munksgaard
92. Spang A, Courtney I, Fackler U, Matzner M, Scheibel E. 1993. *J. Cell Biol.* 123:405–16
93. Baum P, Furlong C, Byers B. 1986. *Proc. Natl. Acad. Sci. USA* 83:5512–16

94. Conde J, Fink GR. 1976. *Proc. Natl. Acad. Sci. USA* 73:3651–55
95. Polaina J, Conde J. 1982. *Mol. Gen. Genet.* 186:253–58
96. Rose MD, Fink GR. 1987. *Cell* 48: 1047–60
97. Winey M, Goetsch L, Baum P, Byers B. 1991. *J. Cell Biol.* 114:745–54
98. Thomas JH, Botstein D. 1986. *Cell* 44:65–76
99. Winey M, Hoyt MA, Chan C, Goetsch L, Botstein D, Byers B. 1993. *J. Cell Biol.* 122:743–51
100. McGrew JT, Goetsch L, Byers B, Baum P. 1992. *Mol. Biol. Cell* 3:1443–54
101. Sluder G, Miller FJ, Rieder CL. 1986. *J. Cell Biol.* 103:1873–81
102. Sluder G, Lewis K. 1987. *J. Exp. Zool.* 244:89–100
103. Picard A, Harricane M-C, Labbe J-C, Doree M. 1988. *Dev. Biol.* 128:121–28
104. Raff JW, Glover DM. 1988. *J. Cell Biol.* 107:2009–19
105. Sluder G, Miller FJ, Cole R, Rieder CL. 1990. *J. Cell Biol.* 110:2025–32
106. Gard DL, Hafezi S, Zhang T, Doxsey SJ. 1990. *J. Cell Biol.* 110:2033–42
107. Reed SI. 1992. *Annu. Rev. Cell Biol.* 8:529–61
108. Murray A, Hunt T. 1993. *The Cell Cycle.* New York: Freeman
109. Nasmyth K. 1993. *Curr. Opin. Cell Biol.* 5:166–79
110. Pringle JR, Hartwell LH. 1981. In *The Molecular Biology of the Yeast* Saccharomyces: *Life Cycle and Inheritance,* ed. J Strathern, EW Jones, JR Broach, p. 97. Cold Spring Harbor, NY: Cold Spring Harbor Lab.
111. Schild D, Ananthaswamy HN, Mortimer RK. 1981. *Genetics* 97:552–62
112. Bender A, Pringle JR. 1989. *Proc. Natl. Acad. Sci. USA* 86:9976–80
113. Hollingsworth RE, Sclafani RA. 1990. *Proc. Natl. Acad. Sci. USA* 87:6272–76
114. Murray AW. 1992. *Nature* 359:599–604
115. Li J, Deshaies R. 1993. *Cell* 74:223–26
116. Kochanski RS, Borisy GG. 1990. *J. Cell Biol.* 110:1599–605
117. Jarrett JT, Lansbury PT. 1993. *Cell* 73:1055–58
118. Brinkley BR, Cox SM, Pepper DA, Wible L, Brenner SL, Pardue RL. 1981. *J. Cell Biol.* 90:554–62
119. Kuriyama R. 1984. *J. Cell Sci.* 66:277–95
120. Evans L, Mitchison T, Kirschner M. 1985. *J. Cell Biol.* 100:1185–91
121. Langford GM. 1980. *J. Cell Biol.* 87: 521–26
122. McEwan B, Edelstein SJ. 1980. *J. Med. Biol.* 139:123–45
123. Pierson GB, Burton PR, Himes RH. 1979. *J. Cell Sci.* 39:89–99
124. Vandre DD, Davis FM, Rao PN, Borisy GG. 1984. *Proc. Natl. Acad. Sci. USA* 81:4439–43
125. Bailly E, Doree M, Nurse P, Bornens M. 1989. *EMBO J.* 8:3985–95
126. Gallant P, Nigg EA. 1992. *J. Cell Biol.* 117:213–14
127. Alfa CE, Ducommun B, Beach D, Hyams JS. 1990. *Nature* 347:680–82
128. Riabowol K, Draetta G, Brizuela L, Vandre D, Beach D. 1989. *Cell* 57: 393–401
129. Maldonado-Codina G, Glover DM. 1992. *J. Cell Biol.* 116:967–76
130. Masuda H, Sevik M, Cande WZ. 1992. *J. Cell Biol.* 117:1055–66
131. Joshi HC, Palacios MJ, McNamara L, Cleveland DW. 1992. *Nature* 356: 80–83
132. Baas PW, Joshi HJ. 1992. *J. Cell Biol.* 119:171–78
133. Muresan V, Joshi HC, Besharse JC. 1993. *J. Cell Sci.* 104:1229–37
134. Palacios MJ, Joshi HC, Simerly C, Schatten G. 1993. *J. Cell Sci.* 104:383–89
135. Rizzolo LJ, Joshi HC. 1993. *Dev. Biol.* 157:147–56
136. Moudjou M, Paintrand M, Vigues B, Bornens M. 1991. *J. Cell Biol.* 115: 129–40
137. Centonze VE, Borisy GG. 1990. *J. Cell Sci.* 95:405–11
138. Mitchison TJ, Salmon ED. 1992. *J. Cell Biol.* 119:569–82
138a. Theurkauf WE, Hawley RS. 1992. *J. Cell Biol.* 116:1167–80
139. Hawley RS, Theurkauf WE. 1993. *Trends Genet.* 9:310–17
140. Sawin KE, Endow SA. 1993. *BioEssays* 15:399–407
141. Sawin KE, Mitchison TJ. 1991. *J. Cell Biol.* 112:941–54
142. Mitchison TJ. 1989. *J Cell Biol.* 109: 637–52
143. Kitanishi-Yumura T, Fukui Y. 1987. *Cell Motil. Cytoskel.* 8:106–17
144. Belmont LD, Hyman AA, Sawin KE, Mitchison TJ. 1990. *Cell* 62:579–89
144a. Mastronarde DN, McDonald KL, Ding R, McIntosh JR. 1993. *J. Cell Biol.* 123:1475–89
145. McNiven M, Wang M, Porter K. 1984. *Cell* 37:753–65
146. McNiven M, Porter K. 1986. *J. Cell Biol.* 103:1547–55
146a. Foe VE, Odell GM. 1993. In *The Development of Drosophila melanogaster,* ed. M Bate, A Martinez Arias,

pp. 149–300. Cold Spring Harbor, NY: Cold Spring Harbor Lab.
147. Vale RD. 1990. *Curr. Opin. Cell Biol.* 2:15–22
148. McIntosh JR, Porter ME. 1989. *J. Biol. Chem.* 265:6001–4
149. Kelly RB. 1990. *Cell* 61:5–7
150. Schroer TA, Scheetz MP. 1991. *Annu. Rev. Cell Physiol.* 53:1309–18
151. Hirokawa N. 1982. *J. Cell Biol.* 94: 129–42
152. Ball EH Singer SJ. 1982. *Proc. Natl. Acad. Sci. USA* 79:123–26
153. Walker RA, Sheetz MP. 1993. *Annu. Rev. Biochem.* 62:429–51
154. Corthesy-Theulaz I, Pauloin A, Pfeffer SR. 1992. *J. Cell Biol.* 118:1333–45
155. Allan VJ, Kreis TE. 1986. *J. Cell Biol.* 103:2229–39
156. Kreis T. 1990. *Cell Motil. Cytoskel.* 15:67–70
157. Rogalski A, Singer SJ. 1984. *J. Cell Biol.* 99:1092–100
158. Wehland J, Henkart M, Klausner R, Sandoval IV. 1983. *Proc. Natl. Acad. Sci. USA* 80:4286–90
159. Vogl AW, Linck RW, Dym M. 1983. *Am. J. Anat.* 168:99–108
160. Louvard D, Reggio H, Warren G. 1982. *J. Cell Biol.* 92:92–107
161. Terasaki M, Chen LB, Fujiwara K. 1986. *J. Cell Biol.* 103:1557–68
162. Sulston JE, Schierenberg E, White JG, Thomson JN. 1983. *Dev. Biol.* 100:64–119
163. Hyman AA, White JG. 1987. *J. Cell Biol.* 105:2123–35
164. Hyman AA. 1989. *J. Cell Biol.* 109: 1185–93
165. Lutz DA, Hamaguchi Y, Inoue S. 1988. *Cell Motil.* 11:83–96
166. Inoue S. 1990. *Ann. NY Acad. Sci.* 582: 1–14
167. Huffaker TC, Thomas JH, Botstein D. 1988. *J. Cell Biol.* 106:1997–2010
168. Jacobs CW, Adams AEM, Staniszlo PJ, Pringle JR. 1988. *J. Cell Biol.* 107:1409–26
169. Sullivan DS, Huffaker TC. 1992. *J. Cell Biol.* 119:379–88
170. Li YY, Yeh E, Hays T, Bloom K. 1993. *Proc. Natl. Acad. Sci. USA* 90:10096–100
170a. Eshel D, Urrestarazu LA, Vissers S, Jauniaux J-C, van Vliet-Reedijk JC, et al. 1993. *Proc. Natl. Acad. Sci. USA* 90:11172–76
171. Palmer RE, Sullivan DS, Huffaker T, Koshland D. 1992. *J. Cell Biol.* 119: 583–93
172. Singer SJ, Kupfer A. 1986. *Annu. Rev. Cell Biol.* 2:337–65
173. Kupfer A, Singer SJ. 1989. *Annu. Rev. Immunol.* 7:309–37
174. Kupfer A, Singer SJ. 1989. *J. Exp. Med.* 170:1697–713
175. Rappaport R. 1986. *Int. Rev. Cytol.* 105:245–81
176. Warn RM, Flegg L, Warn A. 1987. *J. Cell Biol.* 105:1721–30
177. Bray D, White JG. 1988. *Science* 239: 883–88
178. Sullivan W, Fogarty P, Theurkauf W. 1993. *Development* 118:1245–54
179. Postner MA, Miller KG, Wieschaus EF. 1992. *J. Cell Biol.* 119:1205–18
180. Herskowitz I. 1987. *Nature* 329:219–22

Annu. Rev. Biochem. 1994. 63:675–716

ENERGY TRANSDUCTION BY CYTOCHROME COMPLEXES IN MITOCHONDRIAL AND BACTERIAL RESPIRATION: The Enzymology of Coupling Electron Transfer Reactions to Transmembrane Proton Translocation

Bernard L. Trumpower
Department of Biochemistry, Dartmouth Medical School, Hanover, New Hampshire 03755

Robert B. Gennis
School of Chemical Sciences, University of Illinois, Urbana, Illinois 61801

KEY WORDS: respiration, proton translocation, bc_1 complex, cytochrome c oxidase, quinol oxidase

CONTENTS

0066-4154/94/0701-0675$05.00

INTRODUCTION

Mitochondria and a variety of bacteria respire, using oxygen as a terminal electron acceptor. The enzymes that catalyze respiration are oligomeric, cytochrome-containing complexes[1] inserted into the inner mitochondrial membrane or bacterial plasma membrane. These respiratory enzymes convert the free energy from oxidation of hydrogen-containing substrates into a transmembrane proton electrochemical gradient ($\Delta\mu_H$). This gradient, alternatively referred to as protonmotive force (pmf), consists of two components, a proton concentration gradient (ΔpH) and a transmembrane electrical potential ($\Delta\psi$). The protonmotive force is used for energy-requiring reactions such as synthesis of ATP, transport of ions and metabolites, and cell motility.

How the respiratory cytochrome complexes link the oxidation-reduction reactions that they catalyze to the electrogenic (voltage-generating) translocation of protons across the membrane in which they reside is a central theme of this review. Related reviews have been written on the cytochrome bc_1 complexes in microorganisms (1, 2), the protonmotive Q cycle (3–5), cytochrome b (6, 7), and the quinol and cytochrome c oxidases (8–10).

THE PROTONMOTIVE ENZYME COMPLEXES OF RESPIRATION

In mitochondria and respiring bacteria, ubiquinol and its analogs are a confluence point for the collection of reducing equivalents from multiple dehydrogenases, linking metabolism to cellular energetics, and the universal substrates for respiration. These quinol substrates are soluble and mobile in the inner mitochondrial membrane or bacterial plasma membrane, and are

[1]Textbooks commonly refer to a "respiratory chain" consisting of NADH dehydrogenase, succinate dehydrogenase, the cytochrome bc_1 complex, and the cytochrome c oxidase complex. This view is parochial in the context of microorganisms, which use a diverse collection of dehydrogenases to support respiration and frequently lack an energy-transducing form of NADH dehydrogenase, and misleading in the instance of mammalian mitochondria, in which ETF dehydrogenase is quantitatively as important a source of reducing equivalents for respiration in most tissues as is succinate dehydrogenase. Oxidation of quinol is universal in respiring cells. For this reason we have defined "respiratory enzymes" as the cytochrome complexes that catalyze quinol oxidation by molecular oxygen.

oxidized by membranous cytochrome complexes in the process of respiration. Two types of cytochrome complexes, the cytochrome bc_1 complex and the terminal oxidases, participate in respiration, and link electron transfer to proton translocation. The bc_1 complex does not transfer electrons to oxygen, whereas the terminal oxidases do.

A COMPARISON OF MITOCHONDRIAL AND BACTERIAL RESPIRATORY SYSTEMS

In discussing the enzymology of the cytochrome bc_1 complex, the aa_3-type cytochrome c oxidase, the bo-type ubiquinol oxidase, and the bd-type ubiquinol oxidase, emphasis is placed on similarities among the phylogenetically diverse bacterial and mitochondrial enzymes, which have become apparent from amino acid sequence comparisons of subunits, and biochemical and biophysical characterizations (10–13). Randomly generated and site-directed mutants have been particularly valuable in deducing aspects of the structures of the bc_1 complexes and the terminal oxidases. This analysis has been aided by the fact that these complexes contain metalloprotein redox centers amenable to spectroscopic techniques, which are valuable in deducing the nature of perturbations resulting from mutations.

The relationships between the cytochrome complexes in the electron transfer chains of mitochondria and two representative bacteria are shown in Figure 1. In mitochondria of most species, oxidation of ubiquinol by molecular oxygen is catalyzed by the sequential action of two enzyme complexes, the cytochrome bc_1 complex, which oxidizes quinol and reduces cytochrome c, and the cytochrome c oxidase complex, which oxidizes cytochrome c and reduces oxygen to water, as shown in Figure 1*A*.

There are exceptions to this linear arrangement in mitochondria. *Saccharomyces cerevisiae* contain a lactate-cytochrome b_2 oxidoreductase, which transfers electrons directly from L-lactate to cytochrome c (14), and protozoae possess a quinol oxidase resembling that in plant mitochondria, which transfers electrons from ubiquinol to oxygen without the intermediate oxidation-reduction of the bc_1 complex (15). In yeast the bypass of the bc_1 complex may provide a selective advantage over other fungi, which coinhabit decaying vegetation in which some microorganisms produce toxins that act on the cytochrome bc_1 complex. Whether resistance to cytotoxins has selected for the alternate oxidase in protozoans is a matter of speculation.

Bacteria that oxidize quinol through a cytochrome bc_1 complex also possess at least one alternate quinol oxidase, which transfers electrons directly from the quinol to oxygen. A paradigm of the relationship between the cytochrome bc_1 complex and the terminal oxidases in bacteria exists in photosynthetic bacteria, shown in Figure 1*B*, which possess a quinol oxidase

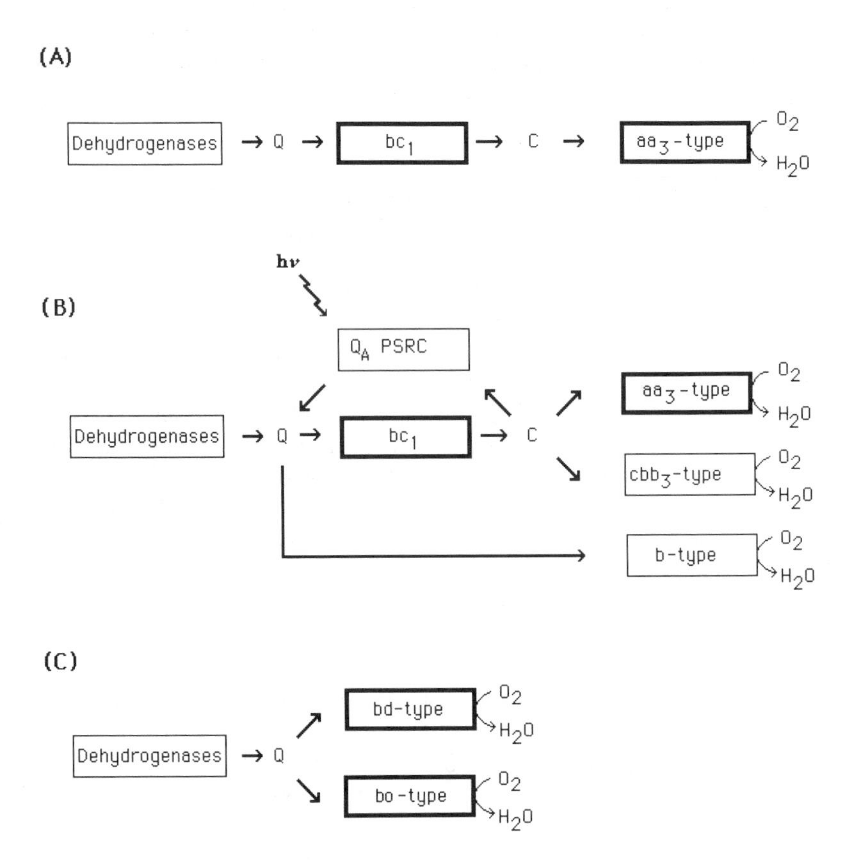

Figure 1 Energy-transducing cytochrome complexes in mitochondrial and bacterial respiration. The diagram summarizes the relationships between the cytochrome complexes, dehydrogenases, quinols (Q), and cytochromes c (C) involved in respiration in (*A*) mitochondria, (*B*) purple nonsulfur photosynthetic bacteria, and (*C*) *Escherichia coli*. The energy-transducing cytochrome complexes discussed in this review are enclosed in bold boxes. The figure also shows the light-harvesting photosynthetic reaction center (PSRC), which drives cyclic electron transfer through the bc_1 complex in photosynthetic bacteria.

that oxidizes ubiquinol without the intermediate participation of the bc_1 complex.

Oxidation of quinol by the sequential action of the cytochrome bc_1 and cytochrome c oxidase complexes translocates a greater number of protons across the mitochondrial membrane or bacterial membrane than does the oxidation of quinol by the singular action of a quinol oxidase complex. Thus, in addition to allowing bacteria to escape toxins that act on the bc_1 complex, these quinol oxidases may also balance the energy requirements of the cell against the requirements for oxidation of NADH and similar hydrogen-containing metabolites.

In addition to participating in respiration, the cytochrome bc_1 complex participates in light-driven cyclic electron transfer when photosynthetic bacteria such as *Rhodobacter sphaeroides* grow phototrophically (16), as shown in Figure 1*B*. In these bacteria there is only one operon for the bc_1 complex, and it appears to be expressed constitutively (17), as is the bc_1 complex in *Paracoccus denitrificans* when they are grown aerobically or under anaerobic, denitrifying conditions (18).

The simplest arrangement of respiratory cytochrome complexes is illustrated by *Escherichia coli,* which lack a bc_1 complex and oxidize quinol directly with molecular oxygen by either of two quinol oxidases as illustrated in Figure 1*C*. The similarities between one of the *E. coli* terminal oxidases and the cytochrome c oxidase of mitochondria (10–13) represent a striking example of how highly analogous enzymes are used as the terminal oxidases in phylogenetically distant species.

PROTON TRANSLOCATION SCHEMES

A question of fundamental interest is how the electron transfer reactions catalyzed by the cytochrome complexes are coupled to the generation of a voltage across the membrane and a transmembrane pH gradient. Nature has devised different mechanisms to accomplish this, and the enzymology and structures of the four best-understood enzymes that illustrate these different mechanisms are discussed and compared.

Cytochrome bd: Formation of a Proton Gradient by Substrate Protons without a Transmembrane Channel

The simplest means of forming a proton gradient is that used by the cytochrome bd ubiquinol oxidase from *E. coli,* illustrated in Figure 2*A*. This enzyme has two spatially separated, active sites. The ubiquinol oxidation site is located near the periplasmic, outer surface of the membrane, and the oxygen reduction site appears to be near the inner, cytoplasmic surface of the membrane. The two-electron oxidation of ubiquinol releases two substrate-derived protons to the periplasm, and the electrons are passed to the second active center, where oxygen is reduced, utilizing protons from the cytoplasm.

For each electron used to form water, one proton appears in the periplasm and one disappears from the cytoplasm, yielding a H^+/e^- ratio of 1. The charge separation that generates a transmembrane voltage is likely associated with electron transfer between the two active centers. It is also possible that movement of protons could be electrogenic, either from the quinol-binding site to the periplasm or from the cytoplasm to the oxygen reduction site.

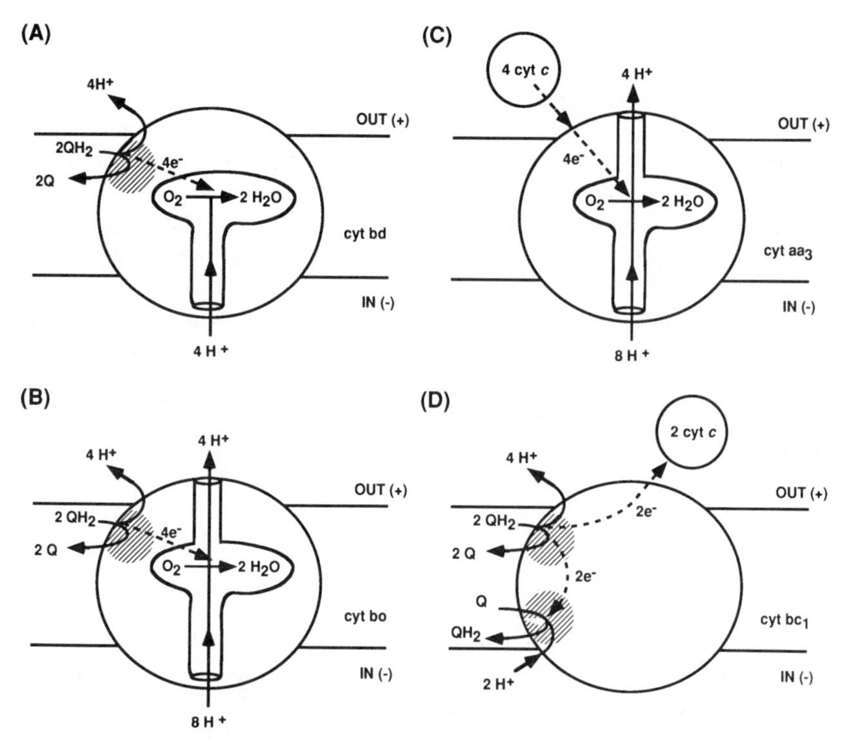

Figure 2 Schemes by which respiratory cytochrome complexes generate a transmembrane proton electrochemical gradient. In each case the enzyme is shown within a membrane. The electrochemically positive, outer side corresponds to the mitochondrial intermembrane space or the bacterial periplasm, for gram-negative species. Protons are taken up from either the mitochondrial matrix or the bacterial cytoplasm. (*A*) Cytochrome bd ubiquinol oxidase from *E. coli*. The location of the site where oxygen reacts is not known and may be close to the inner surface. Protons that appear on the outside are derived from the substrate. (*B*) Cytochrome bo ubiquinol oxidase from *E. coli*. Half of the protons that appear on the outside are transported through a proton-conducting channel through the enzyme. (*C*) Cytochrome c oxidases from prokaryotes and mitochondria. The protons that appear on the outside must be actively transported through a transmembrane channel. (*D*) Cytochrome bc_1 complex from prokaryotes and mitochondria. The two "Q sites" are the heart of the Q cycle, illustrated in Figure 3 below.

However, there is no proton channel through the protein per se, and the alkalinization and acidification reactions are entirely due to substrate protons in the chemistry catalyzed at the active sites. Therefore this enzyme is not referred to as a proton pump.

Cytochrome bo: Proton Translocation by Substrate Protons Plus a Transmembrane Channel

The mechanism by which *E. coli* cytochrome bo ubiquinol oxidase generates a proton electrochemical gradient across the membrane is shown in Figure 2*B*.

Cytochrome bo is a member of the heme-copper oxidase superfamily, and related to the mitochondrial cytochrome c oxidase, but unrelated to the *E. coli* cytochrome bd quinol oxidase. As does cytochrome bd, cytochrome bo deposits one proton per electron into the periplasm from oxidation of the quinol, and consumes protons from the cytoplasm in forming water. However, cytochrome bo pumps one additional proton per electron through a proton-conducting channel. Hence, these two mechanisms combined yield a H^+/e^- ratio of 2.

Cytochrome c Oxidase: A Proton Pump

The eukaryotic cytochrome c oxidases and several prokaryotic homologs pump one proton per electron. There are no substrate protons released at the outer surface concomitant with the oxidation of cytochrome c, but protons from the inside are required both for the pump and for the formation of water, as illustrated in Figure 2*C*.

Cytochrome bc_1 Complex: Proton Translocation By Substrate Protons

This enzyme oxidizes ubiquinol by a unique mechanism that deposits two protons outside of the membrane for each one electron transferred to cytochrome c, without a transmembrane proton-conducting channel. This is accomplished by recycling electrons through separate quinol oxidation and quinone reduction sites within the same enzyme, but on opposite sides of the membrane as shown in Figure 2*D*. Proton translocation involves only substrate protons from oxidation of ubiquinol or reduction of ubiquinone. Ubiquinol is solely responsible for carrying protons through the membrane barrier.

THE CYTOCHROME bc_1 COMPLEX

Energy Transduction Linked to Electron Transfer by the Protonmotive Q Cycle

The cytochrome bc_1 complex[2] contains a di-heme cytochrome b, cytochrome c_1, and an iron-sulfur protein known as the Rieske protein (1, 25). All bc_1

[2]The cytochrome bc_1 complex is a member of the larger family of bc-type complexes, which includes the cytochrome bf complex found in chloroplasts (19), algae (20), and some gram-positive bacteria (21). The bc_1 and bf complexes differ in that the equivalent of the monomeric cytochrome b in the bc_1 complex is an α,β dimer in the bf complex, with the two hemes located in the larger, α subunit (22). Cytochromes c_1 and f differ in that the axial ligands to the covalently attached c-type heme are histidine and methionine in cytochrome c_1 (23), and two nitrogenous ligands in cytochrome f (24). Interestingly, the distal heme ligand of cytochrome f is the α-amino group of the N-terminal tyrosine residue (see SE Martinez, D Huang, A Szczepaniak, WA Cramer, JL Smith, manuscript in preparation). This implies that ligation of the heme cannot be completed until the leader peptide and the section of the polypeptide containing the heme domain have been translocated across the membrane and the leader peptide has been removed by the lumen-side leader peptidase.

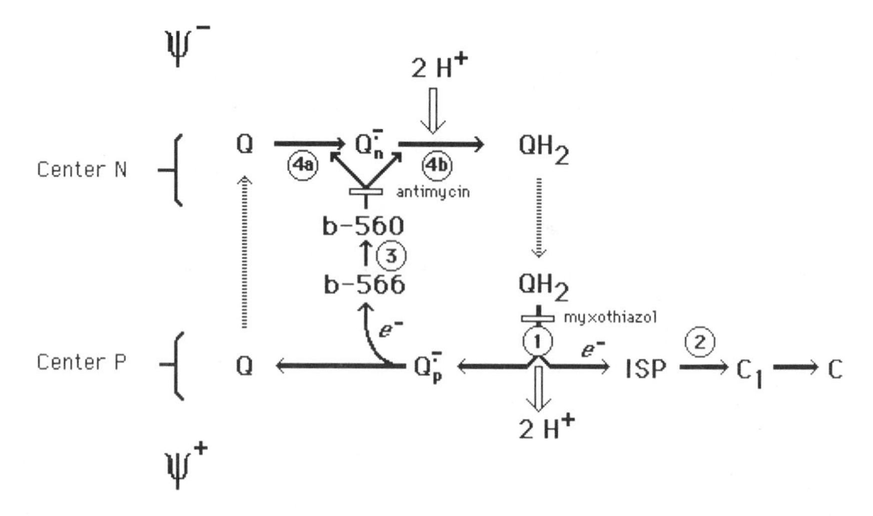

Figure 3 The protonmotive Q cycle path of electron transfer from ubiquinol to cytochrome c through the redox prosthetic groups of the cytochrome bc_1 complex is depicted as a series of numbered reactions, shown by solid arrows. Dashed arrows represent movement of ubiquinol and ubiquinone in the membrane, between the site where ubiquinol is oxidized at the positive side of the membrane ("center P") and the site where ubiquinone and ubisemiquinone are reduced at the negative side of the membrane ("center N"). Open arrows show the reactions in which protons are released and taken up during oxidation of ubiquinol and reduction of ubiquinone. Open rectangles show the reactions that are blocked by myxothiazol and antimycin.

complexes characterized to date contain these three redox proteins, and there is no bc_1 complex known that contains any additional redox prosthetic groups.

The path of electron transfer from ubiquinol to cytochrome c through the bc_1 complex is the protonmotive Q cycle, depicted in Figure 3 as numbered electron transfer reactions, with the redox groups arranged topographically in the complex, as illustrated in Figure 4. Cytochrome b is a transmembrane protein (26, 27) in which the two heme groups form an electrical circuit across the membrane (28), with the low-potential b-566 heme near the positive side of the membrane and the higher-potential b-560 heme more central in the membrane. The 2 Fe:2 S cluster of the iron-sulfur protein and the heme of cytochrome c_1 are near the electropositive surface of the membrane.

Ubiquinol is oxidized by the bc_1 complex by a concerted reaction in which the first electron from ubiquinol is transferred to the 2 Fe:2 S cluster

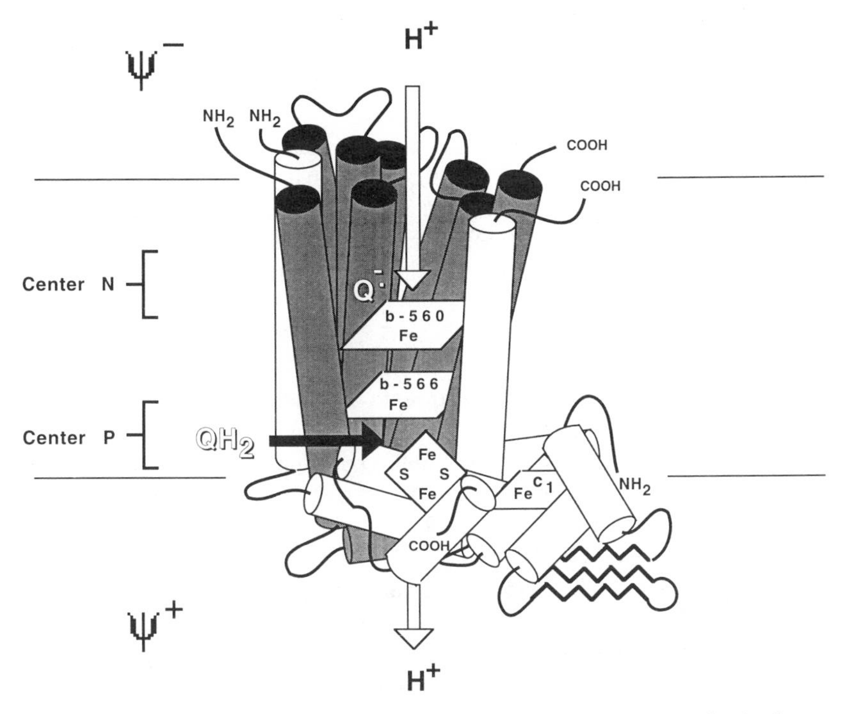

Figure 4 The cytochrome bc_1 complex, showing the three redox proteins in the inner mitochondrial or bacterial cytoplasmic membrane. The electronegative side corresponds to the matrix side of the mitochondrial membrane or the cytoplasmic side of the bacterial membrane. The figure also depicts ubiquinol (QH_2) entering the ubiquinol oxidase center P, and the stable ubisemiquinone anion ($Q^{\cdot -}$) near the b-560 heme at the ubiquinone reductase center N. Protons are shown entering center N and exiting center P. The topography of the proteins is discussed in the text.

of the iron-sulfur protein to form a ubisemiquinone,[3] which immediately reduces the b-566 heme. This two-electron oxidation of ubiquinol (reaction 1 in Figure 3) releases two protons at the positive surface of the membrane, and is catalyzed by a reaction site referred to as center P (Figures 3 and 4). The two electrons removed from ubiquinol at center P diverge and follow separate pathways. The electron transferred to the iron-sulfur protein is then transferred to cytochrome c_1 (reaction 2 in Figure 3) and then to cytochrome c. The second electron from ubiquinol, transferred from $Q_p^{\cdot -}$

[3]Ubisemiquinone is ionized at physiological pH, and thus depicted as the anion in the figures. The ionization state of the semiquinone does not affect the proton translocation stoichiometry of the Q cycle.

to the b-566 heme, recycles through the bc_1 complex from the low-potential b-566 heme to the higher-potential b-560 heme in reaction 3. The b-560 heme then reduces ubiquinone to form a stable ubisemiquinone ($Q_n^{\dot{-}}$), depicted as reaction 4a in Figure 3.

These electron transfer reactions convert ubiquinol to ubisemiquinone. One electron is transferred to cytochrome c, and two protons are released to the outer, P side of the membrane. The protonmotive Q cycle is only half complete in that only one electron from ubiquinol has been transferred to cytochrome c. The result of this first half of the Q cycle is:

$$QH_2 + \text{cyt } c_{ox} \rightarrow Q_n^{\dot{-}} + \text{cyt } c_{red} + 2\ H^+{}_p \qquad 1.$$

In the second half of the Q cycle, a second molecule of ubiquinol is oxidized by transferring one electron to iron-sulfur protein, which then reduces cytochrome c_1, and the resulting $Q_p^{\dot{-}}$ again reduces the heme of b-566, which reduces the heme of b-560. Oxidation of the ubiquinol releases an additional two protons at the outer surface of the membrane. These reactions are repetitive of reactions 1, 2, and 3, which occurred during the first half of the Q cycle. However, during the second half of the Q cycle, the heme of b-560 reduces the ubisemiquinone ($Q_n^{\dot{-}}$), which was formed in reaction 4a during the first half of the cycle, to ubiquinol in a reaction (4b in Figure 3) that consumes two protons from the inner side of the membrane. The site that catalyzes the two-step rereduction of ubiquinone at the negative side of the membrane is referred to as center N. The result of the second half of the Q cycle is:

$$QH_2 + \text{cyt } c_{ox} + Q_n^{\dot{-}} + 2\ H^+{}_n \rightarrow \text{cyt } c_{red} + QH_2 + Q + 2\ H^+{}_p \qquad 2.$$

During one complete Q cycle, two molecules of ubiquinol are oxidized to ubiquinones, but one molecule of ubiquinone is rereduced to ubiquinol. The iron-sulfur protein, cytochrome c_1, and the two hemes of cytochrome b are reduced and reoxidized twice. The b-560 heme reduces ubiquinone to ubisemiquinone during the first half of the Q cycle (reaction 4a), and reduces the ubisemiquinone to ubiquinol during the second half of the cycle (reaction 4b). By summing the two reactions above, one can see that as one molecule of ubiquinol is oxidized to ubiquinone, two molecules of cytochrome c are reduced, two protons are consumed on the negative (N) side of the membrane, and four protons are deposited on the positive (P) side of the membrane.

$$QH_2 + 2\ \text{cyt } c_{ox} + 2\ H^+{}_n \rightarrow Q + 2\ \text{cyt } c_{red} + 4\ H^+{}_p \qquad 3.$$

The concerted oxidation of ubiquinol that initiates the Q cycle is a second-order, diffusion-controlled reaction when the ratio of ubiquinol to bc_1 complex is 3 or less, with a half-time that varies from 300 μs to 7 ms, depending on the amount of quinol in the quinone pool (29). As the quinone pool becomes more reduced and the ratio of ubiquinol to bc_1 complex increases, the rate of the Q cycle becomes limited by the rate at which cytochrome c_1 and the iron-sulfur protein are reoxidized by cytochrome c and cytochrome c oxidase. The electron transfer reactions that intervene between oxidation of ubiquinol and reduction of the b-560 heme, and between ubiquinol oxidation and reduction of cytochrome c_1, are so rapid that they are not temporally resolved under normal experimental conditions. Electron transfer from the iron-sulfur protein to cytochrome c_1 has a half-time $< 200\ \mu s$, and that between the b-566 and b-560 hemes has a half-time $< 300\ \mu s$ (29, 30).

The two-step rereduction of quinone to ubisemiquinone and ubiquinol at center N, which resembles the gated reduction of ubiquinone by the photosynthetic reaction center, occurs with a half-time of 1.7 to 7 msec (29). Like the oxidation of ubiquinol, the rate of this reaction is controlled by the redox poise of the quinone pool, but increases as the ratio of quinone acceptor increases in relation to the ubiquinol product. The net effect of the rates of reactions 1–4 (Figure 3) is that the amount of ubisemiquinone at the b-560 reaction site is determined by the ratio of ubiquinol to ubiquinone in the quinone pool, which paces ubiquinol oxidation at center P and ubiquinone reduction at center N (29).

The Q cycle is supported by extensive experimentation (4, 5). Twenty years before the formulation of the Q cycle, it was observed that when a pulse of oxygen is added to mitochondria it causes an expected oxidation of cytochromes c and c_1, but this oxidation is accompanied by an unexpected transient reduction of cytochrome b (31). This "oxidant-induced reduction" involves both b hemes and is increased by the addition of antimycin, an inhibitor of the bc_1 complex that causes electrons to accumulate in ubiquinol and the b cytochromes when added to respiring mitochondria (32).

Wikström & Berden pointed out that divergent oxidation of ubiquinol would link formation of a cytochrome b reductant to oxidation of the c-cytochromes (33). This interpretation was seminal to Mitchell's subsequent formulation of the Q cycle (34). The Q cycle accounts for oxidant-induced b reduction, since rapid oxidation of the c-type cytochromes and iron-sulfur protein accelerates ubiquinol oxidation, leading to a transient increase in Q_p^-, the reductant for cytochrome b. Antimycin blocks reoxidation of b-560, enhancing the oxidant-induced reduction.

Evidence establishing the Q cycle mechanism came from examining the pre-steady-state reduction of the cytochromes b and c_1 when the Rieske iron-sulfur protein was reversibly resolved from the isolated complex (25, 35–38). This approach was extended by the discovery of a second class of inhibitors specific for the bc_1 complex (39), acting at a site different than antimycin as evidenced by their differential effects on pre-steady-state reduction of the cytochromes (40, 41).

If iron-sulfur protein is extracted from the bc_1 complex, cytochrome c_1 cannot be reduced by ubiquinol, but the b cytochromes can be reduced (37, 38). If iron-sulfur protein is extracted and antimycin is added, reduction of both cytochromes b and c_1 is blocked. This was the first piece of evidence for two pathways of cytochrome b reduction. One pathway, through center P, is iron-sulfur protein dependent. The second pathway, through center N, is iron-sulfur protein independent, but inhibited by antimycin (Figure 3). If antimycin blocks b reduction only through center N, one would predict that it would not inhibit reduction of cytochromes b or c_1 through the alternative, center P pathway. This prediction was confirmed (42).

The second piece of evidence for two pathways of b reduction is that myxothiazol inhibits electron transfer to the b cytochromes through an antimycin-insensitive pathway (39–41). When myxothiazol is added alone it blocks reduction of cytochrome c_1, but does not block reduction of the b cytochromes through center N (Figure 3). When myxothiazol is added with antimycin, the two inhibitors block reduction of both cytochromes b and c_1. Myxothiazol thus mimics removal of the iron-sulfur protein from the bc_1 complex, blocking oxidation of ubiquinol at center P (Figure 3). If myxothiazol blocks oxidation of ubiquinol at center P and the iron-sulfur protein catalyzes this reaction, then myxothiazol should inhibit—and iron-sulfur protein should be required for—oxidant-induced reduction of cytochrome b. These predictions were confirmed (41, 43).

If cytochrome b-560 reduces both ubiquinone to ubisemiquinone (step 4a in Figure 3) and ubisemiquinone to ubiquinol (step 4b), the potentials of the $Q/Q^{\cdot-}$ and $Q^{\cdot-}/QH_2$ couples at center N must be similar. From the equilibrium constant for formation of ubisemiquinone and the Nernst equations for the ubiquinone/ubiquinol couple and the two ubisemiquinone couples, one can relate the oxidation-reduction potentials of the two ubisemiquinone couples and the concentration of ubisemiquinone (3, 44).

$$E_m(Q/Q^{\cdot-}) - E_m(Q^{\cdot-}/QH_2) = \frac{RT}{nF} \ln \frac{(Q^{\cdot-})^2}{(Q)(QH_2)} = 60 \log \frac{(Q^{\cdot-})^2}{(Q)(QH_2)} \qquad 4.$$

From this equation it can be seen that when the potentials of the two semiquinone couples are similar, the equilibrium constant for semiquinone formation approaches unity, and its concentration approaches those of ubiquinol and ubiquinone. This predicts that the ubisemiquinone at center N, $Q_n^{\cdot-}$, is relatively stable. A stable ubisemiquinone was detected in isolated mitochondrial bc_1 complex (45), in mitochondrial membranes (46), and in bacterial membranes (47, 48), and the electron paramagnetic resonance

(EPR) signal from this semiquinone was eliminated upon addition of antimycin.

If ubiquinol is oxidized in a concerted reaction by the iron-sulfur protein (E_m = + 280 mV) and the heme of b-566 (E_m = −90 mV), the potentials of the $Q/Q^{\cdot-}$ and $Q^{\cdot-}/QH_2$ couples at center P must be very different, and the intermediate ubisemiquinone $Q_p^{\cdot-}$ should be unstable. De Vries and coworkers detected a transient, unstable ubisemiquinone during oxidant-induced reduction of cytochrome b, and the EPR signal from this semiquinone was not eliminated by antimycin (49).

If a rate-limiting amount of ubiquinol is added to the bc_1 complex, reduction of cytochrome b is a triphasic reaction, whereas reduction of cytochrome c_1 is monophasic (50–52). Within the first 50 msec there is a rapid partial reduction of cytochrome b, and c_1 reduction parallels this rapid phase of b reduction. Cytochrome b then undergoes reoxidation, while c_1 reduction continues. After approximately 100 msec, c_1 reduction is complete, reoxidation of cytochrome b stops, and cytochrome b is slowly rereduced. Formation of $Q_n^{\cdot-}$ is delayed until reduction of c_1 nears completion (53, 54).

These redox responses are consistent with the Q cycle mechanism (Figure 3). When ubiquinol is added to fully oxidized bc_1 complex it is oxidized at center P, due to the high potential of the iron-sulfur protein (+280 mV) in comparison to the b-560 heme (+40 mV), and this reduction is linked to c_1 reduction. As $Q_n^{\cdot-}$ is formed, it catalyzes reoxidation of cytochrome b through center N, and does not accumulate to detectable levels until c_1 and iron-sulfur protein are reduced, at which point further oxidation of ubiquinol through center P is blocked by lack of an electron acceptor. The b reoxidation phase then ends as ubiquinol oxidation "switches" to center N, and the b hemes are rereduced by reversal of reactions 4b and 4a in Figure 3. Myxothiazol inhibits the initial, rapid phase of b reduction and reduction of c_1, but allows the slower reduction of b to proceed. Antimycin does not inhibit the initial phase of b reduction, but eliminates the subsequent reoxidation and slow rereduction (52).

Structure-Function Relationships in the Cytochrome bc_1 Complex

As the Q cycle gained acceptance, attention focused on elucidating the structure of the cytochrome bc_1 complex as it relates to the mechanism. Three-dimensional crystals of the mitochondrial bc_1 complex have been obtained, and the structure of this oligomeric protein is being solved to high resolution (55–57). Present knowledge of the structure of the complex is based on electron microscopy of membrane crystals (58), as well as a number of other approaches, such as chemical modifications that localize proteins at the membrane surface or interior (26, 27, 59), spectroscopic probes that measure distances of the redox prosthetic groups from the membrane surface (60), proteolytic digestions that release water-soluble domains of the iron-sulfur protein (61) and cytochrome c_1 (62), genetic methods that identify domains of cytochrome b involved in quinol oxidation and quinone reduction (63–75), predictions of secondary structure by empirical algorithms (76, 77), and models that reconcile the location of the redox prosthetic groups with the Q cycle mechanism (1, 4, 74).

CYTOCHROME b Sequences of cytochrome bs from approximately 900 species show a high degree of conservation of secondary structure, and identify the four conserved histidine heme ligands (7). Cytochrome b is folded into an octet of transmembrane helices, and a ninth, extramembranous amphipathic helix, as in Figure 5. Colson, Di Rago, Lemesle-Meunier, and coworkers have extended the elegant mitochondrial cytochrome b genetics developed in Slonimski's laboratory to elucidate structure-function relationships in the cytochrome b of *S. cerevisiae* (for a review see Ref. 75). Similar studies have been initiated with the cytochrome b of photosynthetic bacteria, which is amenable to site-directed mutagenesis (for a review see Ref. 74). The centers N and P domains in cytochrome b have been identified by point mutations, which confer resistance to inhibitors that block Q cycle reactions at these sites, and neighboring relationships between the helices have been deduced from intragenic complementations.

The two b heme groups are ligated through four conserved histidines (7), and bridge the helices B–D (Figure 5). The predicted folding of the protein (Figure 5) and oxidation-reduction of the two hemes in response to a membrane potential (28) imply that His82 and 183 ligate b-566 near the P side of the membrane, and His96 and 197 ligate b-560 in a more electronegative location, near the center of the membrane. These assignments are supported by a point mutation, Cys133-Tyr, which alters the spectrum of b-566, and a second mutation, Ser206-Leu, which affects heme b-560 (72).

Two additional inhibitors, strobilurin and stigmatellin, act at center P in a manner similar to that of myxothiazol, as indicated by their effects on pre-steady-state reactions of the Q cycle (39, 40) and cross-resistance in some of the inhibitor-resistant cytochrome b mutants. Point mutations conferring resistance to these center P inhibitors have been identified at Phe129, Cys133, Gly137, Trp142, Ile147, Thr148, Asn256, Leu275, and Val291 (65, 70, 71, 75). Mutation of Cys133 also modified heme b-566, and a mutation at Gly143 blocked oxidation of ubiquinol at center P, while leaving reactions through center N intact (65). From these mutations it can be inferred that center P is formed from the amphipathic, extramembranous cd helix, the loop between helices E and F, and the distal regions of helices C and F, and that the b-566 heme is near center P.

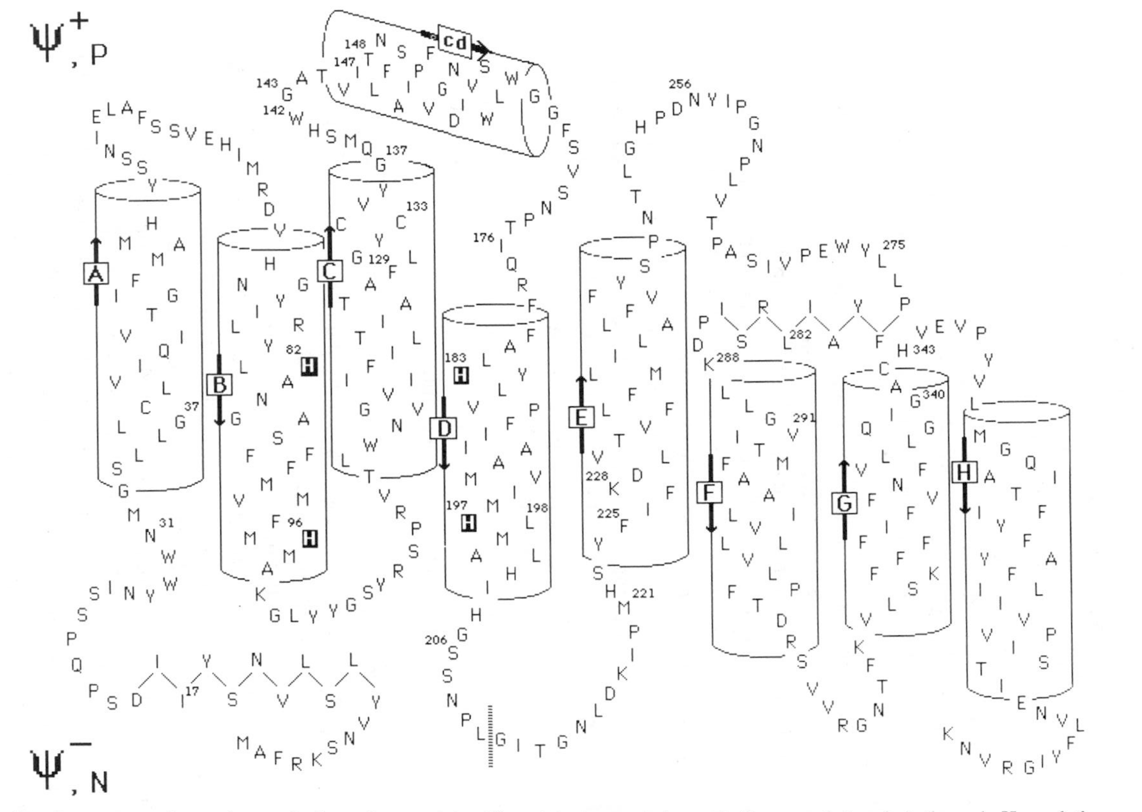

Figure 5 Predicted structure of cytochrome b from *S. cerevisiae* The eight transmembrane helices are lettered A through H, and the amphipathic extramembranous helix at center P is lettered cd. The dashed line in the loop connecting helices D and E depicts the approximate position at which cytochrome b is split into an α/β dimer in the cytochrome bf complexes. Numbers identify amino acids, which are discussed in the text. The four conserved histidine ligands to the hemes are enclosed in bold boxes.

Mutations conferring resistance to antimycin, funiculosin, and diuron have been identified at Ile17, Asn31, Gly37, Leu198, Phe225, and Lys228 (63, 66, 68, 69, 72). These place the amino terminus of the protein, the connecting loop between helices D and E, the end of helix E, and the end of helix D, which includes one of the b-560 histidine ligands, at the site that catalyzes oxidation-reduction of ubiquinol and ubiquinone at center N (reactions 4a and 4b in Figure 3). Since antimycin destabilizes $Q_n^{\bar{}}$ (45–48), and some of the antimycin-resistant mutants exhibit low rates of electron transfer, it would be interesting to determine whether the amino acid changes in these mutants affect the stability of the EPR-detectable semiquinone.

The locations of the resistance mutations at centers N and P provide strong support for the topographical disposition of transmembrane helices A through F and the extramembranous helix cd in cytochrome b, as shown in Figure 5. Analysis of second-site, intragenic pseudorevertants has provided additional evidence for the topology of the transmembrane helices, and has identified interacting amino acids within adjacent helices (70, 78). A mutation at Gly340 that leads to a nonfunctional cytochrome b (72) is complemented by an intragenic reversion mutant at Lys288, indicating that the C terminus of helix G and the N terminus of helix F interact and are on the same side of the membrane (79). A mutation of Cys133 is complemented by reversion mutations at His343 or Ile176, indicating that helix C interacts with helices G and D (Figure 5). Complementing mutations between Gly137 and Asn256 imply that the conserved "PEWY" loop connecting helices E and F interacts with the C terminus of helix C.

RIESKE IRON-SULFUR PROTEIN Sequences of the Rieske iron-sulfur proteins have been deduced from 15 species, including mammals, plants, fungi, and oxygenic and photosynthetic bacteria (for a review see Ref. 80). The 2 Fe:2 S cluster is noncovalently held in the protein by two cysteines and two histidines (81–83). Since histidine is less strongly electron donating than sulfur, it is likely that the two histidine ligands account for the fact that the midpoint potential (+280 mV) of this 2 Fe:2 S protein is 400 mV more positive than those of other ferredoxins (84), which contain four sulfur ligands. The only conserved sequences in the Rieske proteins are two hexapeptides, corresponding to Cys159-THLGC and Cys178-PCHGS in the yeast sequence in Figure 6. These two sequences include the only conserved histidines, His161 and His181, which must be two of the ligands to the 2 Fe:2 S cluster, and four conserved cysteines. Which two cysteines are ligands is still under investigation. It seems unlikely, for steric reasons, that an iron could be liganded to Cys180, if His181 is a ligand. However, site-directed mutagenesis of Cys159, Cys164, Cys178, and Cys180 has

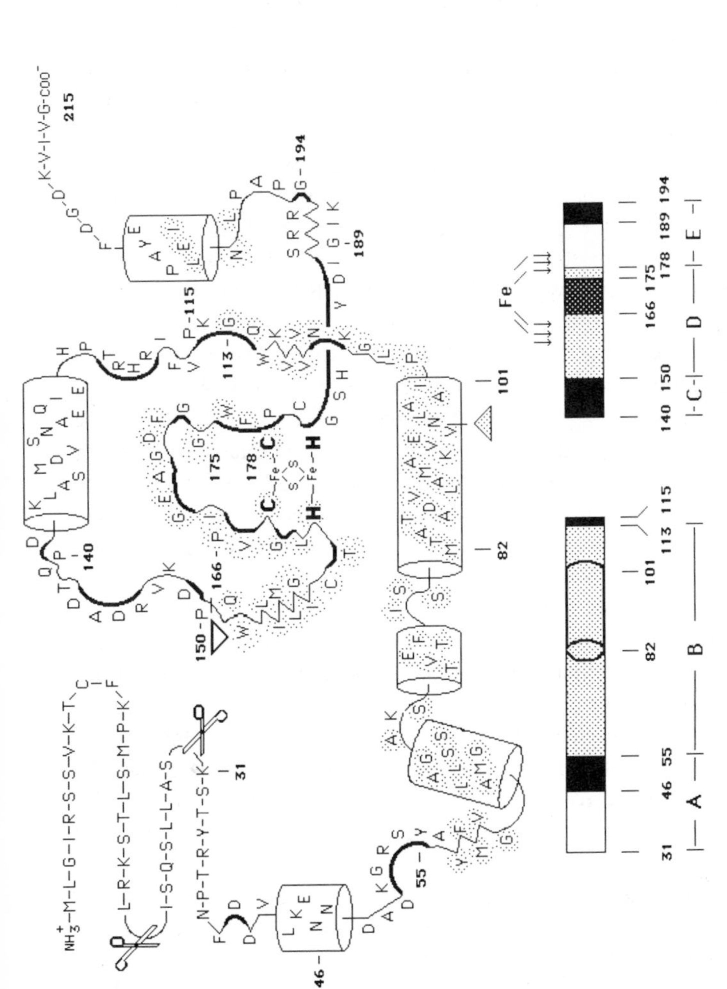

Figure 6 Predicted structure of the Rieske iron-sulfur protein from *S. cerevisiae* showing domains conserved among species. The scissors indicate the two sites of posttranslational processing by mitochondrial proteases. Hydrophobic regions of the protein are indicated by stippling around the amino acids. The stippled triangle indicates the region in the yeast protein where there is a deletion of approximately 15 amino acids in the iron-sulfur proteins from spinach, *Nostoc*, and *Synechococcus*. The inverted open triangle indicates where there is an insertion of 10–15 amino acids in the iron-sulfur proteins from *P. denitrificans* and *R. capsulatus* (80). Below the predicted structure is a map of the five structural domains (A–E), which are conserved in the Rieske iron-sulfur proteins characterized to date. Numbers below the map correspond to amino acids in the yeast protein. Open rectangular areas on the map indicate randomly coiled hydrophilic domains; solid black areas indicate flexible hydrophilic domains; stippled areas on a clear background indicate hydrophobic domains; and stippled black areas indicate flexible hydrophobic domains. Arrows point to locations of the possible ligands to the iron–sulfur cluster.

resulted in each case in a nonfunctional iron-sulfur protein, apparently lacking the Fe:S cluster (80, 85, 86).

Although extensive regions of amino acid sequence are not conserved, the structure of the Rieske iron-sulfur protein is. The diagram in Figure 6 depicts five structural domains in the yeast protein that are conserved across species. The N terminus of the iron-sulfur protein is distinguished by a bipartite structure, in which a hydrophilic domain (A) is followed by a hydrophobic domain (B), which includes one or more extended α helices and is delimited by two highly flexible sequences.

The N terminus of domain B appears to bind the iron-sulfur protein to the bc_1 complex hydrophobically. A synthetic peptide corresponding to Ser54-Lys72 in the yeast protein (Figure 6) prevented the full-length protein from rebinding to the bc_1 complex from which it was extracted (87), and trypsin releases a water-soluble form of the protein by cleavage at Lys72 (61). Whether this hydrophobic anchor spans the membrane or is adsorbed to cytochrome b is not clear. The iron-sulfur protein binds to the bacterial plasma membrane in the absence of cytochromes b or c_1, indicating that its anchorage is not dependent on other subunits of the complex (88). However, extraction of the protein by guanidine (89) or sodium carbonate (90) is difficult to reconcile with a transmembrane anchor, and the protein was not labeled by hydrophobic probes that labeled phospholipids and cytochrome b (27, 59), but was labeled by extrinsic, hydrophilic reagents (91–93).

The Fe:S cluster is held in a flexible, hydrophobic pocket in the C terminus of the protein, flanked by sequences enriched in glycines and prolines (Figure 6), and that are sites of temperature-sensitive mutations (80). Domain C is thought to be at the aqueous surface of the bc_1 complex. The hydrophilic, flexible character of this region is retained in the various Rieske proteins, and mutations that eliminate the acidic charges render the protein temperature sensitive. The hydrophobic C terminus of the protein is thought to cap the Fe:S cluster, creating an environment in which one edge of the cluster, including His161 and one of the cysteines, is in a hydrophobic environment, while the other edge, including His181 and a second cysteine, is at an interface between hydrophobic domain D and hydrophilic domain E (80).

The divergent oxidation of ubiquinol at center P requires that the Fe:S cluster and the b-566 and c_1 hemes must have approximate relationships to each other and to the hydrophobic phase of the membrane implied in Figure 4. The Fe:S cluster must be reducible by ubiquinol from within the membrane, and must generate $Q_p^{\bar{\cdot}}$ within a few angstroms of b-566. The Fe:S cluster must additionally transfer electrons to the heme of c_1, which must be inaccessible as a direct oxidant for either ubiquinol or ubisemiquinone, even when the iron-sulfur protein is removed (37, 38).

These considerations suggest that one edge of the Fe:S cluster is in a hydrophobic environment near the b-566 heme, while the other must be electronically connected to the c_1 heme, in a more hydrophilic region. Since electron transfer from ubiquinol to c_1 occurs on a microsecond time-scale (29, 30), it is likely the electron is transferred from the quinol across the Fe:S cluster to the c_1 heme in an essentially solid-state relay, in which one or both of the histidine ligands may also participate. To prevent spurious reactivities of the strongly reducing $Q_p^{\overline{\cdot}}$, water and oxygen must be excluded from center P, which must conduct protons from ubiquinol to the surface of the membrane. Investigation of which amino acids form this proton-conduction pathway on the iron-sulfur protein or cytochrome b is currently under way (94).

CYTOCHROME c_1 Cytochrome c_1 is anchored to the P surface of the bc_1 complex by a hydrophobic helix at its C terminus (95), and released as a soluble heme-protein by proteolysis (62, 96). The c-type heme is usually covalently attached through thio-ether linkages to two conserved cysteines (97), although protozoan c_1s have only a single thio-ether linkage (98). A histidine adjacent to a cysteine ligand in the sequence CxyCH, and a methionine corresponding to Met164 in the mature yeast c_1, are thought to form the fifth and sixth axial ligands to the heme. These assignments are consistent with site-directed mutagenesis of these conserved residues (99, 100), and optical (101) and magnetic circular dichroism (102) properties of the proteins.

Cytochrome c_1 is the exit point for electrons from the bc_1 complex, and is more accessible to proteases at the surface of the membrane (96) than is the iron-sulfur protein (Figure 4). Mitochondrial bc_1 complexes are dimeric (103), and contain numerous subunits of unknown function, in addition to the three redox proteins (1). One of these supernumerary subunits, an acidic protein, resides on the surface of the bc_1 complex in association with cytochrome c_1, and shuts down electron transfer from c_1 to c in half of the dimeric bc_1 complex in response to high protonmotive force (104).

Cytochrome c_1 is a diffusionally constrained, membrane protein and conducts electrons from a hydrophobic environment to a hydrophilic, soluble protein, cytochrome c. Electron transfer to and from the c_1 heme may follow different routes, requiring an electronic circuit through the protein. It was previously predicted that cytochrome c_1 differs from the other cytochromes c in having a high content of β-pleated sheet structure (1). Well-diffracting crystals of cytochrome f have been obtained (105), the crystal structure of the 252-residue extrinsic domain of the 285-residue cytochrome f polypeptide has been solved to 2.3 Å, and the polypeptide chain has been completely traced (SE Martinez, D Huang, A Szczepaniak, WA Cramer, JL Smith, manuscript in preparation). Cytochrome f is an approximately 75 × 30 × 20 Å elongate structure, which is unique in several aspects. Unlike the single-domain, predominantly α-helical c-type cytochromes, cytochrome f has two domains, and the predominant secondary structure is β sheet. Identifying the electron pathway within these integral membrane cytochromes will be a focus of future investigations.

RESPIRATORY OXIDASE COMPLEXES

A Multitude of Prokaryotic Respiratory Oxidases

Bacteria have modular, branched respiratory systems, designed for optimal growth, depending on physiological demands and environmental conditions (106). Numerous dehydrogenases feed electrons into the respiratory chain, and there are multiple oxidases, as shown in Figure 1. In addition to bioenergetic efficiency, as measured by the H^+/e^- ratio, the rate of respiration can be a critical factor for optimal growth, since respiration maintains redox balance within the cell by eliminating excess reducing equivalents. When the energy needs of the cell can be met by fermentation or photosynthesis, without oxidative phosphorylation, respiration need not be optimized to maximize the H^+/e^-ratio, but rather to maintain redox balance in the cytoplasm.

The affinity of the respiratory oxidase for oxygen is important in maintaining redox balance. Under microaerophilic conditions, the terminal oxidase should have a high affinity for molecular oxygen. This explains why cytochrome bd predominates in *E. coli* under low oxygen tension (Figure 1), whereas cytochrome bo, which has a lower affinity for oxygen, predominates when the oxygen tension is high (107). Cytochrome bd is also present when *E. coli* is grown anaerobically, and may scavenge oxygen to protect oxygen-sensitive enzymes required for anaerobic growth (108). Furthermore, when grown microaerophilically, *E. coli* has a relatively high concentration of menaquinol compared to ubiquinol. Menaquinol is efficiently oxidized by cytochrome bd, but not by cytochrome bo (109). Under such circumstances, the H^+/e^- ratio may be less important and, indeed, cytochrome bd is less efficient ($H^+/e^- = 1$) than is cytochrome bo ($H^+/e^- = 2$, Ref. 110).

It has become clear in the past several years that the majority of the bacterial oxidases are members of a superfamily of heme-copper oxidases closely related to the mitochondrial oxidase (8). One exception is the alternate cytochrome bd quinol oxidase from *E. coli*. This enzyme, found in various bacteria, has been well characterized and is unrelated to the heme-copper oxidases.

Cytochrome bd from E. coli

Cytochrome bd is a heterodimer (111, 112) and subunits I and II are encoded by the *cyd* operon (113). Cytochrome bd translocates protons with a H^+/e^- ratio of 1 (110, 114, 115), consistent with the scheme in Figure 2*A*, in which the enzyme is postulated to translocate protons by separating the sites for oxidation of quinol on the outside and reduction of oxygen on the inside. Controlled proteolysis of the purified oxidase cleaves a unique site within subunit I and specifically eliminates oxidation of quinol (116), while oxidation of tetramethylphenylenediamine is not affected, demonstrating that the site where oxygen is reduced to water is not influenced by the proteolytic cleavage. Subunits I and II of the oxidase are transmembranous, with seven and eight putative transmembrane helices, respectively, demonstrated by gene fusions (117, 118). The proteolytic cleavage site has been located on the periplasmic side of the membrane of *E. coli* sphaeroplasts, in a hydrophilic region connecting helices E and F within subunit I (116), where monoclonal antibodies bind to an epitope and similarly inhibit quinol oxidation (119). This region of subunit I is called the Q loop (quinol site), and its presence on the outer surface suggests that the quinol-binding site is near the periplasmic side of the membrane, as in Figure 2*A*. The release of protons from quinol oxidation at the periplasmic surface accounts for the measured H^+/e^- ratio.

Cytochrome bd contains one heme D (120, 121) and two heme Bs (Figure 7), which constitute the three redox centers in the enzyme (122, 123). Heme b-558 is a six-coordinate, low-spin heme, and the initial site reduced by quinol. The heme b-558 ligands have been identified as His186 (124) and (probably) Met393 in subunit I, and are located at the periplasmic ends of helices D and F, respectively, consistent with placement of the quinol oxidation site near the periplasmic surface. Heme b-595 and heme d appear to be near the cytoplasmic side of the membrane, although this has not been directly shown. Mutagenesis of His19 in subunit I, on the cytoplasmic side of the membrane, eliminates both of these hemes, but leaves heme b-558 intact (124). Heme b-595 and heme d share a common binding pocket in the oxidase (125), and both are high spin, but only heme d binds oxygen or CO (125, 126). The role of heme b-595 is not known. Heme b-595 may simply conduct electrons to the oxygenated intermediates bound to heme d or, perhaps, is directly involved in catalysis.

Cytochrome bd forms stable oxygen adducts (126–128), and is isolated as a mixture of oxygenated complexes. Two different oxygenated forms

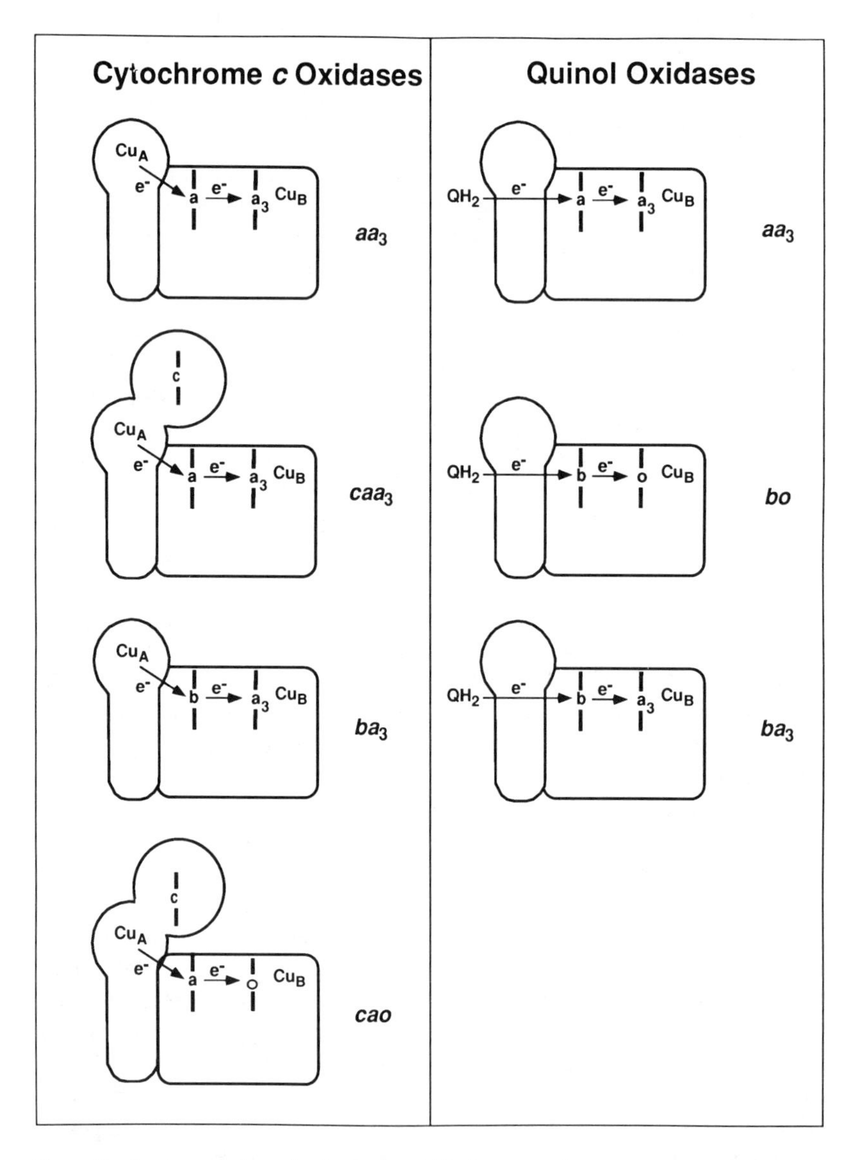

Figure 7 A representation of subunits I and II of some well-characterized members of the heme-copper superfamily of oxidases. The cytochrome c oxidases include cytochrome aa_3 from mitochondria and prokaryotes such as *R. sphaeroides* (150, 153) or *P. denitrificans* (141); cytochrome caa_3 from *Thermus thermophilus* (149) or *Bacillus* PS3 (143); cytochrome ba_3 from *T. thermophilus* (187); and cytochrome cao from *Bacillus PS3* (252). The quinol oxidases include cytochrome aa_3 from *Bacillus subtilis* (148); cytochrome bo from *E. coli* (161); and cytochrome ba_3 (also called cytochrome a_1) from *Acetobacter aceti* (139).

have been characterized by resonance Raman spectroscopy. One form is the oxy complex, equivalent to the oxygen adduct of myoglobin, in which oxygen is reversibly bound to ferrous heme d (126). The second form is the oxoferryl heme d, $Fe(IV){=}O^{-2}$ (127). Both of the stable oxy and oxoferryl forms of cytochrome bd are predicted intermediates during oxidase turnover, though this has yet to be demonstrated. However, both the oxy and oxoferryl forms of cytochrome c oxidase have been identified as reaction intermediates (129–132, see next section), so it is likely that the oxygen chemistry catalyzed by cytochrome bd and the heme-copper oxidases is similar.

In summary, the data suggest that cytochrome bd is a relatively simple proton-translocating device that releases and takes up substrate protons from opposite sides of the membrane. In contrast, the heme-copper oxidases require a proton-conducting channel or pathway across the membrane and a mechanism coupling electron transfer to the vectorial movement of protons through this channel. The subunits of the heme-copper oxidases have no sequence similarity with the subunits of cytochrome bd.

The Superfamily of Heme-Copper Oxidases

Most of the bacterial oxidases characterized are members of a superfamily of proton-pumping oxidases, which includes the mitochondrial cytochrome c oxidase (8, 133, 134). This realization has been exploited over the past few years and has resulted in significant progress in our understanding of these enzymes.

Mammalian cytochrome c oxidase contains 13 different subunits, 10 encoded in the nuclear genome and three that are mitochondrially encoded (135–137). The genes encoding the subunits of several prokaryotic respiratory oxidases have been sequenced, and have revealed homologs of the mitochondrially encoded eukaryotic subunits I, II, and III (8, 138–151, 151a). There are no reports of prokaryotic genes corresponding to any of the nuclear-encoded subunits of the eukaryotic oxidases. Cytochrome c oxidases from *P. denitrificans* (152) and *R. sphaeroides* (153) have been isolated, for example, as three-subunit species that have prosthetic groups identical to the mitochondrial enzyme, that have high cytochrome c oxidase activity, and that pump protons as efficiently as the mitochondrial oxidase. Additionally, studies on mammalian and *P. denitrificans* oxidases have shown that subunit III is not essential for either cytochrome c oxidation or proton pumping (154–158, 158a, 158b). Hence, the minimal unit required for electron transfer–linked proton pumping consists of subunits I and II. The functions of the other 11 subunits of the eukaryotic oxidase are not known, although some roles in regulation of oxidase activity are indicated (159, 160). Some prokaryotic oxidases have one or more subunits that are not homologous to any of the subunits of the mitochondrial oxidase, and whose functions are not known (161, 162).

The prokaryotic oxidases that are related to the mitochondrial oxidase are diverse with respect to the substrate oxidized and heme and copper content. The mitochondrial oxidase utilizes cytochrome c as substrate and has four redox-active metal centers, Cu_A, Cu_B, heme a, and heme a_3 (for reviews see Refs. 8, 134, 163). The heme a and a_3 centers contain the same heme A prosthetic group, but have different properties due to the ways in which they are bound to the polypeptides. Heme a is a low-spin, six-coordinate heme and is responsible for the visible absorbance and green color of the enzyme. Heme a_3 is a high-spin, five-coordinate heme and has an open coordination site where oxygen binds. Its iron is within 5 Å of Cu_B and, together, these constitute the heme a_3-Cu_B binuclear center where oxygen is reduced to water. The presence of a binuclear center is diagnostic for all members of the oxidase superfamily.

Although there is evidence for multiple electron transfer pathways within the oxidase (164, 165), there is substantial support for a linear scheme of electron flux from cytochrome c to oxygen at the binuclear center (166, 167).

$$\text{cyt c} \rightarrow Cu_A \rightarrow \text{heme a} \rightarrow \text{heme } a_3\text{-}Cu_B \qquad 5.$$

The binding site for cytochrome c and the Cu_A center are in subunit II (for a review see Ref. 8). Heme a and the a_3-Cu_B binuclear center are in subunit I, as illustrated in Figure 7, which also displays some of the prokaryotic variants that have been characterized.

The heme-copper superfamily is divided into two branches, the cytochrome c oxidases and the quinol oxidases. The enzymes in Figure 7 are classified as members of the superfamily on the basis of the deduced amino acid sequences of the subunits. Subunit I is particularly highly conserved. For example, the sequences of subunits I from the *R. sphaeroides* aa_3-type and the *E. coli* cytochrome bo ubiquinol oxidase are 50% and 40% identical to subunit I of bovine oxidase, respectively (168). The primary differences between the cytochrome c and quinol oxidases reside in subunit II, which is not as well conserved throughout the superfamily as is subunit I. However, all of the cytochrome c oxidases have two conserved sets of residues that have been implicated in binding of cytochrome c and in ligating Cu_A (8, 168a, 168b). These residues are not conserved in the members of the superfamily that are quinol oxidases (148, 168), and the quinol oxidases neither interact with cytochrome c nor do they have the Cu_A center (161, 169). The quinol-binding site has not been localized.

A variety of heme combinations are found in the prokaryotic oxidases. Cytochrome bo from *E. coli* contains heme B at the low-spin locus (equivalent to heme a) and heme O (equivalent to heme a_3) as a component of the binuclear center (161, 170). Heme O and heme A have a farnesyl side chain that is not present in heme B (171). Heme A differs from heme O in the presence of a formyl group. The hemes in subunit I of a particular oxidase do not correlate with the substrate oxidized by the enzyme. For example, *Bacillus subtilis* contains an aa_3-type quinol oxidase in addition to an aa_3-type cytochrome c oxidase (169, 172). Cu_A is absent from the quinol oxidase but present in the cytochrome c oxidase. The *B. subtilis* cytochrome c oxidase illustrates another variation found in a number of prokaryotic oxidases, in which heme C is attached covalently to a carboxy-terminal extension of subunit II (173). Hence, this enzyme is called a caa_3-type oxidase.

Traditionally, bacterial respiratory oxidases have been classified either by their heme content or substrate oxidized (i.e. cytochrome c or quinol). Data accumulated over the past several years shows that such classifications are superficial, and that many enzymes previously thought to be unrelated are, in fact, very close relatives.

Two of the best-characterized members of this superfamily are cytochrome aa_3 from bovine heart mitochondria and cytochrome bo from *E. coli*. Despite their different hemes and different substrates oxidized, their metal centers show a high degree of similarity, and reinforce the belief that the enzymes in the superfamily share a common structural framework and similar catalytic mechanisms.

SPECTROSCOPIC CHARACTERIZATION OF THE HEME AND CU_B LIGANDS Spectroscopic studies have helped to identify amino acids bound to the prosthetic groups within subunit I. Electron paramagnetic resonance (EPR) and magnetic circular dichroism (MCD) of the mitochondrial oxidase have identified the two axial ligands to heme a as histidines (174–176). The corresponding heme in cytochrome bo is heme b-562 (Figure 7), and the same techniques indicate bis-histidine ligation of this heme as well (177, 178). The oxygen-binding heme a_3 and heme o are coordinated to only a single amino acid, and EPR and resonance Raman spectroscopy indicate a histidine ligand in both cases (179–181).

The ligands to Cu_B are less well defined. Electron nuclear double resonance (ENDOR) studies have suggested two or three histidines ligated to Cu_B of the mitochondrial oxidase (182). Extended X-ray absorption fine structure (EXAFS) analysis indicates, in addition to these histidines, a sulfur or chloride also bound to Cu_B (183, 184). Similar results have been reported for cytochrome bo (185). Biochemical studies (186) make it likely that chloride is responsible for the observed EXAFS. ENDOR studies on an unusual CN^- adduct of cytochrome ba_3 from *Thermus thermophilus* indicated that Cu_B has four equivalent histidine ligands (187). Due to the limited number of conserved histidines within subunit I (133), this interpretation, if correct, cannot apply to all members of the superfamily.

THE Cu_A CENTER Cu_A is ligated to residues within a hydrophilic domain of subunit II that has limited sequence homology to the copper-binding domain of blue copper proteins such as azurin (8). The spectroscopic properties of Cu_A, however, are very different from those of the blue copper proteins. Until recently this was thought to be entirely due to ligation of copper in the Cu_A site to two cysteine residues, in addition to two histidines, whereas the blue copper proteins have only a single cysteine copper ligand. However, there is a growing body of evidence that the Cu_A center is composed of two copper atoms (188–192). This is consistent with some of the unusual spectroscopic properties of Cu_A, and explains the stoichiometry of three copper atoms per enzyme often found in prokaryotic and eukaryotic cytochrome c oxidases (one for Cu_B, and two at the Cu_A center, Refs. 193, 194).

Cytochrome bo and the other quinol oxidases do not have Cu_A, and the putative copper ligands within subunit II are absent (142, 161, 169). These putative ligands have been returned to the hydrophilic domain of subunit II from cytochrome bo by site-directed mutagenesis (188, 195), creating a copper-binding site that accommodates two coppers, with spectroscopic properties similar to those of Cu_A.

METAL DISTANCES AND ELECTROCHEMICAL INTERACTIONS EPR techniques have been used to approximate distances between Cu_A and heme a and between heme a and heme a_3 in the bovine oxidase. The results indicate that Cu_A is only about 8 to 13 Å from heme a (196), and that the distance between the two heme irons is 12 to 19 Å (197, 198). EXAFS analysis indicates that heme a_3 and Cu_B are 3–5 Å apart in the bovine enzyme (183, 199), and a preliminary EXAFS study of cytochrome bo also indicates that heme o and Cu_B may be as close as 3 Å (185) in the oxidized enzyme.

Electrochemical studies of bovine cytochrome c oxidase and cytochrome bo show complicated behavior, because each of the metal centers can be influenced by the redox state of the others (177, 200, 201). For example, the midpoint potential of heme a is decreased by as much as 50 mV when heme a_3 is reduced, and vice versa. The structural model (133, 134) provides a rationale for cooperativity between the two hemes, since they appear to be ligated to histidine residues on the same transmembrane helix.

THE HEME-COPPER BINUCLEAR CENTER Oxygen is reduced to water at the binuclear center, where cyanide, azide, and carbon monoxide bind and inhibit the enzyme. The heme and copper of the binuclear center are spin-coupled, altering the EPR signal that would arise from the oxidized form of either component in isolation (177, 202, 203). In the bovine oxidase, under most conditions no EPR signals attributable to the heme or copper of the binuclear center are observed. However, incubation of the oxidized enzyme under mild acidic conditions reversibly converts the enzyme to a "slow form" (203a), which has a distinct EPR spectrum with a $g = 12$ component (190, 202, 203b). This form of the oxidase differs from the "fast form" in numerous properties, most notably a 300-fold slower rate of reaction with cyanide, from which the name derives (186, 203a). The differences between these forms of the enzyme indicate that conversion from the fast to the slow form of the oxidase involves a change in an endogenous ligand bound to one or both of the metals in the binuclear center. Cytochrome bo does not convert spontaneously to a slow form, suggesting some subtle difference at the binuclear center between these two enzymes (203, 205).

Upon reduction, the heme of the binuclear center can bind CO with high affinity, and the properties of the CO adducts have been useful in characterizing the heme-copper oxidases. The kinetics of CO binding at 10 Kelvin indicate that CO first forms a transient complex with Cu_B (206, 207). The CO stretching frequency observed by Fourier transform infrared (FTIR) spectroscopy is sensitive to the metal to which CO is bound (208), and has been valuable in evaluating mutants that perturb the binuclear center (133).

The binding of cyanide, azide, and formate to the binuclear center of cytochromes aa_3 and bo, examined by various spectroscopic methods (181, 199, 203, 209–213), confirms that the bovine and *E. coli* enzymes are remarkably similar, and shows that Cu_B and the heme Fe can bind exogenous ligands which, in some cases, may bridge the two metals. It is becoming clear that Cu_B acts as a gate, controlling access of the heme Fe to certain exogenous ligands (207, 214). This may also be true for oxygen, which appears to form a transient complex with Cu_B prior to interacting with the heme Fe to initiate catalysis (215).

SUBUNIT TOPOLOGY Figure 8 shows the proposed topology of subunit I of cytochrome bo (216). It is assumed that the membrane-spanning regions are helical, and the hydropathy profile is consistent with the number and approximate location of the transmembrane spans, and with gene fusions that have identified regions of the polypeptide exposed to the periplasmic or cytoplasmic surfaces of the *E. coli* membrane (216). Sequence alignments suggest that subunit I of most of the heme-copper oxidases has 12 transmembrane helices (8), corresponding to helices I to XII in Figure 8.

Subunit II from the *E. coli* oxidase and from most of the heme-copper oxidases has two transmembrane spans and a large hydrophilic domain

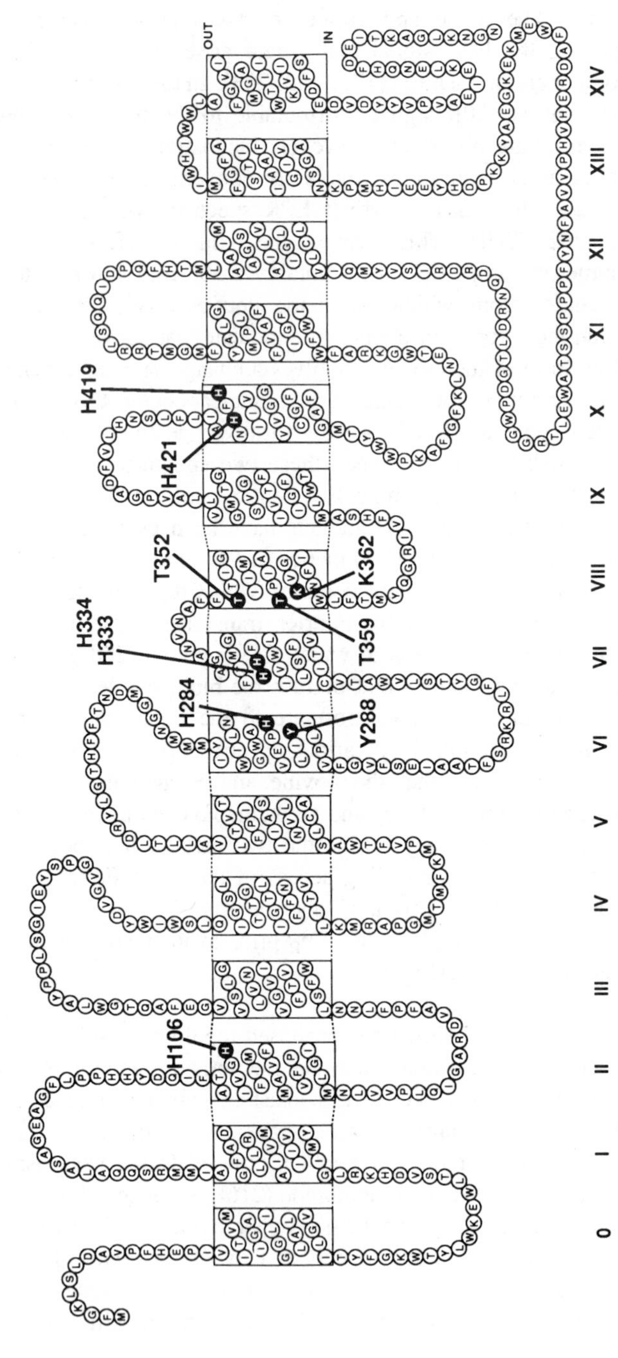

Figure 8 A two-dimensional representation of the 15 putative transmembrane spans of subunit I from cytochrome bo from *E. coli*, with the periplasmic surface of the membrane at the top of the figure. Helices I through XII appear to be common to the entire superfamily of heme-copper oxidases. Residues referred to in the text are numbered.

facing the periplasm or intermembrane space (135, 216). One exception to this topology is subunit II from the quinol oxidase from *Sulfolobus acidocaldarius,* which has only a single putative transmembrane span (142).

Subunit III from *E. coli* cytochrome bo has five putative transmembrane spans (216), but in most of the other heme-copper oxidases has two additional transmembrane spans at the N terminus (8). The *E. coli* and several other bacterial oxidases that have the "short version" of subunit III also have "long versions" of subunit I, with two extra transmembrane spans at the C terminus (helices XIII and XIV in Figure 8, Refs. 143, 148). This suggests that these two helices can be associated with either the C terminus of subunit I or the N terminus of subunit III without major structural consequences. Indeed, in the caa_3 oxidase from *Thermus thermophilus,* the C terminus of subunit I and the N terminus of subunit III are fused to form a single subunit (149). Subunits I and III in cytochrome bo have similarly been fused by mutagenesis without adverse functional consequences (217).

ASSIGNMENT OF THE METAL LIGANDS IN SUBUNIT I Spectroscopic studies on the mitochondrial cytochrome c oxidases and *E. coli* cytochrome bo indicate five or six histidines within subunit I are metal ligands; one for heme a_3 (heme o), two for heme a (heme b-562), and two or three for Cu_B. Alignments of approximately 75 subunit I sequences reveals only six conserved histidines (8, 133, 134). These are in four of the putative transmembrane helices (Figure 8): helix II (His106), helix VI (His284), helix VII (His333 and His334), and helix X (His419 and His421).[4] Substitutions for any of the six conserved histidines by site-directed mutagenesis eliminate oxidase activity, but in most cases the enzyme is assembled in the membrane. UV/visible, Raman, FTIR, and EPR spectroscopies from both of these oxidases present a consistent picture (133, 218–220):

1. His106 in helix II and His421 in helix X ligate the low-spin, six-coordinate heme a or heme b-562.
2. His333 and His334 are ligands for Cu_B.
3. His419 is the ligand to the high-spin, oxygen-binding heme a_3 or heme o.
4. His284 is either a Cu_B ligand or in the immediate vicinity of Cu_B.

This analysis allows a model of subunit I by defining the relative positions of helices II, VI, VII, and X as shown in Figure 9, which is also consistent with biophysical measurements of the metal distances. Several additional mutations perturb the binuclear center, allowing the model to be elaborated beyond the four putative transmembrane helices containing the metal ligands.

[4]The sequence numbering refers to cytochrome bo.

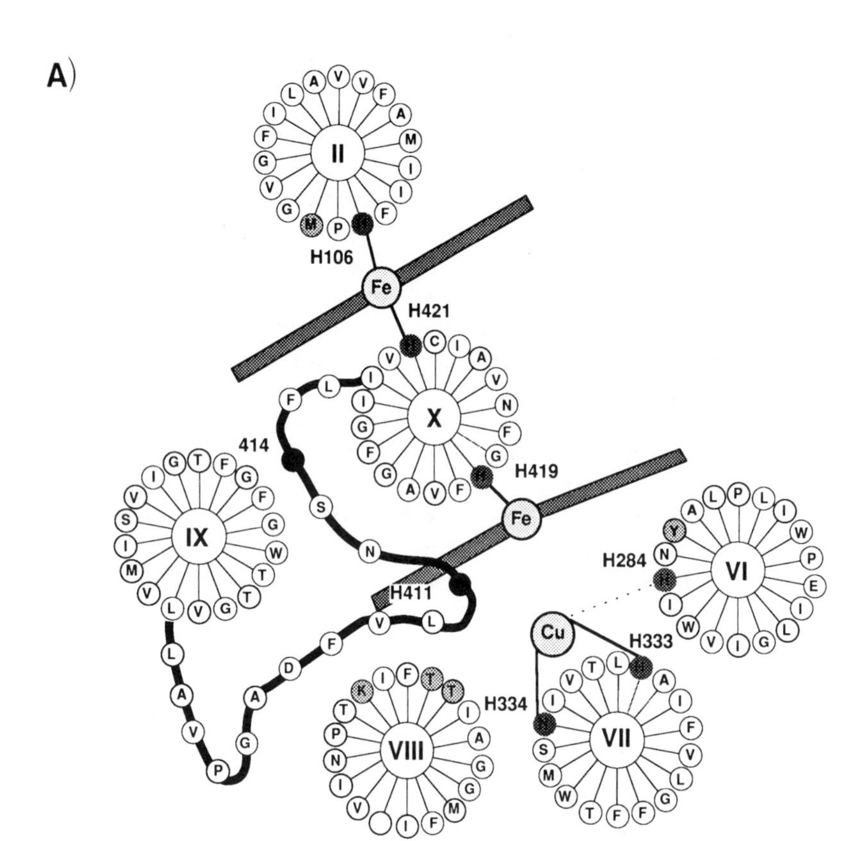
A)
II
H106
Fe
H421
X
414
H419
Fe
IX
VI
H284
H411
Cu
H333
H334
VII
VIII

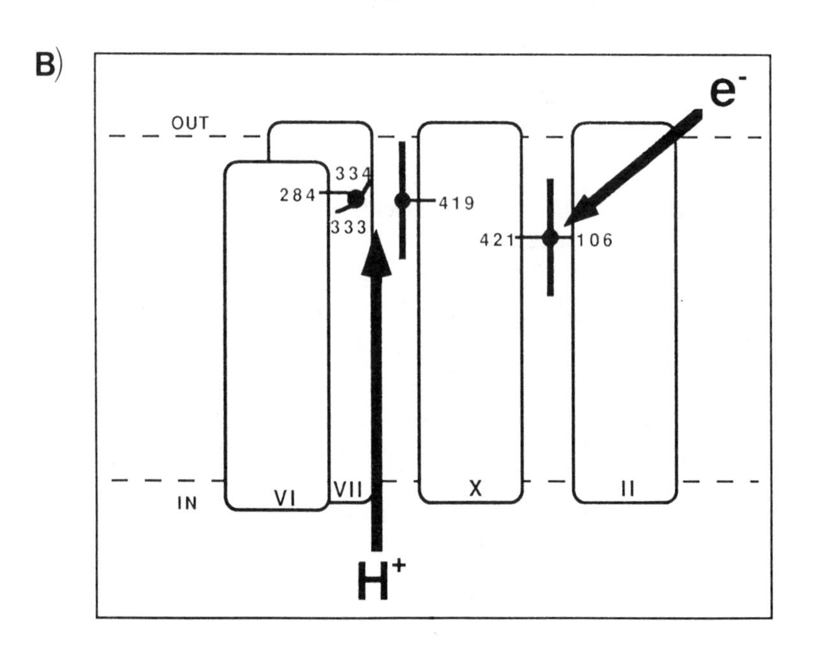
B)
e-
OUT
334
284
333
419
421
106
IN
VI
VII
X
II
H+

THE IX-X LOOP The connection between helix IX and helix X appears to cap the three metal centers within subunit I (Figure 9). In the *R. sphaeroides* oxidase, for example, changing Tyr414 (equivalent to Asn414 in cytochrome bo) to phenylalanine results in an active enzyme, with an altered heme a optical spectrum (133), suggesting that Tyr414 may be in the immediate vicinity of heme a. On the other hand, changing His411 (also His411 in cytochrome bo) to an asparagine disrupts the binuclear center, as does changing His411 to leucine in cytochrome bo. Since substitutions to alanine have no effect, a role for histidine at this position can be ruled out. The IX-X loop is located directly over both the low-spin heme and the binuclear center (Figure 9), consistent with the locations of second-site revertants to point mutations in yeast cytochrome c oxidase (221). An oxidase-defective mutant was isolated in which a residue in the IX-X loop was altered. Two revertants that restore oxidase activity were isolated in which residues predicted to be near the Cu_B-binding site are altered, suggesting the proximity shown in Figure 9.

Residues within the IX-X loop also form part of a Mg/Mn-binding site (222) in the *R. sphaeroides* oxidase (133). The function of this metal is not known, but it is possible that this portion of the protein may interact with the hydrophilic domain of subunit II via a bridging metal.

HELIX VIII The short connection between helix VII and helix VIII (Figure 8) suggests that the two helices are adjacent, and that some of the residues near the periplasmic side of helix VIII may be near the binuclear center, as shown in Figure 9. This was examined by mutating Thr352 (Figures 8 and 9) in the *E. coli* cytochrome bo (223) and the *R. sphaeroides* oxidases (133). In both cases, some mutations inactivate the oxidases, and the binuclear centers are severely perturbed. Helix VIII is proposed to form part of the pocket containing Cu_B.

Thr352, Thr359, and Lys362 are among a set of highly conserved polar residues located on the same face of helix VIII, which have been mutated in the *E. coli* and *R. sphaeroides* oxidases (Figures 8 and 9, Ref. 133). Oxidase activity is impaired or eliminated, but spectroscopy reveals either

←

Figure 9 A model of a portion of subunit I of cytochrome bo and, by extension, of other members of the heme-copper oxidase superfamily. *(A)* A view from the periplasmic side of the membrane in which the putative transmembrane helices are represented as helical wheels. The low-spin heme b-562 is ligated between His106 (helix II) and His421 (helix X). The high-spin heme o is ligated to His419 (helix X). Cu_B is ligated to or near His333 and His334 (helix VII) and His284 (helix VI). The postulated position of the IX-X loop is also indicated. *(B)* A side view in which the relative positions of the metals and several helices are shown. The metals are near the periplasmic side of the membrane, requiring a proton-conducting channel.

no or modest perturbations to the metal centers in some of the Thr359 and Lys362 mutants. Perhaps these residues form part of a proton- and/or water-conducting channel from the cytoplasm to the site where oxygen is reduced, since electron transfer cannot proceed if protons are not delivered to the active site, or water cannot exit.

THE II-III INTERHELICAL LOOP This interhelical connection contains Asp135, one of the most highly conserved acidic residues in subunit I. Changing this residue in cytochrome bo to an asparagine results in a modest decrease in the oxidase activity, but eliminates proton pumping (224). Proton translocation resulting from substrate protons (Figure *2B*) is still observed, but electron transfer is decoupled from the pump mechanism. Asp135 may direct protons into the mouth of the putative proton-conducting channel, but the possibility of nonspecific conformational changes induced by this mutation cannot be excluded. Asparagine substitution at position 135 causes no spectroscopic perturbation of the metal centers, consistent with its proposed location distant from the binuclear center.

IMPLICATIONS OF THE MODEL Site-directed mutagenesis has resulted in a tentative model involving 6 of the 12 putative transmembrane helices that are common to all subunit I variants (133, 134). The two most salient features of the model that have functional consequences concern the locations of the metal redox centers.

1. Heme a, heme a_3, and Cu_B are located near the bacterial periplasm or mitochondrial intermembrane space. Location of the binuclear center near the outer surface of the membrane implies a proton-conducting pathway from the bacterial cytoplasm or mitochondrial matrix to the binuclear center, since protons used to synthesize water from oxygen must originate from the opposite side of the membrane. The polar residues in helix VIII may contribute to this proposed pathway.

2. Heme a and heme a_3 are ligated on opposite sides of helix X, and Cu_B is not located between the two hemes. This is consistent with rapid electron transfer between hemes a and a_3 (225), and with the observation that a mutant that is thought to eliminate Cu_B in cytochrome bo maintains a high rate of electron transfer between the two hemes (226). The model suggests that electron transfer from heme a to oxygenated intermediates on heme a_3 need not first pass through Cu_B, but may go directly between the two hemes. This has implications concerning proton-pumping mechanisms.

THE REACTION MECHANISM Four electrons are required to reduce molecular oxygen to water. Techniques have been developed to observe each of the steps in the chemical sequence and to characterize the products, by resonance

Raman spectroscopy with bovine heart oxidase (129–132), or by UV/visible spectroscopy with bovine cytochrome c oxidase and *E. coli* cytochrome bo (225, 227–230). These experiments usually start with CO bound to the fully reduced form of the oxidase. A flash of light expels the CO and initiates a single-turnover reaction with oxygen. In cytochrome c oxidase each of the four redox-active metals is initially reduced, providing electrons to reduce molecular oxygen to water fully. In cytochrome bo, the lack of Cu_A allows only three electrons for reaction with oxygen, and the reaction sequence leading to two water molecules is incomplete.

Under physiological conditions, fully reduced oxidase is not present (231). The first electron to reduce the fully oxidized enzyme is shared by the three metal centers in subunit I, with Cu_B partially reduced (200). This one-electron reduced, E form of the enzyme probably does not interact rapidly with oxygen (231, 231a, 232). Oxygen binds rapidly to the two-electron reduced form of the enzyme, but if oxygen does not react with this form of the enzyme, it has been argued that negative cooperativity between the hemes (200, 231) makes it thermodynamically unfavorable to reduce the enzyme further. It has also been argued that physiologically it is the three-electron reduced form of the oxidase that reacts with oxygen (231a). Figure 10 shows a feasible reaction sequence beginning with the fully oxidized form

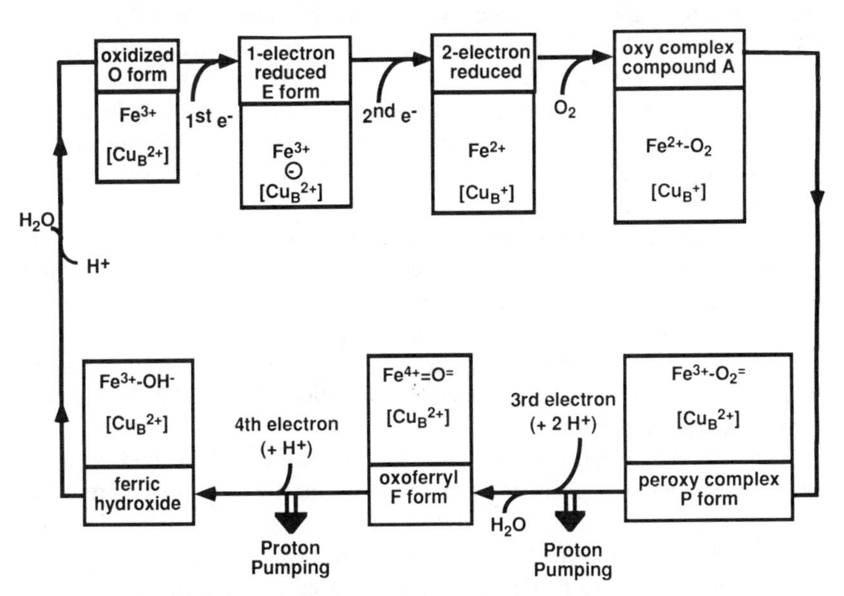

Figure 10 A representation of the sequence of species generated as oxygen is reduced to water at the binuclear center. The peroxy (P) form is not well characterized either as regards a potential bridging structure to Cu_B or its protonation state. All the proton pumping is thought to occur during the delivery of the last two electrons to the binuclear center (P to F and F to O).

of cytochrome c oxidase, focussing on the oxygenated intermediates at the binuclear center.

The first intermediate observed is the oxy complex with heme a_3. This has the same spectroscopic features as the oxygen complex with ferrous myoglobin, and indicates considerable superoxide character of the bound oxygen. Cu_B has no apparent effect on the properties of this first intermediate. The second intermediate observed is presumably a two-electron reduced form of oxygen, called the peroxy intermediate or P form. The structure of this intermediate is controversial and not well defined. Arrival of the third electron coincides with heterolytic cleavage of the oxygen-oxygen bond, yielding water and the heme a_3 oxoferryl intermediate, or F form, defined by resonance Raman spectroscopy. The fourth electron converts the oxoferryl form to the heme a_3 ferric hydroxide. Approximate rate constants for formation of the peroxy, oxoferryl, and fully oxidized forms of the enzyme are 2.5×10^4, 1.0×10^4, and 800 s^{-1}, respectively (233). The latter approximates the maximal turnover number of the oxidase.

PROTON TRANSFER REACTIONS For each molecule of oxygen reduced, eight protons are taken up from the bacterial cytoplasm or mitochondrial matrix, four of these are deposited in the periplasm or mitochondrial intermembrane space, and four are used to form water at the binuclear center (Figure 2*B*, *C*). Protons linked to electron transfer are taken up from the aqueous phase on one side of the membrane (234, 235), transferred within the oxidase (235), and expelled to the aqueous phase on the opposite side of the membrane (236–238). For the mitochondrial oxidase, the stoichiometry of one proton pumped per electron appears to be likely, although lower values have been reported (238, 239).

Cytochrome bo in sphaeroplasts pumps one proton per electron (110). As shown in Figure 2, cytochrome bo also delivers one substrate proton per electron to the periplasm, so under most circumstances the H^+/e^- ratio is 2. One unusual feature of cytochrome bo is that at alkaline pH, the oxidase becomes decoupled, and the H^+/e^- ratio drops to 1, presumably reflecting only the substrate protons (240). This is not observed with the mitochondrial oxidase.

Although the average H^+/e^- ratio for the mitochondrial oxidase is 1, proton pumping in each of the four electron-transfer steps in Figure 10 is unequal. Wikström (239) has deduced that no protons are pumped during the steps leading from the fully oxidized form of the oxidase to the peroxy form (O to E and E to P, Figure 10), and that each of the last two-electron transfers (P to F and F to O, Figure 10) is associated with pumping of two protons. These studies were based on measurements of the distribution of species of the bovine oxidase (P, F, and O) in intact energized mitochondria.

In flash photolysis experiments with purified bovine oxidase in phospholipid vesicles, proton pumping has been observed (236, 237) with a rate equivalent to the final step in the reoxidation of the enzyme (F to O, Figure 10). This and the preceding electron transfer step (P to F, Figure 10) are pH-dependent and show solvent deuterium isotope effects (233, 235), suggesting that proton transfer reactions are partially rate limiting for at least the last two electron transfer steps in the reoxidation of the bovine oxidase. This is not observed in reactions during the reoxidation of cytochrome bo (229) where, presumably, the last step is the conversion of the peroxy to the oxoferryl form (P to F, Figure 10). It is not clear whether this indicates a difference in mechanism or a subtle shift in the rate-limiting steps. Given the similarity between cytochrome bo and the cytochrome c oxidases, it is logical to assume that protons are pumped across the membrane by similar mechanisms. For an alternative view, see Ref. 241.

There are many questions and few hard answers concerning how protons are delivered to the oxygenated intermediates at the binuclear center, as well as how protons are pumped across the membrane. Mutagenesis data indicates that a limited number of amino acid residues are essential for these proton transfer reactions. Tyr288, His284, and Trp280 in helix VI and Thr352 in helix VIII are candidates for residues involved in proton transfer at the binuclear center. In the bacterial photosynthetic reaction center and cytochrome P450, analogous situations exist in which protons must be conveyed from the bulk solvent to a substrate sequestered within these proteins. In the reaction center, protons are delivered to Q_B via acidic residues to serine and histidine residues in contact with the bound quinone (242, 243). In the cytochrome P450s, it appears that protons are delivered via acidic residues to a threonine in contact with the oxygenated heme intermediate (244, 244a).

If the number of residues required for proton transfer reactions is small, then models requiring global conformational changes coupled to proton pumping are less likely to be correct. Although conformational changes in cytochrome c oxidase have been indicated, these have since been shown to be unlikely (232). Any model for proton pumping must minimally involve an input state where protons are taken up from one side of the membrane, and an output state where protons are delivered to the aqueous compartment on the opposite side of the membrane (245, 246). It is likely that the switch from the input to output states involves amino acid residues in the vicinity of the heme-copper binuclear center, connected via hydrogen bond networks to the aqueous compartments. A role for bound water(s) is very likely, as is the case for the internal proton transfer reactions in the bacterial photosynthetic reaction center (242, 243), in cytochrome P450 (244), and in bacteriorhodopsin (247).

Numerous models have been elaborated to describe proton pumping. Models involving Cu_A (163, 245) and the formyl group of heme a (248) are probably incorrect, since cytochrome bo, which lacks these prosthetic groups, pumps protons as efficiently as the mitochondrial oxidase. Proton pumping is probably directly coupled to the chemistry at the binuclear center, and Cu_B is likely to be a critical component in directing the vectorial movement of protons. Some current suggestions involve movement of protonated oxygen intermediates between Cu_B coordination sites facing different sides of the enzyme, or changes in the amino acids ligating the metals at the binuclear center (206, 249–251). Some of these models postulate changes in the coordination of Cu_B triggered by changes in the redox state of Cu_B. However, the current oxidase model (Figure 9) suggests that Cu_B need not function as an obligate intermediate in electron transfer from heme a to heme a_3. Hence, changes in the redox state of Cu_B during the proton-pumping steps (P to F and F to O, Figure 10) cannot be assumed, since direct electron transfer from heme a to heme a_3 appears more probable.

Clearly, additional data are needed to probe the proton transfer reactions. A first priority is to establish firmly the H^+/e^- stoichiometry associated with each step in the reaction. A study of why some of the mutants in the bacterial oxidases do not pump protons, or are inactive, despite having intact metal centers, should be revealing.

ACKNOWLEDGMENTS

We wish to thank Melissa W. Calhoun for help in preparing figures. Research from our laboratories has been supported by NIH grants GM 20379 (BLT) and HL16101 (RBG) and DOE grant DEFG02-87ER13716 (RBG).

Literature Cited

1. Trumpower BL. 1990. *Microbiol. Rev.* 54:101–29
2. Gennis RB, Barquera B, Hacker B, Van Doren SR, Arnaud S, et al. 1993. *J. Bioenerg. Biomembr.* 25:195–209
3. Mitchell P. 1976. *J. Theor. Biol.* 62:327–67
4. Trumpower BL. 1990. *J. Biol. Chem.* 265:11409–12
5. Crofts AR. 1985. In *The Enzymes of Biological Membranes,* ed. A Martonosi, 1:347–82. New York: Plenum
6. Howell N. 1989. *J. Mol. Evol.* 29:157–69
7. Esposti MD, De Vries S, Crimi M, Ghelli A, Patarnello T, Meyer A. 1993. *Biochim. Biophys. Acta* 1143: 243–71
8. Saraste M. 1990. *Q. Rev. Biophys.* 23:331–66
9. Rich PR, Moody J. 1994. In *Treatise on Bioelectrochemistry,* ed. P Gräber, Vol. 3. Basel: Birkhäuser Verlag. In press
10. Hosler JP, Ferguson Miller S, Calhoun

MW, Thomas JW, Hill J, et al. 1993. *J. Bioenerg. Biomembr.* 25:121–36
11. Hill J, Goswitz VC, Calhoun M, Garcia Horsman JA, Lemieux, L, et al. 1992. *Biochemistry* 31:11435–40
12. Puustinen A, Finel M, Haltia T, Gennis RB, Wikström M. 1991. *Biochemistry* 30:3936–42
13. Hill JJ, Alben JO, Gennis RB. 1993. *Proc. Natl. Acad. Sci. USA* 90:5863–67
14. De Vries S, Marres CAM. 1987. *Biochim. Biophys. Acta* 895:205–39
15. Clarkson AB, Bienen EJ, Pollakis G, Grady RW. 1989. *J. Biol. Chem.* 264:17770–76
16. Dutton PL. 1986. In *Encylcopedia of Plant Physiology, NS,* ed. LA Staehelin, CJ Arntzen, 19:197–237. Berlin: Springer-Verlag
17. Gabellini N, Sebald W. 1986. *Eur. J. Biochem.* 154:569–79
18. Yang X. 1988. *The purification and characterization of a three subunit ubiquinol-cytochrome c oxidoreductase complex from* Paracoccus denitrificans. PhD thesis. Dartmouth College, Hanover, NH
19. Widger WR, Cramer WA. 1991. In *Cell Culture and Somatic Cell Genetics of Plants: The Molecular Biology of Plastids and the Photosynthetic Apparatus,* ed. IK Vasil, L Bogorad, 7B: 149–76. Orlando: Academic
20. Malkin R. 1992. *Photosynth. Res.* 33: 121–36
21. Kutoh E, Sone N. 1988. *J. Biol. Chem.* 263:9020–26
22. Widger WR, Cramer WA, Herrmann RG, Trebst A. 1984. *Proc. Natl. Acad. Sci. USA* 81:674–78
23. Simpkin D, Palmer G, Devlin FJ, McKenna MC, Jensen GM, Stephens PJ. 1989. *Biochemistry* 28: 8033–39
24. Rigby SEJ, Moore GR, Gray JC, Gadsby PMA, George SJ, Thomson AJ. 1988. *Biochem. J.* 256:571–77
25. Trumpower BL. 1981. *Biochim. Biophys. Acta* 639:129–55
26. Beattie DS, Clejan L, Chen Y, Lin CP, Sidhu A. 1981. *J. Bioenerg. Bioemembr.* 13:357–72
27. Gutweniger H, Bisson R, Montecucco C. 1981. *J. Biol. Chem.* 256:11132–36
28. West IC, Mitchell P, Rich PR. 1988. *Biochim. Biophys. Acta.* 933:35–41
29. Crofts AR, Meinhardt SW, Jones KR, Snozzi M. 1983. *Biochim. Biophys. Acta.* 723:202–18
30. T'sai AL, Olson JS, Palmer G. 1983. *J. Biol. Chem.* 258:2122–25
31. Chance B. 1974. In *Dynamics of Energy-Transducing Membranes,* ed. L Ernster RW Estabrook, EC Slater, pp. 553–78. Amsterdam: Elsevier Science
32. Slater EC. 1973. *Biochim. Biophys. Acta* 301:129–54
33. Wikström MKF, Berden J. 1972. *Biochim. Biphys. Acta* 283:403–20
34. Mitchell P. 1975. *FEBS Lett.* 56:1–6
35. Trumpower BL, Edwards CA. 1979. *J. Biol. Chem.* 254:8697–706
36. Trumpower BL, Edwards CA, Ohnishi T. 1980. *J. Biol. Chem.* 255:7487–93
37. Edwards CA, Bowyer JR, Trumpower BL. 1982. *J. Biol. Chem.* 257:3705–13
38. Trumpower BL. 1976. *Biochem. Biophys. Res. Commun.* 70:73–79
39. Von Jagow G, Link TA. 1986. *Methods Enzymol.* 126:253–71
40. Von Jagow G, Gribble GW, Trumpower BL. 1986. *Biochemistry* 25:775–80
41. Von Jagow G, Ljungdahl PO, Graf P, Ohnishi T, Trumpower BL. 1984. *J. Biol. Chem.* 259:6318–26
42. Bowyer JR, Trumpower BL. 1981. *J. Biol. Chem.* 256:2245–51
43. Bowyer JR, Edwards CA, Trumpower BL. 1981. *FEBS Lett.* 126:93–97
44. Trumpower BL. 1981. *J. Bioenerg. Biomembr.* 13:1–24
45. Ohnishi T, Trumpower BL. 1980. *J. Biol. Chem.* 255:3278–84
46. De Vries S, Berden JA, Slater EC. 1980. *FEBS Lett.* 122:143–48
47. Dutton PL, Gunner MR, Robertson DE, Prince RC. 1983. *Photochem. Photobiol.* 37:566–67
48. Meinhardt SW, Yang X, Trumpower BL, Ohnishi T. 1987. *J. Biol. Chem.* 262:8702–6
49. De Vries S, Albracht SPJ, Berden JA, Slater EC. 1981. *J. Biol. Chem.* 256: 11996–98
50. Jin YZ, Tang HL, Li SL, Tsou CL. 1981. *Biochim. Biophys. Acta* 637:551–54
51. Tang HL, Jin YZ, Tsou CL. 1981. *Biochem. Int.* 3:327–32
52. Tang HL, Trumpower BL. 1986. *J. Biol. Chem.* 261:6209–19
53. Yu CA, Nagoaka S, Yu L, King TE. 1980. *Arch. Biochem. Biophys.* 204: 59–70
54. De Vries S, Albracht SPJ, Berden JA, Slater EC. 1982. *Biochim. Biophys. Acta* 681:41–53
55. Yue WH, Zou YP, Yu L, Yu CA. 1991. *Biochemistry* 30:2303–6
56. Berry EA, Huang LS, Earnest TN, Jap BK. 1992. *J. Mol. Biol.* 224:1161–66
57. Yu CA, Yu L. 1993. *J. Bioenerg. Biomembr.* 25:259–73
58. Karlsson BS, Hovmöller S, Weiss H,

Leonard K. 1983. *J. Mol. Biol.* 165:287–302
59. Gonzalez-Halphen DM, Lindorfer MA, Capaldi RA. 1988. *Biochemistry* 27:7021–31
60. Ohnishi TH, Schägger H, Meinhardt SW, Lobrutto R, Link TA, Von Jagow G. 1989. *J. Biol. Chem.* 264:735–44
61. Li Y, De Vries S, Leonard K, Weiss H. 1981. *FEBS Lett.* 135:277–80
62. Li Y, Leonard K, Weiss H. 1981. *Eur. J. Biochem.* 116:199–205
63. Di Rago JP, Perea J, Colson AM. 1986. *FEBS Lett.* 208:208–10
64. Brivet-Chevillotte P, Di Rago JP. 1989. *FEBS Lett.* 255:5–9
65. Daldal F, Tokito MK, Davidson E, Faham M. 1989. *EMBO J.* 8:3951–61
66. Coria RO, Garcia MC, Brunner AL. 1989. *Mol. Microbiol.* 3:1599–604
67. Howell N, Appel J, Cook JP, Howell B, Hauswirth WW. 1987. *J. Biol. Chem.* 262:2411–14
68. Di Rago JP, Perea J, Colson AM. 1990. *FEBS Lett.* 263:93–98
69. Di Rago JP, Colson AM. 1988. *J. Biol. Chem.* 263:12564–70
70. Di Rago JP, Netter P, Slonimski PP. 1990. *J. Biol. Chem.* 265:15750–57
71. Tron T, Lemesle-Meunier D. 1990. *Curr. Genet.* 18:413–19
72. Lemesle-Meunier D, Brivet-Chevillotte P, Di Rago JP, Slonimski PP, Bruel C, et al. 1993. *J. Biol. Chem.* 268:15626–32
73. Tokito MK, Daldal F. 1993. *Mol. Microbiol.* 9:965–78
74. Gennis RB, Barquera B, Hacker B, Van Doren SR, Arnaud S, et al. 1993. *J. Bioenerg. Biomembr.* 3:195–209
75. Colson AM. 1993. *J. Bioenerg. Biomembr.* 3:211–20
76. Crofts A, Robinson H, Andrews K, Van Doren S, Berry E. 1987. In *Cytochrome Systems: Molecular Biology and Bioenergetics*, ed. S Papa, B Chance, L Ernster, pp. 617–24. New York: Plenum
77. Brasseur R. 1988. *J. Biol. Chem.* 263:12571–75
78. Di Rago JP, Netter P, Slonimski PP. 1990. *J. Biol. Chem.* 265:3332–39
79. Di Rago JP, Hermann-LeDenmant S, Paques F, Netter P, Slonimski PP. 1994. *J. Mol. Biol.* Submitted
80. Graham LA, Brandt U, Sargent JS, Trumpower BL. 1993. *J. Bioenerg. Biomembr.* 25:245–57
81. Kuila D, Fee JA, Schoonover JR, Woodruff WH, Batie C, Ballou DP. 1987. *J. Am. Chem. Soc.* 109:1559–61
82. Gurbiel RJ, Batie CJ, Sivaraja M, True AE, Fee JA, et al. 1989. *Biochemistry* 28:4861–71
83. Britt RD, Sauer K, Klein MP, Knaff DB, Kriauciunas A, et al. 1991. *Biochemistry* 30:1892–901
84. Blumberg WE, Peisach J. 1974. *Arch. Biochem. Biophys.* 162:502–12
85. Graham LA, Trumpower BL. 1991. *J. Biol. Chem.* 266:22485–92
86. Davidson E, Ohnishi T, Atta-Asafo-Adjei E, Daldal F. 1992. *Biochemistry* 31:3342–51
87. Gonzalez-Halphen D, Vazquez Acevedo M, Garcia Ponce B. 1991. *J. Biol. Chem.* 266:3870–76
88. Van Doren SR, Yun CH, Crofts AR, Gennis RB. 1993. *Biochemistry* 32:628–36
89. Trumpower BL, Edwards CA. 1979. *J. Biol. Chem.* 254:8697–706
90. Hartl FU, Schmidt B, Wachter E, Weiss H, Neupert W. 1986. *Cell* 47:939–51
91. Bell RL, Sweetland J, Ludwig B, Capaldi RA. 1979. *Proc. Natl. Acad. Sci. USA.* 76:741–45
92. Gellerfors P, Nelson BD. 1977. *Eur. J. Biochem.* 80:275–82
93. D'Souza MP, Wilson DF. 1982. *J. Biol. Chem.* 257:11760–66
94. Beattie DS. 1993. *J. Bioenerg. Biomembr.* 25:233–44
95. Konishi K, Van Doren SR, Dramer DM, Crofts AR, Gennis RB. 1991. *J. Biol. Chem.* 266:14270–76
96. Theiler R, Niederman RA. 1991. *J. Biol. Chem.* 266:23163–68
97. Yu L, Chiang YL, Yu CA, King TE. 1975. *Biochim. Biophys. Acta* 379:33–42
98. Mukai K, Yoshida M, Yao Y, Wakabayashi Y, Matsubara H. 1988. *Proc. Jpn. Acad.* 64B:41–44
99. Nakai M, Ishiwatar H, Asad A, Bogati M, Kawai K, et al. 1990. *J. Biochem.* 108:793–803
100. Gray KA, Davidson E, Daldal F. 1992. *Biochemistry* 31:11864–73
101. Yu CA, Yu L, King TE. 1972. *J. Biol. Chem.* 247:1012–19
102. Simpkin D, Palmer G, Devlin FJ, McKenna MC, Jensen GM, Stephens PJ. 1989. *Biochemistry* 28:8033–39
103. Leonard K, Wingfield P, Arad T, Weiss H. 1981. *J. Mol. Biol.* 149:259–74
104. Schmitt ME, Trumpower BL. 1990. *J. Biol. Chem.* 265:17005–11
105. Cramer WA, Martinez SE, Huang D, Tae GS, Everly RM, et al. 1994. *J. Bioenerg. Biomembr.* 26:31–47
106. Anraku Y. 1988. *Annu. Rev. Biochem.* 57:101–32

107. Rice CW, Hempfling WP. 1978. *J. Bacteriol.* 134:115–24
108. Hill S, Viollet S, Smith AT, Anthony C. 1990. *J. Bacteriol.* 172:2071–78
109. Carter K, Gennis RB. 1985. *J. Biol. Chem.* 260:10986–90
110. Puustinen A, Finel M, Haltia T, Gennis RB, Wikström M. 1991. *Biochemistry* 30:3936–42
111. Anraku Y, Gennis RB. 1987. *Trends Biochem. Sci.* 12:262–66
112. Miller MJ, Hermodson M, Gennis RB. 1988. *J. Biol. Chem.* 263:5235–40
113. Green NG, Fang H, Lin R-J, Newton G, Mather M, et al. 1988. *J. Biol. Chem.* 263:13138–43
114. Calhoun MW, Oden KL, Gennis RB, Teixeira de Mattos MJ, Neijssel OM. 1993. *J. Bacteriol.* 175:3020–25
115. Miller MJ, Gennis RB. 1985. *J. Biol. Chem.* 260:14003–8
116. Dueweke TJ, Gennis RB. 1991. *Biochemistry* 30:3401–6
117. Georgiou CD, Dueweke TJ, Gennis RB. 1988. *J. Biol. Chem.* 263:13130–37
118. Newton G, Yun C-H, Gennis RB. 1991. *Mol. Microbiol.* 5:2511–18
119. Dueweke TJ, Gennis RB. 1990. *J. Biol. Chem.* 265:4273–77
120. Timkovich R, Cork MS, Gennis RB, Johnson PY. 1985. *J. Am. Chem. Soc.* 107:6069–75
121. Sotiriou C, Chang CK. 1988. *J. Am. Chem. Soc.* 110:2264–70
122. Meinhardt SW, Gennis RB, Ohnishi T. 1989. *Biochim. Biophys. Acta* 975: 175–84
123. Rothery R, Ingledew WJ. 1989. *Biochem. J.* 262:437–43
124. Fang H, Lin R-J, Gennis RB. 1989. *J. Biol. Chem.* 264:8026–32
125. Hill JJ, Alben JO, Gennis RB. 1993. *Proc. Natl. Acad. Sci. USA* 90:5863–67
126. Kahlow MA, Loehr TM, Zuberi TM, Gennis RB. 1993. *J. Am. Chem. Soc.* 115:5845–46
127. Kahlow MA, Zuberi TM, Gennis RB, Loehr TM. 1991. *Biochemistry* 30: 11485–89
128. Lorence RM, Gennis RB. 1989. *J. Biol. Chem.* 264:7135–40
129. Varotsis C, Zhang Y, Appelman EH, Babcock GT. 1993. *Proc. Natl. Acad. Sci. USA* 90:237–41
130. Han S, Ching Y-c, Rousseau DL. 1990. *J. Am. Chem. Soc.* 112:9445–551
131. Han S, Ching Y-c, Rousseau DL. 1990. *Nature* 348:89–90
132. Babcock GT, Varotsis C. 1993. *J. Bioenerg. Biomembr.* 25:71–80
133. Hosler JP, Ferguson-Miller S, Calhoun MW, Thomas JW, Hill J, et al. 1993. *J. Bioenerg. Biomembr.* 25: 121–36
134. Brown S, Moody AJ, Mitchell R, Rich PR. 1993. *FEBS Lett.* 316:216–23
135. Capaldi RA. 1990. *Annu. Rev. Biochem.* 59:569–96
136. Takamiya S, Lindorfer MA, Capaldi RA. 1987. *FEBS Lett.* 218:277–82
137. Capaldi RA. 1990. *Arch. Biochem. Biophys.* 280:252–62
138. Quirk PG, Hicks DB, Krulwich TA. 1993. *J. Biol. Chem.* 268:678–85
139. Fukaya M, Tayama K, Tamaki T, Ebisuya H, Okumura H, et al. 1993. *J. Bacteriol.* 175:4307–14
140. Wachenfeldt CV, Hederstedt L. 1992. *FEMS Microbiol. Lett.* 100:91–100
141. Raitio M, Jalli T, Saraste M. 1987. *EMBO J.* 6:2825–33
142. Lübben M, Kolmerer B, Saraste M. 1992. *EMBO J.* 11:805–12
143. Ishizuka M, Machida K, Shimada S, Mogi A, Tsuchiya T, et al. 1990. *J. Biochem.* 108:866–73
144. Bott M, Preisig O, Hennecke H. 1992. *Arch. Microbiol.* 158:335–43
145. Bott M, Bolliger M, Hennecke H. 1990. *Mol. Microbiol.* 4:2147–57
146. Preisig O, Anthamatten D, Hennecke H. 1993. *Proc. Natl. Acad. Sci. USA* 90:3309–13
147. Gabel C, Maier RJ. 1990. *Nucleic Acids Res.* 18:6143
148. Santana M, Kunst F, Hullo MF, Rapoport G, Danchin A, Glaser P. 1992. *J. Biol. Chem.* 267:10225–31
149. Mather MW, Springer P, Hensel S, Buse G, Fee JA. 1993. *J. Biol. Chem.* 268:5395–408
150. Shapleigh JP, Gennis RB. 1992. *Mol. Microbiol.* 6:635–42
151. Cao J, Hosler J, Shapleigh J, Revzin A, Ferguson-Miller S, 1992. *J. Biol. Chem.* 267:24273–78
151a. Cao J, Shapleigh J, Gennis R, Revzin A, Ferguson-Miller S. 1991. *Gene* 101:133–37
152. Haltia T, Puustinen A, Finel M. 1988. *Eur. J. Biochem.* 172:543–46
153. Hosler JP, Fetter J, Tecklenburg MMJ, Espe M, Lerma C, Ferguson-Miller S. 1992. *J. Biol. Chem.* 267:24264–72
154. Haltia T, Saraste M, Wikström M. 1991. *EMBO J.* 10:2015–21
155. Hendler RW, Pardhasaradhi K, Reynafarje B, Ludwig B. 1991. *Biophys. J.* 60:415–23
156. Gregory LC, Ferguson-Miller S. 1988. *Biochemistry* 27:6307–14
157. Finel M, Wikström M. 1986. *Biochim. Biophys. Acta* 851:99–108

158. Brunori M, Antonini G, Malatesta F, Sarti P, Wilson MT. 1987. *Eur. J. Biochem.* 169:1–8
158a. Thompson D, Ferguson-Miller S. 1983. *Biochemistry* 22:3178–87
158b. Thompson D, Gregory L, Ferguson-Miller S. 1985. *J. Inorg. Biochem.* 23:357–64
159. Weishaupt A, Kadenbach B. 1992. *Biochemistry* 31:11477–81
160. Lightowlers R, Chrzanowska-Lightowlers Z, Marusich M, Capaldi RA. 1991. *J. Biol. Chem.* 266:7688–93
161. Minghetti KC, Goswitz VC, Gabriel NE, Hill JJ, Barassi C, et al. 1992. *Biochemistry* 31:6917–24
162. Sone N, Shimada S-I, Ohmori T, Souma Y, Gonda M, Ishizuka M. 1990. *FEBS Lett.* 262:249–52
163. Chan SI, Li PM. 1990. *Biochemistry* 29:1–12
164. Nicholls P. 1993. *FEBS Lett.* 327:194–98
165. Nicholls P, Butko P. 1993. *J. Bioenerg. Biomembr.* 25:137–43
166. Hill BC. 1993. *J. Bioenerg. Biomembr.* 25:115–20
167. Pan L-P, Hazzard JT, Lin J, Tollin G, Chan SI. 1991. *J. Am. Chem. Soc.* 113:5908–10
168. Chepuri V, Lemieux L, Au DC-T, Gennis RB. 1990. *J. Biol. Chem.* 265:11185–92
168a. Millett F, De Jong K, Paulson L, Capaldi R. 1983. *Biochemistry* 22:546–52
168b. Taha T, Ferguson-Miller S. 1992. *Biochemistry* 31:9090–97
169. Lauraeus M, Haltia T, Saraste M, Wikström, M. 1991. *Eur. J. Biochem.* 197:699–705
170. Puustinen A, Wikström M. 1991. *Proc. Natl. Acad. Sci. USA* 88: 6122–26
171. Wu W, Chang CK, Varotsis C, Babcock GT, Puustinen A, Wikström M. 1992. *J. Am. Chem. Soc.* 114: 1182–87
172. Lauraeus M, Wikström M. 1993. *J. Biol. Chem.* 268:11470–73
173. Saraste M, Metso T, Nakari T, Jalli T, Lauraeus M, van der Oost J. 1991. *Eur. J. Biochem.* 195:517–25
174. Gadsby PMA, Thomson AJ. 1990. *J. Am. Chem. Soc.* 112:5003–11
175. Carter K, Palmer G. 1982. *J. Biol. Chem.* 257:13507–14
176. Martin CT, Scholes CP, Chan SI. 1985. *J. Biol. Chem.* 260:2857–61
177. Salerno JC, Bolgiano B, Poole RK, Gennis RB, Ingledew WJ. 1990. *J. Biol. Chem.* 265:4364–68
178. Cheesman MR, Watmough NJ, Pires CA, Turner R, Brittain T, et al. 1993. *Biochem. J.* 289:709–18
179. Stevens TH, Chan SI. 1981. *J. Biol. Chem.* 256:1069–71
180. Ogura T, Hon-Nami K, Oshima T, Yoshikawa S, Kitagawa T. 1983. *J. Am. Chem. Soc.* 105:7781–82
181. Ingledew WJ, Horrocks J, Salerno JC. 1994. *Eur. J. Biochem.* In press
182. Cline J, Reinhammar B, Jensen P, Venters R, Hoffman BM. 1983. *J. Biol. Chem.* 258:5124–28
183. Powers L, Chance B, Ching Y, Angiolillo P. 1981. *Biophys. J.* 34:465–98
184. Li PM, Gelles J, Chan SI, Sullivan RJ, Scott RA. 1987. *Biochemistry* 26: 2091–95
185. Ingledew WJ, Bacon M. 1991. *Biochem. Soc. Trans.* 19:613–16
186. Moody AJ, Cooper CE, Rich PR. 1991. *Biochim. Biophys. Acta* 1059: 189–207
187. Surerus KK, Oertling WA, Fan C, Gurbiel RJ, Einarsdóttir O, et al. 1992. *Proc. Natl. Acad. Sci. USA* 89:3195–99
188. Kelly M, Lappalainen P, Talbo G, Haltia T, van der Oost J, Saraste M. 1993. *J. Biol. Chem.* 268:16781–87
189. Kroneck PMH, Antholine WE, Kastrau DHW, Buse G, Steffens GCM, Zumft WG. 1990. *FEBS Lett.* 268:274–76
190. Palmer G. 1993. *J. Bioenerg. Biomembr.* 25:145–51
191. Malmström BG, Aasa R. 1993. *FEBS Lett.* 325:49–52
192. Antholine WE, Kastrau DHW, Steffens GCM, Buse G, Zumft WG, Kroneck PMH. 1992. *Eur. J. Biochem.* 209: 875–81
193. Bombelka E, Richter F-W, Stroh A, Kadenbach B. 1986. *Biochem. Biophys. Res. Commun.* 140:1007–14
194. Steffens GCM, Biewald R, Buse G. 1987. *Eur. J. Biochem.* 164:295–300
195. van der Oost J, Pappalainen P, Musacchio A, Warne A, Lemieux L, et al. 1992. *EMBO J.* 11:3209–17
196. Goodman G, Leigh JS Jr. 1985. *Biochemistry* 24:2310–17
197. Goodman G, Leigh JS Jr. 1987. *Biochim. Biophys. Acta* 890:360–67
198. Ohnishi T, LoBrutto R, Salerno JC, Bruckner RC, Frey TG. 1982. *J. Biol. Chem.* 257:14821–25
199. Thomson AJ, Greenwood C, Gadsby PMA, Peterson J, Eglinton DG, et al. 1985. *J. Inorg. Biochem.* 23:187–97
200. Nicholls P, Wrigglesworth JM. 1988. *Ann. NY Acad. Sci.* 550:59–67
201. Blair DF, Ellis J, Walther R, Wang

H, Gray HB, Chan SI. 1986. *J. Biol. Chem.* 261:11524–37
202. Cooper CE, Salerno JC. 1992. *J. Biol. Chem.* 267:280–85
203. Watmough NJ, Cheesman MR, Gennis RB, Greenwood C, Thomson AJ. 1993. *FEBS Lett.* 319:151–54
203a. Baker G, Noguchi M, Palmer G. 1987. *J. Biol. Chem.* 262:595–604
203b. Brudvig G, Stevens T, Morse R, Chan S. 1981. *Biochemistry* 20:3912–21
204. Day EP, Peterson J, Sendova MS, Schoonover J, Palmer G. 1993. *Biochemistry* 32:7855–60
205. Moody AJ, Rumbley JN, Gennis RB, Ingledew WJ, Rich PR. 1993. *Biochim. Biophys. Acta* 1141:321–29
206. Woodruff WH. 1993. *J. Bioenerg. Biomembr.* 25:177–88
207. Woodruff WH, Einarsdóttir O, Dyer RB, Bagley KA, Palmer G, et al. 1991. *Proc. Natl. Acad. Sci. USA* 88:2588–92
208. Fiamingo FG, Altschuld RA, Alben JO. 1986. *J. Biol. Chem.* 261:12976–87
209. Li W, Palmer G. 1993. *Biochemistry* 32:1833–43
210. Schoonover JR, Palmer G. 1991. *Biochemistry* 30:7541–50
211. Tsubaki M, Yoshikawa S. 1993. *Biochemistry* 32:164–73
212. Tsubaki M, Mogi T, Anraku Y, Hori H. 1993. *Biochemistry* 32:6065–72
213. Calhoun MW, Gennis RB, Salerno JC. 1992. *FEBS Lett.* 309:127–29
214. Berka V, Vygodina T, Musatov A, Nicholls P, Konstantinov AA. 1993. *FEBS Lett.* 315:237–41
215. Blackmore RS, Greenwood C, Gibson QH. 1991. *J. Biol. Chem.* 266:19245–49
216. Chepuri V, Gennis RB. 1990. *J. Biol. Chem.* 265:12978–86
217. Ma J, Lemieux L, Gennis RB. 1993. *Biochemistry* 32:7692–97
218. Lemieux LJ, Calhoun MW, Thomas JW, Ingledew WJ, Gennis RB. 1992. *J. Biol. Chem.* 267:2105–13
219. Minagawa J, Mogi T, Gennis RB, Anraku Y. 1992. *J. Biol. Chem.* 267:2096–104
220. Shapleigh JP, Hosler JP, Tecklenburg MMJ, Kim Y, Babcock GT, et al. 1992. *Proc. Natl. Acad. Sci. USA* 89:4786–90
221. Meunier B, Coster F, Lemarre P, Colson AM. 1993. *FEBS Lett.* 321:159–62
222. Haltia T. 1992. *Biochim. Biophys. Acta* 1098:343–50
223. Thomas JW, Lemieux LJ, Alben JO, Gennis RB. 1993. *Biochemistry* 32:11173–80
224. Thomas JW, Puustinen A, Alben JO, Gennis RB, Wikström M. 1993. *Biochemistry* 32:10923–28
225. Oliveberg M, Malmström BG. 1991. *Biochemistry* 30:7053–57
226. Brown S, Rumbley JN, Moody AJ, Thomas JW, Gennis RB, Rich PR. 1993. *Biochim. Biophys. Acta.* In press
227. Oliveberg M, Malmström BG. 1992. *Biochemistry* 31:3560–63
228. Svensson M, Nilsson T. 1993. *Biochemistry* 32:5442–47
229. Hallén S, Svensson M, Nilsson T. 1993. *FEBS. Lett.* 325:299–302
230. Hill BC, Greenwood C. 1984. *Biochem. J.* 218:913–21
231. Babcock GT, Wikström M. 1992. *Nature* 301–9
231a. Nicholls P. 1992. *Biochem. J.* 288:1070–72
232. Mitchell R, Brown S, Mitchell P, Rich PR. 1992. *Biochim. Biophys. Acta* 1100:40–48
233. Oliveberg M, Brzezinski P, Malmström BG. 1989. *Biochem. Biophys. Acta* 977:322–28
234. Mitchell R, Mitchell P, Rich PR. 1992. *Biochim. Biophys. Acta* 1101:188–91
235. Hallén S, Nilsson T. 1992. *Biochemistry* 31:11853–59
236. Nilsson T, Hallén S, Oliveberg M. 1990. *FEBS Lett.* 260:45–47
237. Oliveberg M, Hallén S, Nilsson T. 1991. *Biochemistry* 30:436–40
238. Antonini G, Malatesta F, Sarti P, Brunori M. 1993. *Proc. Natl. Acad. Sci. USA* 90:5949–53
239. Wikström M. 1989. *Nature* 338:776–78
240. Verkhovskaya M, Verkhovsky M, Wikström M. 1992. *J. Biol. Chem.* 267:14559–62
241. Musser SM, Stowell MHB, Chan SI. 1993. *FEBS Lett.* 327:131–36
242. Takahashi E, Wraight CA. 1991. *Biochemistry* 31:855–66
243. Rongey SH, Paddock ML, Feher G, Okamura MY. 1993. *Proc. Natl. Acad. Sci. USA* 90:1325–29
244. Ravichandran K, Boddupalli SS, Hasemann CA, Peterson JA, Deisenhofer J. 1993. *Science* 261:731–36
244a. Gerber N, Sligar S. 1992. *J. Am. Chem. Soc.* 114:8742–43
245. Gelles J, Blair DF, Chan SI. 1987. *Biochim. Biophys. Acta* 853:205–36
246. Malmström BG. 1993. *Acc. Chem. Res.* 26:332–38
247. Papadopoulos G, Dencher NA, Zaccai

G, Büldt G. 1990. *J. Mol. Biol.* 214:15–19

248. Babcock GT, Callahan PM. 1983. *Biochemistry* 22:2314–19
249. Larsen RW, Pan L-P, Musser SM, Li Z, Chan SI. 1992. *Proc. Natl. Acad. Sci. USA* 89:723–27
250. Rich PR. 1991. *Biosci. Rep.* 11:539–71
251. Rousseau DL, Ching Y-c, Wang J. 1993. *J. Bioenerg. Biomembr.* 25:165–76
252. Sone N, Fujiwara Y. 1991. *FEBS Lett.* 288:154–58

Annu. Rev. Biochem. 1994. 63:717–43

CONTROL OF RNA INITIATION AND ELONGATION AT THE HIV-1 PROMOTER

Katherine A. Jones
The Salk Institute for Biological Studies, Regulatory Biology Laboratory and Molecular Biology & Virology Laboratory, La Jolla, California 92037

B. Matija Peterlin
Howard Hughes Medical Institute, Department of Medicine and Department of Microbiology and Immunology, University of California, San Francisco, California 94143

KEY WORDS: HIV-1 promoter, transcription elongation, TAT, TAR, RNA-protein interactions

---

CONTENTS

INTRODUCTION

The human immunodeficiency virus (HIV) is a highly pathogenic lentivirus that causes acquired immune deficiency syndrome (AIDS) (1–3). Since the first sequence analyses of HIV-1 were reported in 1985 (4–6), a wealth of

0066-4154/94/0701-0717$05.00

information has accrued on the steps that regulate viral gene expression in infected cells. HIV-1 infection is characterized by a lengthy period of proviral latency during which few viral transcripts or regulatory proteins are produced. Much of this regulation is transcriptional, exerted through the viral promoter located in the leftward long terminal repeat (LTR). Most notably, RNA polymerase II (RNA pol II) transcription complexes that are initiated from the viral promoter early in infection terminate RNA synthesis prematurely through a process of attenuation or abortive elongation. HIV-1 encodes six regulatory proteins that control different aspects of the viral life cycle, from virus entry and reverse transcription to transcription of the integrated provirus and nuclear transport of unspliced viral RNAs. One of the earliest regulatory proteins, Tat (7–9), is an essential nuclear protein that is targeted to the viral promoter through a nascent RNA signal to activate transcription. Tat strongly increases the efficiency or processivity of elongation by RNA polymerase II, and can also affect the RNA initiation rate (10–13). In this article, we consider the effects of host cell DNA- and RNA-binding proteins on the assembly of transcription complexes at the viral promoter, and the mechanism of *trans*-activation by Tat. Additional relevant information can be obtained from recent reviews on HIV-1 transcription and *trans*-activation by Tat (14–25), the general RNA pol II transcription factors (26–30), and transcription termination and anti-termination in HIV as well as other systems (31–34).

THE HIV-1 PROMOTER

Distal Enhancer Sequences

The HIV-1 promoter, located in the U3 region of the viral long terminal repeat (LTR), is highly responsive to mitogenic stimulation and to the HIV *trans*-activator, Tat. The core promoter and enhancer, containing the TATA box and binding sites for Sp1 and NF-κB, spans a region of only 250 nucleotides (Figure 1). Although sequences upstream of the NF-κB sites contribute only marginally to HIV-1 promoter activity in vitro or in transfected cells, a recent study suggests that upstream sequences between positions −130 and −201 are nevertheless important for viral replication (or proviral promoter activity) in peripheral blood lymphocytes (PBLs) and in some T cell lines (35). This region of the promoter includes the binding sites for two activators that are highly enriched in T cells: LEF, a lymphocyte-specific high mobility group (HMG) protein that is found in immature B and T cells and in mature T cells (36, 37); and the thymocyte-enriched Ets-1 protein (C Sheline, M Voz, K Jones, unpublished). In addition, this region contains an E-box motif bound by USF (20: see Figure 1). As with

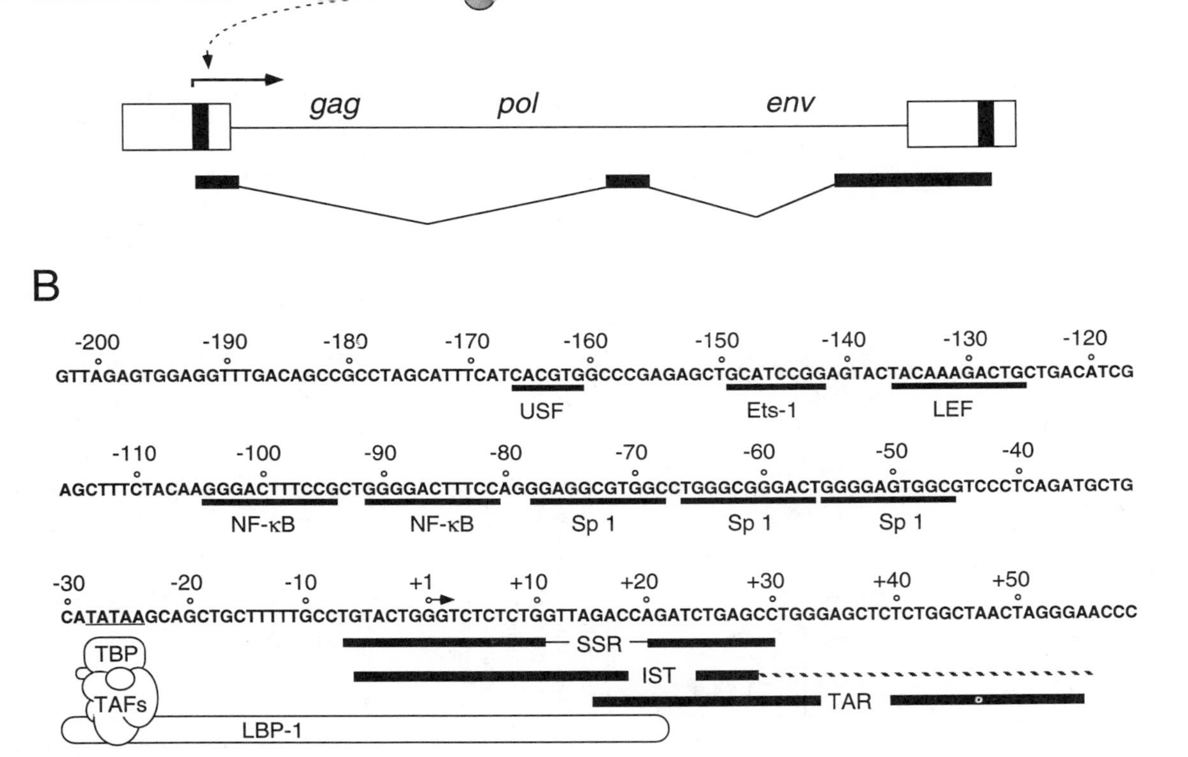

Figure 1 The HIV-1 genome and promoter structure. (*A*) Schematic representation of the 9-kb HIV-1 genome, which encodes Gag, Pol, and Env, as well as six regulatory proteins Tat, Rev, Nef, Vif, Vpr, and Vpu. Tat-1 is synthesized from several multiply spliced viral transcripts, one of which is depicted beneath the genome map. The first exon of *tat*-1, located between the *pol* and *env* genes, is most important for function. Because of the redundancy in the LTR, the TAR RNA structure is present at both ends of all viral transcripts, although Tat-1 functions only through the leftward TAR element (black box below the arrow in the LTR). (*B*) Nucleotide sequence of the HIV-1 LTR from positions -203 to +59. This region contains the enhancer (underlined sites for DNA-binding proteins USF, Ets-1, LEF, and NF-κB), the promoter (underlined Sp1 and TATA box), and the 3' elements (bars denote SSR or initiator element, IST or initiator of short transcripts, TAR element, and LBP-1 site). TBP and TAFs assemble transcription complexes with the help of upstream and downstream DNA-binding proteins.

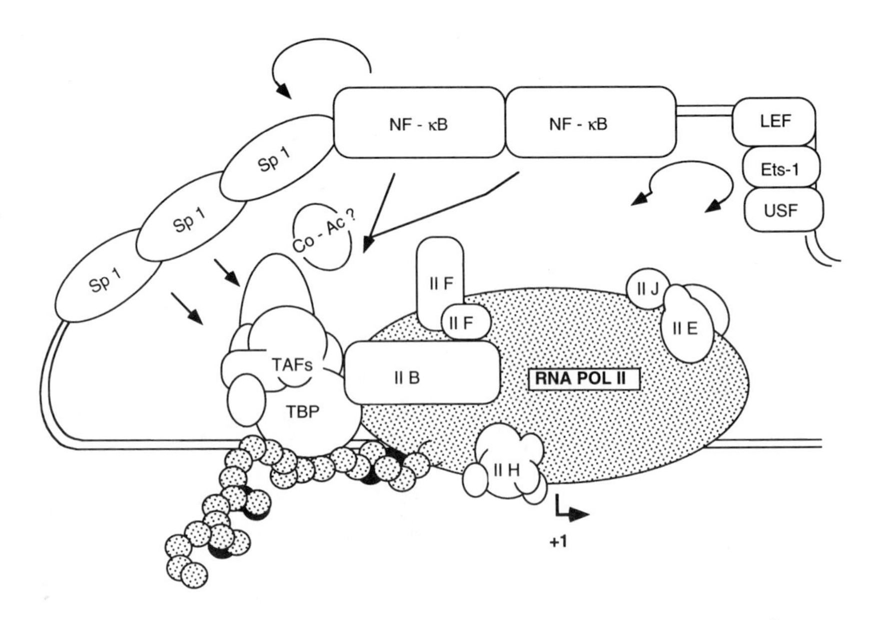

Figure 2 Model for transcriptional activation of the HIV-1 promoter by upstream factors. Initiation of HIV-1 transcription depends upon TBP and TAFs, and promoter factors Sp1 and NF-κB. The LEF, Ets-1, and USF proteins are thought to be important for promoter activity of the integrated provirus. Upstream factors and Sp1 help recruit basal transcription factors (TFIIA-J and RNA polymerase II) to the promoter and stabilize the preinitiation complex. The TATA box and SSR (or initiator) element dictate the site of initiation of HIV-1 transcription. This model for promoter activation is based on that of Drapkin et al (28).

many HMG proteins, LEF generates a strong (130°) bend in the DNA that may influence the local structure of the promoter dramatically (38; Figure 2). LEF has a potent transcription activation domain that can function independently of its HMG box (39). Interestingly, the LEF activation domain is preferentially active in T cells and is strongly influenced by the context of its binding site, indicating that it may act in concert with other T cell–specific proteins such as Ets-1. Transient expression assays indicate that overexpression of LEF strongly activates the HIV-1 promoter (C Sheline, T Mayall, K Jones, unpublished data). This part of the HIV-1 promoter is reminiscent of the enhancer for the alpha chain of the T cell receptor, which also binds LEF and Ets-1 proteins (38).

It is striking that a strong contribution of the LEF-, Ets-1-, and USF-binding sites is evident only in replication assays that monitor transcriptional activity of the integrated provirus. These proteins may function together in a complex to ensure that the promoter is retained in an open configuration, or to counter the repressive effects of sequences flanking the integration site. Further studies are needed to ascertain whether other transcription factors that bind to the remaining upstream sequences in the HIV-1 promoter (such as NFAT, AP-1, COUP, and c-Myb; 20) are also important for proviral promoter activity or virus replication in T cells.

T Cell Activation Signals and NF-κB

The HIV-1 enhancer contains tandem binding sites for NF-κB, an important activator of the cellular acute-phase and immune responses (for recent reviews, see Refs. 43–46). The HIV-1 NF-κB sites can be essential for virus replication (47, 48). NF-κB-binding activity is conferred by a heterodimer of p50 and p65 subunits that are related to the c-Rel proto-oncogene (Figure 3). The p50 subunit arises from proteolytic processing of a larger precursor molecule, p105. In many cells, NF-κB activity is tightly controlled

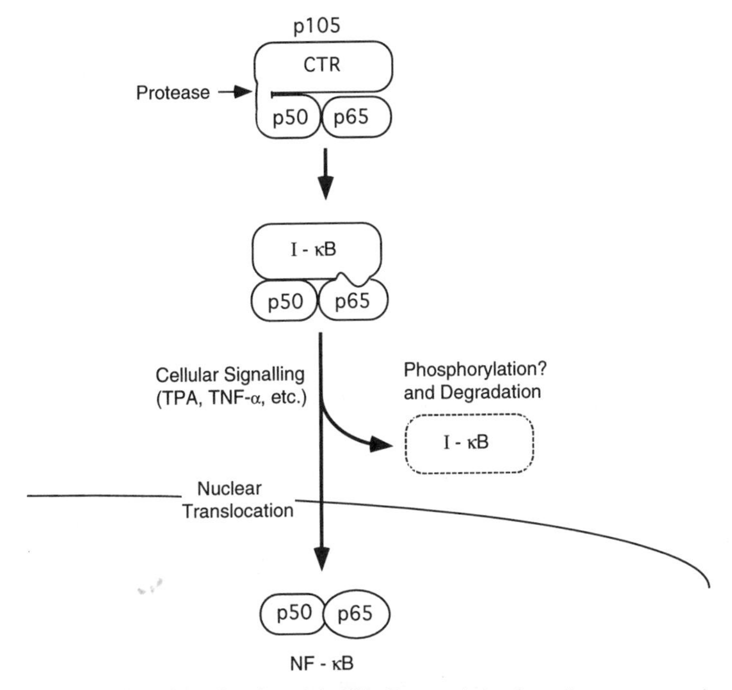

Figure 3 Assembly and translocation of the NF-κB transcription factor in response to mitogenic stimuli. The precursor of the p50 subunit (p105) contains several ankyrin repeats in its C-terminal region (CTR), and associates with p65 in the cytoplasm. An as-yet-unknown protease cleaves p105 into p50 and p70 subunits, and until it is further degraded, p70 (the CTR) also keeps the p50:p65 heterodimer in the cytoplasm. Upon proteolysis, p70 is replaced by I-κB, which binds p65 and masks its nuclear localization signal. Upon cellular activation, I-κB is phosphorylated, dissociates from p50:p65, and is rapidly degraded. For review, see (46).

through specific binding of p65 to an inhibitory molecule, I-κB. The I-κB:NF-κB interaction masks the nuclear localization signal of p65, causing NF-κB to be retained in an inactive protein complex in the cytoplasm. Formation of the I-κB:NF-κB complex is mediated by interactions between the Rel homology domain of NF-κB, which contains the nuclear localization signal and the sequences that mediate DNA-binding and dimerization of the NF-κB subunits, and multiple ankyrin repeats that are found on I-κB.

While NF-κB activity is constitutively high in mature B cells and in some T cell lines, most cells lack constitutive NF-κB-binding activity. In these cells, NF-κB is strongly induced by mitogens, cytokines, oxygen radicals, or other agents. T cell activation causes dissociation of the I-κB:NF-κB complex and subsequent rapid, albeit transient, degradation of I-κB (49, 50). Dissociation of the I-κB:NF-κB complex seems to be initiated through phosphorylation of I-κB (and possibly also NF-κB) by mitogen-stimulated kinases. Once released from I-κB, the NF-κB subunits are free to move into the nucleus and activate transcription.

The response of any given κB element depends not only on the source of the stimulus and the relative abundance of different I-κB molecules, but also on different members of the NF-κB/Rel family, which vary in their transcriptional activities. The HIV-1 κB elements are bound by heterodimers of p50/p65 and p50/RelB, as well as homodimers of p50, which form a relatively low-activity complex called KBF-1 (46). The stronger activity of the p50/p65 heterodimer is due to the presence of a C-terminal transcription activation domain on the p65 subunit that is not found on p50. In addition, the amino terminus of the p65 Rel-homology domain can bind directly to the TATA-binding protein (TBP) present in the holo-TBP:TAF complex (51). Therefore NF-κB might simultaneously contact Sp1 and the basal transcription machinery on the HIV-1 promoter (Figure 2).

NF-κB activity is strongly induced by HIV-1 infection (52). This induction is sustained and is not sensitive to protein kinase C inhibitors, indicating that the mechanism is distinct from that induced by mitogens. High levels of NF-κB could arise from degradation of p105 by viral proteases (53, 54), changes in the redox potential (55–58), decreased expression of cell surface receptor tyrosine phosphatases (CD45) (59), and the expression of the viral Nef protein in infected cells (60). In addition, the p50 promoter is activated directly by NF-κB (61, 62), which could establish a self-sustaining loop to maintain high levels of NF-κB. Although all of these mechanisms are viable, to date only reducing agents and free radical scavengers have been shown to decrease virus replication in infected cells (58, 63, 64).

The Basal Promoter: Sp1, TATA, and the Initiator

The core region of the HIV-1 promoter contains three tandem Sp1-binding sites located upstream of a canonical RNA polymerase II TATA box sequence and a nonconventional initiator element (called SSR, or start-site region—Figure 1). These three elements are essential for minimal promoter activity in vitro and in vivo (65–67). The TATA box is recognized by the TATA-binding protein (TBP) acting in conjunction with 8–12 cellular TAFs (TBP-associated factors; Refs. 68–70). Sp1 activity in reconstituted transcription systems requires the TBP:TAF complex (68), and Sp1 binds directly to the TAF-110 protein (71, 72). In addition, Sp1 can interact with other DNA-bound proteins, including distal Sp1 monomers (73), YY-1 (74, 75), and NF-κB (G Nabel, personal communication). These Sp1-mediated protein interactions may stabilize and enhance the formation of preinitiation complexes.

The TATA box and Sp1-binding sites are essential for HIV-1 promoter activity, and insertions that separate these two elements inhibit *trans*-activation (76). The TATA box appears to be especially important for Tat *trans*-activation, since these insertions do not affect basal promoter activity in vivo. Likewise, some mutations or replacements of the HIV-1 TATA box have been observed to inhibit Tat *trans*-activation without affecting basal promoter activity (77–79). With the caveat that basal promoter activity in vivo is low and difficult to quantitate, these studies suggest that proteins associated with the HIV-1 TATA box are particularly important for *trans*-activation by Tat.

The HIV-1 initiator or SSR element, which is important for basal transcription, cannot functionally substitute for the better-characterized initiator element of the TdT promoter in vitro (67). Whereas the TdT initiator is recognized specifically by the holo-TBP:TAF complex (S Smale, personal communication), the proteins responsible for HIV-1 initiator activity have not yet been identified unambiguously. Several different proteins bind to the HIV-1 TATA and initiator region in vitro, including LBP-1 (80–82), LBP-2 (83), USF/TFII-I (84, 85), YY-1 (74, 75), HIP116 (86), and TDP (87). Although LBP-1 is the predominant protein that binds to the SSR, its binding specificity does not correlate with initiator activity (67). Whether the protein responsible for SSR activity is another promoter-specific factor or a component of the basal transcription complex remains an open question.

Downstream Elements: LBP-1, IST, and TAR

Sequences near the viral RNA start site contain at least three different elements: SSR, IST (inducer of short transcripts) (88, 89), and TAR (*trans*-activation response element; Figure 1). The TAR element is a nascent RNA transcript that directs Tat to the promoter. The IST element (88, 89) is a DNA sequence closely overlapping TAR (−5 to +80), which mediates synthesis of short attenuated RNAs that accumulate in transfected or infected cells in the absence of Tat (11–13, 90, 91). These short viral RNAs range in length from 55–60 nucleotides (HIV-1) to 125 nucleotides (HIV-2), and include the TAR structure. Mutations in IST interfere specifically with the synthesis of short RNAs and do not affect full-length RNA transcripts (89), and fusion of IST to heterologous promoters confers the ability to synthesize short RNAs in vivo (88). Thus a protein bound to IST might recruit or stabilize a form of RNA pol II that elongates inefficiently.

One proposed role of the IST protein is to promote the synthesis of TAR-containing transcripts that recruit Tat to the promoter. However, accumulation of short TAR-containing RNAs also eventually dampens *trans*-activation, by binding Tat and sequestering it from the promoter. LBP-1 does not appear to mediate IST activity (89), and the responsible factor has yet to be identified. Another potential candidate for the IST factor is a recently cloned 43-kDa protein, TDP, which binds to CT-rich sequences overlapping the LBP-1 site. However, TDP does not appear to affect Tat *trans*-activation in vivo, although it does inhibit basal promoter activity (R Gaynor, personal communication). The CT-rich elements bear intriguing similarity to sequences in the hsp70 promoter that are bound by the *Drosophila* GAGA factor, which plays a role in attenuation of hsp70 transcription (92), and so it will be interesting to learn whether the IST-binding protein also binds to cellular promoters that are controlled by attenuation.

Although LBP-1 binds very strongly to downstream sequences on the HIV-1 and HIV-2 promoters, its role in viral transcription is not clearly defined. Whereas binding of LBP-1 to a weak site that overlaps the HIV-1 TATA box is inhibitory to transcription (81), binding of LBP-1 to higher-affinity downstream sites can activate transcription in vitro (80). Since the LBP-1 site is not required for *trans*-activation by Tat, it may act earlier in the virus life cycle. Interestingly, a significant change in the local chromatin structure near the LBP-1 site occurs upon activation of cells with TPA (93, 94). This change does not require ongoing transcription (94), and may reflect a reconfiguration of proteins that bind to this part of the promoter. The role of LBP-1 may therefore be more clearly revealed by in vitro studies that use chromatin-assembled viral DNA templates.

TRANSCRIPTION ACTIVATION BY TAT THROUGH TAR

Domain Structure of Tat

Of the known lentiviruses, HIV-1, HIV-2, SIV (simian immunodeficiency virus), and EIAV (equine infectious anemia virus), but not visna virus or CAEV (caprine arthritis and encephalitis virus), encode Tat proteins that activate transcription through leader TAR RNA structures found in all viral

A

```
EIAV   MADRRIPGTAEENLQKSSGGVPGQNTGGQEARPN..........................YHCQLC
HIV-1  MEPVDPRLEPWKHPGSQPKTA............................CTT.CYCKKCCFHCQVC
HIV-2  METPLKAPESSLKSCNEPFSRTSEQDVATQELARQGEEILSQLYRPLETCNNSCYCKRCCYHCQMC
       ————————— N - TERM ——————————————————————   ——— CYS-RICH ———

EIAV   FL.RSLGIDYLDASLRKKNKQRLKAIQQGRQPQYLL
HIV-1  FITKALGISYG....RKKRRQRRR....PPQGSQTHQVSLSKQPTSQPRGDPTGPKE
HIV-2  FLNKGLGICYE....RKGRRRRTPKKTKTHPSPTPDKSISTRTGDSQPTKKQKKTVEATVETDTGPGR
       — CORE —  ——— BASIC ——  ——————— C-TERM ———————————————
```

B

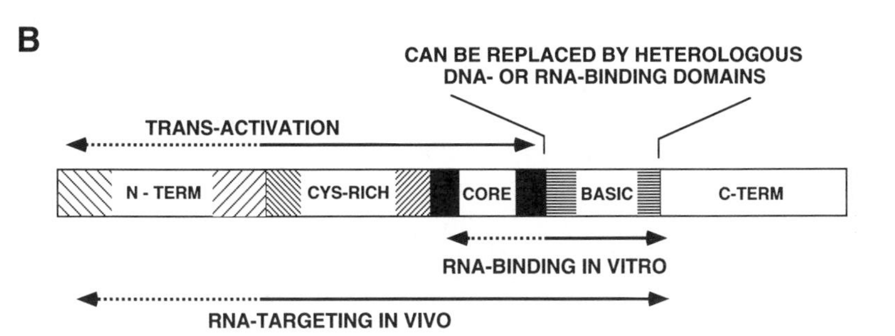

Figure 4 Sequence and structural domains of the lentiviral Tat proteins. (*A*) Amino acid sequences of E-Tat, Tat-1 (HIV1.Bru), and Tat-2 (HIV2.Rod) are aligned for greatest similarities. (*B*) Structural domains of Tat-1, organized as proposed previously (201, 202). Essential and most conserved domains are represented with darker boxes. Sequence from positions +1 to +48, which contain N-terminal, cysteine-rich, and core domains, form the Tat-1 activation domain. Activation domain mutations have been described previously (119–124). The N-terminal 72 amino acids are sufficient for optimal *trans*-activation, and substantial levels of *trans*-activation are observed with shorter fragments of Tat that contain the first 60 amino acids (202, 203).

transcripts (for review, see Refs. 20, 21, 95, 96). The HIV-1 Tat protein (Tat-1) is expressed early in infection from three different multiply spliced viral mRNAs (97). In addition, Tat-1 is expressed as a hybrid Tat-Rev molecule, called Tev, which is encoded by a transcript that splices the first exon of Tat to the middle of the *env* gene and the second exon of Rev (98–101). The various Tat proteins range in size from 130 amino acids (Tat-2), to 82–101 amino acids (Tat-1), to only 75 amino acids (for EIAV Tat; Figure 4). Interestingly, the HIV and SIV Tats are localized to the nucleus or nucleolus, whereas the EIAV Tat protein (E-Tat) is found predominantly in the cytoplasm (102).

Trans-activation by the different Tat proteins is highly species-specific. Thus, although Tat-1 and Tat-2 activate well in primate cells, they work very poorly in rodent cells (103, 104). Likewise, E-Tat, which activates transcription in canine cell lines, cannot *trans*-activate in primate cells (105–107). Moreover, *trans*-activation through the different TAR elements is not always reciprocal. Thus, although the full-length Tat-1 and Tat-2 proteins can *trans*-activate through either TAR-1 or TAR-2 RNA (108), a truncated Tat-2 protein that lacks the second exon (C-terminal domain) functions well only through TAR-2 RNA (109–111). Neither Tat-1 nor Tat-2 can *trans*-activate through the EIAV TAR element (E-TAR), and, conversely, E-Tat does not recognize the HIV TAR structures (102, 105–107, 112).

Domain-swapping experiments suggest that both nuclear localization of the HIV Tat proteins and their ability to recognize a given TAR element are specified by basic domain sequences that also mediate binding of Tat to TAR RNA in vitro. First, the basic domain of Tat-1 contains a functional nuclear localization signal (NLS) (113–115). Second, *trans*-activation by Tat-1 is inhibited by conservative amino acid changes in the basic domain that interfere with its ability to bind TAR in vitro but do not affect its nuclear localization (116–118). Finally, fusion of the basic domain of Tat-1 to E-Tat confers the ability to activate transcription through TAR-1 (105). E-Tat contains only six of the nine residues that are found in the basic domains of Tat-1 or Tat-2, as well as four additional amino acids that separate the basic and core domains (Figure 4). These differences in E-Tat could be responsible for its cytoplasmic retention, or, alternatively, its NLS might be masked in the cell. A small amount of E-Tat probably enters the nucleus by diffusion to mediate *trans*-activation. However, *trans*-activation by E-Tat is not yet characterized in great detail, and it is possible that the mechanisms of *trans*-activation by the HIV and EIAV Tat proteins may differ significantly.

The region most highly conserved among the various Tat proteins is the 10-amino-acid "core" domain (FITKALGISY) and a flanking cysteine-rich domain, both of which are critical for activity (Figure 4). By contrast, E-Tat contains only two of the seven Cys residues found in Tat-1 and Tat-2. Nevertheless, a minimal 25-amino-acid lentiviral Tat (containing the E-Tat core domain and Tat-1 basic domain) weakly activates the HIV-1 promoter (112), indicating that the core and basic domains are sufficient for a minimal response through TAR-1 in vivo.

Mutagenesis and domain-swapping experiments indicate that the core, Cys-rich, and amino-terminal regions of Tat constitute an independent *trans*-activation domain (119, 120). Thus, fusion of the upstream region of Tat-1 to the RNA-binding domain of the bacteriophage MS2 coat protein generates a chimeric Tat:MS2 activator that is capable of activating a hybrid HIV-1 promoter in which TAR has been replaced with the MS2 coat protein RNA-binding site (121). Similarly, hybrid Tat:Rev proteins *trans*-activate through the Rev RNA-binding domain (122, 123), and Tat:GAL4 fusion proteins activate transcription through upstream DNA-binding sites, although activation in the latter case requires at least six upstream GAL4 DNA-binding sites as well as the Sp1 sites in the promoter (124–126). In all of these cases, mutations that affect *trans*-activation by the wild-type Tat protein inhibit activation by the chimeric Tat proteins, indicating that the underlying mechanisms are similar.

Structure of TAR RNA

Transfer of the TAR element (−17 to +81) to heterologous promoters confers responsiveness of these promoters to Tat, although the extent of induction varies depending on the level of basal promoter activity (66, 127, 128). In this respect, TAR behaves as a *cis*-acting enhancer element whose principal role is to recruit Tat to the promoter. As would be expected for a sequence-specific RNA element, TAR is active only in the sense orientation. However, unlike conventional DNA enhancer elements that function well from distal downstream locations, TAR becomes inactivated when moved more than 88 nucleotides downstream of its normal location in the leader. Moreover, multimerization of the TAR sequence does not lead to synergistic activation by Tat (128). Expression of exogenous TAR-containing RNAs (TAR decoys) interferes with *trans*-activation by Tat, and the TAR decoy RNAs cannot act in *trans* to restore the activity of a promoter deleted of TAR (129–132). Therefore the TAR element cannot function in *cis* from locations distal to the promoter, nor can it act in *trans* to recruit Tat to the promoter.

Several independent lines of evidence suggest that TAR is recognized as an RNA element. First, mutations that affect the folding of the TAR hairpin inhibit *trans*-activation, and TAR function can be restored by compensatory changes that repair the TAR stem-loop structure. Second, *trans*-activation by Tat is not affected by competition with TAR DNA, whereas it is specifically inhibited by synthetic TAR RNAs. The TAR decoy RNAs inhibit *trans*-activation by Tat and diminish viral replication in infected cells (129–132). Third, *trans*-activation by Tat is selectively inhibited by insertion 5′ of TAR of an "antisense" sequence that does not affect the primary sequence of TAR but interferes with its folding in the RNA (133, 134). Similarly, *trans*-activation is inhibited by insertion of a self-cleaving ribozyme sequence that cuts the RNA immediately downstream of TAR (135). Finally, Tat does not bind to DNA but can bind specifically to a highly conserved residue (U23) in the bulge of TAR RNA (17–19). Taken together, these studies suggest that Tat sees TAR as a nascent RNA structure.

The HIV-1 TAR RNA hairpin structure spans nucleotides +1 to +59, and deletion studies have established that the region from +19 to +42 incorporates the minimal domain that is both necessary and sufficient to support significant levels of *trans*-activation by Tat in vivo (66, 127, 128). This region includes the upper stem, a tetranucleotide bulge, and a structured, hexanucleotide loop (Figure 5*A*). The sequence of the RNA appears to be

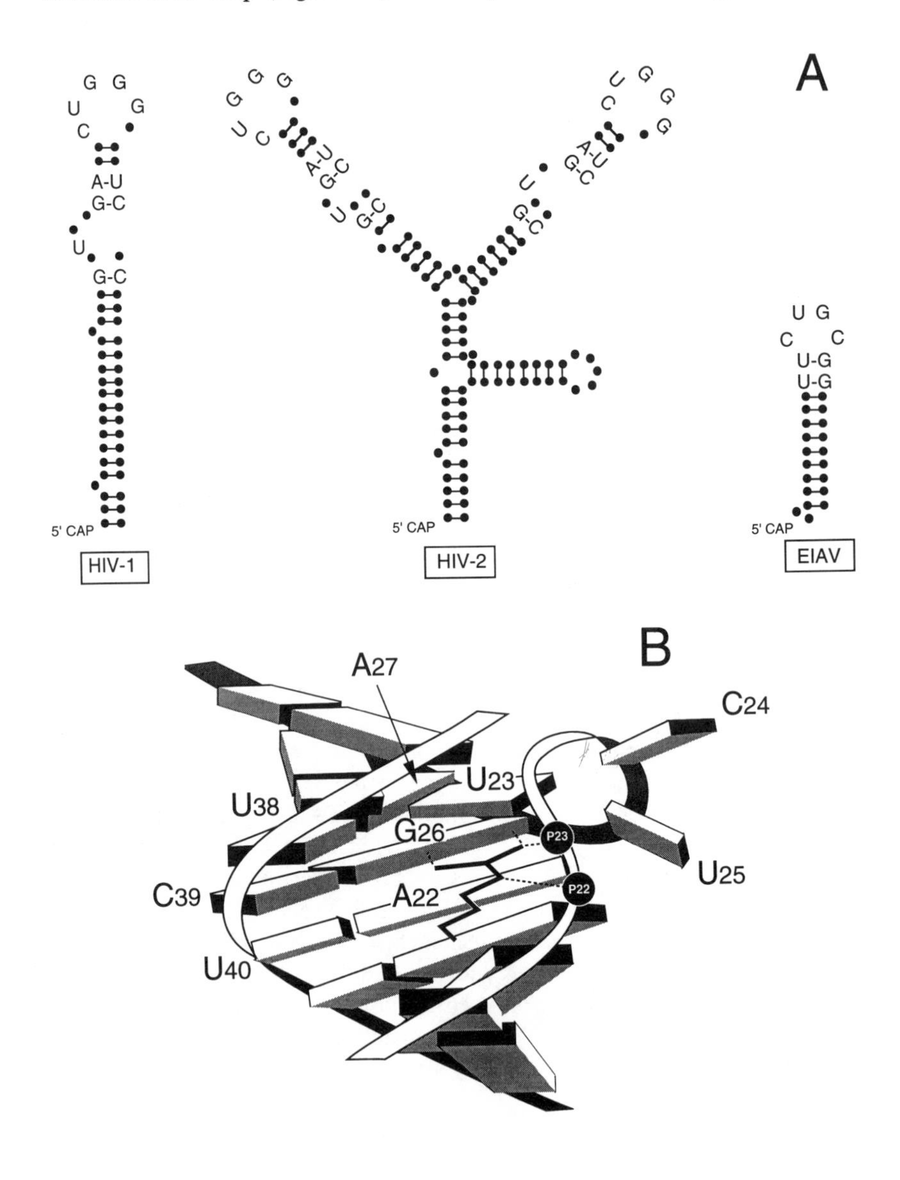

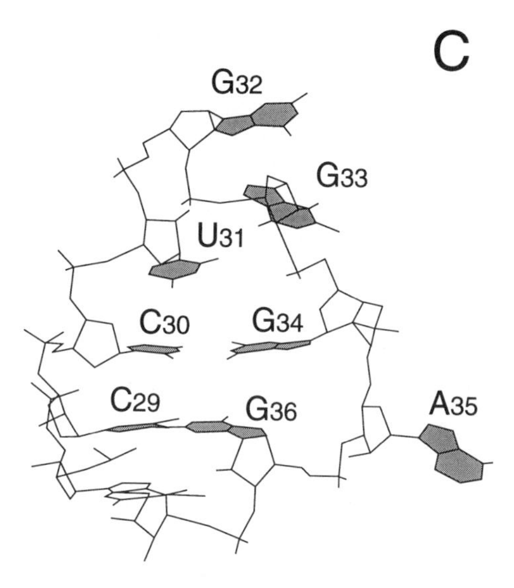

Figure 5 Secondary and tertiary structures of lentiviral TARs. (*A*) Predicted secondary structures of TARs of HIV-1, HIV-2, and EIAV (TAR-1, TAR-2, and E-TAR) according to the method of Zuker & Stiegler (205). The HIV TARs form very stable structures: TAR-1 DG = −37.6 kcal/mol; TAR-2 DG = −80.5 kcal/mol. E-TAR is the smallest TAR (DG = −16 kcal/mol) and does not contain a 5′ bulge. Important contacts for binding of Tat are shown in bold. (*B*) Proposed tertiary structure of the bulge of HIV-1 TAR RNA upon binding of arginine; the structure shown is adapted from Puglisi et al (158). (*C*) Proposed tertiary structure of the loop of TAR-1, including the putative C30-G34 basepair and A35 bulged residue. The structure shown is derived from Critchley et al (153).

as important as its structure, since point mutations in the loop of TAR and, to a lesser extent, in the bulge, inhibit *trans*-activation in vivo (128, 133, 134, 136–141). Although certain point mutations in the loop (e.g. C30-G) are highly detrimental to *trans*-activation, several variations in the loop sequence occur naturally among different HIV isolates, including C30-U (CDC41/RF), U31-C (MAL/OY1/ANT-70), and G33-A (Z2Z6; 142). By contrast, residue U23 in the bulge seems to be universally conserved (77, 142). While the relative activity of these different TAR structures has not been determined directly, these studies suggest that certain loop mutations are tolerated by Tat (143).

The TAR element of the HIV-2 promoter is larger than that of HIV-1 and is composed of tandem hairpin structures (Figure 5*A*). The TAR-2 loop sequence is identical to that of TAR-1, and each TAR-2 stem includes a bulge structure with an unpaired uridine. As mentioned above, Tat-1 functions effectively through either TAR-1 or TAR-2, whereas a version of

Tat-2 that lacks the second exon works well only on its cognate TAR-2 structure (109–111). This nonreciprocity has been attributed to a relatively low affinity of the truncated Tat-2 protein for TAR-1 in vitro (144–146). E-TAR, which is not recognized by either Tat-1 or Tat-2, forms a short hairpin that includes a four-residue loop but lacks a bulge (147). Point mutations affecting G14 in the E-TAR loop, or the two UG basepairs in the stem, dramatically reduce *trans*-activation by E-Tat in vivo (106). Presently it is unclear whether E-Tat binds directly to its TAR RNA, or whether it is targeted to the promoter solely through interactions with a cellular TAR RNA-binding protein.

Studies of the solution structure of TAR-1 indicate that the bulge induces a bend in the RNA helix that distorts the local structure and widens the major groove of the RNA to expose hydrogen-bonding contacts that are important for binding of Tat (138, 148–150; Figure 5*B*). Bending is diminished by the addition of extra bases in the stem opposing the bulge (151, 152). Chemical reactivity studies and NMR data indicate that residues A22 and U40 in the lower stem are unpaired, and consequently the bulge actually contains four, rather than three, unpaired nucleotides (148, 153). NMR and circular dichroism studies further suggest that U23 is stacked on A22 in the absence of Tat, and that the conformation of the bulge changes upon binding of arginine (used to model the Tat:TAR interaction; 154–157). Binding causes unstacking of the bulge nucleotides and repositioning of U23 near A27, potentially forming a base triple with AU27 in the upper stem (Figure 5*B*). The upper and lower stems are then proposed to stack coaxially to create a binding pocket for arginine (156, 158).

Further analysis of the TAR structure suggests that loop residues C30 and G34 may be basepaired (148, 153). This would generate a compact loop containing only three unpaired nucleotides (U31, G32, and G33) and a single bulged residue (A35; Figure 5*C*). However, NMR data suggest that C30 and G34 do not basepair (159), but instead are involved in base-stacking with the upper stem (159). The possibility of a C30:G34 basepair is especially important because point mutations affecting either residue strongly inhibit *trans*-activation by Tat (128, 133, 134, 141). Additional NMR studies to examine the structure of the loop of TAR RNA, currently under way in several laboratories, should help to further elaborate the structure of this critical region of TAR.

Recognition of Tat by TAR in vivo and in vitro

Tat-1 forms a 1:1 complex with HIV-1 TAR RNA in vitro. Although Tat does not bind to TAR DNA, it can bind RNA-DNA hybrids provided that the bulge and flanking 5′ residues are presented as RNA (160). Tat-1 binds TAR-1 principally through electrostatic interactions between the basic domain of Tat and negatively charged phosphates surrounding the RNA bulge, and binding is further stabilized by hydrogen-bonding interactions in the major groove of the RNA (18, 19). Mutations affecting U23 are most detrimental to binding (136–138, 154, 155, 161–163). The number of nucleotides in the bulge can be increased without affecting binding (137, 164). Removal of the N7 position of G26 and A27 reduces Tat binding, as does methylation of U23, confirming that Tat-1 forms essential hydrogen bond contacts in the major groove of TAR RNA (164, 165).

Tat-1 binds to TAR-1 in vitro through its basic domain (136–138, 154, 155, 163), which can be functionally replaced with "scrambled" peptides or basic peptides of unrelated sequence (149, 154, 155). Nevertheless, significant differences are observed among the TAR RNA-binding affinities and specificities of arginine, short basic peptides, and larger segments of the Tat protein (166, 167; reviewed in Ref. 19). In particular, the core domain also contributes significantly to the specificity of the Tat-TAR interaction (167). Most importantly, the specificity of targeting of Tat to TAR in vivo may depend in large part upon a cellular factor that binds to the loop of TAR (see below). Thus a definitive picture of the Tat:TAR recognition complex must await further studies that use both the intact Tat protein and the cellular TAR RNA-binding protein.

Binding of Cellular Factors to TAR RNA

That a cellular factor must bind to the loop of TAR RNA to promote TAR recognition by Tat is supported both by genetic and biochemical experiments. The strongest genetic evidence derives from studies on the species-specificity of *trans*-activation. Tat-1 *trans*-activates well in primate cells but only poorly in rodent or Chinese hamster ovary (CHO) cells. *Trans*-activation by Tat-1 in CHO cells is at least partially restored in human-hamster hybrid cell lines containing human chromosome 12 (104, 168). Two lines of evidence suggest that this species-specific protein facilitates the binding of Tat to TAR. First, the low level of activation by Tat seen in CHO cells is not affected by mutations in the loop of TAR, whereas the higher level of activity in cells containing chromosome 12 is loop-dependent (169; A Alonso, BM Peterlin, unpublished). Second, the Tat:MS2 fusion protein is active in CHO cells even in the absence of chromosome 12 (170). Taken together, these data suggest that chromosome 12 encodes a protein(s) necessary for Tat to function through TAR. Such a factor could bind directly to the loop of TAR and stabilize the Tat-TAR interaction, or it could modify Tat or cellular factors bound to TAR RNA.

Most importantly, neither the bulge sequences of TAR nor the basic domain of Tat is sufficient to target Tat-1 to TAR RNA in vivo. The involvement of a cellular factor was first suggested by the observation that binding of Tat does not require sequences in the loop of TAR, despite the fact that these are critical for *trans*-activation in vivo. Two recent studies provide complementary genetic evidence on this point. First, amino acid insertions that separate the basic domain of Tat from the rest of the activation domain inhibit *trans*-activation in vivo, indicating that the basic region does not act as an independent RNA-binding domain (171). In the second study, TAR was used to direct a Tat:Rev fusion protein to a Rev-dependent reporter system (123). Targeting of the Tat:Rev fusion protein to TAR in vivo required not only the basic domain of Tat-1, but also the Cys-rich, the core, and the amino-terminal sequences (123). Consequently, the nine-amino-acid basic region of Tat-1 is not an independent promoter-targeting domain. Instead, all 60 amino-terminal residues of Tat-1, which contain the entire transcription activation domain, are necessary for a stable interaction between Tat and TAR in the cell. The activation domain could form additional direct contacts with TAR RNA, or it could mediate protein-protein interactions with a cellular factor bound at the loop.

Direct evidence for a cellular loop-binding protein comes from biochemical studies of *trans*-activation in HeLa nuclear extracts (172–175). *Trans*-activation by Tat is TAR dependent in vitro, and can be competed specifically by synthetic wild-type TAR, but not by mutant TARs containing base substitutions in the loop (173, 174). Depletion of nuclear extracts with TAR RNA affects *trans*-activation without affecting basal transcription, indicating that TAR is not bound by a general transcription factor, but rather by a nuclear protein specifically required for *trans*-activation by Tat. Most importantly, *trans*-activation in these TAR-depleted extracts is not restored by Tat but is restored by the addition of nuclear proteins (173, 174) (M Garber, K Jones, unpublished). That TAR decoys diminish *trans*-activation by Tat and viral infectivity, but do not appear to affect cellular viability, suggests that the relevant cellular TAR RNA-binding protein is not a general transcription factor or a protein essential for cell growth.

A number of different cellular proteins bind to synthetic TAR RNA in gel mobility shift assays. These proteins include nonspecific duplex RNA-binding proteins as well as factors that can recognize specifically the lower stem of TAR, or specific residues in either the bulge or the loop (139, 174, 176–182). One of these proteins, called TRP-185 (or TRP-1), binds specifically to the loop of TAR-1 RNA in vitro (139, 174). Binding of TRP-185 is affected by mutations in the loop or by mutations that increase the distance between the loop and the bulge, both of which inhibit Tat *trans*-activation in vivo. However, a protein similar to TRP-185 is present in CHO cells (139, 174; A Alonso, BM Peterlin, unpublished data). The recent isolation of cDNA clones encoding TRP-185 (R Gaynor, personal communication) should reveal whether it is the factor that recruits Tat to

TAR. If not, functional assays based on complementation of TAR-depleted transcription extracts will be required to identify the cellular RNA loop-binding activity.

Cellular Factors that Interact with Tat

Overexpression of the Tat-1 activation domain (either as the wild-type Tat, Tat fusion proteins, or truncated Tat proteins that lack the basic domain) can inhibit or squelch *trans*-activation by either the wild-type Tat-1 or by the hybrid Tat:MS2 and Tat:Rev proteins. Similarly, overexpression of the activation domain of E-Tat squelches *trans*-activation by Tat-1. These findings suggest that the Tat-1 activation domain may bind a cellular transcription factor (or a "coactivator") in the absence of TAR RNA. Several cellular proteins have been identified that bind to Tat in vitro, including TBP (the TATA-binding protein) (183), Sp1 (184), a putative ATPase and DNA helicase called TBP-1 (185), an unknown 36-kDa nuclear factor that binds avidly to Tat-1 affinity columns (186), and a 42-kDa kinase (187). A specific interaction of Tat with TBP and Sp1 might explain its effects on RNA initiation, although loading of Tat onto nascent RNA would require it to "reach-back" to preinitiation complexes at the promoter to stimulate RNA initiation. Moreover, overexpression of the Tat activation domain inhibits *trans*-activation but does not seem to affect basal promoter activity, suggesting that the putative "coactivator" is not a general transcription factor. The potential involvement of a DNA helicase (TBP-1) is of interest, particularly since TBP-1:GAL4 fusion proteins are strong activators of certain promoters in vivo, and since an ATPase and DNA helicase could explain the sensitivity of *trans*-activation to 5,6-dichloro-1-beta-D-ribofuranosyl benzimidazole (DRB) (188). However, overexpression of TBP-1 suppresses rather than enhances Tat-1 *trans*-activation in vivo, and although a human homolog of TBP-1 (called MSS-1) could modestly stimulate Tat activation in P19 embryonic carcinoma cells (189), to date no biochemical evidence has been presented to suggest that TBP-1 or MSS-1 binds specifically to the Tat activation domain in solution or is essential for *trans*-activation by Tat.

The possible involvement of a kinase in Tat activation (187) is especially intriguing because of earlier observations highlighting the importance of protein kinase C in Tat-1 *trans*-activation in vivo (190), and because *trans*-activation by Tat in vitro is highly sensitive to DRB (172), which is a potent kinase inhibitor (191). Mutations in any part of the Tat-1 activation domain interfered with binding of the 42-kDa kinase, and the kinase was also shown to bind to the activation domain of Tat-2 (187). Paradoxically, however, the kinase does not appear to recognize the full-length Tat-1 protein, which should associate with the coactivator since it can squelch *trans*-activation in vivo. Additional studies are needed to determine whether the kinase affects *trans*-activation by Tat in vivo or in vitro, and, if so, to identify potential targets for the kinase on the RNA pol II transcription complex.

Mechanism of Trans-*activation by Tat*

Taken together, the studies cited above suggest a two-step model for the binding of Tat-1 to TAR-1 RNA (Figure 6). Because high levels of the Tat activation domain inhibit or squelch *trans*-activation in vivo, even in the absence of TAR, it appears that Tat may associate with a limiting cellular coactivator (or cofactor) prior to its binding to TAR RNA (for review, see Refs. 20–23). Binding of Tat-1 to the coactivator is mediated by sequences in the activation domain, and the squelching data further suggest that the activation domains of both Tat-1 and Tat-2 interact with the same cellular coactivator. In the model shown in Figure 6, the Tat:Coactivator complex is recruited to nascent TAR RNA through its affinity for the cellular loop-binding protein. This ternary ribonucleoprotein complex can be stabilized by contacts between the cellular factor and the loop of TAR RNA as well as by interactions between the basic domain of Tat-1 and the bulge of TAR. The ternary complex may also be stabilized by interactions between the coactivator and the cellular loop-binding factor, which would explain

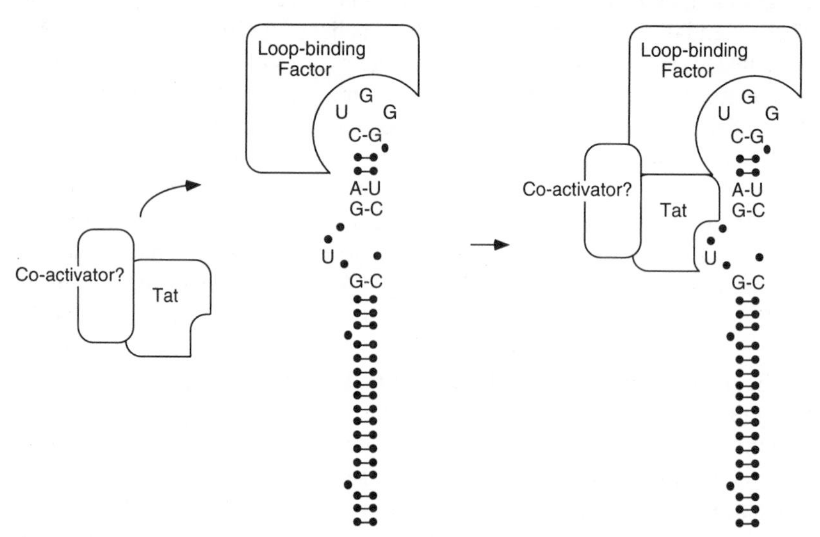

Figure 6 Hypothetical ribonucleoprotein complex of Tat-1, TAR-1, and cellular factors. The model that best fits the genetic and biochemical data suggests that Tat-1 interacts with TAR-1 with the help of a primate cell-specific TAR loop-binding protein. Tat-1 brings to this ribonucleoprotein complex a cellular coactivator.

why targeting of Tat-1 to TAR RNA in vivo requires the entire Tat-1 activation domain in addition to the basic domain. Although Tat-1 may further stabilize the ternary complex by interacting directly with the cellular loop-binding protein, it is equally possible that contact between these two proteins is indirect and bridged by the coactivator. In the latter case, the loop-binding protein would not interact directly with Tat and could not recruit Tat to the promoter in the absence of the coactivator; instead, all three proteins would be needed to create a functional TAR RNA complex.

An alternative model for TAR recognition invokes the idea that a single protein carries out the activities of both the loop-binding factor and the coactivator (122). For a single cellular factor to carry out both TAR recognition and transcription activation, two restrictions must be placed on its properties. First, to explain the species-specificity of *trans*-activation, the human, rodent, and canine or equine versions of this protein would each need to have distinct RNA-binding specificities but identical transcription activation domains. Second, to explain the need for Tat, it would be necessary to invoke that the coactivator binds to TAR-1 RNA only in the presence of the Tat-1 protein. In this scenario, binding of Tat and the coactivator to TAR RNA is highly cooperative, and neither protein should have significant affinity for TAR RNA in the absence of the other. However, in vitro experiments suggest that this may not be the case, because TAR decoys compete for *trans*-activation (172–174), and the TAR loop-binding protein can be removed efficiently from HeLa nuclear transcription extracts in the absence of Tat (M Garber, K Jones, unpublished data). Thus the cellular TAR RNA-binding and transcriptional coactivator proteins are probably distinct. As such, the genetic data predict that the cellular RNA-binding protein is species-specific and encoded or modified by a gene product expressed from human chromosome 12, but that the coactivator protein is ubiquitously distributed and is able to interact with each of the Tat proteins. Because the TAR loop-binding factor does not participate in *trans*-activation by the Tat:MS2, Tat:Rev, or Tat:GAL4 fusion proteins, its role appears to be to recruit Tat (and the presumptive coactivator) to the promoter. Transcriptional activation would then carried out by the coactivator or the Tat:Coactivator protein complex.

Nuclear run-on experiments have established that the HIV-1 promoter is regulated by attenuation (11–13, 90, 91; Figure 7). The accumulation of paused polymerases that contain nascent TAR transcripts may promote rapid binding and *trans*-activation by Tat-1. However, in the absence of Tat, these transcripts can accumulate and down-regulate *trans*-activation by binding and sequestering Tat and the cellular loop-binding factor (Figure 7*A*). Regulation by attenuation is similar in many respects to that of other highly inducible eukaryotic genes such as the hsp70 promoter, which contains

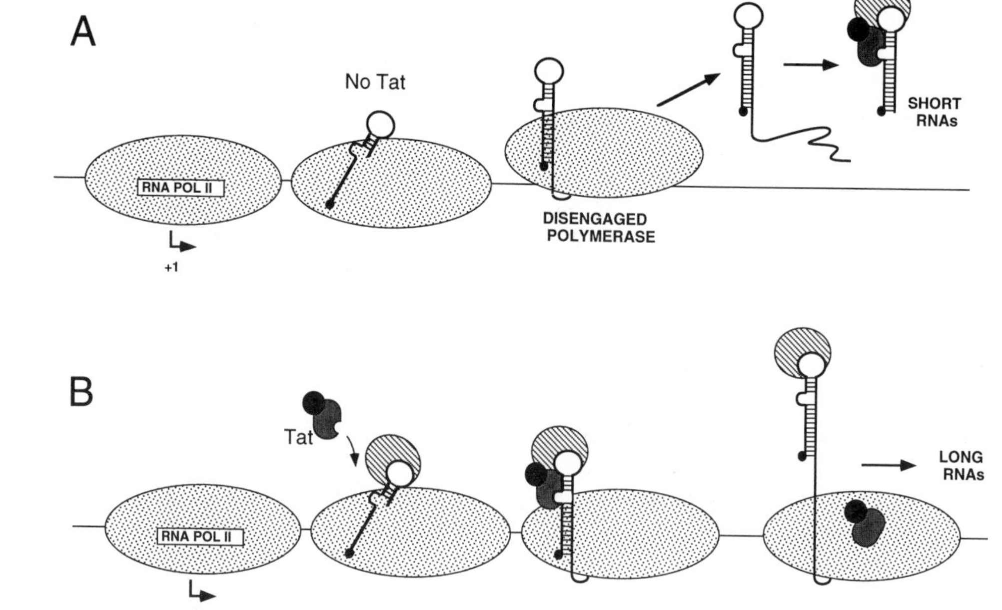

Figure 7 Attenuation of transcription and *trans*-activation by Tat. (*A*) The simplest model of *trans*-activation by Tat-1 allows for the assembly of transcription complexes on the HIV LTR in the absence of Tat. Many transcription complexes pause and terminate prematurely, releasing short viral transcripts that contain TAR. (*B*) In the presence of Tat, nonprocessive transcription complexes are modified by Tat or the coactivator (or both) assembled at TAR. Processive transcription complexes are released from the promoter. Since there is greater clearance 3′ to the promoter, more transcripts can be initiated, resulting in apparent increases in the initiation and elongation rates.

initiated but paused RNA polymerase II molecules that are rapidly released into a productive elongation mode after induction by heat shock (92). Escape of the released RNA polymerase II complex could be tightly coupled to the recruitment of the next transcription complex, which could explain effects of Tat-1 on RNA initiation. The IST-binding protein could play a role in controlling the extent of attenuation or pausing that occurs in the absence of Tat (88, 89).

Trans-activation by Tat-1 in vitro is synergistic with elongation factor TFIIS but not with the initiation/elongation transcription factor, TFIIF (175). Therefore, Tat might act in a manner similar to TFIIF. Transcription factors IIE, IIH, IIF, and IIJ are thought to remain bound to RNA polymerase for at least a short distance following initiation, and thus are also potential

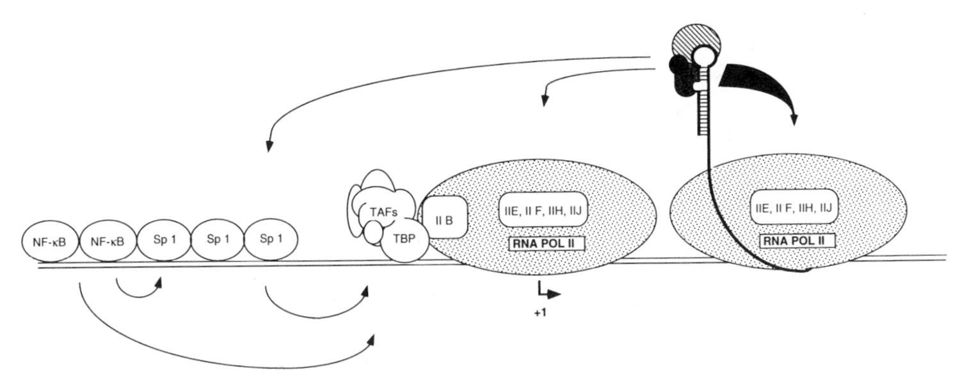

Figure 8 Possible mechanism of *trans*-activation by Tat. The ribonucleoprotein complex on TAR most likely modifies nonprocessive transcription complexes that are paused at TAR. These transcription complexes contain RNA polymerase II as well as a subset of the general transcription factors, possibly including TFIIE-J. In addition, Tat or the coactivator may "reach back" to activate preinitiation complexes assembled at the HIV-1 promoter, perhaps by forming contacts with Sp1 and the TBP:TAF complex, or any of the general factors (TFIIA-J) that are present on the initiating RNA polymerase. Dual functions in both initiation and elongation may also be possible for strong transcriptional activators, such as the herpes virus VP16 activator or the adenovirus E1A protein. Arrows point to possible interactions, and the width of the line indicates relative effects of different interactions.

targets for Tat or its associated coactivator (Figure 8). If the coactivator is a kinase (187), it might regulate any of these transcription factors by phosphorylation. Analysis of the kinetics of *trans*-activation and DRB inhibition in vitro further suggests that two different types of transcription complexes are assembled on the HIV-1 LTR (78, 172). One of these is highly processive, is not sensitive to inhibition by DRB, does not require the TATA box (although it does need the NF-κB and Sp1-binding sites), and cannot respond to Tat. The other transcription complex requires the TATA box, is weakly processive, is readily inhibited by DRB, and is highly responsive to Tat. These two complexes may differ in the types of general transcription factors that are associated with RNA polymerase II.

Tat principally affects RNA elongation in vitro (81, 172, 192–194). That Tat-1 can act on paused RNA polymerase molecules is suggested by in vitro experiments showing that it can modify transcription complexes that have been stalled artificially at position +13 in TAR-1 (175). Tat might also affect RNA initiation, either through its ability to interact with promoter-bound transcription factors such as TBP and Sp1 (183, 184) (Figure 8), or through clearing the promoter of paused RNA polymerase II complexes. It remains to be determined whether Tat travels with the elongating

RNA polymerase II throughout transcription or whether it acts early and then dissociates from the elongation complex. While it is not known whether or not Tat binds avidly to the RNA polymerase or instead modifies some component of the transcription complex, it nevertheless is capable of altering RNA polymerase II to read-through distal termination sites in vitro (192).

Since Tat can also activate transcription from upstream DNA-binding sites (124), it will be interesting to compare its mode of action with that of other DNA-binding proteins that can affect RNA chain elongation, such as those that bind to the hsp70 promoter (92, 195). Alternatively, Tat may preferentially activate RNA initiation when bound to upstream DNA-binding sites. Interestingly, the herpes simplex virus VP-16 activator, which can strongly enhance RNA initiation, can also activate transcription when tethered to a promoter-proximal nascent RNA site (196). While it has not been determined whether the RNA-bound VP-16 molecule affects RNA initiation or elongation, this finding raises the possibility that strong transcriptional activators such as VP-16 may interact with multiple transcription factors at a promoter to affect both RNA initiation and elongation.

FUTURE DIRECTIONS

Tat is the first eukaryotic *trans*-activator to be described that acts via specific binding to a site that forms on nascent RNA. As such, regulation of the HIV-1 promoter by Tat-1 can serve as a useful model for studies on the control of transcription elongation in eukaryotes. In many respects, *trans*-activation by Tat is reminiscent of the bacteriophage lambda Q and N anti-terminator proteins, which modify the *Escherichia coli* core RNA polymerase subunit directly to read-through distal transcription termination sites (31, 32, 197). The lambda N protein recognizes an RNA stem-loop structure (NutB) via the bacterial NusA protein, which facilitates binding of N to not only the RNA but also the RNA polymerase. The N protein remains associated with the transcription complex to read-through distal termination sites. To better characterize the mechanism of *trans*-activation by Tat, the current focus is to isolate, clone, and characterize the cellular proteins that interact with Tat and TAR (as well as with IST, SSR, and the transcription complex) and to reproduce Tat *trans*-activation in reconstituted in vitro transcription systems. These studies should not only further elaborate the mechanism of *trans*-activation, but may also reveal unifying concepts of eukaryotic attenuation that may be shared with the promoters for hsp70, c-Myc, c-Fos, and c-Myb. Since insertion of TAR 3′ to the P2 promoter of the human c-Myc gene leads to anti-attenuation of c-Myc transcription by Tat (S Wright, personal communication), the mechanism of c-Myc attenuation may be similar to that of HIV-1. Because high levels of expression of proto-oncogenes such as c-Myc, c-Fos, and c-Myb have been implicated in several human malignancies, finding cellular counterparts to Tat should also become a high priority.

As for Tat, it is hoped that specific strategies aimed at blocking *trans*-activation will render HIV-1 harmless and help in the treatment of AIDS. Presently, efforts are under way to evaluate whether the use of TAR decoy RNAs (129–132), *trans*-dominant inhibitory Tat proteins (102, 122, 198, 199), and a drug called Ro 5-3335, which interferes with the *trans*-activation by Tat in infected cells (200), will have future clinical and therapeutic importance. A better understanding of the interactions between Tat and cellular transcription factors should suggest additional approaches for future targeted intervention in the *trans*-activation process.

ACKNOWLEDGMENTS

We are very grateful to our many colleagues who sent reprints or preprints, communicated unpublished results from their laboratories, or provided comments on the manuscript, especially John Brady, Alan Frankel, Richard Gaynor, Mark Groudine, Nouria Hernandez, K.T. Jeang, Susan M. Kingsman, and Andy Rice. We are also indebted to the various members of our laboratories, past and present, who contributed to our understanding of HIV transcription and especially those whose unpublished results are cited here. Finally, we thank Lorie Mells and Jamie Simon for assistance in preparing the manuscript and Figure 5, respectively. This article is dedicated to our families, Steven and Kyle Dallas Koerber (K.A. Jones), and Anne, Anton Alexander, and Sebastian Bogomir (B.M. Peterlin) for their encouragement and support.

Literature Cited

1. Barre-Sinoussi T, Chermann JC, Rey F, Nugeyre MT, Chamaret S, et al. 1983. *Science* 220:868–71
2. Gallo RC, Salahuddin SZ, Popovic M, Shearer GM, Kaplan M, et al. 1984. *Science* 224:500–3
3. Levy JA, Hoffman AD, Kramer SM, Landis JA, Shimabukuro JM. 1984. *Science* 225:840–42
4. Ratner L, Haseltine W, Patarca R, Livak KJ, Starcich B, et al. 1985. *Nature* 313:277–84
5. Sanchez-Pescador R, Power MD, Barr PJ, Steimer KS, Stempien MM, et al. 1984. *Science* 227:484–92
6. Wain-Hobson S, Sonigo P, Danos O, Cole S, Alizon M. 1985. *Cell* 40:9–17
7. Sodroski J, Rosen C, Wong-Staal F, Salahuddin SZ, Popovic M, et al. 1985. *Science* 227:171–73
8. Fisher AG, Feinberg MB, Josephs SF, Harper ME, Marselle LM, et al. 1986. *Nature* 320:2367–71
9. Dayton AI, Sodroski JG, Rosen CA, Goh WC, Haseltine WA. 1986. *Cell* 44:941–47

10. Kao S, Calman A, Luciw P, Peterlin B. 1987. *Nature* 330:489–93
11. Laspia MF, Rice AP, Mathews MB. 1989. *Cell* 59:283–92
12. Laspia MF, Rice AP, Mathews MB. 1990. *Genes Dev.* 4:2397–408
13. Feinberg MB, Baltimore D, Frankel AD. 1991. *Proc. Natl. Acad. Sci. USA* 88:4045–49
14. Karn J, Graeble M. 1992. *Trends Genet.* 8:365–68
15. Cullen BR. 1993. *Cell* 73:417–20
16. Cullen BR. 1990. *Cell* 63:655–57
17. Frankel AD, Mattaj IW, Rio DC. 1991. *Cell* 67:1041–46
18. Frankel AD. 1992. *Curr. Opin. Genet. Dev.* 2:293–98
19. Gait M, Karn J. 1993. *Trends Biochem.* 18:255–59
20. Gaynor R. 1992. *AIDS* 6:347–63
21. Jeang K-T, Gatignol A. 1993. *Curr. Top. Microbiol. Immunol.* In press
22. Jones K. 1993. *Curr. Opin. Cell Biol.* 5:461–68
23. Sharp PA, Marciniak RA. 1989. *Cell* 59:229–30
24. Vaishnav YN, Wong-Staal F. 1991. *Annu. Rev. Biochem.* 60:577–630
25. Karn J, Graeble MA. 1992. *Trends Genet.* 8:365–68
26. Conaway RC, Conaway JW. 1993. *Annu. Rev. Biochem.* 62:161–90
27. Zawel L, Reinberg D. 1992. *Curr. Opin. Cell Biol.* 4:488–95
28. Drapkin R, Merino A, Reinberg D. 1993. *Curr. Opin. Cell Biol.* 5:469–76
29. Greenblatt J. 1991. *Trends Biochem. Sci.* 16:408–11
30. Roeder RG. 1991. *Trends Biochem. Sci.* 16:402–8
31. Das A. 1993. *Annu. Rev. Biochem.* 62:893–930
32. Greenblatt J, Nodwell JR, Mason SW. 1993. *Nature* 364:401–6
33. Proudfoot N. 1989. *Trends Biochem. Sci.* 14:105–10
34. Spencer CA, Groudine M. 1990. *Oncogene* 5:777–85
35. Kim JYH, Gonzalez-Scarano F, Zeichner SL, Alwine JC. 1993. *J. Virol.* 67:1658–62
36. Waterman ML, Fischer WH, Jones KA. 1991. *Genes Dev.* 5:656–69
37. Waterman ML, Jones KA. 1990. *New Biol.* 2:621–36
38. Giese K, Cox J, Grosschedl R. 1992. *Cell* 69:185–96
39. Carlsson P, Waterman M, Jones K. 1993. *Genes Dev.* 7:2418–30
40. Li C, Lusis AJ, Sparkes R, Tran SM, Gaynor R. 1992. *Genomics* 13:658–64
41. Yang Z, Engel J. 1993. *Nucleic Acids Res.* 21:2831–36
42. Dasgupta P, Saikumar P, Reddy CD, Reddy EP. 1990. *Proc. Natl. Acad. Sci. USA* 87:8090–94
43. Gilmore TD. 1990. *Cell* 62:841–43
44. Gilmore TD. 1991. *Trends Genet.* 7:318–22
45. Nolan G, Baltimore D. 1992. *Curr. Opin. Genet. Dev.* 2:211–20
46. Liou H-C, Baltimore D. 1993. *Curr. Opin. Cell Biol.* 5:477–87
47. Ross EK, Buckler-White AJ, Rabson AB, Englund G, Martin MA. 1991. *J. Virol.* 65:4350–58
48. Parrott C, Seidner T, Duh E, Leonard J, Theodore TS, et al. 1991. *J. Virol.* 65:1414–19
49. Sun S-C, Ganchi PA, Ballard DW, Greene WC. 1993. *Science* 259:1912–15
50. Brown K, Park S, Kanno T, Franzoso G, Siebenlist U. 1993. *Proc. Natl. Acad. Sci. USA* 90:2532–36
51. Kerr LD, Ransone LJ, Wamsley P, Schmitt MJ, Boyer TG, et al. 1993. *Nature* 365:412–19
52. Bachelerie F, Alcami J, Arenzana-Seisdedos F, Virelizier J-L. 1991. *Nature* 350:709–12
53. Kieran M, Blank V, Logeat F, Vandekerckhove J, Lottspeich F, et al. 1990. *Cell* 62:1007–18
54. Riviere Y, Blank V, Kourilsky P, Israel A. 1991. *Nature* 350:625–26
55. Hayashi T, Ueno Y, Okamoto T. 1993. *J. Biol. Chem.* 268:11380–88
56. Israel N, Gougerot-Pocidalo MA, Aillet F, Virelizier JL. 1992. *J. Immunol.* 149:3386–93
57. Schreck R, Albermann K, Baeuerle PA. 1992. *Free Radic. Res. Commun.* 17:221–37
58. Staal FJ, Roederer M, Herzenberg LA, Herzenberg LA. 1990. *Proc. Natl. Acad. Sci. USA* 87:9943–47
59. Baur A, Peterlin BM. 1994. *J. Immunol.* In press
60. Baur AS, Sawai ET, Cheng-Mayer C, Peterlin BM. 1994. *Cell.* Submitted
61. Ten RM, Paya CV, Israel N, Le Bail O, Mattei M-G, et al. 1992. *EMBO J.* 11:195–203
62. Paya CV, Ten RM, Bessia C, Alcami J, Hay RT, Virelizer J-L. 1992. *Proc. Natl. Acad. Sci. USA* 89:7826–30
63. Harakeh S, Jariwalla RJ, Pauling L. 1990. *Proc. Natl. Acad. Sci. USA* 87:7245–49
64. Staal FJ, Roederer M, Raju PA, Anderson MT, Ela SW, et al. 1993. *AIDS Res. Hum. Retroviruses* 9:299–306
65. Jones KA, Kadonaga JT, Luciw PA, Tjian R. 1986. *Science* 232:755–59

66. Garcia JA, Harrich D, Soultanakis E, Wu F, Mitsuyasu R, Gaynor RB. 1989. *EMBO J.* 8:765–78
67. Zenzie-Gregory B, Sheridan P, Jones KA, Smale ST. 1993. *J. Biol. Chem.* 268:15823–32
68. Dynlacht BD, Hoey T, Tjian R. 1991. *Cell* 55:563–76
69. Tanese N, Pugh BF, Tjian R. 1991. *Genes Dev.* 5:2212–24
70. Zhou Q, Lieberman PM, Boyer TG, Berk AJ. 1992. *Genes Dev.* 6:1964–74
71. Gill G, Pascal E, Tseng Z, Tjian R. 1993. *Proc. Natl. Acad. Sci. USA* 91:192–96
72. Hoey T, Weinzierl R, Gill G, Chen J-L, Dynlacht B, Tjian R. 1993. *Cell* 72:247–60
73. Pascal E, Tjian R. 1991. *Genes Dev.* 7:1521–34
74. Seto E, Lewis B, Shenk T. 1993. *Nature* 365:462–64
75. Seto E, Shi Y, Shenk T. 1991. *Nature* 354:241–45
76. Huang L-M, Jeang K-T. 1993. *J. Virol.* 67:6937–44
77. Berkhout B, Jeang K-T. 1992. *J. Virol.* 66:139–49
78. Lu X, Welsh TM, Peterlin BM. 1993. *J. Virol.* 67:1752–60
79. Olsen HS, Rosen CA. 1992. *J. Virol.* 66:5594–97
80. Jones KA, Luciw PA, Duchange N. 1988. *Genes Dev.* 2:1101–14
81. Kato H, Horikoshi M, Roeder RG. 1991. *Science* 251:1476–79
82. Wu FK, Garcia JA, Harrich D, Gaynor RB. 1988. *EMBO J.* 7:2117–30
83. Margolis D, Ostrove J, Straus S. 1993. *Virology* 192:370–74
84. Du H, Roy AL, Roeder RG. 1993. *EMBO J.* 12:501–11
85. Roy AL, Meisterernst M, Pognonec P, Roeder RG. 1991. *Nature* 354:245–48
86. Sheridan P, Schorpp M, Voz M, Jones K. 1994. *Mol. Cell. Biol.* Submitted
87. Garcia JA, Ou SH, Wu F, Lusis AJ, Sparkes RS, Gaynor RB. 1992. *Proc. Natl. Acad. Sci. USA* 89:9372–76
88. Ratnasabapathy R, Sheldon M, Johal L, Hernandez N. 1990. *Genes Dev.* 4:2061–74
89. Sheldon M, Ratnasbapathy R, Hernandez N. 1993. *Mol. Cell Biol.* 13:1251–63
90. Kessler M, Mathews MB. 1992. *J. Virol.* 66:4488–96
91. Adams M, Romeo J, Kimpton J, Garcia J, Peterlin B, Emerman M. 1994. *Proc. Natl. Acad. Sci. USA.* In press
92. Lee H-S, Kraus K, Wolfner M, Lis J. 1992. *Genes Dev.* 6:284–95
93. Verdin E. 1991. *J. Virol.* 65:6790–99
94. Verdin E, Paras JP, Van Lint C. 1993. *EMBO J.* 12:3249–59
95. Peterlin BM, Adams M, Alonso A, Baur A, Ghosh S, et al. 1993. *Molecular Biology of Human Retroviruses*, ed. BR Cullen, pp. 75–100. Oxford: ILR/Oxford Univ. Press
96. Cullen BR, Garrett ED. 1992. *AIDS Res. Hum. Retroviruses* 8:387–93
97. Schwartz S, Felber BK, Benko DM, Fenyo EM, Pavlakis GN. 1990. *J. Virol.* 64:2519–29
98. Benko DM, Schwartz S, Pavlakis GN, Felber BK. 1990. *J. Virol.* 64:2505–18
99. Furtado MR, Balachandran R, Gupta P, Solinsky SM. 1991. *Virology* 185:258–70
100. Gottlinger HG, Dorfman T, Cohen EA, Haseltine WA. 1992. *Virology* 189:618–28
101. Salfeld J, Gottlinger HG, Sia RA, Park RE, Sodroski JG, Haseltine WA. 1990. *EMBO J.* 9:965–70
102. Carroll R, Peterlin BM, Derse D. 1992. *J. Virol.* 66:2000–7
103. Winslow BJ, Trono D. 1993. *J. Virol.* 67:2349–54
104. Newstein M, Stanbridge EJ, Casey G, Shank PR. 1990. *J. Virol.* 64:4565–67
105. Carroll R, Martarano L, Derse D. 1991. *J. Virol.* 65:3460–67
106. Carvalho M, Derse D. 1991. *J. Virol.* 65:3468–74
107. Dorn P, DaSilva L, Martarano L, Derse D. 1990. *J. Virol.* 64:1616–24
108. Tong-Starksen S, Baur A, Lu X, Peck E, Peterlin B. 1993. *Virology* 195:826–30
109. Emerman M, Guyader M, Montagnier L, Baltimore D, Muesing M. 1987. *EMBO J.* 6:3755–60
110. Fenrick R, Malim MH, Hauber J, Le SY, Maizel J, Cullen BR. 1989. *J. Virol.* 63:5006–12
111. Berkhout B, Gatignol A, Silver J, Jeang KT. 1990. *Nucleic Acids Res.* 18:1839–46
112. Derse D, Carvalho M, Carroll R, Peterlin BM. 1991. *J. Virol.* 65:7012–15
113. Endo S, Kubota S, Siomi H, Adachi A, Oroszlan S, et al. 1989. *Virus Genes* 3:99–100
114. Hauber J, Malim MH, Cullen BR. 1989. *J. Virol.* 63:1181–87
115. Siomi H, Shida H, Maki M, Hatanaka M. 1990. *J. Virol.* 64:1803–7
116. Subramanian T, Kuppuswamy M, Venkatesh L, Srinivasan A, Chinnadurai G. 1990. *Virology* 176:178–83
117. Subramanian T, Govindarajan R,

Chinnadurai G. 1991. *EMBO J.* 10: 2311–18
118. Ruben S, Perkins A, Purcell R, Joung K, Sia R, et al. 1989. *J. Virol.* 63(1):1–8
119. Rice AP, Carlotti F. 1990. *J. Virol.* 64:6018–26
120. Rice AP, Carlotti F. 1990. *J. Virol.* 64:1864–68
121. Selby MJ, Peterlin BM. 1990. *Cell* 62:769–76
122. Madore SJ, Cullen BR. 1993. *J. Virol.* 67:3703–11
123. Luo Y, Madore S, Parslow T, Cullen B, Peterlin B. 1993. *J. Virol.* 67:5617–22
124. Southgate CD, Green MR. 1991. *Genes Dev.* 5:2496–507
125. Kamine J, Loewenstein P, Green M. 1991. *Virology* 182:570–77
126. Ghosh S, Selby M, Peterlin B. 1993. *J. Mol. Biol.* 234:610–19
127. Jakobovits A, Smith DH, Jakobovits EB, Capon D. 1988. *Mol. Cell. Biol.* 8:2555–61
128. Selby MJ, Bain ES, Luciw PA, Peterlin BM. 1989. *Genes Dev.* 3:547–58
129. Lisziewicz J, Rappaport J, Dhar R. 1991. *New Biol.* 3:82–89
130. Sullenger BA, Gallardo HF, Ungers GE, Gilboa E. 1991. *J. Virol.* 65:6811–16
131. Sullenger BA, Gallardo HF, Ungers GE, Gilboa E. 1990. *Cell* 63:601–8
132. Graham GJ, Maio JJ. 1990. *Proc. Natl. Acad. Sci. USA* 87:5817–21
133. Berkhout B, Silverman RH, Jeang KT. 1989. *Cell* 59:273–82
134. Berkhout B, Jeang KT. 1989. *J. Virol.* 63:5501–4
135. Jeang KT, Berkhout B. 1992. *J. Biol. Chem.* 267:17891–99
136. Roy S, Parkin NT, Rosen C, Itovitch J, Sonenberg N. 1990. *J. Virol.* 64: 1402–6
137. Roy S, Delling U, Chen CH, Rosen CA, Sonenberg N. 1990. *Genes Dev.* 4:1365–73
138. Weeks KM, Ampe C, Schultz SC, Steitz TA, Crothers DM. 1990. *Science* 249:1281–85
139. Wu F, Garcia J, Sigman D, Gaynor R. 1991. *Genes Dev.* 5:2128–40
140. Feng S, Holland EC. 1988. *Nature* 334:165–67
141. Dingwall C, Ernberg I, Gait MJ, Green SM, Heaphy S, et al. 1990. *EMBO J.* 9:4145–53
142. Berkhout B. 1992. *Nucleic Acids Res.* 20:27–31
143. LeGuern M, Shioda T, Levy JA, Cheng-Mayer C. 1993. *Virology* 195: 441–47
144. Chang YN, Jeang KT. 1992. *Nucleic Acids Res.* 20:5465–72
145. Elangovan B, Subramanian T, Chinnadurai G. 1992. *J. Virol.* 66: 2031–36
146. Rhim H, Rice AP. 1993. *J. Virol.* 67:1110–21
147. Hoffman DW, Colvin RA, Garcia-Blanco MA, White SW. 1993. *Biochemistry* 32:1096–104
148. Colvin RA, Garcia BM. 1992. *J. Virol.* 66:930–35
149. Delling U, Reid LS, Barnett RW, Ma MY, Climie S, et al. 1992. *J. Virol.* 66:3018–25
150. Weeks KM, Crothers DM. 1991. *Cell* 66:577–88
151. Berkhout B, Jeang KT. 1991. *Nucleic Acids Res.* 19:6169–76
152. Riordan FA, Bhattacharyya A, McAteer S, Lilley DM. 1992. *J. Mol. Biol.* 226:305–10
153. Critchley A, Haneef I, Cousens D, Stockley P. 1993. *J. Mol. Graphics* 11:92–97
154. Calnan BJ, Tidor B, Biancalana S, Hudson D, Frankel AD. 1991. *Science* 252:1167–71
155. Calnan BJ, Biancalana S, Hudson D, Frankel AD. 1991. *Genes Dev.* 5:201–10
156. Puglisi JD, Tan R, Calnan BJ, Frankel AD, Williamson JR. 1992. *Science* 257:76–80
157. Tan R, Frankel AD. 1992. *Biochemistry* 31:10288–94
158. Puglisi JD, Chen L, Frankel AD, Williamson JR. 1993. *Proc. Natl. Acad. Sci. USA* 90:3680–84
159. Colvin RA, White SW, Garcia-Blanco MA, Hoffman DW. 1993. *Biochemistry* 32:1105–12
160. Barnett R, Delling U, Kuperman R, Sonenberg N, Sumner-Smith M. 1993. *Nucleic Acids Res.* 21:151–54
161. Dingwall C, Ernberg I, Gait MJ, Green SM, Heaphy S, et al. 1989. *Proc. Natl. Acad. Sci. USA* 86:6925–29
162. Frankel AD, Biancalana S, Hudson D. 1989. *Proc. Natl. Acad. Sci. USA* 86:7397–401
163. Cordingley MG, LaFemina RL, Callahan PL, Condra JH, Sardana VV, et al. 1990. *Proc. Natl. Acad. Sci. USA* 87:8985–89
164. Sumner-Smith M, Roy S, Barnett R, Reid LS, Kuperman R, et al. 1991. *J. Virol.* 65:5196–202
165. Hamy F, Asseline U, Grasby J, Iwai S, Pritchard C, et al. 1993. *J. Mol. Biol.* 230:111–23
166. Tao J, Frankel AD. 1993. *Proc. Natl. Acad. Sci. USA* 90:1571–75

167. Churcher MJ, Lamont C, Hamy F, Dingwall C, Green SM, et al. 1993. *J. Mol. Biol.* 230:90–110
168. Hart CE, Ou CY, Galphin JC, Moore J, Bacheler LT, et al. 1989. *Science* 246:488–91
169. Hart CE, Galphin JC, Westhafer MA, Schochetman G. 1993. *J. Virol.* 67: 5020–24
170. Alonso A, Derse D, Peterlin BM. 1992. *J. Virol.* 66:4617–21
171. Luo Y, Peterlin B. 1993. *J. Virol.* 67:3441–45
172. Marciniak RA, Sharp PA. 1991. *EMBO J.* 10:4189–96
173. Marciniak RA, Calnan BJ, Frankel AD, Sharp PA. 1990. *Cell* 63:791–802
174. Sheline CT, Milocco LH, Jones KA. 1991. *Genes Dev.* 5:2508–20
175. Kato H, Sumimoto H, Pognonec P, Chen CH, Rosen CA, Roeder RG. 1992. *Genes Dev.* 6:655–66
176. Braddock M, Powell R, Blanchard AD, Kingsman AJ, Kingsman SM. 1993. *FASEB J.* 7:214–22
177. Muckenthaler M, Blanchard A, Vives E, Nacken W, Braddock M, et al. 1994. *J. Virol.* Submitted
178. Marciniak RA, Garcia BM, Sharp PA. 1990. *Proc. Natl. Acad. Sci. USA* 87:3624–28
179. Gatignol A, Buckler-White A, Berkhout B, Jeang KT. 1991. *Science* 251:1597–600
180. Gatignol A, Kumar A, Rabson A, Jeang KT. 1989. *Proc. Natl. Acad. Sci. USA* 86:7828–32
181. Gatignol A, Buckler C, Jeang KT. 1993. *Mol. Cell Biol.* 13:2193–202
182. Gaynor R, Soultanakis E, Kuwabara M, Garcia J, Sigman DS. 1989. *Proc. Natl. Acad. Sci. USA* 86:4858–62
183. Kashanchi F, Piras G, Radonovich M, Duvall J, Fattaey A, et al. 1994. *Nature* 367:295–99
184. Jeang K-T, Chun R, Lin N, Gatignol A, Glabe C, Fan H. 1993. *J. Virol.* 67:6224–33
185. Nelbock P, Dillon PJ, Perkins A, Rosen CA. 1990. *Science* 248:1650–53
186. Desai K, Loewenstein PM, Green M. 1991. *Proc. Natl. Acad. Sci. USA* 88:8875–79
187. Hermann C, Rice A. 1993. *Virology* 197:601–8
188. Ohana B, Moore PA, Ruben SM, Southgate CD, Green MR, Rosen CA. 1993. *Proc. Natl. Acad. Sci. USA* 90:138–42
189. Shibuya H, Irie K, Ninomiya-Tsuji J, Goebl M, Taniguchi T, Matsumoto K. 1992. *Nature* 357:700–2
190. Jakobovits A, Rosenthal A, Capon DJ. 1990. *EMBO J.* 9:1165–70
191. Payne JM, Dahmus ME. 1993. *J. Biol. Chem.* 268:80–87
192. Graeble MA, Churcher MJ, Lowe AD, Gait MJ, Karn J. 1993. *Proc. Natl. Acad. Sci. USA* 90:6184–88
193. Bohan C, Kashanchi F, Ensoli B, Buonaguro L, Boris-Lawrie K, Brady J. 1992. *Gene Expr.* 2:391–407
194. Laspia M, Wendel P, Mathews M. 1993. *J. Mol. Biol.* 232:732–46
195. O'Brien T, Lis JT. 1991. *Mol. Cell. Biol.* 11:5285–90
196. Tiley LS, Madore SJ, Malim MH, Cullen BR. 1992. *Genes Dev.* 6:2077–87
197. Roberts JW. 1993. *Cell* 72:653–55
198. Pearson L, Garcia J, Wu F, Modesti N, Nelson J, Gaynor R. 1990. *Proc. Natl. Acad. Sci. USA* 87:5079–83
199. Modesti N, Garcia J, Debouck C, Peterlin M, Gaynor R. 1991. *New Biol.* 3:759–68
200. Hsu MC, Schutt AD, Holly M, Slice LW, Sherman MI, et al. 1991. *Science* 254:1799–802
201. Tiley LS, Brown PH, Cullen BR. 1990. *Virology* 178:560–67
202. Kuppuswamy M, Subramanian T, Srinivasan A, Chinnadurai G. 1989. *Nucleic Acids Res.* 17:3551–61
203. Garcia JA, Harrich D, Soultanakis E, Wu F, Mitsuyasu R, Gaynor RB. 1988. *EMBO J.* 8:765–78
204. Sadaie MR, Benter T, Wong-Staal F. 1988. *Science* 239:910–13
205. Zuker M, Stiegler P. 1981. *Nucleic Acids Res.* 9:133–48

Annu. Rev. Biochem. 1994. 63:745–76

REGULATION OF EUKARYOTIC DNA REPLICATION

Dawn Coverley and Ronald A. Laskey
Wellcome/CRC Institute of Cancer and Developmental Biology, Tennis Court Road, Cambridge CB2 1QR, England and Department of Zoology, University of Cambridge

KEY WORDS: replication, regulation, origins, cell cycle, S phase

---

CONTENTS

INTRODUCTION

The complexity of eukaryotic chromosomes makes precise duplication a formidable task. It is not surprising, therefore, that eukaryotic DNA repli-

0066-4154/94/0701-0745$05.00

cation is tightly controlled by a range of complex mechanisms. We know that DNA synthesis normally initiates at many specific origins of replication on each chromosome, probably in response to activation by cyclin-dependent kinases. We know that different origins may initiate at different times and that replication forks are spatially organized into clusters within the nucleus. We also know that replication produces exactly one copy and that it normally avoids reinitiation within a single cell cycle. The progress of DNA replication is also monitored so that progression to subsequent cell-cycle phases can be delayed pending completion of S phase.

While this simple picture may appear comfortingly clear, there are many crucial questions that cannot be answered yet.

If replication initiates at specific sites, why has it been so difficult to find DNA sequences that function as replication origins in plasmid replication assays in animal cells? The problem here is not that too few sequences will replicate, but that too many will. We review several aspects of this paradox, and propose that it may be explained if patterns of replication in higher eukaryotes are imposed by chromosome structure and therefore only indirectly by DNA sequence.

The signals that specify entry into S phase are gradually becoming clearer as emphasis in the cell-cycle field is shifting away from mitosis towards the G1- and S-phase events. However, at present there are many cyclin-dependent kinases implicated in replication but no precisely defined roles.

The mechanism that prevents multiple rounds of DNA replication within a single cell cycle can be overcome by certain mutations and a number of chemical treatments. The implications of these observations are discussed in relation to previously proposed models for replication control.

In this review we focus on regulation of eukaryotic DNA replication rather than on its mechanism. We also focus on cellular rather than viral systems, but even within these limitations it has been necessary to omit much interesting work. We apologize to those authors whose work has not been cited.

I. ORIGINS OF REPLICATION

Origins of Replication in S. cerevisiae

One of the most intensively studied areas in eukaryotic DNA replication is the search for DNA sequences that specify its initiation. The term "origins of replication" is used to describe the DNA sequences at which replication initiates as well as *cis*-acting sequences that promote initiation. By analogy with transcription, the replication equivalents of transcription factor–binding sites and transcription start sites would all be included within the term "origins of replication."

Evidence is now overwhelming that eukaryotic DNA replication normally initiates at specific sites. However, with the exception of *S. cerevisiae,* it is more difficult to decide whether patterns of initiation are specified directly, by sequence-specific interaction with replication proteins, or indirectly, as a consequence of structural features of the chromosome. Here we shall argue that the second possibility may explain some of the many paradoxes encountered studying the replication origins of higher eukaryotes.

We start by briefly summarizing knowledge of replication origins in *S. cerevisiae*. This subject has been reviewed more fully elsewhere (1, 2). Identification of yeast replication origins was made possible by an assay for sequences that allow autonomous replication of plasmids (3). The failure of an equivalent system to yield specific DNA sequences has held back understanding of higher eukaryotic replication origins. Two-dimensional (2D) gel mapping methods have greatly reinforced and extended results of autonomous replication assays (4). They have shown that all yeast origin sequences can support autonomous replication of plasmids, but that not all autonomously replicating sequences act as origins of replication in the yeast genome (reviewed in Ref. 4). Furthermore, not all yeast origins initiate replication at the same time; some initiate later than others (2). Timing of origin activity is determined by chromosomal context rather than by autonomously replicating sequence elements themselves, as changing an origin's location can change the time at which it initiates (2).

Systematic deletion of yeast replication origins results in replication of their surrounding DNA by forks that extend from neighboring origins (5). Carried to extremes, this approach has surprisingly revealed that deletion of all the origins from the left arm and part of the right arm of chromosome III is not lethal. The chromosome is still replicated, but by a fork progressing from the right arm. Although these origins are not essential, their deletion results in marginally higher rates of chromosome loss. These observations show that patterns of origin usage are flexible, but that yeast does not respond to origin deletion by initiating at "non-origin" sequences. Instead it retains a stringent definition of origin sequence and replicates the region around the deletion by forks from another site (5).

By contrast with yeast, replication origins in higher eukaryotes have been more elusive. The field has suffered from the lack of efficient assays for specific DNA sequences that confer autonomous replication of plasmids. Ironically, the problem has not been that too few plasmids replicate in higher eukaryotic cells, but that too many replicate. Therefore, knowledge of replication origins in higher eukaryotes comes largely from mapping studies, rather than from assays for origin function.

Evidence for Specific Initiation in the Chromosomes of Higher Eukaryotes

There is growing evidence that replication usually initiates at specific sites within the chromosomes of higher eukaryotes. Evidence in favor of this conclusion comes from many approaches, studying many genes in a wide range of species. It has been reviewed extensively, including in last year's volume of this series (6–11).

By far the most intensively studied region of higher eukaryotic DNA is that which surrounds the dihydrofolate reductase (*DHFR*) gene in Chinese hamster ovary cells. Most of the available origin mapping methods have focused on this region, because it exists as highly amplified tandem repeats. These repeats give stronger origin-mapping signals than do single-copy genes. Although the methods differ in the precision with which they map the point where bidirectional replication initiates, several methods support the preferential use of a specific site called "ori β" (6–10, 12, 13). In fact, all mapping methods provide evidence that replication of this region does not initiate randomly; however, one discrepancy continues to cause confusion. It is the discrepancy between 2D gel electrophoretic studies and other methods. 2D gel mapping shows initiation at precise sites in the chromosomes of *S. cerevisiae* (4), but it consistently shows initiation occurring at multiple sites within broad initiation zones in the chromosomes of higher eukaryotes, including in a broad region of 55 kb around *DHFR* ori β. So, initiation is constrained to a defined part of the amplified repeat, but this part is much larger than suggested by other methods. Other examples of 2D gel mapping in higher eukaryotes also suggest broad initiation zones, for example ribosomal RNA genes in human or frog cells (14–17) or *Drosophila* histone genes (18, 19).

Attempts have been made to reconcile the wide initiation zones seen by 2D mapping studies with the narrower initiation zones seen by alternative origin mapping methods. Suggested explanations include (*a*) the unwinding of large areas of DNA duplex followed by priming at multiple sites before initiation occurs at a specific site (11), (*b*) priming at many sites, but only unidirectional synthesis towards a fixed "origin" from which replication becomes bidirectional (10), and (*c*) initiation at many sites but efficient elongation from only one of these (the Jesuit model—many are called, but few are chosen) (6–9). We suggest that features of these models might be combined in a fourth "quality control" or "reformation" model (Figure 1). In this model, initiation would be mediated at many sites in a broad zone by polymerase α, but the product of this synthesis might be destroyed and replaced by new synthesis at one of these sites by DNA polymerases δ and ϵ. Elongation would occur only within an immobile replication focus, which would form the basis for site selection. This would have the merit that the inaccurate product of polymerase α, which lacks proofreading activity, would be erased and replaced by the more accurate synthesis catalyzed by polymerases δ and ε, both of which have 3′-5′ exonuclease proofreading activity and therefore greater fidelity. Although this model is clearly speculative, it can accommodate most relevant observations. Whether or not it is correct may emerge when the respective roles of the three essential replicative polymerases are clarified.

Are Specific Origin Sequences Required for Replication of Plasmids in Cells of Higher Eukaryotes?

A crucial problem, which has held back studies of replication origins in mammalian cells, is the lack of an assay for the function of candidate sequences. Many attempts have been made to develop such assays and at least one report still deserves further investigation (20), but in general plasmid DNA transfection assays have failed to identify mammalian autonomously replicating sequences.

Although this is widely assumed to be due to inefficient replication of plasmid DNA in mammalian cells, a careful and undervalued series of experiments by Calos and colleagues (21–23) strongly suggests an alternative, though unfashionable, interpretation. Initially they devised an origin-assay strategy using Epstein-Barr virus (EBV) vectors that lacked origin function to isolate human DNA fragments that could replicate in human cells. The rationale of this first generation of experiments was that the EBV DNA would provide a plasmid maintenance function to prevent plasmid loss during cell division. The surprising result was that any large human DNA sequence allowed replication (21, 23). The initial interpretation was that functional origin sequences are frequent in human DNA and are therefore present in any large fragment. However, subcloning did not identify shorter functional origins. This is consistent with the alternative heretical view that size itself is more important than sequence. The one inconsistency with this alternative view was the apparent failure of bacterial DNA to replicate in the same EBV vectors; however, this later turned out to be an oversimplification of the true situation.

Reviews of this work have frequently stopped at this point, but the experiments go much further. Two principles have emerged. First, the length of the DNA template is crucial (22). Tandem repeats of the same sequence replicate much more efficiently than single copies. When long templates are used, 100% of the template can replicate without any viral sequences present. Second, when long sequences are assayed, even DNA from *Escherichia coli* replicates at 66% efficiency in spite of its much lower A + T

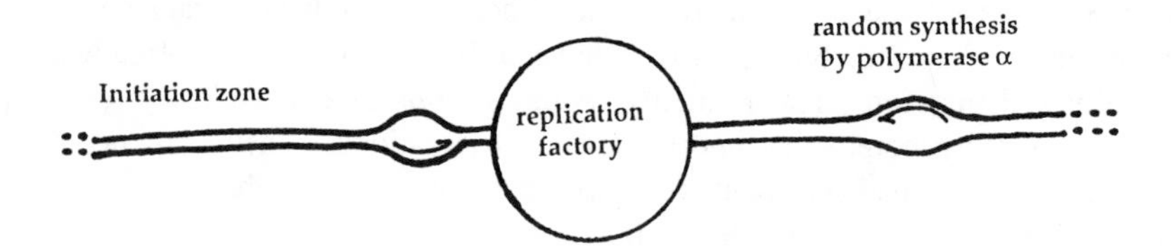
A
Initiation zone
replication factory
random synthesis by polymerase α

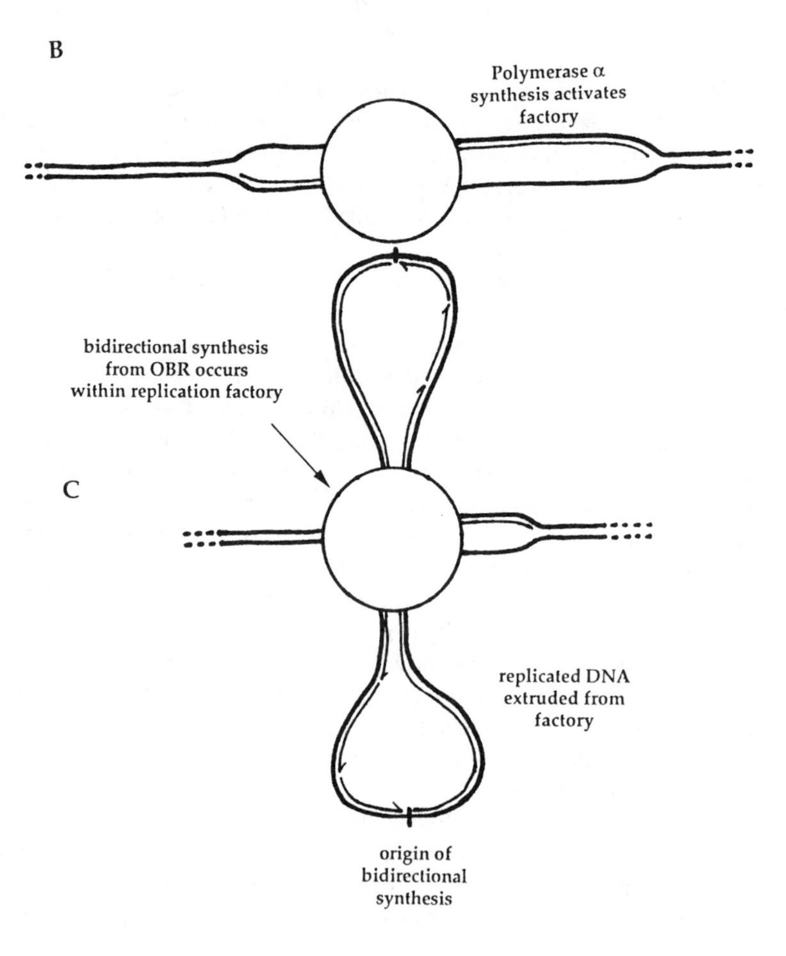
B
Polymerase α synthesis activates factory
bidirectional synthesis from OBR occurs within replication factory
C
replicated DNA extruded from factory
origin of bidirectional synthesis

content (22). It is important to note that these experiments were performed without any viral sequences.

Whether or not there is a 1.5-fold replication advantage for human DNA over bacterial DNA should not obscure two key points. First, in these studies it has not been possible to identify a source of DNA that fails to replicate when long molecules are transfected into human cells. Second, it has not been possible to identify short human DNA sequences that substantially increase replication compared to vector DNA. A popular but inadequate explanation of the length dependence is that higher eukaryotic origins might consist of short elements dispersed over long distances. In this case, only long molecules would contain all the sequence elements necessary to serve as origins. This explanation is difficult to reconcile with the observations that tandem repeats of the same sequence are replicated more efficiently than is a single copy, even for DNA from *E. coli*.

One of the earliest attempts to identify nonviral sequences that allow replication of plasmid DNA in higher eukaryotic cells involved injecting DNA molecules into *Xenopus* eggs. All injected DNA molecules replicated efficiently, including a wide range of bacterial plasmids and bacteriophage genomes (24–26). Subsequent 2D gel mapping has shown that initiation takes place at multiple sites on plasmids in frog egg extracts or injected eggs (15, 17). Furthermore, in these experiments replication occurred only once per cell cycle (24, 27, 28). Therefore, in this system, specific DNA sequences are not essential either for initiation or for the mechanisms that prevent repeated replication within a single cell cycle. In addition, these experiments demonstrated a marked effect of size. Large molecules replicated more efficiently than small (24, 26). This was subsequently explained by showing that formation of exogenous DNA into pseudonuclei is necessary for efficient replication in the *Xenopus* system and that large molecules form pseudonuclei more efficiently than do small molecules (28–30).

←———

Figure 1 Structural selection of eukaryotic origins: a "quality control" or "reformation" model. As a basis for further discussion and experiment, we propose a speculative model for origin selection that appears to reconcile various types of origin-mapping data. It borrows heavily from previously proposed models (6–11), but places more emphasis on the role of immobilized replication "foci" or "factories" (52) and the three types of replicative DNA polymerase. Polymerase α would activate replication enzymes within a factory by catalyzing unidirectional synthesis towards the factory *(B)*, starting anywhere within a broad initiation zone *(A)*. However, synthesis from the origin of bidirectional replication (OBR), with coordinated leading and lagging strands, would occur only within replication factories *(C)*. OBRs may be determined to a greater or lesser extent by DNA sequence, but in this model would be defined by their presence within a replication factory. Once activated, polymerases δ and ϵ and possibly the original polymerase α molecule would catalyze coordinated high-fidelity synthesis from the OBR situated within the factory. This would erase and replace the original low-fidelity synthesis catalyzed by polymerase α outside the factory and would result in extrusion of loops of replicated DNA from the immobile factory.

It is important to understand that these studies do not show that initiation on the endogenous chromosomal DNA is independent of sequence. They show only that exogenous plasmid DNA does not require specific DNA sequences for DNA replication or its coupling to the cell cycle.

Does Replication Initiate at Random Positions in Early Embryos?

Early embryonic cells of *Drosophila* or amphibia initiate replication at closer intervals than do their adult counterparts, allowing them to complete S phase in only 20 minutes in *Xenopus* or only 4 minutes in *Drosophila*. The extreme interpretation of this might be random initiation in embryonic cells. We have argued previously (31) that the data of Blumenthal, Kriegstein, and Hogness (32) for replication in early embryos excludes random initiation. They showed that a maximum of 21 kb DNA can be synthesized in a 4-minute S phase by two forks diverging at 2.6 kb per minute. They also showed an average spacing between consecutive initiations of 7.9 kb. In a Poisson distribution around a mean of 7.9 kb, 11% of all initiations would be too far apart to complete replication within the rigidly determined S phase. Four percent of the template would be unreplicated at the end of S, resulting in chromosome breakage at mitosis. Therefore initiation in early *Drosophila* embryos cannot be random. It must be regular (31).

Resolving the Paradox of Eukaryotic Replication Origins

If initiation in higher eukaryotes is not random, why do assays for autonomous replication of purified DNA in higher eukaryotes fail to find sequences that increase replication? Furthermore, if replication can initiate apparently randomly on long DNA molecules in human cells or frog's eggs, then why isn't replication initiation random in endogenous chromosomes? These questions lead to another more basic question about origins. What is the real function of fixed sites of initiation if not just to allow replication enzymes to initiate or to prevent over-replication in the cell cycle?

The lack of DNA sequences that stimulate plasmid replication significantly in the cells of higher eukaryotes has been frustrating, but it could be explained if patterns of initiation in higher eukaryotes are imposed by features of chromosome structure, rather than by direct interaction of replication proteins with specific DNA sequences. Although chromosome structure is ultimately specified by DNA sequence, in this case DNA sequence would have only indirect effects on patterns of DNA replication.

Evidence that supports this view has recently been reported for mammalian *DHFR* genes replicated in *Xenopus* egg extracts (33). When naked DNA encoding the *DHFR* gene region is injected into intact eggs or incubated in egg extracts, replication initiates at many sites (33), as reported before for other DNA templates (24–26). However, when nuclei from CHOC 400

Chinese hamster ovary cells are added to the same egg extracts, replication of the *DHFR* gene appears to initiate at the same initiation sites as in the intact cell. In these experiments, nuclei in which initiation had occurred were transferred back into hamster cell extracts for elongation (33). These experiments indicate that some feature of nuclear structure can impose selective initiation sites on DNA that could otherwise initiate randomly. So far these experiments have only described the result obtained shortly after nuclei are added to the extract. It would be interesting to know if the same pattern is retained in subsequent cell cycles, or if the egg imposes its own initiation pattern on somatic nuclei, as implied by old nuclear transplantation studies (34) and by studies of clustered replication foci (35–37).

The observation that initiation near the *DHFR* gene is specific when nuclei are added to egg extract, but nonspecific when pure DNA is added, supports the view that chromosome structure imposes specificity. This is also consistent with experiments in which yeast artificial chromosomes containing mammalian genes are introduced into mammalian cells. When introduced as intact chromosomes, they initiate replication at similar sites to those used in intact mammalian cells (38). The concept that patterns of replication initiation are imposed by chromosome structure may seem unconventional, but it is not new: It was clearly suggested 20 years ago by Blumenthal, Kriegstein, and Hogness (32).

Perhaps the most puzzling and challenging question about replication origins is their true function. There is now abundant evidence that specific DNA sequences are not essential for efficient replication of exogenous DNA in a variety of animal cell types, ranging from cultured human cells to *Xenopus* eggs or *Paramecium* (21, 22, 24–30, 39). Nevertheless, there is equally abundant evidence that initiation is not random, but occurs at preferred sites (reviewed in Refs. 1–13); why should this be so? If replication origins are not essential for the replication machinery itself, a more satisfactory explanation of their role is that they coordinate replication and transcription. For example, it may be important to prevent initiation of replication within a gene, particularly if that gene is expressed highly. This would be consistent with the fact that little, if any, transcription occurs in the very short S phases of early embryos of *Drosophila* and *Xenopus*. Transcription starts only when replication slows and when intervals between adjacent replication initiations increase (40–42). Perhaps patterns of DNA replication are imposed by the need to use the same DNA template for both replication and transcription.

Spatial Patterns of Replication Forks within the Nucleus

Replication is spatially regulated within the nucleus so that replication forks are tightly clustered in foci. This is seen clearly when cells are pulse labelled with bromodeoxyuridine (BrdU) or biotinylated deoxyuridine and stained with fluorescent antibodies or streptavidin. Instead of uniform general fluorescence, about 100–300 bright foci of incorporation are seen, consisting of clustered replication forks (43–46).

Similar structures are seen in nuclei replicating in *Xenopus* egg extracts in vitro. Here, there must be at least 300 replication forks in each cluster to account for the rapid rate of replication (47), compared to estimates of around 20 in somatic cultured cells. It is not yet known if replication foci correspond to the chromosomal domains that replicate at different times in S phase, but this seems likely.

Discrete replication foci can also be seen by staining nuclei with antibodies raised against replication proteins such as PCNA or RP-A (35, 48, 49). Perhaps the most interesting proteins to be found in replication foci are the S phase kinase CDK2 and cyclin A (50). Other cyclins and cyclin-dependent kinases were not found there. Consistent with this observation, a cyclin-dependent kinase reported at the time to be cdc2, as well as cyclin A, have been found associated with replicated SV40 DNA (51).

The finding that replication forks are tightly clustered into foci raises important topological questions. Is the pattern of spots constant throughout S phase or does it vary? Do forks remain in the same clusters throughout S phase? If so, it would add direct support for the concept of DNA spooling through fixed sites. If DNA is replicated during passage through fixed sites, how is replication of the DNA that lies between the foci completed? What specifies the pattern of clustered replication foci, and what amino acid sequences target proteins to them? Not all of these questions can be answered yet, but answers are emerging for some.

First, in somatic cells patterns of replication foci change during the course of S phase. Early foci disappear and are replaced by new ones (43–46). This is not observed during the rapid replication of *Xenopus* sperm nuclei in egg extract, where a single pattern of foci persists throughout S phase, with no evidence for addition of new foci later (35, 47). Furthermore when human fibroblast nuclei are incubated in egg extract, their foci change from the programmed sequence of patterns seen in culture, to the single, persistent pattern of egg extracts (35).

The changing patterns of foci in cultured somatic cells can be explained with or without DNA spooling through fixed replication sites. Different regions might simply replicate at different times, producing more and more replicated foci until all the DNA has replicated. However, this argument is much harder to reconcile with the single constant pattern in egg extracts. This pattern is much more easily explained by spooling DNA through immobile replication sites (47). Further evidence for spooling through fixed sites comes from electron microscopy of replicating nuclei, from which most of the chromatin has been removed by nuclease digestion (52). The sites of DNA synthesis as revealed by incorporation of labelled nucleotides are seen as ovoid replication factories. Furthermore, labelled DNA appears to move away from the "factories" during a chase.

A DNA spooling reaction, related to replication, has been seen directly with the electron microscope. During the initial DNA-unwinding reaction of SV40 DNA replication, the two T antigen complexes at the replication forks remain together while loops of unwound DNA are extruded (53). Perhaps this provides an accurate model for the dynamics of eukaryotic DNA replication.

How are replication forks organized into such tight clusters? Clues come from several types of experiments. First, clusters are formed before replication initiates (49). Second, a domain of DNA methyl transferase has been identified that targets the enzyme to the clusters (54). It is not a short targeting signal, like a nuclear localization sequence, but it is more likely to represent a binding domain for interaction with other proteins. Third, characteristic patterns of replication foci can be formed de novo in pseudonuclei assembled from *Xenopus* eggs, even when the template is DNA from bacteriophage λ (Figure 2 and Ref. 36). Clearly, patterns of foci can be

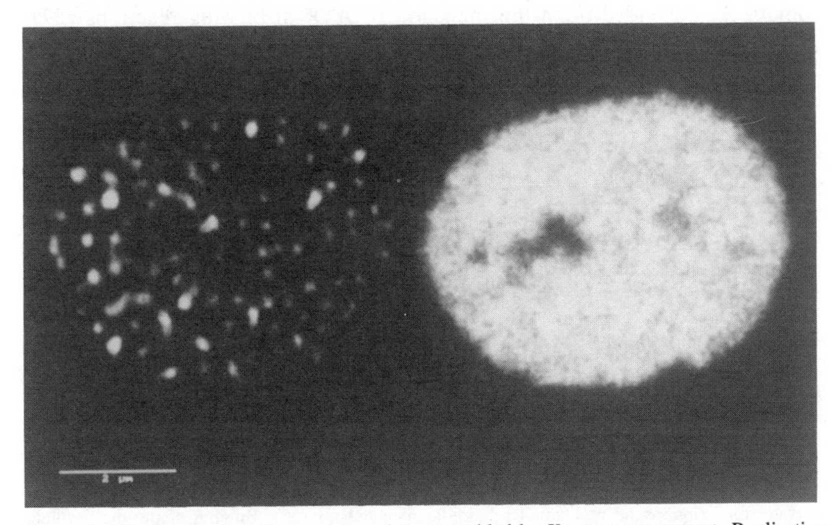

Figure 2 Replication foci in a pseudonucleus assembled by *Xenopus* egg extract. Replication forks are clustered into factories in nuclei assembled from bacteriophage λ DNA. Two images of a pseudonucleus are shown in confocal section after incubation in *Xenopus* egg extract with biotin-dUTP to label nascent DNA. *Left panel,* biotin incorporation is shown by fluorescent streptavidin staining, revealing a similar pattern of replication fork clusters to those seen in normal nuclei. *Right panel,* DNA stained by propidium iodide. Scale bar is 2 μM. For details see Ref. 36.

assembled independently of specific eukaryotic sequences. It will be interesting to know what proteins are responsible for organizing DNA into these remarkable structures.

II. INITIATION MOLECULES AND MECHANISMS

Proteins Involved in Initiation at Saccharomyces cerevisiae *ARS Sequences*

Comparison of DNA sequences that function as replication origins in the yeast genome indicates that they are all A/T rich and that they contain an 11-bp consensus sequence (3, 55). In addition, flanking sequences that vary between autonomously replicating sequences (ARSs) play auxiliary roles in initiation. Identification of the sequence requirements for ARS activity has provided a powerful means of identifying proteins that are involved in ARS function in vivo.

A summary of protein-DNA interactions at *S. cerevisiae* ARS1 is shown in Figure 3. The first protein to be identified, ARS-binding factor 1 (ABF1), is a transcription factor that binds to the B3 auxiliary sequence element of ARS1 (56). Identification of a protein that binds to the core sequence, domain A, depended upon the inclusion of ATP in binding reactions (57). This is the multiprotein origin recognition complex (ORC). ORC consists

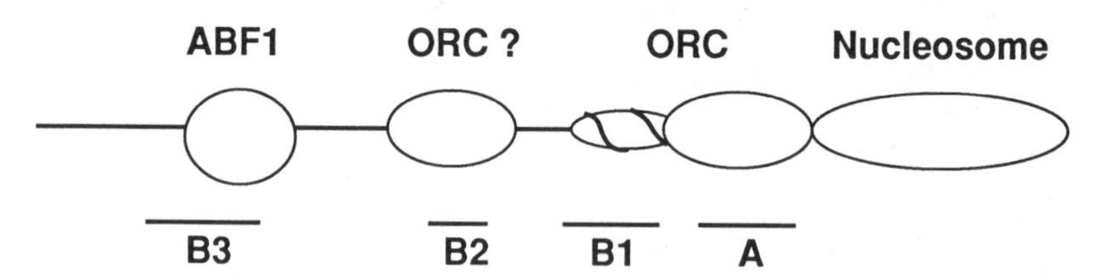

Figure 3 Summary of proposed protein-DNA interactions at ARS1 based on in vivo and in vitro footprinting (57, 59). Mutational analysis of *S. cerevisiae* ARS sequences reveals an organization that has some similarity to that of transcriptional promoters. Generally, *S. cerevisiae* ARS sequences contain an essential AT-rich core sequence (domain A), as well as proximal and distal elements, which are not essential but which stimulate ARS function (domain B). Analysis of domain B of ARS1 has identified three subdomains called B1, B2, and B3 (55). ARS-binding factor 1 (ABF1) is a yeast transcription factor that binds to the B3 element of ARS1 and may activate transcription and replication by similar mechanisms (55). Binding of the multiprotein origin recognition complex (ORC) to the core sequence (domain A) is strongly correlated with ARS1 function, as point mutations that affect binding also influence replication activity (57). DNase I footprinting and analysis of chemical cleavage patterns at the ORC-binding site suggest that binding extends into region B1 and that DNA is wound around the outside of ORC (57, 59). At higher concentrations, ORC binding may extend into region B2, which contains a degenerate match to the ARS consensus of domain A. This representation of protein-DNA interactions at ARS1 is redrawn from Ref. 60.

of six polypeptides, one of which has been shown by genetic analysis to be involved in transcriptional silencing at the yeast mating-type locus (58, 58a, 58b). ORC is believed to associate with ARS sequences throughout most or all of the yeast cell cycle, so initiation is not likely to be regulated at the level of binding (57, 59, 60). It remains to be seen whether the activity of this complex in replication initiation is modulated in a cell cycle–dependent manner.

By selecting for mutants that are defective in the maintenance of ARS-containing minichromosomes (61), a number of other genes have been identified that are also believed to be involved in the initiation of DNA replication in *S. cerevisiae*. Several of these mutant complementation groups show an ARS-specific phenotype, implying that the proteins coded for by these genes interact with ARSs in a sequence-specific manner, either directly or indirectly via other DNA-binding proteins. A number of minichromosome-maintenance (MCM)–defective mutants have been the subject of further analysis.

The product of the *MCM1* gene is a multifunctional protein. It is a transcription factor necessary for the expression of mating-type genes, as well as the maintenance of certain ARS-containing minichromosomes (62–64). *MCM3* is an essential gene that encodes a protein with striking homology to *MCM5* and *MCM2* (65). Overproduction of the protein product of *MCM3* results in instability of ARS minichromosomes, implying that this protein may act as part of a stoichiometrically balanced multiprotein complex. *MCM5* is identical to the previously characterized *CDC46* gene (66). Mutants in CDC46 arrest at the beginning of S phase with unreplicated DNA, implying a role at the G1–S phase boundary, which is likely to be directly linked to the initiation of DNA replication. Members of this multiprotein family have been identified in a broad range of organisms, including fission yeast (67), human cells (68), and amphibians (C-Y Khoo, J Gall, personal communication). The mammalian P1 protein, which has extensive homology to the *MCM3* gene product, was originally identified by its association with complex forms of DNA polymerase α (68), further implicating this protein family in DNA replication. Under nonpermissive conditions, a temperature-sensitive *Mcm3* mutant strain arrests with an almost fully replicated nucleus (69). Incomplete replication has been interpreted in terms of inefficient initiation from a subset of genomic ARS sequences. This would imply that individual members of this protein family contribute to the function of different ARS sequences in vivo, and would lead to a number of obvious questions. Do MCM proteins act sequentially, giving rise to a temporal pattern of replication within S phase? Do they have any developmental significance in terms of changing the order of initiation of replication and possibly also transcriptional activity? Do all members function in all cell types? Analysis of the amino acid sequence of this protein family suggests that they may function as ATP-dependent helicases (70).

The Multifunctional Replication Protein A

Replication protein A (RPA) is a single-stranded DNA-binding protein, originally identified as an essential component of in vitro viral DNA replication systems (71–73). This protein has also been implicated in other processes central to the maintenance of DNA, such as recombination (74) and DNA excision repair (75). Increasingly, efforts to understand DNA metabolism have lead to studies of this protein and its influence on the activities of other proteins. The 34-kDa subunit of human and yeast RPA is phosphorylated in a cell cycle–dependent manner at the start of S phase (76), suggesting a role in the control of initiation. Analysis of the products of SV40 replication suggest that phosphorylation occurs within the replication initiation complex itself (77) by a cyclin-dependent kinase (51, 78, 79). Dephosphorylation at mitosis may then reset the phosphorylation cycle, perhaps allowing nucleation of new replication complexes. Consistent with this, RPA has been found in prereplication foci at the end of mitosis (49).

Stimulation of Initiation by Transcription Factors

Initiation of DNA synthesis is carried out by complicated protein complexes that are known to assemble at specific DNA sequences in viral genomes, and genomes of simple eukaryotes such as yeast. Replication origins in simple genomes often share a common core structure, which usually consists of a recognition sequence for specific initiator proteins, a DNA-unwinding element or DUE (80, 81), and an A/T-rich element, as well as one or more additional sequences that bind transcription factors, reviewed in (6, 82). Transcriptional activators influence the efficiency with which complexes initiate DNA replication by a number of diverse mechanisms, such as by facilitating replication complex assembly or by stabilizing it against repression by chromatin formation, reviewed in (83, 84).

The transcriptional activator NF1 specifically prevents repression of SV40 DNA replication by chromatin assembly. The presence of a single NF1-binding site adjacent to the SV40 origin sequence stimulates DNA replication 20-fold in vivo, but has little effect on DNA synthesis in a cell-free system. However, if synthesis is monitored on SV40 minichromosomes in vitro, incubation with NF1 before chromatin assembly prevents repression of DNA synthesis (85). Disruption of the normal distribution of nucleosomes by the presence of NF1 therefore appears to be responsible for an increased efficiency of initiation, presumably by increasing accessibility of the origin region.

In another example of the influence of transcription factors on initiation, direct interaction with the replication machinery has been demonstrated. The acidic transcriptional activation domains of VP16, GAL4, and p53 bind to the large subunit of replication protein A in cellular extracts (86, 87). Mutations that weaken the ability of chimeric VP16-GAL4 to bind RPA reduce the ability of this transactivator to stimulate polyomavirus DNA replication. Enhanced replication via transcription factor binding to RPA is likely to reflect an influence on a very early stage of the initiation process, such as initiation complex assembly. Other experiments that insert binding sites for a range of cellular transcription factors near the SV40 origin (88) have shown that in these cases, the ability to stimulate replication does not correlate with transcriptional activity, therefore the mechanism by which replication is effected by transcription factor binding is not necessarily a consequence of the transcriptional event itself.

Many more examples exist of the influence of known transcriptional activators on the initiation of DNA replication at sequence-specific origins (84). It will be interesting to see if the principals arising from these observations apply to genomes where origin selection may be based only indirectly on DNA sequence. In complex genomes such as these, transcription factors could play an even more important role in origin selection.

Replication Timing and its Relationship to Transcription

Thousands of initiation events take place in the eukaryotic nucleus during every S phase. These events are spatially (Section I) and temporally coordinated. A correlation between transcriptional activity and early replication within S phase has been reported for many sequences (see below), although there are a number of notable exceptions to this generalization (89, 90).

What is the association between early replication and gene expression? The transient receptiveness of newly replicated DNA to transcription complex assembly could suggest that patterns of RNA synthesis might be partly specified by patterns of DNA synthesis (reviewed in Refs. 91–93), but this simple relationship would shed no light upon the mechanism by which certain replicons are selected for early replication. Do transcriptional activators assist replication complex assembly, either by acting as nucleation centers or by maintaining an open chromatin structure to allow easy access of the replication machinery (94)?

Replication timing of specific regions varies between cell types. This has been demonstrated for the tissue-specific expression of the cystic fibrosis (*CF)* gene locus and flanking DNA by fluorescent in situ hybridization (95). Unreplicated regions are visualized as singlet hybridization signals, while replicated loci are seen as doublets. In cells that express *CF,* a 500-kb domain (containing the *CF* gene) is replicated early in S phase, but it is late replicating in other cell types. Furthermore, in a series of cell fusion experiments, replication timing of the β-globin gene has been manipulated experimentally and coordinately with transcription. In hybrids between mouse hepatoma cells and mouse erythroleukemia (MEL) cells, which normally express β-globin, inactivation of the β-globin gene from the MEL parent cell coincided with a shift to later replication. Similarly, in fusions between MEL cells and human fibroblasts, the human gene becomes activated in conjunction with a shift to earlier replication of the whole of the β-globin domain. As the timing of replication is altered in these hybrids according to transcriptional status, order of replication does not appear to be fixed during the process of differentiation (96). Furthermore, this set of experiments demonstrates that both transcriptional activation and transcriptional repression are linked to replication timing in a predictable way. However, the central question of cause and effect has not been answered.

Replication Timing of Imprinted Genes

A further level of complexity that may have a bearing upon the relationship between transcription and replication arises in the case of differentially regulated alleles of imprinted genes. Hybridization with probes directed to several different chromosomal regions of the mouse genome that are known to contain imprinted genes has demonstrated that the two homologous alleles replicate asynchronously (97). Surprisingly, the paternal copy of each gene is always the early replicating one regardless of its transcriptional status. This raises the possibility that maintenance of the signal that identifies imprinted genes as maternal or paternal in origin is linked to the time at which they replicate. In this example, therefore, temporal control of replication does not appear to be imposed by transcriptional status. Rather, parentally determined patterns of gene expression may by maintained via temporally regulated replication.

III. CELL-CYCLE CONTROL OF DNA REPLICATION

Cyclin-Dependent Kinases and Entry into S Phase

Transitions from one cell-cycle phase to the next are controlled by activation and inactivation of cyclin-dependent protein kinases (cdks). In yeasts, a single cdk encoded by *CDC28* in *Saccharomyces cerevisiae* and *cdc2* in *Schizosaccharomyces pombe* plays a central role in the transitions into both S phase and mitosis. In mammalian cells, induction of these events is achieved by a growing family of *cdc2*-related proteins, in association with a range of cyclin subunits that activate the kinase and confer specificity. Cyclin binding has been shown to control both the timing of kinase activation and in some instances the subcellular location. Cdc2, in association with cyclins A and B, governs the post–S phase part of the cell cycle and is responsible for mitotic induction, but the situation for entry into S phase is more confusing. Cdk2, in association with E or A (and possibly C) type cyclins, appears to regulate G1 as well as the initiation and progression of S phase. This subject has been reviewed extensively (98, 99), so here we only briefly outline G1- and S-phase activities.

Experiments with *Xenopus* egg extracts have shown that *cdk2* but not *cdc2* is required for DNA replication (100). In mammalian cells *cdk2* associates with cyclins A and E. Association of both of these cyclins with the Rb-related protein p107 as well as the E2F transcription factor implies that *cdk2* might act to regulate gene expression. However, interpretation of these in vitro data is not straightforward (for review see Ref. 99). Additional evidence implicating cyclin A and *cdk2* in DNA replication has emerged from microscopy studies of replication fork clusters (Section I). *cdk2* and cyclin A are found to colocalize within these discrete sites of replication (50).

In yeast the decision to begin a new cell cycle is taken at a point in G1 called START. Once cells have passed this point, they are committed to proceed through S phase and mitosis. Accumulation of cyclins during G1 is rate limiting for passage through START. This also appears to be true for cultured mammalian cells, as overexpression of cyclin E in rat fibroblasts results in a shorter G1 coupled with longer S and G2 phases (101). In these experiments, cells overexpressing cyclin E were smaller than control cells even though the doubling times were equivalent. This implies that controls that maintain cell size are dependent on cyclin E levels and are functional in G1. Observations similar to these have also been made with respect to overexpression of D-type cyclins (102).

Other cyclins may also function in G1 based on their ability to act at START in yeast. Cyclin D has been implicated on this basis, but the significance of these observations is not clear, as the mitotic cyclins A and B can also function here. In higher eukaryotes, mitogenic signals are required for passage through the G1 restriction point, which appears to be analogous to START in yeast. Induction of D-type cyclins in quiescent cells upon addition of serum suggests that they may be involved in reentry to the cell cycle, and numerous reports of deregulation associated with malignancy suggest that cyclin D may play a central role in growth control. In mammalian cells, complexes of cyclin D1 and its associated kinase, cdk4, accumulate towards the end of G1 then fall in early S phase, implying a role in the G1–S phase transition (for review see Ref. 98).

Two relatively new additions to the list of B-type cyclins, *CLB5* and *CLB6,* are involved in the progression of S phase in *S. cerevisiae* (103,

104). *CLB5* and *CLB6* are transcribed towards the end of G1 under the control of the START-dependent transcription factor MBF at the same time as many other genes required for DNA replication; for review see Refs. 105, 106. Deletion of *CLB5* and *CLB6* disrupts the initiation of DNA replication, implying that their associated kinase activity may be involved in controlling S-phase entry.

In general, there appear to be more cyclins and cdks than there are identifiable functions. This situation resembles that for eukaryotic DNA polymerases, where there are presently too many replicative enzymes chasing too few defined roles. The relative roles of cyclins and their dependent kinases are subjects of intense study, so their functions are likely to clarify rapidly. Several key questions remain. Is phosphorylation by cdks involved in assembling active replication factories? Are there cascades of different cyclin-dependent kinases and if so, how do they activate and sustain DNA replication? Do cascades of activity relate to the temporal pattern of origin selection?

In addition to the activation of S phase by cyclin-dependent kinase activity, there is evidence that the tumor suppressor protein p53 may inhibit both entry into and progression through S phase (107–109). p53 was discovered by its ability to bind the SV40 initiator protein, T-antigen (110, 111), and it is now clear that it also binds RPA. In both cases p53 inhibits the replicative activity of the bound protein (112–114), suggesting that it might directly inhibit DNA replication.

Why Does DNA Replicate Only Once per Cell Cycle?

Cell fusion experiments by Rao & Johnson (115) found a clear difference in the replication capacity of G1- and G2-phase cells. Fusion with S-phase cells induced G1 nuclei but not G2 nuclei to begin DNA synthesis. This observation could reflect the absence of an essential replication activity from G2 nuclei or alternatively the presence of an inhibitory activity, which does not allow G2 nuclei to go through a second round of DNA synthesis. In either case, it is during mitosis that the G2 nucleus becomes altered to a replication-competent state.

Experiments in which DNA is microinjected into *Xenopus* eggs or incubated in egg extracts show that the inability of replicated DNA to replicate a second time does not depend on specific DNA sequences (24, 27, 28). Subsequent experiments with isolated nuclei replicated in *Xenopus* egg extract showed that this regulatory mechanism depends on the nuclear membrane. *Xenopus* sperm nuclei replicate their DNA only once in egg extract (28), but if they are treated with agents that repermeabilize the nuclear membrane after replication they become competent for a second round of DNA synthesis (116). Therefore nuclear membrane breakdown is the only feature of mitosis that is required to convert a replicated nucleus into a state suitable for new initiations. These observations were confirmed and extended using nuclei isolated from synchronized mammalian cells. G1 HeLa nuclei replicated in egg extract, while G2 nuclei did not, but permeabilization of G2 nuclei allowed reinitiation (117). These observations as well as the cell fusion experiments of Rao & Johnson can be explained by a simple model (116), in which an activity essential for DNA synthesis cannot enter the nucleus from the cytoplasm in interphase because it lacks a nuclear localization sequence. This activity, called "licensing factor," would bind to DNA when the nuclear membrane breaks down at mitosis. A second signal would allow DNA synthesis to begin at the start of S phase, leading to destruction of licensing activity.

Permeabilization of the nuclear membrane could allow a diffusible activator to enter nuclei or a diffusible inhibitor to escape. These possibilities were tested directly by reversibly permeabilizing the nuclear membrane of G2 nuclei (118). If permeabilizing the nuclear membrane allows a diffusible inhibitor to be lost from the nucleus, then subsequently repairing the membrane should not prevent replication, because the inhibitor will have had the opportunity to escape. On the other hand, if permeabilization allows a positive factor to enter the nucleus, then a repaired nucleus will only replicate if the factor was present before repair. The result observed was that membrane repair successfully reimposed the block to DNA synthesis that was originally present in intact G2 nuclei. This argues strongly for the existence of at least one positively acting activity within *Xenopus* egg extract, which enters the nucleus when it is in a permeable state and "licenses" DNA for another round of DNA synthesis. Furthermore, incubation of permeable G2 nuclei in a soluble fraction from the extract before membrane repair allowed the nuclei to replicate (118).

Another study has produced evidence in support of a diffusible positive factor (119). Metaphase-arrested *Xenopus* egg extracts were treated with the protein kinase inhibitor 6-dimethylaminopurine (6-DMAP) before being induced to undergo the transition into S phase. This prevented the generation of an activity whose behavior is consistent with that proposed for licensing factor. This activity must therefore exist in two states, only one of which is active for replication of sperm chromatin ("activated" S-phase extract). This appears to be a contradiction, as any activity present only in S-phase cytoplasm (activated *Xenopus* egg extract) would not be present at the right time (mitosis) to gain access to the nucleus via nuclear membrane breakdown. In order to reconcile these observations with the licensing factor model, it appears necessary to suggest either that activation of licensing factor occurs within the nucleus of cycling cells or that a critical window exists between cytoplasmic activation and the reformation of the nuclear membrane at the end of mitosis. It is not clear at present whether the positive activity reported by Blow (119), and more recently in a similar study by Kubata & Takisawa (119a), is the same as the positive activity that allows new replication in replicated somatic nuclei. If these two activities are the same, then DMAP-treated extracts provide a powerful assay for the purification of licensing factor.

Mutations and Drugs that Cause Repeated Replication

Multiple rounds of replication without an intervening mitosis are not easy to reconcile with a simple licensing factor that acts as part of initiation complexes at the beginning of S phase. Table 1 lists several investigations that describe mutant phenotypes and the actions of specific inhibitors that uncouple S phase from mitosis. We consider some of these examples here and then consider how they can be reconciled with the model in the next section.

A role for *cdc2* kinase in the mechanism that prevents over-replication has been suggested by the identification of *S. pombe cdc2* mutants that can be induced to diploidize (120). The *cdc2* gene product is required twice during the *S. pombe* cell cycle, once in late G2 for the promotion of mitosis and once in G1 for the promotion of S phase. Certain mutations in *cdc2* allow the cell to bypass mitosis, as a shift to the restrictive temperature at the right time results in degradation of the mitotic kinase. It has been proposed that one form of *cdc2* kinase identifies cells as being post–S phase, and that in its absence G2 memory is lost so cells are free to enter G1 and eventually another round of synthesis (reviewed in Ref. 121). G2 memory could be achieved via modification of replicated chromatin or by association

Table 1 Escape from once per cell cycle control[a]

Mutation/treatment	System	Extent of re-replication	Ref.
K252a (protein kinase inhibitor)	rat fibroblasts	complete rounds of DNA synthesis up to 32C	122
nuclear membrane permeabilization	*Xenopus* egg extract and replicated nuclei	up to a single round of re-replication	116, 117
tscdc2-M26 tscdc2-33	*S. pombe*	conversion to diploid state	120
ts41	CHO cells	complete rounds of DNA synthesis up to 16C	123
tshdf	*S. cerevisiae*	partial re-replication	152

[a] Re-replication in the absence of an intervening mitosis has been described in a range of biological systems. In almost all the cases, over-replication takes the form of complete or nearly complete new rounds of DNA synthesis, implying that a block to re-replication of the whole nucleus is bypassed. In the last example, deregulation at the level of individual replicons or parts of replicons may have resulted in partially over-replicated cells, which have variable DNA contents of 2–4C.

of *cdc2* kinase with the nucleus, resulting in a block to re-replication. Secondly, the staurosporine analog, K252a, can also induce repeated rounds of DNA synthesis in mammalian cells without an intervening mitosis, generating cells with DNA contents of 16 or 32C (122). This suggests that a protein kinase sensitive to K252a is involved in the mechanism that normally prevents over-replication in this system.

Thirdly, the *ts41* mutation of Chinese hamster ovary cells allows repeated S phases without an intervening mitosis (123). Unlike cells treated with K252a, which have an inter-S period of approximately eight hours, the altered cell cycle in *ts41* cells occurs without G2, M, and G1 and bypasses both the G2 checkpoint and the G1 restriction point. Temperature-shift experiments indicate that one execution point for the activity defined by the *ts41* mutation is in late G1 and may influence entry into mitosis, and a second is in G2 and is required to inhibit re-entry into S phase. Handeli & Weintraub (123) propose that the *ts41* gene product exists in two alternate configurations. The one that inhibits entry into S phase would be inactivated at the G1 restriction point by conversion into its other form, which would promote mitosis (and be inactivated as a consequence). The complimentary and probably connected cycles of *cdc2* and *ts41* may therefore together define the order of cell-cycle events.

The studies considered in this section could all be explained by failure or inhibition of a negative regulatory mechanism rather than in terms of the simple form of the positive licensing factor model. We consider this in the following section.

Possible Mechanisms for Preventing Repeated Replication within One Cell Cycle

Ideally an explanation of the regulatory mechanisms that prevent repeated replication within a single cell cycle should account for the following observations. First, intact replicated nuclei are unable to replicate again when exposed to S phase cytoplasm (115–118). Second, permeabilizing the nuclear membrane of a replicated nucleus allows it to replicate again (116–118). Third, repairing the membrane of a permeable G2 nucleus restores it to the G2 state, preventing replication (118). Fourth, specific DNA sequences are not required to prevent over-replication; once per cell cycle control can be imposed on prokaryotic DNA (24, 27, 28). Fifth, nuclear membrane breakdown does not occur in lower eukaryotes such as yeast, therefore it cannot be universally essential to allow subsequent replication cycles. Finally, certain mutations and drugs can allow repeated cycles of DNA replication without intervening mitoses (120, 122, 123).

It is difficult to reconcile all of these observations in a single model. Although the first four observations are accommodated by the licensing factor model of Blow & Laskey (116), it is more difficult to reconcile the lack of nuclear envelope breakdown in yeast or the existence of drugs and mutations that allow new cycles of replication without mitosis and therefore presumably without nuclear membrane breakdown. The closed mitosis in yeast could be reconciled with the licensing factor model if regulated nuclear transport allows timed access of a positive factor into the nucleus only during mitosis. There are precedents for proteins that behave in this way, notably the *cdc46/mcm3* family of *S. cerevisiae* (66, 69). These genes encode proteins that are required for initiation of DNA replication, and that enter the yeast nucleus only during mitosis. Although the proteins of this family are strictly conserved throughout eukaryotes (124), early indications of their location during the cell cycle suggest that they do not show the same pattern of timed nuclear entry that is seen in yeast (68).

The existence of drugs and mutations that allow repeated replication without mitosis or apparent nuclear envelope breakdown can be reconciled by a modification of the positive licensing factor model. It is necessary to invoke one additional component, namely a nondiffusible inhibitory factor present in the nucleus of replicated cells, as has been suggested before by experiments using viral origins (125, 126). A diffusible licensing factor would act catalytically to cancel this bound inhibitor during mitosis (Figure 4). A requirement for a catalytic licensing factor could be bypassed in mutant cells if its putative target, the inhibitory activity, were absent. If such a model is correct, then the cyclin-dependent kinases themselves might be candidates for a catalytic licensing role.

Completion of S Phase

In most eukaryotic cells a complicated regulatory system monitors the progress of S phase. By detecting some feature of unreplicated chromatin, it is able to delay mitosis until DNA synthesis is completed, ensuring duplication of the entire genome during each cell cycle. This can be observed when actively replicating S-phase cells are fused to cells in G2 (115). The G2 nucleus is delayed from entering mitosis until the S-phase nucleus has completed DNA synthesis. This observation argues strongly for a diffusible inhibitory signal that is either generated as a consequence of the process of DNA synthesis or activated by something associated with an unreplicated genome. As mitosis is delayed by inhibitors of elongation, the primary signal does not seem to arise as a by-product of synthesis itself.

Biochemical studies have clearly demonstrated the dependency of mitosis upon the completion of S phase and have begun to dissect the mechanism that ensures that these events remain separated in time. Early embryonic cell cycles are not normally subject to mitotic checkpoint control; nevertheless, Dasso & Newport (127) were able to observe this dependency in

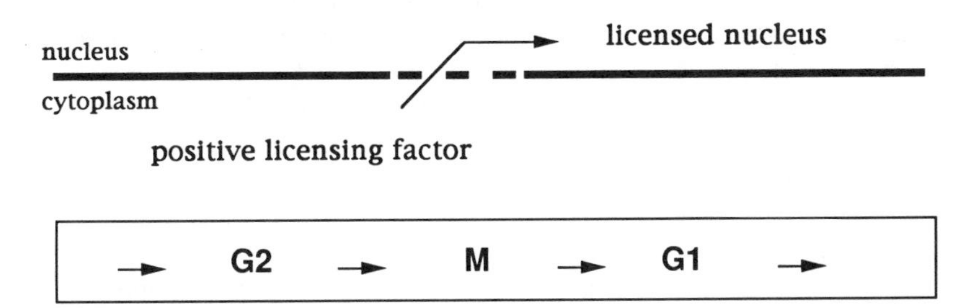

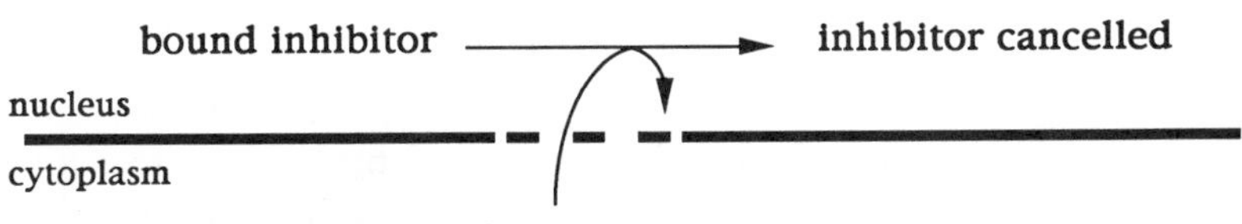

B. Catalytic licensing factor model

bound inhibitor

inhibitor cancelled

nucleus

cytoplasm

positive licensing factor

Figure 4 Variations of a licensing factor model. Licensing factor represents a positive activity, which can convert replicated G2 nuclei into a state suitable for a new round of replication, but only when the nuclear membrane breaks down or is permeabilized by chemical treatments. (*A*) Licensing factor was originally proposed to be an essential DNA-binding protein, which entered the nucleus at mitosis and bound to sites of initiation. Inactivation of the factor as a consequence of DNA synthesis would ensure a single round of replication. (*B*) In addition to this possibility, the available data on licensing factor can be interpreted in terms of a catalytic activity that enters the nucleus at mitosis and carries out a "hit and run" function. Interaction of licensing factor with inhibitory components of a G2 nucleus would lead to the generation of a "licensed" G1 nucleus. It is possible that licensing factor acts by a combination of these two mechanisms and remains in the nucleus after mitosis in order to carry out a later function. The hit and run licensing model is attractive because it can incorporate negative mechanisms that could prevent new replication on already replicated DNA, and could therefore reconcile reports of mutations and drugs that induce re-replication without an intervening mitosis with data on nuclear membrane permeabilization (see Table 1).

cycling *Xenopus* egg extracts that alternate between S phase and mitosis. By adding large numbers of exogenous nuclei, they showed that the presence of unreplicated chromatin prevented the initiation of mitosis by blocking the activation of the mitotic inducer maturation-promoting factor (MPF) (127). Restriction of the activity of this key regulator of cell cycle progression can be thought of as the last of three phases in the process that couples S phase to mitosis. The first is detection of an incompletely replicated genome, and the second defines the process by which the signal is transmitted to the effector, MPF (cdc2 kinase). Signal transmission is as yet relatively undefined, but the major G2 checkpoint that can block cell-cycle progression entirely is believed to lie somewhere in this pathway. In a screen for *S. pombe* mutants that enter mitosis prematurely when DNA synthesis is partially inhibited by hydroxyurea, five new genes (called *hus1–5*) have been identified (128). None of these are members of previously identified linkage groups and all are candidates for components of the system that links the primary signal generated in response to unreplicated DNA to the control of mitotic induction by cdc2 kinase. A list of known mutations that affect the dependence of mitosis upon S phase as well as detailed discussion of those that are thought to influence detection of incomplete replication is given in a review by Li & Deshaies (129). Three genes that may function at the detection end of the coupling mechanism are discussed below.

Detection of Incompletely Replicated DNA

Mitosis in the tsBN2 mutant rodent cell line can be uncoupled from the completion of DNA synthesis, as a shift to the restrictive temperature during S phase results in premature mitotic entry (130). A key component, necessary for the generation, maintenance, or transmission of the signal that restrains the onset of mitosis, must therefore be altered in this cell line. Complementation of tsBN2 allowed identification of the *RCC1* gene (Regulator of Chromosome Condensation) (131), which codes for a nuclear protein with DNA-binding activity (132). RCC1 associates with the Ras-like GTP-binding protein, Ran (133, 134), to form a chromatin-bound protein complex. RCC1 can catalyze guanine nucleotide exchange on Ran, leading to dissociation of Ran-GTP from RCC1. Ran-GTP, in association with a GAP protein (GTPase-activating protein), may possibly act to transmit information on the status of DNA replication to downstream targets, which ultimately influence the activity of MPF. Loss of RCC1 in S phase leads to the activation of cdc2 kinase (130; discussed in 135).

RCC1 may be responsible for sensing the presence of unreplicated DNA and conveying this information to the machinery responsible for mitotic changes, but the central question of how RCC1 activity is modulated in response to the changing replication status of the genome remains to be answered. Presumably the progression of S phase would downregulate RCC1's ability to catalyze nucleotide exchange on Ran, resulting in a progressively weaker signal to restrain mitotic onset.

Analysis of *pim1* mutants (which are defective in the *S. pombe* homolog of RCC1) in a *cdc10* background, suggests that the wild-type protein is active in mitotic restraint even before S phase has begun (134). The dependence of mitosis on the completion of DNA replication does not therefore depend upon events that take place as a consequence of initiation.

In addition to the phenotype seen when tsBN2 cells are shifted to the restrictive temperature during S phase, they also display a G1-S block if shifted during G1. This is consistent with immunodepletion experiments that remove RCC1 from *Xenopus* egg extracts, resulting in a defect in DNA replication at a very early stage, possibly related to initiation or replication complex assembly (136). Surprisingly, at the restrictive temperature, both transcription and translation are also inhibited in tsBN2 cells, which may suggest roles for RCC1 distinct from its involvement in mitotic regulation. The finding that small nucleolar RNAs are processed out of the *RCC1* premessenger RNA or spliced out introns may have a bearing on this observation (137). Discussion of the possible functions of RCC1 in this system is covered in a review on the pleiotropic effects of *RCC1* mutations (138).

A number of other genes that are essential for maintaining the dependence of mitosis upon the completion of DNA synthesis are also required at the start of S phase. This implies that structures that are important for the initiation of DNA synthesis are involved in signalling the presence of unreplicated chromatin. The initiation deficiency in *Xenopus* extracts depleted of RCC1 has been described above, but in addition, the products of the *cut5* and *cdc18* genes of *S. pombe* have recently been reported to be involved in events at the beginning of S phase as well as mitotic restraint. The *cut5* gene appears to play a positive role in S-phase progression, as *cut5* mutants accumulate with an unreplicated DNA content (139). It is not known whether DNA synthesis is completely blocked or whether it is working inefficiently in these mutants. *cut5* mutants also fail to recognize the incomplete S phase and do not arrest in G2, leading to aberrant mitosis and eventual cell death. Failure to restrain mitosis is also evident in *cdc10-cut5* double mutants, as reported for the yeast *RCC1* mutant, implying that *cut5* plays a role in mitotic restraint before the onset of S phase.

Cloning of the *cut5* gene revealed that it is identical to the *S. pombe rad4* gene that is involved in the cellular response to radiation-induced DNA damage. Saka & Yanagida (139) propose that the absence of *cut5*/*rad4* from unreplicated chromatin may lead to failure to initiate DNA synthesis at the start of S phase as well as the inability to generate a signal that restrains mitotic inducers. In addition, they suggest that its absence from duplicated chromatin may cause instability, leading to the *rad* phenotype and a high frequency of sister-chromatid exchange. An alternative and more speculative scenario sees the *cut5*/*rad4* protein as an early part of the signal transmission mechanism that restrains mitotic inducers. Primary signals generated by both unreplicated and damaged DNA may converge on this pathway, explaining the *rad* phenotype as well as the lack of dependence of mitosis on the completion of S phase. The initiation defect observed in *cut5* (and *RCC1*) mutants could reflect the activity of a G1-S checkpoint that monitors the assembly of the mitotic checkpoint detection/signal transmission apparatus, or alternatively a structural requirement for its assembly before replication complexes can be activated.

A dual role in S-phase initiation and completion may also apply to the *cdc18* gene of *S. pombe,* which is expressed at the G1-S boundary under the control of the START gene *cdc10* (140). Deletion of *cdc18* prevents cells from entering S phase, but they still enter mitosis, leading to the appearance of cells with less than a 1C DNA content. The temperature-sensitive *cdc18-k46* mutation displays a different phenotype, causing cells to arrest with 2C DNA content but with chromosomes that are structurally different from those in cells that are arrested late in G2. So the early S-phase function remains intact in this mutant, whereas the completion of replication events and the subsequent dissipation of signals that restrain mitotic onset do not occur. Kelly et al (140) propose that completion of the function of *cdc18* in DNA synthesis would lead to its inability to function as a negative regulator of mitosis. This implies that structures specific to an unreplicated genome, such as unused or unresolved replication complexes, may be the primary signal, leading to the generation of signals that inhibit mitosis. *cdc18, RCC1,* and possibly *cut5* would be an intrinsic part of these structures.

Signal Transmission and Mitotic Restraint

Uncoupling of M phase from the completion of S phase can also be achieved by directly deregulating cdc2 kinase activity. For example, certain alleles of *S. pombe cdc2*, such as *cdc2-3w* mutations, advance mitosis even in the presence of DNA synthesis inhibitors such as hydroxyurea (141). In addition, alterations in the activity of regulators of cdc2 predictably influence mitotic onset. When overexpressed, the *cdc25* gene product can cause premature entry into mitosis by activating cdc2 kinase at the wrong time. Dephosphorylation of a single tyrosine residue in the ATP-binding site of cdc2 by the phosphatase encoded by *cdc25* causes activation in *S. pombe*. Phosphorylation of this tyrosine is required for the inhibition of mitosis occurring in response to the detection of unreplicated DNA, as mutations that substitute the tyrosine abolish the dependence of M on S phase (142). A second independent mitotic inducer, *NIMA,* has been identified in *Aspergillus nidulans*. Overexpression of this gene can induce cells treated with DNA synthesis inhibitors to enter mitosis (reviewed in Ref. 135). The relationship of this pathway to mitotic checkpoints remains to be seen.

In budding yeast the checkpoint that delays mitosis until inhibitory signals have dissipated is defined by mutations in the *RAD9* gene (143). In its absence the dependence of mitosis on the completion of S phase is relieved, so that mutations or some drug treatments that normally delay mitosis via an effect on DNA synthesis do not influence mitosis in *rad9* mutant cells. In addition to incomplete DNA synthesis, the presence of DNA damage can influence the onset of mitosis by modulating the activity of MPF via the *RAD9* checkpoint. A number of genes other than *RAD9* are involved in the mechanisms that monitor both DNA damage and DNA replication, although several features suggest that these pathways are distinct (144, 145). Analysis of double mutant phenotypes suggests that *RAD9* encodes a repressor of mitosis that is activated by the presence of chromosome breaks (143). The *S. pombe chk1* protein kinase, which was isolated as a suppressor of a cold-sensitive *cdc2* allele, may provide a link with the *rad* checkpoint, as multiple copies partially rescue the ultraviolet sensitivity of the checkpoint-deficient mutant *rad1-1* (146). Overexpression of *chk1* in normal cells results in a delay in the onset of mitosis, consistent with the suggestion that *chk1* acts to inhibit *cdc2* kinase activity.

CONCLUSIONS AND PROSPECTS

In the first section of this review we have deliberately chosen to question the widespread view that replication origins in prokaryotes, viruses, and budding yeast are appropriate models for higher eukaryotes. Although it is clear that replication initiates at preferred, nonrandom sites within the chromosomes of higher eukaryotes, there is growing evidence that these sites are not essential for replication of purified DNA in animal cells. Although the report of preferential initiation on plasmids containing *c-myc* upstream sequences deserves to be pursued further (20), the general lack of efficient plasmid assays for replication origins in higher eukaryotes has held back progress.

Even if it becomes possible to identify sequences that specify origins in higher eukaryotes, the question of why replication initiates at preferred sites will remain. The conventional explanation that a specific sequence is necessary to allow replication enzymes to bind DNA has been inadequate for a decade. So why are replication origins really needed? Any explanations must take account of the facts that origin deletions from long tracts of yeast DNA are viable (5), and that patterns of replication vary at different developmental stages within the same organism (32, 147, 148). Two explanations for the existence of preferred sites appear to be consistent with all of the observations. First, the positions of replication origins might be selected simply to ensure that all forks can meet within a limited S phase. This facile explanation is consistent with the small but detectable increase in chromosome loss when several contiguous origins are deleted in yeast

(5). Alternatively, specific replication origins might provide a means of coordinating replication and transcription. Restricting initiation of replication to regions outside transcribed genes might help to coordinate polymerase traffic. For example, some eukaryotic replication forks pause when they approach transcribed regions (14, 19, 149–151). This is difficult to envisage if replication initiates within an active gene. We favor a dual role for replication origins, achieving both the saturation of the genome with evenly spaced replication forks as well as coordination with patterns of transcription.

Evidence that replication forks are clustered into "foci" or "factories" raises important questions about the dynamics of interphase chromatin. The recent evidence for DNA spooling is particularly challenging (52). It will be important to determine how these patterns are organized. The imposition of a typical eukaryotic pattern on pseudonuclei assembled from purified prokaryotic DNA invites the conclusion that a protein framework can measure off DNA at regular intervals, similar to the mechanism of phage packaging. However there must also be a sequence-specific component in somatic cells to account for the results of origin-mapping studies. Several proteins that occupy foci have been identified, but it will be important to identify those proteins that organize the structural pattern.

Our understanding of the molecules and mechanisms of initiation of eukaryotic DNA replication is poised for a major advance following the breakthrough of identification of the origin recognition complex (ORC) in *S. cerevisiae*. Identification of similar complexes in other organisms may provide the key to resolving the enigmatic properties of replication origins in higher eukaryotes. The origin recognition complex is an obvious target for the machinery that triggers entry into S phase. Evidence that cyclin-dependent kinases, especially *cdk2*, are involved is strong, focusing increasing attention on the search for their protein substrates. It will be equally important to identify the pathways that control activation of the cdks, and to assign individual functions to the growing family of cyclins and their associated kinases.

It is not clear yet if cdks are directly involved in preventing repeated rounds of DNA replication within a single cell cycle. Increased understanding of this mechanism should come from identifying the positive replication factor(s) from frog egg extracts that is excluded by the nuclear membrane. It will be important to understand how this relates to the phenotypes of mutants that over-replicate their DNA, apparently without undergoing nuclear envelope breakdown.

Not only are entry into, and progression through, S phase regulated, but exit from S phase is also controlled. The role of *RCC1* is becoming clearer, but many questions remain about how the cell monitors completion of DNA replication.

In conclusion, there can be little doubt that our knowledge of the control of eukaryotic DNA replication would advance much more rapidly if efficient cell-free replication systems that initiate replication in vitro were available from yeast and mammalian cells.

ACKNOWLEDGMENTS

We are grateful to John Pines, Mark Madine, and Mark Jackman for critical reading of the manuscript. Figure 2 was kindly provided by Lynne Cox. The authors are supported by grant SP1961 from the Cancer Research Campaign.

Literature Cited

1. Fangman WL, Brewer BJ. 1991. *Annu. Rev. Cell Biol.* 7:375–402
2. Fangman WL, Brewer BJ. 1992. Cell 71:363–66
3. Stinchcomb DT, Struhl K, Davis RW. 1979. *Nature* 282:39–43
4. Brewer BJ, Fangman WL. 1991. *BioEssays* 13:317–22
5. Newlon CS, Collins I, Dershowitz A, Deshpande AM, Greenfeder SA, et al. 1994. *Cold Spring Harbor Symp. Quant. Biol.* In press
6. DePamphilis ML. 1993. *Annu. Rev. Biochem.* 62:29–63
7. DePamphilis ML. 1993. *Curr. Opin. Cell Biol.* 5:434–41
8. Vassilev LT, DePamphilis ML. 1992. *Crit. Rev. Biochem. Mol. Biol.* 27:445–72
9. DePamphilis ML. 1993. *J. Biol. Chem.* 268:1–4
10. Linskens MHK, Huberman JA. 1990. *Cell* 62:845–47
11. Benbow RM, Zhao J, Larson DD. 1992. *BioEssays* 14:661–70
12. Hamlin JL. 1992. *Crit. Rev. Eukaryot. Gene Expr.* 2:359–81
13. Hamlin JL, Vaughn JP, Dijkwel PA, Leu T-H, Ma C. 1991. *Curr. Opin. Cell Biol.* 3:414–21
14. Little RD, Platt THK, Schildkraut CL. 1993. *Mol. Cell Biol.* 13:6600–13
15. Hyrien O, Mechali M. 1992. *Nucleic Acids Res.* 201:1463–69
16. Hyrien O, Mechali M. 1993. *EMBO J.* 12:4511–20
17. Mahbubani HM, Paull T, Elder JK, Blow JJ. 1992. *Nucleic Acids Res.* 20:1457–62
18. Shinomiya T, Ina S. 1991. *Nucleic Acids Res.* 19:3935–41
19. Shinomiya T, Ina S. 1993. *Mol. Cell Biol.* 13:4098–106
20. McWhinney C, Leffak M. 1990. *Nucleic Acids Res.* 18:1233–42
21. Krysan PJ, Haase SB, Calos MP. 1989. *Mol. Cell Biol.* 9:1026-33
22. Krysan PJ, Smith JG, Calos MP. 1993. *Mol. Cell Biol.* 13:2688–96
23. Krysan PJ, Calos MP. 1991. *Mol. Cell Biol.* 11:1464–72
24. Harland RM, Laskey RA. 1980. *Cell* 21:761–71
25. Laskey RA, Harland RM, Earnshaw WC, Dingwall C. 1981. *International Cell Biology 1980,* ed. HG Schweiger, pp. 162–67. Heidelberg: Springer-Verlag
26. Mechali M, Kearsey S. 1984. *Cell* 38:55–64
27. Mechali M, Mechali F, Laskey RA. 1983. *Cell* 35:63–69
28. Blow JJ, Laskey RA. 1986. *Cell* 47: 577–87
29. Newport J. 1987. *Cell* 48:205–17
30. Blow JJ, Sleeman AM. 1990. *J. Cell Sci.* 95:383–91
31. Laskey RA. 1985. *J. Embryol. Exp. Morphol.* 89:285–96
32. Blumenthal AB, Kriegstein HJ, Hogness DS. 1974. *Cold Spring Harbor Symp. Quant. Biol.* 38:205–23
33. Gilbert DM, Miyazawa H, Nallaspeth FS, Ortega JM, Blow JJ, DePamphilis

ML. 1994. *Cold Spring Harbor Symp. Quant. Biol.* In press
34. Gurdon JB, Laskey RA. 1970. *J. Embryol. Exp. Morphol.* 24:227–48
35. Kill IR, Bridges JM, Campbell KHS, Maldonado-Codina G, Hutchison CJ. 1991. *J. Cell Sci.* 100:869–76
36. Cox LS, Laskey RA. 1991. *Cell* 66: 271–75
37. Sleeman AM, Leno GH, Mills AD, Fairman MP, Laskey RA. 1992. *J. Cell Sci.* 101:509–15
38. Nonet GH, Wahl GM. 1993. *Somat. Cell Mol. Genet.* 19:171–92
39. Gilley D, Preer JRJ, Aufderheide KJ, Polisky B. 1988. *Mol. Cell Biol.* 8: 4765–72
40. Edgar BA, Schubiger G. 1986. *Cell* 44:871–77
41. Newport J, Kirschner M. 1982. *Cell* 30:675–86
42. Newport J, Kirschner M. 1984. *Cell* 37:731–42
43. Nakamura H, Morita T, Sato C. 1986. *Exp. Cell Res.* 165:291–97
44. Nakayasu H, Berezney R. 1989. *J. Cell Biol.* 108:1–11
45. Fox MH, Arndt-Jovin DJ, Jovin TM, Baumann PH, Robert-Nicoud M. 1991. *J. Cell Sci.* 99:247–53
46. O'Keefe RT, Henderson SC, Spector DL. 1992. *J. Cell Biol.* 116:1095–110
47. Mills AD, Blow JJ, White JG, Amos WB, Wilcock D, Laskey RA. 1989. *J. Cell Sci.* 94:471–77
48. Wilcock D, Lane DP. 1991. *Nature* 349:429–31
49. Adachi Y, Laemmli U. 1992. *J. Cell Biol.* 119:1–15
50. Cardoso MC, Leonhardt H, Nadal-Ginard B. 1993. *Cell* 74:1–20
51. Fotedar R, Roberts JM. 1991. *Cold Spring Harbor Symp. Quant. Biol.* 56:325–33
52. Hozak P, Hassan AB, Jackson DA, Cook PR. 1993. *Cell* 73:361–73
53. Wessel R, Schweizer J, Stahl H. 1992. *J. Virol.* 66:804–15
54. Leonhardt H, Page AW, Weier H.-U, Bestor T. 1992. *Cell* 71:865–73
55. Marahrens Y, Stillman B. 1992. *Science* 255:817–23
56. Diffley JFX, Stillman B. 1990. *Trends Genet.* 6:427–32
57. Bell SP, Stillman B. 1992. *Nature* 357:128–34
58. Micklem G, Rowley A, Harwood J, Nasmyth K, Diffley J. 1993. *Nature* 366:87–89
58a. Foss M, McNally FJ, Laurenson P, Rine J. 1993. *Science* 262:1838–44
58b. Bell SP, Kobayashi R, Stillman B. 1993. *Science* 262:1844–49
59. Diffley JFX, Cocker JH. 1992. *Nature* 357:169–72
60. Diffley JFX. 1992. *Trends Cell Biol.* 2:298–303
61. Maine GT, Sinha P, Tye B-K. 1984. *Genetics* 106:365–85
62. Passmore S, Elble R, Tye B-K. 1989. *Genes Dev.* 3:921–35
63. Elble R, Tye B-K. 1991. *Proc. Natl. Acad. Sci. USA* 88:10966–70
64. Christ C, Tye B-K. 1991. *Genes Dev.* 5:751–63
65. Yan H, Gibson S, Tye B-K. 1991. *Genes Dev.* 5:944–57
66. Hennessy KM, Lee A, Chen E, Botstein D. 1991. *Genes Dev.* 5:958–69
67. Coxon A, Maundrell K, Kearsey SE. 1992. *Nucleic Acids Res.* 20:5571–77
68. Thommes P, Fett R, Schray B, Burkhart R, Barnes C, et al. 1992. *Nucleic Acids Res.* 20:1069–74
69. Gibson SI, Surosky RT, Tye B-K. 1990. *Mol. Cell. Biol.* 10:5707–20
70. Koonin EV. 1993. *Nucleic Acids Res.* 21:2541–47
71. Wobbe CR, Weissbach L, Borowiec JA, Dean FB, Murakami Y, et al. 1987. *Proc. Natl. Acad. Sci. USA* 84:1834–38
72. Fairman MP, Stillman B. 1988. *EMBO J.* 7:1211–18
73. Wold MS, Kelly T. 1988. *Proc. Natl. Acad. Sci. USA* 85:2523–27
74. Heyer W-D, Kolodner RD. 1989. *Biochemistry* 28:2856–62
75. Coverley D, Kenny MK, Munn M, Rupp WD, Lane DP, Wood RD. 1991. *Nature* 349:538–41
76. Din S, Brill SJ, Fairman MP, Stillman B. 1990. *Genes Dev.* 4:968–77
77. Fotedar R, Roberts JM. 1992. *EMBO J.* 11:2177–87
78. Dutta A, Din S, Brill SJ, Stillman B. 1991. *Cold Spring Harbor Symp. Quant. Biol.* 56:315–24
79. Dutta A, Stillman B. 1992. *EMBO J.* 11:2189–99
80. Kowalski D, Eddy MJ. 1989. *EMBO J.* 8:4335–44
81. Umek RM, Kowalski D. 1990. *Proc. Natl. Acad. Sci. USA* 87:2486–90
82. Kornberg A, Baker T. 1992. *DNA Replication.* New York: Freeman
83. Heintz NH. 1992. *Curr. Opin. Cell Biol.* 4:459–67
84. DePamphilis ML. 1993. *Trends Cell Biol.* 3:161–67
85. Cheng L, Kelly TJ. 1989. *Cell* 59:541–51
86. He Z, Brinton BT, Greenblatt J, Hassell JA, Ingles CJ. 1993. *Cell* 73:1223–32
87. Li R, Botchan MR. 1993. *Cell* 73: 1207–21

88. Hoang AT, Wang W, Gralla JD. 1992. *Mol. Cell Biol.* 12:3087–93
89. Benard M, Pallotta D, Pierron G. 1992. *Exp. Cell Res.* 201:506–13
90. Pierron G, Benard M, Puvion E, Flanagan R, Sauer HW, Pallotta D. 1989. *Nucleic Acids Res.* 17:553–66
91. Wolffe AP. 1991. *J. Cell Sci.* 99:201–6
92. Almouzni G, Wolffe AP. 1993. *Exp. Cell Res.* 205:1–15
93. Felsenfeld G. 1992. *Nature* 355:219–24
94. Holmquist GP. 1987. *Am. J. Hum. Genet.* 40:151–73
95. Selig S, Okumura K, Ward DC, Cedar H. 1992. *EMBO J.* 11:1217–25
96. Dhar V, Skoultchi AI, Schildkraut CL. 1989. *Mol. Cell Biol.* 9:3524–32
97. Kitsberg D, Selig S, Brandeis M, Simon I, Keshet I, et al. 1993. *Nature* 364:459–63
98. Pines J. 1993. *Trends Biochem. Sci.* 18:195–97
99. Sherr CJ. 1993. *Cell* 73:1059–65
100. Fang F, Newport JW. 1991. *Cell* 66:731–42
101. Ohtsubo M, Roberts JM. 1993. *Science* 259:1908–12
102. Quelle DE, Ashmun RA, Shurtleff SA, Kato J-y, Bar-Sagi D, et al. 1993. *Genes Dev.* 7:1559–71
103. Epstein CB, Cross FR. 1992. *Genes Dev.* 6:1695–706
104. Schwob E, Nasmyth K. 1993. *Genes Dev.* 7:1160–75
105. Johnston LH, Lowndes NF. 1992. *Nucleic Acids Res.* 20:2403–10
106. Johnston LH. 1992. *Trends Cell Biol.* 2:353–57
107. Levine AJ, Momand J, Finlay CA. 1991. *Nature* 351:453–56
108. Lane DP. 1992. *Nature* 358:15–16
109. Pietenpol JA, Vogelstein B. 1993. *Nature* 365:17–18
110. Lane DP, Crawford LV. 1979. *Nature* 278:261–63
111. Linzer DIH, Levine AJ. 1979. *Cell* 17:43–52
112. Braithwaite AW, Sturzbecher H-W, Addison C, Palmer C, Rudge K, Jenkins JR. 1987. *Nature* 329:458–60
113. Wang EH, Friedman PN, Prives C. 1989. *Cell* 57:379–92
114. Dutta A, Ruppert JM, Aster JC, Winchester E. 1993. *Nature* 365:79–82
115. Rao PN, Johnson RT. 1970. *Nature* 225:159–64
116. Blow JJ, Laskey RA. 1988. *Nature* 332:546–48
117. Leno GH, Downes CS, Laskey RA. 1992. *Cell* 69:151–58
118. Coverley D, Downes CS, Romanowski P, Laskey RA. 1993. *J. Cell Biol.* 122:985–92
119. Blow JJ. 1993. *J. Cell Biol.* 122:993–1002
119a. Kubota Y, Takisawa H. 1993. *J. Cell Biol.* 122:1321–31
120. Broek D, Bartlett R, Crawford K, Nurse P. 1991. *Nature* 349:388–93
121. Roberts JM. 1993. *Curr. Opin. Cell Biol.* 5:201–6
122. Usui T, Yoshida M, Abe K, Osada H, Isono K, Beppu T. 1991. *J. Cell Biol.* 115:1275–82
123. Handeli S, Weintraub H. 1992. *Cell* 71:599–611
124. Hu B, Burkhart R, Schulte D, Musahl C, Knippers R. 1994. *Nucleic Acids Res.* In press
125. Roberts JM, Weintraub H. 1986. *Cell* 46:741–52
126. Roberts JM, Weintraub H. 1988. *Cell* 52:397–404
127. Dasso M, Newport JW. 1990. *Cell* 61:811–23
128. Enoch T, Carr AM, Nurse P. 1992. *Genes Dev.* 6:2035–46
129. Li JJ, Deshaies RJ. 1993. *Cell* 74:223–26
130. Nishitani H, Ohtsubo M, Yamashita K, Lida H, Pines J, et al. 1991. *EMBO J.* 10:1555–64
131. Ohtsubo M, Kai R, Furuno N, Sekiguchi T, Sekiguchi M, et al. 1987. *Genes Dev.* 1:585–93
132. Ohtsubo M, Okazaki H, Nishimoto T. 1989. *J. Cell Biol.* 109:1389–97
133. Bischoff FR, Maier G, Tilz G, Ponstingl H. 1990. *Proc. Natl. Acad. Sci. USA* 87:8617–21
134. Matsumoto T, Beach D. 1991. *Cell* 66:347–60
135. Roberge M. 1992. *Trends Genet.* 2: 277–81
136. Dasso M, Nishitani H, Kornbluth S, Nishimoto T, Newport JW. 1992. *Mol. Cell. Biol.* 12:3337–45
137. Kiss T, Filipowicz W. 1993. *EMBO J.* 12:2913–20
138. Dasso M. 1993. *Trends Biochem. Sci.* 18:96–101
139. Saka Y, Yanagida M. 1993. *Cell* 74:383–93
140. Kelly TJ, Martin GS, Forsburg SL, Stephen RJ, Russo A, Nurse P. 1993. *Cell* 74:371–82
141. Enoch T, Nurse P. 1990. *Cell* 60:665–73
142. Gould K, Nurse P. 1989. *Nature* 342: 39–45
143. Hartwell LH, Weinert TA. 1989. *Science* 246:629–34
144. Rowley R, Subramani S, Young PG. 1992. *EMBO J.* 11:1335–42

145. Al-Khodairy F, Carr AM. 1992. *EMBO J.* 11:1343–50
146. Walworth N, Davey S, Beach D. 1993. *Nature* 363:368–71
147. Callan HG. 1972. *Proc. R. Soc. London Ser. B* 181:19–41
148. Callan HG. 1974. *Cold Spring Harbor Symp. Quant. Biol.* 38:195–203
149. Brewer BJ, Fangman WL. 1988. *Cell* 55:637–43
150. Dhar V, Schildkraut CL. 1991. *Mol. Cell Biol.* 11:6268–78
151. Linskens MHK, Huberman JA. 1988. *Mol. Cell Biol.* 8:4927–35
152. Feldman H, Winnacker EL. 1993. *J. Biol. Chem.* 268:12895–900

Annu. Rev. Biochem. 1994. 63:777–822

FUNCTION AND STRUCTURE RELATIONSHIPS IN DNA POLYMERASES

Catherine M. Joyce and Thomas A. Steitz[1]

Department of Molecular Biophysics and Biochemistry, Yale University, Bass Center for Molecular and Structural Biology, 266 Whitney Avenue, New Haven, Connecticut 06520

KEY WORDS: protein structure, X-ray crystallography, DNA-protein interactions, exonuclease, ribonuclease H

CONTENTS

INTRODUCTION

Of fundamental importance to all living organisms is the ability to synthesize DNA efficiently and accurately, thus ensuring the faithful transmission of genetic information from parent to offspring. To achieve this objective, all

[1]and Howard Hughes Medical Institute

0066-4154/94/0701-0777$05.00

free-living organisms encode several DNA polymerases that fulfill the various replicative and repair functions within the cell. Additionally, many viruses bypass the host replicative machinery and encode their own polymerases. At first glance, the DNA polymerases seem a bewilderingly disparate group of enzymes, ranging from the small mammalian repair polymerase β, a single subunit of 39 kDa, to the huge multisubunit replicative polymerases, exemplified by DNA polymerase III holoenzyme of *Escherichia coli*, which has at least 20 subunits and a combined molecular mass close to 900 kDa (1). At an intermediate stage of complexity are polymerases such as *E. coli* DNA polymerase I, having multiple enzymatic activities within a single, fairly large, polypeptide chain.

Despite the rich variety within the DNA polymerase family, it seems likely that, at heart, all DNA polymerases are variations on a single theme. In every case the function of the core polymerase activity is to add deoxynucleotides onto the growing end of a DNA primer strand; the difference between repairing a short patch of DNA in a bacterium and replicating the 46 chromosomes of *Homo sapiens* is one of scale, not of chemistry. The underlying similarity between DNA polymerases is suggested by complementation in vivo between polymerases of quite different types (2–4). Studies of polymerase accuracy, an important attribute of enzymes of this type, also imply that many (though not all) of the physicochemical mechanisms used to discriminate between correct and incorrect basepairs have been preserved throughout this family of enzymes (5, and references cited therein).

Perhaps the most compelling argument for an underlying similarity among polymerases is provided by the analysis of protein sequences; although there is little discernible resemblance overall between the sequences of distantly related polymerases, it appears that a small number of crucial active-site residues are conserved (summarized in Figure 1). The sequence conservation not only encompasses the currently identified families of DNA polymerases (6), but can also be extended to reverse transcriptases, RNA replicases, and DNA-dependent RNA polymerases (7), implying that the basic mechanism of phosphoryl transfer required for polynucleotide synthesis is preserved, but that different enzyme families possess variable features that differentiate between utilization of ribo- or deoxyribo-substrates. That such features are relatively subtle and not central to the overall reaction mechanism is suggested by the observations that polymerases can utilize their "unnatural substrate," albeit at suboptimal efficiency or when minor adjustments are made to the reaction conditions (1, 8), and that the retroviral life-cycle requires that reverse transcriptases use both RNA and DNA templates.

A valuable consequence of the probable similarity between polymerases is that structure-function studies pursued on those polymerases that are

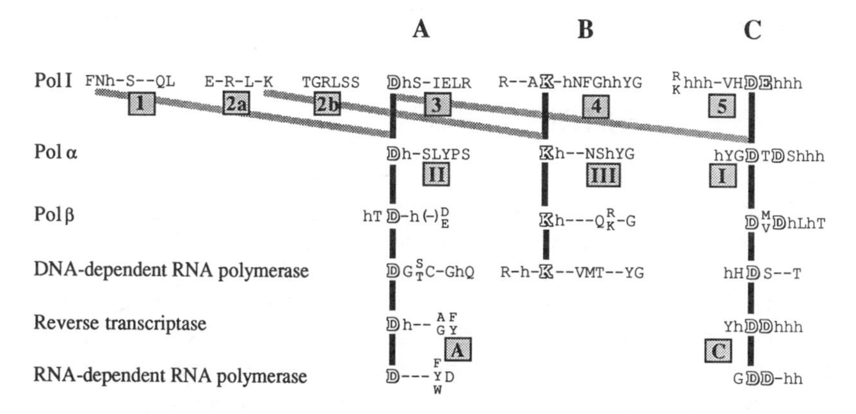

Figure 1 Alignment of the major conserved sequence regions of the polymerase families. The listed motifs are based on published compilations: Ref. 6 for the Pol I and Pol α families, Ref. 104 for the Pol β family and DNA-dependent RNA polymerases, and Ref. 182 for the RNA-dependent polymerases. Positions that are almost invariably occupied by a hydrophobic amino acid are indicated by "h." Hyphens denote nonconserved positions; parentheses are used to indicate length variations within a motif. Following published conventions, conserved sequence blocks in the Pol I family are numbered 1 through 5 according to Delarue et al (7), with the addition of motif 2a according to Blanco et al (183), the Pol α sequences are labeled according to Wong et al (184), and the RNA-dependent polymerase sequences according to Poch et al (182). For clarity, some of the motifs that are conserved within individual families are not shown. The black lines indicate a proposed alignment (7) that gives two motifs, A and C (labeled at the top of the figure), common to the entire polymerase family, containing two invariant aspartates and another highly conserved acidic residue (in motif C). Motif B, containing an invariant lysine, is common to DNA-dependent polymerases (7, 104). As discussed later, mutagenesis experiments provide evidence of the importance of these invariant amino acids (highlighted in the figure). An alternative alignment between the Pol I and Pol α families indicated by shaded lines (183) seems to us to be less satisfactory, since it gives less good conservation of the carboxylates and more variable spacing in regions corresponding to α-helices in the Klenow fragment structure.

simpler and more amenable to such studies are likely to provide information of relevance to all polymerases. Currently, the high-resolution structures of two DNA polymerases have been published (9–11). They are the Klenow fragment of *E. coli* DNA polymerase I and the reverse transcriptase from the human immunodeficiency virus HIV-1. In this review we explore the inferences that can be drawn from a detailed examination of these two structures, together with the available biochemical and genetic data, and the extent to which these deductions can be applied to other polymerases. The focus of our review is confined to structural and mechanistic aspects of the reactions catalyzed by DNA polymerases; to place this information in its biological context, the reader is referred to the book by Kornberg & Baker (1), and to recent reviews in this series on prokaryotic DNA replication

(12), eukaryotic DNA polymerases (13), and DNA replication fidelity (14, 15).

POLYMERASE STRUCTURES

The three-dimensional structures of Klenow fragment and HIV-1 reverse transcriptase will be described in detail. Additionally, though not strictly within the scope of this review, the structure of the DNA-dependent RNA polymerase from bacteriophage T7 (16) will be considered briefly, since its similarity to the other two structures lends additional weight to the idea of a fundamental similarity among all polymerases. Several other polymerase structures are in progress so that, in the not too distant future, it should be possible to evaluate the ideas presented here in the context of a wider range of polymerase structures.

Klenow fragment and HIV-1 reverse transcriptase, representing two distinct classes of DNA polymerase, show substantial differences in their overall molecular properties. Klenow fragment normally uses a DNA template (although it can use RNA, Ref. 8), while reverse transcriptase uses both RNA and DNA. Both enzymes have an associated nuclease domain, but the nucleases serve different functions in the two cases. In Klenow fragment, the associated 3′-5′ exonuclease functions to edit polymerase errors by removing incorrectly incorporated nucleotides from the primer terminus (1). The RNase H activity of reverse transcriptase cleaves the RNA template strand of the RNA-DNA duplex product (17). There are also differences in quaternary structure; Klenow fragment is a monomer, while HIV-1 reverse transcriptase is a dimer of different-sized subunits derived from the same sequence. In contrast to these rather obvious global differences between the two enzymes, there are equally striking similarities between the two when the polymerase active-site regions are examined in detail.

Structure of Klenow Fragment

Three different crystalline complexes have been solved from a tetragonal crystal form grown from high salt. The complex of Klenow fragment with deoxynucleoside monophosphate at the exonuclease active site has been refined at 2.5 Å resolution (LS Beese, TA Steitz, unpublished). The complex with oligo$(dT)_4$ at the exonuclease site has been refined at 2.6 Å resolution (18). Finally, an editing complex containing 11 basepairs of duplex DNA has been refined at 3.2 Å resolution (19). A trigonal crystal form grown at low ionic strength from polyethylene glycol diffracts anisotropically to 2.8–3.5 Å resolution and has been partially refined (JM Friedman, TA Steitz, unpublished). It should be noted that the model-built structures derived by Modak and colleagues (20) are incorrect in many details and should not be considered appropriate substitutes for the experimentally derived structures.

The overall structure of Klenow fragment, initially determined at 3.3 Å resolution, showed very clearly that the 68-kDa molecule consists of two domains (Figure 2*a*): a larger C-terminal domain with a very prominent cleft, and a smaller globular N-terminal domain (9). A large number of experiments have shown conclusively that the 3′-5′ exonuclease active site is located on the small N-terminal domain of Klenow fragment, while the polymerase active site is located in the cleft of the larger C-terminal domain. Thus, in the initial crystallographic studies, deoxynucleoside monophosphate, the product of the exonuclease reaction, was found to bind to the small domain (9); more recently, a binding site for deoxynucleoside triphosphate, the substrate for the polymerase reaction, has been observed on the large domain (21). As described in more detail below, the proposed locations for the exonuclease active site (around the deoxynucleoside monophosphate binding site) and the polymerase active site (within the cleft of the large domain) are strongly supported by site-directed mutagenesis data and protein sequence alignments. Moreover, expression of the DNA encoding the large C-terminal domain gave a protein product that had DNA polymerase activity but no exonuclease activity (22).

The polymerase domain of Klenow fragment can be divided into three subdomains, named "palm," "fingers," and "thumb" for their anatomical analogy to a right hand (Figure 3). The palm subdomain contains a β-sheet that forms the base of the cleft and contains the catalytic residues (see below); the fingers subdomain is virtually all α-helix and forms one wall of the cleft, while the thumb subdomain has two long antiparallel α-helices that interact as coiled coils forming the other side of the cleft. A 50-amino-acid region at the tip of the thumb subdomain was partially disordered in the original description of the structure (9). This region is now seen, both in the low-salt trigonal crystal form and in the complex with duplex DNA, to contain two additional short α-helices and some connecting strands (19). The tips of the fingers and thumb subdomains are in contact so that the cleft is, in fact, more accurately described as a tunnel.

Binding sites for divalent metal ions (magnesium, manganese, and zinc) have been located crystallographically on both the 3′-5′ exonuclease and the polymerase domains. Two metal ions are bound to carboxylate ligands at the 3′-5′ exonuclease active site in the presence of deoxynucleoside monophosphate (18, 23) and have been proposed to play a pivotal role in catalysis, as described in a later section. Magnesium, manganese, or zinc ions bind to the polymerase domain at the bottom of the cleft if the high-salt crystals are transferred from ammonium sulfate into lithium sulfate (LS Beese, TA Steitz, unpublished). No crystallographic information is available

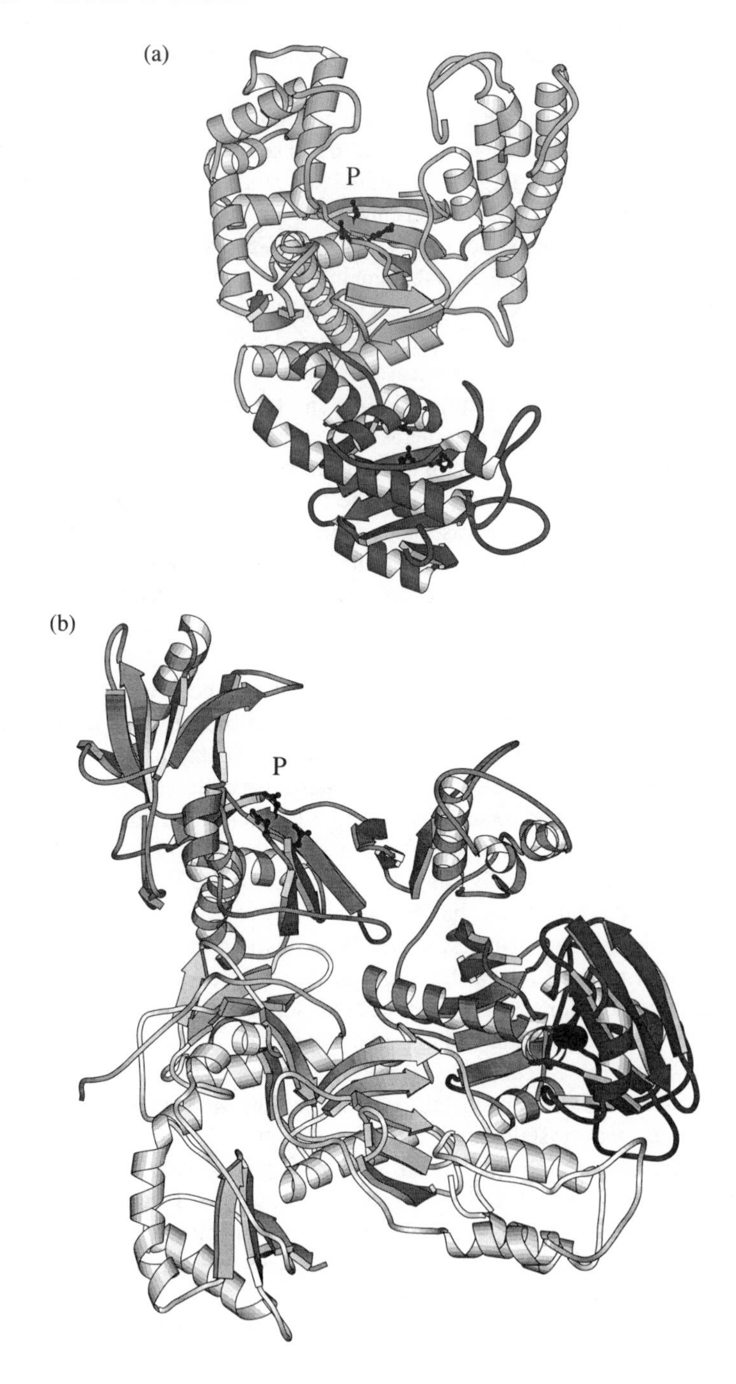
(a)
P
(b)
P

on the binding of metal ions to the polymerase active site in the presence of DNA and deoxynucleoside triphosphate substrates.

Structure of HIV-1 Reverse Transcriptase

HIV-1 reverse transcriptase can employ either RNA or DNA as a template, yielding either RNA-DNA hybrid or duplex DNA products (24). DNA synthesis from the viral RNA as template is initiated in vivo by human $tRNA_3^{Lys}$, whose 3′ end partially unfolds and forms 18 basepairs of duplex with the viral RNA primer-binding site (24, 25). The HIV-1 reverse transcriptase is processed initially from the pol gene products as a 66-kDa polypeptide having an N-terminal polymerase and a C-terminal RNase H domain (26, 27). Subsequent proteolytic cleavage of the homodimer of the 66-kDa subunits removes the RNase H domain from one subunit, leaving a heterodimer containing one 66-kDa subunit (p66) and one 51-kDa subunit (p51) (28, 29). In the p66-p51 heterodimer, only the p66 subunit has polymerase activity, as shown by the observation that mutations of essential active-site residues eliminate polymerase activity only when present on the p66 subunit (30, 31). The p66 subunit also contains the single binding site for non-nucleoside inhibitors such as Nevirapine (32). The heterodimer, but not the isolated subunits, interacts in 1:1 stoichiometry with the tRNA primer (33). A dimeric structure may be typical of most retroviral reverse transcriptases (24); recent results (34) suggest that even those enzymes, such as Moloney murine leukemia virus reverse transcriptase, which are monomeric in solution, may dimerize when bound to their substrate nucleic acid (although in this case the enzymatically active form would be a homodimer).

Since HIV-1 reverse transcriptase is the target of currently approved anti-AIDS drugs, extensive efforts have been directed towards establishing its structure. Initial crystallization experiments led to numerous crystal forms (35, 36), none of which diffracted beyond 6 Å resolution. Two strategies aimed at reducing the inherent flexibility of this polymerase, and consequently improving the resolution to which crystals diffract, have been successful: cocrystallization with an inhibitor, and cocrystallization with a monoclonal antibody F_{ab} fragment and DNA. High-resolution structures have

←

Figure 2 (*a*) Schematic representation of the Klenow fragment structure, with α-helices shown as spiral ribbons and β-strands as arrows. The extent of the separate 3′-5′ exonuclease domain is indicated by darker shading. The catalytically important carboxylate side chains at the polymerase (P) and exonuclease sites are shown. (*b*) A similar representation of HIV-1 reverse transcriptase. The p51 subunit of the heterodimer is shown with the lightest shading. Within the p66 subunit, the RNase H domain is more darkly shaded to distinguish it from the polymerase and connection regions. The catalytically important carboxylate side chains at the polymerase site (P) are shown; the two metal ions at the RNase H site are shown as black spheres. These figures were made by Joe Jäger.

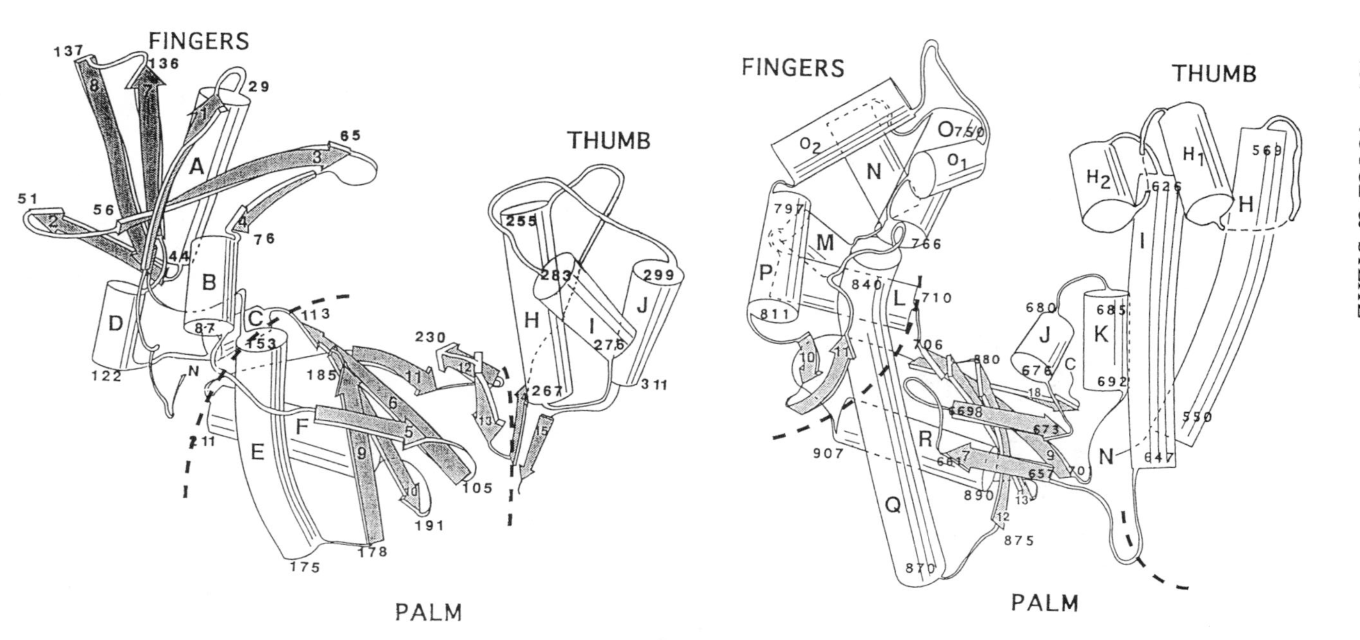

Figure 3 Schematic representation of the polymerase domain of HIV-1 reverse transcriptase (*lefthand panel*) and Klenow fragment (*righthand panel*) with α-helices shown as cylinders and β-strands as arrows (reproduced from Ref. 37). The molecules are positioned with their palm subdomains oriented identically. The amino acid sequence numbers give an approximate indication of the boundaries of secondary structural features. Note that further refinement of the Klenow fragment structure (LS Beese, TA Steitz, unpublished) has allowed identification of four additional helices (H_1, H_2, O_1, and O_2) that were not apparent on the original 3.3 Å structure (9). As previously noted (37), the similarities observed in the two polymerase structures do not exactly parallel the sequence alignments proposed by Delarue et al (7). Thus, the two structures diverge immediately following the invariant aspartate of motif A (although the extended sequence alignment may well be structurally justified within individual polymerase families). Conversely, although there is no apparent primary sequence similarity, one can use the structural similarities to extend motif C, which encompasses the sequences of β-strands 12 and 13 in Klenow fragment and 9 and 10 in reverse transcriptase, so as to overlap helices Q (Klenow fragment) with E (reverse transcriptase) and helices R (Klenow fragment) with F (reverse trancriptase).

been reported for the complex of HIV-1 reverse transcriptase with the non-nucleoside inhibitor Nevirapine (initially at 3.5 Å, and now partially refined at 2.9 Å resolution, Refs. 10, 37), and for the complex with an antibody F_{ab} fragment and 18 basepairs of double-stranded DNA (at 3.0 Å resolution, Ref. 11). Additionally, the structure of the isolated RNase H domain of HIV-1 reverse transcriptase has been solved at 2.4 Å (38), and two separate determinations of the structure of the homologous *E. coli* RNase H, at 2.0 and 1.5 Å resolution, have been described (39, 40).

The structure of the polymerase-proficient p66 subunit of HIV-1 reverse transcriptase is analogous to that of Klenow fragment in several ways. It has a very pronounced domain structure (Figure 2*b*), which reflects its enzymatic activities. The RNase H activity resides on a separate C-terminal domain joined via the "connection" subdomain to the three subdomains that form the polymerase region (10). The three polymerase subdomains can be described as fingers, palm, and thumb of a right hand, and together they form a large cleft, reminiscent of the cleft in the polymerase domain of Klenow fragment (Figure 3). The binding sites for two divalent metal ions on the RNase H domain have been located crystallographically (10, 38), but the binding sites for divalent metal ions on the polymerase domain have yet to be determined. Although comparison of the two reverse transcriptase complexes (one with Nevirapine, the other with DNA and an F_{ab} fragment) shows that the overall structure of the protein is similar in the two complexes, there are significant differences in the relative orientations of the subdomains, particularly the RNase H (see Figure 8, in a later section). It is unclear which of the three ligands is responsible for the differences in the structure or whether crystal packing influences the structure.

The structure provides an obvious rationale for the lack of polymerase activity in the p51 subunit of the heterodimer (30, 31). Although the two subunits have the same amino acid sequence throughout the polymerase region, and similar tertiary structures within each subdomain, the overall conformation of the polymerase domain of p66 is astonishingly different from that of p51, due to differences in the relative positioning of subdomains (10). If the palm subdomains of the two subunits are identically oriented, then the p51 subunit has the fingers closer to the palm and the thumb subdomain further from the fingers and palm. The connection subdomain lies within and fills the expanded cleft between thumb and palm in p51, so that the p51 subunit has no cleft and hence no binding site for the primer-template. The polymerase domain of p66 is not related to p51 by a two-fold rotation axis; rather, the two domains interact in a more head-to-tail arrangement that results from a 16 Å translation and an approximately 74° rotation of one palm subdomain relative to the other.

Comparison of Polymerase Domain Structures in Reverse Transcriptase and Klenow Fragment

A comparison of the crystal structures of the two polymerase domains (10, 37, 41) shows that the three subdomains (palm, fingers, and thumb) that form the cleft (Figure 3) appear to be functionally related. The palm subdomains that lie at the bottom of the polymerase cleft in both enzymes show substantial structural similarity (10), such that it is possible to superimpose 45 Cα atoms from two α-helices and three β-strands with an rms deviation of 1.6 Å (Figure 4). Particularly significant is the virtually identical positioning of three carboxylate residues (Asp705, Asp882, and Glu883 of Klenow fragment; Asp110, Asp185, and Asp186 of reverse transcriptase), which have been shown by mutagenesis studies to be crucial for polymerase activity (42–46). In the alignment of all polymerase sequences proposed by Delarue et al (7), two of these residues are the invariant aspartates of sequence motifs A and C; the third (the Asp186/Glu883 residue) is highly conserved (Figure 1). Multiple magnesium or manganese ions have been observed to bind to this trio of carboxyl groups in Klenow fragment (LS Beese, TA Steitz, unpublished; 10, 19). Thus it appears that these carboxyl groups with associated divalent metal ions form the catalytic center of these two polymerases and are a vital feature that is common to all polymerases.

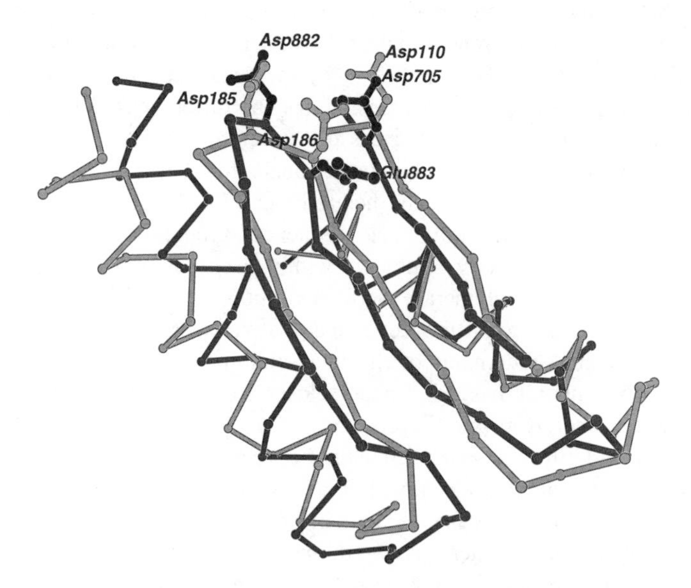

Figure 4 Superposition of a portion of the palm subdomains of Klenow fragment (in darker shading) and HIV-1 reverse transcriptase, showing the α-carbon backbone and the three conserved carboxylate residues. Reproduced from Ref. 37.

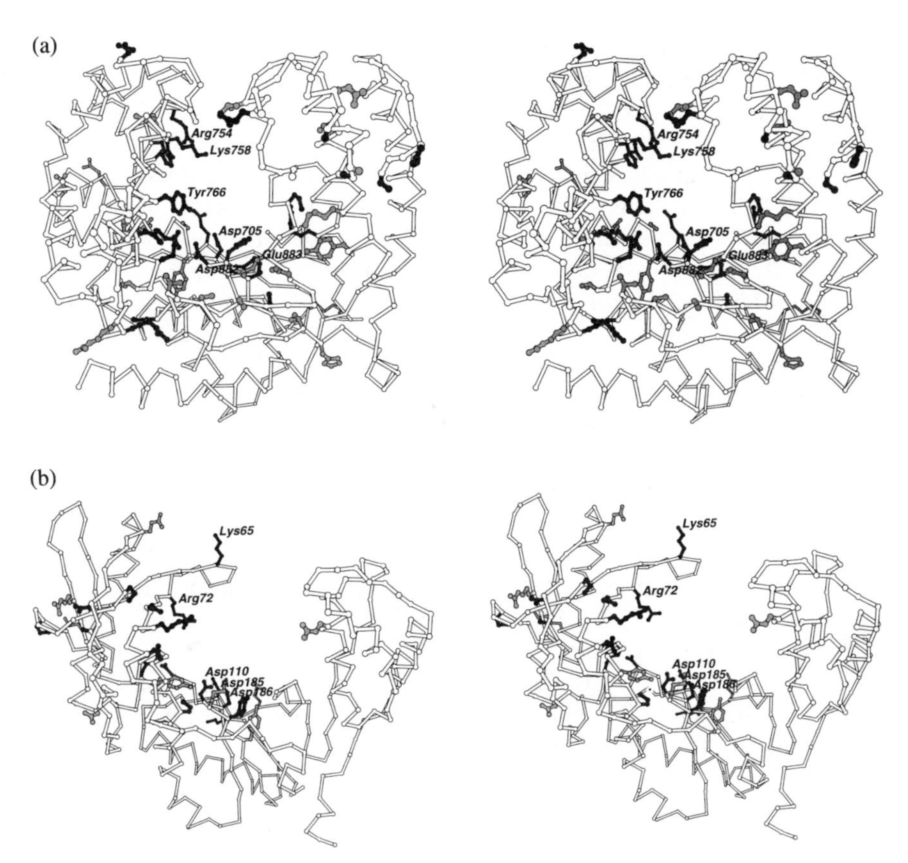

Figure 5 (*a*) Stereo representation of the α-carbon backbone of the polymerase domain of Klenow fragment showing side chains that are conserved in DNA polymerases of the Pol I family (6). The darker-shaded residues are identical in at least eight of the nine sequences; those with lighter shading are conserved in at least six of the sequences. (*b*) A similar representation of the polymerase domain of HIV-1 reverse transcriptase showing side chains that are conserved in the alignment of five reverse transcriptase sequences (26). The darker-shaded side chains are identical in the five sequences; lighter shading indicates side chains that are conserved in four out of the five sequences. In both (*a*) and (*b*), only solvent-accessible side chains are shown. Reproduced from Ref. 37.

In both Klenow fragment and reverse transcriptase, the functional importance of the cleft region could also be inferred from the clustering, particularly at the base of the cleft, of side chains that are invariant or highly conserved in the respective polymerase families (Figure 5).

The thumb subdomains of the two polymerase structures appear to be functionally analogous rather than truly structurally homologous. In reverse transcriptase the thumb consists of three α-helices and one extended strand, which had previously been tentatively assigned as an α-helix (10, 11, 37). In Klenow fragment the thumb is made up of two longer antiparallel α-helices with a small globular domain at its tip consisting of two short helices and short stretches of random coil (19). Not only do the structures of the thumb regions differ in detail, but their relative location in the gene sequence is also different in the two enzymes. The thumb sequence occurs before the palm and fingers sequences in Klenow fragment, but after these sequences in reverse transcriptase. In both polymerases, however, the thumb subdomain appears to serve the same purpose, interacting with the minor groove of the duplex nucleic acid product of DNA synthesis, and moving in response to the binding of DNA (11, 19, 47) (see below).

The fingers subdomains of Klenow fragment and reverse transcriptase bear no structural resemblance to one another. In Klenow fragment this region is predominantly α-helical, whereas in reverse transcriptase it is mixed α-helix and β-sheet. Consistent with the lack of any structural relatedness, the fingers subdomain of Klenow fragment contains the B sequence motif (Figure 1), which is present only in DNA-dependent polymerases and is not seen in RNA-dependent polymerases such as the reverse transcriptase family (7). This motif is located on the O helix of Klenow fragment (Figure 3), which has no structural counterpart in reverse transcriptase. Nevertheless, as discussed below, it is possible that the fingers region fulfills the same role in both polymerases, that of interacting with the template strand close to the site of synthesis.

Comparison with the Structure of T7 RNA Polymerase

The recently reported structure of T7 RNA polymerase (16) shows a striking similarity to Klenow fragment. Not only do the conserved A, B, and C motifs occur in similar spatial relationships, with conserved residues similarly located, but nearly all the secondary structure elements within the polymerase domain are equivalently located. Mutational studies have demonstrated the importance of Asp537 and Asp812, corresponding to the invariant carboxylates of motifs A and C respectively, and Lys631, an invariant residue of motif B in DNA-directed polymerases (48, 49). Like the two structures described above, T7 RNA polymerase also seems to be built in a modular fashion, having an N-terminal domain that is structurally unrelated to any of the domains of Klenow fragment or reverse transcriptase, and that appears to carry out functions specific to RNA synthesis.

POLYMERASE ACTIVE SITE

A complete structural description of the enzymatic mechanism of the polymerase reaction and the enzyme's role in maintaining accuracy will require a high-resolution structure of a stable ternary complex containing both substrates, the deoxynucleoside triphosphate and the primer-template. While such information is not yet available, structural data from binary complexes with either deoxynucleoside triphosphate or duplex DNA, combined with extensive site-directed mutagenesis and kinetic studies, provide some useful insights into active-site location and the role of particular side chains in the polymerase reaction.

As indicated above, distinct functions in the polymerase reaction have been proposed for each of the three subdomains of the polymerase region (10). The palm subdomain contains the catalytic center, the binding site for the 3′ terminus of the primer strand, and contributes to the dNTP-binding site. From the available structural and sequence data, it seems likely to be the most conserved part among all polymerases. The fingers subdomain binds and orients the template strand across from the primer terminus and may also form part of the dNTP-binding site. The three published polymerase structures (9–11, 16), together with protein sequence alignments (7), suggest that there are at least two families of structures for the fingers: The mixed α-helix and β-sheet structure found in reverse transcriptase is hypothesized to occur in polymerases that use RNA as a template; the α-helical structure found in Klenow fragment and T7 RNA polymerase is predicted to be typical of DNA-dependent polymerases. The helical thumb, which is attached flexibly to the rest of the polymerase domain, contacts the minor groove of the product duplex (11, 19). The flexibility of this subdomain may be important in allowing access of the primer-template to the binding site, and in translocation of the product after dNTP incorporation.

Binding of Primer-Template

Essential to understanding the structural basis of polymerase function is a detailed knowledge of the complex between the primer-template DNA and the enzyme. Although there is as yet no high-resolution structure of a complex between Klenow fragment and duplex DNA in which the 3′ end of the primer strand is bound in the polymerase active site, a complex with an 11-basepair duplex DNA having a 3-nucleotide 3′ overhang bound at the 3′-5′ exonuclease site has been partially refined at 3.2 Å resolution (19). While this structure corresponds to an editing complex, the location of the duplex portion of the DNA may provide useful insights into the polymerase primer-template complex. Recently, the structure of HIV-1 reverse transcriptase complexed with both a monoclonal F_{ab} and an 18-basepair duplex DNA with a single-nucleotide 5′ overhang bound to the polymerase active site has been solved at 3 Å resolution. These structures, taken together with the results of model-building, mutagenesis, and chemical modification,

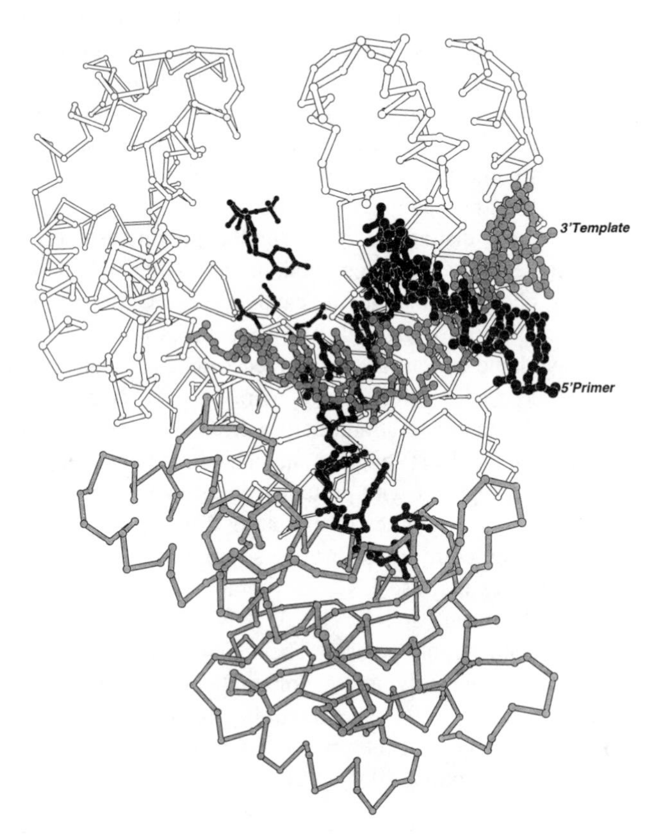

Figure 6 Structure of Klenow fragment complexed with DNA. The α-carbon backbone of Klenow fragment is shown, with the 3′-5′ exonuclease domain lightly shaded. The DNA, shown as a full-atom representation, was derived from the cocrystal structure of an editing complex (19). The position of the dCTP molecule shown was derived from the structure of the enzyme-dNTP binary complex (21). The three conserved carboxylates at the polymerase active site are shown. Reproduced from Ref. 37.

provide a useful, if incomplete, picture of primer-template binding to the polymerase active site.

Although only one cleft was apparent in the initial Klenow fragment structure (9), the cocrystal structure of Klenow fragment with duplex DNA (19) shows the existence of two deep clefts that lie nearly at right angles to each other (Figure 6). The cleft within the polymerase domain was identified as the probable location of the polymerase active site initially because of the cluster of residues that are conserved between T7 DNA polymerase and Klenow fragment (9, 50). Its size and extensive positive electrostatic potential (51) made it a plausible binding site for the primer-template, and this inference was supported by the location within the cleft of two mutations that cause defects in DNA binding (52). Although the binding of primer-template to the polymerase cleft is now established, the accumulating evidence has recently been recognized to indicate that the initially modeled direction of DNA synthesis (9), in which the primer strand entered the cleft from the end furthest from the 3′-5′ exonuclease domain, is not correct (10, 19). The initial model was inconsistent with the increasingly precise localization of the polymerase active site provided by subsequent mutagenesis studies (45, 46). Moreover, the orientation of DNA in the polymerase cleft of Klenow fragment was opposite to that deduced for HIV-1 reverse transcriptase (10), even though the high degree of structural homology in the polymerase clefts of these two proteins would require a similar mode of binding if the active sites were to function analogously. The cocrystal structure of Klenow fragment with duplex DNA indicated how DNA could be bound to Klenow fragment so as to approach the polymerase active site from the same direction as in reverse transcriptase (19). In the cocrystal, the 11-basepair duplex DNA was bound in a second cleft whose axis is roughly orthogonal to the cleft that contains the polymerase active site (Figures 6 and 7). Four single-stranded nucleotides at the 3′ end of the primer strand bound to the exonuclease active site identically to the previously described complexes with single-stranded DNA (18, 53). This second cleft is in part formed by a significant movement of the thumb towards the DNA with which it is interacting and by stabilization of the subdomain at the tip of the thumb that is highly disordered in the apo-structure (9). Extensive interactions are seen across the minor groove between the backbone phosphates of the duplex DNA and amino acid side chains from the thumb subdomain that are highly conserved in the Pol I family (6). These include Arg631 and Lys635 (region 2a, Figure 1), Asn675 and Asn678 (just beyond region 2b), and a conserved motif (region 1) at the tip of the thumb. The conservation of residues that interact with the duplex DNA implies that the binding site observed in the editing complex is biochemically relevant, and supports the contention that it may also be the duplex DNA-binding site employed when DNA is bound to the polymerase active site. In this case, the primer strand would approach the polymerase catalytic site from the direction of the 3′-5′ exonuclease domain and the large cleft in the polymerase domain would bind the single-stranded template strand beyond the site of DNA synthesis (Figure 7).

The location of the polymerase active site centered on the three carboxylates, Asp705, Asp882, and Glu883 (Figure 6), together with the position of the duplex DNA observed in the cocrystal structure, implies that the duplex DNA upstream of the primer terminus is severely bent when the

(a)

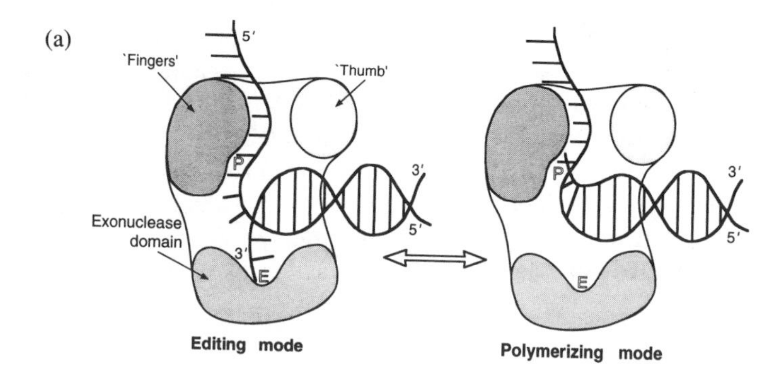

(b)

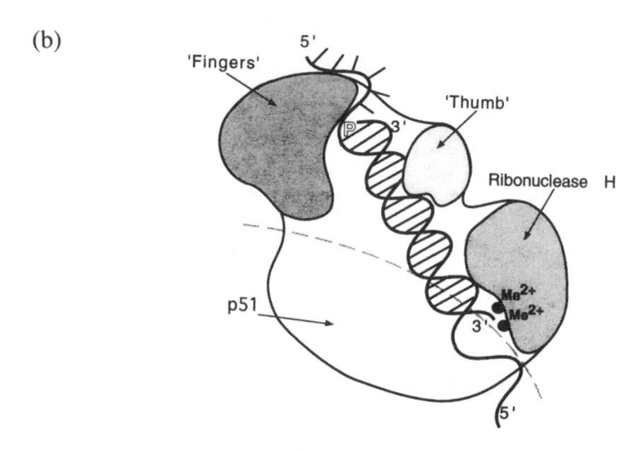

Figure 7 Direction of DNA synthesis and relative location of polymerase ("P") and nuclease active sites in Klenow fragment (*a*) and HIV-1 reverse transcriptase (*b*). For Klenow fragment, "E" indicates the 3′-5′ exonuclease active site; in reverse transcriptase, the two metal ions mark the location of the RNase H active site. Note that the direction of DNA synthesis relative to the conserved polymerase catalytic subdomain appears likely to be the same for both polymerases. Reproduced from Ref. 41.

primer terminus is in the polymerase active site. While this predicted DNA bend and the narrowness of the polymerase cleft discouraged detailed model building, placement of the primer terminus near the catalytic carboxylates necessarily positions the template strand in contact with the fingers subdomain (Figure 7). Therefore, in addition to any role the fingers domain might have in binding the deoxynucleoside triphosphate substrate (see below), it must also play a significant role in binding the template strand (19).

The model deduced for binding of DNA to the polymerase site of Klenow fragment is broadly consistent with a number of biochemical studies. It is important to realize, however, that, because a large area of the protein is proposed to be in contact with the DNA, and because of limitations in the resolution of the techniques used, very few of these experiments provide information at a sufficient level of detail to constrain the model-building process. Thus, the observation that mutation or chemical modification of residues at many different locations within the polymerase cleft has an effect on DNA binding (45, 46, 54–56) is as expected; indeed, it would be surprising if this were not so. The labeling of Tyr766 by a photoaffinity probe at the primer terminus is likewise consistent with the general location of the polymerase catalytic site (57). Chemical footprinting (58), fluorescence (59, 60), and photocrosslinking (57) experiments together indicate that 5–8 basepairs of duplex DNA are covered by Klenow fragment when the primer terminus is at the polymerase site. The range of values probably reflects, on the one hand, the particular requirements for access of the footprinting reagent and, on the other, uncertainties as to the precise orientation of fluorescent or photoactivatable probes attached to DNA; again, the data are consistent with the proposed model but are insufficiently discriminating to rule out alternatives that could be proposed. Using time-resolved fluorescence spectroscopy, it has been concluded that more of the duplex DNA upstream of the primer terminus is drawn into the binding site when the 3′ terminus is at the 3′-5′ exonuclease active site (60). The proposed model (19) suggests a difference of 2–3 basepairs between the polymerase and exonuclease binding modes; it does not appear able to accommodate the 9-nucleotide difference inferred from experiments that compared the effect of a bulky DNA substituent on the two enzymatic reactions (61). This latter experimental result may, however, be reconcilable with the model if, for example, there were a region at the junction of the two clefts where a bulky substituent on the DNA could be tolerated.

When the first high-resolution crystal structure of HIV-1 reverse transcriptase was determined, a model of a binary complex with an A-form DNA-RNA hybrid was constructed by assuming that the 3′ end of the primer strand should be placed near to the catalytic carboxylates, Asp110, Asp185, and Asp186, and that a phosphate of the RNA template strand should lie adjacent to the metal ions at the RNase H active site (10). Employing a rise per residue of 3 Å, 19–20 nucleotides were estimated to lie between these two active sites (Figure 8*a*). Whereas earlier biochemical studies had suggested a distance of 15–16 nucleotides between the polymerase and RNase H active sites (62), more recent experiments, using short incubation times and single–turnover conditions, gave a distance of 18–19 nucleotides, in good agreement with the structural model (63, 64).

More recently, the structure of HIV-1 reverse transcriptase complexed with a duplex DNA oligonucleotide and the F_{ab} fragment of a non-inhibiting antibody has been described at 3.0 Å resolution (11). The positioning of

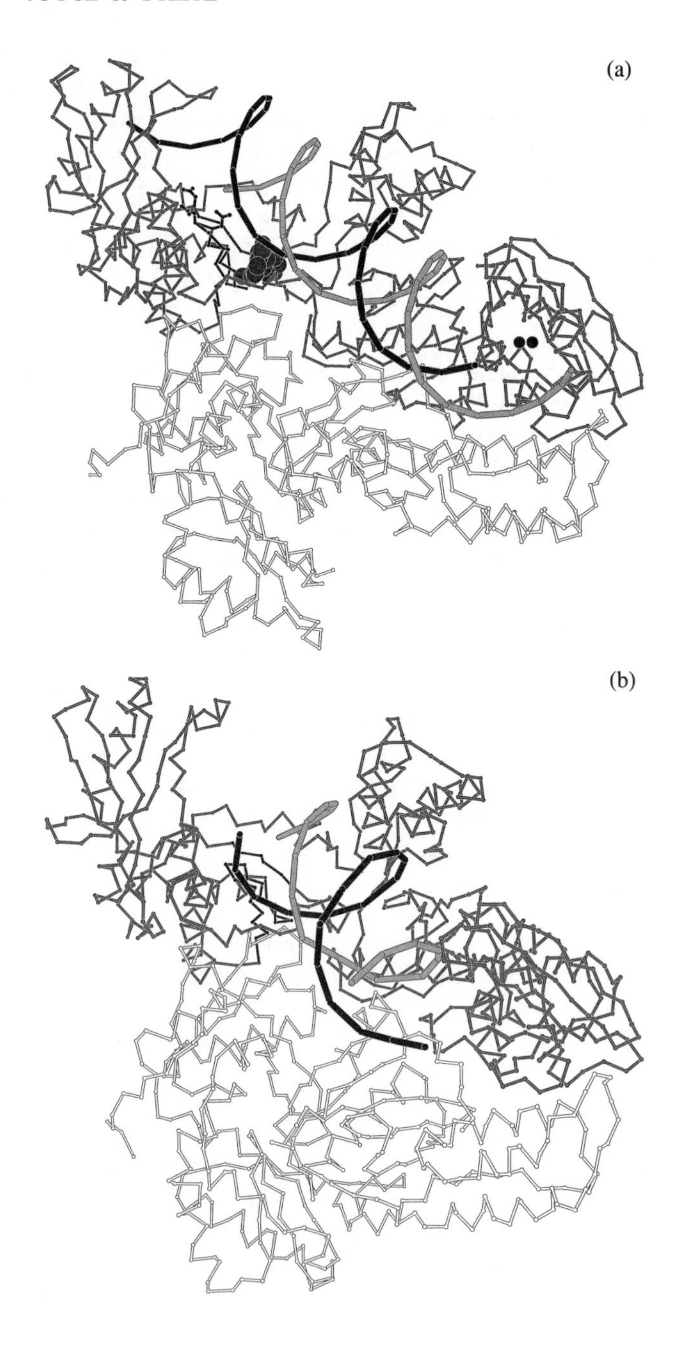
(a)
(b)

the duplex DNA observed experimentally is nearly identical to the model-built A-form DNA-RNA hybrid. However, the duplex DNA is more nearly B-form except in the vicinity of the polymerase active site (where it is A-form), and is significantly distorted (Figure 8*b*), with a bend at the junction of the A-form and B-form segments. This bend is not as large as the one predicted to occur in the Klenow fragment complex and is in the opposite direction.

The structure of the DNA complex raises the intriguing possibility that reverse transcriptases may constrain their template-primer to adopt an A-form conformation at the polymerase site (11); this would permit the use of either RNA or DNA templates, as required during retroviral replication, since both can be A-form, whereas RNA does not adopt a B-form conformation. Model-building the template strand, either in the RNA-DNA hybrid (10) or by extending from the DNA complex (11) (assuming, in either case, that the template continues as an A-form helix beyond the primer terminus) indicates contacts with the fingers subdomain, with the antiparallel β-ribbon formed by strands 3 and 4 (Figure 3) fitting the minor groove side of a template in A-form (Figure 9). By contrast, DNA polymerases, such as Klenow fragment, might favor binding of a B-form template-primer, substantially reducing the efficiency of RNA-templated reactions (8). Thus, it is tempting to speculate that the fingers subdomain (whose structure, as noted earlier, is very different in Klenow fragment and HIV-1 reverse transcriptase) plays a significant role in determining the template specificity of a polymerase, perhaps by constraining the substrate into the appropriate A-form or B-form conformation at the polymerase active site.

The binding of a DNA primer-template to crystals of reverse transcriptase results in substantial movement of the thumb subdomain towards the DNA (11, 47), as was observed with Klenow fragment. Just as in Klenow fragment, the thumb interacts with the phosphate backbone across the minor groove of the product duplex (10, 11). Consistent with the structural data, UV irradiation of short oligonucleotides (acting as model primers) bound to HIV-1 reverse transcriptase resulted in crosslinking to amino acid residues

←

Figure 8 (*a*) The α-carbon backbone of the HIV-1 reverse transcriptase heterodimer with a model of an A-form RNA-DNA hybrid duplex. The p66 subunit is shaded to distinguish it from p51. The bound Nevirapine inhibitor is shown in space-filling representation. The 3′ end of the primer (DNA) strand (gray) is positioned adjacent to the carboxylate side chains of Asp110, Asp185, and Asp186 that define the catalytic center of the polymerase active site; these side chains and the three β-strands on which they are located are shown in black. The two spheres indicate the positions of the two divalent metal ions at the RNase H active site and are seen to be adjacent to the backbone of the RNA template strand (black). (*b*) A similar representation to (*a*), showing the experimentally determined position of a B-form DNA duplex (11). The location of the polymerase active site is marked by the three β-strands shown in black. These figures were made by Steve Smerdon.

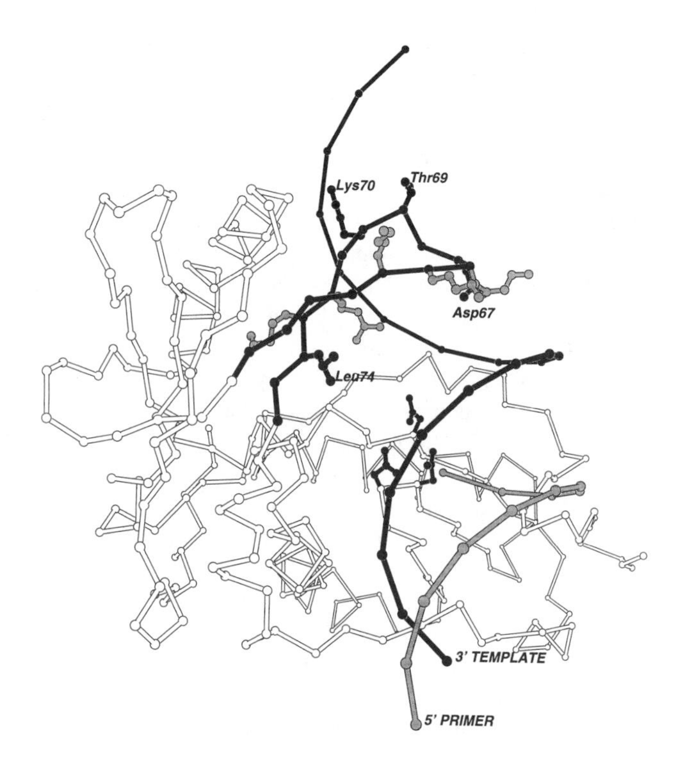

Figure 9 The palm and fingers subdomain of HIV-1 reverse transcriptase, showing a possible interaction between the template strand beyond the site of synthesis and residues in the loop between β-strands 3 and 4 in the fingers subdomain. The model-built primer strand (lightly shaded) has its 3′ terminus close to the three conserved carboxylate side chains, Asp110, Asp185, and Asp186, which are shown in black. The darker template strand of the A-form duplex is suitably placed to interact with the loop, whose backbone atoms are indicated by shading. Several residues whose mutation renders the virus resistant to the nucleoside analogs, AZT, ddI, and ddC, are located in this loop (Asp67, Thr69, Lys70, and Leu74). The approximate positions of the corresponding side chains are shown by dark shading. The remaining positively charged residues (Lys64, Lys65, Lys66, Arg72, and Lys73) are lightly shaded. Reproduced from Ref. 37.

located in the thumb, with Leu289-Thr290 and Leu295-Thr296 as the probable sites of labeling (65, 66).

Unlike other retroviral reverse transcriptases, HIV-1 reverse transcriptase is inhibited noncompetitively by several distinct classes of non-nucleoside compounds whose chemical structures appear largely unrelated to each other, although the isolation of cross-resistant mutants suggests that these compounds share a common binding site (67). When HIV-1 reverse transcriptase was cocrystallized with the non-nucleoside inhibitor Nevirapine (68), the molecule was observed to bind to a hydrophobic pocket that lies at the base of the thumb at the junction between the thumb and palm subdomains (10) (Figure 8*a*). Consistent with its behavior as a noncompetitive inhibitor (68), the Nevirapine-binding site does not overlap the observed or expected binding sites for either substrate of the polymerase reaction. The structure suggests that inhibition by this molecule may be a consequence of a reduction in the mobility of the thumb subdomain (10), since the binding of Nevirapine may immobilize the thumb in one orientation relative to the palm. It is possible that this fixed orientation may not be precisely correct for positioning the DNA substrate in a manner appropriate for catalysis. Alternatively (or additionally) the reduced ability of the thumb to undergo necessary conformational changes (a sort of molecular arthritis) may inhibit translocation of the duplex product.

The template-directed synthesis of DNA by HIV-1 reverse transcriptase is initiated by the 3′ end of human $tRNA^{Lys}_{3}$. Eighteen nucleotides at the 3′ end of the tRNA pair with a specific complementary sequence in the viral template (the primer-binding site). It has been hypothesized that the D-stem and anticodon stem of the tRNA, which remain folded, interact with the p51 subunit (10); this is consistent with the observed binding of the tRNA primer to the heterodimer of reverse transcriptase, but not to the isolated subunits (33). Chemical crosslinking suggests that the anticodon region is in close proximity to the protein (33). Moreover, comparisons of the binding of unmodified tRNAs made in vitro and $tRNA^{Lys}_{3}$ modified in vivo (69, 70) indicate that the modifications on the anticodon loop contribute to the specificity of the interaction of HIV-1 reverse transcriptase with its primer. Other interactions between the tRNA and the protein are independent of RNA sequence (69, 70). Additional selectivity for the primer tRNA is provided by complementary basepairing interactions with the viral RNA template, involving the unraveled 18 nucleotides at the 3′ end of the tRNA, and 6 nucleotides of the anticodon loop (69; LA Kohlstaedt, TA Steitz, unpublished). In the case of HIV-2 RNA, an additional interaction of about 6 nucleotides of the T-stem with the RNA template is predicted (71). Since reverse transcriptase alone can initiate synthesis from the tRNA primer (69), it must facilitate the partial unfolding of the tRNA and formation of a complex structure with the template.

Nucleotide Binding

Our understanding of the polymerase reaction will not be complete without the identification of the side chains responsible for positioning the dNTP substrate, particularly as it is likely that a subset of these interactions will play important roles as the reaction proceeds. Information on polymerase-dNTP contacts can, in principle, be derived from a combination of crystallographic studies of complexes containing the dNTP, affinity labeling and chemical modification experiments, and the study of mutations that affect the binding of nucleotides or their analogs. With all of these approaches, however, there are caveats that preclude an unambiguous interpretation of the results, so that our understanding of dNTP-binding remains at a rather primitive level. In the first place, although a polymerase-dNTP binary complex can be formed, such a complex is not catalytically competent (72). This makes sense since the requirement for complementarity between the incoming dNTP and the DNA template dictates a substantial degree of contact between the nucleotide base and the template-primer. However, it means that any structural or labeling data obtained with the binary complex must be interpreted with caution. Secondly, a clear-cut interpretation of mutational data is difficult, since a mutation could influence dNTP binding either via its effect on dNTP contact residues, or via an effect on template contacts, resulting in a subtle change in position of the opposing template base at the dNTP-binding site. With these caveats in mind, we will examine the available experimental data on dNTP binding.

High-resolution crystallographic studies have been carried out on a binary complex formed between Klenow fragment and dNTP after the tetragonal crystals grown in high salt were transferred to low ionic strength (21). Although a complex between 5-Hg UTP, primer-template DNA, and HIV-1 reverse transcriptase has also been reported (47), the only information presented was the position of the Hg substituent at 7 Å resolution. In the Klenow fragment-dNTP complex, the multiple transfers of the crystals resulted in some loss of resolution; however, a plausible model for the binary complex could be constructed (Figure 10). Consistent with the ideas discussed in the preceding paragraph, the conformation of the nucleotide base was not uniquely defined, but varied depending on the identity of the dNTP. In each case the base made van der Waals and hydrophobic interactions with residues at the bottom of the polymerase cleft, interactions that cannot occur when the base is hydrogen-bonded to the opposing template base. The location of the nucleotide base is consistent with chemical crosslinking experiments that indicated proximity to Tyr766 and His881 in the binary polymerase-dNTP complex (73, 74), implying that the crystalline and solution binary complexes are the same.

Although no useful conclusions could be drawn from the positions of the nucleotide base or sugar, it is possible that the crystalline binary complex may be informative in identifying contacts between Klenow fragment and the dNTP phosphate groups. The β and γ phosphates appear to make interactions with three positively charged side chains within the cleft (Figure 10): Arg754 and Lys758 on helix O of the fingers subdomain, and Arg682 on the thumb. Both Arg682 and Lys758 had previously been suggested as dNTP contacts on the basis of affinity-labeling experiments (75, 76); the

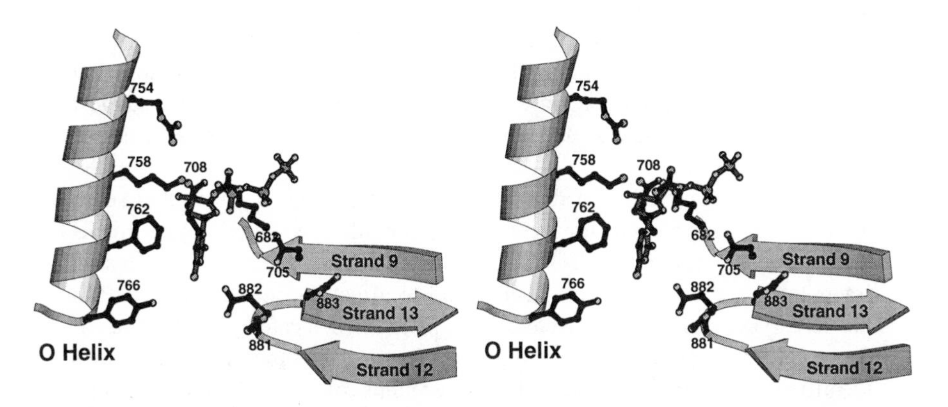

Figure 10 Stereo drawing of the dCTP-binding site of Klenow fragment (21). The protein is viewed looking down into the polymerase cleft, approximately perpendicular to the view shown in Figure 3. The side chains of the trio of carboxylates, and of other residues that contact the dCTP molecule, are shown in black and numbered. The dCTP molecule is shown in gray. Relevant secondary structural features are indicated. This figure was made by Joe Jäger.

data for Arg682 seem particularly persuasive since labeling was carried out in the presence of primer-template DNA, so as to probe dNTP binding in the ternary complex (75). However, it is disturbing that mutation of either of these residues had little or no effect on $K_{m(dNTP)}$ (77; M Astatke, CM Joyce, unpublished work). The positioning of the β and γ phosphates inferred from the Klenow fragment-dNTP complex is consistent with data from a binary complex of Klenow fragment with pyrophosphate (21), and would place the deoxyribose close to Phe762 and the α-phosphate about 6 Å from the catalytically important side chains of Asp705 and Asp882.

While biochemical experiments with HIV-1 reverse transcriptase have identified the general region of the polymerase active site, a detailed interpretation of the role of particular residues cannot be made with the methods used. Photoaffinity labeling of a binary complex with dTTP resulted in crosslinking to Lys73 (78), and an antibody inhibition study suggested that dNTP binds in the vicinity of residues 65–73 (79). However, the residues implicated by these studies are located on the β-hairpin that has been proposed, from crystallographic work, to bind to the template strand (Figure 9), and Lys73 is more than 15 Å from the primer terminus (10, 11, 37; J Jäger, TA Steitz, unpublished). It is possible that in the binary complex the thymine base is not in a fixed orientation (as is the case in the dNTP complex with Klenow fragment), allowing crosslinking with Lys73 to occur. Labeling experiments using pyridoxal phosphate were interpreted as indicating a role in dNTP binding for Lys263 (80), which is, however, located on the thumb subdomain, also at a considerable distance from the position of the primer terminus (37). Furthermore, a subsequent mutational study (81) likewise argues against this interpretation, and kinetic data suggest that pyridoxal phosphate is not a good probe for the dNTP site, at least for this enzyme (82).

A different approach aimed at pinpointing active-site residues that serve as important enzyme-substrate contacts is provided by kinetic experiments that probe the binding of dNTPs to mutant DNA polymerases. In Klenow fragment, mutations that have been found to affect the binding of dNTP in the ternary complex (as reflected in $K_{m(dNTP)}$) are located on one side of the polymerase cleft within or close to the fingers subdomain. Positions identified thus far encompass the N terminus of helix Q (Arg841 and Asn845), the exposed face of helix O (Tyr766, Phe762, and Arg754), and neighboring residues closer to the catalytic center (Asp705 and Glu710) (45, 46; M Astatke, CM Joyce, unpublished). Since these regions include the conserved sequence motifs A (around Asp705) and B (helix O), it is gratifying to note that mutational analyses of the corresponding regions of human DNA-polymerase α and ϕ29 DNA polymerase have also identified side chains that may play a role in dNTP binding (83–85). An advantage of the kinetic approach is that it probes the ternary complex; however, as discussed above, it is impossible, in the absence of other structural evidence, to distinguish direct effects from those mediated via template interactions. Moreover, the side chains listed above encompass an area much larger than the dNTP molecule and therefore cannot all be in direct contact with it. Since the region of Klenow fragment implicated by these studies is thought to make extensive contacts with the template strand, a reasonable interpretation is that a subset of the residues mentioned above are in direct contact with the dNTP, while the remainder bind the template DNA.

In a similar way, studies on mutations that influence the binding of DNA polymerase inhibitors have the potential to provide information on polymerase-dNTP contacts. However, the location of mutations in HIV-1 reverse transcriptase that result in resistance of the virus to one or more of the chain-terminating nucleoside analogs AZT, ddI, and ddC (86–89) is entirely consistent with the conclusion that the mutated side chains are involved in positioning the template strand (10). Among the residues whose mutation confers drug resistance, Asp67, Thr69, and Lys70 are on the antiparallel β-hairpin that is proposed to bind and orient the template strand (Figure 9). Residues Thr215 and Lys217, at the junction between fingers and palm, also appear to interact with the model-built template strand (10), while Leu74 and Met41 could stabilize and position the antiparallel β-hairpin. An intriguing parallel may also exist in Klenow fragment: Mutation of Tyr766, which is located on the fingers domain in the vicinity of the model-built template strand, affects the discrimination between deoxy and dideoxy nucleotide substrates (CM Joyce, unpublished) and also influences polymerase fidelity at the level of dNTP insertion (90).

In conclusion, it is clear that in the catalytically competent ternary complex the dNTP α-phosphate must be close to the carboxylates at the polymerase catalytic center. The precise position of the rest of the molecule remains to be established, although the available data are suggestive of contacts with the neighboring part of the palm subdomain and perhaps with the fingers subdomain. Contacts with the dNTP base must be provided by the primer terminus and the template strand, whose position also seems likely to be determined by contacts with the fingers region.

Catalysis of the Polymerase Reaction

The polymerase active site catalyzes a nucleophilic attack by the 3′ hydroxyl of the primer terminus on the dNTP α-phosphate, with release of pyrophosphate. Allowing for relatively subtle variations to accommodate ribo- or deoxyribo-derivatives, it seems reasonable to predict an essentially similar chemical pathway for all polymerases, regardless of whether synthesis is templated by DNA or RNA or whether the final product is DNA or RNA. The stereochemical outcome of the reaction, determined for several DNA polymerases, is consistent with an "in-line" nucleophilic displacement proceeding via a pentacovalent transition state or intermediate (91–96). The reaction pathway has been extensively characterized for several polymerases, including Klenow fragment (97), the DNA polymerases of bacteriophages T7 (98) and T4 (99), and HIV-1 reverse transcriptase (63, 100, 101). In every case there is an obligatory order of substrate binding, with initial formation of the enzyme–nucleic acid binary complex being required for productive binding of the dNTP to form a catalytically competent ternary complex. For at least some of the enzymes studied, the reaction pathway includes, in addition to the chemical step, one or more nonchemical transformations (63, 97, 98). Thus, for Klenow fragment, the chemical step is both preceded and followed by slow nonchemical processes that may play a part in maintaining polymerase accuracy (102, 103). These nonchemical steps are frequently described as conformational changes, although the precise nature of such changes is unknown at present. It is important to bear in mind that such a change could involve either enzyme or substrate molecules (or both), and need not necessarily be large in structural terms in order to be kinetically significant.

An active-site residue that plays a role in catalysis may do so by accelerating any one of the steps of the reaction. Depending on the nature of the side chain, it could intervene directly in the chemical step, e.g. as a general acid or base, it could accelerate either the chemical or nonchemical steps by preferential binding to the relevant transition state, or it could be binding a catalytically essential cofactor such as a metal ion. It is probably unrealistic to categorize active-site residues as involved exclusively in either substrate binding or catalysis, since some overlap between these functions is inevitable; moreover, an unambiguous assignment of mechanistic roles is virtually impossible in the absence of detailed structural information on the relevant substrate complexes. Thus far, most of the experimental evidence for the participation of particular residues in catalysis has been provided by mutagenesis studies. Removal of a putative catalytic residue by mutation should cause a substantial decrease in the reaction rate; further studies on the mutant protein can then dissect further the involvement (if any) of the side chain in substrate binding, and identify the particular step of the reaction that is affected. In some cases, the study of a series of substitutions at a particular position may lead to inferences about the mechanistic role of the mutated side chain.

In the complex of HIV-1 reverse transcriptase with an oligonucleotide (11), the DNA primer terminus interacts with the palm region of the polymerase domain, placing the 3′-hydroxyl, the attacking nucleophile in the polymerase reaction, close to the three proposed catalytic aspartate side chains (positions 110, 185, and 186) (Figure 11*a*). The structurally and functionally equivalent residues in Klenow fragment are Asp705, Asp882, and Glu883. Substitutions of Asp705 or Asp882 are among the most deleterious mutations isolated to date in Klenow fragment, while the effect of mutations at Glu883 is more modest (45, 46). The role of the three carboxylate residues in Klenow fragment has been examined further by determining the elemental effect for incorporation of an α-thio-substituted dNTP. The results indicated that Asp882 and Glu883 may play some role in the chemical step of the polymerase reaction, with the side chain of Asp882 being located extremely close to the dNTP α-phosphate at which nucleophilic substitution takes place (46). By contrast, the kinetically important contribution of Asp705 is felt at a different step of the reaction, perhaps the preceding conformational change (46; CM Joyce, unpublished work).

Other residues in Klenow fragment that are important in accelerating the reaction include Gln849 (implicated by elemental effect measurements in the chemical step), Arg668, and Lys758 (46; M Astatke, CM Joyce, unpublished). Although widely separated on the primary sequence, the side chains of these residues and the three important carboxylates are sufficiently close in the tertiary structure to define an active site. Gln849 is located on helix Q of Klenow fragment, which is structurally analogous to helix E of the p66 polymerase domain of reverse transcriptase (10, 37) (Figure 3).

Arg668, a residue that is invariant in the Pol I family of polymerases, is on a neighboring β-hairpin that has no obvious counterpart in the reverse transcriptase structure. Lys758, an invariant residue in the sequence motif B of DNA-dependent polymerases (Figure 1; Refs. 7, 104), is located on helix O in the fingers subdomain of Klenow fragment. Although this region shows no structural resemblance to reverse transcriptase (10), mutagenesis data raise the possibility that the fingers subdomain of reverse transcriptase may also contribute a catalytically important lysine residue, Lys65 (44, 105).

The presence of a few highly conserved residues within the entire polymerase family could reflect a common active-site architecture, and mutagenesis experiments on polymerases for which no structural data are available lend additional support to this idea. Within the two-carboxylate motif C (Figure 1), even conservative substitutions at the carboxylate positions cause a dramatic loss of activity in a variety of polymerases: several members of the DNA polymerase α family (e.g. Refs. 106–109), mammalian DNA polymerase β (110), and the RNA replicase of encephalomyocarditis virus (111). The effects of substitutions at the noncarboxylate positions in motif C are more variable (106, 107, 110) and may, in at least some cases, reflect the degree of structural disruption caused by the mutation (e.g. Ref. 112).

In a similar way, mutagenesis studies on φ29 and PRD1 DNA polymerases (108, 113) and encephalomyocarditis virus RNA replicase (111) support the importance of the invariant aspartate of motif A. Surprisingly, however, mutation of the corresponding residue had little effect on the polymerase activity of human DNA polymerase α, the prototype sequence for the φ29 and PRD1 polymerases (83). Mutagenesis of the invariant lysine of motif B in PRD1 DNA polymerase and T7 RNA polymerase suggests an important role for this side chain (49, 108). Other mutations within motifs A and B generally had less dramatic overall effects on polymerase activity, but frequently affected the binding of dNTPs or nucleotide analogs (83–85, 113), consistent with the idea, discussed above, that some of the residues in motifs A and B may be involved—either directly or indirectly—in positioning the dNTP.

In summary, the available structural data together with mutagenesis studies in a variety of polymerases suggest a common polymerase active-site structure containing three crucial carboxylate side chains and probably also a lysine. Additional polar residues, e.g. Arg668 and Gln849 in Klenow fragment, may well be important, but their degree of conservation within the polymerase superfamily is at present unclear. There are many ways in which a constellation of polar side chains of this type could catalyze the polymerase reaction. One possibility, given the large number of active-site

(a)

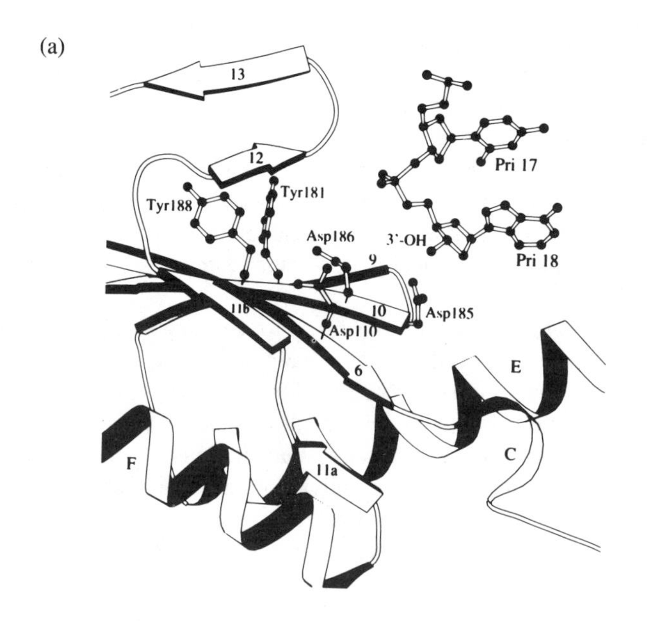
13
12
Tyr181
Tyr188
Pri 17
Asp186
3'-OH
Pri 18
9
10
11b
Asp185
Asp110
6
E
C
F
11a

(b)

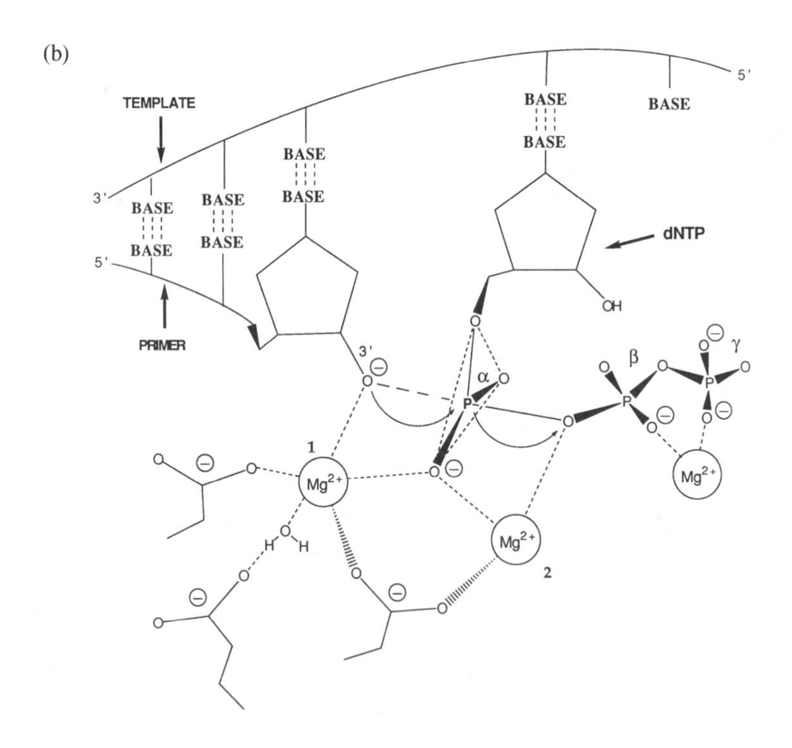
TEMPLATE
3'
5'
BASE
BASE
PRIMER
dNTP
OH
3'
α
β
γ
P
Mg2+
1
2

carboxylates, is that these side chains serve to anchor a pair of divalent metal ions that promote catalysis by a mechanism (Figure 11*b*) analogous to that described below for the 3′-5′ exonuclease reaction (18, 41, 53, 114). In this mechanism, one Mg^{2+} (number 1) would promote the deprotonation of the 3′ hydroxyl of the primer strand, analogous to the role of metal A in the exonuclease active site. The other Mg^{2+} (number 2) would facilitate the formation of the pentacovalent transition state at the α-phosphate of the dNTP and the loss of pyrophosphate. Consistent with the idea of catalysis mediated by metal ions, crystallographic experiments have shown binding of Mg^{2+} and Mn^{2+} to Asp705 and Asp882 of Klenow fragment (LS Beese, TA Steitz, unpublished; Refs. 10, 19). Despite the attractiveness of the two-metal-ion mechanism, the available evidence does not exclude related mechanisms in which the necessary deprotonation of the attacking 3′ hydroxyl is achieved by one of the active-site carboxylates acting as a general base. Clearly, a full understanding of the mechanism of catalysis is unlikely in the absence of structural data for a complex containing both substrates at the polymerase active site.

In addition to understanding the mechanistic pathway of the polymerase reaction, it has long been a goal to describe polymerase fidelity in molecular terms. The effect on fidelity of mutations that alter the relative activities of the polymerase and editing functions is well understood (14), but less is known about the two fidelity-determining processes that take place at the polymerase site, namely the discrimination against incorrect dNTPs in the insertion step and against further extension of a mismatched primer terminus. Biochemical studies on polymerase fidelity have suggested a model based on geometric criteria; the polymerase tolerates an incorrect substrate to a degree related to its resemblance to normal DNA geometry (14). From this model one would predict that at least some of the residues that determine accuracy would be those responsible for positioning the dNTP or the primer terminus in the active site and that mutations in these residues would have a mutator or antimutator phenotype. Mutations that affect the fidelity of the polymerase reaction have been described for several DNA polymerases (e.g. 83, 90, 115–119). However, the absence of a large mutational database, particularly for polymerases whose structure is known, precludes a coherent molecular description of polymerase fidelity at present. Comparison of HIV-1

Figure 11 (*a*) The polymerase active-site region in HIV-1 reverse transcriptase, showing the position of the 3′ terminal dinucleotide of the primer strand, the trio of catalytic carboxylates (Asp110, Asp185, and Asp186), and Tyr181 and Tyr188, which are involved in binding non-nucleoside inhibitors, such as Nevirapine. Reproduced from Ref. 11. (*b*) A possible mechanism for the polymerase reaction, involving catalysis mediated by two divalent metal ions. For details see the text. Based on the conclusions of Burgers & Eckstein (91), a third Mg^{2+} is shown chelated by the dNTP β and γ phosphates. Reproduced from Ref. 41.

reverse transcriptase with Klenow fragment suggests one structural feature that may play a role in polymerase fidelity. HIV-1 reverse transcriptase has an exceptionally high rate of synthesis errors that appear to result from template-primer misalignments (120). It is tempting to speculate that the reverse transcriptase molecule with its more open DNA-binding cleft (Figure 2) may be better able to accommodate the postulated misaligned intermediates (37).

OTHER ACTIVITIES PRESENT IN DNA POLYMERASES

Many DNA polymerases also have nuclease activities, which are frequently part of the same polypeptide chain as the polymerase region. The available evidence suggests that the arrangement seen in the Klenow fragment and HIV-1 reverse transcriptase structures, in which the nuclease active site is located on a separate structural domain, is widespread. Polymerases can therefore be thought of as made up of separate enzymatically active modules that appear in various combinations in the different polymerase families.

3′-5′ Exonuclease

Many DNA polymerases (exemplified by Klenow fragment) have an associated 3′-5′ exonuclease that acts in opposition to the direction of DNA synthesis and serves to remove, or proofread, polymerase errors (1). In most cases the 3′-5′ exonuclease is part of the same polypeptide chain as the DNA polymerase, although DNA polymerase III of *E. coli* has its editing exonuclease as a separate subunit (ϵ) within the core polymerase (121). The preferred substrate for the exonuclease is single-stranded DNA (122), and crystallographic studies have demonstrated the molecular basis for this preference. Thus, when an eight-basepair duplex DNA was cocrystallized with Klenow fragment, four single-stranded nucleotides resulting from partial melting of the duplex were seen bound to the exonuclease active site in a binding pocket that is capable of binding single-stranded (but not duplex) DNA (53). Biochemical experiments using covalently crosslinked DNA duplexes have also shown a requirement for local melting at the 3′ terminus (61). It has therefore been concluded that the DNA primer terminus is bound as a "frayed" or single-stranded end at the 3′-5′ exonuclease active site, even when the enzyme is working on a duplex DNA substrate (53, 58).

The mechanism of the 3′-5′ exonuclease reaction is probably the best understood of the reactions catalyzed by DNA polymerases, thanks to the availability of high-resolution structural data for crystalline complexes of Klenow fragment with the product (dNMP) or the substrate (a single-stranded DNA oligomer) bound at the exonuclease active site (9, 18, 53). The amino acid side chains that bind the metal ions and the deoxynucleotides, and thus form the active site, were identified from these structures (Figure 12*a*), thereby providing the basis for a detailed mutagenesis study of their roles in the reaction (23, 123).

The structural and mutagenesis studies demonstrated that the major catalytic role in the 3′-5′ exonuclease reaction is played by a pair of divalent metal ions, A and B (Figure 12*a*), 4 Å apart, that are coordinated to the oxygens of the phosphodiester bond that is to be cleaved (18). In addition to the phosphate oxygen, metal A is bound to the protein in distorted tetrahedral geometry by the carboxylate groups of Asp355, Glu357, and Asp501. Metal B has octahedral coordination; it shares the Asp355 ligand with the metal A site and is coordinated via bridging water molecules to Asp424. Crystallographic studies have indicated that both sites can be filled by Mg^{2+}, Zn^{2+}, or Mn^{2+} in the presence of dNMP (18); likewise, biochemical studies have shown that any of these three metal ions alone can support the 3′-5′ exonuclease reaction (124). When Klenow fragment crystals were placed in a solution containing Mg^{2+} and Zn^{2+}, site A bound Zn^{2+} and site B bound Mg^{2+}, as expected from the tetrahedral geometry of site A and the octahedral geometry of site B. This may be indicative of the preferred combination of metal ions in vivo. Although spectroscopic data suggest the binding of a third metal ion at or close to the active site (125), no additional binding site has been observed crystallographically in either the substrate or the product complex using Mg^{2+} at concentrations up to 10 mM (18).

Mutations of the aspartic acid side chains that serve as ligands to the metal ions (primarily Asp355, Asp424, and Asp501) were shown crystallographically to disrupt metal binding, and also caused dramatic decreases in 3′-5′ exonuclease activity (23, 123). Since the mutations included one (D424A) that abolished binding of metal B (but not A) (23), and one (D355A) that caused loss of metal A (but not B) (19), it is clear that both metal ions are required for the reaction. Moreover, since mutations in the three aspartate ligands caused the largest decreases ($>10^4$ fold) in exonuclease activity of any active-site mutation that has been tested (123), these two metal ions alone appear likely to play the central role in the chemistry of catalysis.

Refinement of the structures of complexes having single-stranded DNA at the 3′-5′ exonuclease active site gave a detailed view of substrate interactions and suggested a mechanism for catalysis of the exonuclease reaction (Figure 12*b*) (18). Since the stereochemistry of the reaction implies an associative in-line displacement (126), the attacking nucleophile must approach the phosphodiester bond from the direction of Tyr497. At 2.6 Å resolution, there is electron density consistent with an appropriately posi-

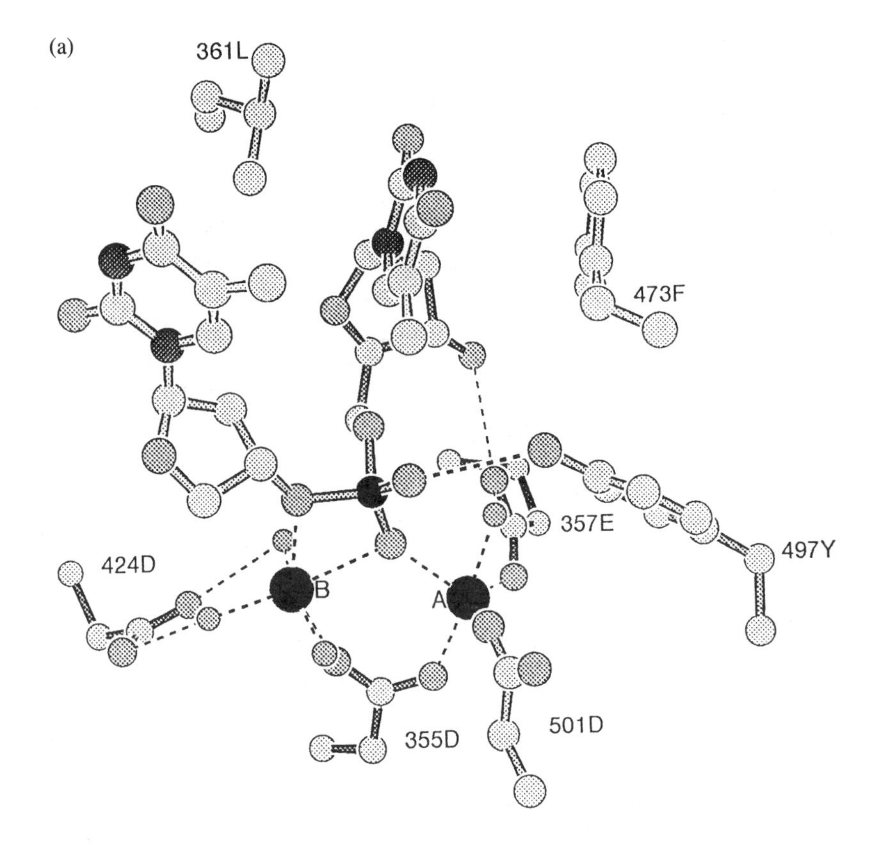

tioned water molecule (or hydroxide ion) as the fifth ligand of the distorted tetrahedral coordination to metal A. Evidence from pH-dependence studies suggests that metal A, and not one of the protein side chains, serves to deprotonate the water molecule and facilitate nucleophilic attack (123). In addition to promoting formation of the nucleophile, the two metal ions probably accelerate the reaction by stabilizing the developing negative charge on the pentacovalent transition state or intermediate and by facilitating departure of the leaving group. Two-metal-ion catalysis of the type proposed for the 3′-5′ exonuclease reaction may be a recurrent theme in phosphoryl transfer reactions. In addition to the likely parallels with the phosphoryl transfer reactions catalyzed by alkaline phosphatase (127), ribonuclease H (38, 39), and by ribozymes (18, 53, 128, 129), one cannot ignore the possibility (discussed above) that a similar catalytic mechanism may be used at the polymerase active site (18).

The mutagenesis results suggest that active-site residues other than the

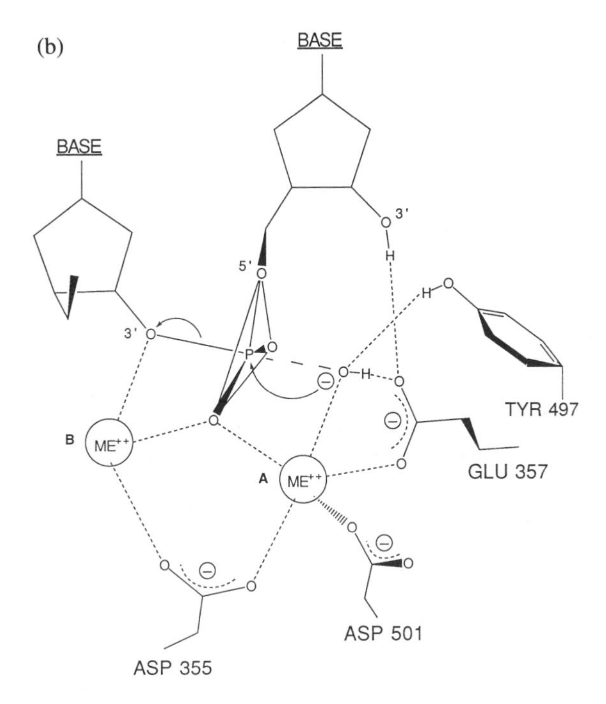

Figure 12 (*a*) Structure of the 3′-5′ exonuclease active site containing a bound dinucleotide. The catalytically essential metal ions, A and B, are shown as large black balls. Other atoms are represented as smaller balls, with phosphorus black, and carbon, oxygen, and nitrogen represented by increasingly darker shades of gray. Water molecules are shown as smaller gray spheres. Reproduced from Ref. 18. (*b*) The proposed transition state for the 3′-5′ exonuclease reaction. The mechanism is thought to involve catalysis mediated by the two bound divalent metal ions. Metal ion A facilitates the formation of the attacking hydroxide ion, whose lone-pair electrons are oriented towards the phosphorus by interactions with metal A, Tyr497, and Glu357. Metal ion B stabilizes the geometry and charge of the pentacovalent transition state and facilitates the departure of the 3′ hydroxyl group. Reproduced from Ref. 18.

aspartates at 355, 424, and 501 play an important but secondary role in the exonuclease reaction (123). From the structural data, a probable role for these side chains is that of holding the DNA substrate and the attacking nucleophile in the correct orientation for efficient catalysis. In the complex with single-stranded DNA, Leu361 and Phe473 anchor the 3′-terminal two residues by stacking with the bases; Glu357 is hydrogen-bonded to the 3′-hydroxyl as well as binding metal ion A, while Tyr497 and Glu357 may serve to orient the attacking nucleophile (18). A combination of polar interactions with the sugar-phosphate backbone and hydrophobic interactions with the nucleotide bases defines a binding pocket capable of binding 3–4 nucleotides of single-stranded DNA in a sequence-independent manner (18, 53). This size is in good agreement with biochemical data indicating that the 3′-5′ exonuclease of Klenow fragment requires the melting of 4 or 5 terminal basepairs (61), but raises questions about the energetics of such extensive melting. It has been suggested (19) that the bent configuration modeled for duplex DNA bound to the polymerase site (Figure 7) may serve to destabilize the duplex terminus and facilitate the melting necessary for the exonuclease reaction. Direct measurement by time-resolved fluorescence spectroscopy indicates that the energetics of melting is indeed in an appropriate range, since a perfectly basepaired duplex DNA molecule partitions about 7:1 in favor of the polymerase site (60).

Protein sequence alignments suggest strongly that a very similar 3′-5′ exonuclease active site will be found in all DNA polymerases that have an editing function, and preliminary structural data for the N-terminal 45-kDa domain of T4 DNA polymerase confirm this prediction (J Wang, P Yu, WH Konigsberg, TA Steitz, unpublished). The important active-site residues identified in the Klenow fragment studies are conserved in all proofreading polymerases, while the rest of the protein scaffold within this region is very divergent. Although some homologies had previously been noted, Bernad et al (130) were the first to recognize that three sequence motifs (Exo I, II, and III) were present in the majority of DNA polymerase sequences. More recently, several groups have arrived independently at essentially identical adjustments to the alignment proposed by Bernad et al (6, 104, 131, 132). The three Exo sequence motifs parallel almost exactly the active-site residues described above. The Exo I motif contains the core sequence DXE, where the two acidic residues correspond to Asp355 and Glu357 of Klenow fragment. Exo II has the sequence NX_{2-3}(F/Y)D; in Klenow fragment the last residue is Asp424, and the motif is immediately preceded by Gln419, whose side chain interacts with the penultimate phosphodiester bond of the DNA substrate (18, 53). The Exo III motif has the sequence YX_3D, containing the active-site residues Tyr497 and Asp501 in Klenow fragment. There is no obvious conservation of the active-site residues Leu361 (potentially part of the Exo I region) and Phe473, although in the latter case several candidate aromatic residues could be proposed in the sequences between Exo II and Exo III. Perhaps the problem of anchoring the terminal base has been solved in different ways by different polymerases.

In the absence of structural data, the validity of the above sequence alignments in predicting active-site residues has been borne out by site-directed mutagenesis studies; the majority of these (exemplified by references 98, 130, 131, 133–135) involved mutation of one or both of the invariant carboxylate residues in the Exo I motif, although a few studies have addressed the importance of residues in Exo II and/or Exo III (130, 133–137).

Despite the overall similarity implied by the conservation of active-site residues, individual polymerases have very different rates for the 3′-5′ exonuclease reaction (e.g. Refs. 99, 123, 138). This is true even when comparing the rates of degradation of single-stranded DNA and thus avoiding complications due to the need for melting at a duplex terminus. Since the available kinetic data do not establish with certainty that the observed rate in every case indicates the rate of chemical catalysis, it remains to be seen whether the differences in rate reflect structural differences at the exonuclease active site.

In Klenow fragment, the polymerase and exonuclease active sites are located about 30 Å apart on separate structural domains. Moreover, in Klenow fragment and a number of other polymerases, point mutations at one active site usually have a negligible effect on the other (23, 98, 107, 108, 123, 130, 137, 139), implying that this arrangement of separate and independent active sites is widespread. Exceptions, such as the DNA polymerase from herpes simplex virus, where the two activities cannot be cleanly separated by mutation (140), are rare, and the underlying structural causes are unclear at present. Although the two catalytic sites usually behave independently of one another, the available evidence suggests that DNA-binding functions are shared between polymerase and exonuclease domains. Even in *E. coli* DNA polymerase III, where the ϵ subunit is independently active as an exonuclease (121), the activity of ϵ is enhanced, particularly on duplex DNA, by assembly with the polymerase subunit, α (141). In Klenow fragment, the structural data place duplex DNA in a cleft between polymerase and exonuclease domains (19); moreover, the polymerase domain contributes the contact residue His660, which interacts with single-stranded DNA bound at the exonuclease site (18, 53). Experiments with the separate domains of Klenow fragment are consistent with idea of a mutual reliance of one domain on the other for DNA-binding functions (although other explanations have not been ruled out). Thus, removal of the exonuclease domain weakens duplex DNA binding to the polymerase (22, 142), while removal of the polymerase domain leaves an exonuclease domain with no detectable enzymatic or single-stranded DNA-binding activity (124).

Despite the physical separation of polymerase and 3′-5′ exonuclease active sites, these two activities must cooperate functionally to proofread polymerase errors. The interplay between the single-stranded DNA-binding properties of the exonuclease region and the duplex DNA binding associated with the polymerase site provides a straightforward structural rationale for the editing process (53, 143). Thus, editing by the 3′-5′ exonuclease is determined by the single-stranded character of the primer terminus, which favors its binding to the associated single-stranded DNA-binding site. Detailed studies of the exonuclease activity of T4 and T7 polymerases demonstrate how an increasing number of terminal mismatches favor the exonuclease reaction both by increasing the proportion of termini in a "frayed" configuration and by accelerating the melting process that converts a duplex molecule into a substrate for the exonuclease (15, 99, 138). At the same time, the polymerase active site plays a role in enhancing the effectiveness of the editing process. Not only is a mispaired terminus preferred over a correctly paired terminus as a substrate for the exonuclease, it is also a very poor substrate for further rounds of dNTP addition at the polymerase site (15, 144, 145). This delay increases the time window for editing of a mismatch and allows even the slow 3′-5′ exonuclease of Klenow fragment to serve as an effective proofreader. Because the efficiency of editing is determined by a physical property (notably, the tendency to melt or fray) of the DNA, effective editing of a polymerase error can be achieved even if the DNA product travels from polymerase to exonuclease site via dissociation. Consistent with this prediction, both intermolecular and intramolecular transfers between active sites have been observed (e.g. 99, 138, 146, 147), as a natural consequence of the balance between the rates of DNA dissociation and of the reactions under consideration (which, in turn, may be related to the in vivo function of a particular enzyme).

Ribonuclease H

The ribonuclease H activity associated with retroviral reverse transcriptases acts as an endonuclease (148, 149) to degrade the viral RNA template during (−) strand DNA synthesis. Additionally, though beyond the scope of this review, RNase H participates in a series of strand-transfer reactions that are necessary for the complete cycle of retroviral replication (150–152). The structure of the RNase H domain of HIV-1 reverse transcriptase has been solved at 2.4 Å resolution from the isolated domain (38) and at 2.9 Å resolution in the context of the heterodimer (10, 37). The isolated RNase H domain, from Tyr427 to the C-terminus, has exactly the same structure as in the heterodimer except for a small loop at the active site containing His539, which is disordered only in the isolated domain (10). A comparison (38) of the structure of the HIV-1 RNase H domain with that of *E. coli* RNase H (39, 40) shows that they are very similar in terms of the overall fold and location of secondary structure elements, as well as the position of active-site residues (see below). An important difference is that the *E. coli* enzyme has an insertion of some 15 residues, 6 of which are basic, which form an additional α helix at the surface of the protein. It has been suggested (40, 153) that this basic protrusion on the *E. coli* enzyme forms part of the nucleic acid–binding site, contacting the RNA template strand about 10 residues 5′ to the catalytic site. This location is similar to the contacts proposed between the HIV-1 reverse transcriptase heterodimer and a DNA-RNA duplex (10), although in reverse transcriptase the contacts are provided by residues from the polymerase domain. The structural comparisons therefore suggest that the polymerase domain of HIV-1 reverse transcriptase provides an important part of the nucleic acid–binding site for the RNase H reaction and also stabilizes the active-site loop containing His539. Either or both of these functions could be the reason why the isolated p15 RNase H domain is only weakly active (27) unless reconstituted with p51 (154). Interestingly, the RNase H region of Moloney murine leukemia virus contains sequences similar to the additional basic region in *E. coli* RNase H and, in contrast to the situation in HIV-1 reverse transcriptase, the murine retroviral RNase H is active as an isolated domain (155).

The RNase H domain consists of a five-stranded β-sheet flanked on each side by α-helices. On this β-sheet are four carboxylate residues (three Asp, one Glu) that are conserved among all RNase H sequences (156). Mutagenesis studies of HIV-1 reverse transcriptase and *E. coli* RNase H indicate that at least three of these carboxylates are crucial for RNase H activity (157–159). In the structure of the HIV-1 RNase H domain, the conserved carboxylates are seen to bind two Mn^{2+} ions spaced 4 Å apart, inviting comparison with the 3′-5′ exonuclease active site of Klenow fragment. In fact, there is substantial structural similarity in the active-site regions of these two nucleases, although the overall topology of folding implies that they do not share a common evolutionary ancestor. As shown in Figure 13, it is possible to superimpose three of the conserved carboxylates, the two metal ions, and a substantial region of the surrounding β-strand scaffold in the two structures (37). The superposition implies a correspondence between Asp443, Glu478, and Asp549 of HIV-1 reverse transcriptase and, respectively, Asp355, Asp424, and Asp501 of Klenow fragment. The suggested equivalence between Asp549 (reverse transcriptase) and Asp501 (Klenow fragment) may explain why the homologous residue in *E. coli* RNase H, Asp134, seems to be dispensable for enzymatic activity even though this residue is conserved in all RNase H sequences (156). The analysis of the role of Asp134 in *E. coli* RNase H made use of an asparagine substitution that had no effect on the enzymatic reaction (159); the same substitution at position 501 of Klenow fragment was likewise without effect, although alanine or glutamate substitutions caused dramatic decreases in exonuclease activity (123). Thus, for RNase H, just as for the Klenow fragment 3′-5′ exonuclease, the requirement at this position may be merely for a metal ligand of the correct stereochemistry.

As in the Klenow fragment 3′-5′ exonuclease, mutational analysis of the RNase H active site suggests that the most important side chains are the carboxylates, implying a central role in catalysis for the metal ions to which they serve as ligands. In addition to the carboxylates, there are three residues (Ser, His, and Asn) that are conserved in all retroviral and bacterial RNase H sequences (156). Although these side chains are located at the RNase H active site, mutations at these positions have much less effect than mutations in the carboxylates, and it has been suggested that these residues may play some role in substrate binding (159–161). Indeed, model building of an RNA-DNA substrate suggests that His539 may interact with the phosphate backbone (J Jäger, TA Steitz, unpublished). The apparent structural similarity between the 3′-5′ exonuclease and RNase H active sites, together with the necessity for a similar reaction pathway (both cleave so as to leave a 5′ phosphate product), suggests that RNase H also may use a two-metal-ion catalytic mechanism (38, 39), as described for the 3′-5′ exonuclease (Figure 12*b*). While the observation of two bound metal ions, with the appropriate spacing, in the HIV-1 RNase H domain structure (38) provides good evidence for such a mechanism, the presence of only a single magnesium ion in the *E. coli* RNase H structure (40) has been taken as evidence for the related reaction mechanism in which the necessary deprotonation of the attacking water molecule is facilitated by one of the carboxylates acting as a general base (162). However, it is also possible that the putative second metal site in the *E. coli* enzyme could be obstructed by basic residues from a neighboring molecule (39, 40). Alternatively, as is the case with the 3′-5′ exonuclease, binding of substrate or product may be required for occupancy of the second metal site by Mg^{2+} (37). Only Mn^{2+} binds to both metal sites of the 3′-5′ exonuclease in the absence of a bound phosphate at the cleavage site (18).

The recently determined NMR structure of an RNA-DNA hybrid (163) has provided a possible mechanism whereby RNase H may discriminate among RNA-RNA, RNA-DNA, and DNA-DNA duplexes in its cleavage specificity. The minor groove of the A-form RNA-DNA duplex is narrower than that of RNA-RNA, though wider than the B-form DNA-DNA minor groove. Fedoroff et al (163) suggest that only this narrower A-form phosphate backbone spacing will fit into the RNase H active site.

The structure of HIV-1 reverse transcriptase shows polymerase and RNase H activities located on separate structural domains, with about 60 Å distance between the two active sites. Although inhibition studies (164) and the properties of many point mutants (e.g. 44, 157) suggest that the two active sites are functionally distinct, there is clearly a substantial degree of communication between the domains. Thus, attempts to separate the two activities genetically using deletion and insertion mutations were largely unsuccessful (165–167). By contrast, the polymerase and RNase H activities of the reverse transcriptase of Moloney murine leukemia virus seem much more independent since they can be cleanly separated both by subcloning

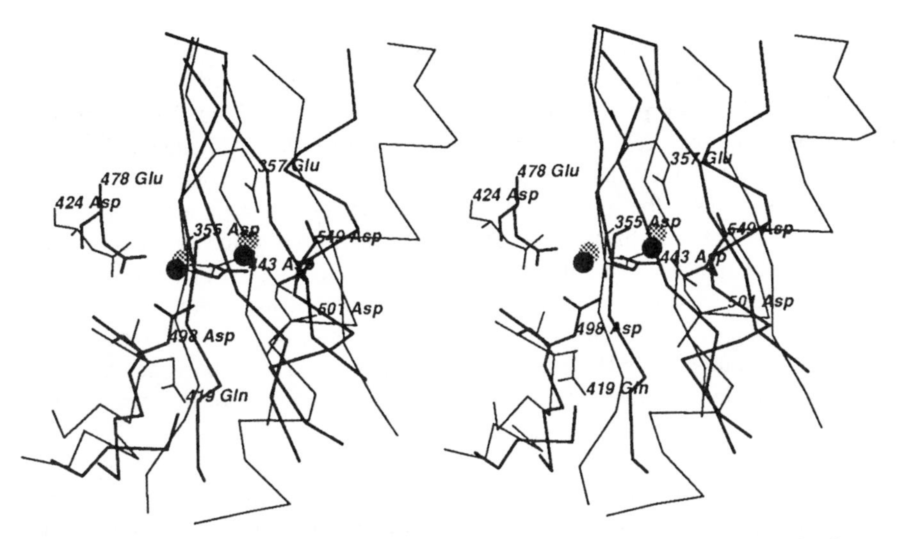

Figure 13 Stereo representation of a least-squares overlap of the 3′-5′ exonuclease domain of Klenow fragment (thin lines) with the RNase H domain of HIV-1 reverse transcriptase (thick lines). The superposition was carried out by first overlapping the $C\gamma$ atoms of the three pairs of matching carboxylic acid residues and the two pairs of metal ions, and was then extended to include 37 $C\alpha$ atoms from each structure in the three-stranded β-sheet and a short section of the α-helix. The RNase H metal ions are shown as black circles and those of the 3′-5′ exonuclease as lightly shaded circles. Reproduced from Ref. 37.

and by linker-insertion mutagenesis (155). As already discussed, some of the differences between these two reverse transcriptases may be due to a greater requirement in the HIV-1 RNase H reaction for substrate contacts provided by the polymerase domain. Additionally, a large area of the reverse transcriptase molecule is involved in the dimerization interface, and mutations that affect these interactions would be likely to influence reactions at both active sites. Consistent with this idea, recent experiments with the reverse transcriptase of Moloney murine leukemia virus have shown that, in spite of the apparent independence of the two activities in this system, some RNase H–defective mutations exert subtle effects on the polymerase reaction by interfering with dimerization (34).

The mechanism of retroviral reverse transcription requires that the two separate active sites cooperate so as to achieve degradation of the original viral RNA template as the (−) strand DNA is synthesized, and raises questions about the temporal and spatial relationship of the two activities. As noted in an earlier section, the 19-nucleotide distance between the RNase H cleavage position and the primer terminus is in good agreement with model-building of an A-form DNA-RNA hybrid onto the reverse transcriptase structure (Figure 8*a*). Moreover, the demonstration that both reactions can occur during a single processive cycle of DNA synthesis (64) is consistent with this model, which allows simultaneous placement of the DNA primer terminus at the polymerase active site and the RNA template phosphodiester backbone at the RNase H active site. Although it is clear that both activities can take place during a single polymerase–nucleic acid encounter (64), the available evidence argues against a tight or obligatory coupling between them (63, 168). More likely, the sequence of reactions that takes place will be determined by the relative rates of the polymerase and RNase H reactions in relation to dissociation of the complex (169). An elegant experiment, showing complementation between a polymerase-defective and an RNase H–defective reverse transcriptase, argues against any strict requirement that the two activities be physically coupled, even in vivo (170).

5′-3′ Exonuclease

Most of the Pol I family of enzymes have an associated 5′-3′ exonuclease activity specific for double-stranded DNA. In the bacterial DNA polymerase I enzymes, this activity is part of the same polypeptide chain as the polymerase. By contrast, bacteriophages T4, T5, and T7 encode essentially the same enzymatic activity in a separate polypeptide (171–173). In several cases, these enzymes have been shown to have RNase H activity with the same 5′-3′ polarity (171, 174–176), an observation that fits well with the in vivo requirement for removal of RNA primers during lagging-strand DNA replication. It seems likely that some eukaryotic and viral polymerases may have similar 5′-3′ exonuclease activities, either covalently attached to the polymerase (177) or as an associated subunit (178).

The conventionally used term "5′-3′ exonuclease" is an inaccurate description of the activity of these enzymes. As was originally shown for *E. coli* DNA polymerase I (179), and more recently for several thermophilic DNA polymerases (175), the preferred substrate is a displaced 5′ end of the type that could be generated by strand-displacement synthesis. Cleavage takes place at or close to the junction between the single-stranded and duplex DNA, though it seems likely that the enzyme requires a free 5′ end for access to the substrate (175). The term "structure-specific endonuclease" is therefore a better description of this enzymatic activity. Specificity for a particular substrate structure seems to be intrinsic to the nuclease and not to require presentation of the substrate by an associated polymerase, since both the isolated exonuclease domain of *E. coli* DNA polymerase I and the T7 exonuclease showed similar rates and positions of cleavage on model substrates as did a polymerase-associated 5′-3′ exonuclease (175, 179). A strong inference from both these model studies is that the well-known stimulation of the 5′-3′ exonuclease by polymerase action may be due to production of the preferred substrate and need not imply any physical or temporal coupling between the two activities.

The lack of structural data for any 5′-3′ exonuclease precludes any definitive conclusions on structure-function relationships, but protein sequence alignments and mutational studies provide some tantalizing clues. Characterization of 5′-3′ exonuclease-proficient amber fragments of *E. coli* DNA polymerase I indicates that the exonuclease active site must be within the first 297 residues of the protein (180). Alignment of the protein sequences of six 5′-3′ exonuclease regions from bacterial polymerases and four bacteriophage 5′-3′ exonucleases indicates six highly conserved sequence motifs containing 14 invariant amino acids (181). Nine of these invariant residues are acidic, inviting the speculation that the 5′-3′ exonuclease, like the other enzymatic activities so far described, may require the coordination of divalent metal ions at the active site. A conservative mutation, D13N, at one of the invariant residues in *E. coli* DNA polymerase I results in a dramatic decrease in 5′-3′ exonuclease activity (CM Joyce, unpublished data). Intriguingly, the sequence surrounding Asp13 (ILVDGSSY) is quite similar to the sequence in *E. coli* RNase H that contains the active-site residue Asp10 (IFTDGSCL). Inspection of the 5′-3′ exonuclease sequence alignment does not, however, suggest any obvious counterparts to the other carboxylates, Glu48 and Asp70, that have been shown to be important for RNase H activity (159).

An entirely different structural question concerns the way in which the 5′-3′ exonuclease domain fits together with the Klenow fragment structure; this question should be resolved by the ongoing structural studies on *Taq* polymerase (Y-S Kim, TA Steitz, unpublished data). The polymerase active site, at the base of a cleft, and the 5′-3′ exonuclease active site, on an independent structural domain, must necessarily be separated in space, and yet the two must work together in vivo so as to leave a ligatable nick. The way in which this cooperation is achieved poses some intriguing mechanistic questions.

CONCLUDING REMARKS

The available evidence leads us to predict that the common polymerase active-site architecture observed in the three published polymerase structures will be widespread, perhaps universal, within the polymerase superfamily. The major global differences between individual polymerases can be attributed to additional functions that are assembled onto the polymerase core, frequently in a modular fashion. Obvious examples are the various nuclease modules (described above) that are present in particular polymerase families. Additionally, some polymerases contain subdomains responsible for the protein-protein interactions that allow assembly of large multisubunit replicative complexes, whose overall molecular properties appear very different from those of the simple polymerase modules that are their functional core. Finally, there is suggestive, though by no means conclusive, evidence that the phosphoryl transfer reactions catalyzed by polymerases and their associated nucleases may share a common mechanism of catalysis mediated by divalent metal ions.

ACKNOWLEDGMENTS

We are grateful to the many coworkers who have shared our investigations of polymerase structure and function over the years. We are particularly grateful to Joe Jäger and Steve Smerdon for invaluable help with the figures, to Nigel Grindley for comments on the manuscript, and to Kate Tatham for her patience in preparing the manuscript. We also thank the large number of colleagues who sent us preprints of their work. Polymerase research in the authors' laboratories is supported by NIH grants GM-28550 and GM-39546, ACS grant BE-52, and HHMI funding.

This paper is dedicated to the memory of Hatch Echols.

Literature Cited

1. Kornberg A, Baker TA. 1992. *DNA Replication*. San Francisco: Freeman. 931 pp. 2nd ed.
2. Bryan S, Chen H, Sun Y, Moses RE. 1988. *Biochim. Biophys. Acta* 951:249–54
3. Witkin EM, Roegner-Maniscalco V. 1992. *J. Bacteriol.* 174:4166–68
4. Sweasy JB, Loeb LA. 1992. *J. Biol. Chem.* 267:1407–10
5. Joyce CM, Sun XC, Grindley NDF. 1992 *J. Biol. Chem.* 267: 24485–500
6. Braithwaite DK, Ito J. 1993. *Nucleic Acids Res.* 21:787–802
7. Delarue M, Poch O, Tordo N, Moras D, Argos P. 1990. *Protein Eng.* 3:461–67
8. Richetti M, Buc H. 1993. *EMBO J.* 12:387–96
9. Ollis DL, Brick P, Hamlin R, Xuong NG, Steitz TA. 1985. *Nature* 313:762–66
10. Kohlstaedt LA, Wang J, Friedman JM, Rice PA, Steitz TA. 1992. *Science* 256:1783–90
11. Jacobo-Molina A, Ding J, Nanni RG, Clark AD Jr, Lu X, et al. 1993. *Proc. Natl. Acad. Sci. USA* 90:6320–24
12. Marians KJ. 1992. *Annu. Rev. Biochem.* 61:673–719
13. Wang TS-F. 1991. *Annu. Rev. Biochem.* 60:513–52
14. Echols H, Goodman MF. 1991. *Annu. Rev. Biochem.* 60:477–511
15. Johnson KA. 1993. *Annu. Rev. Biochem.* 62:685–713
16. Sousa R, Chung YJ, Rose JP, Wang B-C. 1993. *Nature* 364:593–99
17. Crouch RJ. 1990. *New Biol.* 2:771–77
18. Beese LS, Steitz TA. 1991. *EMBO J.* 10:25–33
19. Beese LS, Derbyshire V, Steitz TA. 1993. *Science* 260:352–55
20. Yadav PNS, Yadav JS, Modak MJ. 1992. *Biochemistry* 31:2879–86
21. Beese LS, Friedman JM, Steitz TA. 1993. *Biochemistry* 32:14095–101

22. Freemont PS, Ollis DL, Steitz TA, Joyce CM. 1986. *Proteins* 1:66–73
23. Derbyshire V, Freemont PS, Sanderson MR, Beese L, Friedman JM, et al. 1988. *Science* 240:199–201
24. Goff SP. 1990. *J. Acquired Immune Defic. Syndr.* 3:817–31
25. Weiss R, Teich N, Varmus HE, Coffin J, eds. 1985. *Molecular Biology of Tumor Viruses: RNA Tumor Viruses,* Parts 1 & 2. Cold Spring Harbor, NY: Cold Spring Harbor Lab. 2nd ed.
26. Johnson MS, McClure MA, Feng D-F, Gray J, Doolittle RF. 1986. *Proc. Natl. Acad. Sci. USA* 83:7648–52
27. Hansen J, Schulze T, Mellert W, Moelling K. 1988. *EMBO J.* 7:239–43
28. Di Marzo Veronese F, Copeland TD, DeVico AL, Rahman R, Oroszlan S, et al. 1986. *Science* 231:1289–91
29. Lightfoote MM, Coligan JE, Folks TM, Fauci AS, Martin MA, Venkatesan S. 1986. *J. Virol.* 60:771–75
30. LeGrice SFJ, Naas T, Wohlgensinger B, Schatz O. 1991. *EMBO J.* 10:3905–11
31. Hostomsky Z, Hostomska Z, Fu T-B, Taylor J. 1992. *J. Virol.* 66:3179–82
32. Wu JC, Warren TC, Adams J, Proudfoot J, Skiles J, et al. 1991. *Biochemistry* 30:2022–26
33. Barat C, Lullien V, Schatz O, Keith G, Nugeyre MT, et al. 1989. *EMBO J.* 8:3279–88
34. Telesnitsky A, Goff SP. 1993. *Proc. Natl. Acad. Sci. USA* 90:1276–80
35. Lowe DM, Aitken A, Bradley C, Darby GK, Larder BA, et al. 1988. *Biochemistry* 27:8884–89
36. Lloyd LF, Brick P, Mei-Zhen L, Chayen NE, Blow DM. 1991. *J. Mol. Biol.* 217:19–22
37. Steitz TA, Smerdon S, Jäger J, Wang J, Kohlstaedt LA, et al. 1993. *Cold Spring Harbor Symp. Quant. Biol.* 58:In press
38. Davies JF, Hostomska Z, Homstomsky Z, Jordan SR, Matthews DA. 1991. *Science* 252:88–95
39. Yang W, Hendrickson WA, Crouch RJ, Satow Y. 1990. *Science* 249:1398–405
40. Katayanagi K, Miyagawa M, Matsushima M, Ishikawa M, Kanaya S, et al. 1992. *J. Mol. Biol.* 223: 1029–52
41. Steitz TA. 1993. *Curr. Opin. Struct. Biol.* 3:31–38
42. Larder BA, Purifoy DJM, Powell KL, Darby G. 1987. *Nature* 327:716–17
43. Larder BA, Kemp SD, Purifoy DJM. 1989. *Proc. Natl. Acad. Sci. USA* 86:4803–7
44. Boyer PL, Ferris AL, Hughes SH. 1992. *J. Virol.* 66:1031–39
45. Polesky AH, Steitz TA, Grindley NDF, Joyce CM. 1990. *J. Biol. Chem.* 265: 14579–91
46. Polesky AH, Dahlberg ME, Benkovic SJ, Grindley NDF, Joyce CM. 1992. *J. Biol. Chem.* 267:8417–28
47. Arnold E, Jacobo-Molina A, Nanni RG, Williams RL, Lu X, et al. 1992. *Nature* 357:85–89
48. Bonner G, Patra D, Lafer EM, Sousa R. 1992. *EMBO J.* 11:3767–75
49. Osumi-Davis PA, de Aguilera MC, Woody RW, Woody AY. 1992. *J. Mol. Biol.* 226:37–45
50. Ollis DL, Kline C, Steitz TA. 1985. *Nature* 313:818–19
51. Warwicker J, Ollis D, Richards FM, Steitz TA. 1985. *J. Mol. Biol.* 186: 645–49
52. Joyce CM, Fujii DM, Laks HS, Hughes CM, Grindley NDF. 1985. *J. Mol. Biol.* 186:283–93
53. Freemont PS, Friedman JM, Beese LS, Sanderson MR, Steitz TA. 1988. *Proc. Natl. Acad. Sci. USA* 85:8924–28
54. Basu A, Williams KR, Modak MJ. 1987. *J. Biol. Chem.* 262:9601–07
55. Basu S, Basu A, Modak MJ. 1988. *Biochemistry* 27:6710–16
56. Mohan PM, Basu A, Basu S, Abraham KI, Modak MJ. 1988. *Biochemistry* 27:226–33
57. Catalano CE, Allen DJ, Benkovic SJ. 1990. *Biochemistry* 29:3612–21
58. Joyce CM, Steitz TA. 1987. *Trends Biochem. Sci.* 12:288–92
59. Allen DJ, Darke PL, Benkovic SJ. 1989. *Biochemistry* 28:4601–7
60. Guest CR, Hochstrasser RA, Dupuy CG, Allen DJ, Benkovic SJ, Millar DP. 1991. *Biochemistry* 30:8759–70
61. Cowart M, Gibson KJ, Allen DJ, Benkovic SJ. 1989. *Biochemistry* 28: 1975–83
62. Furfine ES, Reardon JE. 1991. *J. Biol. Chem.* 266:406–12
63. Kati WA, Johnson KA, Jerva LF, Anderson KS. 1992. *J. Biol. Chem.* 267:25988–97
64. Gopalakrishnan V, Peliska JA, Benkovic SJ. 1992. *Proc. Natl. Acad. Sci. USA* 89:10763–67
65. Basu A, Ahluwalia KK, Basu S, Modak MJ. 1992. *Biochemistry* 31: 616–23
66. Sobol RW, Suhadolnik RJ, Kumar A, Lee BJ, Hatfield DL, Wilson SH. 1991. *Biochemistry* 30:10623–31
67. Sardana VV, Emini EA, Gotlib L, Graham DJ, Lineberger DW, et al. 1992. *J. Biol. Chem.* 267:17526–30

68. Merluzzi VJ, Hargrave KD, Labadia M, Grozinger K, Skoog M, et al. 1990. *Science* 250:1411–13
69. Kohlstaedt LA, Steitz TA. 1992. *Proc. Natl. Acad. Sci. USA* 89:9652–56
70. Barat C, LeGrice SFJ, Darlix J-L. 1991. *Nucleic Acids Res.* 19:751–57
71. Leis J, Aiyar A, Cobrinik D. 1993. In *Reverse Transcriptase,* ed. AM Skalka, SP Goff, pp. 33–47. Cold Spring Harbor, NY: Cold Spring Harbor Lab.
72. Bryant FR, Johnson KA, Benkovic SJ. 1983. *Biochemistry* 22:3537–46
73. Rush J, Konigsberg WH. 1990. *J. Biol. Chem.* 265:4821–27
74. Pandey VN, Williams KR, Stone KL, Modak MJ. 1987. *Biochemistry* 26: 7744–48
75. Pandey VN, Kaushik NA, Pradhan DS, Modak MJ. 1990. *J. Biol. Chem.* 265:3679–84
76. Basu A, Modak MJ. 1987. *Biochemistry* 26:1704–9
77. Pandey VN, Kaushik N, Sanzgiri RP, Patil MS, Modak MJ, Barik S. 1993. *Eur. J. Biochem.* 214:59–65
78. Cheng N, Merrill BM, Painter GR, Frick LW, Furman PA. 1993. *Biochemistry* 32:7630–34
79. Wu J, Amandoron E, Li X, Wainberg MA, Parniak MA. 1993. *J. Biol. Chem.* 268:9980–85
80. Basu A, Tirumalai RS, Modak MJ. 1989. *J. Biol. Chem.* 264:8746–52
81. Martin JL, Wilson JE, Furfine ES, Hopkins SE, Furman PA. 1993. *J. Biol. Chem.* 268:2565–70
82. Mitchell LLW, Cooperman BS. 1992. *Biochemistry* 31:7707–13
83. Dong Q, Copeland WC, Wang TS-F. 1993. *J. Biol. Chem.* 268:24163–74
84. Blasco MA, Lázaro JM, Bernad A, Blanco L, Salas M. 1992. *J. Biol. Chem.* 267:19427–34
85. Blasco MA, Lázaro JM, Blanco L, Salas M. 1993. *J. Biol. Chem.* 268: 16763–70
86. Larder BA, Kemp SD. 1989. *Science* 246:1155–58
87. Fitzgibbon JE, Howell RM, Haberzettl CA, Sperber SJ, Gocke DJ, Dubin DT. 1992. *Antimicrob. Agents Chemother.* 36:153–57
88. St. Clair MH, Martin JL, Tudor-Williams G, Bach MC, Vavro CL, et al. 1991. *Science* 253:1557–59
89. Kellam P, Boucher CAB, Larder BA. 1992. *Proc. Natl. Acad. Sci. USA* 89:1934–38
90. Carroll SS, Cowart M, Benkovic SJ. 1991. *Biochemistry* 30:804–13
91. Burgers PMJ, Eckstein F. 1979. *J. Biol. Chem.* 254:6889–93
92. Brody RS, Frey PA. 1981. *Biochemistry* 20:1245–52
93. Romaniuk PJ, Eckstein F. 1982. *J. Biol. Chem.* 257:7684–88
94. Bartlett PA, Eckstein F. 1982. *J. Biol. Chem.* 257:8879–84
95. Brody RS, Adler S, Modrich P, Stec WJ, Leznikowski ZL, Frey PA. 1982. *Biochemistry* 21:2570–72
96. Hopkins S, Furman PA, Painter GR. 1989. *Biochem. Biophys. Res. Commun.* 163:106–10
97. Kuchta RD, Mizrahi V, Benkovic PA, Johnson KA, Benkovic SJ. 1987. *Biochemistry* 26:8410–17
98. Patel SS, Wong I, Johnson KA. 1991. *Biochemistry* 30:511–25
99. Capson TL, Peliska JA, Kaboord BF, Frey MW, Lively C, et al. 1992. *Biochemistry* 31:10984–94
100. Reardon JE. 1992. *Biochemistry* 31: 4473–79
101. Reardon JE. 1993. *J. Biol. Chem.* 268:8743–51
102. Dahlberg ME, Benkovic SJ. 1991. *Biochemistry* 30:4835–43
103. Eger BT, Benkovic SJ. 1992. *Biochemistry* 31:9227–36
104. Heringa J, Argos P. 1994. *Evolutionary Biology of Viruses,* ed. SS Morse. New York: Raven. In press
105. Boyer PL, Ferris AL, Hughes SH. 1992. *J. Virol.* 66:7533–37
106. Copeland WC, Wang TS-F. 1993. *J. Biol. Chem.* 268:11028–40
107. Bernad A, Lázaro JM, Salas M, Blanco L. 1990. *Proc. Natl. Acad. Sci. USA* 87:4610–14
108. Jung G, Leavitt MC, Schultz M, Ito J. 1990. *Biochem. Biophys. Res. Commun.* 170:1294–300
109. Joung I, Horwitz MS, Engler JA. 1991. *Virology* 184:235–41
110. Date T, Yamamoto S, Tanihara K, Nishimoto Y, Matsukage A. 1991. *Biochemistry* 30:5286–92
111. Sankar S, Porter AG. 1992. *J. Biol. Chem.* 267:10168–76
112. Wakefield JK, Jablonski SA, Morrow CD. 1992. *J. Virol.* 66:6806–12
113. Blasco MA, Lázaro JM, Blanco L, Salas M. 1993. *J. Biol. Chem.* 268: 24106–13
114. Joyce CM. 1991. *Curr. Opin. Struct. Biol.* 1:123–29
115. Dong Q, Copeland WC, Wang TS-F. 1993. *J. Biol. Chem.* 268:24175–82
116. Reha-Krantz LJ, Nonay RL, Stocki S. 1993. *J. Virol.* 67:60–66
117. Copeland WC, Lam NK, Wang TS-F. 1993. *J. Biol. Chem.* 268:11041–49

118. Reha-Krantz LJ. 1988. *J. Mol. Biol.* 202:711–24
119. Hall JD. 1988. *Trends Genet.* 4:42–46
120. Bebenek K, Abbotts J, Roberts JD, Wilson SH, Kunkel TA. 1989. *J. Biol. Chem.* 264:16948–56
121. Scheuermann RH, Echols H. 1984. *Proc. Natl. Acad. Sci. USA* 81:7747–51
122. Brutlag D, Kornberg A. 1972. *J. Biol. Chem.* 247:241–48
123. Derbyshire V, Grindley NDF, Joyce CM. 1991. *EMBO J.* 10:17–24
124. Derbyshire V. 1990. *Studies of the 3′ to 5′ exonuclease of DNA polymerase I of Escherichia coli.* PhD thesis. Yale Univ., New Haven. 111 pp.
125. Han H, Rifkind JM, Mildvan AS. 1991. *Biochemistry* 30:11104–8
126. Gupta AP, Benkovic SJ. 1984. *Biochemistry* 23:5874–81
127. Kim EE, Wykoff HW. 1991. *J. Mol. Biol.* 218:449–64
128. Piccirilli JA, Vyle JS, Caruthers MH, Cech TR. 1993. *Nature* 361:85–88
129. Steitz TA, Steitz JA. 1993. *Proc. Natl. Acad. Sci. USA* 90:6498–502
130. Bernad A, Blanco L, Lázaro JM, Martin G, Salas M. 1989. *Cell* 59:219–28
131. Morrison A, Bell JB, Kunkel TA, Sugino A. 1991. *Proc. Natl. Acad. Sci. USA* 88:9473–77
132. Blanco L, Bernad A, Salas M. 1992. *Gene* 112:139–44
133. Foury F, Vanderstraeten S. 1992. *EMBO J.* 11:2717–26
134. Reha-Krantz LJ, Nonay RL. 1993. *J. Biol. Chem.* 268:27100–8
135. Simon M, Giot L, Faye G. 1991. *EMBO J.* 10:2165–70
136. Soengas MS, Esteban JA, Lázaro JM, Bernad A, Blasco MA, et al. 1992. *EMBO J.* 11:4227–37
137. Frey MW, Nossal NG, Capson TL, Benkovic SJ. 1993. *Proc. Natl. Acad. Sci. USA* 90:2579–83
138. Donlin MJ, Patel SS, Johnson KA. 1991. *Biochemistry* 30:538–46
139. Eger BT, Kuchta RD, Carroll SS, Benkovic PA, Dahlberg ME, et al. 1991. *Biochemistry* 30:1441–48
140. Gibbs JS, Weisshart K, Digard P, de Bruynkops A, Knipe DM, Coen DM. 1991. *Mol. Cell. Biol.* 11:4786–95
141. Maki H, Kornberg A. 1987. *Proc. Natl. Acad. Sci. USA* 84:4389–92
142. Derbyshire V, Astatke M, Joyce CM. 1994. *Nucleic Acids Res.* 21:5439–48
143. Joyce CM, Friedman JM, Beese L, Freemont PS, Steitz TA. 1988. *DNA Replication and Mutagenesis,* ed. RE Moses, WC Summers, pp. 220–26. Washington, DC: Am. Soc. Microbiol.
144. Kuchta RD, Benkovic P, Benkovic SJ. 1988. *Biochemistry* 27:6716–25
145. Wong I, Patel SS, Johnson KA. 1991. *Biochemistry* 30:526–37
146. Joyce CM. 1989. *J. Biol. Chem.* 264:10858–66
147. Reddy MK, Weitzel SE, von Hippel PH. 1992. *J. Biol. Chem.* 267:14157–66
148. Oyama F, Kikuchi R, Crouch RJ, Uchida T. 1989. *J. Biol. Chem.* 264:18808–17
149. Krug MS, Berger SL. 1989. *Proc. Natl. Acad. Sci. USA* 86:3539–43
150. Luo GX, Taylor J. 1990. *J. Virol.* 64:4321–28
151. Champoux JJ. 1993. In *Reverse Transcriptase,* ed. AM Skalka, SP Goff, pp. 103–17. Cold Spring Harbor, NY: Cold Spring Harbor Lab.
152. Peliska JA, Benkovic SJ. 1992. *Science* 258:1112–18
153. Kanaya S, Katsuda-Nakai C, Ikehara M. 1991. *J. Biol. Chem.* 266:11621–27
154. Hostomsky Z, Hostomska Z, Hudson GO, Moomaw EW, Nodes BR. 1991. *Proc. Natl. Acad. Sci. USA* 88:1148–52
155. Tanese N, Goff SP. 1988. *Proc. Natl. Acad. Sci. USA* 85:1777–81
156. Doolittle RF, Feng D-F, Johnson MS, McClure MA. 1989. *Q. Rev. Biol.* 64:1–29
157. Mizrahi V, Usdin MT, Harington A, Dudding LR. 1990. *Nucleic Acids Res.* 18:5359–63
158. Schatz O, Cromme FV, Grüninger-Leitch F, LeGrice SFJ. 1989. *FEBS Lett.*257:311–14
159. Kanaya S, Kohara A, Miura Y, Sekiguchi A, Iwai S, et al. 1990. *J. Biol. Chem.* 265:4615–21
160. Tisdale M, Schulze T, Larder BA, Moelling K. 1991. *J. Gen. Virol.* 72:59–66
161. Wöhrl BM, Volkmann S, Moelling K. 1991. *J. Mol. Biol.* 220:801–18
162. Nakamura H, Oda Y, Iwai S, Inoue H, Ohtsuka E, et al. 1991. *Proc. Natl. Acad. Sci. USA* 88:11535–39
163. Fedoroff OY, Salazar M, Reid BR. 1993. *J. Mol. Biol.* 233:509–23
164. Tan C-K, Zhang J, Li Z-Y, Tarpley WG, Downey KM, So AG. 1991. *Biochemistry* 30:2651–55
165. Prasad VR, Goff SP. 1989. *Proc. Natl. Acad. Sci. USA* 86:3104–08
166. Hizi A, Barber A, Hughes SH. 1989. *Virology* 170:326–29
167. Hizi A, Hughes SH, Shaharabany M. 1990. *Virology* 175:575–80
168. DeStefano JJ, Buiser RG, Mallaber

LM, Myers TW, Bambara RA, Fay PJ. 1991. *J. Biol. Chem.* 266:7423–31
169. Dudding LR, Mizrahi V. 1993. *Biochemistry* 32:6116–20
170. Telesnitsky A, Goff SP. 1993. *EMBO J.* 12:4433–38
171. Hollingsworth HC, Nossal NG. 1991. *J. Biol. Chem.* 266:1888–97
172. Sayers JR, Eckstein F. 1990. *J. Biol. Chem.* 265:18311–17
173. Kerr C, Sadowski PD. 1972. *J. Biol. Chem.* 247:311–18
174. Berkower I, Leis J, Hurwitz J. 1973. *J. Biol. Chem.* 248:5914–21
175. Lyamichev V, Brow MAD, Dahlberg JE. 1993. *Science* 260:778–83
176. Shinozaki K, Okazaki T. 1978. *Nucleic Acids Res.* 5:4245–61
177. Crute JJ, Lehman IR. 1989. *J. Biol. Chem.* 264:19266–70
178. Siegal G, Turchi JJ, Myers TW, Bambara RA. 1992. *Proc. Natl. Acad. Sci. USA* 89:9377–81
179. Lundquist RC, Olivera BM. 1982. *Cell* 31:53–60
180. Kelley WS, Joyce CM. 1983. *J. Mol. Biol.* 164:529–60
181. Gutman PD, Minton KW. 1993. *Nucleic Acids Res.* 21:4406–7
182. Poch O, Sauvaget I, Delarue M, Tordo N. 1989. *EMBO J.* 8:3867–74
183. Blanco L, Bernad A, Blasco MA, Salas M. 1991. *Gene* 100:27–38
184. Wong SW, Wahl AF, Yuan P-M, Arai N, Pearson BE, et al. 1988. *EMBO J.* 7:37–47

Annu. Rev. Biochem. 1994. 63:823–67

CALCIUM CHANNEL DIVERSITY AND NEUROTRANSMITTER RELEASE: The ω-Conotoxins and ω-Agatoxins[1]

Baldomero M. Olivera
Department of Biology, University of Utah, Salt Lake City, Utah 84112

George P. Miljanich and J. Ramachandran
Neurex Corporation, 3760 Haven Avenue, Menlo Park, California 94025

Michael E. Adams
Departments of Entomology and Neuroscience, 5419 Boyce Hall, University of California, Riverside, California 92521

KEY WORDS: Ca channels, neurotransmitter release, ω-conotoxins, ω-agatoxins, ligand:receptor interactions

---

CONTENTS

[1]Abbreviations used: GVIA, ω-conotoxin GVIA from *Conus geographus*; MVIIC, ω-conotoxin MVIIC from *Conus magus*; MVIIA, ω-conotoxin MVIIA from *Conus magus*; ω-AgaIIIA or AgaIIIA, ω-agatoxin-IIIA from *Agelenopsis aperta*; ω-AgaIVA or AgaIVA, ω-agatoxin-IVA from *Agelenopsis aperta*; DHP, the dihydropyridine drugs.

0066-4154/94/0701-0823$05.00

I. INTRODUCTION

The level of intracellular Ca regulates many cellular processes, including neurotransmitter and hormone secretion, the activity of ion channels and enzymes, cytoskeletal function, cell proliferation, and gene expression. Voltage-sensitive Ca channels are among the most heterogeneous of ion channels. In neurons, Ca channels differ in cellular location, biophysical and pharmacological properties, and modulation. A single neuron generally contains multiple types of Ca channels, and such channels are central to the integration and expression of activity in the nervous system. It is clearly important to understand the functional significance of Ca channel diversity; a major research effort that is under way has made clear that different calcium channel types are all part of a family of multisubunit ion channels.

Two classes of polypeptide ligands, collectively referred to as ω-toxins, have been instrumental in defining various classes of Ca channels (1–3a). A number of ω-conotoxins, which are 24–29 amino acids in length, have been isolated from *Conus* snails, and somewhat longer ω-agatoxins have been isolated from the funnel-web spider *Agelenopsis aperta*. In both cases the calcium channel ligands are produced in venoms, presumably to capture prey. It is clear that a wealth of similar peptides are available from venoms of related snails and spiders.

Over many millions of years, in two different phyla, the ω-toxins have evolved to interact with critical target regions in the Ca channels of their prey (which represent at least six different phyla). The molecular mechanisms of this evolution include familiar and unfamiliar elements. In tracking the variations among prey species, the predators have incidentally explored a range of structures that correspond to the various Ca channels in our own central nervous system. Hence the extraordinary interest in these toxins as tools for exploring the functional diversity of this important class of neuronal signaling molecules.

The classical experiments of Katz & Miledi (4–6) showed that for neurotransmitter release to occur, calcium entry into the presynaptic termini is required. It has since become clear that voltage-activated calcium channels are responsible for this entry. Selective inhibition of the different Ca channel types at the presynaptic termini of functional synaptic preparations became possible for the first time with the discovery of ω-toxins. These toxins, with their high affinity and selectivity for specific Ca channel types, are proving to be powerful pharmacological tools for blocking neurotransmitter release.

This article surveys how the various ω-toxins from snail and spider venoms are being used to explore voltage-sensitive calcium channel diversity, and provides an overview of the effects of the ω-toxins on neurotransmitter release.

II. THE ω-CONOTOXINS

A. *Introduction and Biological Background*

Predatory marine snails of the genus *Conus* ("cone snails") produce a set of small, disulfide-rich, calcium channel–targeted peptides, the ω-conotoxins (7–9). In the literature, ω-conotoxin often refers to one specific peptide, ω-conotoxin GVIA (10). This peptide, the first to be described, is commercially available and used in hundreds of laboratories as a standard ligand for characterizing voltage-sensitive Ca channels. However, in this review, ω-conotoxin is used as a generic term for all homologous Ca channel–targeted peptides from the venom of cone snails, not only ω-conotoxin GVIA. The specificity for blocking Ca channels of many of these peptides differs greatly from that of ω-conotoxin GVIA.

A number of features are common to all ω-conotoxins. The natural peptides are found in *Conus* venoms, and specifically target voltage-sensitive calcium channel subtypes. The characteristic arrangement of cysteine residues—C. . . .C. . . . . .CC. . .C. . . . . .C—is called the "four-loop Cys scaffold" of the conotoxins (1). While all ω-conotoxins described have a four-loop scaffold, the converse is not true. Many *Conus* peptides of the four-loop class are not ω-conotoxins. At least one other large family of *Conus* peptides belongs to this class, the "King-Kong peptides" from mollusk-hunting cones (which have no direct effect on voltage-sensitive calcium channels) (11, 12).

All ω-conotoxins characterized so far are from fish-hunting *Conus* species. Of the ca. 500 *Conus* species, it seems likely that 40–100 prey primarily,

Figure 1 Shells of six fish-hunting *Conus* species from which ω-conotoxins have been purified. Top row (left to right): *Conus stercusmuscarum,* the fly-speck cone (ω-conotoxin SmVIIA); *Conus magus,* the magus cone (ω-conotoxins MVIIA, MVIIB, MVIIC, MVIID); *Conus radiatus,* the radial cone (ω-conotoxin RVIA). Bottom row (left to right): *Conus striatus,* the striated cone (ω-conotoxins SVIA, SVIB); *Conus geographus,* the geography cone (ω-conotoxins GVIA, GVIIA); *Conus tulipa,* the tulip cone (ω-conotoxin TVIA). The sequences of ω-conotoxins from *Conus magus, geographus* and *striatus* are shown in Figure 2.

if not exclusively, on fish. ω-Conotoxins have been purified and sequenced from six different fish-hunting *Conus* (see Figure 1) (7, 10, 13, 14). The venoms of cone snails are produced in a long tubular venom duct, and ejected from the duct by the contraction of a muscular venom bulb. The venom is injected into the prey through a hollow, harpoon-like tooth. For fish-hunting *Conus*, the specialized tooth plays a dual role: to tether the prey, and to serve as an injection needle. An average fish-hunting species 3–5 cm in length injects approximately 5 μl of venom (8, 15, 16).

An obvious problem for the cone snails that feed primarily on fish is to get close enough to the prey to strike. A wide variety of behavioral strategies are employed; most *Conus* are nocturnal, and many piscivorous cone snails probably stalk sleeping fish that are relatively stationary, or that hide in crevices within a coral reef or rocky area. Some cone snails engulf the prey before stinging, others bait the fish and sting them in the mouth, and still others are poised on top of small caves or tunnels, harpooning fish from above when they come to hide.

The ω-conotoxins are not the only paralytic components in a fish-hunting *Conus* venom. In *Conus geographus,* at least two other families of paralytic peptides are found: the α-conotoxins, which target nicotinic acetylcholine receptors, and the μ-conotoxins, which inhibit voltage-sensitive sodium channels. In *Conus magus* venom, three other paralytic families of *Conus* peptides have been identified besides the ω-conotoxins. Although ω-conotoxins are found in almost all fish-hunting species examined, they may be absent in a few. A detailed examination of *Conus purpurascens,* a fish hunter, has so far failed to identify any ω-conotoxins (it is noteworthy that all other fish-hunting species analyzed are from the Indo-Pacific, while *C. purpurascens* is a Panamic *Conus*).

It is not yet established whether ω-conotoxins are restricted to fish-hunting *Conus*; many more *Conus* species hunt mollusks or polychaete worms. It seems quite possible, and even likely, that these venoms will have peptides that target to presynaptic calcium channels to effect rapid prey paralysis. Such peptides could either be independently evolved and unrelated to the ω-conotoxins or belong to the ω-conotoxin family as defined above, with potentially novel target specificities.

B. Purification, Sequencing, and Chemical Synthesis

The first ω-conotoxins were purified directly from the venom of several fish-hunting *Conus* species—the geography cone, *Conus geographus*; the striated cone, *Conus striatus*; and the magus cone, *Conus magus*. This initial characterization was labor intensive: Large numbers of snails were collected, venom ducts dissected, venom extracted, and purification of ω-conotoxins from crude *Conus* venom carried out by a series of chromatographic steps. The ω-conotoxins could be identified using fish paralysis as one assay; in much of the earlier work, mice were also used. The ω-conotoxins targeted to the N-type Ca channel such as ω-conotoxin GVIA (see discussion below) typically elicit a highly characteristic “shaking” syndrome upon intracranial injection into mice; however, ω-conotoxins with different Ca channel subtype specificities elicit different symptoms.

Amino acid sequences of purified ω-conotoxins were determined using standard Edman procedures, with alkylation of Cys residues and the determination of hydroxyproline being the only nonroutine features of the analysis. The only remaining question regarding primary structure was whether the C terminus was blocked; in all natural ω-conotoxins so far, the C terminus has proven to be amidated. This is generally determined by mass spectroscopy and verified by chemical synthesis of the peptide.

More recently, ω-conotoxin sequences have been elucidated indirectly, by a molecular genetic strategy (13). By using homologous sequences either in the mature toxin region itself, in the precursor, or in conserved untranslated regions of the messenger RNA, ω-conotoxin-encoding clones can be identified from a venom duct cDNA library. Once such clones have been sequenced, the amino acid sequence of the mature ω-conotoxin can be predicted. Thus, ω-conotoxins MVIIC and MVIID have never actually been purified from *Conus magus* venom, but their primary structures were deduced by sequencing cDNA clones. The nucleic acid sequence typically encodes the amino acid codons -X-Gly-STOP to specify the C terminus of the peptide; this is presumably processed to -X-NH_2 in the mature toxin. In addition, all proline residues in ω-conotoxins appear to be processed to 4-*trans*-hydroxyproline in the natural peptides.

Whether an ω-conotoxin sequence is obtained by direct amino acid sequencing of purified peptide or through cloning, chemical synthesis is required to make a comprehensive biochemical and physiological characterization possible. The linear peptide is generally synthesized by a Merrifield solid-phase synthesis (8, 16, 17); some successful large commercial syntheses of ω-conotoxins employ liquid-phase synthetic methods (18). The major problem is to form correct disulfide bonds. Optimal cyclization conditions have had to be developed for each ω-conotoxin in order to provide a sufficient yield of the product with the correct disulfide bonds. Folding into the correct disulfide-bonded form has recently been facilitated by using a different protective group on one of the pairs of Cys residues that is to form a disulfide bond in the mature toxin (19); if fewer Cys residues are unblocked at one time, fewer isomers can be formed upon oxidation. Thus, there are theoretically 15 different monomeric isomers of a peptide with six cysteine residues (such as ω-conotoxins). However, if one pair of Cys residues is synthesized with a protective group resistant to HF, allowing the other four Cys residues to be unblocked first, only three monomeric disulfide-bonded forms would be formed initially. The three possible forms can be rapidly folded, separated from each other, and the remaining Cys pair unblocked and oxidized to form the corresponding three-disulfide-bonded isomers. This allows a much more rapid synthesis of the biologically active form, and obviates the need for developing special conditions for each peptide.

Once synthetic material is available, an ω-conotoxin can be derivatized for specific experimental purposes. Various reporter groups have been covalently attached to these peptides without prohibitive loss of biological activity. ω-Conotoxins have been successfully iodinated with ^{125}I at Tyr residues (20, 21); such radiolabeled derivatives have been widely used for binding assays (20, 22, 23, 23a), autoradiographic investigation to identify the anatomical location of calcium channel subtypes (24–28), as well as chemical and photoactivatable crosslinking studies to identify the biochemical targets of these peptides (22, 29–34). In addition, both biotinylated and fluorescent derivatives of ω-conotoxins have been described (35–40). These ω-conotoxin derivatives should greatly increase the utility of these peptides as biochemical reagents for investigating different calcium channel types.

C. Biosynthesis of ω-Conotoxins; Hypervariability

In contrast to many marine natural products in the same general size range, the conotoxins are directly translated from genes. The sequence of messenger RNA encoding several ω-conotoxins has been determined from *Conus* venom duct cDNA libraries. Although mature ω-conotoxins are 24–29 amino acids in length, the initial translation product is a prepropeptide precursor of ca. 70 amino acids. This precursor is then processed to the mature peptide, typically with additional posttranslational modification.

The prepropeptides for ω-conotoxins have not yet been directly purified and characterized from *Conus* venoms. However, toxin precursors are believed to be stored as ellipsoidal granules present in the venom duct of *Conus*. Although the biochemistry of processing has not been experimentally investigated, the primary structure of the precursors strongly suggests a specific processing scenario.

For example, ω-conotoxin GVIA from *Conus geographus,* which is a 27-amino-acid peptide, is initially translated as a 73-amino-acid precursor (see Figure 2) (41). The prepropeptide must undergo a minimum of three posttranslational processing events before the mature peptide can be generated: (*a*) the C-terminal glycine must be converted to an amide residue by the standard C-terminal amidation pathway (to yield a blocked Cys-Tyr amide), (*b*) the proline residues are hydroxylated to yield 4-*trans*-hydroxyproline, and (*c*) the peptide bond between Arg45 and Cys46 of the prepropeptide precursor must be cleaved; Cys46 of the prepropeptide precursor is thus converted to Cys1 of the mature peptide. A protease that recognizes the sequence around Arg45 to cleave specifically at this locus is presumably present in *Conus geographus* venom ducts.

A. Mature Toxin Sequences

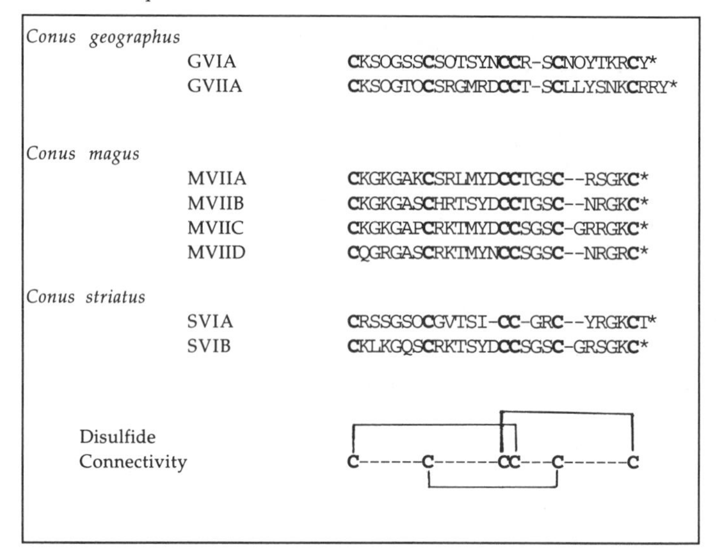

Conus geographus
GVIA CKSOGSSCSOTSYNCCR-SCNOYTKRCY*
GVIIA CKSOGTOCSRGMRDCCT-SCLLYSNKCRRY*
Conus magus
MVIIA CKGKGAKCSRLMYDCCTGSC--RSGKC*
MVIIB CKGKGASCHRTSYDCCTGSC--NRGKC*
MVIIC CKGKGAPCRKTMYDCCSGSC-GRRGKC*
MVIID CQGRGASCRKTMYNCCSGSC--NRGRC*
Conus striatus
SVIA CRSSGSOCGVTSI-CC-GRC--YRGKCT*
SVIB CKLKGQSCRKTSYDCCSGSC-GRSGKC*
Disulfide Connectivity
C------C------CC---C------C

B. Predicted Sequence of ω-conotoxin GVIA precursor

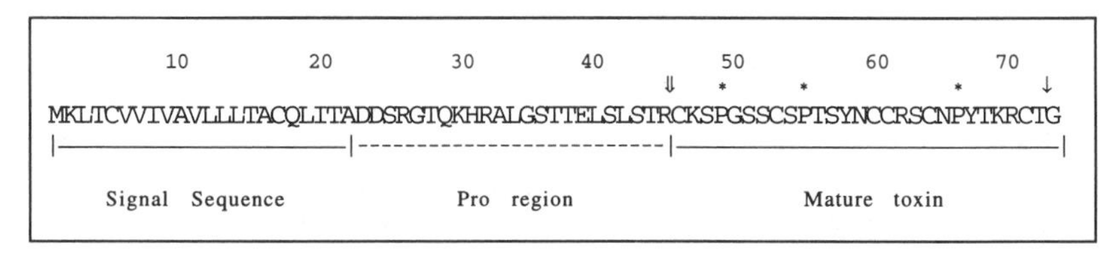

Figure 2 Sequences of ω-conotoxins. *A*. All sequences given were verified by chemical synthesis. Sequences for ω-conotoxins from *C. radiatus* and *C. stercusmuscarum* (B Olivera, M Grilley, F Abogadie, LJ Cruz, unpublished) as well as for *C. tulipa* (J Haack, J Bell, L Nadasdi, J Ramachandran, LJ Cruz, B Olivera, unpublished) have also been determined. The asterisk indicates an amidated C terminus. O is 4-*trans*-hydroxyproline. *B*. The predicted amino acid sequence of ω-conotoxin GVIA prepropeptide is shown, with the three major regions indicated (41). The open arrow indicates a protease processing site, while the regular arrow indicates processing of glycine to a C-terminal amide moiety. The asterisks show proline residues that are posttranslationally modified to 4-*trans*-hydroxyproline.

All ω-conotoxin sequences published in the literature are summarized in Figure 2. Two features of this figure are noteworthy. First is the multiplicity of ω-conotoxins in the venom of each *Conus* species. The three *Conus* venoms that contain ω-conotoxins that have been systematically examined have yielded more than one molecular form. Furthermore, different ω-conotoxins from the same venom are not merely minor sequence variants that might arise from neutral polymorphism. The ω-conotoxin isoforms within the same venom are likely to be functionally selected, and the considerable sequence divergence observed between isoforms is presumably relevant for differential targeting. Preliminary evidence for different target specificity has been obtained for the toxin isoforms from *Conus magus* (see Section IV below).

The sequences in Figure 2 also exhibit remarkable interspecies sequence variation. ω-Conotoxins from two different venoms may have less than 30% sequence identity (if Cys residues are excluded). In many cases, highly divergent sequences from two different *Conus* venoms compete for binding to the same site on a calcium channel. Indeed, although all of the peptides have the same conserved four-loop Cys scaffold, the only conserved non-Cys residue is Gly5. However, a comparison of ω-conotoxin precursor sequences from different *Conus* species deduced from cDNA clones reveals that the N-terminal regions of the precursor (i.e. both the signal sequence and the pro region) show the high degree of conservation expected between homologous proteins derived from species within the same genus. Thus, the hypervariability is restricted to the loops between Cys residues in the mature toxin region.

The unexpected interspecies variation in mature ω-conotoxin sequences raises two questions. First, how is this interspecific hypervariability generated? Second, how is the extreme divergence in peptide sequence compatible with targeting to the same binding sites on calcium channels? It has been suggested that the hypervariability observed in conotoxin sequence may be the result of a specialized genetic strategy of the cone snails, which has allowed them to quickly evolve novel peptide ligands with altered target specificity. Different individuals of the same species do not exhibit unusual hypervariability; the peptide sequences from one individual to another are identical. Even populations of the same species widely separated in habitat, with significant differences in ecological parameters, nevertheless largely retain the same peptide sequences. On the other hand, homologous peptides from different *Conus* species that have the same function, although they exhibit sequence identity in the signal sequence and propeptide regions, show great hypervariability in the blocks of amino acids between cysteine residues of the mature peptide. An antibody-like strategy has been suggested for the generation of novel *Conus* peptide sequences: conserved sequences to specify a structural scaffold, and short regions that become hypermutable during speciation, to determine binding specificity (8). However, the precise mechanism of the observed hypermutation in *Conus* peptide sequences has not been elucidated.

The extreme interspecific hypervariability of *Conus* sequences demonstrates that many different sequence solutions bind and inhibit the same target. If the ω-conotoxins were conventional peptides, and their sequences aligned, one might postulate that only conserved residues were critical for biological activity and that the variable residues were not functionally important. This does not appear to be the case. The six cysteine residues and one glycine that are conserved in all ω-conotoxin sequences are clearly insufficient to specify a functional ω-conotoxin. This conclusion can be stated unequivocally, because the six Cys residues and conserved glycine are also found in the King-Kong peptide from *Conus textile,* which is clearly not a voltage-sensitive calcium channel antagonist [recent physiological experiments suggest that the King-Kong peptide changes the time course of inactivation of sodium channels in molluskan systems, with no direct effects on Ca channels at all (12)]. Thus, the hypervariable loops between Cys residues must provide specificity determinants that are essential for ω-conotoxin function.

If the non-Cys amino acids present in loops between disulfide bonds are essential for ω-conotoxin binding and activity, why are such different sequences all permissible solutions for binding to the same sites on voltage-sensitive calcium channels? To take one example that has been intensively investigated, ω-conotoxins GVIA and MVIIA—which share only 27% of all non-Cys amino acids—nevertheless compete for binding to the vast majority of their high-affinity sites in mammalian brain. Furthermore, they have the same physiological effects on fish, i.e. an inhibition of presynaptic calcium channels at the neuromuscular junction, resulting in paralysis. How can the same ligand sites exhibit extremely high affinity for two peptides with such divergent sequences? We will attempt to explain this dilemma by invoking the "macrosite" hypothesis described in more detail below (see Section VI). A key insight that has been obtained from examining different ω-conotoxins is that extreme degeneracy is permissible in the sequence solutions that target the same ligand pocket of the voltage-sensitive calcium channel complex.

III. SPIDER TOXINS TARGETING Ca CHANNELS: THE ω-AGATOXINS

A. Biological and Physiological Background

A largely unexplored resource for ligands affecting receptors and ion channels are the venoms of spiders. Some 30,000 spider species have been described, and this represents only about 20% of the estimated total (42). The principal prey for spiders are insects, but a number of spider venom polypeptides have been found to bind with high affinity and selectivity to vertebrate Ca channels. This is not surprising, given the high degree of

Figure 3 The American funnel-web spider, *Agelenopsis aperta*. The venom of this spider is the source of the ω-agatoxins discussed in this review.

sequence similarity already demonstrated between membrane proteins in vertebrates and insects, many of which serve as targets for toxins. There is little doubt that spiders and other arthropods provide a vast reservoir of toxins with potential usefulness in the discrimination of ion channels and receptors in the brain. Thus far, venoms of only a few spider species have been examined, and only one, that of the American funnel-web spider, *Agelenopsis aperta* (see Figure 3), has been studied in comprehensive biochemical detail.

As is the case for *Conus,* spider venom contains a mixture of toxins directed toward multiple molecular targets for prey capture. The ω-agatoxins, selective peptide antagonists of voltage-activated calcium channels, are only one of three groups of neurotoxins present in *Agelenopsis aperta* venom (3). Two additional neurotoxin classes in this venom are the acylpolyamines, most of which target glutamate receptors (the α-agatoxins), and the μ-agatoxins, peptide activators of neuronal sodium channels.

The coexistence of multiple toxins acting at different targets appears to be a strategy for synergistic action. Indeed, measurable synergism between classes of agatoxins has been demonstrated (43). The α-agatoxins, which occur at high levels (> 30 nM) in *A. aperta* venom, are use-dependent blockers of glutamate receptors (43–45). Channel block caused by exposure to α-agatoxins occurs upon motor nerve activation, or after addition of exogenous glutamate. These toxins apparently require channel opening in order to gain access to their binding sites. The presence of glutamate also appears to slow recovery of synaptic potentials following removal of the α-agatoxins (44). The μ-agatoxins increase spontaneous transmitter release by activation of sodium channels, thereby elevating the concentration of glutamate in the synaptic area. Synergism between these two classes of toxins was demonstrated by coinjection of α- and μ-agatoxins into insects, resulting in higher than additive rates of paralysis as compared with injection of either type of toxin alone (43).

Acylpolyamine toxins occur in a variety of spider groups, including the Agelenids (*Agelenopsis aperta, Hololena curta*) (43, 46–48), orb weavers (various *Argiope* species, including *A. lobata, A. aurantia, A. trifasciata*) (49–51), trapdoor spiders (*Hebestatis theveniti*) (52), and tarantulas (*Aphonopelma chalcodes*) (52). The orb weaver species, though lacking μ-agatoxin-like peptides, appear to synergize the actions of the polyamine toxins by including high concentrations of free glutamate in their venoms (53). Toxins similar to the μ-agatoxins have been identified in venom of *Hololena curta* ["curtatoxins," cf. (54)].

The presence of the ω-agatoxins in *Agelenopsis* venom appears at first glance difficult to reconcile with the synergistic action observed between the α-agatoxins and μ-agatoxins. The ω-agatoxins inhibit transmitter release by blocking presynaptic Ca channels, which would tend to counteract the increased spontaneous transmitter release caused by the μ-agatoxins. It can be pointed out that the paralysis associated with the α-agatoxins is extremely rapid and reversible, whereas that caused by the ω-agatoxins may be of slower onset in vivo and, based on in vitro experiments, would certainly be expected to be irreversible. Thus, the different components of the venom may be involved in different stages of paralysis.

B. The ω-Agatoxins: Discovery and Characterization

The ω-agatoxins of *Agelenopsis aperta* venom are a heterogeneous group of polypeptides varying in molecular mass from 5 to 10 kDa. The earliest accounts of the ω-agatoxins described their block of presynaptic Ca channels at the insect neuromuscular junction by two groups of toxins (55–57). Type I ω-agatoxins (ω-AgaIA, ω-AgaIB, and ω-AgaIC) are similar in size (ca. 7 kDa) and amino acid sequence; all appear to be heterodimers with 10 cysteine residues (58). Somewhat larger "Type II" (ca. 9 kDa) ω-agatoxins, ω-AgaIIA and ω-AgaIIB (~10 kDa), showed little sequence similarity to the Type I toxins. Both of these toxin types are potent blockers of insect neuromuscular transmission, but block junctional potentials only partially. This block is additive when the toxins are applied jointly, suggesting that

Table 1 ω-Toxins: Bioassays

	Inhibition of:		
	Insect neuromuscular junction	ω-Conotoxin GVIA binding	^{45}Ca influx into rat brain synaptosomes[a]
ω-Agatoxin			
IA	+	−	−
IIA	+	+	+/−
IIIA	−	+	+/−
IVA	−	−	+/−
ω-Conotoxin			
GVIA/MVIIA	−	+	−
MVIIC/MVIID	−	+	+

[a] +/− signifies partial block of ^{45}Ca influx.

they bind to distinct sites. This was confirmed by later experiments showing different pharmacological selectivities for the two toxins (56).

Additional ω-agatoxins were discovered upon testing of *A. aperta* venom fractions for antagonism of vertebrate Ca channels. Many fractions from the venom inhibited the binding of [^{125}I]ω-conotoxin GVIA to chick brain synaptosomal membranes. This facile assay led to the isolation of the first Type III ω-agatoxin, ω-AgaIIIA (59), which went undetected in earlier studies, being inactive against presynaptic Ca channels at the fly neuromuscular junction (57). Both Type II and Type III ω-agatoxins block the binding of [^{125}I]ω-conotoxin GVIA to synaptosomal membranes prepared from chick or rat brain, while the Type I ω-agatoxins do not. Therefore, a subset of ω-agatoxins binds to sites that overlap with GVIA-binding sites. It was further shown that inhibition of [^{125}I]ω-conotoxin GVIA binding correlates with functional block of Ca channels: Voltage-dependent (potassium-stimulated) ^{45}Ca entry into chick brain synaptosomes is blocked by ω-AgaIIA and ω-AgaIIIA, but not by ω-AgaIA (59).

ω-AgaIIIA and GVIA, though potent blockers of chick synaptosomal Ca channels, proved to be poor blockers of rat synaptosomal Ca channels. This suggested a clear pharmacological difference between avian and mammalian nerve terminal Ca channels. With the aim of isolating a mammalian Ca channel antagonist, fractions from *A. aperta* venom were systematically tested for antagonism of potassium-stimulated ^{45}Ca entry into both chick and rat brain synaptosomes. This approach yielded the "Type IV" Ca channel antagonists ω-AgaIVA (60) and ω-AgaIVB (61; see also 62, 63), which have amino acid sequences unrelated to Types I-III ω-agatoxins. These two toxins, 48 amino acids in length, are the smallest of the ω-toxins isolated from *A. aperta* venom. The Type IV toxins showed a pharmacology opposite that of GVIA, i.e. strong block of mammalian, but weak block of avian Ca channels. The properties of the different ω-agatoxin types are summarized in Table 1.

C. ω-Agatoxin Chemistry and Biosynthesis

Complete amino acid sequences of Types I, III, and IV ω-agatoxins have been determined and are shown in Figure 4. Only a partial amino acid sequence for ω-AgaIIA is available at this time.

The original published sequence of ω-AgaIA was determined by direct Edman degradation of the reduced, alkylated toxin to be 66 amino acids, including 9 cysteine residues (56). Further investigation was prompted by two curious aspects of the toxin structure. The molecular mass of the native toxin was measured at 7791 kDa, while that predicted from the 66-amino-acid sequence was only 7495. Secondly, the odd number of cysteines—9—was unprecedented. The discrepancy in molecular mass was resolved when the ω-AgaIA precursor was deduced from a cDNA clone; it was predicted that the mature toxin consisted of a major chain that contains 66 amino acids (corresponding to the polypeptide originally sequenced), and a minor chain of 3 amino acids. The two chains are attached to each other by a fifth disulfide linkage (see Figure 4). The cloned sequence also revealed that the precursor polypeptide contains a signal sequence, followed by an N-terminal Glu-rich region (with 8/17 amino acids being Glu), and a shorter C-terminal Glu-rich region (3/7 amino acids are Glu), which was located between the major and minor chains. Arg residues flanking Glu-rich regions are likely to be substrates for proteases; proteolytic cleavage at these sites would generate the two-chain structure of ω-AgaIA having a molecular mass consistent with that determined for the native toxin (7791 kDa). Edman sequencing of the native toxin (without reduction and alkylation) produced two amino acids for the first two cycles (A/S, D/P), consistent with the amino acids shown in the cDNA precursor. It is noteworthy that the entire prepropeptide is only slightly larger than the processed, biologically active product. In this respect, the spider toxin prepropeptides differ from their cone snail counterparts and resemble more closely precursors of the snake and scorpion toxins.

ω-AgaIIIA is a monomeric toxin containing 76 amino acids, including 12 cysteines that form six disulfide bonds (59). Although ω-AgaIIIA is the only published Type III ω-agatoxin thus far, it is clear that this group is very diverse. Preliminary evidence from peptide isolation, cDNA cloning, and mass spectrometry indicate the presence of many closely related toxins

```
                                      (Major Chain)                                       (Minor Chain)
ω-Aga-IA     AKALPPGSVCDGNESDCKC YGKWHKCRCPWKWHFTGEGPCTCEKGMKHTCITKLHCPNKAEWGLDW       SPC

ω-Aga-IIA    GCIEIGGDCDGYQEKSYCQCCRNNGFCS...

ω-Aga-IIIA   SCIDIGGDCDG--EKDDCQCCRRNGYCSCYSLFGYLKSGCKCVVGTSAEFQGICRRKARQCYNSDPDKCESHNKPKRR*

ω-Aga-IVA    KKKCIAKDYGRCKWGGTPCCRGRGCICSIMGTNCECKPRLIMEGLGLA

ω-Aga-IVB    EDNCIAEDYGKCTWGGTKCCRGRPCRCSMIGTNCECTPRLIMEGLSFA

Disulfide    ---C-------C------CC----CXC------CEC------------
connectivity
(IVA/IVB)
```

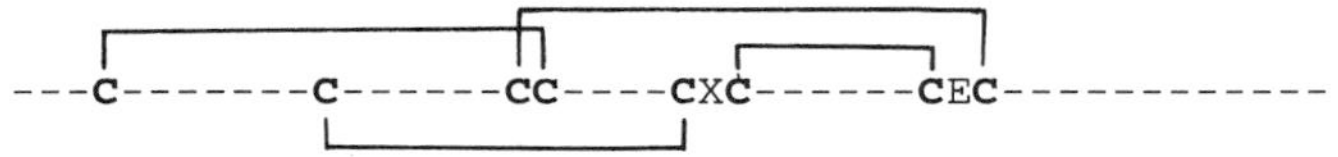

Figure 4 ω-Agatoxin sequences. ω-Agatoxins IIA and IIIA have been aligned to maximize sequence identity; note the extensive sequence homology between these two classes. The asterisk at the end of the IIIA sequences indicates amidation.

in the venom (64, 65). All of these variant forms have amidated C termini and closely related cysteine placements.

The best-characterized of the Ca channel–targeted peptides in *Agelenopsis* venom are ω-AgaIVA and ω-AgaIVB. These closely related toxins are unblocked 48-amino-acid peptides, including 8 cysteines having identical placement. Of all the ω-agatoxins, only ω-AgaIVA has been successfully chemically synthesized in a biologically active form (66–69). The disulfide linkage configuration of ω-AgaIVA also has been determined by chemical methods (68). Preliminary structural studies of ω-AgaIVB using multidimensional NMR have yielded a proposed disulfide bonding configuration that is identical to that of ω-AgaIVA (61). The disulfide loop configuration of the Type IV ω-agatoxins bears the same relationship to the four-loop configuration described above for the ω-conotoxins as do the "long" to the "short" snake α-neurotoxins (i.e. a small additional disulfide-bonded loop nested within one of the larger loops).

D. Other Venom Toxins Targeting Ca Channels

FTX (funnel-web toxin), a polyamine-containing fraction from *Agelenopsis aperta* venom, was reported to be a specific blocker of P-type Ca channels (70–73). However, when tested in isolated rat neocortical neurons, FTX was found to block P-type channels as well as additional components of Ca current (67). The chemical structure of FTX has not yet been determined, but a synthetic analog ("synthetic FTX," or "sFTX") has been prepared. sFTX has been shown to be a selective blocker of T-type current in rat dorsal root ganglion cells at 10 nM concentration, and appears to be a nonspecific blocker of other calcium currents at higher concentrations (74, 75).

Another peptide fraction from *Agelenopsis aperta* venom (AgaGI), reported to be distinctly different from the ω-agatoxins, has the novel property of partially blocking a non-inactivating plateau of synaptosomal calcium flux in rat brain synaptosomes; this partial block was reported to inhibit glutamate release completely (76). The chemical nature of this interesting factor has not yet been reported.

Spider neurotoxins that target Ca channels are not unique to *Agelenopsis aperta* venom. Venom fractions from *Hololena curta* (77–79) and *Plectreurys tristis* (80, 81) showed antagonism of insect and vertebrate calcium channels. Some fractions appeared to be specific for insect calcium channels (77, 80, 82, 83), while others were active on vertebrates (78, 79, 81). One toxin from *Plectreurys,* PLTX II, has been purified and shown to have a unique posttranslational modification, palmitoylation of threonine (84); the fatty acid substitution has been confirmed by chemical synthesis (85).

Many of the peptides isolated from *A. aperta* venom are active against vertebrate calcium channels (56, 60, 86), and some are quite selective for subtypes of vertebrate Ca channels. Recently, calcium channel–targeted toxins from other venoms active on vertebrate systems have been described. One is a 37-amino-acid peptide called "ω-grammotoxin SIA" from venom of the tarantula *Grammostola spatulata*. ω-Grammotoxin SIA completely inhibits Ca entry into rat brain synaptosomes and appears to define a site common to a broad range of vertebrate high-threshold Ca channels (87).

The first selective peptide antagonist of L-type channels has been isolated from venom of the black mamba snake (88). The peptide, called calciseptine, is 60 amino acids long; its sequence has been verified by preparation of the synthetic product, which shows properties identical to the native material (89). Calciseptine is a competitive blocker of [^{3}H]nitrendipine binding, and allosterically inhibits the binding of diltiazem (90).

IV. Ca CHANNEL DIVERSITY: MULTIPLE ω-TOXIN TARGETS

A. Phylogenetic Considerations: Nonmammalian Ca Channels

This section—which focuses on ω-toxin interactions with calcium channel types—is largely restricted to the mammalian central nervous system (almost all studies use rat brain). All five classes of α_1 subunits characterized from rat brain (α_{1A}, α_{1B}, α_{1C}, α_{1D}, and α_{1E}) will almost certainly also be found in the CNS of nonmammalian vertebrates; indeed several of these α_1 classes have been described in an elasmobranch system (91, 92). However, the use of toxins to define specific Ca channel types (e.g. ω-conotoxin GVIA for N-type channels) probably cannot be routinely extended to nonmammalian systems. The assumption that the subtype selectivity of ω-toxin interactions in mammalian systems will be unchanged in lower vertebrates is unwarranted.

For example, ω-conotoxin GVIA is highly specific for the N-type calcium channels in mammalian systems (see below), but also has been used to define "N-type" in a variety of nonmammalian systems. However, the specificity of the toxin in nonmammalian systems may be much broader. For instance, significant overlap between dihydropyridine-sensitive and GVIA-sensitive Ca currents in avian neurons has been reported (93, 94). In contrast, in mammalian neurons tested, GVIA-sensitive current did not overlap with dihydropyridine-sensitive current (95–97).

Similarly, GVIA blocks calcium entry into chick brain synaptosomes (59, 98, 99) and abolishes neurotransmitter release at the frog neuromuscular junction (100), leading to a widespread assumption that these avian and amphibian calcium channels are predominantly "N-type." In contrast, neither synaptosomal calcium entry (99) nor neuromuscular transmission (101, 102) is inhibited by this toxin in mammalian systems, and consequently corresponding mammalian Ca channels are "non-N-type." These conclusions necessarily require that an abrupt switch from N-type to non-N-type calcium channels occurred at both the neuromuscular junction and at CNS presynaptic termini during the evolution of mammals. An alternative and perhaps more parsimonious hypothesis would hold that the toxin is simply a less discriminating ligand in the lower vertebrates (much as ω-conotoxin MVIIC is against mammalian calcium channels—see below). With this in mind, it seems unwise at this stage to conclude that all calcium channels in the lower vertebrates that exhibit ω-conotoxin GVIA sensitivity must be of the "N-type."

Until these issues have been definitively settled, it is suggested that exclusive use of pharmacological agents as diagnostic tools for determining calcium channel types across broader taxonomic boundaries may be misleading. In addition to pharmacological criteria, a corroborating electrophysiological or molecular characterization is desirable before conclusions are made regarding the nature of nonmammalian Ca channel subtypes.

B. Subunit Composition of Ca Channels

One approach to the classification of voltage-sensitive calcium channels is to construct a molecular genealogy based on cloned sequences, and to match the electrophysiological and pharmacological properties of calcium channels to the clones, in effect assigning phenotypes to each genotype. Since the cloning of the first Ca channel from skeletal muscle (103), great progress has been made towards elucidating the family tree. At the same time, the

Table 2 Subunits of purified Ca channel complexes

	DHP receptor (L-type)	ω-Conotoxin GVIA receptor (N-type)	
	Skeletal muscle[a]	Rabbit brain[b]	Rat brain[c]
Subunit	Molecular mass (kDa)		
α_1	175	230	230
$\alpha_2\delta$	160	160	170
β	52	57	60
γ	32		
Others		95	110
			70

[a] Campbell et al (110)
[b] Witcher et al (112)
[c] McEnery et al (111)

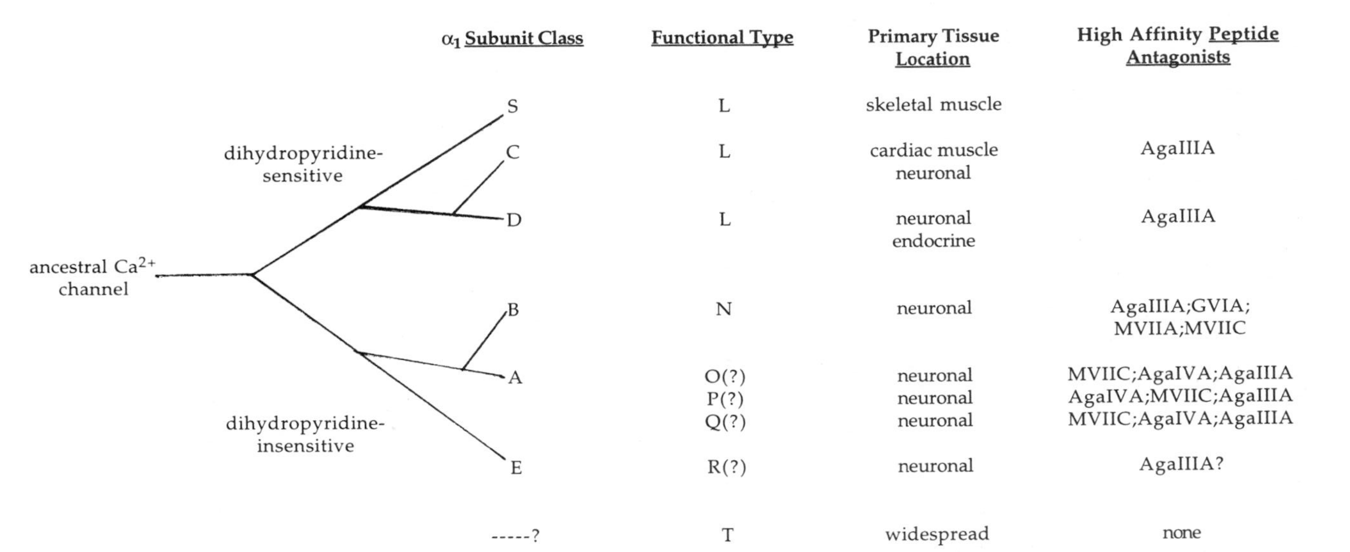

α_1 Subunit Class
Functional Type
Primary Tissue Location
High Affinity Peptide Antagonists
ancestral Ca^{2+} channel
dihydropyridine-sensitive
dihydropyridine-insensitive
S
L
skeletal muscle
C
L
cardiac muscle
neuronal
AgaIIIA
D
L
neuronal
endocrine
AgaIIIA
B
N
neuronal
AgaIIIA;GVIA;
MVIIA;MVIIC
A
O(?)
P(?)
Q(?)
neuronal
neuronal
neuronal
MVIIC;AgaIVA;AgaIIIA
AgaIVA;MVIIC;AgaIIIA
MVIIC;AgaIVA;AgaIIIA
E
R(?)
neuronal
AgaIIIA?
-----?
T
widespread
none

Figure 5 **Correlating α_1 subunits, Ca channel types, and ω-toxin ligands. This figure is adapted from Tsien et al (109) and Fujita et al (129). The nomenclature of Snutch and coworkers (113) for α_1 subunits is adopted in this figure. In rabbit, cloned α_1 subunits are referred to as BI, BII, and BIII; these are equivalent to α_{1A}, α_{1E}, and α_{1B}, respectively.**

functional and pharmacological characterization of calcium channels proceeds apace (for general reviews see Refs. 104–109).

There is a fundamental "body plan" that is largely shared by all voltage-sensitive calcium channel complexes. Two Ca channel complexes that have been purified and analyzed are shown in Table 2 (110–112); both are multisubunit protein complexes. Although the physiological roles of these two complexes are as disparate as is possible within the broad functional spectrum of Ca channels, three of the heteromeric subunits are clearly homologous, α_1, $\alpha_2\delta$, and β. The key subunit appears to be α_1, which possesses the essential machinery of a calcium channel: a central calcium-conducting pore, and a voltage-sensitive gate. For native Ca channels, it is likely that each α_1 subunit is part of a larger heteromeric complex, with the other subunits necessary for proper function, modulation, and localization of the channel. One source of functional diversity of the Ca channels clearly resides in the existence of multiple α_1 isoforms; indeed, the α_1 subunit appears to be a dominant determinant of the functional properties of a Ca channel subtype. The current state of understanding of the correlation of each known α_1 class with the corresponding Ca channel type is summarized in Figure 5. The nomenclature of the α_1 gene class originally proposed by Snutch and coworkers (113) is adopted in this review.

C. LVA and HVA Ca Channels: T and L Subtypes

Ca channels are activated, or opened, by a depolarization of the cell membrane voltage. Subsequently, the channels may inactivate and become nonconducting; both activation and inactivation occur at characteristic rates and membrane potentials. During the activated period, the channels conduct a current of calcium ions. The kinetics of activation, the subsequent inactivation, and the character of the calcium current all can be measured experimentally and have been used to classify Ca channel subtypes. Three parameters traditionally used for identification of Ca channel types are (*a*) the magnitude of the depolarizing voltage step required for activation, (*b*) the time course of inactivation, and (*c*) the magnitude of the conductance of a single calcium channel. Based on these parameters, Ca channels were divided into two general groups: (*a*) low voltage–activated (LVA), rapidly inactivating Ca channels with smaller conductances; and (*b*) high voltage–activated (HVA) Ca channels (114). The first HVA Ca channels described were moderately-to-slowly inactivating with larger conductances.

LVA Ca channels in cardiac muscle and neurons have been referred to as T-type channels in the widely used nomenclature of Tsien and colleagues (108), because of the shorter or "transient" nature of their activation and their small or "tiny" unitary conductances. The electrical properties of T-type channels (activation at very negative potentials and rapid inactivation) reflect their functional roles: generation of rhythmic pacemaker activity in cardiac muscle and neurons. Importantly for this review, no known ω-toxin blocks any T-type Ca channel with high potency, although the polyamine sFTX reportedly showed selective block of T-type current in rat dorsal root ganglion (DRG) neurons (74, 75). Recently, an α_{1E} subunit from rat brain has been cloned and expressed that exhibits some properties of LVA Ca channels (activation at negative potentials, rapid rate of inactivation), but differs significantly from the standard T-channels in several respects [sensitivity to Cd^{2+}, insensitivity to amiloride and octanol, peak current at relatively positive potentials (115)]. Unlike typical T-type channels, the α_{1E} channel is mildly sensitive to the spider toxin, ω-AgaIVA (see below).

The HVA calcium channels (non-T-type) originally were distinguished by electrophysiological and pharmacological criteria (see Refs. 94, 116), a division now supported by molecular data. The availability of small organic molecules that block calcium channels, exemplified by the 1,4-dihydropyridines (DHPs), provided a criterion for dividing HVA calcium channels into two classes, DHP-sensitive and -insensitive. The DHP-sensitive class was called the L-type Ca channel because the first of these channels described typically had long inactivation kinetics and large unitary conductances that distinguished them from most other HVA Ca channels.

As shown in Figure 5, at least three distinct α_1 subunits in the L-type Ca channel subfamily have been cloned. The α_{1S} subunit, the first one cloned, so far only has been found in skeletal muscle and was originally purified from T-tubules. Two additional α_1 subunits of L-type Ca channels have been cloned. The α_{1C} subunit of Snutch and coworkers was originally identified in a brain cDNA library by homology to the α_{1S} sequence; this form is also found in cardiac muscle. A second clone derived from a rat brain cDNA library first defined the α_{1D} class, which is found not only in neuronal tissue, but also in endocrine tissues. When expressed in oocytes, these α_1 subunits were sensitive to the dihydropyridine drugs. The L-type calcium channels also are sensitive to ω-AgaIIIA (117, 118). Early reports of high-affinity L-type calcium channel block by ω-conotoxin GVIA involved currents recorded in avian sensory neurons and have proved to be generally inapplicable to L-type calcium channels in mammalian systems. However, high concentrations of GVIA reversibly block a human α_{1D} clone expressed in oocytes (119, 120).

The sequences of the three L-type α_1 subunits, α_{1S}, α_{1C}, and α_{1D}, are evolutionarily more closely related to each other than to the other α_1 subunits that have been identified by molecular cloning so far. Thus, the L-type calcium channels appear to form a related set that have in common their sensitivity to the dihydropyridine drugs, and this sensitivity resides in the α_1 subunits (121–123). The L-type calcium channel oligomers with the best understood physiological functions are those present in cardiac and vascular smooth muscle, where the key role played by calcium currents is well defined. While L-type calcium channel complexes with both α_{1C} and α_{1D} subunits occur in neuronal cell bodies, the precise role of L-type channels in neurons is not yet well understood.

Recent work in chromaffin cells (124) suggests that Ca channel regulation may be analogous to a metabolic situation where two isozymes are present in a cell, one constitutive and the other inducible. In certain cells, L-type Ca channels may be the electrical equivalent of the inducible isozyme. In chromaffin cells there is a "constitutive" Ca current (non-L-type), but with large priming depolarization pre-pulses, a substantial "facilitation" L-type calcium current is induced (123a, 124, 124a, 125).

The L-type calcium channel is a well-understood pharmacological entity, defined by its sensitivity to dihydropyridines; these channels are resistant to all ω-toxins except ω-AgaIIIA (117, 118) and sensitive to calciseptine (88). Despite being clearly defined pharmacologically, the L-type channel is clearly extremely heterogeneous at the molecular level. The genealogical tree of L-type α_1 subunits reveals that α_{1S}, α_{1C}, and α_{1D} belong to an entirely distinct evolutionary lineage from the other α_1 subunits. Furthermore, there is further heterogeneity in each L-type α_1 branch; for example, in α_{1D} at least four distinct regions exhibit alternatively spliced variants (119, 126). In addition, multiple forms of the α_2 and β subunits have been identified. Thus there are potentially hundreds of different isoforms that could fit the tightly defined pharmacological entity called the L-type Ca channel.

D. N-type and P-type Calcium Channels

The N-type Ca channel was originally identified by its characteristic electrophysiological properties in chick sensory neurons, and named N because its intermediate inactivation kinetics made it neither transient (T) nor long-lasting (L) (116, 127). In mammalian neurons, ω-conotoxin GVIA from *Conus geographus* produces virtually irreversible block of N-type calcium channels (95, 96). It is now well established that Ca channel complexes that contain an α_{1B} subunit have the characteristic properties of an N-type calcium channel. Three α_{1B}-like genes have been cloned: human (120), rat (128), and rabbit (129). Functional N-type channels from both rabbit and human have been expressed in oocytes using a cloned calcium channel α_{1B} subunit; both the defining electrophysiological and pharmacological properties are found in such expressed clones (i.e. N-like channel kinetics and potent block by ω-conotoxin GVIA). Whether expression of any other α_1 genes apart from the α_{1B} subunits already cloned would result in an ω-conotoxin GVIA–sensitive Ca channel remains to be established; however, two α_{1B} subunit isoforms were found in the human system (120).

High sensitivity to ω-conotoxin GVIA is a definitive pharmacological hallmark for the presence of the N-type calcium channel in mammalian systems. N-type channels are restricted almost entirely to neurons in both peripheral and central nervous systems. Although N-type currents were originally characterized from neuronal cell body recordings, their major role may be to mediate neurotransmitter release at synaptic endings, in contrast to T-type and L-type currents. Indeed, $[^{125}I]$ω-conotoxin GVIA-binding sites are particularly abundant in synaptic regions of layered structures, such as the hippocampus and cerebellum in the rat brain (25, 28, 130). In addition, binding of fluorescent GVIA derivatives has been localized to synaptic contact sites in mammalian hippocampal neurons (40), as well as at the frog neuromuscular junction (37, 118). Direct measurement of nerve terminal currents has been made in the chick ciliary ganglion neurons, where GVIA irreversibly blocks presynaptic Ca currents (131).

It should be noted that all ω-conotoxins isolated to date inhibit the N-type calcium channel, with binding affinities ranging from subpicomolar to micromolar (14, 132). Apart from ω-conotoxin-GVIA, some conotoxins with strikingly different sequences appear to be highly selective for this Ca channel type, notably ω-conotoxin MVIIA from *Conus magus*. This peptide is a high-affinity but reversible blocker of N-type calcium channels (133). Other ω-conotoxins, such as ω-conotoxin SVIB, also bind and block N-type Ca channels, while retaining significant affinity to other calcium channel types (14). Newly discovered ω-conotoxins MVIIC and MVIID show higher affinity for, and preferential block of, non-N-type channels, but still retain measurable affinity for N-type channels as well (13, 19, 23a).

Another DHP-insensitive Ca channel is the P-type channel, originally described in cerebellar Purkinje neuron cell bodies, hence the name, P-type (70, 95). In mammalian systems, this Ca channel is blocked potently ($K_d \sim 2$ nM) by the spider toxin, ω-AgaIVA (60, 86). MVIIC also blocks P-type channels in Purkinje neurons, but the on-rate kinetics governing formation of the toxin-channel complex are considerably slower under physiological conditions (134). Currents that are insensitive to DHPs and GVIA, yet sensitive to ω-AgaIVA, have been identified in a variety of central and peripheral neurons (86); these are often referred to as "P-type" or "P-like" currents, but the designation may be overly simplistic.

The first clearly non-L, non-N-type Ca channel α_1 subunit to be cloned from the mammalian brain was an α_{1A} subunit derived from a rabbit brain cDNA library ("the BI gene") (135). This channel expressed in *Xenopus* oocytes was blocked by crude *A. aperta* venom, and its message was found to be enriched in the cerebellum. Because the rabbit α_{1A} channel was blocked by *A. aperta* venom, and the venom subsequently was shown to contain the first potent blockers of P-type Ca channels (ω-AgaIVA and FTX), it was suggested that the rabbit α_{1A} might be a subunit of the P-type channel. The sequence for the rabbit BI gene is nearly identical to that of the rbA rat gene, which encodes the α_{1A} subunit (136). However, the contrast between the high-affinity ω-AgaIVA block of P-type current in cerebellar Purkinje neurons ($K_d \sim 2$ nM) and the toxin's lower-affinity block of rabbit α_{1A} currents expressed in oocytes ($IC_{50} \sim 200$ nM), suggested that the currents elicited by the cloned α_{1A} subunits might not be precisely identical to native P-type calcium channels on Purkinje cell bodies (137, 138). Thus, whether α_{1A} subunits are present in the P-type Ca channel complexes as defined in Purkinje cells needs to be definitively established.

E. Other Ca Channel Subtypes: An "OPQ" Subfamily?

The four Ca channel subtypes described above, T, L, N, and P, are widely accepted in the literature. This is partly due to clear-cut criteria for identifying these subtypes: The T-type Ca channels have very distinctive electrophysiological characteristics, while facile identification of the three other subtypes can be achieved by using key pharmacological reagents, i.e. dihydropyridines for L-type, ω-conotoxins GVIA or MVIIA for N-type, and ω-AgaIVA for P-type Ca channels. It is clear, however, that other calcium channel types and ω-toxin-binding sites are present in the nervous system that do not precisely fit into the classification described above. We summarize some recent proposals for additional subtypes that can be differentiated from T-, L-, N- and P-type channels using either pharmacological or electrophysiological criteria.

The binding of MVIIC to a class of sites in rat brain has been characterized, and a range of affinities (from 40 pM to about 1 nM, depending on conditions) has been reported (13, 23a, 132, 139). These affinities are considerably higher than both the binding affinity of MVIIC for N-type channels and the potency of MVIIC for blocking P-type channels in cerebellar Purkinje cells under most assay conditions (13). The high-affinity target has been proposed to represent an O-type channel (139). This has also been referred to as "site 2 to distinguish it from N-type channels, or "site 1 (132). A substantial fraction of norepinephrine release in the hippocampus is inhibited by MVIIC at subnanomolar concentrations, consistent with the high-affinity binding observed (132, 140). At this time, the O-type channel is primarily defined by binding experiments, and may, in fact, be closely related to the MVIIC-sensitive current observed in cerebellar granule cells (see below) and to the P-type channels on Purkinje cells. An estimate of the number of O-type binding sites on crude rat brain suggests that there may be significantly more O-type than N-type binding sites. If

O-type Ca channels were found almost exclusively at presynaptic termini (and not neuronal cell bodies), this might explain why Ca currents with an O-type pharmacology have not yet been observed by electrophysiological methods, but only through a neurotransmitter release assay.

A Ca current present in cerebellar granule cells appears to be pharmacologically and electrically distinct from the P-type Ca channel, and has been designated "Q-type" (138). In contrast to the P-type current, which is effectively blocked by low concentrations of ω-AgaIVA, but is only inhibited with very slow kinetics by ω-conotoxin MVIIC under most physiological conditions, MVIIC and AgaIVA have been reported to be roughly equipotent at blocking a Q-like current in oocytes expressing the rabbit equivalent of the α_{1A} subunit (137), and the Q current in cerebellar granule cells (ca. 30–300 nM IC_{50}). The Q-type and O-type channels appear to differ in their affinity for MVIIC, with the O-type channel having significantly higher affinity.

The proposed O-type and Q-type channels are clearly closely related to P-type channels in that both are also blocked by AgaIVA and MVIIC, albeit with different apparent affinities. In addition, all three are GVIA- and dihydropyridine-resistant. One possibility is that there is a pharmacologically distinct subfamily of Ca channels, a situation analogous to L-type channels, which exist as a subfamily of several distinct molecular entities (e.g. α_{1S}, α_{1C}, α_{1D}, and variants thereof). According to such a scheme, the channel blocked with moderate potency by MVIIC and AgaIVA in neurotransmitter release and calcium influx experiments using rat brain synaptosomes would also belong to this subfamily. If this turns out to be the case, the "OPQ" subfamily is widely distributed in the nervous system and may mediate a substantial fraction of neurotransmitter release in the brain.

Thus, the Ca channels that are members of the proposed subfamily are dihydropyridine- and GVIA-resistant, but AgaIVA- and MVIIC-sensitive, and may all have an α_1 subunit identical or very closely related to α_{1A}. Members of the family differ in the relative inhibitory potency and kinetics of block by MVIIC: For P-type channels, MVIIC has very slow on and off times; the O-type channels are the highest-affinity site; for Q-type channels, AgaIVA $\approx$ MVIIC, with IC_{50}s in the 10^{-7} M range. These differences may arise because of heterogeneity in α_{1A} subunits—in rabbit brain, α_{1A} splice variants have been reported (135). Alternatively, the properties of identical α_{1A} subunits may be altered by differences in the other subunits making up the channel complex. This hypothesis predicts that molecular heterogeneity should be found between the channel complexes underlying the P-type currents of Purkinje cells, the Q-type currents of cerebellar granule cells, and the O-type channels regulating norepinephrine release in hippocampus. However, because binding assays and the different electrophysiological experiments were carried out under different conditions, the possibility that some of the observed differences are due to variations in the conditions used cannot be eliminated at this time.

Finally, there is a proposal for an R-type Ca channel (92, 138, 141). When dihydropyridine, GVIA, AgaIVA, and MVIIC are applied to cultured cerebellar granule cells to block the L, N, P, Q, and O channels, a residual calcium current with distinct electrical properties remains. This current and the Ca channel that gives rise to it have been termed the R-type channel. The electrical properties of this current and the current resulting from expression of the cloned α_{1E} gene in oocytes are similar. Furthermore, the oocyte current also is resistant to block by DHP, GVIA, AgaIVA, and MVIIC. In situ hybridization experiments suggest that R-type Ca channels are widely distributed in the brain, but their functional roles are not known (115). In cerebellar granule cell bodies, the R-type channel, in concert with L-, N-, P-, and Q-types, presumably regulates complex electrical activities. A recent study suggests that there may be two distinct currents that have R-type pharmacology in cerebellar granule cells (142). A specific antagonist for the proposed R-type Ca channels would clearly be extremely useful in understanding the functional role(s) of this channel type.

V. NEUROTRANSMITTER RELEASE AND ω-TOXIN INHIBITION

A key step in excitation-secretion coupling in presynaptic nerve terminals and in many neuroendocrine cells is the influx of calcium through voltage-sensitive calcium channels. It is clear that different calcium channel subtypes contribute differentially to secretion of the various neurotransmitters and of many hormones (143, 144). Due to the small size of most presynaptic nerve terminals, it is generally not possible to identify the channel types contributing to transmitter release using a standard electrophysiological approach. Therefore, in general, the pharmacological sensitivity of the release process is used as the major criterion for identifying the calcium channel subtype(s) involved. Since the amount of transmitter released is estimated to be a function of $[Ca]^3$ or $[Ca]^4$ (6, 145, 146), it must be kept in mind that the observed inhibition of synaptic responses by ω-toxins is not directly proportional to the number of calcium channels blocked.

The ω-toxins are the major pharmacological agents used to block neurotransmitter release by inhibiting Ca entry through voltage-sensitive Ca channels. The effects of the ω-toxins, especially ω-conotoxin GVIA, on the release of neurotransmitters and hormones in numerous physiological preparations have now been reported in hundreds of publications, and an enormous literature is accumulating on the effects of ω-AgaIVA and other ω-conotoxins (such as

MVIIC). A selective survey is given here to highlight critical issues in this area, with emphasis on evidence for Ca channel diversity in neurotransmitter release. This section is organized by tissue type, with supporting data such as results from binding, Ca influx, electrophysiological, and physiological studies presented in each subsection, as necessary. Only mammalian systems are considered, with an emphasis on the rat.

A. Central Nervous System Calcium Channels

BRAIN The present consensus is that little or no transmitter release in the brain is mediated by T-type or L-type calcium channels. A substantial fraction of release is sensitive to various ω-toxins, however, and therefore is likely to be mediated by the N-type calcium channels, or the "OPQ-subfamily" of channels that are sensitive to these ligands (147–151).

In some studies, using either slices or synaptosomes prepared from various brain regions, and employing either electrical stimulation or elevated K^+ to effect depolarization, GVIA has been shown to inhibit potently but partially the release of glutamate (99, 147, 151), acetylcholine (152), dopamine (151, 153), and norepinephrine (154). Thus, N-type calcium channels mediate a substantial fraction, but not all, of transmitter release in the brain. These results are consistent with autoradiographic studies in brain slices that show radioiodinated GVIA binding enriched in synaptic layers (25, 26, 28).

The effects of GVIA on release of glutamate in the hippocampus have been measured indirectly in hippocampal slices by examining electrically induced synaptic responses electrophysiologically (155–158). Excitatory postsynaptic potentials (EPSPs) resulting from electrically evoked Schaffer collateral input to the CA1 region of the hippocampus were 70–80% blocked by 100 nM GVIA (156). Similar results were obtained in the CA3 region upon stimulation of hippocampal mossy fibers (155).

Transmitter release and synaptic transmission also are inhibited by AgaIVA. Glutamate release from whole rat brain synaptosomes was reported to be 60% inhibited by saturating concentrations of AgaIVA, but unaffected by GVIA under these conditions (150). However, both toxins are partial antagonists of glutamate release from hippocampal synaptosomes. These studies, which directly measure glutamate release from nerve terminals, are in reasonable agreement with electrophysiological studies on slice preparations from various regions of the brain. Excitatory transmission in hippocampus (147, 159), as well as inhibitory transmission in cerebellum and spinal cord (159), is partially blocked by AgaIVA and GVIA. Interestingly, strong stimuli overcome the inhibition of glutamate release by each toxin alone, but not by the toxins applied jointly (147). Thus, substantial evidence is emerging to implicate both AgaIVA- and GVIA-sensitive channels in neurotransmitter release in the brain. However, Ca channels that are both GVIA- and AgaIVA-resistant were estimated to mediate 17–41% of synaptic transmission in central synapses (159). Microdialysis studies have shown that potassium-induced glutamate release in the hippocampus is potently blocked with ω-conotoxin MVIIC, whereas the N-type blocker, MVIIA, was three orders of magnitude less potent in this regard (184). These results imply a predominant role for non-N-type, non-L-type Ca channels in glutamate release.

The calcium channel types coupled to the release of different neurotransmitters need to be identified precisely. In particular, what types of Ca channel mediate the DHP-, GVIA-resistant release of norepinephrine, dopamine, acetylcholine, and serotonin? GVIA and AgaIVA each inhibit roughly 50% of K^+-evoked norepinephrine release from rat hippocampal brain slices (149). Joint application of the two toxins was found to be additive, indicating that two distinct toxin-sensitive components of release are present. Similarly, AgaIVA and GVIA inhibit what appear to be separate components of K^+-induced dopamine release from rat striatal synaptosomes (151). Inhibition by each toxin is maximal at low K^+ concentrations and markedly diminished at high concentrations. Significantly, the two toxins are synergistic when applied jointly at high-stimulus intensities. These results suggest that the AgaIVA- and GVIA-sensitive Ca channels may coexist in the same nerve terminals.

NEUROHYPOPHYSIS Calcium entry into nerve endings of hypothalamic neurosecretory cells terminating in the neurohypophysis triggers the release of oxytocin and vasopressin. DHPs and GVIA each partially block release of these peptide hormones, suggesting the involvement of two pharmacologically distinct calcium channels: L-type and N-type. Direct examination of the calcium currents in isolated neurohypophyseal terminals indeed revealed evidence for at least two calcium channel types coexpressed in these terminals on both biophysical and pharmacological grounds (160, 161). One component is DHP sensitive and has electrical properties characteristic of L-type channels. The other is maximally blocked up to about 80% by GVIA, with an IC_{50} of about 50 nM. The electrical properties of this component are qualitatively similar to the N-type currents first observed in neuronal cell bodies. The quantitative differences are large enough, however, for the investigators to designate the underlying channel subtype as Nt (for terminal N-type) to differentiate it from the neuronal cell body N-type. The investigators further suggest that this component may contain a P-like current as well. Moreover, in contrast to the L-type channels in neuronal cell bodies, the L-type current is blocked by GVIA, albeit with considerably lower potency (IC_{50} of about 500 nM) than the Nt current. This is consistent with the presence of Ca channel complexes that contain an α_{1D} subunit in these cells.

SPINAL CORD Reports on the effects of the ω-toxins on transmitter release in the spinal cord are limited. It has been shown that nearly all release of the neuropeptide, CGRP, from terminals of peripheral afferent neurons in the rat spinal cord is potently blocked by GVIA and not by DHP and is, therefore almost entirely mediated by N-type calcium channels (162). These nerve terminals relay sensory signals from peripheral receptive fields of the sensory neurons. In fact, intrathecal spinal administration of GVIA blocks the reflex response activated by peripheral stimulation of afferent neurons. CGRP release from efferent nerves in the periphery is largely GVIA insensitive, suggesting that the calcium channels mediating this release are not N-type and providing evidence that presynaptic channel subtype is not necessarily correlated with the type of transmitter released.

B. Peripheral Calcium Channels

NEUROMUSCULAR JUNCTION The primary natural target of the ω-conotoxins is very likely the presynaptic calcium channels at the neuromuscular junction of the fish prey of the cone snails. Likewise the primary target of the ω-agatoxins is thought to be the presynaptic calcium channels of the motor neurons of the spiders' insect prey. The sensitivity of fish neuromuscular junction calcium channels to ω-conotoxins is shared by other lower vertebrates, including reptiles, amphibians, and birds (100, 163), but not by mammals. Thus, GVIA has little or no effect on evoked acetylcholine release at the mammalian neuromuscular junction (102, 152, 164). These presynaptic calcium channels are not L-type, because DHPs also have little or no effect on evoked release (165). It has been proposed recently that mammalian neuromuscular junction calcium channels are P-type due to the sensitivity of these channels to a nonpeptide toxin obtained from *A. aperta* venom, FTX (164). However, the uncertain chemical nature of the toxin and conflicting reports on its selectivity (67, 75) have left this proposal less than certain. More recently, MVIIC and AgaIVA have been shown to interfere with mouse neuromuscular transmission; both toxins block endplate potential amplitude without diminishing miniature endplate potential amplitude (166). Several ω-conopeptides with a range of affinities for the GVIA-binding site in brain (i.e. N-type channels) and to the MVIIC-binding site in brain were tested for block of neuromuscular transmission. The rank order potency for inhibition of electrically stimulated muscle contraction correlates well with the rank order potency for binding to the MVIIC-binding site, while there is no correlation with the potency for binding to N-type channels. Thus, the mammalian neuromuscular junction calcium channel has about equal sensitivity to AgaIVA and MVIIC, and, with respect to ω-conotoxins, has a similar pharmacological profile to that of the central channel characterized by high-affinity binding to MVIIC (166). Further work is needed to assign this channel definitively to the O, P, Q, or perhaps some other type.

SYMPATHETIC NEURONS In all the sympathetic systems studied, the dominant channel mediating noradrenergic release is the N-type. GVIA and MVIIA have potent hypotensive effects due to inhibition of nonepinephrine release from sympathetic innervation of the vascular system (167, 168). The effects of GVIA have also been studied in numerous isolated tissue preparations receiving sympathetic innervation (e.g. Refs. 169, 170). In cultured rat superior cervical ganglion cells, release of noradrenaline is completely blocked by GVIA, but not by DHP (171). In rat vas deferens, electrically evoked muscle responses were inhibited by GVIA (172). The results suggest that GVIA-sensitive (N-type) channels mediate release at lower-frequency stimulation, but that a non-N-type channel may mediate release at high-frequency stimulation. A similar frequency-dependent inhibition of electrically stimulated contraction by GVIA in rat bladder has been reported (173). How the peripheral N-type channels compare structurally, pharmacologically, and with respect to modulation to the central N-type channels remains to be determined.

PARASYMPATHETIC SYSTEMS GVIA blocks electrically evoked contraction of ileal muscle (e.g. Refs. 172, 173). In this cholinergic system, exogenously applied acetylcholine partially overcomes this blockade, indicating that GVIA acts by inhibiting transmitter release through block of presynaptic N-type calcium channels. It is interesting that the major calcium channel type mediating release at this cholinergic synapse is clearly different from that mediating release at another cholinergic synapse, the skeletal neuromuscular junction (see above). Here again is evidence that a simple correlation between Ca channel type and transmitter does not appear to be generally applicable.

CHROMAFFIN CELLS Chromaffin cells of the adrenal medulla secrete noradrenaline and adrenaline in response to a variety of secretagogues such as acetylcholine. DHP- and GVIA-binding sites are present on chromaffin cell membranes (174, 175). Several studies have shown that chromaffin cell Ca currents are mediated by DHP-sensitive and, to a lesser extent, GVIA-sensitive calcium channels, but that the bulk of release is mediated by L-type channels (e.g. Refs. 176, 177). As described in Section IV above, investigations using whole-cell patch-clamp electrophysiological methods indicate that the whole-cell calcium current is composed of a DHP-insensitive "standard" component that is activated by moderate stimulation, and a DHP-sensitive "facilitated" component that is recruited by stronger stimulation (178). Two distinct channels contribute to the standard component, one of which is blocked by GVIA, and the other by AgaIVA (178a). The investigators suggest that the GVIA-insensitive current represents a non-N-channel, perhaps a P-type channel, while the GVIA-sensitive current represents an N-type channel variant with electrical properties that differ from those of N-type channels in neuronal cell bodies. Molecular cloning will be required to classify the three currents definitively.

VI. INTERACTIONS BETWEEN ω-TOXINS AND CALCIUM CHANNELS

A. General Considerations

Studies of several toxins interacting with a multitude of Ca channel subtypes have the potential to yield a very complex data set. In order to simplify the analysis of these interactions, we use the key parameters that determine interaction between a ligand and an ensemble of different calcium channels subtypes as a framework for discussion. These critical functional determinants, which would be expected to vary from one peptide ligand to another, are:

1. *Channel specificity:* the specificity or lack of specificity of the peptide ligand for different calcium channel types. A particular peptide could be very narrowly targeted to just one molecular form, may bind a small spectrum of types, or be active on a wide variety of voltage-sensitive calcium channels.
2 *Affinity:* the binding affinity of the peptide for each of its channel targets, and the kinetic components that determine affinity (on- and off-times).
3 *Blocking efficacy:* the intrinsic antagonist activity of the peptide when it is bound to a particular calcium channel target. This parameter may vary independently of binding affinity; a peptide bound to a Ca channel complex may either completely block channel function, partially block so some Ca conductance is observed at saturating peptide concentrations, or as an extreme, cause no block. A peptide that bound a Ca channel but did not block could conceivably act as a protective "antitoxin," i.e. would inhibit the toxins that blocked that Ca channel (179).

The potential complexity in using toxins as biochemical tools to study Ca channels arises from the fact that each ω-conotoxin and ω-agatoxin has its characteristic Ca channel specificity, and the binding affinity and blocking efficacy vary for each channel type targeted by the toxin. For example, both physiological and binding data suggest that the spider toxin ω-AgaIIIA broadly targets high-threshold calcium channels, but does not have the same blocking efficacy on different channel types. ω-AgaIIIA blocks 100% of L-type current in cardiac muscle, but this decreases in the order L > N > P (117, 179a). For the OPQ subfamily of Ca channels, ω-AgaIIIA appears to act as a high-affinity partial antagonist, typically blocking less than 50% of calcium conductance (see, for example, Ref. 179a). Thus, affinity and binding efficacy can be varied independently.

B. The Macrosite Hypothesis

Several lines of evidence suggest that ω-conotoxins and many spider toxins may be targeted to homologous ligand-binding pockets on different voltage-sensitive calcium channel subtypes. It was previously proposed that ω-conotoxins bind to a relatively large site exposed to the extracellular medium on the ion channel complex, a "macrosite" (1). The macrosite comprises multiple microsites that are potential points of focal contact with peptide ligands. The extreme degeneracy observed in ω-conotoxin sequences competing for the same binding site was explained by postulating that each ω-conotoxin interacted with only a small subset of all potential microsites within a particular macrosite. Peptides with very different sequences probably interact with different subsets of microsites, but occupancy of the macrosite by one ω-conotoxin would exclude another.

If two homologous macrosites on different voltage-sensitive calcium channel subtypes are compared, many microsites would be conserved, but a subset would be different. If all microsites with which a peptide ligand interacts in a given channel subtype are conserved in a second subtype, that ligand would be expected to bind the two subtypes equally well. However, if a significant number of the microsites interacting with the ligand were altered in the second subtype, then the peptide ligand should show strong discrimination in its effects on the two channels.

It has been suggested that AgaIIIA from *Agelenopsis aperta* (see Section III above) can be used to define a macrosite on Ca channels operationally (139). If AgaIIIA either prevents binding or interferes with the activity of a particular ligand, that ligand can be presumed to be targeted to at least a portion of the macrosite. Using this operational test, it can be concluded that all ω-conotoxins tested, as well as some other ω-agatoxins (e.g. AgaIIA) target the macrosite.

The complex effects of ω-toxins on Ca channels can be understood in the framework of two interacting sets. The first is a set of polypeptide toxins comprising the individual ω-toxins varying in channel type specificity,

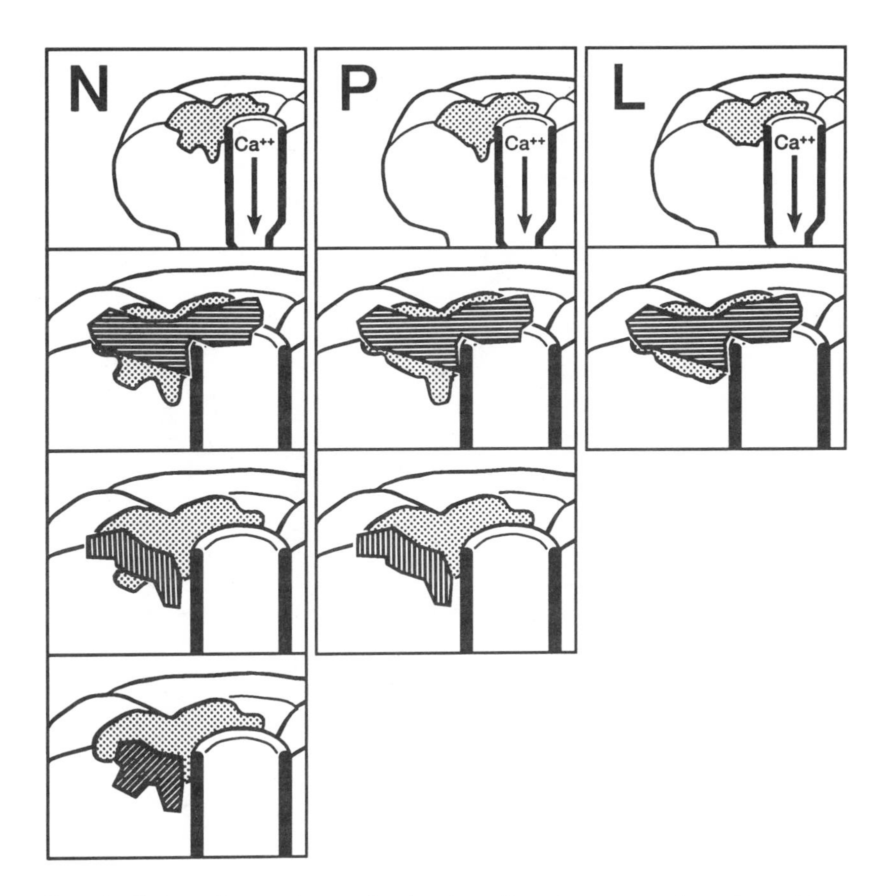

Figure 6 Ca channel types and macrosites. A cartoon of homologous macrosites on three calcium channel subtypes. Shown are a diagrammatic representation of the N-, P-, and L-type calcium channels with the macrosite represented as a ligand-binding pocket oriented towards the extracellular surface of the channel. Although macrosites on different channel subtypes are homologous, they are not identical, and different toxins show different selectivities. In the second row the macrosite is occupied by ω-Aga-IIIA, in the third row by ω-conotoxin MVIIC, and in the fourth row by ω-conotoxin GVIA. Note that ω-conotoxin GVIA can only bind the N-type calcium channel. Changes in the shape of the macrosite preclude binding to P-type and L-type calcium channels. On the other hand, ω-Aga-IIIA has the capacity to bind all three subtypes. Furthermore, the affinity of the ω-toxins changes from one subtype to the next, as does the blocking efficacy.

affinity, and blocking efficacy. The second set consists of the various calcium channel isoforms with homologous macrosites. The picture that emerges (see Figure 6) is that all ω-conotoxins target the homologous macrosite, but each individual conotoxin differs in affinity, blocking effi-

Table 3 Interactions of ω-toxins with Ca channel subtypes

Toxin	Channel subtype	Binding affinity[a]	Antagonist activity[b]
GVIA/MVIIA	L	−(+)[c]	
	N	+++	total
	O	−	
	P	−	
	Q	−	
	R	−	
MVIIC	L	−	
	N	++	[e]
	O	+++	total
	P	++[d]	total
	Q	++[d]	total
	R	−	
ω-AgaIVA	L	−	
	N	−	
	O	++	n.d.
	P	+++	total
	Q	++[d]	n.d.
	R	−	
ω-AgaIIIA	L	+++	total
	N	+++	partial
	O	+++	n.d.
	P	+++	partial
	Q	n.d.	n.d.
	R	n.d.	

[a] +++ < 10 nM; + > 1 μM; n.d., not determined.
[b] n.d., not determined.
[c] For L-channels, GVIA is active only on α_{1D}-containing complexes.
[d] Binding affinity inferred from electrophysiological or calcium flux measurements.
[e] MVIIC blocks almost totally under most conditions, but a small conductance remains at high [Ba^{2+}] (K Swartz, unpublished).

cacy, and subtype specificity. Binding by ω-conotoxins can be inhibited by preincubation with AgaIIIA, no matter what ω-conotoxin:subtype pair is being tested. Furthermore, in those cases where AgaIIIA does not have high blocking efficacy, preexposure to AgaIIIA prevents the more complete blocking effects of ω-conotoxins. Thus, AgaIIIA and the ω-conotoxins can be thought of as all targeting homologous macrosites, with each toxin being different in the three key functional determinants.

All ω-conotoxins in Figure 2 are capable of inhibiting N-type calcium channels, but there is considerable variation in binding affinity (GVIA >

MVIIA >> SVIA). A subset of ω-conotoxins in Figure 2 clearly has broader subtype activity than is exhibited by GVIA and MVIIA, which are highly specific for N-type channels. Of these, ω-conotoxin MVIIC is the best characterized: In addition to inhibiting the N-type calcium channel, ω-conotoxin MVIIC inhibits the "OPQ" family of calcium channels, but does not block L-type calcium channels (134). Interestingly, ω-conotoxin MVIIC blocks the N-type calcium channel reversibly, and with a faster on-time than for the P-type calcium channel, which is blocked essentially irreversibly. These functional determinants for the four key ω-toxins are summarized in Table 3. We should emphasize that while the homologous macrosite hypothesis satisfactorily explains all the data so far, it has not been rigorously proven. The possibility that the spider toxins and conotoxins interact allosterically, instead of competing for the same macrosite, has not been rigorously eliminated.

The relationship of the binding site of ω-agatoxin-IVA to the proposed macrosite is one unsettled issue. AgaIVA has been characterized as a high-affinity ligand for P-type calcium channels (60); several lines of evidence suggest that this toxin binds to a site at least somewhat distinct from that identified by the ω-conotoxins or AgaIIIA. AgaIIIA does not inhibit binding of AgaIVA to its high-affinity site (139), and preincubation of Purkinje cells with AgaIIIA did not prevent the complete inhibition of P-type currents by AgaIVA (179a). Furthermore, the interactions between MVIIC and AgaIVA are nonreciprocal. Although AgaIVA does not displace high-affinity MVIIC binding (suggesting that the AgaIVA-binding site on O-type channels does not overlap with the MVIIC-binding site), preincubation with rather high concentrations of ω-conotoxin MVIIC (≥ 1 mM) will prevent AgaIVA from binding its high-affinity sites (presumably on the P-type channel) (139). One explanation is that the two toxins bind spatially separate sites, and occupancy of the macrosite by MVIIC affects the conformational state of the AgaIVA-binding site. However, it remains a formal possibility that the MVIIC sites overlap both spider toxin–binding sites on the P-channel, although the two spider toxin sites do not overlap each other.

We might add parenthetically that another class of Ca channel ligands that may not target the homologous macrosite are the dihydropyridines; ω-AgaIIIA does not block dihydropyridine binding, and vice versa, though both inhibit L-type channels. Clearly, there are nonmacrosite ligand-binding sites on some channels, which may affect channel function. However, since both spiders and cone snails independently evolved macrosite-targeted ligands, a promising strategy for novel Ca channel–targeted drugs would appear to be the development of subtype-discriminating macrosite peptidomimetic ligands (see Section VI.D below).

C. ω-Toxins as Probes for Ca Channels; Structure:Function Experiments

A number of experiments using ω-toxins as probes to further characterize their target calcium channels have been carried out. With the availability of radiolabeled ω-conotoxin GVIA, both chemical crosslinking and photoactivatable crosslinking experiments were performed (22, 29–34). These earlier crosslinking experiments can be reinterpreted in the light of the purification of the N-type calcium channel complex, presumably the substrate for crosslinking to the toxin. Although it is generally believed that the ω-toxins primarily affect the function of the α_1 subunit, and that most binding determinants are on the α_1 subunit, the crosslinking data indicate that ω-toxins may be useful probes for mapping the orientation of the other subunits in the Ca channel complex.

Surprisingly, chemical crosslinking using ^{125}I-labeled ω-conotoxin GVIA yields primarily a covalent adduct to the α_2 subunit of the channel complex, not to the α_1 subunit. The molecular mass of the adduct formed upon the addition of the bivalent chemical crosslinker, disuccinimidyl suberate, was consistent with an intact α_2 subunit (~170 kDa); upon reduction, the decreased molecular mass observed (~140 kDa) is consistent with removal of the δ subunit, believed to be disulfide-bonded to the α_2 subunit. However, photoactivatable crosslinking caused labeling of a different subset of subunits in the complex; specifically labeled bands with molecular mass > 230 kDa (presumably the α_1 subunit) and 95 kDa were observed. The last may correspond to the 95-kDa subunit of Campbell and coworkers (see Table 2). Thus, although the photoactivatable group on the toxin reacted with the α_1 subunit, it also had access to the α_2 and 95-kDa subunits in the channel complex.

In addition, it has been found that ω-AgaIIIA binding is abolished during purification of the N-type channel complex from rat brain, although ω-conotoxin GVIA binding to the N-type calcium channel remains (111). It is possible that a critical determinant for ω-AgaIIIA binding was lost in the purification procedure. Thus, although we have focused on α_1 subunits in this review, there is evidence of proximity of other subunits, and possibly even some binding determinants for ω-toxins being on non-α_1 components of the Ca channel complex.

The three-dimensional conformation of several ω-toxins in solution has been solved by multidimensional NMR techniques. Several groups have published proposed structures for ω-conotoxin GVIA. The conformation of the peptide backbone is highly constrained by three disulfide bonds and several hydrogen bonds; this is the smallest peptide known to have a triple-stranded β-sheet (180–183). One ω-agatoxin, ω-AgaIVB, has been studied by NMR as well, and the disulfide bonding of the toxin inferred from the NMR structure (61).

The availability of NMR structures makes the interpretation of structure:function studies much more incisive. A derivatization study adding biotinyl groups to various amino groups on ω-conotoxins is an example (35). When the α-amino group and two ϵ-amino groups of ω-conotoxin GVIA were biotinylated, it was found that each biotinyl moiety increased the apparent IC_{50} 10-fold, with each modification having an additive effect. In contrast, in ω-conotoxin MVIID, biotinylation of the α-amino group caused the same 10-fold increase in IC_{50}, but biotinylation of the single ϵ-amino group of MVIID caused a 500-fold increase in IC_{50}. The NMR structure of ω-conotoxin GVIA reveals that all of the free amino groups in ω-conotoxin GVIA are on the same side of the molecule. However, the ϵ-amino group of MVIID would be predicted to be on the opposite side of the peptide. Thus, the ϵ-amino group of MVIID is on a toxin surface that is presumably more critical for channel interaction. A study supporting this conclusion was carried out with ω-conotoxin SVIB, which has a broader subtype specificity than the highly specific N-type blockers, GVIA and MVIIA. SVIB has approximately equal affinity for GVIA/MVIIA- and MVIIC-binding sites in rat brain (132, 140). Synthesis and characterization in both binding and norepinephrine release assays of two hybrid peptides in which four amino acids, 9 through 12, from SVIB and MVIIA were interchanged showed this region is a major determinant of subtype specificity (132); this is the same region with the ϵ-amino group of MVIID. Clearly, as more NMR structures are solved, there will be new opportunities for increasingly refined structure:function analyses using native, chemically modified, and amino acid–substituted peptides.

D. Therapeutic and Diagnostic Potential for Peptide Blockers of Ca Channels

Classical calcium channel antagonists such as the dihydropyridines, which selectively block L-type calcium channels in muscle, have been highly successful in treating hypertension and other cardiovascular disorders. The availability of several synthetic conopeptides has provided the opportunity to evaluate the therapeutic potential of selectively blocking N-type calcium channels in a variety of pathological conditions, including cerebral ischemia, in which these channels have been implicated. Synthetic MVIIA (also referred to as SNX-111) protected the pyramidal neurons of the CA1 subfield of the hippocampus from damage caused by transient, global forebrain ischemia in the rat. A single-bolus intravenous administration of MVIIA was sufficient to provide highly significant neuroprotection even when the peptide was given 24 hours after the ischemic insult (184). On the other hand, the synthetic peptide MVIIC (also referred to as SNX-230), which blocks other types of neuronal calcium channels more potently than N-type channels (13, 132), did not provide any neuroprotection in the rat model of global ischemia (184). The efficacy of MVIIA in preventing damage caused by severe forebrain ischemia even when administered 6–24 hours after ischemia has been confirmed in two additional studies (185, 186).

MVIIA (SNX-111) has also been found to be highly effective in reducing neocortical infarct volume in rat models of focal ischemia, both when administered during occlusion (187) and after the ischemic episode (188). Calcium accumulation in the cerebral cortex and the hippocampus caused by traumatic brain injury in a rat model was significantly reduced by the administration of MVIIA one hour after the insult (189). These results with MVIIA are highly interesting. They suggest that the window of opportunity for therapeutic intervention after ischemia may be much longer than previously thought, and point to the potential use of conopeptides and their derivatives in the prevention of neuronal damage resulting from ischemic episodes due to cardiac arrest, head trauma, or stroke.

Selective blockade of N-type calcium channels may also be beneficial in treatment of specific pain syndromes. Intrathecal administration of as little as 0.3 μg MVIIA completely suppressed the nociceptive responses in the rat hindpaw formalin test (190, 191). Tactile allodynia was also selectively abolished in a rat neuropathic pain model by intrathecal administration of MVIIA at doses that did not impair motor function. MVIIA was found to be 100 times more potent than morphine (192).

ω-Toxins have potential for use as diagnostic indicators of neural degenerative syndromes. Patients exhibiting symptoms of Lambert-Eaton myasthenic syndrome (LEMS) have been shown to possess serum autoantibodies that recognize voltage-sensitive calcium channels. Serum from LEMS-positive patients precipitates calcium channels from neuroblastoma cell lines (193) or small cell carcinoma (194) that have been prebound with $[^{125}I]$ω-conotoxin GVIA. Such types of immunoprecipitation assays appear to provide a facile means of diagnosis, which may eliminate the need for muscle biopsies to distinguish between LEMS and symptomatically similar syndromes such as myasthenia gravis.

VII. DISCUSSION AND PERSPECTIVES

We conclude with a brief overview of some general issues that need to be resolved. In this article, we have outlined the role of ω-toxins in identifying voltage-sensitive calcium channel subtypes, and reviewed their use in inhibiting neurotransmitter release. It is obvious from what is presented above that this is a field in which rapid progress is being made. A continuing correlation between the electrophysiological characterization of Ca currents on the one hand, and the rapidly progressing molecular definition by cloning on the other hand, is clearly a major goal in understanding Ca channel subtypes; the ability of ω-toxins to target specific channel subtypes provides a synergistic linkage between cloning and electrophysiology.

Some electrophysiological characteristics that might have been predicted to be most important in distinguishing calcium channel types have proven not to be critical criteria for classifying calcium channels at the molecular level. Some of the electrophysiological parameters described in section IV.A do not correlate tightly to the major classes of calcium channel subunits cloned so far. In particular, it has been shown that a molecularly homogeneous cloned channel can show quite strikingly different inactivation characteristics (129, 195), indicating that inactivation cannot be used as a reliable parameter for gauging broader relationships between channel types. Somewhat surprisingly, pharmacological criteria seem to do better.

Another fundamental problem involves assignment of calcium channel types to specific functions in neurons, such as neurotransmitter release. Essentially all of the highly discriminating electrophysiological data on calcium channel pharmacology has been obtained from native channels expressed in neuronal cell bodies (60, 86, 95, 96, 117, 196) or cloned channels expressed in a heterologous system such as the *Xenopus* oocyte (115, 120, 137). Although these studies illuminate the properties of calcium channels extremely well, questions remain as to the relationships between cell body channels and those that control neurotransmitter release. Since direct measurements of nerve terminal calcium current have been possible only in rare instances (131, 160, 197), this question has been difficult to answer with any certainty.

We note that in many cases a toxin interacting with a calcium channel does not cause a complete block of conductance. When partial inhibition is found in living cells, this immediately creates an ambiguity. Is the residual calcium current after applying the toxin a different subtype, elicited by a calcium channel resistant to the toxin, or is it the same current resulting from partial toxin block? The problem is that such questions are not necessarily resolved by analyzing the voltage and current characteristics of the residual current, since a toxin binding a particular ion channel and partially blocking it could also change the fundamental parameters of activation, inactivation, and conductance. Thus, it is important to establish whether or not a particular ω-toxin completely blocks a channel subtype to which it is bound.

A final issue relates to how widely diverse calcium channels ultimately will prove to be. The data in Figure 5 indicate two main evolutionary trunks in the Ca channel α_1 subunit family, one of which can be neatly sorted out pharmacologically into the dihydropyridine-sensitive L-type channel family, and the other into a non-L-type channel family. Do additional trunks exist that have not yet been discovered? For instance, will the T-type channel ultimately prove to have an α_1 subunit closely related to α_{1E} (115), or will it be completely unrelated?

Enough molecular data has been collected for the α_1 subunits that are DHP sensitive to reveal the remarkable potential for molecular heterogeneity in the L-type family. Not only do several genes encode different α_1 subunits, but there are multiple splice variants of each gene. The number of combinatorial possibilities expands rapidly if heterogeneity in the other Ca channel subunits is taken into account. Since it appears that the non-L-type family controls the major fraction of neurotransmitter release from neurons, as well as many of the electrical properties of cell bodies, it seems likely that this other major trunk of the calcium channel family tree will show a degree of molecular heterogeneity that is at least comparable to the L-type family. The ω-toxins are clearly most useful for sorting out and further understanding the diversity of DHP-resistant calcium channel types.

Since spiders and cone snails operate in vastly different ecological settings, and have distinctive evolutionary histories, the peptide toxins they have evolved have been subject to different selective pressures. Thus, it is not surprising that the spider and *Conus* toxins that both target to N-type calcium channels (e.g. ω-conotoxin GVIA and ω-agatoxin-IIIA) exhibit a striking contrast in channel subtype range for mammalian Ca channels. A cross-comparison of the subtype selectivity of *Conus* ω-conotoxins and spider ω-agatoxins should continue to be illuminating. Even if a spider and *Conus* toxin both target largely the same spectrum of subtypes, such as AgaIVA and ω-conotoxin MVIIC, it has proven insightful to evaluate the effect of both toxins on each particular Ca channel. Such studies have revealed that channel subtypes that are AgaIVA sensitive can be further subdivided if the effects of MVIIC are taken into account. This is the basis for our tentative suggestion that a molecularly heterogeneous "OPQ subfamily" of Ca channels, together with the N-type channels, may be the major players in neurotransmitter release in the mammalian central nervous system.

A key question that needs to be addressed aggressively is how many functionally distinguishable Ca channels control neurotransmitter release. The present data suggest considerable diversity; in Section V, examples of apparently different Ca channel subtypes controlling release of the same neurotransmitter were cited. Do the different presynaptic Ca channel subtypes correspond to different postsynaptic receptor subtypes? For example, would presynaptic termini releasing dopamine to postsynaptic termini with D_1, D_2, D_3, and D_4 receptors each have a different, characteristic spectrum of Ca channel isoforms at the corresponding presynaptic termini? The number of

Ca channel isoforms described in the literature will undoubtedly increase. Although each new Ca channel type elucidated often seems to cause only more confusion and complexity, it may be reassuring to remember that the greater the molecular diversity of Ca channels, the greater the potential therapeutic possibilities.

ACKNOWLEDGMENTS

The work of the authors was supported by National Institutes of Health Grant No. PO1 GM48677 (to B.M.O.) and Grant No. NS24473 (to M.E.A).

Literature Cited

1. Olivera BM, Rivier J, Scott JK, Hillyard DR, Cruz LJ. 1991. *J. Biol. Chem.* 266:22067–70
2. Olivera BM, Imperial JS, Cruz LJ, Bindokas VP, Venema VJ, Adams ME. 1991. *Ann. NY Acad. Sci.* 635:114–22
3. Adams ME, Bindokas VP, Venema VJ. 1992. In *Neurotox '91 The Molecular Basis of Drug and Pesticide Action,* ed. IR Duce, pp. 33–44. Amsterdam: Elsevier Sci.

3a. Sher E, Clementi F. 1991. *Neuroscience* 42:301–7

4. Katz B, Miledi R. 1967. *Proc. R. Soc. London Ser. B* 167:23
5. Katz B, Miledi R. 1967. *J. Physiol.* 189:535–44
6. Katz B. 1969. *The Release of Neural Transmitter Substances.* Liverpool: Liverpool Univ. Press
7. Olivera BM, Gray WR, Zeikus R, McIntosh JM, Varga J, et al. 1985. *Science* 230:1338–43
8. Olivera BM, Rivier J, Clark C, Ramilo CA, Corpuz GP, et al. 1990. *Science* 249:257–63
9. Gray WR, Olivera BM, Cruz LJ. 1988. *Annu. Rev. Biochem.* 57:665–700
10. Olivera BM, McIntosh JM, Cruz LJ, Luque FA, Gray WR. 1984. *Biochemistry* 23:5087–90
11. Hillyard DR, Olivera BM, Woodward S, Corpuz GP, Gray WR, et al. 1989. *Biochemistry* 28:358–61
12. Hasson A, Fainzilber M, Gordon D, Zlotkin E, Spira ME. 1993. *Eur. J. Neurosci.* 5:56–64
13. Hillyard DR, Monje VD, Mintz IM, Bean BP, Nadasdi L, et al. 1992. *Neuron* 9:69–77
14. Ramilo CA, Zafaralla GC, Nadasdi L, Hammerland LG, Yoshikami D, et al. 1992. *Biochemistry* 31:9919–26
15. Kohn AJ, Saunders PR, Wiener S. 1960. *Ann. NY Acad. Sci.* 90:706–25
16. Olivera BM, Gray WR, Cruz LJ. 1988. In *Handbook of Natural Toxins,* ed. AT Tu, pp. 327–52. New York: Dekker
17. Rivier J, Galyean R, Gray WR, Azimi-Zonooz A, McIntosh JM, et al. 1987. *J. Biol. Chem.* 262:1194–98
18. Nishiuchi Y, Kumagaye K, Noda Y, Watanabe TX, Sakakibara S. 1986. *Biopolymers* 25:561–68
19. Monje VD, Haack JA, Naisbitt SR, Miljanich G, Ramachandran J, et al. 1993. *Neuropharmacology* 32:1141–49
20. Cruz LJ, Olivera BM. 1986. *J. Biol. Chem.* 261:6230–33
21. Olivera BM, Cruz LJ, De Santos V, LeCheminant GW, Griffin D, et al. 1987. *Biochemistry* 26:2086
22. Abe T, Koyano K, Saisu H, Nishiuchi Y, Sakakibara S. 1986. *Neurosci. Lett.* 71:203–8
23. Wagner JA, Snowman AM, Biswas A, Olivera BM, Snyder SH. 1988. *J. Neurosci.* 8:3354–59

23a. Kristipati R, Nadasdi L, Tarczy-Hornoch K, Lau K, Miljanich GP, et al. 1994. *Mol. Cell. Neurosci.* In press

24. Albensi BC, Ryujin KT, McIntosh JM, Naisbitt SR, Olivera BM, Fillous F. 1993. *NeuroReport* 4:1331–34

25. Kerr LM, Filloux F, Olivera BM, Jackson H, Wamsley JK. 1988. *Eur. J. Pharmacol.* 146:181–83
26. Takemura M, Kiyama H, Fukui H, Tohyama M, Wada H. 1988. *Brain Res.* 451:386–89
27. McIntosh JM, Adams ME, Olivera BM, Filloux F. 1992. *Brain Res.* 594:109–14
28. McIntosh JM, Schapper A, Naisbitt SR, Olivera BM, Filloux F. 1993. *Soc. Neurosci. Abstr.* 19:1717
29. Cruz LJ, Johnson DS, Olivera BM. 1987. *Biochemistry* 26:820–24
30. Myers RA, McIntosh JM, Imperial J, Williams RW, Oas T, Haack JA. 1990. *J. Toxicol.-Toxin Rev.* 9:179–202
31. Yamaguchi T, Saisu H, Mitsui H, Abe T. 1988. *J. Biol. Chem.* 263:9491–98
32. Glossmann H, Striessnig J. 1988. *ISI Atlas Pharmacol.* 2:202–10
33. Barhanin J, Schmid A, Lazdunski M. 1988. *Biochem. Biophys. Res. Commun.* 150:1051–62
34. Marqueze B, Martin-Moutot N, Leveque C, Couraud F. 1988. *Mol. Pharmacol.* 34:87–90
35. Haack J, Kinser P, Yoshikami D, Olivera B. 1993. *Neuropharmacology* 32:1151–59
36. Sugiura Y, Ko C-P, Woppmann A, Miljanich G. 1993. *Soc. Neurosci. Abstr.* 19:1756
37. Robitaille R, Adler EM, Charlton MP. 1990. *Neuron* 5:773–79
38. Cohen MW, Jones OT, Angelides KJ. 1991. *J. Neurosci.* 11:1032–39
39. Tarelli FT, Passafaro M, Clementi F, Sher E. 1991. *Brain Res.* 547:331–34
40. Jones OT, Kunze DL, Angelides KJ. 1989. *Science* 244:1189–93
41. Colledge CJ, Hunsperger JP, Imperial JS, Hillyard DR. 1992. *Toxicon* 30: 1111–16
42. Coddington JA, Levi HW. 1991. *Annu. Rev. Ecol. Syst.* 22:565–92
43. Adams ME, Herold EE, Venema VJ. 1989. *J. Comp. Physiol. A* 164:333–42
44. Adams ME, Bindokas VP, Zlotkin E. 1989. In *Insecticide Action From Molecule to Organism,* ed. T Narahashi, JE Chambers, pp. 189–203. New York: Plenum
45. Parks TN, Mueller AL, Artman LD, Albensi BC, Nemeth EF, et al. 1991. *J. Biol. Chem.* 266:21523–29
46. Jasys VJ, Kelbaugh PR, Nason DM, Phillips D, Rosnack KJ, et al. 1990. *J. Am. Chem. Soc.* 112:6696–704
47. Jasys VJ, Kelbaugh PR, Nason DM, Phillips D. 1992. *J. Org. Chem.* 57: 1814–20
48. Quistad GB, Reuter CC, Skinner WS, Dennis PA, Suwanrumpha S, Fu EW. 1991. *Toxicon* 29:329–36
49. Bateman A, Boden P, Dell A, Duce IR, Quicke DLJ, Usherwood PNR. 1985. *Brain Res.* 339:237–44
50. Grishin EV, Volkova TM, Arseniev AS, Reshetova OS, Onoprienko VV, et al. 1986. *Bioorg. Khim.* 12:110–12
51. Adams ME, Carney RL, Enderlin FE, Fu ET, Jarema MA, et al. 1987. *Biochem. Biophys. Res. Commun.* 148: 678–83
52. Skinner WS, Dennis PA, Lui A, Carney RL, Quistad GB. 1990. *Toxicon* 28:541–46
53. Early SL, Michaelis EK. 1987. *Toxicon* 25:433–42
54. Stapleton A, Blankenship DT, Ackermann BL, Chen TM, Gorder GW, et al. 1990. *J. Biol. Chem.* 265:2054–59
55. Bindokas VP, Adams ME. 1989. *J. Neurobiol.* 20:171–88
56. Adams ME, Bindokas VP, Hasegawa L, Venema VJ. 1990. *J. Biol. Chem.* 265:861–67
57. Bindokas VP, Venema VJ, Adams ME. 1991. *J. Neurophysiol.* 66:590–601
58. Santos AD, Imperial JS, Chaudhary T, Beavis RC, Chait BT, et al. 1992. *J. Biol. Chem.* 267:20701–5
59. Venema VJ, Swiderek KM, Lee TD, Hathaway GM, Adams ME. 1992. *J. Biol. Chem.* 267:2610–15
60. Mintz IM, Venema VJ, Swiderek KM, Lee TD, Bean BP, Adams ME. 1992. *Nature* 355:827–29
61. Adams ME, Mintz IM, Reily MD, Thanabal V, Bean BP. 1993. *Mol. Pharmacol.* 44:681–88
62. Hirning LD, Mueller AL, Alasti N, Artman LD, Delmar EG, et al. 1993. *Neurosci. Abstr.* 19:1753
63. Teramoto T, Kuwada M, Nidome T, Sawada K. 1993. *Biochem. Biophys. Res. Commun.* 196:134–40
64. Ertel EA, Warren VA, Adams ME, Griffin PR, Cohen CJ, Smith MM. 1994. *Biochemistry.* In press
65. Hillyard DR, Santos AD, Monje VD, Marsh M, Adams ME. 1994. *Biochemistry.* Submitted
66. Ahlijanian M, Andrews G, Guarino B, Saccomano NA, Hirning LD, et al. 1993. *Soc. Neurosci. Abstr.* 19: 1753
67. Brown AM, Schwindt PC, Crill WE. 1993. *J. Physiol.* In press
68. Nishio H, Kumagaye KY, Chen Y-n, Kimura T, Sakakibara S. 1993. *Am. Peptide Symp. Abstr.* 13:2-1

69. Mintz IM, Bean BP. 1993. *Soc. Neurosci. Abstr.* 19:1478
70. Llinas R, Sugimori M, Lin JW, Cherksey B. 1989. *Proc. Natl. Acad. Sci. USA* 86:1689–93
71. Lin JW, Rudy B, Llinas R. 1990. *Proc. Natl. Acad. Sci. USA* 87:4538–42
72. Llinas R, Sugimori M, Hillman DE, Cherksey B. 1992. *Trends Neurosci.* 15:351–55
73. Usowicz MM, Sugimori M, Cherksey B, Llinas R. 1992. *Neuron* 9:1185–99
74. Scott RH, Sweeney MI, Kobrinsky EM, Pearson HA, Timms GH, et al. 1992. *Br. J. Pharmacol.* 106:199–207
75. Dolphin AC, Huston E, Pearson H, Menon JA, Sweeney MI, et al. 1991. *Ann. NY Acad. Sci.* 635:139–52
76. Pocock JM, Nicholls DG. 1992. *Eur. J. Pharmacol.* 226:343–50
77. Bowers CW, Phillips HS, Lee P, Jan YN, Jan LY. 1987. *Proc. Natl. Acad. Sci. USA* 84:3506–10
78. Lundy PM, Hong A, Frew R. 1992. *Eur. J. Pharmacol.* 225:51–56
79. Lundy PM, Frew R. 1993. *Eur. J. Pharmacol.* 231:197–202
80. Branton WD, Kolton L, Jan YN, Jan LY. 1987. *J. Neurosci.* 7:4195–200
81. Newman EA, Zhou Y, Rudnick MS, Branton WD. 1992. *Soc. Neurosci. Abstr.* 18:971
82. Leung HT, Branton WD, Phillips HS, Jan L, Byerly L. 1989. *Neuron* 3:767–72
83. Leung HT, Byerly L. 1991. *J. Neurosci.* 11:3047–59
84. Branton WD, Rudnick MS, Zhou Y, Eccleston ED, Fields GB, Bowers LD. 1993. *Nature* 365:496–97
85. Branton WD, Fields CG, Vandrisse VL, Fields GB. 1993. *Tetrahedron Lett.* 34:4885–88
86. Mintz IM, Adams ME, Bean BP. 1992. *Neuron* 9:85–95
87. Lampe RA, DeFeo PA, Davison MD, Young J, Herman JL, et al. 1993. *Mol. Pharmacol.* 44:451–59
88. de Weille JR, Schweitz H, Maes P, Tartar A, Lazdunski M. 1991. *Proc. Natl. Acad. Sci. USA* 88:2437–40
89. Kuroda H, Chen YN, Watanabe TX, Kimura T, Sakakibara S. 1992. *Pept. Res.* 5:265–68
90. Yasuda O, Morimoto S, Chen YH, Jiang BB. 1993. *Biochem. Biophys. Res. Commun.* 194:587–94
91. Horne WA, Ellinor PT, Inman I, Zhou M, Tsien RW, Schwarz TL. 1993. *Proc. Natl. Acad. Sci. USA* 90:3787–91
92. Ellinor PT, Zhang JF, Randall AD, Zhou M, Schwarz TL, et al. 1993. *Nature* 363:455–58
93. Aosaki T, Kasai H. 1989. *Pflugers Arch.* 414:150–56
94. McCleskey EW, Fox AP, Feldman DH, Cruz LJ, Olivera BM, et al. 1987. *Proc. Natl. Acad. Sci. USA* 84:4327–31
95. Regan LJ, Sah DW, Bean BP. 1991. *Neuron* 6:269–80
96. Mogul DJ, Fox AP. 1991. *J. Physiol.* 433:259–81
97. Kasai H, Neher E. 1992. *J. Physiol.* 448:161–88
98. Rivier J, Galyean R, Simon L, Cruz LJ, Olivera BM, Gray WR. 1987. *Biochemistry* 26:8508–12
99. Suszkiw JB, Murawsky MM, Fortner RC. 1987. *Biochem. Biophys. Res. Commun.* 145:1283–86
100. Kerr LM, Yoshikami D. 1984. *Nature* 308:282–84
101. Wessler I, Dooley DJ, Werhand J, Schlemmer F. 1990. *Naunyn Schmiedebergs Arch. Pharmacol.* 341: 288–94
102. Sano K, Enomoto K-I, Maeno T. 1987. *Eur. J. Pharmacol.* 141:235–41
103. Tanabe T, Takeshima H, Mikami A, Flockerzi V, Takahashi H, et al. 1987. *Nature* 328:313–18
104. Bean BP. 1989. *Annu. Rev. Physiol.* 51:367–84
105. Hess P. 1990. *Annu. Rev. Neurosci.* 13:337–56
106. Miller RJ. 1987. *Science* 235:46–52
107. Snutch TP. 1992. *Curr. Biol.* 2:247–53
108. Tsien RW, Lipscombe D, Madison DV, Bley KR, Fox AP. 1988. *Trends Neurosci.* 11:431–38
109. Tsien RW, Ellinor PT, Horne WA. 1991. *Trends Pharmacol. Sci.* 12:349–54
110. Campbell KP, Leung AT, Sharp AH. 1988. *Trends Neurosci.* 11:425–30
111. McEnery MW, Snowman AM, Sharp AH, Adams ME, Snyder SH. 1991. *Proc. Natl. Acad. Sci. USA* 88:11095–99
112. Witcher DR, De WM, Sakamoto J, Franzini AC, Pragnell M, et al. 1993. *Science* 261:486–89
113. Snutch TP, Leonard JP, Gilbert MM, Lester HA, Davidson N. 1990. *Proc. Natl. Acad. Sci. USA* 87:3391–95
114. Llinas R, Yarom Y. 1981. *J. Physiol.* 315:549–67
115. Soong TW, Stea A, Hodson CD, Dubel SJ, Vincent SR, Snutch TP. 1993. *Science* 260:1133–36
116. Nowycky MC, Fox AP, Tsien RW. 1985. *Nature* 316:440–43
117. Mintz IM, Venema VJ, Adams ME, Bean BP. 1991. *Proc. Natl. Acad. Sci. USA* 88:6628–31
118. Cohen CJ, Ertel EA, Smith MM, Venema VJ, Adams ME, Leibowitz

MD. 1992. *Mol. Pharmacol.* 42:947–51
119. Williams ME, Feldman DH, McCue AF, Brenner R, Velicelebi G, et al. 1992. *Neuron* 8:71–84
120. Williams ME, Brust PF, Feldman DH, Patthi S, Simerson S, et al. 1992. *Science* 257:389–95
121. Regulla S, Schneider T, Nastainczyk W, Meyer HE, Hofmann F. 1991. *EMBO J.* 10:45–49
122. Nakayama H, Taki M, Striessnig J, Glossmann H, Catterall WA, Kanaoka Y. 1991. *Proc. Natl. Acad. Sci. USA* 88:9203–7
123. Striessnig J, Murphy BJ, Catterall WA. 1991. *Proc. Natl. Acad. Sci. USA* 88:10769–73
123a. Hoshi T, Rothlein J, Smith SJ. 1984. *Proc. Natl. Acad. Sci. USA* 81:5871–75
124. Artalejo CR, Mogul DJ, Perlman RL, Fox AP. 1991. *J. Physiol.* 444:213–40
124a. Hoshi T, Smith SJ. 1987. *J. Neurosci.* 7:571–80
125. Artalejo CR, Ariano MA, Perlman RL, Fox AP. 1990. *Nature* 348:239–42
126. Perez-Reyes E, Wei XY, Castellano A, Birnbaumer L. 1990. *J. Biol. Chem.* 265:20430–36
127. Fox AP, Nowycky MC, Tsien RW. 1987. *J. Physiol.* 394:149–72
128. Dubel SJ, Starr TVB, Hell J, Ahlijanian MK, Enyeart JJ, et al. 1992. *Proc. Natl. Acad. Sci. USA* 89:5058–62
129. Fujita Y, Mynlieff M, Dirksen RT, Kim M-S, Niidome T, et al. 1993. *Neuron* 10:585–98
130. Albensi BC, Ryujin KT, McIntosh JM, Naisbitt SR, Olivera BM, Filloux F. 1993. *Soc. Neurosci. Abstr.* 19:1755
131. Stanley EF, Goping G. 1991. *J. Neurosci.* 11:985–93
132. Ramachandran J, Nadasdi L, Gohil K, Kristipati R, Tarczy-Hornoch K, et al. 1993. In *Perspectives in Medicinal Chemistry,* ed. B Testa, E Kyburz, W Fuhrer, R Giger, pp. 374–88. Basel: Verlag Helvetica Chimica Acta
133. Yoshikami D, Bagabaldo Z, Olivera BM. 1989. *Ann. NY Acad. Sci.* 560:230–48
134. Swartz KJ, Mintz IM, Boland LM, Bean BP. 1993. *Soc. Neurosci. Abstr.* 19:1478
135. Mori Y, Friedrich T, Kim M-S, Mikami A, Nakai J, et al. 1991. *Nature* 350:398–402
136. Starr TVB, Prystay W, Snutch TP. 1991. *Proc. Natl. Acad. Sci. USA* 88:5621–25
137. Sather WA, Tanabe T, Zhang J-F, Mori Y, Adams ME, Tsien RW. 1993. *Neuron* 11:291–303
138. Randall AD, Wendland B, Schweizer F, Miljanich G, Adams ME, Tsien RW. 1993. *Soc. Neurosci. Abstr.* 19:1478
139. Adams ME, Myers RA, Imperial JS, Olivera BM. 1993. *Biochemistry* 32:12566
140. Gaur S, Nadasdi L, Bell J, Ramachandran J, Miljanich G. 1992. *Soc. Neurosci. Abstr.* 18:972
141. Zhang J-F, Randall AD, Ellinor PT, Horne WA, Sather WA, et al. 1993. *Neuropharmacology* 32:1075–88
142. Pietrobon D, Forti L, Tottene A, Moretti A. 1993. *Soc. Neurosci. Abstr.* 19:1479
143. Augustine GJ, Charlton MP, Smith SJ. 1987. *Annu. Rev. Neurosci.* 10:633–93
144. Sher E, Clementi F. 1991. *Neuroscience* 42:301–7
145. Augustine GJ, Charlton MP. 1986. *J. Physiol.* 381:619–40
146. Dodge FAJ, Rahamimoff R. 1967. *J. Physiol.* 193:419–32
147. Luebke JL, Dunlap K, Turner TJ. 1993. *Neuron* 11:1–20
148. Mangano TJ, Patel J, Salama AI, Keith RA. 1991. *Eur. J. Pharmacol.* 192:9–17
149. DeFeo PA, Mangano TJ, Adams ME, Keith RA. 1992. *Pharmacol. Commun.* 1:273–78
150. Turner TJ, Adams ME, Dunlap K. 1992. *Science* 258:310–13
151. Turner TJ, Adams ME, Dunlap K. 1993. *Proc. Natl. Acad. Sci. USA* 90:9518–22
152. Wessler I, Dooley DJ, Osswald H, Schlemmer F. 1990. *Neurosci. Lett.* 108:173–78
153. Woodward JJ, Rezazadeh SM, Leslie SW. 1988. *Brain Res.* 475:141–45
154. Dooley DJ, Lupp A, Hertting G, Osswald H. 1988. *Eur. J. Pharmacol.* 148:261–67
155. Kamiya H, Sawada S, Yamamoto C. 1988. *Neurosci. Lett.* 91:84–88
156. Krishtal OA, Petrov AV, Smirnov SV, Nowycky MC. 1989. *Neurosci. Lett.* 102:197–204
157. Dutar P, Rascol O, Lamour Y. 1989. *Eur. J. Pharmacol.* 174:261–66
158. Horne AL, Kemp JA. 1991. *Br. J. Pharmacol.* 103:1733–39
159. Takahashi T, Momiyama A. 1993. *Nature* 366:156–58
160. Lemos JR, Nowycky MC. 1989. *Neuron* 2:1419–26
161. Wang X, Treistman SN, Lemos JR. 1992. *J. Physiol.* 445:181–99
162. Santicoli P, Del Bianco E, Tramontana

M, Geppetti P, Maggi CA. 1992. *Neurosci. Lett.* 136:161–64
163. Lindgren CA, Moore JW. 1989. *J. Physiol.* 414:201–22
164. Uchitel OD, Protti DA, Sanchez V, Cherksey BD, Sugimori M, Llinas R. 1992. *Proc. Natl. Acad. Sci. USA* 89:3330–33
165. Anderson AJ, Harvey AL. 1987. *Neurosci. Lett.* 82:177–80
166. Bowersox S, Ko C-P, Sugiura Y, Li CZ, Fox J, et al. 1993. *Soc. Neurosci. Abstr.* 19:1478
167. Pruneau D, Angus JA. 1990. *J. Cardiovasc. Pharmacol.* 16:675–80
168. Bowersox SS, Singh T, Nadasdi L, Zukowska GZ, Valentino K, Hoffman BB. 1992. *J. Cardiovasc. Pharmacol.* 20:756–64
169. Mohy El-Din MM, Malik KV. 1988. *Br. J. Pharmacol.* 94:355–62
170. Clasbrummel B, Osswald H, Illes P. 1989. *Br. J. Pharmacol.* 96:101–10
171. Hirning LD, Fox AP, McCleskey EW, Olivera BM, Thayer SA, et al. 1988. *Science* 239:57–61
172. Keith R, LaMonte D, Salama A. 1990. *J. Auton. Pharmacol.* 10:139–51
173. Maggi CA. 1991. *J. Auton. Pharmacol.* 11:295–304
174. Ballesta JJ, Palmero M, Hidalgo MJ, Gutierrez LM, Reig JA, et al. 1989. *J. Neurochem.* 53:1050–56
175. Jan CR, Titeler M, Schneider AS. 1990. *J. Neurochem.* 54:355–58
176. Owen PJ, Marriott DB, Boarder MR. 1989. *Br. J. Pharmacol.* 97:133–38
177. Rosario LM, Soria B, Feuerstein G, Pollard HB. 1989. *Neuroscience* 29: 735–47
178. Artalejo CR, Perlman RL, Fox AP. 1992. *Neuron* 8:85–95
178a. Artalejo CR, Adams ME, Fox AP. 1994. *Nature* 367:72–76
179. Myers RA, Cruz LJ, Rivier J, Olivera BM. 1993. *Chem. Rev.* 93: 1923–36
179a. Mintz IM. 1994. *J. Neurosci.* In press
180. Nishiuchi Y, Kumagaye K, Noda Y, Watanabe T, Sakakibara S. 1986. *Biopolymers* 25:561–68
181. Davis J, Bradley E, Miljanich G, Nadasdi L, Ramachandran J, Basus V. 1993. *Biochemistry* 32:7396–405
182. Sevilla P, Bruix M, Santoro J, Gago F, Garcia AG, Rico M. 1993. *Biochem. Biophys. Res. Commun.* 192:1238–44
183. Skalicky JJ, Metzler WJ, Ciesla DJ, Galdes A, Pardi A. 1993. *Protein Sci.* 2:1591–603
184. Valentino K, Newcomb R, Gadbois T, Singh T, Bowersox S, et al. 1993. *Proc. Natl. Acad. Sci. USA* 90:7894–97
185. Smith M-L, Siesjo B. 1992. In *Pharmacology of Cerebral Ischemia,* ed. J Krieglstein, H Oberpichler-Schwenk, pp. 161–66. Stuttgart: Wissenschaftlische Velagsgesellschaft
186. Xue D, Huang ZG, Barnes K, Lesiuk H, Smith KE, Buchan AM. 1993. *Soc. Neurosci. Abstr.* 19:1643
187. Takizawa S, Matsushima K, Shinohara Y, Fujita H, Nanri K, Ogawa S. 1994. *J. Cereb. Blood Flow.* In press
188. Buchan A, Lesiuk H, Xue D, Li H, Huang Z-G, et al. 1994. *J. Cerebral Blood Flow.* In press
189. Badie H, Smith ML, Hovda DA, Fu K, Pinanong P, et al. 1993. *Soc. Neurosci. Abstr.* 19:1485
190. Gohil K, Bowersox S, Singh T, Ramachandran J, Miljanich G. 1993. *Soc. Neurosci. Abstr.* 19:235
191. Singh T, Malmberg AB, Yaksh TL. 1993. *7th World Congr. Pain,* 225:225. Seattle: IASP
192. Chaplan SR, Pogrel J, Yaksh TL. 1993. *7th World Congr. Pain,* 225:37. Seattle: IASP
193. Sher E, Gotti C, Canal N, Scoppetta C, Piccolo G, et al. 1989. *Lancet* 2:640–43
194. Lennon VA, Lambert EH. 1989. *Mayo Clin. Proc.* 64:1498–504
195. Plummer MR, Hess P. 1991. *Nature* 351:657–59
196. Regan LJ. 1991. *J. Neurosci.* 11:2259–69
197. Stanley EF. 1991. *Neuron* 7:585–91

Annu. Rev. Biochem. 1994. 63:869–914

GENETIC AND BIOCHEMICAL STUDIES OF PROTEIN N-MYRISTOYLATION

D. Russell Johnson, Rajiv S. Bhatnagar, Laura J. Knoll, and Jeffrey I. Gordon
Department of Molecular Biology and Pharmacology, Washington University School of Medicine, St. Louis, Missouri 63110

KEY WORDS: enzyme kinetics, titration calorimetry, acylCoA metabolism, yeast genetics, protein-protein/protein-membrane interactions

---

CONTENTS

INTRODUCTION

Protein N-myristoylation refers to the covalent attachment of myristate (C14:0), via an amide bond, to the amino-terminal Gly residue of a nascent polypeptide. This cotranslational modification (1, 2) has only been observed in eukaryotes and appears to be irreversible (3, 4). The first N-myristoylproteins were described in 1982 (5, 6). Since that time, a large number of cellular N-myristoylproteins have been identified. These proteins have diverse functions and include serine/threonine kinases, tyrosine kinases,

0066-4154/94/0701-0869$05.00

kinase substrates, phosphoprotein phosphatases, other types of proteins involved in signal transduction cascades (e.g. the α subunits of heterotrimeric G proteins, the constitutively expressed endothelial nitric oxide synthase), and mediators of protein and vesicular transport (e.g. ADP ribosylation factors) (reviewed in Ref. 7).

A myristoyl group appears to mean different things to different N-myristoylproteins. N-myristoylproteins have diverse intracellular destinations. Myristate appears to be critical for mediating protein-protein and/or protein-membrane interactions for some N-myristoylproteins. For others, interactions with cell membranes requires accessory factors or other covalent modification: e.g. calcium ions in the case of recoverin (8), GTP in the case of ADP ribosylation factors (9–11), or phosphorylation in the case of the myristoylated alanine-rich C kinase substrate (MARCKS) (12–14). For some N-myristoylproteins, "deleting" myristate has no obvious functional consequences, as with *Saccharomyces cerevisiae* Cnb1p, the regulatory subunit of the type IIB calcium phosphatase calcineurin, which is involved in adaptation of haploid cells to pheromone (M Cyert, unpublished observations). For other N-myristoylproteins, the effects of removing the acyl chain are subtle. For example, deleting the myristoyl moiety from the catalytic (C_α) subunit of bovine cAMP-dependent protein kinase (PK-A) does not appear to alter its activity or its ability to interact with other proteins (15). However, the non-acylated protein is more susceptible to cycles of thermal denaturation-renaturation, revealing a role for the fatty acid in conferring structural stability to C_α, both in its monomeric form and as part of the holoenzyme complex (16–18).

Current interest in protein N-myristoylation arises in part from the fact that the enzyme that catalyzes this covalent protein modification, myristoyl-CoA:protein N-myristoyltransferase (Nmt; EC 2.1.3.97), is a potential target for antiviral, antifungal, and/or antineoplastic therapy. For example, a variety of structural and nonstructural proteins encoded by retroviruses, hepadnaviruses, papovaviruses, picornaviruses, and reoviruses are substrates for human Nmt (reviewed in Ref. 19). Myristoylation of some of these proteins is essential for viral assembly (e.g. the $Pr55^{gag}$ polyprotein precursor encoded by human immunodeficiency virus I; Refs. 20, 21) or the production of infectious virions (e.g. the L protein of hepatitis B virus; Refs. 22–24). A number of transforming tyrosine kinases are N-myristoylated. These include src and members of its family: fyn, fgr, hck, lck, blk, yes, and yrk (25–32; reviewed in 32a). For some, such as src, myristate is necessary for stable association with the plasma membrane and for cellular transformation, although C14:0 does not appear to be required for src's tyrosine kinase activity (33–35a). *Candida albicans* is the most common cause of systemic fungal infections in immunocompromised humans. It produces a small number of N-myristoylproteins (36, 37). The peptide substrate specificity of its Nmt is distinct from that of the human enzyme, and the organism requires Nmt for vegetative growth (38), suggesting that the acyltransferase may be a good target for development of fungicidal drugs.

This review focuses on several model systems that have been useful for assessing how protein N-myristoylation is regulated and for examining its functional significance.

REGULATION OF PROTEIN N-MYRISTOYLATION IN *SACCHAROMYCES CEREVISIAE*

S. cerevisiae represents a very attractive model system for analyzing protein N-myristoylation. It is genetically manipulatable, and there is a large body of information about the location and regulation of the organism's lipid biosynthetic and degradative pathways (reviewed in Ref. 39). When wild-type *S. cerevisiae* strains are incubated with tritiated myristate during exponential growth in rich media, at least 12 cellular proteins incorporate the fatty acid in a form that resists cleavage by 1 N hydroxylamine, pH 10—conditions known to cleave thioester and oxyester but not amide bonds (40–42). Seven *S. cerevisiae* N-myristoylproteins have been identified to date: Gpa1p, the α subunit of a heterotrimeric G protein involved in mating factor signal transduction (4, 43–46); vacuolar sorting protein 15 (Vps15p; Refs. 47, 48); two functionally interchangeable ADP ribosylation factors (Arf1p and Arf2p; Refs. 49–51); Cnb1p (52); and two functionally interchangeable type 1–related protein phosphatases (Ppz1p and Ppz2p; Ref. 53).

Gpa1p, Ste4p, and Ste18p are the α, β, and γ subunits, respectively, of *S. cerevisiae*'s pheromone-responsive heterotrimeric G protein. When the α- and **a**-factor pheromones bind to their receptors, Ste2p and Ste3p, the receptors are activated and catalyze the exchange of GDP for GTP on Gpa1p. This produces a conformational change in Gpa1p that allows it to dissociate from the Ste4p/Ste18p (βγ) dimer. The dimer then initiates a kinase cascade that results in phosphorylation of the transcription factor Ste12p which, in turn, activates transcription of mating-specific genes such as *FUS1*, which is required for cell fusion (54, 55). The signal transduction pathway produces arrest in the G1 phase of the cell cycle and induces morphologic changes resulting in projection formation (reviewed in Ref. 56). Myristoylation of Gpa1p can be blocked by a Gly2→Ala mutation (45, 46). Cells that undermyristoylate or fail to myristoylate Gpa1p exhibit constitutive activation of the pheromone response pathway and they arrest in G1. Loss of C14:0 does not affect targeting of the protein to the cell membrane but likely perturbs its interaction with the βγ dimer (45).

Vps15p is a 1455-residue protein kinase homolog required for sorting of soluble hydrolases to the yeast vacuole (47, 48). Cells that only produce Vps15p$^{Gly2 \rightarrow Ala}$ grow normally at 38°C and are able to deliver carboxypeptidase Y to the vacuole at 26°C (48). The non-acylated Vps15p^{Ala2} mutant is less efficiently phosphorylated in vivo than its wild-type counterpart. It is unclear whether Vps15p is responsible for its own phosphorylation and, if so, whether C14:0 influences its kinase activity. Loss of the myristoyl group does not affect the protein's ability to associate with cellular membranes (48). Deleting 30–214 residues from the wild-type protein's carboxy terminus markedly decreases its degree of phosphorylation and results in mild sorting defects at 26°C, which worsen at 38°C but do not produce growth arrest. Double mutants with carboxy-terminal truncations and Gly2→Ala substitutions produce severe temperature-sensitive growth defects and mis-sorting via an apparent default transit pathway to the cell surface (48). This finding suggests that myristoylation and phosphorylation may together affect the function of Vps15p.

Arf1p and Arf2p have 96% sequence identity (51). Arf1p is the dominant Arf species in *S. cerevisiae*—accounting for ~90% of total cellular Arfs. Disruption of *ARF1* produces slow growth, cold sensitivity, and sensitivity to fluoride ion (51). Disruption of *ARF2* has no phenotypic effect. Overexpression of wild type Arf2p but not Arf2p$^{Gly2 \rightarrow Ala}$ can suppress the effects of *ARF1* disruption. Strains with disruption of *ARF1* and of *ARF2* are not viable and cannot be rescued by overexpression of Arf1p$^{Gly2 \rightarrow Ala}$ or of Arf2p$^{Gly2 \rightarrow Ala}$ (51; R Kahn, personal communication). Current models suggest that phospholipids are required to induce a conformational change in *S. cerevisiae* Arfs that affects their amino terminus so that GDP dissociation is favored. Release of GDP does not require myristate. However, once GDP is released and the protein binds GTP, its amino terminus is locked in an exposed conformation that allows it to interact with cell membranes. This latter interaction is dependent upon a covalently bound myristoyl group (R Kahn, personal communication; and see below).

Comparative sequence analyses predict that Ppz1p and Ppz2p are type-1 protein phosphatases (53). They were identified based on their ability to suppress cell lysis produced by a null allele of the protein kinase C gene (*PKC*) (57). Deletion of either *PPZ1* or *PPZ2* in strains with wild-type *PKC* has no phenotypic effect at 24 or 37°C (53). However, strains with deletions of both *PPZ1* and *PPZ2* fail to grow at 37°C due to cell lysis (53).

NMT1 *Is Essential for Vegetative Growth*

S. cerevisiae Nmt1p is a 455-residue monomeric protein (58, 59). The purified protein lacks methionylaminopeptidase activity (58), indicating that the initiator Met residues of nascent N-myristoylproteins must be removed by cellular methionylaminopeptidase(s) so that their Gly^2 residue can be exposed. The *NMT1* gene[1] is located on chromosome XII (59). Hybridization

[1]Note that *nmt1* is also a name given to a *Schizosaccharomyces pombe* gene that is completely repressed by thiamine (no message in thiamine) (61, 62). *nmt1* encodes a 346-residue protein with no homology to *S. cerevisiae* Nmt1p.

studies, using a yeast cosmid clone grid that includes overlapping fragments of >95% of the yeast genome, have revealed that the methionylamino-peptidase-1 gene (*MAP1*) is located ~80 kb downstream from *NMT1* (F Li, R Johnson, J Gordon, unpublished observations). Another gene is present <1 kb upstream of *NMT1*. It encodes a 576-residue protein, Pwp1p, which belongs to the β-transducin superfamily (60). *PWP1* is divergently transcribed from *NMT1,* and their initiator Met codons are only separated by 664 nucleotides. Strains containing *pwp1* null alleles are viable, but growth is severely retarded and the steady-state levels of at least two N-myristoyl-proteins are markedly reduced (60).

Given that several yeast N-myristoylproteins are essential for vegetative growth and that some of these proteins are entirely dependent upon myristate for full expression of their biological activity, it is not surprising that insertional mutagenesis or deletion of *NMT1* causes recessive lethality (42, 59). Spores lacking a functional *NMT1* locus can germinate, but stop dividing after passing through the cell cycle 1–4 times. Growth arrest occurs at different stages of the cell cycle (59). This ability to undergo 1–4 cell divisions prior to growth arrest likely reflects spore inheritance of maternal Nmt1p and/or essential N-myristoylproteins, which are subsequently diluted or degraded so that they fall below a critical concentration. This notion is supported by studies of the diploid *S. cerevisiae* strain SK-1, a homothallic prototroph that can be induced to sporulate synchronously (63). The steady-state levels of Nmt1p mRNA rise abruptly during meiosis at the time of spore formation (60). (Hybridization studies of RNA prepared from mitotically synchronized cell cultures have also shown that Nmt1p mRNA levels do not change appreciably during the vegetative cell cycle; Ref 60.)

The *nmt1Δ* recessive lethal allele can be fully complemented with an episome containing the open reading frame of *NMT1* plus 185 nucleotides of sequence 5′ of its initiator Met codon (59). All attempts to isolate extragenic suppressors *nmt1* null alleles have been unsuccessful (see Ref. 42). This failure strongly suggests that Nmt activity is uniquely derived from *NMT1* and that the enzyme activity cannot be replaced by alteration of any other single genetic locus.

The Kinetic Mechanism and Substrate Specificities of Nmt1p

A first step in understanding how protein N-myristoylation is regulated in *S. cerevisiae* was to purify Nmt1p and characterize the enzyme's kinetic mechanism and substrate specificities. These studies used a discontinuous in vitro assay system: (*a*) myristoylCoA is produced from [^{3}H]myristate and CoA using the relatively nonspecific *Pseudomonas* acylCoA synthetase (64, 65); (*b*) a source of Nmt activity is then added to the reaction mixture together with an octapeptide derived from residues 2–9 of a known N-myristoylprotein; and (*c*) the resulting labeled myristoylpeptide is recovered and quantitated by reverse-phase high-performance liquid chromatography (HPLC) (40, 66). Variations of the assay include the use of labeled CoA or labeled peptide (when unlabeled fatty acid is being tested) and various methods for separating the acylpeptide from other components in the reaction mixture (e.g. see Refs. 67–69). A readily available source of *S. cerevisiae* Nmt1p was identified: Nmt1p can be efficiently synthesized in *Escherichia coli,* a bacterium with no endogenous Nmt activity. Production of Nmt1p does not have any apparent effect on bacterial growth at 24–37°C (70). Moreover, the recombinant enzyme can be rapidly purified to apparent homogeneity using P11 phosphocellulose and MonoS chromatography (71, 66). The activities of the authentic *S. cerevisiae*–derived enzyme (58) and *E. coli*–derived Nmt1p are indistinguishable (70, 72).

Nmt1p is a monomeric enzyme at concentrations up to 20 μM. It has a pH optimum of 7.5–8.0 and is inactive below pH 6 (58, 73). The enzyme does not appear to require any cofactors to sustain activity in vitro (7).

A large body of evidence indicates that Nmt1p has a sequential ordered mechanism. The apo-enzyme (E) first forms a high-affinity myristoyl-CoA:Nmt binary complex. Peptide then binds to form a ternary complex. This is followed by catalytic transfer of myristate from CoA to peptide, release of CoA, and subsequently release of myristoylpeptide:

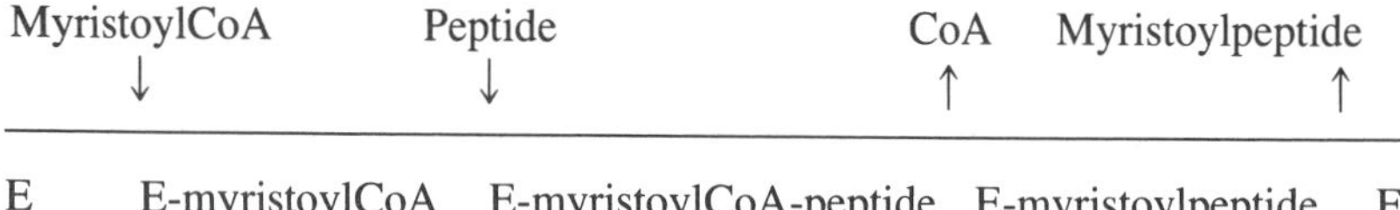

The ordered Bi Bi reaction mechanism was deduced from kinetic studies of product inhibition (CoA is a noncompetitive inhibitor vs myristoylCoA and a mixed inhibitor vs peptide, while myristoylpeptide is competitive vs myristoylCoA), and kinetic studies of dead-end inhibitors [the nonhydrolyzable derivative of myristoylCoA, S-(2-oxo)pentadecylCoA (74), is competitive vs myristoylCoA, while an octapeptide derivative of the N-terminal sequence of Arf1p, ALYASKLS-NH2, is competitive vs peptide and uncompetitive vs myristoylCoA (75, 76)]. In addition, binding assays using isoelectric focusing and native polyacrylamide gel electrophoresis, fluorescence spectroscopy, and isothermal titration calorimetry have shown that Nmt1p is not able to form a detectable complex with peptide in the absence of bound acylCoA (71, 75, 77). A similar conclusion was made after crosslinking studies with photoactivatable octapeptide substrates (76).

These findings indicate that binding of acylCoA to apo-Nmt1p is required for the formation of a functionally competent peptide-binding site. This heterotropic cooperativity is also evident when the physicochemical properties of the acyl chain of acylCoAs are varied. In vitro kinetic studies have shown that adding or subtracting one methylene from the acyl chain of myristoylCoA has little effect on binding affinity or catalytic efficiency: The activities of tridecanoylCoA and pentadecanoylCoA are similar to those of tetradecanoylCoA (78). However, C13:0 and C15:0 are either present in trace concentrations or are undetectable in vivo (79), eliminating any selective pressure to engineer the acylCoA-binding site of Nmt1p so that it has to discriminate against these ligands. PalmitoylCoA binds to Nmt1p with an affinity (K_m) similar to that of myristoylCoA, but it has a greatly reduced V_{max} (78). Isothermal titration calorimetric analysis of the binding of myristoylCoA to apo-Nmt1p (77) reveal a 1:1 stoichiometry and a K_d of 15 nM—corresponding to a binding free energy (ΔG) of -10.9 kcal/mol (Table 1). This free energy is composed of a very large enthalpic term (ΔH), favoring binding by -24 kcal/mol, and a large unfavorable entropic term ($T\Delta S = -13.4$ kcal/mol; see Table 1). Although there is little variation in binding affinity between myristoylCoA and palmitoylCoA, there is a large difference in the enthalpic and entropic components: Addition of two methylenes to the acyl chain of myristoylCoA reduces the favorable enthalpy of binding by approximately 9 kcal/mol with a nearly equal compensating change in the enthalpic contribution to binding free energy (Table 1). This finding indicates palmitoylCoA:Nmt1p binary complexes exist in a broader, shallower potential energy well than do myristoylCoA:Nmt1p complexes.

The large difference in the energetics of ligation of myristoylCoA and palmitoylCoA likely reflects a significant difference in the conformation of the corresponding binary complexes. One possibility is that when Nmt1p binds palmitoylCoA, it does not undergo the putative conformational change necessary to generate a functionally competent peptide-binding site (as mandated by the sequential ordered reaction mechanism). However, this does not appear to be the case, based on the results of isothermal titration calorimetric studies with nonhydrolyzable derivatives of myristoylCoA and palmitoylCoA: i.e. S-(2-oxo)pentadecylCoA and S-(2-oxo)heptadecylCoA. Although these analogs are one carbon atom longer than the hydrolyzable compounds, they have the same chain length when measured from the carbonyl carbon to the ω-terminal methyl of the acyl chain (74; see Figure 1). The binding of the analogs to apoNmt1p is energetically similar to the binding of the corresponding hydrolyzable forms to apoNmt1p (Table 1). Moreover, the analogs allow generation of stable Nmt1p:alkyl(thioether) CoA complexes, which are not catalytically productive. This makes the

Table 1 Thermodynamic parameters for Nmt1p binary and ternary complex formation[a]

Binding reaction	K_d (kcal/mol)	ΔG (kcal/mol)	ΔH (kcal/mol)	$T\Delta S$
MyristoylCoA·Nmt1p binary complex formation	15 ± 16 nM	−10.9 ± 0.5	−24.4 ± 1.7	−13.5 ± 2.0
PalmitoylCoA·Nmt1p binary complex formation	37 ± 25 nM	−10.3 ± 0.4	−15.8 ± 1.4	−5.5 ± 1.6
S-(2-oxo)pentadecylCoA (nonhydrolyzable (nh) myristoylCoA)·Nmt1p binary complex formation	4.9 ± 2.2 nM	−11.5 ± 0.3	−25.3 ± 1.0	−13.9 ± 1.1
S-(2-oxo)heptadecylCoA (nonhydrolyzable palmitoylCoA)·Nmt1p binary complex formation	45 ± 12 nM	−10.1 ± 0.2	−13.2 ± 0.06	−3.1 ± 0.1
GAAPSKIV-NH_2·Nmt1p·nhMyrCoA ternary complex formation	3.6 ± 0.3 μM	−7.47 ± 0.05	−6.54 ± 0.4	0.92 ± 0.4
GAAPSKIV-NH_2·Nmt1p·nhPalmCoA ternary complex formation	0.67 ± .06 μM	−8.47 ± 0.05	−8.77 ± 0.5	−0.30 ± 0.46

Chain length dependence of the energetics of ligand binding to apo-Nmt1p

	$\Delta\Delta G$	$\Delta\Delta H$ (kcal/mol)	$T\Delta\Delta S$
AcylCoAs			
MyristoylCoA	0	0	0
PalmitoylCoA	0.6	8.5	7.9
ThioetherCoAs			
S-(2-oxo)pentadecylCoA	−0.6	−1.0	−0.5
S-(2-oxo)heptadecylCoA	0.8	11.1	10.3

[a] Values are at 300 K.

analogs very useful for characterizing the binding thermodynamics of peptide ligands (77). Table 1 demonstrates that an octapeptide derived from the amino-terminal sequence of Cnb1p (GAAPSKIV-NH_2) binds to a complex of Nmt1p with S-(2-oxo)*hepta*decylCoA with higher affinity than it does to a complex formed with the nonhydrolyzable myristoylCoA analog [S-(2-oxo)*penta*decyl CoA]. The altered energetics of binding the off-length acylCoA allows generation of a high-affinity peptide-binding pocket but a less productive active site, as indicated by the reduced catalytic efficiency of palmitoylCoA compared to myristoylCoA. This situation could arise from small differences in the backbone conformation of the protein and acylCoA and significant differences in the specific contacts between protein and ligand, resulting in a failure to position catalytically important functional groups properly. Thus, one model for explaining Nmt1p's observed chain length specificity is that binding of myristoylCoA but not palmitoylCoA allows formation of a precise set of "tight" electrostatic contacts involving CoA and the enzyme. The result is a much greater immobilization of the ligand and the interacting protein residues, giving rise to the observed favorable enthalpy and unfavorable entropy while making critical intermolecular contacts that provide catalytic destabilization. Failure of the CoA moiety of palmitoylCoA to form the necessary contacts with the enzyme could result in either unpaired polar functional groups in the CoA or formation of intramolecular contacts in the acylCoA ligand (77). (The less favorable binding of the Cnb1p peptide substrate to the Nmt1p:S-(2-oxo)pentadecylCoA complex may represent binding into a strained conformation, indicating catalytic destabilization.)

This model for chain length discrimination by Nmt1p, in which "off-length" acylCoAs are able to bind to the enzyme with high affinity, but fail to make the necessary contacts to facilitate catalysis, is consistent with the known structures of protein-CoA complexes. Five such structures have been reported (see Ref. 77 for summary, and Refs. 79a–h). Each structure shows a large number of specific contacts between CoA and protein functional groups, making use of the CoA molecule's great potential for binding energy. The binding of CoA shares several common features among these proteins, including a similar bent conformation, certain hydrogen bonds involving the adenine moiety of CoA, and hydrophobic and pi stacking interactions involving the adenine and the pantetheine and cysteamine amide bonds. However, there are also significant differences in how the five proteins bind CoA, and no conserved pattern of protein residues interacting with CoA has been found. Thus, these proteins have evolved overlapping yet distinct ways of forming low-energy complexes with CoA.

Fierke & Jencks (79i) have suggested that CoA can be considered as two

S-(2-oxo)pentadecylCoA

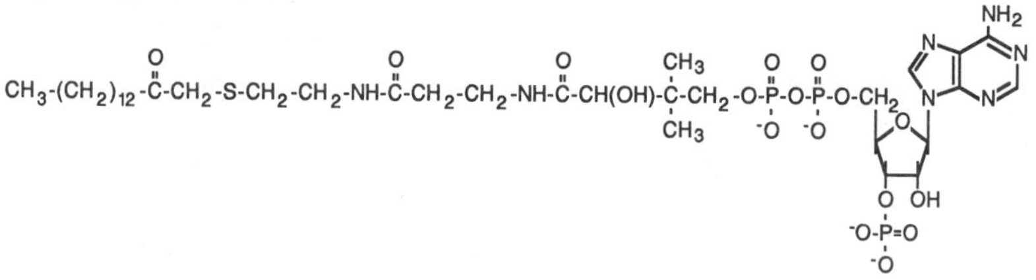

CH_3-$(CH_2)_{12}$-C(=O)-CH_2-S-CH_2-CH_2-NH-C(=O)-CH_2-CH_2-NH-C(=O)-CH(OH)-C(CH_3)(CH_3)-CH_2-O-P(=O)($^-$O)-O-P(=O)($^-$O)-O-CH_2 (adenosine; 3′-O-P(=O)($^-$O)$_2$, 2′-OH; NH_2)

MyristoylCoA

CH_3-$(CH_2)_{12}$-C(=O)-S-CH_2-CH_2-NH-C(=O)-CH_2-CH_2-NH-C(=O)-CH(OH)-C(CH_3)(CH_3)-CH_2-O-P(=O)($^-$O)-O-P(=O)($^-$O)-O-CH_2 (adenosine; 3′-O-P(=O)($^-$O)$_2$, 2′-OH; NH_2)

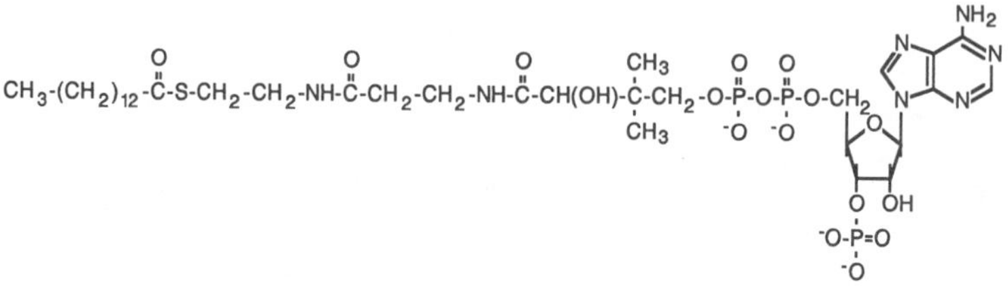

Figure 1 Structures of myristoylCoA and S-(2-oxo)-pentadecanoylCoA. Although the non-hydrolyzable analog of myristoylCoA is one carbon atom longer than the hydrolyzable compounds, it has the same chain length when measured from the carbonyl carbon to the ω-terminal methyl of the acyl chain.

functional domains: the pantetheine-cysteamine arm, which makes contacts responsible for transition-state stabilization, and the adenosine-3′-monophosphate-5′-diphosphate moiety, which provides the binding energy necessary to position the molecule. Computer modelling of the catalytic subunit of the dihydrolipoyl transacetylate (Ep2, Ref. 79f) and chloramphenicol acetyltransferase type III (79j) support this theory. In both proteins, a residue critical to stabilization of a putative transition state is positioned by a hydrogen bond with the cysteamine nitrogen of the bound CoA. The alternative conformation of E2p in which the position of the adenosine head is unchanged but the cysteamine arm is no longer in the vicinity of the active site (79f) provides additional support. The formulation of Fierke & Jencks is consistent with a model for the chain-length specificity of Nmt1p in which acylCoAs outside the catalytic specificity of the enzyme are able to bind with high affinity due to the anchoring of the CoA adenosine triphosphate moiety, but the pantetheine arm is displaced from its catalytically relevant position by the improper length of the acyl chain. The formation of a high-affinity peptide-binding site in Nmt1p when palmitoyl-CoA binds indicates that the binding site of this "off-length" acylCoA is sufficient to induce the cooperative transition to allow peptide binding, but not to generate a highly efficient active site.

The kinetic and thermodynamic analyses described above define an apparent dilemma that Nmt1p may face in vivo, i.e. how does the enzyme avoid inhibition through formation of catalytically less productive complexes with palmitoylCoA? In *S. cerevisiae* as well as in several higher eukaryotic cell lineages where acylCoA levels have been measured, it appears that palmitoylCoA is 5–20-fold more abundant than myristoylCoA (46, 80–82). Moreover, purified Nmt1p does not possess a detectable, intrinsic, peptide-independent thioesterase activity that can serve as an editing function to remove bound palmitoylCoA (77). These considerations suggest that there must be functional segregation of cellular myristoylCoA and palmitoylCoA pools so that Nmt1p does not acquire palmitoylCoA. The work of Peitzsch & McLaughlin (83) provides insights about how this compartmentalization may be achieved in vivo. They measured the free energy of association of myristoylCoA and acylpeptides with model membrane systems. The free energy of association of acylpeptides with uncharged membranes depends only on the length of the acyl chain. Each methylene contributes 0.8 kcal/mol. Affinity is independent of the chemical nature of the polar head group of the amphipathic molecule. These findings predict that myristoylCoA associates with membranes with free energy 1.6 kcal/mol less favorable than that of palmitoylCoA. This is equivalent to a $\sim$15-fold greater partitioning of palmitoylCoA into membranes at 300 K compared to myristoyl-CoA (cf Ref. 84).

More than 250 fatty acid analogs of myristate and palmitate have been utilized to determine the structural features in myristoylCoA that are recognized by Nmt1p's acylCoA-binding site (64, 65, 85–91). Surveys of the CoA derivatives of C12-C18 fatty acids containing triple bonds, *cis* and *trans* double bonds, *para*-phenylene, or a 2,5 furyl group have suggested that Nmt1p's acylCoA-binding site requires the acyl chain of active substrates to adopt a bent conformation, with the principal bend occurring in the vicinity of C5 (64, 78). For example, kinetic studies of 12 tetradecynoylCoAs (Y-3–Y-13) have demonstrated that all are active except the analog with a triple bond between C5 and C6 (Y-5-tetradecynoylCoA), which is not bound (64). Similarly, placement of a *trans* double bond between C5 and C6 of myristoylCoA (yielding E-5-tetradecenoylCoA) results in marked reductions in activity, while E-6- and E-7-tetradecenoylCoAs are as active as tetradecanoylCoA (64). The distance between carboxylate (C1) and this bend at C5-C6 appears to be a very important determinant for optimal positioning of acylCoA substrates so that peptides can subsequently bind to the binary acylCoA:Nmt1p complex. PalmitoylCoA can be "converted" to a more active Nmt1p substrate in vitro by introducing a *cis* double bond between C5 and C6 (yielding Z-5-hexadecenoylCoA) or a triple bond between C4 and C5, or C6 and C7 (Y-4- and Y-6-hexadecynoylCoAs). [The activities of these analogCoAs approach that of myristoylCoA (78), affording an opportunity to use them to explore the question of whether N-myristoylproteins can retain their biological function(s) with a covalently bound C16 fatty acid (see below).]

Studies of other members of this analog panel (e.g. ω-phenyl-substituted fatty acids with chain lengths equivalent to C11–C15, and azidophenylalkyl acids of varying chain length with the linear azide unit attached either *meta* or *para* to phenyl) have indicated that Nmt1p's myristoylCoA-binding site contains a conical receptor that interacts with the ω-terminus of the bound acyl chain. The acuteness of the cone determines the enzyme's sensitivity to the distance between the bend at C5-C6 and the ω-terminus of an acyl chain, and the enzyme's sensitivity to steric bulk at the ω-terminus (64, 91).

The peptide substrate specificity of Nmt1p has been examined using the discontinuous in vitro assay together with >100 synthetic peptides (40–58, 73, 92–94; summarized in Ref. 7). Nmt1p appears to be able to sense the physicochemical properties of the amino acids occupying at least the first eight residues of nascent polypeptides (58, 94–96). *S. cerevisiae* Nmt1p has an absolute requirement for an amino-terminal Gly (40, 73, 93). Its small side chain and primary amine are critical for ligand binding. For example, β-alanine substitution at the amino terminus of substrate peptides blocks binding to the enzyme (58). Nmt1p is not able to attach myristate to the ε amino group of Lys (cf Ref. 97) or to form ester linkages with Ser or Cys (cf Ref. 98). Uncharged residues are generally preferred at position 2 (e.g. Cys, Ala, Leu), while Nmt1p can accommodate a spectrum of amino acids at positions 3 and 4 (neutral preferred over basic which, in turn, are preferred over acidic residues). A Ser is present at position 5 in all known yeast N-myristoylproteins (Table 2). Substitution of this Ser with an Ala (i.e. replacement of hydroxyl with a proton) decreases affinity (K_m) for Nmt1p by 2–3 orders of magnitude in several sequence contexts. Threonine, a β-methyl-substituted "analog" of serine, is not able to promote high-affinity binding (58, 92). Thr, Lys, and Leu are found at position 6 of the seven known myristoylproteins described in Table 2. Nmt1p prefers basic amino acids at positions 7 and 8 over neutral and acidic residues.

Based on these observations, the sequence MGXXXSXX can be used to search current databases that encompass the protein products of ~40% of the yeast genome and recover all known *S. cerevisiae* N-myristoylproteins. Several potential substrates for Nmt1p can also be identified from such a search (Table 2) and evaluated in at least two ways: (*a*) by characterizing the kinetic properties of octapeptides representing their amino-terminal sequence in the in vitro Nmt enzyme assay or (*b*) by sequentially expressing Nmt1p and then the candidate protein in *E. coli,* a bacterium which, as noted above, lacks endogenous Nmt activity. Two plasmids are used in this

Table 2 Known and putative *S. cerevisiae* N-myristoylproteins

	Amino-terminal sequence	Ref.
Known[a]		
Arf1p	GLFASKLFSNLFGNKEMRILMVGL	49
Arf2p	GLYASKLFSNLFGNKEMRILMVGL	51
Cnb1p	FAAPSKIVDGLLEDTNFDRDEIERL	52
Gpa1p	GCTVSTQTIGDESDPFLQNKRAND	43, 45
Vps15p	GAQLSLVVQASPSIAIFSYIDVLEE	47, 48
Ppz1p	GNSSSKSSKKDSHSNSSSRNPRP	53
Ppz2p	GNSGSKQHTKHNSKKDDHDGDRK	53
Putative[b]		
Gpa2p	GLCASSEKNGSTPDTQTASAGSD (nonessential G protein α subunit homolog)	99
Pho4p	GRTTSEGIHGFVDDLEPKSSILDKV (positive regulator of phosphatase gene expression)	100, 101
Ef1αp	GKEKSHINVVVIGHVDSDKSTTT (polypeptide chain elongation factor)	102, 103
Sip2p	GTTTSHPAQKKQTTKKCRAPIMS (protein identified by two-hybrid system as capable of interacting with Snf1p, a kinase required for release of genes from glucose repression; cf Ref. 104)	X Yang and M Carlson, personal communication

[a] Based on the results of metabolic labeling studies of *S. cerevisiae* with tritiated myristate, expression of the protein with Nmt1p in the *E. coli* coexpression system, and/or analysis of the kinetic properties of octapeptides, representing the amino-terminal sequence of these proteins, in the in vitro Nmt assay system.

[b] Based on the results of a search of a yeast protein database conducted by M. Goebl (University of Indiana) using MGXXXSXX as a query sequence. These sequences contain residues at position 1–8 which, based on an analysis of >100 synthetic peptides, should be tolerated by Nmt1p (see text for further discussion).

bacterial coexpression system, each with a different inducible promoter, a different antibiotic resistance gene, and a different but compatible origin of replication (70, 72). [^{3}H]Myristate is added to the media at the time of initiation of production of the potential substrate protein. Myristate can permeate across the bacterial inner membrane by simple diffusion (105) and is subsequently activated by FadD, a 580-residue acylCoA synthetase (106) loosely associated with the cytoplasmic side of the inner membrane (107). Nmt1p-dependent transfer of [^{3}H]myristate from CoA to the amino-terminal Gly residue of the protein is then evaluated by examining autoradiographs of SDS-polyacrylamide gels containing total bacterial lysates. These lysates are prepared from a series of strains that contain one of the following: (*a*) plasmids encoding Nmt1p and the protein of interest, (*b*) a single plasmid encoding Nmt1p or a single plasmid specifying the protein of interest (negative controls), or (*c*) two plasmids, one specifying Nmt1p and the other a mutant derivative of the protein of interest with a Gly2→Ala substitution (another negative control). Studies of strains that contain a mutation in *fadE* (108) that inhibits β-oxidation of fatty acids have confirmed that Nmt1p retains its specificity for myristoylCoA over palmitoylCoA in *E. coli* (105).[2]

Determining the Minimal Catalytic Domain of Nmt1p

The dual plasmid expression system can be used not only to assess whether a protein is a substrate for Nmt1p, but also to define the effects of mutating Nmt1p on its ability to acylate a known substrate. Deletion of the amino-terminal 35 or 59 residues of Nmt1p (i.e. Met1→Ala34 or Met1→Ile59) has no apparent effect on its activity in the dual plasmid expression system at 24°C or 37°C or in the discontinuous in vitro enzyme assay (109). nmtΔ35p and nmtΔ59p can also rescue the lethal phenotype of a *nmt1* null allele. Haploid strains of *S. cerevisiae* containing a chromosomal *nmt1Δ* allele and episomes encoding either of the mutant enzymes grow on synthetic or rich (yeast/peptone/dextrose, YPD) media at a rate comparable to that of an isogenic strain that produces similar amounts of wild-type Nmt1p. Moreover, metabolic labeling studies indicate that the pattern of incorporation of labeled myristate into cellular proteins is similar in the three isogenic *NMT1*, *nmt1-Δ35*, and *nmt1-Δ59* strains (109). In contrast, removal of Met1→Phe96 or the carboxy-terminal five residues (Gly451→Leu455) destroys activity in all three functional assay systems, suggesting that the minimal catalytic domain of Nmt1p is located between Ile59→Phe96 and Gly451→Leu455 (109).

The rate of growth of a diploid *NMT1/nmt1Δ* strain of *S. cerevisiae* is not affected when Nmt1p, nmtΔ35p, nmtΔ59p, nmtΔ96p, or nmt-stop451p is overexpressed using an episome containing the mutant allele under the control of the galactose-inducible *GAL1-10* promoter (109). This finding suggests that they do not function as dominant negative mutants—i.e. the truncated nmts do not form complexes with the (monomeric) wild-type enzyme that affect its activity in vivo (nor for that matter are they apparently able to bind a functionally significant fraction of cellular myristoylCoA pools or compete for Nmt1p's nascent polypeptide substrates).

[2]Isogenic strains with inactivating mutations of *fadD* do not incorporate exogenous tritiated myristate into known substrates such as Arf1p, indicating that myristoyl-acyl carrier protein (myristoyl-ACP) produced by the bacterial inner-membrane acyl-ACP synthetase cannot be used by Nmt1p (105).

Mutant Alleles of NMT1 Are Powerful Tools for Examining Regulation of Protein N-Myristoylation

Analyses of *S. cerevisiae* strains containing mutant *nmt1* alleles have provided important insights, about not only structure/activity relationships in Nmt1p but also myristoylCoA metabolism—i.e. what are the sources of cellular myristoylCoA pools and what determines the ability of Nmt1p to gain access to them?

STUDIES OF *nmt1-Δ59* Subcellular fractionation studies as well as electron microscopic immunocytochemical analysis indicate that Nmt1p is a cytosolic enzyme (110). The components of *S. cerevisiae*'s fatty acid biosynthetic pathway that produce long-chain acylCoAs from acetate are also located in the cytoplasm. AcetylCoA carboxylase (Acc1p, also known as Fas3p; E.C. 6.4.1.2) is an essential cytosolic enzyme that catalyzes carboxylation of acetylCoA, yielding malonylCoA (111, 112). Fatty acid synthetase (EC 2.3.1.85) is a large, cytosolic 2.4×10^6 kDa complex of six trifunctional α subunits and six pentafunctional β subunits encoded by the *FAS2* and *FAS1* genes, respectively (113–116). *S. cerevisiae* Fas produces long-chain acylCoAs from acetylCoA and malonylCoA in the presence of NADPH. PalmitoylCoA and stearoylCoA are its principal products. MyristoylCoA represents 3–5% of the total acylCoAs produced by this de novo pathway (117, 118). Stearoyl-CoA production increases relative to palmitoylCoA as the yeast is grown at higher temperatures (e.g. 35–37°C). This effect appears to be mediated by temperature-dependent changes in the activity of Acc1p: At higher temperatures, Acc1p activity is greater, malonylCoA concentrations are higher, and Fas production of longer-chain acylCoAs is favored (119). However, myristoylCoA production is not affected by these temperature changes (119, 120). The de novo pathway for acylCoA biosynthesis can be blocked by cerulenin ((2*R*,3*S*)-2,3-epoxy-4-oxo-7,10-*trans,trans*-dodecanienamide). This compound, which is obtained from the fungus *Cephalosporium caerulens,* covalently modifies the SH group of a Cys present in the β-ketoacyl-(ACP)synthase activity present in Fas′ α subunits (121). Cerulenin (CER) does not appear to inhibit fatty acid elongation or desaturation systems present in *S. cerevisiae* (79).

Isogenic haploid strains of *S. cerevisiae* that produce comparable levels of Nmt1p or nmtΔ35p are unable to grow at 24 or 37°C in YPD media containing 25 μM cerulenin (CER) unless the media is supplemented with myristate (MYR).[3] The exogenous myristate presumably rescues growth

[3]The growth inhibition of wild-type strains of *S. cerevisiae* produced by CER can be readily reversed by addition of the saturated fatty acids C14:0, C15:0, or C16:0. Awaya et al (79) noted that when pentadecanoic acid is added to rescue growth, more than 90% of the native even-numbered cellular fatty acids are substituted by odd-numbered fatty acids.

because it can be taken up by *S. cerevisiae* and activated by cellular acylCoA synthetases—thereby maintaining cellular acylCoA pools that would otherwise be depleted by CER-mediated inactivation of the de novo pathway. The viability of Nmt1p- and Δnmt35p-producing cells on YPD/MYR/CER suggests that Nmt1p and nmtΔ35p are able to gain access to and effectively utilize myristoylCoA produced by activation of exogenously derived C14:0. In contrast, a strain that only produces nmtΔ59p cannot grow in YPD/CER/-MYR at any of these temperatures (109). This phenomenon cannot be ascribed to any apparent differences in the stability of the three Nmts, to any apparent differences in their specific activities in vitro or in the *E. coli* coexpression system (see above), or to any appreciable differences in their substrate specificities (as judged by the similarities in the pattern of metabolic labeling of cellular N-myristoylproteins when the haploid strains are grown in YPD alone). Thus, it appears that Met1→Ile59 of Nmt1p may function as a noncatalytic targeting signal that allows the acyltransferase to gain access to and effectively utilize myristoylCoA produced by cellular long-chain acylCoA synthetases and/or deposited in cellular membranes. These experiments do not resolve the question as to whether myristoylCoA derived from the Fas pathway and myristoylCoA produced by acylCoA synthetases are located in similar or distinct compartments in the cell or how myristoyl-CoA is "delivered" to Nmt1p. For example, are specific acylCoA carrier proteins involved? Does Nmt interact directly with one or more cellular acylCoA synthetases or with Fas itself? Some of these questions can be addressed in strains containing conditional lethal *nmt1* alleles (see below) and/or through use of the two-hybrid system, which provides a genetic approach for detecting protein-protein interactions in vivo (122, 123).[4]

STUDIES OF *nmt1-181*

A Gly451→Asp substitution in Nmt1p produces marked temperature-dependent reductions in its affinity for myristoylCoA and global defects in cellular protein N-myristoylation More than 20 years ago, mutations were identified in *S. cerevisiae* that produce temperature-sensitive growth arrest, which can be relieved by addition of C14:0, C16:0, or C18:0 to the media (117, 124, 125). Most of the mutations responsible for producing this saturated fatty acid auxotrophy were mapped to *FAS1* or *FAS2*. Meyer & Schweizer (126) characterized a strain (LK181) with a mutation that did not map to either

[4]The two-hybrid system takes advantage of the distinct DNA-binding and activation domains (GBD and GAD) of Gal4p, which are required for its activation of the *GAL1-10* promoter. Transcription of a *GAL1-10-REPORTER* gene will occur even if the GBD and GAD domains are contained on separate polypeptides as long as they are able to interact with one another: i.e. GBD-Nmt1p could activate *GAL1-10-REPORTER* if it interacts with a GAD-containing fusion protein that possesses elements that bind to Nmt1p.

FAS1 or *FAS2*. This recessive mutation results in specific auxotrophy for myristate at 37°C. Cells can only grow at the restrictive temperature if 500 μM myristate is added to the medium: C12:0, C15:0, and C16:0 cannot replace C14:0. Assays of Fas activity in vitro and in vivo revealed that there were no impairments in de novo fatty acid biosynthesis in this strain (126). Meyer & Schweizer (126) predicted that "exogenously supplied fatty acids act as allosteric effectors for a mutationally altered cellular protein to restore its biological function at elevated temperatures." This prediction was corroborated 17 years later. The phenotype in LK181 cells arises from a single G-to-A transition that changes codon 451 of *NMT1* from Gly (GGT) to Asp (GAT) (42). Overexpression of nmt181p rescues the temperature-sensitive growth arrest and myristic acid auxotrophy produced by *nmt1-181*. Comparison of the patterns of incorporation of [^{3}H]myristate into cellular proteins synthesized by isogenic haploid *NMT1* and *nmt1-181* strains suggests that the efficiency of acylation of several N-myristoylproteins is reduced in *nmt1-181* cells after a 1–4 h incubation at the nonpermissive temperature (42). Defects in protein N-myristoylation are also apparent at the permissive temperature. These reductions in protein N-myristoylation are not due to degradation of a thermolabile enzyme: The steady-state levels of Nmt1p and nmt181p are equivalent in cells grown at 24°C and in cells incubated at 37°C for 1–4 h.

The notion that several essential cellular N-myristoylproteins are undermyristoylated by nmt181p is supported by several other observations. Overexpression of Arf1p fails not only to rescue growth, but also to worsen the phenotype of *nmt1-181* strains at 24, 30, or 37°C (46). The latter finding is significant, because overexpression of Arf1p in *NMT1* strains adversely affects vegetative growth (46, 50, 51). Overexpressing Gpa1p alone does not rescue *nmt1-181* strains. Epistasis tests indicate that blocking induction of the mating pathway by deletion of *STE4* (the nonessential gene encoding the β subunit of the heterotrimeric G protein involved in the mating cascade) does not restore growth of *nmt1-181* cells. Such a restoration would be expected if nmt181p produced its effects only by undermyristoylating Gpa1p (46). The time course to death of *nmt1-181* cells is long (up to 24 h in YPD), and likely reflects the cumulative effects of undermyristoylating a variety of proteins. Microscopic studies of *nmt1-181* cells recovered after a 1–2 h incubation at the nonpermissive temperature indicate that they exhibit growth arrest at all stages of the cell cycle (42), and that they are abnormally large with prominent vacuoles. The transcriptional activity of a chimeric gene containing the *NMT1* promoter linked to a reporter remains constant for at least 1 h after shifting to the nonpermissive temperature of 37°C, yet cell division ceases almost immediately (127). One interpretation of these results is that shortly after a shift to 37°C, cellular pools and/or levels of acylation of an N-myristoylprotein involved in cell-cycle regulation fall below some critical concentration. Prolonged incubation at 37°C eventually yields more global reductions in the level of acylation and/or concentrations of other essential cellular N-myristoylproteins, leading to the demise of the organism.

nmt181p has been expressed in *E. coli,* purified to apparent homogeneity, and had its kinetic properties analyzed in vitro. The wild-type and mutant enzymes have a similar K_m for myristoylCoA at 24°C. However, at 37°C the myristoylCoA K_m of nmt181p is 10-fold higher and the V_m/K_m 200-fold lower than for Nmt1p (42). This reduced affinity for myristoylCoA was independently confirmed by fluorescence spectroscopic studies. Temperature-dependent decreases in nmt181p's specific activity can be readily appreciated by coexpressing it with several known substrates of Nmt1p in *E. coli* (e.g. Gpa1p and Arf1p; Ref. 46).

The systematic variations in conformations, steric bulk, polarity, and other physicochemical properties incorporated into the panel of >250 fatty acid analogs described above can be used to probe the nature of the acylCoA-binding sites of mutant Nmts. Kinetic analyses of C14-C16 acylCoAs containing single triple bonds, *para*-phenylene, as well as *cis* and *trans* double bonds, indicate that the Gly451→Asp substitution in nmt181p does not grossly perturb the overall geometry of the enzyme's myristoylCoA-binding site (78).

The temperature-dependent reduction in nmt181p's affinity for myristoylCoA undoubtedly accounts for the fact that several essential cellular N-myristoylproteins are undermyristoylated when strains are grown at the nonpermissive temperature. A chimeric gene consisting of the 5′ nontranscribed domain of *NMT1* linked to a reporter is transcribed with the same efficiency in isogenic *NMT1* and *nmt1-181* strains during exponential growth on YPD at 24 and 30°C (or after a 1 h incubation at the nonpermissive temperature). This suggests that the level of protein N-myristoylation in yeast does not serve as a feedback mechanism to control the transcriptional activity of *NMT1* (127)—an adaptive response that, based on the overexpression experiments with *nmt1-181* described above, would help rescue the growth arrest produced by the mutant enzyme.

Analyses of strains containing nmt1-181 *provide insights about the contributions of fatty acid synthetase and acylCoA synthetases to cellular myristoylCoA metabolism* The temperature-dependent reduction in nmt181p's affinity for myristoylCoA makes the enzyme a useful reporter of cellular myristoylCoA pool size and the pathways that contribute to maintenance of these pools: If accessible cellular pools cannot be increased to a level that can overcome the enzyme's catalytic defects, then global defects in protein acylation will lead to growth arrest at the restrictive temperature.

Strains containing either *NMT1* or *nmt1-181* require de novo fatty acid synthesis. They can only grow on YPD/CER media at 24 and 37°C if myristate is added to the media (42). Unlike *NMT1* strains, the growth arrest produced by CER treatment of isogenic *nmt1-181* strains cannot be reversed by adding palmitate, suggesting that palmitate cannot augment cellular myristoylCoA pools to a level that supports growth. The failure to convert palmitate to myristoylCoA perhaps is not surprising, given the fact that yeast mitochondria do not contain enzymes that catalyze β-oxidation, whereas peroxisomes only acquire this metabolic capability when cells are grown in the presence of oleate as their sole carbon source (reviewed in Ref. 39).

Although studies of the growth of *S. cerevisiae* strains containing *NMT1* or *nmt1-181,* in the presence or absence of an active de novo fatty acid biosynthetic pathway, establish that Nmt uses myristoylCoA derived directly or indirectly from the cytoplasmic Fas complex, they raise another question: i.e. How much of the myristoylCoA used by Nmt1p or nmt181p is contributed by cellular acylCoA synthetases? The principal fatty acids of *S. cerevisiae* are palmitic acid, palmitoleic acid (Z-9-hexadecenoic acid, $C16{:}1^{\Delta 9}$), stearic acid (octadecanoic acid, C18:0), and oleic acid (Z-9-octadecenoic acid; $C18{:}1^{\Delta 9}$) (128). Myristic acid represents 2–3% of total cellular fatty acids. AcylCoA synthetases catalyze thioester bond formation between CoA and free fatty acids in a two-step reaction:

$$\text{Free fatty acid} + \text{ATP} \xrightarrow{Mg^{2+}} \text{acyl-AMP} + PP_i \qquad 1.$$

$$\text{Acyl-AMP} + \text{CoA} \rightarrow \text{acylCoA} + \text{AMP} \qquad 2.$$

The *S. cerevisiae* genome contains at least three genes encoding acylCoA synthetases. The *FAA1* (fatty acid activation-1) gene is located on yeast chromosome XV and encodes a 700-residue acylCoA synthetase (129). It was isolated by genetic complementation of a strain of yeast whose acetyl-CoA carboxylase activity cannot be repressed by treatment with long-chain fatty acids (such repression requires acylCoAs), and that is not able to use exogenous palmitate as the sole source of fatty acids when the de novo pathway for fatty acid synthesis is blocked (130, 131). *FAA2* encodes a 744-residue protein and is located on chromosome V. It was recovered using the polymerase chain reaction and oligonucleotide primers derived from conserved domains represented in Faa1p, FadD, and several mammalian acylCoA synthetases (132). *FAA3* specifies a 694-residue protein and was identified during an analysis of open reading frames contained in yeast chromosome IX (132). Alignments of the primary structures of Faa1p (129), Faa2p (132), Faa3p (132), human acylCoA synthetase (133), two rat long-chain acylCoA synthetases (134, 135), FadD (106), and alkK, a long-chain acylCoA synthetase produced by *Pseudomonas oleovorans* (136), indicate that Faa2p has a greater degree of sequence similarity to mammalian acylCoA synthetases than to either Faa1p or Faa3p (132, 137).

S. cerevisiae Faas have been expressed in, and purified from, *fadD*$^-$ *E. coli* strains (137). In vitro biochemical studies of C3:0-C24:0 fatty acids indicate that these enzymes have distinct chain-length specificities. Faa1p prefers saturated fatty acids containing 12–16 carbon atoms, with myristate and pentadecanoate having the greatest activities. C3:0-C:10 and C20:0-C24:0 are not substrates. Faa2p prefers saturated fatty acids with 9–12 carbons, although it tolerates C7 through C17 fatty acid substrates with no more than a twofold variation in activity. The acylCoA synthetase activity of Faa3p is 2–40-fold lower than that of Faa1p and Faa2p for C8:0-C18:0. However, it is the only one of the three Faas that can convert C24:0 (lignoceric acid) to its CoA derivative. Ole1p is a cis-Δ^9 desaturase, which accounts for all de novo unsaturated fatty acid production from saturated acylCoAs in *S. cerevisiae*. Palmitoleate (Z-9-hexadecenoate) and oleate (Z-9-octadecenoate) are the two most abundant fatty acids in wild-type yeast strains when they are grown at 15–30°C. Faa3p prefers C16 and C18 fatty acids with a *cis* double bond at C9-C10. Moreover, myristoleate is a more active substrate than myristate for Faa1p and Faa2p. Faa1p has a higher temperature optimum (30°C) than either Faa2p or Faa3p (25°C), although Faa2p appears to be more tolerant of changes in temperature than either of the other two acylCoA synthetases. Faa1p and Faa3p show only modest (less than twofold) changes in activity between pH 6.5 and 7.9. In contrast, Faa2p is very sensitive to changes in pH: Its optimum pH is 7.1. Activity is reduced 10-fold when the pH is raised to 7.9. The subcellular locations of Faa1p, Faa2p, and Faa3p have not been determined.

In vitro biochemical assays using purified *E. coli*–derived Faas indicate that the myristoylCoA synthetase activity of Faa1p is equal to that of Faa2p and ~70-fold greater than that of Faa3p. These relative myristoylCoA synthetase activities can be independently confirmed in a triple plasmid expression system, which allows production of Nmt1p, a substrate protein (e.g. Arf1p or Arf2p) and an Faa in an *fadD*$^-$ strain of *E. coli*. Both Faa1p and Faa2p can activate exogenous radiolabeled myristate to myristoylCoA, which can be subsequently used by Nmt1p to acylate either Arf protein. In contrast, Arf acylation is undetectable in the presence of Faa3p (137).

The relative contributions of Faa1p, Faa2p, and Faa3p to myristoylCoA metabolism in *S. cerevisiae* can be assessed by growing isogenic strains with *NMT1* or *nmt1-181* alleles and *faa1, faa2,* and/or *faa3* null alleles at 24, 30, and 37°C in YPD, YPD supplemented with myristate or palmitate, and YPD supplemented with cerulenin plus myristate or palmitate. The results can be summarized as follows (cf Ref. 132):

(i) Faa1p is the acylCoA synthetase most responsible for activation of exogenous fatty acids. Deletion of *FAA1, FAA2,* or *FAA3* in an *NMT1* strain, either singly or in all possible combinations, has no observable effect on growth in YPD at 24, 30, or 37°C when the de novo fatty acid biosynthetic pathway is active. As noted above, strains that produce Nmt1p are able to grow at 24, 30, or 37°C when de novo fatty acid synthesis is inhibited by CER if the media is supplemented with C14:0, C16:0, or C18:1$^{\Delta 9}$. Deletion of *FAA1,* but not *FAA2* or *FAA3,* eliminates the ability to grow in YPD-CER plus fatty acids at 37°C, although growth at 24 and 30°C is largely unaffected. Deletion of both *FAA1* and *FAA2,* or *FAA1* and *FAA3,* or all three *FAA*s, does not create any additional growth deficit when compared to deleting *FAA1* alone. The failure of *NMT1 Δfaa1* strains to grow at 37°C on YPD/CER plus fatty acids can only be overcome by overexpressing Faa1p: Faa2p and Faa3p do not complement the phenotype and therefore appear to have no role in the activation of exogenous fatty acids.

(ii) Several observations indicate that Faa2p and Faa3p are able to activate endogenous pools of free fatty acids. First, overexpression of Faa2p or Faa3p, but not Faa1p, reduces the amount of exogenous fatty acid required to rescue an *NMT1* strain when de novo fatty acid synthesis is blocked with CER. Second, overexpression of Faa2p, but not Faa1p or Faa3p, rescues the moderate growth retardation seen when an *nmt1-181* strain is incubated at 30°C on YPD media containing no exogenous fatty acids. This rescue by Faa2p is only partial—overexpression Faa2p does not support growth of an *nmt1-181* strain on YPD alone at 37°C. Third, overexpression of Faa2p and Faa3p does not rescue the ability of an *nmt1-181/faa1Δ* strain to grow on YPD/CER/MYR at 37°C. As noted above, only overexpression of Faa1p is able to complement the *faa1* deletion.

(iii) There are no appreciable phenotypic differences between *NMT1* or *nmt1-181* strains expressing wild-type levels of Faa1p and strains that overexpress Faa1p when they are grown on YPD, YPD plus CER, or YPD/CER/MYR at 24–37°C. This suggests that there is a functional reserve of Faa1p: i.e. utilization of exogenous fatty acids is not normally limited by their rate of activation by Faa1p. In contrast, strains that overexpress Faa2p or Faa3p have phenotypes that are distinct from those of isogenic strains that express wild-type levels of the corresponding acylCoA synthetases. This suggests that utilization of fatty acids from endogenous pools may be limited (regulated) by their rate of activation by Faa2p or Faa3p.

(iv) Faa-mediated activation of endogenous fatty acids appears to be required, together with Fas, to maintain myristoylCoA pool size so that nmt181p can acylate essential N-myristoylproteins at an efficiency required for sustaining vegetative growth. This conclusion is based on the finding that *nmt1-181* strains with deletions of all three FAAs are not viable at the permissive temperature (24°C) on YPD alone, even when the de novo pathway for fatty acid biosynthesis is intact.

(v) *S. cerevisiae* has more than three long-chain acylCoA synthetases. A *nmt1-181faa1ΔFAA2FAA3* strain is not able to grow on YPD alone at 30°C, but is viable on YPD supplemented with myristate. Moreover, a *NMT1* strain with *faa1, faa2,* and *faa3* null alleles can grow on YPD/CER supplemented with C14:0, C16:0, or C18:$1^{\Delta 9}$, indicating that there is at least one other Faa activity (Faa4p) that can activate exogenously derived fatty acids.

(vi) *NMT1* strains containing all possible combinations of *faa1, faa2, faa3* null alleles are able to grow on media containing C14:0, C16:0, or C18:$1^{\Delta 9}$ as the sole carbon source. Utilization of these exogenous fatty acids by a *NMT1faa1Δfaa2Δfaa3Δ* strain indicates that they are imported, activated by another faa, and acetylCoA is generated from the acylCoAs in peroxisomes via β-oxidation. This finding does not rule out the possibility that Faa1p, Faa2p, and/or Faa3p are able to activate fatty acids destined for β-oxidation.

Searching for suppressors of the temperature-sensitive growth arrest and myristic acid auxotrophy of nmt1-181 strains: an approach for studying the factors that regulate protein N-myristoylation in vivo The studies with overexpression or deletion of *FAA1* and *FAA2* in *nmt1-181* strains illustrate how the contributions of a specific gene to the regulation of protein N-myristoylation can be audited by determining whether its overexpression or deletion relieves or worsens the phenotype produced by this conditional lethal *nmt1* allele. For example, searches for suppressors of *nmt1-181* would be expected to yield several general classes of genes: (*a*) genes that help regulate the size and availability of myristoylCoA pools (e.g. a fatty acid transporter or proteins that might bind myristoylCoA and facilitate its delivery to Nmt1p); (*b*) genes encoding factors that regulate the transcriptional activity of *NMT1* and/or the steady-state levels of its mRNA and protein products; and (*c*) genes that encode substrate proteins or determine their availability as substrates for Nmt1p (e.g. methionylaminopeptidases, proteins that might help target Nmt1p to nascent polypeptides) or (*d*) general factors that affect the biological activities of N-myristoylproteins (an N-myristoylprotein receptor, specific chaperones?). Searches for suppressors could involve introduction of yeast genomic libraries into an *nmt1-181* strain and subsequent identification of those products that can partially rescue or fully rescue the phenotype. Alternatively, a directed analysis of the effects of a known gene product on the strain's phenotype could be carried out. The value of these approaches is illustrated by the results of four recent studies.

(i) Since purified Nmt1p has no associated methionylaminopeptidase (Map) activity, cellular Map(s) must remove the initiator Met residue from nascent Nmt1p substrates to reveal the required Gly2 and its primary amino group. Therefore, does methionylaminopeptidase (Map) function to limit (regulate) the overall efficiency of protein N-myristoylation in vivo? Are there cellular methionylaminopeptidases that are specifically responsible for removing the initiator Met of nascent Nmt1p substrates? Chang et al (138) purified a Map from *S. cerevisiae*. In vitro studies of its substrate specificities indicate a strict requirement for a Met at position 1 and small uncharged residues (e.g. Ala, Gly) at position 2. This indicates that the Met-Gly bond at the amino terminus of nascent Nmt1p substrates should be readily cleaved by this Map. Chang et al (139) subsequently isolated a *S. cerevisiae MAP* gene. Its 377-residue protein product contains a carboxy-terminal domain with sequence similarities to other prokaryotic Maps, and a unique amino-terminal domain with two predicted zinc finger motifs, which may mediate its interaction with ribosomes. Although *E. coli* and *Salmonella typhimurium MAP* genes are essential for cell growth (140–144), disruption of the yeast *MAP1* gene is not lethal, although it does reduce the rate of cell growth (139). Map1p has been overexpressed in *S. cerevisiae* by placing *MAP1* under the control of the *GAL1-10* promoter and growing cells in the presence of galactose. Overexpression produces no phenotypic effects in isogenic strains containing *NMT1* or *nmt1-181* (F Li, DR Johnson, JI Gordon, unpublished observations). The failure of Map1p overexpression to relieve the temperature-sensitive growth arrest and myristic acid auxotrophy of the *nmt1-181* strain, together with the results of the *MAP1* disruption experiments, suggests—but does not prove—that *S. cerevisiae* may contain other *MAP* gene(s) involved in removing Met residues from Nmt1p substrates.

(ii) *S. cerevisiae* also contains another important N-terminal modifying enzyme activity that catalyzes the transfer of an acetyl moiety from acetylCoA to the α-amino group of nascent proteins (145, 146). The principal amino-terminal residues of N^{α}-acetylated proteins are Ser, Ala, Met, Gly, Thr, and Asp (147–150). Two genes are needed for expression of *S. cerevisiae* N^{α}-acetyltransferase activity (151–153). *NAT1* encodes the catalytic subunit of this enzyme, while *ARD1* specifies an essential subunit of the heterodimeric acetyltransferase complex (154). Loss of acetyltransferase activity due to disruption of *NAT1* and *ARD1* results in an **a**-factor-specific mating defect, abnormalities in sporulation, and failure to survive during stationary phase (151, 152, 155, 156). Does this enzyme activity compete with Nmt1p for common substrates? The answer appears to be no. Overexpression of *NAT1* and/or *ARD1* produces no detectable phenotypic alterations in isogenic strains containing *NMT1* or *nmt1-181*. *nmt1-181* strains with *nat1* and/or *ard1* null alleles exhibit similar growth patterns on YPD, YPD/CER, YPD/CER/MYR at 24, 30, and 37°C as isogenic strains with *NMT1*. There is no change in the temperature-sensitivity of the growth arrest or the requirements for myristate at the nonpermissive temperature (F Li, DR Johnson, JI Gordon, unpublished observations). Moreover, in vitro studies of purified Nmt1p and octapeptides representing the amino-terminal sequences of several known N-acetylproteins (e.g. GIPETQKG from alcohol dehydrogenase-3) have shown that they do not bind to Nmt1p (E Jackson-Machelski, JI Gordon, unpublished observations).

(iii) The increased requirements of nmt181p for myristoylCoA also make it useful for assessing whether cellular acylCoA-binding proteins participate in the delivery of myristoylCoA to the acyltransferase and/or to the sequestration of this substrate or palmitoylCoA. One candidate binding protein is Acb1p (acylCoA-binding protein-1). *S. cerevisiae ACB1* encodes an 87-residue protein (157).[5] Studies by Mandrup et al (158) have shown that overproduction of bovine acylCoA-binding protein in *S. cerevisiae* results in an increase in the size of the total cellular acylCoA pool, with a proportional increase in palmitoylCoA and relative decreases in stearoylCoA and oleoylCoA. The size of the myristoylCoA pool was not documented in their study. Overexpression of Acb1p in *NMT1* and *nmt1-181* strains has no effect on their phenotypes, whether they are grown at 24, 30, or 37°C, in the presence or absence of an active de novo pathway for fatty acid biosynthesis. Similarly, deletion of *ACB1* neither ameliorates nor worsens the temperature sensitivity and myristic acid auxotrophy of *nmt1-181* strains (DR Johnson, TM Rose, JI Gordon, unpublished observations). Together, these findings suggest that this acylCoA-binding protein does not play a significant role in regulating myristoylCoA pool size or in the metabolic trafficking of myristoylCoA to Nmt1p.

[5]Sequence comparisons of Acb1p with orthologous Acbs indicate that the primary structure of this protein has been highly conserved during the course of eukaryotic evolution. The three-dimensional structure of the 86-residue bovine acylCoA-binding protein has been solved with and without bound palmitoylCoA using multidimensional ^{1}H, ^{13}C, and ^{15}N nuclear magnetic resonance spectroscopy (159, 160), making it an attractive model for studying the molecular details of acylCoA-protein interaction (cf Ref. 162).

(iv) A search for high-copy suppressors of the growth arrest and myristic acid auxotrophy exhibited by *nmt1-181* strains at 37°C only yielded genomic copies of *NMT1*—indicating that only Nmt1p can correct the defect in protein N-myristoylation at the nonpermissive temperature (127). However, searches can be initiated for high-copy suppressors of the growth retardation and myristic acid auxotrophy observed when *nmt1-181* strains are grown on YPD media at 30°C. Once suppressors are recovered, their mechanism of action can be examined using a series of secondary assays. As noted above, mating pathway signaling is exquisitely sensitive to the level of myristoyl-Gpa1p in a cell. Suppressors that function by increasing myristoyl-CoA pool size or that affect the abundance or activity of nmt181p should lower *FUS1* transcription (i.e. myristoylGpa1p leads to suppression of *FUS1* transcription). Genes that rescue *nmt1-181* strains by increasing expression of some critical protein substrate of Nmt1p other than Gpa1p should not affect expression of *FUS1*. Genes that are able to increase transcription of *nmt1-181* can be identified based on their ability to change the activity of chimeric gene, consisting of *NMT1*'s promoter linked to a reporter such as LacZ or the bacterial gene encoding chloramphenicol acetyltransferase (CAT) (the chimeric gene is integrated into the *S. cerevisiae* genome). One screen for suppressors of the myristoylation defect (*SMD*) produced by *nmt1-181* at 30°C yielded six different genes (127). The increased *NMT1*-CAT activity in strains expressing *SMD4, SMD5,* and *SMD6* suggested that these plasmids contain genomic DNAs that rescue growth by increasing transcription of *nmt1-181*. *SMD1* and *SMD2* decreased *FUS1* expression, but did not affect transcription of *nmt1-181*; consistent with a role for these genes in modulating the size or accessibility of myristoylCoA pools, or the amount or activity of nmt181p independent of increased transcription of *nmt1-181*. *SMD3* rescued growth without altering *FUS1* or *nmt1-181* expression, implying that rescue was accomplished without a global increase in protein N-myristoylation. One would predict that *SMD3* encodes a critical N-myristoylprotein or a factor that affects N-myristoylprotein-mediated process(es). The validity of these assumptions is supported by the revelation that *SMD2* and *SMD6* are allelic with *FAS1* and *CDC39,* respectively. Overexpression of FAS1 can induce overexpression of *FAS2* (162), leading to increased activity in the de novo pathway and presumably greater production of myristoylCoA. *CDC39* has recently been shown to negatively regulate transcription of multiple genes (163). A mutant allele, *cdc39-1,* was originally isolated from a strain that arrested in G1 of the cell cycle due to constitutive activation of the mating response (164, 165). Extensive genetic analysis of this mutant allele localized the signaling defect to perturbed interactions between Gpa1p and Ste4p/Ste18p [i.e. the β and γ subunits of the G protein (165)]. This is the same defect observed in strains with mutant *nmt1* alleles (45, 46). Thus, decreased expression of *NMT1* is a plausible explanation for the G1 arrest phenotype observed in *cdc39-1* strains.

The ability of fatty acid analogs to rescue the growth arrest of nmt1-181 cells provides insights about the physicochemical properties of myristate that are necessary for proper functioning of essential cellular N-myristoylproteins The requirement of an *nmt1-181* strain for exogenous myristate to grow at the nonpermissive temperature makes it an attractive model system for examining whether myristic acid analogs with alterations in conformation, steric bulk, or polarity can substitute for the naturally occurring fatty acid. Rescue by an analog can be taken as prima facie evidence that the compound is efficiently imported into *S. cerevisiae,* that it is activated by cellular Faas, that it is incorporated into essential N-myristoylproteins, and that it is able to support expression of the biological function of these proteins at the nonpermissive temperature. A number of analogs have been identified as rescuers of the temperature-sensitive growth arrest produced by *nmt1-181* (42, 78). Surveys of oxatetradecanoic and thiatetradecanoic acids reveal that 7-thia- and 9-thiatetradecanoic acids can fully rescue growth at the nonpermissive temperature. These compounds have polarities that are significantly less than that of myristate (i.e. equivalent to C12:0), although they have comparable bond lengths and bond angles (64). Interestingly, movement of the sulfur-for-methylene substitution by one carbon atom, either towards carboxyl (yielding 6-thia and 8-thiatetradecanoic acids) or towards the ω-terminus (yielding 8-thia- and 10-thiatetradecanoic acids), is associated with loss of the ability to rescue growth (DR Johnson, JI Gordon, unpublished observations). Y-6-tetradecynoic acid, an analog with conformational restriction owning to its triple bond (78), can also rescue growth. In vitro studies have confirmed that the analogs that rescue are excellent substrates for cellular Faas as well as Nmt1p (64, 137; LJ Knoll, GW Gokel, JI Gordon, manuscript in preparation). Other compounds that are also good substrates, specifically C16 fatty acids with bends in the vicinity of C5 (Z-5-hexadecenoic acid and Y-6-hexadecynoic acids), fail to rescue, just as is the case with the parental C16:0 fatty acid.

Unlike hexadecanoic acid (palmitate), Z-5-hexadecenoic acid suppresses growth of an *nmt1-181* (but not an *NMT1*) strain at the permissive temperature of 24°C (78). A similar effect was noted with 6-oxatetradecanoic acid, but not the analogous sulfur-containing derivative (6-thiatetradecanoic acid; Ref. 42). The uptake of tritiated 6-oxatetradecanoic acid into *S. cerevisiae* is $<1\%$ that of myristate and it is not incorporated into cellular N-myristoylproteins at detectable levels (42). Analog-induced growth arrest could reflect a number of possible mechanisms, including an ability to produce absolute or relative reductions in the size of myristoylCoA pools used by nmt181p. If so, these analogs could represent starting points for identifying genes that regulate acylCoA metabolism and/or availability—e.g. genes that encode proteins that might be involved in the transport of myristate across cellular membranes.[6] These points are illustrated by a recent search for high-copy suppressors of 6-oxatetradecanoic acid's toxicity to *nmt1-181* strains. Overexpression of Faa1p in *nmt1-181* strains partially suppresses the growth defect caused by 6-oxatetradecanoic acid, even though the analog is neither an inhibitor of, nor a substrate for, the acylCoA synthetase (DR Johnson, LJ Knoll, JI Gordon, unpublished observations). Overexpression of *nmt1-181* or *FAS1* also overcomes the suppression produced at 24°C, as does adding myristate to the media. These results are consistent with the notion that 6-oxatetradecanoic acid's suppressive effect can be overcome by increasing myristoylCoA size.

Isolation of orthologous Nmts cDNA by complementation of nmt1-181 The inability to find any extragenic suppressors when *nmt1-181* strains are grown on YPD at 37°C can be used to isolate cDNAs encoding orthologous Nmts—assuming of course that their peptide substrate specificities have been sufficiently conserved so they can acylate essential *S. cerevisiae* N-myristoylproteins at levels sufficient to support its vegetative growth. A human Nmt cDNA was recovered in this fashion using an expression library of liver cDNAs placed under the control of a strong constitutive promoter (from the glyceraldehyde-3-phosphate dehydrogenase gene) in a high-copy plasmid (171). The 416-residue human acyltransferase has 44% identity with Nmt1p, including a Gly residue located 5 residues from its carboxy terminus (Figure 2). Southern blot analysis suggests that there is a single *NMT* gene in *Homo sapiens* (171).

Comparison of the kinetic mechanisms and substrate specificities of human and yeast Nmts reveals important similarities and differences. Kinetic studies of product inhibition, as well as an examination of the effects of covariation of subsaturating concentrations of myristoylCoA and peptide substrates on initial velocity, indicate that human Nmt has a preferred ordered mechanism, with myristoylCoA binding before peptide followed by release of CoA and

[6]There is little information about how myristate is transported across cell membranes. Studies using the triple plasmid expression system in *E. coli* strains with various *fad* mutants showed that permeation of myristate across the bacterial inner membrane can occur by simple diffusion and does not need a functional bacterial long-chain fatty acid transporter encoded by the *fadL* gene (105; cf Refs. 167–169). Studies in *S. uvarum* and *S. lipolytica* suggest that these yeasts contain two systems for fatty acid uptake: Passive diffusion predominates at high fatty acid concentrations and a saturable high-affinity system is responsible for fatty acid uptake at low (< 10 μM) concentrations (170).

```
H. sapiens      .......... .......... .......... .......... .......... .......... .......... ........MN SLPAERIQEI QKAIELFSVG
C. albicans     .......... .......... .......... .......... .......... .......... .......... .MASGDNTGN KSNSAPSKS! EELLKLLAMG
S. cerevisiae   .......... .......... .......... .......... .......... .......... .......... .......... MSEEDKAKKL ENLLKLLQLN

C. neoformans   .......... .......... .......... .......... .......... .......... .......... ...MDSSDNK AATDEEIRRA LKAADLMKIL
H. capsulatum   MSEQEGNQSE HQSEHVGESE GKLPNETPTT SQSTNASTGT AGKGEKKSSD GDPAAANPAT KLTPSMAESL LELNPALRSE LAGMDKEKAT EALRQMNISD

H. sapiens      QGPAKTMEEA SKRS...YQF WKTQPVPKLG EVVNT....H GPVEPDK..D NIRQEPYTLP QGFTWDALDL GDRGVLKELY TLLNENYVED DDNMFRFDYS
C. albicans     QE....LSPA QQKEMKDYKF WDTQPVPSLS ETVTE....E GPIDKLKTPE DVPNDPLPLI SDFEWSTLDI DDNLQLDELY KLLYDNYVED IDATFRFKYS
S. cerevisiae   NDDTSKFTQE QKKAMKDHKF WRTQPVKDFD EKVVE....E GPIDKPKTPE DISDKPLPLL SSFEWCSIDV DNKKQLEDVF VLLNENYVED RDAGFRFNYT

C. neoformans   DGKMALGNKS GTKNLGEHKF WKTQPVPQIT GSGASAPMEE GPIDDPKTPA DVKQEPGVLP AGFEWSTIDI NDEEQSKEVY VLLCENYVED DDAMFRFNYS
H. capsulatum   LLTGLSVNPK NQKDMASFKF WQTQPVIRFD DRESESP..D GPI.KIVELD QVSREPIPLV DGFEWVTLDI DDEADVKEFY ELLANHYVED GSAMFRFNYS

H. sapiens      PEFLLWALRP PGWLPQWHCG VRVVSSRKLV GFISAIPANI HI..YDTEKK MVEINFLCVH KKLRSKRVAP VLIREITRRV HLEGIFQAVY TAGVVLPKPV
C. albicans     HEFFQWALKP PGWRKDWHVG VRVKSTGKLV AFIAATPVTF KLNKSNKVID SVEINFLCIH KKLRNKRLAP VLIKEITRRV NKQNIWQALY TGGSILPTPL
S. cerevisiae   KEFFNWALKS PGWKKDWHIG VRVKETQKLV AFISAIPVTL GV..RGKQVP SVEINFLCVH KQLRSKRLTP VLIKEITRRV NKCDIWHALY TAGIVLPAPV

C. neoformans   REFLLWALTA PGYLPDWHIG VRVQKTKKLV AFISGIKIDI RV..RAKTFP AAEINFLCVH KKLRSKRLAP VLIKEVTRRV NLTNIWQAIY TAGVILPTPI
H. capsulatum   PAFLNWALKA PGWKREWHVG VRASKSGKLV ASICGVPAEI AV..RGKSLK VTEINFLCVH KKLRSKRLTP VLIKEITRRC YLNGIYQAIY TVGIMLPTPV

H. sapiens      GTCRYWHRSL NPRKLIEVKF SHLSRNMTMQ RTMKLYRLPE TPKTAGLRPM ETKDIPVVHQ LLTRYLKQFH LTPVM.SQEE VEHWFYPQ.. ........EN
C. albicans     TTCRYQHRPI NWSKLHDVGF SHLPPNQTKS SMVASYTLPN NPKLKGLRPM TGKDVSTVLS LLYKYQERFD IVQLF.TEEE FKHWMLGHDE NS.D....SN
S. cerevisiae   STCRYTHRPL NWKKLYEVDF TGLPDGHTEE DMIAENALPA KTKTAGLRKL KKEDIDQVFE LFKRYQSRFE LIQIF.TKEE FEHNFIGEES LPLD....KQ

C. neoformans   GTCRYFHRNL NPPKLVDIGF SPLPRGSTIA RLVQQYSVPS HPRIPGFREM KKEDVPQVGA LLRRYLDRFD VAQAFKDDDE VEHWFLSGQG KEVGGRRVEQ
H. capsulatum   SACRYYHRAL DWLKLHEVGF SPLPIGSTKS RQVTRNHLPG HTSTPGLRPM QSKDIDAVQD LLNRYLKRFD LSQIF.SRKE VDHLLL.HKE KPGA....EQ

H. sapiens      IIDTFVVENA NGE.VTDFLS FYTLPSTIMN HPTHKSLKAA YSFYNVHT.. .......... .......... .......... .......QTP LLDLMSDALV
C. albicans     VVKSYVVEDE NG.IITDYFS YYLLPFTVLD NAQHDELGIA YLFYYAS... .......... .......... .........D SFEKPNYKKR LNELITDALI
S. cerevisiae   VIFSYVVEQP DG.KITDFFS FYSLPFTILN NTKYKDLGIG YLYYYATDAD FQ........ .......... ....FKDRFD PKATKALKTR LCELIYDACI

C. neoformans   VVWAYVVEDP TTHRITDLIS FYALPSTIMK HPKHNLLNAA YMFYYATDVI FPPPSSSAHS DADVNAGESS VAAVGTGGED AKTKKKLETR LNALTADLLI
H. capsulatum   IVWSYVAEEP GTHRITDFAA FYSLESSVLQ NSKHKNVKAA YLYYYATETA FA........ .......... ....EKE... ....KGLKER LLMLINDVLI

H. sapiens      LAKMKGFDVF NALDLMENKT FLEKLKFGIG DGNLQYYLYN WKCPSMGAE. .......... ...KVGLVLQ  416
C. albicans     TSKKFGVDVF NCLTCQDNTY FLKDCKFGSG DGFLNYYLFN YRTFPMDGGI DKKTKEVVED QTSGIGVVLL  451
S. cerevisiae   LAKNANMDVF NALTSQDNTL FLDDLKFGPG DGFLNFYLFN YRAKPITGGL NPDNSNDIKR R.SNVGVVML  455

C. neoformans   IAKQAGFDVF NALTLLDNNM FLQEQKFGPG DGYLNYYLYN WNCAPIDGGQ HSTT....AK QGSKIGVVML  491
H. capsulatum   LAKKERFDVM NALTLHDNPL FLEQLKFGAG DGQLHYYLFN YRTAPIAGGV NDKNLPD.ER KRGGVGVILV  529
```

↓ = Intron

Figure 2 Alignment of the primary structures of orthologous Nmts. This multiple sequence alignment was generated using the PILEUP program contained in the GCG sequence analysis package and taken from Ref. 173. The locations of introns are indicated by closed arrows. *S. cerevisiae NMT1* and *C. albicans NMT* genes are not interrupted (37, 59). The two introns of *Histoplasma capsulatum NMT* (from strain G217B, cf. Ref. 173) contain 66 and 73 bp, obey the dinucleotide rule (i.e. begin with GT and terminate with AG), and are located at positions identical to 2 of the 10 introns in *Cryptococcus neoformans* NMT. The 10 introns of *C. neoformans* var. *neoformans* strain L210425 *NMT* range in length from 71 to 118 bp. All have the consensus GT ... AG sequence at their 5′ and 3′ junctions (173). *H. sapiens NMT* contains several introns, but the position of only one of them has been determined to date (173). Southern blot studies indicate that each of these organisms has a single copy of the *NMT* gene/haploid genome (37, 59, 171, 173). The vertical black stripes indicate residues that genetic studies have shown to be important for Nmt1p activity. A previous search for intragenic suppressors of the temperature-sensitive myristic acid auxotrophy conferred by *nmt1-181* (nmt$^{Gly451 \rightarrow Asp}$) yielded several pseudorevertant alleles, two of which retained an Asp451 codon but had a Glu167→Lys substitution or a Glu293→Lys substitution. These Glu residues are retained in the orthologous Nmts. *S. cerevisiae nmt1*-72 encodes a mutant enzyme with a Leu99→Pro substitution (46). This temperature-sensitive mutant causes arrest in G1 of the cell cycle due to reduced acylation of Gpa1p (45). The temperature-sensitive growth arrest observed at 37–39°C in minimal or rich media can be rescued by addition of 500 μM myristate to the media. nmt72p has a reduced affinity for myristoylCoA, but its reduction is less than that of nmt181p and is associated with more restricted defects in acylation of essential cellular N-myristoylproteins (and by a more restricted pattern of temperature sensitivity, cf. Ref. 46).

then myristoylpeptide (172). Surveys of a panel of C7-C17-saturated fatty acids plus >70 myristic acid analogs that include oxa- and thiatetradecanoic, oxotetradecanoic, tetradecenoic, and tetradecynoic acids suggest that the acylCoA-binding sites of the yeast and human enzymes have been highly conserved (65). Moreover, substitution of Gly412 in human Nmt with an Asp (comparable to the Gly451→Asp substitution in nmt181p) produces a marked temperature-dependent reduction in affinity for myristoylCoA as judged by the in vitro enzyme assay system, the *E. coli* coexpression system, and by complementation studies in *S. cerevisiae* (e.g. see Ref. 171). Despite this conservation of acylCoA-binding sites and kinetic mechanism, there are differences in the peptide substrate specificities of *H. sapiens* and *S. cerevisiae* Nmts. Analysis of strains containing episomes with human Nmt cDNA under the control of the galactose-inducible *GAL1-10* promoter indicates that overexpression of the human enzyme is sufficient to rescue growth of *nmt1-181* strains on YPD at 37°C or the lethal *nmt1* null strain at 24–37°C. In contrast to Nmt1p, the human enzyme cannot complement either mutant *nmt1* allele when its levels are markedly reduced by glucose repression of the *GAL1-10* promoter (173). Moreover, coexpression of the wild-type yeast or human acyltransferases with several G protein α subunits in *E. coli* indicates that both enzymes can transfer myristate to yeast Gpa1p (amino-terminal sequence = GCTVSTQT) and rat G_o (GCTLSAEE), but that only human Nmt can acylate human G_z (GCRQSSEE) (109, 129, 171). In vitro studies using a panel of octapeptide substrates reveal distinct differences in their specificities for amino acids occupying positions 3, 4, 7, and 8: e.g. the human enzyme can tolerate acidic residues at positions 7 and 8 better than the yeast enzyme can (172). Analysis of human/yeast Nmt chimeras in *E. coli* and in *S. cerevisiae* suggest that the apparent structural variations in the peptide-binding sites of human Nmt and *S. cerevisiae* Nmt1p arise from differences present in the amino- and carboxy-terminal halves of the enzymes (109).

The experiments with human Nmt emphasize the potential utility of using *S. cerevisiae* as a genetically manipulatable surrogate environment for studying regulation of protein N-myristoylation in higher eukaryotes. For example, the ability of cDNAs encoding human or other mammalian long-chain acylCoA synthetases (133–135) to complement *faa* null alleles in NMT1 and/or *nmt1-181* strains could provide insights about the contributions of these enzymes to cellular myristoylCoA metabolism. Strains of *S. cerevisiae* expressing human $Nmt^{Gly412 \rightarrow Asp}$ can be used to search for high-copy suppressors of their temperature sensitivity and myristic acid auxotrophy. This suppressor screen could use a library of *S. cerevisiae* genomic DNA and lead to identification of gene products that regulate protein N-myristoylation in this organism. Alternatively, the suppressor screen could use cDNA libraries prepared from various human tissues and yield essential proteins that function as substrates for the human Nmt in vivo and/or gene products that modulate the levels of myristoylCoA.

S. CEREVISIAE AS A SURROGATE ENVIRONMENT FOR STUDYING PROTEIN N-MYRISTOYLATION IN PATHOGENIC FUNGI *Candida albicans* is a dimorphic asexual fungus and the most common cause of systemic fungal infections in immunocompromised patients. *Cryptococcus neoformans* is a monomorphic haploid yeast (174), the most common cause of fungal meningitis, and a leading cause of death among patients with acquired immunodeficiency syndrome (AIDS). Metabolic labeling studies using [^{3}H]myristate indicate that both *C. albicans* and *C. neoformans* synthesize <10 N-myristoylproteins during exponential growth in rich media (36, 37, 173). *C. albicans* contains two *ARF* genes, while *C. neoformans* only has one (36, 173, 175). Studies in the *E. coli* coexpression system confirm that these Arf proteins are substrates for their respective Nmts (173). Even though *C. albicans* does not have a known sexual pathway, it produces a protein, Cag1, which is more homologous to *S. cerevisiae* Gpa1p than any other reported G protein α subunit (176). The N-terminal sequence of this G protein α subunit homolog (GCGASVPVDD) makes it a likely substrate for *C. albicans* Nmt (37). Moreover, *CAG1* can complement the growth arrest and mating defects found in strains of *S. cerevisiae* with *gpa1* null alleles (176), indicating that it is a substrate for *S. cerevisiae* Nmt1p.

Alignments of the primary structures of *S. cerevisiae, C. albicans, C. neoformans,* and *H. sapiens* Nmts reveal comparable degrees of sequence identity (Figure 2). The regions of amino acid sequence identity are distributed throughout their primary structures, a result that is not surprising given the fact that both acylCoA and peptide recognition requires elements from both the amino- and carboxy-terminal halves of Nmt1p and human Nmt (109). It is also not surprising that the amino-terminal sequences of the orthologous Nmts are the most divergent regions of the orthologous acyltransferases: As noted above, the amino-terminal 59 residues of Nmt1p have no apparent function in binding substrates or in catalysis. In vitro studies have shown that the peptide substrate specificities of purified *C. albicans* and *H. sapiens* Nmts have diverged (37, 172), although surveys using members of the panel of myristic acid analogs indicate that the enzymes' acylCoA-binding sites are quite similar (N Kishore, JI Gordon, unpublished observations). This apparent divergence in the peptide- but not acylCoA-binding sites likely reflects both enzymes' requirements for myristoylCoA and the marked differences in the number of protein substrates they must recognize (metabolic labeling studies in a variety of mammalian cell lines reveal >50 N-myristoylproteins; cf Refs. 41, 177).

These observations suggest that it may be possible to design peptide-based, species-specific inhibitors of fungal Nmts that could function as antifungal agents. Of course, support for this strategy requires information about whether Nmt is essential for vegetative growth of *C. albicans* or *C. neoformans*. One obvious choice for addressing this question is to introduce a mutation into either *C. albicans* or *C. neoformans NMT* that is analogous to the mutation in *nmt1-181* and genetically engineer conditional lethal fungal *nmt* alleles. When the conserved Gly residue located five residues from the carboxy terminus of *C. albicans* Nmt (Gly447) or *C. neoformans* Nmt (Gly487) is mutated to an Asp, each of the mutant fungal enzymes displays a temperature-dependent reduction in their ability to acylate Arf proteins in the *E. coli* coexpression system (173). The functional properties of the wild-type and mutant Nmts can be assessed further based on their ability to complement the lethal phenotype of an *S. cerevisiae nmt1* null allele (173). Wild-type *C. albicans* and *C. neoformans NMT,* placed under the control of the *GAL1-10* promoter and introduced into an *S. cerevisiae nmt1*Δ strain, can each sustain growth when the yeast strains are grown in YP-galactose media (YP-GAL) at 24 or 37°C or on YP-GAL-supplemented cerulenin and myristate. These results indicate that (*a*) when either wild-type fungal Nmt is expressed at high concentrations, it is able to acylate essential cellular N-myristoylproteins at levels sufficient to support vegetative growth and (*b*) both wild-type enzymes can access *S. cerevisiae* myristoylCoA pools derived through Faa-mediated activation of exogenous (and/or endogenous) myristate. In contrast, only wild-type *C. albicans NMT* can support growth when the *GAL1-10* promoter is repressed by glucose. An isogenic strain producing *C. neoformans* Nmt is unable to grow on YP-GLU at either 24 or 37°C even when the media is supplemented with 500 μM myristate. The *mutant Candida* and the *mutant Cryptococcus* nmts are both able to support growth of *S. cerevisiae* strains with *nmt1* null alleles on YP-GAL at 24°C but not at 37°C, even though the steady-state levels of wild-type and mutant enzymes are similar after the cells have been grown on YP-GAL at 24°C and then shifted to the nonpermissive temperature for 1 h (173). Addition of 500 μM myristate to YP-GAL media rescues growth of yeast strains producing *C. albicans* $\text{nmt}^{\text{Gly447}\rightarrow\text{Asp}}$ at 37°C just as it does with isogenic strains containing *nmt1-181*. Yeast strains producing *C. neoformans* $\text{nmt}^{\text{Gly487}\rightarrow\text{Asp}}$ fail to grow on YP-GAL-MYR at 37°C.

These results predict that a Gly447→Asp mutation in *C. albicans* NMT may yield an allele that produces temperature-sensitive growth arrest and myristic acid auxotrophy in this pathogenic fungus. Subsequent studies, conducted in a laboratory strain of *C. albicans,* revealed that when this diploid organism contains one chromosomal copy of an *nmt1* null allele and one copy of the wild-type *NMT1* allele, it is able to grow on YPD at 24 and 37°C at rates similar to an isogenic strain that is homozygous for *NMT* (38). However, a strain with an *nmt* null allele and an *nmt* allele containing the Gly447→Asp substitution is unable to grow at either 24 or 37°C unless the YPD media is supplemented with 500 μM myristate (38). C16:0 cannot substitute for C14:0. Neither isogenic *NMT/NMT*, *nmtΔ/NMT*, nor *nmtΔ/nmtG447D* strains grow on YPD-CER at 24 or 37°C. All can grow at either temperature on YPD-CER supplemented with 500 μM MYR. The ability of the mutant *C. albicans nmtG447D* allele to support growth of *S. cerevisiae nmt1Δ* strains on YP-GAL without myristate at 24°C, but not to support growth of the *nmt1Δ* strain of *C. albicans* itself under these conditions, can be ascribed to one or more of the following: differences in the steady-state levels of the mutant fungal acyltransferase in *S. cerevisiae* and in *C. albicans*, differences in the import of myristate by these two organisms, differences in the size or availability of their myristoylCoA pools, or differences in the levels of acylation of essential cellular N-myristoylproteins required to sustain growth of these organisms.

A final experiment using these isogenic *C. albicans nmtΔ/NMT* and *nmtΔ/nmtG447D* strains established that Nmt is essential for vegetative growth. Cells were grown at 24°C in YPD/MYR and then switched to YPD without myristate. A 10,000-fold reduction in the number of viable *nmtΔ/ nmtG447D* cells occurs after a 24 h incubation in YPD media. In contrast, the isogenic *nmtΔ/NMT* strain is able to grow at the same rate in YPD compared to YPD/MYR (38). These results not only suggest that potent inhibitors of *C. albicans* Nmt will be fungicidal, but also underscore the utility of using the *E. coli* coexpression system together with *S. cerevisiae* to examine the functional consequences of mutations in fungal *NMT* genes prior to their introduction into *C. albicans*. Moreover, strains of *S. cerevisiae* that express these conditional-lethal fungal *nmt* genes can be used as starting points for suppressor screens—screens that may yield critical insights about gene products that regulate protein N-myristoylation in these fungal organisms.

HETEROGENEOUS ACYLATION OF N-MYRISTOYLPROTEINS

Heterogeneous Acylation of Proteins with C12:0, C14:0, C14:1$^{\Delta 5}$, and C14:2$^{\Delta 5,8}$

Transducin (Tr) is a heterotrimeric GTP-binding protein located in the rod outer segment (ROS) membrane. It is composed of α, β, and γ subunits. Transducin couples photolysis of rhodopsin to activation of cGMP phosphodiesterase in the vertebrate visual signal transduction cascade (178). Two groups of investigators (179, 180) used a variety of methods, including electrospray and tandem mass spectrometry, to demonstrate that bovine ROS

$T\alpha_r$ is heterogeneously acylated at its amino-terminal Gly with C12:0, C14:0, C14:$1^{\Delta 5}$, and C14:$2^{\Delta 5,8}$. Although these two groups report differences in the relative abundance of the different acyltransducins isolated from rod outer segments, both note that the N-myristoylated form of $T\alpha_r$ does not constitute the majority or even the most abundant isoform.

Retinal recoverin mediates the Ca^{2+} sensitivity of vertebrate guanylate cyclase in photoreceptors. Following light-induced depletion of intracellular cGMP and Ca^{2+}, recoverin stimulates photoreceptor guanylate cyclase to synthesize cGMP (178, 181, 182). Dizhoor et al (8) found that recoverin isolated from bovine ROS is also heterogeneously acylated with C12:0, C14:0, C14:$1^{\Delta 5}$, and C14:$2^{\Delta 5,8}$. The Z-5-tetradecenoyl group is the most abundantly represented acyl chain in both bovine ROS recoverin and $T\alpha_r$.

These analyses of the acylation of retinal transducin and retinal recoverin provide startling evidence that acyl moieties other than C14:0 can be covalently bound to the amino-terminal Gly residue of Nmt substrates. A critical control experiment performed by Johnson et al (RS Johnson, KA Walsh, JB Hurley, TA Neubert, manuscript in preparation) showed that the C_α subunit of PK-A isolated from bovine heart and brain only contains covalently bound C14:0, while the protein isolated from bovine retina is heterogeneously acylated with C14:0, C14:$1^{\Delta 5}$, C14:$2^{\Delta 5,8}$, and C12:0. In addition, heterogeneous acylation of the catalytic subunit is not a feature unique to the bovine retina: C_α purified from *Xenopus laevis* retina contains covalently bound C14:$2^{\Delta 5,8}$ (RS Johnson, KA Walsh, JB Hurley, TA Neubert, manuscript in preparation). These data suggest that heterogeneous amino-terminal acylation depends either on cell lineage–specific differences in acylCoA metabolism or on the presence of cell lineage–specific Nmts with varying acyl-CoA substrate specificities. The former possibility seems more likely than the latter. As noted above, in vitro studies of the acyl-CoA substrate specificities of *S. cerevisiae* and human Nmts demonstrated that Z-5-tetradecenoylCoA is a better substrate than myristoylCoA and that both enzymes can readily accommodate a *cis* double bond between C8 and C9 of C14 fatty acids (64, 65). Moreover, in vitro studies of a partially purified myristoylCoA synthetase activity recovered from a human erythroleukemia cell line indicate that the enzyme activity can use C12:0 as well as C14 fatty acids with *cis* double bonds between C5 and C6, or C8 and C9 (65). These observations suggest that heterogeneous acylation of N-myristoyl-proteins is regulated by the availability of "alternative" acyl chains, underscoring the importance of cellular acylCoA metabolism in regulating the specificity of this covalent protein modification. This emphasizes the necessity of using varied approaches in the future to understand how myristoyl-CoA metabolism is regulated in vivo, e.g. by employing currently available conditional lethal *S. cerevisiae nmt1* alleles for suppressor and/or synthetic lethality screens (cf. Refs. 184, 185) or by studying myristoyl CoA metabolism in various mutant mammalian cell lines (cf. Ref. 186).

Quantitative Heterogeneity

In addition to the qualitative heterogeneous acylation described above, there have been two recently reported examples of quantitative heterogeneity in protein N-myristoylation. Dohlman et al (4) noted that α-factor treatment of *MAT* **a** strains of *S. cerevisiae* promotes replacement of a pool of nonmyristoylated Gpa1p with newly synthesized myristoyl-Gpa1p, providing a mechanism for long-term regulation of the pheromone signal transduction pathway. Their data suggest that both the myristoylated and nonmyristoylated forms of Gpa1p are rapidly degraded, that they require continuous replacement to be maintained, and that Gpa1p synthesized after α-factor treatment is more efficiently acylated than is the pre-existing protein. Myristoylated and nonmyristoylated pools of the MARCKS protein (myristoylated alanine-rich C kinase substrate) have also been described by McIlhinney & McGlone (187) and Manenti et al (188) in rat and bovine brain, respectively. MARCKS binds calcium/calmodulin and actin (3, 189, 190; reviewed in Ref. 191). Elegant studies by Aderem and coworkers have suggested that this protein may function as a PKC-sensitive, reversible bridge between actin filaments and membranes, thereby participating in the regulation of cell motility and the endocytic pathway (13, 192, 193). Analysis of a MARCKS Gly2→Ala mutant indicated that myristoylation is required for membrane binding and for efficient phosphorylation: i.e. the myristoyl moiety targets the protein to the membrane where it is in close apposition to activated PKC. Subsequent phosphorylation of MARCKS is required for its displacement from the membrane (13). Nonmyristoylated bovine brain MARCKS represents up to 30% of cytosolic MARCKS and has a markedly lower affinity for calmodulin, suggesting a role for the myristoyl moiety in promoting MARCKS-calmodulin interactions (188).

It is unclear how nonmyristoylated pools of Gpa1p or MARCKS are created. McIlhinney & McGlone (187) have provided evidence that non-myristoylated MARCKS does not arise from deacylation. They observed continued N-myristoylation of MARCKS in rat brain synaptosomes up to 12 h after inhibition of protein synthesis (suggesting that some proteins may be acylated posttranslationally; cf Ref. 194). Dohlman et al (4) used isogenic strains of yeast containing wild-type *NMT1* or a temperature-sensitive *nmt1* allele that produces specific defects in Gpa1p acylation (*nmt1*-72: Refs. 45, 46 and Figure 2) to examine whether the size of the nonmyristoylated pool of Gpa1p could be influenced by the activity of Nmt. Overexpression of

Nmt1p or nmt72p produces no change in the fractional representation of nonmyristoylated Gpa1p (Ref. 4; DR Johnson, H Dohlman, JI Gordon, unpublished observations). The mechanism by which α factor can increase the fraction of myristoylGpa1p in yeast is unknown, but this phenomenon would appear to represent a very interesting model for identifying mechanisms that control protein N-myristoylation in this organism. Gpa1p is efficiently myristoylated by Nmt1p in the *E. coli* coexpression system: A nonmyristoylated species is not detectable (46). The kinetic properties of an octapeptide derived from the amino-terminal sequence of Gpa1p are comparable to that of an octapeptide representing the amino terminus of Arf1p, as judged by the in vitro Nmt assay (E Jackson-Machelski, JI Gordon, unpublished observations). Moreover, there is no evidence for demyristoylation of myristoylGpa1p in *S. cerevisiae* (4). Replacement of the amino-terminal sequence of Gpa1p with the amino-terminal sequence of another Nmt1 substrate (e.g. Arf1p or Arf2p) would be one way of determining whether the signals that control the levels of Gpa1p myristoylation are specified by Gpa1p itself (i.e. are intrinsic to the protein). In addition, since several G protein α subunits have been found to be palmitoylated at their Cys3 residue (see below), it will be interesting to determine whether C16:0 is added to Gpa1p, whether the presence of an amino-terminal C14:0 moiety influences the level of Gpa1p palmitoylation, and/or whether palmitoylation influences the level of Gpa1p N-myristoylation.

THE FUNCTIONAL SIGNIFICANCE OF THE MYRISTOYL MOIETY IN N-MYRISTOYLPROTEINS

An obvious and critical question to workers in this field is "why myristate"? Why has myristate been selected over more abundant cellular fatty acids such as palmitate (195)? One potential answer is provided by the work of Peitzsch & McLaughlin (83). As noted above, these workers showed that the free energy of association of acylpeptides with uncharged model membranes depends on the length of the acyl chain. For a series of myristoylpeptides, affinity appears to be independent of the chemical nature of the polar head of the amphipathic molecule. Based on their findings, they suggest that myristate has been chosen because the hydrophobicity of a tetradecanoyl group may favor reversible membrane association of N-myristoylproteins as opposed to longer acyl chains, which could serve as more permanent membrane anchors. A C14 fatty acid may also reversibly interact with other proteins or with domains present in its own "attached" protein.

Expression of wild-type N-myristoylproteins and their mutant derivatives containing Gly→Ala substitutions in mammalian cells, and/or functional studies of myristoylated and nonmyristoylated forms of a protein produced in *E. coli,* indicate that different N-myristoylproteins have different dependencies on their acyl chains for protein-membrane and/or protein-protein interactions. For example, the myristoyl group of $p60^{v\text{-}src}$ is necessary but not sufficient for stable association with the plasma membrane (35a, 196–198). NADH-cytochrome b_5 reductase (199), endothelial cell nitric oxide synthase (200, 201), *S. purpuratus* sperm flagellar creatine kinase (202), and a 43-kDa protein closely associated with the cytoplasmic face of the nicotinic acetylcholine receptor-rich postsynaptic plasma membrane (203) all appear to require myristate for their association with cellular membranes. X-ray studies demonstrate that the myristoyl group of poliovirus's VP4 capsid protein functions as a hydrophobic anchor to stabilize interactions between polypeptide subunits on the viral coat surface (204, 205). N-myristoylation of G protein α subunits is required for their association with membranes, for their binding to $\beta\gamma$ subunits (45, 206–208), and/or for signaling and transformation functions (as in the case of $G_{\alpha i2}$ mutants known as gip2 oncoproteins, Ref. 209).

Cotranslational myristoylation of the amino-terminal Gly of certain G protein α subunits may be a requirement for subsequent recognition and posttranslational acylation by cellular palmitoyltransferases (210). Palmitoylation is a reversible protein modification. Proteins that have juxtaposed myristoyl and palmitoyl groups near their amino terminus, such as $G\alpha_o$, $G\alpha_i$, $G\alpha_z$ (210), and the protein tyrosine kinase $p56^{lck}$ (211, 212), may be able to use the reversibility of palmitoylation to create a titratable, reversible interaction with cellular membranes and/or with other proteins (213).

Heterogeneous acylation could allow generation of subsets of an acylprotein with different partitioning characteristics. This might allow subsets of a protein to localize to different cellular membrane systems or to receptor proteins based on differences in the hydrophobicity or conformation of their covalently bound acyl chain. For example, Neubert et al (180) suggest that the heterogeneous acyl modifications of bovine ROS $T\alpha_r$ might yield subsets that can function under different physiological conditions, e.g. varying light intensities. Heterogeneous acylation in the ROS could titrate the hydrophobicity of a protein's amino terminus. Rather than considering each subset separately, the bulk effect would be to generate a range of hydrophobicity, averaging less than C14:0, thereby satisfying two requirements not made by other cell lineages: (*a*) the need for extremely rapid turnover and response times in the visual signal transduction pathway, and (*b*) the unusually high concentration of membranes in the ROS that will draw membrane association equilibria toward the bound state. The idea of shifting the membrane association equilibrium toward the unbound state is supported by the work of Kokame et al (179), which showed weaker association of lauroyl-Tα_r than of myristoyl-Tα_r with membranes.

Studies with myristic acid analogs also support the notion that different N-myristoylproteins have different dependencies on the physicochemical properties of their acyl chain for expression of their biological activities. 4-Oxa, 6-oxa, 11-oxa-, and 13-oxatetradecanoic acids have chain lengths equivalent to that of myristate, but their single oxygen-for-methylene substitutions reduce hydrophobicity so that they have polarities equivalent to that of dodecanoic acid (C12:0). Metabolic labeling studies using tritiated myristate and the four oxatetradecanoic acids plus a variety of cultured mammalian cell lines and/or strains of *S. cerevisiae, C. neoformans,* and *C. albicans* reveal that these analogs are selectively incorporated into overlapping yet distinct subsets of cellular N-myristoylproteins (36, 41, 177). (Selective incorporation appears to arise from the cooperative interactions between Nmt's acylCoA- and peptide-binding sites: The analogCoA:-Nmt binary complex produces a peptide-binding (and/or catalytic) site with some alterations in its conformation compared to the site synthesized by a myristoylCoA:Nmt binary complex. These perturbations affect the enzyme's ability to interact productively with some but not all protein substrates; Refs. 41, 64, 85.) The biological consequences of analog incorporation are protein- and analog-specific (41, 177). Incorporation of an oxatetradecanoic acid whose hydrophobicity is equivalent to C12:0 has no detectable effects on the subcellular location of most analog-substituted proteins. For some proteins such as p60^{src}, incorporation can produce redistribution from membrane to cytosolic fractions (41). The extent of redistribution is dependent upon the location of the oxygen-for-methylene substitution.[7]

The apparent sensitivity of some N-myristoylproteins to acyl chain length and/or hydrophobicity for proper expression of their biological function may reflect their need for "switches" that allow conformational heterogeneity. X-ray studies of nonmyristoylated, *E. coli*-derived, mouse C_α subunit of

[7]The selective incorporation of analogs into subsets of cellular N-myristoylproteins and the fact that only some members of the subset of analog-substituted proteins exhibit perturbations in their function likely accounts for the fact that many of these compounds are not toxic to cells at concentrations up to 200 μM. This allows some myristic acid analogs that function as alternate substrates for cellular acylCoA synthetases and Nmt to be used as antiviral or antifungal agents (36, 214, 215). For example, 13-oxatetradecanoic acid inhibits replication of human immunodeficiency virus 1 in acutely and chronically infected T-lymphocytes at doses that are not toxic to the cells. The antiviral effect appears to be due, at least in part, to selective incorporation of the analog into the virus's Pr55^{gag} polyprotein precursor. Myristoylated Pr55^{gag} is associated with the cell membrane. The analog-substituted protein is primarily cytosolic and demonstrates reduced proteolytic processing by viral protease, perhaps because less protease is released owing to reductions in the amount of dimer formation (protease is released from myristoyl-gag-pol dimers that form at the cell membrane; Refs. 21, 215, 216).

PK-A indicate that the protein's amino-terminal 14 residues are unstructured (217, 218). In contrast, the myristoylated catalytic subunit purified from porcine heart has a well-ordered amino-terminal domain. This is due, at least in part, to the fact that the myristoyl group binds to a hydrophobic pocket located within the protein (18). These findings prompted a proposal that cotranslational addition of myristate to C_α may facilitate its proper folding (18). As noted above, studies of the MARCKS protein indicate that addition or subtraction of a cofactor (a phosphate group) may affect protein conformation so that the acyl chain is presented in a way that allows it to modulate function (and thereby create a myristoyl-phosphoryl switch[8]). Taniguchi & Manenti (219) suggested a model where nonphosphorylated MARCKS interacts with negatively charged membranes using two of its domains, the amino-terminal myristoyl group and the basic side of an amphipathic α-helix. Phosphorylation of as many as three serine residues in this amphipathic phosphorylation domain adds significant negative charge to the region and results in dissociation of the protein from negatively charged membranes. Phosphorylation does not affect binding to neutral membranes. In fact, the MARCKS protein may possess the most complex switching system described to date for any N-myristoylprotein. In addition to reversible membrane association (which is phosphorylation- and myristoylation-dependent), MARCKS has Ca^{2+}- and phosphate-dependent associations with calmodulin and actin, which are also affected by the presence or absence of its myristate moiety. The finding that nonmyristoylated MARCKS does not bind to calmodulin in the presence of Ca^{2+}, as the acylated protein does, "unifies" the membrane association and Ca^{2+}/calmodulin-binding switches: i.e. both involve C14:0. The simple models proposed in which there are spatially separated domains for membrane association, phosphorylation, calmodulin binding, and actin binding may give way to a more allosteric model in which the covalent phosphate switch modifies the conformation of the myristoylated amino terminus, resulting in altered affinities for membranes and other proteins.

Not all myristoyl-cofactor switches involve covalent modifications. The binding of GTP to Arf proteins appears to affect the conformation of their amino terminus, allowing presentation of the myristoyl moiety to membranes [a myristoyl-GTP switch; see Refs. 10 and 221 for data that support this model; see Franco et al (222), who report that myristoylation is not required for GTP-dependent binding of Arf to phospholipids]. Binding of calcium to retinal recoverin induces a conformational change in the protein, which apparently allows the protein to insert itself into rod outer segment mem-

[8]The notion of myristoyl-cofactor switches was presented in a paper by Zozulya & Stryer (220).

branes (a myristoyl-calcium switch; Ref. 220). Zozulya & Stryer (220) have suggested that several other members of the EF-hand superfamily of calcium-binding proteins may use a Ca^{2+}-myristoyl switch. These members include rat hippocalcin (223), chicken visinin (224), bovine neurocalcin (225), rat 21-kDa calcium-binding protein (226), bovine calcineurin B (227), and *Drosophila melanogaster* frequenin. Qualitative heterogeneous acylation of a protein such as recoverin could represent a way of adding sensitivity (complexity) to a Ca^{2+}-myristoyl conformational switch, thereby expanding the protein's repertoire of potential functions.

Zozulya & Stryer (220) state that "it should be interesting to learn whether covalently attached prenyl and palmitoyl groups are also dynamically switched in signal transduction processes." Although palmitoylation is a posttranslational and reversible modification, more hydrophobic lipid modifications may not permit dynamic conformational switches in the same sense as proposed for the myristoyl moiety of recoverin. The studies of Pietzsch & McLaughlin (83) suggest that the enhanced reversibility of association of C14:0 (or C12:0, $14{:}1^{\Delta5}$, and $14{:}2^{\Delta5,8}$) with membrane systems should also apply to hydrophobic binding sites on receptor proteins. In order to have a reversible conformational switch, it is necessary that the acyl group be easily removed from interactions with membranes or other proteins. The quantitative heterogeneity of acylation found for MARCKS and Gpa1p could represent a way of generating a subset of a protein that does not have a switch, or else a longer time course of switching, involving the increased or decreased acylation of newly synthesized protein.

ACKNOWLEDGMENTS

We thank our colleagues Jennifer Lodge, David Rudnick, Emily Jackson-Machelski, Frank Li, Robert Duronio, Robert Heuckeroth, Dwight Towler, George Gokel, Tianbao Lu, Charles McWherter, Robin Weinberg, and Nandini Kishore for their contributions to work cited in this review. We are grateful to Alan Aderem, Thomas Neupert, James Hurley, Richard Kahn, Henrik Dohlman, Martha Cyert, Xiaolu Yang, and Marian Carlson for providing information about their unpublished work. Mark Goebl kindly searched his yeast protein database to provide the data presented in Table 2. D.R.J. and R.S.B. are participants in the Medical Scientist Training Program. Work from our laboratory was supported by grants from the National Institutes of Health (AI27179 and AI30188) and Monsanto.

Literature Cited

1. Wilcox C, Hu J-S, Olson EN. 1987. *Science* 238:1275–78
2. Deichaite I, Casson LP, Ling H-P, Resh MD. 1988. *Mol. Cell Biol.* 8: 4295–301
3. James G, Olson EN. 1989. *J. Biol. Chem.* 264:20928–33
4. Dohlman HG, Goldsmith P, Spiegel AM, Thorner J. 1993. *Proc. Natl. Acad. Sci. USA* 90:9688–92
5. Carr SA, Biemann K, Shoji S, Parmelee DC, Titani K. 1982. *Proc. Natl. Acad. Sci. USA* 79:6128–31
6. Aitken A, Cohen P, Santikarn S, Williams DH, Calder AG, et al. 1982. *FEBS Lett.* 150:314–18
7. Rudnick DA, McWherter CA, Gokel GW, Gordon JI. 1993. *Adv. Enzymol.* 67:375–430
8. Dizhoor AM, Ericsson LH, Johnson RS, Kumar S, Olshevskaya E, et al. 1992. *J. Biol. Chem.* 267:16033–36
9. Serafini T, Orci L, Amherdt M, Brunner M, Kahn RA, Rothman JI. 1991. *Cell* 67:239–53
10. Haun R, Tsai S-C, Adamik R, Moss J, Vaughan M. 1993. *J. Biol. Chem.* 268:7064–68
11. Randazzo PA, Yang YC, Rulka C, Kahn RA. 1993. *J. Biol. Chem.* 268: 9555–63
12. Wang JK, Walaas SI, Sihra TS, Aderem A, Greengard P. 1989. *Proc. Natl. Acad. Sci. USA* 86:2253–56
13. Thelen M, Rosen A, Nairn AC, Aderem A. 1991. *Nature* 351:320–22
14. Sawai T, Negishi M, Nishigaki N, Ohno T, Ichikawa A. 1993. *J. Biol. Chem.* 268:1995–2000
15. Clegg CH, Ran W, Uhler MD, McKnight GS. 1989. *J. Biol. Chem.* 264:20140–46
16. Slice LW, Taylor SS. 1989. *J. Biol. Chem.* 264:20940–46
17. Yonemoto W, McGlone ML, Taylor SS. 1993. *J. Biol. Chem.* 268:2348–52
18. Zheng J, Knighton DR, Xuong N-H, Taylor SS, Sowadski JM, Ten Eyck LF. 1993. *Protein Sci.* 2:1559–73
19. Chow M, Moscufo N. 1992. In *Lipid Modifications of Proteins,* ed. MJ Schlesinger, pp. 59–81. Boca Raton, Fla: CRC Press
20. Gottlinger HG, Sodroski JG, Haseltine WA. 1989. *Proc. Natl. Acad. Sci. USA* 86:5781–85
21. Bryant ML, Ratner L. 1990. *Proc. Natl. Acad. Sci. USA* 87:523–27
22. Persing DH, Varmus HE, Ganem D. 1987. *J. Virol.* 61:1672–77
23. Bruss V, Ganem D. 1991. *Proc. Natl. Acad. Sci. USA* 88:1059–63
24. Macrae DR, Bruss V, Ganem D. 1991. *Virology* 181:359–63
25. Marchildon GA, Casnellie JE, Walsh KA, Krebs EG. 1984. *Proc. Natl. Acad. Sci. USA* 81:7679–82
26. Schultz AM, Henderson LE, Oroszlan S, Garber EA, Hanafusa H. 1985. *Science* 227:427–29
27. Buss JE, Sefton BM. 1985. *J. Virol.* 53:7–12
28. Semba K, Nishizawa M, Miyajima N, Yoshida MC, Sukegawa J, et al. 1986. *Proc. Natl. Acad. Sci. USA* 83:5459–63
29. Sukegawa J, Semba K, Yamanashi Y, Nishizawa M, Miyajima N, et al. 1987. *Mol. Cell Biol.* 7:41–47
30. Kypta RM, Hemming A, Courtneidge SA. 1988. *EMBO J.* 7:3837–44
31. Cheng SH, Harvey R, Espino PC, Semba K, Yamamoto T, et al. 1988. *EMBO J.* 7:3845–55
32. Lock P, Ralph S, Stanley E, Boulet I, Ramsay R, Dunn AR. 1991. *Mol. Cell Biol.* 11:4363–70

32a. Resh MD. 1993. *Biochem. Biophys. Acta* 1115:307–22

33. Kamps MP, Buss JE, Sefton BM. 1985. *Proc. Natl. Acad. Sci. USA* 82:4625–28
34. Buss JE, Kamps MP, Gould K, Sefton BM. 1986. *J. Virol.* 58:468–74
35. Linder ME, Burr JG. 1988. *Proc. Natl. Acad. Sci. USA* 85:2608–12

35a. Resh MD. 1994. *Cell* 76:411–13

36. Langner CA, Lodge J, Travis S, Caldwell JE, Lu T, et al. 1992. *J. Biol. Chem.* 267:17159–69
37. Wiegand RC, Carr C, Minnerly JC, Pauley AM, Carron CP, et al. 1992. *J. Biol. Chem.* 267:8591–98
38. Weinberg R, McWherter C, Freeman SK, Wood, DC, Gordon JI, Lee S. 1994. *J. Biol. Chem.* Submitted
39. Paltauf F, Kohlwein SD, Henry SA. 1992. In *The Molecular and Cellular Biology of the Yeast Saccharomyces: Gene Expression,* ed. EW Jones, JR Pringle, JR Broach, II:415–500. Cold Spring Harbor, NY: Cold Spring Harbor Lab.
40. Towler D, Glaser L. 1986. *Proc. Natl. Acad. Sci. USA* 83:812–2816
41. Heuckeroth RO, Gordon JI. 1989. *Proc. Natl. Acad. Sci. USA* 86:5262–66
42. Duronio RJ, Rudnick DA, Johnson RJ, Johnson DR, Gordon JI. 1991. *J. Cell Biol.* 113:1313–30

43. Dietzel C, Kurjan J. 1987. *Cell* 50: 1001–10
44. Miyajima I, Nakafuku M, Nakayama N, Brenner C, Miyajima A, et al. 1987. *Cell* 50:1011–19
45. Stone DE, Cole GM, Lopes MD, Goebl M, Reed SI. 1991. *Genes Dev.* 5:1969–81
46. Johnson DR, Duronio RJ, Langner CA, Rudnick DA, Gordon JI. 1993. *J. Biol. Chem.* 268:483–94
47. Herman PK, Stack JH, DeModena JA, Emr SD. 1991. *Cell* 64:425–37
48. Herman PK, Stack JH, Emr SD. 1991. *EMBO J.* 10:4049–60
49. Sewell JL, Kahn RA. 1988. *Proc. Natl. Acad. Sci. USA* 85:4620–24
50. Stearns T, Willingham MC, Botstein D, Kahn RA. 1990. *Proc. Natl. Acad. Sci. USA* 87:1238–42
51. Stearns T, Kahn RA, Botstein D, Hoyt MA. 1990. *Mol. Cell Biol.* 10:6690–99
52. Cyert MS, Thorner J. 1992. *Mol. Cell Biol.* 12:3460–69
53. Lee KS, Hines LK, Levin DE. 1993. *Mol. Cell Biol.* 13:5843–53
54. Trueheart J, Boeke JD, Fink GR. 1987. *Mol. Cell Biol.* 7:2316–28
55. McCaffrey G, Clay FJ, Kelsay K, Sprague GF Jr. 1987. *Mol. Cell Biol.* 7:2680–90
56. Blumer KJ, Thorner J. 1991. *Annu. Rev. Physiol.* 53:37–57
57. Levin DE, Bartlett-Heubusch E. 1992. *J. Cell Biol.* 116:1221–29
58. Towler DA, Adams SP, Eubanks SR, Towery DS, Jackson-Machelski E, et al. 1987. *Proc. Natl. Acad. Sci. USA* 84:2708–12
59. Duronio RJ, Towler DA, Heuckeroth RO, Gordon JI. 1989. *Science* 243: 796–800
60. Duronio RJ, Gordon JI, Boguski MS. 1992. *Proteins Struct. Funct. Genet.* 13:41–56
61. Maundrell K. 1990. *J. Biol. Chem.* 265:10857–64
62. Tommasino M, Maundrell K. 1991. *Curr. Genet.* 20:63–66
63. Johnston LH, Williamson DH, Johnson AL, Fennell DJ. 1982. *Exp. Cell Res.* 141:53–62
64. Kishore NS, Lu T, Knoll LJ, Katoh A, Rudnick DA, et al. 1991. *J. Biol. Chem.* 266:8835–53
65. Kishore NS, Wood DC, Mehta PP, Wade AC, Lu T, et al. 1993. *J. Biol. Chem.* 268:4889–902
66. Rudnick DA, Duronio RJ, Gordon JI. 1992. In *Lipid Modification of Proteins: A Practical Approach,* ed. NM Hooper, AJ Turner, pp. 37–61. New York: IRL Press at Oxford Univ. Press
67. Paige LA, Chafin DR, Cassady JM, Geahlen RL. 1989. *Anal. Biochem.* 181:254–58
68. Wagner AP, Retey J. 1990. *Anal. Biochem.* 188:356–58
69. McIlhinney RAJ, McGlone K. 1989. *Biochem. J.* 263:387–91
70. Duronio RJ, Jackson-Machelski E, Heuckeroth RO, Olins PO, Devine CS, et al. 1990. *Proc. Natl. Acad. Sci. USA* 87:1506–10
71. Rudnick DA, McWherter CA, Adams SP, Ropson IJ, Duronio RJ, Gordon JI. 1990. *J. Biol. Chem.* 265:13370–78
72. Duronio RJ, Rudnick DA, Johnson RL, Linder ME, Gordon JI. 1990. *Methods Enzymol.* 1:253–63
73. Towler DA, Eubanks SR, Towery DS, Adams SP, Glaser L. 1987. *J. Biol. Chem.* 262:1030–36
74. Paige LA, Zheng G-Q, DeFrees SA, Cassady JM, Geahlen RL. 1989. *J. Med. Chem.* 32:1667–73
75. Rudnick DA, McWherter CA, Rocque WJ, Lennon PJ, Getman DP, Gordon JI. 1991. *J. Biol. Chem.* 266:9732–39
76. Rudnick DA, Rocque WJ, McWherter CA, Toth MV, Jackson-Machelski E, Gordon JI. 1993. *Proc. Natl. Acad. Sci. USA* 90:1087–91
77. Bhatnagar R, Jackson-Machelski E, McWherter C, Gordon JI. 1994. *J. Biol. Chem.* 269:In press
78. Rudnick DA, Lu T, Jackson-Machelski E, Hernandez JC, Li Q, et al. 1992. *Proc. Natl. Acad. Sci. USA* 89:10507–11
79. Awaya J, Ohno T, Ohno H, Omura S. 1975. *Biochim. Biophys. Acta* 409: 267–73
79a. Remington S, Weigand G, Huber R. 1982. *J. Mol. Biol.* 158:111–52
79b. Leslie AGW, Moody PCE, Shaw WV. 1988. *Proc. Natl. Acad. Sci. USA* 8:4133–37
79c. Kim JP, Wang M, Paschke R. 1993. *Proc. Natl. Acad. Sci. USA* 90:7523–27
79d. Kragelund BB, Andersen KV, Madsen JC, Poulsen FM. 1993. *J. Mol. Biol.* 230:1260–77
79e. Mattevi A, Obmolova G, Kalk KH, Wesphal AH, de Kok A, Hol WGJ. 1993. *J. Mol. Biol.* 230:1183–99
79f. Mattevi A, Obmolova G, Kalk KH, Teplyakov A, Hol WGJ. 1993. *Biochemistry* 32:3887–901
79g. Day PJ, Shaw WV. 1992. *J. Biol. Chem.* 267:5122–27
79h. Day PJ, Shaw WV, Gibbs MR, Leslie AGW. 1992. *Biochemistry* 31:4198–205

79i. Fierke CA, Jencks WP. 1986. *J. Biol. Chem.* 261:7603–6
79j. Lewendon A, Murray IA, Shaw WV, Gibbs MR, Leslie AGW. 1990. *Biochemistry* 29:2075–80
80. Woldegiorgis G, Spennetta T, Corkey BE, Williamson JR, Shrago E. 1985. *Anal. Biochem.* 150:8–12
81. Corkey BE. 1988. *Methods Enzymol.* 166:55–70
82. Corkey BE, Deeney JT. 1990. In *Fatty Acid Oxidation: Clinical, Biochemical, and Molecular Aspects*, pp. 217–32. Liss
83. Peitzsch RM, McLaughlin S. 1993. *Biochemistry* 32:10436–43
84. Smith RH, Powell GL. 1986. *Arch. Biochem. Biophys.* 244:357–60
85. Heuckeroth RO, Glaser L, Gordon JI. 1988. *J. Biol. Chem.* 85:8795–99
86. Heuckeroth RO, Jackson-Machelski E, Adams SP, Kishore NS, Huhn M, et al. 1990. *J. Lipid Res.* 31:1121–29
87. Devadas B, Lu T, Katoh A, Kishore NS, Wade AC, et al. 1992. *J. Biol. Chem.* 267:7224–39
88. Devadas B, Kishore NS, Adams SP, Gordon JI. 1993. *Bioorgan. Med. Chem. Lett.* 31:779–84
89. Gokel GW, Lu T, Rudnick DA, Jackson-Machelski E, Gordon JI. 1992. *Isr. J. Chem.* 32:127–33
90. Peseckis SM, Deichaite I, Resh MD. 1993. *J. Biol. Chem.* 268:5107–14
91. Lu T, Li Q, Katoh A, Hernandez J, Duffin K, et al. 1994. *J. Biol. Chem.* 269:5349–57
92. Towler DA, Adams SP, Eubanks SR, Towery DS, Jackson-Machelski E, et al. 1988. *J. Biol. Chem.* 263:1784–90
93. Towler DA, Gordon JI, Adams SP, Glaser L. 1988. *Annu. Rev. Biochem.* 57:69–99
94. Duronio RJ, Rudnick DA, Adams SP, Towler DA, Gordon JI. 1991. *J. Biol. Chem.* 266:10498–504
95. Pellman D, Garber EA, Cross FR, Hanafusa H. 1985. *Nature* 314:374–77
96. Kaplan JM, Mardon G, Bishop JM, Varmus HE. 1988. *Mol. Cell Biol.* 8:2435–41
97. Stevenson FT, Bursten SL, Fanton C, Licksley RM, Lovett DH. 1993. *Proc. Natl. Acad. Sci. USA* 90:7245–49
98. Muszbek L, Laposata M. 1993. *J. Biol. Chem.* 268:8251–55
99. Nakafuku M, Obara T, Kaibuchi K, Miyajima K, Miyajima A, et al. 1988. *Proc. Natl. Acad. Sci. USA* 85:1374–78
100. Koren R, LeVitre JA, Bostian KA. 1986. *Gene* 41:271–80
101. Legrain M, DeWilde M, Hilger F. 1986. *Nucleic Acids Res.* 14:3059–73
102. Nagata S, Nagashima K, Tsunetsugu-Yokota Y, Fujimura K, Mizazaki M, Kaziro Y. 1984. *EMBO J.* 3:1825–30
103. Schirmaier F, Philippsen P. 1984. *EMBO J.* 3:3311–15
104. Yang X, Hubbard EJA, Carlson M. 1992. *Science* 257:680–82
105. Knoll LJ, Gordon JI. 1993. *J. Biol. Chem.* 268:4281–90
106. Black PN, DiRusso CC, Metzger AK, Heimert TL. 1992. *J. Biol. Chem.* 267:25513–20
107. Kameda K, Nunn WD. 1981. *J. Biol. Chem.* 256:5702–7
108. Spratt SK, Black PN, Ragozzino M, Nunn WD. 1984. *J. Bacteriol.* 158: 535–42
109. Rudnick DA, Johnson RL, Gordon JI. 1992. *J. Biol. Chem.* 267:23852–61
110. Knoll LJ, Levy MA, Stahl PD, Gordon JI. 1992. *J. Biol. Chem.* 267:5366–73
111. Mishina M, Roggenkamp R, Schweizer E. 1980. *Eur. J. Biochem.* 111:79–87
112. Al-Feel W, Chirala SS, Wakil SJ. 1992. *Proc. Natl. Acad. Sci. USA* 89:4534–38
113. Schweizer M, Roberts LM, Holtke H-J, Takabayashi K, Hollerer E, et al. 1986. *Mol. Gen. Genet.* 203:479–86
114. Schweizer E, Muller G, Roberts LM, Schweizer M, Rosch J, et al. 1987. *Fet. Wiss. Technol.* 89:570–77
115. Chirala SS, Kuziora MA, Spector DM, Wakil SJ. 1987. *J. Biol. Chem.* 262: 4231–40
116. Mohamed AH, Chirala SS, Mody NH, Huang W-Y, Wakil SJ. 1988. *J. Biol. Chem.* 263:12315–25
117. Schweizer E, Bolling H. 1970. *Proc. Natl. Acad. Sci. USA* 67:660–66
118. Singh N, Wakil SJ, Stoops JK. 1985. *Biochemistry* 24:6598–602
119. Hori T, Nakamura N, Okuyama H. 1987. *J. Biochem.* 101:949–56
120. Okuyama H, Saito M, Joshi VC, Gunsberg S, Wakil SJ. 1979. *J. Biol. Chem.* 254:12281–84
121. Funabashi H, Kawaguchi A, Tomoda H, Omura S, Okuda S, Iwasaki S. 1989. *J. Biochem.* 105:751–55
122. Fields S, Song O-K. 1989. *Nature* 340:245–46
123. Chien C-T, Bartel PL, Sternglanz R, Fields S. 1991. *Proc. Natl. Acad. Sci. USA* 88:9578–82
124. Henry SA, Fogel S. 1971. *Mol. Gen. Genet.* 113:1–19
125. Kuhn L, Castorph H, Schweizer E. 1972. *Eur. J. Biochem.* 24:492–97
126. Meyer KH, Schweizer E. 1974. *J. Bacteriol.* 117:345–50

127. Johnson DR, Coks S, Feldman H, Gordon JI. 1994. *Proc. Natl. Acad. Sci. USA*. Submitted
128. Cottrell M, Viljoen BC, Kock JLF, Lategan PM. 1986. *J. Gen. Microbiol.* 132:2401–3
129. Duronio RJ, Knoll LJ, Gordon JI. 1992. *J. Cell Biol.* 117:515–29
130. Kamiryo T, Parthasarathy S, Numa S. 1976. *Proc. Natl. Acad. Sci. USA* 73:386–90
131. Kamiryo T, Parthasarathy S, Mishina M, Iida Y, Numa S. 1977. *Agric. Biol. Chem.* 41:1295–301
132. Johnson DR, Knoll LJ, Rowley N, Gordon JI. 1994. *J. Cell Biol.* Submitted
133. Abe T, Fujino T, Fukuyama R, Minoshima S, Shimizu N, et al. 1992. *J. Biochem.* 111:123–28
134. Suzuki H, Kawarabayasi Y, Kondo J, Abe T, Nishikawa K, et al. 1990. *J. Biol. Chem.* 265:8681–85
135. Fujino T, Yamamoto T. 1992. *J. Biochem.* 111:197–203
136. van Beilen JB, Eggink G, Enequist H, Bos R, Witholt B. 1992. *Mol. Microbiol.* 6:3121–36
137. Knoll LJ, Johnson DR, Gordon JI. 1994. *J. Biol. Chem.* Submitted
138. Chang Y-H, Teichert U, Smith JA. 1990. *J. Biol. Chem.* 265: 19892–97
139. Chang Y-H, Teichert U, Smith JA. 1992. *J. Biol. Chem.* 267:8007–11
140. Ben-Bassat A, Bauer K, Chang S-Y, Myambo K, Boosman A, Chang S. 1987. *J. Bacteriol.* 169:751–57
141. Miller CG, Strauch KL, Kukral AM, Miller JL, Wingfield PT, et al. 1987. *Proc. Natl. Acad. Sci. USA* 84:2718–22
142. Miller CG, Kukral AM, Miller JL, Movva NR. 1989. *J. Bacteriol.* 171: 5215–17
143. Nakamura K, Nakamura A, Takamatsu H, Yoshikawa H, Yamane K. 1990. *J. Biochem.* 107:603–7
144. Chang S-YP, McGary EC, Chang S. 1989. *J. Bacteriol.* 171:4071–72
145. Lee F-JS, Lin L-W, Smith JA. 1988. *J. Biol. Chem.* 263:14948–55
146. Lee F-JS, Lin L-W, Smith JA. 1990. *J. Biol. Chem.* 265:11576–80
147. Driessen HPC, deJong WW, Tesser GI, Bloemendal H. 1985. *CRC Crit. Rev. Biochem.* 18:281–325
148. Persson B, Flinta C, von Heijne G, Jornvall H. 1985. *Eur. J. Biochem.* 152:523–27
149. Augen J, Wold F. 1986. *Trends Biochem. Sci.* 11:494–97
150. Tsunasawa S, Stewart JW, Sherman F. 1985. *J. Biol. Chem.* 260:5382–91
151. Lee F-JS, Lin L-W, Smith JA. 1989. *J. Bacteriol.* 171:5795–802
152. Mullen JR, Kayne PS, Moerschell RP, Tsunasawa S, Gribskov M, et al. 1989. *EMBO J.* 8:2067–75
153. Park E-C, Szostak JW. 1990. *Mol. Cell Biol.* 10:4932–34
154. Park E-C, Szostak JW. 1992. *EMBO J.* 11:2087–93
155. Whiteway M, Szostak JW. 1985. *Cell* 43:483–92
156. Whiteway M, Freedman R, Van Arsdell S, Szostak JW, Thorner J. 1987. *Mol. Cell Biol.* 7:3713–22
157. Rose TM, Schultz ER, Todaro GJ. 1992. *Proc. Natl. Acad. Sci. USA* 89:11287–91
158. Mandrup S, Jepsen R, Skøtt H, Rosendal J, Højrup P, et al. 1993. *Biochem. J.* 290:369–74
159. Andersen KV, Poulsen FM. 1992. *J. Mol. Biol.* 226:1131–41
160. Kragelund BB, Andersen KV, Madsen JC, Knudsen J, Poulsen FM. 1993. *J. Mol. Biol.* 230:1260–77
161. Rosendal J, Ertbjerg P, Knudsen J. 1993. *Biochem. J.* 290:321–26
162. Chirala SS. 1992. *Proc. Natl. Acad. Sci. USA* 89:10232–36
163. Collart MA, Struhl K. 1993. *EMBO J.* 12:177–86
164. Reed SI. 1980. *Genetics* 95:561–77
165. Neiman AM, Chang F, Komachi K, Herskowitz I. 1990. *Cell Regul.* 1:391–401
166. Deleted in proof
167. Nunn WD, Colburn RW, Black PN. 1986. *J. Biol. Chem.* 261:167–71
168. Black PN. 1991. *J. Bacteriol.* 173:435–42
169. Kumar GB, Black PN. 1993. *J. Biol. Chem.* 268:15469–76
170. Kohlwein SD, Paltauf F. 1983. *Biochim. Biophys. Acta* 792:310–17
171. Duronio RJ, Reed SI, Gordon JI. 1992. *Proc. Natl. Acad. Sci. USA* 89:4129–33
172. Rocque W, McWherter CA, Wood DC, Gordon JI. 1993. *J. Biol. Chem.* 268:9964–71
173. Lodge JK, Johnson RL, Weinberg RA, Gordon JI. 1994. *J. Biol. Chem.* 269: 2996–3009
174. Kwon-Chung KJ. 1976. *Mycologia* 68:-821–33
175. Denich KT, Malloy PJ, Feldman D. 1992. *Gene* 110:123–28
176. Sadhu C, Hoekstra D, McEachern MJ, Reed SI, Hicks JB. 1992. *Mol. Cell Biol.* 12:1977–85

177. Johnson DR, Cox AD, Solski PA, Devadas B, Adams SP, et al. 1990. *Proc. Natl. Acad. Sci. USA* 87:8511–15
178. Stryer L. 1991. *J. Biol. Chem.* 266: 10711–14
179. Kokame K, Fukada Y, Yoshizawa T, Takao T, Shimonishi Y. 1992. *Nature* 359:749–52
180. Neubert TA, Johnson RS, Hurley JB, Walsh KA. 1992. *J. Biol. Chem.* 267:18274–77
181. Dizhoor AM, Ray S, Kumar S, Niemi G, Spencer M, et al. 1991. *Science* 251:915–18
182. Lambrecht HG, Koch KW. 1991. *EMBO J.* 10:793–98
183. Deleted in proof
184. Huffaker TC, Hoyt MA, Botstein D. 1987. *Annu. Rev. Genet.* 21:259–84
185. Bender A, Pringle JR. 1991. *Mol. Cell Biol.* 11:1295–305
186. Wang L, Yerram NR, Kaduce TL, Specto AA. 1992. *J. Biol. Chem.* 267:18983–90
187. McIlhinney RAJ, McGlone K. 1990. *Biochem. J.* 271:681–85
188. Manenti S, Sorokine O, Van Dorsselaer A, Taniguchi H. 1993. *J. Biol. Chem.* 268:6878–81
189. Aderem AA, Albert KA, Keum MM, Wang JKT, Greengard P, Cohn ZA. 1988. *Nature* 332:362–64
190. Stumpo DJ, Graff JM, Albert KA, Greengard P, Blackshear PJ. 1989. *Proc. Natl. Acad. Sci. USA* 86:4012–16
191. Aderem A. 1992. *Cell* 71:713–16
192. Hartwig JH, Thelen M, Rosen A, Janmey PA, Nairn AC, Aderem A. 1992. *Nature* 356:618–22
193. Allen L-AH, Aderem A. 1994. *J. Cell Biol.* Submitted
194. da Silva AM, Klein C. 1990. *J. Cell Biol.* 111:401–7
195. Boyle JJ, Ludwig EH. 1962. *Nature* 196:893–94
196. Cross FR, Garber EA, Pellman D, Hanafusa H. 1984. *Mol. Cell Biol.* 4:1834–42
197. Kaplan JM, Varmus HE, Bishop JM. 1990. *Mol. Cell Biol.* 10:1000–9
198. Silverman L, Resh MD. 1992. *J. Cell Biol.* 119:415–25
199. Strittmatter P, Kittler JM, Coghill JE, Ozols J. 1993. *J. Biol. Chem.* 268: 23168–71
200. Sessa WC, Harrison JK, Barber CM, Zeng D, Durieux ME, et al. 1992. *J. Biol. Chem.* 267:15274–76
201. Sessa WC, Barber CM, Lynch KR. 1993. *Circ. Res.* 72:921–24
202. Quest AFG, Chadwick JK, Wothe DD, McIlhinney RAJ, Shapiro BM. 1992. *J. Biol. Chem.* 267:15080–85
203. Phillips WD, Maimone MM, Merlie JP. 1991. *J. Cell Biol.* 115:1713–23
204. Chow M, Newman JFE, Filman D, Hogle JM, Rowlands DJ, Brown F. 1987. *Nature* 327:482–86
205. Marc D, Drugeon G, Haenni A-L, Girard M, van der Werf S. 1989. *EMBO J.* 8:2661–68
206. Mumby SM, Heuckeroth RO, Gordon JI, Gilman AG. 1990. *Proc. Natl. Acad. Sci. USA* 87:728–32
207. Jones TLZ, Simonds WF, Merendino JJ Jr, Brann MR, Spiegel AM. 1990. *Proc. Natl. Acad. Sci. USA* 87:568–72
208. Linder ME, Pang I-H, Duronio RJ, Gordon JI, Sternweis PC, Gilman AG. 1991. *J. Biol. Chem.* 266:4654–59
209. Gallego C, Gupta SK, Winitz S, Eisfelder BJ, Johnson GL. 1992. *Proc. Natl. Acad. Sci. USA* 89:9695–99
210. Linder ME, Middleton P, Hepler JR, Taussig R, Gilman AG, Mumby SM. 1993. *Proc. Natl. Acad. Sci. USA* 90:3675–79
211. Paige LA, Nadler MJS, Harrison ML, Cassady JM, Geahlen RL. 1993. *J. Biol. Chem.* 268:8669–74
212. Nadler MJS, Harrison ML, Ashendel CL, Cassady JM, Geahlen RL. 1993. *Biochemistry* 32:9250–55
213. Shenoy-Scaria AM, Timson Gauen LK, Kwong J, Shaw AS, Lublin DM. 1993. *Mol. Cell Biol.* 13:6385–92
214. Bryant ML, Heuckeroth RO, Kimata JT, Ratner L, Gordon JI. 1989. *Proc. Natl. Acad. Sci. USA* 86:8655–59
215. Bryant ML, Ratner L, Duronio RJ, Kishore NS, Devadas B, et al. 1991. *Proc. Natl. Acad. Sci. USA* 88:2055–59
216. Navia MA, Fitzgerald PMD, McKeever BM, Leu C, Heimbach JC, et al. 1989. *Nature* 337:615–20
217. Knighton DR, Zheng JH, Ten Eyck LF, Ashford VA, Xuong N-h, et al. 1991. *Science* 253:407–14
218. Knighton DR, Zheng J, Ten Eyck LF, Xuong N-h, Taylor SS, Sowadski JM. 1991. *Science* 253:414–20
219. Taniguchi H, Manenti S. 1993. *J. Biol. Chem.* 268:9960–63
220. Zozulya S, Stryer L. 1992. *Proc. Natl. Acad. Sci. USA* 89:11569–73
221. Kahn RA, Randazzo P, Serafini T, Weiss O, Rulka C, et al. 1992. *J. Biol. Chem.* 267:13039–46

222. Franco M, Chardin P, Chabre M, Paris S. 1993. *J. Biol. Chem.* 268: 24531–34
223. Kobayashi M, Takamatsu K, Saitoh S, Noguchi T. 1993. *J. Biol. Chem.* 268:18898–904
224. Yamagata K, Goto K, Kuo CH, Kondo H, Miki N. 1990. *Neuron* 4:469–76
225. Okazaki K, Watanabe M, Ando Y, Hagiwara M, Terasawa M, Hikada H. 1992. *Biochem. Biophys. Res. Commun.* 185:147–53
226. Kuno T, Kajimoto Y, Hashimoto T, Mukai H, Shirai Y, et al. 1992. *Biochem. Biophys. Res. Commun.* 184: 1219–25
227. Minta A, Kao JP, Tsien RY. 1989. *J. Biol. Chem.* 264:8171–78

Annu. Rev. Biochem. 1994. 63:915–48

REPAIR OF OXIDATIVE DAMAGE TO DNA: Enzymology and Biology

Bruce Demple and Lynn Harrison
Department of Molecular and Cellular Toxicology, Harvard School of Public Health, Boston, Massachusetts 02115

KEY WORDS: free radicals, DNA glycosylases, nucleases, apurinic endonucleases, mutagenesis

CONTENTS

INTRODUCTION

Considerable interest has arisen in recent years in the formation and consequences of oxidative damage to DNA. This interest derives in large part from the realization that the use of oxygen by aerobic organisms is

0066-4154/94/0701-0915$05.00

accompanied by the formation of reactive by-products: free-radical forms of oxygen that can damage most cellular components. Many toxic agents also generate intracellular oxygen radicals. Oxidative damage refers to the damages formed by these reactive oxygen species. Repair enzymes that mediate the removal of oxidative damages from DNA help counteract the potential cytotoxic, mutagenic, and carcinogenic effects of these damages.

Oxygen radicals are formed by many pathways in biology, which will be summarized here only briefly; recent comprehensive reviews are available (1–3). Components of the chain of electron carriers that bring reducing equivalents to cytochrome oxidase are subject to oxidation by O_2 to yield superoxide radical ($O_2^{\bullet -}$); a substantial portion of the respiratory electrons seem to be diverted in this way (1, 3). Evolution counteracted this problem by generating defensive activities such as superoxide dismutases, which catalyze the conversion of $O_2^{\bullet -}$ to hydrogen peroxide (H_2O_2), and catalases and peroxidases that destroy hydrogen peroxide. H_2O_2 that escapes destruction can react with reduced transition metals (especially Fe^{2+} or Cu^{1+}) to form hydroxyl radical ($^{\bullet}OH$), the proximal agent of much oxidative damage to DNA (4–6). Hydroxyl radical reacts avidly with all macromolecules, but "sacrificial" defenses exist in the form of small molecules such as glutathione (7) and α-tocopherol (2, 3), which break free-radical chain reactions (8).

Free radicals formed from the radiolysis of H_2O in aqueous solution are the agents of most ionizing radiation damage, which is further potentiated by the presence of O_2 (5). Numerous other toxic agents depend directly on oxygen to kill cells or cause mutations. These include redox-cycling agents, such as paraquat, which catalytically divert electrons from NAD(P)H to O_2 to generate intracellular superoxide (8), and inhibitors of electron transport that cause the accumulation of reduced components of the electron transport chain (1).

Free radicals are not solely the accidents of aerobic metabolism or the agents of environmental toxins. Superoxide is generated deliberately by activated macrophages and some other immune system cells as a cytotoxic weapon (9). Mammalian macrophages can also generate another free radical, nitric oxide ($NO^{\bullet}$), which contributes significantly to the killing of bacteria and tumor cells (10, 11). $NO^{\bullet}$ can combine with $O_2^{\bullet -}$ to yield another potent damaging agent, peroxynitrite ($ONOO^-$), which is as reactive as $^{\bullet}OH$ (12, 13). Since $NO^{\bullet}$ is also produced for intercellular signaling by various other cell types (14–16), peroxynitrite might be a common biological agent of radical damage.

The load of oxidative damage from endogenous processes is evidently significant. As a measure of this load, specific oxidative damages may accumulate with time in the DNA of animal cells (3), although the precise numbers are subject to debate (17). Numerous inverse correlations have been made between the metabolic rate of various mammalian species and their maximum lifespan (18), and used along with other observations to argue that oxidative damage is a cause of aging and age-related diseases such as cancer and atherosclerosis (2, 3, 18).

Connections to oxidative damage have also turned up unexpectedly in other areas of medicine. For example, mutations in the gene encoding the cytoplasmic superoxide dismutase underlie an inherited form of the neurodegenerative disease amyotrophic lateral sclerosis (19). The cumulative genetic (and other) lesions arising from incessant oxidative damage could be the degenerative mechanism in this and other cases. DNA repair systems are key lines of defense that prevent or limit the biological effects of this damage.

Oxidative DNA Damage

Not only are oxidative DNA damages formed by diverse pathways, they constitute probably the most varied class of DNA damages: Nearly 100 different free radical damages have been identified (5, 20). Much of the current knowledge about such damages derives from studies of the radiation chemistry of DNA and nucleotides. In this context, it should be noted that the identified oxidation products include both primary damages, which are often unstable, and the breakdown products of these damages arising from hydrolysis and rearrangement reactions.

BASE DAMAGES As a rule, compared to single-stranded polynucleotides, bases in duplex DNA are relatively protected from free-radical attack (5). A major family of thymine oxidation products—collectively called "thymine glycol"—consists of four isomers of 5,6-dihydroxy-5,6-dihydrothymine (Figure 1), in which the vicinal hydroxyl groups may be either *cis* or *trans* to each other, and the 5-carbon may exist in either the *R* or *S* configuration (5, 20). The *cis*-glycol forms predominate under most conditions, and the *R* form is the main product in native DNA exposed to γ-radiation or chemical oxidants (20).

Thymine glycol undergoes alkali-catalyzed decomposition to form various fragmentation products, ultimately to yield urea residues N-linked to deoxyribose. This property of thymine glycols was the basis for assays that demonstrated the production of this lesion in DNA following chemical oxidation or ionizing radiation treatment in vitro, or in the DNA of cells irradiated with 313-nm UV light (21). Intact thymine glycol released from DNA was eventually assayed by high performance liquid chromatography (HPLC) methods (22), and has also been measured in mammalian urine as an indicator of endogenous oxidative DNA damage (23).

Ionizing radiation also produces the ring-saturated product 5,6-dihydro-

DAMAGE	FORMATION	PROPERTIES
Thymine glycol 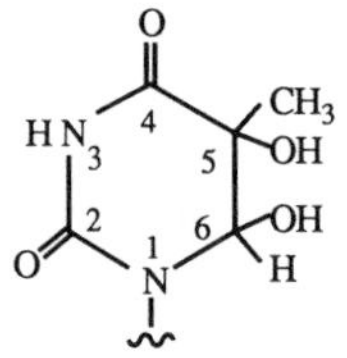	$^{\bullet}OH$; ionizing radiation; ultraviolet radiation	Replicative block; poorly mutagenic (?)
8-Oxoguanine 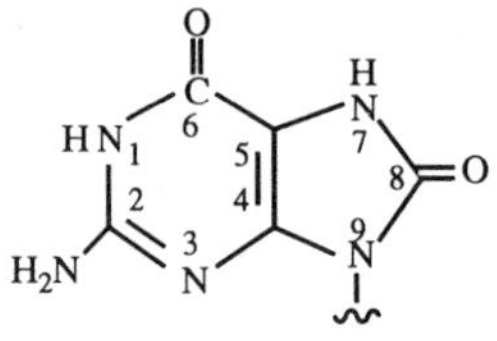	$^{\bullet}OH$; ionizing radiation; singlet oxygen	Miscoding; mutagenic
Formamidopyrimidine 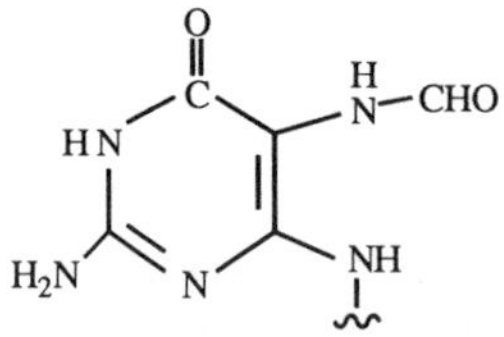	$^{\bullet}OH$; ionizing radiation	Replicative block; cytotoxic (?)

Figure 1 Representative oxidative base lesions. See text for discussion of the formation and repair of individual damages.

thymine; two stereoisomers of this product are possible (5, 20). Urea and other fragmented derivatives of thymine are direct radiation products; these ruptured bases also include methyltartronylurea, and the cyclized product 5-hydroxy-5-methylhydantoin (5, 20). Although none of the lesions thus far mentioned seems to be strongly mutagenic, both urea and thymine glycol in the template block DNA synthesis in vitro (24–26).

The exocyclic methyl group of thymine does not escape damage: 5-Hydroxymethyluracil (5-hydroxythymine) is a significant product of radiation damage or the chemical oxidants generated by activated polymorphonuclear neutrophils (5, 20, 27). 5-Methylcytosine can be oxidized to the analogous 5-hydroxymethyl derivative (28). 5-Hydroxymethyluracil does not seem to have inherent genetic instability, because at least one genome (that of the

Bacillus subtilis phage SP8) has a complete substitution of thymine by this modified base (29). One might therefore consider these damages as mere markers for free-radical damage to DNA, but some evidence for cytotoxic effects of 5-hydroxymethyluracil in mammalian DNA has appeared (30, 31). The mechanism that might account for such differences in different contexts is unclear.

Most oxidation products of cytosine are analogous to those detected for thymine. 5,6-Hydrated cytosines in DNA even predominate over thymine glycols after H_2O_2 treatment of cultured mammalian cells (20). However, cytosine glycols can undergo a secondary reaction not available to thymines: deamination. The slow, acid-catalyzed hydrolytic deamination of cytosine is substantially accelerated upon saturation of the 5,6-bond (5, 32). Deamination of cytosine hydrates produces uracil derivatives that form basepairs preferentially with adenine instead of guanine, so this process probably enhances the mutagenic effects of cytosine hydrates relative to thymine glycols.

Purines can undergo oxidation of the ring atoms to form various products. Of these, 8-oxo-7,8-dihydroguanine (8oxoG; Figure 1) [sometimes cited as 8-hydroxyguanine, the less common tautomeric form (33, 34)] has become the species of greatest current interest. 8OxoG was identified as a radiation damage initially (35), but has now been demonstrated as the product of an array of agents that generate reactive oxygen, including numerous chemical oxidants (35), photosensitizers (such as methylene blue) that generate the reactive species singlet oxygen ($^{\Delta 1}O_2$) (36), and biological sources such as activated polymorphonuclear leukocytes (37). 8-Oxoadenine is also formed, in much smaller amounts (20, 35), although the biological effects of the adenine lesion appear to be limited (38).

Fragmented purines, with ruptured imidazole rings joined to an intact pyrimidine ring, are significant radiation products (5, 20). These formamidopyrimidine (FAPy) residues (Figure 1) are derived from both adenine and guanine, and can arise as secondary products of N7-alkylated guanines or adenines: Base-catalyzed hydrolysis generates N-methyl FAPy from N7-methylguanine (39). This mechanism, and the identification of a common repair activity for radiation- and alkylation-generated FAPy lesions (see below), shows that DNA damages cannot always be neatly categorized as "oxidative," "alkylation-induced," etc, and that repair of damages of different origins can proceed through common steps.

Monomeric purine damages can also have bonding rearrangements. Abstraction of the C1′-hydrogen in monomeric dAMP and fixation of the α-anomer of this nucleotide generates a product with an inverted (α) glycosylic bond (40). The steric constraints of double-stranded DNA probably counteract this reaction. Radical attack at C5′ can lead to cyclization of dAMP (40), but this product is also expected to be rare in duplex DNA. For both of these lesions, single-stranded DNA would presumably be a more likely target, and the possibility of specific repair systems for lesions of this type has not been much addressed.

DEOXYRIBOSE DAMAGES The phosphodiester backbone is exposed to solvent even in duplex DNA and is readily attacked by free radicals. The attraction by this polyanion of metals such as Fe^{2+} or Cu^{1+} may also target localized formation of $^{\bullet}OH$ by H_2O_2 (6). The resulting damages can displace bases, oxidize deoxyribose, and fragment the sugar. The resulting strand breaks are the hallmarks of oxidative damaging agents such as X-rays or hydrogen peroxide, and all of these damages cause a complete loss of genetic information in the affected strand.

Radical attack at C1′ can eliminate the apparently undamaged base and oxidize C1′ to the carboxylate to form deoxyribonic acid (41). Base displacement also occurs after attack at C4′, which can form 4-keto-deoxyribonate (41, 42). Both of these lesions leave the DNA chain intact, but the lability of these abasic sites in alkali allows their release from oxidized DNA (41, 42). Oxidized abasic sites are quantitatively the major products of the important antitumor drug bleomycin (43) and ene-diyne agents such as neocarzinostatin (44). Abasic sites are also formed by the action of repair enzymes called DNA N-glycosylases, which remove damaged or unconventional bases, including some oxidative damages (see below). Collectively, these are called AP sites (for apurinic or apyrimidic sites).

C4′ oxidation also leads to the fragmentation of deoxyribose through a series of reactions not yet fully characterized. This fragmentation produces a strand break bracketed by a normal nucleotide 5′-terminus and a 3′ end that bears either the two-carbon glycolate (3′-phosphoglycolate esters) or simple 3′-phosphate monoesters. The formation of 3′-phosphoglycolates was demonstrated first for bleomycin, for which it was also shown that the base is lost as a base-propenal adduct (45). Subsequent work confirmed the formation of 3′-phosphoglycolate esters by ionizing radiation (46).

Agents such as bleomycin and neocarzinostatin also produce complex lesions that consist of an oxidized abasic site in one strand accompanied by a 3′-phosphoglycolate ester or an abasic site nearby in the opposite strand. These and other complex lesions are also produced by ionizing radiation (5). The double-strand breaks that characterize both bleomycin and ionizing radiation are evidently the result of the cleavage of an abasic site closely opposed to a strand break, or the fragmentation of deoxyribose in both strands within a short distance (5, 47, 48). The presence of damages in both stands probably presents particular problems for DNA repair.

CROSSLINKS Ionizing radiation generates crosslinks between DNA and proteins (49, 50), for which a few linkage sites (e.g. thymine 5-methyl to tyrosine 4-oxygen) have been identified (20, 51). Such lesions are not likely to be innocuous, but pathways mediating their repair have not been identified.

Small amounts of interstrand crosslinks might also be formed by radical attack; the repair of such lesions would presumably recruit recombination machinery, as is the case for some chemical or photochemical crosslinks (52, 53).

Crosslinks between groups in the same chain (intrastrand) could be cytotoxic or mutagenic, as in the case of *cis*-platinum or even UV-induced pyrimidine dimers (52). Recent studies with some model compounds suggest the oxidative formation of such damages between nearby purines (54). Whether these lesions are also formed significantly in DNA (duplex or single-stranded) has not been established.

DNA GLYCOSYLASES

Enzymes of this class initiate repair by hydrolyzing the base-sugar (N-C *glycosylic*) bond of modified or incorrect bases to generate abasic (AP) sites (32, 52). The resulting pathway of "base excision repair" acts on oxidative and non-oxidative damages, determined by the specificity of individual DNA glycosylase enzymes.

Thymine Glycol Glycosylases

ENDONUCLEASE III OF *E. COLI* Endonuclease III (M_r 27,000) specifically cleaves duplex DNA damaged by X-rays (55), UV light (56), osmium tetroxide, or acid pH (57). This range of substrates is due to the combination of a broad-specificity DNA glycosylase and an activity in the same protein that cleaves AP sites (22, 58, 59). Thymine glycol and dihydrothymine were the first identified substrates for the endonuclease III glycosylase (22), but a more complete listing now includes a range of 5,6-saturated and fragmented pyrimidine derivatives, such as *cis*- and *trans*-thymine glycols, methyltartronylurea, 5-hydroxy-5-methylhydantoin, 6-hydroxy-5,6-dihydrothymine, and 6-hydroxy-5,6-dihydrocytosine (58–60). Endonuclease III also accounts for the urea-DNA glycosylase identified by Breimer & Lindahl (61). The range of base damages recognized by endonuclease III illustrates how evolution generated broad-specificity enzymes to cope with oxidative damage, rather than many individual activities of high specificity.

The "nicking activity" by which endonuclease III was first identified (55–57) is due to its AP lyase function (22). The enzyme cleaves abasic

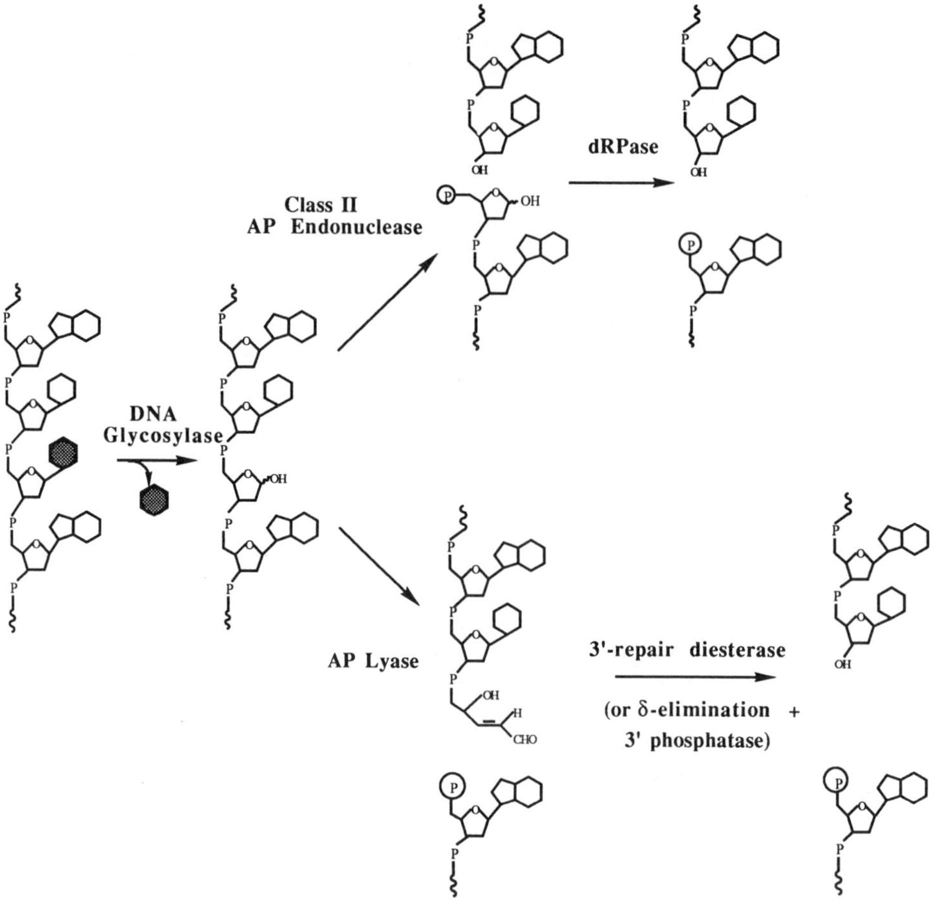

Figure 2 Base excision repair pathways. For simplicity, only one strand of the duplex is shown; the enzymes involved usually have specificity for double-stranded DNA. Upper pathway: Release of a damaged base (filled hexagon) is followed by incision by a class II AP endonuclease, the abasic deoxyribose-5-phosphate released by a dRPase (5′-deoxyribophosphodiesterase; Refs. 89, 226) or 5′-exonuclease, and DNA synthesis and ligation (not shown) fill the gap. Lower pathway: In some cases, class I AP endonuclease associated with the glycosylase may form a strand break that requires 3′-processing to generate a suitable primer terminus for DNA polymerase. See text for details.

sites that are generated either by hydrolytic base loss (17, 32) or as products of endonuclease III or other DNA glycosylases (22, 58, 59). The phosphodiester cleavage by endonuclease III occurs via β-elimination to generate 5′ termini with 5′-phosphate nucleotides and 3′ termini bearing the 2,3-unsaturated abasic residue 4-hydroxy-2-pentenal (Figure 2) (62, 63). These blocked 3′ termini require further processing by other enzymes to generate 3′-OH primers, which can be used for repair synthesis (see section on REPAIR OF DEOXYRIBOSE DAMAGES below).

E. coli endonuclease III has no apparent cofactor requirements, and the activity is not significantly stimulated in the presence of metals or other small molecules (56–59). However, the protein does contain a $[4Fe\text{-}4S]^{2+}$ iron-sulfur cluster that is not particularly redox-sensitive (64). Although iron-sulfur centers can be involved in hydrolytic reactions, as in aconitase (65), there is no reason to expect oxidation and reduction in the reactions catalyzed by this enzyme (66). The structure of the cluster is altered only slightly by the binding to endonuclease III of either thymine glycol, a competitive inhibitor of the enzyme, or an oligonucleotide containing an abasic site (67). These studies indicate a primarily structural role for the metal center in this enzyme (67), although a role in the catalytic chemistry has not been ruled out entirely.

The three-dimensional structure of endonuclease III shows that the iron-sulfur center is anchored by a $Cys\text{-}X_6\text{-}Cys\text{-}X_2\text{-}Cys\text{-}X_5\text{-}Cys$ sequence (68), which also occurs in a 181-amino-acid segment of the MutY DNA-glycosylase (see below) with 66% similarity and 24% identity to endonuclease III (69). Recent experimental evidence suggests that MutY is also an FeS protein (70). Perhaps a family of such iron-sulfur DNA glycosylase/AP lyase enzymes exists.

Bacteria deficient in endonuclease III were engineered using the cloned *nth* gene (71). Null mutants containing insertions in the *nth* gene were not hypersensitive to any genotoxic agent tested, including hydrogen peroxide or γ-rays (71), which generate significant numbers of oxidized pyrimidines in DNA (5, 20). These lesions might be expected to be cytotoxic, because thymine glycols strongly block DNA synthesis in vitro (24–26). These observations suggest either (*a*) that the production or the effects of thymine glycol and other endonuclease III–sensitive lesions have been overestimated significantly, or (*b*) that other repair activities might substitute for endonuclease III in *E. coli*. Such a secondary role is proposed for an activity called endonuclease VIII, which attacks thymine glycol, dihydrothymine, and abasic sites in DNA in vitro (72). Such redundancy could also be provided by the nucleotide excision repair system (see below).

The elimination of endonuclease III from bacteria does produce one modest phenotype: a ~2-fold increase in the spontaneous rate of certain reversions (71). This mutator phenotype is not enhanced by the additional elimination of exonuclease III or endonuclease IV (71), which suggests that the effect does not involve AP sites. Although the mutator effect of endonuclease III deficiency appears small, it could involve a relatively large increase in the rate of some otherwise rare mutation. Of course, it remains possible that endonuclease III has an important in vivo substrate that has yet to be identified.

Expression of the *nth* gene in X-ray-sensitive Chinese hamster cells (xrs7; Ref. 73) increased cellular resistance to H_2O_2, consistent with a role for the enzyme in repairing oxidized bases, although the cells were sensitized to bleomycin (74). Endonuclease III in this context was proposed to incise abasic sites closely opposed to single-strand breaks to generate double-strand breaks, as shown to occur with bleomycin-damaged DNA in vitro (75).

EUKARYOTIC THYMINE GLYCOL GLYCOSYLASES Possible counterparts to *E. coli* endonuclease III have been found in other organisms. Like endonuclease III, these enzymes operate without the aid of cofactors and show strong specificity for damages in double-stranded DNA. Eukaryotic cells contain enzymes with demonstrated DNA glycosylase activity against thymine glycol. Such activities have been partially purified from the baker's yeast *Saccharomyces cerevisiae* (76, 77), calf thymus (78, 78a), and cultured mouse and human cells (79, 80, 81). Whether the eukaryotic enzymes are structurally related to *E. coli* endonuclease III has not yet been established. Table 1 summarizes some key features of endonuclease III and other *E. coli* enzymes for oxidative DNA damage, and lists some eukaryotic counterparts.

Formamidopyrimidine Glycosylase (Fpg/MutM)

E. coli Fpg is a 31-kDa enzyme, purified originally as a DNA glycosylase activity that releases fragmented purine lesions (formamidopyrimidines, or FAPy; Figure 1) from methylated, alkali-treated DNA (82). Guanine adducts of acetylaminofluorene to C8 can undergo a different imidazole ring cleavage to generate a ruptured species that Fpg can liberate from DNA (83). Fpg also releases the unmethylated FAPy derivatives of both adenine and guanine generated by ionizing radiation (84).

The *E. coli* gene (*fpg*) encoding this glycosylase was cloned by large-scale screening for overproduction of FAPy glycosylase activity among strains expressing individual recombinant plasmids (85). Overexpressing the cloned *fpg* gene allowed the purification of large amounts of Fpg for detailed characterization of the protein (86). The purified glycosylase displays strand-cleaving activity that acts at AP sites in DNA (87). Unlike AP lyases or class II AP endonucleases, the products of the Fpg reaction are 3′ and 5′-phosphate termini. The enzyme is hypothesized to catalyze a β-elimination reaction to yield a 3′-terminal unsaturated deoxyribose (as for other AP lyases), followed by a δ-elimination reaction to release the modified sugar (88). Fpg acts by β-elimination to remove deoxyribose-5-phosphate moieties

Table 1 Repair enzymes for oxidative DNA damage[a]

E. coli enzyme	Specificity	Copy no. in *E. coli*	Eukaryotic counterpart
Exonuclease III	AP endo (II); 3'-repair Inducible (*katF/rpos*)	1000 (146) 5000–10,000 (225)	Ape (human, 159–161) Apex (mouse, 163) Bapl (bovine, 162)
Endonuclease IV	AP endo (II); 3'-repair Inducible (*soxRS*)	50 (193) 500–1000 (193)	Apnl (yeast, 200)
Endonuclease III	TG glycosylase; β-lyase	500 (59)	UV endonuclease I & II (mouse, human, 79–81); Activity found in yeast (76, 77) & bovine (78, 78a) cells
FAPy glycosylase (MutM)	FAPy/8oxoG glycosylase β-lyase; dRPase	400 (calculated from Ref. 86)	Activity detected in HeLa cells (103)
MutY protein	Adenine glycosylase (8oxoG : A preference)	30 (calculated from Ref. 108)	G:A mismatch endonuclease activity detected in human cells (116)

[a] Abbreviations used: AP endo (II), class II (hydrolytic) AP endonuclease; 3'-repair, 3'-PGA diesterase/3'-phosphatase; TG glycosylase, thymine glycol glycosylase; β-lyase, class I AP endonuclease; FAPy, formamidopyrimidine; 8oxoG: 8-oxo-7,8-dihydroguanine (8-hydroxyguanine); dRPase, 5'-deoxyribophosphodiesterase.

from DNA 5′-termini (89), which are the products of class II AP endonuclease incision. Therefore, the concerted actions of enzymes such as endonuclease IV and Fpg might completely remove oxidative damage from DNA to produce undamaged 3′-OH and 5′-phosphate termini that require only DNA polymerase and ligase to complete the repair process (Figure 2).

FAPy lesions strongly block DNA synthesis in vitro (90), and so have cytotoxic potential, although their mutagenic effect could be limited (17). Thus, the significance of an enzyme specific for FAPy lesions was open to debate. However, Fpg was found to act as an efficient 8oxoG glycosylase (91), which accounted for an activity previously termed 8-hydroxyguanine endonuclease (92). Hence, Fpg removes purines with either ruptured (FAPy) or intact (but oxidized) imidazole rings (8oxoG).

8oxoG in DNA has significant miscoding potential in vitro: Replicative DNA polymerases of both bacteria and mammalian cells insert dAMP preferentially opposite 8oxoG in the template, whereas polymerases associated with repair inserted dCMP (the "correct" nucleotide) preferentially (93, 94). This in vitro miscoding specificity is reflected in results of transfection studies with single-stranded DNA vectors containing a single 8oxoG residue, which gave rise to G→T transversions at a frequency of ~1% after replication in *E. coli* (95). Grollman (96) has discussed the possible physical basis of 8oxoG mispairing. The ease with which

8oxoG is formed by numerous oxidizing agents (35) suggests that 8oxoG could make a significant contribution to mutagenesis and genetic instability in some circumstances.

The biological importance of 8oxoG removal by Fpg protein came suddenly into focus with the realization that the *mutM* gene (97) of *E. coli* is identical to *fpg* (98). Strains with *mutM* mutations have a ~5-fold increase in the spontaneous rate of GC→TA transversions (97). It was not easy to rationalize this mutational specificity as a result of unrepaired FAPy lesions, which strongly block DNA synthesis (90). The propensity of 8oxoG to mispair with A during DNA synthesis (93) suggested that the mutator effect of *mutM* (*fpg*) mutations arises from replication of unrepaired 8oxoG residues in the genome of enzyme-deficient cells. Indeed, Fpg-deficient strains were recently reported to contain a steady-state level of 8oxoG in their DNA six times higher than measured for wild-type cells (99). How this difference relates to mutagenesis is not entirely clear, because the small fraction of 8oxoG residues present in the DNA template ahead of a replication fork would constitute most of the mutagenic threat at any given time, compared to those positioned elsewhere.

Fpg contains one zinc atom per molecule (86). Zinc is tightly bound to the protein through the carboxy-terminal sequence Cys-X_2-Cys-X_{16}-Cys-X_2-Cys (100). Substitution of any of the cysteines in this segment eliminates zinc from the protein and inactivates the DNA glycosylase and deoxyribophosphodiesterase activities of Fpg (100, 102). Since the mutant proteins do not bind to oligonucleotides containing 8oxoG (101, 102), the zinc in Fpg may help to form a recognition domain for stable interactions with DNA.

If 8oxoG is so readily formed in DNA and effective in miscoding, one would expect repair of this lesion to be universal in aerobic organisms (see Table 1). Indeed, mammalian activities against 8oxoG-containing DNA have been reported. Separate 8oxoG-DNA glycosylase and endonuclease activities from HeLa cells have been partially purified (103). The HeLa 8oxoG-DNA endonuclease activity was shown to cleave the 8oxoG-containing strand, which distinguishes the human enzyme from the bacterial MutY protein (see below). The possible roles of separate glycosylase and endonuclease enzymes have not been established. Such biochemical complexity in repair pathways for 8oxoG could recapitulate the multilevel system found in *E. coli* (94, 104), and is certainly consistent with the proposed biological significance of 8oxoG in DNA. Surprisingly, a recent report (105) suggests removal of 8oxoG by mammalian N-methylpurine-DNA glycosylases expressed in *E. coli,* with the human enzyme much less active than the murine protein. Whether these enzymes contribute to 8oxoG repair in mammalian cells has not yet been addressed.

MutY: A DNA Mismatch Glycosylase for Oxidative Damage

Mutations in the *mutY* gene of *E. coli* elevate the spontaneous frequency of GC→TA transversions substantially (106). The MutY protein was initially characterized as an adenine-specific DNA glycosylase acting at A:G mismatches (107, 108). More recently, MutY has been implicated in the *E. coli* defense system against mutagenesis from 8oxoG. MutY protein efficiently removes adenine from 8oxoG:A basepairs (109), which are generated during replication of 8oxoG-containing DNA templates (93). By contrast, Fpg/MutM protein acts poorly on 8oxoG:A pairs (91), which would avoid production of a T:A basepair at the site (94, 104).

Mutations in *mutY* act synergistically with *mutM* mutations to elevate the GC→TA transversion frequency by 2–3 orders of magnitude, and the *mutY* mutator phenotype can be nearly fully suppressed by overexpression of Fpg activity (110). Thus, if Fpg (MutM) does not remove 8oxoG from DNA before replication, MutY can remove an A incorporated opposite 8oxoG; DNA repair synthesis then has a high probability of forming an 8oxoG:C pair (93, 94), which is a good substrate for repair by Fpg (91). MutT, a dGTPase (111) that preferentially hydrolyses 8oxodGTP (112), also apparently forms part of the *E. coli* defense system against 8oxoG by eliminating this mutagenic precursor from the nucleotide pool. A similar activity has recently been described in human cells (113, 114).

The specificity of MutY may also include 8oxoA:A and C:A mismatches as weak substrates (70, 109, 115). A corresponding repair system for mismatches opposite 8oxoadenine may not be necessary: dTMP is inserted preferentially opposite 8oxoA during in vitro DNA synthesis (93), and 8oxoA in single-stranded phage DNA was not detectably mutagenic upon transfection into *E. coli* (38).

There has been some dispute as to whether MutY has an associated AP endonuclease: Au et al (108) did not detect AP endonuclease activity in their MutY preparation, in contrast to Lu & Chang (115). Elution and renaturation (in the presence of ferrous ion and inorganic sulfide) of MutY from an SDS/polyacrylamide gel produced an enzyme with both adenine-DNA glycosylase and class I AP endonuclease activities, although it is not known whether the DNA was cleaved by a β-elimination reaction (70).

An activity that cleaves both 5′ and 3′ to G:A mismatches has been detected in human nuclear extracts (116). However, it is not known whether this enzyme has an associated DNA glycosylase activity, acts on 8oxoG:A mismatches, or is related to MutY protein.

Hypoxanthine-DNA Glycosylase

Deamination of adenine to generate hypoxanthine (the nucleotide dIMP in DNA) produces damage with mutagenic potential (117). Hypoxanthine (Hx) glycosylase releases deaminated adenine from DNA and has been detected in *E. coli* (118), bovine (119), and human cells (120, 121). Nitrosative deamination of adenine by the free radical nitric oxide can form significant amounts of Hx in DNA (122, 123). Partial purification of the *E. coli* and bovine Hx glycosylases has allowed initial characterization of this enzyme activity. Hx DNA glycosylases (M_r ~30,000) act preferentially and without any known cofactor on double-stranded DNA to release free Hx and produce AP sites (118, 119).

Hx-DNA glycosylase does not recognize various mismatched basepairs (G:T, A:G, or A:C), but specifically acts on dIMP in DNA (124). The bovine enzyme releases Hx basepaired with T or C in a synthetic substrate, although the reaction is 15–20 times slower with the latter substrate than with the former (124).

Recent experiments have unexpectedly identified Hx glycolsylase activity as a function of both bacterial and mammalian 3-methyladenine-DNA glycoylase enzymes (124a). In *E. coli,* only the AlkA protein releases Hx, although with a specificity constant nearly five orders of magnitude lower than that for 3-methyladenine. Purified mammalian alkylpurine glycosylases also released Hx, with activities significantly higher than measured for *E. coli* AlkA (124a). These observations now provide for a genetic analysis of the biological significance of Hx glycosylase, and further emphasize that single repair activities might act in multiple pathways.

5-Hydroxymethyluracil and 5-Hydroxymethylcytosine DNA Glycosylases

A DNA glycosylase activity that liberates HMU (hydroxymethyluracil) from DNA was partly purified from mouse plasmacytoma cells (125). By various indirect criteria, this activity seemed to be distinct from other DNA glycosylases. An HMU glycosylase activity of M_r ~38,000 was extensively purified from calf thymus and exhibited an apparent K_m for HMU of 0.7 μM (126). This glycosylase, which was active on both duplex and single-stranded DNA, was not inhibited by HMU up to 30 mM. A Chinese hamster cell line deficient in the HMU glycosylase activity was resistant to 5-hydroxymethyl-2′-deoxyuridine present in the culture medium (30), the toxicity of which has been suggested to be a consequence of DNA degradation due to excessive HMU repair (127).

HMU glycosylase activity has been detected in extracts of a wide variety of eukaryotes (excepting *Drosophila melanogaster* and the yeasts *Schizosaccharomyces pombe* and *Saccharomyces cerevisiae*) but not in bacteria (128). A distinct DNA glycosylase activity for the corresponding oxidation product of 5-methylcytosine, 5-hydroxymethylcytosine (HMC), was detected in calf thymus, but also was absent from bacterial extracts (129). It has been suggested that HMU and HMC glycosylases exist to maintain 5-methylcytosine in the DNA of vertebrate cells in the face of free radical damage (126).

In line with the above suggestion, a limited amount of evidence connects HMU glycosylase activity to genetic stability. HMU glycosylase activity may decline with the population doubling of human fibroblasts (130). Measured HMU glycosylase activity was also lower in a cell line derived from a patient with Werner's syndrome than in cells from age-matched individuals (131). Werner's syndrome is an autosomal recessive disease, in which features similar to accelerated aging and chromosomal instability are seen (32).

UV Endonucleases

Three "UV endonucleases" purified from mouse plasmacytoma cells (79) cleave heavily UV-irradiated DNA; two of these enzymes (UV endonucleases I and II) are thymine glycol glycosylases with accompanying β-lyase activity for AP sites. Similar enzymes have been found in mammalian mitochondria (132), indicative of DNA repair processes in that organelle (133). The UV endonuclease III activity was substantially diminished in fibroblasts derived from patients of the D-complementation group of the inherited disease xeroderma pigmentosum (XP-D; Refs. 132, 134). The known molecular defect in XP is in the excision repair of UV photoproducts, specifically cyclobutane pyrimidine dimers and 6,4-pyrimidine photoproducts, as well as other "bulky" lesions (32, 52). However, DNA substrates for UV endonuclease III require heavy UV doses (79, 132), more consistent with a specificity for minor photoproducts, such as thymine glycol.

Genetic analysis has actually obscured the link between the fibroblast enzyme and the XP syndromes and reveals the second surprise connected with this enzyme. Polypeptide sequence data revealed that the fibroblast enzyme is identical to S3, a mammalian ribosomal protein (135). Indeed the cloned human S3 cDNA (136) directs the synthesis in bacteria of a protein with all the enzymatic properties of the fibroblast UV endonuclease III (135). The same ribosomal protein from *Drosophila melanogaster,* expressed from the cloned *S3* gene of that organism, also has AP lyase activity (137), which may therefore be a general property of these eukaryotic ribosomal proteins.

A likely candidate gene (*ERCC2*) affected in XP-D has been cloned using genetic complementation (138). This gene maps to chromosome 19q13 (139), but it is unclear whether this locus includes an *S3* gene. In fact, it is likely that there are multiple copies of the *S3* gene in vertebrates (140).

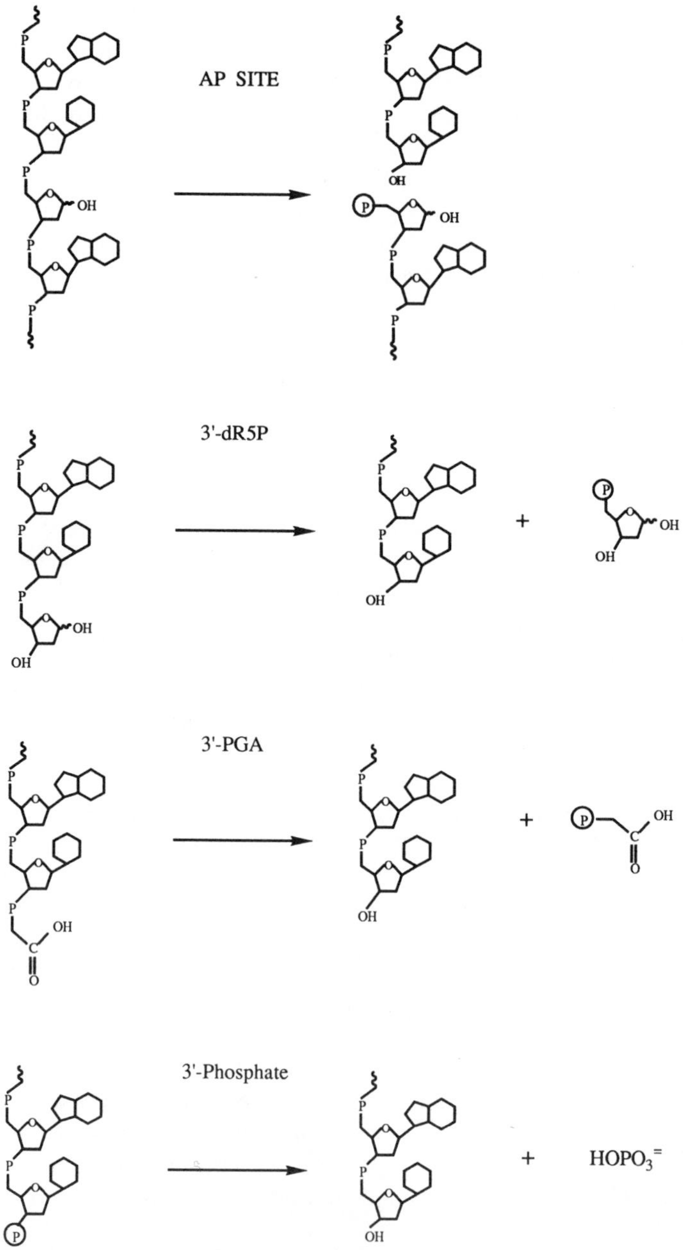

Figure 3 Activities of Class II (Hydrolytic) AP Endonucleases. Active primer termini for DNA repair synthesis are generated by the action of class II AP endonucleases on AP sites or 3′-damages. 3′-Repair by these enzymes can release intact deoxyribose-5-phosphate (dR5P), fragments such as 3′-phosphoglycolate (PG), or free phosphate. See text for details.

It also remains possible that the expression of S3 protein as a "UV endonuclease" is in some way affected by mutations at the XP-D locus.

REPAIR OF DEOXYRIBOSE DAMAGES: AP ENDONUCLEASES AND 3′-TRIMMING

The same enzymes that cleave AP sites arising from all DNA glycosylases also initiate the repair of oxidized abasic sites and deoxyribose fragments formed by free radical attack. These enzymes often combine phosphodiesterase and phosphomonesterase activities (Figure 3), and their broad specificity involves them as a focal point in multiple repair pathways. Molecular analysis has shown that these enzymes fall into two families based on sequence homology.

Exonuclease III of E. coli

Exonuclease III (M_r 30,000) was first discovered as a combined 3′-phosphatase-exonuclease separated from DNA polymerase I during a late stage of purification (141, 142). By 1976, exonuclease III was also established as the major AP endonuclease of *E. coli* (143, 144), despite some earlier confusion over this point (145). Exonuclease III also has a ribonuclease H activity of unknown biological significance (146). The activities of exonuclease III show a near-absolute preference for double-stranded substrates, inhibition by metal chelators such as ethylenediaminetetraacetic acid (EDTA), and stimulation by 1–10 mM Mg^{2+} (141–143, 146). The 3′-phosphatase is highly specific for DNA and accounts for >99% of this activity in extracts of wild-type *E. coli* (141, 146). The 3′→5′ exonuclease degrades linear DNA from both 3′ ends until ~50% is digested, at which point insufficient basepairing remains to support the enzyme (142); exonuclease III can also digest from internal nicks (142, 146). The AP endonuclease function cleaves hydrolytically on the 5′-side of base-free deoxyribose (146), which leaves an abasic 5′-terminal deoxyribose-5-phosphate (a so-called "Class II" AP endonuclease; Figure 2). The 3′-termini generated by exonuclease III acting on all its substrates are normal nucleotides with 3′-hydroxyl groups that are effective primers for DNA polymerases (141, 146–148) (Figure 3).

The properties of exonuclease III have made the enzyme useful as a reagent for molecular biology manipulations (146, 149, 150). The exonuclease can be blocked by phosphorothioate esters to "cap" the 3′-ends of DNA (151). The AP endonuclease is also strongly inhibited by a site-specifically engineered phosphorothioate immediately 5′ to an AP site, with one of the two stereoisomers of the substituted phosphate causing a much greater inhibition of incision than the other (151a). Such a difference would be in keeping with a stereospecific attack of exonuclease III on its target phosphodiester (151).

The need for a powerful DNA 3′-phosphatase/exonuclease activity in *E. coli* was unclear, since no enzymes were known that generate DNA 3′-phosphates in this organism (but see section on Fpg protein). Mutants lacking exonuclease III (*xth* strains) were isolated (152) to address the enzyme's biological function. The *xth* mutants were modestly sensitive to simple alkylating agents, such as methyl methane sulfonate, but not to UV light, ionizing radiation, and many other DNA-damaging agents (152). The sensitivity of *xth* mutants to alkylating agents was consistent with a role for exonuclease III in the repair of AP sites, and the mildness of the phenotype could be explained by the identification of another AP endonuclease in *E. coli,* endonuclease IV, present at lower levels (153). Overwhelming cells with AP sites by eliminating dUTPase (to cause incorporation of uracil into DNA and its subsequent removal by uracil glycosylase) also supported a role for exonuclease III in AP site repair (146).

The key function of exonuclease III in the repair of oxidative damage was revealed with the unexpected discovery that *xth* mutants have extreme hypersensitivity to H_2O_2 (154). Confirmation of the generality of this role was provided by the subsequent observation that oxidative damage generated in vivo by near-UV light also requires exonuclease III for its efficient repair (155).

The specific role of exonuclease III in repair of oxidative damage was established by analysis of chromosomal DNA from bacteria challenged with H_2O_2, which revealed that the enzyme removes deoxyribose fragments from the 3′-termini of DNA strand breaks generated in vivo by free-radical attack (148). Single-strand breaks persisted in the DNA of cells lacking the enzyme (*xth* mutants), while such damages were efficiently eliminated from wild-type bacteria. Most importantly, after isolation of the chromosomal DNA, the persistent strand breaks from *xth* mutants did not support the activity of DNA polymerase I in vitro, but could be activated by a prior treatment with exonuclease III; endonuclease IV (see below) also effected this primer activation (148). The strand breaks in DNA from wild-type cells treated with even very high amounts of H_2O_2 did not exhibit such a requirement for pretreatment. These data suggested that exonuclease III and endonuclease IV are required to remove oxidative damage that blocks DNA polymerase (148).

Likely candidates for such blocking groups were the 3′-phosphoglycolate esters and 3′-phosphates generated by free radical agents (45, 46). We developed a synthetic DNA substrate containing the analog 3′-phosphoglycoaldehyde (3′-PGA), radiolabeled and present at each 3′-end as the sole oxidative damage in this molecule (148). Exonuclease III actively released

PGA from this substrate, with k_{cat}/K_m about equal to that measured for AP sites; a similar result was obtained for endonuclease IV. Several other enzymes have now been identified that are capable of hydrolyzing 3′-PGA and/or activating DNA synthesis primers in oxidatively damaged DNA. In each case, the enzyme also displays hydrolytic AP endonuclease activity as described for exonuclease III. These enzymes are discussed individually below.

Combined exonuclease/AP endonuclease enzymes similar to exonuclease III have been described in other bacteria. These include the Gram-positive species *Haemophilus influenzae* (156) and *Diploccocus pneumoniae* (157). The cloned *exoA* gene from the latter organism encodes a polypeptide clearly homologous to *E. coli* exonuclease III (158). This conservation of proteins related to exonuclease III extends to eukaryotic homologs that have AP endonuclease activity (see below), which suggests a general biological role for this enzyme.

Eukaryotic AP Endonucleases Related to Exonuclease III

Recent efforts have shown that a family of enzymes related to exonuclease III includes proteins from both prokaryotes and eukaryotes (Table 1). As mentioned above, exonuclease III has a counterpart (ExoA) in Gram-positive cells; other members of this family include proteins from *Drosophila* and several mammalian species (Figure 4). Considerable information is now available about the mammalian enzymes, which will be summarized here. The *Drosophila* protein is discussed in the next section.

THE APE-ENCODED ENZYME OF MAMMALIAN CELLS In the past several years, molecular cloning efforts have yielded the cDNAs encoding the major AP endonucleases of mammalian cells. These cDNAs from human (159–161), bovine (162), and murine (163) sources predict proteins of molecular mass 35,500 that are closely related to one another (91–93% identity) and, more distantly, to *E. coli* exonuclease III and *S. pneumoniae* ExoA protein. The homology of the mammalian polypeptides is in keeping with their powerful hydrolytic AP endonuclease activities in vitro: The human enzyme has k_{cat} ~10-fold higher than found for exonuclease III (164, 165). Although the preferred name for this gene is *APE* (159), it has also been called *HAP1* (160), *BAP1* (162), *APEX* (161, 163), and *REF1* (166). The *APE* gene is relatively compact (~3 kb total) and has been mapped to the q11.2-12 region of human chromosome 14 (167–169). This genomic site has not been associated with any known syndrome of DNA repair deficiency, although recent experiments show that *APE* exhibits significant regulation in vivo (L Harrison et al, unpublished data).

The connection of these enzymes to oxidative damage arose from the

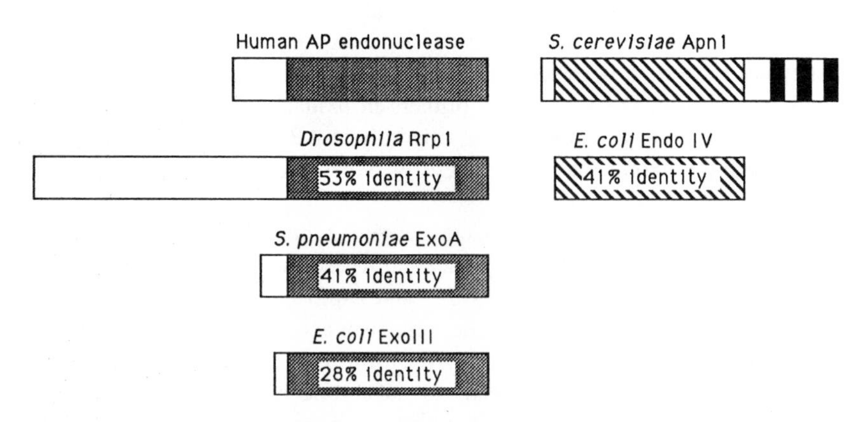

Figure 4 Families of Class II AP Endonucleases. The core regions of homology shared by the Human AP endonuclease (Ape) family (stippled) and Apn1/endonuclease IV family (hatched) are highlighted. The "% identity" in polypeptide refers to the protein shown at the top in each case. The other mammalian AP endonucleases have essentially the same structure as Ape, with 94% (mouse) and 93% (bovine) identities. For Apn1, the three clusters of basic residues in the C terminus are indicated in black. Reproduced with permission from (4).

identification of their 3′-PGA diesterase (3′-repair) activity, shown first for the HeLa cells AP endonuclease (170) and mouse endonuclease (171). Eventually, the HeLa, mouse, and bovine thymus enzymes were purified by monitoring this activity using the synthetic 3′-PGA substrate (165) or primer activation with bleomycin-damaged DNA (162, 171). However, it is not certain that this 3′-repair activity is biologically significant for these enzymes: The catalytic rate of the HeLa enzyme acting on 3′-PGA and 3′-phosphates is nearly 200-fold lower than its AP endonuclease function (165).

One group has reported significant 3′→5′ exonuclease activity of the mouse and human enzymes (hence the name Apex, for AP endonuclease-exonuclease) (161, 163, 171, 172). Several other groups found AP endonucleases isolated from mouse (173), bovine (162, 174), and human (164, 165) to be free of detectable endo- and exonuclease activity against undamaged DNA. It is possible that the detection of exonuclease activity depends on the specific substrate to an unusual degree: The Apex activity has been assayed monitoring the release of labeled nucleotides from DNA polymer containing 3′-terminal label (172) that has not generally been employed by others. However, our laboratory has not detected exonuclease activity in recombinant Ape protein using this approach (DM Wilson et al, unpublished data).

The homology between the Ape family of proteins and exonuclease III is not confined to a single region, but extends nearly the length of both proteins (Figure 4) and includes some highly conserved blocks of sequence (159–163). The N-terminal ~60 residues of the mammalian proteins probably encodes the signal for nuclear targeting (159, 166). The homology shared among the prokaryotic and eukaryotic proteins suggests a significant conservation of function during evolution from a common ancestor. Such conservation is reflected in the potent hydrolytic (Class II) AP endonuclease activity of all the members of this family (146, 158–165, 172–174). However, the powerful 3′-repair activities found in the bacterial prototype (exonuclease III) either did not evolve in the mammalian lineage or were lost. Among these enzymes, the levels of 3′-phosphatase, 3′-PGA diesterase, and 3′→5′ exonuclease activity are approximately correlated with one another, and a relatively limited number of amino acid changes might account for the different specificities (159).

The in vitro properties of the human AP endonuclease are reflected in the *trans*-complementation specificity of the *APE* cDNA in *E. coli*. Expression of Ape protein in AP endonuclease-deficient *E. coli* strains restores some resistance to monofunctional alkylating agents (159, 160) or dUTPase deficiency (160), which shows that Ape (and by extension its counterparts in other mammalian species) can initiate AP site repair in vivo. However, because resistance to oxidative damage is not provided by the human enzyme (159, 160, 175), the relatively weak 3′-repair diesterase activity of Ape is evidently insufficient to replace the loss of this activity in exonuclease III–deficient strains. Other observations suggest that this 3′-repair requirement might be fulfilled by another enzyme in human cells (165) (see below). A modest amount of resistance to γ-rays is conferred by Ape protein expressed in AP endonuclease-deficient *E. coli* (175); perhaps this rescuing effect is due to repair of oxidized abasic sites generated by radiation.

The Ape protein has recently appeared in another context, as a protein (called Ref1) isolated based on its ability to restore DNA binding to Jun•Fos heterodimers or Jun•Jun homodimers damaged in vitro by oxidation (166). The nature of this damage is uncertain, but does depend on a conserved cysteine residue in the DNA-binding region of these proteins and is thought not to involve disulfide bonding (176). The reaction of Jun/Fos with Ape/Ref1 is probably stoichiometric, and depends on the N-terminal segment of the protein not related to exonuclease III (177, 178); exonuclease III itself does not replace the mammalian protein in this reaction (177).

The significance of the Ref1 activity remains to be established. Treatment of cells in culture with oxidizing agents has a modest activating effect on AP1-binding activity (containing Jun and Fos) assayed in vitro, while radical-scavenging or reducing agents seem to stimulate it strongly (179). The extent to which these in vivo effects depend on redox reactions involving the DNA-binding domain cysteines has not been tested. It is suggestive that *fos* genes with the critical cysteine changed to serine, which should render Fos protein more stable toward oxidation, have enhanced transforming ability (cited in Ref. 178). Both the possibility of significant redox regulation exerted by Ape, and the specific role of the protein in various types of DNA repair, will have to be resolved by modulating the levels and activity of this protein in vivo.

DROSOPHILA RRP1 PROTEIN A protein related to exonuclease III has also been isolated from *Drosophila melanogaster,* and its cDNA (*RRP1*) cloned and sequenced (180–182). The predicted Rrp1 polypeptide has a 250-residue C terminus clearly belonging to the exonuclease III family of proteins (Figure 4), while the N-terminal ~400 residues are not significantly related to known sequences (181). Rrp1 was isolated initially from *Drosophila* Schneider cells as an activity that forms joint DNA molecules in an in vitro recombination reaction (180). After the relation to exonuclease III was noted, it was verified that Rrp1 protein is both a 3′→5′ exonuclease against undamaged, double-stranded DNA and a class II (hydrolytic) AP endonuclease (182, 183).

Although the role of this enzyme in *Drosophila* and its possible relation to oxidative damage has not yet been established, recent work confirms that the exonuclease III–related segment of the protein contains the 3′ exonuclease activity and probably the AP endonuclease (184). It is noteworthy that the Rrp1 exonuclease activity is only 10^{-2} to 10^{-3} of its AP endonuclease activity. Thus, Rrp1 may lie midway on a continuum of multifunctional proteins ranging from those that combine robust exonuclease and AP endonuclease activities (e.g. exonuclease III and ExoA), to those that favor the AP endonuclease (e.g. Ape, with exonuclease $\sim10^{-5}$ of AP endonuclease; Ref. 165).

Rrp1 [predicted mass 75,000 (181); M_r 105,000 on SDS-polyacrylamide gels (180)] is much larger than Ape (35,500) or exonuclease III (31,000). The N-terminal region of Rrp1 unrelated to the Ape/exonuclease III family (Figure 4) is responsible for the unusual electrophoretic mobility of Rrp1, and harbors the strand transferase activity (184) detected originally (180). Mammalian cells also contain a relatively large (M_r 65,000) class II AP endonuclease, but that enzyme is of mitochondrial origin (185). The mitochondrial AP endonuclease is evidently a member of the Ape/exonuclease III family, because the protein crossreacts with anti-Ape antibodies in immunoblots (185).

Like other eukaryotic members of this family (159, 160), Rrp1 complements the sensitivity of AP endonuclease-deficient *E. coli* to alkylating agents (186). Unlike the mammalian proteins, however, Rrp1 also restores resistance to H_2O_2 in these cells (186), in keeping with the higher level of 3′-specific activity in the *Drosophila* enzyme.

E. coli *Endonuclease IV*

Endonuclease IV was discovered as an EDTA-resistant AP endonuclease activity present in *xth* mutants of *E. coli* (153). Although the enzyme accounts for only ~5% of the total AP endonuclease in wild-type cells, endonuclease IV is induced up to 10-fold in response to certain superoxide-generating agents (187) or by nitric oxide (188). This oxidative stress response is under the transcriptional control of the *soxRS* system (189–191). Along with the phenotypes of endonuclease IV–deficient strains, this regulation has provided insights into the biological function of the enzyme (189–191).

Purified endonuclease IV (M_r 30,000) is a very active and heat-stable class II AP endonuclease (192). The enzyme is also a potent 3′-PGA diesterase and DNA-specific 3′-phosphatase (193). These 3′-repair activities probably account for the ability of endonuclease IV to activate DNA synthesis primers in chromosomal DNA isolated from H_2O_2-treated *xth* mutants (148). Exonuclease III and endonuclease IV account for virtually all of the DNA-specific 3′-phosphatase activity in *E. coli*.

Genetic studies show that endonuclease IV functions in the repair of oxidative DNA damages in vivo. Enzyme-deficient mutants were constructed by targeted disruption of the *nfo* structural gene, which was isolated in a mass screening for EDTA-resistant AP endonuclease activity (194). Bacterial strains lacking endonuclease IV are hypersensitive to killing by the oxidative agents bleomycin and *t*-butyl hydroperoxide; *nfo* mutations also exacerbate the sensitivity of *xth* mutants to H_2O_2 and to the simple alkylating agent methyl methane sulfonate (MMS), and only *nfo xth* double mutants show abnormal sensitivity to γ-radiation (194). A mutant endonuclease IV that failed to provide H_2O_2 resistance to an *nfo xth* double mutant strain had lost the 3′-phosphatase activity but retained AP endonuclease activity; this mutant protein still conferred resistance to alkylation damage (195). Thus, wild-type endonuclease IV operates both to repair specific classes of oxidative damage and as a backup activity for exonuclease III in repairing other oxidative damages and AP sites. The inducible expression of endonuclease IV probably functions in both these capacities (191).

The specific DNA damages whose repair in vivo requires endonuclease IV have not yet been established. It seems likely that these include the oxidized abasic sites produced by agents such as bleomycin, neocarzinostatin, and ionizing radiation (41–44). These damages are cleaved more efficiently in vitro by endonuclease IV than by exonuclease III (B Epe, B Demple, unpublished data), and chromosomal DNA isolated from bleomycin-treated *nfo* mutants contains sites that are cleaved preferentially by endonuclease IV in vitro (196). Conversely, an endonuclease isolated from *E. coli* in an assay for cleavage of bleomycin-treated DNA turned out to be endonuclease IV (197).

Complex damages, with oxidized abasic sites closely opposed to strand breaks in bleomycin- or neocarzinostatin-treated DNA, are better substrates for endonuclease IV than for exonuclease III (47). Such damages are still relatively poor substrates even for endonuclease IV, which may account for mutational hotspots associated with neocarzinostatin (198).

Endonuclease IV as a component of the oxidative stress-inducible *soxRS* regulon (189–191) has a distinct role in repair. For example, the *soxRS*-dependent adaptive resistance to the killing effects of activated macrophages depends strongly on endonuclease IV in a manner that parallels the synthesis of $NO^{\bullet}$ by the macrophages (188). This observation implies that nitric oxide, perhaps in combination with the other toxic agents generated by macrophages, produces a specific class of oxidative damages to the sugar-phosphate backbone that cannot be handled effectively by exonuclease III.

S. cerevisiae *Apn1 Protein*

Apn1 (M_r 40,000) is the major AP endonuclease in baker's yeast, and is present at ~7000 molecules per cell (199). The in vitro activities of the purified yeast protein are quite similar to those of *E. coli* endonuclease IV, to which Apn1 is structurally related (Table 1; see below): combined class II AP endonuclease, 3′-PGA diesterase, and 3′-phosphatase functions specific for duplex DNA (200). Both Apn1 and endonuclease IV have apparent K_m's in the 1–10 nM range for all these substrates (193, 200).

Specific antisera generated against purified Apn1 were used to identify recombinant phage bearing the AP endonuclease structural gene, and polypeptide sequence data confirmed the identity of the cloned *APN1* gene (201). Yeast Apn1 protein (367 residues) is clearly homologous to *E. coli* endonuclease IV, with 41% amino acid identity through a 280-residue segment that corresponds to nearly the entire length of the bacterial enzyme (Figure 4). The occurrence of this protein in both prokaryotic and eukaryotic cells suggests that the enzyme may be more widely distributed, although other members of this family have not yet been confirmed by sequence analysis. Activities with enzymatic properties similar to endonuclease IV have been identified in thermophilic bacteria (202), radioresistant bacteria (203), and plant tissues (204).

The C-terminal segment of Apn1 functions in the localization of the protein to the yeast nucleus, where virtually all of Apn1 normally resides (205). This C terminus is especially rich in basic residues (of 80 amino acids, 24 are lysines or arginines; Ref. 201), and three lysine/arginine clusters can be identified. The most C-terminal of these clusters (KKRKTKK) is essential for nuclear targeting; its removal does not change the stability or enzymatic activity of Apn1, but leaves >90% of the protein in the cytoplasm (205). Surprisingly, nuclear localization of this truncated Apn1 was not restored by attaching the nuclear targeting signal of SV40 T-antigen (205). It is not known whether the C terminus of Apn1 has additional functions (e.g. protein-protein interactions).

Like its bacterial counterpart, Apn1 functions in the repair of both oxidative damages and alkylation-induced AP sites in yeast. Apn1-deficient strains are hypersensitive to H_2O_2, *t*-butyl hydroperoxide, or alkylating agents (201, 206). This hypersensitivity results from the buildup in yeast chromosomal DNA of unrepaired oxidative strand breaks or apurinic sites, detectable in the isolated DNA (206). Apn1 can operate on the same oxidative damages as endonuclease IV: Expression of Apn1 in *E. coli* corrects the oxidant-sensitive phenotype of *nfo* mutants, but not the H_2O_2 hypersensitivity of *xth* mutants (207). Unlike endonuclease IV, the level of Apn1 activity is not detectably regulated in response to oxidative stress or other toxic insults, nor is the level of the active enzyme or of the *APN1* transcript modulated by genetoxic challenges or during the cell cycle (201, 206).

Apn1 performs an important role in genetic stability in yeast. Apn1-deficient yeast strains have ~5-fold increases in the rate of spontaneous mutation, depending on the locus examined (206). This increase is largely due to a nearly 70-fold increase in the rate of AT to CG transversions (206a). Evidently, endogenous DNA damages that accumulate in the repair-deficient strains are targets for mutagenesis. At least some of the mutator phenotype of *apn1* mutants persists during anaerobic growth (206), and a portion of the aerobic mutator effect depends on the presence in yeast of an active DNA glycosylase for alkylated DNA purines (208). It is therefore likely that, in Apn1-deficient strains, DNA glycosylases act on endogenous damages (17) to generate AP sites that can cause miscoding during DNA replication (209); in the absence of the alkylpurine glycosylase, some of those AP sites would not be formed. The demonstrable role of Apn1 in the genetic stability of yeast underscores the steady erosion of DNA integrity that would occur in the absence of DNA repair.

AP Endonucleases as Metalloproteins

Although it is resistant to heat inactivation, endonuclease IV is quite sensitive to metal-chelating agents such as EDTA or 1,10-phenanthroline; the activity is restored by addition of Mn^{2+}, Co^{2+}, or Ni^{2+}, but not by Zn^{2+} (210). Atomic absorption spectroscopy revealed an unusual content of transition metals in the active enzyme, with average stoichiometries of 2.4 atoms of

Zn and 0.6 atom of Mn per endonuclease IV monomer (210). These and other analyses suggest the presence of three distinct metal-binding sites in endonuclease IV: one site that retains Zn in the presence of chelators; a second site that binds Zn but loses the metal to chelators; and a third, chelator-sensitive site that binds either Zn or Mn, but that requires Mn to give enzymatic activity. The Mn available in vivo may be limiting when endonuclease IV is overproduced (193), which could account for the non-integral copy number of this metal in the protein. Endonuclease IV activity is apparently also satisfied by Co or Ni in the third site. Roles for two metals may be envisioned in coordination to both sides of the phosphodiester undergoing hydrolytic attack and in the activation of water, as seen in the 3′→5′ (proofreading) exonuclease of *E. coli* DNA polymerase I (211). How the third site would activate the enzyme with Mn but not Zn is unknown.

Apn1 also appears to contain three metals, all Zn atoms. Unlike endonuclease IV, after inactivation by chelators, Apn1 activity can be restored by addition of Zn alone (210), or by addition of either Co^{2+} or Mn^{2+} (199). It is noteworthy that both endonuclease IV (192, 193) and active Apn1 (201) are resistant to chelators in the presence of DNA; the critical metal sites might therefore be protected in endonuclease-DNA complexes.

The foregoing shows that endonuclease IV and Apn1 are probably *metallo-enzymes,* i.e. enzymes in which one or more metals play a role in the catalytic chemistry. The full metal complement is not required to maintain structure in Apn1, because inactive protein purified in the presence of metal chelators is activated by certain metals in the assay buffer (199). In contrast, endonuclease III (Fe; Ref. 64) and Fpg protein (Zn; Ref. 86) contain metals that may be involved only in maintaining the overall protein structure.

GENERAL REPAIR: NUCLEOTIDE EXCISION AND RECOMBINATION

Excision Repair

General repair systems that recognize diverse DNA lesions also act on oxidative damages. The nucleotide excision repair system of *E. coli* is based on the UvrA, UvrB, and UvrC proteins, which act in a complex series of reactions to locate and form incisions on both sides of DNA lesions (52, 212–214). DNA helicase, polymerase, and ligase reactions then eliminate the damaged 12- or 13-mer oligonucleotide and complete the repair. For at least some DNA damages (particularly those that block transcription), UvrABC-dependent repair in *E. coli* is actively recruited to the transcribed strand of active genes by the Mfd protein (transcription-repair coupling factor; Ref. 215).

The *E. coli* UvrABC system is not pivotal for handling oxidative damage, in contrast to its crucial role in the repair of structurally distorting lesions such as pyrimidine photodimers, some carcinogen adducts, or *cis*-platinum intrastrand crosslinks (52, 212–214). Nevertheless, both in vivo and in vitro evidence indicates a demonstrable function for UvrABC as a secondary defense against oxidative damages. For example, *uvr* mutations combined with *fpg* mutations decrease the transforming efficiency and increase the mutagenesis of plasmid DNA treated in vitro with methylene blue plus light, which generates large amounts of 8oxoG (216). In vitro, UvrABC incises DNA containing thymine glycol, AP sites, or 8oxoG, although these activities are relatively weak compared to the incision rates for some damages such as pyrimidine photodimers (212).

Nucleotide excision repair may also play a secondary role in repairing oxidative damages in the yeast *S. cerevisiae* (217, 218). In humans, DNA photodamages account for much of the cellular and genetic deterioration in individuals with the inherited repair-deficiency disease xeroderma pigmentosum (32). The damage-specificity of the mammalian excision repair system, which is defective in this disease, is reminiscent of the UvrABC system of *E. coli,* and the mammalian repair mechanism also includes many components (212, 218). A recent report indicates that the human excision repair system can act in vitro on an unspecified DNA damage produced by γ-rays or by H_2O_2/Cu^{2+}, at lesions distinct from those recognized by endonuclease III (AP sites, thymine glycols, etc) and Fpg/MutM protein (8-oxoguanines and FAPy residues) (219). It was speculated that these damages might be dihydrothymine crosslinks (20), 8,5′-cyclodeoxyadenosines (40), or purine-purine intrastrand crosslinks (54).

Recombinational Repair

Recombination mechanisms can in principle repair any DNA lesion, provided that an intact copy of the affected region resides in the same cell (53, 220). Recombination machinery can be recruited when strand breaks remain open at a lesion site, or particularly when unrepaired damage blocks the progress of a DNA replication fork to produce a daughter-strand gap. Regarding lesions for which other, specific repair activities exist, it seems likely that only a small fraction would undergo recombinational repair, provided that the lesions remain in double-stranded DNA.

Oxidative damages include at least two classes for which recombination is likely to be crucial: interstrand crosslinks and double-strand breaks. The former are only poorly characterized, but are demonstrated substrates for the recombinational machinery in *E. coli* (52, 53). Repair of double-strand breaks has a strong requirement for recombination proteins for obvious reasons (53, 220): The two free ends must be realigned to restore chromosomal integrity. Such breaks are not produced in significant amounts by most chemical oxidants, but they are among the major products formed by ionizing radiations (5), bleomycin (43, 47), and neocarzinostatin (47, 48). As a consequence, recombination-deficient bacteria and yeast are strongly hypersensitive to all these agents (52, 53, 217, 220).

The details of recombinational repair in mammalian cells are unknown. One candidate gene involved in this process for radiation damages has been identified, however. *XRCC1* was cloned by molecular complementation of the X-ray sensitivity of CHO cells deficient in repairing strand breaks (221). This cell line (EM9) shows a high frequency of sister chromatid exchange that is also suppressed by human *XRCC1* (221). Efforts to purify recombinant XRCC1 protein have revealed a possible interaction with DNA ligase III (222).

SUMMARY AND PERSPECTIVES

In keeping with the conclusion that free radical damage to DNA constitutes a constant and significant threat for aerobic organisms, enzymes that attack specific oxidative damages are widespread among oxygen-using bacteria and yeast, and among mammalian cells. Although free radical attack produces a multitude of DNA damages, the ability of at least some of these repair activities to recognize a range of substrates allows relatively few enzymes to repair the bulk of base and sugar lesions; general repair mechanisms act as secondary repair pathways and can handle more complex lesions.

Despite the abundant information flowing from the facile combination of biochemical and genetic approaches in *E. coli* and, to a lesser extent, *S. cerevisiae,* the function of mammalian counterparts of the microbial enzymes (Table 1) is only hypothesized or extrapolated from microbial models. Moreover, superimposed on the primary reactions of base excision, AP site incision, and fragment release are cellular responses that control the access of enzymes to damages in chromatin. For example, poly(ADP-ribose) polymerase may bind strand breaks before or during repair, and require automodification to induce its release from DNA in order that repair continues (223).

A more complete understanding of oxidative damage to DNA under physiological conditions is desirable, and the repair enzymes identified so far can be used as gentle probes for such studies (224). Key experimental approaches for the near future will also focus on using molecular manipulation of cultured cells and transgenic animal technology to establish the specific role of DNA repair systems as the various genes become available. Among the fundamental issues that such work can address is the role of oxidative damage to DNA in the normal aging process and in age-related diseases such as cancer (3). However, given the redundancy already observed for repair mechanisms in microbes, such a relationship may be difficult to establish in the milieu of multiple repair systems.

ACKNOWLEDGMENTS

We thank the many colleagues who supplied reprints and preprints of their work. We are indebted to Dr. D. M. Wilson III for a criticism of the manuscript, and to Ms. K. Silva for her indefatigable help in preparing it. Work in the authors' laboratory has been supported by the National Institutes of Health (National Cancer Insitute, National Institute of General Medical Sciences, and National Institute of Environmental Health Sciences) and the Camille and Henry Dreyfus Foundation.

Literature Cited

1. Cadenas E. 1989. *Annu. Rev. Biochem.* 58:79–110
2. Sies H. 1991. In *Oxidative Stress: Oxidants and Antioxidants,* ed. H Sies. New York: Academic
3. Ames BN, Shigenaga MK, Hagen TM. 1993. *Proc. Natl. Acad. Sci. USA* 90:7915–22
4. Halliwell B, Aruoma OI. 1993. *DNA and Free Radicals.* London: Horwood
5. von Sonntag C. 1987. *The Chemical Basis of Radiation Biology.* London: Taylor & Francis
6. Imlay JA, Linn S. 1988. *Science* 240: 1302–9
7. Meister A, Anderson ME. 1983. *Annu. Rev. Biochem.* 52:711–60
8. Kappus H, Sies H. 1981. *Experientia* 37:1233–41
9. Babior BM. 1987. *Trends Biochem. Sci.* 12:241–43
10. Hibbs JB Jr, Taintor RR, Vavrin Z, Rachlin EM. 1988. *Biochem. Biophys. Res. Commun.* 157:87–94
11. Drapier JC, Hibbs JB Jr. 1988. *J. Immunol.* 140:2829–38
12. Beckman JS, Beckman TW, Chen J, Marshall PA. 1990. *Proc. Natl. Acad. Sci. USA* 87:1620–24
13. Koppenol WH, Moreno JJ, Pryor WA, Ischiropoulos H, Beckman JS. 1992. *Chem. Res. Toxicol.* 5:834–42
14. Knowles RG, Moncada S. 1992. *Trends Biochem. Sci.* 17:399–402
15. Ignarro LJ. 1990. *Pharmacol. Toxicol.* 67:1–7
16. Bredt DS, Snyder SH. 1992. *Neuron* 8:3–11
17. Lindahl T. 1993. *Nature* 362:709–15
18. Cutler R. 1991. *Ann. NY Acad. Sci.* 621:1–28
19. Rosen DR, Siddique T, Patterson D, Figleqicz DA, Sapp P, et al. 1993. *Nature* 362:59–62
20. Dizdaroglu M. 1992. *Mutat. Res.* 275: 331–42
21. Hariharan PV, Cerutti PA. 1977. *Biochemistry* 16:2791–95
22. Demple B, Linn S. 1980. *Nature* 287:203–8
23. Cathcart R, Schweirs E, Saul RL, Ames BN. 1984. *Proc. Natl. Acad. Sci. USA* 81: 5633–73
24. Clark JM, Pattabiraman M, Jarvis W, Beardsley GP. 1987. *Biochemistry* 26: 5404–9
25. Hayes RC, Petrullo LA, Huang H, Wallace SS, LeClerc JE. 1988. *J. Mol. Biol.* 201:239–46
26. Basu AK, Loechler EL, Leadon SA, Essigmann JM. 1989. *Proc. Natl. Acad. Sci. USA* 86:7677–81
27. Frenkel K, Chrzan K, Troll W, Teebor GW, Steinberg JJ. 1986. *Cancer Res.* 46:5533–40
28. Frenkel K, Cummings A, Solomon J, Cadet J, Steinberg JJ, Teebor GW. 1985. *Biochemistry* 24:4527–33

29. Kallen RG, Simon M, Marmur J. 1962. *J. Mol. Biol.* 5:248–50
30. Boorstein RJ, Chiu LN, Teebor GW. 1992. *Mol. Cell. Biol.* 12:5536–40
31. Kasai H, Iida A, Yamaizumi Z, Nishimura S, Tanooka H. 1990. *Mutat. Res.* 243:249–53
32. Friedberg EC. 1985. *DNA Repair*. New York: Freeman
33. Kouchakdjian M, Bodepudi V, Shibutani S, Eisenberg M, Johnson F, et al. 1991. *Biochemistry* 30: 1403–12
34. Oda Y, Uesugi S, Ikehara M, Nishimura S, Kawase Y, et al. 1991. *Nucleic Acids Res.* 19:1407–12
35. Kasai H, Nishimura S. 1991. See Ref. 2, pp. 99–116
36. Boiteux S, Gajewski E, Laval J, Dizdaroglu M. 1992. *Biochemistry* 31: 106–10
37. Dizdaroglu M, Olinski R, Doroshow JH, Akman SA. 1993. *Cancer Res.* 53:1269–72
38. Wood ML, Esteve A, Morningstar ML, Kuziemko GM, Essigmann JM. 1992. *Nucleic Acids Res.* 20:6023–32
39. Haines JA, Reese CB, Todd L. 1962. *J. Chem. Soc.* 5281–88
40. Raleigh JA, Fuciarelli AF, Kulatunga CR. 1987. In *Anticarcinogenesis & Radiation Protection,* ed. PA Cerutti, OF Nygaard, MG Simic, pp. 33–39. New York: Plenum
41. Dizdaroglu M, Schulte-Frohlinde D, von Sonntag C. 1977. *Int. J. Rad. Biol.* 32:481–83
42. Beesk F, Dizdaroglu M, Shulte-Frohlinde D, von Sonntag C. 1979. *Int. J. Rad. Biol.* 36:565–76
43. Rabow L, Stubbe J, Kozarich JW, Gerlt JA. 1986. *J. Am. Chem. Soc.* 108:7130–31
44. Kappen LS, Goldberg IH. 1989. *Biochemistry* 28:1027–32
45. Giloni L, Takeshita M, Johnson F, Iden C, Grollman AP. 1980. *J. Biol. Chem.* 256:8606–15
46. Henner WD, Rodriguez LO, Hecht SM, Haseltine WA. 1983. *J. Biol. Chem.* 258:711–13
47. Povirk LF, Houlgrave CW, Han Y-H. 1988. *J. Biol. Chem.* 263:19263–66
48. Deadon PC, Jiang ZW, Goldberg IH. 1992. *Biochemistry* 31:1917–27
49. Mee LK, Adelstein SJ. 1981. *Proc. Natl. Acad. Sci. USA* 78:2194–98
50. Oleinick NL, Chiu S, Friedman LR, Xue L, Ramakrishnan N. 1986. In *Mechanisms of DNA Damage and Repair,* ed. MG Simic, L Grossman, AC Upton, pp. 181–92. New York: Plenum
51. Margolis SA, Coxon B, Gajewski E, Dizdaroglu M. 1988. *Biochemistry* 27: 6353–59
52. Sancar A, Sancar GB. 1988. *Annu. Rev. Biochem.* 57:29–67
53. West SC. 1992. *Annu. Rev. Biochem.* 61:603–40
54. Carmichael PL, Nishé M, Phillips DH. 1992. *Carcinogenesis* 13:1127–35
55. Strniste GF, Wallace SS. 1975. *Proc. Natl. Acad. Sci. USA* 72:1997–2001
56. Radman M. 1976. *J. Biol. Chem.* 251:1438–45
57. Gates FT, Linn S. 1977. *J. Biol. Chem.* 252:2802–7
58. Katcher HL, Wallace SS. 1983. *Biochemistry* 22:4071–81
59. Breimer LH, Lindahl T. 1984. *J. Biol. Chem.* 259:5543–48
60. Boorstein RJ, Hilbert TP, Cadet J, Cunningham RP, Teebor GW. 1989. *Biochemistry* 28:6164–70
61. Breimer LH, Lindahl T. 1980. *Nucleic Acids Res.* 8:6199–210
62. Bailly V, Verly WG. 1987. *Biochem. J.* 242:565–72
63. Kim J, Linn S. 1988. *Nucleic Acids Res.* 16:1135–41
64. Cunningham RP, Asahara H, Bank JF, Scholes CP, Salerno JC, et al. 1989. *Biochemistry* 28:4450–55
65. Beinert H. 1990. *FASEB J.* 4:2483–91
66. Kow YW, Wallace SS. 1987. *Biochemistry* 26:8200–6
67. Fu WG, O'Handley S, Cunningham RP, Johnson MK. 1992. *J. Biol. Chem.* 267:16135–37
68. Kuo CF, McRee D, Fisher CL, O'Handley SF, Cunningham RP, Tainer JA. 1992. *Science* 258:434–40
69. Michaels ML, Pham L, Nghiem Y, Cruz C, Miller JH. 1990. *Nucleic Acids Res.* 18:3841–45
70. Tsai-Wu JJ, Liu HF, Lu AL. 1992. *Proc. Natl. Acad. Sci. USA* 89:8779–83
71. Cunningham RP, Weiss B. 1985. *Proc. Natl. Acad. Sci. USA* 82:474–78
72. Wallace SS. 1988. *Environ. Mol. Mutagen.* 12:431–77
73. Jeggo PA, Kemp LM. 1983. *Mutat. Res.* 112:313–27
74. Harrison L, Skorvaga M, Cunningham RP, Hendry JH, Margison GP. 1992. *Radiat. Res.* 132:30–39
75. Povirk LF, Houlgrave CW. 1988. *Biochemistry* 27:3850–57
76. Demple B, Daikh Y, Greenberg JT, Johnson AW. 1986. In *Antimutagenesis and Anticarcinogenesis: Mechanisms,* ed. DM Shankel, P Hartman, T Kada, A Hollander, pp. 205–17. New York: Plenum
77. Gossett J, Lee K, Cunningham RP,

Doetsch PW. 1988. *Biochemistry* 27: 2629–34
78. Helland DE, Doetsch PW, Haseltine WA. 1986. *Mol. Cell. Biol.* 6:1983–90
78a. Huq I, Haukanes B-I, Helland DE. 1992. *Eur. J. Biochem.* 206:833–39
79. Kim J, Linn S. 1989. *J. Biol. Chem.* 264:2739–45
80. Doetsch PW, Henner WD, Cunningham RP, Toney JH, Helland DE. 1987. *Mol. Cell. Biol.* 7:26–32
81. Lee K, McCray WH Jr, Doetsch PW. 1987. *Biochem. Biophys. Res. Commun.* 149:93–101
82. Chetsanga CJ, Lindahl T. 1979. *Nucleic Acids Res.* 6:3673–83
83. Boiteux S, Bichara M, Fuchs RPP, Laval J. 1989. *Carcinogenesis* 10: 1905–9
84. Breimer LH. 1984. *Nucleic Acids Res.* 12:6359–67
85. Boiteux S, O'Connor TR, Laval J. 1987. *EMBO J.* 6:3177–83
86. Boiteux S, O'Connor TR, Lederer F, Gouyette A, Laval J. 1990. *J. Biol. Chem.* 265:3916–22
87. O'Connor TR, Laval J. 1989. *Proc. Natl. Acad. Sci. USA* 86:5222–26
88. Bailly V, Verly WG, O'Connor T, Laval J. 1989. *Biochem. J.* 262:581–89
89. Graves RJ, Felzenszwalb I, Laval J, O'Connor TR. 1992. *J. Biol. Chem.* 267:14429–35
90. Boiteux S, Laval J. 1983. *Biochem. Biophys. Res. Commun.* 110:552–58
91. Tchou J, Kasai H, Shibutani S, Chung MH, Laval J, et al. 1991. *Proc. Natl. Acad. Sci. USA* 88:4690–94
92. Chung MH, Kasai H, Jones D, Inoue H, Ishikawa H, Nishimura S. 1991. *Mutat. Res.* 254:1–12
93. Shibutani S, Takeshita M, Grollman AP. 1991. *Nature* 349:431–34
94. Grollman AP, Moriya M. 1993. *Trends Genet.* 9:246–49
95. Wood ML, Dizdaroglu M, Gajewski E, Essigmann JM. 1990. *Biochemistry* 29:7024–32
96. Grollman AP. 1992. In *Structure and Function,* Vol 1: *Nucleic Acids,* ed. RH Sarma, MH Sarma, pp. 165–70. Guilderland, NY: Adenine
97. Cabrera M, Nghiem Y, Miller JH. 1988. *J. Bacteriol.* 170:5405–7
98. Michaels ML, Pham L, Cruz C, Miller JH. 1991. *Nucleic Acids Res.* 18:3841–45
99. Bessho T, Tano K, Kasai H, Nishimura S. 1992. *Biochem. Biophys. Res. Commun.* 188:372–78
100. O'Connor TR, Graves RJ, de Murcia G, Castaing B, Laval J. 1993. *J. Biol. Chem.* 268:9063–70
101. Castaing B, Geiger A, Seliger H, Nehls P, Laval J, et al. 1993. *Nucleic Acids Res.* 21:2899–905
102. Tchou J, Miller JH, Michaels ML, Grollman AP. 1993. *J. Biol. Chem.* 268:26738–44
103. Bessho T, Tano K, Kasai H, Ohtsuka E, Nishimura S. 1993. *J. Biol. Chem.* 268:19416–21
104. Michaels ML, Miller JH. 1992. *J. Bacteriol.* 174:6321–25
105. Bessho T, Roy R, Yamamoto K, Kasai H, Nishimura S, et al. 1993. *Proc. Natl. Acad. Sci. USA* 90:8901–4
106. Nghiem Y, Cabrera M, Cupples CG, Miller JH. 1988. *Proc. Natl. Acad. Sci. USA* 85:2709–13
107. Au KG, Cabrera M, Miller JH, Modrich P. 1988. *Proc. Natl. Acad. Sci. USA* 85:9163–66
108. Au KG, Clark S, Miller JH, Modrich P. 1989. *Proc. Natl. Acad. Sci. USA* 86:8877–81
109. Michaels ML, Tchou J, Grollman AP, Miller JH. 1992. *Biochemistry* 31: 10964–68
110. Michaels ML, Cruz C, Grollman AP, Miller JH. 1992. *Proc. Natl. Acad. Sci. USA* 89:7022–25
111. Bhatnagar SK, Bullions LC, Bessman MJ. 1991. *J. Biol. Chem.* 266:9050–54
112. Maki H, Sekiguchi M. 1992. *Nature* 355:273–75
113. Mo JY, Maki H, Sekiguchi M. 1992. *Proc. Natl. Acad. Sci. USA* 89:11021–25
114. Sakumi K, Furuichi M, Tsuzuki T, Kakuma T, Kawabata S, et al. 1993. *J. Biol. Chem.* 268:23524–30
115. Lu AL, Chang DY. 1988. *Cell* 54:805–12
116. Yeh YC, Chang DY, Masin J, Lu AL. 1991. *J. Biol. Chem.* 266:6480–84
117. Hill-Perkins M, Jones MD, Karran P. 1986. *Mutat. Res.* 162:153–63
118. Karran P, Lindahl T. 1978. *J. Biol. Chem.* 253:5877–79
119. Karran P, Lindahl T. 1980. *Biochemistry* 19:6005–11
120. Myrnes B, Guddal PH, Krokan H. 1982. *Nucleic Acids Res.* 10:3693–701
121. Dehayzya P, Sirover MA. 1986. *Cancer Res.* 46:3756–61
122. Wink DA, Kasprzak KS, Maragos CM, Elespuru RK, Misra M, et al. 1991. *Science* 254:1001–3
123. Nguyen TT, Brunson DC, Crespi CL, Penman BW, Wishnok JS, Tannenbaum SR. 1992. *Proc. Natl. Acad. Sci. USA* 89:3030–34
124. Dianov GT, Lindahl T. 1991. *Nucleic Acids Res.* 19:3829–33

124a. Saparbaev M, Laval J. 1994. *Proc. Natl. Acad. Sci. USA*. Submitted
125. Hollstein MC, Brooks P, Linn S, Ames BN. 1984. *Proc. Natl. Acad. Sci. USA* 81:4003–7
126. Cannon-Carlson SV, Gokhale H, Teebor GW. 1989. *J. Biol. Chem.* 264:13306–12
127. Boorstein RJ, Levy DD, Teebor GW. 1987. *Cancer Res.* 47:4372–77
128. Boorstein RJ, Chiu LN, Teebor GW. 1989. *Nucleic Acids Res.* 17:7653–61
129. Cannon SV, Cummings A, Teebor GW. 1988. *Biochem. Biophys. Res. Commun.* 151:1173–9
130. Gangly T, Duker NJ. 1990. *Mutat. Res.* 237:107–15
131. Gangly T, Duker NJ. 1992. *Mutat. Res.* 275:87–96
132. Tomkinson AE, Bonk RT, Kim J, Bartfield N, Linn S. 1990. *Nucleic Acids Res.* 18:929–35
133. Driggers WJ, LeDoux SP, Wilson GL. 1993. *J. Biol. Chem.* 268:22042–45
134. Kuhnlein U, Penhoet EE, Linn S. 1976. *Proc. Natl. Acad. Sci. USA* 73:1169–73
135. Kim J, Chubatsu LS, Admon A, Stahl J, Fellous R, Linn S. 1994. *Cell*. Submitted
136. Zhang XT, Tana YM, Tane YH. 1990. *Nucleic Acids Res.* 18:6689
137. Wilson DM III, Deutsch WA, Kelley MR. 1993. *Nucleic Acids Res.* 21:2516
138. Flejter WL, McDaniel LD, Johns D, Friedberg EC, Schultz RA. 1992. *Proc. Natl. Acad. Sci. USA* 89:261–65
139. Smeets H, Bachinski L, Coerwinkel M, Schepens J, Hoeijmakers J, et al. 1990. *Am. J. Hum. Genet.* 46:492–501
140. Chan YL, Devi KR, Olvera J, Wool IG. 1990. *Arch. Biochem. Biophys.* 283:546–50
141. Richardson CC, Kornberg A. 1964. *J. Biol. Chem.* 239:242–50
142. Richardson CC, Lehman IR, Kornberg A. 1964. *J. Biol. Chem.* 239:251–58
143. Weiss B. 1976. *J. Biol. Chem.* 251:1896–901
144. Gossard F, Verly WG. 1978. *Eur. J. Biochem.* 82:321–32
145. Hadi SM, Goldthwait DA. 1971. *Biochemistry* 10:4986–94
146. Rogers S, Weiss B. 1980. *Methods Enzymol.* 65:201–11
147. Warner HR, Demple B, Deutsch WA, Kane CM, Linn S. 1980. *Proc. Natl. Acad. Sci. USA* 77:4602–6
148. Demple B, Johnson AW, Fung D. 1986. *Proc. Natl. Acad. Sci. USA* 83:7731–35
149. Sambrook J, Fritsch EF, Maniatis T. 1989. *Molecular Cloning—A Laboratory Manual*. Cold Spring Harbor, NY: Cold Spring Harbor Lab. 2nd ed.
150. Ausubel M, Brent R, Kingston RE, Moore DD, Smith JA, et al. 1988. *Current Protocols in Molecular Biology*. New York: Wiley
151. Eckstein F. 1985. *Annu. Rev. Biochem.* 54:367–402
151a. Takeuchi M, Lillis R, Demple B, Takeshita M. 1994. *J. Biol. Chem.* Submitted
152. Milcarek C, Weiss B. 1972. *J. Mol. Biol.* 68:303–18
153. Ljungquist S, Lindahl T, Howard-Flanders P. 1976. *J. Bacteriol.* 126:646–53
154. Demple B, Halbrook JH, Linn S. 1983. *J. Bacteriol.* 153: 1079–82
155. Sammartano LJ, Tuveson RW. 1983. *J. Bacteriol.* 156:904–6
156. Clements JE, Rogers SG, Weiss B. 1978. *J. Biol. Chem.* 253:2990–99
157. Lacks S. 1970. *J. Bacteriol.* 101:373–83
158. Puyet A, Greenberg B, Lacks S. 1989. *J. Bacteriol.* 171:2278–86
159. Demple B, Herman T, Chen DS. 1991. *Proc. Natl. Acad. Sci. USA* 88:10450–54
160. Robson CN, Hickson ID. 1991. *Nucleic Acids Res.* 19:5519–23
161. Seki S, Hatsushika M, Watanabe S, Akiyama K, Nagao K, Tsutsui K. 1992. *Biochim. Biophys. Acta* 1131:287–99
162. Robson CN, Milne AM, Pappin DJC, Hickson ID. 1991. *Nucleic Acids Res.* 19:1087–92
163. Seki S, Akiyama K, Watanabe S, Hatsushika M, Ikeda S, Tsutsui K. 1991. *J. Biol. Chem.* 266:20797–802
164. Kane CM, Linn S. 1981. *J. Biol. Chem.* 256:3405–14
165. Chen D, Herman V, Demple B. 1991. *Nucleic Acids Res.* 19:5907–14
166. Xanthoudakis S, Miao G, Wang F, Pan YCE, Curran T. 1992. *EMBO J.* 11:3323–35
167. Zhao B, Grandy DK, Hagerup JM, Magenis RE, Smith L, et al. 1992. *Nucleic Acids Res.* 20:4097–98
168. Robson CN, Hochhauser D, Craig R, Rack K, Buckle VJ, Hickson ID. 1992. *Nucleic Acids Res.* 20:4417–21
169. Harrison L, Ascion G, Menninger JC, Ward DC, Demple B. 1992. *Hum. Mol. Genet.* 1:677–80
170. Demple B, Greenberg JT, Johnson A, Levin JD. 1988. In *Mechanisms & Consequences of DNA Damage Processing*, ed. EC Friedberg, PC Hanawalt. *UCLA Symp. Mol. Cell. Biol.* NS, 83:151–55. New York: Liss

171. Seki S, Oda T. 1988. *Carcinogenesis* 9:2239–44
172. Seki S, Ikeda S, Watanabe S, Hatsushika M, Tsutsui K, et al. 1991. *Biochim. Biophys. Acta* 1079:57–64
173. Nes IF. 1980. *Nucleic Acids Res.* 8:1575–89
174. Sanderson BJS, Chang CN, Grollman AP, Henner WD. 1989. *Biochemistry* 28:3892–901
175. Chen DS, Law C, Keng P. 1993. *Radiat. Res.* 135:405–10
176. Abate C, Patel L, Rauscher FJ III, Curran T. 1990. *Science* 249:1157–61
177. Walker LJ, Robson CN, Black E, Gillespie D, Hickson ID. 1993. *Mol. Cell. Biol.* 13:5370–76
178. Xanthoudakis S, Miao G, Curran T. 1994. *Proc. Natl. Acad. Sci. USA* 91:23–27
179. Meyer M, Schreck R, Bauerle PA. 1993. *EMBO J.* 12:2005–15
180. Lowenhaupt K, Sander M, Hauser C, Rich A. 1989. *J. Biol. Chem.* 264: 20568–75
181. Sander M, Lowenhaupt K, Lane WS, Rich A. 1991. *Nucleic Acids Res.* 19:4523
182. Sander M, Lowenhaupt K, Rich A. 1991. *Proc. Natl. Acad. Sci. USA* 88:6780–84
183. Nugent M, Huang SM, Sander M. 1993. *Biochemistry* 32:11445–52
184. Sander M, Carter M, Huang SM. 1993. *J. Biol. Chem.* 268:2075–82
185. Tomkinson AE, Bonk RT, Linn S. 1988. *J. Biol. Chem.* 263:12532–37
186. Gu L, Huang SM, Sander M. 1993. *Nucleic Acids Res.* 21:4788–95
187. Chan E, Weiss B. 1987. *Proc. Natl. Acad. Sci. USA* 84:3189–93
188. Nunoshiba T, deRojas-Walker T, Wishnok JS, Tannenbaum SR, Demple B. 1993. *Proc. Natl. Acad. Sci. USA* 90:9993–97
189. Tsaneva IR, Weiss B. 1990. *J. Bacteriol.* 172:4197–205
190. Greenberg JT, Monach PA, Chou J, Josephy PD, Demple B. 1990. *Proc. Natl. Acad. Sci. USA* 87:6181–85
191. Demple B. 1991. *Annu. Rev. Genet.* 25:315–37
192. Ljungquist S. 1977. *J. Biol. Chem.* 252:2808–14
193. Levin JD, Johnson AW, Demple B. 1988. *J. Biol. Chem.* 263: 8066–71
194. Cunningham RP, Saporito S, Spitzer SG, Weiss B. 1986. *J. Bacteriol.* 168:1120–27
195. Izumi T, Ishizaki K, Ikenaga M, Yonei S. 1992. *J. Bacteriol.* 174:7711–16
196. Levin JD. 1991. *Repair of free radical damage to DNA deoxyribose: Biochemical and cellular studies on E. coli endonuclease IV*. PhD thesis. Harvard Univ.
197. Hagnesee ME, Moses RE. 1990. *Biochim. Biophys. Acta* 1048:19–23
198. Povirk L, Golberg IH. 1985. *Proc. Natl. Acad. Sci. USA* 82:3182–86
199. Johnson AW, Demple B. 1988. *J. Biol. Chem.* 263:18009–16
200. Johnson AW, Demple B. 1988. *J. Biol. Chem.* 263:18017–22
201. Popoff SC, Spira AI, Johnson AW, Demple B. 1990. *Proc. Natl. Acad. Sci. USA* 87:4193–97
202. Kaboev OK, Luchkina LA, Kuziakina TI. 1985. *J. Bacteriol.* 164:878–81
203. Masters CI, Moseley BE, Minton KW. 1991. *Mutat. Res.* 254:263–72
204. Thibodeau L, Verly WG. 1977. *J. Biol. Chem.* 252:3304–9
205. Ramotar D, Kim C, Lillis R, Demple B. 1993. *J. Biol. Chem.* 268:20533–39
206. Ramotar D, Popoff SC, Gralla EB, Demple B. 1991. *Mol. Cell. Biol.* 11:4537–44
206a. Kunz BA, Henson E, Roche H, Ramotar D, Nunoshiba T, Demple B. 1994. *Proc. Natl. Acad. Sci. USA*. Submitted
207. Ramotar DS, Popoff SC, Demple B. 1990. *Mol. Microbiol.* 5:149–55
208. Xiao W, Samson L. 1993. *Proc. Natl. Acad. Sci. USA* 90:2117–21
209. Loeb LA, Preston B. 1986. *Annu. Rev. Genet.* 20:201–30
210. Levin J, Shapiro R, Demple B. 1991. *J. Biol. Chem.* 266: 22893–98
211. Beese LS, Steitz TA. 1991. *EMBO J.* 10:25–33
212. Sancar A, Tang MS. 1993. *Photochem. Photobiol.* 57:905–21
213. Grossman L, Thiagalingam S. 1993. *J. Biol. Chem.* 268:16871–74
214. Hoeijmakers JHJ. 1993. *Trends Genet.* 9:173–77
215. Selby CP, Sancar A. 1993. *J. Bacteriol.* 175:7509–14
216. Czeczot H, Tudek B, Lambert B, Laval J, Boiteux S. 1991. *J. Bacteriol.* 173: 3419–24
217. Friedberg EC, Siede W, Cooper AJ. 1991. In *The Molecular and Cellular Biology of the Yeast Saccharomyces*, ed. JR Broach, JR Pringle, EW Jones, pp. 147–92. Cold Spring Harbor, NY: Cold Spring Harbor Lab.
218. Hoeijmakers JHJ. 1993. *Trends Genet.*. 9:211–17
219. Satoh MS, Jones CJ, Wood RD, Lindahl T. 1993. *Proc. Natl. Acad. Sci. USA* 90:6335–39
220. Petes TD, Malone J, Jones E. 1991. See Ref. 217, pp. 407–521

221. Thompson LH, Brookman KW, Jones NL, Allen SA, Carrano AV. 1990. *Mol. Cell. Biol.* 10:6160–71
222. Caldecott KW, McKeown, Tucker JD, Ljungquist S, Thompson LH. 1993. *Mol. Cell. Biol.* 14:68–76
223. Satoh MS, Poirier GG, Lindahl T. 1993. *J. Biol. Chem.* 268:5480–87
224. Epe B, Hegler J. 1993. *Methods Enzymol.* In press
225. Sak BD, Eisenstark A, Touati D. 1989. *Proc. Natl. Acad. Sci. USA* 86:3271–75
226. Price A, Lindahl T. 1991. *Biochemistry* 30:8631–37

Annu. Rev. Biochem. 1994. 63:949–90

GTPases: Multifunctional Molecular Switches Regulating Vesicular Traffic

Claude Nuoffer and William E. Balch

Departments of Cell and Molecular Biology, The Scripps Research Institute, 10666 N. Torrey Pines Road, La Jolla, California 92037

KEY WORDS: Rab, Arf, Sar1; dynamin; G proteins; exocytic and endocytic pathways; signal transduction

---

CONTENTS

0066-4154/94/0701-0949$05.00

SUMMARY AND PERSPECTIVES

The vectorial movement of proteins between compartments of the exocytic and endocytic pathways of eukaryotic cells is mediated by carrier vesicles that bud from a donor organelle, and are targetted to and fuse with the appropriate acceptor organelle. GTPases of a remarkable diversity are now recognized to play key roles in the regulation of these complex events. By alternating between conformational states in a fashion controlled by guanine nucleotide exchange and GTP hydrolysis, GTPases act as molecular switches to control a wide variety of cellular functions. Like their well-characterized counterparts involved in translation (Ef-Tu) and signal transduction (Ras and heterotrimeric G proteins), GTPases participating in membrane traffic are likely to control the assembly and disassembly of protein complexes driving the formation, targetting, and fusion of carrier vesicles. In so doing, they may confer vectoriality and fidelity to the transport process, and play a critical role in maintaining the structural and functional integrity of subcellular compartments.

The seven major groups of GTPases implicated in vesicular traffic include the members of the Ras-like Arf, Sar1, Rab/*YPT*, Rac/*CDC42*, and Rho families, as well as members of the heterotrimeric G protein ($G_{\alpha\beta\gamma}$) and dynamin families. These proteins share sequence motifs that are essential for guanine nucleotide binding and GTP hydrolysis. The GTP-binding domain encoded by these motifs is a hallmark of all GTPases examined to date. The members of different families are related to each other by variable degrees of homology in regions outside the conserved GTP-binding motifs. Such divergent sequences determine functionality by conferring specificity to interactions with upstream/downstream effectors, and accessory factors that control guanine nucleotide exchange [guanine nucleotide exchange proteins (GEPs)] or stimulate GTP hydrolysis [GTPase activating proteins (GAPs)]. While the Sar1 family contains a single member at this time, more than 30 members of the Rab family have been identified.

Given the exceptional structural diversity of these GTPases, one must anticipate a correspondingly diverse range of functions. In this review, we examine the functions of members of the different families with respect to their potential roles in vesicular transport. While members of the Arf, Sar1, $G_{\alpha\beta\gamma}$, and dynamin families appear to integrate the biochemical machinery involved in vesicle budding, a preliminary consensus suggests that members of the Rab/*YPT* family monitor protein-protein interactions during vesicle targetting/fusion. These proteins are likely to function as key regulatory components in a larger cast of proteins that constitute specific fusion "machines" operating at different stages along the exocytic and endocytic pathways. Each of these GTPases is thought to interact in a coordinated fashion with a specific set of upstream and downstream effectors and accessory factors that modulate the GTPase cycle, as well as proteins responsible for the recycling of all these components for use in multiple, successive rounds of transport. Thus, GTPase function is critical not only for the timing, vectoriality, and fidelity of the vesicle budding, targetting, and fusion processes per se, but also for maintaining critical steady-state pools of the proteins or protein complexes that constitute the transport machinery.

While it is clear that GTPases play a fundamental role in vesicular traffic, analysis of their function is also providing new insight into the close relationship between membrane transport and organelle structure. Alterations in the guanine nucleotide exchange or hydrolytic properties of distinct GTPases have dramatic effects on the integrity of transport compartments. Distinct GTPases may act in a coordinated fashion to provide continuity between processes that must be tuned to each other to ensure efficient operation of vesicular traffic and maintenance of organelle structure/ function.

In contrast to the Rab, Arf, Sar1, and dynamin families, which play a direct role in transport, members of the Rac/*CDC42* and Rho families participate in the organization of the cytoskeleton. These GTPases may indirectly serve to link vesicular traffic to the cytoskeleton and are likely to play a role in the establishment and maintenance of cell polarity. In addition, members of the heterotrimeric family of GTPases may integrate vesicular traffic with the processes of cell proliferation and differentiation through control of coat assembly and protein phosphorylation. Given the many excellent recent reviews on GTPases (see Refs. 1–7, 7a, 7b), we focus on new insights relevant to their role as regulators of the multiplicity of biochemical interactions governing the secretory and endocytic pathways.

MEMBRANE TRAFFIC AND GTPases

The mechanics of the membrane-membrane interactions along vesicular trafficking pathways can be divided conceptually into steps involved in vesicle budding, targetting, and fusion. In the first step, coat components involved in the sequestration of membrane into carriers must be recruited to the site of vesicle formation. This process involves both membrane-associated proteins and numerous soluble cytosolic factors. The second step, targetting or recognition of the downstream compartment, is expected to require receptors on the incoming vesicle and the acceptor membrane, which confer exquisite specificity to vesicle docking. The last step, vesicle fusion, is likely to include a biochemical machinery distinct from that involved in vesicle budding and should be activated only under rigorously defined circumstances following docking. The absence of such specificity would have catastrophic consequences on the distribution of cargo and the organization of subcellular compartments.

Insight into the mechanisms governing vesicular transport has come through a combination of genetic and biochemical approaches, which have led to the identification of a number of components that may be involved in these complex events (see Refs. 3, 8, 8a). The development of versatile cell-free systems that faithfully reconstitute transport steps between compartments of the endocytic and exocytic pathways in vitro has allowed rapid progress in assigning potential roles to some of these proteins in different aspects of the transport process. In particular, the involvement of various types of GTPases stems from the recognition that GTPγS (guanosine-5′-O-thiotriphosphate), an unspecific, nonhydrolyzable analog of GTP, and AlF_4^-, which acts as a phosphate analog and specifically activates members of the heterotrimeric G protein family (9), both inhibit intracellular transport between nearly all compartments.

Transport of soluble and membrane-bound proteins through the exocytic and endocytic pathways is accomplished by at least two classes of vesicular carriers that differ in coat structure. Clathrin-coated vesicles mediate transport from the cell surface through compartments of the endocytic pathway, as well as the export of proteins from the trans Golgi network (TGN) to lysosomal/vacuolar compartments and terminal compartments of the regulated secretory pathway (see Ref. 10). In contrast, non-clathrin-coated vesicles are responsible for traffic between compartments of the early secretory pathway and possibly traffic from the TGN to the cell surface along the constitutive pathway (see Refs. 3, 8, 8a). The coats of non-clathrin-coated vesicles contain a spectrum of COPs (coat proteins), which may be functionally analogous to the clathrin/adaptor complexes of clathrin-coated vesicles (see Refs. 8, 11).

While the identification and characterization of components of the biochemical machines involved in vesicle budding, targetting, and fusion is critical to understanding vesicular traffic, it is now evident that the formation of both clathrin- and non-clathrin-coated carriers involves sorting and concentration of cargo. The clathrin/adaptor coats underlying the plasma membrane act as molecular filters to sort and concentrate protein up to 10-fold during receptor-mediated endocytosis (see Ref. 10). Vesicular stomatitis virus glycoprotein, a type I integral membrane protein, is also sorted and concentrated nearly 10-fold into COP-coated carriers budding from the endoplasmic reticulum (ER) (12). These results suggest the existence of selective mechanisms for the packaging of proteins in both the endocytic and exocytic pathways. Thus, it will be important to consider the possibility that GTPases may play a role in coordinating coat assembly with the recruitment of cargo from the cell surface, or from the membrane or lumen of different intracellular compartments.

FUNDAMENTALS OF GTPase DESIGN

Conserved Structural Features

All GTPases examined to date share, in most cases, four highly conserved sequence motifs (G1–G4), which are required for guanine nucleotide binding and GTP hydrolysis (Figure 1) (see Refs. 13–16). The first motif, $Gx_4GK(S/T)$, forms a loop in which main-chain amide hydrogens of several amino acids and the ϵ-amino group of lysine interact with the α- and β-phosphates of GDP and GTP. A second motif corresponding to the putative "effector" region of Ras and Rab proteins contains a highly conserved threonine that interacts with a Mg^{2+} ion coordinated to oxygens of the β- and γ-phosphates of GTP. This Mg^{2+}, which is essential for GTP hydrolysis, also interacts with serine/threonine in the first motif and, through an intervening H_2O molecule, with an invariant aspartate in the third motif, Dx_2G. The fourth motif, (N/T)(K/Q)xD, interacts with the guanine nucleotide ring. A fifth motif, which is less conserved but typical of Ras and Ras-related proteins, is located towards the carboxyl terminus and is involved indirectly

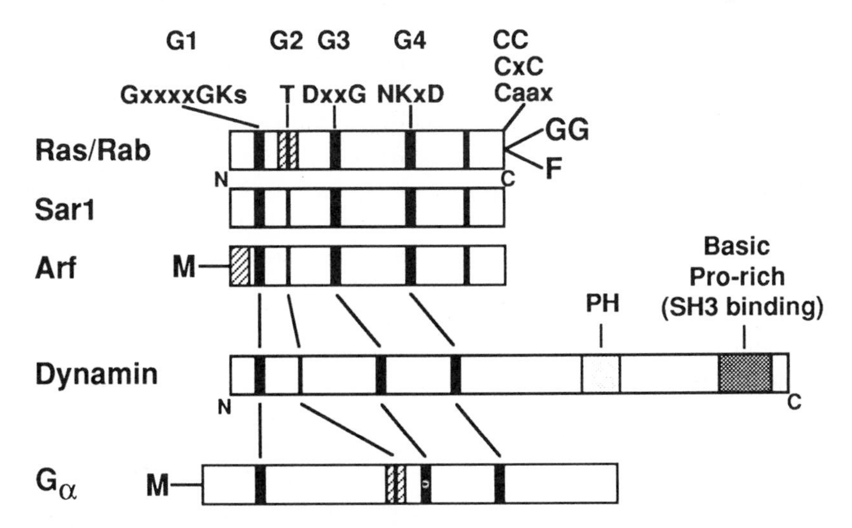

Figure 1 Common and unique structural features of distinct GTPase families. The open boxes represent the linear sequences of the specified GTPases. The black bars show the relative locations of conserved motifs responsible for guanine nucleotide binding and GTP hydrolysis. Hatched areas illustrate the positions of putative effector domains. a, aliphatic residue; F, farnesyl chain; GG, geranylgeranyl chain(s); M, myristyl chain. The proline-rich (SH3-binding) and pleckstrin homology (PH) domains are indicated for dynamin.

in guanine nucleotide binding through hydrogen bonds that stabilize the side chains of residues in the third motif. Despite the divergence of Ras, Ef-Tu, and G_α, the guanine nucleotide-binding domain in the three-dimensional crystal structures of these proteins are nearly superimposable (16a,b,c). These structural data, in combination with mutational studies on Ras (17, 18), Ef-Tu, and G_α, have provided direct evidence for the roles of these motifs in guanine nucleotide binding and GTP hydrolysis, and their importance in interactions with effectors and accessory proteins. Thus, mutations in these highly conserved regions can be used to investigate the function of other, less extensively characterized GTPases.

Distinctive Characteristics of Individual Families

GTPases involved in vesicular traffic can be divided into two groups based on their subunit composition. The heterotrimeric G proteins contain an α-subunit (~40 kDa), which binds guanine nucleotides, as well as two additional subunits, β and γ (19). The GTPases of the Ras superfamily (including the members of the Sar1, Arf, Rab/*YPT*, Rac/*CDC42*, and Rho families) are 20–30 kDa in size, whereas those of the dynamin family are considerably larger, normally in the 60–80 kDa range. While Ras-related GTPases are generally considered to be monomeric, evidence now indicates that, with the exception of Arf, all of them form complexes with other cytosolic and/or membrane-bound proteins, which may serve functions analogous to those of the βγ subunits of heterotrimeric G proteins and their cognate cell surface receptors in controlling activation and recycling.

Each of these GTPases contains stretches of homologous sequence characteristic of all members of a particular family, as well as variable regions specifying unique function. In some cases, such as Arf, a slight variation in the GTP-binding motifs serves as a hallmark of a particular family. An effector domain located between the first and second GTP-binding motifs (Figure 1) has been proposed to participate in interactions with specific upstream and downstream effector molecules (see Refs. 7a, 17, 18) and is a useful marker to delineate subgroups within a family, as this stretch of sequence is homologous among closely related members, but diverges considerably among distinct subgroups (16). More variable domains include the amino- and carboxyl-terminal regions. One currently recognized role of the carboxyl-terminal variable domain of members of the Rab/*YPT* family is membrane localization (Ref. 7; see below). Rab, Rho, Rac, and the γ subunit of heterotrimeric G proteins are posttranslationally modified at carboxyl-terminal and/or subterminal cysteine residues by the addition of one or two prenyl moieties that are essential for function and membrane association (Figure 1) (see Refs. 20–22). In contrast, members of the Arf and G_α families are myristylated at an amino-terminal glycine residue (Figure

1) (see Ref. 23). Palmitylation of cysteine residues in amino- or carboxyl-terminal regions variably occurs on different members of the Ras and G_α families. Apparently, Sar1 and dynamin lack any such posttranslational modifications, although dynamin is subject to cyclical phosphorylation/dephosphorylation as are some Rab proteins. The sequential interaction of individual GTPases with various components of the transport machinery is likely to involve multiple specific protein-protein contacts, consistent with the notion that the sequence and structure of each of these proteins are precisely tuned to ensure optimal function.

The Guanine Nucleotide Cycle

GTPases alternate between distinct conformations in response to the phosphorylation state of the bound nucleotide, which varies as the proteins go through a cycle of guanine nucleotide exchange and GTP hydrolysis. The binding of GTP switches the protein to an "active" form, in which it may interact with a downstream effector to trigger an appropriate response. GTP hydrolysis returns the GTPase to the "inactive" GDP-bound form. Regeneration of the active GTP-bound state requires exchange of the bound GDP by an exogenous GTP molecule, a reaction stimulated by specific guanine nucleotide exchange proteins (GEPs). In addition, members of Rab/*YPT* and Rho families are known to interact with guanine nucleotide dissociation inhibitions (GDIs), which can bind the GDP-bound form and prevent exchange. Since in many cases the transition from the inactive to the active state is likely to represent a rate-limiting step, these regulatory factors may play a key role in GTPase function. The role of the nucleotide-free state is unknown, but, by analogy to EF-tu or G_α, is likely to be critical for function.

While the α subunits of heterotrimeric G proteins are able to hydrolyze GTP rapidly, most monomeric GTPases have low intrinsic hydrolytic activities. Therefore, members of the Ras superfamily of GTPases require a class of proteins—referred to as GTPase activating proteins (GAPs)—that facilitate GTP hydrolysis (7a, 17). GEPs and GAPs, which most probably represent integral components of the machineries mediating transport between distinct compartments, are likely to be as diverse in structure and function as are their cognate GTPases. Their fundamental role is to control the transitions between the various conformational states by regulating the rates of guanine nucleotide exchange and GTP hydrolysis to levels appropriate for a given physiological process. GAPs may function either upstream as negative regulators or downstream as effectors. GDIs may play a role analogous to the $\beta\gamma$ subunits of heterotrimeric G proteins, which sequester the α subunit in the GDP-bound form until activation occurs, while GEPs are likely to be functionally equivalent to G protein–coupled cell surface receptors containing agonist. The importance of GAPs, GEPs, and GDIs emphasizes that the GTPases active in vesicular traffic do not function independently, but are continuously involved in interactions with a variety of effectors and accessory proteins, which respond to their action and/or control their function.

Thus, the GTPase cycle can be divided into two parts—a "charging" step, which activates the switch and promotes interactions with specific effectors, and a "discharging" step, which controls the function of this complex via GTP hydrolysis. Both steps may regulate in either a negative or a positive fashion the assembly or disassembly of protein complexes involved in coat assembly and vesicle budding, or vesicle docking and fusion.

SMALL GTPases INVOLVED IN VESICLE BUDDING

A multiplicity of GTPases have now been demonstrated to participate in vesicular traffic. Below we emphasize new insights into the function of small GTPases in vesicle formation.

Sar1 Regulation of Vesicle Budding from the Endoplasmic Reticulum (ER)

SAR1 was first identified in yeast as a multicopy suppressor of the temperature-sensitive *sec12-4* mutation, which blocks ER-to-Golgi transport at the restrictive temperature (24, 25). Homologs have recently been identified in *Schizosaccharomyces pombe*, *Arabidobsis thaliana* (26), and mammalian cells (27). While the homology of Sar1 to Ras-like GTPases is low, it is most closely related to the members of the Arf family (~30–35% identical). Like Arf, Sar1 requires detergent or phospholipid for efficient guanine nucleotide exchange in vitro (28). However, Sar1 differs from Arf and other Ras-like GTPases by the absence of consensus motifs for posttranslational processing by either amino-terminal myristyl or carboxyl-terminal prenyl groups (Figure 1). The intrinsic rate of GTP hydrolysis is comparable to the low rates of other monomeric GTPases.

Sar1 has been localized in pancreatic acinar cells using immunoelectron microscopy (27). It is found on transitional elements of the ER, which are believed to be specialized sites for the export of newly synthesized secretory proteins. Moreover, Sar1 is highly enriched (20–40-fold) on ER-to-Golgi carrier vesicles compared to the bulk of the ER. It is also present on the *cis* face of the Golgi complex, but is absent from the *medial* and *trans* cisternae, suggesting that it may only be required for ER-to-Golgi transport. Yeast Sar1p has been localized by indirect immunofluorescence to the perinuclear ER and punctate structures that are likely to represent pre-Golgi carriers (29). It has also been detected in ER structures that accumulate in *sec12* and *sec18* mutants at the restrictive temperature, but not in the Golgi-like compartments that accumulate in *sec1* or *sec7* strains (29).

Sar1p is an essential component of the machinery involved in the formation of transport vesicles from the ER in both yeast (25, 28, 29, 31–33) and mammalian cells (27). Two proteins with Sar1p-specific GEP and GAP activities have been isolated and characterized in yeast. These are encoded by the *SEC12* and *SEC23* genes, respectively. Both proteins are essential for export from the ER. Sec12p is a 70-kDa type II integral membrane protein with a 40-kDa cytoplasmic domain that contains the GEP activity (32, 34). This domain can be dissociated from the transmembrane and luminal portions using genetic or biochemical approaches without apparent loss of biological function (32, 35). Homologs of Sec12p have been detected in *S. pombe* and *A. thaliana* (26). Sec23p is an 84-kDa peripheral membrane protein that has been isolated in a soluble form as part of multimeric complex that also contains Sec24p, a 105-kDa protein of unknown function (36, 37). Both monomeric and multimeric forms of Sec23p have GAP activity towards Sar1p (38). Sar1p-GEP and Sar1p-GAP are inactive towards representative members of the Ras, Rab/*YPT*, and Arf families. A mammalian protein crossreacting with Sec23p-specific antibodies has been detected in the cytoplasm surrounding transitional elements in insulin-secreting cells (40), and has recently been cloned (JE Rothman, personal communication).

Sec23p stimulates the intrinsic rate of GTP hydrolysis of Sar1p by ~10-fold in vitro (38). A 50-fold stimulation can be achieved in the presence of a combination of Sec23p and Sec12p, suggesting that guanine nucleotide exchange is a rate-limiting step in Sar1p function. However, even a 50-fold stimulation corresponds to the hydrolysis of only ~1 molecule of GTP per molecule of Sar1p in 20 min. In contrast, Ras-GAP stimulates the intrinsic rate of hydrolysis of Ras by nearly five orders of magnitude, resulting in a turnover of ~1 molecule per sec. Given that yeast secretory proteins are transported to the cell surface in less than 5 min, it is clear that important factors involved in the GTPase cycle of Sar1p remain to be identified. These results also illustrate that characterization of the exchange and hydrolytic properties of purified GTPases in the presence or absence of accessory proteins may be somewhat artificial, as these conditions are unlikely to reflect all of the physiological properties of these components in the context of the transport reaction.

Additional lines of evidence support a role for Sar1 in vesicle budding from the ER, but not later compartments in mammalian cells. First, Sar1-specific antibodies inhibit ER-to-Golgi but not intra-Golgi transport in vitro (27). Second, a Sar1 mutant with a substitution in the $Gx_4GK(T/S)$ motif, Sar1(T39N), inhibits export from the ER, but not transport between Golgi compartments (27). Based on the three-dimensional structure of

Ras, the Sar1 (T39N) mutation is predicted to disrupt the interaction between the hydroxyl group of threonine-39 and the Mg^{2+} ion that is coordinated to the β- and γ-phosphates of GTP. The equivalent Ras mutant [H-Ras(S17N)] inhibits cell proliferation (41). In Ras, this mutation results in a protein with a preferential affinity for GDP and appears to restrict its conformation to a constitutively inactive GDP-bound state (42). The mechanism by which H-Ras(S17N) interferes with normal Ras function remains unknown, but it has been proposed to act as a competitive inhibitor of a Ras-specific GEP (T39N) (42). By analogy to the Ras model, the Sar1 mutant, which also has a markedly reduced affinity for GTP (27), may sequester the function of Sar1-GEP, thereby interfering with recruitment and activation of the wild-type protein (Figure 2*A*). This interpretation is consistent with the phenotype of the temperature-sensitive yeast sec12-*4* mutant, which accumulates ER structures at the restrictive temperature (24, 29). It is further supported by the observation that overexpression of Sec12p sequesters soluble Sar1p to membranes, resulting in a Sar1-depleted cytosolic fraction that requires the addition of exogenous Sar1p to initiate budding in vitro (35, 28). Thus, interaction with Sar1-GEP is likely to be a critical early step for the function of Sar1 in export of protein from the ER (Figure 2*A*).

Sar1p, the Sec23p/Sec24p complex, as well as another complex containing Sec13p and a 150-kDa protein of unknown function represent all of the cytosolic components required to form carrier vesicles from the yeast ER in vitro (28, 36, 45, 46) (Figure 2*A*). In contrast, as described below, Arf1 and its machinery apparently are sufficient to drive vesicle budding from the Golgi compartments (Figure 2*B;* see below). How can two different GTPases and their unrelated machineries promote the formation of vesicles from the ER and the Golgi, respectively? One possibility is that the cell has invented multiple mechanisms to form carrier vesicles from exocytic compartments, similar to the observation that import from the cell surface occurs via receptor-mediated and bulk mechanisms. The different mechanisms could be related to the transport of different proteins. A second possibility is that the need to concentrate cargo, as occurs during export from the ER but not during transit between Golgi compartments (12), imposes an additional level of specificity to the budding process. Although the mechanics of vesicle budding per se can be driven by a Sar1 or Arf1 GTPase cycle in the ER and the Golgi, respectively, Sar1 may be utilized to establish an initial concentration of cargo, which is maintained by Arf during transit through the Golgi stack. In support of the latter model, transport between the ER and the Golgi in yeast also requires Ypt1p and COP-related components involved in Arf function (3, 5). Similarly, export from the ER in mammalian cells requires, in addition to Sar1, Rab1, β-COP, and Arf1 (47–51; see below). In any event, to date, Sar1 appears to be the first example of a GTPase specific for vesicle formation from a single compartment (27), although there is a distinct possibility that additional members of a larger Sar1 family will emerge from future investigations.

Arf Control of Coat Assembly

Arf (ADP-ribosylation factor) was originally discovered as a cofactor required for the ADP-ribosylation of the heterotrimeric G protein G_s (55, 56). The Arf family now includes at least six mammalian and two yeast members, although recent studies suggest the existence of more than 15 distinct Arf proteins in mammalian cells (57). Based on sequence homology, the known mammalian Arf proteins (80% identical) fall into three classes (Class 1, Arf1–3; Class 2, Arf4–5; Class 3, Arf6) (58, 59). Arf proteins are highly conserved across phylogenetic lines (74% identity between yeast and bovine sequences) and differ from other members of the Ras superfamily by the absence of a carboxyl-terminal prenylation motif (Figure 1). Instead, they undergo posttranslational myristylation on an amino-terminal glycine residue, a modification essential for membrane association and function (60). In addition, the association of Arf proteins with membranes depends on an amino-terminal sequence that forms an amphiphilic α-helix and may be involved in interactions with a putative effector (60).

Initial evidence for a role of Arf in vesicular transport came from genetic studies in yeast. While the deletion of both *ARF* genes (*ARF1* and *ARF2*) is lethal, deletion of *ARF1* alone results in a cold-sensitive phenotype and partial secretory defects (61). Consistent with these results, Arf proteins have been localized to the cytoplasmic face of Golgi cisternae (61) and Golgi-derived carrier vesicles (54, 62, 63). Arf has been best characterized as an essential component of COP-coated vesicles mediating transport between Golgi compartments. Incubation of Golgi stacks in the presence of GTPγS in vitro inhibits transport, and results in the accumulation of coated buds and vesicles (52, 53, 64). These contain both Arf and "coatomer," a soluble coat precursor with an S value of 14S, which contains five COPs (α, β, β′, γ, and δ) and three smaller polypeptides (p36, p35, and p20) (65–67). Two distinct Arf proteins that are immunologically related to Arf1 are sufficient to confer sensitivity of intra-Golgi traffic to GTPγS (68). Interestingly, in the absence of Arf, COP-coated vesicles fail to form in vitro, and a new form of vesicle-independent fusion occurs between tubules extending from Golgi cisternae (54, 68a). These results demonstrate that Arf and coat components play a crucial role in both vesicular traffic and Golgi integrity.

Further insight into the role of Arf in vesicular transport has come from the observation that Arf is essential for the recruitment of coatomer to Golgi membranes (69, 70, 70a, 71). It has now been established that the formation of functional COP-coated vesicles from Golgi stacks can be reconstituted in vitro by supplementing the membranes with two fractions—one containing purified Arf and the other purified coatomer (54, 63, 71). These results suggest that Arf and coatomer represent the only cytosolic components required for vesicle budding from Golgi cisternae. A mutant form of Arf1 likely to be defective for GTP hydrolysis [Arf1(Q71L); see below] promotes vesicle budding and coat assembly but prevents uncoating, indicating that hydrolysis is essential for later targetting/fusion steps (71a). The phosphorylation state of the bound nucleotide is critical for membrane association, as myristylated Arf readily binds to Golgi membranes in the GTP-, but not the GDP-bound form (72, 73). Moreover, at least two pools of Arf1 (GTP-bound) can be detected on Golgi membranes—a loosely bound, nonsaturable pool extractable with liposomes, and a tightly bound, saturable pool that is resistant to liposome extraction and implicates an Arf receptor (73). One possible explanation for these results is that the binding of GTP triggers a conformational change(s) that exposes the amino-terminal domain to the lipid bilayer, and subsequently, to a specific membrane receptor. Consistent with this hypothesis, a synthetic peptide derived from the amino terminus of Arf1 prevents membrane association and inhibits intra-Golgi transport in vitro (60).

The mechanism by which Arf is recruited to membranes is believed to involve an Arf-specific GEP (Figure 2*B*). Initial evidence for the involvement of an Arf-GEP came from studies with the fungal metabolite brefeldin A (BFA). Treatment of most mammalian cells with BFA triggers the rapid release of Arf and β-COP from Golgi membranes and results in the reversible collapse of the *cis/medial* Golgi compartments into the ER (see Ref. 74). The drug has been suggested to inhibit either directly or indirectly an Arf-GEP activity associated with Golgi membranes (75–77). Recent mutational studies on Arf function in vivo have led to similar conclusions (Ref. 48; see below). Thus, similar to Sar1/Sar1-GEP (Sec12p) in export from the ER, Arf/Arf-GEP appears to play an important role in the formation of carrier vesicles from Golgi compartments (Figure 2).

Is Arf function only required for transport between Golgi compartments? Apparently not. In fact, although the studies discussed above have focused on Arf/coatomer function in intra-Golgi traffic, several lines of evidence strongly suggest a role of Arf in transport from the ER to the Golgi complex in mammalian cells. As observed for intra-Golgi transport, the amino-terminal Arf peptide is a potent inhibitor of ER-to-Golgi transport in vitro, preventing the budding of carrier vesicles from the ER (78). While the

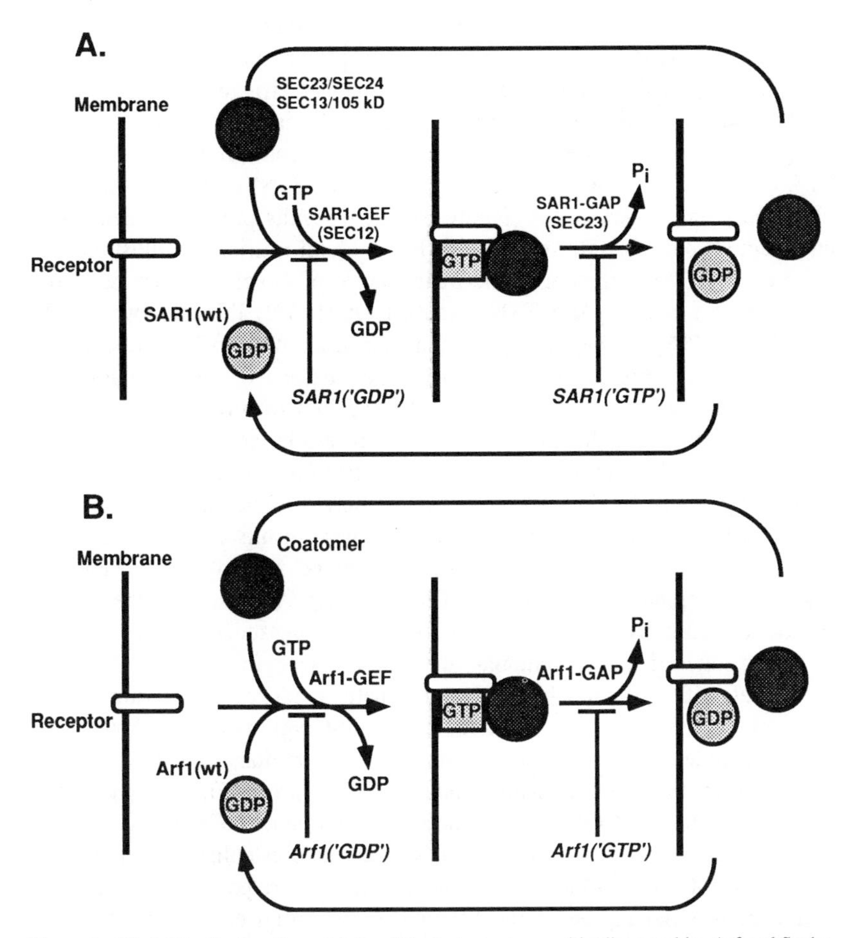

Figure 2 Model for the function of Arf and Sar1p in coat assembly/disassembly. Arf and Sar1p are recruited to the membrane in the GDP-bound state via a putative membrane receptor. Interactions with coat precursors result in coat assembly, which is coupled to guanine nucleotide exchange through the action of specific GEFs. In the GTP-bound form, the coat is stabilized until interactions with specific GAPs trigger GTP hydrolysis and initiate coat disassembly. Mutants restricted to the inactive or activated GTP-bound states (italics) are likely to interfere with the functions of the GEF and GAP effectors, respectively.

specific target of this amphipathic peptide is unknown, excess levels of myristylated Arf1 also block export from the ER, suggesting a critical balance between the GTPase and a limiting pool of cytosolic or membrane-bound transport factors. More direct experiments involving mutant forms of Arf1 support a role of Arf in both ER-to-Golgi and intra-Golgi transport in vivo (48). Mutants in three of the guanine nucleotide–binding motifs

(Figure 1) inhibit ER-to-Golgi transport. Overexpression of Arf1(T31N) [analogous to H-Ras(S17N), a mutant with preferential affinity for GDP (41, 42)] markedly reduces export from the ER, triggers the release of β-COP from intracellular membranes, and results in the collapse of Golgi compartments into the ER in a fashion indistinguishable from the effects of BFA (48). Given the likely possibility that Arf1(T31N) sequesters the function of a putative Arf1-GEP (Figure 2), these results provide evidence for an essential role of Arf1/Arf1-GEP in ER-to-Golgi traffic, and illustrate the importance of Arf function in the maintenance of Golgi structure in vivo.

In contrast to the Arf1(T31N) mutant, Arf1(Q71L) [analogous to H-Ras(Q61L) (79)], a mutation predicted to restrict Arf1 to an active GTP-bound form by preventing GTP hydrolysis, causes the accumulation of ER-to-Golgi carrier vesicles in vivo (48) and in vitro (WE Balch, unpublished), and promotes the binding of β-COP to vesicles and Golgi membranes (48, 71a). The Arf1(Q71L) mutant most likely forms a stable complex with coat components by interfering with the normal function of a putative Arf1-GAP in coat disassembly (Figure 2). These results emphasize a cycle like that observed for Sar1p, in which guanine nucleotide exchange is essential during recruitment whereas GTP hydrolysis is critical for the release of coat proteins prior to or during vesicle targetting and fusion (Figure 2).

The results of the mutational analysis of Arf1 function in vivo are consistent with both biochemical and genetic evidence suggesting that a form of coatomer is also required for ER-to-Golgi transport. Antibodies against synthetic peptides derived from different regions of β-COP inhibit export from the ER in mammalian cells in vitro (49). Likewise, the microinjection of β-COP-specific antibodies into intact cells inhibits ER-to-Golgi transport in vivo (80). Several studies have established that β-COP is present not only on Golgi cisternae and Golgi-derived vesicles, but also in the transitional region of the ER and on pre-Golgi carrier vesicles (50, 80–83, 83a). Surprisingly, when coatomer was tested for its ability to promote vesicle budding from the ER in vitro, it was found to be inactive (49). However, a high-molecular-weight fraction (S value of 18–19S) was active in vesicle budding. This complex was subsequently found to contain β-COP, Rab1, and a mammalian homolog of Sec23p (49). These data suggest that only a subfraction of the total COP pool is active in ER export. Consistent with these results, the yeast homolog of mammalian γ-COP, Sec21p, is essential for ER-to-Golgi transport (84, 85). Moreover, genetic studies suggest that Sec21p interacts with the yeast Rab1 homolog Ypt1p through the *SLY1* gene product (86).

While the role for Arf1 has been extensively studied with respect to COP recruitment, the formation of clathrin-coated buds on the TGN is also sensitive to GTPγS and BFA (87, 87a, 87b), raising the possibility that

Arf may also be involved in the assembly of clathrin coats. Interestingly, β-COP shares sequence homology with the adaptor subunits β- and β′-adaptin (66, 81), components of adaptor complexes believed to link clathrin to transmembrane receptors (10). Consistent with this possibility, Arf1 was recently shown to promote association of the TGN-specific AP-1 adaptor complex to the Golgi stack (88, 88a). Taken together, these observations suggest mechanistic parallels between the two different types of carrier vesicles. Although similar experiments have yet to demonstrate a role for Arf in the recruitment of the plasma membrane–specific AP-2 adaptor complex, the selective uptake of plasma membrane proteins into clathrin-coated vesicles is sensitive to GTPγS (89), and Arf has been implicated in endosome function (90). Other functions attributed to Arf include organelle acidification (91) and activation of phospholipase D (PLD) (92, 92a). How these latter observations relate to the mechanism by which Arf initiates coat assembly and vesicle budding remains unknown (92), although a link to heterotrimeric G protein function has been established by several investigators (see below).

DYNAMINS: A NEW CLASS OF GTPases INVOLVED IN VESICLE FORMATION

Dynamin is a member of a novel group of GTPases that share high sequence homology (43–66% identity) in an amino-terminal GTPase domain responsible for guanine nucleotide binding and GTP hydrolysis, but are highly divergent (10–30% identity) in their carboxyl-terminal regions (Figure 1) (see Ref. 93). Functionally unrelated members of this family include the mammalian Mx proteins involved in viral resistance and the yeast *MGM1* gene product, the latter being required for the maintenance of mitochondrial DNA (see Ref. 93). Based on sequence homology, dynamin is most closely related to a third member, the yeast *VPS1* gene product. *VPS1* encodes a non-essential protein that is involved in vacuolar protein sorting from the Golgi (94). Dynamin contains in the carboxyl-terminal region a long, basic, proline-rich extension, which distinguishes it from all other GTPases (Figure 1). Vps1p and dynamin also differ strikingly from Ras-like GTPases by their higher (~10 to 1000-fold, respectively) intrinsic GTP hydrolysis rates. In the case of dynamin, this rate can be further stimulated 10–20-fold by interaction with many apparently unrelated factors, including acidic phospholipids (95), microtubules, GRB2, and many other proteins containing SH3 domains (96). The proline-rich region that contains multiple predicted SH3 domain–binding sites is essential for these effects (96; see Ref. 97). The significance of these observations for the function of dynamin in vesicular transport is unknown.

Initial evidence for a role of dynamin in the endocytic pathway came from the recognition that the *shibire* locus in *Drosophila* encodes a homolog of mammalian dynamin (98, 99). Flies with temperature-sensitive alleles of *shibire* exhibit rapid paralysis at elevated temperatures (100). This correlates with a depletion of synaptic vesicles from nerve terminals and the accumulation of membrane invaginations and coated pits on the cell surface (101). Similar morphological changes are observed in cells of other tissues of the fly, suggesting that *shibire* mutations lead to a pleiotropic defect in the endocytic pathway. Recently, dynamin has been found to be identical to the neuronal phosphoprotein dephosphin (102), originally identified by its rapid stimulus-dependent dephosphorylation in nerve terminals. Dynamin is now recognized to be a substrate for protein kinase C (PKC) and casein kinase II and contains a domain with weak homology to pleckstrin (Figure 1, PH), which is a major PKC substrate of platelets (103). Phosphorylation by PKC enhances its GTPase activity nearly 12-fold (102). These results have led to the hypothesis that the regulation of the dynamin GTPase activity by phosphorylation in the nerve terminal may play a critical role in the rapid endocytosis of synaptic vesicles after exocytosis (102).

Dynamin mutants with substitutions in the highly conserved motifs essential for guanine nucleotide binding and GTP hydrolysis block receptor-mediated endocytosis in mammalian cells in vivo (104, 105). Using in vitro assays that distinguish between a series of sequential steps involved in the formation of clathrin-coated vesicles, one of these mutants predicted to have a lower affinity for guanine nucleotides has been shown to inhibit vesicle formation after the initiation of coat assembly, but prior to the sequestration of ligands into deeply invaginated coated pits (104). A truncated form of dynamin lacking the entire amino-terminal GTPase domain also acts as a potent inhibitor of endocytic transport, apparently by preventing the recruitment of coat components (105). Select carboxyl-terminal deletions reverse the inhibitory phenotypes of these mutants, suggesting that this region interacts with a putative effector(s). The involvement of dynamin in endocytosis is also consistent with the requirement for at least two different GTPases in the formation of clathrin-coated vesicles in vitro (89). Since dynamin mutants have no effect on transport between the ER and the Golgi compartments (104) (S Schmid, personal communication), dynamin per se appears not to be implicated in the early secretory pathway, although we cannot exclude the possibility that unknown members of this family may be involved.

Mutations in the conserved GTP-binding motifs of Vps1p cause dominant defects in vacuolar protein sorting, while endocytosis appears to be normal (106). Consistent with the mutational analysis of dynamin, deletion of the entire amino-terminal GTPase domain inhibits sorting, while select carboxyl-terminal deletions reverse the mutant phenotypes (106). The *VPS1* gene product has also been implicated in the mechanisms responsible for the retention of resident membrane proteins in Golgi compartments (107). Since both vacuolar protein sorting and the retention of Golgi membrane proteins are sensitive to mutations in yeast clathrin genes (see Ref. 3), it is possible that Vps1p (and most likely a mammalian homolog) participates in the assembly of Golgi-derived clathrin-coated vesicles, while dynamin is involved in the assembly of plasma membrane–derived clathrin-coated vesicles.

THE RAB FAMILY: THE TARGETTING/FUSION PARADIGM

Diversity

The Rab/*YPT* gene family is the largest and most diverse group of GTPases involved in vesicular traffic. To date, more than 30 members have been identified in mammalian cells (see Refs. 6, 7, 76). The sequences of these proteins share between 35% and 95% identity, indicating a broad range of functional specificities. Consistent with a role for these GTPases in intracellular transport, individual Rab proteins and their many *YPT* yeast counterparts (5) have been localized to distinct compartments of both the endocytic and exocytic pathways (Figure 3). This distribution strongly suggests that distinct Rab proteins have specific roles in controlling traffic between different subcellular compartments. The complexity of the Rab family is increased by the fact that several members form subfamilies comprising multiple, closely related isoforms sharing more than 90–95% identity. While there is evidence supporting the possibility that these isoforms may regulate different aspects of a particular transport event, other studies suggest that they may be functionally interchangeable.

A given organelle may contain several species of Rab proteins (Figure 3). For instance, the vesicles mediating ER-to-Golgi transport contain at least two Rab proteins, Rab1 and Rab2 (50, 51, 83a, 108, 109); Rab4 and Rab5 have both been localized to early endosomes (109, 110); and Rab7 and Rab9 have been localized to late endosomes (109, 111). Given the low levels of similarity (<55% identical) within each pair, each of these proteins is likely to serve a different function. This may reflect the fact that many organelles of the exocytic and endocytic pathways are dynamic structures at the crossroads of multiple anterograde and retrograde transport routes. Furthermore, during development, mammalian cells elaborate cell-type-specific transport pathways, which may be expected to require a repertoire of specific Rab proteins. A number of cell-type- or tissue-specific Rab proteins have been identified. For instance, the four isoforms of Rab3 (>80% identity) are largely restricted to cell lineages containing regulated secretory

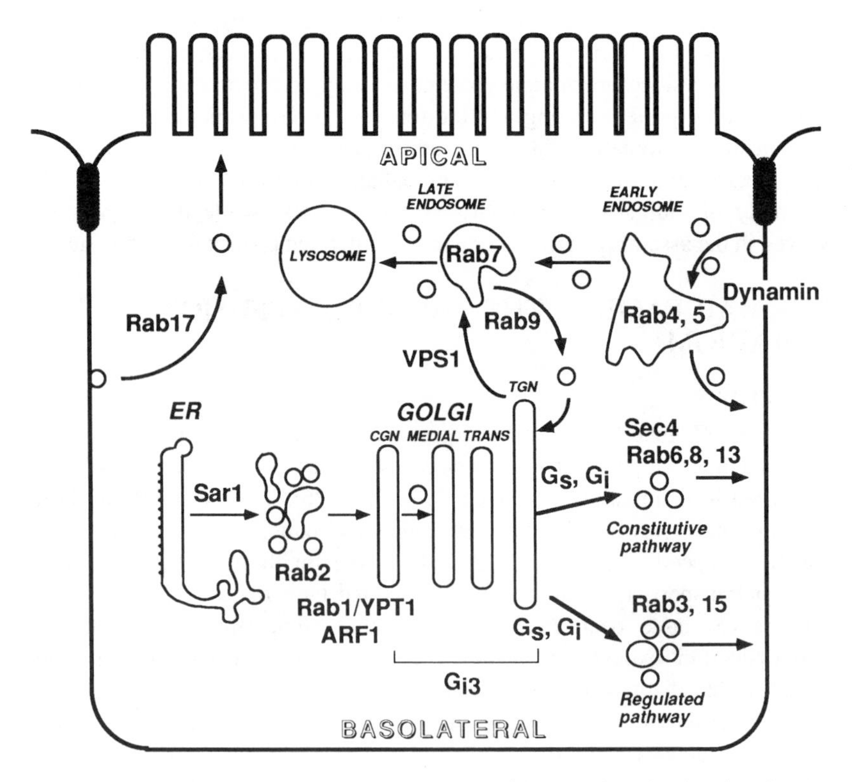

Figure 3 Distribution of GTPases along compartments of the exocytic and endocytic pathways.

pathways, such as neurons, endocrine, and exocrine cells (112–114). In contrast, Rab17 is found in epithelial cells, which contain distinct apical, basolateral, and transcytotic transport pathways (115).

Structural Organization

Based on the striking similarity of Rab proteins with Ras throughout most of their sequence, the structure of Ras can be used as a paradigm to investigate the function of individual domains. The alignment of Rab protein sequences identifies the amino- and carboxyl-terminal regions as the most divergent domains. These are likely to confer functional specificity to individual proteins (116). Molecular modeling predicts that three separate regions of Ras are involved in the conformational changes accompanying guanine nucleotide exchange and GTP hydrolysis (16a, 118a). The corresponding domains of Rab proteins are likely to play a key role in interactions with other components of the transport machinery. These "switch" domains

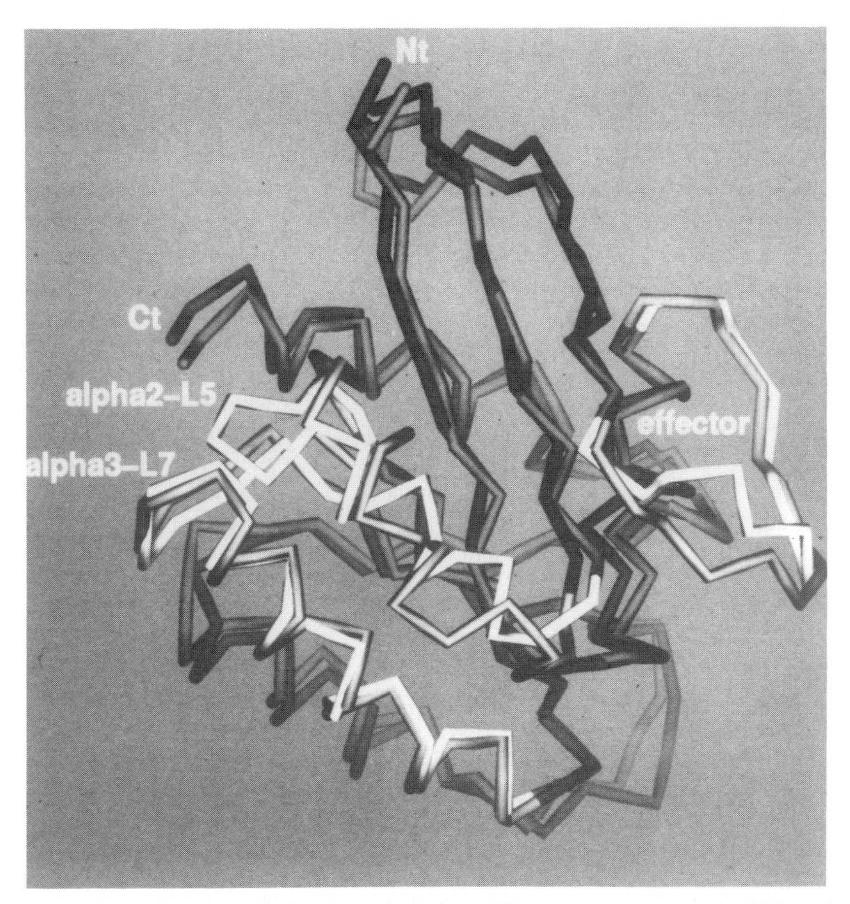

Figure 4 Schematic representation of the three-dimensional structure of Ras in the GDP- and GTP-bound states. Note the conformational differences in the putative effector (L2) domain, and the alpha2L5 and alpha3L7 "switch" regions (GDP-bound form in white). The Ras GDP L4-alpha2 region of the GDP-bound form was modelled (118a) using the structure of the GTP-bound form (16a).

include many of the residues that discriminate between different families of Ras-like GTPases and subfamilies of Rab proteins. They include the L2 effector loop, helix α2/loop 5 (α2L5), and helix α3/loop 7 (α3L7) (Figure 4). Mutational studies on Ras (see Refs. 17, 18) and Rab3a (118) have demonstrated that both L2 and/or α2L5 are involved in interactions with specific GEPs and/or GAPs. All of these regions are located on one surface of the protein opposite to the entry site for guanine nucleotides (Figure 4).

To address the role of individual domains in Rab function, a series of chimeras (117) have been generated between Rab5, a protein involved in early endosome function (119, 120), and Rab6, which has been localized to the Golgi complex (121) and may be involved in vesicle budding from the TGN (122). The rationale for the generation of chimeras was to determine systematically which domains would be required to confer function of the endocytic GTPase Rab5 to the exocytic GTPase Rab6. Replacing the carboxyl-terminal variable region of Rab6 with that of Rab5 led to the targetting of the chimera to early endosomes, confirming previous studies implicating the importance of this region in membrane localization (116). However, as this hybrid could not substitute for Rab5 function in endocytosis, a series of additional domains were swapped. Adding both the α2-L5 and the α3-L7 switch region domains predicted to interact with each other based on modeling studies of H-Ras (Figure 4) inhibited Rab5 function (117). These results suggested that only a limited subset of functional interactions with Rab5 effector proteins was generated. However, a hybrid with partial Rab5 function was obtained when the amino-terminal sequence of Rab5 was added to the chimera (117). This is consistent with recent studies indicating that the amino-terminal regions of several Rab proteins play an important role. For example, it has been shown that a synthetic peptide derived from the amino-terminal domain of Rab8 inhibits function (123). Thus, it appears that determinants in four separate domains (amino terminus, α2/L5, α3/L7, and carboxyl terminus) are required for a basal Rab5 function. Surprisingly, introducing the Rab5 L2 effector region (Figure 4) into the Rab6/Rab5 chimera disrupted function (117), indicating that even more elements are important for proper effector interaction. These results illustrate the remarkable evolutionary fine-tuning of Rab proteins in the context of an invariant core essential for guanine nucleotide binding and GTP hydrolysis.

Related studies using chimeras between two *YPT* proteins, *SEC4* and *YPT1*, involved at consecutive steps in the yeast exocytic pathway, have led, for the most part, to similar conclusions (124, 125; see Ref. 5). However, one construct was particularly informative. In this case, a single hybrid containing both *SEC4* and *YPT1* domains was able to complement a yeast strain lacking both *SEC4* and *YPT1* function (125). In cells expressing only the hybrid protein, transport through the secretory pathway was normal with no indication of missorting. These results indicate that Rab proteins are unlikely to be sufficient to specify proper targetting of carrier vesicles as proposed previously (13), but are consistent with the notion that these GTPases coordinate the function of multiple components constituting targetting/fusion machineries between distinct compartments.

Regulatory Factors Controlling Rab Function

A general class of molecules critical for Rab function are those involved in posttranslational modification of the carboxyl-terminal CC or CXC motifs with geranylgeranyl moieties. These have been extensively reviewed elsewhere (see Refs. 21, 22). One of the three subunits (component A) of the Rab-specific geranylgeranyl transferase II (GGTase II) is the rat counterpart of the human choroideremia (CHM) gene product (125a, 126), which is responsible for a late-onset, recessive, X-linked disease resulting in retinal degeneration. The sequence of the CHM gene has been shown to be related to Rab-GDI (see below). Thus, GGTase II appears to be a member of a larger family of components that recognize nascent Rab proteins, control their posttranslational prenylation, and deliver the mature proteins to the transport machinery.

Given that Rab proteins exhibit low intrinsic rates of guanine nucleotide exchange and GTP hydrolysis, the transport machinery is likely to require a set of accessory factors that modulate these reactions. At present, our knowledge on the role of GAPs in Rab function is quite limited. GAP activities specific for Rab3a (127), Rab1/Ypt1p (128), and Sec4p (129, 130) have been identified in crude extracts, but to date, none of these factors have been purified or cloned. The stimulatory activities of these Rab/Yptp-specific GAPs is weak (10- to 100-fold stimulation) compared to that of Ras-GAP (10,000-fold). Each of these proteins is likely to be specific for a particular member of the Rab/*YPT* family and tuned to promote GTP hydrolysis in the context of a distinct machinery operating under control of a given GTPase. Recently, a GAP (*GYP6*) specific for Ypt6p, a protein of unknown function, has been cloned from yeast (131). The sequence of *GYP6* shows no similarity to any of the known GAPs specific for members of the Ras superfamily (see Refs. 7a, 17), emphasizing its unique role in the biology of vesicular traffic.

More is known about Rab/Yptp-specific GEPs. A dominant suppressor of the temperature-sensitive *sec4–8* mutation in yeast, *DSS4* (132), and a putative mammalian homolog, *Mss4* (133), have been cloned. These genes encode hydrophilic proteins of 17 kDa and 14 kDa, respectively, which stimulate the rate of guanine nucleotide dissociation from Sec4p. In vivo, Dss4p appears to be a component of a high-molecular-weight complex (132). The sequences of these GEPs have no homology to exchange factors specific for other members of the Ras superfamily. Both proteins also exhibit weak exchange activities towards Ypt1p and Rab3a (132, 133). Since these GEP measurements are rather artificial in design and execution, it remains unclear whether Dss4p (or other Rab-GEPs) is active on divergent Rab proteins or, in a context more relevant to the physiological function of these components, will prove to be specific for a particular or a limited set of Rab proteins.

GDIs represent another class of regulatory molecules, which appear to oppose the function of GEPs. GDIs are functionally defined as proteins that prevent guanine nucleotide dissociation from Ras-like GTPases. A Rab-specific GDI was first identified by its ability to inhibit the dissociation of GDP from Rab3a (134, 135). Unlike GAPs and GEPs, which appear to be relatively specific for individual Rab proteins, Rab3a-GDI is active towards a wide range of Rab proteins, including Rab1, 2, 3, 11, and Sec4p (136). Thus, Rab3a-GDI can actually be considered a Rab-GDI. Further studies have revealed another function of Rab-GDI—it is able to extract Rab proteins (Rab1, 2, 5, 7, 8, 9, and 11) from membranes in vitro (136–139). The solubilizing activity of Rab-GDI towards Rab proteins seems to depend on the phosphorylation state of the bound guanine nucleotide. For instance, Rab-GDI solubilizes Rab3a and Rab9 in the GDP-, but not the GTP-bound form (134, 137). Furthermore, all interactions between Rab-GDI and Rab proteins require the presence of carboxyl-terminal prenyl groups (137).

An intriguing new aspect of GDI function concerns the possibility that its distribution between the cytoplasm and intracellular membranes may be regulated by phosphorylation. In fact, mutations in the *quartet* locus of *Drosophila,* which cause developmental defects at the larval stage, are associated with a decrease in the phosphorylation state of a *Drosophila* homolog of Rab-GDI (140). Although at present there is no evidence that *quartet* encodes a kinase, it is interesting that cytosolic Rab-GDI is highly phosphorylated in mammalian cells (141), while the small membrane-bound pool, which can be detected, is not. Recently, a gene encoding a yeast Rab-GDI homolog has been cloned and shown to be essential for cell viability. Depletion of this protein causes transport defects at multiple stages of the secretory pathway (139b). Purified GDI has dominant inhibitory effects on intra-Golgi transport (139a) and prevents export from the ER (WE Balch, unpublished). Taken together, these observations argue for an important physiological role for Rab-GDI in the recycling of Rab proteins for use in multiple rounds of transport. Several studies have demonstrated that the cytosolic pools of numerous Rab proteins, including Rab3a (142), Rab5 (141), Rab9 (137), and Rab1 (WE Balch, unpublished), form stable complexes with Rab-GDI. Moreover, Rab-GDI complexes can be used to deliver Rab5 (143) and Rab9 (137) to membranes in vitro. Since delivery is closely linked to guanine nucleotide exchange (143), membrane association may be coupled with transition from the GDP- to the GTP-bound form, a reaction that is likely to be facilitated by a putative Rab5-GEP.

Role of Rab Proteins in the Constitutive Pathway

In the following we review progress in identifying the function of specific Rab proteins in vesicular traffic.

ER AND GOLGI TRANSPORT (RAB1/YPT1/RAB2) The Rab1a and Rab1b isoforms (92% identical) are highly homologous to the yeast *YPT1* gene product (75% and 66% identity, respectively) (144–147) (Ypt1p function has been reviewed in Refs. 3, 5). Morphological, genetic, and biochemical approaches support an essential role for Rab1 in transport through early compartments of the secretory pathway in mammalian cells (47, 49–51). Rab1-specific antibodies inhibit export from the ER (51). This indicates either that Rab1 plays an active role in budding or, more likely, that it is recruited during vesicle formation to regulate interactions with late-acting effectors involved in vesicle targetting or fusion. Both Ypt1p (148) and Rab1 (50) are associated with 40–80-nm vesicles mediating ER-to-Golgi traffic in yeast and mammalian cells, respectively. In cytosol, at least one functional form of Rab1 serves as a component of a β-COP-containing complex that is essential for vesicle formation from the ER in vitro (49). Rab1 is required not only for ER-to-Golgi traffic, but also for transport between the *cis* and *medial* Golgi compartments (149). This is consistent with the notion that Rab proteins per se are not sufficient for vesicle targetting/fusion, but serve to regulate the function of components involved in these processes.

Further insight into the function of Rab1 stems from the characterization of mutants with altered guanine nucleotide–binding properties. Substitutions in the $Gx_4GK(T/S)$ [Rab1a(S25N)] or NKxD [Rab1a(N124I)] motifs result in proteins that are potent inhibitors of ER-to-Golgi transport in vivo (150) and in vitro (47, 50). The Rab1a(S25N) mutant [analogous to H-Ras(S17N) (41, 42)], has multiple effects on transport. It markedly reduces vesicle budding from the ER and prevents the fusion of carrier vesicles with the *cis* Golgi compartment (47). Biochemical data indicate that this substitution results in a protein with preferential affinity for GDP (47), suggesting that it may be largely restricted to the inactive state. Consistent with a competitive mode of action, the inhibitory phenotype of the S25N mutant can be antagonized by equivalent levels of wild-type Rab1 (47). This observation has been used to demonstrate that the Rab1a and Rab1b isoforms are functionally interchangeable (47). Thus, the Rab1a(S25N) mutant may inhibit export from the ER by competing for a putative Rab1-GEP facilitating recruitment of the wild-type protein during vesicle budding. In contrast to GDP-bound mutant, the Rab1a(N124I) mutant has an exceptionally high exchange rate and inhibits a late, vesicle fusion step. In the presence of this mutant, pre-Golgi vesicular carriers accumulate (47, 50). Since the

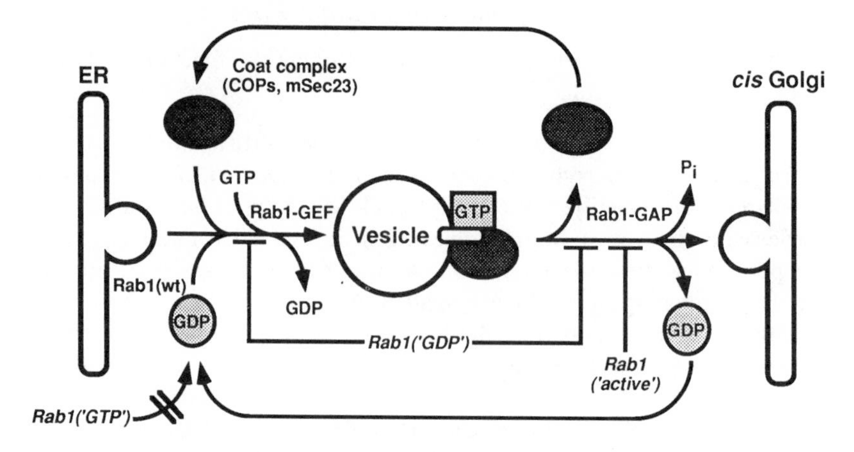

Figure 5 Model for the function of Rab1 in transport. Rab1 is recruited in the GDP-bound form during vesicle budding along with a coat precursor containing β-COP. This is coupled to guanine nucleotide exchange facilitated by a putative Rab1-GEF. The GTP-bound form remains associated with the transport vesicles until interaction with a putative Rab1-GAP triggers GTP hydrolysis during a subsequent targetting and/or fusion step. A mutant (italics) restricted to the inactive (GDP-bound) state reduces the rate of export from the ER. Vesicles that form may also be incompetent for fusion. A mutant with a high exchange rate that may mimic an active configuration (Rab1 "active") inhibits fusion by interfering with the function of Rab1-GAP or another downstream effector, while mutants that cannot hydrolyze GTP (Rab1 "GTP") fail to integrate into the transport machinery. The Rab1 cycle may need to operate in conjunction with the Sar1p and/or Arf1 cycle (Figure 2) to initiate vesicle budding.

equivalent H-Ras(N116I) mutant is transforming (151), the N124I mutant may block fusion by adopting a constitutively active conformation. Thus, in addition to conformational changes associated with nucleotide exchange during recruitment, a change in protein conformation reflecting GTP hydrolysis may be required during a late targetting/fusion step. A similar model applies to Sec4p (5, 7b).

In contrast to the S25N or N124I mutants, Rab1 effector domain mutants have no detectable consequence on transport in vivo or in vitro, consistent with the recessive character of similar mutations in Ras and *YPT1* (152). However, synthetic peptides derived from this region have been shown to arrest ER-to-Golgi traffic at a late, prefusion step in vitro (153, 154). While the target for these peptides remains to be established, one possibility is that they may prevent the interaction of Rab1 with a downstream effector(s) critical for fusion.

The cumulative evidence suggests a tentative working model for the function of Rab1 in ER-to-Golgi transport and traffic between Golgi compartments (Figure 5). A putative Rab1-GEP controls recruitment of Rab1 in the GDP-bound form. Consistent with this notion, a mutation that restricts the protein to the GTP-bound form by preventing GTP hydrolysis neither inhibits nor stimulates transport (150; WE Balch, unpublished results) (Figure 5). Hydrolysis, presumably facilitated by a putative Rab1-GAP, is likely to be required for vesicle targetting and/or fusion. The altered conformation of the Rab1a(N124I) mutant may inhibit these steps by sequestering the function of a putative downstream Rab1-GAP or another effector molecule necessary for the delivery of vesicles to the *cis* Golgi compartment. Alternatively, recent evidence indicates that the H-Ras(N116I) mutant forms a stable complex with a Ras-GEP in vitro (155, 156). These results raise the possibility that the S25N and N124I mutants of Rab1 share the same target, namely, Rab1-GEP. While the GDP-restricted form of Rab1 would markedly attenuate GEP function, the N124I mutant, unlike the GTP-restricted mutant, may be recruited efficiently but would be unable to undergo a conformational change essential to initiate vesicle fusion (50).

In addition to Rab1, Rab2, which shares less than 40% identity with Rab1, has been localized to pre-Golgi intermediates (83a, 109). Substitutions equivalent to those discussed above result in proteins that are potent *trans* dominant inhibitors of ER-to-Golgi transport in vivo (150). The precise role of Rab2, as well as its functional relationship to Rab1, is presently obscure.

TRAFFIC TO THE CELL SURFACE (SEC4/RAB6/RAB8) In mammalian cells, several Rab proteins are associated with late Golgi compartments and post-Golgi transport vesicles, and therefore represent strong candidates for a function analogous to that of Sec4p, a small GTPase that is required for transport between the TGN and the plasma membrane in yeast (see Refs. 3, 5, 7b).

Rab6 involvement in budding from the TGN Rab6, which shares less than 40% identity with Sec4p, can be isolated from Golgi membranes in a complex with TGN38/41, a type I integral membrane protein that recycles between the TGN and the cell surface (122). A larger pool of Rab6, however, forms a cytosolic complex with a 62-kDa protein whose association with Golgi membranes may be modulated by phosphorylation (122). Given the striking similarity between the properties of p62 and Rab-GDI (136), it seems likely that p62 is a Rab-GDI-like protein. Immunodepletion of the cytosolic complex using either Rab6-specific or p62-specific antibodies inhibited the budding of carrier vesicles containing a marker protein destined for the plasma membrane by >70%, suggesting that Rab6 may be involved in the function of vesicles mediating the constitutive pathway (122). The prominent position of the TGN at the crossroads of multiple transport pathways, however, does not rule out a possible involvement of Rab6 in the endocytic, lysosomal, or regulated pathways. A Rab6 homolog, Ypt6p, as well as a Ypt6p-GAP, have been identified in yeast (131). Their functions remain to be characterized.

Membrane traffic in polarized epithelial cells (Rab8/Rab13/Rab17) A subset of Rab proteins is likely to be required for the regulation of transport pathways involved in the establishment and maintenance of cell polarity in epithelial cells and other polarized cell-types. Rab8, which shares ~60% identity with Sec4p, has been implicated in vesicular traffic between the TGN and basolateral domain in polarized MDCK cells (123). The protein has been localized to Golgi membranes, vesicular-tubular structures adjacent to the Golgi complex, and the basolateral surface, where it appears to be restricted to regions involved in the formation of tight junctions. Consistent with a role in transport from the TGN to the basolateral domain, immunoisolation studies have revealed that Rab8 is found on carrier vesicles destined for the basolateral, but not the apical, surface (123). Moreover, a synthetic peptide derived from the amino-terminal region of Rab8 inhibits basolateral but not apical traffic (123). Thus, the evidence suggests that Rab8 mediates transport of a particular subset of vesicles between the TGN and the basolateral domain in polarized cells. Consistent with this hypothesis, Rab8 is restricted to the somatodendritic surface in polarized hippocampal neurons (158). Incubation of cultured neurons with antisense oligonucleotides, which reduce the expression of Rab8, inhibits transport to the dendrites but has no effect on axonal traffic.

Rab13 has recently been cloned from a human intestinal cDNA library (158a). The Rab13 sequence is 61% identical to human Rab8 and 56% identical to Sec4p. It shows less than 47% identity with other Rab proteins, suggesting a function in post-Golgi traffic. Interestingly, Rab13 colocalizes with the tight junction marker ZO-1 in Caco-2 cells as shown by confocal microscopy. This indicates that the protein may be associated with the cytoplasmic face of junctional complexes and/or carrier vesicles destined for this region. This particular distribution of Rab13 depends on the integrity of tight junctions, as their dispersal by incubation in the presence of low Ca^{2+} concentrations triggers the redistribution of the protein to a diffuse cytoplasmic staining pattern. The establishment of cell polarity in higher eukaryotes involves the selective delivery of proteins to the differentiating apical and basolateral domains via specific exocytic routes from the TGN and through transcytotic pathways. An interesting possibility is that junctional complexes may serve as targets for the delivery of vesicles destined for the apical domain (158a). If this is the case, Rab13, like Rab8, may be involved in controlling the function of surface-specific transport machineries.

In adult kidney epithelia, Rab17 is associated with the basolateral surface and apical endocytic tubules (115). While it is absent from early mesenchymal precursors, induction of Rab17 expression during differentiation into epithelial cells is essential for proper development of the tissue (115). Although the function of Rab17 remains unknown, it has been speculated to be involved in a tissue-specific event controlling transcytosis (115).

Bud site selection in yeast (BUD1/CDC42/Rac/Rho) The potential role of Rab8 and Rab13 in the establishment or maintenance of cell polarity in epithelial tissues may be analogous to the process of bud site selection in *S. cerevisiae,* which represents a special form of cell surface polarization. Cell division by budding involves a series of events, including bud site selection, bud growth, and cytokinesis at the boundary between the bud and the mother cell. Genetic analysis of these processes has revealed the importance of a number of Ras-like GTPases (159–162). Bud formation depends directly on Bud1p, a protein 57% identical to human Ras, and Cdc42p, which shares 52–80% identity with members of the human Rho family (160). Bud1p and Cdc42 function in the context of protein complexes controlling bud site selection (including Bud1p, 2p, 3p, 4p, and 5p) and bud growth (Cdc42p, 24p, 43p, and Bem1p) (160, 161). These proteins further interact with two additional GTPases, Rho3p and Rho4p (162). Rho proteins might participate in the organization of the actin cytoskeleton, as previous results have implicated actin in the polarized movement of secretory vesicles towards the bud (164). Current models suggest that Bud1p regulates the recruitment of proteins required for bud site selection, while Cdc42p, Rho3p, and Rho4p may control the function of complexes responsible for bud growth. Interestingly, *BUD5* shows similarity to *CDC25,* which encodes a Ras-GEP (165), raising the possibility that Bud5p may act on Bud1p or Cdc42p to promote guanine nucleotide exchange. *BUD2* encodes a *BUD1*-specific GAP (166). Thus, the polarized movement of secretory vesicles to the cell surface of *S. cerevisiae* involves the function of at least four Ras-like GTPases—Sec4p for vesicle targetting/fusion and Bud1p, Cdc42p, and Rho3/4p for bud site selection and bud growth.

The role of Cdc42p and Rho3/4p in bud site selection may be similar to the newly recognized involvement of mammalian Rho proteins in the organization of focal adhesion sites, stress fibers, and possibly tight junctions in response to growth factors (167–169). The expression of dominant inhibitory Rho mutants causes dramatic changes in the organization of the actin cytoskeleton, blocking both focal adhesion (169) and stress fiber (168) formation in mammalian cells. Perhaps a similar spectrum of GTPases involving Ras, Rab (i.e. Rab8/13), and Rho family members is required in higher eukaryotes for transport pathways involved in (*a*) the differentiation of apical and basolateral domains in epithelial cells, and (*b*) the establishment and maintenance of highly specialized structures as found, for example, in the synapse. Taken together, these observations illustrate the exceptional versatility of the GTPase design. Moreover, they emphasize the importance of vesicular traffic in development and proliferation.

Control of the Regulated Secretory Pathway by Rab3

Another form of developmental specialization involves cells having a regulated secretory pathway, including cells of endocrine and exocrine origin, as well as neuronal cells involved in synaptic transmission. These express a number of cell-type-specific Rab proteins of which Rab3 is the best characterized. Four closely related isoforms have been identified to date (112, 113, 170, 172, 173). While Rab3a is largely restricted to synaptic vesicles in neurosecretory cells (114), Rab3b and 3c have been localized to anterior pituitary (174) and insulin-secreting cells, respectively (142). Rab3d is found in adipocytes and is likely to be involved in regulating the delivery of the GLUT4 glucose permease to the cell surface in response to insulin (173). The Rab3 subfamily is likely to expand as new members are discovered in cells containing other specialized regulated pathways.

While Rab3a-GEP and Rab3a-GAP activities have been identified and partially purified (118, 127), their roles in the regulation of Rab3a function remain unknown. Stimulation of neurotransmitter release is accompanied by the dissociation of Rab3a from synaptic vesicles (113, 114), implicating a role for Rab-GDI in this process. Recently, a membrane-bound protein that interacts with the GTP-, but not the GDP-bound form of Rab3a has been purified and cloned from bovine brain (175). The properties of this protein, termed rabphilin 3a, are consistent with its proposed role as a downstream effector of Rab3a. Rabphilin 3a shares homology to synaptotagmin through conserved C2 phospholipid- and Ca^{2+}-binding domains (175a). Synaptotagmin has been shown to bind syntaxins in vitro, molecules that may associate with Ca^{2+} channels in the active release zone and thereby promote vesicle fusion (see Refs. 176, 177, 177a).

Several lines of evidence indicate that Rab3 participates in vesicle targetting/fusion. The introduction of a synthetic peptide homologous to the Rab3 effector domain (L2, see Figure 4) into mast cells via a patch pipette triggers rapid and complete degranulation (178). It is important to note that the effector domain peptide as well as GTPγS appear to stimulate regulated secretion, in contrast to the inhibitory effects of these reagents in the constitutive secretory pathway. The Rab3 effector peptide also stimulates regulated secretion in a number of other cell types, including insulin release from HIT-T15 cells (179) and amylase release from pancreatic acinar cells (180, 181), and triggers degranulation of chromaffin cells (182). Microinjection of the peptide into cultured neurons stimulates the release of neurotransmitter (183). One possible explanation for these results is that the peptide bypasses a tightly regulated upstream signaling cascade involving a Rab3 effector to initiate fusion. An alternative explanation stems from the observation that an analogous Ras-derived, effector-domain peptide prevents the interaction of Ras with Ras-GAP (184). Perhaps the role of Rab3-GAP in the synapse is to maintain Rab3 in an inactive GDP-bound form—in this case, as a negative regulator. Only the GTP-bound form is able to initiate fusion. Addition of effector peptide may prevent interaction with Rab3-Gap, stabilizing the GTP-bound form and triggering vesicle fusion. In addition to peptides, the introduction of antisense oligonucleotides to Rab3b into rat pituitary cells using whole-cell patch clamp techniques specifically blocks expression of Rab3b and markedly attenuates regulated secretion without interfering with the constitutive pathway or endocytosis (186). Attesting to the specific function of Rab3b, no effect is observed using antisense oligonucleotides to Rab3a. Transient expression of wild-type and mutant Rab3a proteins in nondividing bovine chromaffin cells inhibits nicotinic agonist-stimulated exocytosis in intact and permeabilized cells (186a). These results provide the first evidence indicating that highly related isoforms of a given Rab protein have unique functions.

The function of Rab3a has been analyzed using site-directed mutagenesis (118, 185). A series of mutations equivalent to those described previously for Rab1, Sar1, and Arf1 [analogous to oncogenic Ras mutants] demonstrated that while some of the basal properties of exchange and hydrolysis were similar to those found for Ras, others were different (118). These results emphasize that specific amino acid residues in individual members of the Ras superfamily may differ in function. Furthermore, analysis of a number of effector domain mutations on the interaction of Rab3a with partially purified GEP and GAP in vitro led to the suggestion that multiple factors may interact through this site (185).

Control of the Endocytic Pathway (Rab4, 5, 7, and 9)

Compartments of the endocytic pathway are at the crossroads of exocytic and endocytic transport routes. Early endosomes serve as a site for the sorting of internalized proteins destined for recycling to the plasma membrane, transport to the late endosome, and transcytosis to the opposite surface domain in epithelial cells. Late endosomes, in addition to mediating the delivery of proteins from both the endocytic and exocytic pathways to lysosomes, are involved in the recycling of membrane receptors to the TGN. Given the functional diversity of these compartments, it is not surprising that at least four Rab proteins—Rab4, 5, 7, and 9—have been localized to endocytic organelles. Rab4 and 5 are found on early endosomes (109, 187) and regulate early steps in endocytosis (119). Rab7 and 9 are associated with late endosomes (109, 111). Rab7 is believed to function in the delivery of proteins to lysosomes, Rab9 in the recycling of membrane receptors to the TGN (109, 111).

Members of a Rab5/*YPT5* subgroup have been identified in both yeast and mammalian cells (7, 186a, 187b). These proteins are 48–54% identical, which is low compared to the high level of conservation across phylogenetic lines between Ypt1p and the mammalian Rab1 isoforms (66–75% identity). However, these proteins are considered to be related based on identical effector domains (L2) and sequence similarities in the α2L5 and other regions (see Figure 4). Three isoforms (Rab5a, b, and c; ~95% identical) and three potential yeast homologs have been cloned (see Refs. 6, 187a). Rab5b is essential for the fusion of early endosomes in vitro (120). Rab5b-specific antibodies inhibit fusion in vitro, whereas cytosol prepared from cells overexpressing wild-type Rab5b but not Rab5b(N133I), a mutant with a high guanine nucleotide exchange rate [equivalent to (H-Ras (N116I)], stimulates the reaction. Analysis of Rab5b function in vivo has revealed that overexpression of the wild-type protein in BHK cells stimulates internalization of endocytic markers and causes the appearance of large endosomal structures (119). In contrast, the N133I mutant inhibits endocytosis in vivo and results in extensive fragmentation of endosomes. In yeast, mutations in the Rab5-like GTPase (Vps21p) lead to a pronounced defect in vacuolar sorting (187b). In the endocytic pathway, no effect was detected in α-factor uptake, although α-factor turnover was impaired (187a).

Rab4 also plays an essential role in the early endocytic pathway. In contrast to Rab5, overexpression of Rab4 reduces the intracellular steady-state level of the transferrin receptor (110, 188). However, this is not due to a reduction in the rate of internalization, but rather to an increase in the rate of recycling from early endosomes to the plasma membrane (188). Overexpression of Rab4 also prevents the delivery of transferrin to an acidic compartment essential for iron release, resulting in its accumulation in non-acidic, perinuclear tubular-vesicular structures. Thus, at least two GTPases appear to regulate opposing pathways relevant to early endosome function. Rab4 redistribution has also been linked to insulin stimulation in adipocytes, suggesting it plays a critical role in the translocation of the glucose transporter (GLUT4) to the cell surface (188a).

At present, little is known about the function of mammalian Rab7. A

Rab7 homolog, Ypt7p (63% identical), has been identified in yeast (189, 190). While the internalization of α-factor from the cell surface is normal in strains lacking *YPT7* function, degradation of the pheromone is severely inhibited due to a delay in delivery to the vacuole. Furthermore, these strains show extensive fragmentation of the vacuole. Ypt7p has been proposed to be involved in transport between early and late endosomes (189), although it may also function between late endosomes and the vacuole. The correlation between the function of Ypt7p and vacuolar biogenesis in yeast is consistent with the role of early/late endosomes in the lysosomal pathway in mammalian cells.

Rab9 has been localized to late endosomes (111). Through use of a cell-free system that reconstitutes the movement of the mannose 6-phosphate receptor (M6PR) between late endosomes and the TGN in vitro, Rab9 has been found to play an important role in the delivery of M6PR to the TGN (111). Rab9-specific antibodies inhibit transport, while the reaction is stimulated by the presence of excess Rab9. Since Rab9 is found in the cytosol as a GDI complex (137) and delivery to the membrane is accompanied by guanine nucleotide exchange, a putative Rab9-GEP is likely to be involved in an early recruitment step (S Pfeffer, personal communication).

Summary: Rab Proteins and the Targetting/Fusion Paradigm

While Rab proteins and other small GTPases are likely to be recruited to membranes from a recycling, GDI-bound pool via a variety of distinct GEPs, a growing consensus points to the possibility that Rab proteins and Ca^{2+} sensors may play a prominent role in vesicle targetting/fusion (Figure 6). Vesicles involved in ER-to-Golgi transport (39), intra-Golgi transport (46, 50, 108, 149), neurotransmitter release (see Ref. 177), and regulated secretion by exocrine and endocrine cells (see Ref. 191) require Ca^{2+} for fusion with their respective target membranes. These processes may represent variations on a common theme. For example, genetic studies in yeast have provided evidence for a link between the requirement for Ca^{2+} in traffic through the early secretory pathway and *YPT1* function (192). Mutations in *PMR1*, a protein that encodes a Ca^{2+}-dependent adenosine triphosphatase associated with early Golgi compartments (193), suppress the transport defects of *ypt1-1* mutants (194). Other suppressors of *ypt1* mutations include *BET1(SLY12)* and *SEC22(SLY2)* (86, 195). Interestingly, these proteins show homology to the synaptic vesicle proteins synaptobrevin 1 and 2 (also known as VAMPs) (see Ref. 177a), Snc1p (196), a protein involved in the late secretory pathway in yeast, and Pep12p, involved in vacuolar biogenesis (197) (Figure 6). Homologs of the synaptic protein syntaxin are also found in both the early and late secretory pathways in yeast (197a) and mammalian

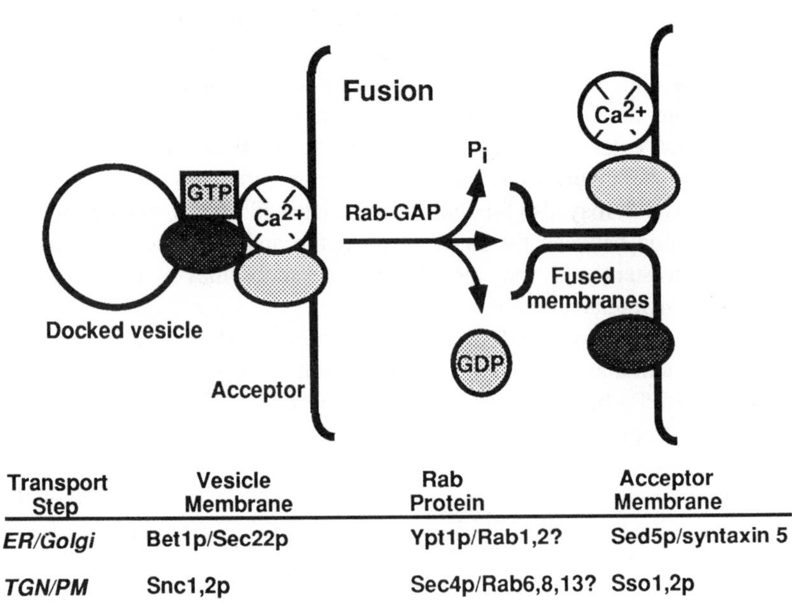

Transport Step	Vesicle Membrane	Rab Protein	Acceptor Membrane
ER/Golgi	Bet1p/Sec22p	Ypt1p/Rab1,2?	Sed5p/syntaxin 5
TGN/PM	Snc1,2p	Sec4p/Rab6,8,13?	Sso1,2p
Synapse	Synaptobrevin/VAMP	Rab3	Syntaxin
Vaculolar	?	Ypt5 (Vps21),7p?	Pep12p

Figure 6 The targetting/fusion paradigm. Rab proteins may control the targetting and/or fusion of transport vesicles in cooperation with other components functioning at different stages of the exocytic pathway. In this particular model, the Rab protein serves as a negative regulator. A conformational change associated with the hydrolysis of GTP (presumably involving Rab-GAP) is responsible for initiating the fusion process. Other models may also be applicable.

cells (198) (Figure 6). It appears that synaptobrevin and syntaxin are part of a growing family of related components involved in conferring specificity to fusion between incoming vesicles and their target membranes at distinct stages of the secretory pathway (199, 200). This model has been referred to as the SNARE hypothesis (199a, 200; see 7b, 8a). While the role of individual Rab proteins in the function of such fusion complexes is presently unknown, they may monitor the fidelity of these protein-protein interactions through changes in conformation associated with the GTPase cycle.

HETEROTRIMERIC G PROTEINS

Classical heterotrimeric G proteins, unlike Ras-like GTPases, consist of α, β, and γ subunits (19). In their well-established role as signal transducers, the binding of ligand to a cell surface receptor triggers guanine nucleotide exchange on the α subunit (G_α). This activates G_α, which dissociates from the βγ dimer and, with few exceptions, acts to stimulate or inhibit ion channels or the production of second messengers by interacting with downstream effector molecules. Subsequently, G_α inactivates itself by GTP hydrolysis, which allows reassociation with the βγ dimer.

Initial evidence for a role of heterotrimeric G proteins in vesicular traffic came from observations that degranulation of mast cells, chromaffin cells, neutrophils, and a range of other cells is stimulated by extracellular G protein agonists (see Refs. 4, 191; 43, 201–203). In this case, the function of the G protein(s) involved is in keeping with a more classical view, whereby second messengers trigger an intracellular response leading to vesicle fusion. Although progress in identifying the G proteins involved in these events has been slow, recent indirect evidence suggests that G_o may regulate granule discharge from chromaffin cells (204), while $G_{i\alpha3}$ may promote degranulation of mast cells (205).

The possibility that heterotrimeric G proteins regulate earlier steps in the secretory pathway stems from the observation that AlF_4^- [a phosphate analog that binds to G_α and activates heterotrimeric G proteins, but not monomeric GTPases (9)], inhibits ER-to-Golgi (108) and intra-Golgi (64) transport, as well as vesicle budding from the TGN (203, 206, 212). Two distinct G_α subunits cooperate antagonistically to control a rate-limiting step in the formation of both constitutive and regulated secretory vesicles from the TGN in PC12 and MDCK cells. An inhibitory $G_{\alpha i}$ sensitive to ADP-ribosylation by pertussis toxin (PtX) suppresses vesicle budding, whereas a stimulatory $G_{\alpha s}$ sensitive to cholera toxin (CtX) promotes vesicle formation (1, 206, 212). $G_{i\alpha3}$ is a candidate for the $G_{\alpha i}$ involved, as it has been localized to Golgi compartments (207), and overexpression of $G_{i\alpha3}$ reduces the flux of heteroglycosaminoglycans through the Golgi complex and to the cell surface (207). This inhibition can be overcome by treating the cells with PtX, which prevents guanine nucleotide exchange and release from βγ subunits. An alternatively spliced form of $G_{i\alpha2}$ has also been localized to the Golgi (207a). Its potential role in transport is unknown.

Export from the ER is also sensitive to a range of agonists and antagonists of G_α function, including mastoparan and βγ dimers (108). Mastoparan is a peptide that mimics ligand-induced receptor coupling, stimulates guanine nucleotide exchange, and thereby activates G_α (208). βγ subunits, on the other hand, directly antagonize activation by sequestering G_α, or might modulate the activity of an unknown downstream effector(s) (see Refs. 209, 210). Surprisingly, the putative G_α protein(s) implicated in export from the ER is insensitive to both PtX and CtX (108), suggesting the involvement of a family member(s) different from those implicated in Golgi function. While there is no evidence for G protein function in vesicular traffic in yeast, Sar1-GAP (Sec13p) contains sequence motifs reminiscent of the WD40 repeats common to β subunits of heterotrimeric G proteins (45). This motif is found in a wide range of proteins with diverse functions. Thus, the significance of this observation remains obscure. Heterotrimeric G proteins may also be involved in endosome function, as AlF_4^-, PtX, CtX, and mastoparan have been demonstrated to perturb the flow of proteins through the endocytic pathway (211), in transcytosis of the polymeric immunoglobulin receptor (211a, 211b) and in vacuolar biogenesis (211c, 211d).

What are the potential mechanisms by which heterotrimeric G proteins may regulate transport through the constitutive pathway? Given the possible importance of Ca^{2+} and/or pH in the function of exocytic compartments, one possibility is that proton and ion channels are regulated through a G protein–mediated pathway, although no direct evidence currently indicates this to be the case. The apparent role of Arf in compartment acidification or stimulation of PLD may be related to these events. On the other hand, a number of studies have provided direct evidence for the involvement of heterotrimeric G proteins in coat assembly. For example, the association of Arf, β-COP, and clathrin with Golgi membranes is sensitive to a number of reagents that perturb G protein function (see Refs. 1, 4). There is a second possibility that has received scant attention—a potential role of cargo in modulating vesicular traffic through G proteins. As mentioned previously, both the exocytic and endocytic pathways transport a diverse range of soluble and membrane-bound proteins destined for distinct compartments. In many cases cargo must be sorted from the bulk of resident proteins and concentrated for transport through divergent routes. Sorting signals in the cytoplasmic domains of transmembrane proteins promote their concentration into clathrin-coated carriers. In contrast, sorting signals triggering recruitment to non-clathrin-coated vesicles mediating traffic through the early exocytic pathway have not been identified, even though it is now clear that proteins can be sorted and concentrated nearly 10-fold during export from the ER (12). Given the ability of G proteins to serve as a communication link between opposite sides of a membrane bilayer, this raises the possibility that G protein–coupled receptors may, at least in some cases, be involved in a novel form of signal transduction to regulate coat assembly in response to cargo (108, 201).

ROLE OF PHOSPHORYLATION IN CONTROL OF GTPase FUNCTION

Recently, treatment of rat basal leukemia (RBL) cells with the extracellular agonist IgE or with phorbol esters was found to stimulate constitutive secretion and promote Arf/β-COP association with membranes in a protein kinase C (PKC)–sensitive fashion (213, 215). These observations are consistent with earlier results that documented the role of kinases and phosphatases in regulating vesicular budding from the ER and Golgi compartments (149). Cytosolic PKC substrates have previously been suggested by a number of investigators to modulate the regulation of fusion during regulated secretion from endocrine and exocrine cells, and at the synapse (see Ref. 213a). In addition, phosphorylation of Rab proteins via extracellular agonists has been detected in human platelets (188a). Phosphorylation of Rab-GDI (141) and possibly other GTPase regulatory factors may be critical for function. These observations raise the possibility that both the constitutive and regulated pathways may be subject to global regulation by extracellular stimuli via second messenger pathways involving G proteins and protein tyrosine kinases. While the role of phosphorylation is largely unknown, these results have important implications not only on the analysis of GTPase function, but also in understanding the role of these GTPases during cell proliferation, differentiation, and in metastatic diseases.

GTPase CONTROL OF ORGANELLE STRUCTURE

While the regulation of vesicular traffic by multiple GTPases is now clear, an important and emerging aspect of GTPase function concerns the dynamic organization of endocytic and exocytic compartments. As alluded to previously, alterations in the balance of traffic to and from a given compartment can have striking consequences on organelle integrity. Overexpression of an Arf1 (T31N) mutant restricted to the inactive GDP-bound state triggered the collapse of Golgi cisternae into the ER in a manner that is indistinguishable from the well-documented effects of BFA (48) (Figure 7). Overexpression of Rab5 results in the appearance of large endosomal structures, while the Rab5(N133I) mutant triggers fragmentation of early endosomes (119) (Figure 7). Similarly, the *shibire* mutation in *Drosophila* dynamin has profound effects on the morphology of compartments of the early endocytic pathway in the synapse (101), and depletion of Vps21p (Rab5-like) and Ypt7p leads to extensive fragmentation of the vacuole in yeast (187b, 189).

Rab1 is also critical for Golgi integrity. Overexpression of either the Rab1a(S25N) or Rab1a(N124I) mutants results in the disassembly of the Golgi stack without collapse into the ER (47, 216). Unlike the effects of the Arf1 (T31N) mutant (48), the Golgi was reduced to 100–300-nm remnants containing resident Golgi proteins and a large population of 60–80-nm carrier vesicles devoid of Golgi markers (216) (Figure 7). Since individual compartments of the Golgi stack appeared to shed most of their

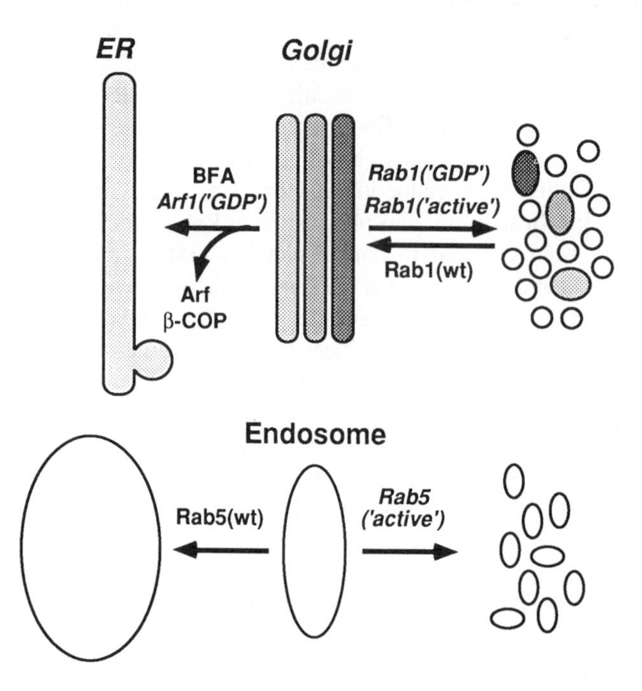

Figure 7 GTPase control of organelle structure. In the exocytic pathway, an Arf1 mutant restricted to the inactive (GDP-bound) (italics) state causes the release of Arf and β-COP from Golgi membranes, leading to its fusion with the ER. This phenotype is identical to that triggered by the drug BFA. In contrast, the equivalent Rab1 (GDP-bound) mutant (italics) or a mutant with a high exchange rate (Rab1 "active") (italics) triggers disassembly of the Golgi stack into remnants and carrier vesicles. In the endosomal pathway, overexpression of Rab5 results in the formation of large endosomal structures, while a mutant with a high exchange rate (Rab5 "active") (italics) leads to extensive fragmentation of endosomes.

membrane to carrier vesicles, this raises the possibility that much of the membrane of the Golgi stack is in transit. If so, this observation could account for a number of results, including (*a*) the striking accumulation of coat proteins on Golgi membranes in the presence of GTPγS (62) or constitutively active forms of Arf1 (48, 73), (*b*) the dramatic effects of BFA, which, by promoting the release of coat proteins (74, 217), may unmask and activate components of an abundant recognition/fusion machinery whose function is normally restricted to carrier vesicles (68a), and (*c*) the inability to detect further concentration of protein during migration from the *cis* to *trans* faces of the Golgi stack (12, 218). In addition, at the onset of mitosis, intracellular traffic is arrested and the Golgi disassembles into remnants and carrier vesicles similar to those observed in the presence of the Rab1 mutants (216). It has previously been proposed that dismantling of the Golgi complex during mitosis might be a direct consequence of a regulatory phosphorylation step that does not interfere with vesicle budding but blocks fusion (219; see Ref. 219a). The dramatic consequences of Rab1 dysfunction on Golgi integrity provide at least one line of biochemical evidence to support this model, and suggest that the vesicle budding and fusion machinery may be differentially sensitive to phosphorylation/dephosphorylation at different stages of the cell cycle (149, 220).

OUTLOOK

Many problems remain to be solved in order to understand how GTPases control vesicular traffic and organelle structure. Although a variety of components have been identified, few of them have been characterized in detail, and most of the accessory factors and effector molecules that are likely to play important roles in GTPase function are unknown or poorly understood. The combination of genetic approaches in vivo and biochemical approaches in vitro will continue to provide clues towards unraveling the precise function of each of these components and understanding how this cast of players cooperates at the molecular level to promote in a specific manner the lipid rearrangements underlying the mechanics of vesicle budding and fusion. The potential role of GTPases in signal-mediated sorting mechanisms for the recruitment of cargo during coat assembly, as well as the function of heterotrimeric G proteins and protein kinases in the global control of vesicular traffic in response to extracellular signals and during different stages of the cell cycle, remain to be elucidated. Whether non-clathrin-coated and clathrin-coated vesicles function in similar or dissimilar ways has not been directly addressed. The growing list of GTPases and components now recognized to govern membrane traffic and organelle structure/function will undoubtedly continue to surprise us, both by their exquisite specificity and their common themes of action.

ACKNOWLEDGMENTS

We acknowledge S. Schmid for many helpful comments and criticisms during preparation of the manuscript. We thank T. Macke for the computer graphics illustration of the H-ras GDP/GTP forms.

Literature Cited

1. Leyte A, Barr FA, Kehlenbach RH, Huttner WB. 1992. *Trends Cell Biol.* 2:91–94
2. Pfeffer SR. 1992. *Trends Cell Biol.* 2:41–46
3. Pryer NK, Wuestehube LJ, Schekman R. 1992. *Annu. Rev. Biochem.* 61:471–516
4. Bomsel M, Mostov K. 1993. *Mol. Biol. Cell* 3:1317–28
5. Ferro-Novick S, Novick P. 1993. *Annu. Rev. Cell Biol.* 9:575–99
6. Simons K, Zerial M. 1993. *Neuron* 11:789–99
7. Zerial M, Stenmark H. 1993. *Curr. Biol.* 5:613–20

7a. Boguski MS, McCormick F. 1993. *Nature* 16:643–54

7b. Novick P, Brennwald P. 1993. *Cell* 75:597–601

8. Rothman JE, Orci L. 1992. *Nature* 355:409–15

8a. Takizawa PA, Malhotra V. 1993. *Cell* 75:593–96

9. Kahn RA. 1991. *J. Biol. Chem.* 266: 15595–97
10. Schmid SL. 1993. *Trends Cell Biol.* 3:145–48
11. Kreis TE. 1992. *Curr. Opin. Cell Biol.* 4:609–15
12. Balch WE, McCaffery JM, Plutner H, Farquhar MG. 1994. *Cell.* In press
13. Bourne HR. 1988. *Cell* 53:669–71
14. Bourne HR, Sanders DA, McCormick F. 1990. *Nature* 348:125–32
15. Wittinghofer A, Pai EF. 1991. *Trends Biochem. Sci.* 16:67–80
16. Valencia A, Chardin P, Wittinghofer A, Sander C. 1991. *Biochemistry* 30: 4638–48

16a. Pai EF, Kabsch W, Krengel U, Holmes KC, John J, Wittinghofer A. 1989. *Nature* 341:209–14

16b. Noel JP, Hamm HE, Sigler PB. 1994. *Nature* 366:654–63

16c. Bechthold H. 1993. *Nature* 365:126–32

17. Bokoch GM, Der CJ. 1993. *FASEB* 7:750–59
18. Lowy DR, Willumsen BM. 1993. *Annu. Rev. Biochem.* 62:851–91
19. Simon MI, Strathmann MP, Gautam N. 1991. *Science* 252:802–8
20. Newman CM, Magee AI. 1993. *Biochim. Biophys. Acta.* 1155:79–96
21. Clarke S. 1992. *Annu. Rev. Biochem.* 61:355–86
22. Casey PJ. 1992. *J. Lipid Res.* 33:1–23
23. Gordon JI, Duronio RJ, Rudnick DA, Adams SP, Gokel GW. 1991. *J. Biol. Chem.* 266:8647–50
24. Nakano A, Brada D, Schekman R. 1988. *J. Cell Biol.* 107:851–63
25. Nakano A, Muramatsu M. 1989. *J. Cell Biol.* 109:2677–91
26. d'Enfert C, Gensse M, Gaillardin C. 1992. *EMBO J.* 11:4205–11
27. Kuge O, Dascher C, Orci L, Amherdt M, Rowe T, et al. 1994. *J. Cell Biol.* In press
28. Barlowe C, d'Enfert C, Schekman R. 1993. *J. Biol. Chem.* 268:873–79
29. Nishikawa S, Nakano A. 1991. *Biochem. Biophys. Acta* 103:135–43
30. Deleted in proof
31. Rexach MF, Schekman RW. 1991. *J. Cell Biol.* 114:219–29
32. d'Enfert C, Wuestehube LJ, Lila T, Schekman R. 1991. *J. Cell Biol.* 114: 663–70
33. Oka T, Nishikawa S-I, Nakano A. 1991. *J. Cell Biol.* 114:671–79
34. Barlowe C, Schekman R. 1993. *Nature* 365:347–49
35. d'Enfert C, Barlowe C, Nishikawa S-I, Nakano A, Schekman R. 1991. *Mol. Cell. Biol.* 11:5727–34
36. Hicke L, Yoshihisa T, Schekman R. 1992. *Mol. Biol. Cell* 3:667–76
37. Hicke L, Schekman R. 1989. *EMBO J.* 8:1677–84
38. Yoshihisa T, Barlowe C, Schekman R. 1993. *Science* 259:1466–68
39. Beckers CJM, Balch WE. 1989. *J. Cell Biol.* 108:1245–56
40. Orci L, Ravazzola M, Meda P, Holcomb C, Moore H-P, et al. 1991. *Proc. Natl. Acad. Sci. USA* 88:8611–15
41. Feig LA, Cooper GM. 1988. *Mol. Cell. Biol.* 8:3235–43
42. Farnsworth CL, Marshall MS, Gibbs JB, Stacey DW, Feig LA. 1991. *Cell* 64:625–33
43. Aridor M, Sagi-Eisenberg R. 1990. *J. Cell Biol.* 111:2885–91
44. Deleted in proof
45. Pryer NK, Salama NR, Schekman R, Kaiser CA. 1993. *J. Cell Biol.* 120: 865–75
46. Salama NR, Yeung T, Schekman RW. 1993. *EMBO J.* 12:4073–82
47. Nuoffer C, Davidson HW, Matteson J, Meinkoth J, Balch WE. 1994. *J. Cell Biol.* In press
48. Dascher C, Balch WE. 1994. *J. Biol. Chem.* 269:1437–48
49. Peter F, Plutner H, Kreis T, Balch WE. 1993. *J. Cell Biol.* 122:1155–68
50. Pind S, Nuoffer C, McCaffery JM, Plutner H, Davidson HW, et al. 1993. *J. Cell Biol.* In press

51. Plutner H, Cox AD, Pind S, Khosravi-Far R, Bourne JR, et al. 1991. *J. Cell Biol.* 115:31–43
52. Orci L, Glick BS, Rothman JE. 1986. *Cell* 46:171–84
53. Orci L, Malhotra V, Amherdt M, Serafini T, Rothman JE. 1989. *Cell* 56:357–68
54. Orci L, Palmer DJ, Ravazzola M, Perrelet A, Amherdt M, Rothman JE. 1993. *Nature* 362:648–52
55. Kahn RA, Gilman AG. 1984. *J. Biol. Chem.* 259:6228–34
56. Kahn RA. 1994. In *Handbook of Experimental Pharmacology: GTPases in Biology,* ed. B Dickey, L Birnbaumer. Berlin: Springer-Verlag. In press
57. Clark J, Moore L, Krasinskas A, Way J, Battey J, Tamkun J, Kahn RA. 1993. *Proc. Natl. Acad. Sci. USA* 90:8952–56
58. Kahn RA, Kern FG, Clark J, Gelmann EP, Rulka C. 1991. *J. Biol. Chem.* 266:2606–14
59. Tsuchiya M, Price SR, Tsai S-C, Moss J, Vaughan M. 1991. *J. Biol. Chem.* 266:2772–77
60. Kahn RA, Weiss O, Rulka C, Clark J, Amherdt M, et al. 1992. *J. Biol. Chem.* 267:13039–46
61. Stearns T, Willingham MC, Botstein D, Kahn RA. 1990. *Proc. Natl. Acad. Sci. USA* 87:1238–42
62. Serafini T, Amherdt M, Brunner M, Kahn RA, Rothman JE. 1991. *Cell* 18:239–53
63. Orci L, Palmer DJ, Amherdt M, Rothman JE. 1993. *Nature* 364:732–34
64. Melancon P, Glick BS, Malhotra V, Weidman PJ, Serafini T, et al. 1987. *Cell* 51:1053–62
65. Duden R, Allan V, Kreis T. 1991. *Trends Cell Biol.* 1:14–19
66. Serafini T, Stenbeck G, Brecht A, Lottspeich F, Orci L, et al. 1991. *Nature* 349:215–19
67. Waters MG, Serafini T, Rothman JE. 1991. *Nature* 349:248–51
68. Taylor TC, Kahn RA, Melancon P. 1992. *Cell* 70:69–79
68a. Elazar Z, Orci L, Ostermann J, Amherdt M, Tanigawa G, Rothman JE. 1994. *J. Cell Biol.* 124:415–24
69. Palmer DJ, Helms JB, Beckers CJM, Orci L, Rothman JE. 1993. *J. Biol. Chem.* 268:12083–89
70. Donaldson JG, Cassel D, Kahn RA, Klausner RD. 1992. *Proc. Natl. Acad. Sci. USA* 89:6408–12
70a. Hara-Kuge S, Kuge O, Orci L, Amherdt M, Ravazzola M, et al. 1994. *J. Cell Biol.* In press
71. Ostermann J, Orci L, Tani K, Amherdt M, Ravazzola M, et al. 1993. *Cell* 75:1015–25
71a. Tanigawa G, Orci L, Amherdt M, Ravazzola M, Helms JB, Rothman JE. 1993. *J. Cell Biol.* 123:1365–71
72. Kahn RA, Rulka C. 1993. *J. Biol. Chem.* 268:9555–63
73. Helms BJ, Palmer DJ, Rothman JE. 1993. *J. Cell Biol.* 121:751–60
74. Lippincott-Schwartz J. 1993. *Trends Cell. Biol.* 3:81–87
75. Donaldson JG, Finazzi D, Klausner RD. 1992. *Nature* 360:350–52
76. Helms JB, Rothman JE. 1992. *Nature* 360:352–54
77. Randazzo PA, Yang YC, Rulka C, Kahn RA. 1993. *J. Biol. Chem.* 268: 9555–63
78. Balch WE, Kahn RA, Schwaninger R. 1992. *J. Biol. Chem.* 267:13053–61
79. Haubruck H, McCormick F. 1991. *Biochim. Biophys. Acta* 1072:215–29
80. Pepperkok R, Scheel J, Hauri HP, Horstmann H, Griffiths G, Kreis TE. 1993. *Cell* 74:71–82
81. Duden R, Griffiths G, Frank R, Argos P, Kreis TE. 1991. *Cell* 64:649–65
82. Oprins A, Duden R, Kreis TE, Geuze HJ, Slot JW. 1993. *J. Cell Biol.* 121:49–59
83. Hendricks LC, McCaffery M, Palade GE, Farquhar MG. 1993. *Mol. Biol. Cell* 4:413–24
83a. Krijnse-Locker J, Ericsson M, Rottier PJM, Griffiths G. 1994. *J. Cell Biol.* 124:55–70
84. Stenbeck G, Schreiner R, Herrmann D, Auerbach S, Lottspeich F, et al. 1992. *FEBS Lett.* 314:195–98
85. Hosobuchi M, Kreis T, Schekman R. 1992. *Nature* 360:603–5
86. Dascher C, Ossig R, Gallwitz D, Schmitt HD. 1991. *Mol. Cell. Biol.* 11:872–85
87. Robinson MS, Kreis TE. 1992. *Cell* 69:129–38
87a. Wong DH, Brodsky FM. 1992. *J. Cell Biol.* 117:1171–79
87b. Seaman MNJ, Ball CL, Robinson MS. 1993. *J. Cell Biol.* 123:1093–105
88. Stamnes MA, Rothman JE. 1993. *Cell* 73:999–1005
88a. Traub LM, Ostrom JA, Kornfeld S. 1993. *J. Cell Biol.* 123:561–73
89. Carter LL, Redelmeier TE, Woolenweber LA, Schmid SL. 1993. *J. Cell Biol.* 120:37–45
90. Lenhard JM, Kahn RA, Stahl PD. 1992. *J. Biol. Chem.* 13047–52
91. Zeuzem S, Feick P, Zimmermann P, Haase W, Kahn RA. 1992. *Proc. Natl. Acad. Sci. USA* 89:6619–23
92. Brown HA, Gutowski S, Moomaw

CR, Slaughter C, Sternweis PC. 1993. *Cell* 75:1137–44
92a. Kahn RA, Yucel JK, Malhotra V. 1993. *Cell* 75:1045–48
93. Collins CA. 1991. *Trends Cell Biol.* 1:57–60
94. Rothman JH, Raymond CK, Gilbert T, O'Hara PJ, Stevens TH. 1990. *Cell* 61:1063–74
95. Tuma PL, Stachniak MC, Collins CA. 1993. *J. Biol. Chem.* 268:17240–46
96. Gout I, Dhand R, Hiles ID, Fry MJ, Panayotou G, et al. 1993. *Cell* 75:25–36
97. Vallee RB. 1992. *J. Muscle Res. Cell Motil.* 13:493–96
98. Chen MS, Obar RA, Schroeder CC, Austin TW, Poodry CA, et al. 1991. *Nature* 351:583–86
99. van der Bliek AM, Meyerowitz EM. 1991. *Nature* 351:411–14
100. Grigliatti TA, Hall L, Rosenbluth R, Suzuki DT. 1973. *Mol. Gen. Genet.* 120:107–14
101. Kosaka T, Ikeda K. 1983. *J. Neurobiol.* 14:207–25
102. Robinson PJ, Sontag J-M, Liu J-P, Fykse EM, Slaughter C, et al. 1993. *Nature* 365:163–66
103. Mayer BJ, Ren R, Clark KL, Baltimore D. 1993. *Cell* 73:629–30
104. van der Bliek AM, Redelmeier TE, Damke H, Tisdale EJ, Meyerowitz EM, Schmid SL. 1993. *J. Cell Biol.* 122:553–63
105. Herskovits JS, Burgess CC, Obar RA, Vallee RB. 1993. *J. Cell Biol.* 122: 565–78
106. Vater CA, Raymond CK, Ekena K, Howald-Stevenson I, Stevens TH. 1992. *J. Cell Biol.* 119:773–86
107. Wilsbach K, Payne GS. 1993. *EMBO J.* 12:3049–59
108. Schwaninger R, Plutner H, Bokoch GM, Balch WE. 1992. *J. Cell Biol.* 119:1077–96
109. Chavrier P, Parton RG, Hauri HP, Simons K, Zerial M. 1990. *Cell* 62: 317–29
110. Van Der Sluijs P, Hull M, Zahraoui A, Tavitian A, Goud B, Mellman I. 1991. *Proc. Natl. Acad. Sci. USA* 88:6313–17
111. Lombardi D, Soldati T, Riederer MA, Goda Y, Zerial M, Pfeffer SR. 1993. *EMBO J.* 12:677–82
112. Darchen F, Zahraoui A, Hammel F, Monteils M-P, Tavitian A, Scherman D. 1990. *Proc. Natl. Acad. Sci. USA* 87:5692–96
113. Fischer von Mollard G, Mignery G, Baumert M, Perin M, Hanson T, et al. 1990. *Proc. Natl. Acad. Sci. USA* 87:1988–92
114. Matteoli M, Takei K, Cameron R, Hurlbut P, Johnston PA, et al. 1991. *J. Cell Biol.* 115:625–33
115. Lutcke A, Jansson S, Parton RG, Chavrier P, Valencia A, et al. 1993. *J. Cell Biol.* 121:553–64
116. Chavrier P, Gorvel JP, Steizer E, Simons K, Gruenberg J, Zerial M. 1991. *Nature* 353:769–72
117. Stenmark H, Valencia A, Martinez O, Ullrich O, Goud B, Zerial M. 1993. *EMBO J.* 13:575–83
118. Burstein ES, Brondyk WH, Macara IG. 1992. *J. Biol. Chem.* 267:22715–18
118a. Stouten PFW, Sander C, Wittinghofer A, Valencia A. 1993. *FEBS Lett.* 320:1–6
119. Bucci C, Parton RG, Mather IH, Stunnenberg H, Simons K, et al. 1992. *Cell* 70:715–28
120. Gorvel JP, Chavrier P, Zerial M, Gruenberg J. 1991. *Cell* 64:915–25
121. Goud B, Zahraoui A, Tavitian A, Saraste J. 1990. *Nature* 345:553–56
122. Jones SM, Crosby JR, Salamero J, Howell KE. 1993. *J. Cell Biol.* 122: 775–88
123. Huber LA, Pimplikar S, Parton RG, Virta H, Zerial M, Simons K. 1993. *J. Cell Biol.* 123:35–45
124. Dunn B, Stearns T, Botstein D. 1993. *Nature* 362:563–65
125. Brennwald P, Novick P. 1993. *Nature* 362:560–63
125a. Seabra MC, Brown MS, Goldstein JL. 1993. *Science* 259:377–81
126. Andres DA, Seabra MC, Brown MS, Armstrong SA, Smeland TE, et al. 1993. *Cell* 73:1091–99
127. Burstein ES, Linko-Stentz K, Lu Z, Macara IG. 1991. *J. Biol. Chem.* 266:2689–92
128. Tan TJ, Vollmer P, Gallwitz D. 1991. *FEBS Lett.* 291:322–26
129. Walworth NC, Brennwald P, Kabcenell AK, Garrett M, Novick P. 1992. *Mol. Cell. Biol.* 12:2017–28
130. Jena BP, Brennwald P, Garrett MD, Novick P, Jamieson JD. 1992. *FEBS Lett.* 309:5–9
131. Strom M, Vollmer P, Tan TJ, Gallwitz D. 1993. *Nature* 361:736–39
132. Moya M, Roberts D, Novick P. 1993. *Nature* 361:460–63
133. Burton J, Roberts D, Montaldi M, Novick P, DeCamilli P. 1993. *Nature* 361:464–67
134. Araki S, Kikuchi A, Hata Y, Isomura M, Takai Y. 1990. *J. Biol. Chem.* 265:13007–15

135. Sasaki T, Kikuchi A, Araki S, Hata Y, Isomura M, et al. 1990. *J. Biol. Chem.* 265:2333–37
136. Ullrich O, Stenmark H, Alexandrov K, Hubert L, Kaibuchi K, et al. 1993. *J. Biol. Chem.* 268:18143–50
137. Soldati T, Riederer MA, Pfeffer SR. 1993. *Mol. Biol. Cell* 4:425–34
138. Sasaki T, Kaibuchi K, Kabacenell AK, Novick PJ, Takai Y. 1991. *Mol. Cell. Biol.* 11:2909–12
139. Ueda T, Kikuchi A, Ohga N, Yamamoto J, Takai Y. 1990. *J. Biol. Chem.* 265:9373–80
139a. Elazar Z, Mayer T, Rothman JE. 1994. *J. Biol. Chem.* 269:794–97
139b. Garrett MD, Zahner JE, Cheney CM, Novick PJ. 1994. *EMBO J.* In press
140. Zahner JE, Cheney CM. 1993. *Mol Cell Biol* 13:217–27
141. Steele-Mortimer O, Gruenberg J, Clague MJ. 1993. *FEBS Lett.* 329:313–18
142. Regazzi R, Kikuchi A, Takai Y, Wollheim CB. 1992. *J. Biol. Chem.* 267:17512–19
143. Ullrich O, Horiuchi H, Bucci C, Zerial M. 1994. *Nature*. In press
144. Haubruck H, Disela C, Wagner P, Gallwitz D. 1987. *EMBO J.* 6:4049–53
145. Haubruck H, Prange R, Vorgias C, Gallwitz D. 1989. *EMBO J.* 8:1427–32
146. Touchot N, Chardin P, Tavitian A. 1987. *Proc. Natl. Acad. Sci. USA* 84:8210–14
147. Zahraoui A, Touchot N, Chardin P, Tavitian A. 1989. *J. Biol. Chem.* 264:12394–401
148. Lian JP, Ferro-Novick S. 1993. *Cell* 73:735–45
149. Davidson HW, Balch WE. 1993. *J. Biol. Chem.* 268:4216–26
150. Tisdale EJ, Bourne JR, Khosravi-Far R, Davidson HW, Der CJ, Balch WE. 1992. *J. Cell Biol.* 119:749–61
151. Walter M, Clark SG, Levinson AD. 1986. *Science* 233:649–52
152. Becker J, Tan TJ, Trepte H, Gallwitz D. 1991. *EMBO J.* 10:785–92
153. Plutner H, Schwaninger R, Pind S, Balch WE. 1990. *EMBO J.* 9:2375–83
154. Balch WE, Fernandez J, Plutner H. 1993. *Methods: A Companion to Methods in Enzymology.* 5:258–63
155. Duronio RJ, Jackson-Machelski EJ, Heuckeroth RO, Olins PO, Devine CS, et al. 1990. *Proc. Natl. Acad. Sci. USA* 87:1506–10
156. Hwang Y-W, Zhong J-M, Poullet P, Parmeggiani A. 1993. *J. Biol. Chem.* 268:24692–98
157. Deleted in proof
158. Huber LA, de Hoop MJ, Dupree P, Zerial M, Simons K, Dotti C. 1993. *J. Cell Biol.* 123:47–55
158a. Zahraoui A, Joberty G, Arpin M, Fontaine JJ, Hellio R, et al. 1994. *J. Cell Biol.* 124:101–15
159. Chant J, Corrado K, Pringle JR, Herskowitz I. 1991. *Cell* 65:1213–24
160. Chant J, Herskowitz I. 1991. *Cell* 65:1203–12
161. Adams AEM, Johnson DI, Longnecker RM, Sloat BF, Pringle JR. 1990. *J. Cell Biol.* 111:131–42
162. Matsui Y, Toh-E A. 1992. *Mol. Cell Biol.* 12:5690–99
163. Deleted in proof
164. Novick P, Botstein D. 1985. *Cell* 40:405–16
165. Jones S, Vignais ML, Broach JR. 1991. *Mol. Cell. Biol.* 11:2647–55
166. Park H-O, Chant J, Herskowitz I. 1993. *Nature* 365:269–74
167. Ridley AJ, Peterson HF, Johnston CL, Diekmann D, Hall A. 1992. *Cell* 70:401–10
168. Ridley AJ, Hall A. 1992. *Cell* 70:389–99
169. Hall A. 1992. *Mol. Biol. Cell* 3:475–79
170. Matteoli M, Takei K, Cameron R, Hurlbut P, Johnston PA, et al. 1991. *J. Cell Biol.* 115:625–33
171. Deleted in proof
172. Mizoguchi A, Kim S, Ueda T, Kikuchi A, Yorifuji H, et al. 1990. *J. Biol. Chem.* 265:11872–79
173. Baldini G, Hohl T, Lin HY, Lodish HF. 1993. *Proc. Natl. Acad. Sci. USA* 89:5049–52
174. Liedo P, Vernier P, Vincent J, Mason WT, Zorec R. 1993. *Nature* 364:540–44
175. Shiratakis H, Kaibuchi K, Sakoda T, Kishida S, Yamaguchi T, et al. 1993. *Mol. Cell. Biol.* 13:2061–68
175a. Yamaguchi T, Shirataki H, Kishida S, Miyazaki M, Nishikawa J, et al. 1993. *J. Biol. Chem.* 268:27164–70
176. Elferink LA, Peterson MR, Scheller RH. 1993. *Cell* 72:153–59
177. Bennett MK, Scheller RH. 1994. *Annu. Rev. Biochem.* 63:63–100
177a. Südhof TC, De Camilli P, Niemann H, Jahn R. 1993. *Cell* 75:1–4
178. Oberhauser AF, Monck JR, Balch WE, Fernandez JM. 1992. *Nature* 360:270–73
179. Li G, Regazzi R, Balch WE, Wolheim CB. 1993. *FEBS Lett.* 327:145–49
180. Padfield PJ, Balch WE, Jamieson JD. 1992. *Proc. Natl. Acad. Sci. USA* 89:1656–60
181. Edwardson JM, MacLean CM, Law GJ. 1993. *FEBS Lett.* 320:52–56

182. Senyshyn J, Balch WE, Holz RW. 1992. *FEBS Lett.* 309:41–46
183. Richmond J, Haydon PG. 1993. *FEBS Lett.* 326:124–30
184. Schaber MD, Garsky VM, Boylan D, Hill WS, Scolnick EM, et al. 1989. *Proteins: Struct., Funct. Genet.* 6:306–15
185. McKiernan CJ, Brondyk WH, Macara IG. 1993. *J. Biol. Chem.* 268:2444–52
186. Liedo P, Vernier P, Vincent J, Mason WT, Zorec R. 1993. *Nature* 364:540–44
186a. Holz RW, Brondyk WH, Senter RA, Kuizon L, Macara IG. 1994. *J. Biol. Chem.* In press
187. van der Sluijs P, Hull M, Zahraoui A, Tavitian A, Goud B. 1991. *Proc. Natl. Acad. Sci. USA* 88:6313–17
187b. Horazdovsky BF, Busch GR, Emr SD. 1994. *EMBO J.* In press
188. van der Sluijs P, Hull M, Webster P, Male P, Goud B, Mellman I. 1992. *Cell* 70:729–40
188a. Cormont M, Tanti J-F, Zahraoui A, Obberghen EV, Tavitian A, Marchand-Brustel YL. 1993. *J. Biol. Chem.* 268:19491–97
189. Wichmann H, Hengst L, Gallwitz D. 1992. *Cell* 71:1131–42
190. Schimoller F, Riezman H. 1993. *J. Cell Sci.* 106:823–30
191. Holz RW, Senyshyn J, Bittner MA. 1991. *Ann. NY Aacd. Sci.* 635:382–92
192. Schmitt HD, Puzicha M, Gallwitz D. 1988. *Cell* 53:635–47
193. Antebi A, Fink GR. 1992. *Mol. Biol. Cell* 3:1–22
194. Rudolph HK, Antebi A, Fink GR, Buckley CM, Dorman TE, et al. 1989. *Cell* 58:133–45
195. Newman AP, Shim J, Ferro-Novick S. 1990. *Mol. Cell. Biol.* 10:3405–14
196. Protopopov V, Govindan B, Novick P, Gerst JE. 1993. *Cell* 74:855–61
197. Jones EW. 1976. *Genetics* 85:23–33
197a. Hardwick KG, Pelham HRB. 1993. *J. Cell Biol.* 119:513–21
198. Bennett MK, Garcia-Arraras JE, Elferink LA, Peterson K, Fleming AM, et al. 1993. *Cell* 74:863–73
199. Bennett MK, Scheller RH. 1993. *Proc. Natl. Acad. Sci. USA* 90:2559–63
199a. Söllner T, Whiteheart SW, Brunner M, Erdjument-Bromage H, Geromanos S, et al. 1993. *Nature* 362:318–24
200. Söllner T, Bennett MK, Whiteheart SW, Scheller RH, Rothman JE. 1993. *Cell* 75:409–18
201. Balch WE. 1992. *Curr. Biol.* 2:157–69
202. Burgoyne RD. 1987. *Nature* 328:112–13
203. Leyte A, Barr FA, Kehlenbach RH, Huttner WB. 1992. *EMBO J.* 11:4795–804
204. Vitale N, Mukai H, Rouot B, Thierse D, Aunis D, Bader M. 1993. *J. Biol. Chem.* 268:14715–23
205. Aridor M, Rajmilevich G, Beaven MA, Sagi-Eisenberg R. 1993. *Science* 262: 1569–72
206. Pimplikar SW, Simons K. 1993. *Nature* 362:456–58
207. Stow JL, de Almeida JB, Narula N, Holtzman KJ, Ercolani L, Ausiello DA. 1991. *J. Cell Biol.* 114: 1113–24
207a. Montmayeur J-P, Borrelli E. 1994. *Science* 263:95–98
208. Higashijima T, Uzu S, Nakajima T, Ross EM. 1988. *J. Biol. Chem.* 263: 6491–94
209. Iniguez-Lluhi J, Kleuss C, Gilman AG. 1993. *Trends Cell Biol.* 3:230–35
210. Clapham DE, Neer EJ. 1993. *Nature* 365:403–6
211. Colombo MI, Mayorga LS, Casey PJ, Stahl PD. 1992. *Science* 255:1695–97
211a. Bomsel M, Mostov KE. 1993. *J. Biol. Chem.* 268:25824–35
211b. Barroso M, Sztul ES. 1994. *J. Cell Biol.* 124:83–100
211c. Haas A, Conradt B, Wickner W. 1994. *J. Cell Biol.* In press
211d. Conradt B, Haas A, Wickner W. 1994. *J. Cell Biol.* In press
212. Barr FA, Leyte A, Mollner S, Pfeuffer T, Tooze SA, Huttner WB. 1991. *FEBS Lett.* 294:239–43
213. de Matteis MA, Santini G, Kahn RA, Tullio GD, Luini A. 1993. *Nature* 364:818–21
213a. Nishizuka Y. 1992. *Science* 258:607–14
214. Deleted in proof
215. Luini A, De Matteis MA. 1993. *Trends Cell Biol.* 3:291–93
216. Wilson BS, Nuoffer C, Meinkoth JL, McCaffery M, Feramisco JR, et al. 1993. *J. Cell Biol.* In press
217. Klausner RD, Donaldson JG, Lippincott-Schwartz J. 1992. *J. Cell. Biol.* 116:1071–80
218. Orci L, Glick BS, Rothman JE. 1986. *Cell* 46:171–84
219. Warren G. 1989. *Nature* 342:857
219a. Warren G. 1993. *Annu. Rev. Biochem.* 263:323–48
220. Lucocq J, Warren G, Pryde J. 1991. *J. Cell Sci.* 100:753–59

Annu. Rev. Biochem. 1994. 63:991–1043

HOMOLOGOUS PAIRING AND DNA STRAND-EXCHANGE PROTEINS

Stephen C. Kowalczykowski and Angela K. Eggleston
Division of Biological Sciences, Sections of Microbiology and of Molecular and Cellular Biology, University of California, Davis, California 95616-8665

KEY WORDS: genetic recombination, DNA renaturation, DNA-binding proteins, DNA-dependent ATPase, nuclease

---

CONTENTS

0066-4154/94/0701-0991$05.00

PERSPECTIVES AND SUMMARY

The ability to pair two DNA chromosomes homologously and to exchange DNA between them lies at the heart of all models for general recombination. This process requires that sequence similarity between two DNA molecules is searched, homology is recognized, and individual DNA strands are mutually exchanged. The complexity of this molecular recognition process has hampered mechanistic analysis, but recent concerted effort has resulted in significant understanding of this elaborate series of events. The biochemical features of this process and of the proteins that promote it are reviewed here.

Major insight came with the discovery that the *Escherichia coli* RecA protein, known from genetic analysis to be crucial to recombination (1), promoted the homologous pairing and exchange of DNA strands (2–6). It is, perhaps, surprising that a single protein can carry out such a complicated biochemical process, but RecA protein is a remarkably complex entity (see Figure 1; details are explained below). RecA protein binds both ATP and DNA, and acts not as a monomer or a limited assemblage of monomers, but rather as a helical filament of indefinite length polymerized on DNA. This nucleoprotein complex, the presynaptic complex, requires ATP binding to attain its striking functional form and is the active species during the homology search and DNA strand exchange. Despite the need for ATP binding in filament assembly and in the homologous alignment of DNA, neither the homology search nor DNA strand exchange requires ATP hydrolysis (see below), further highlighting the unusual nature of this reaction.

The ubiquity of RecA-like proteins in eubacteria (7, 8) argues for conservation of the mechanism for homologous pairing and DNA strand exchange. The extension of this mechanism to eukaryotes is supported by a growing list of proteins that are structurally similar to the *E. coli* RecA protein (see below). Thus, the paradigms established from studies of RecA protein can be tested for their generality.

A hallmark of RecA protein–promoted DNA strand exchange is its ATP dependence. However, a class of eukaryotic pairing proteins can function in the absence of ATP. Although they were initially thought to be a limited case of the RecA paradigm [reviewed in (9)], recent evidence argues that these proteins promote pairing and apparent DNA strand exchange by a distinct reaction mechanism. Most, if not all, of these ATP-independent proteins require nucleolytic degradation of one strand of a duplex molecule

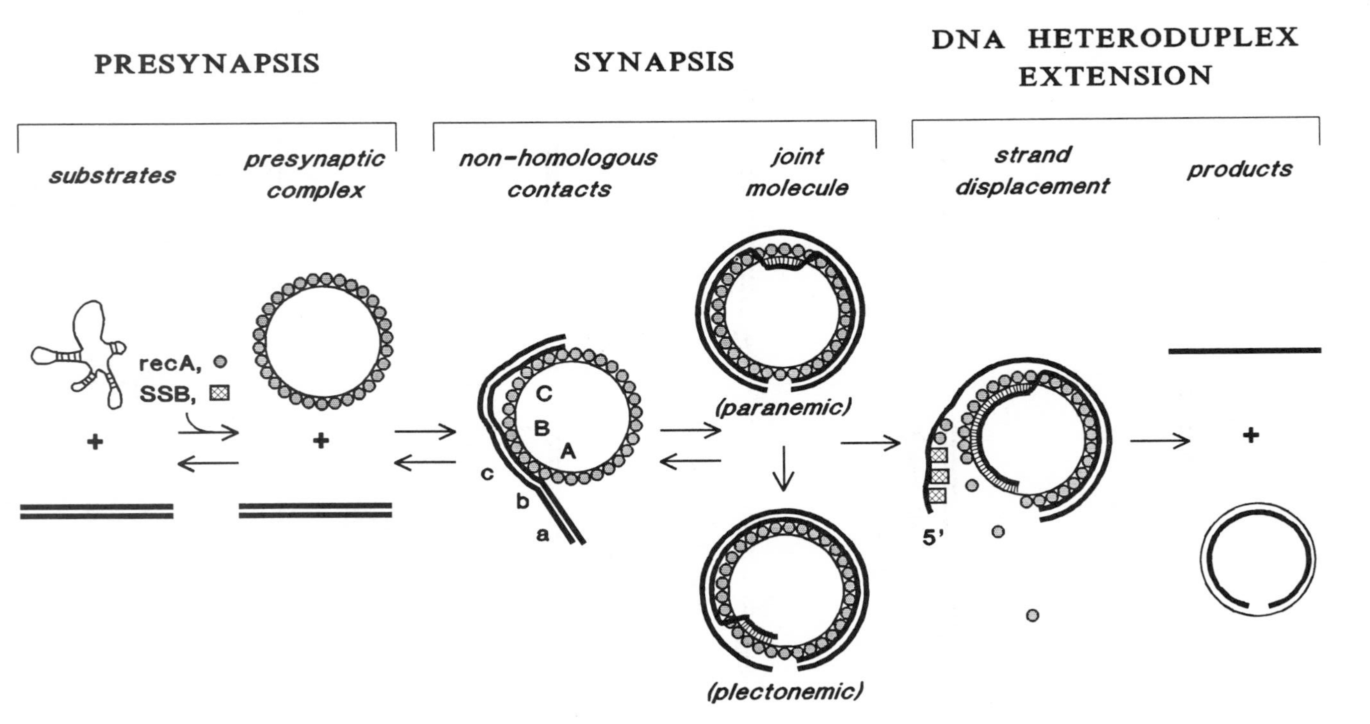

Figure 1 Model for DNA strand exchange promoted by RecA protein. See text for details. Modified from (16).

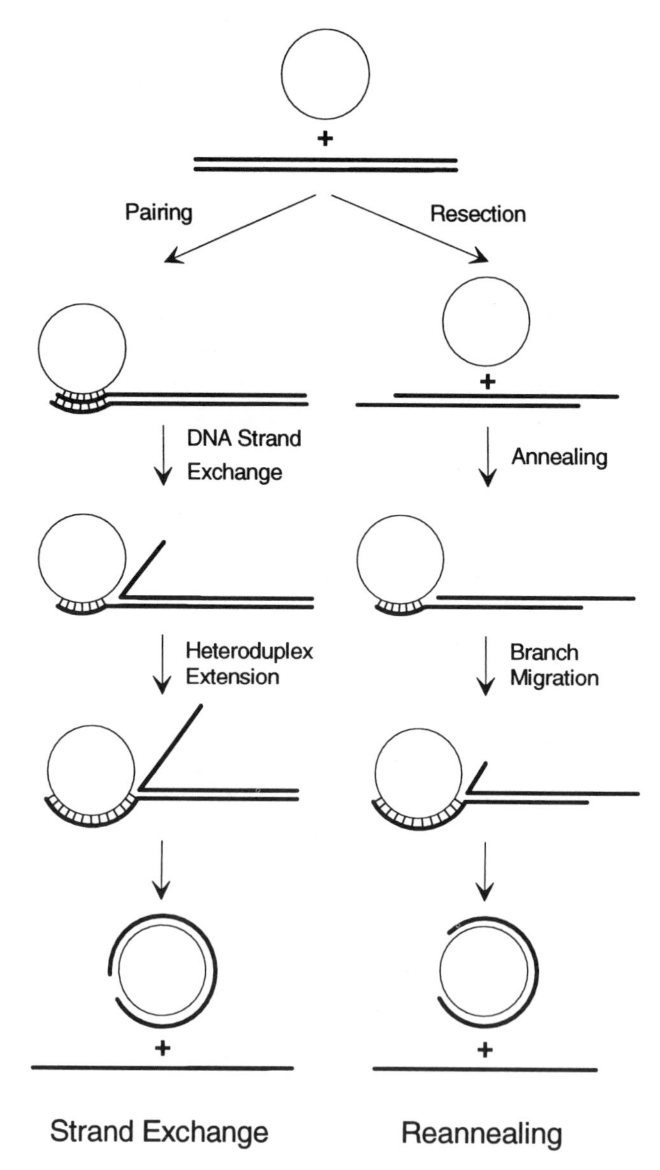

Figure 2 Models for the generation of heteroduplex DNA. *(Left)* DNA strand-exchange mechanism involving initial strand invasion of dsDNA by ssDNA followed by DNA heteroduplex extension. *(Right)* Reannealing mechanism involving renaturation of ssDNA between resected dsDNA and ssDNA molecules, followed by either thermal or protein-mediated branch migration.

as a first step (Figure 2; see below). The annealing of complementary regions of single-stranded DNA (ssDNA), rather than the invasion of double-stranded DNA (dsDNA) by ssDNA, is responsible for the observed homologous pairing. The ensuing extension of DNA heteroduplex may or may not be protein-promoted.

The existence of two different biochemical mechanisms for effecting DNA strand exchange (ATP-dependent and -independent) raises the question: Is the net input of free energy, or even the participation of proteins, necessary? While the complexity of the reaction seems to favor protein-mediated catalysis, neither condition is essential in vitro. DNA strand exchange between identical sequences is isoenergetic (i.e. an equal number of basepairs are disrupted and reformed), so DNA strand exchange is not restricted thermodynamically. This fact argues that the major mechanistic need for proteins is kinetic. Since catalysis involves lowering the activation energy of a rate-limiting step, DNA strand-exchange proteins must facilitate the formation or stabilization of a normally unstable transition-state structure, which many lines of evidence suggest is a three-stranded intermediate (10, 11; see below). In contrast, when DNA sequences are not identical (due to mismatches), then a need for energy input arises; this consideration predicts the involvement of an ATP-dependent step when sequence similarity is imperfect.

This introductory perspective has raised issues that will be elaborated below. The discussion first addresses structural, energetic, and experimental aspects of the homologous pairing of DNA molecules. This groundwork is followed by a discussion of the ATP-dependent class of pairing proteins and the mechanism by which they promote DNA strand exchange. Next, the ATP-independent class of pairing proteins and their mechanism of action are examined. A brief description of eukaryotic structural homologs of RecA protein and their potential as DNA strand-exchange proteins follows. Finally, protein-independent renaturation, pairing, and strand exchange are compared to the protein-promoted reactions. Table 1 summarizes pertinent information about the proteins that are discussed. Other perspectives on DNA strand-exchange proteins and homologous recombination are found in (7, 12–22). This article is an elaboration of a previous overview of this topic (9).

PRINCIPLES OF HOMOLOGOUS PAIRING AND DNA STRAND EXCHANGE

Homologous Pairing

The problem of homologous recognition between DNA molecules is, in principle, no different than that of site-specific recognition by DNA-binding proteins. There are typically few appropriate targets in the entire genome,

Table 1 Properties of homologous pairing and DNA strand-exchange proteins[a]

ATP-dependent proteins: Protein	Organism	M_r[b]	ATPase activity	Stoich.[c]	Mode[d]	Pairing end bias[e]	Joint extension[f]	Renat.[g]	Aggreg./coagg.[h]	Accessory factors[i]	Assays[j]
RecA	*E. coli*	38	Y	3	S	none[k]	5'→3'	Y	Y/Y	SSB	em, fb, jm, ns, se
RecA	*P. mirabilis*	38	Y	1.5	S	—	—	—	—		fb, jm, se
RecA (RecE)	*B. subtilis*	42	Y[l]	—	S	—	—	—	—	SSB	jm, se
RecA	*T. aquaticus*	36	Y	—	—	—	—	—	—	SSB?	jm
UvsX	phage T4	44	Y	3–5	S	5'	5'→3'	Y	Y/Y	G32P; UvsY	em, fb, jm
RecA-like	*P. sativum*	40	?[m]	—	S?	—	—	—	—		em, jm
Rec1	*U. maydis*	70?	Y	200	C?	3'	3'→5'	Y	—		fb, jm, ns

ATP-independent proteins: Protein	Organism	M_r	ATP binding	Stoich.	Mode	Nuclease activity[n]	Extension polarity[o]	Renat.	Aggreg./coagg.	Accessory factors	Assays
RecT	*E. coli*	33	N	13	S	N[p]	none[q]	Y	—		em, jm
β	phage λ	28	N	6[r]	S	N[p]	—	Y	—		—[r]
Sep1/Xrn1	*S. cerevisiae*	175	N	35–40	S	I (5'→3')	none[q]	Y	Y/Y	yRPA; SF1	em, jm
DPA/EF3	*S. cerevisiae*	120	—	~20	S	N[p]	none[q]	Y	—		em, jm
STPα/TFIIS	*S. cerevisiae*	38	—	—	—	Y (<0.5%)	—	Y	N/—	ySSBs	em, jm
p^{140}/ExoII	*S. pombe*	148	N	40	S	I (5'→3')	none[q]	Y	—	FAS	em, jm, ns
Rrp1	*D. melanogaster*	105	N	400	C	I (3'→5')	none[q]	Y	Y/Y		em, jm, se
HPP-1	human T-cell	130	Y[s]	25[t]	S	Y (3.3%)	—	—	—	hRPA, others	em, jm, se
v-SEP	vaccinia	—	—	—	—	Y (0.9%)	5'→3'[u]	Y	—		em, jm
ICP8	HSV-1	138	—	10	S	N (<0.01%)	—	—	—		em, jm
p53	human	53	Y	—	—	—	—	Y	—		se[v]

[a] Adapted from (9).
[b] Expressed as kDa.
[c] Optimal stoichiometry of homologous pairing protein (nt ssDNA per protein monomer) required for strand exchange in the absence of any other protein.
[d] Mode of action (stoichiometric or catalytic).
[e] Preferred end (5' or 3') of displacement of the noncomplementary strand during initial pairing.
[f] Polarity of extension of the DNA heteroduplex joint relative to the displaced, noncomplementary strand.
[g] Ability to renature complementary ssDNA.
[h] Ability to either aggregate ssDNA or coaggregate ssDNA and dsDNA.
[i] Accessory factor(s) that stimulate the homologous pairing protein.
[j] Assay used to determine activities: em = electron microscopy with visualization of the displaced strand; fb = filter-binding assay with retention of joint molecules; jm = gel assay with production of joint molecule intermediates; ns = nuclease sensitivity assay; se = gel assay with the production of form II molecules.
[k] Differing results have been obtained (see Refs. 38, 59, 92a, 93, 226–228). The discrepancies may be due to variations in substrates, reaction conditions, and experimental technique; Refs. 93 and 226 employed filter-binding assays and EM, whereas the other five references used gel assays.
[l] Although both ATP and dATP are bound by the protein, only dATP activates nucleotide hydrolysis and DNA strand exchange.
[m] The addition of exogenous ATP is required for DNA strand exchange.
[n] Presence of nuclease activity in the protein preparation: I = intrinsic to strand exchange protein (with polarity of nuclease activity indicated); Y = nuclease activity reported in protein preparation (level of detection indicated); N = no nuclease activity reported.
[o] Polarity of branch migration.
[p] Requires ssDNA tails to initiate pairing.
[q] Observed polarity is dependent on the polarity of the ssDNA overhang used to initiate pairing.
[r] Optimal stoichiometry of the homologous pairing protein required for ssDNA renaturation; no strand-exchange assay has been reported for this protein.
[s] Based on UV crosslinking by the ATP analog, 8-azido ATP.
[t] Expressed as bp dsDNA per protein monomer.
[u] This observed polarity may be dictated by the nuclease contaminant in this preparation.
[v] The gel assay for strand transfer involved linear molecules <100 nt in length.

making the mechanics of the search process seem insurmountable. Yet despite the vast excess of inappropriate sites or alignments, specific sites and homologous sequences are nevertheless located. Clearly, both recognition elements and a mechanism for identifying them exist. In contrast to site-specific DNA binding, however, protein-promoted DNA homology recognition must involve proteins sufficiently nonspecific so that they can interact with any DNA sequence, yet specific enough so that only the homologous counterpart is recognized. This is possible only if nonspecific DNA-binding proteins utilize the sequence-specific information inherent to DNA. Both the major and minor grooves of dsDNA are sources of potential recognition elements that might permit homology to be detected; once pairing is achieved, the proteins involved are imagined to provide stability to the aligned structures.

MAJOR GROOVE PAIRING The possibility of specifically paired four-stranded DNA structures involving homologous dsDNA molecules was recognized by McGavin (23, 24). It was proposed that non-Watson-Crick hydrogen bonding involving atoms in the major groove of dsDNA could provide the required specificity. The pairing scheme is specific, in that any given basepair can bond only with its homolog and not with another basepair. The bases of the resulting tetraplex structure form a near square, with the basepairs related by a dyad axis perpendicular to their common plane (Figure 3). The corners of the rectangle are ~10–11 Å apart, and the diameter of the tetraplex is essentially unchanged from that of dsDNA. Model building affirms that the structure is feasible. In the absence of charge neutralization, this DNA structure would have twice the charge density of B-form DNA, indicating that without the participation of protein or other stabilizing components, this structure would be less stable than dsDNA. The experimental observation that RecA protein binds to the minor groove of dsDNA (25, 26) suggests that major groove pairing is utilized in the RecA protein–dependent homology search.

MINOR GROOVE PAIRING Minor groove pairing schemes lack the specificity of those invoking major groove pairing (27). In addition, intercoiling of two duplexes along their minor grooves requires untwisting and unstacking of dsDNA, both of which are unfavorable processes. Interestingly, however, such perturbations are features typical of RecA protein–dsDNA complexes [see (21)]. Though interactions between the minor grooves of DNA lack specificity, it was proposed that high specificity arises from the major groove contacts that result after DNA strands are exchanged, providing a means for stabilizing the desired products (27).

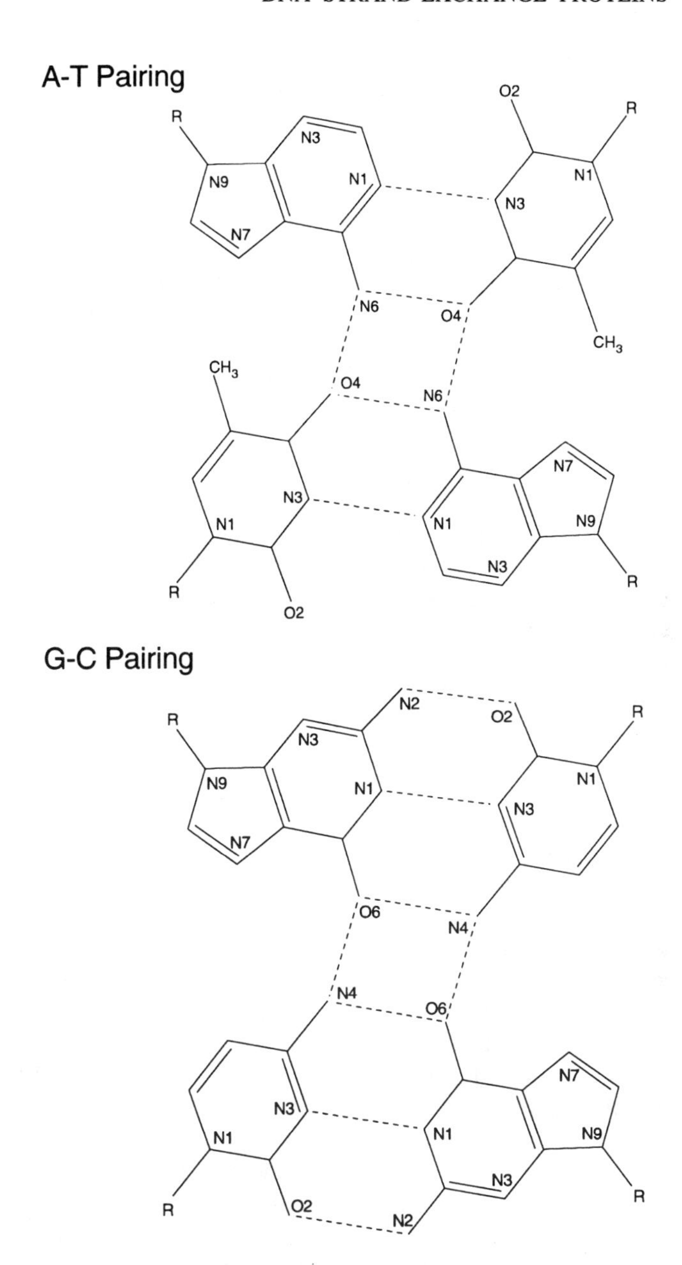

Figure 3 Tetraplex (four-stranded) DNA structure proposed by McGavin (23, 24).

TRIPLEX DNA STRUCTURES Pairing reactions promoted by RecA protein require that one of the substrate molecules be partially single stranded, which suggests that recognition involves contacts between ssDNA and its dsDNA homolog. Since only one of the molecules is basepaired, it is possible that the dsDNA is disrupted before homologous contacts are established, so that conventional Watson-Crick basepairing between the ssDNA and its complement from the dsDNA can be made. Though plausible, no evidence supports this hypothetical "dsDNA opening before pairing" scheme. Instead, experimental evidence argues for a pairing intermediate involving three juxtaposed strands of DNA [see (10, 11, 28) for a critical appraisal of this topic].

The notion of a triple-stranded intermediate in the RecA protein–promoted DNA strand-exchange reaction was advanced by Howard-Flanders and colleagues (29). Although many studies provide compelling evidence for a close association of three strands within the RecA protein–DNA filament (30–36), the precise nature of this structure remains elusive. A three-stranded structure is likely to exist, at least transiently; this transition-state complex was featured in a model to explain DNA strand exchange without the need for ATP hydrolysis (37, 38). Other studies, however, suggest that this structure (or an analogous triplex structure) persists after removal of RecA protein (33, 34, 36).

Several unique triplex DNA structures have been proposed as the recognition intermediates (Figure 4). In contrast to triplex structures formed non-enzymatically (where the two identical strands assume an antiparallel orientation), the identical strands of recombination triplexes must be in a parallel orientation (10, 11, 39). One of the first pairing schemes for recombination intermediates envisioned pairing of the ssDNA only with purine residues in the dsDNA (33) (Figure 4). Subsequently, it was found that replacement of the N7 guanine by a carbon atom in 7-deazaguanine had no effect on DNA strand exchange; this result suggested that the interactions at the N7 position were not crucial to the rate-limiting step of the pairing reaction (40). In agreement, neither was the N7 guanine protected from dimethyl sulfate under conditions that promoted pairing nor did methylation affect pairing (41, 42). Methylation of N6 adenine and N4 cytosine did lower the T_m of a possible three-stranded structure, however, leading to a proposal for an alternative structure that retained some of the characteristics of the McGavin pairing schemes and that did not involve bonding with the N7 position (42) (Figure 4). Finally, energy minimization analysis led to yet a third structure that differs somewhat from the second model (43); the authors' calculations suggest that the triplex structure results from an electrostatic recognition code, although one of the basepairing arrangements involves interaction with an N7 atom (Figure 4). At this time it is not clear which, if any, of these triplex pairing schemes represent stable intermediates of homologous pairing reactions and, hence, these structures

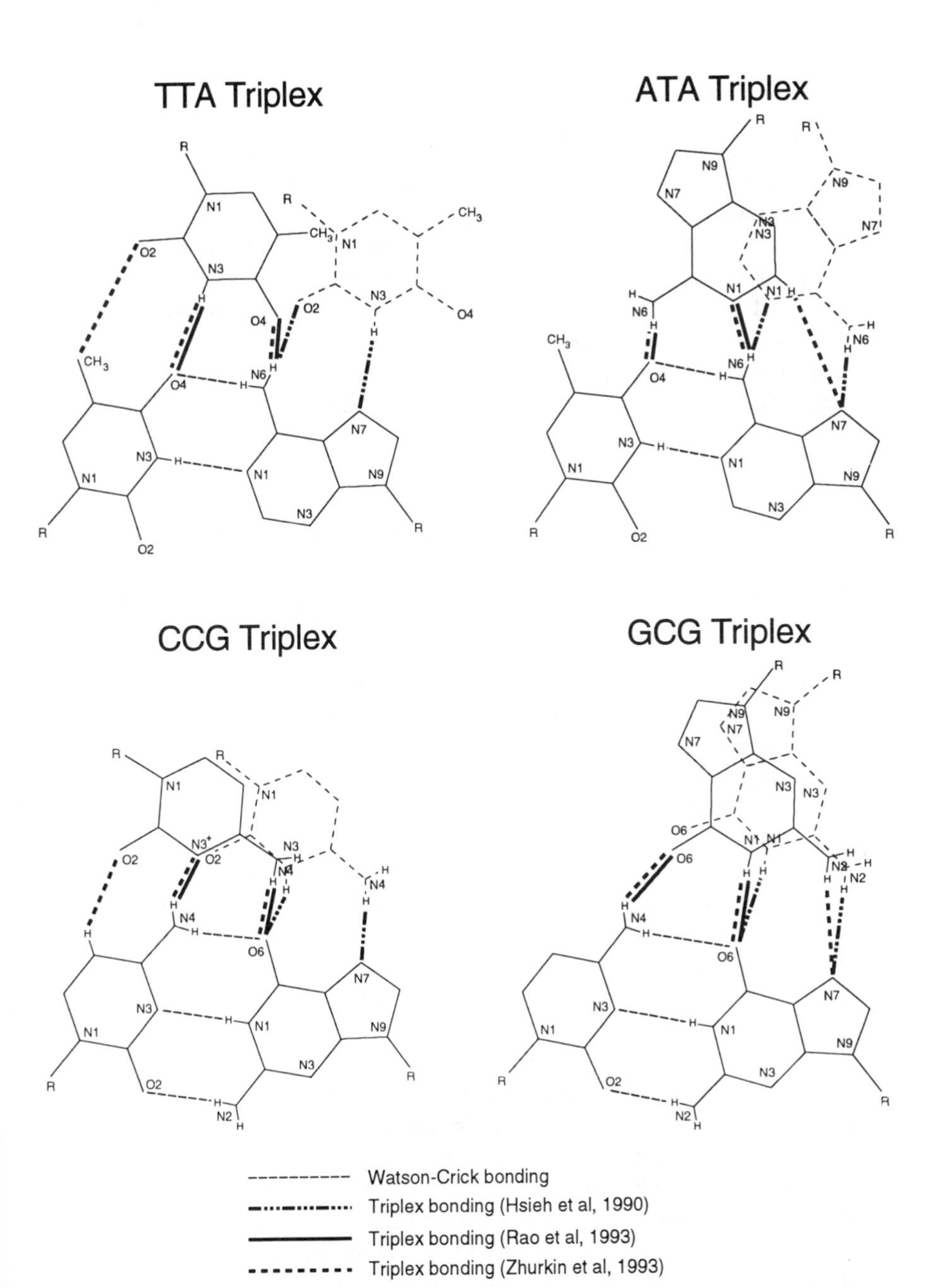

Figure 4 Proposed pairing schemes for RecA protein–mediated triplex (three-stranded) DNA structures. The two bases at the bottom of each structure are bonded by typical Watson-Crick pairing. The third base at the top of each structure is bonded by non-Watson-Crick pairing. Three models of non-Watson-Crick pairing are shown, as indicated in the legend. For clarity, the third base in the Hsieh et al model (33) is composed of a dashed line, while that for the Rao et al (42) and Zhurkin et al (43) models is a solid line. These diagrams are for comparison only and are not meant to imply specific bond angles or lengths.

should be viewed as hypothetical. In fact, a recent examination of the disposition of the three homologous strands of DNA within the RecA protein filament reveals that they are nearly identical to that expected for products of the reaction rather than substrates or intermediates (41). This result implies that the three-stranded DNA intermediate of the enzymatic reaction is short lived, and that upon homologous recognition, it is rapidly converted to a structure with exchanged DNA strands (41). The reason for the detection of apparently stably paired three-stranded species after deproteinization of joint molecules formed at the ends of the linear dsDNA remains unknown.

DNA Strand Exchange

After pairing is achieved, the resultant joint molecules can exchange homologous strands. Conceptually, homologous pairing and DNA strand exchange are separable events, but experimentally, strand exchange may be instantaneous. The act of DNA strand exchange can lead to two types of structures—plectonemic and paranemic—that possess different stabilities.

PLECTONEMIC JOINT MOLECULES In a plectonemic structure, the DNA strands of the heteroduplex are intertwined. Consequently, a free homologous end must be present on one of the DNA molecules involved in plectonemic joint molecule formation. This normally requires that one of the molecules be linear or that a topoisomerase be present to introduce a transient break. Figure 5*A* shows DNA substrate pairs capable of plectonemic joint molecule formation. Once formed, plectonemic joint molecules are stable in the absence of protein and, being stabilized by conventional basepairing, have a T_m characteristic of dsDNA. In the absence of topological constraints, the length of the DNA heteroduplex region is unlimited; for typical in vitro substrates (6–7 kilobases, or kb, in length), a fully displaced DNA strand is readily detected.

PARANEMIC JOINT MOLECULES In a paranemic structure, net intertwining of DNA strands is prevented, which for covalently closed molecules results in no net change in the linking number for the joint molecule. Paranemic joint molecules result when the invading DNA strand is unable to rotate freely around its complement (i.e. when the DNA molecules are topologically constrained). An example of this is pairing that initiates away from the ends of the linear DNA or between two circular DNA molecules (Figure 5*B*). Experimentally, homologous pairing can be limited to regions internal to the dsDNA by introducing heterologous DNA sequences on either the ssDNA or the dsDNA. Despite being basepaired, the topological strain imposed on paranemic joint molecules makes them wholly dependent on the binding of protein for stability (44). Paranemic joints are kinetically convertible to

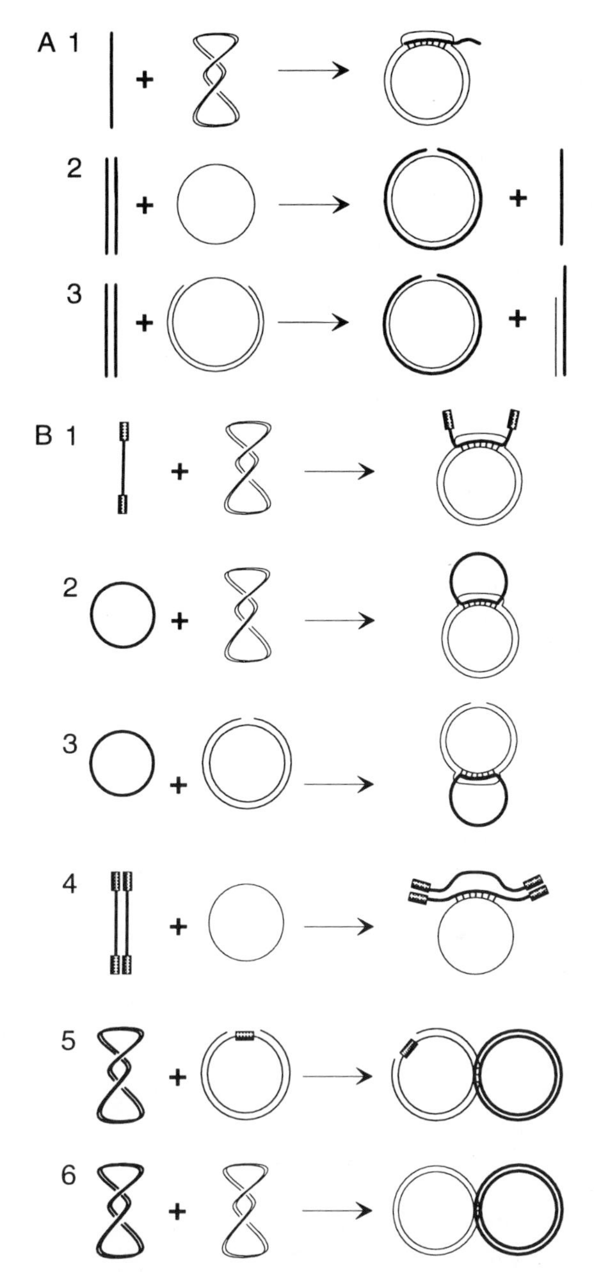

Figure 5 Substrates used to characterize homologous pairing and DNA strand-exchange proteins. Reactions that result in the production of plectonemically intertwined product molecules are shown in (*A*); those that are restricted to forming paranemic, non-intertwined molecules are shown in (*B*). Regions of nonhomology are indicated by shaded cylinders.

plectonemic joints when pairing reaches the end of a linear DNA molecule, or when they are acted upon by a topoisomerase (45).

Energetic Considerations

Because DNA strand exchange between identical sequences is isoenergetic, there is no thermodynamic requirement for energy input in the protein-catalyzed reaction. This rather obvious statement seemed to fly in the face of observations that anywhere from 1 to 1000 ATP molecules were hydrolyzed per basepair (bp) of DNA exchanged [see (16) for discussion]. However, various studies showed that both homologous pairing and DNA strand exchange (resulting in formation of up to 3.4 kb of DNA heteroduplex) could occur in the presence of the essentially nonhydrolyzable ATP analog, ATPγS (37, 38, 46–48). In fact, a nucleoside triphosphate is not needed for limited (800–900 bp) DNA strand exchange (SC Kowalczykowski, RA Krupp, in preparation), and a mutant RecA protein (RecA K72R) that reduces NTP hydrolysis by more than 600-fold is nevertheless capable of 1.5 kb of DNA strand exchange (50). Collectively, these results demonstrate that the free energy derived from ATP hydrolysis is not linked to the physical exchange of DNA strands.

Instead, ATP hydrolysis is important for the dissociation of RecA protein upon completion of strand exchange. This seemingly trivial role for ATP hydrolysis is readily explained by considering ATP as an allosteric effector (16). Binding of ATP induces a functional state of RecA protein that has high affinity for DNA, whereas ADP, the product of ATP hydrolysis, induces a nonfunctional state that has low affinity for DNA (51). The ATP hydrolytic cycle therefore serves an important function from an enzymatic perspective: It allows alternation between high- and low-affinity states, thus enabling successive rounds of protein binding and dissociation. Since the substrates and products of DNA strand exchange are nearly identical, such modulation permits RecA protein to bind the substrates with a sufficiently high affinity needed for DNA strand exchange and, at the same time, prevents product dissociation from becoming rate limiting. Thus, even though ATP hydrolysis and dissociation are not directly coupled for RecA protein (see Refs. 7, 16), ATP hydrolysis resolves the "tight-binding dilemma" faced by enzymes that must act on DNA yet dissociate with sufficient rapidity.

Physiological substrates contain regions of DNA sequence nonhomology that may be as small a single basepair mismatch or as large as several kilobasepairs. RecA protein can promote DNA strand exchange across such heterologies, but the reaction requires continual ATP hydrolysis (52, 53). A heterologous region as short as six nucleotides (nt) is sufficient to block the ATPγS-dependent reaction (52), whereas the ATP-dependent reaction can traverse heterologies in either ssDNA or dsDNA as large as 1308 nt, albeit inefficiently when present in the dsDNA (54). Consequently, the existence of heterologies introduces a thermodynamic need for energy input. It is noteworthy that in the four-stranded reaction, DNA heteroduplex extension also does not occur in the presence of ATPγS, despite the absence of any heterology (55); the requirement for ATP hydrolysis in this case may reflect either a need for rotation of the two dsDNA molecules or an inability to bind another molecule of dsDNA when dsDNA is bound irreversibly to the RecA protein filament.

These observations define additional roles for ATP hydrolysis beyond dissociation of RecA protein from DNA: to bypass heterologies, to promote DNA heteroduplex extension between DNA duplexes, and to impart a directionality to the DNA strand-exchange process (see below). The ability to bypass heterologies and to exchange DNA between duplex molecules seemed a reasonable justification for extensive ATP hydrolysis by RecA protein when it was known to be the only protein capable of promoting branch migration. Recently, however, two proteins essential to homologous recombination, the RuvAB and RecG proteins, have been shown to promote ATP-dependent branch migration [see (18, 56)]. This raises the question of which protein promotes DNA heteroduplex extension in vivo and whether this particular property of RecA protein is essential to cellular function.

Experimental Assays

Identification of homologous pairing and DNA strand-exchange proteins requires reliable in vitro assays. Several assays have been developed, chiefly to examine the properties of RecA protein–promoted DNA strand exchange. However, although RecA protein can both homologously pair and exchange DNA strands, other proteins may conceivably only pair DNA molecules, with responsibility for DNA strand exchange being relegated to a second factor. Hence, it is worth noting explicitly what each of these assays measures (i.e. only homologous pairing or both pairing and strand exchange).

NITROCELLULOSE FILTER–BINDING ASSAY The first assay used to detect joint molecule formation was the nitrocellulose filter–binding (displacement loop or D-loop) assay (57). Using conditions that minimize retention of DNA on the filter by protein, it is possible to selectively retain DNA with single-stranded character. Experimentally, the reaction contains unlabeled ssDNA and homologous labeled dsDNA (supercoiled or linear). When pairing occurs (producing joint molecules having either ssDNA tails or ssDNA in the D-loop), the labeled DNA is retained on the filter. A recent variation of this assay uses ssDNA immobilized on a filter to detect pairing (58). Although generally reliable, these assays assume that the DNA is completely deproteinized, which depends on the treatment used (44, 48). The assays are subject to artifacts introduced by nucleases or helicases, which can convert part or all of the labeled dsDNA to ssDNA; this ssDNA is retained, either independently of homologous ssDNA or by renaturation with the ssDNA. Another potential problem is retention of labeled DNA as part of a large aggregate, but this can also be detected by the absence of a requirement for homologous ssDNA. The filter-binding assay measures homologous pairing and not necessarily the exchange of DNA strands since, unless short oligonucleotides are used, the "tail" of the homologously paired ssDNA can account for retention on the filter.

AGAROSE GEL ASSAY Perhaps the most informative assay is the agarose gel assay (46). In this assay, the reaction products are deproteinized and analyzed by gel electrophoresis. Joint molecule intermediates appear as species with lower mobility than either of the DNA substrates. In the absence of topological constraints, the DNA strands can be completely exchanged, resulting in the formation of discrete product molecules. Thus, two potentially distinct phases of the reaction, initial pairing and DNA heteroduplex extension, can be simultaneously analyzed.

The favored substrate pair employed in this assay consists of circular ssDNA and linear dsDNA, because these substrates are easily obtained from ssDNA phages, and because the substrates, intermediates (joint molecules), and products are easily discerned. However, as with any other assay, this assay is subject to potential artifacts, the most prominent of which is nucleolytic digestion of the linear dsDNA by a strand-specific exonuclease. Should this occur, any protein or treatment capable of renaturing ssDNA will produce intermediates with a mobility comparable to that of joint molecules, but that in fact are reannealed molecules. Processive degradation by an exonuclease can result in the production of a species indistinguishable from the gapped circular product, but with complete loss of the displaced strand. This artifact is easily controlled for by individually labeling the ends of the DNA strand that is to be displaced (i.e. the strand in the dsDNA that is identical to the invading strand) and verifying that each end is intact in the joint molecules; it is highly recommended that this control become an absolute requirement in the characterization of any new pairing protein.

DNA helicases can contribute to artifactual pairing if the unwinding of the dsDNA is coupled to the action of a DNA renaturation protein. This artifact would not be revealed by the labeling experiment described above. However, both this artifact and that introduced by nucleases can be avoided by using covalently closed, supercoiled DNA. Joint molecule formation between supercoiled DNA and homologous ssDNA is readily detected, although extension of the DNA heteroduplex joint is limited by the topological constraint of using covalently closed DNA (59).

NUCLEASE SENSITIVITY ASSAY The most direct assay for DNA strand exchange involves measuring the displacement of ssDNA from a linear dsDNA molecule (46). Using uniformly labeled dsDNA, the existence of a displaced DNA strand can be assayed by adding a ssDNA-specific nuclease (e.g. S_1 or P_1) to deproteinized samples. This assay is perhaps the easiest to quantify and generally yields the most accurate kinetic data and measurements of DNA heteroduplex length. In addition, if the specific activity of the labeled DNA is sufficiently high, the presence of contaminating DNA exonuclease activity can be monitored by intentionally omitting the nuclease. If the nuclease assay conditions do not perturb the joint molecules, this assay is a direct measure of DNA strand exchange.

ELECTRON MICROSCOPY Electron microscopy (EM) provides the most visual evidence of pairing and DNA strand exchange. Micrographs of DNA strand exchange taken before removal of protein can provide striking displays of the pairing process (30, 60), while micrographs of deproteinized samples can demonstrate the presence of a displaced DNA strand [see (14, 21)]. Provided that the spreading procedures do not select for a specific subclass of molecules, EM is a direct assay for both pairing and DNA strand exchange. However, it is neither the most convenient nor the most accessible assay and is subject to the same nuclease and helicase artifacts.

THREE-STRANDED VS FOUR-STRANDED REACTIONS The most common substrates are a ssDNA molecule and a homologous, fully dsDNA molecule (Figure 5*A*, *reactions 1 and 2*); this is the three-stranded reaction. Because of the enzymatic requirements imposed by the properties of RecA protein, pairing between intact duplex substrates does not occur. However, DNA strand exchange can occur between duplex DNA pairs if one molecule has a homologous ssDNA region 37–52 nt in length (61–63); this is the four-stranded reaction (Figure 5*A*, *reaction 3*) (64, 65). Pairing and exchange initiate in the ssDNA region, and DNA heteroduplex then extends into the double-stranded region. The characteristics of both reactions are similar, but some differences exist. Most notably, ssDNA-binding protein (SSB protein) is not needed for joint molecule formation when the regions of ssDNA are short (<162 nt) or for exchange between regions of dsDNA (65). In addition, an intermediate in the four-stranded reaction is a bona fide Holliday junction rather than a D-loop joint molecule (66). These particular characteristics may prove useful in the identification of novel pairing proteins.

DETECTION OF PARANEMIC JOINT MOLECULES Every pair of DNA substrates designed to detect plectonemic joint molecule formation is also capable of forming paranemic joint molecules (32, 44, 61–63, 67–69). Paranemic pairing can be studied directly using DNA substrates that either are covalently closed or are prevented from pairing at DNA ends by the presence of heterologous DNA sequences (typical substrate pairs are shown in Figure 5*B*). Since paranemic joint molecules are unstable when deproteinized, any of the assays described above can be used as long as bound proteins are not removed. Alternatively, paranemic pairing can be detected by treating the closed circular molecule containing the paranemic joint with a topoisomerase; a homology-dependent perturbation of the linking number confirms the presence of pairing (61–63, 69). Since the presence of bound protein can affect the accuracy of all of the aforementioned assays, dependence on DNA sequence homology must be absolute, even though transient interactions with heterologous DNA may result in a much smaller but still detectable unwinding of supercoiled DNA (70).

POTENTIAL ARTIFACTS All pairing reactions are susceptible to artifacts, because any activity that generates ssDNA can yield a positive result due to renaturation of complementary regions. Although a potentially interesting reaction in itself, this does not constitute DNA strand exchange. Biochemical activities that contribute to such artifacts include strand-specific dsDNA exonucleases, helix-destabilizing proteins, and helicases. Trace amounts of strand-specific dsDNA nuclease activity can generate sufficient ssDNA in the dsDNA substrate to permit reannealing with the ssDNA; results with ATP-independent pairing proteins demonstrate that 20 nt or less of homologous ssDNA are sufficient (see below). Thus, nucleolytic degradation corresponding to as little as 0.3% of a 6 kb dsDNA substrate would suffice to produce pairing by DNA renaturation rather than by DNA strand exchange. Helix-destabilizing proteins (e.g. *E. coli* SSB protein) can potentially lower the T_m of dsDNA below the assay temperature. Upon deproteinization of the assay mixture prior to analysis, the free DNA strands can spontaneously renature to give the appearance of ATP-independent DNA strand exchange. The presence of helicase activity is particularly misleading because unwinding requires ATP hydrolysis; thus, helicases also introduce artifactual ATP dependence to an apparent DNA strand-exchange reaction. The artifactual results caused by these activities are compounded when deproteinizing conditions that enhance renaturation are used. For example, drying of the DNA following ethanol precipitation led to an incorrect assignment of DNA strand-exchange activity to histone H1 (71, 72). Phenol extraction in the presence of salt is another example that led to the detection of an artifactual pairing activity in *S. pombe* nuclei that resulted from nuclease activity (73).

For these reasons, DNA strand-exchange assays of partially purified fractions, particularly those using linear dsDNA, can be notoriously unreliable. Most, but not all, of these artifacts can be minimized by using covalently closed dsDNA as one of the substrates. Since DNA shorter than ~400 bp is particularly susceptible to denaturation (73a), DNA substrates greater than this length should be used.

ATP-DEPENDENT DNA STRAND-EXCHANGE PROTEINS

The *E. coli* RecA protein was the first DNA strand-exchange protein discovered; consequently, its properties have served as a benchmark against which all newly discovered proteins are compared. Genes encoding proteins with high degrees of similarity to RecA protein have been identified in every prokaryote examined. Thus, it is likely that the biochemical properties of RecA protein are characteristic of a broad and ubiquitous family of DNA strand-exchange proteins.

Escherichia coli *RecA Protein*

The RecA protein (M_r 37,842) was discovered as a DNA-dependent ATPase and as a DNA- and ATP-dependent coprotease (74, 75, 75a). Subsequently, the RecA protein was found to possess ATP-stimulated DNA renaturation and ATP-dependent DNA strand-exchange activities (2–6). The unique DNA strand-exchange activity almost certainly reflects the protein's intracellular recombination function, although a role for its DNA renaturation activity in vivo cannot be eliminated [see (76)]. The DNA strand-exchange activity of RecA protein consists of three major phases: presynapsis, synapsis, and DNA heteroduplex extension [Figure 1; see (7, 12, 14, 16, 17, 19, 21, 22)].

PRESYNAPSIS The first step of DNA strand exchange is the assembly of RecA protein on ssDNA to form a right-handed helical structure known as the presynaptic complex. The assembly of RecA protein on ssDNA is polar, with association and dissociation occurring in the 5′→3′ direction (77, 78). This structure has 6.2 monomers per turn, a pitch of ~95 Å, and a diameter of ~100 Å [(14, 21, 79, 80) and references therein]. The most unusual characteristic of this complex (as well as the one formed with dsDNA) is that the DNA is extended ~50% relative to B-form DNA, increasing the axial spacing between basepairs to 5.1 Å, and unwinding the DNA to 18.6 bp per turn. Assembly into the active form requires ATP, dATP, or ATPγS and a saturated complex (one monomer per 3–4 nt). Thus, for the typical ssDNA substrate used in vitro, the functional form of RecA protein in the homology search is a filament of approximately 2000 protein monomers. This complex is capable of hydrolyzing ATP at a modest rate (k_{cat}) of 25–30 min^{-1} (81–83). ATP hydrolysis—though it accompanies this and the subsequent steps—is not required for presynaptic complex formation, the homology search, or DNA strand exchange (37, 38, 46–48, 53; SC Kowalczykowski, RA Krupp, in preparation).

SYNAPSIS The presynaptic filament is capable of rapidly searching for DNA sequence homology. Although the details of the homology search remain unclear [see (16) for discussion of the limitations of existing data regarding the kinetics of this process], it is certain that the first step involves the formation of random nonhomologous contacts. These interactions typically result in large, easily sedimented complexes of nonhomologously paired ssDNA and dsDNA called coaggregates (84, 85); they are detected under many, but not all, conditions that support DNA strand exchange (86–89). The heterologous contacts are promiscuous, being independent of orientation of the DNA strands and capable of recognizing either complementary or identical sequences (90), and they lead to a transient unwinding of the dsDNA (70). The minimum length of homology required for recognition in vitro can be as low as 8 nt (91). Iteration of these random collisions is envisioned ultimately to align a region of homology; thereafter, the two DNA molecules pair homologously along their length. Since this process occurs with equal efficiency and rate in the absence of ATP hydrolysis, the mechanism of the homology search must be completely passive. The recognition of DNA sequence homology results in formation of a region of nascent DNA heteroduplex estimated to range from 100 to 300 bp in length (32, 35, 92). Plectonemic joint molecule formation occurs at the homologous ends of the DNA substrate pairs (Figure 5*A*). Pairing between circular ssDNA and linear dsDNA occurs at either end of the dsDNA; however, pairing between linear ssDNA and supercoiled dsDNA occurs preferentially at the 3′ end of the ssDNA (59), primarily as a consequence of the polarity of RecA protein assembly/disassembly (see Ref. 92a for discussion).

DNA HETERODUPLEX EXTENSION The region of DNA heteroduplex formed in the synaptic phase can enlarge, provided there is no topological constraint. In the RecA protein–promoted reaction, this process is not random but, instead, is protein-mediated. The direction of DNA heteroduplex formation is 5′→3′ relative to the displaced ssDNA (or the invading ssDNA) (6, 93–95), which is the same direction as RecA protein polymerization (77). This phase of DNA strand exchange requires ATP hydrolysis (37, 46), and introduces torsional strain into the dsDNA (96, 97). Under typical reaction conditions, RecA protein–promoted DNA heteroduplex extension occurs at a rate of 2–10 bp s^{-1} (46, 93, 94), leading to the complete exchange of DNA strands between substrates 7 kb in length.

STIMULATORY FACTOR: SSB PROTEIN DNA strand exchange is stimulated by the *E. coli* SSB protein (65, 98); other ssDNA-binding proteins from both prokaryotic and eukaryotic sources function similarly (82, 99–102). Each of these stimulatory proteins binds cooperatively and preferentially to ssDNA (103). Thus, the stimulatory effects of SSB protein are mediated through its interaction with ssDNA rather than through specific protein-protein interactions. The binding of RecA and SSB proteins to ssDNA is competitive, with the outcome determined by reaction conditions (82, 104–106).

The stimulatory effects of SSB protein are manifest both pre- and postsynaptically (Figure 6). Presynaptic complex formation is impeded by the presence of DNA secondary structure, to which RecA protein cannot bind (82, 104, 105). SSB protein removes this impediment to complete presynaptic complex formation by removing the secondary structure. About one SSB protein monomer per 15 nt ssDNA is required for optimal presynaptic complex formation. In addition, SSB protein eliminates aggregation of ssDNA caused by RecA protein binding (85). Thus, joint molecule formation is stimulated by eliminating DNA secondary structure, which permits formation of a continuous filament, and by preventing nonproductive aggregation. Excess SSB protein often inhibits RecA protein–dependent activities (98).

Postsynaptically, SSB protein serves two functions. The first is to prevent formation of homologously paired networks of DNA that result from intermolecular reinvasion events (107). The single strand displaced from one DNA molecule can reinvade another dsDNA molecule in an infinite pattern, causing formation of extensive, basepaired DNA networks. Because RecA protein polymerizes $5' \rightarrow 3'$ (77) and nucleates binding randomly, SSB protein can bind to the $5'$ end of the displaced linear ssDNA and hinder its utilization by RecA protein [although under conditions where displacement of SSB protein from ssDNA by RecA protein is enhanced, network formation still occurs (87, 108)]. Beyond this sequestration role, SSB protein plays a direct role in joint molecule formation. This is detected under conditions where the presynaptic role is completely bypassed through the inclusion of volume-excluding agents (e.g. polyethylene glycol or polyvinyl alcohol) (109, 110). Under these conditions, the requirement for SSB protein is directly proportional to the amount of ssDNA produced by DNA strand exchange, and not to the amount of ssDNA initially present in the reaction. The binding of SSB protein to the displaced ssDNA directly

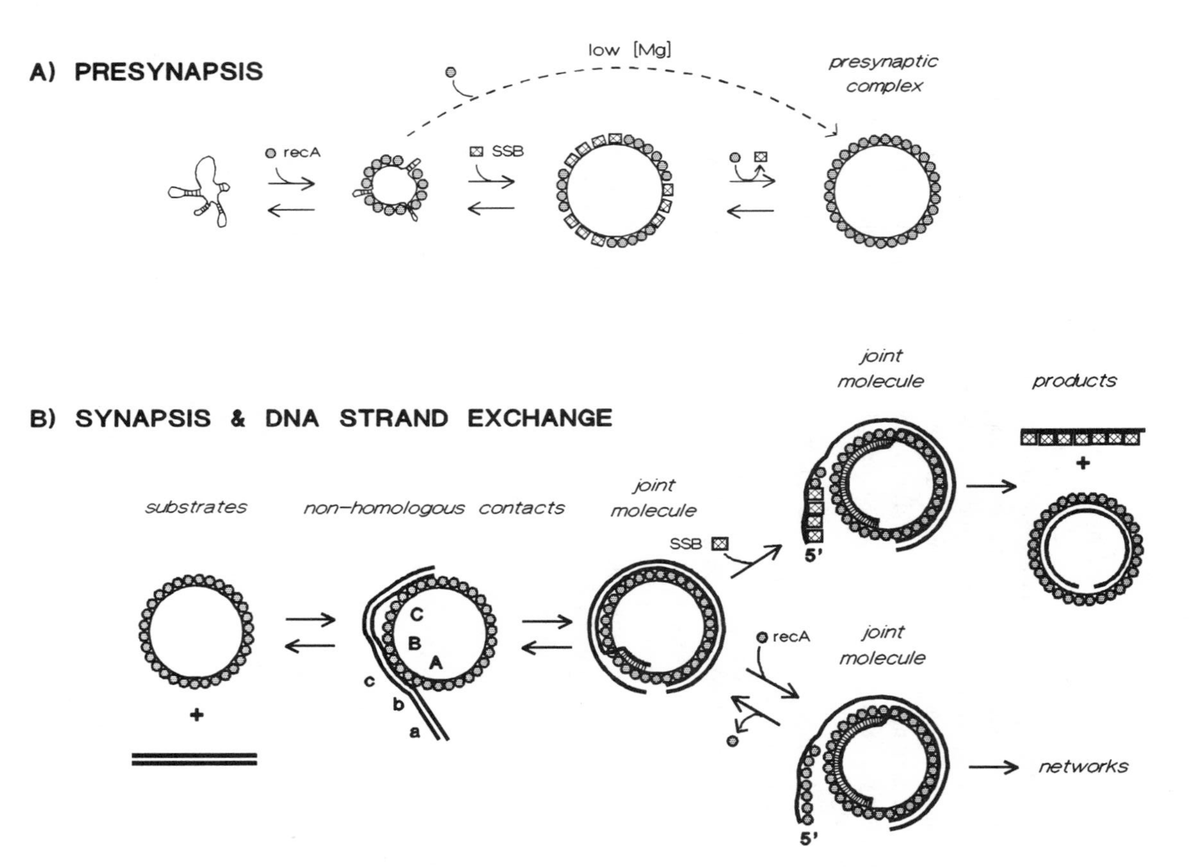

Figure 6 Models for the stimulatory effects of SSB protein on RecA protein–mediated DNA strand exchange during both presynaptic (*A*) and postsynaptic (*B*) steps. See text for details. From (110).

stimulates, by almost 10-fold, the observed rate of joint molecule formation, presumably by preventing the reverse reaction.

DNA RENATURATION ACTIVITY The ability of RecA protein to renature ssDNA was recognized before its DNA strand-exchange activity. Renaturation is optimal at lower molar ratios of RecA protein to ssDNA (1 monomer per 30 nt), where most of the ssDNA is devoid of protein (2, 111, 112). This renaturation is unusual because it is stimulated 2–3-fold by ATP, but it retains the characteristics of the ATP-independent reaction (112). In contrast to many protein-promoted renaturation reactions [e.g. by SSB and T4 gene 32 proteins (G32P) (113, 114)], the reaction is first order rather than second order in DNA concentration. DNA renaturation coincides with conditions that promote extensive aggregation of ssDNA, and both aggregation and renaturation activities of RecA protein are inhibited by SSB protein (85, 111). Thus, the mechanism of DNA renaturation most likely involves a rapid condensation of ssDNA into aggregates; the bimolecular reaction becomes unimolecular because of the high effective DNA concentration. The products of the reaction are normally large intermolecularly basepaired networks, but in the absence of ATP, simple unit-length dsDNA can be obtained (115). The biological role of RecA protein–promoted renaturation remains an open question, because there currently exists no mutant RecA protein differentially affected in its DNA strand-exchange and DNA renaturation activities (76).

Proteus mirabilis *RecA Protein*

The RecA protein from *P. mirabilis* (M_r 38,176) is 73% identical to that of *E. coli* (116). Not surprisingly, it complements an *E. coli recA* mutation (117) and has all the activities of the *E. coli* protein: ssDNA-dependent ATPase (118), DNA strand exchange (118), and LexA repressor cleavage (119) activities. The protein can promote DNA strand exchange using many of the substrate pairs used by the *E. coli* protein, including ssDNA and supercoiled DNA, circular ssDNA and linear dsDNA, and gapped and linear dsDNA substrates. In contrast, *E. coli* SSB protein reduces the ATPase activity of *P. mirabilis* RecA protein by about 80%, which suggests that *P. mirabilis* RecA protein can only partially resist displacement by SSB protein from ssDNA. Consistent with this inhibitory effect of SSB protein, *P. mirabilis* RecA protein does not complete the exchange of DNA strands between circular ssDNA and linear dsDNA when *E. coli* SSB protein is present, limiting the reaction to formation of intermediate joint molecules (118). The behavior of *P. mirabilis* SSB protein in these reactions is untested.

Bacillus subtilis *RecA (RecE) Protein*

The *B. subtilis* RecA (RecE) protein (M_r 38,300), which displays 60% identity with the *E. coli* protein (120), is encoded by the gene previously known as *recE* (121). RecA protein has both DNA strand-exchange and *E. coli* LexA protein cleavage activities (122). Despite these parallels, the *B. subtilis* protein is unable to hydrolyze ATP; it can, however, hydrolyze dATP at a rate about 65% that of *E. coli* RecA protein. DNA strand exchange requires dATP, and ATP is an inhibitor of both the dATPase and the DNA strand-exchange activities. In the presence of dATP and *E. coli* SSB protein, 60% of the linear dsDNA and circular ssDNA is converted to complete DNA strand-exchange product. The requirement for dATP is unusual, but this characteristic is mimicked by a mutant *E. coli* RecA protein, RecA K72R protein, which also requires dATP for DNA strand-exchange activity (50), of which ATP is also a competitive inhibitor (WM Rehrauer, SC Kowalczykowski, unpublished observation).

Thermus aquaticus *RecA Protein*

A RecA protein homolog was isolated from the thermophilic eubacteria, *Thermus aquaticus* (50a; JG Wetmur, DM Wong, B Ortiz, J Tong, F Reichert, DH Gelfand, personal communication). The protein has 59% identity and 78% similarity to the *E. coli* RecA protein. Binding to ssDNA requires ATPγS for detection, is optimal at about 55°C, and is detectable to about 70°C. The protein possesses an optimum for DNA-dependent ATPase activity above 70°C. Joint molecule formation occurs at the optimal temperature of 65°C. The yield of joint molecules formed is about 4–5-fold less efficient than that promoted by *E. coli* RecA protein, and DNA heteroduplex formation is limited (i.e. the complete exchange of DNA strands between M13 DNA substrates does not occur) (50a). The failure to promote extensive DNA heteroduplex formation may have been due to the use of *E. coli* SSB protein in the reactions. RecA proteins were also isolated from three other thermophiles: *Thermus thermophilus, Thermotoga maritima,* and *Aquifex pyrophilus* (JG Wetmur et al, personal communication); the cognate SSB proteins are yet to be isolated.

Bacteriophage T4 UvsX Protein

The bacteriophage T4 analog of *E. coli* RecA protein is encoded by the *uvsX* gene. UvsX protein (M_r 43,760) bears many biochemical similarities to RecA protein, despite being the most divergent of the prokaryotic RecA-like proteins; it has only 23% identical and 15% similar residues (123). Most of the identities cluster in the ATP-binding site, with the remaining conserved residues being primarily hydrophobic amino acids important to tertiary or quaternary structure (124). Next to *E. coli* RecA protein, UvsX protein is the best characterized DNA strand-exchange protein. The mechanism of UvsX protein–promoted DNA strand exchange, while different in some important details from that of RecA protein, is the same globally as that of RecA protein: Presynaptic complex formation results in a helical UvsX protein–ssDNA filament; synapsis results in both paranemic and plectonemic joint molecules; and DNA heteroduplex extension results in the complete exchange of 6–7 kb of DNA. However, UvsX protein appears to be more dynamic in its kinetic behavior than RecA protein and, perhaps most significantly, interacts directly with a novel auxiliary factor, the UvsY protein (see below).

UvsX protein has ssDNA-dependent NTPase, DNA renaturation, and DNA strand-exchange activities (125–128). The ATPase activity of UvsX protein is distinctive among RecA-like proteins, producing both ADP and AMP, and its rate (k_{cat}) is ~15-fold greater than that of RecA protein (240 ADP and 145 AMP min^{-1}) (127). UvsX protein binds both ssDNA and dsDNA cooperatively with a stoichiometry of one monomer per 3–5 nt, and forms a presynaptic filament with a structure similar to that made by RecA protein (129). ATP or ATPγS binding stabilizes the UvsX protein–ssDNA filament (125), suggesting that the appropriate nucleoside triphosphate induces a transition to a higher-affinity state. Joint molecule formation, which is poor in the absence of stimulatory factors (see below), yields both paranemic and plectonemic molecules (130). The DNA strand-exchange reaction displays an optimum in UvsX protein concentration: lower concentrations are suboptimal for presynaptic complex formation, and higher concentrations (which exceed saturation of the ssDNA) reduce joint molecule formation due to binding of UvsX protein to dsDNA (130). Branch migration by UvsX protein (15 bp s^{-1}) is somewhat faster than that promoted by RecA protein (131, 132). Because of the higher rate of ATP turnover and the concomitant increased rate of protein dissociation, the UvsX protein–ssDNA filament is more dynamic than the RecA protein–ssDNA filament (132, 133).

STIMULATORY FACTOR: GENE 32 PROTEIN DNA strand exchange promoted by UvsX protein is enhanced by the T4 phage–encoded ssDNA-binding (helix-destabilizing) protein, G32P. This protein is a 33.5-kDa analog of SSB protein that binds preferentially and cooperatively ($\omega \approx 10^3$) to ssDNA (103). G32P increases both the rate and the yield of joint molecule formation (134). The optimal concentration of G32P (one monomer per 8–10 nt) needed for DNA strand exchange, when UvsX protein is present at suboptimal concentrations, represents the amount needed to saturate the ssDNA. Maximal joint molecule formation in the presence of G32P can occur at subsaturating concentrations of UvsX protein (one monomer per 8 nt) (125, 134, 135), but excess G32P inhibits both the ATPase and the joint molecule formation activities of UvsX protein. SSB protein can substitute for G32P, although the rates of joint molecule formation and of complete exchange between DNA substrates are more than 5- and 15-fold slower, respectively (127, 134). In contrast to the *E. coli* system, where SSB protein typically acts in both the pre- and postsynaptic phases, G32P is not needed in presynapsis when UvsX protein is in large excess over the DNA concentration (135). Under these conditions, G32P seems to function only in the postsynaptic phase by stabilizing the displaced strand (135), in much the same way that SSB protein functions (110).

The mechanism of the homology search is not well understood. Like RecA protein, UvsX protein can coaggregate nonhomologous ssDNA and dsDNA, but these coaggregates are not detected under optimal conditions in the presence of G32P (134). The polarity of pairing and DNA strand displacement is also $5' \rightarrow 3'$ relative to the displaced strand (125, 132), and is stimulated 5–10-fold by G32P (132). DNA strand exchange promoted by UvsX protein between circular ssDNA and linear dsDNA commonly results in the formation of homology-dependent DNA networks that fail to enter an agarose gel; though not typically reported for RecA protein, homology-dependent networks are the major product under conditions that enhance the ability of RecA protein to displace SSB protein (87, 108, 109). This parallel in network formation argues that UvsX protein can more effectively remove G32P from the displaced strand of DNA and utilize this ssDNA in subsequent invasion events. As with RecA protein, DNA heteroduplex extension is stopped by the addition of ATPγS to an ongoing reaction, but, unlike RecA protein, a brief acceleration in the rate is seen (127, 132). This transient increase was interpreted to mean that DNA strand exchange could not only occur in the presence of ATPγS, but that the stabilizing effect of ATPγS actually provided a burst of enhanced exchange; due to the inability of UvsX protein to redistribute itself in the presence of ATPγS, further DNA strand exchange was prohibited. A similar conclusion was reached for RecA protein, based on very different experiments [see (16)].

STIMULATORY FACTOR: UVSY PROTEIN The T4 phage recombination system is unique in that both genetic and biochemical data demonstrate the need for an accessory protein that, as yet, has no counterpart in any other system. This protein, UvsY protein (M_r 16,000), binds cooperatively to both ssDNA and dsDNA (136). It interacts directly with UvsX protein in a 1:1 molar ratio (137). UvsY protein increases the rate of UvsX protein–dependent ATP hydrolysis by 2–3-fold under suboptimal conditions and DNA strand exchange by ~3-fold (133, 136, 137). Because UvsY protein increases the apparent affinity of UvsX protein for ssDNA, UvsX protein has increased resistance to displacement from ssDNA by G32P (136, 137). All of these important stimulatory effects are specific to UvsX protein and are not observed with RecA protein.

A complex series of interactions are proposed for the reaction containing all three proteins (138). UvsY protein interacts with the carboxyl terminus of G32P, and this interaction is necessary to load UvsY protein on the ssDNA. Once bound to the DNA, UvsY protein promotes the binding of UvsX protein to the DNA, presumably through direct protein-protein interactions. It appears that not all of the G32P is displaced by the binding of UvsX and UvsY proteins; it has been suggested that G32P remains associated with the DNA-bound UvsX-UvsY complex via protein-protein interactions (139). The T4 phage system represents the best example of functionally important specific protein-protein interactions in a DNA strand-exchange reaction; if there is any counterpart to the UvsX-UvsY protein interaction, it may be that of the *S. cerevisiae* Rad51 and Rad52 proteins (140) (see below).

Pisum sativum *RecA Protein*

A 39-kDa protein that is immunologically related to *E. coli* RecA protein was identified in pea (*Pisum sativum* L.) chloroplasts (141) and, consistent with this observation, genomic (but not chloroplast) DNA of pea hybridizes to a *Synechococcus recA* probe (142). Extracts of these chloroplasts were subsequently shown to possess DNA strand-exchange activity (142a), which is both ATP- and Mg^{2+}-dependent. Both linear dsDNA by circular ssDNA and linear ssDNA by supercoiled DNA substrate pairs form joint molecules. Because the linear DNA in each case was end-labeled, these assays were controlled for the potential occurrence of exonuclease-dependent renaturation. EM analysis confirms the existence of a displaced strand (142a).

Ustilago maydis *Rec1 Activity*

The first, and so far only, eukaryotic ATP-dependent homologous pairing activity to be purified is from *U. maydis*. It was initially called Rec1 protein because its activity was not detectable in preparations from a *rec1* mutant strain (143, 144); it is now clear that this protein is not the product of the *REC1* gene (145–147), which encodes a 3′→5′ exonuclease lacking pairing activity (148). The protein that encodes this pairing activity is not known, but recently it was found that the *U. maydis* Rec2 protein (M_r 84,000) bears similarity to RecA protein (BP Rubin, DO Ferguson, WK Holloman, personal communication). Homology resides in the regions required for nucleotide binding and, in this respect, it is similar to other eukaryotic structural homologs of RecA protein (see below). Until the genetic identity of the Rec1 activity is determined, the term "Rec1 is maintained to indicate the first pairing activity identified in *U. maydis* (149).

The Rec1 activity (estimated molecular weight of 70,000) catalyzes ssDNA-dependent ATP hydrolysis, renaturation, and DNA strand exchange (143, 149, 150). Somewhat unexpectedly, the protein neither binds ssDNA cooperatively nor forms filaments, although it does bind Z-DNA with greater affinity [2–6-fold (151) to 20–75-fold (152)] than it binds B-DNA. The ATPase activity is cooperative in protein concentration (Hill coefficient of 1.8), and the specific activity ($\sim$225 min^{-1}) is comparable to that of UvsX protein but is $\sim$10-fold greater than that of RecA protein (143, 149). Like the case of RecA protein, the DNA renaturation activity is first order in DNA concentration, is stimulated by ATP (10–15-fold), and is optimal at substoichiometric concentrations (1 monomer per 300 nt); unlike the case of RecA protein, the amount of renatured DNA product is proportional to the amount of protein present, suggesting that the protein does not turnover in this reaction (149).

Rec1 activity homologously pairs the same kinds of substrates used by RecA protein (143, 150) and, in addition, pairs fully duplex DNA substrates (143, 150, 153, 154). Unlike the reaction with the prokaryotic proteins, a fraction (40%) of the joint molecules detected using ssDNA fragments and supercoiled DNA are formed independently of ATP, but DNA heteroduplex extension is fully ATP dependent (143). In the presence of the nonhydrolyzable ATP analog, AMP-PNP, the joint molecules formed are apparently paranemic (150) and contain unwound dsDNA. Joint molecule formation occurs preferentially, displacing the 3′ end of the linear duplex DNA (150); this polarity is opposite that of both RecA and UvsX proteins. Unlike either RecA or UvsX protein, Rec1 activity can pair two supercoiled DNA molecules provided that a topoisomerase and either homologous ssDNA fragments or actively transcribing RNA polymerase are present (153). Thus, generation of a displaced strand in one of the DNA molecules is sufficient for pairing between intact duplex DNA molecules. Since as little as one protein monomer per 200 nt is sufficient for pairing (143), it appears that this activity is not required in stoichiometric amounts. In sum, although there are similarities to the *E. coli* model, the Rec1 pairing activity displays some notable differences.

ATP-INDEPENDENT DNA STRAND-EXCHANGE PROTEINS

The discovery of DNA strand-exchange activity in *E. coli* stimulated a search for similar activity in other organisms. Given the importance of recombination, it was not surprising that such an activity was found in many species. But the discovery of ATP-independent DNA strand exchange, promoted by activities primarily from eukaryotic sources, seemed to controvert the well-established properties of the RecA protein–promoted reaction. The subsequent realization that ATP binding, and not hydrolysis, was sufficient for DNA strand exchange by RecA protein blunted this criticism, because it could be argued that the ATP-independent proteins represented a class of proteins that were equivalent to the active form of RecA protein that results from ATP binding [see (9)].

This encompassing explanation proved not to be accurate. Careful characterization of these ATP-independent reactions uncovered the presence of nuclease activity that, in at least a few cases, was intrinsic to the purified pairing protein (see below). Most of the assays conducted with this class of putative DNA strand-exchange proteins involved the typical circular ssDNA and linear dsDNA substrates. In several well-documented cases, it is now clear that the observed DNA pairing and/or strand-exchange activity does not occur between ssDNA and an intact dsDNA molecule, but only with digested dsDNA containing a ssDNA tail (Figure 2). Thus, pairing is initiated by reannealing of two ssDNA regions rather than by DNA strand invasion and displacement.

The model in Figure 2 illustrates a mechanism for homologous pairing and strand exchange that describes the reactions promoted by the ATP-independent class of activities. The first step requires resection by a strand-specific dsDNA exonuclease; it appears that a nuclease of either polarity will suffice and that the polarity of this degradation step determines the observed apparent polarity in the subsequent DNA strand-exchange step. The nuclease activity is intrinsic to some, but not all, pairing proteins (see below). The second step involves the protein-mediated renaturation of the complementary ssDNA. These two steps are sufficient to be interpreted as DNA strand-exchange activity in nearly any in vitro assay even though no exchange of strands has occurred. The next step is displacement of ssDNA by a process that can be referred to as DNA strand exchange but is mechanistically more akin to the DNA heteroduplex extension phase of the RecA protein–promoted reaction. This step has an inherent asymmetry attributable to the DNA degradation step. Thermodynamic considerations dictate that the reaction can proceed in only one direction so as to maximize both the number of basepairs formed and the entropy of the products. Consequently, random thermal branch migration would appear unidirectional, with a displaced strand being liberated only if exchange occurs in one direction (to the right in Figure 2). The process may be accelerated by the pairing protein, as appears to be the case for the *E. coli* RecT protein (see below) (155). This model is biochemically distinct from the model for DNA strand exchange promoted by RecA protein.

The biological role of the ATP-independent DNA strand-exchange proteins is difficult to assess because genetic analysis is either non-existent or complex. Notable exceptions include the *E. coli* RecT and λ phage β proteins, which are important to certain types of recombination events, and the *Drosophila* Rrp1 protein, which is important in DNA repair (see below).

It is too early to say whether all ATP-independent proteins act by a similar mechanism, but they obviously do not imitate the behavior of RecA protein. It is also unclear how many of these proteins actually function in recombination, precluding their description as recombination proteins. The realization that most, if not all, of these proteins act on nuclease-digested DNA calls for a re-examination of their characteristics. Until then, it would be prudent to refer to such proteins as either reannealing or homologous pairing proteins, and not DNA strand-exchange proteins, unless the displacement of ssDNA by strand exchange is demonstrated to be protein dependent.

Escherichia coli *RecT Protein*

The RecE and RecT proteins are encoded by a genetic locus that was originally designated *recE*. This locus is part of a cryptic lambdoid prophage, *rac,* and is composed of two genes, *recE* and *recT,* which encode a nuclease (exonuclease VIII) and a DNA-binding protein, respectively (156, 157). These genes bear functional, but not sequence, similarity to the bacteriophage λ recombination genes, *redα* and *redβ* (see below). The *recT* gene can complement a *recA* defect in plasmid recombination, arguing that the RecT protein must possess an activity that either alone or in concert with other proteins is functionally equivalent to one possessed by RecA protein (156, 158).

The 33-kDa RecT protein is a tetramer in solution (157). It binds to ssDNA, but not to dsDNA; half-maximal binding, as monitored by a nitrocellulose filter-binding assay, occurs at a stoichiometry of one tetramer per 80 nt. Like the β protein of bacteriophage λ, RecT protein promotes ATP-independent, partially (75%) Mg^{2+}-dependent renaturation of ssDNA (157).

In addition to renaturation, RecT protein can promote homologous pairing between circular ssDNA and linear dsDNA, provided that one strand of the linear dsDNA is digested to produce a homologous ssDNA tail (155). This pairing activity depends absolutely on prior nucleolytic function, but accepts both 3′ and 5′ ssDNA tails. Any nuclease (e.g. exoIII, exoVIII, or T7 gene 6 protein) can serve in this capacity. Thus, the reaction must initiate by renaturation of the complementary regions of ssDNA. The pairing reaction is ATP independent, is Mg^{2+} dependent, and requires at least one RecT protein monomer per 13 nt. The region of DNA heteroduplex is not restricted to the region of resected DNA, but extends into the region of intact dsDNA;

displacement of ssDNA can be detected by EM. This displacement appears to depend on a unique RecT protein activity, because another protein (histone H1) capable of promoting the initial DNA renaturation step cannot displace ssDNA (155).

Bacteriophage λ β Protein

Bacteriophage λ encodes two proteins, λ exonuclease and β, that are essential for phage-specific recombination. The β protein (M_r 28,000) has biochemical properties that partly resemble those of RecA protein and partly those of SSB protein (99, 159). β protein has neither nuclease nor D-loop formation activities, but like SSB protein, it stimulates the activity of RecA protein (99). Stimulation is particularly evident at suboptimal concentrations of RecA protein, and requires a stoichiometric amount of β protein (one monomer per 4 nt).

Under conditions of one protein monomer per 6 nt, β protein also possesses a first-order DNA renaturation activity (99, 159). It is not known whether β protein can promote a pairing reaction using resected dsDNA. The genetic and biochemical similarities between the *recET* system and the phage λ *red* system suggest that the combined actions of a strand-specific dsDNA exonuclease and an annealing protein constitute a biochemical alternative to the type of pairing reaction promoted by RecA protein. This parallel leaves open the possibility that other functionally similar proteins exist (see following sections).

Saccharomyces cerevisiae *Sep1/STPβ*

Sep1 (Strand-exchange protein 1) and STPβ (Strand Transfer Protein β) are independent isolates of a yeast protein that effects transfer between circular ssDNA and linear dsDNA (160, 161). Initially, Sep1 was purified as a 130-kDa proteolytic fragment, which accounted for the initial size discrepancy between it and STPβ (180 kDa). The gene (*SEP1/DST2*) encoding Sep1/STPβ has been identified, and it encodes a protein of 175 kDa (162, 163).

Unlike for RecA protein, the homologous pairing and DNA strand-exchange reaction promoted by Sep1/STPβ (hereafter referred to simply as Sep1) is slightly inhibited by, rather than dependent on, ATP (160). Sep1 has both DNA renaturation (164) and exonuclease activities (165). The latter activity is responsible for resection of the linear dsDNA to reveal ssDNA that is utilized by the DNA reannealing activity to produce paired complexes. As for *E. coli* RecT protein, a ssDNA tail of at least 20 nt of either polarity is required for pairing (165a); in the absence of nuclease activity (e.g. in the presence of Ca^{2+}), no plectonemic pairing occurs (165). Consequently, the initiation phase of pairing for Sep1 proceeds by an annealing, rather than a strand invasion, reaction. Maximal pairing was reported to require about one Sep1 fragment (p^{130}) monomer per 12–14 nt or one intact Sep1 monomer per 35–40 nt (165), but recent work finds that optimal joint molecule formation occurs at about 1 Sep1 monomer per 100 nt (165b). This value is in agreement with direct ssDNA-binding studies that yield a binding site size of 70–100 nt and coincides with the amount needed to aggregate the DNA (165b). The stoichiometric requirement for Sep1 protein can be alleviated more than 10–30-fold by loading the protein onto the ssDNA ends of resected dsDNA (165b). Sep1 displays no end-bias in joint molecule formation, and strand displacement proceeds $5'\rightarrow3'$ (relative to the displaced strand), consistent with the polarity of strand degradation. Based on EM observations, Sep1 action can result in the net displacement of 4.1 kb (165); it is not known whether Sep1 directly promotes this exchange step. Recently, Sep1 was found to promote paranemic joint formation, in a reaction that required at least 41 bp of homology (J Chen, R Kanaar, NR Cozzarelli, personal communication). Paranemic joints between ssDNA and dsDNA were detected by both filter-binding and EM assays and, interestingly, pairing between supercoiled DNA and linear dsDNA was also observed.

The fragment of Sep1 protein (p^{130}) binds noncooperatively to both ssDNA and dsDNA (164). Its affinity for ssDNA is higher than that for dsDNA. p^{130}-ssDNA complexes are stable to 200 mM NaCl, but both ssDNA renaturation and DNA strand exchange are inhibited well below this salt concentration, arguing that a step succeeding ssDNA binding must be responsible for the salt sensitivity. As for RecA protein, optimal renaturation of ssDNA occurs at a substoichiometric concentration of p^{130} ($\sim$one monomer per 100 nt) (164).

The exonuclease activity of Sep1 has been extensively characterized. This protein was initially identified as an exoribonuclease, Xrn1, which processively degrades both poly(A)$^+$-tailed RNA and rRNA in a $5'\rightarrow3'$ manner (166). In addition, the protein has RNaseH activity (167). Although the early experiments did not detect nuclease activity on DNA substrates, subsequent work showed that Sep1 degrades both ds- and ssDNA (at rates of 20 and 70 mol nt per min per mol protein, respectively), although ssRNA is preferred (165, 165a, 166). The nuclease activity requires Mg^{2+}, is inhibited by Ca^{2+}, displays a pH optimum of 8.5, and has an average processivity of 45 nt. The polarity of this intrinsic exonuclease activity dictates the apparent polarity of DNA strand exchange ($5'\rightarrow3'$) promoted by Sep1.

Mutations in the gene encoding Sep1 exhibit pleiotropic effects. In addition to its identification as an exoribonuclease (*XRN1*) (168), this gene was also identified as being involved in nuclear fusion (*KEM1*) (169) and in the maintenance of plasmids containing a defective ARS (autonomously replicating sequence) (*RAR5*) (170). An essential gene (*RAT1/TAP1/HKE1*) with homology to *SEP1* was also isolated (171, 172, 172a). It encodes a 116-kDa 5′→3′ exoribonuclease that is implicated in mRNA trafficking and transcriptional activation. The meiotic *S. pombe* homolog of Sep1, exoII, was initially purified as a ssDNA nuclease (173). Mutations in *SEP1/DST2* have a slight defect (2–3-fold) for intragenic mitotic recombination (162, 163), but no intergenic defect (169); they do not sporulate, arrest in pachytene, and show certain defects in some, but not all, recombination assays (15, 15a, 173a). Formation and processing of dsDNA breaks occurs in *sep1* mutants, but the level of recombination is reduced. It appears that the absence of a striking recombination phenotype is at least partially due to redundant functions. The *sep1Δdmc1Δ* or *sep1Δrad51Δ* double mutants display more severe defects than any single mutation; meiotic intrachomosomal recombination was reduced more than 20-fold and meiotic interchromosomal recombination was partially reduced (D Tishkoff, B Rockmill, GS Roeder, RD Kolodner, personal communication). This complicated behavior potentially argues for a direct role for Sep1 in meiotic recombination and, in addition or alternatively, these phenotypes are indirect consequences of the pleiotropic physiological defects of *sep1* mutants.

STIMULATORY FACTOR: yRPA A number of yeast ssDNA-binding (ySSB) proteins stimulate the pairing activity of Sep1. One of these is the large subunit of a heterotrimeric protein, known as yeast replication protein A (yRPA) (174). Though normally isolated as a complex consisting of 69-, 36-, and 13-kDa polypeptides, a 34-kDa proteolytic fragment of the large subunit, encompassing the central portion of the polypeptide and containing the Zn^{2+} finger DNA-binding domain, can by itself stimulate the activity of Sep1 (101).

Addition of the 34-kDa fragment results in an 18-fold increase in the initial rate, primarily by reducing a kinetic lag in the formation of joint molecules (101). This ySSB protein does not change the sigmoid dependence on Sep1 concentration, but it does reduce (by ~2–3-fold) the amount of Sep1 required for optimal levels of DNA pairing. There is little enhancement of DNA pairing by ySSB protein at saturating concentrations of Sep1. Likewise, a variety of other ssDNA-binding proteins stimulate joint molecule formation by Sep1 (161). A 50-fold stimulation is observed at optimal concentrations, and up to 1.5 kb of heteroduplex DNA is formed. Only 2–3 molecules of Sep1 per linear dsDNA molecule are required when optimal concentrations of these ySSB proteins are present (161).

The trimeric yRPA holoprotein also stimulates Sep1, with maximal stimulation occurring at saturating concentrations of yRPA (102). yRPA binds ssDNA with a stoichiometry of one molecule per 90 nt, forming a beaded structure similar to that formed by *E. coli* SSB protein. It binds with both high affinity ($>10^9$ M^{-1}) and cooperativity ($\omega = 10^4$–10^5). Excess yRPA inhibits pairing, but not nuclease, activity. It has been proposed that this ySSB protein stimulates the activity of Sep1 by inhibiting the aggregation of ssDNA and promoting the coaggregation of ssDNA and dsDNA molecules (102).

STIMULATORY FACTOR: SF1 SF1 (Stimulatory Factor 1) is a 55-kDa protein (originally described as a 33-kDa protein) that substantially reduces the amount of Sep1 required for DNA strand exchange (15, 175). When an optimal amount of SF1 (one monomer per 20 nt) is present, the amount of Sep1 is reduced to only one molecule per 5800 nt. The rate of DNA strand exchange is increased by at least 3–4-fold and, rather than simple joint molecules, large DNA networks are formed. SF1 can aggregate both ssDNA and dsDNA, but this property is not the basis of its stimulatory effect, since SF1 is effective under conditions that reduce its aggregation activity.

Perhaps the most significant property of SF1 is its ssDNA renaturation activity (175, 176). Since it is now apparent that pairing in the Sep1-dependent reaction initiates by reannealing of complementary ssDNA, the ability of SF1 to alleviate the amount of Sep1 required is easily understood: SF1, itself, must catalyze the renaturation step, and the few molecules of Sep1 needed provide the nucleolytic activity necessary to resect the ends of the linear dsDNA. Whether SF1 or Sep1 promotes the strand displacement step is unclear. Furthermore, the identity of SF1 is unknown; given that fatty acid synthase can stimulate the activity of the *S. pombe* homolog of Sep1 (177), the prospect of nonspecific stimulation remains open.

Saccharomyces cerevisiae *DPA Protein*

The second ATP-independent DNA strand-exchange activity (M_r 120,000) isolated from mitotic yeast cells was called DNA pairing activity (DPA) (178). Although its biochemical properties are similar to those of Sep1, it is a distinct protein (15, 179). The sequence of the gene for DPA reveals that it is identical to translation elongation factor 3 (EF3) (K McEntee, personal communication); the likelihood that EF3 is directly involved in recombination is low, and since EF3 is an essential protein, establishing a role for EF3 in recombination will be difficult.

DPA possesses ssDNA-binding, DNA aggregation, and ATP-independent DNA renaturation activities, but no nuclease activity (178). DNA renaturation is extremely rapid (<1 min), and is optimal at stoichiometric amounts of protein (one monomer per 50 nt). The yield shows a sigmoid dependence on protein. Anticipating the need for nucleolytic processing by the ATP-independent homologous pairing protein preparations, Halbrook & McEntee appreciated that DPA required dsDNA substrates with either 5′ or 3′ tails approximately 50 nt in length in order to initiate homologous pairing via its renaturation activity; DNA substrates with only a 4-nt overhang failed to pair (178). DPA forms up to 3–5 kb of heteroduplex DNA and appears to require the continued presence of protein, but this phase of the reaction displays no preferred polarity. Thus, DPA-promoted homologous pairing initiates by renaturation and is followed by a DNA heteroduplex extension phase.

Saccharomyces cerevisiae *STPα*

STPα (Strand Transfer Protein α) (38 kDa) was isolated from meiotic yeast cells (180). It apparently catalyzes DNA strand transfer between linear dsDNA and circular ssDNA as well as DNA renaturation. Its biochemical properties bear a similarity to those of Sep1, and the presence of trace levels of nuclease activity appear to explain the apparent DNA strand transfer activity. STPα acts catalytically (2–3 molecules per dsDNA molecule) in the presence of a saturating amount of 26-kDa ySSB protein (one monomer per 6–8 nt). Pairing activity is stimulated by nonspecific agents (e.g. histone H1 and spermidine), with optimal stimulation occurring at concentrations that aggregate 50% of the DNA. The gene encoding STPα, *DST1,* has been cloned (181), and was found to be a previously identified gene (*PPR2*) encoding the transcription elongation factor, TFIIS (182, 183). A role for TFIIS in genetic recombination appears unlikely.

Schizosaccharomyces pombe *p^{140}/ExoII Protein*

The p^{140} protein (M_r 140,000) was purified from vegetative *S. pombe* cells (184). It promotes homologous pairing and DNA strand exchange (as observed by EM) between circular ssDNA and linear dsDNA, and it has an intrinsic nuclease activity. A monomer in solution, the p^{140} protein degrades ssDNA, dsDNA, and RNA in a $5' \rightarrow 3'$ direction; activity requires Mg^{2+}, is inhibited by Ca^{2+}, and degrades these nucleic acids at rates of 180, 5, and 0.07 nt per min per molecule, respectively. These nucleolytic properties are similar to those of the *S. cerevisiae* Sep1 protein. Based on its biochemical properties, protein sequence, and antigenic behavior, this protein is identical to a polypeptide, exoII, that was purified from meiotic cells as a ssDNA exonuclease (173). As for *E. coli* RecT protein and *S. cerevisiae* Sep1, the DNA pairing activity of p^{140}/exoII requires resection of the linear dsDNA by a nuclease to reveal complementary ssDNA and is enhanced by 6% polyethylene glycol. Consistent with the expectation that pairing initiates in the ssDNA regions, p^{140}/exoII is also capable of renaturing ssDNA.

Independently, a multicomponent pairing system was partially purified from mitotic *S. pombe* cells (185). The predominant species in the active fraction have molecular weights (100, 65, and 30 kDa) different from that of p^{140}/exoII. The 65-kDa polypeptide fraction alone promotes limited pairing and DNA renaturation, but the reaction is stimulated by the addition of fractions containing the other two proteins. It is not clear how the apparent pairing activity of this complex initiates; while it was reported that the fractions contained no nuclease activity (185), the gel assay used was not sensitive enough to detect the small amounts of nuclease activity that are sufficient to activate other ATP-independent pairing proteins.

STIMULATORY FACTOR: FATTY ACID SYNTHASE A factor ($p^{190/210}$) that stimulates the activity of p^{140}/exoII was also isolated (177). Remarkably, protein sequencing identifies $p^{190/210}$ as fatty acid synthase (FAS). This protein binds both ss- and dsDNA and is capable of renaturing DNA. Renaturation activity presumably stems from the ability of FAS to aggregate DNA when present at a ratio of about one molecule per 250 nt or bp of DNA. FAS has no nuclease activity and does not stimulate the nuclease activity of p^{140}/exoII. Furthermore, it can promote homologous pairing of resected dsDNA molecules with complementary ssDNA, as well as the subsequent displacement of ssDNA, at least as effectively as p^{140}/exoII. The amount of protein needed for this activity coincides with the amount required for aggregation. The mechanistic basis for this reaction is unclear, but must be a consequence of the ability of FAS to bind and aggregate DNA nonspecifically. As it is unlikely that FAS has any role in homologous recombination, this observation should be taken as an indication that, although renaturation and DNA strand exchange as measured by this assay may represent a genuine activity of recombination proteins, the existence of such an activity does not automatically identify a recombination protein.

Drosophila melanogaster *Rrp1*

Two ATP-independent DNA strand transfer activities, which may be promoted by the same 105-kDa protein, were isolated from *D. melanogaster* embryo nuclear extracts (186–188). Rrp1 (Recombination repair protein 1) has ssDNA and dsDNA aggregation, ssDNA renaturation, and DNA strand-exchange activities. Most significantly, the protein possesses 3′-strand specific dsDNA exonuclease and apurinic endonuclease activities (188a). In agreement, sequencing of the gene (*rrp1*) revealed strong homology to both *E. coli* exonuclease III (a 3′ exonuclease) and apurinic/lapyrimidinic endonucleases (189–191). Since the *RRP1* gene can complement *E. coli* cells defective in exonuclease III (*xth*$^-$) and endonuclease IV (*nfo*$^-$), it is likely that one in vivo role of Rrp1 is as an endonuclease (191a).

The reaction promoted by Rrp1 appears catalytic, and demonstrates a pairing bias for the 5′ complementary DNA strand, observations that are explained by its nucleolytic activity (186, 187, 191). The maximum extent of DNA heteroduplex formation is ~600 bp (186), and joint molecule formation requires as little as 13 bp of homology (33). This result suggests that as few as 13 nt need to be resected for pairing activity, a value that is in good agreement with the *S. cerevisiae* Sep1 and DPA data. The pairing activity also correlates with the ability of the purified protein to aggregate ssDNA, conditions that would encourage renaturation of ssDNA with the resected dsDNA. Deletion of the carboxyl terminus results in a 452-amino-acid polypeptide that retains renaturation, but not nuclease, activity; this truncated protein cannot promote homologous pairing between ssDNA and linear dsDNA substrates unless ssDNA tails at least 35 nt long and of either polarity are produced by an exogenous nuclease (191). Complete strand displacement could occur if the displaced strand is <400 nt long; the observation that histone H1 did not promote DNA strand displacement suggests that Rrp1 may facilitate this branch migration phase.

Human HPP-1

A homologous DNA pairing protein (HPP-1; M_r 130,000) was purified from human T cells (192, 193). Its pairing activity is ATP independent, apparently requires a 5′ complementary strand in the dsDNA, and proceeds 3′→5′ relative to the displaced strand; all three observations can be explained by the presence of a trace 3′→5′ exonuclease activity. Joint molecule formation between circular ssDNA and linear dsDNA occurs within a few minutes, but only ~6% of the DNA is converted into a product with more than 7 kb of heteroduplex DNA. Extension of DNA heteroduplex proceeds at a rate of 2 nt s^{-1}. Optimal pairing requires as little as one monomer per 25 bp dsDNA, but both the rate and extent are dependent on the concentrations of protein and DNA (192, 193). Even though HPP-1 appears to bind ssDNA cooperatively, it does not form extensive nucleoprotein filaments.

HPP-1 is found associated in a 500-kDa recombination complex at earlier steps in the purification (192). This complex is also proficient in DNA strand exchange, but the reaction promoted by the complex fraction is both ATP-dependent and catalytic. Interestingly, HPP-1 binds the photoaffinity analog, 8-azido ATP. Thus, it is possible that, in vivo, HPP-1 functions as part of a complex that utilizes ATP (possibly that bound by HPP-1) to promote DNA strand exchange. The presumptive ATP-dependent factor may assist in HPP-1 turnover, since the rate of strand exchange in crude extracts is 10-fold faster.

One component of the 500-kDa complex has been identified as the human SSB protein, hRPA (194). Although normally isolated as a heterotrimer of 70-, 32-, and 14-kDa subunits, only the large and small subunits are present in the 500-kDa complex. Addition of stoichiometric amounts of this protein decreases the amount of HPP-1 required for strand exchange 10-fold and increases the rate of the reaction >50-fold. The 70-kDa subunit alone stimulates the reaction, although the stimulation is somewhat greater when purified hRPA is used. Since this effect is specific (neither SSB protein, G32P, nor *S. cerevisiae* yRPA substitutes for hRPA), it is likely that direct interactions between the hSSB protein and HPP-1 facilitate the strand-exchange reaction.

Another protein that both renatures complementary ssDNA and forms joint molecules between circular ssDNA and linear dsDNA was partially purified from human B cells (195, 195a). Nonhydrolyzable ATP analogs inhibit the reaction by up to 50%. The paired DNA molecules contain limited regions of heteroduplex DNA and initiate at the 5′ end of the complementary strand in the linear dsDNA. Initial preparations degraded 30–45% of the 3′-end label, raising the possibility that the nuclease-reannealing mechanism for pairing applies to this activity as well (195); however, a recent preparation promotes homologous pairing without detectable nuclease or ATPase activity (33).

Vaccinia v-SEP

An extract from vaccinia virus–infected HeLa cells was partially purified and shown to promote ATP-independent DNA strand exchange (196). The active fraction (v-SEP) contains three predominant polypeptides with apparent molecular weights of 110, 52, and 32 kDa, but it is not known which protein(s) promote(s) the reaction. v-SEP possesses DNA renaturation and dsDNA exonuclease activities, consistent with the basic requirements common to the ATP-independent DNA pairing reactions. Presumably as a result of the polarity of the exonuclease activity, pairing proceeds unidirectionally (5′→3′). A displaced strand as long as 3 kb was detected by EM. Whether this DNA strand displacement step is protein promoted is unknown.

Herpes Simplex ICP8

The 138-kDa ICP8 was purified from herpes simplex virus-1 (HSV-1)-infected cells (197). It binds ssDNA, but not dsDNA, with a stoichiometry of ~10 nt per monomer (197, 198), and forms protein filaments in the absence of DNA (197). When dsDNA is added to ICP8-ssDNA complexes, homology-dependent pairing occurs. Strand exchange is less efficient than that promoted by RecA protein; transferred fragments of 1 kb are detected, but exchange of molecules the size of full-length M13 is not detected. Direct measurement of nuclease activity detected low levels of activity (~0.3 nt per 5′ end), but other evidence suggests that pairing proceeds by an annealing mechanism. First, blunt-ended DNA is fourfold less efficient than molecules with 1–2-nt overhangs at forming joint molecules. Second, a maximum of ~35% of the substrate forms joint molecules. Finally, ICP8 does not form D-loops with supercoiled DNA.

Human p53 Protein

The p53 tumor-suppressor protein binds preferentially to the ends of ssDNA, and it promotes both DNA renaturation (198a, 198b) and DNA strand transfer (198b). In the presence of Mg^{2+}, p53 protein renatures both DNA and RNA. DNA renaturation is inhibited by all of the NTPs examined, suggesting that they either are allosteric effectors or occupy a DNA-binding site. DNA strand transfer between a variety of oligonucleotides up to 70 nt in length was examined in the absence of Mg^{2+} (198b). Transfer of DNA strands occurred in a protein concentration–dependent reaction that was not inhibited by DNA mismatches as long as 4 contiguous nt. As with DNA renaturation, GTP inhibited the DNA strand transfer reaction, and mutant p53 proteins that bind ssDNA poorly fail to promote DNA strand transfer. The DNA strand-exchange characteristics of this protein, particularly the need to omit Mg^{2+}, are unlike those of any ATP-independent DNA strand-exchange protein summarized herein, but the short length of oligonucleotides used combined with the preferential binding of p53 protein to ssDNA suggest the possibility that p53 protein is a helix-destabilizing protein and is simply melting the dsDNA; reannealing could occur upon deproteinization without the need for a concerted strand transfer event (see discussion in section on *Experimental Assays*). The only other report of DNA strand transfer in the absence of Mg^{2+} was by histone H1 protein, which was ultimately traced to an artifactual denaturation event (71, 72).

STRUCTURAL HOMOLOGS OF RecA PROTEIN

As indicated in the previous section, pairing activities isolated from eukaryotic cells promote homologous pairing and joint molecule formation in the absence of either ATP binding or hydrolysis. However, joint molecule formation by nearly all ATP-independent proteins is absolutely dependent on either the presence of an exonuclease activity or the resection of the ends of the duplex DNA, suggesting that the mechanism of pairing is based upon reannealing of ssDNA regions rather than upon strand invasion. This observation raises the question of whether homologous pairing and DNA strand-exchange proteins exist in eukaryotic organisms that function by a mechanism of DNA strand exchange as defined by *E. coli* RecA and T4 phage UvsX proteins.

Recently, eukaryotic proteins having sequence and structural homology to RecA protein (see below) have been identified. Their sequence similarity

Table 2 Structural homologs of RecA protein

Protein	Organism	M_r[a]	A.a.[b]	Sequence comparison Region[c]	Cons.[d]	Ident.[e]
Rad51	*S. cerevisiae*	43	400	33–240	59	33
Rad55	*S. cerevisiae*	46	406	39–229	35	17
Rad57	*S. cerevisiae*	52	460	39–229	36	19
Dmc1	*S. cerevisiae*	37	334	24–262	67	26
				33–240	50	27
Rad51-like	*S. pombe*	—	365	33–240	53	30
Mei3	*N. crassa*	27[f]	266[f]	33–256	54	25
Rec2	*U. maydis*	84	781	40–85	51	40
				140–157	61	33
Lim15	*L. longiflorum*	—	349	33–240	—	29
RecA-like	*A. thaliana*	42[g]	387[g]	1–352	53	20
Rad51-like	chicken	38	339	33–240	—	26
Rad51-like	mouse	38	339	1–303	55	30
				33–240	51	28
Rad51-like	human	38	339	31–260	56	30
				33–240	51	28

[a] Expressed in kDa.
[b] Number of amino acid residues.
[c] Region of RecA protein used in sequence comparison analysis.
[d] Percent conserved (similar and identical) amino acid residues between indicated protein and RecA protein within the stated region of RecA protein.
[e] Percent identical amino acid residues between indicated protein and RecA protein within the stated region of RecA protein.
[f] These values are based on the open reading frame identified by Cheng et al (210). It has been noted that sequences upstream of the reported coding region also contain homology to Rad51 protein (R Rothstein, personal communication); therefore, it remains unclear whether the true size of Mei3 protein is larger than that reported.
[g] Estimated size of the mature product after cleavage of the chloroplast transit peptide.

to RecA protein is summarized in Table 2. While limited biochemical data is available for these proteins and none have yet been shown to possess DNA strand-exchange activity, their characteristics allude to the universality of the prokaryotic paradigm.

Saccharomyces cerevisiae *Rad51 Protein*

Mutations in *RAD51* display recombination and repair defects consistent with the involvement of this protein in DNA strand exchange [reviewed in (199)]. In a strain containing a *rad51* null mutation, the double-strand breaks with processed ssDNA tails indicative of the earliest steps of recombination are formed, but these intermediates are not processed to produce recombinants (140).

The central portion of the Rad51 protein (M_r 42,961) is 30% identical and 24% similar to that of RecA protein. Included in this region is the nucleotide-binding fold of RecA protein, a motif that is conserved in Rad51 protein (124). Mutation of the conserved lysine residue (Lys191) in the polyphosphate-binding loop of Rad51 protein results in a null phenotype (140), as it does for *E. coli* RecA protein (50, 200) (WM Rehrauer, SC Kowalczykowski, unpublished observation).

Rad51 protein was first purified from *E. coli,* although it has subsequently been purified from yeast with no apparent difference in biochemical activity (140, 201, 201a). Like RecA protein, this protein binds both ss- and dsDNA, and the protein concentration dependence displays sigmoid behavior, which indicates that the active form of this protein may be a RecA protein–like filament. DNA binding is stimulated by the presence of ATP and saturates at a stoichiometry of 2 nt per monomer, and the structure of the Rad51 protein–ssDNA complex undergoes a conformational transition upon ATP binding (140), properties that are characteristic of RecA protein–ssDNA complexes (51). Despite these indications that Rad51 protein might form filaments on ssDNA, EM imaging showed only filaments formed on dsDNA. Image reconstruction from electron micrographs of these Rad51 protein–dsDNA complexes finds that the structures of the prokaryotic and eukaryotic filaments are highly conserved (201). The RecA protein–ATP–dsDNA filament is right handed, has a pitch of 92–97 Å, a 5.1 Å axial rise, and 18.6 bp per turn, parameters that are nearly identical for the Rad51 protein–ATP–dsDNA filament (a pitch of 99 Å, and the same 1.5-fold-increased axial rise and number of bp per turn). No structure was observed for the ~120 amino acids at the amino terminus of Rad51 protein that are not present in RecA protein, suggesting that this region of unknown function is disordered. Genetic analysis suggests that the functional form of Rad51 protein in vivo is a filament, as many mutations in *RAD51* display codominant behavior (201b), a behavior manifest by *recA* mutations (see Ref. 76). Yet, despite these similarities to RecA protein, studies have failed to identify other activities, such as DNA renaturation and (co)aggregation, that are normally associated with RecA protein function. Initially, Rad51 protein displayed no detectable ATPase activity (140); however, it was recently reported that Rad51 protein hydrolyzes ATP in a ssDNA-, but not a dsDNA-, dependent manner at a rate that is about one-fourth that of RecA protein (201a). No conditions have yet been found that allow Rad51 protein to promote DNA strand exchange (140, 201a).

A possible reason for the inability to detect Rad51 protein–mediated DNA strand-exchange activity is that the protein functions as one necessary but insufficient component of a complex. The ability of Rad51 protein to act in the presence of ssDNA-binding proteins such as yRPA has not been reported, but other studies suggest that Rad51 protein may comprise only part of a recombination-promoting complex, perhaps similar to the T4 system. Using affinity chromatography, Shinohara et al demonstrated that Rad51 and Rad52 proteins interact (140). Genetic evidence confirms a direct functional interaction between these proteins (202) (C Bendixen, R Rothstein, personal communication; JH New, SC Kowalczykowski, unpublished observations). The *RAD52* gene lies in the same epistasis group as *RAD51* and was reported initially to have no biochemical activity (140). But recently it was found that Rad52 protein binds both ssDNA and dsDNA, and that it can promote ATP-independent renaturation of ssDNA (201a). Perhaps somewhat surprisingly, it can also promote DNA strand exchange between circular ssDNA and linear dsDNA molecules, although the efficiency is only 5% of the reaction promoted by RecA protein; the Rad52 protein–promoted reaction is not enhanced by ATP. It is possible that the interaction of one or more of the other proteins in this epistasis group (Rad50, Rad54, Rad55, Rad57) with Rad51 and/or Rad52 proteins may be required for efficient DNA strand exchange to occur. In fact, both Rad55 and Rad57 proteins (52 and 53 kDa, respectively) have homology to RecA protein (203, 203a). As with Rad51 and Dmc1 proteins (see below), the most highly conserved region of these proteins is in the nucleotide-binding fold, a region referred to as Domain II (15a, 201a). Outside of this region, Rad55 and Rad57 proteins share no additional similarity to other RecA-like proteins. No biochemistry has been reported for either protein.

Saccharomyces cerevisiae *Dmc1 Protein*

The *DMC1* gene was identified as a cDNA clone derived from a transcript that was meiosis specific and that, when disrupted, was essential for meiosis (204). It was also identified by cross-hybridization to a *Lilium longiflorum* cDNA clone, *LIM15*, which is specifically transcribed at meiotic prophase (204a). Genetic experiments indicate that *DMC1* is involved in the early stages of recombination. As in *rad51* cells, exonucleolytically processed double-strand breaks persist in *dmc1* strains. These and other effects of *dmc1* mutations are only observed in meiotic cells.

The Dmc1 protein sequence has a significant homology to RecA protein, particularly in the hydrophobic core and the nucleotide-binding domain (124, 204). Sequence similarity diverges at both the amino and carboxyl termini of the proteins. Greater similarity is found between Dmc1 and Rad51 proteins; their sequences are 45% identical, again primarily in the central region of the proteins. Even greater similarity exists between Dmc1 and Lim15 proteins (48%); similarities are found not only in the central core region [Domain II (15a, 201a)], but also in the amino-terminal Domain I, suggesting that these two proteins represent a subclass of the RecA-like proteins (15a, 201a).

Analysis of the sequence similarity between Dmc1, RecA, and UvsX proteins makes a case for the conservation of function as well (124). A 288-amino-acid region of Dmc1 protein shares 26% identity and 41% similarity with the central portion of RecA protein (204). Modeling of the Dmc1 protein sequence on the crystallographic structure of RecA protein reveals several classes of identical or conserved residues (124). This analysis suggests that the fundamental properties of Dmc1 and RecA proteins are similar, but until demonstrated biochemically, such a proposal must be viewed as a hypothesis.

Arabidopsis thaliana *RecA-like Proteins*

Two RecA protein homologs have been identified in *Arabidopsis thaliana*. One was identified on the basis of both its immunological crossreactivity to the *E. coli* RecA protein and hybridization to the cyanobacterial *recA* gene (142), while the other (Drt100) was isolated on the basis of its ability to complement partially the recombination and repair defects of *recA*$^-$ uvrC$^-$ phr$^-$ *E. coli* cells (205). Both proteins appear to be targeted to the chloroplast, but they are distinct in sequence. The former shows 61% sequence identity to the *Synechococcus* RecA protein and 52–57% identity with 20 other prokaryotic RecA proteins; the sequence conservation is primarily in the central core, with poor conservation at the amino and carboxyl termini (142). Drt100 shows no identity to RecA proteins but does possess a consensus ATP-binding site (205).

Other Rad51-like Proteins

Homologs of Rad51 protein have been identified in *S. pombe* (201a, 206, 206a), chicken (207), mouse (206, 208), and human (206, 209) cells using probes derived from *RAD51* sequences. The fission yeast protein is 30% identical and 53% similar to RecA protein, and 69% identical to Rad51 protein. The chicken protein is 68% and 49% identical to Rad51 and Dmc1 proteins, respectively, and 95% identical to its mammalian counterpart. It is found in lymphoid tissue and germ cells, suggesting that it is involved in not only DNA repair but also recombination. The mouse protein is 83% and ~55% homologous to Rad51 and RecA proteins, respectively. It is expressed in spleen and intestine as well as lymphoid tissue and germ cells, suggesting that it is involved in both immunoglobulin and general recombination. When expressed in *S. cerevisiae,* this protein partially suppresses *rad51* defects (208). The human protein is 83% homologous (67% identical) to Rad51 protein and 56% homologous (30% identical) to RecA protein. The region encompassing the nucleotide-binding fold (Domain II) of each of these homologs displays the greatest amino acid conservation. However, in addition, these proteins show conservation of their amino-terminal Domain

I sequences that is distinct from the sequences of both the Dmc1 and Lim15 protein subclass and the Rad55 and Rad57 proteins (15a, 201a). Use of the mouse *RAD51* cDNA probe reveals a single genetic homolog in sources as diverse as human, chicken, rabbit, pig, snake, turtle, frog, swellfish, sea urchin, mussel, lamprey, fruit fly, and tobacco (206, 208).

A RecA protein structural homolog has also been identified in *Neurospora crassa* (210). Mutations in the gene encoding this protein, *mei3,* have recombination and repair defects. This protein is smaller than the other RecA protein homologs (27 kDa), but retains the central hydrophobic core and nucleotide-binding domains. Mei3 protein has 27% identity to RecA protein over a 214-amino-acid region, but has much greater identity to Rad51 protein (73% over 260 amino acids).

PROTEIN-FREE DNA PAIRING AND DNA STRAND EXCHANGE

The complexity of the protein-promoted DNA strand-exchange reactions belie their physicochemical simplicity. All of the processes promoted in vivo by proteins occur in vitro in the complete absence of proteins. As for all protein-promoted reactions, the intracellular process benefits from both increased rates and control of the reactions. However, analysis of the protein-free reactions offers the benefits of simplicity and of physical insight into the underlying mechanism.

DNA Renaturation

Nearly every protein discussed above possesses the ability to renature complementary DNA strands. Protein-free renaturation of DNA is usually a second-order kinetic process that is limited by at least two important characteristics of DNA: electrostatic repulsion between charged phosphates and secondary structure [see (211)]. Proteins could mitigate either or both limitations. Charge repulsion can be minimized by reagents as simple as inorganic salts, but positively charged alkyl detergents (e.g. dodecyl- and cetyltrimethyl-ammonium bromide) are far more effective reagents. Though DNA renaturation in the presence of these detergents is still a second-order process, the rate of annealing is more than 2000-fold faster than the reaction in 1 M NaCl due to weak favorable interactions between the detergent-coated DNA molecules (212). Condensation of DNA into aggregates by agents such as poly(ethylene oxide), sodium dextran sulfate, phenol-salt emulsions, NaCl in the presence of ethanol, spermine, spermidine, and hexaminecobalt (III) ion greatly increases renaturation (213). DNA-binding proteins not only reduce charge repulsion, but eliminate secondary structure as well; the rapid second-order reactions catalyzed by SSB protein and G32P are notable examples (113, 114). Other proteins that bind DNA [transcription factor IIIA (TFIIIA) (214), histones (215), ribonucleoprotein A1 (216), fatty acid synthase (177), and the "model" protein polylysine (215)] also renature ssDNA. Thus, agents that increase the local concentration of DNA by either miminizing repulsive interactions or by introducing weak favorable interactions enhance DNA renaturation; hence it should be no surprise that many proteins promote DNA annealing.

Since DNA renaturation is promoted by many reagents, both protein and nonprotein, it is reasonable to ask whether a special characteristic distinguishes the renaturation activity of a protein as being exclusive to (and perhaps important to) homologous recombination. Two proteins known to be essential to recombination, *E. coli* RecA protein and λ phage β protein, promote DNA renaturation by a first-order process, but this feature is not unique: TFIIIA-dependent DNA renaturation is also first order. From examination of the many proteins described in this review, the only property unique to recombination proteins is the ATP stimulation of renaturation displayed by RecA-like proteins.

D-loop Formation

The uptake of ssDNA fragments by homologous supercoiled DNA can occur independent of protein (57). This process requires an optimal temperature (75–78°C) about 5°C below the T_m. At 37°C, the rate of D-loop formation is ~100-fold slower but is nevertheless detectable. The apparent equilibrium constant for D-loop formation is favorable and has a value of about 10^6 M^{-1} (57). Thermodynamic analysis shows that the reaction is driven by the entropy increase associated with loss of superhelical turns. The rate-limiting step exhibits positive entropy and enthalpy of activation, suggesting that this step involves the unstacking of a few basepairs in the dsDNA (57). Knowledge that the rate-limiting step requires dsDNA opening suggests that this step is a candidate for acceleration by a catalyst; this prediction is consistent with the ability of RecA protein to unwind dsDNA in anticipation of the strand-exchange step.

DNA Strand Exchange

Like renaturation, DNA strand exchange requires that the impediment to bringing two DNA molecules into proximity be overcome; but unlike DNA renaturation, DNA strand exchange requires destabilization of the dsDNA. Ostensibly, DNA strand exchange would seem too concerted a process to be catalyzed at room temperature without the intervention of proteins; this, too, is not the case. Condensation of DNA into aggregates by 15% poly(ethylene oxide) and 0.3 M NaCl results in DNA strand exchange between circular viral ssDNA and duplex DNA fragments as large 2748 bp

(213). Condensation is proposed to bring DNA molecules into proximity and to cause destabilization of the dsDNA. DNA strand exchange of a 240-bp fragment is less demanding, occurring at or above either 1% polyethylene glycol or 1 M NaCl (217). It was suggested that RecA protein–promoted DNA strand exchange proceeds by a similar mechanism [i.e. RecA protein both increases the local concentration of the DNA partners and destabilizes the dsDNA, permitting exchange to occur (213)]. In agreement with this physical chemical analogy, RecA protein is highly proficient at pairing DNA molecules, regardless of sequence, within the confines of the presynaptic filament, and is also able to unwind and unstack dsDNA, distortion that clearly destabilizes duplex structure (see section on *E. coli* RecA protein).

Exchange of DNA strands as measured by branch migration of DNA joined by reannealing of ssDNA can also be promoted by a passive ingredient such as bovine serum albumin, suggesting that volume-excluding reagents may promote this kind of exchange as well (217).

Part of the problem encountered when bringing two DNA molecules into close proximity is electrostatic repulsion. Certainly one function of a DNA strand-exchange protein must be to minimize this repulsive interaction. This condition is also required for protein-independent DNA collapse. A DNA analog, polyamide nucleic acid (PNA), that has an uncharged polyamide backbone is useful in studying the effects of charge repulsion. ssPNA not only recognizes its complementary sequence in dsDNA, but it also spontaneously displaces the identical strand in the dsDNA to form a D-loop structure (218). This structure is stabilized by the high stability of PNA-DNA hybrids, and the enhanced rate of formation argues that the electrostatic repulsion encountered by normally charged ssDNA is no longer rate limiting. This result, which agrees with experimental conditions that promote DNA condensation, argues that an important function of DNA strand-exchange proteins is to facilitate an increase in local DNA concentration by masking the strong electrostatic repulsion that occurs when DNA molecules are brought within 10 Å of one another.

CONCLUSIONS

The universal prevalence of proteins that can promote homologous pairing and the exchange of DNA strands argues for their definition as a new class of proteins. The ease with which these proteins locate DNA sequence homologies is both unique and remarkable. At present, these proteins comprise two classes that are distinguished by the need for or independence from ATP in their action; it is likely that most of the as-yet-uncharacterized

RecA-like proteins will prove to be members of the first class. The ATP-dependent proteins have, as a distinguishing characteristic, the ability to initiate pairing and exchange between ssDNA and dsDNA molecules, whereas the ATP-independent proteins are limited to the initiation of pairing between two ssDNA molecules. Both classes of protein promote a DNA heteroduplex extension phase, but only the ATP-dependent proteins have an absolute intrinsic polarity for this exchange. Interestingly, proteins that permit pairing between two fully intact and unperturbed duplex DNAs have not yet been described.

Although it is convenient from an organizational view to group these proteins into just two classes, it is perhaps more interesting to recognize that within each group, unique variations exist. For example, within the ATP-dependent class, in contrast to the *E. coli* RecA protein, we find a protein (T4 phage UvsX protein) that utilizes a third factor (UvsY protein) as an essential component of the complete functional apparatus and a protein that requires dATP instead of rATP (*B. subtilis* RecA protein). Within the ATP-independent class, we find some proteins (*S. cerevisiae* Sep1 and *S. pombe* p^{140}/exoII proteins) that possess an intrinsic nuclease activity that is essential for initiation via DNA annealing, and others (*E. coli* RecT and λ phage β proteins) that recruit a second protein (RecE protein and λ exonuclease, respectively) to provide nuclease function. The mechanistic bases and functional reasons for these intriguing differences are unknown.

Despite the extensive study of these proteins, a number of very significant questions regarding their function remain, including: What is the mechanism of the homology search? What is the precise structure of the homologously paired DNA molecules? What role does ATP hydrolysis play in DNA heteroduplex extension? What is the biological function of most of the eukaryotic ATP-independent proteins? Is the mechanism by which some of the ATP-independent proteins promote DNA heteroduplex extension active or passive? What are the biochemical activities of the apparently ubiquitous eukaryotic RecA-like proteins?

It is already clear, however, that these interesting proteins will be very useful to those wishing to locate and manipulate specific DNA sites in genomes. *E. coli* RecA protein has been used to enrich for homologous DNA sequences in a genomic pool by more than 10^4–10^5-fold (219), and has been used to target unique DNA sequences for enzymatic modification (220, 221) and identification (222–224). Applications for homologous pairing proteins will continue to evolve, bolstered by further appreciation of both their enzymatic characteristics and biological behavior, ultimately permitting their use in applications as far reaching as gene replacement strategies [see (225)]. These proteins do, indeed, represent a fascinating and important group.

ACKNOWLEDGMENTS

The authors are thankful for the helpful comments or preprints provided by Dan Camerini-Otero, John Clark, Mike Cox, Nick Cozzarelli, Ed Egelman, Jack Griffith, Wolf Heyer (in particular for his careful reading and remarks), Bill Holloman, Andre Jagendorf, Arlen Johnson, Richard Kolodner, Boyana Konforti, Sue Lovett, Kevin McEntee, Tomoko Ogawa, Rod Rothstein, Miriam Sander, Dan Tishkoff, Dave Weaver, and Victor Zhurkin, as well as the members of the laboratory—Jim New, Cliff Ng, Bill Rehrauer, Bob Tracy, and Eugene Zaitsev. The studies on RecA protein in this laboratory are supported by NIH grant AI-18987.

Literature Cited

1. Clark AJ, Margulies AD. 1965. *Proc. Natl. Acad. Sci. USA* 53:451–59
2. Weinstock GM, McEntee K, Lehman IR. 1979. *Proc. Natl. Acad. Sci. USA* 76:126–30
3. McEntee K, Weinstock GM, Lehman IR. 1979. *Proc. Natl. Acad. Sci. USA* 76:2615–19
4. Shibata T, Cunningham RP, DasGupta C, Radding CM. 1979. *Proc. Natl. Acad. Sci. USA* 76:5100–4
5. Cassuto E, West SC, Mursalim J, Conlon S, Howard-Flanders P. 1980. *Proc. Natl. Acad. Sci. USA* 77:3962–66
6. Cox MM, Lehman IR. 1981. *Proc. Natl. Acad. Sci. USA* 78:6018–22
7. Roca AI, Cox MM. 1990. *CRC Crit. Rev. Biochem. Mol. Biol.* 25:415–56
8. Miller RV, Kokjohn TA. 1990. *Annu. Rev. Microbiol.* 44:365–94
9. Eggleston AK, Kowalczykowski SC. 1991. *Biochimie* 73:163–76
10. Camerini-Otero RD, Hsieh P. 1993. *Cell* 73:217–23
11. Stasiak A. 1992. *Mol. Microbiol.* 6: 3267–76
12. Cox MM, Lehman IR. 1987. *Annu. Rev. Biochem.* 56:229–62
13. Fishel R. 1991. In *Modern Microbial Genetics,* pp. 91–121. New York: Wiley-Liss
14. Griffith JD, Harris LD. 1988. *CRC Crit. Rev. Biochem.* 23(Suppl. 1):S43–S86
15. Heyer W-D, Kolodner RD. 1993. *Prog. Nucleic Acids Res. Mol. Biol.* 46:221–71
15a. Heyer W-D. 1994. *Experientia* 50:In press
16. Kowalczykowski SC. 1991. *Annu. Rev. Biophys. Biophys. Chem.* 20: 539–75
17. Kowalczykowski SC, Dixon DA, Eggleston AK, Lauder SD, Rehrauer WM. 1994. *Microbiol. Rev.* In press
18. Lloyd RG, Sharples GJ. 1992. *Curr. Opin. Genet. Dev.* 2:683–90
19. Radding CM. 1989. *Biochim. Biophys. Acta* 1008:131–45
20. Radding CM. 1991. *J. Biol. Chem.* 266:5355–58
21. Stasiak A, Egelman EH. 1988. In *Genetic Recombination,* ed. R Kucherlapati, GR Smith, pp. 265–308. Washington, DC: Am. Soc. Microbiol.
22. West SC. 1992. *Annu. Rev. Biochem.* 61:603–40
23. McGavin S. 1971. *J. Mol. Biol.* 55: 293–98
24. McGavin S. 1977. *Heredity* 39:15–25
25. Di Capua E, Müller B. 1987. *EMBO J.* 6:2493–98
26. Dombroski DF, Scraba DG, Bradley RD, Morgan AR. 1983. *Nucleic Acids Res.* 11:7487–504
27. Wilson JH. 1979. *Proc. Natl. Acad. Sci. USA* 76:3641–45
28. West SC. 1991. *BioEssays* 13:37–38
29. Howard-Flanders P, West SC, Stasiak A. 1983. *Nature* 309:1–29
30. Stasiak A, Stasiak AZ, Koller T. 1984.

Cold Spring Harbor Symp. Quant. Biol. 49:561–70
31. Griffith J, Bortner C, Christiansen G, Register J, Thresher R. 1989. In *Molecular Mechanisms in DNA Replication and Recombination,* ed. CC Richardson, IR Lehman, pp. 105–14. New York: Wiley-Liss
32. Bortner C, Griffith J. 1990. *J. Mol. Biol.* 215:623–34
33. Hsieh P, Camerini-Otero CS, Camerini-Otero RD. 1990. *Genes Dev.* 4: 1951–63
34. Rao BJ, Jwang B, Radding CM. 1990. *J. Mol. Biol.* 213:789–809
35. Umlauf SW, Cox MM, Inman RB. 1990. *J. Biol. Chem.* 265:16898–912
36. Rao BJ, Dutreix M, Radding CM. 1991. *Proc. Natl. Acad. Sci. USA* 88:2984–88
37. Menetski JP, Bear DG, Kowalczykowski SC. 1990. *Proc. Natl. Acad. Sci. USA* 87:21–25
38. Rosselli W, Stasiak A. 1990. *J. Mol. Biol.* 216:335–52
39. Hélène C. 1991. *Anti-Cancer Drug Design* 6:569–84
40. Jain SK, Inman RB, Cox MM. 1992. *J. Biol. Chem.* 267:4215–22
41. Adzuma K. 1992. *Genes Dev.* 6:1679–94
42. Rao BJ, Chiu SK, Radding CM. 1993. *J. Mol. Biol.* 229:328–43
43. Zhurkin VB, Raghunathan G, Ulyanov NB, Camerini-Otero RD, Jernigan RL. 1994. *J. Mol. Biol.* Submitted
44. Riddles PW, Lehman IR. 1985. *J. Biol. Chem.* 260:165–69
45. Cunningham RP, Wu AM, Shibata T, DasGupta C, Radding CM. 1981. *Cell* 24:213–23
46. Cox MM, Lehman IR. 1981. *Proc. Natl. Acad. Sci. USA* 78:3433–37
47. Honigberg SM, Gonda DK, Flory J, Radding CM. 1985. *J. Biol. Chem.* 260:11845–51
48. Riddles PW, Lehman IR. 1985. *J. Biol. Chem.* 260:170–73
49. Deleted in proof
50. Rehrauer WM, Kowalczykowski SC. 1993. *J. Biol. Chem.* 268:1292–97
50a. Angov E, Camerini-Otero RD. 1994. *J. Bacteriol.* 176:1405–12
51. Menetski JP, Kowalczykowski SC. 1985. *J. Mol. Biol.* 181:281–95
52. Rosselli W, Stasiak A. 1991. *EMBO J.* 10:4391–96
53. Kim JI, Cox MM, Inman RB. 1992. *J. Biol. Chem.* 267:16438–43
54. Bianchi ME, Radding CM. 1983. *Cell* 35:511–20
55. Kim JI, Cox MM, Inman RB. 1992. *J. Biol. Chem.* 267:16444–49
56. West SC, Connolly B. 1992. *Mol. Microbiol.* 6:2755–59
57. Beattie KL, Wiegand RC, Radding CM. 1977. *J. Mol. Biol.* 116:783–803
58. Bertrand P, Corteggiani E, Dutreix M, Coppey J, Lopez BS. 1993. *Nucleic Acids Res.* 21:3653–57
59. Konforti BB, Davis RW. 1987. *Proc. Natl. Acad. Sci. USA* 84:690–94
60. Register JC III, Christiansen G, Griffith J. 1987. *J. Biol. Chem.* 262:12812–20
61. Conley EC, West SC. 1989. *Cell* 56:987–95
62. Conley EC, West SC. 1990. *J. Biol. Chem.* 265:10156–63
63. Lindsley JE, Cox MM. 1990. *J. Biol. Chem.* 265:10164–71
64. West SC, Cassuto E, Howard-Flanders P. 1981. *Proc. Natl. Acad. Sci. USA* 78:2100–4
65. West SC, Cassuto E, Howard-Flanders P. 1982. *Mol. Gen. Genet.* 186:333–38
66. West SC, Countryman JK, Howard-Flanders P. 1983. *Cell* 32:817–29
67. Bianchi M, DasGupta C, Radding CM. 1983. *Cell* 34:931–39
68. Christiansen G, Griffith J. 1986. *Proc. Natl. Acad. Sci. USA* 83:2066–70
69. Chiu SK, Wong BC, Chow SA. 1990. *J. Biol. Chem.* 265:21262–68
70. Rould E, Muniyappa K, Radding CM. 1992. *J. Mol. Biol.* 226:127–39
71. Kawasaki I, Sugano S, Ikeda H. 1989. *Proc. Natl. Acad. Sci. USA* 86:5281–85
72. Kawasaki I, Sugano S, Ikeda H. 1990. *Proc. Natl. Acad. Sci. USA* 87:1628
73. Arai N, Kawasaki K, Iwabuchi M, Shibata T. 1992. *Nucleic Acids Res.* 20:3679–84
73a. Suaren J, Chalkley R. 1987. *Nucleic Acids Res.* 15:8739–54
74. Roberts JW, Roberts CW, Craig NL. 1978. *Proc. Natl. Acad. Sci. USA* 75:4714–18
75. Roberts JW, Roberts CW, Craig NL, Phizicky EM. 1978. *Cold Spring Harbor Symp. Quant. Biol.* 43:917–20
75a. Ogawa T, Wabiko H, Tsurimoto T, Horii T, Masukata H, Ogawa H. 1978. *Cold Spring Harbor Symp. Quant. Biol.* 43:909–14
76. Kowalczykowski SC. 1991. *Biochimie* 73:289–304
77. Register JC III, Griffith J. 1985. *J. Biol. Chem.* 260:12308–12
78. Lindsley JE, Cox MM. 1989. *J. Mol. Biol.* 205:695–711
79. Stasiak A, DiCapua E. 1982. *Nature* 299:185–86
80. DiCapua E, Schnarr M, Ruigrok RW, Lindner P, Timmins PA. 1990. *J. Mol. Biol.* 214:557–70
81. Brenner SL, Mitchell RS, Morrical

SW, Neuendorf SK, Schutte BC, Cox MM. 1987. *J. Biol. Chem.* 262:4011–16
82. Kowalczykowski SC, Krupp RA. 1987. *J. Mol. Biol.* 193:97–113
83. Menetski JP, Varghese A, Kowalczykowski SC. 1988. *Biochemistry* 27: 1205–12
84. Chow SA, Radding CM. 1985. *Proc. Natl. Acad. Sci. USA* 82:5646–50
85. Tsang SS, Chow SA, Radding CM. 1985. *Biochemistry* 24:3226–32
86. Kowalczykowski SC, Krupp RA. 1989. *J. Mol. Biol.* 207:735–47
87. Lavery PE, Kowalczykowski SC. 1990. *J. Biol. Chem.* 265:4004–10
88. Ramdas J, Mythili E, Muniyappa K. 1991. *Proc. Natl. Acad. Sci. USA* 88:1344–48
89. Pinsince JM, Griffith JD. 1992. *J. Mol. Biol.* 228:409–20
90. Rao BJ, Radding CM. 1993. *Proc. Natl. Acad. Sci. USA* 90:6646–50
91. Hsieh P, Camerini-Otero CS, Camerini-Otero RD. 1992. *Proc. Natl. Acad. Sci. USA* 89:6492–96
92. Kahn R, Radding CM. 1984. *J. Biol. Chem.* 259:7495–503
92a. Konforti BB, Davis RW. 1992. *J. Mol. Biol.* 227:38–53
93. Kahn R, Cunningham RP, DasGupta C, Radding CM. 1981. *Proc. Natl. Acad. Sci. USA* 78:4786–90
94. West SC, Cassuto E, Howard-Flanders P. 1981. *Proc. Natl. Acad. Sci. USA* 78:6149–53
95. West SC, Cassuto E, Howard-Flanders P. 1982. *Mol. Gen. Genet.* 187:209–17
96. Honigberg SM, Radding CM. 1988. *Cell* 54:525–32
97. Jwang B, Radding CM. 1992. *Proc. Natl. Acad. Sci. USA* 89:7596–600
98. Shibata T, DasGupta C, Cunningham RP, Radding CM. 1980. *Proc. Natl. Acad. Sci. USA* 77:2606–10
99. Muniyappa K, Radding CM. 1986. *J. Biol. Chem.* 261:7472–8
100. Egner C, Azhderian E, Tsang SS, Radding CM, Chase JW. 1987. *J. Bacteriol.* 169:3422–28
101. Heyer W-D, Kolodner RD. 1989. *Biochemistry* 28:2856–62
102. Alani E, Thresher R, Griffith JD, Kolodner RD. 1992. *J. Mol. Biol.* 227:54–71
103. Kowalczykowski SC, Bear DG, von Hippel PH. 1981. *Enzymes* 14:373–442
104. Muniyappa K, Shaner SL, Tsang SS, Radding CM. 1984. *Proc. Natl. Acad. Sci. USA* 81:2757–61
105. Tsang SS, Muniyappa K, Azhderian E, Gonda DK, Radding CM, et al. 1985. *J. Mol. Biol.* 185:295–309
106. Kowalczykowski SC, Clow JC, Somani R, Varghese A. 1987. *J. Mol. Biol.* 193:81–95
107. Chow SA, Rao BJ, Radding CM. 1988. *J. Biol. Chem.* 263:200–9
108. Menetski JP, Kowalczykowski SC. 1989. *Biochemistry* 28:5871–81
109. Lavery PE, Kowalczykowski SC. 1992. *J. Biol. Chem.* 267:9307–14
110. Lavery PE, Kowalczykowski SC. 1992. *J. Biol. Chem.* 267:9315–20
111. McEntee K. 1985. *Biochemistry* 24: 4345–51
112. Bryant FR, Lehman IR. 1985. *Proc. Natl. Acad. Sci. USA* 82:297–301
113. Alberts BM, Frey L. 1970. *Nature* 227:1313–18
114. Christiansen C, Baldwin RL. 1977. *J. Mol. Biol.* 115:441–54
115. Bryant FR, Menge KL, Nguyen TT. 1989. *Biochemistry* 28:1062–69
116. Akaboshi E, Yip ML, Flanders PH. 1989. *Nucleic Acids Res.* 17:4390
117. Eitner G, Adler B, Lanzov VA, Hofemeister J. 1982. *Mol. Gen. Genet.* 185:481–86
118. West SC, Countryman JK, Howard-Flanders P. 1983. *J. Biol. Chem.* 258:4648–54
119. West SC, Little JW. 1984. *Mol. Gen. Genet.* 194:111–13
120. Stranathan MC, Bayles KW, Yasbin RE. 1990. *Nucleic Acids Res.* 18: 4249
121. Yasbin RE, Stranathan M, Bayles KW. 1991. *Biochimie* 73:245–50
122. Lovett CM, Roberts JW. 1985. *J. Biol. Chem.* 260:3305–13
123. Fujisawa H, Yonesaki T, Minagawa T. 1985. *Nucleic Acids Res.* 13:7473–81
124. Story RM, Bishop DK, Kleckner N, Steitz TA. 1993. *Science* 259:1892–96
125. Yonesaki T, Minagawa T. 1985. *EMBO J.* 4:3321–28
126. Hinton DM, Nossal NG. 1986. *J. Biol. Chem.* 261:5663–73
127. Formosa T, Alberts BM. 1986. *J. Biol. Chem.* 261:6107–18
128. Griffith J, Formosa T. 1985. *J. Biol. Chem.* 260:4484–91
129. Yu X, Egelman EH. 1993. *J. Mol. Biol.* 232:1–4
130. Harris LD, Griffith J. 1987. *J. Biol. Chem.* 262:9285–92
131. Kodadek T, Alberts BM. 1987. *Nature* 326:312–14
132. Kodadek T, Wong ML, Alberts BM. 1988. *J. Biol. Chem.* 263:9427–36
133. Harris LD, Griffith JD. 1989. *J. Mol. Biol.* 206:19–27
134. Harris LD, Griffith JD. 1988. *Biochemistry* 27:6954–59

135. Kodadek T. 1990. *J. Biol. Chem.* 265:20966–69
136. Yonesaki T, Minagawa T. 1989. *J. Biol. Chem.* 264:7814–20
137. Kodadek T, Gan DC, Stemke-Hale K. 1989. *J. Biol. Chem.* 264:16451–57
138. Jiang H, Giedroc D, Kodadek T. 1993. *J. Biol. Chem.* 268:7904–11
139. Hashimoto K, Yonesaki T. 1991. *J. Biol. Chem.* 266:4883–88
140. Shinohara A, Ogawa H, Ogawa T. 1992. *Cell* 69:457–70
141. Cerutti H, Ibrahim H, Jagendorf AT. 1993. *Plant Physiol.* 102:155–63
142. Cerutti H, Osman M, Grandoni P, Jagendorf AT. 1992. *Proc. Natl. Acad. Sci. USA* 89:8068–72
142a. Cerutti H, Jagendorf AT. 1993. *Plant Physiol.* 102:145–53
143. Kmiec E, Holloman WK. 1982. *Cell* 29:367–74
144. Holliday R, Taylor SY, Kmiec EB, Holloman WK. 1984. *Cold Spring Harbor Symp. Quant. Biol.* 49:669–73
145. Tsukuda T, Bauchwitz R, Holloman WK. 1989. *Gene* 85:335–41
146. Holden DW, Spanos A, Banks GR. 1989. *Nucleic Acids Res.* 17:10489
147. Holden DW, Spanos A, Kanuga N, Banks GR. 1991. *Curr. Genet.* 20:145–50
148. Thelen MP, Onel K, Holloman WK. 1994. *J. Biol. Chem.* 269:747–54
149. Kmiec EB, Holloman WK. 1994. *Eur. J. Biochem.* 219:865–75
150. Kmiec EB, Holloman WK. 1983. *Cell* 33:857–64
151. Blaho JA, Wells RD. 1987. *J. Biol. Chem.* 262:6082–88
152. Kmiec EB, Angelides KJ, Holloman WK. 1985. *Cell* 40:139–45
153. Kmiec EB, Kroeger PE, Brougham MJ, Holloman WK. 1983. *Cell* 34:919–29
154. Kmiec EB, Holloman WK. 1986. *Cell* 44:545–54
155. Hall SD, Kolodner RD. 1994. *Proc. Natl. Acad. Sci. USA.* 91:In press
156. Clark AJ, Sharma V, Brenowitz S, Chu CC, Sandler S, et al. 1993. *J. Bacteriol.* 175:7673–82
157. Hall SD, Kane MF, Kolodner RD. 1993. *J. Bacteriol.* 175:277–87
158. DeLuca CL, Lovett ST, Kolodner RD. 1989. *Genetics* 122:269–78
159. Kmiec E, Holloman WK. 1981. *J. Biol. Chem.* 256:12636–39
160. Kolodner R, Evans DH, Morrison PT. 1987. *Proc. Natl. Acad. Sci. USA* 84:5560–64
161. Dykstra CC, Hamatake RK, Sugino A. 1990. *J. Biol. Chem.* 265:10968–73
162. Dykstra CC, Kitada K, Clark AB, Hamatake RK, Sugino A. 1991. *Mol. Cell. Biol.* 11:2583–92
163. Tishkoff DX, Johnson AW, Kolodner RD. 1991. *Mol. Cell. Biol.* 11:2593–608
164. Heyer W-D, Evans DH, Kolodner RD. 1988. *J. Biol. Chem.* 263:15189–95
165. Johnson AW, Kolodner RD. 1991. *J. Biol. Chem.* 266:14046–54
165a. Johnson AW, Kolodner RD. 1994. *J. Biol. Chem.* 269:In press
165b. Johnson AW, Kolodner RD. 1994. *J. Biol. Chem.* 269:In press
166. Stevens A. 1980. *J. Biol. Chem.* 255: 3080–85
167. Stevens A, Maupin MK. 1987. *Arch. Biochem. Biophys.* 252:339–47
168. Larimer FW, Hsu CL, Maupin MK, Stevens A. 1992. *Gene* 120:51–7
169. Kim J, Ljungdahl PO, Fink GR. 1990. *Genetics* 126:799–812
170. Kipling D, Tambini C, Kearsey SE. 1991. *Nucleic Acids Res.* 19:1385–91
171. Amberg DC, Goldstein AL, Cole CN. 1992. *Genes Dev.* 6:1173–89
172. Aldrich TL, Di Segni G, McConaughy BL, Keen NJ, Whelen S, Hall BD. 1993. *Mol. Cell. Biol.* 13:3434–44
172a. Kenna M, Stevens A, McCammon M, Douglas MG. 1993. *Mol. Cell. Biol.* 13:341–50
173. Szankasi P, Smith GR. 1992. *Biochemistry* 31:6769–73
173a. Bähler J, Hagens G, Holzinger G, Scherthan H, Heyer W-D. 1994. *Chromosoma.* In press
174. Heyer W-D, Rao MRS, Erdile LF, Kelly TJ, Kolodner RD. 1990. *EMBO J.* 9:2321–29
175. Norris D, Kolodner R. 1990. *Biochemistry* 29:7903–11
176. Norris D, Kolodner R. 1990. *Biochemistry* 29:7911–17
177. Käslin E, Heyer W-D. 1994. *J. Biol. Chem.* In press
178. Halbrook J, McEntee K. 1989. *J. Biol. Chem.* 264:21403–12
179. Heyer W-D, Johnson AW, Norris DN, Tishkoff D, Kolodner RD. 1991. *Biochimie* 73:269–76
180. Sugino A, Nitiss J, Resnick MA. 1988. *Proc. Natl. Acad. Sci. USA* 85:3683–87
181. Clark AB, Dykstra CC, Sugino A. 1991. *Mol. Cell. Biol.* 11:2576–82
182. Kipling D, Kearsey SE. 1991. *Nature* 353:509
183. Kipling D, Kearsey SE. 1993. *Cell* 72:12
184. Käslin E, Heyer W-D. 1994. *J. Biol. Chem.* 269:In press
185. Arai N, Kawasaki K, Shibata T. 1992. *J. Biol. Chem.* 267:3514–22

186. Eisen A, Camerini-Otero RD. 1988. *Proc. Natl. Acad. Sci. USA* 85:7481–85
187. Lowenhaupt K, Sander M, Hauser C, Rich A. 1989. *J. Biol. Chem.* 264: 20568–75
188. McCarthy JG, Sander M, Lowenhaupt K, Rich A. 1988. *Proc. Natl. Acad. Sci. USA* 85:5854–58
188a. Nugent M, Huang S-M, Sander M. 1993. *Biochemistry* 32:11445–52
189. Sander M, Lowenhaupt K, Lane WS, Rich A. 1991. *Nucleic Acids Res.* 19:4523–29
190. Sander M, Lowenhaupt K, Rich A. 1991. *Proc. Natl. Acad. Sci. USA* 88:6780–84
191. Sander M, Carter M, Huang S-M. 1993. *J. Biol. Chem.* 268:2075–82
191a. Gu L, Huang S-M, Sander M. 1993. *Nucleic Acids Res.* 21:4788–95
192. Moore SP, Rich A, Fishel R. 1989. *Genome* 31:45–52
193. Moore SP, Fishel R. 1990. *J. Biol. Chem.* 265:11108–17
194. Moore SP, Erdile L, Kelly T, Fishel R. 1991. *Proc. Natl. Acad. Sci. USA* 88:9067–71
195. Hsieh P, Meyn MS, Camerini-Otero RD. 1986. *Cell* 44:885–94
195a. Hsieh P, Meyn MS, Camerini-Otero RD. 1987. In *DNA Replication and Recombination,* ed. R McMacken, TJ Kelly, pp. 671–77. New York: Liss
196. Zhang W, Evans DE. 1993. *J. Virol.* 67:204–12
197. Bortner C, Hernandez TR, Lehman IR, Griffith J. 1993. *J. Mol. Biol.* 231:241–50
198. O'Donnell ME, Elias P, Funnell BE, Lehman IR. 1987. *J. Biol. Chem.* 262:4260–66
198a. Oberoster P, Hloch P, Ramsperger U, Stahl H. 1993. *EMBO J.* 12:2389–96
198b. Bakalkin G, Yakovleva T, Selivanova G, Magnusson KP, Szekeley L, et al. 1994. *Proc. Natl. Acad. Sci. USA* 91:413–17
199. Petes TD, Malone RE, Symington LS. 1991. In *The Molecular and Cellular Biology of the Yeast Saccharomyces: Genome Dynamics, Protein Synthesis, and Energetics,* ed. JR Broach, E Jones, J Pringle, pp. 407–521. New York: Cold Spring Harbor Lab.
200. Logan KM, Knight KL. 1993. *J. Mol. Biol.* 232:1048–59
201. Ogawa T, Yu X, Shinohara A, Egelman EH. 1993. *Science* 259: 1896–99
201a. Ogawa T, Shinohara A, Nabetani A, Ikeyu T, Yu X, et al. 1993. *Cold Spring Harbor Symp. Quant. Biol.* 58:In press
201b. Aboussekhra A, Chanet R, Adjiri A, Fabré F. 1992. *Mol. Cell Biol.* 12: 3224–34
202. Milne GT, Weaver DT. 1993. *Genes Dev.* 7:1755–65
203. Kans JA, Mortimer RK. 1991. *Gene* 105:139–40
203a. Lovett ST. 1994. *Gene.* In press
204. Bishop DK, Park D, Xu L, Kleckner N. 1992. *Cell* 69:439–56
204a. Kobayashi T, Hotta Y, Tabata S. 1993. *Mol. Gen. Genet.* 237:225–32
205. Pang Q, Hays JB, Rajagopal I. 1992. *Proc. Natl. Acad. Sci. USA* 89:8073–77
206. Shinohara A, Ogawa H, Matsuda Y, Ushio N, Ikeo K, Ogawa T. 1993. *Nature Genet.* 4:239–43
206a. Muris DF, Vreeken K, Carr AM, Broughton BC, Lehman AR, et al. 1993. *Nucleic Acids Res.* 21:4586–91
207. Bezzubova O, Shinohara A, Mueller RG, Ogawa H, Buerstedde JM. 1993. *Nucleic Acids Res.* 21:1577–80
208. Morita T, Yoshimura Y, Yamamoto A, Murata K, Mori M, et al. 1993. *Proc. Natl. Acad. Sci. USA* 90:6577–80
209. Yoshimura Y, Morita T, Yamamoto A, Matsushiro A. 1993. *Nucleic Acids Res.* 21:1665
210. Cheng R, Baker TI, Cords CE, Radloff RJ. 1993. *Mutat. Res.* 294:223–34
211. Wetmur JG. 1991. *CRC Crit. Rev. Biochem. Mol. Biol.* 26:227–59
212. Pontius BW, Berg P. 1991. *Proc. Natl. Acad. Sci. USA* 88:8237–41
213. Sikorav J-L, Church GM. 1991. *J. Mol. Biol.* 222:1085–108
214. Fiser-Littell RM, Hanas JS. 1988. *J. Biol. Chem.* 263:17136–41
215. Cox MM, Lehman IR. 1981. *Nucleic Acids Res.* 9:389–400
216. Pontius BW, Berg P. 1990. *Proc. Natl. Acad. Sci. USA* 87:8403–7
217. Kmiec EB, Holloman WK. 1994. *J. Biol. Chem.* 269:In press
218. Nielson PE, Egholm M, Berg RH, Buchart O. 1991. *Science* 254:1497–500
219. Rigas B, Welcher AA, Ward DC, Weissman SM. 1986. *Proc. Natl. Acad. Sci. USA* 83:9591–95
220. Ferrin LJ, Camerini-Otero RD. 1991. *Science* 254:1494–97
221. Koob M, Burkiewicz A, Kur J, Szybalski W. 1992. *Nucleic Acids Res.* 20:5831–36
222. Sena EP, Zarling DA. 1993. *Nature Genet.* 3:365–72
223. Jayasena VK, Johnston BH. 1993. *J. Mol. Biol.* 230:1015–24
224. Revet BM, Sena EP, Zarling DA. 1993. *J. Mol. Biol.* 232:779–91

225. Kowalczykowski SC, Zarling DA. 1994. In *Gene Targeting*, ed. MA Vega. Boca Raton: CRC. In press
226. Wu AM, Kahn R, DasGupta C, Radding CM. 1982. *Cell* 30:37–44
227. Konforti BB, Davis RW. 1990. *J. Biol. Chem.* 265:6916–20
228. Konforti BB, Davis RW. 1991. *J. Biol. Chem.* 266:10112–21

Annu. Rev. Biochem. 1994. 63:1045–83

SIGNAL TRANSMISSION BETWEEN THE PLASMA MEMBRANE AND NUCLEUS OF T LYMPHOCYTES

Gerald R. Crabtree and Neil A. Clipstone
Beckman Center, Stanford University, Howard Hughes Institute, Stanford California 94305

KEYWORDS: signal transduction, transcription, T cell activation, immunology, immunosuppression

CONTENTS

0066-4154/94/0701-1045$05.00

INTRODUCTION

Perhaps one of the greatest attractions of the T lymphocyte to those interested in signal transduction is the diverse roles of the T lymphocyte antigen receptor during development. Early in the life of a T cell, after it has migrated to the thymus, signals initiated by interaction with self-antigen begin a pathway leading to the death and elimination of autoreactive T cells, a process known as negative selection. Also in the thymus, the antigen receptor initiates a signalling pathway that results in the survival of cells having receptors capable of interacting with antigen bound in the antigen-presenting groove of MHC (major histocompatibility complex) molecules expressed on the surface of cells of the host animal—a process referred to as positive selection. Later in the life of a T cell, after it has migrated to the peripheral lymph nodes, interactions with foreign antigen through the antigen receptor initiate a pathway of differentiation, leading to immunologic activation, proliferation, and differentiation of other cells required for the immune response. How are signals from a single receptor redirected to initiate these different responses? At a descriptive level, present data indicate that the outcome of signalling through the receptor is dictated by the cellular

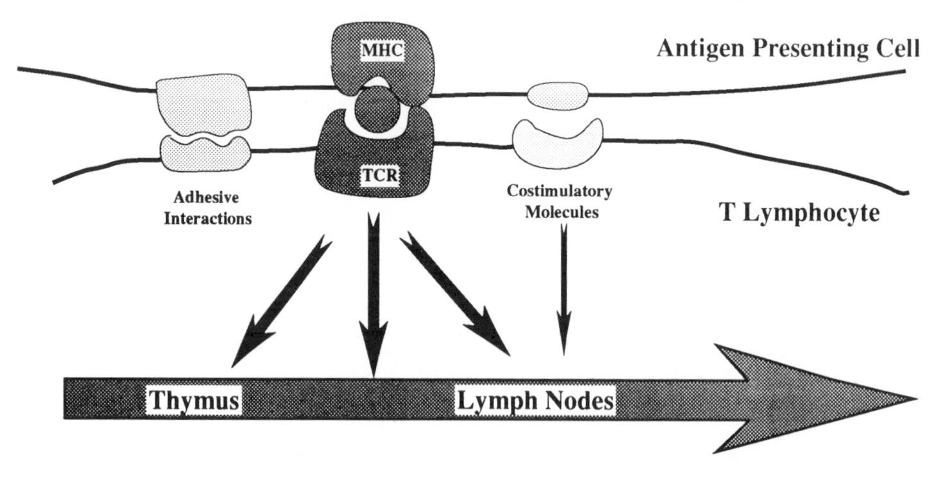

Figure 1 Cell interactive and cell autonomous pathways intersect to determine T cell fates. The interactions between an antigen-presenting cell and a T lymphocyte are depicted as intersecting with a cell autonomous regulatory pathway shown as the time line at the bottom of the figure. Outcomes of the interactions, which may have different costimulatory requirements, are shown in the boxes. The costimulatory signal is shown as coming from the antigen-presenting cell, but can also come from a different adjacent cell.

context in which the antigen receptor is activated, often referred to as costimulatory requirements, by the antigen itself, and by developmental windows resulting from cell autonomous processes (Figure 1). Other examples of a receptor determining different cell fates have been described and are instructive by analogy (1). This chapter attempts to place the recent elucidation of steps in the signalling pathway in mature T cells in the context of these more mysterious early developmental events.

T Lymphocyte Activation and its Role in Coordinating the Immune Response

The encounter between a mature lymphocyte and foreign antigen bound to the MHC molecule initiates the sequence of events leading to immunologic activation. However, the outcome of this event depends on the history of the cell. Lymphocytes that have differentiated in the thymus to bear the CD4 marker on their surface respond only to antigen bound to the MHC class II molecule. These cells are often referred to as helper cells, and in response to antigen undergo a sequence of events that results in proliferation, production of cytokines, and the exchange of cell-surface homing receptors. In contrast, cells that have differentiated in the thymus to bear the CD8 molecule on their surface respond only to antigen bound to the MHC class I molecule, and develop specialized functions such as cytotoxicity to the cell bearing the presented antigen. Each of these cell populations is essential, and the severity of diseases such as AIDS, which eliminates the CD4 cell population, illustrates the contribution of each cell type to the full immune response. Thus, the signalling pathways in mature T cells are in part determined by prior developmental events. As yet, the signal transmission/transduction pathways used in these two cell populations have not been differentiated. However, most work has been carried out using cytokine gene induction in CD4 cells as the paradigm for T cell activation, so this is the main focus of this review.

Analogy to Developmental Induction

Although the response to antigen is commonly called activation, this term belies the complexity of the process and its similarity to developmental inductive events. For example, Table 1 lists some of the molecules that are expressed de novo during T cell activation, grouped by function. What emerges from this analysis is a sense of the similarities with inductive events, in which the same general categories of molecules are expressed, and with directed morphogenic events. These similarities invite the speculation that lymphocyte activation arose by the full-blown diversion of some developmental inductive event from a process that was no longer needed at that point in development, for example some tissue or organ that

Table 1 Examples of T cell activation molecules

Transcription factors
NF-AT
NF-κB
AP-1
Nur77
c-myc
Bcl-3
Growth factors and cytokines
IL-2,3,4,5,6,10,11, and γ Interferon
Encephalin
FGF
TGFβ
Homing receptors and integrins
MEL-14
VLA-1 to VLA-5
Mediators of cell migration
IL-8
Rantes
Cell-cell contact signalling molecules
CD-40 ligand
Putative effectors of programmed cell death
Bcl-2
Fas-1
Nur77
TNFβ
Molecules associated with immunologic memory
CD-45 RO
CD-44
Cell division control molecules
Cyclins D1, D3, and E
CDC 2
PCN

disappeared near the time of the origin of vertebrates. The term activation refers to the fact that the cells become capable of carrying out immune responses, but in addition, activated T cells acquire the ability to home to particular morphologic sites, migrate in response to gradients of immune mediators such as interleukin-8 (IL-8), engage in cell-cell interactions by virtue of induced cell-surface molecules such as the CD40 ligand, and acquire certain ill-defined but possibly permanent changes in phenotype referred to as memory. The fact that these general processes are also fundamental to morphogenic events invites the comparison to a developmental inductive event rather than a transient activation. Although aspects of T cell activation are reversible, the differentiated phenotype is far more plastic than commonly expected (2), and indeed even well-differentiated cells are capable of reversion to other phenotypes by the process known as metaplasia.

EVENTS AT THE CELL SURFACE

The Antigen Receptor and its Associated Molecules

The mystery of how different antigens engage a similar receptor to initiate a program of immunologic activation and differentiation was elucidated from the structure of the subunits of the antigen receptor (Figure 2) (3–10). This receptor consists of two polymorphic chains capable of interacting with antigen when bound to self-MHC molecules of the antigen-presenting cell. These polymorphic chains are coupled tightly, but noncovalently, to the CD3 chains that have a large intracellular region and carry the signalling motifs essential for activation. Interaction of the variable chains with MHC/antigen leads to aggregation, or at least physical approximation of CD3 complexes (for reviews see Refs. 11, 12). In addition, as illustrated in Figure 2, antigen binding results in the physical approximation of lck to the CD3 complex. This occurs owing to an extracellular interaction of CD4 with MHC class II or of CD8 with MHC class I on the antigen-presenting cell, and an intracellular interaction between either CD4 or CD8 and lck. A common motif (ARAM or TAM) (13) occurs in each of the CD3 chains as well as several signalling molecules in B cells, mast cells, and macrophages (14). Mutations in this motif block the ability of the CD3 molecules to initiate signals necessary for activation. Furthermore, this region is sufficient to activate cells when fused to other non-activating receptors (15–17). Recent work has shown that this region can signal when totally within a cell, but requires membrane proximity (18). The nature of the activating event is most likely multimerization or aggregation, since dimerization of the receptor or the TAM motifs is apparently not sufficient (18). Additional evidence supporting the importance of aggregation is that TAM motifs are present in each of the CD3 chains in the antigen receptor complex, and while they appear to be somewhat redundant in function (19), the response to aggregation is proportional to the number of repeated TAMs (15, 18, 20). The importance of multimerization and the apparent inadequacy

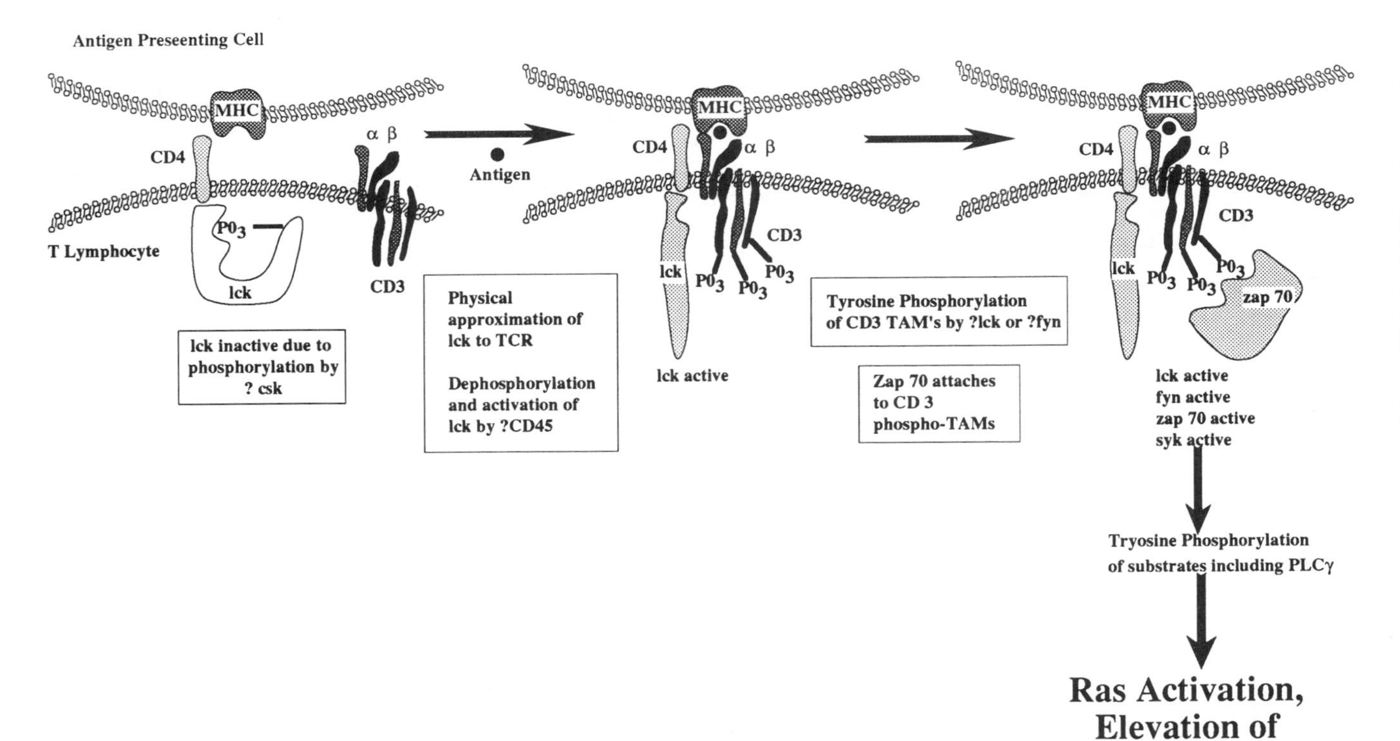

Figure 2 Membrane interactions involved in the initiation of T cell activation (current model). The antigen-presenting cell is shown at the top of the figure and the lymphocyte at the bottom. Antigen is shown as a small black circle. The boxes give interpretations of the experimental data at each stage in the process. The time elapsed in the course of these events is on the order of seconds and all of the phosphorylations shown precede calcium mobilization.

of dimerization suggest that a localization of concentrated substrate rather than the formation of a new activation surface underlies the next step in the signalling process.

The tyrosines of the TAMs in the T cell receptor (TCR) ζ chain and the CD3 γ, δ, and ϵ chains are rapidly phosphorylated preceding the mobilization of calcium and the activation of other second messengers (21). This tyrosine phosphorylation is likely to be the result of the actions of lck or fyn, which are src-like tyrosine kinases largely restricted to lymphocytes and neurons (22). Although relatively little is known about the mechanism by which the aggregated antigen receptor TAMs become subject to the activity of lck or fyn, the observation that the CD4/CD8 cytoplasmic tail binds lck (23–27) indicates that lck could both be activated and brought to a locus containing a high concentration of the unphosphorylated TAMs by the extracellular interaction with MHC (Figure 2).

Phosphorylation of the antigen receptor CD3 chains leads to the binding and the subsequent activation of the ZAP-70 tyrosine kinase (20), which is a homolog of syk, found exclusively in lymphocytes. In addition, a number of other SH2-containing proteins, including Vav, GAP, PLCγ, and GRB-2, may be bound to the TAM phosphotyrosines. However, these latter molecules are not easily seen under the usual assay conditions (20). The result of the binding of ZAP-70 is its apparent activation and phosphorylation of substrates required to continue the signal transduction cascade (Figure 2).

The importance of ZAP-70 in the signal transduction cascade has recently been underscored by the identification of a human deficiency of ZAP-70, which gives rise to a severe combined immunodeficiency phenotype with reduced numbers of $CD8^+$ cells and a profound block in both proximal and distal signalling events initiated through the antigen receptor (A Weiss, personal communication), thereby providing uniquivocal evidence for the involvement of ZAP-70 in signal transduction. A role for the syk tyrosine kinase has also been inferred from the use of chimeric molecules made up of the extracellular domain of CD16 and the intracellular domain of syk or ZAP-70. Chimeras containing syk appeared to function much better in T cells than chimeras containing ZAP-70 (20a). However, since these chimeras restrain either syk or zap, which are normally cytosolic proteins, in nonphysiologic ways near the membrane, the results are difficult to interpret fully.

There is no clear understanding of why there are so many TAMs in the TCR-associated polypeptides or why so many tyrosine kinases are required for T cell activation. However, part of the answer may lie in what is termed partial T cell activation (28). Antigens containing mutations in key contact residues often give T cell responses that are a subset of the entire activation pathway seen with the wild-type antigen (28–31). These responses can include the expression of selective growth factors or cytokines. In some cases, contact with the mutated antigen results in the unresponsiveness to later exposure to the wild-type antigen (32). These observations suggest that the antigen receptor CD3 chains may have several functions controlled by subtleties that are not yet apparent—reminiscent of observations on the roles of the different phosphotyrosines of the PDGF and CSF receptors (33, 34)—and seem inconsistent with a simple on-off switch. As yet, the signalling pathways initiated by these mutant, yet active antigens have not been traced, but such an analysis could be fruitful.

Another protein essential to activation is the tyrosine phosphatase CD45 (35, 36). This protein has been proposed to dephosphorylate lck (37–39) and perhaps fyn at their inhibitory phosphotyrosine, an event that results in the activation of their tyrosine kinase activity on ZAP-70 as discussed above. CD45 coimmunoprecipitates with CD2, TCR, Thy1, and CD4 or CD8, indicating that it is likely to be physically associated with these molecules (40–42). Although the physiologic regulation of CD45 is as yet unclear, studies using a chimeric receptor suggest that its phosphatase activity might be negatively regulated by an as-yet-unknown ligand (43, 44, 44a).

Costimulatory Signals

The interaction of the TCR with appropriately presented antigen, and subsequent clustering of the TCR/CD3 complex, are not sufficient for the initiation of physiological antigen-specific T cell proliferation (45, 46, 48, 49). Schwartz and colleagues have found that stimulation of murine T cell clones via the TCR/CD3 complex alone, in the absence of costimulatory signals, induces a state of long-lived antigen unresponsiveness known as clonal anergy (45, 46, 48, 49). The induction of anergy can be avoided by the inclusion of allogenic accessory cells, such as macrophages or B cells, during the initial stimulation. Indeed, the presence of IL-2 itself is also sufficient to overcome anergy. Although perturbation of the TCR/CD3 complex is sufficient to induce expression of the IL-2 receptor and render the T cell responsive to the effects of IL-2, other signals derived from direct T cell/accessory cell interaction are apparently necessary for the initiation of IL-2 gene expression and concomitant T cell proliferation.

Accessory cell–derived costimulatory signals appear to require intimate contact between the T cell and accessory cell, which has strongly implicated intercellular adhesion molecules with a role in these events (47). In this respect, adhesion molecules, such as LFA-1, enhance intercellular contact and increase both the surface contact area and the time of interaction between a T cell and the apposing cell, which in turn increase the chances of a productive TCR/antigen-MHC interaction. Accordingly, intercellular adhesion acts to increase the level of TCR occupancy and as a result augments the level of TCR/CD3-induced intracellular biochemical second messengers. Certain adhesion molecules are also capable of contributing to costimulatory activity by independently engaging in signal transduction events (50, 51). Thus, T cell surface molecules appear to provide costimulatory signals in one of two ways: either quantitatively by augmenting the second messengers generated by TCR/CD3 complex itself (the CD2 antigen is a good example of this class), or qualitatively, as in the case of CD28 (see below), by generating intracellular biochemical signals distinct from those generated by the TCR/CD3 complex.

In many in vitro culture systems, T cell activation is augmented by antibodies to CD28 (52–55), a cell surface molecule that interacts with a ligand, B7 (56–59), that is an intrinsic cell membrane protein. B7 is present on many cells capable of antigen processing. (The term *processing* refers to the process by which antigen is taken into the cell, proteolytically digested, and passed through the correct cellular compartments to reappear on the cell surface bound to MHC molecules; Ref 60.) CD28 contributes a signal that appears to be independent of the T cell activation pathway elicited by antigen (61, 62), and overcomes the induction of T cell anergy (45, 62a). This signalling pathway is often said to function in *trans,* in that CD28 can activate a cell even if its ligand is not exposed on the antigen-presenting cell itself but rather on an adjacent cell. Recently, the gene for CD28 was ablated in mice, and the cells from the animals found to have reduced activation in response to antigen presentation or treatment with plant lectins that aggregate cell membrane receptors (63). Otherwise, the animals were normal, suggesting that some pathway of compensation can be brought into play that can bypass the requirement for CD28. Presently, relatively little is known of the signalling pathway used by the CD28 molecule; however, it might involve tyrosine kinases (64, 65), and is insensitive to the immunosuppressive drugs cyclosporin A and FK506 (61). CD28 also appears to have a distinct response element on the IL-2 gene (66).

The reason that concurrent interactions between multiple cell membrane proteins are required to activate T lymphocytes probably relates to the deleterious consequences of indiscriminate activation, the result of which is the production of many toxic cytokines such as IL-2 and TNF and the production of cytotoxic T cells that may kill indiscriminately if not properly localized. The requirement for the delivery of two independent signals could provide a proofreading mechanism. Despite speculations on the logic underlying the requirement for several concurrently delivered signals, an analogy to developmental inductive events again emerges in that inductive signals, including spatial and timing cues, must be delivered concurrently to promote organogenesis (for reviews, see Refs. 67a–e).

TYROSINE KINASES AND THEIR SUBSTRATES

Several immediate substrates for tyrosine kinases have been identified and play clear roles in perpetuating signal transduction. PLCγ is phosphorylated on tyrosine immediately after activation (68), and is thought to initiate the breakdown of phosphatidylinositol 4,5 bisphosphate. Other substrates include the product of the *vav* proto-oncogene (69). This protein contains an SH2 domain and two SH3 domains, and is rapidly tyrosine phosphorylated after T cell activation (70–73). Initially, Vav attracted significant attention because of its suggested homology to transcription factors; however, later it became clear that this proposed homology was the result of a sequencing error. Instead, it was found that the bona fide Vav protein sequence exhibited significant homology with guanine nucleotide exchange factors (see below). Recently, the tyrosine kinases required for calcium mobilization were examined by reconstitutions in nonlymphoid cells (74). Surprisingly, $p56fyn^T$ and CD45 were necessary, but ZAP-70 and lck were not, despite the fact that they are T cell–specific. The lack of requirements for ZAP-70 and lck may reflect redundancy with other protein tyrosine kinases, such as syk and src, which are similar in structure to ZAP-70 and lck, respectively. Many other tyrosine-phosphorylated proteins appear on two-dimensional gels after T cell activation and are apparently the result of the tyrosine kinases listed above. Whether any of these molecules have functions other than leading to the mobilization of Ca^{2+}, the activation of ras, and the activation of protein kinase C (PKC) is not presently known. The possibility remains that some molecule conveys signals directly to the nucleus and thereby preserves the specificity of the signalling process.

SECOND MESSENGERS

The early observations that agents that activate PKC and that increase intracellular Ca^{2+} could lead to essentially all of the events of T cell activation, including induction of cytotoxicity, late gene functions, and the expression of markers of memory, suggested a critical role for these second messengers in relaying signals to the nucleus and initiating the program of early gene activation (75–77). More recent evidence supports the contention that an increase in intracellular Ca^{2+} and the activation of ras can initiate most of the early gene activations in T cells (78).

ras Activation in T Lymphocytes

The first demonstration that ras could be physiologically activated was in T lymphocytes (79), and led to the speculation that ras functioned downstream of PKC. However, more recent studies indicate that antigen receptor occupancy can lead to the activation of ras without PKC activation by a mechanism that likely involves protein tyrosine phosphatases (80). The mechanisms leading to ras activation have been recently reviewed (81–83). An elegant picture is emerging in which a family of linking proteins such as GRB-2 (84, 85), sem 5 (86), and Drk (87) function to bridge phosphotyrosine residues on activated receptors to a group of guanine nucleotide–releasing exchange factors or GRFs, capable of activating ras (for review see Ref. 83). These linking molecules are small proteins that bind to the tyrosine-phosphorylated receptor through their SH2 domains and that also bind proline-rich sequences in the GRF through their SH3 domain. Their function appears to be to localize the GRF to the cell membrane, where it can have access to ras. Although this mechanism seems to be used for the activation of ras by a variety of receptors in many different species, the possibility that this mechanism may be specialized is suggested from some recent studies on the Vav protein. Vav appears to be an in vitro substrate for lck (88), and also has regions that are similar to GRFs, including CDC 24, human DBL, Bcr, and CDC 24Hs, that regulate ras and ras-related proteins. Recently, the Vav protein was reported to have guanine nucleotide–releasing activity for ras (88), suggesting that it could serve the role of a more cell-type-specific activator of ras action, since it is largely restricted to hematopoietic cells. It is also possible that GRF (89) or hSos (90, 91), both of which have GRF activity and bind GRB-2, could serve this function (92, 93). The recent finding that Shc protein is both tyrosine phosphorylated and interacts with GRB-2 and the TCR ζ chain suggests that Shc functions to link ζ phosphorylation to ras activation in T lymphocytes (94).

An effector of ras action essential to development was originally identified from genetic studies in *Drosophila* and *Caenorhabditis elegans*, and found to be the homolog of c-raf (95–97). Recently, a direct interaction between activated ras and raf has been demonstrated (93, 98–100), presumably leading to the activation of raf by mechanisms that are yet to be understood. raf is both phosphorylated and activated after antigen receptor interactions (101). raf, in turn, has been shown to phosphorylate map kinase-kinase and lead to its activation (102). This signalling pathway is perpetuated to map kinase, which has been shown in one study to phosphorylate c-jun in vitro (103) and in another study to have no effect on c-jun in vitro (104). In yet another study, several protein kinases were equally able to phosphorylate c-jun, and unlike in other studies appeared not to be associated with enhanced activity (105). Other possible substrates of map kinase appear to be Elk-1, Sap-1, or p62TCF (106). Recently, another kinase implicated in the regulation of c-jun has been identified, namely c-jun kinase (Jnk). Jnk, which has been found to associate physically with the c-jun N terminus and is activated by either the ras-signalling pathway or UV radiation, phosphorylates c-jun on important regulatory N-terminal sites (106a), and appears to be a likely candidate as the physiological regulator of c-jun activity. Although the ras-activated pathway clearly controls a group of transcription factors, the details remain to be clarified.

Calcium

Activation of T lymphocytes with either antigen or TCR/CD3 mAbs results in a rapid increase in the intracellular calcium concentration (76, 107–109). This increase can be separated into two components: an initial transient peak, which lasts about 1 min, and a lower sustained plateau phase. The initial increase in Ca^{2+} is probably caused by the inositol-1,4,5-triphosphate (Ins 1,4,5 P_3)-mediated release of calcium from intracellular stores. Ins 1,4,5 P_3, which is generated from the hydrolysis of phosphatidylinositol-4,5-bisphosphate, binds to its specific receptor on the endoplasmic reticulum, eliciting a conformational change that triggers the transport of calcium from the intracellular stores into the cytoplasm (110). Unlike the release of Ca^{2+} from intracellular stores, the mechanism by which Ca^{2+} enters T lymphocytes from extracellular sources is not well understood (109). Unlike excitable cells, such as nerve and muscle cells, T cells do not express classical voltage-gated Ca^{2+} channels. In most cells, Ca^{2+} influx across the plasma membrane is regulated by the status of the intracellular calcium stores, by a process known as capacitance calcium entry (111). Potential mediators of this phenomenon have recently been described (112).

Calcium clearly plays a pivotal role in the regulation of T cell activation. This is exemplified by the inhibitory effects of depleting extracellular calcium or the use of calcium channel blockers on TCR/CD3-induced T cell proliferation (113). Although the increase in intracellular Ca^{2+} per se is apparently required for IL-2 gene expression, the component derived from intracellular stores is insufficient for activation (114). Further insight into the role of Ca^{2+} in T cell activation has been provided by studies utilizing somatic cell mutants of the Jurkat cell line, which are partially deficient in TCR/CD3-mediated signal transduction (115). The mutant cells manifest a severely attenuated increase in intracellular Ca^{2+} after TCR/CD3 stimulation, and are unable to elicit IL-2 gene expression. Further analysis of this mutant cell line revealed a deficiency in the lck tyrosine kinase (116). An important finding of this study was that the concentration of intracellular Ca^{2+} must be maintained at an elevated level for 1–2 hours for IL-2 production (115). Since the IL-2 gene is initially activated and transcription detected within 40 minutes of contact with antigen (117), these results imply that the sustained elevated calcium concentration is essential for both the initiation and maintenance of transcription (115). Thus, it appears that prolonged elevation of intracellular Ca^{2+}, rather than the initial transient increase in

Ca^{2+}, is the critical component of the Ca^{2+} signal, and that prolonged elevation of intracellular Ca^{2+} may be causally linked to IL-2 gene expression and the commitment to DNA synthesis. There are obviously many potential effectors of the Ca^{2+} signal; foremost among these, however, is the Ca^{2+}/calmodulin-regulated serine/threonine phosphatase, calcineurin. The importance of this enzyme in T cell activation is underscored by two observations. Firstly, inhibition of calcineurin by cyclosporin A (CsA) or FK506 completely inhibits IL-2 gene expression and subsequent T cell proliferation. Secondly, overexpression of calcineurins A and B dramatically increases the sensitivity of T cells to activating stimuli, and moreover, a Ca^{2+}-independent constitutively active mutant calcineurin can synergize with phorbol myristate acetate (PMA; 12-O-tetradecanoyl-phorbol-13 acetate) to activate IL-2 gene expression, thereby replacing the normal requirement for an increase in intracellular Ca^{2+} (see below).

INITIATING THE GENETIC PROGRAM LEADING TO T CELL ACTIVATION

During the 10–14-day period following the interaction of a T cell with antigen, hundreds of genes are activated, inactivated, or modified in a precisely ordered sequence (118). This process is critically dependent on the initial interactions at the cell membrane, but once established is perpetuated by largely cell autonomous events, such as the cytokine-dependent autocrine pathway that controls cell proliferation (119–121). The role of IL-2 in controlling T cell proliferation, as well as many of the late events of T cell activation, has led a number of investigators to study the control of the IL-2 gene to gain an understanding how T lymphocytes come to be committed to the program of T cell activation. Also, since initiation of transcription of the IL-2 gene is rapid, rigorously under the control of the antigen receptor, and restricted to T lymphocytes, its activation has served as a useful monitor for the successful transmission of signals from the antigen receptor to the nucleus of T lymphocytes.

Integration of the Diverse Signals Required for T Cell Activation on the Regulatory Regions of Early Genes

Initial studies employing an enhancer trap strategy found that sequences from the 5′ flanking region of the IL-2 gene were able to activate an unrelated indicator gene when transfected into murine or human T cells (122, 123). The regulatory regions were rather compact, occupying only about 275 bp of 5′ flanking sequence. Surprisingly, this region appears to contain response elements for essentially all of the known regulatory influences over IL-2 production. Three regions within the enhancer, when multimerized, activate transcription in response to antigen receptor cross-linking or to agents that both activate PKC and increase intracellular Ca^{2+} (117, 123, 124). Two regions activate transcription in response to agents that activate PKC and perhaps to cytokines such as TNF or IL-6 (123, 125). Finally, one element has been shown to be required for the enhancement of stimulation by antibodies to CD28 (66). Since deletion of any of these regions within the IL-2 enhancer greatly impairs its ability to be activated by signals from the cell membrane, the concept emerged that each of the regulatory regions must be occupied to give rise to full activation (118). This lack of redundancy in the regulatory region of the gene led to the speculation that the enhancer integrated the complex requirements for IL-2 gene activation such that several stimuli must occur nearly simultaneously to achieve activation (118). Similar conclusions have been reached from the analysis of other genes involved in controlling differentiation and development in species as diverse as yeast, fruitflies, and mammals.

The NF-AT Transcriptional Complex

Two of the regions within the IL-2 enhancer that respond to signals initiated at the antigen receptor bind a protein complex called NF-AT (nuclear factor–activated T cells) (117, 126). The binding site for the NF-AT transcription complex (Figure 3) is found within the enhancer of several early activation genes, and is required for their regulation by the antigen receptor (127–129) or by agents that both mobilize Ca^{2+} and activate PKC (117, 123). Interest in this protein complex has derived largely from a series of observations that indicate that it is a specific target for signals from the antigen receptor and appears to be responsible for the cell-type specificity of expression of the genes for IL-2 (130), IL-3 (131), IL-4 (132, 133), GMCSF, and TNF (131, 134). As the name implies, this complex is largely restricted to activated T lymphocytes and perhaps to subpopulations of B lymphocytes (130). NF-AT is a multisubunit protein consisting of a constitutive, cytosolic fraction (NF-ATc) (135) that is a very distant Dorsal/rel homolog (136), and a nuclear subunit (NF-ATn) that is rapidly induced by agents that activate PKC (135). In studies of transgenic mice, the NF-AT-binding sequence can restrict expression of a linked gene to activated T cells and to a subpopulation of B cells as well as an unknown cell type present in the dermis (130). The binding site for NF-AT has some of the characteristics of a dominant control locus (137), in that it appears to be relatively independent of its site of insertion in the genome in both T cells of transgenic mice and T cell lines (130, 138). The binding site for this protein does not activate transcription in transgenic mice in non-T cells, but seems to repress the initiation of transcription, suggesting that some negative influence functions at this site in other tissues. A silencing activity

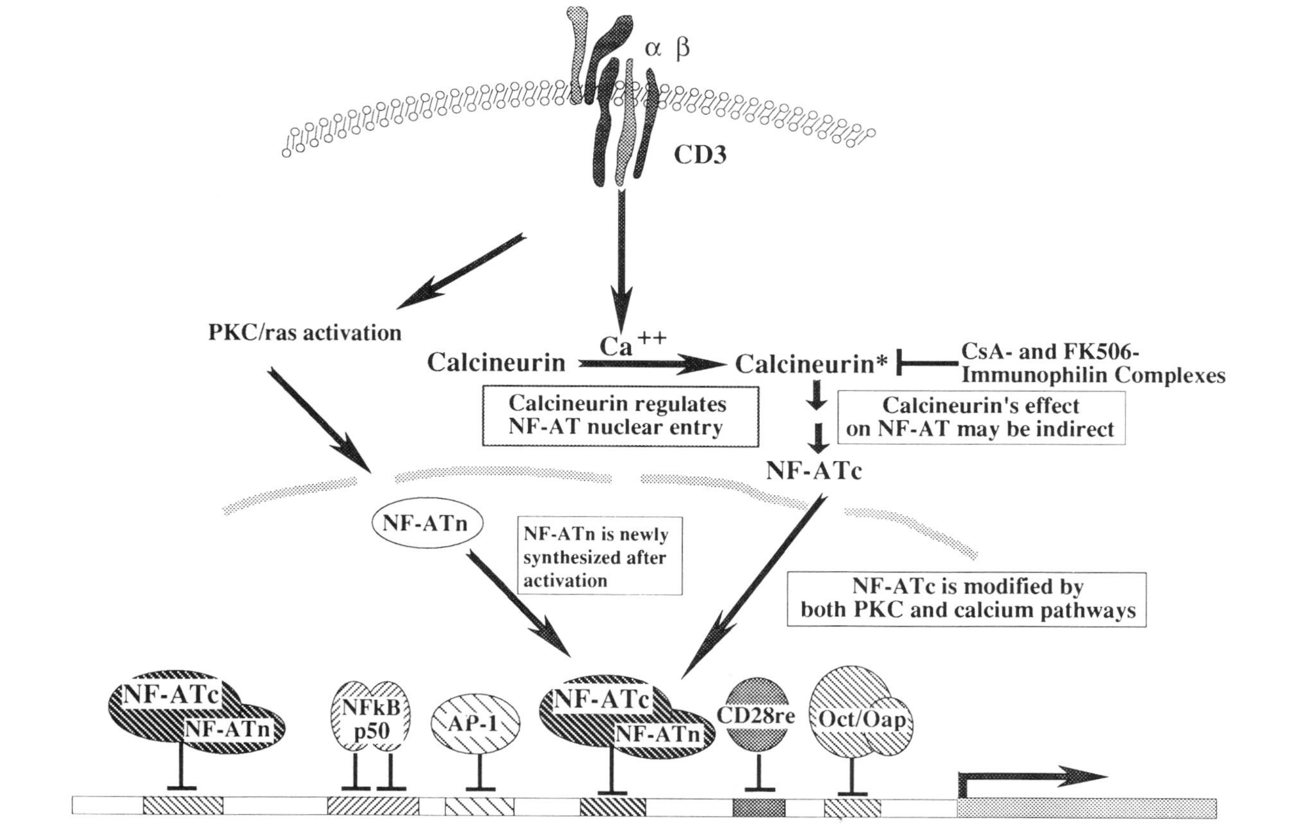

Figure 3 Signal transmission from the cytoplasm to the nucleus in T cell activation. The enhancer of the interleukin 2 gene is shown at the bottom of the figure to illustrate a typical early T cell activation gene. The ovals represent different transcription factors. p50 refers to a NFκB p50 homodimer. The events leading to the activation of NF-AT are shown. The boxes give interpretations of the experimental data. Since the activation of NFκB and AP-1 have been extensively reviewed, their activation is not illustrated.

operating at the NF-AT site has been detected in oocyte transcription assays (139, 140). However, to date, there is no molecular description of this negative influence acting at the NF-AT site. Since NF-ATc shows limited similarity to Dorsal/rel proteins, the recent discovery that Dorsal appears to acquire repressive activity by means of a spatially limited association with a repressor (141) may be instructive and point to the possibility of similar proteins in mammals that act to repress NF-AT-dependent transcription. Interestingly, γδ cells, which recognize antigen more in the manner of an immunoglobulin molecule, also activate transcription through the NF-AT site in response to correct antigen presentation (141a).

A possible silencing protein, NIL-2, acting on a site within the IL-2 enhancer distinct from the NF-AT site, was recently cloned by screening an expression library with a sequence of DNA that in some assays is associated with a silencing influence (142). However, the mechanism underlying the actions of this protein, which is present in all cell types including lymphoid cells, has not been elucidated. One possibility is that it is a universal silencer, playing a role like that of the polycomb group proteins in limiting the activity of homeotic genes during *Drosophila* development (143–146). However, NIL-2 has no obvious similarity to the polycomb group.

The subunit of NF-AT that gives the complex its tissue specificity normally resides in the cytoplasm, and hence is called NF-ATc (135) (or NF-ATp) (147) (see below). An increase in the concentration of intracellular Ca^{2+} somehow leads to the nuclear association of this subunit, which is now known to consist of at least two separate proteins, NF-ATc and NF-ATp, the products of different genes (136) (see below). As indicated below, calcineurin appears to be the effector of the nuclear import of NF-ATc; however, it is not presently known if the phosphatase activity of calcineurin is directly exerted on NF-ATc. The second subunit, NF-ATn, is rapidly synthesized de novo after treatment with agents that activate PKC (135). This nuclear subunit has not been purified or identified definitively to date, but can be replaced by overexpression of either c-jun, jun D, or c-fos proteins (148, 149). When an antibody to one of the various jun family members is used, a fraction of the total DNA-protein complex appears bound by the antibody (i.e. supershifted) (148–150), suggesting that at least a fraction of the total complex involves jun and fos proteins. Whether the NF-ATc/AP-1 complex is functional in vivo has been difficult to assess and requires additional study. To date, fos- or jun-related peptides have not been found upon sequencing preparations of NF-AT (136, 155).

Recently, a possible genetic defect in NF-AT was reported (151) in a child with severe combined immunodeficiency. The lymphocytes of the affected individual are unable to produce IL-2 and other cytokines in response to antigen receptor signalling; however, PKC is activated and [Ca^{2+}]i increases normally. Most of the immunologic defect can be overcome by administering IL-2, and hence the defect is thought to be proximal to the activation of the IL-2 gene. The IL-2 gene itself is normal. An unstable NF-AT protein appears to be at fault; however, to date the basis of the abnormality has not been demonstrated. Another individual with a defect in the signalling cascade distal to Ca^{2+} mobilization has been reported (151a), suggesting that defects, possibly involving NF-AT, may be relatively common.

NF-AT undergoes changes in expression during thymocyte maturation, suggesting that it might play a role in regulating transitions between thymic T cell subsets (152–154). It is expressed at high levels in the immature $CD4^-$, $CD8^-$ cells, disappears in $CD4^+$, $CD8^+$ cells that are the next stage of thymic maturation, and reappears in mature $CD4^+$ or $CD8^+$ cells (152, 153) as they leave the thymus. While these changes could reflect a requirement for NF-AT to direct these transitions, they could relate to changes in the proliferative status of the cells (154).

Two cDNAs for the cytosolic component of the NF-AT transcription factor have been isolated and found to be distantly related to the Dorsal/rel family of DNA-binding proteins (136). While most of the signature features of the Dorsal/rel group are present, there are significant differences: a number of charge reversals within the DNA-binding and dimerization domain and a lack of conservation of key cysteine residues, indicating that the protein is probably the most distant member of the Dorsal/rel family isolated to date. An additional difference is that the rel domain of NF-ATc is at the carboxy terminus, while it is at the amino terminus of all other Dorsal/rel proteins. The NF-ATc mRNA is present almost exclusively in T lymphocytes, consistent with its role in the tissue-specific expression of the IL-2 gene. In addition, the protein appears to be heavily posttranslationally modified, such that it migrates as a 120-kDa band on SDS gels compared with a 87-kDa predicted size. Western blots of the protein demonstrate that agents that activate PKC and agents that increase intracellular Ca^{2+} produce independent modifications of the protein, suggesting that it is a point of convergence for these signalling pathways (136). The mRNA for NF-ATc is rapidly induced in T cells and is largely T cell–specific, while the mRNA for a similar protein, NF-ATp (155), copurified with NF-ATc (136), is present in many cells and is not effected by activating signals (136). Compelling evidence that NF-ATc is active in T cells and is likely to play an essential role in IL-2 gene regulation comes from data indicating that a dominant negative of NF-ATc can be produced that blocks the activation of the IL-2 promoter after transfection into T cells (136). Expression of NF-ATc will activate the IL-2 promoter in non-T cells; however, not to the extent that it will in the Jurkat T cell line, suggesting that additional components of the signalling pathway may play a role in regulating the tissue specificity of the IL-2 promoter (136).

The finding that NF-ATc is a Dorsal/rel homolog suggests that NF-ATn may also be a Dorsal/rel homolog, since these proteins often form heterodimers. NF-ATp is about 65% identical to NF-ATc through the DNA-binding region; however, similarity in other regions cannot be fully assessed because of the fragmentary nature of the NF-ATp cDNA reported. Thus, there appears to be a subfamily of Dorsal/rel-related proteins that can bind to the NF-AT site, and that have a requirement for a ubiquitous nuclear component that well may be another member of this family or AP-1 or related protein(s).

An Association Between jun D and Oct1 Appears to Play an Important Role in Activating the IL-2 Gene

The most essential site in the IL-2 enhancer, the A site or ARRE-1 (126), was shown to bind the Oct1 transcription factor, and upon induction, with calcium ionophore and PMA bound a second protein (OAP or Octamer-associated protein) (124, 156). OAP is present in all cell types examined, and the formation of the complex between OAP and Oct1 correlated with the ability of the isolated sequence, to which it bound, to activate transcription. When the 28-bp region protected from DNase 1 digestion was ligated to a basal promoter, the sequence was found to activate transcription in response to PMA and ionomycin in several different cell types (126). The presence of OAP reduces the rate of dissociation of Oct1 from this site by about 10-fold, suggesting that either these proteins directly interact, or OAP alters the DNA configuration to favor Oct1 binding (124). OAP was purified and found to be a mixture of jun D and c-jun (157), giving the first definitive evidence for the involvement of jun proteins in T cell activation. Transfection of jun D or c-jun activates expression from plasmids that contained the Oct/OAP sequence (157). Furthermore, activation was sensitive to cyclosporin A, suggesting that calcineurin somehow regulated transcription dependent upon this site (see below). Since transcription from conventional AP-1 or Oct1 sites is not cyclosporin sensitive (158), these results suggest that either other, as-yet-unrecognized proteins are involved in the Oct/OAP complex, or that a DNA-dependent modification of jun proteins somehow selects the Oct/OAP site for activity. In this regard, Curran and colleagues have recently presented evidence that cellular protein kinases are able to distinguish DNA-bound and DNA-unbound fos/jun dimers (159). An additional difference between this composite Oct/OAP site and AP-1 sites is that transcription directed by oligomers of the sequence requires a calcium stimulus, whereas AP-1-directed transcription does not (157). A concern suggested by genetic experiments in yeast (160) is the possibility that since jun and fos are both present at high concentrations in nuclear extracts, they may bind to the transfected regulatory regions and can be easily purified, but do not actually function at this site in vivo. While transfection experiments help resolve this issue, they are subject to the same artifacts, since transfected fos and jun may bind (to sites that they do not interact with in vivo) because of their high cellular concentrations. However, consistent with a role for c-jun or jun D at this site is the observation that the inhibition of IL-2 production by glucocorticoids maps to this sequence (161). In the past, the ability of glucocorticoids to inhibit the activation of several genes has mapped to AP-1 sequences, and the glucocorticoid receptor has been shown to form a complex with c-jun that negates the ability of c-jun to activate transcription (162–164).

The Roles of NF-κB and AP-1 in T Cell Activation

NF-κB and AP-1 play important roles in activating a wide variety of genes in many different cell types. Both have been the objects of several recent reviews (165–168). In general, they probably play permissive roles, allowing a gene to be activated if other limiting cell type–specific factors are available and active. As mentioned above, the fos-jun proteins appear to account for a part of the transcriptionally active complexes functioning at both the NF-AT and OAP sites of the IL-2 promoter. In addition, a conventional AP-1 site is present within the IL-2 promoter (125) (Figure 3).

NF-κB has been found to play an important role in controlling several early T cell activation genes, including IL-2r and others (169–171). NF-κB's contribution to the activation of the IL-2 gene is not resolved, despite a number of studies (123, 126, 172, 173). Apparently, the NF-IL2D site (126) binds a p50 homodimer (172, 174), the function of which is unclear (126). p50 homodimers appear to occupy the site shortly after activation. Recently, Bcl-3, which is induced about 1 hour after activation of T lymphocytes (175, 176), has been shown to activate transcription from sites that bind p50, which normally acts as a repressor (174). Surprisingly, activation of T lymphocytes by crosslinking the ζ chain of the T cell receptor or by using antibodies to the entire T cell receptor does not activate transcription from the immunoglobulin κB sites (18), while TNF activates transcription quite well from the immunoglobulin or IL-2 receptor κB sites (177). Because of the large number of NF-κB-related proteins and their variation in concentration from one cell line to another, the analysis of several cell lines as well as normal T lymphocytes, as has been conducted for the NF-AT site (130), will probably be essential to understand their function in T lymphocyte activation.

CALCINEURIN: AN ESSENTIAL SIGNALLING INTERMEDIATE

Efforts to trace the pathways by which the immediate consequences of antigen receptor occupancy are relayed to the nucleus have been greatly facilitated by the immunosuppressive drugs, cyclosporin A and FK506. These drugs have proven to be extremely powerful therapeutic agents in both the prevention of allograft rejection after solid organ engraftment and in the prophylactic treatment of graft-versus-host disease following bone marrow transplantation (178–180). The immunosuppresive properties of CsA and FK506 can largely be ascribed to their potent inhibition of the T cell–activation-dependent transcription of T cell growth factor (interleukin 2) and other immunologically important cytokine genes, including IL-3, IL-4, GMCSF, and γ-IFN (181–184). Although structurally unrelated, both CsA and FK506 inhibit lymphokine gene expression in a similar fashion. Thus, CsA and FK506 do not inhibit the initial generation of biochemical second messengers that result from perturbation of the TCR/CD3 complex (185–189). Neither do they directly inhibit the action of the specific transcription factors that are involved in lymphokine gene transcription (158). Rather, both CsA and FK506 act at an intermediate step in the signal transduction cascade, to specifically interfere with a Ca^{2+}-sensitive T cell signal transduction pathway (190–193), thereby preventing the activation of specific transcription factors (such as NF-AT and NF-IL2A or Oct/OAP) involved in the regulation of lymphokine gene expression (135, 158, 190, 194). Consequently, these drugs have attracted much attention, since their point of action defines a critical step in the transmission of information from primary biological second messengers generated at the plasma membrane to the activation of gene expression in the nucleus.

CsA and FK506 appear to act via interaction with their cognate intracellular receptors (195–197), cyclophilin and FKBP (FK506-binding protein) respectively, collectively known as immunophilins (198). In accord with the lack of structural similarity between CsA and FK506, molecular cloning and structural studies have failed to detect any significant similarities between these two receptor families (199–201). Despite these structural differences, cyclophilin and FKBP both possess peptidyl prolyl-*cis-trans* isomerase activity, an enzymatic activity involved in the catalysis of the *cis-trans* isomerization of proline residues in polypeptide substrates (198, 202–205). Intriguingly, both CsA and FK506 specifically inhibit the isomerase activity of their cognate receptors (198, 202–205). This latter observation led initially to the notion that peptidyl prolyl-*cis-trans* isomerases play a direct role in the T cell signal transduction cascade, perhaps by catalyzing the folding of an inactive signalling intermediate into an active form, and that the inhibition of isomerase activity mediated by CsA and FK506 underlies their immunosuppressive properties. However, a number of independent observations argued strongly against the veracity of this simple loss of function model. Firstly, since the immunophilins are extremely abundant, each accounting for approximately 0.5% of total cellular protein, the concentration of drug required to inhibit T cell activation completely is significantly less than that required to saturate binding of the relevant immunophilin and hence inhibit cellular isomerase activity completely (198, 206). Secondly, Schreiber and colleagues demonstrated that each of these drugs binds to its respective immunophilin, inhibiting its activity by virtue of a twisted amide surrogate structure that acts as a transition-state mimic of a peptidyl prolyl bond undergoing isomerization (198, 207). This observation led to clear predictions about the expected behavior of other molecules containing the transition-state mimic, i.e. that they should be powerful inhibitors of signal transduction and the activation of NF-AT in the nucleus. The surprising observation that certain structural analogs of FK506, which also contained the twisted amide surrogate and potently inhibited FKBP's isomerase activity, were not inhibitors of signal transmission, but rather competitive antagonists of FK506 (195, 196, 208), led to the conclusion that inhibition of prolyl isomerase activity was not relevant to the inhibitory effects of these drugs. Thirdly, the yeast homologs of cyclophilin and FKBP are non-essential genes, and their disruption results in yeast strains insensitive to the effects of CsA and FK506, while expression of the relevant immunophilin in the mutant yeast strains once again confers sensitivity to these drugs (197, 209). Taken together, these studies strongly suggest that rather than a loss of function, CsA and FK506 impose a gain of function on their cognate receptors, resulting in the formation of an inhibitory complex that acts dominantly to interfere with a step in the signal transduction cascade.

The formulation of this inhibitory complex model was greatly aided by an analysis of the structure of the drug/immunophilin complex. This structure revealed a composite surface created by the drug bound in the active site of the enzyme (201, 210–214). This new surface composed of drug and protein was readily available for an interaction with an essential signalling molecule.

The view that the complex of drug and immunophilin per se directly interacted with a component of the signal transduction pathway and thereby inhibited signalling led to a search for a cellular factor(s) that bound specifically to drug-immunophilin complexes. The result of these endeavors was the identification of calcineurin (protein phosphatase 2B, PP2B) as the major cellular target of both CsA-cyclophilin and FK506-FKBP (215, 216). Calcineurin is a calcium/calmodulin-regulated serine/threonine phosphatase that is ubiquitously expressed, but is most abundant in brain and is highly conserved throughout evolution (217, 218). It comprises two subunits, a calmodulin-binding 59-kDa catalytic subunit (CNA) and an intrinsic Ca^{2+}-binding 19-kDa regulatory subunit (CNB), which appears to be required for enzymatic activity (217, 218). Although calcineurin appears to be the predominant calmodulin-binding protein in T lymphocytes (219), it had not previously been implicated in T cell activation and the regulation of lymphokine gene expression. Intriguingly, therefore, in vitro studies revealed that FK506-FKBP and CsA-cyclophilin complexes potently inhibited calcineurin's activity towards a well-characterized model phosphopeptide substrate, whereas either FKBP and cyclophilin alone or FKBP and non-immunosuppressive FK506 analogs were without effect (215, 220). This suggested the possibility that calcineurin might play an integral role in the T cell signalling pathway and that the immunosuppressive properties of CsA and FK506 could be explained by their inhibitory effects on calcineurin. In support of this were two findings: that the concentrations of CsA and FK506 required to inhibit calcineurin in T cell extracts were similar to the immunosuppressive doses that block lymphokine gene expression (221), and that the immunosuppressive potential of a panel of CsA/FK506 analogs correlated with their ability to inhibit calcineurin's phosphatase activity (222).

Although highly suggestive, the above in vitro studies did not definitively establish calcineurin as the physiologically relevant target of the immunosuppressive drugs. Evidence of this was provided by a series of in vivo expression studies. Firstly, overexpression of calcineurin in the CsA/FK506-sensitive Jurkat human T leukemia cell line results in a marked resistance of these cells to the inhibitory effects of CsA and FK506 (223–225). Secondly, expression of wild-type calcineurin was found to synergize with PMA-mediated signals to significantly enhance Ca^{2+}-dependent signalling in response to the calcium ionophore, ionomycin (223). Furthermore, a Ca^{2+}-independent constitutively active mutant of calcineurin was able to replace the normal requirement for ionomycin in the activation of transcription from NF-AT, OAP/Oct1 (NF-IL2A), and IL-2 promoter constructs (224, 225). Together, these results provided strong in vivo biological evidence to support the notion that the inhibition of calcineurin's enzymatic activity by drug-isomerase complexes is the molecular basis of CsA and FK506 action. More importantly, however, these data identified calcineurin as a downstream effector of the calcium signal in the T cell signal transduction cascade and provided direct evidence linking calcineurin to the activation of at least two transcription factor complexes that play essential roles in IL-2 gene expression (NF-AT and OAP/Oct1) and possibly the expression of other lymphokine genes (126, 194, 226). This established calcineurin as a nodal point in the transduction of signals from the early biochemical events that occur at the T cell membrane to gene regulatory events in the nucleus.

How exactly does calcineurin regulate the activity of NF-AT and OAP/Oct1? Although little is known about the involvement of calcineurin in the regulation of OAP/Oct1, recent results have afforded insights into the role of calcineurin in the regulation of NF-AT function. NF-AT is composed of at least two components, a pre-existing T cell–specific cytosolic component and an inducible, ubiquitously expressed nuclear component. The NF-AT cytosolic component is translocated to the nucleus in response to an increase in intracellular calcium, whereupon it interacts with the nuclear component and binds to its cognate DNA-binding site. It is this calcium-dependent nuclear translocation event that is inhibited by CsA and FK506 (135). Therefore, calcineurin probably regulates NF-AT function by controlling the nuclear translocation of NF-ATc in response to T cell activation.

The implication of calcineurin in the regulation of NF-ATc nuclear translocation suggests several independent models. Firstly, in the simplest-case scenario, calcineurin could regulate NF-ATc nuclear translocation via direct dephosphorylation of the NF-AT cytoplasmic component. Here, phosphorylation of NF-ATc would in some way preclude its entry into the nucleus; removal of these inhibitory phosphates by calcineurin following T cell activation would then allow NF-ATc to target directly to the nucleus. In this regard, the yeast transcription factor SWI5 represents a well-characterized precedent, since its cell cycle–dependent nuclear translocation is regulated by the dephosphorylation of critical serine residues and the subsequent dissolution of salt bridges with its basic nuclear localization signal (227). NF-ATc is heavily phosphorylated (NA Clipstone, GR Crabtree, unpublished observation), and its mobility on SDS-PAGE appears to be modified by calcineurin (128, 136, 147; NA Clipstone, GR Crabtree, unpublished observation). Secondly, since NF-ATc is a distant Dorsal/rel homolog (155), its nuclear translocation may be prevented in nonstimulated cells by a cytoplasmic anchor protein, whose function is regulated by its phosphorylation state. In this scheme, following T cell activation and the consequent increase in intracellular calcium, calcineurin could dephosphorylate the putative anchor protein, thereby inhibiting its function, causing it to release NF-ATc, which is then free to translocate directly to the nucleus. In addition to these two alternatives, there obviously exist more complex, less direct models of regulation; which model is operative in vivo remains to be seen. These observations have led to a conception of the mechanism by which signals from the antigen receptor activate early genes such as IL-2 in the nucleus (Figure 3).

The studies with CsA and FK506 have provided valuable insights into the T cell signal transduction cascade and the regulation of IL-2 gene expression in particular; however, they have also raised significant questions. First and foremost, do cellular analogs of CsA and FK506 exist that might provide an additional means of regulating and controlling signal transduction pathways? Or does the usurping of the immunophilins by these microbial metabolites simply reflect a quirk of nature or perhaps the evolution of a sophisticated means by which yeast can avoid the immune systems of more advanced eukaryotes? The latter is unlikely, since the microbes that produce cyclosporin of FK506 are nonpathogenic and have not been found to produce mammalian infections. Related to this is the question: What is the normal function of the immunophilins in the cell? There is compelling evidence that members of this enzyme class play a role in protein folding and might also fulfill a role as molecular chaperones (228, 229). In addition, certain provocative results—e.g (*a*) the colocalization of FKBP-12 with calcineurin in the brain (230), (*b*) the association of FKBP-12 with the calcium-release channel, ryanodine receptor (231), and (*c*) the association of immunophilins with steroid hormone receptors (232–235)—suggest that the immunophilins perhaps play a direct role in signal transduction events. In this respect, the identification of any possible cellular analogs of CsA and FK506 is clearly of the utmost importance.

Since calcineurin, cyclophilin, and FKBP are all ubiquitous proteins, how do they mediate an in vivo effect that is relatively T cell specific? Overexpression of calcineurin in the CsA/FK506-sensitive cell line Jurkat renders the cells markedly resistant to the inhibitory effects of the immunosuppressive drugs CsA and FK506 (223, 224). A priori, these findings suggest that one of the primary determinants of cellular sensitivity to these immunosuppressive drugs is likely to be the intracellular concentration of calcineurin. For example, the low levels of calcineurin expressed by T lymphocytes (219, 236) may in part explain the exquisite sensitivity of this cell lineage to CsA and FK506. Other factors that are likely to be important determinants of cellular sensitivity to these drugs include the concentration and availability of immunophilins, and the central, apparently nonredundant, role of calcineurin in T cell activation.

How does one rationalize the unseemly coincidence that two different inhibitors of prolyl isomerase activity, which are structurally very different, are both immunosuppressive? One possibility is a curious unity of purpose between yeast evolution and the pharmaceutical industry. These molecules both evolved in soil microorganisms to allow them to compete effectively against their fellow soil microbes. To do this, the microorganisms had to synthesize a molecule meeting two stringent requirements: (*a*) biologic specificity, i.e. the molecule should not be harmful to the cell that makes it, and (*b*) hydrophobicity, to enable the molecule to penetrate the cell membrane of its competitor. These two properties are also requirements of a good drug, so the pharmaceutical industry screened large numbers of yeast and microorganisms for the production of the same type of molecule selected by evolution in microorganisms. The physical chemical correlate of biologic specificity is a large surface area of interaction, necessitating a very large molecule that is unlikely to be produced by a single biosynthetic pathway. However, one solution to this problem, as exemplified by the interaction of antigen with MHC molecules, is to combine a small molecule with a larger molecule (in this case an immunophilin) and make use of the composite surface created by this interaction (198, 214, 237). The enzymatic site of either class of immunophilin is among the most hydrophobic sites within a cell (206, 212, 213). Thus, this site could readily bind small diffusible hydrophobic molecules, and also help create the composite surfaces needed to give the specificity for competition between different microbes for nutrients (i.e. it must specifically destroy its neighbors based on small differences in a protein between species) or needed by the phamaceutical industry for highly specific inhibitors. Hence, the coincidence of two structurally different immunosuppressive molecules both having the ability to bind to the hydrophobic enzymatic site of a prolyl isomerase may not be at all unlikely, in light of the large numbers of compounds screened for the ability to produce specific immunosuppression.

Calcineurin in Other Cells and Species

The establishment of CsA and FK506 as specific inhibitors of calcineurin has proven invaluable to understanding the roles of this enzyme in a variety of cellular processes. In addition to T cell activation, calcineurin plays roles in mast cell degranulation (238, 239); the regulation of sodium/potassium transport in the kidney (240), which may contribute to the significant nephrotoxicity experienced with the clinical use of these drugs; the regulation of plant potassium channels (241); and leukocyte chemotaxis (242).

Calcineurin has also been implicated in control of yeast growth. The budding yeast *Saccharomyces cerevisiae* undergoes a transient G1 cell cycle arrest in response to mating pheromones (243). Recovery from this arrested state and reentry into the cell cycle requires an increase in cytosolic calcium (244). Interestingly, disruption of yeast calcineurin genes or the addition of CsA and FK506 prevents the recovery from this mating factor–induced cell cycle arrest, indicating a role for calcineurin in the reentry of yeast to the cell cycle (245–247). In contrast to this positive role, calcineurin has recently been found also to have a negative role in yeast growth control. In this regard, Cunningham & Fink have found that mutations in the *pmc1* gene, a Ca^{2+} ATPase homolog that regulates intracellular calcium in yeast by pumping cytosolic calcium into vacuoles, when coupled with high levels of calcium in the growth media, result in growth arrest (248). Interestingly, this effect is mitigated by either CsA/FK506 or mutations in calcineurin, and requires the presence of a functional calmodulin gene. These results suggest that prolonged activation of calcineurin, as a consequence of elevated intracellular calcium, is extremely deleterious and results in a pathway leading to cell death or at least inhibition of growth. This latter result also raises caution in the interpretation of results obtained with the unregulated expression of constitutively active calcineurin, which may directly or indirectly result in cellular toxicity. Whether calcineurin plays similar roles in growth regulation in mammalian cells to those identified above in yeast is not known. However, these observations underscore an interesting parallel between the calcineurin-dependent cell death or growth inhibition observed in yeast and activation-induced cell death in T cells, which is thought to be a model of negative selection events in the thymus, and is also associated with elevated intracellular calcium and is blocked by CsA/FK506 (249–251).

Clearly, the continued use of CsA and FK506 as inhibitors of calcineurin should lead to our further understanding of the role of this important enzyme. Of particular interest, given the high level of expression of calcineurin in the brain, is the potential role that calcineurin plays in neuronal signal transduction and development.

Is Calcineurin a Rate-Limiting Step in T Cell Activation?

The observation that screening programs developed by the pharmaceutical industry for molecules that block the mixed lymphocyte reaction have repeatedly found calcineurin inhibitors suggests that calcineurin activation is a particularly critical step in the activation pathway. This notion is supported by studies demonstrating that overexpression of calcineurin leads to an increase in the sensitivity of the pathway to activating agents (223, 224). A quantitative analysis of the inhibition of calcineurin by the drug-isomerase complexes suggests that concentrations of drug that would be predicted to reduce calcineurin activity by 50% lead to a substantial reduction in signalling (221). Such predictions may in part be tested by the ablation of the calcineurin genes. However, selective and compensatory mechanisms both at the cellular and molecular levels, brought into play over the several-week period of mammalian development, may blur the interpretation of such experiments. This is likely to be the case, since calcineurin is probably required for the activity of several developmental processes prior to its role in the immune system.

The observation that expression of a constitutively active, Ca^{2+}-independent calcineurin mutant can replace the normal requirement for the increase in intracellular calcium in IL-2 gene expression indicates that calcineurin is most likely the principal effector of the calcium signal in this signal transduction pathway (224, 225). Indeed, coexpression of this calcineurin mutant together with an oncogenic ras molecule, which can replace the requirement for activation of the PKC pathway (252–254), can activate both NF-AT- and IL-2 promoter–dependent transcription in the absence of any pharmacological stimuli (78). These results indicate that the elevation in intracellular calcium and activation of PKC, which are thought essential to T cell activation and IL-2 gene expression, are largely mediated through the effects of calcineurin and of the ras pathway, respectively. This apparently nonredundant role of calcineurin provides a further explanation for the exquisite sensitivity of T lymphocytes to CsA and FK506.

Clearly, the concentration of intracellular calcium plays a pivotal role in the regulation of IL-2 gene expression and T cell activation, and its principal target, in this respect, appears to be calcineurin, whose phosphatase activity is regulated by the calcium-binding protein, calmodulin. Interestingly, analysis of the activation of calcineurin by calcium/calmodulin in vitro has shown this process to be highly cooperative. Calmodulin can exist in an inactive enzymatic complex with calcineurin. Under these conditions, only two of calmodulin's four calcium-binding sites are occupied. The interaction of calcium with the two remaining binding sites is highly cooperative (255); thus in this state, only a small incremental increase in calcium concentration is required for a large increase in calcineurin's phosphatase activity. If this scheme were operative in vivo, diffusion of calcium, rather than the interaction of calmodulin with calcineurin, would be rate limiting for the activation of calcineurin, and hence IL-2 gene expression. In this regard, it is of interest to note that at physiological calcium concentrations, because of the effects of immobile calcium-binding proteins that buffer calcium, calcium acts in tightly restricted microdomains (256). Thus, the spatial relationship of calcineurin to the site of calcium influx is likely to be very important. Interestingly, in this regard, recent data indicates that while all cyclophilins–cyclosporin A complexes can bind and inactivate calcineurin in vitro, only cyclophilins A and B—but not cyclophilin C—can mediate a biologic response (257). Preliminary studies reveal that the abilities of the different cyclophilins to mediate a differential inhibitory response may be related to their cellular locations, and hence the proximity of the drug isomerase complex to the signal transmission pathway.

SIGNALLING DEATH THROUGH THE T CELL ANTIGEN RECEPTOR: NEGATIVE SELECTION

While the pathways responsible for T cell activation are nearing a molecule-by-molecule description of how information obtained at the cell surface is transmitted to the nucleus, little is known of the pathways that initiate cell death and result in the elimination of clones of T cells responsive to self. The antigen receptor itself must play the role of determining the specificity of this process. However, two hypotheses presently coexist concerning the general logic underlying the developmental redirection of antigen receptor signals from death to activation. One view that has been in favor for many years is that the signalling pathway used by the antigen receptor is modulated by costimulatory molecules on the surface of cells found only in the thymus. In its simplest form, this theory now seems untenable, since virtually any cell can present antigen and induce death of thymus cells or antigen-specific hybridoma cells (258). This also appears to be the case for positive selection (259). Furthermore, there is now evidence that deletion of clones of antigen-specific T cells can occur outside the thymus (260).

An alternative view, which is not inconsistent with some aspects of the costimulatory model, is that lymphocytes are on a developmental pathway that allows them to become susceptible to death at a specific stage in their development in response to antigen receptor occupancy. This view proposes a largely cell autonomous pathway (conditioned by the past history of the cell) that interprets signals from the antigen receptor by virtue of the signal-transducing molecules or transcription factors expressed at different times in the life of the cell. The observation that cell death is largely independent of the nature of the costimulatory cell and that cell lines have been established that either die or reproduce the early events of activation suggests that cell autonomous pathways in part govern the outcome of signalling. However, many observations now support the role of costimulation as an important and perhaps essential feature of activation or death. A synthesis of the costimulatory model and the cell autonomous model is illustrated in Figure 1. The central feature of this synthesis is simply that the cell autonomous pathway delineated by the time line at the bottom of the figure must mesh with events at the surface of the cell, costimulatory signals and the like, to bring about a specific cell fate.

fas-Induced Cell Death

Complicating these two models is the recent realization that two different pathways for signalling cell death are used in T cells. One pathway occurs in mature cells that have been stimulated through their antigen receptor. This involves the induction of the fas or Apo-1 antigen on the surface of cells several hours after contact with antigen (261, 262). The antigen-induced expression of fas or Apo-1 leads T cells to become susceptible to death mediated by the hypothetical fas ligand. The fas "receptor" is an intrinsic cell membrane protein with similarity in the signalling domain to the TNF receptor and CD40 (263). Additional information on this pathway is provided by the two known mutations in the fas antigen and the fas signalling pathway. The *lpr* mutant alleles eliminate the function of the fas antigen by virtue of a point mutation or a truncation of the protein (for review see Refs. 264, 265). A second mutation, *gld,* believed to be in the fas ligand or a signal transduction molecule, has phenotype nearly identical to that of the *lpr* mutation (264). In each case the phenotype appears to result in the accumulation of $CD4^-$,$CD8^-$,Thy-1^+ cells in the periphery. However, there is conflicting data on the question of whether these mutations influence immunologic negative selection (266, 267). The phenotypic characteristics of these mutations indicate that loss of cells in the periphery after normal activation may be mediated by the fas antigen. While cell death due to fas is ultimately mediated by the antigen receptor, which induces expression of the fas antigen, fas may have little to do with negative selection in the thymus (267). fas has been proposed to play some role in the normal activation process, since most investigators have found that cells taken from *lpr* or *gld* mice are unable to activate the IL-2 gene in response to signals through the antigen receptor (264). However, these changes may simply be secondary effects of a mutation in a receptor that drastically alters late phases in lymphocyte development. Although the fas receptor mediates cell death when crosslinked (261), there is not clear evidence that this is its normal function. The cell death that is seen in many cell lines could be due to some fatal miscoordination of metabolic events induced by the receptor. Until the fas ligand is identified and studied more thoroughly, the normal function of the fas antigen cannot be easily inferred from the phenotype of the mutant.

Possible Signalling Intermediates in Negative Selection

Several studies have supported a role for calcineurin in negative selection. The administration of cyclosporin to animals or to humans in utero results in a high incidence of autoimmune manifestations (268–270). In addition, the elimination of clones of antigen-specific T cells is reduced drastically in mice treated in utero (271, 272). The interpretation of these experiments is clouded by the observation that cyclosporin also blocks earlier developmental steps in the thymus, leading to a reduced population of cells that are susceptible to cell death (271). Recently, tissue-culture conditions have been perfected that permit many of the aspects of negative selection to be examined in tissue culture. Surprisingly, cyclosporin has little or no effect on negative selection in vitro (273, 274). However, this may be due to the fact that cyclosporin blocks antigen-driven steps prior to negative selection.

To date, none of the genes in the signal transduction pathway that have been genetically inactivated by homologous recombination have given a phenotype indicative of a selective defect in negative selection or programmed cell death in the thymus. Mutations in the *lck* gene result in an early block in thymic development (275–277), a failure of allelic exclusion

(278), but this defect is not associated with autoimmunity. In the absence of the product of the *fyn* gene, T cells pass through thymic development normally but are defective in their ability to be activated in the peripheral lymphatic organs (279, 280) and may show a selective defect in the elimination of certain clones of T cells responsive to self-antigen. The mutations in the ζ chain of the antigen receptor result in abnormal thymic development, but T cells from these animals are still able to be activated (19, 281). Probably because of related functions performed by the TAMs on the γ, δ, and ε chains, the defect is not as severe as might be predicted from the role of TCR ζ in recruiting the ZAP-70 tyrosine kinase.

The pathway leading to negative selection does not appear to involve *bcl-2* (282, 283), which clearly protects against other types of cell death (284–286), possibly by protecting against the deleterious effects of reactive oxygen species (287). *bcl-2* is a distant homolog of the *ced9* gene (288) that protects against cell deaths in *C. elegans,* and the *bcl-2* gene will protect against certain cell deaths when put back into *C. elegans* (289). If these results can be extrapolated, it may mean that the homologs of the ced3 or ced4 proteins (290), which function in a contingent series with ced9 protein (288, 291), may not be on the pathway leading to antigen receptor–induced cell death but could quite likely be involved in other developmentally important cell deaths. However, fas-induced cell death is prevented by *bcl-2* (292), suggesting that ced3 and ced4 could also be on a fas-induced pathway leading to cell death.

Recently, dominant negative mutations of the *nur77* gene have been shown to protect against cell death resulting from antigen receptor activation of a T cell hybridoma formed between antigen-activated lymphocytes and a malignant thymoma (293). The Nur77 protein is a zinc finger protein that is rapidly induced by antigen receptor activation, and its induction is blocked by cyclosporin A and FK506. Interestingly, the promoter of the *nur77* gene contains a putative binding site for the NF-AT transcription factor (A Winoto, personal communication), suggesting the following sequence of events leading to antigen-induced cell death: TCR→calcineurin→NF-AT→Nur77→cell death.

THE DILEMMA OF SPECIFICITY: HOW DO SPECIFIC RECEPTORS ACTIVATE SPECIFIC RESPONSES?

Several cell membrane receptors seem to use the same second messengers to elicit a number of different cellular responses. The release of intracellular stores of calcium, activation of PKC, and activation of ras are the result of many receptor-induced events that have distinct outcomes often in the same cell (294). A number of models can be invoked to rationalize this dilemma, the major and perhaps most credible of which is that not all of the signalling molecules have been found and that more specific ones will come to light with additional study. For example, the present methods for identifying transcriptional regulatory molecules are so crude that investigators may simply be finding the same factors, again and again.

One possible explanation for signalling specificity is that the outcomes of signal transduction are highly restricted by prior developmental events, and that the outcome of ubiquitous mediators is far less diverse than would be expected. This possibility is suggested by the observation that in a differentiated T cell line, the ligand for a transfected muscarinic acid receptor, which increases intracellular Ca^{2+}, can activate the IL-2 gene (295). These observations suggests that a genetic locus might be preconditioned, perhaps by remodeling chromatin, to respond to ubiquitous second messenger molecules. Such preconditioning may occur because the chromatin structure of the IL-2 gene has undergone remodeling by the time that it has become committed to the hematopoietic lineage (122). In yeast and *Drosophila*, chromatin remodelling results at least in part from the actions of opposing sets of positively and negatively acting genes. The positively acting members include *SWI1,2,3*, *snf5*, and *snf6* in yeast (296–300) and the members of the trithorax group including *Brahma* in *Drosophila* (143, 301, 302). These molecules appear to counteract the effects of chromatin in yeast (296) and the polycomb group in *Drosophila* (143) that oppose transcriptional activators. The first information on these molecules in vertebrates (146, 303) indicates that, as in yeast and *Drosophila*, they are dedicated to the activity of specific sets of genes, and hence might somehow be involved in remodeling chromatin in response to cell interactive or cell autonomous developmental pathways that specify cell type and condition genetic loci for activation. In invertebrates and yeasts, these classes of proteins such as Brahma are major regulators of developmental events and give rise to homeotic transitions (143, 301); exploration of their function in vertebrates may yield additional components of specificity that have not yet been discerned.

More specific mechanisms involved in signal transduction may remain to be elucidated. One of the most exciting developments in signal transduction in recent years is the discovery of a direct and specific mechanism of signal transduction for the activation of the interferon genes. Here, a 91-kDa protein (stat 91) is rapidly tyrosine phosphorylated by the JAK tyrosine kinase (304) immediately after binding of interferon (305–307). This protein enters the nucleus directly and participates in the activation of interferon-responsive genes. The possibility that this protein itself could account for the specificity of the interferon-signalling pathway was recently dashed when it was discovered that this protein also was phosphorylated and translocated to the nucleus in response to numerous other cytokines and growth factors (308–311). However more complex models for the generation of specificity, involving stat 91 in combination with other proteins, are still viable.

CONCLUDING REMARKS

A molecule-by-molecule description of the signal transduction pathway leading to immunologic activation of T lymphocytes will probably be attained in the near future. Beginning with the antigen receptor and its associated tyrosine activation motifs, a linear pathway of informational transfer from the cell membrane to the nucleus can roughly be drawn (Figure 3). Although gaps remain in the regions of the mechanism of activation of ras and the step or steps by which calcineurin controls the nuclear import of NF-ATc, this signalling pathway is probably as complete as any in present-day cell biology. These developments open up the possibility of studying the more complex mechanism by which the antigen receptor sets in motion pathways that will lead to cell death, anergy, or positive selection. Since some of the signal transfer steps appear to be shared between the activation and the other pathways, the next several years should provide substantial insight into the pathways to all three fates.

Literature Cited

1. Sternberg PW, Horvitz HR. 1989. *Cell* 58:679–93
2. Blau HM, Pavlath GK, Hardeman EC, Chiu CP, Silberstein L, et al. 1985. *Science* 230(4727):758–66
3. Davis MM, Chien YH, Gascoigne NR, Hedrick SM. 1984. *Immunol. Rev.* 81:235–58
4. Chien Y, Becker DM, Lindsten T, Okamura M, Cohen DI, Davis MM. 1984. *Nature* 312(5989):31–35
5. Yanagi Y, Yoshikai Y, Leggett K, Clark SP, Aleksander I, Mak TW. 1984. *Nature* 308:145–49
6. Yoshikai Y, Anatoniou D, Clark SP, Yanagi Y, Sangster R, et al. 1984. *Nature* 312:521–24
7. Allison JP, Lanier LL. 1985. *Nature* 314(6006):107–9
8. McIntyre BW, Allison JP. 1983. *Cell* 34:739–46
9. Samelson LE, Harford JB, Klausner RD. 1985. *Cell* 43:223–31
10. Weissman AM, Baniyash M, Hou D, Samelson LE, Burgess WH, Klausner RD. 1988. *Science* 239:1018–21
11. Weiss A. 1993. *Cell* 73:209–12
12. Samelson LE, Klausner RD. 1992. *J. Biol. Chem.* 267:24913–16
13. Reth M. 1989. *Nature* 338:383–84
14. Reth M, Hombach J, Wienands J, Campbell KS, Chien N, et al. 1991. *Immunol. Today* 12:196–201
15. Irving BA, Weiss A. 1991. *Cell* 64:891–901
16. Romeo C, Amiot M, Seed B. 1992. *Cell* 68:889–97
17. Letourneur F, Klausner RD. 1992. *Science* 255:79–82
18. Spencer DM, Wandless TJ, Schreiber SL, Crabtree GR. 1993. *Science*. In press
19. Love PE, Shores EW, Johnson MD, Tremblay ML, Lee EJ, et al. 1993. *Science* 261:918–21
20. Chan AC, Iwashima M, Turck CW, Weiss A. 1992. *Cell* 71:649–62

20a. Kolanus W, Romeo C, Seed B. 1993. *Cell* 74:171–83
21. Mustelin T, Coggeshall KM, Isakov N, Altman A. 1990. *Science* 247:1584–87
22. Sefton BM, Campbell M.-A. 1991. *Annu. Rev. Cell Biol.* 7:257–74
23. Shaw AS, Amrein KE, Hammond C, Stern DF, Sefton BM, Rose JK. 1989. *Cell* 59:627–36
24. Veillette A, Bookman MA, Horak EM, Bolen JB. 1988. *Cell* 55:301–8
25. Rudd CE, Trevillyan JM, Dasgupta JD, Wong LL, Schlossman SF. 1988. *Proc. Natl. Acad. Sci. USA* 85:5190–94
26. Turner JM, Brodsky MH, Irving BA, Levin SD, Perlmutter RM, Littman DR. 1990. *Cell* 60:755–65
27. Haughn L, Gratton S, Caron L, Sekaly R-P., Veillette A, Julius M. 1992. *Nature* 358:328–31
28. Evavold BD, Allen PM. 1991. *Science* 252:1308–10
29. Racioppi L, Ronchese F, Matis LA, Germain RN. 1993. *J. Exp. Med.* 177:1047–60
30. Jameson SC, Carbone FR, Bevan MJ. 1993. *J. Exp. Med.* 177:1541–50
31. Ostrov D, Krieger J, Sidney J, Sette A, Concannon P. 1993. *J. Immunol.* 150:4277–83
32. Sloan-Lancaster J, Evavold BD, Allen PM. 1993. *Nature* 363:156–59
33. Roussel MF, Shurtleff SA, Downing JR, Sherr CJ. 1990. *Proc. Natl. Acad. Sci. USA* 87:6738–42
34. Fantl WJ, Escobedo JA, Martin GA, Turck CW, del Rosario M, et al. 1992. *Cell* 69:413–23
35. Koretzky GA, Picus J, Thomas ML, Weiss A. 1990. *Nature* 346:66–68
36. Pingel JT, Thomas ML. 1989. *Cell* 58:1055–65
37. Xu H, Littman DR. 1993. *Cell* 74:633–43
38. Ostergaard HL, Shackelford DA, Hurley TR, Johnson P, Hyman R, et al. 1989. *Proc. Natl. Acad. Sci. USA* 86:8959–63
39. Mustelin T, Coggeshall KM, Altman A. 1989. *Proc. Natl. Acad. Sci. USA* 86:6302–6
40. Schraven B, Samstag Y, Altevogt P, Meuer SC. 1990. *Nature* 345:71–74
41. Volarevic S, Burns CM, Sussman JJ, Ashwell JD. 1990. *Proc. Natl. Acad. Sci. USA* 87:7085–89
42. Mittler RS, Rankin BM, Kiener PA. 1991. *J. Immunol.* 147:3434–40
43. Desai DM, Sap J, Schlessinger J, Weiss A. 1993. *Cell* 73:541–54
44. Volarevic S, Niklinska BB, Burns CM, June CH, Weissman AM, Ashwell JD. 1993. *Science* 260:541–44
44a. Hovis RR, Donovan JA, Musci MA, Motto DG, Goldman FD, et al. 1993. *Science* 260:544–46
45. Schwartz RH. 1992. *Cell* 71:1065–68
46. Mueller DL, Jenkins MK, Schwartz RH. 1989. *Annu. Rev. Immunol.* 7:445–80
47. Springer TA. 1990. *Nature* 346:425–34
48. Jenkins MK, Pardoll DM, Mizuguchi J, Chused TM, Schwartz RH. 1987. *Proc. Natl. Acad. Sci. USA* 84:5409–13
49. Quill H, Schwartz RH. 1987. *J. Immunol.* 138:3704–12
50. Meuer SC, Hussey RE, Fabbi M, Fox D, Acuto O, et al. 1984. *Cell* 36:397–406
51. Hunig T, Tiefenthaler G, zum Buschenfelde KH, Meuer SC. 1987. *Nature* 326:298–300
52. Linsley PS, Ledbetter JA. 1993. *Annu. Rev. Immunol.* 11:191–212
53. Turka LA, Ledbetter JA, Lee K, June CH, Thompson CB. 1990. *J. Immunol.* 144:1646–53
54. Linsley PS, Brady W, Grosmaire L, Aruffo A, Damle NK, Ledbetter JA. 1991. *J. Exp. Med.* 173:721–30
55. Linsley PS, Clark EA, Ledbetter JA. 1990. *Proc. Natl. Acad. Sci. USA* 87:5031–35
56. Freeman GJ, Freedman AS, Segil JM, Lee G, Whitman JF, Nadler LM. 1989. *J. Immunol.* 143:2714–22
57. Freedman AS, Freeman G, Horowitz JC, Daley J, Nadler LM. 1987. *J. Immunol.* 139:3260–67
58. Hathcock KS, Laszlo G, Dickler HB, Bradshaw J, Linsley P, Hodes RJ. 1993. *Science* 262:905–7
59. Freeman GJ, Gribben JG, Boussiotis VA, Ng JW, Restivo VA Jr. et al. 1993. *Science* 262:909–11
60. Guagliardi LE, Koppelman B, Blum JS, Marks MS, Cresswell P, Brodsky FM. 1990. *Nature* 343:133–39
61. June CH, Ledbetter JA, Gillespie MM, Lindsten T, Thompson CB. 1987. *Mol. Cell. Biol.* 7:4472–81
62. Thompson CB, Lindsten T, Ledbetter JA, Kunkel SL, Young HA, et al. 1989. *Proc. Natl. Acad. Sci. USA* 86:1333–37
62a. Harding FA, McArthur JG, Gross JA, Raulet DH, Allison JP. 1992. *Nature* 356:607–9
63. Shahinian A, Pfeffer K, Lee KP, Kundig TM, Kishihara K, et al. 1993. *Science* 261:609–12
64. Ledbetter JA, Linsley PS. 1992. *Adv. Exp. Med. Biol.* 323:23–27
65. Vandenberghe P, Freeman GJ, Nadler

LM, Fletcher MC, Kamoun M, et al. 1992. *J. Exp. Med.*. 175:951–60
66. Fraser JD, Irving BA, Crabtree GR, Weiss A. 1991. *Science* 251:313–16
67. Horvitz HR, Sternberg PW. 1991. *Nature* 351:535–41
67a. Rubin GM. 1989. *Cell* 57:519–20
67b. Smith JC. 1989. *Development* 105:665–77
67c. Gurdon JB. 1989. *Development* 105: 27–33
67d. Gilbert SF. 1991. *Developmental Biology*. Sunderland, MA: Sinauer Assoc.
67e. Dixon JE, Kintner CR. 1989. *Development* 106:749–57
68. Weiss A, Koretzky G, Schatzman RC, Kadlecek T. 1991. *Proc. Natl. Acad. Sci. USA* 88:5484–88
69. Katzav S, Martin-Zanca D, Barbacid M. 1989. *EMBO J.* 8:2283–90
70. Margolis B, Hu P, Katzav S, Li W, Oliver JM, et al. 1992. *Nature* 356:71–74
71. Bustelo XR, Ledbetter JA, Barbacid M. 1992. *Nature* 356:68–71
72. Alai M, Mui AL, Cutler RL, Bustelo XR, Barbacid M, Krystal G. 1992. *J. Biol. Chem.* 267:18021–25
73. Bustelo XR, Barbacid M. 1992. *Science* 256:1196–99
74. Hall CG, Sancho J, Terhorst C. 1993. *Science* 261:915–18
75. Truneh A, Albert F, Golstein P, Schmitt-Verhulst AM. 1985. *Nature* 313:318–21
76. Imboden JB, Weiss A, Stobo JD. 1985. *J. Immunol.* 134:663–65
77. Freedman MH, Raff MC, Gomperts B. 1975. *Nature* 255:378–82
78. Woodrow M, Clipstone NA, Cantrell DA. 1993. *J. Exp. Med.* 178:1517–22
79. Downward J, Graves JD, Warne PH, Rayter S, Cantrell DA. 1990. *Nature* 346:719–23
80. Izquierdo M, Downward J, Graves JD, Cantrell DA. 1992. *Mol. Cell Biol.* 12:3305–12
81. Lowy DR, Willumsen BM. 1993. *Annu. Rev. Biochem.* 62:851–91
82. Blenis J. 1993. *Proc. Natl. Acad. Sci. USA* 90:5889–92
83. McCormick F. 1993. *Nature* 363:15–16
84. Gale NW, Kaplan S, Lowenstein EJ, Schlessinger J, Bar-Sagi D. 1993. *Nature* 363:88–92
85. Lowenstein EJ, Daly RJ, Batzer AG, Li W, Margolis B, et al. 1992. *Cell* 70:431–42
86. Clark SG, Stern MJ, Horvitz HR. 1992. *Nature* 356:340–44
87. Simon MA, Dodson GS, Rubin GM. 1993. *Cell* 73:169–77
88. Gulbins E, Coggeshall KM, Baier G, Katzav S, Burn P, Altman A. 1993. *Science* 260:822–25
89. Shou C, Farnsworth CL, Neel BG, Feig LA. 1992. *Nature* 358:351–54
90. Li N, Batzer A, Daly R, Yajnik V, Skolnik E, et al. 1993. *Nature* 363:85–88
91. Chardin P, Camonis JH, Gale NW, van Aelst L, Schlessinger J, et al. 1993. *Science* 260:1338–43
92. Vojtek AB, Hollenberg SM, Cooper JA. 1993. *Cell* 74:205–14
93. Zhang XF, Settleman J, Kyriakis JM, Takeuchi-Suzuki E, Elledge SJ, et al. 1993. *Nature* 364:308–13
94. Ravichandran KS, Lee KK, Songyang Z, Cantley LC, Burn P, Burakoff SJ. 1993. *Science* 262:902–5
95. Dickson B, Sprenger F, Morrison D, Hafen E. 1992. *Nature* 360:600–3
96. Lu X, Chou TB, Williams NG, Roberts T, Perrimon N. 1993. *Genes Dev.* 7:621–32
97. Han M, Golden A, Han Y, Sternberg PW. 1993. *Nature* 363:133–40
98. van Aelst L, Barr M, Marcus S, Polverino A, Wigler M. 1993. *Proc. Natl. Acad. Sci. USA* 90:6213–17
99. Warne PH, Viciana PR, Downward J. 1993. *Nature* 364:352–55
100. Moodie SA, Willumsen BM, Weber MJ, Wolfman A. 1993. *Science* 260: 1658–61
101. Siegel JN, Klausner RD, Rapp UR, Samelson LE. 1990. *J. Biol. Chem.* 265:18472–80
102. Kyriakis JM, App H, Zhang XF, Banerjee P, Brautigan DL, et al. 1992. *Nature* 358:417–21
103. Pulverer BJ, Kyriakis JM, Avruch J, Nikolakaki E, Woodgett JR. 1991. *Nature* 353:670–74
104. Alvarez E, Northwood IC, Gonzalez FA, Latour DA, Seth A, et al. 1991. *J. Biol. Chem.* 266:15277–85
105. Baker SJ, Kerppola TK, Luk D, Vandenberg MT, Marshak DR, et al. 1992. *Mol. Cell. Biol.* 12:4694–705
106. Marais R, Wynne J, Treisman R. 1993. *Cell* 73:381–93
106a. Hibi M, Lin A, Smeal T, Minden A, Karin M. 1993. *Genes Dev.* 7:2135–48
107. Weiss A, Imboden J, Shoback D, Stobo J. 1984. *Proc. Natl. Acad. Sci. USA* 81:4169–73
108. Nisbet-Brown E, Cheung RK, Lee JW, Gelfand EW. 1985. *Nature* 316: 545–47
109. Gardner P. 1989. *Cell* 59:15–20
110. Berridge MJ, Irvine RF. 1989. *Nature* 341:197–205
111. Putney JW, Bird GSJ. 1993. *Cell* 75:199–201

112. Randriamampita C, Tsien RY. 1993. *Nature* 364:809–14
113. Gelfand EW, Cheung RK, Grinstein S, Mills GB. 1986. *Eur. J. Immunol.* 16:907–12
114. Gelfand EW, Cheung RK, Mills GB, Grinstein S. 1988. *Eur. J. Immunol.* 18:917–22
115. Goldsmith MA, Weiss A. 1988. *Science* 240:1029–31
116. Straus DB, Weiss A. 1992. *Cell* 70: 585–93
117. Shaw J.-P., Utz PJ, Durand DB, Toole JJ, Emmel EA, Crabtree GR. 1988. *Science* 241:202–5
118. Crabtree GR. 1989. *Science* 243:355–61
119. Cantrell DA, Smith KA. 1984. *Science* 224:1312–16
120. Smith KA, Cantrell DA. 1985. *Proc. Natl. Acad. Sci. USA* 82:864–68
121. Fernandez-Botran R, Sanders VM, Oliver KG, Chen YW, Krammer PH, et al. 1986. *Proc. Natl. Acad. Sci. USA* 83:9689–93
122. Siebenlist U, Durand DB, Bressler P, Holbrook NJ, Norris CA, et al. 1986. *Mol. Cell Biol.* 6:3042–49
123. Durand DB, Bush MR, Morgan JG, Weiss A, Crabtree GR. 1987. *J. Exp. Med.* 165:395–407
124. Ullman KS, Flanagan WM, Edwards CA, Crabtree GR. 1991. *Science* 254: 558–62
125. Serfling E, Barthelmäs R, Pfeuffer I, Schenk B, Zarius S, et al. 1989. *EMBO J.* 8:465–73
126. Durand DB, Shaw JP, Bush MR, Replogle RE, Belageje R, Crabtree GR. 1988. *Mol. Cell Biol.* 8:1715–24
127. Karttunen J, Shastri N. 1991. *Proc. Natl. Acad. Sci. USA* 88:3972–76
128. Jain J, McCaffrey PG, Miner Z, Kerppola TK, Lambert JN, et al. 1993. *Nature* 365:352–55
129. Hivroz-Burgaud C, Clipstone NA, Cantrell DA. 1991. *Eur. J. Immunol.* 21:2811–19
130. Verweij CL, Guidos C, Crabtree GR. 1990. *J. Biol. Chem.* 265:15788–95
131. Cockerill PN, Shannon MF, Bert AG, Ryan GR, Vadas MA. 1993. *Proc. Natl. Acad. Sci. USA* 90:2466–70
132. Todd MD, Grusby MJ, Lederer JA, Lacy E, Lichtman AH, Glimcher LH. 1993. *J. Exp. Med..* 177:1663–74
133. Szabo SJ, Gold JS, Murphy TL, Murphy KM. 1993. *Mol. Cell Biol.* 13: 4793–805
134. Goldfeld AE, McCaffrey PG, Strominger JL, Rao A. 1993. *J. Exp. Med..* 178:1365–79
135. Flanagan WM, Corthesy B, Bram RJ, Crabtree GR. 1991. *Nature* 352:803–7
136. Northrop JP, Ho SN, Timmermann LA, Thomas DJ, Chen L, et al. 1993. *Nature*. In press
137. Grosveld F, van Assendelft GB, Greaves DR, Kollias G. 1987. *Cell* 51:975–85
138. Fiering S, Northrop JP, Nolan GP, Mattila PS, Crabtree GR, Herzenberg LA. 1990. *Genes Dev.* 4:1823–34
139. Mouzaki A, Rungger D, Tucci A, Doucet A, Zubler RH. 1993. *Eur. J. Immunol.* 23:1469–74
140. Mouzaki A, Weil R, Muster L, Rungger D. 1991. *EMBO J.* 10:1399–406
141. Kirov N, Zhelnin L, Shah J, Rushlow C. 1993. *EMBO J.* 8:3193–99
141a. Schild H, Mavaddat N, Litzenberger C, Ehrlich EW, Davis MM, et al. 1994. *Cell* 57:519–20
142. Williams TM, Moolten D, Burlein J, Romano J, Bhaerman R, et al. 1991. *Science* 254:1791–94
143. Paro R, Hogness DS. 1991. *Proc. Natl. Acad. Sci. USA* 88:263–67
144. Simon J, Chiang A, Bender W. 1992. *Development* 114:493–505
145. Zink B, Engstrom Y, Gehring WJ, Paro R. 1991. *EMBO J.* 10:153–62
146. Tkachuk DC, Kohler S, Cleary ML. 1992. *Cell* 71:691–700
147. McCaffrey PG, Perrino BA, Soderling TR, Rao A. 1993. *J. Biol. Chem.* 268:3747–52
148. Northrop JP, Ullman KS, Crabtree GR. 1993. *J. Biol. Chem.* 268:2917–23
149. Jain J, McCaffrey PG, Valge-Archer VE, Rao A. 1992. *Nature* 356:801–4
150. Castigli E, Chatila TA, Geha RS. 1993. *J. Immunol.* 150:3284–90
151. Castigli E, Pahwa R, Good RA, Geha RS, Chatila TA. 1993. *Proc. Natl. Acad. Sci. USA* 90:4728–32
151a. Weinberg K, Parkman R. 1990. *New Engl. J. Med.* 322:1718–23
152. Riegel JS, Richie ER, Allison JP. 1990. *J. Immunol.* 144:3611–18
153. Chen D, Rothenberg EV. 1993. *Mol. Cell. Biol.* 13:228–37
154. Zuniga-Pflucker JC, Schwartz HL, Lenardo MJ. 1993. *J. Exp. Med..* 178: 1139–49
155. McCaffrey PG, Luo C, Kerppola TK, Jain J, Badalian TM, et al. 1993. *Science* 262:750–54
156. Kamps MP, Corcoran L, LeBowitz JH, Baltimore D. 1990. *Mol. Cell Biol.* 10:5464–72
157. Ullman KS, Northrop JP, Admon A, Crabtree GR. 1993. *Genes Dev.* 7:188–96
158. Emmel EA, Verweij CL, Durand DB,

Higgins KM, Lacy E, Crabtree GR. 1989. *Science* 246:1617–20
159. Abate C, Baker SJ, Lees-Miller SP, Anderson CW, Marshak DR, Curran T. 1993. *Proc. Natl. Acad. Sci. USA* 90:6766–70
160. Dohrmann PR, Butler G, Tamai K, Dorland S, Greene JR, et al. 1992. *Genes Dev.* 6:93–104
161. Northrop JP, Crabtree GR, Mattila PS. 1992. *J. Exp. Med.* 175:1235–45
162. Yang-Yen H-F, Chambard J-C, Sun Y-L, Smeal T, Schmidt TJ, et al. 1990. *Cell* 62:1205–15
163. Schüle R, Rangarajan P, Kliewer S, Ransone LJ, Bolado J, et al. 1990. *Cell* 62:1217–26
164. Jonat C, Rahmsdorf HJ, Park K-K, Cato ACB, Gebel S, et al. 1990. *Cell* 62:1189–204
165. Liou HC, Baltimore D. 1993. *Curr. Opin. Cell Biol.* 5:477–87
166. Kerr LD, Inoue J, Verma IM. 1992. *Curr. Opin. Cell Biol.* 4:496–501
167. Karin M, Smeal T. 1992. *Trends Biochem. Sci.* 17:418–22
168. Hunter T, Karin M. 1992. *Cell* 70:375–87
169. Böhnlein E, Lowenthal JW, Siekevitz M, Ballard DW, Franza BR, Greene WC. 1988. *Cell* 53:827–36
170. Cross SL, Halden NF, Lenardo MJ, Leonard WJ. 1989. *Science* 244:466–69
171. Shibuya H, Yoneyama M, Taniguchi T. 1989. *Int. Immunol.* 1:43–50
172. Hoyos B, Ballard DW, Böhnlein E, Siekevitz M, Greene WC. 1989. *Science* 244:457–60
173. Williams TM, Eisenberg L, Burlein JE, Norris CA, Pancer S, et al. 1988. *J. Immunol.* 141:662–66
174. Kang SM, Tran AC, Grilli M, Lenardo MJ. 1992. *Science* 256:1452–56
175. Ohno H, Takimoto G, McKeithan TW. 1990. *Cell* 60:991–97
176. Reed JC, Tsujimoto Y, Alpers JD, Croce CM, Nowell PC. 1987. *Science* 236:1295–99
177. Lowenthal JW, Ballard DW, Böhnlein E, Greene WC. 1989. *Proc. Natl. Acad. Sci. USA* 86:2331–35
178. Kahan BD. 1989. *New Engl. J. Med.* 321:1725–38
179. Thompson AW. 1989. *Immunol. Today* 10:6–9
180. Sigal NH, Dumont FJ. 1992. *Annu. Rev. Immunol.* 10:519–60
181. Kronke M, Leonard WJ, Depper JM, Arya SK, Wong-Staal F, et al. 1984. *Proc. Natl. Acad. Sci. USA* 81:5214–18
182. Granelli-Piperno A, Andrus L, Steinman RM. 1986. *J. Exp. Med..* 163:922–37
183. Tocci MJ, Matkocich DA, Collier KA, Kwok P, Dumont F, et al. 1989. *J. Immunol.* 143:718–26
184. Bickel M, Tsuda H, Amstad P, Evequoz V, Mergenhagen SE, et al. 1987. *Proc. Natl. Acad. Sci. USA* 84:3274–77
185. Shevach EM. 1985. *Annu. Rev. Immunol.* 3:397–423
186. Bijsterbosch MK, Klaus GG. 1985. *Immunology* 56:435–40
187. Isakov N, Mally MI, Scholz W, Altman A. 1987. *Immunol. Rev.* 95:89–111
188. Trenn G, Taffs R, Hohman R, Kincaid R, Shevach EM, Sitkovsky M. 1989. *J. Immunol.* 142:3796–802
189. Mizushima Y, Kosaka H, Sakuma S, Kanda K, Itoh K, et al. 1987. *J. Biochem.* 102:1193–201
190. Mattila PS, Ullman KS, Fiering S, McCutcheon M, Crabtree GR, Herzenberg LA. 1990. *EMBO J.* 9:4425–33
191. Lin CS, Boltz RC, Siekierka JJ, Sigal NH. 1991. *Cell. Immunol.* 133:269–84
192. Kay JE, Doe SEA, Benzie CR. 1989. *Cell. Immunol.* 124:175–81
193. Gunter KC, Irving SG, Zipfel PF, Siebenlist U, Kelly K. 1989. *J. Immunol.* 142:3286–91
194. Randak C, Brabletz T, Hergenröther M, Sobotta I, Serfling E. 1990. *EMBO J.* 9:2529–36
195. Bierer BE, Somers PK, Wandless TJ, Burakoff SJ, Schreiber SL. 1990. *Science* 250:556–59
196. Dumont FJ, Melino MR, Staruch MJ, Koprak SL, Fischer PA, Sigal NH. 1990. *J. Immunol.* 144:1418–24
197. Tropschug M, Barthelmess IB, Neupert W. 1989. *Nature* 342:953–55
198. Schreiber SL. 1991. *Science* 251:283–87
199. Standaert RF, Galat A, Verdine GL, Schreiber SL. 1990. *Nature* 346:671–74
200. Maki N, Sekiguchi F, Nishimaki J, Miwa K, Hayano T, et al. 1990. *Proc. Natl. Acad. Sci. USA* 87:5440–43
201. Schreiber SL. 1992. *Cell* 70:365–68
202. Takahashi N, Hayano T, Suzuki M. 1989. *Nature* 337:473–75
203. Fischer G, Wittmann-Liebold B, Lang K, Kiefhaber T, Schmid FX. 1989. *Nature* 337:476–78
204. Harding MW, Galat A, Uehling DE, Schreiber SL. 1989. *Nature* 341:758–60
205. Siekierka JJ, Hung SHY, Poe M, Lin CS, Sigal NH. 1989. *Nature* 341:755–57
206. Schreiber SL, Crabtree GR. 1992. *Immunol. Today* 13:136–42

207. Rosen MK, Standaert RF, Galat A, Nakatsuka M, Schreiber SL. 1990. *Science* 248:863–66
208. Bierer BE, Mattila PS, Standaert RF, Herzenberg LA, Burakoff SJ, et al. 1990. *Proc. Natl. Acad. Sci. USA* 87:9231–35
209. Heitman J, Movva NR, Hiestand PC, Hall MN. 1991. *Proc. Natl. Acad. Sci. USA* 88:1948–52
210. Theriault Y, Logan TM, Meadows R, Yu L, Olejniczak ET, et al. 1993. *Nature* 361:88–91
211. Pflugl G, Kallen J, Schirmer T, Jansanius JN, Zurinum G, Walkinshaw MD. 1993. *Nature* 361:91–94
212. Van Duyne GD, Standaert RF, Karplus PA, Schreiber SL, Clardy J. 1991. *Science* 252:839–42
213. Michnick SW, Rosen MK, Wandless TJ, Karplus M, Schreiber SL. 1991. *Science* 252:836–39
214. Van Duyne GD, Standaert RF, Karplus PA, Schreiber SL, Clardy J. 1993. *J. Mol. Biol.* 229:105–24
215. Liu J, Farmer JD, Lane WS, Friedman J, Weissman I, Schreiber SL. 1991. *Cell* 66:807–15
216. Friedman J, Weissman I. 1991. *Cell* 66:799–806
217. Klee CB, Draetta GF, Hubbard MJ. 1988. In *Advances in Ezymology and Related Areas of Molecular Biology*, ed. A. Meister, pp. 149–209. New York: Wiley
218. Kincaid R. 1993. *Adv. 2nd Messenger Phosphoprotein Res.* 27:1–23
219. Kincaid RL, Takayama H, Billingsley ML, Sitkovsky MV. 1987. *Nature* 330:176–78
220. Swanson SK, Born T, Zydowsky LD, Cho H, Chang HY, et al. 1992. *Proc. Natl. Acad. Sci. USA* 89:3741–45
221. Fruman DA, Klee CB, Bierer BE, Burakoff SJ. 1992. *Proc. Natl. Acad. Sci. USA* 89:3686–90
222. Liu J, Albers MW, Wandless TJ, Luan S, Alberg DG, et al. 1992. *Biochemistry* 31:3896–901
223. Clipstone NA, Crabtree GR. 1992. *Nature* 357:695–97
224. O'Keefe SJ, Tamura J, Kincaid RL, Tocci MJ, O'Neill EA. 1992. *Nature* 357:692–95
225. Clipstone NA, Crabtree GR. 1993. *Ann. NY Acad. Sci.* 696:20–30
226. Ullman KS, Northrop JP, Verweij CL, Crabtree GR. 1990. *Annu. Rev. Immunol.* 8:421–52
227. Moll T, Tebb G, Surana U, Robitsch H, Nasmyth K. 1991. *Cell* 66:743–58
228. Shieh B-H, Stamnes MA, Seavello S, Harris GL, Zuker CS. 1989. *Nature* 338:67–70
229. Colley NJ, Baker EK, Stamnes MA, Zuker CS. 1991. *Cell* 67:255–63
230. Steiner JP, Dawson TM, Fotuhi M, Glatt CE, Snowman AM, et al. 1992. *Nature* 358:584–87
231. Jayaraman T, Brillantes AM, Timerman AP, Fleischer S, Erdjument-Bromage H, et al. 1992. *J. Biol. Chem.* 267:9474–77
232. Tai PK, Albers MW, Chang H, Faber LE, Schreiber SL. 1992. *Science* 256: 1315–18
233. Yem AW, Tomasselli AG, Heinrikson RL, Zurcher-Neely H, Ruff VA, et al. 1992. *J. Biol. Chem.* 267:2868–71
234. Peattie DA, Harding MW, Fleming MA, DeCenzo MT, Lippke JA, et al. 1992. *Proc. Natl. Acad. Sci. USA* 89:10974–78
235. Ratajczak T, Carrello A, Mark PJ, Warner BJ, Simpson RJ, et al. 1993. *J. Biol. Chem.* 268:13187–92
236. Kincaid RL, Giri PR, Higuchi S, Tamura J, Dixon SC, et al. 1990. *J. Biol. Chem.* 265:11312–19
237. Yang D, Rosen MK, Schreiber SL. 1993. *J. Am. Chem. Soc.* 115:819–20
238. Hultsch T, Rodriguez JL, Kaliner MA, Hohman RJ. 1990. *J. Immunol.* 144: 2659–64
239. Hultsch T, Albers MW, Schreiber SL, Hohman RJ. 1991. *Proc. Natl. Acad. Sci. USA* 88:6229–33
240. Aperia A, Ibarra F, Svensson LB, Klee C, Greengard P. 1992. *Proc. Natl. Acad. Sci. USA* 89:7394–97
241. Luan S, Li W, Rusnak F, Assmann SM, Schreiber SL. 1993. *Proc. Natl. Acad. Sci. USA* 90:2202–6
242. Hendey B, Klee CB, Maxfield FR. 1992. *Science* 258:296–99
243. Marsh L, Neiman AM, Herskowitz I. 1991. *Annu. Rev. Cell Biol.* 7:699–728
244. Iida H, Yagawa Y, Anraku Y. 1990. *J. Biol. Chem.* 265:13391–99
245. Cyert MS, Thorner J. 1992. *Mol. Cell Biol.* 12:3460–69
246. Cyert MS, Kunisawa R, Kaim D, Thorner J. 1991. *Proc. Natl. Acad. Sci. USA* 88:7376–80
247. Foor F, Parent SA, Morin N, Dahl AM, Ramadan N, et al. 1992. *Nature* 360:682–84
248. Cunningham KW, Fink GR. 1994. *J. Cell Biol.* In press
249. Shi YF, Sahai BM, Green DR. 1989. *Nature* 339:625–26
250. Kizaki H, Tadakuma T, Odaka C, Muramatsu J, Ishimura Y. 1989. *J. Immunol.* 143:1790–94
251. McConkey DJ, Hartzell P, Amador-

Perez JF, Orrenius S, Jondal M. 1989. *J. Immunol.* 143:1801–6
252. Woodrow MA, Rayter S, Downward J, Cantrell DA. 1993. *J. Immunol.* 150:3853–61
253. Rayter SI, Woodrow M, Lucas SC, Cantrell DA, Downward J. 1992. *EMBO J.* 11:4549–56
254. Baldari CT, Heguy A, Telford JL. 1993. *J. Biol. Chem.* 268:2693–98
255. Kincaid RL, Vaughan M. 1986. *Proc. Natl. Acad. Sci. USA* 83:1193–97
256. Allbritton NL, Meyer T, Stryer L. 1992. *Science* 258:1812–15
257. Bram RJ, Hung DT, Martin PK, Schreiber SL, Crabtree G. 1993. *Mol. Cell. Biol.* 13:4760–69
258. Iwabuchi K, Nakayama K, McCoy RL, Wang F, Nishimura T, et al. 1992. *Proc. Natl. Acad. Sci. USA* 89:9000–4
259. Pawlowski T, Elliott JD, Loh DY, Staerz UD. 1993. *Nature* 364:642–45
260. Russell JH, White CL, Loh DY, Meleedy-Rey P. 1991. *Proc. Natl. Acad. Sci. USA* 88:2151–55
261. Trauth BC, Klas C, Peters AM, Matzku S, Moller P, et al. 1989. *Science* 245:301–5
262. Itoh N, Yonehara S, Ishii A, Yonehara M, Mizushima S, et al. 1991. *Cell* 66:233–43
263. Itoh N, Nagata S. 1993. *J. Biol. Chem.* 268:10932–37
264. Cohen PL, Eisenberg RA. 1991. *Annu. Rev. Immunol.* 9:243–69
265. Watanabe-Fukunaga R, Brannan CI, Copeland NG, Jenkins NA, Nagata S. 1992. *Nature* 356:314–17
266. Herron LR, Eisenberg RA, Roper E, Kakkanaiah VN, Cohen PL, Kotzin BL. 1993. *J. Immunol.* 151: 3450–59
267. Zhou T, Bluethmann H, Zhang J, Edwards CK, Mountz JD. 1992. *J. Exp. Med.* 176:1063–72
268. Hess AD, Horwitz L, Beschorner WE, Santos GW. 1985. *J. Exp. Med.* 161: 718–30
269. Glazier A, Tutschka PJ, Farmer ER, Santos GW. 1983. *J. Exp. Med.* 158:1–8
270. Sakaguchi S, Sakaguchi N. 1988. *J. Exp. Med.* 167:1479–85
271. Gao E-K, Lo D, Cheney R, Kanagawa O, Sprent J. 1988. *Nature* 336:176–79
272. Jenkins MK, Schwartz RH, Pardoll DM. 1988. *Science* 241:1655–58
273. Page DM, Kane LP, Allison JP, Hedrick SM. 1993. *J. Immunol.* 151: 1868–80
274. Vasquez NJ, Kaye J, Hedrick SM. 1992. *J. Exp. Med.* 175:1307–16
275. Molina TJ, Kishihara K, Siderovski DP, Van Ewijk W, Narendran A, et al. 1992. *Nature* 357:161–64
276. Abraham KM, Levin SD, Marth JD, Forbush KA, Perlmutter RM. 1991. *J. Exp. Med.*. 173:1421–32
277. Levin SD, Anderson SJ, Forbush KA, Perlmutter RM. 1993. *EMBO J.* 12: 1671–80
278. Anderson SJ, Levin SD, Perlmutter RM. 1993. *Nature* 365:552–54
279. Appleby MW, Gross JA, Cooke MP, Levin SD, Qian X, Perlmutter RM. 1992. *Cell* 70:751–63
280. Stein PL, Lee HM, Rich S, Soriano P. 1992. *Cell* 70:741–50
281. Malissen M, Gillet A, Rocha B, Trucy J, Vivier E, et al. 1993. *EMBO J.* 12:4347–55
282. Sentman CL, Shutter JR, Hockenbery D, Kanagawa O, Korsmeyer SJ. 1991. *Cell* 67:879–88
283. Strasser A, Harris AW, Cory S. 1991. *Cell* 67:889–99
284. Hartley SB, Cooke MP, Fulcher DA, Harris AW, Cory S, et al. 1993. *Cell* 72:325–35
285. Hockenbery D, Nuñez G, Milliman C, Schreiber RD, Korsmeyer SJ. 1990. *Nature* 348:334–36
286. Veis DJ, Sorenson CM, Shutter JR, Korsmeyer SJ. 1993. *Cell* 75:229–40
287. Hockenbery DM, Oltvai ZN, Yin X-M, Milliman CL, Korsmeyer SJ. 1993. *Cell* 75:241–51
288. Hengartner MO, Ellis RE, Horvitz HR. 1992. *Nature* 356:494–99
289. Vaux DL, Weissman IL, Kim SK. 1992. *Science* 258:1955–57
290. Yuan JY, Horvitz HR. 1990. *Dev. Biol.* 138:33–41
291. Ellis RE, Yuan JY, Horvitz HR. 1991. *Annu. Rev. Cell Biol.* 7:663–98
292. Itoh N, Tsujimoto Y, Nagata S. 1993. *J. Immunol.* 151:621–27
293. Woronicz JD, Calnan B, Ngo V, Winoto A. 1993. *Nature*. In press
294. Chao MV. 1992. *Cell* 68:995–97
295. Goldsmith MA, Desai DM, Schultz T, Weiss A. 1989. *J. Biol. Chem.* 264: 17190–97
296. Stern M, Jensen R, Herskowitz I. 1984. *J. Mol. Biol.* 178:853–68
297. Laurent BC, Carlson M. 1992. *Genes Dev.* 6:1707–15
298. Laurent BC, Treitel MA, Carlson M. 1991. *Proc. Natl. Acad. Sci. USA* 88:2687–91
299. Laurent BC, Treitel MA, Carlson M. 1990. *Mol. Cell Biol.* 10:5616–25
300. Estruch F, Carlson M. 1990. *Mol. Cell Biol* 10:2544–53
301. Tamkun JW, Deuring R, Scott MP,

Kissinger M, Pattatucci AM, et al. 1992. *Cell* 68:561–72
302. Yoshinaga SK, Peterson CL, Herskowitz I, Yamamoto KR. 1992. *Science* 258:1598–604
303. Khavari PA, Peterson CL, Tamkun JW, Crabtree GR. 1993. *Nature*. In press
304. Velazquez L, Fellous M, Stark GR, Pellegrini S. 1992. *Cell* 70:313–22
305. Shuai K, Schindler C, Prezioso VR, Darnell JE Jr. 1992. *Science* 258:1808–12
306. Shuai K, Stark GR, Kerr IM, Darnell JE Jr. 1993. *Science* 261:1744–46
307. Schindler C, Fu XY, Improta T, Aebersold R, Darnell JE Jr. 1992. *Proc. Natl. Acad. Sci. USA* 89:7836–39
308. Sadowski HB, Shuai K, Darnell JE Jr, Gilman MZ. 1993. *Science* 261:1739–44
309. Larner AC, David M, Feldman GM, Igarashi K-i, Hackett RH, et al. 1993. *Science* 261:1730–33
310. Ruff-Jamison S, Chen K, Cohen S. 1993. *Science* 261:1733–36
311. Silvennoinen O, Schindler C, Schlessinger J, Levy D. 1993. *Science* 261:1736–39

AUTHOR INDEX

C

D

F

G

N

P

T

Z

SUBJECT INDEX

U

CUMULATIVE INDEXES

CONTRIBUTING AUTHORS, VOLUMES 59–63

CHAPTER TITLES, VOLUMES 59–63

PROTEINS

Structure

ANNUAL REVIEWS

a nonprofit scientific publisher
4139 El Camino Way
P.O. Box 10139
Palo Alto, CA 94303-0139 • USA

ORDER FORM

ORDER TOLL FREE
1.800.523.8635
from USA and Canada

Fax: 1.415.855.9815

Annual Reviews publications may be ordered directly from our office; through booksellers and subscription agents, worldwide; and through participating professional societies. **Prices are subject to change without notice. We do not ship on approval.**

- **Individuals:** Prepayment required on new accounts. in US dollars, checks drawn on a US bank.
- **Institutional Buyers:** Include purchase order. Calif. Corp. #161041 • ARI Fed. I.D. #94-1156476
- **Students / Recent Graduates: $10.00 discount from retail price, per volume.** ***Requirements:*** **1.** be a degree candidate at, or a graduate within the past three years from, an accredited institution; **2.** present proof of status (photocopy of your student I.D. or proof of date of graduation); **3.** Order direct from Annual Reviews; **4.** prepay. This discount **does not** apply to standing orders, *Index on Diskette,* Special Publications, ARPR, or institutional buyers.
- **Professional Society Members:** Many Societies offer *Annual Reviews* to members at reduced rates. Check with your society or contact our office for a list of participating societies.
- **California orders** add applicable sales tax. • **Canadian orders** add 7% GST. Registration #R 121 449-029.
- **Postage paid** by Annual Reviews (4th class bookrate/surface mail). UPS ground service is available at $2.00 extra per book within the contiguous 48 states only. UPS air service or US airmail is available to any location at actual cost. UPS requires a street address. P.O. Box, APO, FPO, not acceptable.
- **Standing Orders:** Set up a standing order and the new volume in series is sent automatically each year upon publication. Each year you can save 10% by prepayment of prerelease invoices sent 90 days prior to the publication date. Cancellation may be made at any time.
- **Prepublication Orders:** Advance orders may be placed for any volume and will be charged to your account upon receipt. Volumes not yet published will be shipped during month of publication indicated.

NOTE: For copies of individual articles from any *Annual Review,* or copies of any article cited in an *Annual Review,* call **Annual Reviews Preprints and Reprints (ARPR)** toll free 1-800-347-8007 (fax toll free 1-800-347-8008) from the USA or Canada. From elsewhere call 1-415-259-5017.

ANNUAL REVIEWS SERIES *Volumes not listed are no longer in print*		**Prices, postpaid, per volume. USA/other countries**	Regular Order Please send Volume(s):	Standing Order Begin with Volume:
❑ *Annual Review of*	**ANTHROPOLOGY**			
Vols. 1-20	(1972-91)	$41 / $46		
Vols. 21-22	(1992-93)	$44 / $49		
Vol. 23	(avail. Oct. 1994)	$47 / $52	Vol(s). ______	Vol. ______
❑ *Annual Review of*	**ASTRONOMY AND ASTROPHYSICS**			
Vols. 1, 5-14, 16-29	(1963, 67-76, 78-91)	$53 / $58		
Vols. 30-31	(1992-93)	$57 / $62		
Vol. 32	(avail. Sept. 1994)	$60 / $65	Vol(s). ______	Vol. ______
❑ *Annual Review of*	**BIOCHEMISTRY**			
Vols. 31-34, 36-60	(1962-65,67-91)	$41 / $47		
Vols. 61-62	(1992-93)	$46 / $52		
Vol. 63	(avail. July 1994)	$49 / $55	Vol(s). ______	Vol. ______
❑ *Annual Review of*	**BIOPHYSICS AND BIOMOLECULAR STRUCTURE**			
Vols. 1-20	(1972-91)	$55 / $60		
Vols. 21-22	(1992-93)	$59 / $64		
Vol. 23	(avail. June 1994)	$62 / $67	Vol(s). ______	Vol. ______